Lecture Notes in Computer Science 8087

Commenced Publication in 1973
Founding and Former Series Editors:
Gerhard Goos, Juris Hartmanis, and Jan van Leeuwen

Advanced Research in Computing and Software Science
Subline of Lectures Notes in Computer Science

Krishnendu Chatterjee   Jirí Sgall (Eds.)

Mathematical Foundations of Computer Science 2013

38th International Symposium, MFCS 2013
Klosterneuburg, Austria, August 26-30, 2013
Proceedings

 Springer

Volume Editors

Krishnendu Chatterjee
Institute of Science and Technology
3400 Klosterneuburg, Austria
E-mail: krishnendu.chatterjee@ist.ac.at

Jiří Sgall
Charles University
11800 Prague, Czech Republic
E-mail: sgall@iuuk.mff.cuni.cz

ISSN 0302-9743   e-ISSN 1611-3349
ISBN 978-3-642-40312-5   e-ISBN 978-3-642-40313-2
DOI 10.1007/978-3-642-40313-2
Springer Heidelberg New York Dordrecht London

Library of Congress Control Number: 2013945881

CR Subject Classification (1998):
F.2, G.2, G.1, F.1, F.4.1, E.1, D.2.4, F.4.3, I.2.3-4, G.3

LNCS Sublibrary: SL 1 – Theoretical Computer Science and General Issues

Typesetting: Camera-ready by author, data conversion by Scientific Publishing Services, Chennai, India

Printed on acid-free paper

Springer is part of Springer Science+Business Media (www.springer.com)

Preface

This volume contains the papers that were presented at MFCS 2013: The 38th International Symposium on Mathematical Foundations of Computer Science held during August 26–30, 2013, at IST Austria, in Klosterneuburg, Austria. It contains six invited and 67 contributed papers presented at the symposium. The contributed papers were selected by the Program Committee (PC) out of a total of 191 submissions. All submitted papers were peer reviewed and evaluated on the basis of originality, quality, significance, and presentation. Each paper was reviewed by at least three PC members with the help of external experts. The PC also selected to give the Best Paper Awards, sponsored by European Association of Theoretical Computer Science (EATCS), jointly to "Improved bounds for reduction to depth 4 and depth 3" by Sébastien Tavenas; and "Minimal indices for successor search" by Sarel Cohen, Amos Fiat, Moshik Hershcovitch, and Haim Kaplan. In addition, the paper by Sébastien Tavenas was also selected for the Best Student Paper Award.

The program included six invited talks by:

- Sam Buss, University of California, San Diego, USA
- Leah Epstein, University of Haifa, Israel
- Jean Goubault-Larrecq, LSV, ENS Cachan, CNRS, INRIA, France
- Martin Grohe, RWTH Aachen University, Germany
- Elias Koutsoupias, University of Oxford, UK
- Nir Piterman, University of Leicester, UK

We thank all invited speakers for accepting our invitation and for their excellent presentation at the symposium.

We thank all authors who submitted their work to MFCS 2013 for consideration. We wish to thank all PC members and external reviewers for their competent and timely handling of the submissions. The success of the scientific program is due to their hard work. During the selection process and for preparing this volume, we used the EasyChair conference management system, which provided excellent support.

The series of MFCS symposia has a well-established tradition since 1972, and has been organized on a rotating basis in Poland, Czech Republic, and Slovakia. The 2013 meeting added a new country to this history, and we are happy that it could be Austria with its lasting and fruitful relationship to the original host countries.

We gratefully acknowledge the support of IST Austria and EATCS. Special thanks for local organization are due to: Martin Chmelík, Sebastian Nozzi, Andreas Pavlogiannis, Johannes Reiter, and Marie Trappl.

June 2013

Krishnendu Chatterjee
Jiří Sgall

Conference Organization

Program Committee Chairs

Krishnendu Chatterjee
Jiří Sgall

Program Committee

Parosh Abdulla
Eli Ben-Sasson
Nathalie Bertrand
Markus Bläser
Tomáš Brázdil
Ioannis Caragiannis
Thomas Colcombet
Anuj Dawar
Giorgio Delzanno
Martin Dietzfelbinger
Krzysztof Diks
Zoltan Esik
Sándor Fekete
Eldar Fischer
Dmitry Gavinsky
Andrew Goldberg
Kristoffer Arnsfelt Hansen
Tao Jiang
Barbara König
Pascal Koiran
Rastislav Královič

Erik Jan Van Leeuwen
Stefano Leonardi
Christof Löding
Zvi Lotker
Jerzy Marcinkowski
Dániel Marx
Peter Bro Miltersen
Madhavan Mukund
Rasmus Pagh
Madhusudan Parthasarathy
Daniel Paulusma
Holger Petersen
Alex Rabinovich
Rahul Santhanam
Martin Strauss
Ola Svensson
Maxim Sviridenko
Pavel Valtr
Peter Widmayer
Gerhard Woeginger
James Worrell

Steering Committee

Juraj Hromkovič
Antonín Kučera
Jerzy Marcinkowski

Damian Niwinski
Branislav Rovan
Jiří Sgall

Local Organization

Krishnendu Chatterjee, Martin Chmelík, Sebastian Nozzi, Andreas Pavlogiannis, Johannes Reiter, Marie Trappl (IST Austria, Klosterneuburg, Austria)

External Reviewers

Adamczyk, Marek
An, Hyung-Chan
Anagnostopoulos, Aris
Aronov, Boris
Arun-Kumar, S
Atig, Mohamed Faouzi
Atserias, Albert
Aubrun, Nathalie
Bader, David A.
Bansal, Nikhil
Barvinok, Alexander
Bekos, Michael
Björklund, Andreas
Björklund, Henrik
Blume, Christoph
Böhmova, Katerina
Bonnet, Remi
Borokhovich, Michael
Bozga, Marius
Bredereck, Robert
van Breugel, Franck
Brihaye, Thomas
Bruggink, Harrie Jan
   Sander
Bundala, Daniel
Cabello, Sergio
Carayol, Arnaud
Cardinal, Jean
Chan, Timothy
Chattopadhyay, Arkadev
Chaturvedi, Namit
Chen, Hubie
Cibulka, Josef
Cohen, Edith
Colini Baldeschi,
   Riccardo
Cording, Patrick Hagge
Curticapean, Radu
Cygan, Marek
Czeizler, Eugen
D'Souza, Deepak
Dassow, Jürgen
Daviaud, Laure

de Nivelle'S, Hans
de Souza, Rodrigo
de Weerdt, Mathijs
Demri, Stéphane
Dennunzio, Alberto
Dinitz, Yefim
Duan, Ran
Edmonds, Jeff
Eliáš, Marek
Elkind, Edith
Even, Guy
Fanelli, Angelo
Fearnley, John
Fenner, Stephen
Fernau, Henning
Figueira, Diego
Fijalkow, Nathanaël
Forejt, Vojtech
Friedman, Arik
Friedrich, Tobias
Friggstad, Zachary
Fulek, Radoslav
Ganty, Pierre
Garg, Pranav
Genest, Blaise
Gilboa, Niv
Gogacz, Tomasz
Goldhirsh, Yonatan
Gottlob, Georg
Grabowski, Szymon
Grandoni, Fabrizio
Gravin, Nikolai
Grenet, Bruno
Guillemot, Sylvain
Gurevich, Yuri
Göller, Stefan
Haddad, Axel
Hague, Matthew
Hajgato, Tamas
Halava, Vesa
Halevi, Shai
Harju, Tero
Harrenstein, Paul

Heindel, Tobias
Hidalgo-Herrero,
   Mercedes
Ho, Hsi-Ming
Holík, Lukáš
Holzer, Markus
Horn, Florian
Huang, Chien-Chung
Hunter, Paul
Hüffner, Falk
Ibsen-Jensen, Rasmus
Immerman, Neil
Ingolfsdottir, Anna
Ito, Tsuyoshi
Ivan, Szabolcs
Jeż, Lukasz
Jeż, Artur
Jones, Andrew
Jones, Neil
Jurdziński, Tomasz
Kabanets, Valentine
Kaiser, Lukasz
Kalampakas, Antonis
Kambites, Mark
Kanellopoulos,
   Panagiotis
Kanza, Yaron
Kaplan, Haim
Kaporis, Alexis
Kartzow, Alexander
Katz, Matthew
Kerenidis, Iordanis
Kerstan, Henning
Ketema, Jeroen
Kiefer, Stefan
Kim, Eun Jung
Kitaev, Sergey
Klauck, Hartmut
Klazar, Martin
Kleszowski, Wojciech
Klimann, Ines
Klin, Bartek
Koivisto, Mikko

Kolliopoulos, Stavros
Kontchakov, Roman
Kopczynski, Eryk
Kotrbcik, Michal
Koucký, Michal
Kowalik, Lukasz
Krajíček, Jan
Kratsch, Dieter
Krčál, Jan
Křetínský, Jan
Krumke, Sven
Kučera, Antonín
Kuperberg, Denis
Kyropoulou, Maria
La Torre, Salvatore
Lachish, Oded
Lasota, Slawomir
Lauze, Francois
Leupold, Peter
Ligett, Katrina
Limaye, Nutan
Lodya, Kamal
Lohrey, Markus
Lombardy, Sylvain
Lovett, Shachar
Lozin, Vadim
Lu, Songjian
Łącki, Jakub
Magnin, Loïck
Makowsky, Johann
Maletti, Andreas
Malinowski, Adam
Manea, Florin
Manuel, Amaldev
Marcovici, Irène
Markakis, Evangelos
Markey, Nicolas
Mayr, Richard
McCann, Mark
McKenzie, Pierre
Meir, Or
Mercas, Robert
Mestre, Julian
Michaliszyn, Jakub
Mihalak, Matus

Mitchell, Joe
Moore, Christopher
Morgenstern, Gila
Morvan, Christophe
Mosses, Peter
Mucha, Marcin
Muscholl, Anca
Naehrig, Michael
Nagaj, Daniel
Nagarajan, Viswanath
Nomikos, Christos
Nordstrom, Jakob
Novotný, Petr
Nöllenburg, Martin
Obdržálek, Jan
Ogihara, Mitsunori
Ong, Luke
Ouaknine, Joel
Oualhadj, Youssouf
Padrubská, Dana
Parys, Paweł
Pasechnik, Dmitrii
Penna, Paolo
Persiano, Giuseppe
Pighizzini, Giovanni
Pilipczuk, Marcin
Pilipczuk, Michal
Pin, Jean-Eric
Pinchasi, Rom
Polák, Libor
Popa, Alex
Pratt-Hartmann, Ian
Praveen, M.
Radoszewski, Jakub
Rahmann, Sven
Ramanujam, R.
Randour, Mickael
Rao, Michael
Reidl, Felix
Reyzin, Lev
Richerby, David
Riordato, Matteo
Rossmanith, Peter
Russo, Alejandro
Rutter, Ignaz

Rytter, Wojciech
Röglin, Heiko
Safernová, Zuzana
Samorodnitsky, Alex
Sanchez Villaamil,
    Fernando
Sanders, Peter
Sangnier, Arnaud
Sankur, Ocan
Sau, Ignasi
Saumell, Maria
Saurabh, Saket
Savickyý, Petr
Sawa, Zdeněk
Schlotter, Ildikó
Schmid, Markus L.
Schmitz, Sylvain
Schönhage, Arnold
Seki, Shinnosuke
Serre, Olivier
Shah, Nisarg
Sitters, René
Skopalik, Alexander
Soulignac, Francisco
Srinivasan, Srikanth
van Stee, Rob
Štefankovič,
    Daniel
Straubing, Howard
Strozecki, Yann
Stückrath, Jan
Sundarararaman,
    Akshay
Suresh, S.P.
Sznajder, Nathalie
Talebanfard, Navid
Tan, Tony
Tancer, Martin
Thielecke, Hayo
Tiedemann, Peter
Toft, Tomas
Tóth, Csaba D.
Tzameret, Iddo
Tzevelekos, Nikos
Uehara, Ryuhei

Upadhyay, Sarvagya
Vagvolgyi, Sandor
Verschae, Jose
Villanger, Yngve
von Falkenhausen,
    Philipp

Vrt'o, Imrich
Waleń, Tomasz
Ward, Justin
Wassermann, Alfred
Weil, Pascal
Wieczorek, Piotr

Yaroslavtsev, Grigory
Yildiz, Hakan
Zeitoun, Marc
Zhou, Chunlai
Živný, Stanislav
Zych, Anna

Table of Contents

Erratum

Alternation Trading Proofs and Their Limitations

Sam Buss⋆

Department of Mathematics
University of California, San Diego
La Jolla, CA 92130-0112, USA
sbuss@math.ucsd.edu

Abstract. Alternation trading proofs are motivated by the goal of separating NP from complexity classes such as LOGSPACE or NL; they have been used to give super-linear runtime bounds for deterministic and co-nondeterministic sublinear space algorithms which solve the Satisfiability problem. For algorithms which use $n^{o(1)}$ space, alternation trading proofs can show that deterministic algorithms for Satisfiability require time greater than n^{cn} for $c < 2\cos(\pi/7)$ (as shown by Williams [21,19]), and that co-nondeterministic algorithms require time greater than n^{cn} for $c < \sqrt[3]{4}$ (as shown by Diehl, van Melkebeek and Williams [5]). It is open whether these values of c are optimal, but Buss and Williams [2] have shown that for deterministic algorithms, $c < 2\cos(\pi/7)$ is the best that can obtained using present-day known techniques of alternation trading.

This talk will survey alternation trading proofs, and discuss the optimality of the unlikely value of $2\cos(\pi/7)$.

Keywords: Satisfiability, alternation trading, indirect diagonalization, lower bounds.

1 Introduction

A central open problem in computer science is the question of whether nondeterministic polynomial time (NP) is more powerful than ostensibly weaker computational classes such as polynomial time (P) or logarithmic space (LOGSPACE). These are famously important and difficult questions, and unfortunately, in spite of over 40 years of concerted efforts to prove that NP $\neq$ P or NP $\neq$ LOGSPACE, it is generally felt that minimal progress has been made on resolving them.

Alternation trading proofs are a method aimed at separating NP from smaller complexity classes, by using "indirect" diagonalization to prove separations. A typical alternation trading proof begins with a simulation assumption, for instance the assumption that the NP-complete problem of Satisfiability (SAT) can be recognized by an algorithm which uses time n^c and space $n^{o(1)}$. Iterated

---

⋆ Supported in part by NSF grant DMS-1101228.

K. Chatterjee and J. Sgall (Eds.): MFCS 2013, LNCS 8087, pp. 1–7, 2013.

application of the simulation assumption allows it to be amplified into an assertion which can be refuted by diagonalization. This yields a proof that the simulation assumption is false.

One of the strongest alternation trading separations known to date is that SAT cannot be recognized by a deterministic algorithm which uses time n^c and space $n^{o(1)}$ for c a constant $< 2\cos(\pi/7) \approx 1.8109$ (see Theorem 8 below). The bound of $2\cos(\pi/7)$ on the runtime exponent might seem unlikely; however, it has recently been shown that this bound on the exponent is *optimal* in the sense that present-day techniques of alternation trading proofs cannot establish any better runtime bound. This is stated as Theorem 10 below, and thus gives an upper bound on the lower bounds that can be achieved with alternation trading proofs — at least using currently known techniques. In short, we provably need better techniques — or better ways to apply known techniques — in order to get improved separation results via alternation trading proofs.

The next section outlines these results in more detail. However, many details of the definitions and proofs are omitted. These details and additional background information can be found in [19,2]. The earlier survey [12] provides an excellent introduction to alternation trading proofs, but does not include the upper bounds on lower bounds of Theorem 10.

2   Definitions and Preliminaries

We adopt the convention that time- and space-bounded algorithms are run on Turing machines with random access tapes, as this permits robust definitions for subquadratic time and sublinear space computational classes. Specifically, Turing machines are assumed to be multitape machines that have random access (indexed) tapes. This means that the Turing machine's tapes come in pairs. Each pair consists of a sequential access tape and a random access tape. The sequential access tape is accessed as usual in the Turing machine model with a tape head that can move at most one tape cell left or right per step. The random access tape is indexed by the sequential access tape, so that the Turing machine has access to the symbol written in the tape cell whose index is written on the sequential access tape. The input string is stored on a read-only random access tape.

Random access Turing machines form a very robust model of computation; for instance, [9] shows their equivalence to more general random access computers up to logarithmic factors on runtime and space.

The *space* used by the Turing machine is the number of cells which are accessed on either kind of tape, except that the contents of the (read-only) input tape do not count towards the space used by the Turing machine. For t a time-constructible function, the complexity classes DTIME(t) and NTIME(t) contain the languages L which can be recognized by deterministic, respectively nondeterministic, algorithms which use time $O(t)$.

We will work primarily with algorithms for Satisfiability that use sublinear space of only $n^{o(1)}$ or $n^{e+o(1)}$ for some constant $e < 1$. Note these sublinear space algorithms do not even have sufficient space to store a single truth assignment for an instance of Satisfiability.

Definition 1. *Let $c, e \geq 0$. The complexity class $\mathrm{DTISP}(n^c, n^e)$ is the set of decision problems L such that L can be recognized by a deterministic algorithm which uses time $n^{c+o(1)}$ and space $n^{e+o(1)}$. The complexity class $\mathrm{NTISP}(n^c, n^e)$ is defined similarly but allowing nondeterministic algorithms instead of deterministic algorithms.*

$\mathrm{DTS}(n^c)$ *is equal to* $\mathrm{DTISP}(n^c, n^0)$. *And* $\mathrm{NTS}(n^c)$ *is* $\mathrm{NTISP}(n^c, n^0)$.

It is a little unusual for the definitions of DTISP and NTISP to include the "$o(1)$" terms in the exponents, but the advantage is that it gives extra $n^{o(1)}$ factors which can absorb polylogarithmic factors in time or space bounds.

The Cook-Levin theorem states that SAT is NP-complete. In fact, SAT is NP-complete in a very strong way. An algorithm is called "quasilinear time" provided it has runtime $n(\log n)^{O(1)}$, and "polylogarithmic time" provided it has runtime $(\log n)^{O(1)}$.

Theorem 2. *Let $L \in \mathrm{NTIME}(n)$. Then there is a quasilinear time many-one reduction f from L to SAT such that there is a polylogarithmic time algorithm, which given x and j, produces the j-th symbol of $f(x)$.*

The point of Theorem 2 is that the computational complexity of SAT is as strong as any language in $\mathrm{NTIME}(n)$. In particular:

Corollary 3. *Fix $c \geq 0$. $\mathrm{NTIME}(t) \subseteq \mathrm{DTS}(n^c)$ if and only if $\mathrm{SAT} \in \mathrm{DTS}(n^c)$.*

Proofs of Theorem 2 and its precursors were given by [14,17,15,3,16,18,7,12]. For the most direct proof of Theorem 2 as stated see [12], which uses much the same methods as [17,16].

Corollary 3 provides the justification for "slowdown" steps in alternation trading proofs. Alternation trading proofs also contain "speedup" steps which allow sublinear space computations to be speeded up, at the cost of introducing alternations. Speedup steps are based on the following theorem which states that runtime can be speeded up by alternation. The theorem is based on techniques independently developed by Bennett [1], Nepomnjaščiĭ [13], and Kannan [10]. We state it only for the special case where the space is $n^{o(1)}$, but it can be generalized to space n^e for constants $e < 1$.

Theorem 4. *Suppose $a > b > 0$ and that $L \in \mathrm{DTS}(n^a)$. Then membership in L can be expressed as*

$$x \in L \iff (\exists y, |y| \leq |x|^{b+o(1)})(\forall z, |z| \leq d \log |x|)(\langle x, g(y, z)\rangle \in L')$$

for some constant $d > 0$, some $L' \in \mathrm{DTS}(n^{a-b})$, and some function $g \in \mathrm{DTS}(n^0)$ such that $|(g(y, z)| = |x|^{o(1)}$.

3   Separation Results with Alternation Trading

The first separation results using alternation trading were established by Kannan [10] and Fortnow [6], who were motivated by problems such as proving that NP is not equal to NL. Theorem 5 states a simplified version of Fortnow's results.

Theorem 5. *Let $\epsilon > 0$. Then* SAT $\notin$ DTISP$(n^1, n^{1-\epsilon})$. *In fact, we have* $\overline{\text{SAT}} \notin$ NTISP$(n^1, n^{1-\epsilon})$. *Consequently,* NTIME$(n) \not\subseteq$ coNTISP$(n^1, n^{1-\epsilon})$.

Fortnow's theorem was quickly extended to better runtime lower bounds. Lipton and Viglas [11] improved the n^1 time bound to n^c for all $c < \sqrt{2}$, but with polylogarithmic space instead of $n^{1-\epsilon}$. Their methods give the following theorem:

Theorem 6. *Let $c < \sqrt{2} \approx 1.414$. Then* SAT $\notin$ DTS(n^c).

This bound was improved by Fortnow and van Melkebeek [8,7] to use $c < \phi$ where $\phi = (1 + \sqrt{5})/2 \approx 1.618$ is the golden ratio.

Theorem 7. *Let $c < \phi$. Then* SAT $\notin$ DTS(n^c).

The bound $c < \phi$ was improved to $c < \sqrt{3} \approx 1.732$ by Williams [20] and to $c < 1.759$ by Diehl and van Melkebeek [4] (the latter result was a more general result about randomized computation). Finally, these bounds were improved by Williams [21,19] to $c < 2\cos(\pi/7) \approx 1.8109$. His theorem applied to a more general setting of modular counting, but for SAT and NTIME(n) his results were:

Theorem 8. *Let $c < 2\cos(\pi/7)$. Then* SAT $\notin$ DTS(n^c).

Corollary 9. *Let $c < 2\cos(\pi/7)$. Then* NTIME$(n) \not\subseteq$ DTS(n^c).

Subsequently to proving Theorem 8, Williams used a computer-based search (coded in Maple) to search for better alternation trading proofs. For this, Williams formulated a precise set of inference rules that allow the derivation of assertions about inclusions between complexity classes. We do not describe the inference rules here, but they can be found in [19,2]. The essential idea is that the inference rules formalize the "slowdown" and "speedup" principles of Corollary 3 and Theorem 4. This computerized search did not lead to any improved alternation trading proofs beyond those already found for Theorem 8.

The somewhat mysterious value $2\cos(\pi/7)$ arises from its being one of the roots of $x^3 - x^2 - 2x + 1 = 0$.

4   Limits on Alternation Trading Proofs

It had long been informally conjectured that alternation trading proofs should be able to establish Theorems 6-8 for all values of $c < 2$. However, as a result of the computerized search, Williams conjectured that the (admittedly unlikely sounding) value $2\cos(\pi/7)$ is the best that can be achieved with his formalized inference rules. This conjecture was recently proved by Buss and Williams [2]:

Theorem 10. *The alternation trading proof inference system, as described in [19,2], can prove that* SAT $\notin$ DTS(n^c) *if and only if $c < 2\cos(\pi/7)$.*

e	c
0.001	1.80083
0.01	1.79092
0.1	1.69618
0.25	1.55242
0.5	1.34070
0.75	1.15765
0.9	1.06011
0.99	1.00583
0.999	1.00058

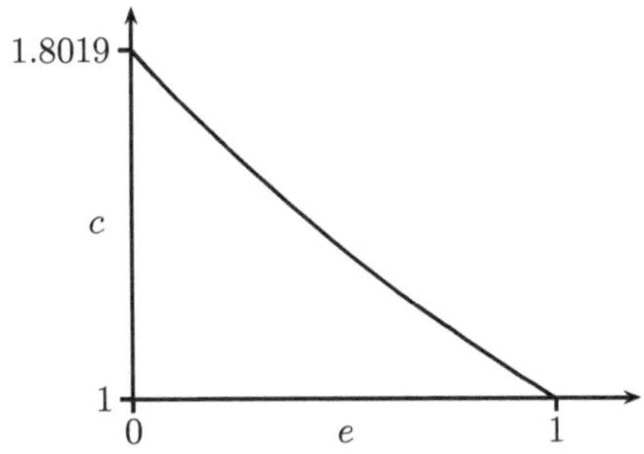

Fig. 1. Showing the maximum value of c, as a function of e, for which alternation trading proofs suffice to show that SAT is not in $DTISP(n^c, n^e)$. The values are accurate to within 10^{-5}. This figure is from [2].

This inference system for alternation trading proofs includes all alternation trading proofs which have been developed so far, and seems to fully capture the power of the Bennett-Nepomnjaščiĭ-Kannan technique of Theorem 4. Thus, Theorem 10 appears to put a meaningful bound on what can be achieved by alternation trading proofs.

Fortnow and van Melkebeek [8] and Williams [19] also used alternation trading proofs to prove results about $NTIME(n) \nsubseteq DTISP(n^c, n^e)$ for values of $c > 1$ and $e < 1$. Already [8] showed that, for any value of $e < 1$, this holds for c sufficiently close to 1; and improved values were given by [19]. The possible values for c and e were further improved, and shown to be optimal by Buss and Williams [2]:

Theorem 11. *The alternation trading proof inference systems described in [19,2] can prove* $SAT \notin DTISP(n^c, n^e)$ *for precisely the values of c and e graphed in Fig. 1.*

Unfortunately, the values shown in Fig. 1 are numerically computed; there is no known formula for describing the values of c and e for which alternation trading proofs exist.

5   Other Directions

So far, we have discussed the question of whether SAT lies in $DTS(n^c)$ or $DTISP(n^c, n^e)$ for constant values of c and e. The alert reader will have noticed that Theorem 5 also discussed whether $\overline{SAT}$ lies in the nondeterministic class $NTISP(n^1, n^{1-\epsilon})$. A number of further such results have been obtained, in particular by [8,7,21,19], culminating in the following theorem proved by Diehl, van Melkebeek, and Williams [5]:

Theorem 12. *Let $c < \sqrt[3]{4}$. Then* $\overline{SAT} \notin NTS(n^c)$. *Consequently,* $NTIME(n) / \subseteq coNTS(n^c)$.

It is tempting to conjecture that the methods of [2] can be extended to prove that the constant $\sqrt[3]{4}$ is optimal for what can be proved with alternation trading proofs. However, to the best of our knowledge, this has not been attempted yet and so it remains an open problem.

References

1. Bennett, J.: On Spectra. Ph.D. thesis, Princeton University (1962)
2. Buss, S., Williams, R.: Limits on alternation-trading proofs for time-space lower bounds, manuscript, submitted for publication. Shorter version appeared in IEEE Conf. on Computational Complexity (CCC), pp. 181–191 (2012)
3. Cook, S.A.: Short propositional formulas represent nondeterministic computations. Information Processing Letters 26, 269–270 (1988)
4. Diehl, S., van Melkebeek, D.: Time-space lower bounds for the polynomial-time hierarchy on randomized machines. SIAM Journal on Computing 36, 563–594 (2006)
5. Diehl, S., van Melkebeek, D., Williams, R.: An improved time-space lower bound for tautologies. Journal of Combinatorial Optimization 22(3), 325–338 (2011), an earlier version appeared in: Ngo, H.Q. (ed.) COCOON 2009. LNCS, vol. 5609, pp. 429–438. Springer, Heidelberg (2009)
6. Fortnow, L.: Nondeterministic polynomial time versus nondeterministic logarithmic space: Time-space tradeoffs for satisfiability. In: Proc. IEEE Conference on Computational Complexity (CCC), pp. 52–60 (1997)
7. Fortnow, L., Lipton, R., van Melkebeek, D., Viglas, A.: Time-space lower bounds for satisfiability. Journal of the ACM 52(6), 835–865 (2005)
8. Fortnow, L., van Melkebeek, D.: Time-space tradeoffs for nondeterministic computation. In: Proc. IEEE Conference on Computational Complexity (CCC), pp. 2–13 (2000)
9. Gurevich, Y., Shelah, S.: Nearly linear time. In: Meyer, A.R., Taitslin, M.A. (eds.) Logic at Botik 1989. LNCS, vol. 363, pp. 108–118. Springer, Heidelberg (1989)
10. Kannan, R.: Towards separating nondeterminism from determinism. Mathematical Systems Theory 17, 29–45 (1984)
11. Lipton, R., Viglas, A.: On the complexity of SAT. In: Proc. 40th Annual IEEE Symposium on Foundations of Computer Science (FOCS), pp. 459–464 (1999)
12. van Melkebeek, D.: Time-space lower bounds for NP-complete problems. In: Current Trends in Theoretical Computer Science, pp. 265–291. World Scientific (2004)
13. Nepomnjaščiǐ, V.A.: Rudimentary predicates and Turing computations. Dokl. Akad. Nauk SSSR 195, 282–284 (1970), English translation in Soviet Math. Dokl. 11, 1462–1465 (1970)
14. Pippenger, N., Fisher, M.J.: Relations among complexity measures. Journal of the ACM 26, 361–381 (1979)
15. Robson, J.M.: A new proof of the NP completeness of satisfiability. In: Proc. 2nd Australian Computer Science Conference, pp. 62–69 (1979)
16. Robson, J.M.: An $O(T \log T)$ reduction from RAM computations to satisfiability. Theoretical Computer Science 81, 141–149 (1991)
17. Schnorr, C.P.: Satisfiability is quasilinear complete in NQL. Journal of the ACM 25, 136–145 (1978)
18. Tourlakis, I.: Time-space tradeoffs for SAT and related problems. Journal of Computer and System Sciences 63(2), 268–287 (2001)

19. Williams, R.: Alternation-trading proofs, linear programming, and lower bounds, to appear. A shorter extended abstract appeared in Proc. 27th Intl. Symp. on Theory of Computings (STACS 2010) (2010), http://stacs-conf.org, doi: 10.4230/LIPIcs.STACS.2010.2494

20. Williams, R.: Algorithms and Resource Requirements for Fundamental Problems. Ph.D. thesis, Carnegie Mellon University (August 2007)

21. Williams, R.: Time-space tradeoffs for counting NP solutions modulo integers. Computational Complexity 17(2), 179–219 (2008)

Bin Packing Games with Selfish Items

Leah Epstein

Department of Mathematics, University of Haifa, 3190501 Haifa, Israel
lea@math.haifa.ac.il

Abstract. We discuss recent work on the subject of selfish bin packing. In these problems, items are packed into bins, such that each item wishes to minimize its own payoff. We survey the known results for a number of variants, focusing on worst-case Nash equilibria and other kinds of equilibria, and mentioning several results regarding issues of complexity and convergence to equilibria.

1 Introduction

We discuss bin packing games with selfish items. In such games, n items are to be packed into (at most) n bins, where each item chooses a bin that it wishes to be packed into. The cost or payoff of an item i of size $0 < s_i \leq 1$ is defined based on its weight $w_i > 0$ and the contents of its bin. Nash equilibria (NE) are defined as solutions where no item can change its choice unilaterally and gain from this change. Bin packing games were inspired by the well-known bin packing problem [49,18,21,17,19,20,65]. In this problem, a set of items, each of size in $(0,1]$, is given. The goal is to partition (or pack) the items into a minimum number of blocks, called bins. Each bin has unit capacity, and the load of a bin is defined to be the total size of items packed into it. That is, the goal is to find a packing of the items into a minimum number of bins, such that the load of each bin is at most 1. The problem is NP-hard in the strong sense, and thus theoretical research has concentrated on the study and development of approximation algorithms, which allow to design nearly optimal solutions. A bin packing algorithm is called *online* if it receives the items one by one, and it must assign each item to a bin immediately and irrevocably without any information on subsequent items. If the input is given as a set, then the problem is called *offline*, in which case an algorithm is typically expected to run in polynomial time. We define the measures for quality of equilibria seeing the suitable solutions as approximation algorithms, possibly resulting from a process of local search.

The approximation ratio. Consider a minimization problem P with a set of instances X. Given a set of solutions $\mathcal{S}$, containing one solution for each $\sigma \in X$, whose cost is denoted by $\mathcal{S}(\sigma)$, assume that $\mathcal{S}(\sigma)$ is a positive integer for any $\sigma \in X$. The set $\mathcal{S}$ can be a set of outputs of a given algorithm, or another class of solutions with specific properties, such as various kinds of equilibria, outputs of an algorithm (or a class of algorithms), etc. Let $OPT(\sigma)$ denote the minimum

K. Chatterjee and J. Sgall (Eds.): MFCS 2013, LNCS 8087, pp. 8–21, 2013.
© Springer-Verlag Berlin Heidelberg 2013

cost for the specific input σ, that is, the cost of an optimal algorithm for this input. We define the (asymptotic) **approximation ratio** of S to be:

$$R(S) = \limsup_{M \to \infty} \max_{\sigma} \{S(\sigma)/OPT(\sigma) \mid OPT(\sigma) = M\}.$$

We use the term (asymptotic) approximation ratio for any type of set of solutions S. Note that in the literature, the term (asymptotic) competitive ratio is usually used for online algorithms and has the same meaning. In what follows we mostly deal with asymptotic approximation ratios and usually omit the word *asymptotic*. The absolute approximation of S is $R_{abs}(S) = \sup_{\sigma}\{S(\sigma)/OPT(\sigma)\}$, and the approximation ratio for the input σ is simply $S(\sigma)/OPT(\sigma)$.

Games.   A game (or a *strategic game*) consists of a finite set of (at least two) players, and a finite, non-empty, set of strategies (or actions) that the set of players can perform. Each player has to choose a strategy, and it has a payoff associated with each one of the possible situations or outcomes (sets of strategies of all players, containing one strategy for each player). Each outcome has a social cost associated with it. A *Nash equilibrium* (an NE), introduced by Nash [62], is a famous solution concept of a game, where no player can gain anything (that is, decrease its payoff) by changing only its own strategy unilaterally. If each player has chosen a strategy, and no player can benefit by changing its strategy while the other players keep their unchanged, then the current set of strategy choices and the corresponding payoffs constitute an NE. If a player chooses to take one action with probability 1, then that player is playing a pure strategy, and otherwise it is playing a mixed strategy. If all players play pure strategies, then the resulting NE is called a pure NE. In what follows we focus on pure NE, and use the term NE for a pure NE.

Measures.   The price of anarchy (PoA) [53] of a game G is the ratio between the maximum social cost of any NE, and the minimum social cost of any solution (with a minimal number of non-empty bins). The price of stability (PoS) [3] is defined analogously, taking into account the NE of the minimum social cost. Note that in order to use the definitions of PoA and PoS, one obviously has to prove first that G admits an NE. Using the definitions above, we can consider a set of the worst NE and a set of the best NE (each containing one solution for each input) as our sets of solutions, and then the PoA and PoS of G are the approximation ratio for it according to the first set and second set, respectively. A bin packing problem, where for each input and packing every item has a payoff associated with it, can be seen as a class of games (details are given below). If every such game admits an NE, then the PoA of the problem (respectively, PoS) is defined as the asymptotic supremum PoA (respectively, PoS) over all games in this class. That is, each one of the measures is the approximation ratio of a class of solutions: the PoA is the approximation ratio of the set of worst NE, and the PoS is the approximation ratio of the set of best NE.

Bin packing games.   We now define classes of games based on bin packing problems, called bin packing games. Every input of standard bin packing induces a standard bin packing game, and all such games are the class of standard bin packing games. Moreover, we can define other classes of bin packing games that are based on other variants of bin packing. In such games, each item corresponds to a selfish player, trying to minimize its own cost, and the strategy of a player is the bin in which it is packed. Changing the strategy of an item means that it moves to be packed in a different bin. Such a deviation is possible only if the empty space in the targeted bin is sufficient, and additional conditions on the packing (if any) are satisfied (in the standard bin packing problem there are no additional conditions). One example of such a variant is bin packing with cardinality constraints [54,55,51,12,26,32,6], introduced by Krause, Shen, and Schwetman [54]. In this problem, in addition to the usual constraint that the total size of items packed into one bin cannot exceed 1, no bin can contain more than k items for a given integer parameter $k \geq 2$. In the corresponding games, an item can move to a bin that has sufficient space for it, and has at most $k - 1$ items. In particular, for both these classes of games, an item can always move to an empty bin.

Recall that every player i corresponding to an item (or simply the item i, since we do not distinguish the player from its item) has a positive weight w_i. We define the cost (or payoff) of an item as follows. If an item of weight v is packed into a bin where the total weight of packed items is η (including this item), then its cost is $\frac{v}{\eta}$. The total payoff for a subset of items packed into one bin is exactly 1. The set of strategies consists of all n bins, and thus we next define the cost of an item that is packed in an invalid way. If an item is packed into a bin whose packing is invalid, then we define its payoff as infinity. Recall that an NE is a packing where no item can benefit from changing its strategy. It is usually assumed that all bins of any considered packing are packed in a valid way, due to the following. A socially optimal solution must be a valid solution of the optimization problem, so all its bins must be packed in a valid way. An NE cannot contain invalid bins either, as an item having an infinite cost would benefit from moving into an empty bin (which must exist, if some bin contains more than one item). Additionally, there is no need to consider the option of an item migrating to a bin that becomes invalid as a result, since this can never decrease the cost of the item. We are interested in the case of arbitrary or general positive weights, but also in the special cases of unit weights (which is equivalent to the case of equal weights), and it proportional weights (where for every item i, we assume that $w_i = s_i$). An NE packing is not necessarily optimal. Consider for example proportional weights or unit weights. Take three items of size 0.51 and three items of size 0.26. The solution that packs the three smallest items in one bin and the larger items in dedicated bins is an NE with respect to both kinds of weights, but an optimal solution packs three bins with one item of each size in each bin (see Figure 1).

This survey. We are interested in the price of anarchy for classes of bin packing games. Such studies originate in the inefficiency of large networks such as the

Internet, where the users act selfishly, and often even in an uncoordinated way. These studies were first done for scheduling and routing [53,64,63,22,58,39] (in some articles, the price of anarchy is called "coordination ratio", and in particular, this is the case in the seminal paper of Koutsoupias and Papadimitriou [53]), but there are applications where the bin packing models are more appropriate [27,1]. Bin packing models are useful for scenarios where resources cannot have arbitrary loads, and can be occupied only up to a certain level. In the case of cardinality constraints, a resource cannot be used by an unlimited number of users. We discuss standard bin packing games and classes of games corresponding to variants of bin packing. The specific game class depends on the types of items, the allowed weights, the calculation of payoffs, and the definitions of valid bins. We also discuss other kinds of equilibria, issues of complexity, and convergence to equilibria.

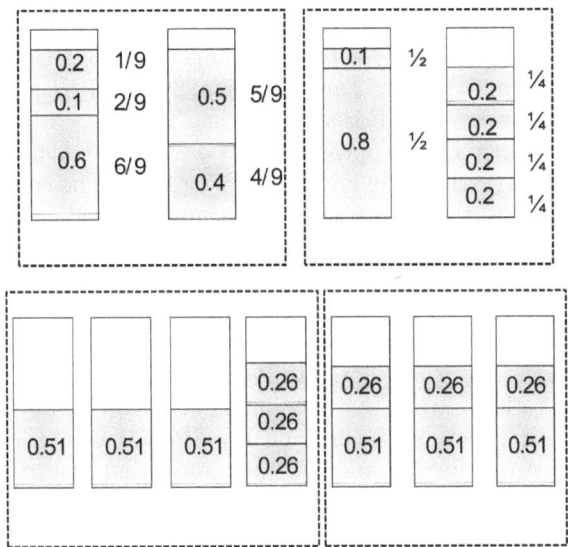

Fig. 1. The top figures are examples of solutions for two games, where the packing on the left hand side is an NE packing for unit weights, but not for proportional weights, and the packing next to it is an NE for proportional weights but not for unit weights. The costs of items are written next to them, where the costs for the figure on the left hand side are for the case of proportional weights, and on the right hand side, for unit weights. In the figure of the left hand size, if the item of size 0.1 migrates to the other bin, its cost will be 0.1. In the figure on the right hand side, if an item this size migrates to the other bin, its cost will become 0.2. The bottom figures are two solutions for one input. The packing on the right hand side is an optimal packing and an NE for any set of weights. It is not, however, an SNE for unit weights and for proportional weights. The packing next to it is an SNE for both these kinds of weights. If we change the size of larger items to 0.53, the packing on the left hand side will no longer be an NE for proportional weights, but it remains an SNE for unit weights.

2   A Discussion on Equilibria for Bin Packing Games

Stronger concepts of stable solutions were defined in order to separate the effect of selfishness from the effect of lack of coordination. A strong equilibrium (an SNE) [4,67,45,2,27,30,39] is a solution concept where not only a single player cannot benefit from changing its strategy but no non-empty subset of players can form a coalition, where a coalition means that a subset of players change their strategies simultaneously, all gaining from the change. Obviously, by this definition, an SNE is an NE. The grand coalition is defined to be a coalition composed of the entire set of players. A solution is called *weakly Pareto optimal* if there is no alternative solution to which the grand coalition can deviate simultaneously and every player benefits from it. A solution is called *strictly Pareto optimal* if there is not alternative solution to which the grand coalition can deviate simultaneously, such that at least one player benefits from it, and no player has a larger cost as a result. The last two concepts are borrowed from welfare economics. The two requirements, that a solution is both (strictly or weakly) Pareto optimal and an NE results in two additional kinds of NE, Strictly Pareto optimal NE (SPNE) and Weakly Pareto optimal NE (WPNE) [25,16,28,5,23]. By these definitions, every WPNE is an NE, every SPNE is a WPNE, and every SNE is a WPNE. Strictly Pareto optimal points are of particular interest in economics, as stated in a textbook, in a chapter by by Luc: "The concept of Pareto optimality originated in the economics equilibrium and welfare theories at the beginning of the past century. The main idea of this concept is that society is enjoying a maximum ophelimity when no one can be made better off without making someone else worse off" [57]. Even though these concepts are stronger than NE, still for many problems a solution which is an SNE, an SPNE, or a WPNE is not necessarily socially optimal. We define the strong price of anarchy (SPoA) [2] as the ratio between the maximum social cost of any SNE and the minimum social cost of any situation, and the SPoS is the ratio between the minimum social cost of any SNE and the minimum social cost of any situation, that is, the SPoA is the approximation ratio of the set of worst SNE, and the SPoS is the approximation ratio of the set of best SNE. For SPNE and WPNE we define the SPPoA and SPPoS, and the WPPoA and WPPoS analogously. See Figure 1 and Figure 2 for some examples for the different concepts.

There are many ways of showing that NE exist for various bin packing games. One way to show that every game has an NE is by showing that all games of a given class of bin packing games belong to a specific class of (weighted) singleton congestion games [45,46], such that it is known that every game in the class has a (pure) NE. Another method is by defining a process that converges into an NE. More accurately, given a packing that is not necessarily an NE (it can be any packing, in particular it can be a packing where each item is packed into a separate bin, or a packing with a relatively small number of bins such as a socially optimal packing), define a process where at each time, one item can migrate to another bin, where its cost will be strictly smaller (the costs of other items packed into the target bin may increase as a result). Such processes must converge since they can reach every possible packing at most once, but they usually require exponential time (see [9,59,60]). Note that games with unit weights are in fact singleton congestion games for which this process converges in polynomial time [45,46], but the number of resources has an exponential size in the number of players, and it is not given explicitly (these are all possible subsets of items that can be packed into a bin). It is known, however, that for standard bin packing games with unit weights, these processes still take polynomial time [44,23]. In [44], using a potential function, it is shown that the process converges using $O(n^2)$ migration steps. In [23], a tight bound of $\Theta(n^{3/2})$ on the worst-case number of migration steps is shown using an additional potential function. The advantage of this approach is that it is possible to use an approximate solution as an initial configuration. For example, it is possible to apply an asymptotic approximation scheme by Fernandez de la Vega and Lueker [38] or a fully polynomial approximation scheme by Karmarkar and Karp [50] as suggested by Han et al. [44], and since no item can benefit from moving into an empty bin, the output cannot contain a larger number of bins that the initial configuration, so the output has both properties that it is an NE, and it is an approximate solution of the same quality as the initial packing. Since the initial configuration can potentially be an optimal one, we find that for any set of weights, the PoS is equal to 1. Another way of showing that an NE exists for every game is to show the stronger property that every game has an SNE. The algorithm *Greedy Set Cover (GSC)* creates such equilibria [27,23], and the set of

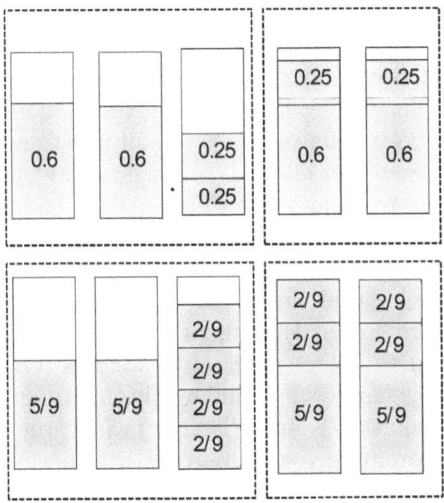

Fig. 2. The two top figures are two solutions for one input. The one on the left hand side is not an NE for proportional weights, but it is an NE for unit weights. For unit weights, it is not an SPNE, since in the packing on the right hand side no item has a larger cost, but the large items have smaller costs. The two bottom figures are two solutions of one input, where the one on the left hand side is an NE for proportional weights and for unit weights, but it not an NE, for example, if the weights of smaller items are equal to 1, and the weights of larger items are equal to 4. For proportional weights, it is not a WPNE, since in the packing on the right hand side every item has a smaller cost (for unit weights there is no such example as every NE is a WPNE).

executions of this algorithm gives exactly the set of SNE. The algorithm acts as follows. Given an input set of items, it repeatedly find a maximum weight subset of unpacked items that can be packed into a bin, packs it, and removes it from the set of unpacked items. This algorithm requires exponential time for arbitrary weights and for proportional weights (because it has to solve a knapsack problem before it packs a bin, and we cannot expect it to run in polynomial time for proportional weight unless P=NP, as otherwise it would solve the 3-PARTITION problem in polynomial time), but using the properties of its output the existence of an SNE (and thus also the existence of an NE) follows.

For proportional weights, this algorithm is actually the algorithm *Subset Sum* [43,13,14,27,30], and for unit weights, a special case of GSC is the greedy algorithm NEXT FIT INCREASING (NFI), which sorts the items by non-decreasing size and applies NEXT FIT (NF) [47,48], that is, uses one active bin at a time, and replaces it with a new active bin if an item does not fit. The running time is $O(n \log n)$. The running time is lower than that of the process described above, and the resulting packing is not only an NE but also an SNE. Interestingly, it turns out that NFI always creates the worst SNE (roughly speaking, the reason for this property is that it always selects the smallest items and thus packs large numbers of items in the few first bins). If an NE of better quality is needed, it is possible to apply a fast heuristic [47,49,42,66] first, and then apply a process that converges to an NE with at most the same number of bins.

For any set of weights, it is strongly NP-hard to find the best SNE, and it is also NP-hard to find the best NE, which can be proved using a simple reduction from 3-PARTITION. However, for the case of proportional weights, finding an arbitrary NE can be done using a polynomial time algorithm that is based on applying FFD [47,49] (see definition below) multiple times in a certain way, as Yu and Zhang [71] showed. For all kinds of weights, any optimal solution is both weakly and strictly Pareto optimal. This property holds since the total cost of all items is equal to the number of non-empty bins, and if in an alternative packing no item increases its cost while some item reduces its cost, we find that the number of bins in the alternative packing is smaller, contradicting optimality. Thus, since there exists a socially optimal solution that is an NE, the SPPoS and WPPoS are equal to 1 for every set of weights. On the other hand, for some games (including games with unit weights and games with proportional weights), no optimal solution is an SNE, and in fact, the SPoS is much higher than 1. Consider the example given above (consisting of six items, also illustrated in Figure 1). The optimal solution described there is unique (up to swapping the positions of identical items), but it is not an SNE packing, as the three smallest items would migrate together to an empty bin. The alternative solution described above is an SNE. Note, however, that not *every* optimal solution is an NE, for example, for an input consisting of four items, two of sizes 0.6 and two of sizes 0.2, a solution with two bins containing each one item of each size is optimal, but both for proportional weights and for unit weights, a small item can benefit from migrating to the other bin.

3   Standard Bin Packing Games

In this section, we survey the known bounds on the measures defined above. Some of the tight bounds are the approximation ratios of well-known algorithms, but some bounds were not found in the past as approximation ratios of algorithms.

General Weights. The algorithm First Fit (FF) for bin packing processes a list of items, and packs each item into the bin of smallest index where it can be packed. The variant First Fit Decreasing (FFD) sorts the items by non-increasing size and applies FF. Every NE can be obtained as an output of FF; sort the bins by non-increasing total weights, and create a list of items according to the ordering of bins. FF will create exactly the bins of the original packing. FF will actually act as NF since no item can be packed into earlier bins (as this would mean that the item benefits from moving there in the original packing, which is an NE packing). Such a relation holds also between outputs of the algorithm Subset Sum and of FF (that every output of Subset Sum can be achieved also by FF) [13,14], and it holds for other classes of bin packing games.

Interestingly, for arbitrary weights, all inefficiency measures (PoA, SPoA, WP-PoA, SPPoA), and even the SPoS are all equal to 1.7 [23], which is both the asymptotic and absolute approximation ratio of First Fit [68,49,41,24]. As mentioned above, the PoS, WPPoS, and SPPoS are equal to 1. Additional properties are revealed in [23], for example, the WPPoA is equal to the PoA for any class of weights.

Proportional Weights. The bin packing game with proportional weights $(w_i = s_i)$ was introduced by Bilò [9], who was the first to study the bin packing problem from this type of game theoretic perspective. He provided the first bounds on the PoA, a lower bound of $\frac{8}{5}$ and an upper bound of $\frac{5}{3}$. The quality of NE solutions was further investigated in [27], where nearly tight bounds for the PoA were given; an upper bound of 1.6428 and a lower bound of 1.6416 (see also [71]). Interestingly, the PoA is not equal to the approximation ratio of any natural algorithm for bin packing. The SPoA and SPoS were also analyzed in [27], and it was shown that these two measures are equal. Moreover, it was shown the set of SNE and outputs of the Subset Sum algorithm [43,13] is the same, which gave bounds on the SPoA and SPoS. In the paper [30], the exact SPoA (which is also the approximation ratio of Subset Sum) was determined, and it was shown that its value is approximately 1.6067. In the same article, the parametric problem where the size of every item is upper bounded by a parameter is studied. Some properties of other measures that were not studied in [9,27,30] (the ones related to Pareto optimal solutions) are mentioned in [23].

Unit Weights. The case of unit weights was introduced and studied by Han et al. [44]. The authors showed that NFI creates an NE for every input. The approximation ratio of NFI is known to be a sum of a series that is equal approximately to 1.69103 [40,7]. In fact, this last value is the approximation ratio of a number of algorithms, it is the approximation ratio of NFI and of NFD (which sorts the items by non-increasing size and applies NF), and it is the limit of the sequence of approximation ratios of a class of online algorithms, called the HARMONIC ALGORITHMS [56], which partition items into classes according to size and pack each class independently using NF. It is also mentioned there that since NE packing is the output of a run of FF, the PoA is at most 1.7. An example is provided where an optimal solution uses 10 bins, while an NE solution uses 17 bins. The problem was studied further in [23], and it turns out that the PoA, WPPoA, and SPPoA are strictly below 1.7 (unlike the case of general weights). The PoA is however at least 1.6966, so it is very close to 1.7. The SPoA is equal to the value 1.69103 that is mentioned above, since NFI creates the worst SNE. However, unlike the case of proportional weights, the SPoS is lower, and its value is approximately 1.611824, a new number in bin packing that is also the sum of a series. The SPPoA is in [1.61678, 1.628113] (and the WPPoA is equal to the PoA, and in fact, in this case every NE is a WPNE).

4   Games for Variants of Bin Packing

In this section we discuss several variants that have been studied. There are additional variants that we do not discuss, since they are very different from standard bin packing games. Two examples are selfish bin covering [10] and selfish bin coloring [31]. Note that the term "bin packing games" is used in the literature for a completely different type of games [35,36,52], and there is recent interest in those games as well.

Bin Packing Game with Cardinality Constraints. These games were studied with respect to the PoA and PoS for the case of proportional weights [1]. By applying GSC (that is, at each step, packing a set of maximum total size that consists of at most k items) it is possible to show that every such game has an SNE. For this class of games, the PoA and PoS were studied for the subclasses of games with fixed k, as functions of k. The PoS is equal to 1 here as well, and a complete analysis with tight bounds of the PoA as a function of k was given. For $k = 2$ any NE is a socially optimal solution (thus the PoA is equal to 1, but not every optimal solution is an NE). For $k = 3$ the PoA is equal to $\frac{11}{7}$, and for any $k \geq 4$ it is equal to $2 - \frac{1}{k}$. The overall PoA (over all values of k) is exactly 2. Interestingly, this property does not depend on the allowed item sizes, even if items are restricted to be smaller than some value, still the overall PoA is 2. For standard bin packing, this is not the case (already the approximation ratio of FF tends to 1 when item sizes are very small) [49,30]. Additionally, one may expect that cardinality constraints would have a different effect on the PoA. Intuitively, it seems that very large values of k should have the same effect as no cardinality constraints, but comparing these results to the results of [27] it can be seen that this is not the case. Another possible expected behavior could be an increase of the approximation ratio by 1 for large k as in [54,26], but this is not the case here either.

Multidimensional Generalizations. There are several kinds of multidimensional bin packing problems. The dimension is usually denoted by $d \geq 2$, and obviously bounds depend on this parameter. In geometric packing [34,11,8], bins are multidimensional cubes (or squares, if $d = 2$), and the items are boxes (or rectangles, if $d = 2$). The items are to be packed in a non-overlapping manner into the bins, such that the sides of the items are parallel to the sides of the bin. In oriented packing, items cannot be rotated, while non-oriented packing is the variant where an item can be rotated arbitrarily. There are variants where items can be rotated only in specific directions [61,33]. The special case (of all these variants) where items are cubes or squares is of particular interest. In vector packing [69,15,8], bins are d-dimensional all-1 vectors, while items are non-zero vectors of the same dimension whose components are in $[0, 1]$. The last problem with $d \geq 2$ generalizes bin packing with cardinality constraints (see [12]). Payoffs of geometric packing games as well as vector packing games are defined similarly to the one-dimensional one, and the proportional weights are defined according to area or volume, where the volume of a d-dimensional box is its actual volume, and the volume of a vector is the sum of its components. GSC can be applied on inputs of these problems as well, showing that such games always have SNE (but papers that studied such games usually prove directly that every such game has at least one NE).

Two-dimensional geometric bin packing games were considered in [37]. It is shown that the PoA for rectangle packing games is unbounded, while the PoA for square packing games is constant. Ye and Chen [70] studied d-dimensional vector packing games. They showed that the PoA for this class of games is $\Theta(d)$. The upper bound shown by [70] once again follows from the relation to FF. Improved bounds were given in [29]. In particular, it is proved in [29] that the PoA for d dimensions exceeds the PoA for one dimension by at least $d - 1$, and thus it is at least $d + 0.6416$.

5 Summary

We surveyed the state-of-the-art for bin packing games with selfish items. Obviously, as there are many variants of bin packing, many other classes of games can be studied. It is also interesting to design polynomial time algorithms for computing NE for those variants for which such algorithms are not known, and to analyze additional kinds of equilibria.

References

1. Adar, R., Epstein, L.: Selfish bin packing with cardinality constraints. Theoretical Computer Science (to appear, 2013)
2. Andelman, N., Feldman, M., Mansour, Y.: Strong price of anarchy. Games and Economic Behavior 65(2), 289–317 (2009)
3. Anshelevich, E., Dasgupta, A., Kleinberg, J.M., Tardos, É., Wexler, T., Roughgarden, T.: The price of stability for network design with fair cost allocation. SIAM Journal on Computing 38(4), 1602–1623 (2008)

4. Aumann, R.J.: Acceptable points in general cooperative n-person games. In: Tucker, A.W., Luce, R.D. (eds.) Contributions to the Theory of Games IV. Annals of Mathematics Study, vol. 40, pp. 287–324. Princeton University Press (1959)

5. Aumann, Y., Dombb, Y.: Pareto efficiency and approximate pareto efficiency in routing and load balancing games. In: Kontogiannis, S., Koutsoupias, E., Spirakis, P.G. (eds.) SAGT 2010. LNCS, vol. 6386, pp. 66–77. Springer, Heidelberg (2010)

6. Babel, L., Chen, B., Kellerer, H., Kotov, V.: Algorithms for on-line bin-packing problems with cardinality constraints. Discrete Applied Mathematics 143(1-3), 238–251 (2004)

7. Baker, B.S., Coffman Jr., E.G.: A tight asymptotic bound for Next-Fit-Decreasing bin-packing. SIAM Journal on Algebraic and Discrete Methods 2(2), 147–152 (1981)

8. Bansal, N., Caprara, A., Sviridenko, M.: A new approximation method for set covering problems, with applications to multidimensional bin packing. SIAM Journal on Computing 39(4), 1256–1278 (2009)

9. Bilò, V.: On the packing of selfish items. In: Proc. of the 20th International Parallel and Distributed Processing Symposium (IPDPS 2006), 9 p. IEEE (2006)

10. Cao, Z., Yang, X.: Selfish bin covering. Theoretical Computer Science 412(50), 7049–7058 (2011)

11. Caprara, A.: Packing d-dimensional bins in d stages. Mathematics of Operations Research 33(1), 203–215 (2008)

12. Caprara, A., Kellerer, H., Pferschy, U.: Approximation schemes for ordered vector packing problems. Naval Research Logistics 50(1), 58–69 (2003)

13. Caprara, A., Pferschy, U.: Worst-case analysis of the subset sum algorithm for bin packing. Operetions Research Letters 32(2), 159–166 (2004)

14. Caprara, A., Pferschy, U.: Modified subset sum heuristics for bin packing. Information Processing Letters 96(1), 18–23 (2005)

15. Chekuri, C., Khanna, S.: On multidimensional packing problems. SIAM Journal on Computing 33(4), 837–851 (2004)

16. Chien, S., Sinclair, A.: Strong and Pareto price of anarchy in congestion games. In: Albers, S., Marchetti-Spaccamela, A., Matias, Y., Nikoletseas, S., Thomas, W. (eds.) ICALP 2009, Part I. LNCS, vol. 5555, pp. 279–291. Springer, Heidelberg (2009)

17. Coffman Jr., E.G., Csirik, J.: Performance guarantees for one-dimensional bin packing. In: Gonzalez, T.F. (ed.) Handbook of Approximation Algorithms and Metaheuristics, ch. 32, 18 p. Chapman & Hall/Crc (2007)

18. Coffman Jr., E.G., Garey, M.R., Johnson, D.S.: Approximation algorithms for bin packing: A survey. In: Hochbaum, D. (ed.) Approximation Algorithms. PWS Publishing Company (1997)

19. Csirik, J., Leung, J.Y.-T.: Variants of classical one-dimensional bin packing. In: Gonzalez, T.F. (ed.) Handbook of Approximation Algorithms and Metaheuristics, ch. 33, 13 p. Chapman & Hall/Crc (2007)

20. Csirik, J., Leung, J.Y.-T.: Variable-sized bin packing and bin covering. In: Gonzalez, T.F. (ed.) Handbook of Approximation Algorithms and Metaheuristics, ch. 34, 11 p. Chapman & Hall/Crc (2007)

21. Csirik, J., Woeginger, G.J.: On-line packing and covering problems. In: Fiat, A., Woeginger, G.J. (eds.) Online Algorithms: The State of the Art, ch. 7, pp. 147–177. Springer (1998)

22. Czumaj, A., Vöcking, B.: Tight bounds for worst-case equilibria. ACM Transactions on Algorithms 3(1) (2007)

23. Dósa, G., Epstein, L.: Generalized selfish bin packing. CoRR, abs/1202.4080 (2012)
24. Dósa, G., Sgall, J.: First Fit bin packing: A tight analysis. In: Proc. of 30th International Symposium on Theoretical Aspects of Computer Science (STACS 2013), pp. 538–549 (2013)
25. Dubey, P.: Inefficiency of Nash equilibria. Mathematics of Operations Research 11(1), 1–8 (1986)
26. Epstein, L.: Online bin packing with cardinality constraints. SIAM Journal on Discrete Mathematics 20(4), 1015–1030 (2006)
27. Epstein, L., Kleiman, E.: Selfish bin packing. Algorithmica 60(2), 368–394 (2011)
28. Epstein, L., Kleiman, E.: On the quality and complexity of Pareto equilibria in the job scheduling game. In: Proc. of the 10th International Conference on Autonomous Agents and Multiagent Systems (AAMAS 2011), pp. 525–532 (2011)
29. Epstein, L., Kleiman, E.: Vector packing and vector scheduling with selfish jobs and items. Work in progress (2013)
30. Epstein, L., Kleiman, E., Mestre, J.: Parametric packing of selfish items and the subset sum algorithm. In: Leonardi, S. (ed.) WINE 2009. LNCS, vol. 5929, pp. 67–78. Springer, Heidelberg (2009)
31. Epstein, L., Krumke, S.O., Levin, A., Sperber, H.: Selfish bin coloring. Journal of Combinatorial Optimization 22(4), 531–548 (2011)
32. Epstein, L., Levin, A.: AFPTAS results for common variants of bin packing: A new method for handling the small items. SIAM Journal on Optimization 20(6), 3121–3145 (2010)
33. Epstein, L., van Stee, R.: This side up! ACM Transactions on Algorithms 2(2), 228–243 (2006)
34. Epstein, L., van Stee, R.: Multidimensional packing problems. In: Gonzalez, T.F. (ed.) Handbook of Approximation Algorithms and Metaheuristics, ch. 35, 15 p. Chapman & Hall/Crc (2007)
35. Faigle, U., Kern, W.: On some approximately balanced combinatorial cooperative games. Mathematical Methods of Operations Research 38(2), 141–152 (1993)
36. Faigle, U., Kern, W.: Approximate core allocation for binpacking games. SIAM Journal on Discrete Mathematics 11(3), 387–399 (1998)
37. Fernandes, C.G., Ferreira, C.E., Miyazawa, F.K., Wakabayashi, Y.: Selfish square packing. Electronic Notes in Discrete Mathematics 37, 369–374 (2011)
38. Fernandez de la Vega, W., Lueker, G.S.: Bin packing can be solved within $1 + \varepsilon$ in linear time. Combinatorica 1(4), 349–355 (1981)
39. Fiat, A., Kaplan, H., Levy, M., Olonetsky, S.: Strong price of anarchy for machine load balancing. In: Arge, L., Cachin, C., Jurdziński, T., Tarlecki, A. (eds.) ICALP 2007. LNCS, vol. 4596, pp. 583–594. Springer, Heidelberg (2007)
40. Fisher, D.C.: Next-fit packs a list and its reverse into the same number of bins. Operations Research Letters 7(6), 291–293 (1988)
41. Garey, M.R., Graham, R.L., Johnson, D.S., Yao, A.C.-C.: Resource constrained scheduling as generalized bin packing. Journal of Combinatorial Theory Series A 21(3), 257–298 (1976)
42. Garey, M.R., Johnson, D.S.: A 71/60 theorem for bin packing. Journal of Complexity 1(1), 65–106 (1985)
43. Graham, R.L.: Bounds on multiprocessing anomalies and related packing algorithms. In: Proceedings of the 1972 Spring Joint Computer Conference, pp. 205–217 (1972)
44. Han, X., Dósa, G., Ting, H.-F., Ye, D., Zhang, Y.: A note on a selfish bin packing problem. Journal of Global Optimization (2012)

45. Holzman, R., Law-Yone, N.: Strong equilibrium in congestion games. Games and Economic Behavior 21(1-2), 85–101 (1997)
46. Ieong, S., McGrew, B., Nudelman, E., Shoham, Y., Sun, Q.: Fast and compact: A simple class of congestion games. In: Proceedings of the 20th National Conference on Artificial Intelligence, pp. 489–494 (2005)
47. Johnson, D.S.: Near-optimal bin packing algorithms. PhD thesis. MIT, Cambridge (1973)
48. Johnson, D.S.: Fast algorithms for bin packing. Journal of Computer and System Sciences 8(3), 272–314 (1974)
49. Johnson, D.S., Demers, A.J., Ullman, J.D., Garey, M.R., Graham, R.L.: Worst-case performance bounds for simple one-dimensional packing algorithms. SIAM Journal on Computing 3(4), 299–325 (1974)
50. Karmarkar, N., Karp, R.M.: An efficient approximation scheme for the one-dimensional bin-packing problem. In: Proceedings of the 23rd Annual Symposium on Foundations of Computer Science (FOCS 1982), pp. 312–320 (1982)
51. Kellerer, H., Pferschy, U.: Cardinality constrained bin-packing problems. Annals of Operations Research 92(1), 335–348 (1999)
52. Kern, W., Qiu, X.: Integrality gap analysis for bin packing games. Operations Research Letters 40(5), 360–363 (2012)
53. Koutsoupias, E., Papadimitriou, C.: Worst-case equilibria. In: Meinel, C., Tison, S. (eds.) STACS 1999. LNCS, vol. 1563, pp. 404–413. Springer, Heidelberg (1999)
54. Krause, K.L., Shen, V.Y., Schwetman, H.D.: Analysis of several task-scheduling algorithms for a model of multiprogramming computer systems. Journal of the ACM 22(4), 522–550 (1975)
55. Krause, K.L., Shen, V.Y., Schwetman, H.D.: Errata: "Analysis of several task-scheduling algorithms for a model of multiprogramming computer systems". Journal of the ACM 24(3), 527–527 (1977)
56. Lee, C.C., Lee, D.T.: A simple online bin packing algorithm. Journal of the ACM 32(3), 562–572 (1985)
57. Luc, D.T.: Pareto optimality. In: Chinchuluun, A., Pardalos, P.M., Migdalas, A., Pitsoulis, L. (eds.) Pareto Optimality, Game Theory and Equilibria, pp. 481–515. Springer (2008)
58. Mavronicolas, M., Spirakis, P.G.: The price of selfish routing. Algorithmica 48(1), 91–126 (2007)
59. Miyazawa, F.K., Vignatti, A.L.: Convergence time to Nash equilibrium in selfish bin packing. Electronic Notes in Discrete Mathematics 35, 151–156 (2009)
60. Miyazawa, F.K., Vignatti, A.L.: Bounds on the convergence time of distributed selfish bin packing. International Journal of Foundations of Computer Science 22(3), 565–582 (2011)
61. Miyazawa, F.K., Wakabayashi, Y.: Approximation algorithms for the orthogonal z-oriented threedimensional packing problem. SIAM Journal on Computing 29(3), 1008–1029 (1999)
62. Nash, J.: Non-cooperative games. Annals of Mathematics 54(2), 286–295 (1951)
63. Roughgarden, T.: Selfish routing and the price of anarchy. MIT Press (2005)
64. Roughgarden, T., Tardos, É.: How bad is selfish routing? Journal of the ACM 49(2), 236–259 (2002)
65. Seiden, S.S.: On the online bin packing problem. Journal of the ACM 49(5), 640–671 (2002)
66. Simchi-Levi, D.: New worst-case results for the bin-packing problem. Naval Research Logistics 41(4), 579–585 (1994)

67. Rozenfeld, O., Tennenholtz, M.: Strong and correlated strong equilibria in monotone congestion games. In: Spirakis, P.G., Mavronicolas, M., Kontogiannis, S.C. (eds.) WINE 2006. LNCS, vol. 4286, pp. 74–86. Springer, Heidelberg (2006)
68. Ullman, J.D.: The performance of a memory allocation algorithm. Technical Report 100, Princeton University, Princeton, NJ (1971)
69. Woeginger, G.J.: There is no asymptotic PTAS for for two-dimensional vector packing. Information Processing Letters 64(6), 293–297 (1997)
70. Ye, D., Chen, J.: Non-cooperative games on multidimensional resource allocation. Future Generation Computer Systems 29(6), 1345–1352 (2013)
71. Yu, G., Zhang, G.: Bin packing of selfish items. In: Papadimitriou, C., Zhang, S. (eds.) WINE 2008. LNCS, vol. 5385, pp. 446–453. Springer, Heidelberg (2008)

A Constructive Proof of the Topological Kruskal Theorem

Jean Goubault-Larrecq

LSV, ENS Cachan, CNRS, INRIA

Abstract. We give a constructive proof of Kruskal's Tree Theorem—precisely, of a topological extension of it. The proof is in the style of a constructive proof of Higman's Lemma due to Murthy and Russell (1990), and illuminates the role of regular expressions there. In the process, we discover an extension of Dershowitz' recursive path ordering to a form of cyclic terms which we call μ-terms. This all came from recent research on Noetherian spaces, and serves as a teaser for their theory.

1 Introduction

Kruskal's Theorem [33] states that the homeomorphic embedding ordering on finite trees is a a well quasi-ordering. This is a deep and fundamental theorem in the theory of well quasi-orderings. The aim of this paper is to give a *constructive*, that is, an intuitionistic proof of this fact[1].

I will explain what all that means in Section 2. I should probably admit right away that I have not *actively* looked for such a proof. It came to me in 2010 as a serendipitous by-product of research I was doing on Noetherian spaces, seen as a generalization of well quasi-ordered spaces. The result is, hopefully, a nice piece of mathematics. It is also an opportunity for me to explain various related developments which I would dare to say have independent interest.

I would like to issue a word of warning, though. The constructive proofs of the topological Higman and Kruskal theorems I am giving here were the first I found. The non-constructive proofs of [29, Section 9.7] came second. These are the ones I chose to publish, for good reason: once cast in formal language, the original constructive proofs are terribly heavy. I have therefore opted for a somewhat lighter presentation here, which stresses the beautiful *core* of the proof, at the cost at being somewhat sketchy in Sections 4 (Higman) and 5 (Kruskal). And this core is: these theorems reduce to questions of *termination* problems, which one can solve by using multiset orderings (Higman), resp. an extension of Dershowitz' multiset path ordering (Kruskal).

---

[1] I will avoid any debate of what intuitionism or constructivism is, and assume the logic of any of the modern proof assistants based on intuitionistic type theory, such as Coq [6]. The full calculus of inductive constructions with universes is definitely not needed, though. I only need first-order intuitionistic logic, plus a few inductively defined predicates and relations, and their associated induction principles.

K. Chatterjee and J. Sgall (Eds.): MFCS 2013, LNCS 8087, pp. 22–41, 2013.

2   Well Quasi-Orderings, Noetherian Spaces

A *quasi-ordering* on a set X is a reflexive and transitive binary relation $\leqslant$ on X. Given a subset A of X, we write $\uparrow A$ for its upward closure $\{y \in X \mid \exists x \in A \cdot x \leqslant y\}$, and call A *upward closed* if and only if $A = \uparrow A$. A *basis* of an upward closed subset E is any set A such that $E = \uparrow A$; E has a *finite basis* if and only if one can take A finite. We define the downward closure $\downarrow A$, and downward closed subsets, similarly. We also write $\geqslant$ for the converse of $\leqslant$, $<$ for the strict part of $\leqslant$ ($x < y$ iff $x \leqslant y$ and not $y \leqslant x$), $>$ for that of $\geqslant$.

There are many equivalent definitions of a *well* quasi-ordering (wqo for short), of which here are a few:

1. every infinite sequence $(x_n)_{n \in \mathbb{N}}$ in X is *good*, namely, there are two indices $m < n$ with $x_m \leqslant x_n$;
2. every infinite sequence $(x_n)_{n \in \mathbb{N}}$ in X is *perfect*, i.e., has an infinite ascending subsequence $x_{n_0} \leqslant x_{n_1} \leqslant \ldots \leqslant x_{n_i} \leqslant \ldots$ (with $n_0 < n_1 < \ldots < n_i < \ldots$);
3. $\leqslant$ is *well-founded* (there is no infinite descending sequence of elements $x_0 > x_1 > \ldots > x_n > \ldots$) and has *no infinite antichain* (an infinite sequence of pairwise incomparable elements);
4. every upward closed subset U has a finite basis;
5. every ascending chain $U_0 \subseteq U_1 \subseteq \ldots \subseteq U_n \subseteq \ldots$ of upward closed subsets is stationary (i.e., all U_ns are equal from some rank n onwards);
6. every descending chain $F_0 \supseteq F_1 \supseteq \ldots \supseteq F_n \supseteq \ldots$ of downward closed subsets is stationary;
7. the *strict inclusion* ordering $\subset$ is well-founded on downward closed subsets, i.e., there is no infinite descending chain $F_0 \supset F_1 \supset \ldots \supset F_n \supset \ldots$ of downward closed subsets.

The latter shows that being a wqo is merely a *termination* property, only one not on words, or on terms, as would be familiar in computer science [13], but rather on downward closed subsets.

There are many useful wqos in nature: $\mathbb{N}$ with its natural ordering $\leqslant$, any finite set, any finite product of wqos (in particular $\mathbb{N}^k$ with its componentwise ordering: this is Dickson's Lemma [18]), any finite coproduct of wqos, the set of finite words X^* over a well-quasi-ordered alphabet X (with the so-called word embedding quasi-ordering: this is Higman's Lemma [30]), the set of finite trees, a.k.a., first-order terms, $\mathcal{T}(X)$ over a well-quasi-ordered signature X (with the so-called tree embedding quasi-ordering: this is Kruskal's Theorem [33]), notably.

There are also more and more applications of wqo theory in computer science.

Termination. An early application is Nachum Dershowitz' discovery of the *multiset path ordering* on terms. This is a strict ordering $<^{\mathrm{mpo}}$ on terms that is well-founded, i.e., such that there is no infinite $>^{\mathrm{mpo}}$-chain $t_0 >^{\mathrm{mpo}} t_1 >^{\mathrm{mpo}} \ldots >^{\mathrm{mpo}} t_n >^{\mathrm{mpo}} \ldots$: to show that a rewrite system $\mathcal{R}$ terminates, it is enough to show that $\ell >^{\mathrm{mpo}} r$ for every rule $\ell \to r$ in $\mathcal{R}$. Dershowitz' initial proof ([12], see also [11]) rested on the remark that $>^{\mathrm{mpo}}$ is a simplification ordering: if t embeds into s, then $t \leqslant^{\mathrm{mpo}} s$. Given any infinite $>^{\mathrm{mpo}}$-descending chain as above, by Kruskal's Theorem one can find $i < j$ such that t_i embeds into t_j. It follows that $t_i \leqslant^{\mathrm{mpo}} t_j$, contradicting $t_i >^{\mathrm{mpo}} t_j$. This uses characterization 1 of wqos.

This simple argument definitely relies on Kruskal's deep result. The realization that Dershowitz' theorem required much less logical clout [26,8] came to me as both a relief and a disappointment : I'll recapitulate the elementary argument in Section 3. I'll also give a slight extension of this elementary argument to a form of cyclic terms I have decided to call μ-terms. This will be instrumental in the rest of the paper, and may even be useful in the rewriting community.

Minimal patterns. A second application arises from characterization 4. Given an upward closed language L of elements in a wqo X, one can test whether $x \in L$ by just checking finitely many equalities $x_1 \leqslant x$, ..., $x_n \leqslant x$. Indeed, property 4 states that one can write L as $\uparrow\{x_1, \ldots, x_n\}$. For example, this is how van der Meyden shows that fixed monadic queries to indefinite databases can be evaluated in linear time in the size of the database [44], where x, x_1, ..., x_n are (encodings of models as) finite sequences of finite sets of logical atoms. The query L defines the minimal patterns x_1, ..., x_n to be checked, in the embedding quasi-ordering on words. That the latter is a wqo is Higman's Lemma, and the fact that its standard proofs are non-constructive implies the curious fact that one cannot *a priori* compute x_1, ..., x_n from L. That is, a linear time algorithm exists for each L... but what is it? Ogawa [40] solves the issue by extracting the computational content of Murthy and Russell's constructive proof of Higman's Lemma [37]. This computes the values x_1, ..., x_n, hence derives a linear-time algorithm for the query L, from L given as input.

WSTS. Another application is in verification of *well-structured transition systems* (WSTS) [1,25]. A WSTS is a (possible infinite-state) transition system $(X, \rightarrow)$, with a *wqo* $\leqslant$ of the set of states X, satisfying a monotonicity property. For simplicity, we shall only consider strong monotonicity: if $s \rightarrow s'$ and $s \leqslant t$, then there is a state t' such that $t \rightarrow t'$ and $s' \leqslant t'$.

Examples of WSTS abound. Petri nets are WSTS whose state space is $\mathbb{N}^k$, where k is the number of places. Affine nets [24] generalize these and many other variants, and are still WSTS on $\mathbb{N}^k$. Lossy channel systems [3] are networks of finite-state automata that communicate over FIFO queues. They are WSTS whose state space is $\prod_{i=1}^m Q_i \times \prod_{j=1}^n \Sigma_j^*$, where Q_i is the finite state space of the ith automaton, and Σ_j is the finite alphabet of the jth queue. Let us also cite data nets [34], BVASS [46,10], and recent developments in the analysis of processes [36,4,47,42], which require tree representations of state.

The simple structure of a WSTS implies that *coverability* is decidable in every effective WSTS. This is the following question: given a state $s \in X$ and an upward closed subset U of X, is there a state $t \in U$ that is reachable from s, i.e., such that $s \rightarrow^* t$, where $\rightarrow^*$ is the reflexive-transitive closure of $\rightarrow$? By *effective* WSTS, we mean that we can represent states on a computer (which implies that every upward closed subset U is representable as well, as a finite set E, by property 4), that $\leqslant$ is decidable, and that the set of one-step predecessors $\mathrm{Pre}(U) = \{s \in X \mid \exists t \in U \cdot s \rightarrow t\}$ of a state t is computable. This is the case of all WSTS mentioned above. Inclusion of upward closed subsets is decidable, since $\uparrow E_1 \subseteq \uparrow E_2$ if and only if for every $x \in E_1$, there is a $y \in E_2$ with $y \leqslant x$. That coverability is decidable is almost trivial: using a while loop, compute the successive sets $U_0 = U$, $U_{n+1} = U_n \cup \mathrm{Pre}(U_n)$, and stop when $U_{n+1} \subseteq U_n$; this must eventually happen by property 5. Then there is a state in U that is reachable from s if and only if $s \in U_n$.

In 1969, Karp and Miller [32] devised another way (historically, the first one) of deciding coverability. They built a so-called coverability tree, and showed that it was finite and effectively constructible by resorting to Dickson's Lemma, plus a few additional tricks. One of the tricks they required was to extend the state space from $\mathbb{N}^k$ to $\mathbb{N}_\omega^k$, where $\mathbb{N}_\omega$ is $\mathbb{N}$ plus a fresh top element ω, the *limit* of any ever growing sequence. Although it would seem natural that the construction would generalize to every WSTS, progress was slow. One of the blocking factors was to define a completion $\hat{X}$ of a well quasi-ordered state space X, so that Karp and Miller's construction would adapt.

By analogy with $\mathbb{N}^k$, $\hat{X}$ should be X with some limit points added, and this naturally calls for topology. Alain Finkel once asked me whether there would be a notion of completion from topology that could serve this purpose. We realized that the *sobrification* of X (see [29, Section 8.2]) was the right candidate, and this led us to a satisfactory extension of Karp and Miller's procedure to all WSTS [20,21,23].

Noetherian spaces. In the process, going to topology begged the question whether there is a topological characterization of wqos. I realized in [27] that this would be the notion of Noetherian space, invented in algebraic geometry in the first half of the 20th century. A *Noetherian* space is a space where every ascending chain of opens is stationary: comparing this with property 5, we have merely replaced "upward closed" by "open".

Every quasi-ordered set can be equipped with the so-called *Alexandroff topology*, whose opens are just the upward closed subsets. Property 5 immediately implies that every wqo is Noetherian, once equipped with its Alexandroff topology. The framework of Noetherian spaces also allows us to extend the WSTS methodology to more kinds of transition systems. I have explained this in [28], applying this to two examples: a certain kind of multi-stack automata, and concurrent polynomial programs manipulating numerical values (in $\mathbb{R}$) that communicate through discrete signals over lossy channels. The decidability results that I'm stating in these settings are far from trivial, but are low-hanging fruit once we have the theory of Noetherian spaces available.

By "theory of Noetherian spaces", I do not mean the one we inherit from algebraic geometry, rather some natural results that arise from cross-fertilization with wqo theory. (See [29, Section 9.7] for a complete treatment.) Of interest to us are the following generalizations of Higman's Lemma and Kruskal's Theorem, respectively:

Topological Higman Lemma [29, Theorem 9.7.33]: if X is Noetherian, then the space of finite words X^* with the word topology is Noetherian, too.

Topological Kruskal Theorem [29, Theorem 9.7.46]: if X is Noetherian, then set space of finite trees $\mathcal{T}(X)$ with symbol functions taken from X is Noetherian under the tree topology.

We define the word and tree topologies as follows. Intuitively, think of an open set U as a test—namely, x passes the test if and only if $x \in U$. In the word topology, we wish the following to be a test: given tests $U_1, \ldots, U_n$ on letters (open subsets of X), the word w passes the test $X^* U_1 X^* \ldots X^* U_n X^*$ if and only if w contains a (not necessarily contiguous) subword $a_1 a_2 \ldots a_n$ with each a_i in U_i. In the tree topology, the basic tests are whether a given tree has an embedded subtree of a given shape, and where each function symbol is in a given open subset of X (possibly different at each node). In each case, these tests form bases for the required topologies, i.e., the opens are all unions of such tests.

The proofs I give of these theorems in [29, Section 9.7] are elegant, yet terribly topological, and rest on many results that require classical logic, and the Axiom of Choice. Instead, we shall use the following remark.

Call a closed subset F *irreducible* if and only if, for every finite family of closed subsets $F_1, \ldots, F_n$, if $F \subseteq F_1 \cup \ldots \cup F_n$, then $F \subseteq F_i$ for some i already. By [29, Theorem 9.7.12], a space X is Noetherian if and only if: ($\downarrow$) the strict inclusion relation $\subset$ is well-founded on the set $\mathcal{S}(X)$ of irreducible closed subsets of X ($\mathcal{S}(X)$ happens to be the sobrification of X we alluded to above), (T) the whole space X can be written as the union of finitely many irreducible closed subsets of X, and (W) given any two irreducible closed subsets F_1, F_2 of X, $F_1 \cap F_2$ can be written as the union of finitely many irreducible closed subsets of X. It follows that every closed subset will be a finite union of irreducible closed subsets, and that the strict inclusion ordering $\subset$ will be well-founded on closed subsets. The latter generalizes property 7, since in a quasi-ordered set, the (Alexandroff) closed sets are exactly the downward closed sets.

This leads us to the following proof plan:

(A) Find concrete representations of all irreducible closed subsets. This programme was initiated in [20] and carried out in [22], where we call the latter *S-representations*. In both the word and tree cases, our S-representations are certain forms of regular expressions, over words, or over trees. On words, this generalizes the products and the semi-linear regular expressions (SRE) of [2]; on trees, no prior work seems to have existed. These are effective representations: we can decide inclusion (in polynomial time, modulo an oracle deciding inclusion of irreducible closed subsets of letters, resp., of function symbols), and we can compute finite intersections of S-representations (in polynomial time again, provided the number of input representations is bounded).

(B) Show directly that strict inclusion is well-founded on S-representations. This will establish property ($\downarrow$). Properties (T) and (W) are mostly obvious, since we even have *algorithms* to compute finite intersections.

In the case of the topological Higman Lemma (on words), we shall obtain a re-reading of Murthy and Russell's celebrated constructive proof of Higman's

Lemma ([37]; see also [40], footnotes 6 and 7, for fixes to the definition of sequential regular expression). Our S-representations will be their *sequential regular expressions*, seen as the result of building SREs (originating in [2]) over a cotopology. Our constructive proof of Kruskal's Theorem, and indeed of its topological generalization, is in the same spirit, and we believe it provides a satisfactory answer to Murthy and Russell's final question [37].

Intuitionism. One difficulty with finding intuitionistic proofs in the theory of wqos is that properties 1–7 are *not* constructively equivalent. Notably, 2 is intuitionistically strictly stronger than 1, as Veldman notes [45, 1.3]. Indeed, 2 fails on $X = \mathbb{N}$ in an intuitionistic setting, while 1 is constructively valid. Similarly, 4 fails on $\mathbb{N}^2$, intuitionistically, even for decidable subsets of $\mathbb{N}^2$ [45, 1.2].

Following Murthy and Russell, a *constructive wqo* is defined by the following reformulation of property 1: (1') the opposite of the prefix ordering on bad finite sequences of words in X^* is well-founded. A finite sequence $x_0, x_1, \ldots, x_n$ is *bad* iff it is not good, that is, if $x_i \not\leqslant x_j$ for no $i < j$. The well-foundedness requirement means that one cannot extend finite sequences (adding x_{n+1}, x_{n+2}, etc.) indefinitely, keeping them all bad.

Murthy and Russell actually proved property 7. They derived (1') from 7, assuming $\leqslant$ decidable in the constructive sense that $\forall x, y \in X \cdot x \leqslant y \vee \neg(x \leqslant y)$ is provable. All the other constructive proofs I know of Higman's Lemma prove (1'), some of them directly [41,7,5]; the latter two do not require $\leqslant$ to be decidable. There are fewer intuitionistic proof of Kruskal's Theorem. One is due to Monika Seisenberger [43], who gives a direct proof of (1') on trees, based on a intuitionistic variant of Nash-Williams' minimal bad sequence argument [38]. She requires the quasi-ordering $\leqslant$ on function symbols to be decidable. Wim Veldman's proof [45] does not make this requirement, but models tree embedding with so-called at-most-ternary relations rather than using a binary relation $\leqslant$. He shows that Kruskal's original proof [33] can be made constructive, replaying the needed part of Ramsey theory in intuitionistic logic. Curiously, our proofs of the *topological* versions of Higman's Lemma and Kruskal's Theorem are entirely constructive, and we only need to assume $\leqslant$ decidable to deduce the *ordinary*, order-theoretic versions of these results from the topological versions.

3 Path Orderings

Path orderings (mpo, lpo, rpo) have been an essential ingredient of termination proofs for rewrite systems since their inception by Nachum Dershowitz in 1982 [12]. We shall concentrate on Dershowitz' original *multiset path ordering* (a.k.a., mpo). He proved that the mpo was well-founded as a consequence of Kruskal's Theorem. We give an elementary, inductive, intuitionistic proof instead. This is based on a paper I wrote in 2001 [26]. Coupet-Grimal and Delobel [8] implemented a similar proof in Coq, with a proof of the Dershowitz-Manna Theorem (which I had not given, but Nipkow had [39]—see below). Dershowitz and Hoot's earlier proof that the general path ordering is well-founded [15] is non-constructive but elementary as well. Even earlier, Lescanne had already given an inductive proof that the mpo was well-founded [35, Theorem 5]; his proof relies on Zorn's Lemma (op.cit., Lemma 5), and ours will be simpler anyway, but his notion of decomposition ordering is illuminating.

Let X be a set with a binary relation $<$ on it. We again write $>$ for the converse of $<$. One thinks of $<$ as a strict ordering, but this is not needed. What will be important is that $<$ is *well-founded*: classically, this means that there is no infinite $>$-chain $x_1 > x_2 > \ldots > x_n > \ldots$ Constructively, it is better to say that $<$ is well-founded iff every element is $<$-*accessible*, where $<$-accessibility is the predicate defined inductively by (i.e., the least predicate such that):

$$\frac{\text{every } y < x \text{ is } <\text{-accessible}}{x \text{ is } <\text{-accessible}}$$

The set of $<$-accessible elements is traditionally called the *well-founded part* of $<$, i.e., the set of elements that cannot start an infinite $>$-chain. Since $<$-accessibility is defined inductively, we obtain the following useful principle of $<$-*induction*: to prove that a property P holds of every $<$-accessible element x, it is enough to show it under the additional assumption that P holds of every $y < x$ (the *induction hypothesis*). Another useful principle is $<$-*inversion*: if x is $<$-accessible, and $x > y$, then y is $<$-accessible as well.

Write $\{x_1, \ldots, x_n\}$ for the (finite) multiset consisting of the elements $x_1, \ldots, x_n \in X$. Let $\emptyset$ be the empty multiset, and $\uplus$ denote multiset union. We use the letters M, M', $\ldots$, for multisets. Intuitionistically, we assume an inductive definition of multisets, e.g., as finite lists, and we will reason up to permutation. (This actually incurs some practical difficulties in proof assistants such as Coq, which we shall merrily gloss over.) On the set $\mathcal{M}(X)$ of multisets of elements of X, we define the *multiset extension* $<_{\mathrm{mul}}$ of $<$, inductively, by:

$$\frac{\text{for every } i \ (1 \leqslant i \leqslant n), \ x > x_i}{M \uplus \{x\} >_{\mathrm{mul}} M \uplus \{x_1, \ldots, x_n\}}$$

That is, we replace some element x by arbitrarily many smaller elements $x_1, \ldots, x_n$. The following *Dershowitz-Manna Theorem* [17] is crucial.

Lemma 1 (Dershowitz-Manna, Nipkow). *For all $<$-accessible elements x_1, $\ldots, x_n \in X$, $\{x_1, \ldots, x_n\}$ is $<_{mul}$-accessible. In particular, if $<$ is well-founded on X, then $<_{mul}$ is well-founded on $\mathcal{M}(X)$.*

Proof. We give Nipkow's intuitionistic proof [39]. Let Acc denote the set of $<_{\mathrm{mul}}$-accessible multisets. We prove that $\{x_1, \ldots, x_n\} \in Acc$ by induction on n. The case $n = 0$ is obvious, while the induction step consists in showing that, for every $<$-accessible x: $(*)$ for every $M \in Acc$, $M \uplus \{x\} \in Acc$. Fix an $<$-accessible x, and use $<$-induction. This provides us with the induction hypothesis: (a) for every $y < x$, for every $M \in Acc$, $M \uplus \{y\} \in Acc$. To prove $(*)$, we show by $<_{\mathrm{mul}}$-induction on $M \in Acc$ that: $(**)$ $M \uplus \{x\} \in Acc$. This gives us the extra induction hypothesis (b): for every $M' <_{\mathrm{mul}} M$, $M' \uplus \{x\} \in Acc$. It now remains to show that (a) and (b) imply $(**)$. By definition of $<_{\mathrm{mul}}$-accessibility, this means showing that every multiset $M_1 <_{\text{mul}} M \uplus \{\!\{x\}\!\}$ is in Acc. There are two cases: either $M_1 = M' \uplus \{\!\{x\}\!\}$ for some $M' <_{\text{mul}} M$, and the claim follows from (b); or $M_1 = M \uplus \{\!\{x_1, \ldots, x_m\}\!\}$ with $x > x_1, \ldots, x_m$, then the claim follows by induction on m, using (b) in the base case and (a) in the induction step.   □

It follows that, under the same assumptions, the transitive closure $<^+_{\text{mul}}$ of $<_{\text{mul}}$ is well-founded: for any relation R, R-accessibility and R^+-accessibility coincide.

Let now Σ be a signature, i.e., just a set whose elements will be understood as function symbols, with arbitrary, finite arity. The *terms* s, t, u, v, $\ldots$, are inductively defined as tuples $f(t_1, \ldots, t_n)$ of an element f of Σ and of finitely many terms t_1, $\ldots$, t_n. The base case is obtained when $n = 0$. There are no variables here, so our terms are the *ground terms* considered in the literature [16]. This is no loss of generality, as one can encode general terms as ground terms over a signature that includes all variables, understanding the variable term x as the application $x()$ to no argument. However, please do not confuse the latter (free) variables with the (μ-bound) variables that we introduce later.

Let $\approx$ be the relation defined inductively by: $f(s_1, \ldots, s_m) \approx g(t_1, \ldots, t_n)$ if and only if $f = g$, $m = n$, and there is a permutation π of $\{1, \ldots, n\}$ such that $s_{\pi(i)} \approx t_i$ for each i, $1 \leqslant i \leqslant n$. This is an equivalence relation, and relates terms that are equal up to permutations of arguments, anywhere in the term.

Call *precedence* any binary relation $<$ on Σ. The *multiset path ordering*, or *mpo*, $<^{\text{mpo}}$ is defined inductively (together with an auxiliary relation $\ll$) by:

$$\frac{\exists i \cdot s_i \gtrsim^{\text{mpo}} t}{f(s_1, \ldots, s_m) >^{\text{mpo}} t} (Sub) \qquad \frac{\begin{array}{c} f(s_1, \ldots, s_m) \gg g(t_1, \ldots, t_n) \\ \forall j \cdot f(s_1, \ldots, s_m) >^{\text{mpo}} t_j \end{array}}{f(s_1, \ldots, s_m) >^{\text{mpo}} g(t_1, \ldots, t_n)} (Gt)$$

where $s \gtrsim^{\text{mpo}} t$ abbreviates $s >^{\text{mpo}} t$ or $s \approx t$, and here are the clauses for $\ll$:

$$\frac{f > g}{f(s_1, \ldots, s_m) \gg g(t_1, \ldots, t_n)} (\gg Fun) \qquad \frac{\{\!\{s_1, \ldots, s_m\}\!\} (>^{\text{mpo}})^+_{\text{mul}} \{\!\{t_1, \ldots, t_n\}\!\}}{f(s_1, \ldots, s_m) \gg f(t_1, \ldots, t_n)} (\gg Args)$$

In other words, $\ll$ is the lexicographic product of $<$ and of $(<^{\text{mpo}})^+_{\text{mul}}$. The relation $\ll$ is a *lifting* (a notion called as such in [19], and which one can trace back to [31]), meaning that it is well-founded on the set $\overline{Acc}$ of terms of the form $f(s_1, \ldots, s_m)$ with f $<$-accessible and s_1, $\ldots$, s_m $<^{\text{mpo}}$-accessible. Beware that this does *not* mean that any $\gg$-chain starting from a term $f(s_1, \ldots, s_m)$ with f $<$-accessible and s_1, $\ldots$, s_m $<^{\text{mpo}}$-accessible is finite. It only means that any infinite such chain must eventually exit $\overline{Acc}$, i.e., reach a term $g(t_1, \ldots, t_n)$ where g is not $<$-accessible, or where some t_j is not $<^{\text{mpo}}$-accessible. Intuitionistically, we define the restriction $\ll_{|\overline{Acc}}$ of $\ll$ to $\overline{Acc}$ by $t \ll_{|\overline{Acc}} s$ iff $t \in \overline{Acc}$ and $s \in \overline{Acc}$ and $t \ll s$; and we note that every term in $\overline{Acc}$ is $\ll_{|\overline{Acc}}$-accessible.

Replacing $\gg$ by other liftings would yield similar orderings: if we compare arguments lexicographically, for example, we would get the lexicographic path ordering (lpo), and mixing the two kinds yields the recursive path ordering (rpo) [13]. The following theorem is intuitionistic.

Proposition 1. *Every term whose function symbols are all $<$-accessible is $<^{mpo}$-accessible. In particular, if $<$ is well-founded, then $<^{mpo}$ is well-founded on terms.*

Proof. In the course of the proof, we shall need to observe that: (∗) for every $<^{mpo}$-accessible term u, for every term t such that $u \approx t$, t is $<^{mpo}$-accessible. This requires us to show first that if $u \approx t$ and $t >^{mpo} s$, then $u >^{mpo} s$, an easy induction on the definition of $<^{mpo}$. We show (∗) by $<^{mpo}$-induction on u, i.e., that for every t such that $u \approx t$, for every $s <^{mpo} t$, s is $<^{mpo}$-accessible; the assumptions imply $s <^{mpo} u$, and the claim follows by induction hypothesis.

Let Acc be the set of $<^{mpo}$-accessible terms, and W be the set of terms whose function symbols are all $<$-accessible. As above, we define $\overline{Acc}$ as the set of terms of the form $f(t_1, \ldots, t_n)$ such that f is $<$-accessible and whose arguments $t_1, \ldots, t_n$ are in Acc. We show that every $t \in W$ is in Acc, by structural induction on t. This means showing that for every $s \in \overline{Acc}$, s is in Acc.

We first give a classical argument, in the hope that it will be clearer. We shall need to use the immediate subterm relation $\lhd$, defined inductively by $g(t_1, \ldots, t_m) \rhd t_j$ for all $g, t_1, \ldots, t_m$ and j. This is a well-founded relation. Assume there is term $s \in \overline{Acc}$ that is not in Acc. In other words, the set $\overline{Acc} \smallsetminus Acc$ is non-empty. Since $\ll$ is a lifting, it is well-founded on $\overline{Acc}$, hence on $\overline{Acc} \smallsetminus Acc$: so there is a $\ll$-minimal element s in $\overline{Acc} \smallsetminus Acc$. Since $s \notin Acc$, it starts an infinite $>^{mpo}$-chain, so $s >^{mpo} t$ for some $t \notin Acc$. Among these terms t we pick one that is $\lhd$-minimal: writing t as $g(t_1, \ldots, t_n)$, this assures us that for every j such that $s >^{mpo} t_j$, $t_j \in Acc$.

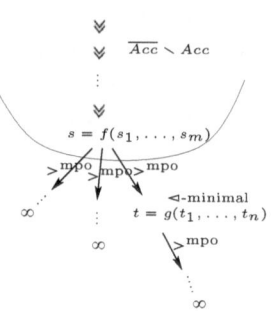

The fact $s >^{mpo} t$ is obtained by rule (Sub) or by rule (Gt). (Sub) is out of the question, though, since that would mean $s = f(s_1, \ldots, s_m)$ with some $s_i \geq^{mpo} t$; but $s \in \overline{Acc}$ implies $s_i \in Acc$, hence $t \in Acc$, either because $s_i \approx t$, using (∗), or because $s_i >^{mpo} t$, using $<^{mpo}$-inversion: contradiction. So rule (Gt) must have been used: $s \gg t = g(t_1, \ldots, t_n)$ with $s >^{mpo} t_j$ for every j. Since s was chosen $\gg$-minimal, t cannot be in $\overline{Acc} \smallsetminus Acc$, and since $t \notin Acc$, t is not in $\overline{Acc}$: so $t_j \notin Acc$ for some j. However, $s >^{mpo} t_j$ together with the fact that t was $\lhd$-minimal implies $t_j \in Acc$, a contradiction.

We obtain an intuitionistic proof by replacing minimal counter-examples by induction principles. We wish to show that for every term $s \in \overline{Acc}$ then $s \in Acc$. Since $\ll$ is a lifting, $s \in \overline{Acc}$ is $\ll_{|\overline{Acc}}$-accessible, so $\ll_{|\overline{Acc}}$-induction applies and we obtain the following induction hypothesis: (a) for every $t \ll s$, if $t \in \overline{Acc}$ then $t \in Acc$. Our goal is to prove that $s \in Acc$, i.e., that every $t <^{mpo} s$ is in Acc. We show this by $\lhd$-induction on $t = g(t_1, \ldots, t_n)$, which means that we have the extra induction hypothesis: (b) for every j, if $t_j <^{mpo} s$ then $t_j \in Acc$. If $t <^{mpo} s$ was obtained by (Sub), then $s = f(s_1, \ldots, s_m)$ with $s_i \geq^{mpo} t$ for some i; since $s \in \overline{Acc}$, $s_i \in Acc$ hence $t \in Acc$, either by (∗) if $s_i \approx t$, or by $<^{mpo}$-inversion if

$s_i >^{\mathrm{mpo}} t$. If $t <^{\mathrm{mpo}} s$ was obtained by (Gt), then $s \gg t$ and $s >^{\mathrm{mpo}} t_j$ for every j. By (b), $t_j \in Acc$ for every j. Also, $s \gg t$ implies $f \gtrsim g$, and since $s \in \overline{Acc}$, f is $<$-accessible, hence also g: so $t = g(t_1, \ldots, t_n)$ is in $\overline{Acc}$. By (a), $t \in Acc$. $\square$

I'm not claiming that the above proof is novel. This is the core of Theorem 1 of [26], later improved by Dawson and Goré [9]. Dershowitz [14] gives a broader perspective on this kind of results. That its proof is constructive is also one argument set forth in [26], and, as I've said already, this was made precise and implemented in Coq by Coupet-Grimal and Delobel [8].

I had also argued that the proof technique of [26] extended to prove abstract termination arguments, some of whose applied to graphs, for example. I'll develop this now for a new relation on a class of so-called μ-*terms*, defined by the following (pseudo-)grammar:

$$s, t, u, v, \ldots ::= x \qquad \text{variables}$$
$$\mid \; f(t_1, \ldots, t_n) \; \text{applications}, \; f \in \Sigma, n \in \mathbb{N}$$
$$\mid \; \mu x = s \cdot t \quad \text{iterators}.$$

The iterator $\mu x = s(x) \cdot t$ should be thought of as some kind of infinite term $\ldots s(\ldots s(s(t)) \ldots)$. The variable x is bound in $\mu x = s \cdot t$, its scope is s. A term t is ground if and only if $\mathrm{fv}(t) = \varnothing$, where the set $\mathrm{fv}(t)$ of free variables of t is defined inductively by $\mathrm{fv}(x) = \{x\}$, $\mathrm{fv}(f(s_1, \ldots, s_m)) = \bigcup_{i=1}^{m} \mathrm{fv}(s_i)$, $\mathrm{fv}(\mu x = s \cdot s') = (\mathrm{fv}(s) \setminus \{x\}) \cup \mathrm{fv}(s')$. For instance, $\mu x = f(x) \cdot g(a)$ is a ground μ-term.

Again, we give ourselves a precedence $<$ on Σ. We extend the definition of $\approx$ by letting $\mu x = s \cdot s' \approx \mu x = t \cdot t'$ if and only if $s \approx t$ and $s' \approx t'$, and $x \approx x$ for every variable x. (We make an abuse of notation here and silently assume a form of α-renaming. A more correct definition would be: $\mu x = s \cdot s' \approx \mu y = t \cdot t'$ iff $s[x := z] \approx t[y := z]$ and $s' \approx t'$, for z a fresh variable. We shall make similar abuses of notation in rules $(\mu Gt\mu)$, $(\mu \ll)$ and $(\mu \ll \mu)$ below, to avoid clutter.) We take the same rules defining $<^{\mathrm{mpo}}$ and $\ll$ as above, and add the following to also compare variables and iterations, either together or with other terms:

$$\frac{(t \text{ ground } \mu\text{-term})}{x >^{\mathrm{mpo}} t} \; (Var) \qquad\qquad \frac{s' \gtrsim^{\mathrm{mpo}} t}{\mu x = s \cdot s' >^{\mathrm{mpo}} t} \; (\mu Sub)$$

$$\frac{\mu x = s \cdot s' \gg g(t_1, \ldots, t_n) \quad \forall j \cdot \mu x = s \cdot s' >^{\mathrm{mpo}} t_j}{\mu x = s \cdot s' >^{\mathrm{mpo}} g(t_1, \ldots, t_n)} \; (\mu Gt) \qquad \frac{\mu x = s \cdot s' \gg \mu x = t \cdot t' \quad \mu x = s \cdot s' >^{\mathrm{mpo}} t'}{\mu x = s \cdot s' >^{\mathrm{mpo}} \mu x = t \cdot t'} \; (\mu Gt\mu)$$

$$\frac{s >^{\mathrm{mpo}} g(t_1, \ldots, t_n)}{\mu x = s \cdot s' \gg g(t_1, \ldots, t_n)} \; (\mu \ll) \qquad\qquad \frac{s >^{\mathrm{mpo}} t}{\mu x = s \cdot s' \gg \mu x = t \cdot t'} \; (\mu \ll \mu)$$

The unusual rule (Var) states that *every* ground μ-term is strictly smaller than any variable. This allows us to check, for example, that $\mu x = f(x) \cdot g(a) >^{\mathrm{mpo}} f(f(f(f(g(a)))))$, where a is a constant: using (μGt) and $(\mu \ll)$, this requires us to check two premises, of which one is $f(x) >^{\mathrm{mpo}} f(f(f(f(g(a)))))$; the latter follows, using (Gt), from $x >^{\text{mpo}} f(f(f(g(a))))$, and this, in turn, is an instance of (Var). We leave the rest of the verification to the reader.

The above rules are probably not the ones one would have imagined. In particular, it would seem natural to consider $\mu x = s(x) \cdot s'$ and $s(\mu x = s \cdot s')$ as equivalent. This would suggest the following alternative to (μGt): to prove $\mu x = s(x) \cdot s' >^{\text{mpo}} t$ (where $t = g(t_1, \ldots, t_n)$, and for simplicity we assume both sides of the inequality to be ground), prove $s(\mu x = s(x) \cdot s') >^{\text{mpo}} t$ and $\forall j \cdot \mu x = s \cdot s' >^{\text{mpo}} t_j$. Instead of proving $s(\mu x = s(x) \cdot s') >^{\text{mpo}} t$, (μGt) (together with $(\mu \ll)$) only requires us to prove $s(x) >^{\text{mpo}} t$, a seemingly much weaker statement, since x is not just greater than or equal to $\mu x = s(x) \cdot s'$, but strictly greater than *any* ground term by (Var). Although they are not what we would imagined at first, these are the rules that arise from our study of the topological Kruskal Theorem (Section 5).

The following is new, and probably useful in other contexts. Our proof is intuitionistic. The proof is similar to Proposition 1, or to Theorem 1 of [26], but we need a few easy additional arguments near the end of the proof.

Theorem 1. *Every μ-term whose function symbols are all $<$-accessible is $<^{\text{mpo}}$-accessible. In particular, if $<$ is well-founded, then $<^{\text{mpo}}$ is well-founded on μ-terms.*

Proof. One might think that Theorem 1 is an easy consequence of Proposition 1: encode $\mu x = s \cdot s'$ as the ordinary term $\mu(s, s')$, and the variable x as $x()$, and extend the precedence appropriately. This strategy does not work, as for example (Var) requires $x >^{\text{mpo}} \mu x = f(x) \cdot g(a)$. In the encoding, this would force $x >^{\text{mpo}} \mu(f(x), g(a))$, which is plainly false, since $\mu(f(x), g(a)) >^{\text{mpo}} x$.

We imitate the proof of Proposition 1. Again, we have: $(*)$ for every $<^{\text{mpo}}$-accessible μ-term u, for every μ-term t such that $u \approx t$, t is $<^{\text{mpo}}$-accessible. Define the *immediate subterms* of a μ-term in the expected way, as follows: the immediate subterms of $g(t_1, \ldots, t_m)$ are $t_1, \ldots, t_n$, the immediate subterms of $\mu x = s \cdot s'$ are s and s', and variables have no immediate subterms. We need to define $\lhd$ slightly differently, inductively, by: (i) $g(t_1, \ldots, t_m) \rhd t_j$ for all $g \in \Sigma$, μ-terms $t_1, \ldots, t_m$ and j; (ii) $x \rhd t$ for every variable x and ground μ-term t; (iii) $\mu x = s \cdot s' \rhd s'$ (not s!).

We first show that $\lhd$ is well-founded. This is done in several steps. We first show that every ground μ-term t is $\lhd$-accessible, by induction on t; crucially, if $\mu x = s \cdot s'$ is ground and $\mu x = s \cdot s' \rhd s'$, then s' is ground and the induction hypothesis applies. We then do a secondary induction to establish that every μ-term is $\lhd$-accessible, using the previous claim in the case of variables.

Let Acc be the set of $<^{\text{mpo}}$-accessible μ-terms, and W be the set of μ-terms whose function symbols are all $<$-accessible. Say that a μ-term is *head accessible* if and only if it is a variable, an iterator $\mu x = s \cdot s'$ with s head accessible, or an application $f(s_1, \ldots, s_m)$ with f $<$-accessible. The point is: $(\dagger)$ if $s \gg g(t_1, \ldots, t_n)$ and s is head accessible, then g is $<$-accessible. This is proved by induction on the proof of $s \gg g(t_1, \ldots, t_n)$; the base case is when s is of the form $f(s_1, \ldots, s_m)$, where necessarily $f \geq g$, and f is $<$-accessible since s is head accessible.

We also define $\overline{Acc}$ as the set of head accessible μ-terms s whose immediate subterms are all in Acc.

Again, $\ll$ is a lifting, namely, every term in $\overline{Acc}$ is $\ll_{|\overline{Acc}}$-accessible. This is proved in two steps. We first show that every variable x is $\ll_{|\overline{Acc}}$-accessible (vacuous: $x \gg t$ for no μ-term t), and that every application $f(s_1, \ldots, s_m)$ in $\overline{Acc}$ is $\ll_{|\overline{Acc}}$-accessible: this is by double induction (<-induction on f, then $(<^{\mathrm{mpo}})^+_{\mathrm{mul}}$-induction on $\{\!|s_1, \ldots, s_m|\!\}$), using the fact that $f(s_1, \ldots, s_m) \gg t$ implies that $t = g(t_1, \ldots, t_n)$ with $f > g$ or $[f = g$ and $\{\!|s_1, \ldots, s_m|\!\} (>^{\mathrm{mpo}})^+_{\mathrm{mul}} \{\!|t_1, \ldots, t_n|\!\}]$. We then show that every iterator $\mu x = s \cdot s'$ in $\overline{Acc}$ is $\ll_{|\overline{Acc}}$-accessible, by $<^{\mathrm{mpo}}$-induction on s. To do so, we consider the μ-terms $t \in \overline{Acc}$ such that $t \ll \mu x = s \cdot s'$. Those obtained by rule $(\mu\ll)$ are $\ll_{|\overline{Acc}}$-accessible by the first step, and those obtained by rule $(\mu\ll\mu)$ are $\ll_{|\overline{Acc}}$-accessible by the induction hypothesis.

Let us pause a minute, and observe the following, called 'Property 1' in [26]. For all μ-terms s, t, if $s >^{\mathrm{mpo}} t$ then either:

(i) $s \rhd u \gtrsim^{\mathrm{mpo}} t$ for some μ-term u, or:
(ii) $s \gg t$ and $s >^{\mathrm{mpo}} u$ for every $u \lhd t$.

Case (i) happens in case $s >^{\mathrm{mpo}} t$ was derived using (Sub), (μSub), or (Var). Case (ii) happens in case it was derived using (Gt), (μGt), or $(\mu Gt\mu)$.

We now show that every $t \in W$ is in Acc, by structural induction on t. This means showing that for every $s \in \overline{Acc}$, s is in Acc. Since $\ll$ is a lifting, $s \in \overline{Acc}$ is $\ll_{|\overline{Acc}}$-accessible, so $\ll_{|\overline{Acc}}$-induction applies and we obtain the following induction hypothesis: (a) for every $t \ll s$, if $t \in \overline{Acc}$ then $t \in Acc$. Our goal is to prove that $s \in Acc$, i.e., that every $t <^{\mathrm{mpo}} s$ is in Acc. We show this by $\lhd$-induction on t, which means that we have the extra induction hypothesis: (b) for every $u \lhd t$, if $u <^{\mathrm{mpo}} s$ then $u \in Acc$. Since $t <^{\mathrm{mpo}} s$, either (i) or (ii) is true. If (i) holds, then $s \rhd u \gtrsim^{\mathrm{mpo}} t$, so $u \in Acc$ since $s \in \overline{Acc}$ and $u \lhd s$; therefore $t \in Acc$, by $(*)$ if $u \approx t$, by $<^{\mathrm{mpo}}$-inversion if $u >^{\mathrm{mpo}} t$. So assume (ii). We claim that t is in $\overline{Acc}$. This is trivial if t is a variable. If t is an application $g(t_1, \ldots, t_n)$ then for each j, $t_j \lhd s$, so by taking $u = t_j$ in (b), we obtain that t_j is in Acc; g is <-accessible since $s \gg g(t_1, \ldots, t_n)$, using $(\dagger)$; so $t \in \overline{Acc}$. If t is an iterator $\mu x = t_1 \cdot t_2$, then (b) only implies that t_2 is in Acc. To obtain $t_1 \in Acc$, we realize that we can only have derived $s \gg t$ by rule $(\mu\ll\mu)$, which implies that s is of the form $\mu x = s_1 \cdot s_2$ with $s_1 >^{\mathrm{mpo}} t_1$: since $s \in \overline{Acc}$, s_1 is in Acc hence t_1 is in Acc by $<^{\mathrm{mpo}}$-inversion. In any case, t is in $\overline{Acc}$. Since also $t \ll s$, (a) applies, so that t is in Acc, as desired. $\square$

4   A Constructive Proof of Higman's Lemma

It is time to apply all this and prove the topological Higman Lemma. Given a set X with a quasi-ordering $\leqslant$, the *embedding* quasi-ordering $\leqslant^*$ on X^* is the smallest relation such that $x_1 \leqslant y_1$, $\ldots$, $x_n \leqslant y_n$ imply $x_1 \ldots x_n \leqslant w_0 y_1 w_1 \ldots w_{n-1} y_n w_n$, where w_0, w_1, $\ldots$, w_{n-1}, w_n are arbitrary words in X^*. In other words, to go down in $\leqslant^*$, remove some letters and replace the others by smaller ones.

Higman's Lemma states that if $\leqslant$ is wqo, then so is $\leqslant^*$. The topological Higman Lemma states that if X is a Noetherian topological space, then X^* with the word topology is Noetherian, too. We have already discussed this in Section 2.

Step (A) of our proof plan consists in discovering an S-representation of X^*, for X Noetherian. (Step (A) is *not* constructive.) In [22], we defined an *S-representation* of a Noetherian space X as a tuple $(S, \mathcal{S}\,[\![\_]\!], \unlhd, \tau, \wedge)$, where S is a set of elements, meant to denote the irreducible closed subsets of X, through the denotation map $[\![\_]\!]$, $\unlhd$ denotes inclusion, τ represents the whole space, and $\wedge$ implements intersection. We change this slightly, and replace $\unlhd$ by its strict part $\sqsubset$ [2]. Hence, call *S-representation* of a Noetherian space X any tuple $(S, \sqsubset, \tau, \wedge)$, where S is a set, $[\![\_]\!] : S \to \mathcal{S}(X)$ is a bijective denotation function, $\sqsubset$ is a binary relation on S denoting strict inclusion (i.e., $a \sqsubset b$ iff $[\![a]\!] \subset [\![b]\!]$), τ is a finite subset of S denoting the whole of X ($[\![\tau]\!] = X$, where we extend the notation $[\![a]\!]$ for $a \in S$ to $[\![A]\!]$ for $A \in \mathbb{P}(S)$, by letting $[\![A]\!] = \bigcup_{a \in A} [\![a]\!]$), and for all $a, b \in S$, $a \wedge b$ is a finite subset of S denoting their intersection ($[\![a \wedge b]\!] = [\![a]\!] \cap [\![b]\!]$). When X is Noetherian, $\sqsubset$ will be well-founded (property ($\downarrow$)), τ will exist by property (T), and $\wedge$ will make sense because of property (W).

Since $[\![a']\!]$ is irreducible for every $a' \in A'$, the inclusion $[\![A]\!] \subseteq [\![A']\!]$ is equivalent to $A \sqsubseteq^\flat A'$, where we write $\sqsubseteq$ for the union of $\sqsubset$ and $=$, and the Hoare quasi-ordering $\sqsubseteq^\flat$ is defined by: for every $a \in A$, there is an $a' \in A'$ such that $a \sqsubseteq a'$. Since A, A' are antichains, one can encode them as multisets. A moment's notice shows that the strict part of $\sqsubseteq^\flat$ is just $\sqsubset_{mul}^+$. This will be used to compare antichains A, A' below.

$$\frac{e\boldsymbol{P} \sqsubseteq^{\mathrm{w}} \boldsymbol{P}'}{e\boldsymbol{P} \sqsubset^{\mathrm{w}} e'\boldsymbol{P}'}\ (\mathrm{w1}) \qquad \frac{a \sqsubset a'\quad \boldsymbol{P} \sqsubseteq^{\mathrm{w}} \boldsymbol{P}'}{a^?\boldsymbol{P} \sqsubset^{\mathrm{w}} a'^?\boldsymbol{P}'}\ (\mathrm{w2}) \qquad \frac{\boldsymbol{P} \sqsubset^{\mathrm{w}} \boldsymbol{P}'}{a^?\boldsymbol{P} \sqsubset^{\mathrm{w}} a^?\boldsymbol{P}'}\ (\mathrm{w3})$$

$$\frac{\forall i \cdot e_i \sqsubset^e A'^*\quad \boldsymbol{P} \sqsubseteq^{\mathrm{w}} \boldsymbol{P}'}{e_1 \ldots e_k \boldsymbol{P} \sqsubset^{\mathrm{w}} A'^* \boldsymbol{P}'}\ (\mathrm{w4}) \qquad \frac{\boldsymbol{P} \sqsubset^{\mathrm{w}} \boldsymbol{P}'}{A^* \boldsymbol{P} \sqsubset^{\mathrm{w}} A^* \boldsymbol{P}'}\ (\mathrm{w5})$$

Fig. 1. Deciding strict inclusion between word-products

Given an S-representation $(S, \mathcal{S}\,[\![\_]\!], \sqsubset, \tau, \wedge)$ of X, Theorem 6.14 of [22] gives us an S-representation $(S^{\mathrm{w}}, \mathcal{S}\,[\![\_]\!]^{\mathrm{w}}, \sqsubset^{\mathrm{w}}, \tau^{\mathrm{w}}, \wedge^{\mathrm{w}})$ of X^*. S^{w} is a set of so-called word-products, first invented in the setting of forward coverability procedures for lossy channel systems [2]. Define the *atomic expressions* as $a^?$ with $a \in S$ (denoting the set of words with at most one letter in $[\![a]\!]$), and A^* with A a non-empty finite antichain of S (denoting the set of words, of arbitrary length, whose

---

[2] In all rigor, we should also include the associated congruence $\equiv$, defined by $a \equiv b$ iff $a \unlhd b$ and $b \unlhd a$. We silently assume we are working in the quotient of the S-representation by $\equiv$. In proof assistants such as Coq, this is not an option, and the standard solution is to use setoid types. In any case, considering $\equiv$ explicitly would make our exposition too complex, and we shall therefore avoid it. We also change the notation from $\unlhd$ to $\sqsubset$ to avoid a conflict with the relations $\rhd$ of Section 3.

letters are all in $[\![A]\!]$). The *word-products* P, P', ..., are the finite sequences $e_1 e_2 \ldots e_n$ of atomic expressions, denoting the concatenations of words in the denotations of e_1, e_2, ..., e_n, and we define S^w as those that are *reduced*, namely those where $[\![e_i e_{i+1}]\!]^w$ is included neither in $[\![e_i]\!]^w$ nor in $[\![e_{i+1}]\!]^w$ for every i. Inclusion between word-products is decidable, using simple formulae given for example in [22, Lemma 6.8, Lemma 6.9], and this allows us to give computable predicates that sieve out the non-reduced word-products. We are more interested in the relation $\sqsubset^w$. Two reduced word-products, that is, two elements of S^w, have equal denotations iff they are equal. One can show that the strict inclusion relation $\sqsubset^w$ on reduced word-products is defined inductively by the rules of Figure 1. We write $P \sqsubseteq^w P'$ for $P \sqsubset^w P'$ or $P = P'$. We also define the auxiliary relation $\sqsubset^e$ (strict inclusion of atomic expressions) by: $a^? \sqsubset^e a'^?$ iff $a \sqsubset a'$; $a^? \sqsubset^e A'^*$ iff $a \sqsubseteq a'$ for some $a' \in A'$; $A^* \sqsubset^e a'^?$ never; and $A^* \sqsubset^e A'^*$ iff $A \sqsubset^+_{mul} A'$. We define τ^w as the antichain $\{\tau^*\}$, and omit the definition of $\wedge^w$ [22, Lemma 6.11].

We now embark on step (B) of our proof plan. Contrarily to step (A), we must pay attention to only invoke *constructive* arguments. So forget everything we have done in step (A), except for the final result. Say that $(S, \sqsubset, \tau, \wedge)$ is a *constructive S-representation* (without reference to X) if and only if S is a set with a strict ordering $\sqsubset$, and where: $(\downarrow)$ $\sqsubset$ is well-founded; (T) $S = \downarrow\tau$; (W) for all $a, b \in S$, $\downarrow a \cap \downarrow b = \downarrow(a \wedge b)$; $\sqsubseteq$ stands for the union of $\sqsubset$ and $=$, $\downarrow A$ for the downward closure of a subset A of S with respect to $\sqsubseteq$, and $\downarrow a$ for $\downarrow\{a\}$.

We now *posit* $(S^w, \sqsubset^w, \tau^w, \wedge^w)$ by the syntax given above, in step (A). S^w is the set of reduced word-products over S, $\sqsubset^w$ is defined inductively by (w1)–(w5), $\tau^w = \{\tau^*\}$, and we define $\wedge^w$ by the recursive formula of [22, Lemma 6.11].

Theorem 2. *If $(S, \sqsubset, \tau, \wedge)$ is a constructive S-representation, then so is $(S^w, \sqsubset^w, \tau^w, \wedge^w)$.*

Proof. (Sketch.) There is a boring part, consisting in checking that $\sqsubset^w$ is a strict ordering, and that properties (T) and (W) hold. We omit it here. The interesting part is checking that $\sqsubset^w$ is well-founded. Define a mapping μ from atomic expressions to pairs $(i, A) \in \{0, 1\} \times \mathcal{M}(S)$ by $\mu(a^?) = (0, \{\!|a|\!\})$, $\mu(A^*) = (1, A)$, and order them by the lexicographical product $\prec$ of the ordering $0 < 1$ and of $\sqsubset^+_{mul}$. Extend μ to word-products by $\mu(e_1 \ldots e_n) = \{\!|\mu(e_1), \ldots, \mu(e_n)|\!\}$. In other words, we look at word-products as though they were multisets of atomic expressions, where the latter as read as multisets of letters from S, plus a tag, 0 or 1. It is fairly easy to show that for all reduced word-products P, P', if $P \sqsubset P'$ then $\mu(P) \prec^+_{mul} \mu(P')$, by induction on the structure of a proof of $P \sqsubset P'$. By Lemma 1, $\prec^+_{mul}$ is well-founded. By $\prec^+_{mul}$-induction on $\mu(P)$, P is then $\sqsubset$-accessible, for every $P \in S^w$.   □

The statement of Theorem 2 seems very far from Higman's Lemma. Call *constructive Noetherian space* any tuple $(X, T, <, \varepsilon)$, where $<$ is a well-founded ordering on the set T (T is the *cotopology*) whose reflexive closure $\leq$ makes T a distributive lattice (this much implies classically that $(T, \leq)$ is the lattice of closed subsets of some Noetherian space, up to isomorphism), and $\varepsilon \subseteq X \times T$

(membership) is a binary relation such that for all $A, B \in T$, $A \preceq B$ iff for every $x \varepsilon A$, $x \varepsilon B$. We observe the following:

(a) Given a constructive S-representation $(S, \sqsubset, \tau, \wedge)$, we think of elements of S as irreducible closed subsets of some Noetherian space X, and we can build all closed sets as finite unions thereof. We encode the latter as finite antichains, hence as multisets. Letting $T = \mathcal{M}(S)$, $\prec = \sqsubset^+_{mul}$ then defines the *canonical cotopology* on $(S, \sqsubset, \tau, \wedge)$. Any subset X of S then gives rise to a constructive Noetherian space $(X, \mathcal{M}(S), \sqsubset^+_{mul}, \varepsilon)$, where $x \varepsilon M$ iff $\{\!|x|\!\} \; (\sqsubset_{mul})^* \; M$.

(b) Conversely, every cotopology $(T, \prec)$ gives rise to a trivial constructive S-representation $(S, \sqsubset, \tau, \wedge)$ where $S = T$, $\sqsubset$ is $\prec$, $\tau = \{\top\}$ where $\top$ is the top element of T, and $A \wedge B = \{A \sqcap B\}$ where $\sqcap$ is meet in T.

Given (a) and (b), Theorem 2 and Lemma 1 then imply:

Corollary 1 (Topological Higman Lemma, Constructively). *For every constructive Noetherian space* $(X, T, \prec, \varepsilon)$, $(X^*, \mathcal{M}(T^w), (\prec^w_{mul})^+, \varepsilon^w)$ *is a constructive Noetherian space, with* $w \; \varepsilon^w \; M$ *iff* $\{\!|\eta^w(w)|\!\} \; (\prec^w_{mul})^* \; M$, *where* $\eta^w(x_1 x_2 \ldots x_m) = x_1^? x_2^? \ldots x_m^?$.

This implies the usual form of Higman's Lemma, by similar arguments as in [37]. Assuming a decidable constructive wqo $\leqslant$ on a set X, one can show, constructively, that the antichains $E = \{x_1, \ldots, x_n\}$ (interpreted as the downward closed set $X \smallsetminus \uparrow E$) are the elements of a cotopology, where $\prec$ is the strict part of $\leqslant$; we let $E \leqslant E'$ iff $X \smallsetminus \uparrow E \subseteq X \smallsetminus \uparrow E'$, iff for every $y \in X'$, there is an $x \in E$ such that $x \leqslant y$; and $x \varepsilon E$ iff $x \in X \smallsetminus \uparrow E$, iff for every $y \in E$, $y \nleqslant x$. Recall that a finite sequence $w_1, \ldots, w_n$ in X^* is *bad* iff $w_i \leqslant^* w_j$ for no $i < j$. Following Murthy and Russell, we show that the converse of the prefix ordering on bad sequences $w_1, \ldots, w_n$ is well-founded, by $(\prec^w_{mul})^+$-induction on the closed subset $X^* \smallsetminus \uparrow\{w_1, \ldots, w_n\}$—this induction principle is given to us by Corollary 1. The set $X^* \smallsetminus \uparrow\{w_1, \ldots, w_n\}$ is represented, constructively, as the finite intersection of the sets $X \smallsetminus \uparrow w_i$, using the $\top$ and $\sqcap$ operations of the cotopology; writing w_i as the word $x_1 x_2 \ldots x_m$, $X \smallsetminus \uparrow w_i$ is the word-product $(X \smallsetminus \uparrow x_1)^* X^? (X \smallsetminus \uparrow x_2)^* X^? \ldots X^? (X \smallsetminus \uparrow x_m)^*$ if $m \geqslant 1$, the empty set otherwise [22, Lemma 6.1]. This is the core of Murthy and Russell's proof:

Theorem 3 (Murthy-Russell). *Let* X *be a set with a decidable constructive wqo* $\leqslant$. *Then* $\leqslant^*$ *is a (decidable) constructive wqo on* X^*.

5   A Constructive Proof of Kruskal's Theorem

We use the same strategy for trees, i.e., first-order terms. Given a set X with a quasi-ordering $\leqslant$, the (tree) *embedding* quasi-ordering $\leqslant_{\leqslant}$ is inductively defined by $s \leqslant_{\leqslant} t$, where $s = f(s_1, \ldots, s_m)$ and $t = g(t_1, \ldots, t_n)$, iff $s \leqslant_{\leqslant} t_j$ for some j, or $f \leqslant g$ and $s_1 \ldots s_m \leqslant^*_{\leqslant} t_1 \ldots t_n$; note the use of the word embedding ordering $\leqslant^*_{\leqslant}$ on lists of immediate subterms, considered as words.

Given a constructive S-representation $(S, \sqsubset, \tau, \wedge)$, we define a set S^t of regular expressions on trees (the *tree-products* $P, Q, \ldots$) inductively, as follows. Let $\square$ be a fresh constant. The elements P of S^t are the *tree steps* $a^{[?]}(\boldsymbol{P})$, where $a \in S$ and $\boldsymbol{P}$ is a reduced word-product over S^t, and the *tree iterators* $(\bigcup_{i=1}^m a_i(\boldsymbol{Q}_i))^{[*]}._{\square}A$, where A is a finite set of elements of S^t, $a_i \in S$, and $\boldsymbol{Q}_i$ is a word-product over $S \cup \{\square\}$, which is either equal to $\{\square\}^*$ (which we shall simply write $\square^*$), or of the form $\boldsymbol{Q}_{i1}\square^? \boldsymbol{Q}_{i2}\square^? \ldots \square^? \boldsymbol{Q}_{ik_i}$ where all $\boldsymbol{Q}_{ij}$s are reduced word-products over S^t [22, Lemma 9.20].

Tree steps are the analogue of $a^?$ for words. Intuitively, $a^{[?]}(e_1 e_2)$ will contain all the terms of the form $f(t_1, t_2)$ with f in (the denotation of) a, t_1 in e_1 and t_2 in e_2, plus all the terms from e_1 and from e_2. Of course, $e_1 e_2$ is not a word-product, but, say, $e_1^? e_2^?$ is, and $a^{[?]}(e_1^? e_2^?)$ will contain not just the terms above, but also the terms of the form $f(t_1)$, $f(t_2)$ and $f()$, with $f \in a$, $t_1 \in e_1$, $t_2 \in e_2$.

Tree iterators $(\bigcup_{i=1}^m a_i(\boldsymbol{Q}_i))^{[*]}._{\square}A$ define the following language L, inductively, by the following two rules. First, A is a set $\{P_1, \ldots, P_n\}$ of tree-products, and every element of any P_i is in L. Second, given any term t in the set denoted by $a_i^{[?]}(\boldsymbol{Q}_i)$, the term obtained from t by replacing each occurrence of $\square$ by a (possibly different) term from L is again in L. For example, if P contains terms t_1, t_2 and t_3, then $(a(\square^? \square^?))^{[*]}._{\square}\{P\}$ will contain $f(t_1, t_1)$, $f(t_2, t_2)$, but also $f(t_1, t_2)$, $f(t_1, f(t_2, t_2))$, $f(f(t_1, f(t_2, t_1)), f(t_1, t_3))$, for f in a, among other terms. As another example, $(a(\square^*))^{[*]}._{\square}\varnothing$ is the set of terms all of whose function symbols are in a.

Much as we only considered reduced word-products in Section 4, we shall restrict to *canonical* tree-products here. The tree-products considered in [22, Section 9] are *normal* tree-products, a closely related notion. Normality requires, for example, that in a tree iterator, $(\bigcup_{i=1}^m a_i(\boldsymbol{Q}_i))^{[*]}._{\square}A$, (i) m is non-zero, (ii) $\square$ occurs in every $\boldsymbol{Q}_i$, and (iii) A contains just one tree-product in case every $\boldsymbol{Q}_i$ is $\square$-*linear*, i.e., does not contain $\square^*$ and only one occurrence of $\square^?$. Here, we need to require that the support supp $\boldsymbol{Q}_i$ of every $\boldsymbol{Q}_i$, namely, the set of ($\square$-free) terms t such that the one-element sequence t is in the denotation of $\boldsymbol{Q}_i$, is entirely contained in the denotation of A. This is easy to ensure, by adding the required tree-products from supp $\boldsymbol{Q}_i$ to A... but breaks (iii). Instead, we define *canonical* tree iterators as those satisfying (i), (ii), (iii'): if every $\boldsymbol{Q}_i$ is $\square$-linear, then A denotes the union of $\bigcup_{i=1}^m$ supp $\boldsymbol{Q}_i$ with at most one tree-product; we also require: (iv) the tree steps $a_i(\boldsymbol{Q}_i)$ are pairwise incomparable, (v) the elements of A are pairwise incomparable, and (vi) $(\bigcup_{i=1}^m a_i(\boldsymbol{Q}_i))^{[*]}._{\square}A$ must not be included in $\bigcup_{i=1}^m$ supp $\boldsymbol{Q}_i$. Similarly, we define *canonical* trees steps as those $a^?(\boldsymbol{P})$ that are not included in supp $\boldsymbol{P}$. Every tree-product can be *canonicalized*, i.e., transformed to a canonical one with the same denotation.

One can decide inclusion of canonical tree-products, in polynomial time, and also compute finite intersections thereof ($\wedge^t$, τ^t), using formulae given in [22, Section 9], plus canonicalization. From these formulae, we deduce the rules for strict inclusion $\sqsubset^t$ on S^t—to be precise, on S^t union $\{\square\}$, where $\square$ will be topmost—given in Figure 2. We again write $\sqsubseteq^t$ for the reflexive closure of $\sqsubset^t$. For a word-product $\boldsymbol{P}$ over $S^t \cup \{\square\}$, define sub($\boldsymbol{P}$) (denoting the support of $\boldsymbol{P}$) by: sub($e_1 \ldots e_n$) = $\bigcup_{i=1}^n$ sub(e_i), sub($P^?$) = $\{P\}$ for $P \in S^t$, sub($\square^?$) = $\varnothing$, sub(A^*) = A for A an antichain in S^t, sub($\square^*$) = $\varnothing$. We also use an auxiliary

$$\frac{(P' \in S^{\mathrm{t}}) \qquad \exists P \in \mathrm{sub}(\boldsymbol{P}) \cdot P' \sqsubseteq^{\mathrm{t}} P}{P' \sqsubset^{\mathrm{t}} a^{[?]}(\boldsymbol{P})} \qquad \frac{(P' \in S^{\mathrm{t}})}{P' \sqsubset^{\mathrm{t}} \Box} \qquad \frac{a'^{[?]}(\boldsymbol{P}') \sqsubseteq^{\mathrm{t}} a^{[?]}(\boldsymbol{P}) \qquad \forall P' \in \mathrm{sub}(\boldsymbol{P}') \cdot P' \sqsubset^{\mathrm{t}} a^{[?]}(\boldsymbol{P})}{a'^{[?]}(\boldsymbol{P}') \sqsubset^{\mathrm{t}} a^{[?]}(\boldsymbol{P})}$$

$$\frac{a' \sqsubset a}{a'^{[?]}(\boldsymbol{P}) \sqsubseteq^{\mathrm{t}} a^{[?]}(\boldsymbol{P})} \qquad \frac{\boldsymbol{P}' \ (\sqsubset^{\mathrm{t}})^{\mathrm{w}} \ \boldsymbol{P}}{a^{[?]}(\boldsymbol{P}') \sqsubseteq^{\mathrm{t}} a^{[?]}(\boldsymbol{P})}$$

$$\frac{(P' \in S^{\mathrm{t}}) \qquad \exists P \in A \cdot P' \sqsubseteq^{\mathrm{t}} P}{P' \sqsubset^{\mathrm{t}} (\bigcup_{i=1}^{m} a_i(\boldsymbol{Q}_i))^{[*]} \cdot_{\Box} A} \qquad \frac{\exists i \cdot a'^{[?]}(\boldsymbol{P}') \sqsubseteq^{\mathrm{t}} a_i^{[?]}(\boldsymbol{Q}_i) \qquad \forall P' \in \mathrm{sub}(\boldsymbol{P}') \cdot P' \sqsubset^{\mathrm{t}} (\bigcup_{i=1}^{m} a_i(\boldsymbol{Q}_i))^{[*]} \cdot_{\Box} A}{a'^{[?]}(\boldsymbol{P}') \sqsubset^{\mathrm{t}} (\bigcup_{i=1}^{m} a_i(\boldsymbol{Q}_i))^{[*]} \cdot_{\Box} A}$$

$$\frac{\{\!|a_j'^{[?]}(\boldsymbol{Q}_j') \mid 1 \leqslant j \leqslant n|\!\} \sqsubseteq^{\mathrm{t}}_{\mathrm{mul}} \{\!|a_i^{[?]}(\boldsymbol{Q}_i) \mid 1 \leqslant i \leqslant n|\!\} \qquad \forall P' \in A' \cdot P' \sqsubset^{\mathrm{t}} (\bigcup_{i=1}^{m} a_i(\boldsymbol{Q}_i))^{[*]} \cdot_{\Box} A}{(\bigcup_{j=1}^{n} a_j'(\boldsymbol{Q}_j'))^{[*]} \cdot_{\Box} A' \sqsubset^{\mathrm{t}} (\bigcup_{i=1}^{m} a_i(\boldsymbol{Q}_i))^{[*]} \cdot_{\Box} A}$$

Fig. 2. Deciding strict inclusion between tree-products

relation $\sqsubseteq^{\mathrm{t}}$, which should be reminiscent of $\ll$. The whole definition should, in fact, remind you of the definition of $<^{\mathrm{mpo}}$ on μ-terms, and this is no accident.

Theorem 4. *If $(S, \sqsubset, \tau, \wedge)$ is a constructive S-representation, then so is $(S^{\mathrm{t}}, \sqsubset^{\mathrm{t}}, \tau^{\mathrm{t}}, \wedge^{\mathrm{t}})$.*

Proof. (Sketch.) Only property ($\downarrow$) deserves attention. Define a syntactic translation from $P \in S^{\mathrm{t}}$ to μ-terms $\langle P \rangle$, as follows. Our signature consists of all elements of S, plus one fresh function symbol $\mathtt{u}$ (union). The following formulae also define $\langle \_ \rangle$ translations of various other syntactic categories, e.g., $\langle \boldsymbol{P} \rangle$ will be a list of μ-terms for every word-product $\boldsymbol{P}$ over S^{t}, so that $\langle a^?(\boldsymbol{P}) \rangle = a \langle \boldsymbol{P} \rangle$ will be the application of the function symbol a to the list of arguments $\langle \boldsymbol{P} \rangle$. We use only *one* μ-bound variable, which we call $\Box$: this serves for tree iterators, which are translated as iterators of the form $\mu \Box = s \cdot t$ (third row below).

$$\langle a^{[?]}(\boldsymbol{P}) \rangle = a\langle \boldsymbol{P} \rangle \qquad \langle e_1 e_2 \ldots e_m \rangle = (\langle e_1 \rangle, \langle e_2 \rangle, \ldots, \langle e_m \rangle) \qquad \langle \Box \rangle = \Box$$
$$\langle P^? \rangle = \langle P \rangle \qquad \langle A^* \rangle = \langle A \rangle = \mathtt{u}(\langle P_1 \rangle, \ldots, \langle P_n \rangle) \text{ where } A = \{\!|P_1, \ldots, P_n|\!\}$$
$$\langle (\textstyle\bigcup_{i=1}^{m} a_i(\boldsymbol{Q}_i))^{[*]} \cdot_{\Box} A \rangle \ = \ \mu \Box = \mathtt{u}(\langle a_1^{[?]}(\boldsymbol{Q}_1) \rangle, \ldots, \langle a_m^{[?]}(\boldsymbol{Q}_m) \rangle) \cdot \langle A \rangle$$

Define the precedence $<$ by $a < b$ iff $a, b \in S$ and $a \sqsubset b$, or $a = \mathtt{u}$ and $b \in S$ ($\mathtt{u}$ is least). We check that $P \sqsubset^{\mathrm{t}} P'$ implies $\langle P \rangle <^{\mathrm{mpo}} \langle P' \rangle$. (This was how $<^{\mathrm{mpo}}$ was found on μ-terms!) Theorem 1 then implies that $\sqsubset^{\mathrm{t}}$ is well-founded on S^{t}. $\qquad \Box$

Corollary 2 (Topological Kruskal Theorem, Constructively). *For every constructive Noetherian space $(X, T, <, \varepsilon)$, $(\mathcal{T}(X), \mathcal{M}(T^{\mathrm{t}}), (<^{\mathrm{t}}_{\mathrm{mul}})^+, \varepsilon^{\mathrm{t}})$ is a constructive Noetherian space, with $t \ \varepsilon^{\mathrm{t}} \ M$ iff $\{\!|\eta^{\mathrm{t}}(t)|\!\} \ (<^{\mathrm{t}}_{\mathrm{mul}})^* \ M$, where $\eta^{\mathrm{t}}(f(t_1, t_2, \ldots, t_n)) = f^{[?]}(\eta^{\mathrm{t}}(t_1)^? \eta^{\mathrm{t}}(t_2)^? \ldots \eta^{\mathrm{t}}(t_n)^?)$.*

As for Higman's Lemma, we obtain the ordinary form of Kruskal's Theorem by assuming a decidable constructive wqo $\leqslant$ on X, and proving that the complement $\mathsf{C}t$ of the upward closure of a single tree t in $\leq_{\leqslant}$ is defined as the following tree-product, built using tree iterators only: letting a abbreviate $X \smallsetminus {\uparrow}f$ and b abbreviate X itself, $\mathsf{C}f() = (a(\square^*))^{[*]}._{\square}\varnothing$, and $\mathsf{C}f(t_1, \ldots, t_n)$ for $n \geqslant 1$ is equal to $(a(\square^*) \cup b(\mathsf{C}t_1\square^? \mathsf{C}t_2\square^? \ldots \square^? \mathsf{C}t_n))^{[*]}._{\square}\varnothing$ (see [22, Lemma 9.8]; then use canonicalization). The following is then constructive.

Theorem 5 (Kruskal, Constructively). *Let X be a set with a decidable constructive wqo $\leqslant$. Then $\leq_{\leqslant}$ is a (decidable) constructive wqo on $\mathcal{T}(X)$.*

6 Conclusion

The main thing one should remember is that proving that a given quasi-ordering is well is just a matter of proving termination—not of the ordering itself, but of strict inclusion between downward-closed subsets. In and of itself, this would be no breakthrough. However, in applying this to Higman's and Kruskal's classical theorems, this exposed a tight coupling between the word embedding ordering and the multiset ordering (on word-products), and between the tree embedding quasi-ordering and Dershowitz' multiset path ordering (on tree-products).

While I have given relatively exhaustive proofs of the termination results of Section 3, I have barely sketched the constructive proofs of the (topological) Higman and Kruskal theorems in Sections 4 and 5. Playing these proofs in a proof assistant such as Coq is in order, but certainly somewhat of an endeavor.

Finally, I would like to stress that although our proof techniques establish the classical, order-theoretic versions of Higman's and Kruskal's theorems under a decidability assumption, the topological versions are *entirely* constructive. It therefore seems that the constructive contents of the order-theoretic and the topological theorems are different—something that should be explored, by investigating into the computational contents of the relevant constructive proofs. We also believe that the notion of constructive Noetherian space, and the related notion of S-representation, should be of some importance, in intuitionistic logic (where it sheds some light on the precise role of the sequential regular expressions of Murthy and Russell, notably), as well as in the field of WSTS model-checking.

Acknowledgments. I must thank David Baelde, who found a mistake in an early version of this paper, and Nachum Dershowitz, who gave me several additional pointers. I have had several interesting discussions with Sylvain Schmitz, Alain Finkel, and Jean-Pierre Jouannaud. All remaining errors are of course mine.

References

1. Abdulla, P.A., Čerāns, K., Jonsson, B., Tsay, Y.-K.: Algorithmic analysis of programs with well quasi-ordered domains. Inf. Comput. 160(1-2), 109–127 (2000)
2. Abdulla, P.A., Collomb-Annichini, A., Bouajjani, A., Jonsson, B.: Using forward reachability analysis for verification of lossy channel systems. Formal Methods in System Design 25(1), 39–65 (2004)

3. Abdulla, P.A., Jonsson, B.: Verifying programs with unreliable channels. In: LICS 1993, pp. 160–170. IEEE Computer Society Press (1993)
4. Acciai, L., Boreale, M.: Deciding safety properties in infinite-state pi-calculus via behavioural types. In: Albers, S., Marchetti-Spaccamela, A., Matias, Y., Nikolet-seas, S., Thomas, W. (eds.) ICALP 2009, Part II. LNCS, vol. 5556, pp. 31–42. Springer, Heidelberg (2009)
5. Berghofer, S.: A constructive proof of Higman's lemma in Isabelle. In: Berardi, S., Coppo, M., Damiani, F. (eds.) TYPES 2003. LNCS, vol. 3085, pp. 66–82. Springer, Heidelberg (2004)
6. Bertot, Y., Castéran, P.: Interactive Theorem Proving and Program Development—Coq'Art: The Calculus of Inductive Constructions. Texts in Theor. Comp. Sci., an EATCS Series, vol. XXV. Springer (2004)
7. Coquand, T., Fridlender, D.: A proof of Higman's lemma by structural induction (November 2003) (unpublished note),
http://www.cse.chalmers.se/~coquand/open1.ps
8. Coupet-Grimal, S., Delobel, W.: An effective proof of the well-foundedness of the mul-tiset path ordering. Appl. Algebra Eng., Commun. Comput. 17(6), 453–469 (2006)
9. Dawson, J.E., Goré, R.: A general theorem on termination of rewriting. In: Marcinkowski, J., Tarlecki, A. (eds.) CSL 2004. LNCS, vol. 3210, pp. 100–114. Springer, Heidelberg (2004)
10. de Groote, P., Guillaume, B., Salvati, S.: Vector addition tree automata. In: LICS 2004, pp. 64–73. IEEE Computer Society Press (2004)
11. Dershowitz, N.: A note on simplification orderings. Inf. Proc. Letters 9(5), 212–215 (1979)
12. Dershowitz, N.: Orderings for term-rewriting systems. Theor. Comp. Sci. 17(3), 279–301 (1982)
13. Dershowitz, N.: Termination of rewriting. J. Symb. Comp. 3, 69–116 (1987)
14. Dershowitz, N.: Jumping and escaping: Modular termination and the abstract path ordering. Theor. Comp. Sci. 464, 35–47 (2012)
15. Dershowitz, N., Hoot, C.: Natural termination. Theor. Comp. Sci. 142(2), 179–207 (1995)
16. Dershowitz, N., Jouannaud, J.-P.: Rewrite systems. In: van Leeuwen, J. (ed.) Hand-book of Theor. Comp. Sci., ch. 6, pp. 243–320. Elsevier (1990)
17. Dershowitz, N., Manna, Z.: Proving termination with multiset orderings. Comm. ACM 22(8), 465–476 (1979)
18. Dickson, L.E.: Finiteness of the odd perfect and primitive abundant numbers with n distinct prime factors. Amer. J. Math. 35(4), 413–422 (1913)
19. Ferreira, M.C.F., Zantema, H.: Well-foundedness of term orderings. In: Linden-strauss, N., Dershowitz, N. (eds.) CTRS 1994. LNCS, vol. 968, pp. 106–123. Springer, Heidelberg (1995)
20. Finkel, A., Goubault-Larrecq, J.: Forward analysis for WSTS, part I: Completions. In: STACS 2009, Freiburg, Germany, pp. 433–444. Leibniz-Zentrum für Informatik, Intl. Proc. in Informatics 3 (2009)
21. Finkel, A., Goubault-Larrecq, J.: Forward analysis for WSTS, part II: Complete WSTS. In: Albers, S., Marchetti-Spaccamela, A., Matias, Y., Nikoletseas, S., Thomas, W. (eds.) ICALP 2009, Part II. LNCS, vol. 5556, pp. 188–199. Springer, Heidelberg (2009)
22. Finkel, A., Goubault-Larrecq, J.: Forward analysis for WSTS, part I: Completions. Logical Methods in Computer Science (2012) (submitted)
23. Finkel, A., Goubault-Larrecq, J.: Forward analysis for WSTS, part II: Complete WSTS. Logical Methods in Computer Science 8(3:28) (2012)
24. Finkel, A., McKenzie, P., Picaronny, C.: A well-structured framework for analysing Petri net extensions. Inf. Comput. 195(1-2), 1–29 (2004)

25. Finkel, A., Schnoebelen, P.: Well-structured transition systems everywhere! Theor. Comp. Sci. 256(1-2), 63–92 (2001)
26. Goubault-Larrecq, J.: Well-founded recursive relations. In: Fribourg, L. (ed.) CSL 2001. LNCS, vol. 2142, pp. 484–497. Springer, Heidelberg (2001)
27. Goubault-Larrecq, J.: On Noetherian spaces. In: LICS 2007, pp. 453–462. IEEE Computer Society Press (2007)
28. Goubault-Larrecq, J.: Noetherian spaces in verification. In: Abramsky, S., Gavoille, C., Kirchner, C., Meyer auf der Heide, F., Spirakis, P.G. (eds.) ICALP 2010. LNCS, vol. 6199, pp. 2–21. Springer, Heidelberg (2010)
29. Goubault-Larrecq, J.: Non-Hausdorff Topology and Domain Theory—Selected Topics in Point-Set Topology. New Mathematical Monographs, vol. 22. Cambridge University Press (2013)
30. Higman, G.: Ordering by divisibility in abstract algebras. Proc. London Math. Soc. 2(7), 326–336 (1952)
31. Kamin, S., Lévy, J.-J.: Attempts for generalizing the recursive path orderings. Unpublished letter to N. Dershowitz (1980), http://nachum.org/term/kamin-levy80spo.pdf
32. Karp, R.M., Miller, R.E.: Parallel program schemata. J. Comp. Sys. Sci. 3(2), 147–195 (1969)
33. Kruskal, J.B.: Well-quasi-ordering, the tree theorem, and Vazsonyi's conjecture. Trans. AMS 95(2), 210–225 (1960)
34. Lazič, R., Newcomb, T., Ouaknine, J., Roscoe, A.W., Worrell, J.: Nets with tokens which carry data. Fund. Informaticae 88(3), 251–274 (2008)
35. Lescanne, P.: Some properties of decomposition ordering, a simplification ordering to prove termination of rewriting systems. RAIRO Theor. Inform. 16(4), 331–347 (1982)
36. Meyer, R.: On boundedness in depth in the π-calculus. In: Ausiello, G., Karhumäki, J., Mauri, G., Ong, L. (eds.) TCS 2008. IFIP, vol. 273, pp. 477–489. Springer, Boston (2008)
37. Murthy, C.R., Russell, J.R.: A constructive proof of Higman's lemma. In: LICS 1990, pp. 257–267. IEEE Computer Society Press (1990)
38. Nash-Williams, C.S.-J.A.: On well-quasi-ordering infinite trees. Proc. Cambridge Phil. Soc. 61, 697–720 (1965)
39. Nipkow, T.: An inductive proof of the wellfoundedness of the multiset order. Technical report, Technische Universität München (October 1998)
40. Ogawa, M.: A linear time algorithm for monadic querying of indefinite data over linearly ordered domains. Inf. Comput. 186(2), 236–259 (2003)
41. Richman, F., Stolzenberg, G.: Well quasi-ordered sets. Technical report, Northeastern University, Boston, MA and Harvard University, Cambridge, MA (1990)
42. Rosa-Velardo, F., Martos-Salgado, M.: Multiset rewriting for the verification of depth-bounded processes with name binding. Inf. Comput. 215, 68–87 (2012)
43. Seisenberger, M.: Kruskal's tree theorem in a constructive theory of inductive definitions. In: Proc. Symp. Reuniting the Antipodes—Constructive and Nonstandard Views of the Continuum. Synthese Library, vol. 306. Kluwer Academic Publishers, Dordrecht (1999, 2001)
44. van der Meyden, R.: The complexity of querying indefinite data about linearly ordered domains. J. Comp. Sys. Sci. 54(1), 113–135 (1997)
45. Veldman, W.: An intuitionistic proof of Kruskal's theorem. Report 0017, Dept. of Mathematics, U. Nijmegen (2000)
46. Verma, K.N., Goubault-Larrecq, J.: Karp-Miller trees for a branching extension of VASS. Discr. Math. & Theor. Comp. Sci. 7(1), 217–230 (2005)
47. Wies, T., Zufferey, D., Henzinger, T.A.: Forward analysis of depth-bounded processes. In: Ong, L. (ed.) FOSSACS 2010. LNCS, vol. 6014, pp. 94–108. Springer, Heidelberg (2010)

Logical and Structural Approaches to the Graph Isomorphism Problem

Martin Grohe

RWTH Aachen University
grohe@informatik.rwth-aachen.de

Abstract. It is a long-standing open question whether there is a polynomial time algorithm deciding if two graphs are isomorphic. Indeed, graph isomorphism is one of the very few natural problems in NP that is neither known to be in P nor known to be NP-complete. The question is still wide open, but a number of deep partial results are known. On the complexity theoretic side, we have good reason to believe that graph isomorphism is not NP-complete: if it was NP-complete, then the polynomial hierarchy would collapse to its second level. On the algorithmic side, we know a nontrivial algorithm with a worst-case running time of $2^{O(\sqrt{n \log n})}$ and polynomial time algorithms for many specific classes of graphs. Many of these algorithmic results have been obtained through a group theoretic approach that dominated the research on the graph isomorphism problem since the early 1980s.

After an introductory survey, in my talk I will focus on approaches to the graph isomorphism problem based on structural graph theory and connections between logical definability, certain combinatorial algorithms, and mathematical programming approaches to the isomorphism problem.

K. Chatterjee and J. Sgall (Eds.): MFCS 2013, LNCS 8087, p. 42, 2013.
© Springer-Verlag Berlin Heidelberg 2013

Prior-Free Auctions of Digital Goods

Elias Koutsoupias

Department of Computer Science
University of Oxford
elias@cs.ox.ac.uk

The study of prior-free auctions brings the Computer Science approach of worst-case analysis to the classical economic problem of designing optimal auctions. In this talk, I will discuss some recent developments on prior-free auctions for digital goods.

A digital good, which can be reproduced without any cost, is to be sold to a set of bidders (potential buyers); each bidder is willing to pay up to a certain price to acquire the item and this price is a private value. The seller wants to maximize the profit without knowing the private values of the bidders. Had the private values been available, the solution would have been trivial—sell to every bidder at their maximum acceptable price—but, the private information of the bidders renders the objective unattainable, even in-approximable, when the values are selected by an adversary. However, all is not lost if we are willing to compare the obtained profit with a weaker optimal objective. A natural such objective is $F^{(2)}$, equal to the optimal value that one can extract by offering the same price to everyone; there is also a technical requirement that the price is low enough to allow at least two bidders to buy the item. The obvious question is to come up with an auction that achieves the best approximation ratio over such an objective. Despite the simplicity of the framework, this question is still open; even specific interesting algorithms, such as the RSOP algorithm, are not completely understood.

In my talk, I will focus on two extensions of this framework:

Online Prior-Free Auctions. In this version, the bidders arrive one-by-one in a random order and the seller must offer each bidder a take-it-or-leave-it price. Because of the random arrival of the bidders, this model lies at the intersection of prior-free auctions and secretary problems. I will discuss specific natural online auctions, as well as the relation between online and offline auctions. As in the case of offline auctions, we know algorithms that achieve constant approximation ratio, but the optimal ratio is still unknown.

Ordered Bidders. In this version, the bidders have a specific order, although they are not processed online. The requirement is that instead of competing against the benchmark $F^{(2)}$, the auction competes against the harder benchmark $M^{(2)}$, equal to the optimal profit from a set of prices that are non-decreasing (with respect to the fixed order of the bidders); as in the case of $F^{(2)}$, there is a technical requirement that the prices should not be higher than the second

K. Chatterjee and J. Sgall (Eds.): MFCS 2013, LNCS 8087, pp. 43–44, 2013.

value. Benchmark $M^{(2)}$ captures the case of asymmetric bidders: an auction with constant approximation ratio against $M^{(2)}$ is within a constant factor of the optimal profit when the bids are drawn from any sequence of probability distributions in which each distribution stochastically dominates the next one. I will discuss a recent result that gives an auction with constant approximation ratio in this framework.

Synthesis from Temporal Specifications: New Applications in Robotics and Model-Driven Development

Nir Piterman

University of Leicester

Synthesis from temporal specifications is the automatic production of adaptable plans (or input enabled programs) from high level descriptions. The assumption underlying this form of synthesis is that we have two interacting reactive agents. The first agent is the system for which the plan / program is being designed. The second agent is the environment with which the system interacts. The exact mode of interaction and the knowledge available to each of the agents depends on the application domain. The high level description of the plan is usually given in some form of temporal logic, where we often distinguish between assumptions and guarantees. As we do not expect the system to function correctly in arbitrary environments, the assumptions detail what the system expects from the environment. The guarantees are what the system is expected to fulfill in such environments. Our algorithms then produce a plan that interacts with the environment and reacts to it so that the tasks assigned to the plan are fulfilled. By definition, the plan is reacting to the moves of the environment and tries to adapt itself to the current condition (as a function of that interaction).

Technically, the interaction between the system and its environment is modeled as a two-player game, where system choices correspond to the execution of the plan and environment choices correspond to the behavior of the environment. The specifications, i.e., the assumptions on the environment and the guarantees of the system, are translated to the winning conditions in the game: the system has to be able to resolve its choices in such a way that it satisfies the specification. The way the system resolves its choices, called *strategy*, is then translated to a design that satisfies the specification. Verifying that such a strategy exists and computing the strategy is referred to as "solving the game". Different types of games arise depending on the exact conditions of the interaction between the agents, and depending on the winning conditions. In order to make synthesis useful we have to come up with algorithms that work well for the games that arise from interesting applications.

The theoretical framework for synthesis from temporal specifications has been known for many years. The question of decidability of this form of synthesis was raised by Church in the late 50's [8]. Independently, Rabin [29] and Büchi and Landweber [7] suggested tree automata and two-player games as a way to reason about the interaction between the program and its environment. These solutions concentrated on decidability and were not concerned with practicality. Pnueli and Rosner cast this question in a modern setting and proved that synthesis

K. Chatterjee and J. Sgall (Eds.): MFCS 2013, LNCS 8087, pp. 45–49, 2013.

from linear-time temporal logic (LTL) specifications is 2EXPTIME-complete [28]. Indeed, this is the framework considered here.

The solution of Pnueli and Rosner called for the translation of the specification to a deterministic Rabin automaton over infinite words [31]. Integrating this automaton with the approach of Rabin produced a Rabin tree automaton accepting winning strategies. Checking emptiness of this automaton corresponds to deciding whether the specification is *realizable*. Finding a tree accepted by this automaton corresponds to extracting a strategy. The two components of this solution proved very hard to implement. Determinization of automata on infinite words proved complicated to implement [18,1]. To the best of our knowledge, emptiness of Rabin automata (equivalently solution of Rabin games) was never implemented [13,28,26]. Improvements to determinization [25] are still challenging to implement effectively [34]. They lead to the slightly simpler parity automata / games, for which no efficient solution is known [17,15,32].

These difficulties led researchers to suggest two ways to bypass the two complicated parts of this approach. One approach is to avoid determinization and reduce synthesis to safety games [24,14,33]. This approach has been implemented in various tools [16,12,5]. The second approach, the one advocated here, is to restrict attention to a subset of LTL that can be solved more efficiently [27,4].

Specifically, we consider LTL formulas over Boolean variables partitioned to sets of *inputs* and *outputs*, $\mathcal{X}$ and $\mathcal{Y}$, respectively. Then, the specification has the format $\varphi_e \to \varphi_s$, where φ_e is a conjunction of assumptions on the behavior of the environment and φ_s is a conjunction of guarantees of the system. Both φ_e and φ_s are restricted to the form $\psi_i^a \wedge \Box \rho_t^a \wedge \bigwedge_{i \in I_g^a} \Box \Diamond J_i^a$, for $a \in \{e, s\}$, where the components of φ_a take the following form.

- ψ_i^e is a Boolean formula over $\mathcal{X}$ and ψ_i^s is a Boolean formula over $\mathcal{X} \cup \mathcal{Y}$.
- ρ_t^e is a Boolean formula over $\mathcal{X} \cup \mathcal{Y}$ and $\bigcirc \mathcal{X}$ and ρ_t^s is a Boolean formula over $\mathcal{X} \cup \mathcal{Y}$ and $\bigcirc \mathcal{X} \cup \bigcirc \mathcal{Y}$. That is, ρ_t^e is allowed to relate to the next values of input variables and ρ_t^s is allowed to relate to the next values of both input and output variables..
- J_i^a is a Boolean formula over $\mathcal{X} \cup \mathcal{Y}$.

That is, the specification takes the following format:

$$\left(\psi_i^e \wedge \Box \rho_t^e \wedge \bigwedge_{i \in I_g^e} \Box \Diamond J_i^e \right) \to \left(\psi_i^s \wedge \Box \rho_t^s \wedge \bigwedge_{i \in I_g^s} \Box \Diamond J_i^s \right)$$

Intuitively, this formula allows the system to update its initial assignment to output variables based on the assignment to the input variables; it allows the system to update output variables based on the way the environment updates the input variables; and it allows the system to fulfill some liveness requirements based on the environment fulfilling its own liveness requirements.[1] We argue

---

[1] We note that presentation of the specification in the form of such an implication depends on the ability of the environment to fulfil its assumptions [19,4].

that this form of specifications arise in practice and are sufficient to specify many interesting designs. Furthermore, we show how to implement the solution to the synthesis problem arising from such specifications using BDDs.

This approach has been adopted by some practitioners and led to applications of synthesis in hardware design [2,3], robot-controller planning [9,20,21,35,37,36], and user programming [22,23]. Adapting our solution to be used in the context of robot-controller required to consider how to combine the discrete controller produced by our approach with continuous controllers for various parts of the robot [30]. Recently, we have adapted this approach to applications in model-driven development [10,11,6]. This required us to adjust the setting to that of games defined by labeled-transition systems, winning conditions defined by fluent linear-temporal logic, and to enumerative representation of games.

Here we will survey the theoretical solution to synthesis proposed by Pnueli and Rosner and some of the difficulties in applying it in practice. We will then present our approach and some of the applications it was used for. We will also cover some of the issues arising from adaptation of our approach to the usage by practitioners in robotics and model-driven development.

Acknowledgements. The work surveyed here is based on joint work with (mostly) Roderick Bloem, Victor Braberman, Nicolas D'Ippolito, Barbara Jobstmann, Hadas Kress-Gazit, Amir Pnueli, Vasu Raman, Yaniv Sa'ar, and Sebastian Uchitel. References to our joint work are mentioned in the paper. I am grateful to them for the great pleasure in working with them on these results.

References

1. Althoff, C.S., Thomas, W., Wallmeier, N.: Observations on determinization of Büchi automata. In: Farré, J., Litovsky, I., Schmitz, S. (eds.) CIAA 2005. LNCS, vol. 3845, pp. 262–272. Springer, Heidelberg (2006)

2. Bloem, R., Galler, S., Jobstmann, B., Piterman, N., Pnueli, A., Weiglhofer, M.: Automatic hardware synthesis from specifications: A case study. In: Design Automation and Test in Europe, pp. 1188–1193 (2007)

3. Bloem, R., Galler, S., Jobstmann, B., Piterman, N., Pnueli, A., Weiglhofer, M.: Specify, compile, run: Hardware from PSL. In: 6th International Workshop on Compiler Optimization Meets Compiler Verification. Electronic Notes in Computer Science, vol. 190, pp. 3–16 (2007)

4. Bloem, R., Jobstmann, B., Piterman, N., Pnueli, A., Sa'ar, Y.: Synthesis of reactive(1) designs. Journal of Computer and Systems Science 78(3), 911–938 (2012)

5. Bohy, A., Bruyère, V., Filiot, E., Jin, N., Raskin, J.-F.: Acacia+, a tool for LTL synthesis. In: Madhusudan, P., Seshia, S.A. (eds.) CAV 2012. LNCS, vol. 7358, pp. 652–657. Springer, Heidelberg (2012)

6. Braberman, V., D'Ippolito, N., Piterman, N., Sykes, D., Uchitel, S.: Controller synthesis: From modelling to enactment. In: 35th International Conference on Software Engineering, San Francisco, USA (2013)

7. Büchi, J., Landweber, L.: Solving sequential conditions by finite-state strategies. Trans. AMS 138, 295–311 (1969)

8. Church, A.: Applications of recursive arithmetic to the problem of circuit synthesis. In: Summaries of the Summer Institute of Symbolic Logic, vol. I, pp. 3–50. Cornell University, Ithaca (1957)
9. Conner, D., Kress-Gazit, H., Choset, H., Rizzi, A., Pappas, G.: Valet parking without a valet. In: Proceedings IEEE/RSJ International Conference on Intelligent Robots and Systems, pp. 572–577. IEEE (2007)
10. D'Ippolito, N., Braberman, V., Piterman, N., Uchitel, S.: Synthesis of live behavior models for fallible domains. In: 33rd International Conference on Software Engineering. ACM, ACM Press (2011)
11. D'Ippolito, N., Braberman, V., Piterman, N., Uchitel, S.: Synthesising non-anomalous event-based controllers for liveness goals. Transactions on Software Engineering and Methodology 22(1) (2012)
12. Ehlers, R.: Symbolic bounded synthesis. Formal Methods in System Design 40(2), 232–262 (2012)
13. Emerson, E., Jutla, C.: The complexity of tree automata and logics of programs. In: Proc. 29th IEEE Symp. on Foundations of Computer Science, pp. 328–337 (1988)
14. Henzinger, T.A., Piterman, N.: Solving games without determinization. In: Ésik, Z. (ed.) CSL 2006. LNCS, vol. 4207, pp. 395–410. Springer, Heidelberg (2006)
15. Vöge, J., Jurdziński, M.: A discrete strategy improvement algorithm for solving parity games. In: Emerson, E.A., Sistla, A.P. (eds.) CAV 2000. LNCS, vol. 1855, pp. 202–215. Springer, Heidelberg (2000)
16. Jobstmann, B., Bloem, R.: Game-based and simulation-based improvements for ltl synthesis. In: 3nd Workshop on Games in Design and Verification (2006)
17. Jurdziński, M.: Deciding the winner in parity games is in up ∩ co-up. Information Processing Letters 68(3), 119–124 (1998)
18. Klein, J., Baier, C.: Experiments with deterministic ω-automata for formulas of linear temporal logic. In: Farré, J., Litovsky, I., Schmitz, S. (eds.) CIAA 2005. LNCS, vol. 3845, pp. 199–212. Springer, Heidelberg (2006)
19. Klein, U., Pnueli, A.: Revisiting synthesis of GR(1) specifications. In: Barner, S., Kroening, D., Raz, O. (eds.) HVC 2010. LNCS, vol. 6504, pp. 161–181. Springer, Heidelberg (2010)
20. Kress-Gazit, H., Fainekos, G., Pappas, G.: From structured english to robot motion. In: Proceedings IEEE/RSJ International Conference on Intelligent Robots and Systems, pp. 2717–2722. IEEE (2007)
21. Kress-Gazit, H., Fainekos, G., Pappas, G.: Where's waldo? sensor-based temporal logic motion planning. In: Proc. IEEE International Conference on Robotics and Automation, pp. 3116–3121. IEEE (2007)
22. Kugler, H., Plock, C., Pnueli, A.: Controller synthesis from LSC requirements. In: Chechik, M., Wirsing, M. (eds.) FASE 2009. LNCS, vol. 5503, pp. 79–93. Springer, Heidelberg (2009)
23. Kugler, H., Segall, I.: Compositional synthesis of reactive systems from live sequence chart specifications. In: Kowalewski, S., Philippou, A. (eds.) TACAS 2009. LNCS, vol. 5505, pp. 77–91. Springer, Heidelberg (2009)
24. Kupferman, O., Piterman, N., Vardi, M.Y.: Safraless compositional synthesis. In: Ball, T., Jones, R.B. (eds.) CAV 2006. LNCS, vol. 4144, pp. 31–44. Springer, Heidelberg (2006)
25. Piterman, N.: From nondeterministic Büchi and Streett automata to deterministic parity automata. Logical Methods in Computer Science 3(3), e5 (2007)
26. Piterman, N., Pnueli, A.: Faster solution of Rabin and Streett games. In: Proc. 21st IEEE Symp. on Logic in Computer Science, pp. 275–284. IEEE, IEEE Computer Society Press (2006)

27. Piterman, N., Pnueli, A., Sa'ar, Y.: Synthesis of reactive(1) designs. In: Emerson, E.A., Namjoshi, K.S. (eds.) VMCAI 2006. LNCS, vol. 3855, pp. 364–380. Springer, Heidelberg (2006)

28. Pnueli, A., Rosner, R.: On the synthesis of a reactive module. In: Proc. 16th ACM Symp. on Principles of Programming Languages, pp. 179–190 (1989)

29. Rabin, M.: Decidability of second order theories and automata on infinite trees. Transaction of the AMS 141, 1–35 (1969)

30. Raman, V., Piterman, N., Kress-Gazit, H.: Provably correct continuous control for high-level robot behaviors with actions of arbitrary execution durations. In: IEEE International Conference on Robotics and Automation. IEEE, IEEE Computer Society Press (2013)

31. Safra, S.: On the complexity of ω-automata. In: Proc. 29th IEEE Symp. on Foundations of Computer Science, pp. 319–327 (1988)

32. Schewe, S.: Solving parity games in big steps. In: Arvind, V., Prasad, S. (eds.) FSTTCS 2007. LNCS, vol. 4855, pp. 449–460. Springer, Heidelberg (2007)

33. Schewe, S., Finkbeiner, B.: Bounded synthesis. In: Namjoshi, K.S., Yoneda, T., Higashino, T., Okamura, Y. (eds.) ATVA 2007. LNCS, vol. 4762, pp. 474–488. Springer, Heidelberg (2007)

34. Tsai, M.-H., Fogarty, S., Vardi, M.Y., Tsay, Y.-K.: State of büchi complementation. In: Domaratzki, M., Salomaa, K. (eds.) CIAA 2010. LNCS, vol. 6482, pp. 261–271. Springer, Heidelberg (2011)

35. Wongpiromsarn, T., Topcu, U., Murray, R.M.: Receding horizon temporal logic planning for dynamical systems. In: IEEE Conference on Decision and Control, pp. 5997–6004. IEEE Computer Society Press (2009)

36. Wongpiromsarn, T., Topcu, U., Murray, R.M.: Automatic synthesis of robust embedded control software. In: AAAI Spring Symposium on Embedded Reasoning: Intelligence in Embedded Systems (2010)

37. Wongpiromsarn, T., Topcu, U., Murray, R.M.: Receding horizon control for temporal logic specifications. In: Johansson, K.H., Yi, W. (eds.) Proceedings of the 13th ACM International Conference on Hybrid Systems: Computation and Control, HSCC 2010, Stockholm, Sweden, April 12-15, pp. 101–110. ACM (2010)

Clustering on k-Edge-Colored Graphs[*]

Eric Angel[1], Evripidis Bampis[2], Alexander Kononov[3], Dimitris Paparas[4],
Emmanouil Pountourakis[5], and Vassilis Zissimopoulos[6]

[1] IBISC, Université d'Evry
Eric.Angel@ibisc.fr
[2] LIP6, Université Pierre et Marie Curie
Evripidis.Bampis@lip6.fr
[3] Sobolev Institute of Mathematics, Novosibirsk
alvenko@math.nsc.ru
[4] Computer Science Department, Columbia University
paparas@cs.columbia.edu
[5] EECS Department, Northwestern University
Emmanouil.Pountourakis@eecs.northwestern.edu
[6] Department of Informatics & Telecommunications,
National and Kapodistrian University of Athens
vassilis@di.uoa.gr

Abstract. We study the Max k-colored clustering problem, where, given
an edge-colored graph with k colors, we seek to color the vertices of the
graph so as to find a clustering of the vertices maximizing the number
(or the weight) of matched edges, i.e. the edges having the same color
as their extremities. We show that the cardinality problem is NP-hard
even for edge-colored bipartite graphs with a chromatic degree equal to
two and $k \geq 3$. Our main result is a constant approximation algorithm
for the weighted version of the Max k-colored clustering problem which
is based on a rounding of a natural linear programming relaxation. For
graphs with chromatic degree equal to two, we improve this ratio by
exploiting the relation of our problem with the MAX 2-AND problem. We
also present a reduction to the maximum-weight independent set (IS)
problem in bipartite graphs which leads to a polynomial time algorithm
for the case of two colors.

1 Introduction

We consider the following problem: we are given an edge-colored graph $G = (V, E)$, where every edge e is labeled with one color among $\{1, 2, \ldots, k\}$ and it
is associated with a weight w_e. We are interested in coloring every vertex of the
graph with one of the k available colors so as to create at most k clusters. Each
cluster corresponds to the subgraph induced by the vertices colored with the

---

[*] This work has been partially supported by the ANR project TODO (09-EMER-
010), and by the project ALGONOW of the research funding program THALIS
(co-financed by the European Social Fund-ESF and Greek national funds).

K. Chatterjee and J. Sgall (Eds.): MFCS 2013, LNCS 8087, pp. 50–61, 2013.

same color. Given a coloring of the vertices, an edge is called *matched* if its color is the same as the color of both its extremities. Our goal is to find a clustering of the vertices maximizing the total weight of the matched edges of the graph. We call this problem the *Max k-colored clustering problem* and we denote it as MAX-k-CC.

Our model has similarities with the centralized version of the information-sharing model introduced by Kleinberg and Ligett [2,7]. In their model, the edges are not colored and two adjacent nodes share information only if they are colored with the same color. As they mention, one interesting extension of their model would be the incorporation of different categories of information. The use of colors in our model goes in this direction. Every edge-color corresponds to a different information category and two adjacent vertices share information if their color is the same as the color of the edge that connects them. While the centralized version of the information-sharing problem of Kleinberg and Ligett is easy to solve, we show that the introduction of colors in the edges of the graph renders the problem NP-hard. In this paper, we focus on the centralized variant of our problem and we study its approximability. Studying our problem from a game theoretic point of view would be an interesting direction for future work. Our problem is also related to the classical correlation clustering problem [1,6].

1.1 Related Works and Our Contribution

In Section 2, we formulate the problem as an integer linear program and we propose a constant-approximation ratio algorithm which is based on a randomized rounding of its linear programming relaxation. Notice here that simpler rounding schemes, apparently do not lead to a constant approximation algorithm. Another observation is that our problem can be formulated as a combinatorial allocation problem [4]. We can consider each color as a player and each vertex as an item, where items have to be allocated to competing players by a central authority, with the goal of maximizing the total utility provided to the players. Every player (each color) has utility functions derived from the different subsets of vertices. Feige and Vondrák [4] consider subadditive, fractional subadditive and submodular functions. It is easy to see that in our case the function is supermodular and hence, their method cannot be directly applied. At the end of Section 2, we show that in the special case where the chromatic degree[1] of the graph is equal to two, our problem is a special case of the MAX 2-AND problem [10]. We show in Section 3 that the cardinality Max k-colored problem is strongly NP-hard by a reduction from MAX-2-SAT, even for bipartite graphs with chromatic degree equal to two, whenever the number of colors is any constant number $k \geq 3$. In Section 4, we present a reduction to the maximum-weight independent set (IS) problem in bipartite graphs which allows us to get an optimal polynomial-time algorithm for the case of two colors. Furthermore, we exploit this idea to get a

---

[1] We define the *chromatic degree* of a vertex as the number of different colors which appear in its incident edges. The *chromatic degree* of an edge-colored graph is the maximum chromatic degree over all its vertices.

$\frac{2}{k}$-approximation algorithm whose approximation ratio is better than the ratio of the constant-approximation algorithm presented in Section 3 for any $k \leq 14$.

2   A Constant Approximation Algorithm

As the problem is strongly NP-hard (see section 3), in the first part of this section, we present a constant-factor approximation algorithm for our problem, while in the second part we focus on graphs with chromatic degree equal to two.

For every vertex i of the graph and for every available color c, we introduce a variable x_{ic} which is equal to one if i is colored with color c and zero otherwise. Also, for every edge $e = [i, j]$, we introduce a variable z_{ij} which is equal to one if both extremities are colored with the same color as e, and zero otherwise. We obtain the following ILP:

$$\max \sum_e w_e z_e$$

$$z_e \leq x_{ic}, \quad \forall\, e = [i, j] \text{ which is } c\text{-colored}$$

$$z_e \leq x_{jc}, \quad \forall\, e = [i, j] \text{ which is } c\text{-colored}$$

$$\sum_c x_{ic} = 1, \qquad\qquad \forall i$$

$$x_{ic}, z_e \in \{0, 1\}, \qquad\qquad \forall i, c, e$$

We consider its linear relaxation, and we denote it by LP.

Our algorithm works in k iterations, by considering each color c, $1 \leq c \leq k$, independently from the others, and so the order in which the colors are considered does not matter. When an edge is chosen, this means that its two extremities get the color of this edge. Since in general a vertex is adjacent to edges of different colors, a vertex may get more than one colors. We want to avoid such situations, and the way the algorithm assigns colors to vertices is designed to minimize the number of such conflicts.

The algorithm is given below.

Algorithm RR

Phase I:

Solve the linear program LP, and let z_e^* be the values of variables z_e.

For each color c

Order, non decreasingly, the c-colored edges $e_1, \ldots, e_{l(c)}$ according to their z_e^* values.

Let us assume that we have $z_{e_1}^* \leq z_{e_2}^* \leq \ldots \leq z_{e_{l(c)}}^*$.

Let r be a random value in $[0, 1]$.

Choose edges e with $z_e^* > r$.

End For

Phase II:
For each vertex v
    If v gets no color or more than two colors, remove it (together with all its
    adjacent edges) from graph G.
End For
Let G' be the obtained graph.
For each vertex v in G'
    If v got one color, then assign this color to it.
    If v got two colors, then choose randomly one of them, each one with
        probability $1/2$.
End For

Notice that the algorithm does not assign colors to all vertices. Indeed, at the end of Phase I some vertex may get no colors, and then in Phase II the algorithm assigns colors only for a subgraph G' of the initial graph G.

The following two Lemmas are straighforward to prove.

Lemma 1. _For any edge e, the probability that e is chosen is z_e^*._

Notice that for a vertex v, it may be the case that none of its adjacent edges are chosen. In that case, v gets no color. But in general, several of its adjacent edges can be chosen, and the vertex v can get more than one colors. We denote by X_{vc} (resp. $\overline{X_{vc}}$) the event that v gets (resp. does not get) color c.

Lemma 2. _For any vertex v, if there exists at least one c-colored edge which is incident to v then one has $Pr(X_{vc}) = z_{e'}^*$, with e' the c-colored edge which has the maximal value of z_e^* among all c-colored edges e which are incident to v._

Lemma 3. _For any vertex v, one has $\sum_c Pr(X_{vc}) \le 1$._

Proof. For any color c, let $e(c)$ be the c-colored edge which is incident to vertex v (if such an edge exists), and with the maximal value of z_e^* among all such edges. From Lemma 2, one has $Pr(X_{vc}) = z_{e(c)}^*$. Therefore, $\sum_c Pr(X_{vc}) \le \sum_c z_{e(c)}^* \le \sum_c x_{vc}^* = 1$.

As stated before, a vertex v can get more than one colors during the execution of the algorithm. However, in general this number will be small. We have the following lemmas.

Lemma 4. _Given a set of independent events such that the sum of their probabilities is less than or equal to 1, the probability of getting at most one of them is greater or equal to $2/e$._

Lemma 5. _At any time during the execution of the algorithm, for any vertex v, the probability that v gets at most one additional color until the end of the Phase I of the algorithm is greater than or equal to $2/e$._

Proof. The events: "v gets color c" for $1 \le c \le k$, are independent, and the sum of their probabilities is less than or equal to 1 according to Lemma 3. Therefore, the result follows from Lemma 4.

Proposition 1. *At any time during the execution of the algorithm, consider any edge $e = [u, v]$, and let us denote by p_u (resp. p_v) the probability that u (resp. v) gets at most one additional color until the end of the Phase I of the algorithm. Let us denote by $p_{u \wedge v}$ the probability that both u and v get each one at most one additional color until the end of the Phase I of the algorithm. Then, one has $p_{u \wedge v} \geq p_u \cdot p_v$.*

Proof. In order to prove that the proposition holds, we consider a sequence of algorithms denoted by $\Sigma_0, \dots, \Sigma_k$, where Σ_0 is our algorithm.

The difference among these algorithms comes from the way in which the vertices get a color. Let us fix a color c. We consider two different procedures for assigning colors to the vertices. The *First procedure*, assigns the colors in the same way as our algorithm does. Let us recall how our algorithm works for just two vertices: Without loss of generality, we assume that there exist an edge e' adjacent to u with color c and an edge e'' adjacent to v with color c (if such an edge e' does not exist, we are in the case $p = 0$ and $y = q$). Moreover e' (resp. e'') is the edge with the maximal value of $z_{e'}^*$ (resp. $z_{e''}^*$) among all c-colored edges incident to u (resp. v). Let us assume that $z_{e'}^* \leq z_{e''}^*$. Let p be the probability that u gets color c in the algorithm (we know that it is $z_{e'}^*$ from Lemma 2), and let q be the probability that v gets color c assuming that u does not get color c. Using the *First procedure*, we color both vertices u and v (with color c) with probability p, and we color only vertex v with probability $(1 - p)q$. The *Second procedure* colors the vertices with color c independently. More precisely, we color vertex u with probability p, and we color vertex v with probability $(1 - p)q + p := y$.

In the algorithm Σ_0, for each color c, $1 \leq c \leq k$, we use the *First procedure* for assigning colors to vertices. In the algorithm Σ_i, $1 \leq i \leq k$, for colors c such that $1 \leq c \leq i$ (resp. $i + 1 \leq c \leq k$) we use the *Second procedure* (resp. *First procedure*) for assigning those colors to vertices. Thus, in algorithm Σ_k, all colors are assigned to vertices using the *Second procedure*.

Let us fix any iteration (color) t, and let us analyze the behavior of those algorithms from iteration t until the end of their execution (at the end of Phase I), i.e. when colors $t, t+1, \dots, k$ are assigned to vertices. Let us also consider any edge $e = [u, v]$. We denote by $p_u(\Sigma_i)$ (resp. $p_v(\Sigma_i)$) the probability that u (resp. v) gets at most one additional color from iteration t until the end of iteration k, for the algorithm Σ_i. Moreover, we denote by $p_{u \wedge v}(\Sigma_i)$ the probability that both u and v get each one at most one additional color from iteration t until the end of iteration k, for the algorithm Σ_i. Notice that one has for any vertex v, $p_v(\Sigma_i) = p_v(\Sigma_0)$ for $1 \leq i \leq k$. Let us now prove that for any $1 \leq i \leq k - 1$, one has $p_{u \wedge v}(\Sigma_i) \geq p_{u \wedge v}(\Sigma_{i+1})$.

If $t \geq i + 2$, since both algorithms Σ_i and Σ_{i+1} use the *First procedure* to assign colors c to vertices, for $i + 2 \leq c \leq k$, they behave in the same way during iterations $t, t + 1, \dots k$, and so $p_{u \wedge v}(\Sigma_i) = p_{u \wedge v}(\Sigma_{i+1})$.

We now assume that $1 \leq t \leq i + 1$. Algorithms Σ_i and Σ_{i+1} only differ in the way they assign color $i+1$ to vertices. If there is no $(i+1)$-colored edge adjacent to either u or v, then again those two algorithms have the same behavior from iteration t to k and so $p_{u \wedge v}(\Sigma_i) = p_{u \wedge v}(\Sigma_{i+1})$. Let us assume now w.l.o.g. that there exists at least one $(i+1)$-colored edge which is adjacent to u. Recall that we denote by X_{vc} (resp. $\overline{X_{vc}}$) the event that v gets (resp. does not get) color c. We have the following probabilities:

	when $\Sigma = \Sigma_i$	when $\Sigma = \Sigma_{i+1}$
$Pr_\Sigma(X_{u,i+1} \wedge \overline{X_{v,i+1}})$	0	$p(1-y)$
$Pr_\Sigma(\overline{X_{u,i+1}} \wedge X_{v,i+1})$	$(1-p)q$	$(1-p)y$
$Pr_\Sigma(\overline{X_{u,i+1}} \wedge \overline{X_{v,i+1}})$	$(1-p)(1-q)$	$(1-p)(1-y)$
$Pr_\Sigma(X_{u,i+1} \wedge X_{v,i+1})$	p	py

Let us denote by A_0 (resp. B_0) the event which corresponds to the situation where vertex u (resp. v) gets no additional color when considering iterations (colors) in $\{t, t+1, \ldots, k\} \setminus \{i+1\}$. Let us also denote by A_1 (resp. B_1) the event which corresponds to the situation where vertex u (resp. v) gets one additional color when considering iterations (colors) in $\{t, t+1, \ldots, k\} \setminus \{i+1\}$. Since these events do not depend on the color $i+1$, they have the same probability for algorithms Σ_i and Σ_{i+1}.

For $\Sigma \in \{\Sigma_0, \ldots, \Sigma_k\}$, one has $p_{u \wedge v}(\Sigma) = Pr(A_0 \wedge B_0) + Pr(A_1 \wedge B_0) \cdot [Pr_\Sigma(\overline{X_{u,i+1}} \wedge X_{v,i+1}) + Pr_\Sigma(\overline{X_{u,i+1}} \wedge \overline{X_{v,i+1}})] + Pr(A_0 \wedge B_1) \cdot [Pr_\Sigma(X_{u,i+1} \wedge \overline{X_{v,i+1}}) + Pr_\Sigma(\overline{X_{u,i+1}} \wedge \overline{X_{v,i+1}})] + Pr(A_1 \wedge B_1) \cdot [Pr_\Sigma(\overline{X_{u,i+1}} \wedge \overline{X_{v,i+1}})]$.

As stated above, $Pr(A_0 \wedge B_0), Pr(A_1 \wedge B_0), Pr(A_0 \wedge B_1), Pr(A_1 \wedge B_1)$ are the same for Σ_i and Σ_{i+1}. In the following table, we give the remaining terms, with $A = Pr_\Sigma(\overline{X_{u,i+1}} \wedge X_{v,i+1}) + Pr_\Sigma(\overline{X_{u,i+1}} \wedge \overline{X_{v,i+1}})$, $B = Pr_\Sigma(X_{u,i+1} \wedge \overline{X_{v,i+1}}) + Pr_\Sigma(\overline{X_{u,i+1}} \wedge \overline{X_{v,i+1}})$, and $C = Pr_\Sigma(\overline{X_{u,i+1}} \wedge \overline{X_{v,i+1}})$.

	$\Sigma = \Sigma_i$	$\Sigma = \Sigma_{i+1}$
A	$(1-p)q + (1-p)(1-q) = (1-p)$	$y(1-p) + (1-p)(1-y) = (1-p)$
B	$(1-p)(1-q)$	$p(1-y) + (1-p)(1-y) = (1-p)(1-q)$
C	$(1-p)(1-q)$	$(1-p)(1-y)$

A term-by-term comparison is sufficient for concluding that $p_{u \wedge v}(\Sigma_i) \geq p_{u \wedge v}(\Sigma_{i+1})$.

Since in algorithm Σ_k, all colors are assigned to vertices using the *Second procedure*, i.e. in an independent way, one has $p_{u \wedge v}(\Sigma_k) = p_u(\Sigma_k) \cdot p_v(\Sigma_k)$. Then $p_{u \wedge v} = p_{u \wedge v}(\Sigma_0) \geq p_{u \wedge v}(\Sigma_1) \geq \ldots \geq p_{u \wedge v}(\Sigma_k) = p_u(\Sigma_k) \cdot p_v(\Sigma_k) = p_u(\Sigma_0) \cdot p_v(\Sigma_0) = p_u \cdot p_v$.

Corollary 1. *At any time during the execution of the algorithm, for any edge $e = [u, v]$, the probability that both u and v get each one at most one additional color until the end of the algorithm is greater than or equal to $4/e^2$.*

Proof. It follows directly from Proposition 1 and Lemma 5.

Definition: An edge $e = [u, v]$ is *safe* if both its extremities u and v are colored with the color of edge e and each one of them gets at most one additional color.

Theorem 1. *The algorithm RR is $1/e^2 \simeq 0.135$-approximate for* MAX-k-CC.

Proof. Let e be any edge of the graph G. We are going to evaluate the probability that edge e is matched in the solution returned by the algorithm. Let OPT be the sum of weights of the matched edges in an optimal solution. Since the linear program LP is a linear relaxation, we have $\sum_{e \in E} w_e z_e^* \geq OPT$. Since the colors are considered in an independent way by the algorithm, we can assume w.l.o.g. that edge e has color 1. This edge needs to be chosen in the first iteration of the algorithm (i.e. when color 1 is considered). This occurs with the probability z_e^* according to Lemma 1. Then during the remaining iterations until the end of the Phase I of the algorithm, i.e. when colors from 2 to k are considered, this edge must remain safe so that it belongs to graph G'. This occurs with a probability greater than or equal to $4/e^2$ according to Corollary 1. Thus, we have proved that each edge $e = [u, v]$ from G belongs to the graph G' with a probability greater than or equal to $4z_e^*/e^2$. We also know that if $e = [u, v]$ belongs to G' then each of the two vertices u and v got either one color, in this case it is the color of edge e, or two colors, and in this case one of them is the color of edge e. So assuming that e belongs to G' the probability that e is matched is at least $1/4$. Overall, this probability is equal to $4z_e^*/e^2 \times 1/4 = z_e^*/e^2$. Thus the cost of the solution returned by the algorithm is in expectation at least equal to $\sum_{e \in E} w_e z_e^*/e^2 \geq OPT/e^2$.

This algorithm can be derandomized by the method of conditional expectations [9]. The algorithm RR can also be used for the case where there are more than one colors on the edges. It is sufficient to create parallel edges, i.e. one edge for each color.

2.1   Graphs with a Chromatic Degree Equal to 2

In this case it is possible to define a quadratic program for this problem and use semi definite relaxations to obtain algorithms with constant approximation ratio.

We associate to each vertex $v \in V$ a variable $y_v \in \{-1, 1\}$. Furthermore, we have an additional variable $x \in \{-1, 1\}$. Now for each couple (v, e), with e an edge adjacent to v, we define a label $l(v, e) \in \{-1, 1\}$. This set of labels must satisfy the following condition: For any vertex v, if e and e' are two edges adjacent to v with different colors, then $l(v, e) \neq l(v, e')$. Since we know that in the graph G', for any vertex v, the set of its adjacent edges are colored with at most 2 colors, it is easy to define such a set of labels (i.e. for all edges e colored with the first (resp. second) color we set $l(v, e) = 1$ (resp. $l(v, e) = -1$). Notice that it is possible to have $l(u, e) \neq l(v, e)$ for an edge $e = [u, v]$. An example is given in Figure 1.

Now, for each edge $e = [u, v]$ we define $f(e) = \frac{1 + l(u,e)y_u x}{2} + \frac{1 + l(v,e)y_v x}{2}$. It is easy to see that the objective function that we need to maximize can be written as

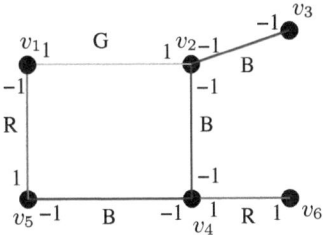

Fig. 1. An example of labeling

$\sum_{e \in E'} w_e f(e)$. Various approximation ratios have been found for such problems like MAX DICUT (see for example [10,8,3]).

In particular, this problem can be seen as a particular case of the problem MAX 2-AND [10]. An instance of MAX 2-AND is composed of a collection of clauses (with non-negative weights assigned to them) such that each clause is either of the form z_i or $z_i \wedge z_j$, where each z_i is either a boolean variable x_k or its negation $\overline{x_k}$. The goal is to find an assignment of the boolean variables $x_1, \ldots x_n$, in order to maximize the weight of the satisfied clauses. It is easy to see that any instance of our problem can be transformed to an equivalent MAX 2-AND instance for which an algorithm with an approximation ratio of 0.859 exists [10]. For example, for the instance given in Figure 1 we obtain the set of clauses: $x_1 \wedge x_2$ (for edge $[v_1, v_2]$), $\overline{x_1} \wedge x_5$ (for edge $[v_1, v_5]$), $\overline{x_5} \wedge \overline{x_4}$ (for edge $[v_5, v_4]$), and so on.

3   Complexity

In this part we show that the problem is NP-complete for bipartite graphs if we allow the initial coloring of the edges to contain three or more colors. Our reduction is from MAX-2-SAT.

Theorem 2. *The* MAX-3-CC *problem is NP-complete even for bipartite graphs with chromatic degree two and* $w_e = 1$, *for every edge* e *of the graph.*

Proof. Clearly the problem is in NP. Let us now give a polynomial time reduction R that maps any instance of the MAX-2-SAT problem $\mathcal{I}_{\text{MAX-2-SAT}} =< \mathcal{X}, \mathcal{C}, B >$ where $\mathcal{X} = \{x_1, \ldots, x_n\}$ is a set of variables, $\mathcal{C} = \{c_1, \ldots, c_m\}$ is a set of disjunctive clauses with exactly two literals, and $B \leq m$ is a positive integer, to an instance $R(\mathcal{I}_{\text{MAX-2-SAT}}) = \mathcal{I}_{\text{MAX-3-CC}} =< G, C, f, 3 >$ for the MAX-3-CC problem. For the rest of the proof we assume that $C = \{\text{R(ed)},\text{B(lue)},\text{G(reen)}\}$. The reduction is based on the gadgets presented in Figures 2 and 3.

The Gadgets. The set V is constructed as follows: For each variable x_i of the MAX-2-SAT formula we create a new node v_i and for each $c \in \mathcal{C}$ we construct *four* nodes v_{up}^c, v_{down}^c, v_{left}^c and v_{right}^c. Then for each clause, we add six edges

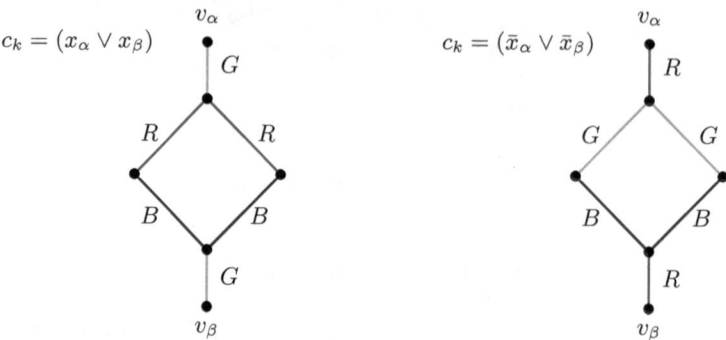

Fig. 2. The case when all literals are positive or all literals are negative

based on whether both of the literals are positive or negative, or one of them is negative and the other positive.

First Case: Assume that both of the literals are positive or both are negative i.e. the clause is either $c_k = (x_\alpha \vee x_\beta)$ or $c_k = (\bar{x}_\alpha \vee \bar{x}_\beta)$. Then we construct the gadgets in Figure 2.

Second Case: Assume that one literal is positive and the other is negative. That is, the clause is of the form $c_k = (x_\alpha \vee \bar{x}_\beta)$ or $c_k = (\bar{x}_\alpha \vee x_\beta)$. Respectively we construct the gadgets in Figure 3.

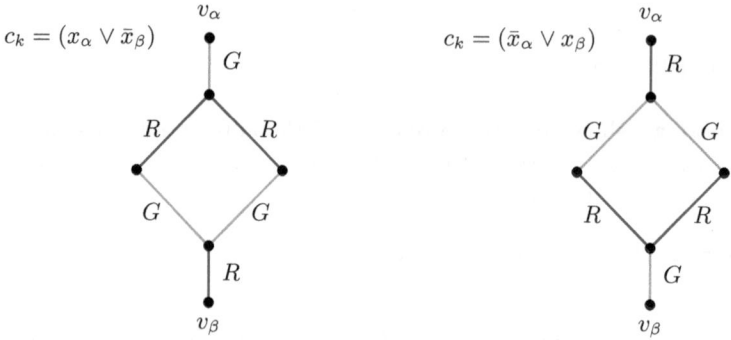

Fig. 3. The case when one literal is positive and one literal is negative

Finally we set $P = 3B + 2(m - B)$, where m is the total number of clauses.

It is not difficult to check that the constructed graph does not contain any odd cycle and so it is bipartite. Also, for every vertex of the constructed graph the edges that are incident to this vertex are colored with at most two different colors, i.e. the chromatic degree of the graph is two.

Lemma 6. *The maximum contribution that any gadget can have is exactly 3 and is obtained when at least one of the nodes v_α and v_β has the same color as*

the edge that connects it with the rest of the gadget. If none of v_α and v_β has the same color with the edge that connects it with the rest of the gadget then the maximum contribution that can be achieved is 2.

Proof. Simple case analysis.

Lemma 7. *For an instance of the MAX-2-SAT problem $\mathcal{I}_{\text{MAX-2-SAT}} =< X, \mathcal{C}, B >$, there is a truth-assignment that satisfies at least B clauses if and only if there is a clustering for the corresponding MAX-3-CC problem with contribution greater than or equal to $3 \cdot B + 2(m - B)$, where $m = |\mathcal{C}|$ is the number of clauses.*

Proof. To prove the if direction, let T be a truth assignment that satisfies at least B clauses of a 2-SAT formula F. In the derived graph, color green all the nodes that correspond to variables that are true and red all the nodes that correspond to false variables. In this way, for each satisfied clause the corresponding gadget in the optimum clustering will have pay-off three.

Since each of the gadgets representing a satisfied clause will contribute three to the pay-off and the satisfied clauses are $L \geq B$ the total optimal contribution of these clauses will be $3 \cdot L$. The gadgets of the rest $m - L$ clauses will each have optimal contribution 2 and the total optimal contribution from the unsatisfied clauses will be $2 \cdot (m - L)$. Hence the total pay-off will be $2 \cdot (m - L) + 3 \cdot L = 2 \cdot m + L \geq 2 \cdot m + B = 3 \cdot B + 2(m - B)$.

For the opposite direction, suppose that the corresponding graph of a formula F has a partition with pay-off at least $3 \cdot B + 2(m - B) = 2 \cdot m + B$. Since each one of the gadgets contributes to the pay-off either 2 or 3, there must exist at least B gadgets with pay-off 3.

Let us assign the value true to the variables with green corresponding nodes and the value false to the rest of the variables. Notice now that each one of the gadgets with pay-off three corresponds to a satisfied clause.

Since the gadgets with pay-off 3 are at least B, there are at least B clauses that are satisfied and the only if direction holds too.

4   A Reduction to the Independent Set problem

In this section we will show that the colored clustering problem can be reduced to the IS problem in bipartite graphs.

Given an instance of the MAX-k-CC problem, we create the line graph G_{line} corresponding to the initial graph. We then construct a new graph G'_{line} by deleting the edges between the vertices of G_{line} that correspond to neighboring edges of the same color in G.

Lemma 8. *The MAX-k-CC problem has a clustering with pay-off P if and only if the graph G'_{line} has an independent set of size P.*

Proof. For the if direction, suppose that the initial problem has a partition with pay-off P. For this to happen there must exist a set $\mathcal{L}$ of P edges with properly colored ends. Each edge $e \in \mathcal{L}$ is either adjacent to some other edges in $\mathcal{L}$ and all have the same color or not adjacent with any other edge in $\mathcal{L}$. In either case, the vertex in G'_{line} that corresponds to e is not adjacent to any vertex corresponding to some other edge in $\mathcal{L}$, because in G'_{line} we have eliminated the edges between vertices corresponding to adjacent edges with the same color. Hence, the nodes of G'_{line} that correspond to edges in $\mathcal{L}$ form an independent set of size P.

To prove the opposite direction, let us examine an instance of the induced problem that has an independent set of size P. The nodes that form the independent set correspond to edges of the initial graph that either are not adjacent or are adjacent and have the same color. Therefore it is possible to color the extremities of these edges with the same color as the edges themselves and hence to produce a solution with pay-off P, because there are P such edges.

For $k = 2$, the constructed graph is always bipartite. Indeed, in G'_{line} we have eliminated the edges between nodes of the line graph G_{line} that correspond to edges of the same color in the initial graph G. So, while traversing any cycle of G'_{line} the color of the corresponding edge must change from node to node. Since there are only two different colors, any cycle must have even length and, therefore, the graph is bipartite. Notice that our reduction holds also for the weighted case.

As a result, given that a weighted independent set can be found in polynomial time in a bipartite graph, we get that the weighted MAX-2-CC is polynomially solvable.

For any $k \geq 3$ we can also derive from Lemma 8 a $\frac{2}{k}$ approximation algorithm for the weighted MAX-k-CC. Although, it is not a constant-approximation algorithm its ratio is better than $1/e^2$ for every $k \leq 14$. We use the following Theorem, from [5]: Let G be a weighted graph with n vertices and m edges; let k be an integer greater than one. If it takes only s steps to color the vertices of G in k colors, then it takes only $s + O(nm \log(n^2/m))$ steps to find an independent set whose weight is at least $2/k$ times the weight of an optimal independent set. In our case we have $s = 0$.

References

1. Bansal, N., Blum, A., Chawla, S.: Correlation Clustering. Machine Learning 56, 89–113 (2004)
2. Ducoffe, G., Mazauric, D., Chaintreau, A.: Convergence of Coloring Games with Collusions. CoRR abs/1212.3782 (2012)
3. Feige, U., Goemans, M.X.: Approximating the value of two prover proof systems, with applications to MAX 2SAT and MAX DICUT. In: Proc. of 3rd Israel Symposium on the Theory of Computing and Systems, pp. 182–189 (1995)
4. Feige, U., Vondrák, J.: Approximation algorithms for allocation problems: Improving the factor of 1 - 1/e. In: FOCS 2006, pp. 667–676 (2006)
5. Hochbaum, D.S.: Approximating Covering and Packing Problems: Set Cover, Vertex Cover, Independent Set, and Related Problems. In: Hochbaum, D.S. (ed.) Approximation Algorithms for NP-hard Problems, PWS Publishing Company (1997)
6. Jain, A.K., Dubes, R.C.: Algorithms for Clustering Data. Prentice-Hall (1981)

7. Kleinberg, J.M., Ligett, K.: Information-Sharing and Privacy in Social Networks. CoRR abs/1003.0469 (2010)
8. Lewin, M., Livnat, D., Zwick, U.: Improved rounding techniques for the MAX 2-SAT and MAX DI-CUT problems. In: Cook, W.J., Schulz, A.S. (eds.) IPCO 2002. LNCS, vol. 2337, pp. 67–82. Springer, Heidelberg (2002)
9. Vazirani, V.: Approximation algorithms. Springer (2004)
10. Zwick, U.: Analyzing the MAX 2-SAT and MAX DI-CUT approximation algorithms of Feige and Goemans, currently available from,
`http://www.cs.tau.ac.il/~zwick/online-papers.html`

How to Pack Your Items When You Have to Buy Your Knapsack

Antonios Antoniadis[1,*], Chien-Chung Huang[2],
Sebastian Ott[3], and José Verschae[4]

[1] Computer Science Department, University of Pittsburgh, Pittsburgh, USA
antoniosantoniadis@gmail.com
[2] Chalmers University, Göteborg, Sweden
villars@gmail.com
[3] Max-Planck-Institut für Informatik, Saarbrücken, Germany
ott@mpi-inf.mpg.de
[4] Departamento de Ingeniería Industrial, Universidad de Chile, Santiago, Chile
jverscha@ing.uchile.cl

Abstract. In this paper we consider a generalization of the classical knapsack problem. While in the standard setting a fixed capacity may not be exceeded by the weight of the chosen items, we replace this hard constraint by a weight-dependent cost function. The objective is to maximize the total profit of the chosen items minus the cost induced by their total weight. We study two natural classes of cost functions, namely convex and concave functions. For the concave case, we show that the problem can be solved in polynomial time; for the convex case we present an FPTAS and a 2-approximation algorithm with the running time of $\mathcal{O}(n \log n)$, where n is the number of items. Before, only a 3-approximation algorithm was known.

We note that our problem with a convex cost function is a special case of maximizing a non-monotone, possibly negative submodular function.

1 Introduction

The knapsack problem is a classical problem of combinatorial optimization [5]. In the standard setting we must choose a subset of items to fit into a given knapsack. Each item comes with a *weight* w_j and a *profit* p_j, and the goal is to maximize the total profit, under the constraint that the total weight of the chosen items should not exceed the knapsack's capacity. A natural generalization of the knapsack problem (and backed up by applications - see discussion below) is to assume that we can "customize" the knapsack in which we pack the selected items. In other words, we can pay more to buy a knapsack of a larger capacity. This motivates our problem.

Consider U to be a set of items j, each with its own positive *weight* w_j and positive *profit* p_j. Let $c(W)$ be a non-decreasing *cost function* $c(W)$,

* The author was supported by a fellowship within the Postdoc-Programme of the German Academic Exchange Service (DAAD).

K. Chatterjee and J. Sgall (Eds.): MFCS 2013, LNCS 8087, pp. 62–73, 2013.

where W is the total weight of the chosen items. The function c is given by an oracle and it is not part of the input. The goal is to maximize the *net value*, which is the total profit of the chosen items minus the cost of their total weight. Precisely, we want to maximize the objective $\pi(A) = \sum_{j \in A} p_j - c(\sum_{j \in A} w_j)$ over all possible subsets $A \subseteq U$ of items.

Clearly, our problem contains the standard knapsack problem as a special case, when the cost function switches from 0 to infinity at the capacity of the knapsack.

If we allow arbitrary non-decreasing functions, the problem is inapproximable within any factor, unless $P = NP$. This follows by a simple reduction from the 2-partition problem [5]; for details see the full version of our paper [11]. Therefore, in this work, we consider two natural special cases of cost functions, namely convex and concave functions.

Our Results. Let $n = |U|$ be the number of items. We assume that each call of the cost function takes constant time. Suppose that the cost function is convex. Then our problem is (weakly) NP-hard, since it contains the original knapsack problem as a special case.

– In Section 3 we present an $O(n \log n)$ time 2-approximation algorithm.
– In Section 4, we present an FPTAS (Fully Polynomial Time Approximation Scheme). With any $\epsilon > 0$, our FPTAS returns a solution with net value at least $1 - \epsilon$ fraction of the optimal solution, in $O(n^3/\epsilon^2)$ time. Our FPTAS is only slower than the well-known FPTAS for the original knapsack problem by a factor of $1/\epsilon$.

Suppose that the cost function is convave.

– In Section 5, we present an exact algorithm with running time $O(n \log n)$.

Applications. Applications for our problem naturally arise in many areas, such as cloud computing or connection management in wireless access points [1]. Consider for example a data center which earns a certain amount for every job it processes. When accepting more and more jobs, the workload increases and the data center must run its processors at higher speed to maintain an acceptable quality of service. This results in a higher power consumption, which typically grows convexly with the speed of the processor (cf. the cube-root rule for CMOS based processors [2]). Consequently, the data center wants to select the most profitable set of jobs, taking into account the energy costs, which depend on the total volume of accepted jobs. This directly translates into our generalized knapsack problem with a convex cost function.

The study of concave cost functions is motivated by the *economies of scale* principle [10]. This principle states that the cost per unit decreases if the total number of units increases, because efficiency increases and fixed costs are spread over more units.

Related Work. The standard knapsack problem is well known to be NP-hard, but one can approximate it within any factor greater than 1 in polynomial time

[8]. Our generalized problem has been studied before under a convex cost function, by Barman et al. [1]. There the authors consider an online setting, where items arrive over time in random order and the algorithm must make an irrevocable accept/reject decision as soon as a new item arrives. They assume a convex cost function and develop an online algorithm which is constant competitive in expectation. In addition they also study the problem under several feasibility constraints. Even though their focus is on the online version, they also present a result for our offline setting, namely a greedy algorithm which achieves a 3-approximation. To the best of our knowledge, this was the best known approximation ratio for this problem so far.

We note that in the convex costs setting, our problem amounts to maximizing a non-monotone, possibly negative submodular function. Submodular function maximization in general is a very active area of research that has recently attracted much attention [3,4,9], and many other special cases (e.g. several graph cut problems [6,7]) have been investigated. Not much is known for other problems where the submodular function is non-monotone and can take negative values. Our results can be regarded as a first step in understanding this type of problems.

Challenges and Techniques. A tricky aspect of our problem is that the objective, i.e. the net value, can be very small and even negative. In this situation, the design and analysis of approximation algorithms often becomes more challenging. Although the usual dynamic program by Ibarra and Kim [8] can be easily adapted for our more general case, the *rounding* technique is troublesome. The reason is that our goal is to maximize the difference between the profit and the cost. It could happen that both values are very large and almost coincide in the optimal solution, resulting in a rather small net value. Thus, a relatively small error in the rounding of profit and/or weight would cascade into a large mistake in the final outcome. The same problem prevents us from guessing the weight of the optimal solution, and then applying known algorithms for the classic knapsack problem.

Our 2-approximation algorithm for convex costs does not resort to any rounding or guessing. Informally speaking, we use the optimal solution of a relaxed version of our problem, where the items can be fractionally added into the collection, to upper bound the net value of the original (non-fractional) problem. However, this idea has to be carefully applied since the gap between the cost of a fractional solution and an integral one can be arbitrarily large. We show that this gap is bounded if the contribution (defined appropriately) of a job to the optimal solution is bounded. Turning these ideas into a greedy algorithm we can achieve the claimed result.

As just said, rounding the profits as for the usual knapsack problem might decrease the objective by an arbitrarily large factor. Therefore our FPTAS cannot rely on this technique. To overcome this difficulty, we use two crucial ideas. First, we observe that the convexity of the cost function implies the following fact: if A is an optimal subset of items, and $A' \subseteq A$, we can assume that A' has minimum weight among all subsets of $(U \setminus A) \cup A'$ with net value at least $\pi(A')$.

This indicates the use of a specially crafted dynamic programming *tableau* that is indexed by the achievable net value. Secondly, instead of rounding the profits/weights, we merge several entries of the tableau into one. Intuitively, this corresponds to rounding the objective function so that it takes polynomially many different values.

2 Preliminaries and Notations

The input is given by a set U of items, each having a positive weight w_j and positive profit p_j, together with a non-decreasing cost function $c = c(W)$ that denotes the cost of selecting items with total weight W. We assume that c is given by an oracle which can be called in constant time, and that $c(0) = 0$. Our goal is to find a set $A \subseteq U$ of items that maximizes the total net value $\pi(A) = \sum_{j \in A} p_j - c(\sum_{j \in A} w_j)$. W.l.o.g. we assume the weights w_j to be positive integers (otherwise we can multiply them with their least common denominator τ and use the modified cost function $c'(W) = c(W/\tau)$).

Throughout the paper we consider a fixed optimal solution OPT. We slightly abuse notation by letting OPT denote both the set of items and the net value of this solution. Furthermore, let W_{OPT} denote the total weight of all items in OPT. More generally we use W_S to denote the total weight of all items in a given set S.

For any item j, we define the *density* d_j as the profit per unit of weight, i.e. $d_j := p_j/w_j$. Furthermore, it will sometimes be helpful to consider *fractional items*, i.e. we pick only $w < w_j$ units of an item j and assume this fraction has profit $w \cdot d_j$. Finally, we define $\delta_j^w(W) := w \cdot d_j - c(W + w) + c(W)$ for every item j and $w \geq -W$. The intuitive meaning of $\delta_j^w(W)$ is the change in the net value when adding w units of item j to a knapsack with current total weight W. Note that we also allow negative w, which then describes the change in the net value when removing $|w|$ units of item j.

It is easy to see, that the function $\delta_j^w(W)$ satisfies the following property if c is convex.

Proposition 1. *Let c be a convex function. Then $\delta_j^w(W) = w \cdot d_j - c(W + w) + c(W)$ is non-increasing in W, and $\delta_j^{-w}(W) = -w \cdot d_j - c(W - w) + c(W)$ is non-decreasing in W, for any item j and $w \geq 0$.*

3 2-Approximation for Convex Costs

In this section we present a 2-approximation with running time $\mathcal{O}(n \log n)$ for the case of a convex cost function c. The main idea of the algorithm is as follows. We try every item individually and compare their net values to the outcome of a greedy procedure, which adds items in order of non-increasing densities. During the procedure, we add an item j only if $\delta_j^1(W_A + w_j - 1) \geq 0$, i.e. adding the last unit of j does not decrease the net value. Thus, we may discard certain items from the sorted list, but possibly continue to add further items with smaller

---

Algorithm:

Initialize a set of candidate solutions $C := \{\emptyset\}$, and an empty solution $A := \emptyset$.

For every item $j \in U$ **do:** $C := C \cup \{\{j\}\}$.

Sort the items in U by non-increasing densities, and let $i_1, i_2, \ldots i_n$ be the items in this sorted order.

For k from 1 to n, **do:**

   If $\delta_{i_k}^1(W_A + w_{i_k} - 1) \geq 0$, **then** $A := A \cup \{i_k\}$.

Set $C := C \cup \{A\}$.

Output a set $B \in C$ with maximum net value $\pi(B)$.

Fig. 1. A 2-approximation with running time $\mathcal{O}(n \log n)$

densities at a later point of time. The details of the algorithm are presented in Figure 1.

It is easy to see that the running time of the proposed algorithm is dominated by the sorting of the items, which can be done in $\mathcal{O}(n \log n)$ time. We now turn to proving the claimed approximation ratio of 2.

Theorem 1. *The set B returned by the algorithm in Figure 1 satisfies $\pi(B) \geq 1/2 \cdot \mathrm{OPT}$.*

The high-level idea of the proof of Theorem 1 is as follows. First, we divide the items into two categories: *heavy* and *light* (see Definition 1). In Lemma 1 we show that each heavy item in the optimal solution, by itself, is a 2-approximation. On the other hand, if there is no heavy item in OPT, in Lemma 3 we show that the greedy procedure results in a 2-approximation. Then Theorem 1 follows easily from the two lemmas.

Definition 1. *An item $j \in U$ is called* heavy *if one of the following is true:*

- $w_j > W_{\mathrm{OPT}}$
- $w_j \leq W_{\mathrm{OPT}}$ *and* $\delta_j^{-w_j}(W_{\mathrm{OPT}}) < -1/2 \cdot \mathrm{OPT}$ *(removing j from a knapsack with total weight W_{OPT} decreases the net value by more than $1/2 \cdot \mathrm{OPT}$)*

Items which are not heavy are called light. *We denote the set of heavy items by $H := \{j | j \text{ is heavy}\}$.*

We remark that this definition is only used in the analysis of our algorithm and never in the algorithm itself.

Lemma 1. *Assume* OPT *contains some heavy item j. Then $\pi(\{j\}) > 1/2 \cdot \mathrm{OPT}$.*

Proof. First of all, observe that any heavy item in OPT satisfies the second condition of Definition 1, because its weight cannot be more than the total weight of OPT. Thus we have $\delta_j^{-w_j}(W_{\mathrm{OPT}}) < -1/2 \cdot \mathrm{OPT}$. Furthermore it holds that $\delta_j^{-w_j}(w_j) \leq \delta_j^{-w_j}(W_{\mathrm{OPT}})$ by Proposition 1. Therefore $\delta_j^{-w_j}(w_j) < -1/2 \cdot \mathrm{OPT}$, and since $\pi(\{j\}) + \delta_j^{-w_j}(w_j) = 0$, we get $\pi(\{j\}) > 1/2 \cdot \mathrm{OPT}$. $\qquad \square$

We now consider the case that OPT does not contain any heavy item. Our first step is to define a family of fractional solutions, which will serve as an upper bound on OPT. To this end, consider the items in $U \setminus D$, where D is a given set of items that were discarded by the algorithm in the second for loop, and sort them by non-increasing densities (breaking ties in the same way as the algorithm does). We regard these items as a continuous "stream", i.e. a sequence of unit-size item fractions.

Definition 2. *We denote by $F_D(W)$ the (fractional) solution obtained by picking the first W units from the described stream.*

Note that $F_D(W)$ contains only complete (non-fractional) items, plus at most one fractional item. The next lemma summarizes a number of useful properties satisfied by $F_D(W)$.

Lemma 2. *The following three statements are true for any set of items D and integer weights W and W':*

(1) $\pi(F_D(W-1)) \leq \pi(F_D(W)) \Rightarrow \pi(F_D(W')) \leq \pi(F_D(W)) \; \forall \, W' \leq W$
(2) $\pi(F_D(W-1)) > \pi(F_D(W)) \Rightarrow \pi(F_D(W')) < \pi(F_D(W)) \; \forall \, W' > W$
(3) $\mathrm{OPT} \cap D = \emptyset \Rightarrow \pi(F_D(W_{\mathrm{OPT}})) \geq \mathrm{OPT}$

Proof. Observe that $\pi(F_D(W)) - \pi(F_D(W-1)) = d - c(W) + c(W-1)$, where d is the density of the Wth unit in the stream. Since the items in the stream are ordered by non-increasing densities, and c is convex, it follows that $\pi(F_D(W)) - \pi(F_D(W-1))$ is non-increasing in W. This immediately proves statements (1) and (2).

For statement (3), note that $F_D(W)$ maximizes the net value among all (possibly fractional) solutions with total weight W, that do not contain any item from D. This is clearly true, because the cost is the same for all those solutions ($= c(W)$), and $F_D(W)$ maximizes the profit by choosing the units with the highest densities. Statement (3) follows.   □

Using these properties, we can now prove the following lemma, which is the main technical contribution of this section.

Lemma 3. *If $\mathrm{OPT} \cap H = \emptyset$, then the candidate A constructed by the greedy procedure achieves $\pi(A) \geq 1/2 \cdot \mathrm{OPT}$.*

Proof. The algorithm starts the construction of A with an empty set $A_0 = \emptyset$, and then iterates through the sorted list of items to augment A_0. During the iteration, some items are added to A_0, while others are discarded. We restrict our attention to the moment t, when it happens for the first time that a light item z gets discarded. If this never happens, t is just the time when the iteration is finished and all items have been considered. Let D_t be the set of items discarded up to this point (excluding z), and let A_t be the solution constructed until now. Then $D_t \subseteq H$, while $\mathrm{OPT} \cap H = \emptyset$. Thus $D_t \cap \mathrm{OPT} = \emptyset$, and hence

$$\pi(F_{D_t}(W_{\mathrm{OPT}})) \geq \mathrm{OPT} \tag{1}$$

by Lemma 2 (3). In the remaining part of the proof, our goal is to show the following inequality:

$$\pi\big(F_{D_t}(W_{A_t})\big) \geq \pi\big(F_{D_t}(W_{\mathrm{OPT}})\big) - 1/2 \cdot \mathrm{OPT}. \tag{2}$$

Together with (1) this proves the lemma for the following reason. First of all, observe that A_t is exactly the same set of items as $F_{D_t}(W_{A_t})$, and therefore (1) and (2) imply $\pi(A_t) \geq 1/2 \cdot \mathrm{OPT}$. As the iteration proceeds, the algorithm may add further jobs to A_t, but for each such job j we know that $\delta_j^1(W_{cur}+w_j-1) \geq 0$, where W_{cur} is the total weight of A before j is added. Therefore adding item j changes the net value by

$$\delta_j^{w_j}(W_{cur}) = \sum_{r=0}^{w_j-1} \delta_j^1(W_{cur}+r) \geq w_j \cdot \delta_j^1(W_{cur}+w_j-1) \geq 0,$$

where the second inequality holds by Proposition 1. Thus every further job which is added to A_t does not decrease the net value, and hence the final candidate A satisfies $\pi(A) \geq \pi(A_t) \geq 1/2 \cdot \mathrm{OPT}$.

For the proof of (2), we distinguish two cases. Let us first assume that $W_{\mathrm{OPT}} \leq W_{A_t}$. Clearly (2) is satisfied when $W_{\mathrm{OPT}} = W_{A_t}$, so we can assume $W_{\mathrm{OPT}} < W_{A_t}$. Then at least one item has been added to A_0 during the iteration up to time t. Let k be the last item that was added on the way from A_0 to A_t. Since the algorithm decided to add k, we have $0 \leq \delta_k^1(W_{A_t}-1)$. Now observe that $\pi\big(F_{D_t}(W_{A_t})\big) - \pi\big(F_{D_t}(W_{A_t}-1)\big) = d_k - c(W_{A_t}) + c(W_{A_t}-1) = \delta_k^1(W_{A_t}-1) \geq 0$. Hence we can apply Lemma 2 (1) to obtain $\pi\big(F_{D_t}(W_{A_t})\big) \geq \pi\big(F_{D_t}(W_{\mathrm{OPT}})\big)$.

We are left with the case that $W_{\mathrm{OPT}} > W_{A_t}$. In this case there must indeed be a light item z which gets discarded at time t, because if the iteration was finished at t and all light items were accepted, we would have $\mathrm{OPT} \subseteq A_t$, implying $W_{\mathrm{OPT}} \leq W_{A_t}$, a contradiction.

The reason for the rejection of z is that $0 > \delta_z^1(W_{A_t}+w_z-1)$, where the latter is equal to $\pi\big(F_{D_t}(W_{A_t}+w_z)\big) - \pi\big(F_{D_t}(W_{A_t}+w_z-1)\big)$. We claim that this implies $W_{\mathrm{OPT}} \leq W_{A_t}+w_z$. To see this, suppose for the sake of contradiction that $W_{\mathrm{OPT}} > W_{A_t}+w_z$. Then Lemma 2 (2) yields $\pi\big(F_{D_t}(W_{\mathrm{OPT}})\big) < \pi\big(F_{D_t}(W_{A_t}+w_z)\big)$, and together with (1) we obtain $\pi\big(F_{D_t}(W_{A_t}+w_z)\big) > \mathrm{OPT}$. However, $F_{D_t}(W_{A_t}+w_z)$ is a feasible (non-fractional) solution, which contradicts the optimality of OPT. Therefore we have

$$W_{A_t} < W_{\mathrm{OPT}} \leq W_{A_t}+w_z. \tag{3}$$

We continue by comparing the solutions $F_{D_t}(W_{A_t})$ and $F_{D_t}(W_{\mathrm{OPT}})$. Using (3), we see that both solutions differ only by some non-zero fraction of z, namely by $\Delta w := W_{\mathrm{OPT}} - W_{A_t}$ units of z. In terms of the net value, we have

$$\pi\big(F_{D_t}(W_{A_t})\big) = \pi\big(F_{D_t}(W_{\mathrm{OPT}})\big) + \delta_z^{-\Delta w}(W_{\mathrm{OPT}}).$$

So all we need to show for completing the proof of (2), is that

$$\delta_z^{-\Delta w}(W_{\mathrm{OPT}}) \geq -1/2 \cdot \mathrm{OPT}.$$

As z is a light item, we already know that $\delta_z^{-w_z}(W_{\text{OPT}}) \geq -1/2 \cdot \text{OPT}$. Now consider the function $f(x) := x \cdot d_z - c(W_{\text{OPT}} + x) + c(W_{\text{OPT}})$ and observe that f is concave. Clearly, $f(0) = 0$ and $f(-w_z) = \delta_z^{-w_z}(W_{\text{OPT}}) \geq -1/2 \cdot \text{OPT}$. By the concavity of f we can conclude that $f(x) \geq -1/2 \cdot \text{OPT}$ for all $-w_z \leq x \leq 0$. Therefore $-1/2 \cdot \text{OPT} \leq f(-\Delta w) = \delta_z^{-\Delta w}(W_{\text{OPT}})$. This completes the proof of (2), and hence the proof of the entire lemma. □

Now Theorem 1 follows from Lemmas 1 and 3.

4   FPTAS for Convex Costs

In this section we describe an FPTAS for our problem if the cost function is convex. The main idea of the FPTAS is to build up a tableau of polynomial size using dynamic programming. Although obtaining a dynamic program of pseudo-polynomial size is a relatively straightforward task, designing one that can be effectively rounded is significantly more involved. To this end, we exploit the convexity of the objective function to obtain a tableau that is indexed by the net value achievable by sub-solutions. We will show that we can reduce the size of this tableau to make it polynomial. In this process we again exploit convexity. For the sake of conciseness we describe directly the tableau with polynomial size.

Before defining the dynamic program, we use the 2-approximation algorithm from the previous section to obtain an upper bound UB on OPT, s.t. OPT $\leq$ UB $\leq$ 2OPT. Further, for all $k \in \{0, \ldots, n\}$ and $N \in [0, \text{UB}]$, we define $T^*(k, N)$ to be the minimum weight achievable by a subset of items $A \subseteq \{1, \ldots, k\}$ whose net value is at least N, i.e.,

$$T^*(k, N) := \min\{W_A \mid \pi(A) \geq N, A \subseteq \{1, \ldots, k\}\}.$$

If no subset $A \subseteq \{1, \ldots, k\}$ with $\pi(A) \geq N$ exists, we let $T^*(k, N) := \infty$.

To elaborate the details of the FPTAS, let us fix an error bound $\epsilon > 0$, and let $\ell_{max} := \lceil (n+1)/\epsilon \rceil$ and $C := \text{UB}/\ell_{max}$. Using dynamic programming, we build up a table $T(k, l)$, where $k \in \{0, \ldots, n\}$ and $\ell \in \{0, \ldots, \ell_{max}\}$. Intuitively, $T(k, \ell)$ compacts all values of $T^*(k, N)$ for $N \in [\ell C, (\ell + 1)C)$ into one entry. We now describe the dynamic program. Initially, we set $T(0, 0) := 0$ and $T(0, \ell) := \infty$ for all $\ell \neq 0$. The remaining entries are computed by the recursive formula

$$T(k, \ell) := \min\big\{T(k-1, \ell), \tag{4}$$
$$\min_{\ell' \geq 0}\{T(k-1, \ell') + w_k \mid \ell'C + \delta_k^{w_k}(T(k-1, \ell')) \geq \ell C\}\big\}.$$

Intuitively, the first term in the recursive formula corresponds to the case in which item k does not belong to the set A in $\arg\min\{W_A : \pi(A) \geq \ell C, A \subseteq \{1, \ldots, k\}\}$. The second term is when k belongs to this set A. In this case we need to check for the best choice among all $T(k-1, \ell')$ for $0 \leq \ell'$.

In the following lemma we show that unless $T(k, \ell) = \infty$, there always exists a set $A \subseteq \{1, \ldots, k\}$ with weight $T(k, \ell)$ and net value at least ℓC.

Lemma 4. *If $T(k, \ell) < \infty$, then there exists $A \subseteq \{1, \ldots, k\}$ such that $W_A = T(k, \ell)$ and $\pi(A) \geq \ell C$.*

Proof. We show this by induction on k. The base case for $k = 0$ holds trivially. If $T(k, \ell) = T(k - 1, \ell)$ then the claim is trivially true. Otherwise, let ℓ' be such that $T(k, \ell) = T(k - 1, \ell') + w_k$ and $\ell'C + \delta_k^{w_k}\left(T(k - 1, \ell')\right) \geq \ell C$. By induction hypothesis, we have that there exists $A' \subseteq \{1, \ldots, k-1\}$ with $W_{A'} = T(k-1, \ell')$ and $\pi(A') \geq \ell'C$. Then taking $A := A' \cup \{k\}$ yields a set satisfying $W_A = T(k, \ell') + w_k = T(k, \ell)$ and $\pi(A) = \pi(A') + \delta_k^{w_k}(W_{A'}) \geq \ell'C + \delta_k^{w_k}\left(T(k-1, \ell')\right) \geq \ell C$. $\qquad \square$

The next lemma shows that the table T provides, in an appropriate fashion, a good approximation for T^*. In the following we denote by $x_+ := \max\{x, 0\}$ for all $x \in \mathbb{R}$.

Lemma 5. *For all $k \in \{0, \ldots, n\}$ and $N \in [0, \mathrm{UB}]$ it holds that*

$$T\left(k, (\ell_N - k)_+\right) \leq T^*(k, N), \text{ where } \ell_N = \left\lfloor \frac{N}{C} \right\rfloor.$$

Before showing this lemma we show why it implies an FPTAS for our problem.

Theorem 2. *If the function c is convex, our problem admits a $(1 - \epsilon)$-approximation algorithm with running time $\mathcal{O}(n^3/\epsilon^2)$ for any $\epsilon > 0$.*

Proof. Using the proposed dynamic program, we fill the table T and search for the largest ℓ such that $T(n, \ell) < \infty$. For proving correctness, take $N := \mathrm{OPT}$ and $k := n$ in our previous lemma. It follows that $T\left(n, (\ell_{\mathrm{OPT}} - n)_+\right) \leq T^*(n, \mathrm{OPT}) < \infty$. Then, by Lemma 4, there exists a set of items A with

$$\pi(A) = (\ell_{\mathrm{OPT}} - n)_+ C \geq \left(\left\lfloor \frac{\mathrm{OPT}}{C} \right\rfloor - n\right)C \geq \left(\frac{\mathrm{OPT}}{C} - n - 1\right)C = \mathrm{OPT} - C(n+1)$$

$$\geq \mathrm{OPT} - \epsilon\mathrm{UB} \geq \mathrm{OPT} - 2\epsilon\mathrm{OPT}.$$

This set A can be easily computed by back-tracking. Thus, redefining ϵ as $\epsilon/2$ yields a $(1 - \epsilon)$-approximation. The table has $\mathcal{O}(n^2/\epsilon)$ many entries, and each entry takes $\mathcal{O}(n/\epsilon)$ time to compute. This implies the claimed running time. $\qquad \square$

It remains to show Lemma 5. In order to do that we need the following technical property.

Proposition 2. *For all k and $\ell_1 \leq \ell_2$ it holds that $T(k, \ell_1) \leq T(k, \ell_2)$.*

Proof. We argue by induction on k. If $k = 0$ then the proposition holds immediately. By induction hypothesis, we assume that the proposition holds for $k - 1$ and all $\ell_1 \leq \ell_2$. Let us fix a value for ℓ_1 and ℓ_2 with $\ell_1 \leq \ell_2$. By (4) we have that either $T(k, \ell_2) = T(k - 1, \ell_2)$, or $T(k, \ell_2) = T(k - 1, \ell') + w_k$ for some $\ell' \geq 0$ such that $\ell'C + \delta_k^{w_k}(T(k - 1, \ell')) \geq \ell_2 C$. If the first case holds, then (4) implies that $T(k, \ell_1) \leq T(k - 1, \ell_1)$, and the induction hypothesis yields that

$T(k-1, \ell_1) \le T(k-1, \ell_2) = T(k, \ell_2)$. We conclude that $T(k, \ell_1) \le T(k, \ell_2)$ in this case.

For the second case, where $T(k, \ell_2) = T(k-1, \ell') + w_k$, it is enough to notice that $T(k, \ell_1) \le T(k-1, \ell') + w_k = T(k, \ell_2)$. Indeed, this follows from (4), by noting that

$$\ell' C + \delta_k^{w_k}(T(k-1, \ell')) \ge \ell_2 C \ge \ell_1 C.$$

$\square$

Proof (Lemma 5). We show the lemma by induction on k. For $k = 0$, the claim follows directly from the initial conditions for T. We now assume that the claim is true for $k-1$ and all $N \in [0, \mathrm{UB}]$. For the induction step, let us fix a value N. If $T^*(k, N) = \infty$, the claim is obviously true. Otherwise there exists a set of items $A \subseteq \{1, \ldots, k\}$, such that $\pi(A) \ge N$ and $W_A = T^*(k, N)$. We consider two cases.

Case 1. $k \notin A$.
In this case we have that $T^*(k-1, N) = T^*(k, N)$. Therefore it holds that

$$T\left(k, (\ell_N - k)_+\right) \le T\left(k-1, (\ell_N - k)_+\right) \le T\left(k-1, (\ell_N - k + 1)_+\right)$$
$$\le T^*(k-1, N) = T^*(k, N).$$

Here, the first inequality follows from (4), the second by the previous proposition, and the third one by the induction hypothesis.

Case 2. $k \in A$.
For this case we consider the value $N' := \pi(A \setminus \{k\})$, and we use the convexity of the cost function to show the following property.

Claim. It holds that $T^*(k-1, N') = W_{A \setminus \{k\}}$.
We show the claim by contradiction. Let us assume that there exists a set $A^* \subseteq \{1, \ldots, k-1\}$ with $\pi(A^*) \ge N'$ and $W_{A^*} < W_{A \setminus \{k\}}$. Then $W_{A^* \cup \{k\}} < W_A$, and

$$\pi(A^* \cup \{k\}) = \pi(A^*) + \delta_k^{w_k}(W_{A^*})$$
$$\ge N' + \delta_k^{w_k}(W_{A^*})$$
$$\ge N' + \delta_k^{w_k}(W_{A \setminus \{k\}}) = \pi(A),$$

where the last inequality follows from Proposition 1 and $W_{A^*} < W_{A \setminus \{k\}}$. This implies that $T^*(k, N) < W_A$, which contradicts the definition of A. The claim follows.

Using this claim we can conclude

$$N' + \delta_k^{w_k}\left(T^*(k-1, N')\right) \ge N. \tag{5}$$

To simplify notation, let us call $r := (\ell_{N'} - k + 1)_+$. Our goal is to prove

$$T(k, (\ell_N - k)_+) \le T(k-1, r) + w_k. \tag{6}$$

Then

$$T(k-1, r) + w_k \le T^*(k-1, N') + w_k = W_{A \setminus \{k\}} + w_k = T^*(k, N),$$

where the first inequality follows from the induction hypothesis, and the subsequent equality holds by the above claim, completes the induction step. To see that (6) holds, note that

$$r \cdot C + \delta_k^{w_k}\left(T(k-1,r)\right) \geq N' - kC + \delta_k^{w_k}\left(T(k-1,r)\right)$$
$$\geq N' - kC + \delta_k^{w_k}\left(T^*(k-1,N')\right)$$
$$\geq N - kC \geq (\ell_N - k)C.$$

In this computation, the first inequality holds by the definition of $\ell_{N'}$. The second one follows by Proposition 1, and since the induction hypothesis implies that $T(k-1,r) \leq T^*(k-1,N')$. The third inequality is implied by (5).

This chain of inequalities implies (6), as long as $\ell_N - k = (\ell_N - k)_+$, since $\ell' = r$ is one possible choice in (4). On the other hand, if $(\ell_N - k)_+ = 0$, then (6) also holds. Indeed, it is enough to notice that $T(0,0) = 0$, and that applying (4) iteratively implies that $T(k,0) = 0$. This completes the proof.   □

5   Concave Costs

In this section we consider the problem under a concave cost function c. The proof of the following lemma can be found in the full version of our paper [11].

Lemma 6. *Let* OPT *be an optimal solution of maximal weight. Then* OPT *is comprised of exactly the items that have density at least* $c(W_{\text{OPT}}+1) - c(W_{\text{OPT}})$.

Lemma 6 directly suggests the algorithm in Figure 2.

Algorithm:

Initialize a solution $A := \emptyset$, and a candidate solution $D := \emptyset$.
Sort the items in U by non-increasing densities, and let $i_1, i_2, \ldots i_n$ be the items in this sorted order.
For j from 1 to n, **do:**
$\quad D := D \cup \{i_j\}$.
$\quad$**If** $\pi(D) \geq \pi(A)$, **then**
$\quad\quad A := D$.
Output set A.

Fig. 2. An optimal algorithm for the concave costs setting

It is easy to see that this algorithm returns an optimal solution, since it selects the best prefix of the sorted items, and Lemma 6 implies that OPT is such a prefix. The running time is dominated by the sorting of the items. Note that by storing some additional information, $\pi(D)$ can be computed in constant time in each iteration, because $\pi(D) = \pi(D') + \delta_{i_j}^{w_{i_j}}(W')$, where D' corresponds to the candidate solution of the previous iteration, and W' to its weight. Therefore we obtain the following theorem.

Theorem 3. *For concave cost functions, an optimal solution can be computed in time $\mathcal{O}(n \log n)$.*

6 Conclusion

In this paper, we study a generalization of the classical knapsack problem, where the "hard" constraint on the capacity is replaced by a "soft" weight-dependent cost function. An interesting direction for further study is to impose certain restrictions on the set of chosen items. In Barman et al. [1], they show that if a matroid constraint is imposed on the items, they can achieve a 4-approximation for the case of a convex cost function. An obvious question is whether this can be further improved.

References

1. Barman, S., Umboh, S., Chawla, S., Malec, D.: Secretary problems with convex costs. In: Czumaj, A., Mehlhorn, K., Pitts, A., Wattenhofer, R. (eds.) ICALP 2012, Part I. LNCS, vol. 7391, pp. 75–87. Springer, Heidelberg (2012)
2. Brooks, D., Bose, P., Schuster, S., Jacobson, H.M., Kudva, P., Buyuktosunoglu, A., Wellman, J.D., Zyuban, V.V., Gupta, M., Cook, P.W.: Power-aware microarchitecture: Design and modeling challenges for next-generation microprocessors. IEEE Micro 20(6), 26–44 (2000)
3. Buchbinder, N., Feldman, M., Naor, J., Schwartz, R.: A tight linear time (1/2)-approximation for unconstrained submodular maximization. In: FOCS, pp. 649–658. IEEE Computer Society (2012)
4. Feige, U., Mirrokni, V.S., Vondrák, J.: Maximizing non-monotone submodular functions. In: FOCS, pp. 461–471. IEEE Computer Society (2007)
5. Garey, M., Johnson, D.: Computers and Intractability: A Guide to the Theory of NP-Completeness. W. H. Freeman (1979)
6. Goemans, M.X., Williamson, D.P.: Improved approximation algorithms for maximum cut and satisfiability problems using semidefinite programming. J. ACM 42(6), 1115–1145 (1995)
7. Halperin, E., Zwick, U.: Combinatorial approximation algorithms for the maximum directed cut problem. In: Kosaraju, S.R. (ed.) SODA, pp. 1–7. ACM/SIAM (2001)
8. Ibarra, O.H., Kim, C.E.: Fast approximation algorithms for the knapsack and sum of subset problems. J. ACM 22(4), 463–468 (1975)
9. Lee, J., Mirrokni, V.S., Nagarajan, V., Sviridenko, M.: Non-monotone submodular maximization under matroid and knapsack constraints. In: Mitzenmacher, M. (ed.) STOC, pp. 323–332. ACM (2009)
10. O'Sullivan, A., Sheffrin, S.: Economics: Principles in Action. Pearson Prentice Hall (2003)
11. http://www.mpi-inf.mpg.de/~ott/download/MFCS2013_FULL.pdf

Computing Behavioral Distances, Compositionally[*]

Giorgio Bacci, Giovanni Bacci, Kim G. Larsen, and Radu Mardare

Department of Computer Science, Aalborg University, Denmark
{grbacci,giovbacci,kgl,mardare}@cs.aau.dk

Abstract. We propose a general definition of composition operator on
Markov Decision Processes with rewards (MDPs) and identify a well
behaved class of operators, called *safe*, that are guaranteed to be *non-
extensive* w.r.t. the bisimilarity pseudometrics of Ferns et al. [10], which
measure behavioral similarities between MDPs. For MDPs built using
safe/non-extensive operators, we present the first method that exploits
the structure of the system for (exactly) computing the bisimilarity distance on MDPs. Experimental results show significant improvements
upon the non-compositional technique.

1 Introduction

Probabilistic bisimulation of Larsen and Skou [13] is the standard equivalence for
analyzing the behaviour of Markov chains. In [12], this notion has been extended
to Markov Decision Processes with rewards (MDPs) with the intent of reducing
the size of large systems to help the computation of optimal policies.

However, when the numerical values of probabilities are based on statistical
sampling or subject to error estimates, any behavioral analysis based on a notion
of equivalence is too fragile, as it only relates processes with identical behaviors.
This is a common issue in applications such as systems biology [15], games [4], or
planning [7]. Such problems motivated the study of *behavioral distances* (pseudometrics) for probabilistic systems, firstly developed for Markov chains [9,17,16]
and later extended to MDPs [10]. These distances support approximate reasoning
on probabilistic systems, providing a way to measure the behavioral similarity
between states. They allow one to analyze models obtained as approximations
of others, more accurate but less manageable, still ensuring that the obtained
solution is close to the real one. For instance, in [2,3] the pseudometric of [10] is
used to compute (approximated) optimal polices for MDPs in applications for
artificial intelligence. These arguments motivate the development of methods to
efficiently compute behavioral distances for MDPs.

Realistic models are usually specified compositionally by means of operators
that describe the interactions between the subcomponents. These specifications
may thus suffer from an exponential growth of the state space, e.g. the parallel

---

[*] Work supported by the VKR Center of Excellence MT-LAB and by the Sino-Danish
Basic Research Center IDEA4CPS.

K. Chatterjee and J. Sgall (Eds.): MFCS 2013, LNCS 8087, pp. 74–85, 2013.

composition of n subsystems with m states may cause the main system to have m^n states. To cope with this problem, algorithms like [10,7,5] that need to investigate the entire state space of the system and even more recent proposals [1], that avoid the entire state space exploration using on-the-fly techniques, are not sufficient: one needs to reason compositionally.

Classically, the exact behavior of systems can be analyzed compositionally if the considered behavioral equivalence (e.g. bisimilarity) is a congruence w.r.t. the composition operators. When the behavior of processes is approximated by means of behavioral distances, congruence is generalized by the notion of *non-extensiveness* of the composition operators, that describes the relation between the distances of the subcomponents to that of the composite system [9].

In this paper we study to which extent compositionality on MDPs can be exploited in the computation of the behavioral pseudometrics of [10], hence how the compositional structure of processes can be used in an approximated analysis of behaviors. To this end we introduce a general notion of composition operator on MDPs and characterize a class of operators, called *safe*, that are guaranteed to be non-extensive. This class is shown to cover a wide range of known operators (e.g. synchronous and asynchronous parallel composition), moreover its defining property provides an easy systematic way to check non-extensiveness.

We provide an algorithm to compute the bisimilarity pseudometric by exploiting both the on-the-fly state space exploration in the spirit of [1], and the compositional structure of MDPs built over safe operators. Experimental results show that the compositional optimization yields a significant additional improvement on top of that obtained by the on-the-fly method. In the best cases, the exploitation of compositionality achieves a reduction of computation time by a factor of 10, and for least significant cases the reduction is that of a factor of 2.

2   Markov Decision Processes and Behavioral Metrics

In this section we recall the definitions of *finite discrete-time Markov Decision Process with rewards* (MDP), and of *bisimulation relation* on MDPs [12]. Then we recall the definition of *bisimilarity pseudometric* introduced in [10], which measures behavioral similarities between states.

We start recalling a few facts related to probability distributions that are essential in what follows. A *probability distribution* over a finite set S is a function $\mu \colon S \to [0,1]$ such that $\sum_{s \in S} \mu(s) = 1$. We denote by $\Delta(S)$ the set of probability distributions over S. Given $\mu, \nu \in \Delta(S)$, a distribution $\omega \in \Delta(S \times S)$ is a *matching* for (μ, ν) if for all $u, v \in S$, $\sum_{s \in S} \omega(u, s) = \mu(u)$ and $\sum_{s \in S} \omega(s, v) = \nu(v)$; we denote by $\Pi(\mu, \nu)$ the set of matchings for (μ, ν). For a (pseudo)metric $d \colon S \times S \to [0, \infty)$ over a finite set S, the *Kantorovich (pseudo)metric* is defined by $\mathcal{T}_d(\mu, \nu) = \min_{\omega \in \Pi(\mu,\nu)} \sum_{u,v \in S} \omega(u, v) d(u, v)$, for arbitrary $\mu, \nu \in \Delta(S)$.[1]

Definition 1 (Markov Decision Process). *A Markov Decision Process is a tuple $\mathcal{M} = (S, A, \tau, \rho)$ consisting of a finite nonempty set S of states, a finite*

---

[1]   Since S is finite, $\Pi(\mu, \nu)$ describes a *bounded* transportation polytope [8], hence the minimum in the definition of $\mathcal{T}_d(\mu, \nu)$ exists and can be achieved at some vertex.

nonempty set A of actions, a transition function $\tau\colon S \times A \to \Delta(S)$, and a reward function $\rho\colon S \times A \to \mathbb{R}$.

The operational behavior of an MDP $\mathcal{M} = (S, A, \tau, \rho)$ is as follows: the process in the state $s_0 \in S$ chooses nondeterministically an action $a \in A$ and it changes the state to $s_1 \in S$, with probability $\tau(s_0, a)(s_1)$. The choice of a in s_0 is rewarded by $\rho(s_0, a)$. The *executions* are transition sequences $w = (s_0, a_0)(s_1, a_1)\ldots$; the challenge is to find *strategies* for choosing the actions in order to maximize the reward $R_\lambda(w) = \lim_{n\to\infty} \sum_{i=0}^{n} \lambda^i \rho(s_i, a_i)$, where $\lambda \in (0,1)$ is a *discount factor*. A *strategy* is given by a function $\pi\colon S \to \Delta(A)$, called *policy*, where $\pi(s_0)(a)$ is the probability of choosing the action a at state s_0. Each policy π induces a probability distribution over executions defined, for an arbitrary $w = (s_0, a_0)(s_1, a_1)\ldots$, by $P^\pi(w) = \lim_{n\to\infty} \prod_{i=0}^{n} \pi(s_i)(a_i)\cdot\tau(s_i, a_i)(s_{i+1})$. The *value of $s \in S$ according to π*, written $V_\lambda^\pi(s)$, is the expected value of R_λ w.r.t. P^π on the measurable cylinder set of the executions starting from s. The mapping $V_\lambda^\pi\colon S \to \mathbb{R}$ is the *value function according to π*. The value functions induce a preorder on policies defined by $\pi \preceq \pi'$ iff $V_\lambda^\pi(s) \leq V_\lambda^{\pi'}(s)$, for all $s \in S$. A policy π^* is *optimal* for an MDP $\mathcal{M}$ if it is maximal w.r.t. $\preceq$ among all policies for $\mathcal{M}$. Given $\mathcal{M}$, there always exists an optimal policy π^*, but it might not be unique; it has a unique value function $V_\lambda^{\pi^*}$ satisfying the following system of equations known as the *Bellman optimality equations*: $V_\lambda^{\pi^*}(s) = \max_{a\in A} \left(\rho(s,a) + \lambda \sum_{t\in S} \tau(s,a)(t) \cdot V_\lambda^{\pi^*}(t)\right)$, for all $s \in S$. As reference on MDPs we recommend to consult [14].

Definition 2 (Stochastic Bisimulation). *Let $\mathcal{M} = (S, A, \tau, \rho)$ be an MDP. An equivalence relation $R \subseteq S \times S$ is a* stochastic bisimulation *if whenever $(s,t) \in R$ then, for all $a \in A$, $\rho(s,a) = \rho(t,a)$ and, for all R-equivalence classes C, $\tau(s,a)(C) = \tau(t,a)(C)$. Two states $s, t \in S$ are* stochastic bisimilar, *written $s \sim_{\mathcal{M}} t$, if they are related by some stochastic bisimulation on $\mathcal{M}$.*

To cope with the problem of measuring how similar two MDPs are, Ferns et al. [10] defined a bisimilarity pseudometrics that measure the behavioural similarity of two non-bisimilar MDPs. This is defined as the least fixed point of a transformation operator on functions in $[0, \infty)^{S\times S}$.

Let $\mathcal{M} = (S, A, \tau, \rho)$ be an MDP and $\lambda \in (0,1)$ be a discount factor. The set $[0,\infty)^{S\times S}$ of $[0,\infty)$-valued maps on $S \times S$ equipped with the point-wise partial order defined by $d \sqsubseteq d'$ iff $d(s,t) \leq d'(s,t)$, for all $s, t \in S$, forms an ω-complete partial order with bottom the constant zero-function $\mathbf{0}$, and greatest lower bound given by $(\bigsqcap_{i\in\mathbb{N}} d_i)(s,t) = \inf_{i\in\mathbb{N}} d_i(s,t)$, for all $s, t \in S$. We define a fixed point operator $F_\lambda^{\mathcal{M}}$ on $[0,\infty)^{S\times S}$, for $d\colon S \times S \to [0,\infty)$ and $s, t \in S$, as follows:

$$F_\lambda^{\mathcal{M}}(d)(s,t) = \max_{a\in A} \left(|\rho(s,a) - \rho(t,a)| + \lambda \cdot \mathcal{T}_d(\tau(s,a), \tau(t,a))\right).$$

$F_\lambda^{\mathcal{M}}$ is monotonic [10], thus, by Tarski's fixed point theorem, it admits a least fixed point. This fixed point is the *bisimilarity pseudometric*.

Definition 3 (Bisimilarity pseudometric). *Let $\mathcal{M}$ be an MDP and $\lambda \in (0,1)$ be a discount factor, then the λ-discounted bisimilarity pseudometric for $\mathcal{M}$, written $\delta_\lambda^{\mathcal{M}}$, is the least fixed point of $F_\lambda^{\mathcal{M}}$.*

The pseudometric $\delta_\lambda^{\mathcal{M}}$ enjoys the property that two states are at zero distance if and only if they are bisimilar. Moreover, in [6] it has been proved, using Banach's fixed point theorem, that for $\lambda \in (0,1)$, $F_\lambda^{\mathcal{M}}$ has a *unique* fixed point.

3 Non-extensiveness and Compositional Reasoning

In this section we give a general definition of composition operator on MDPs that subsumes most of the known composition operators such as the synchronous, asynchronous, and CCS-like parallel compositions. We introduce the notion of *safeness* for an operator and prove that it implies non-extensiveness. Recall that, non-extensiveness corresponds to the quantitative analogue of congruence when one aims to reason with behavioral distances, as advocated e.g. in [11,9].

Definition 4 (Composition Operator). *Let* $\mathcal{M}_i = (S_i, A_i, \tau_i, \rho_i)$, $i = 1..n$, *be MDPs. A composition operator on* $\mathcal{M}_1, \ldots, \mathcal{M}_n$ *is a tuple* $op = (A, op_\tau, op_\rho)$ *consisting of a nonempty set* A *of actions and the following operations*

- **on transitions functions:** $op_\tau \colon \prod_{i=1}^{n} \Delta(S_i)^{S_i \times A_i} \to \Delta(S)^{S \times A}$,
- **on reward functions:** $op_\rho \colon \prod_{i=1}^{n} \mathbb{R}^{S_i \times A_i} \to \mathbb{R}^{S \times A}$.

where, $S = \prod_{i=1}^{n} S_i$ *denotes the cartesian product of* S_i, $i = 1..n$. *We denote by* $op(\mathcal{M}_i, \ldots, \mathcal{M}_n)$ *the composite MDP* $(S, A, op_\tau(\tau_1, \ldots, \tau_n), op_\rho(\rho_1, \ldots, \rho_n))$.

Below we present examples, for two fixed MDPs $\mathcal{M}_X = (X, A_X, \tau_X, \rho_X)$ and $\mathcal{M}_Y = (Y, A_Y, \tau_Y, \rho_Y)$, of some of the known parallel composition operators.

Example 5. **Synchronous Parallel Composition** can be given as a binary composition operator $| = (A_X \cap A_Y, |_\tau, |_\rho)$, where

$$(\tau_X \mid_\tau \tau_Y)((x,y),a)(u,v) = \tau_X(x,a)(u) \cdot \tau_Y(y,a)(v) \,,$$
$$(\rho_X \mid_\rho \rho_Y)((x,y),a) = \rho_X(x,a) + \rho_Y(y,a) \,.$$

The process $\mathcal{M}_X \mid \mathcal{M}_Y$ reacts iff $\mathcal{M}_X$ and $\mathcal{M}_Y$ can react synchronously. Actions are rewarded by summing up the rewards of the components.   ∎

Example 6. **CCS-like Parallel Composition** can be defined by the composition operator $\| = (A_X \cup A_Y, \|_\tau, \|_\rho)$, where

$$(\tau_X \mid\mid_\tau \tau_Y)((x,y),a)(u,v) = \begin{cases} \tau_X(x,a)(u) & \text{if } a \notin A_Y \text{ and } v = y \\ \tau_Y(y,a)(v) & \text{if } a \notin A_X \text{ and } u = x \\ \tau_X(x,a)(u) \cdot \tau_Y(y,a)(v) & \text{if } a \in A_X \cap A_Y \\ 0 & \text{otherwise} \end{cases}$$

$$(\rho_X \mid\mid_\rho \rho_Y)((x,y),a) = \begin{cases} \rho_X(x,a) & \text{if } a \notin A_Y \\ \rho_Y(y,a) & \text{if } a \notin A_X \\ \rho_X(x,a) + \rho_Y(y,a) & \text{if } a \in A_X \cap A_Y \end{cases}$$

In the process $\mathcal{M}_X \parallel \mathcal{M}_Y$, the components synchronize on the same action, otherwise they proceed asynchronously. Asynchronous parallel composition can be defined as above, requiring that the MDPs have disjoint set of actions.   ∎

Before introducing the concept of non-extensiveness for a composition operator, we provide some preliminary notations. Consider the sets X_i, the functions
$d_i \colon X_i \times X_i \to [0, \infty)$, for $i = 1..n$, and $p \in [1, \infty]$. We define the *p-norm* function
$\|d_1, \ldots, d_n\|_p \colon \prod_{i=1}^n X_i \times \prod_{i=1}^n X_i \to [0, \infty)$ as follows:

$$\|d_1, \ldots, d_n\|_p((x_1, \ldots, x_n), (y_1, \ldots, y_n)) = \left(\sum_{i=1}^n d_i(x_i, y_i)^p\right)^{\frac{1}{p}} \quad \text{if } p < \infty,$$
$$\|d_1, \ldots, d_n\|_\infty((x_1, \ldots, x_n), (y_1, \ldots, y_n)) = \max_{1 \le i \le n} d_i(x_i, y_i).$$

Note that, if (X_i, d_i) are (pseudo)metric spaces, $\|d_1, \ldots, d_n\|_p$ is a (pseudo)metric
on $\prod_{i=1}^n X_i$, known in the literature as the *p-product (pseudo)metric*.

Definition 7. *Let $p \in [1, \infty]$. A composition operator op on MDPs $\mathcal{M}_1, \ldots, \mathcal{M}_n$
is p-non-extensive if $\delta_\lambda^{op(\mathcal{M}_1, \ldots, \mathcal{M}_n)} \sqsubseteq \|\delta_\lambda^{\mathcal{M}_1}, \ldots, \delta_\lambda^{\mathcal{M}_n}\|_p$. A composition operator
is* non-extensive *if it is p-non-extensive for some p.*

Non-extensiveness for a composition operator ensures that bisimilarity is a congruence with respect to it —direct consequence of Theorem 4.5 in [10].

Lemma 8. *Let $\mathcal{M}_i = (S_i, A_i, \tau_i, \rho_i)$ be an MDP and $s_i, t_i \in S_i$, for $i = 1..n$,
and op be a p-non-extensive composition operator on $\mathcal{M}_1, \ldots, \mathcal{M}_n$. Then,*

i) *if $p < \infty$, $\delta_\lambda^{op(\mathcal{M}_1, \ldots, \mathcal{M}_n)}((s_1, \ldots, s_n), (t_1, \ldots, t_n)) \le \left(\sum_{i=1}^n \delta_\lambda^{\mathcal{M}_i}(s_i, t_i)^p\right)^{\frac{1}{p}}$*

ii) *if $p = \infty$, $\delta_\lambda^{op(\mathcal{M}_1, \ldots, \mathcal{M}_n)}((s_1, \ldots, s_n), (t_1, \ldots, t_n)) \le \max_{i=1}^n \delta_\lambda^{\mathcal{M}_i}(s_i, t_i)$.*

Corollary 9. *Let $\mathcal{M}_i = (S_i, A_i, \tau_i, \rho_i)$ be an MDP, $s_i, t_i \in S_i$, for $i = 1..n$, and
op be a non-extensive composition operator on $\mathcal{M}_1, \ldots, \mathcal{M}_n$. If $s_i \sim_{\mathcal{M}_i} t_i$ for all
$i = 1..n$, then $(s_1, \ldots, s_n) \sim_{op(\mathcal{M}_1, \ldots, \mathcal{M}_n)} (t_1, \ldots, t_n)$.*

In general, proving non-extensiveness for a composition operator on MDPs is
not a simple task, since one needs to consider the pseudometrics $\delta_\lambda^{\mathcal{M}_i}$ which
are defined as the least fixed point of $F_\lambda^{\mathcal{M}_i}$. A simpler sufficient condition that
ensures non-extensiveness is the following:

Definition 10. *Let $\mathcal{M}_i = (S_i, A_i, \tau_i, \rho_i)$, for $i = 1..n$, be MDPs and $p \in [1, \infty]$.
A composition operator op on $\mathcal{M}_1, \ldots, \mathcal{M}_n$ is p-safe if, for any d_i pseudometric
on S_i, such that $d_i \sqsubseteq F_\lambda^{\mathcal{M}_i}(d_i)$, it holds*

$$F_\lambda^{op(\mathcal{M}_1, \ldots, \mathcal{M}_n)}(\|d_1, \ldots, d_n\|_p) \sqsubseteq \|F_\lambda^{\mathcal{M}_1}(d_1), \ldots, F_\lambda^{\mathcal{M}_n}(d_n)\|_p.$$

A composition operator on MDPs is safe *if it is p-safe for some $p \in [1, \infty]$.*

Theorem 11. *Any safe composition operator on MDPs is non-extensive.*

The examples of compositional operators that we have presented in this section
are all 1-safe, hence non-extensive.

Proposition 12. *The composition operators of Examples 5–6 are 1-safe.*

4   Alternative Characterization of the Pseudometric

In this section we give an alternative characterization of $\delta_\lambda^{\mathcal{M}}$ based on the notion of *coupling* that allows us to transfer the results previously proven for Markov chains in [1,5] to MDPs. Then, we show how to relate this characterization to the concept of non-extensiveness for compositional operators on MDPs.

Definition 13 (Coupling). *Let $\mathcal{M} = (S, A, \tau, \rho)$ be an MDP. A coupling for $\mathcal{M}$ is a pair $\mathcal{C} = (\rho, \omega)$, where $\omega \colon (S \times S) \times A \to \Delta(S \times S)$ is such that, for any $s, t \in S$ and $a \in A$, $\omega((s,t), a) \in \Pi(\tau(s, a), \tau(t, a))$.*

Given a coupling $\mathcal{C} = (\rho, \omega)$ for $\mathcal{M}$ and a discount factor $\lambda \in (0, 1)$, we define the operator $\Gamma_\lambda^{\mathcal{C}} \colon [0, \infty)^{S \times S} \to [0, \infty)^{S \times S}$, for $d \in [0, \infty)^{S \times S}$ and $s, t \in S$, by

$$\Gamma_\lambda^{\mathcal{C}}(d)(s,t) = \max_{a \in A} \left(|\rho(s, a) - \rho(t, a)| + \lambda \sum_{u,v \in S} d(u, v) \cdot \omega((s,t), a)(u, v) \right).$$

Note that, any coupling $\mathcal{C} = (\rho, \omega)$ for $\mathcal{M}$ induces an MDP $\mathcal{C}^* = (S \times S, A, \omega, \rho^*)$, defined for any $s, t \in S$ and $a \in A$ by $\rho^*((s,t), a) = |\rho(s, a) - \rho(t, a)|$, and $\Gamma_\lambda^{\mathcal{C}}$ corresponds to the *Bellman optimality operator* on $\mathcal{C}^*$. This operator is monotonic and has a unique fixed point, hereafter denoted by $\gamma_\lambda^{\mathcal{C}}$, corresponding to the value function for $\mathcal{C}^*$ (see [14, §6.2]).

Next we see that the bisimilarity pseudometric $\delta_\lambda^{\mathcal{M}}$ can be characterized as the minimum $\gamma_\lambda^{\mathcal{C}}$ among all the couplings $\mathcal{C}$ for $\mathcal{M}$.

Theorem 14. *Let $\mathcal{M}$ be an MDP. Then, $\delta_\lambda^{\mathcal{M}} = \min \left\{ \gamma_\lambda^{\mathcal{C}} \mid \mathcal{C} \text{ coupling for } \mathcal{M} \right\}$.*

Theorem 14 allows us to transfer the compositional reasoning on couplings. To this end, we introduce the notion of composition operator on couplings.

Definition 15. *Let $\mathcal{M}_i = (S_i, A_i, \tau_i, \rho_i)$ be MDPs, for $i = 1..n$. A coupling composition operator for $\mathcal{M}_1, \ldots, \mathcal{M}_n$ is a tuple $op^* = (A, op_\rho^*, op_\omega^*)$ consisting of a nonempty set A, and the following operations, where $S = \prod_{i=1}^n S_i$.*
- $op_\rho^* \colon \prod_{i=1}^n \mathbb{R}^{S_i \times A_i} \to \mathbb{R}^{S \times A}$,
- $op_\omega^* \colon \prod_{i=1}^n \Delta(S_i \times S_i)^{S_i \times S_i \times A_i} \to \Delta(S \times S)^{S \times S \times A}$,

Let $\mathcal{C}_i = (\rho_i, \omega_i)$ be a coupling for $\mathcal{M}_i$, for $i = 1..n$, we denote by $op^*(\mathcal{C}_1, \ldots, \mathcal{C}_n)$ the composite coupling $(op_\rho^*(\rho_1, \ldots, \rho_n), op_\omega^*(\omega_1, \ldots, \omega_n))$. Moreover, op^* is called lifting of a composition operator op on $\mathcal{M}_1, \ldots, \mathcal{M}_n$ if, for all $i = 1..n$ and $\mathcal{C}_i$ coupling for $\mathcal{M}_i$, $op^*(\mathcal{C}_1, \ldots, \mathcal{C}_n)$ is a coupling for $op(\mathcal{M}_1, \ldots, \mathcal{M}_n)$.

It is not always possible to find coupling composition operators that lift a composition operator on MDPs. Nevertheless, the composite operators presented in Examples 5–6 can be lifted on couplings. We show in the next example how this can be done for the CCS-like parallel composition. For the other example the construction is similar.

Example 16. The composition operator of Example 6 can be lifted on couplings by the operator $\|^* = (A_X \cup A_Y, \|_\rho, \|_\omega)$

$$(\omega_X \|_\omega \omega_Y)(((x,y),(x',y')),a)((u,v),(u',v')) =$$

$$= \begin{cases} \omega_X((x,x'),a)(u,u') & \text{if } a \notin A_Y, (v,v') = (y,y') \\ \omega_Y((y,y'),a)(v,v') & \text{if } a \notin A_X, (u,u') = (x,x') \\ \omega_X((x,x'),a)(u,u') \cdot \omega_Y((y,y'),a)(v,v') & \text{if } a \in A_X \cap A_Y \\ 0 & \text{otherwise} \end{cases}$$

Note how the definition above mimics the one in Example 6.   ∎

Next we adapt the concept of safeness to coupling composition operators.

Definition 17. *Let $\mathcal{M}_i = (S_i, A_i, \tau_i, \rho_i)$ be MDPs, $i = 1..n$ and $p \in [1, \infty]$. A coupling composition operator op^* on $\mathcal{M}_1, \ldots, \mathcal{M}_n$ is p-safe if, for all $i = 1..n$, $\mathcal{C}_i$ coupling for $\mathcal{M}_i$ and $d_i \colon S_i \times S_i \to [0, \infty)$ such that $d_i \sqsubseteq \Gamma_\lambda^{\mathcal{C}_i}(d_i)$, it holds*

$$\Gamma_\lambda^{op^*(\mathcal{C}_1,\ldots,\mathcal{C}_n)}(\|d_1,\ldots,d_n\|_p) \sqsubseteq \|\Gamma_\lambda^{\mathcal{C}_1}(d_1),\ldots,\Gamma_\lambda^{\mathcal{C}_n}(d_n)\|_p.$$

A coupling composition operator is safe *if it is p-safe for some $p \in [1, \infty]$.*

As done for Proposition 12, the lifting in Example 16 can be shown to be 1-safe.

Non-extensiveness for an operator is ensured if it admits a lifting composition operator on couplings that is safe, as proven by the following theorem.

Theorem 18. *Let op^* be a coupling composition operator that lifts a composition operator op on $\mathcal{M}_1, \ldots, \mathcal{M}_n$. If op^* is safe, then op is non-extensive.*

5   Exact Computation of Bisimilarity Distance

Inspired by the characterization given in Theorem 14, in this section we propose a procedure to exactly compute the bisimilarity pseudometric. This extends to MDPs a method that has been proposed in [1] for Markov chains. We also show how this strategy can be optimized to cope well with composite MDPs.

For a discount factor $\lambda \in (0,1)$, the set of couplings for $\mathcal{M}$ can be endowed with the preorder $\trianglelefteq_\lambda$, defined by $\mathcal{C} \trianglelefteq_\lambda \mathcal{D}$ iff $\gamma_\lambda^{\mathcal{C}} \sqsubseteq \gamma_\lambda^{\mathcal{D}}$. Theorem 14 suggests to look for a coupling for $\mathcal{M}$ which is minimal w.r.t. $\trianglelefteq_\lambda$. The enumeration of all the couplings is clearly unfeasible, therefore it is crucial to provide an efficient search strategy which prevents us to do that.

A Greedy Search Strategy. We provide a greedy strategy that explores the set of couplings until an optimal one is eventually reached.

Let $\mathcal{M} = (S, A, \tau, \rho)$ and $\mathcal{C} = (\rho, \omega)$ be a coupling for $\mathcal{M}$. Given $s, t \in S$, $a \in A$, and $\mu \in \Pi(\tau(s,a), \tau(t,a))$, we denote by $\mathcal{C}[(s,t), a/\mu]$ the coupling (ρ, ω') for $\mathcal{M}$, where ω' is such that $\omega'((s,t),a) = \mu$ and $\omega'((s',t'),a') = \omega((s',t'),a')$ for all $s', t' \in S$ and $a' \in A$ with $((s',t'),a') \neq ((s,t),a)$.

Lemma 19. *Let $\mathcal{M} = (S, A, \tau, \rho)$ be an MDP, $\mathcal{C}$ be a coupling for $\mathcal{M}$, $s, t \in S$, $a \in A$, $\mu \in \Pi(\tau(s,a), \tau(t,a))$, and $\mathcal{D} = \mathcal{C}[(s,t), a/\mu]$. If $\Gamma_\lambda^{\mathcal{D}}(\gamma_\lambda^{\mathcal{C}})(s,t) < \gamma_\lambda^{\mathcal{C}}(s,t)$, then $\gamma_\lambda^{\mathcal{D}} \sqsubseteq \gamma_\lambda^{\mathcal{C}}$.*

The lemma above states that $\mathcal{C}$ can be improved w.r.t. $\unlhd_\lambda$ by locally updating it as $\mathcal{C}[(s,t), a/\mu]$, with a matching $\mu \in \Pi(\tau(s,a), \tau(t,a))$ such that

$$\sum_{u,v \in S} \gamma_\lambda^{\mathcal{C}}(u,v) \cdot \mu(u,v) < \sum_{u,v \in S} \gamma_\lambda^{\mathcal{C}}(u,v) \cdot \omega((s,t), a)(u,v),$$

where $a \in A$ is the action that maximizes $\Gamma^{\mathcal{C}}(\gamma_\lambda^{\mathcal{C}})_\lambda(s,t)$. A matching μ satisfying the condition above can be obtained as a solution of a Transportation Problem [8] with cost matrix $(\gamma_\lambda^{\mathcal{C}}(u,v))_{u,v \in S}$ and marginals $\tau(s,a)$ and $\tau(t,a)$, hereafter denoted by $TP(\gamma_\lambda^{\mathcal{C}}, \tau(s,a), \tau(t,a))$. This gives us a strategy for moving toward $\delta_\lambda^{\mathcal{M}}$ by successive improvements on the couplings.

Now we give a necessary and sufficient condition for termination.

Lemma 20. *Let $\mathcal{M} = (S, A, \tau, \rho)$ be an MDP and $\mathcal{C}$ be a coupling for $\mathcal{M}$. If $\gamma_\lambda^{\mathcal{C}} \neq \delta_\lambda^{\mathcal{M}}$, then there exist $s, t \in S$, $a \in A$, and $\mu \in \Pi(\tau(s,a), \tau(t,a))$ such that $\Gamma_\lambda^{\mathcal{D}}(\gamma_\lambda^{\mathcal{C}})(s,t) < \gamma_\lambda^{\mathcal{C}}(s,t)$, where $\mathcal{D} = \mathcal{C}[(s,t), a/\mu]$.*

The above result ensures that, unless $\mathcal{C}$ is optimal w.r.t $\unlhd_\lambda$, the hypotheses of Lemma 19 are satisfied, so that, we can further improve $\mathcal{C}$ following the same strategy. The next statement proves that this search strategy is correct.

Theorem 21. *$\delta_\lambda^{\mathcal{M}} = \gamma_\lambda^{\mathcal{C}}$ iff there exists no coupling $\mathcal{D}$ for $\mathcal{M}$ s.t. $\Gamma_\lambda^{\mathcal{D}}(\gamma_\lambda^{\mathcal{C}}) \sqsubseteq \gamma_\lambda^{\mathcal{C}}$.*

Remark 22. In general, there could be an infinite number of couplings (ρ, ω). However, for each fixed $d \in [0, \infty)^{S \times S}$, the linear function mapping $\omega((s,t), a)$ to $\sum_{u,v \in S} d(u,v) \cdot \omega((s,t)a)(u,v)$ achieves its minimum at some vertex of the transportation polytope $P = \Pi(\tau(s,a), \tau(t,a))$. Since the number of such vertices is finite, using the optimal transportation schedule (which is a vertex in P) for the update ensures that the search strategy is always terminating. $\blacksquare$

Compositional Heuristic: Assume we want to compute the bisimilarity distance for a composite MDP $\mathcal{M} = op(\mathcal{M}_1, \ldots, \mathcal{M}_n)$. The greedy strategy described above moves toward an optimal coupling for $\mathcal{M}$ starting from an arbitrary one. Clearly, the better is the initial coupling the fewer are the steps to the optimal one. The following result gives a heuristic for choosing such a coupling when op admits a safe lifting coupling composition operator.

Proposition 23. *Let op be a composition operator on $\mathcal{M}_1, \ldots, \mathcal{M}_n$, and op^* be a p-safe coupling composition operator that lifts op. Then,*

(i) *$\gamma_\lambda^{op^*(\mathcal{C}_1, \ldots, \mathcal{C}_n)} \sqsubseteq \|\gamma_\lambda^{\mathcal{C}_1}, \ldots, \gamma_\lambda^{\mathcal{C}_n}\|_p$, for any $\mathcal{C}_i$ coupling for $\mathcal{M}_i$;*

(ii) *$\delta_\lambda^{op(\mathcal{M}_1, \ldots, \mathcal{M}_n)} \sqsubseteq \gamma_\lambda^{op^*(\mathcal{D}_1, \ldots, \mathcal{D}_n)} \sqsubseteq \|\delta_\lambda^{\mathcal{M}_1}, \ldots, \delta_\lambda^{\mathcal{M}_n}\|_p$, where $\mathcal{D}_i$ is a coupling for $\mathcal{M}_i$ which is minimal w.r.t. $\unlhd_\lambda$.*

Proposition 23(ii) suggests to start from the coupling $op^*(\mathcal{D}_1, \ldots, \mathcal{D}_n)$, i.e., the one given as the composite of the optimal couplings $\mathcal{D}_i$ for the subcomponents $\mathcal{M}_i$. This ensures that the first over-approximation of $\delta_\lambda^{\mathcal{M}}$, that is $\gamma_\lambda^{op^*(\mathcal{D}_1, \ldots, \mathcal{D}_n)}$, is at least as good as the upper bound given by non-extensiveness of op.

Algorithm 1. On-the-Fly Bisimilarity Pseudometric

Input: MDP $\mathcal{M} = (S, A, \tau, \rho)$; discount factor $\lambda \in (0, 1)$; query $Q \subseteq S \times S$.
1. $\mathcal{C} \leftarrow (\rho, \mathsf{empty})$; $d \leftarrow \mathsf{empty}$; visited $\leftarrow \emptyset$; exact $\leftarrow \emptyset$; toComp $\leftarrow Q$;   // Initialize
2. **while** $\exists (s, t) \in$ toComp **do**
3.   **for all** $a \in A$ **do** guess $\mu \in \Pi(\tau(s, a), \tau(t, a))$; $UpdateC(\mathcal{M}, (s, t), a, \mu)$
4. $d \leftarrow BellmanOpt(\lambda, \mathcal{C}, d)$ // update the current estimate
5.   **while** $\mathcal{C}[(u, v), a]$ is not optimal for $TP(d, \tau(u, a), \tau(v, a))$ **do**
6. $\mu \leftarrow$ optimal schedule for $TP(d, \tau(u, a), \tau(v, a))$
7. $UpdateC(\mathcal{M}, (u, v), a, \mu)$ // improve the current coupling
8. $d \leftarrow BellmanOpt(\lambda, \mathcal{C}, d)$ // update the current estimate
9.   **end while**
10.   exact $\leftarrow$ exact $\cup$ visited   // add new exact distances
11.   toComp $\leftarrow$ toComp $\setminus$ exact   // remove exactly computed pairs
12. **end while**
13. **return** $d{\upharpoonright}_Q$ // return the distance restricted to the pairs in Q

---

6   A Compositional On-the-Fly Algorithm

In this section we provide an on-the-fly algorithm for computing the bisimilarity distance making full use of the greedy strategy presented in Section 5. Then, we describe how to optimize the computation on composite MDPs.

Let $\mathcal{M} = (S, A, \tau, \rho)$ be an MDP, $Q \subseteq S \times S$, and assume we want to compute $\delta_\lambda^{\mathcal{M}}$ restricted to Q, written $\delta_\lambda^{\mathcal{M}}{\upharpoonright}_Q$. Our strategy has the following features:

- when a coupling $\mathcal{C}$ is considered, $\gamma_\lambda^{\mathcal{C}}$ can be computed solving the Bellman optimality equation system associated with it;
- the current coupling $\mathcal{C}$ can be improved by a *local* update $\mathcal{C}[(u, v), a/\mu]$ that satisfies the hypotheses of Lemma 19.

Note that, $\gamma_\lambda^{\mathcal{C}}{\upharpoonright}_Q$ can be computed considering only the smallest independent subsystem containing the variables associated with the pairs in Q. Therefore, we do not need to store the entire coupling, but we can construct it on-the-fly.

The computation of $\delta_\lambda^{\mathcal{M}}{\upharpoonright}_Q$ is implemented by Algorithm 1. We assume the following global variables to store: $\mathcal{C}$, the current partial coupling; d, the current partial over-approximation of $\delta_\lambda^{\mathcal{M}}$; toComp, the pairs of states for which the distance has to be computed; exact, the pairs of states (s, t) such that $d(s, t) = \delta_\lambda^{\mathcal{M}}(s, t)$; visited, the pair of states considered so far.

At the beginning $\mathcal{C}$ and d are empty, there are no visited states and no exact distances. While there are pairs (s, t) left to be computed we update $\mathcal{C}$ calling the subroutine $UpdateC$ on a matching $\mu \in \Pi(\tau(s, a), \tau(t, a))$, for each $a \in A$. Then, d is updated on all visited pairs with the over-approximation $\gamma_\lambda^{\mathcal{C}}$ by calling $BellmanOpt$. According to the greedy strategy, $\mathcal{C}$ is successively improved and d is consequently updated, until no further improvements are possible. Each improvement is demanded by the existence of a better transportation schedule. When line 10 is reached, $d(u, v) = \delta_\lambda^{\mathcal{M}}(u, v)$ for all $(u, v) \in$ visited, therefore visited is added to exact and removed from toComp. If no more pairs have to be considered, the exact distance on Q is returned.

Algorithm 2. $UpdateC(\mathcal{M}, (s,t), a, \mu)$

Input: MDP $\mathcal{M} = (S, A, \tau, \rho)$; $s, t \in S$; $a \in A$, $\mu \in \Pi(\tau(s,a), \tau(t,a))$
1. $\mathcal{C} \leftarrow \mathcal{C}[(s,t), a/\mu]$ // update the coupling
2. visited $\leftarrow$ visited $\cup \{(s,t)\}$ // set (s,t) as visited
3. **for all** $(u,v) \in \{(u',v') \mid \mu(u',v') > 0\} \setminus$ visited **do**   // for all demanded pairs
4.   visited $\leftarrow$ visited $\cup \{(u,v)\}$
5.   // propagate the construction
6.   **for all** $a \in A$ **do** guess $\mu' \in \Pi(\tau(u,a), \tau(v,a))$; $UpdateC(\mathcal{M}, (u,v), a, \mu')$
7. **end for**

The subroutine $UpdateC$ (Algorithm 2) updates the coupling $\mathcal{C}$ and recursively populates it on all demanded pairs. $BellmanOpt(\lambda, \mathcal{C}, d)$ solves the smallest independent subsystem of the Bellman optimality equation system on the MDP induced by $\mathcal{C}$, that contains all the visited pairs. Notice that, the equation system can be further reduced by Gaussian elimination, substituting the variables associated with pairs $(u,v) \in$ exact with $d(u,v)$.

Compositional Optimizations: Algorithm 1 can be modified to handle composite MDPs efficiently. Assume $\mathcal{M} = op(\mathcal{M}_1, \ldots, \mathcal{M}_n)$ and to have a safe coupling composition operator op^* that lifts op. The compositional heuristic described in Section 5 suggests to start from the coupling $op^*(\mathcal{D}_1, \ldots, \mathcal{D}_n)$ obtained by composing the optimal couplings $\mathcal{D}_i$ for each $\mathcal{M}_i$. This is done running Algorithm 1 in two modalities: master/slave. For each $\mathcal{M}_i$, the master shares the data structures $\mathcal{C}_i, d_i$, visited$_i$, toComp$_i$ and exact$_i$ with the corresponding slave to keep track of the computation of $\delta_\lambda^{\mathcal{M}_i}$. When a new pair $((s_i, \ldots, s_n), (t_1, \ldots, t_n))$ is considered, the master runs (possibly in parallel) n slave threads of Algorithm 1 on the query $\{(s_i, t_i)\}$. At the end of these subcomputations, the couplings $\mathcal{C}_i$ are optimal, and they are composed to obtain a coupling for $\mathcal{M}$. Note that, the master can reuse the values stored by the slaves in their previous computations.

Experimental Results: For Markov chains, in [1] it has already been shown that an on-the-fly strategy yields, on average, significant improvements with respect to the corresponding iterative algorithms.

Here we focus on how the compositional optimization affects the performances. To this end we consider a simple yet meaningful set of experiments performed on a collection of MDPs, parametric in the probabilities, modeling a pipeline. The figure aside specifies an element $E_i(p, q)$ of the pipeline with actions $A_i = \{a_i, a_{i+1}, b_i\}$. Pipelines are modeled as the parallel composition of different processing elements, that are connected in series by means of synchronization on shared actions. Table 1 reports the computation times of the tests[2] we have run both

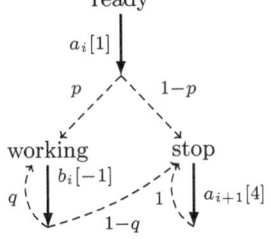

---

[2] The tests have been made using a prototype implementation coded in Mathematica® (available at http://people.cs.aau.dk/~giovbacci/tools.html) running on an Intel Core-i5 2.4 GHz processor with 4GB of RAM.

Table 1. Comparison between the on-the-fly algorithm (OTF) and its compositional optimization (COTF); $E_0 = E_0(0.7, 0.2)$, $E_1 = E_1(0.6, 0.2)$, and $E_2 = E_2(0.5, 0.3)$

Query	Instance	OTF	COTF	# States
All pairs	$E_0 \parallel E_1$	0.654791	0.97248	9
	$E_1 \parallel E_2$	0.702105	0.801121	9
	$E_0 \parallel E_0 \parallel E_1$	48.5982	13.5731	27
	$E_0 \parallel E_1 \parallel E_2$	23.1984	19.9137	27
	$E_0 \parallel E_1 \parallel E_1$	126.335	13.6483	27
	$E_0 \parallel E_0 \parallel E_0$	49.1167	14.1075	27
Single pair	$E_0 \parallel E_0 \parallel E_0 \parallel E_1 \parallel E_1$	16.7027	11.6919	243
	$E_0 \parallel E_1 \parallel E_0 \parallel E_1 \parallel E_1$	20.2666	16.6274	243
	$E_2 \parallel E_1 \parallel E_0 \parallel E_1 \parallel E_1$	22.8357	10.4844	243
	$E_1 \parallel E_2 \parallel E_0 \parallel E_0 \parallel E_2$	11.7968	6.76188	243
	$E_1 \parallel E_2 \parallel E_0 \parallel E_0 \parallel E_2 \parallel E_2$	Time-out	79.902	729

on all-pairs queries and single-pair queries for several pipeline instances; timings are expressed in seconds and, as for the single-pair case, they represent the average of 20 randomly chosen queries. Table 1 shows that the required overhead for maintaining the additional data structure for the subcomponents, affects the performances only on very small systems. In all other cases the compositional optimization yields a significant reduction of the computation time that varies from a factor of 2 up to a factor of 10. Notably, on single-pair queries the compositional version can manage (relatively) large systems whereas the non-compositional one exceeds a time-bound of 3 minutes. Interestingly, we observe better reductions on all-pairs queries than in single-pairs; this may be due to fact that the exact distances collected during the computation are used to further reduce the size of the equation systems that are successively encountered.

7   Conclusions and Future Work

We have proposed a general notion of composition operator on MDPs and identified safeness as a sufficient condition for ensuring non-extensiveness. We showed that the class of safe operators is general enough to cover a wide range of known composition operators. Moreover, we presented an algorithm for computing bisimilarity distances on MDPs, which is able to exploit the compositional structure of the system and applies on MDPs built over any safe operators. This is the first proposal for a compositional algorithm for computing bisimilarity distances; before our contribution, the known tools were based on iterative methods that, by their nature, cannot take advantage of the structure of the systems.

Our work can be extended in several directions. For instance, the notion of safeness can be easily adapted to other contexts where bisimilarity pseudometrics have a fixed point characterization. In the same spirit, one may obtain a sufficient condition that ensures continuity of operators, which is the natural generalization of non-extensiveness.

References

1. Bacci, G., Bacci, G., Larsen, K.G., Mardare, R.: On-the-Fly Exact Computation of Bisimilarity Distances. In: Piterman, N., Smolka, S.A. (eds.) TACAS 2013. LNCS, vol. 7795, pp. 1–15. Springer, Heidelberg (2013)
2. Castro, P.S., Precup, D.: Using bisimulation for policy transfer in MDPs. In: Proceedings of the 9th International Conference on Autonomous Agents and Multiagent Systems, AAMAS 2010, Richland, SC, vol. 1, pp. 1399–1400. International Foundation for Autonomous Agents and Multiagent Systems (2010)
3. Castro, P.S., Precup, D.: Automatic Construction of Temporally Extended Actions for MDPs Using Bisimulation Metrics. In: Sanner, S., Hutter, M. (eds.) EWRL 2011. LNCS, vol. 7188, pp. 140–152. Springer, Heidelberg (2012)
4. Chatterjee, K., de Alfaro, L., Majumdar, R., Raman, V.: Algorithms for Game Metrics. Logical Methods in Computer Science 6(3) (2010)
5. Chen, D., van Breugel, F., Worrell, J.: On the Complexity of Computing Probabilistic Bisimilarity. In: Birkedal, L. (ed.) FOSSACS 2012. LNCS, vol. 7213, pp. 437–451. Springer, Heidelberg (2012)
6. Comanici, G., Panangaden, P., Precup, D.: On-the-Fly Algorithms for Bisimulation Metrics. In: International Conference on Quantitative Evaluation of Systems, pp. 94–103 (2012)
7. Comanici, G., Precup, D.: Basis function discovery using spectral clustering and bisimulation metrics. In: AAMAS 2011, Richland, SC, vol. 3, pp. 1079–1080. International Foundation for Autonomous Agents and Multiagent Systems (2011)
8. Dantzig, G.B.: Application of the Simplex method to a transportation problem. In: Koopmans, T. (ed.) Activity Analysis of Production and Allocation, pp. 359–373. J. Wiley, New York (1951)
9. Desharnais, J., Gupta, V., Jagadeesan, R., Panangaden, P.: Metrics for labelled Markov processes. Theoretical Computer Science 318(3), 323–354 (2004)
10. Ferns, N., Panangaden, P., Precup, D.: Metrics for finite Markov Decision Processes. In: Proceedings of the 20th Conference on Uncertainty in Artificial Intelligence, UAI, pp. 162–169. AUAI Press (2004)
11. Giacalone, A., Jou, C., Smolka, S.A.: Algebraic reasoning for probabilistic concurrent systems. In: Proc. IFIP TC2 Working Conference on Programming Concepts and Methods, pp. 443–458. North-Holland (1990)
12. Givan, R., Dean, T., Greig, M.: Equivalence notions and model minimization in Markov decision processes. Artificial Intelligence 147(1-2), 163–223 (2003)
13. Larsen, K.G., Skou, A.: Bisimulation through probabilistic testing. Information and Computation 94(1), 1–28 (1991)
14. Puterman, M.L.: Markov Decision Processes: Discrete Stochastic Dynamic Programming, 1st edn. John Wiley & Sons, Inc., New York (1994)
15. Thorsley, D., Klavins, E.: Approximating stochastic biochemical processes with Wasserstein pseudometrics. IET Systems Biology 4(3), 193–211 (2010)
16. van Breugel, F., Sharma, B., Worrell, J.: Approximating a Behavioural Pseudometric without Discount for Probabilistic Systems. Logical Methods in Computer Science 4(2), 1–23 (2008)
17. van Breugel, F., Worrell, J.: Approximating and computing behavioural distances in probabilistic transition systems. Theoretical Computer Science 360(1-3), 373–385 (2006)

Which Finitely Ambiguous Automata Recognize Finitely Sequential Functions?*

(Extended Abstract)

Sebastian Bala

Institute of Computer Science
University of Wrocław
Joliot-Curie 20, Wrocław, Poland

Abstract. Weighted automata, especially min-plus automata that operate over the tropical semiring, have both a beautiful theory and important practical applications. In particular, if one could find a sequential or finitely sequential equivalent to a given (or learned) min-plus automaton, one could increase performance in several applications. But this question has long remained open even as a decision problem. We show that existence of a finitely sequential equivalent for a given finitely ambiguous min-plus automaton is decidable.

1 Introduction

Contrary to the classical case, weighted automata, in particular min-plus automata, cannot always be determinized. Therefore, one classifies min-plus automata into a hierarchy, depending on the level of non-determinism. This hierarchy starts with the class of sequential (deterministic) min-plus automata, and includes unambiguous, finitely sequential, finitely ambiguous, and polynomially ambiguous ones. In this context, natural questions arise: (1) given an automaton from one class, is there a criterion that allows to determine if it recognizes a function from a specific sub-class? (2) is the criterion decidable? (3) if the criterion holds, can we effectively construct an equivalent automaton that belongs to the sub-class? In practice (e.g. in speech recognition systems) one often learns automata that are relatively unambiguous. For efficiency, one would like to convert them into a disjoint union of a fixed number of sequential (deterministic) automata - which can then be run very fast in parallel. Abstracting away from many practical constraints, this problem corresponds to the following theoretical question: given a finitely ambiguous min-plus automaton, does there exist an equivalent finitely sequential one?

---

* Supported by NCN grant DEC-2011/01/D/ST6/07164, 2011-2015. The work was done in main part during the author visit at LIAFA laboratory Univeristy Paris Diderot, and in cooperation with the European Union's Seventh Framework Programme (FP7/2007-2013) grant agreement 259454. The visit at LIAFA was sponsored by European Science Foundation, within the project GAMES.

K. Chatterjee and J. Sgall (Eds.): MFCS 2013, LNCS 8087, pp. 86–97, 2013.

In this paper, we provide an algorithm for deciding the above question. ,thereby solving an extended a version of the problem raised in [1] by Kirsten and Lombardy. Our proof is quite technical (the full version is in the appendix), but we hope that the presented ideas and the algorithm can be used in many areas of the theory of min-plus automata. The presented algorithm works in 2-NEXPTIME by deciding the A–Fork Property. When the given automaton is finitely ambiguous, the A-Fork Property can be seen as a generalization of the Twin Property which is a necessary and sufficient condition for the existence of a deterministic counterpart [2] for unambiguous automata and sufficient for an arbitrary min-plus automata. We prove that the negation of the A-Fork Property is a necessary and sufficient condition for the existence of a finitely sequential counterpart of an automaton. We also show how to construct the suitable set of deterministic automata if the A-Fork Property does not hold. Similar properties have been defined for translations of finitely ambiguous [3] and polynomially ambiguous automata [1] into unambigous ones.

2   Preliminaries

Formally a *weighted finite automaton over a semiring* $\mathbb{K} = \langle K, +, \cdot, 1_K, 0_K \rangle$ (wfa) is a quintuple $\mathcal{A} = \langle Q, \Sigma, \lambda, \mu, \gamma \rangle$ where Σ is a finite alphabet, Q is a finite set called states, $\lambda, \gamma \in \mathbb{K}^Q$, and $\mu : \Sigma^* \to \mathbb{K}^{Q \times Q}$ is a homomorphism into the semiring of $Q \times Q$-matrices over $\mathbb{K}$. A state $q \in Q$ is called *initial* if $\lambda[q] \neq 0_K$ and *final* if $\gamma[q] \neq 0_K$.

A quadruple $(p, a, l, q) \in Q \times \Sigma \times \mathbb{K} \times Q$ is a *transition* of the wfa $\mathcal{A}$ if $\mu(a)[p, q] = l$. A path π in $\mathcal{A}$ of length k is a sequence of transitions $t_1 t_2 \ldots t_k$, where $t_i = (q_{i-1}, a_i, l_i, q_i)$. The word $a_1 a_2 \ldots a_k$ is the *label* of π and it is denoted by $label(\pi)$. A path $\pi = t_1 t_2 \ldots t_k$ is *accepting* if the first state of t_1 is an initial one and the last state of t_k is an accepting one. An automaton $\mathcal{A}$ *accepts* a string w if there exists an accepting path $t_1 t_2 \ldots t_k$ labeled by w. By $L(\mathcal{A})$ we denote the set of all strings accepted by $\mathcal{A}$ and we say that $\mathcal{A}$ *recognizes* the language $L(\mathcal{A})$. We call $\pi(i, j) = t_i t_{i+1} \ldots t_{j-1}$ a *subpath* of π. The expression $\pi' \sqsubseteq \pi$ denotes that π' is a subpath of π.

An automaton $\mathcal{A}$ is *deterministic (or sequential)* if for each $a \in \Sigma, q \in Q$ there exists at most one $q \in Q$ such that $\mu(a)[p, q] \neq 0_K$ and the set of initial states I is a singleton. If the second condition is not satisfied, an automaton is called *finitely sequential* (fseq). Let $amb_{\mathcal{A}} : \Sigma^* \mapsto \mathbb{N}$ be a function which for each string $w \in \Sigma^*$ assigns to w number of different accepting paths labeled by w. An automaton $\mathcal{A}$ is *finitely ambiguous* (famb) if there exists a nonnegative integer c such that $amb_{\mathcal{A}}(w) \leq c$ for all $w \in L(\mathcal{A})$. If $c \leq 1$ then finitely ambiguous automata are called *unambiguous* (namb). An automaton $\mathcal{A}$ is *polynomially ambiguous* (pamb) if there exists a polynomial p such that $amb_{\mathcal{A}}(w) < p(|w|)$. The class of sequential automata and class of all automata are denoted respectively by seq and rat.

In this paper we study only automata over the *tropical semiring* $\langle \{\mathbb{R} \cup \infty\} = \mathbb{R}_\infty, \min, +, \infty, 0 \rangle$ earlier called min-plus automata. The constructive part of this paper deals with the semiring with rational domain rather than real numbers, to deal with finite representations of numbers. A wfa $\mathcal{A}$ over the tropical semiring defines a function $S(\mathcal{A}) : \Sigma^* \to \mathbb{R}_\infty$ such that $S(\mathcal{A})(w) = \lambda \cdot \mu(w) \cdot \gamma$ for $w \in \Sigma^*$. Later, the value $S(\mathcal{A})(w)$ will be also denoted by $\langle w, \mathcal{A} \rangle$.

In the tropical semiring, the weight of an accepting path is the sum of the weights of the transitions taken along the path, and the value of a word w is the minimal weight of an accepting path on it. Every not accepted string is mapped into ∞.

By Seq, Namb, Fseq, Famb, Pamb, Rat we denote classes of functions described by automata which respectively belong to seq, namb, fseq, famb, pamb, rat.

In order to make the text more readable we use an arrow notation together with the notation of wfa introduced above. If we write $\xrightarrow{x} q$, it means $\lambda[q] = x$. A sequence $\xrightarrow{v_0} q_0 \xrightarrow{a_1|v_1} q_1 \xrightarrow{a_2|v_2} \cdots \xrightarrow{a_i|v_i} q_i$ denotes a path π being a sequence of transitions $t_1, \ldots, t_i$ such that $label(\pi) = a_1 \ldots a_i$, the transition t_j has weight v_j, and $\lambda[q_0] = v_0$. We use an arrow notation solely to denote paths with weights in $\mathbb{R}$. A state $q \in Q$ is *accessible* if there exists string $w \in \Sigma^*$ such that $\lambda \mu(w)[q] \in \mathbb{R}$ and *co-accessible* if $\mu(w)\gamma[q] \in \mathbb{R}$. An automaton A is *trimmed* if all of its states are accessible and co-accessible. We say that two states p and q are *siblings* if they satisfy $\to q_I \xrightarrow{u} p$ and $\to p_I \xrightarrow{u} q$ for some $u \in \Sigma^*$.

A subpath $\pi(i,j) = t_i \ldots t_j$ is *non-empty* if $0 < i \le j \le k$ and *proper* if weights of all transitions t_i is in $\mathbb{R}$. A subpath $\pi(i,j)$ is a *loop* if the source state of transition t_i and the destination state of t_j are the same, and it is proper. We say also that $\pi(i,j)$ is a z-loop if $z = label(\pi(i,j))$ and $\pi(i,j)$ is a loop. A triple $\beta = (p,q,z) \in Q^2 \times \Sigma^*$ is a z-loop, if $\mu(z)[p,p], \mu(z)[q,q] \in \mathbb{R}$.

States $p, q \in Q$ are twins if for every words u_1, u_2 the following holds

$$\text{if } \xrightarrow{x_0} p_I \xrightarrow{u_1|x_1} p \xrightarrow{u_2|x_2} p \text{ and } \xrightarrow{y_0} q_I \xrightarrow{u_1|y_1} q \xrightarrow{u_2|y_2} q \text{ then } x_2 = y_2.$$

An automaton $\mathcal{A}$ has the *twin property* [2] if all pairs of its states p, q are twins. For a $t \in \Sigma^+$, a triple (q_1, q_2, t), where $q_1, q_2 \in Q$ is a t-*fork* if $q_1 \xrightarrow{t} q_1$ and $q_1 \xrightarrow{t} q_2$.

By the $[k]$ we denote the set $\{1, \ldots, k\}$. An automaton $\mathcal{A} = \bigcup_{i=1}^{k} \mathcal{A}_i$ is a *disjoint sum* of automata $\mathcal{A}_i$ if

1. Q_i are pairwise disjoint sets of states and $Q = \bigcup_{i=1}^{k} Q_i$,
2. for each $i \in [k], q \in Q_i$ it holds that $\lambda_i(q) = \lambda(q)$ and $\gamma_i(q) = \gamma(q)$,
3. for each $i, j \in [k]$, $p \in Q_i$, $q \in Q_j$ if $i = j$ then $\mu(a)[p,q] = \mu_i(a)[p,q]$, else $\mu(a)[p,q] = \infty$.

For given two weighted transitions of the form $t_1 = (p_1, a, v_1, q_1)$, $t_2 = (p_2, a, v_2, q_2)$ we define a product of transitions $t_1 \times t_2$ as a quadruple $((p_1, p_2), a, (v_1, v_2), (q_1, q_2))$. For given two paths $\theta_1 = t_1 t_2 \ldots t_n$ and $\theta_2 = t'_1 t'_2 \ldots t'_n$ by the *product of paths* $\theta_1 \times \theta_2$ we mean the sequence $(t_1 \times t'_1)(t_2 \times t'_2) \ldots (t_n \times t'_n)$. For given functions $f : A \to B$ and $g : B \to C$, by $g \circ f$ we denote a function from A into C which, for all $a \in A$, satisfies $g(f(a)) = (g \circ f)(a)$. A *min-plus transducer* is a min-plus automaton $\langle Q, \Sigma, \lambda, \mu, \gamma \rangle$ which differs from previously defined automata by definition of μ. In case of transducers, μ is a mapping of type $\Sigma^* \times \Delta^* \to \mathbb{K}^{Q \times Q}$, where Δ is a finite alphabet called the *output alphabet*.

Let $\mathcal{T}_1 = \langle Q_1, \Sigma \times \Delta, \lambda^1, \mu^1, \gamma^1 \rangle$ and $\mathcal{T}_2 = \langle Q_2, \Delta \times \Theta, \lambda^2, \mu^2, \gamma^2 \rangle$ be transducers. The *compostion of transducers* $\mathcal{T}_1$ and $\mathcal{T}_1$, denoted by $\mathcal{T}_2 \circ \mathcal{T}_1$, is $\mathcal{T} = \langle Q, \Sigma \times \Theta, \lambda, \mu, \gamma \rangle$, where $Q = Q_1 \times Q_2$, $\lambda((p,q)) = \lambda_1(p) \cdot \lambda_2(q)$, $\gamma(p,q) = \gamma_1(p) \cdot \gamma_2(q)$ and

$$\mu((a,b))[(p_1, p_2),(q_1, q_2)] := \min\{\mu^1((a,c))[p_1, q_1] \cdot \mu^2((c,b))[p_2, q_2] \mid c \in \Delta\}.$$

Later on we compose also a transducer $\mathcal{T}_1$ with an automaton $\mathcal{D}$. The only difference between composition of a transducer with a transducer and a transducer with an automaton is the lack of the output alphabet in the second case. In case of a classical automaton the mappings μ, λ and γ have weights from the set $\{0_K, 1_K\}$ of the semiring $\mathbb{K}$.

Let $\mathcal{A}_1, \mathcal{A}_2$ be classical automata. By $\mathcal{A}_1 \triangleright \mathcal{A}_2$ we denote automaton which accepts all strings w such that $w = uv$, $u \in L(\mathcal{A}_1)$ and $v \in L(\mathcal{A}_2)$.

2.1   Related Work

With respect to forms of nondeterminism, functions recognizable by automata over tropical semirings form a hierarchy [3,4] which can be depicted as follows.

$$\text{Seq} \subsetneq (\text{Namb} \cap \text{Fseq}) \begin{array}{c} \nearrow \subsetneq \text{Fseq} \subsetneq \searrow \\ \searrow \subsetneq \text{Namb} \subsetneq \nearrow \end{array} \text{Famb} \subsetneq \text{Pamb} \subsetneq \text{Rat}$$

The question whether exists equivalent automaton which is deterministic is called the determisation problem. Despite of the fact the problem was studied by several researchers e.g. [1,3,5,2] the determinisation problem is still open. The best known result [1] shows decidability of the determinisation problem for polynomially ambiguous automata.

For unambiguous automata, the determinisation problem has a positive answer if and only if the Twin Property holds [6,2]. The Twin Property is decidable in polynomial time $(O(|Q|^2 + |E|^2)$, where E is the set of transitions) for unambiguous automata [7] and in polynomial space for arbitrary automata [8]. Moreover, the determinisation algorithm [2] returns automata of size at most exponential with respect to the size of an input.

For finitely ambiguous automata the determinisation problem has been solved [3] in two stages: First deciding if a translation to an unambiguous automaton is possible and then deciding the Twin Property over the unambiguous equivalent one. The first stage starts from translation of the given automaton into a finite union of unambiguous ones. The best known translation is exponential in the size of the automaton [9]. Then the Dominance Property is defined over the new representation – the union of unambiguous automata. The Dominance Property separates these unions which can be translated into unambiguous automata from those which have no unambiguous equivalent. The property is decidable in polynomial time with respect to the size of the union of unambiguous automata, hence exponential with respect to the input.

In [1] Kirsten and Lombardy posed question if it is decidable whether finitely ambiguous or finitely sequential equivalent exists for given polynomially ambiguous automaton. The question is especially interesting in one aspect, namely one of the algorithms presented in [1] depends on the equivalence between polynomially ambiguous min-plus automaton and unambiguous min-plus automaton. This problem is known to be decidable. Also equivalence between finitely ambiguous automata is decidable [10,11]. The weaker formulation of the equivalence problem known to be undecidable is between to polynomially ambiguous automata [12]. However, it is not known whether the equivalence is decidable if one of the automata is finitely ambiguous or even finitely sequential.

3   Decomposition and Stronger Fork Property

By a result of Schützenberger and later (in most effective way) from [13], [3], [9], every weighted finitely ambiguous automaton $\mathcal{A}$ can be decomposed into finitely many unambiguous ones, denoted $U(\mathcal{A}) = \bigcup_{i=1}^{k} \mathcal{U}_i$. Thus the problem whether given finitely ambiguous automaton has finitely sequential counterpart can be transformed to the form with finitely many unambiguous automaton on the input. In [13], [3] the outcome unambiguous automata are doubly exponential in the size of the finitely ambiguous input. In [9], the resulting unambiguous automata are at most exponential.

The automaton $\mathcal{U}_i$ is described by a tuple $\langle \Sigma, J_i, \lambda_i, \mu_i, \gamma_i \rangle$. In [13], all $\mathcal{U}_i$ recognizes the same language. The unambiguous $\mathcal{U}_i$ in Sakarovitch's and de Souza's construction [9] do not recognize the same language, what would be useful for our purpose. However, the construction from [9] can be easily extended to one which fulfills this property.

The main object we further consider is the automaton $\mathcal{U} = \langle \Sigma, Q, \lambda, \mu, \gamma \rangle$ which is the product of the $\mathcal{U}_i$ with $Q = \prod_{i \in [k]} J_i$. Morphism μ maps Σ^* into $(\mathbb{R} \cup \{\infty\})^{Q \times Q \times [k]}$. Let the set of all proper paths in $\mathcal{U}_i$ be denoted by $\Pi(\mathcal{U}_i)$. From now on, slightly abusing notation, assume that $\mathcal{U}_i$ is the automaton $\mathcal{U}$ with transitions (p, a, q) weighted by $(\mu(a)[p,q])_i$. The component $(\mu(a)[p,q])_i$ will also be denoted $\mu_i(a)[p,q]$.

In [14] it has been shown that an unambiguous trimmed automaton $\mathcal{A} = \langle Q, \Sigma, \lambda, \mu, \gamma \rangle$ can be translated into a finitely sequential one if and only if it does not satisfy the Fork Property defined in [14] as follows:

Definition 1 (Fork Property). *A trimmed unambiguous automaton $\mathcal{A}$ has the Fork Property (FP) if there exist states q_1, q_2 such that q_1 and q_2 are not twins, but there exists a $t \in \Sigma^+$ such that (q_1, q_2, t) is a t–fork.*

Our main goal is to define a version of the fork property suitable for finitely ambiguous automata. But first we need to restrict the formulation of the current Fork Property, introducing the *Stronger Fork Property*.

If q_1, q_2 are not twins then there exists z such that (q_1, q_2, z) is a z–loop with $\mu(z)[q_1, q_1] \neq \mu(z)[q_2, q_2]$. In the Fork Property, defined above, we have not restricted z and t in any way. In particular z and t do not have to depend on the size of the given unambiguous automata. We are going to change this for our purpose. Given a triple $\beta = (p, q, t')$, a quadruple $\alpha = (p, q, t', t)$ is a

β –*modification* if there exists a way of gradually removing loops from $p \xrightarrow{t'} q$ obtaining as a result a path $p \xrightarrow{t} q$ such that (p, q, t) is a t–fork. Additionally $|t| \leq |Q|^2$ as in Lemma 1 and there are no repeated states in the product path $(p, p) \xrightarrow{t} (p, q)$.

Definition 2 (Stronger Fork Property). *A trimmed unambiguous automaton $\mathcal{A}$ has the* Stronger Fork Property (SFP) *if* (1) *there exists a computable function $c(\mathcal{A})$, a v–loop (q_1, q_2, v) such that $|v| \leq c(\mathcal{A})$ and $q_1 \xrightarrow{v|x_2} q_1$ and $q_2 \xrightarrow{v|y_2} q_2$ and $x_2 \neq y_2$, and* (2) *there exists a $t' \in \Sigma^+$ such that (q_1, q_2, t', t) is a (q_1, q_2, t') –modification and $t \leq |Q|^2$.*

We assume in the next lemma that weights of transitions of $\mathcal{A}$ are rational.

Lemma 1. *An unambiguous automaton $\mathcal{A}$ satisfies the Fork Property if and only if it satisfies the Stronger Fork Property.*

4   A-Fork Property

In this section we assume that each state of $\mathcal{U}$ is accessible and co-accessible. A triple $\beta = (p, q, z) \in Q^2 \times \Sigma^*$ is a z-*loop*, if $\mu(z)[p, p], \mu(z)[q, q] \in \mathbb{R}$. Let $\tau_i(\beta) = \nearrow$ if $\mu_i(z)[p, p] \neq \mu_i(z)[q, q]$ and $\tau_i(\beta) = \rightarrow$ if $\mu_i(z)[p, p] = \mu_i(z)[q, q]$.

We may say that $\tau_i(\beta)$ describes the type of β from a point of view its i'th coordinate. A path π_i of a component $\mathcal{U}_i$ is a β-*witness* for $\beta = (p, q, t')$ if the quadruple (p, q, t', t) is a β-modification for some word t; $\pi = \theta_1 \theta_2 \theta_3$, where p and q are respectively the first and the last state of θ_2, $label(\theta_2) = t'$ and there exists $z \in \Sigma^+$ such that $\tau_i((p, q, z)) = \nearrow$. Then, $\alpha = (p, q, z)$ is a *distinguisher* of β. Assume that (p, q, t', t) is a β-modification such that $|t| \leq |Q|^2$.

Definition 3 (Broken Path). *A path π of the automaton $\mathcal{U}$ is i-broken if for some $i \in [k]$ the path π_i is a β-witness for some $\beta = (p, q, t)$ and there exists a distiguisher $\alpha = (p, q, z)$. Otherwise, π is called i-nonbroken.*

We also say that a projection π_i of a path π is *broken* if a path π is i-broken. Here and subsequently, $\mu_j(\pi)$ stands for the the projection on the weight of j'th coordinate of the path π in $\mathcal{U}$.

Definition 4 (Spoonfulness). *An automaton $\mathcal{U}$ is a* spoonful *if there exists natural number C such that for all accepting $\pi \in \Pi(\mathcal{U})$ and $i \in [k]$ there exists $j \in [k]$ such that if π is i-broken then π is j-nonbroken and $\mu_j(\pi) < \mu_i(\pi) + C$.*

The intention of spoonfulness is to express that if π_i can be ground to infer nonexistence of a finitely sequential counterpart for $\mathcal{U}_i$ then there always exists π_j with $label(\pi_i) = label(\pi_j)$ such that π_j is nonbroken and it's weight is not substantially greater than π_i. We say that an automaton $\mathcal{U}$ satisfies the **A-Fork** property if it is not spoonful.

5   Necessity

Assume now that $\mathcal{U}$ satisfies the A-Fork property. The aim of this section is to give the sketch of the proof of the following theorem.

Theorem 1. *If the automaton $\mathcal{U}$ satisfies the A-Fork property then $S(\mathcal{A}) \notin$ Fseq.*

Assumption 2.   *Suppose that there exists a finitely sequential automaton with deterministic components $\mathcal{D}_1, \ldots, \mathcal{D}_l$ which recognizes the function $S(\mathcal{A})$.*

By $\mathcal{D} = \langle \Sigma, D, \lambda^d, \mu^d, \gamma^d \rangle$ we denote product of automata $\mathcal{D}_1, \ldots, \mathcal{D}_l$ defined in the same way as $\mathcal{U}$ was defined with the respect to $\mathcal{U}_i$. By m_D and n_D we denote respectively the cell of the biggest transition weight in $\mathcal{D}_1, \ldots, \mathcal{D}_l$ and $|D|$. By m_U we denote the biggest transition's weight in $\mathcal{U}_1, \ldots, \mathcal{U}_k$.

Let $\alpha = (p, q, z)$ and $\beta = (p, q, t')$. Assume that $\mathcal{U}_\chi$ is β–witness and α is a distiguisher of β. A component $\mathcal{U}_\chi$ satisfies the Stronger Fork Property. Assume that $\tau_\chi(\beta) = \mathscr{Z}$ and $|z| < c(\mathcal{U}_\chi)$, where $c(\mathcal{U}_\chi)$ is the constants mentioned in the definition of the Stronger Fork Property.

Let $\sigma = \mu_\chi(z)[p, p] - \mu_\chi(z)[q, q]$ and let $\delta = \lceil ((m_D + m_U) \cdot |zt|)/|\sigma| \rceil$. Define $\varsigma(u, m)$, as a sequence of strings $u_{m+1}, u_m, u_{m-1}, \ldots, u_1$ such that

$$u_{m+1} = u; \quad u_i = u_{i+1} t z^{f_i}, \text{ where } f_{i+1} > 4 \cdot \delta \cdot f_i, f_1 \geq 1 \text{ and } m > (k+1) n_D + 1.$$

Consider now a sequence $\varsigma'(u, m)$, of the form $u'_{m+1}, \ldots, u'_1$, where $u'_i = u_i t' z^{f_{i-1}}$. The recursive construction is similar to constructions presented in [14] and [3].

Assumption 3.   *There exist m and $\varsigma'(u, m)$ such that for more than n_D indexes $\kappa \in [m]$ there are v_κ which satisfy $\langle u'_\kappa v_\kappa, \mathcal{A} \rangle = \lambda_\chi \mu_\chi(u'_\kappa v_\kappa) \gamma_\chi = (\lambda_\chi \mu_\chi(u_\kappa))[q] + (\mu_\chi(\rho_\kappa v_\kappa) \gamma_\chi)[q] \in \mathbb{R}^+$.*

Because of the lack of the space we do not present here the proof which is technical. It may not be quite clear for the reader what kind of role Assumption 2 plays. Satisfaction of the Fork Property means that there exists some skeleton of a path which can be a generator of a family of paths which form a witness for nonexistence of a finitely sequential equivalent. The forming process follows by proper alternate pumping of the fork part and the distinguisher part. The technique is described in [14]. In case of the union of unambiguous automata, accepting paths come from different components and we need to describe how the generators behave simultaneously. Precisely, it is crucial whether for some generator and the generated witness consists of paths which are optimal in large enough amount. Optimal means with minimal weight among paths labeled by the same word. This is what Assumption 3 expresses. In the next step, we prove the following:

Lemma 2. *Assumptions 2 and 3 are inconsistent.*

Finally we prove the following lemma which already gives us the main result of this section as a simple conclusion.

Lemma 3. *If $\mathcal{U}$ has the A-Fork property then Assumption 3 holds.*

Because of the lack of the space we describe only the core idea of the proof. We assume that Assumption 2 is satisfied along with the A-Fork Property. For any positive integer C one can choose some path π in the product automaton $\mathcal{U}$ in such a way that there exists a component π_i which can be a generator of witness. Additionally, all components which are nonbroken have weight greater than the weight of π_i and the difference of weights is at least C. It may happen that there are other components which are of weight greater than the weight of π but the difference of weights is smaller than C. However, all such paths are also generators of some witnesses. We proceed by proving nonexistence of fseq equivalent for all mentioned components simultaneously. But we need suitably more pumping. This means that we tend to have u_1 as a substring of the label of the resulting path, but for properly large m. How many pumpings we do depends in principle on the constant x and the maximal weight of loops in distinguisher. The second parameter is also a constant because we can perform pumping for strings $z < c(\mathcal{A})$. There is one parameter of our schema which, during the pumping we do, can have strong variation in the size and form of strings t which form proper t–forks of pumped generators. The most technical part is to choose the size of pumping in such a way that the size guarantees existence of a proper number of strings which are of minimal weight and all come from the same component. This means that for assumption for any l we can choose such big C and the size of pumping such that the number of mentioned paths is bigger than n_D. Thus Assumption 3 is satisfied. Since we have to join the effect of simultaneous proofs in one object, we use in this part some more advance combinatorial analysis than only the pumping technique. The technique deals with colorings of finite hypercubes.

6   Sufficiency

Recall that the automata $\mathcal{U}_i = \langle J_i, \Sigma, \lambda_i, \mu_i, \gamma_i \rangle$ introduced earlier are unambiguous. Let us introduce more definitions from [14].

Definition 5 (Critical Pair). *Let*

$$D_j := \{\langle q_1, q_2 \rangle \in J_j \times J_j \mid q_1 \text{ and } q_2 \text{ are siblings, not twins in } \mathcal{U}_j\}.$$

A pair $\langle q, E \rangle \in J_j \times P(J_j)$ *is said to be critical if* $\exists p \in J_j, \langle q, p \rangle \in D_j \wedge \{q, p\} \subseteq E$.

We define $\hat{\mathcal{U}}_j$ as $\langle \hat{J}, \Sigma, \hat{\lambda}_j, \hat{\mu}_j, \hat{\gamma}_j \rangle$, where $\hat{J}_j := J_j \times P(J_j)$, $\hat{\gamma}_j(\langle q, E \rangle) := \gamma_j(q)$,

$$\hat{\lambda}_j(\langle q, E \rangle) := \begin{cases} \lambda_j(q), & \text{if } E = \{r \in J_j \mid \lambda_j(r) \neq \infty \} \\ \infty & \text{otherwise} \end{cases}$$

$$\hat{\mu}_j(a)(\langle q_1, E_1 \rangle, \langle q_2, E_2 \rangle) := \begin{cases} \mu_j(a)(q_1, q_2), & \text{if } q_1 \in E_1, \text{ and } \langle q_1, E_1 \rangle \text{ is not critical} \\ & \text{and } E_2 = \{q \mid \exists p \in E_1 \quad \mu_j(a)(p, q) \neq \infty\} \\ \mu_j(a)(q_1, q_2), & \text{if } q_1 \in E_1 \quad \wedge \langle q_1, E_1 \rangle \text{ is critical} \\ & \text{and } E_2 = \{q \mid \mu_j(a)(q_1, q) \neq \infty\} \\ \infty & \text{otherwise} \end{cases}$$

Each path $\hat{\pi}_j$ in $\hat{\mathcal{U}}_j$ corresponds to the path π_j in $\mathcal{U}_j$, where consecutive states of π_j are created as projections on the first component of corresponding states in $\hat{\pi}_j$. The hat operation establishes $1-1$ correspondence between paths in $\mathcal{U}_i$ and $\hat{\mathcal{U}}_i$.

Definition 6 (Critical Form of Automaton). *Let* $\hat{\mathcal{U}} = \langle \Sigma, \hat{Q}, \hat{\lambda}, \hat{\mu}, \hat{\gamma} \rangle$ *be the product of* $\hat{\mathcal{U}}_i$ *automata with* $\hat{Q} = \prod_{i \in [k]} \hat{J}_i$. *The automaton* $\hat{\mathcal{U}}$ *is a* critical form *of* $\mathcal{U}$.

Definition 7 (Close Approximation). *The min-plus automaton* $\mathcal{B}$ *closely approximates a min-plus automaton* $\mathcal{A}$ *if they recognize the same language* L *and there exists constants* C *which depends on* $\mathcal{A}$ *such that for all* $w \in L$

$$S(\mathcal{A})(w) - C \leq S(\mathcal{B})(w) \leq S(\mathcal{A})(w) + C.$$

This section shows that if the automaton $\mathcal{U}$ does not satisfy A-fork property then the $\mathcal{A}$ is closely approximated by some finitely sequential automaton. From now on assume that A–Fork Property is not satisfied for the automaton $\mathcal{U}$. In this section we show how to build an finitely sequential automaton, which is close approximation of $\mathcal{A}$, as a union $\bigcup_{j \in [s]} \bigcup_{s \in [|Q|]} \bigcup_{a \in (R_j)^k \cup \{\epsilon\}} \mathcal{L}_j[a]$. It will be define later what means the R_j and m. In [14] the following theorems has been proven:

Theorem 4 ([14]). *Let* $\hat{\pi}_j$ *be a path of* $\hat{\mathcal{U}}_j$. *If* $\hat{\pi}_j$ *contains repeating critical states then* $\hat{\pi}_j$ *is broken.*

Theorem 5 ([14]). *If any accepting path* $\hat{\pi}_j$ *in unambiguous automaton* $\hat{\mathcal{U}}_j$ *contains at most* $|Q|$ *critical states then* $\hat{\mathcal{U}}_j$ *has finitely sequential equivalent.*

In [14] it has been proven a bit more, namely repeating critical states guarantees nonexistence of the equivalent. Automaton $\hat{\mathcal{U}}$ can be seen as directed graph then all cycles without repeating nodes can be determined.

Let us sketch the idea of our proof. Each accepting path in graph $\hat{\mathcal{U}}$ can be decomposed into simple cycles which occurs on them and the part without cycle. The first part is characterized by some sequence of nonnegative integers $a_1, \dots, a_l \in \mathbb{N}$. Particular number a_i denote the number of occurrence of suitable simple cycle θ_i in given accepting path. A cycle θ_i corresponds to vector of weight $\overrightarrow{\theta}_i$. Our approach is to consider linear combinations $\overrightarrow{\theta} = a_1 \overrightarrow{\theta}^1 + \dots + a_l \overrightarrow{\theta}^l$ which correspond to some accepting paths. Along with linear combinations we observe which components of simple cycles contains at least one critical state and which of them have coefficient $a_i > 1$. Such coefficients make the particular component of given π broken. The crucial element of the proof is to observe which coordinates of $\overrightarrow{\theta}$ are minimal if it is the set $L \subseteq [k]$ the $\overrightarrow{\theta}$ is called L–minimal. The key lemma of this section is the following.

Lemma 4. *Let* $\hat{\pi}$ *be an accepting path in* $\hat{\mathcal{U}}$. *If* $\hat{\pi}$ *is* L–*minimal and for all* $j \in L$ *the path* $\hat{\pi}$ *is* j–*broken then* $\mathcal{U}$ *satisfies A-fork property.*

The direct conclusion of this lemma is that

Corollary 1. *For all $L \subseteq [k]$ if $\hat{\pi}$ is L–minimal then there exists $j \in L$ such that π is j–nonbroken.*

Corollary 2. *There exists constant C, depending on the automaton $\hat{\mathcal{U}}$, such that for all accepting paths π there exists nonbroken π_j that for all $i \in [k]$ the following equation hold $\mu_i(\pi) + C > \mu_i(\pi)$.*

In the next part we constitutes a type of path (T, T'), where T contains labels of simple cycles which have only one occurrence in $\hat{\pi}$ and T' contains labels of simple cycles which appears more than one in $\hat{\pi}$.

Using all possible types and L–minimality it is possible to show that existence of $\hat{\pi}$ from Lemma 4, which is j–broken for all $j \in L$ is expressible in existential fragment of Presburger Arithmetic. This fragment is decidable in NP.

Now the automaton $\hat{\mathcal{U}}$ can be covered by several its simulations or rather restrictions. Each of the restrictions is involved with sequence of critical states $\boldsymbol{a}$. The restriction involved with $\boldsymbol{a}$ accepts only this paths which have the sequence $\boldsymbol{a}$ on it. The order of critical states vector $\boldsymbol{a}$ and the path coincide. If precondition in the Lemma 4 is not satisfied for $\hat{\pi}$ labeled by any word w then the weight of the minimal component in the union of coverings can differ only from the best path labeled by w in $\hat{\mathcal{U}}$ by some constant. The constant depends on the part of the $\hat{\pi}$ after removing all cycles. Hence it has to be bounded by some constant, which depends only on $\hat{\mathcal{U}}$. The conclusion of the reasoning above is the following theorem.

Theorem 6. *It is decidable in 2-NEXPTIME if a given finitely ambiguous automaton is closely approximated by a finitely sequential automaton. If $\mathcal{A}$ is closely approximated by some finitely ambiguous automaton then there exists approximating automaton of size at most triply exponential in the size of $\mathcal{A}$.*

The strategy of the previous section rules out all accepting paths of $\hat{\mathcal{U}}_j$ which contains repeating critical states which always make the path broken. Instead of analysis of paths in $\hat{\mathcal{U}}$, this section applies strategy based on detection of suitable properties of paths in $\mathcal{U}_i$ namely except for currently walked path π is accepting we detect if

Property 1. The path π has β–witness for some $\beta = (p, q, t')$ such that $\beta' = (p, q, t', t)$ is a modifier for some t.

Property 2. There exists distinguisher $\alpha = (p, q, z)$ such that we visited a c disjoint copies of both types loops of the distinguisher. Only distinguishers with $|z| < c(\mathcal{A})$ are taken into account.

To be more precise we detect subpaths of the form $(p \xrightarrow{z})^c p \xrightarrow{t'} q (\xrightarrow{z} q)^c$ for fixed positive integer c.

The constant c is bounded by the bigger weight of accepting, cycle free path $\mathcal{U}$. The strategy we presented above is not yet precise. Despite we would like to avoid very formal presentation but presented strategy still needs more detailed exposition. In order to check the first property, for given t–fork $\beta = (p, q, t)$, it is enough to detect if given path π visits the path $p \xrightarrow{t} q$.

The size of distinguishers depends on the automaton $\mathcal{A}$ and it is at most exponential in the size of representation of $\mathcal{A}$ (exponential because we assume that weights are stored in binary representation). In spite of z can be exponentially large, z can be in the form $uv^x w$ where u, v, w and representation of number x are of length polynomially bounded to the size of $\mathcal{A}$. Therefore one can enumerate all pairs (distinguisher ,t–fork) (α, β). A pair (α, β) can be seen also as a type of set of paths (ρ_1, ρ_2, ρ_3) which starts respectively with states p, q, q, ends respectively with p, q, q and are labeled respectively by z, z, t. The number of such triples (ρ_1, ρ_2, ρ_3) is at most exponential with respect to the size of representation of automaton $\mathcal{A}$. By $\mathcal{T}_i$ we denote the set of such triples for $\mathcal{U}_i$. We emphasize that the first two coordinates of any triple denotes the loops of distinguisher.

The detection can be realized by composition of the following automata: Let $\rho \in \mathcal{T}_i$ and $\rho = (\rho_1, \rho_2, \rho_3)$. By Γ_{ρ_1} and Γ_{ρ_2} we denote the classical nondeterministic automata over the alphabet $\mathcal{G} = Q \times \Sigma \times Q$ accepting only these sequences $\pi = t_1 t_2 \cdots t_l$ which belongs to $\Pi(\mathcal{U}_i)$ and π visits respectively ρ_1, ρ_2 at least c times. The automaton Γ_{ρ_3} accepts such proper paths which visits ρ_3. Let Γ_ρ be the composition $\Gamma_{\rho_1} \triangleright \Gamma_{\rho_3} \triangleright \Gamma_{\rho_1}$. By Γ_i we denote the $\bigcup_{\rho \in \mathcal{T}_i} \Gamma_\rho$. By $\overline{\Gamma}_i$ we denote automaton which recognizes $\Sigma^* \setminus L(\Gamma_i)$ an by $D(\overline{\Gamma}_i)$ its minimized deterministic version.

Now extend an $\mathcal{U}_i = \langle J_i, \Sigma, \lambda_i, \mu_i, \gamma_i \rangle$ to transducer $\mathcal{T}_i = \langle J_i, \Sigma \times \Delta, \lambda_i, \overline{\mu}_i, \gamma_i \rangle$, where $\Delta = J_i \times \Sigma \times J_i$, in such a way that

$$\overline{\mu}_i((r, a, s))[p, q] := \begin{cases} \mu_i(a)[p, q] & \text{if } r = p \text{ and } q = s \\ \infty & \text{otherwise.} \end{cases}$$

Let π be the accepting path labeled by word w. The composition $\mathcal{D}_i = D(\overline{\Gamma}_i) \circ \mathcal{T}_i$ is a min-plus automaton with useful properties, namely the weight of accepted word w in $\mathcal{D}_i$ equals exactly the weight of w in $\mathcal{U}_i$ if there is no a pair $(\rho_1, \rho_2, \rho_3) \in \mathcal{T}_i$ such that $(\rho_1)^c p \xrightarrow{t'} q(\rho_3)^c$ is a subpath of π and $(p, q, t', t) \in modif((p, q, t))$.

With this definition one can prove that the min-plus automaton $\mathcal{D}_i$ has no Fork Property. Hence $\mathcal{D}_i$ has a finitely sequential equivalent. Now one can show that $\bigcup_{i=1}^k \mathcal{D}_i \equiv \bigcup_{i=1}^k \mathcal{U}_i \equiv \mathcal{U}$. However this is the most technical part of this section and we left it out because of lack of space. The direct consequence of the last equivalence is the following.

Theorem 7. *For the automaton $\mathcal{U}$ the following three statements are equivalent.* (1) *The automaton $\mathcal{U}$ is spoonful.* (2) *The $\mathcal{U}$ is closely approximated by some finitely sequential automaton.* (3) *$S(\mathcal{A}) \in$ Fseq.*

Theorem 8. *For a given finitely ambiguous automaton $\mathcal{A}$ it is decidable in 2-NEXPTIME whether $S(\mathcal{A}) \in$ Fseq. If $S(\mathcal{A}) \in$ Fseq then there exists a finitely sequential equivalent of the size at most triply exponential in the size of $\mathcal{A}$.*

References

1. Kirsten, D., Lombardy, S.: Deciding unambiguity and sequentiality of polynomially ambiguous min-plus automata. In: Albers, S., Marion, J.Y. (eds.) STACS. LIPIcs, vol. 3, pp. 589–600. Schloss Dagstuhl - Leibniz-Zentrum fuer Informatik, Germany (2009)
2. Mohri, M.: Finite-state transducers in language and speech processing. Computational Linguistics 23(2), 269–311 (1997)
3. Klimann, I., Lombardy, S., Mairesse, J., Prieur, C.: Deciding unambiguity and sequentiality from a finitely ambiguous max-plus automaton. Theor. Comput. Sci. 327(3), 349–373 (2004)
4. Kirsten, D.: A burnside approach to the termination of mohri's algorithm for polynomially ambiguous min-plus-automata. ITA 42(3), 553–581 (2008)
5. Lombardy, S., Sakarovitch, J.: Sequential? Theor. Comput. Sci. 356(1-2), 224–244 (2006)
6. Choffrut, C.: Une caracterisation des fonctions sequentielles et des fonctions sous-sequentielles en tant que relations rationnelles. Theor. Comput. Sci. 5(3), 325–337 (1977)
7. Allauzen, C., Mohri, M.: Efficient algorithms for testing the twins property. Journal of Automata, Languages and Combinatorics 8(2), 117–144 (2003)
8. Kirsten, D.: Decidability, undecidability, and pspace-completeness of the twins property in the tropical semiring. Theor. Comput. Sci. 420, 56–63 (2012)
9. Sakarovitch, J., de Souza, R.: On the decomposition of k-valued rational relations. In: Albers, S., Weil, P. (eds.) STACS. LIPIcs, vol. 1, pp. 621–632. Schloss Dagstuhl - Leibniz-Zentrum fuer Informatik, Germany (2008)
10. Hashiguchi, K., Ishiguro, K., Jimbo, S.: Decidability of the equivalence problem for finitely ambiguous finance automata. IJAC 12(3), 445 (2002)
11. Haumbold, N.: The similarity and equivalence problem for finitely ambiguous automata over tropical semiring. Master Thesis under supervision of Daniel Kirsten, TU Dresden Farichtung Mathematic Institute fur Algebra (2006)
12. Krob, D.: The equality problem for rational series with multiplicities in the tropical semiring is undecidable. In: Kuich, W. (ed.) ICALP 1992. LNCS, vol. 623, pp. 101–112. Springer, Heidelberg (1992)
13. Weber, A.: Decomposing a k-valued transducer into k unambiguous ones. ITA 30(5), 379–413 (1996)
14. Bala, S., Koniński, A.: Unambiguous automata denoting finitely sequential functions. In: Dediu, A.-H., Martín-Vide, C., Truthe, B. (eds.) LATA 2013. LNCS, vol. 7810, pp. 104–115. Springer, Heidelberg (2013)

Rewriting Guarded Negation Queries

Vince Bárány*, Michael Benedikt, and Balder ten Cate**

1 LogicBlox Inc., Atlanta, GA
2 Department of Computer Science, University of Oxford
3 Department of Computer Science, UC-Santa Cruz

Abstract. The Guarded Negation Fragment (GNFO) is a fragment of first-order logic that contains all unions of conjunctive queries, a restricted form of negation that suffices for expressing some common uses of negation in SQL queries, and a large class of integrity constraints. At the same time, as was recently shown, the syntax of GNFO is restrictive enough so that static analysis problems such as query containment are still decidable. This suggests that, in spite of its expressive power, GNFO queries are amenable to novel optimizations. In this paper we provide further evidence for this, establishing that GNFO queries have distinctive features with respect to rewriting. Our results include effective preservation theorems for GNFO, Craig Interpolation and Beth Definability results, and the ability to express the certain answers of queries with respect to GNFO constraints within very restricted logics.

1 Introduction

The guarded negation fragment (GNFO) is a syntactic fragment of first-order logic, introduced in [BtCS11]. On the one hand, GNFO can be seen as a constraint language: it captures classical database referential integrity constraints (that is, inclusion dependencies), specifications of relationships between schemas given in a common schema mapping language (namely that of Local-As-View constraints [Len02, FKMP05]) and the first-order translations of ontologies specified in some of the most popular description logics [BCM+03]. It contains these prior classes by virtue of extending the Guarded Fragment of first-order logic [AvBN98]. On the other hand, GNFO is more suitable than the Guarded Fragment for defining queries: for example, it contains all positive existential queries, corresponding in expressiveness to unions of conjunctive queries. The defining characteristic of GNFO formulas is that a subformula $\psi(\mathbf{x})$ with free variables $\mathbf{x}$ can only be negated when used in conjunction with a positive literal $\alpha(\mathbf{x}, \mathbf{y})$, i.e. a relational atom or an equality, containing all free variables of ψ, as in

$$\alpha(\mathbf{x}, \mathbf{y}) \wedge \neg \psi(\mathbf{x}) ,$$

where order and repetition of variables is irrelevant. One says that the literal $\alpha(\mathbf{x}, \mathbf{y})$ *guards* the negation. Unguarded negations $\neg \phi(x)$ of formulas with at most one free variable are supported through the use of an equality guard $x = x$.

· * Work done while affiliated with TU Darmstadt.
** Supported by NSF Grants IIS-0905276 and IIS-1217869. Benedikt supported by EPSRC grant EP/H017690/1.

K. Chatterjee and J. Sgall (Eds.): MFCS 2013, LNCS 8087, pp. 98–110, 2013.
© Springer-Verlag Berlin Heidelberg 2013

It was shown in [BtCS11] that GNFO possesses a number of desirable static analysis properties. For example, every satisfiable GNFO-formula has a finite model (*finite model property*), as well as a, typically infinite, model of bounded tree-width (*tree-like model property*). It follows that satisfiability and implication (hence, by the finite model property, finite satisfiability and finite implication) of GNFO formulas are decidable.

In [BtCO12] the implications of GNFO for database theory are explored: for example, an SQL-based syntax for GNFO is defined, and an analog of stratified Datalog is also presented. The complexity of query evaluation and "open world query answering" (i.e. computing certain answers) is identified for several GNFO-based languages, and many important static analysis problems for queries (e.g. boundedness for the GNFO-variant of Datalog) are shown to be decidable.

In this work we investigate properties of GNFO related to *rewriting*. We first present results showing that GNFO queries or constraints satisfying additional semantic properties can be rewritten into restricted syntactic forms. For example, we show that every GNFO query that is closed under extensions can be effectively rewritten as an existential GNFO formula.We give an analogous result for queries closed under homomorphisms. We also show that the GNFO sentences that can be expressed in a common constraint language – that of tuple-generating dependencies (TGDs), are precisely those that can be rewritten into a recently-introduced class of TGDs, the frontier-guarded TGDs.

We then turn to the setting where one has views and queries both defined within GNFO, imposing an additional restriction that the free variables in the views and queries are guarded. We show that if the views and queries satisfy the semantic restriction that the views *determine* the query, then we can find a rewriting of the query in terms of the views, with the rewriting belonging again to GNFO. Following ideas of Marx [Mar07], we proceed by showing that an important model theoretic theorem for first-order logic, the Projective Beth Definability theorem, holds in GNFO. We show that, unlike in the case of the Guarded Fragment, the more general Craig Interpolation Theorem of first-order logic holds for GNFO. In contrast, we show that Craig Interpolation and Projective Beth fail for the guarded fragment, contradicting claims made in earlier work.

We also study the existence of rewritings computing the certain answer to conjunctive queries. We show that GNFO sentences that take the form of dependencies have particularly attractive properties from the point of view of open world query answering. We extend and correct results of Baget et. al. [BMRT11b] by showing that the certain answers are expressible in a small fragment of Datalog. Using this, we show that the existence of first-order rewritings can be effectively decided for dependencies in GNFO.

For space reasons, most proofs are deferred to the full version.

2   Definitions and Preliminaries

We will make use of some basic notions of database theory – in particular the notion of schema or signature, relational structures, and the following "classical" query classes: conjunctive queries (CQs), Unions of Conjunctive Queries (UCQs), first-order logic formulas (FO), existential and positive existential FO, and Datalog. Abiteboul, Hull, and Vianu [AHV95] is a good reference for all of these languages. Note that by default we allow constants in our signature (that is, in CQs, UCQs, etc.). In this work, by the *active domain* of a structure I we mean the set of values that occur in some relation of I along with the values named by constants. In our arguments we will often make use of the following basic notions from classical model theory: s a *reduct* of a structure is obtained by restricting the signature), an *expansion* of a structure (obtained by adding additional relations). By a *fact* of a structure $\mathfrak{A}$ we mean an expression $R(a_1,\ldots,a_n)$ where $(a_1,\ldots,a_n)$ is a tuple belonging to a relation $R^{\mathfrak{A}}$. For structures I,J, we write $I \subseteq J$ if the domain of I is contained in the domain of J, every fact of I is also a fact of J, and I and J agree on the interpretation of all constant symbols. In this case, we say that I is a *subinstance* of J and that J is a *super-instance* of I. If, furthermore, every fact of J containing only values from the domain of I belongs to I, then we say that I is an *induced substructure* of J and J is an *extension* of I.

The Basics of GNFO. The Guarded Negation Fragment (GNFO) is built up inductively according to the grammar:

$$\phi ::= R(\mathbf{t}) \mid t_1 = t_2 \mid \exists x \, \phi \mid \phi \vee \phi \mid \phi \wedge \phi \mid R(\mathbf{t}, \mathbf{y}) \wedge \neg \phi(\mathbf{y})$$

where R is either a relation symbol or the equality relation $x = y$, and the t_i represent either variables or constants. Notice that any use of negation must occur conjoined with an atomic relation that contains all the free variables of the negated formula – such an atomic relation is a *guard* of the formula. The purpose of allowing equalities as guards is to ensure that every formula with at most one free variable can be considered guarded, and we often write $\neg\phi$ instead of $(x = x) \wedge \neg\phi$, when ϕ has no free variables besides (possibly) x. If τ is a signature consisting of constants and predicates, GNFO[τ] denotes the GNFO formulas in signature τ.

GNFO should be compared to the *Guarded Fragment*, GFO [AvBN98], typically defined via the grammar:

$$\phi ::= R(\mathbf{x}) \mid \exists \mathbf{x} \, R(\mathbf{x}, \mathbf{y}) \wedge \phi(\mathbf{x}, \mathbf{y}) \mid \phi \vee \phi \mid \phi \wedge \phi \mid \neg\phi(\mathbf{y})$$

It is easy to see that every union of conjunctive queries is expressible in GNFO. It is only slightly more difficult to verify that every GFO sentence can be expressed in GNFO [BtCS11]. Turning to fragments of first-order logic that are common in database theory, consider *guarded tuple-generating dependencies*: that is, sentences of the form

$$\forall \mathbf{x} \, R(\mathbf{x}) \wedge \phi(\mathbf{x}) \rightarrow \exists \mathbf{y} \, \psi(\mathbf{x}, \mathbf{y}) \, .$$

where ϕ, ψ are conjunctions of relational atomic formulas. By simply writing out such a sentence using $\exists, \neg, \wedge$, one sees that it is convertible to a GNFO sentence. In particular, every *inclusion dependency* is expressible in GNFO, and many of the common dependencies used in data integration and and in exchange (e.g. linear-guarded dependencies, also known as Local-As-View (LAV) constraints [Len02, FKMP05]) lie in GNFO.

Looking at constraints that come from Entity-Relationship and other semantic data models, we see that concept subsumption, when translated into relational database terminology, is expressible in GFO, hence in GNFO. Going further, many of the common description logic languages used in the semantic web (e.g. $\mathcal{ALC}$ and $\mathcal{ALCHIO}$ [BCM^{+}03]) are known to admit translations into GFO, and hence into GNFO.

We will frequently make use of the key result from [BtCS11]:

Theorem 1. *A GNFO sentence is satisfiable over all structures iff it is satisfiable over finite structures. Satisfiability and validity are decidable (and* 2ExpTime*-complete).*

We have mentioned before that GNFO can capture important integrity constraints, but in [BtCO12] it is also argued that GN-RA, and hence GNFO, captures many uses of negation in queries in practice.

3   Rewriting Special GNFO Queries

Preservation theorems in model theory are results that syntactically characterize the formulas within a logic that satisfy important semantic properties. Two examples from classical model theory are the Łoś-Tarski theorem, stating that the universal formulas capture all first order properties closed under taking induced substructures, and the Homomorphism Preservation theorem, stating that existential positive sentences capture all first order properties closed under homomorphism [CK90]. It is known that the Łoś-Tarski theorem fails if we consider equivalence only over finite structures [EF99], while Rossman [Ros08] has shown that the Homomorphism Preservation theorem does hold if we restrict attention to finite structures. A well-known preservation theorem from modal logic is Van Benthem's theorem, stating that basic modal logic captures precisely the fragment of first-order logic invariant under bisimulation [vB83]. The analog for finite structures was proven to hold by Rosen [Ros97], cf. also [Ott04].

Here we will investigate the analogous questions for GNFO. We will start by showing analogs of Van Benthem's theorem for GNFO. We will then identify syntactic fragments that capture the intersection of important fragments of first-order logic with GNFO – from these, new semantic characterizations will follow, including analogs of the Łoś-Tarski and Homomorphism Preservation theorems.

Characterizing GNFO within FO. We first look at the question of characterizing GNFO as the set of all first-order formulas that are invariant under certain simulation relations. In [BtCS11], *guarded-negation bisimulation* were introduced, and it was shown that GNFO captures the fragment of first-order logic that is invariant under GN-bisimulations. Here we give a characterization theorem for a simpler kind of simulation relation, which we call a *strong GN-bisimulation*. We will use this characterization as a basic tool throughout the paper – to show that a certain formula is in GNFO, to argue that two structures must agree on all GNFO formulas, and to amalgamate structures that cannot be distinguished by GN-sentences in a subsignature. The many uses of strong GN-bisimulations suggest that it is really the "right" equivalence relation for GNFO.

Recall that a homomorphism from a structure $\mathfrak{A}$ to a structure $\mathfrak{B}$ is a map from the domain of $\mathfrak{A}$ to the domain of $\mathfrak{B}$ that preserves the relations as well as the interpretation of the constant symbols. We say that a set, or tuple, of elements from a structure $\mathfrak{A}$ is *guarded* in $\mathfrak{A}$ if there is a fact of $\mathfrak{A}$ that contains all elements in question except possibly for those that are the interpretation of a constant symbol.

Definition 1 (Strong GN-bisimulations). *A strong GN-bisimulation between structures $\mathfrak{A}$ and $\mathfrak{B}$ is a non-empty collection Z of pairs $(\mathbf{a}, \mathbf{b})$ of guarded tuples of elements of $\mathfrak{A}$ and of $\mathfrak{B}$, respectively, such that for every $(\mathbf{a}, \mathbf{b}) \in Z$:*
- *there is a homomorphism $h : \mathfrak{A} \to \mathfrak{B}$ with $h(\mathbf{a}) = \mathbf{b}$ and such that "h is compatible with Z", meaning that $(\mathbf{c}, h(\mathbf{c})) \in Z$ for every guarded tuple $\mathbf{c}$ in $\mathfrak{A}$.*
- *there is a homomorphism $g : \mathfrak{B} \to \mathfrak{A}$ with $g(\mathbf{b}) = \mathbf{a}$ and such that "g is compatible with Z", meaning that $(g(\mathbf{d}), \mathbf{d}) \in Z$ for every guarded tuple $\mathbf{d}$ in $\mathfrak{B}$.*

We write $(\mathfrak{A}, \boldsymbol{a}) \to_{GN}^{s} (\mathfrak{B}, \boldsymbol{b})$ if the map $\boldsymbol{a} \mapsto \boldsymbol{b}$ extends to a homomorphism from $\mathfrak{A}$ to $\mathfrak{B}$ that is compatible with some strong GN-bisimulation between $\mathfrak{A}$ and $\mathfrak{B}$. Note that, here, $\boldsymbol{a}$ and $\boldsymbol{b}$ are not required to be guarded tuples. We write $(\mathfrak{A}, \boldsymbol{a}) \sim_{GN}^{s} (\mathfrak{B}, \boldsymbol{b})$ if, furthermore, $\boldsymbol{a}$ is a guarded tuple in $\mathfrak{A}$ (in which case we also have that $(\mathfrak{B}, \boldsymbol{b}) \sim_{GN}^{s} (\mathfrak{A}, \boldsymbol{a})$). These notations can also be indexed by a signature σ, in which case they are defined in terms of σ-reducts of the respective structures. A first-order formula $\phi(x)$ is preserved by $\sim_{GN}^{s}$ if, whenever $(\mathfrak{A}, \boldsymbol{a}) \to_{GN}^{s} (\mathfrak{B}, \boldsymbol{b})$ and $\mathfrak{A} \models \phi(\boldsymbol{a})$, then $\mathfrak{A} \models \phi(\boldsymbol{b})$.

The reader may verify as an exercise that if there exists a strong GN-bisimulation between two structures, then the respective induced substructures consisting of the elements designated by constant symbols must be isomorphic.

Our first "expressive completeness" result characterizes GNFO as the fragment of first-order logic that is invariant for strong GN-bisimulations.

Theorem 2. *A first-order formula $\phi(x)$ is preserved by $\to_{GN}^{s}$ (over all structures) iff it is equivalent to a GNFO formula.*

Strong bisimulations will play a key role in our remaining results. When we want to show that a GNFO formula ϕ can be replaced by another simpler ϕ', we will often justify this by showing that an arbitrary model of ϕ can be replaced by a strongly bisimilar structure where ϕ' holds (or vice versa). The proof of the "hard direction" of Theorem 2 uses the technique of *recursively saturated models* [CK90].

Characterizing Fragments of GNFO. We now look at characterizing the intersection of GNFO with smaller fragments of first-order logic. We will start with tuple-generating dependencies (TGDs). Recall that these are sentences of the form:

$$\forall \mathbf{x} \phi(\mathbf{x}) \to \exists \mathbf{y} \rho(\mathbf{x}, \mathbf{y})$$

where ϕ and ρ are conjunctions of relational atoms (not equalities). TGDs capture many classes of integrity constraints used in classical databases, in data exchange, and in ontological reasoning. Static analysis and query answering problems in the latter contexts have in recent years been driving a quest for identifying expressive yet computationally well-behaved classes of TGDs. A guarded TGD (GTGD) is one in which ϕ includes an atom containing all variables occurring in the rule. Guarded TGDs constitute an important class of TGDs at the heart of the Datalog$^{\pm}$ framework [CGL09, BGO10] for which static analysis problems are decidable. More recently, Baget, Leclère, and Mugnier [BLM10] introduced *frontier-guarded TGDs* (FGTGD), defined like guarded TGDs, but where only the variables occurring both in ϕ and in ρ (the *exported* variables) must be guarded by an atom in ϕ. All FGTGDs are equivalent to GNFO sentences, obtained just by writing them out using existential quantification, negation, and conjunction. Theorem 3 below shows that these are *exactly* the TGDs that GNFO can capture.

We need two lemmas, one about GNFO and one about TGDs. For structure I and superinstance J of I, let us denote by $J \ominus I$ the substructure of J obtained by removing all facts containing only values from the active domain of I. We say that J is a *squid-superinstance* of I if (i) every set of elements from the active domain of I that is guarded in J is already guarded in I, and (ii) $J \ominus I$ is a disjoint union of structures J' for which it holds that $(adom(J') \cap adom(I)) \setminus C$ is guarded in I, where C is the set of elements of I named by a constant symbol (intuitively, we can think of J as a squid, where each J' is one of its tentacles). The following lemma, intuitively, allows one to turn an arbitrary superinstance of a structure I into a squid-superinstance of I, modulo strong GN-bisimulation.

Lemma 1. *For every pair of structures I,J with J being a super-instance of I, there is a squid-superinstance J' of I and a homomorphism $h : J' \to J$ whose restriction to I is the identity function, such that $J' \sim^s_{GN} J$ via a strong GN-bisimulation that is compatible with h. Moreover, we can choose J' to be finite if J is.*

We will make use of Lemma 1 as a tool for bringing certain conjunctive queries into a restricted syntactic form, by exploiting the fact that, whenever a tuple from $adom(I)$ satisfies a conjunctive query in a squid-superinstance J of I, then we can partition the atoms of the query into independent subsets that are mapped into different tentacles of J.

The following lemma expresses a general property of TGDs that follows from the fact that TGDs are preserved under taking direct products of structures [Fag82].

Lemma 2. *Let Σ be any set of TGDs and suppose $\Sigma \models \forall \mathbf{x}(\phi(\mathbf{x}) \to \bigvee_{i=1\ldots n} \exists \mathbf{y}_i \psi_i(\mathbf{x}, \mathbf{y}_i))$, where ϕ, ψ_i are conjunctions of atoms. Then $\Sigma \models \forall \mathbf{x}(\phi(\mathbf{x}) \to \exists \mathbf{y}_i \psi_i(\mathbf{x}, \mathbf{y}_i))$ for some $i \leq n$. This holds both over finite structures and over arbitrary structures.*

We now return to describing our characterization of TGDs that lie in GNFO. Consider a TGD $\rho = \forall \mathbf{x} \beta(\mathbf{x}) \to \exists \mathbf{z} \gamma(\mathbf{xz})$. A *specialisation* of ρ is a TGD of the form $\rho^\theta = \forall \mathbf{x} \beta(\mathbf{x}) \to \exists \mathbf{z}' \gamma'(\mathbf{xz}')$ obtained from ρ by applying some substitution θ mapping the variables $\mathbf{z}$ to constant symbols or to variables among $\mathbf{x}$ and $\mathbf{z}$. The following lemma states that as far as strong GN-bisimulation invariant TGDs are concerned, we can replace any TGD by specializations of it that are equivalent to frontier-guarded TGDs. Its proof relies heavily on the two lemmas above.

Lemma 3. *[TGD specialisations] Let Σ be a set of TGDs that is strong GN-bisimulation invariant and let ρ be a TGD such that $\Sigma \models \rho$. Then there exists a specialisation ρ' of ρ such that $\Sigma \models \rho'$, and such that ρ' is logically equivalent to a conjunction of frontier-guarded TGDs. This holds both over finite structures and over arbitrary structures.*

The result above immediately implies our first main characterization:

Theorem 3. *Every GNFO-sentence that is equivalent to a finite set of TGDs on finite structures is equivalent to a finite set of TGDs on arbitrary structures, and such a formula is equivalent (over all structures) to a finite set of FGTGDs.*

In the light of the above result, it may seem tempting to suppose that, similarly, guarded TGDs form the intersection of TGDs and GFO. This is, however, not the case: the TGD $\forall xyz R(x,y) \wedge R(y,z) \to P(x)$ can be equivalently expressed in GFO, but not by means of a guarded TGD; and the guarded TGD $\forall x P(x) \to \exists yz\, E(x,y) \wedge E(y,z) \wedge E(z,x)$ is not expressible in GFO. Instead, we show that the intersection of GFO and TGDs is *acyclic frontier-guarded TGDs*.

Recall from [Yan81] that an *acyclic conjunctive query* is a conjunctive query whose hypergraph is acyclic. There is another equivalent characterization of acyclic conjunctive queries, which is more convenient for our present purposes: a conjunctive query is acyclic if it can be equivalently expressed by a formula of GFO built up from atomic formulas using only conjunction and guarded existential quantification [GLS03]. We say that a TGD $\rho = \forall \mathbf{xy}\beta(\mathbf{x},\mathbf{y}) \rightarrow \exists \mathbf{z}\gamma(\mathbf{x},\mathbf{z})$ is acyclic if the conjunctive queries $\exists \mathbf{y}\beta(\mathbf{x},\mathbf{y})$ and $\exists \mathbf{z}\gamma(\mathbf{x},\mathbf{z})$ are both acyclic. Using Theorem 3 above, plus the "Treeification Lemma" of [BGO10], we can characterize the GFO sentences that are equivalent to TGDs:

Theorem 4. *Every GFO-sentence that is equivalent to a finite set of TGDs over finite structures is equivalent to a finite set of TGDs on arbitrary structures, and such a formula is equivalent (over all structures) to a finite set of acyclic FGTGDs.*

Existential and Positive-Existential Formulas. We turn to characterizing the existential formulas that are in GNFO, establishing an analog of the Łoś-Tarski theorem. We say that a first-order formula $\phi(\mathbf{x})$ is *preserved under extensions* over a given class of structures if for all structure $\mathfrak{A}$ and $\mathfrak{B}$ from the class, such that $\mathfrak{A} \models \phi(\mathbf{a})$ and $\mathfrak{A}$ is an induced substructure of $\mathfrak{B}$, we have that $\mathfrak{B} \models \phi(\mathbf{a})$.

Theorem 5. *Every GNFO formula that is preserved under extensions over finite structures has the same property over all structures, and such a formula is equivalent (over all structures) to an existential formula in GNFO. Furthermore, we can decide whether a formula has this property, and also find the existential GNFO formula effectively.*

The first part of the first statement follows from the fact that the property of preservation of a GNFO formula can be expressed as a GNFO sentence, along with the finite model property for GNFO. The second part uses the classical Łoś-Tarski theorem to show that a sentence is rewritable as an existential, and then uses our previous infrastructure (e.g. strong bisimulations) to show that any unguarded negations in the existential formula can be removed.

Finally, we consider the situation for GNFO formulas that are positive-existential (for short, $\exists^+$), i.e., that do not contain any negation (and hence, also, only existential quantification) Since GNFO contains all $\exists^+$ formulas, Rossman's theorem [Ros08] implies that the $\exists^+$ formulas are exactly the formulas in GNFO preserved by homomorphism, over all structures or (equivalently, by the finite model property for GNFO) over finite structures. In addition, using the proof of Rossman's theorem plus the decidability of GNFO we can decide whether a GNFO formula can be written in $\exists^+$.

Theorem 6. *There is an effective algorithm for testing whether a given GNFO formula is equivalent to a UCQ and, if so, computing such a UCQ.*

4   Determinacy and Rewriting for Queries with Respect to Views

We now investigate properties pertaining to view-based query rewriting for GNFO.

Suppose V is a finite set of relation names, and we have FO formulas $\{\phi_v : v \in V\}$ over a signature S that is disjoint from V. We consider each ϕ_v as defining a view v that is to be made accessible to a user, where given a finite structure I, this view is the set $\phi_v(I)$ of all tuples of elements satisfying ϕ_v in I. Suppose ϕ_Q is another first-order formula over the signature S. We say that the views ϕ_v's *determine* ϕ_Q if: for all finite structures I and I' with $\phi_v(I) = \phi_v(I')$ for all $v \in V$, we have $\phi_Q(I) = \phi_Q(I')$. Determinacy states that the query result can be recovered from the results of the views, via some function. Note that in this paper, when we talk about a set of views determining a query, we will *always* be working only over finite structures. Segoufin and Vianu initiated a study of determinacy for queries, including the question of when the assumption of determinacy implies that the recovery function is realized by a query. A *rewriting of* ϕ_Q *over* $\{\phi_v : v \in V\}$ is a formula ρ over the signature V (where the arity of a relation $v \in V$ is the number of arguments of ϕ_v), such that for every structure I for signature S, ρ applied to the view structure is the same as $\phi_Q(I)$. The view structure is the structure whose domain is the set of all elements occurring in $\phi_v(I)$ for some $v \in V$, and that interprets each $v \in V$ by $\phi_v(I)$. It is known that determinacy for unions of conjunctive queries is undecidable [NSV10], and that for UCQs determinacy does not imply rewritability even in first-order logic.

In contrast, we will show that whenever GNFO $\{\phi_v : v \in V\}$ determines GNFO ϕ_Q, then there is a rewriting, with the additional assumption that both $\{\phi_v : v \in V\}$ and ϕ_Q are *answer-guarded* – for FO formulas, we mean by this that they are of the form $\phi(\mathbf{x}) = R(\mathbf{x}) \wedge \phi'$ for some ϕ' and relation symbol R. That is, we show that determinacy implies rewritability for GNFO queries and views whose free variables are guarded. Note that rewritings, when they exist, can always be taken to be domain-independent queries, since the recovery function is (by definition) dependent only on the view extent.

Nash, Segoufin, and Vianu [NSV10] showed that these notions of determinacy and rewritability are closely related to interpolation and definability theorems in classical model theory. The Craig Interpolation theorem for first-order logic can be stated as follows: given formulas ϕ, ψ such that $\phi \models \psi$, there is a formula χ such that (i) $\phi \models \chi$, and $\chi \models \psi$ (ii) all relations occurring in χ occur in both ϕ and ψ (iii) all constants occurring in χ occur in both ϕ and ψ (iv) all free variables of χ are free variables of both ϕ and ψ.

The Craig Interpolation theorem has a number of important consequences, including the *Projective Beth definability theorem*. Suppose that we have a sentence ϕ over a first-order signature of the form $S \cup \{G\}$, where G is an n-ary predicate, and suppose S' is a subset of S. We say that ϕ *implicitly defines predicate G over* S' if: for every S'-structure I, every expansion to an $S \cup \{G\}$-structure I' satisfying ϕ has the same restriction to G up to isomorphism. Informally, the S' structure and the sentence ϕ determine a unique value for G. We say that an n-ary predicate G is *explicitly definable over* S' *for models of* ϕ if there is another formula $\rho(x_1 \ldots x_n)$ using only predicates from S' such that $\phi \models \forall \mathbf{x}\, \rho(\mathbf{x}) \leftrightarrow G(\mathbf{x})$. It is easy to see that whenever G is explicitly definable over S' for models of ϕ, then ϕ implicitly defines G over S'. The Projective Beth Definability theorem states the converse: if ϕ implicitly defines G over S', then G is explicitly definable over S' for models of ϕ. In the special case where $S' = S$, this is called simply the Beth Definability theorem.

A proof of the Craig Interpolation theorem can be found in any model theory textbook (e.g. [CK90]). The proof is not effective, and it has been shown that it cannot be made effective [Fri76]. The Projective Beth Definability theorem follows from the Craig Interpolation theorem. Both theorems fail when restricted to finite structures.

We say that a fragment of first-order logic has the Craig Interpolation Property (CIP) if for all $\phi \models \psi$ in the fragment, the result above holds relative to the fragment. We similarly say that a fragment satisfies the Projective Beth Definability Property (PBDP) if the Projective Beth Definability theorem holds relativized to the fragment – that is, if ϕ in the hypothesis of the theorem lies in the fragment then there is a corresponding formula ρ lying in the fragment as well. We talk about the Beth Definability Property (BDP) for a fragment in the same way. The argument for first-order logic applies to any fragment with reasonable closure properties [Hoo00] to show that CIP implies PBDP.

As shown by Nash, Segoufin, and Vianu, the PBDP easily implies that whenever an FO query is determined by a set of FO views over all models, it is rewritable in FO. The fact that determinacy of FO queries does not imply FO rewritability over finite structures is related to the fact that CIP, PBDP, and BDP all fail for FO when implication is considered over finite structures [EF99]. Hence it is of particular interest to look at fragments of FO that have the finite model property, since there equivalence over finite structures can be replaced by equivalence over all structures. Hoogland, Marx, and Otto [HMO99] showed that the Guarded Fragment satisfies BDP but lacks CIP. Marx [Mar07] went on to explore determinacy and rewriting for the Guarded Fragment and its extensions. He argues that the PBDP holds for an extension of GFO called the Packed Fragment; using this, he concludes that determinacy implies rewritability for queries and views in the Packed Fragment. The definition of the Packed Fragment is not important for this work, but at the end of this section we show that PBDP fails for GFO, and also (contrary to [Mar07]) for the Packed Fragment. But we will adapt ideas of Marx to show that CIP and PBDP do hold for GNFO. Using this we will conclude that determinacy implies rewritability for answer-guarded GNFO views and queries.

Craig Interpolation and Beth Definability for GNFO. We now present the main technical result of this section. It is proven following a common approach in modal logic (see, in particular, Hoogland, Marx, and Otto [HMO99]), via a result saying that we can take two structures over different signatures, behaving similarly in the common signature, and *amalgamate* them to get a structure that is simultaneously similar to both of them. The amalgamation results in turn rely on the notion of strong GN-bisimulation, and use the proof of Theorem 2 to construct equivalent structures.

Theorem 7 (GNFO has Craig interpolation). *For each pair of GNFO-formulas ϕ, ψ such that $\phi \models \psi$, there is a GNFO-formula χ such that (i) $\phi \models \chi$, and $\chi \models \psi$, (ii) all relations occurring in χ occur in both ϕ and ψ, (iii) all constants occurring in χ occur in ϕ or ψ (or both), (iv) all free variables of χ are free variables of both ϕ and ψ.*

Projective Beth Definability for GNFO follows by standard arguments [Hoo00]:

Theorem 8. *If a GNFO-sentence ϕ in signature σ implicitly defines a relation symbol G in terms of a signature $\tau \subset \sigma$, and τ includes all constants from σ, then there is an explicit definition of G in terms of τ relative to ϕ.*

Observe that in Theorem 7, the interpolant is allowed to contain constant symbols outside of the common language. Indeed, this must be so, for GNFO lacks the stronger version of interpolation where the interpolant can only contain constant symbols occurring both in the antecedent and in the consequent. Recall that, in GNFO, as well as GFO, constant symbols are allowed to occur freely in formulas, and that their occurrence is not governed by guardedness conditions. In particular, for example, the formula $\forall y R(c, y)$ belongs to GFO (and to GNFO), while the formula $\forall y R(x, y)$ does not. Now, consider the valid entailment $(x = c) \wedge \forall y R(c, y) \models (x = d) \rightarrow \forall y R(d, y)$. It is not hard to show that any interpolant $\phi(x)$ not containing the constants c and d must be equivalent to $\forall y R(x,y)$. This shows that there are valid GFO-implications for which interpolants cannot be found in GNFO, if the interpolants are required to contain only constant symbols occurring both in the antecedent and the consequent. In fact, in [tC05] it was shown that, in a precise sense, every extension of GFO with this strong form of interpolation has full first-order expressive power and is undecidable for satisfiability.

Applications to Rewriting. We can now state the consequence of the PBDP for determinacy-and-rewriting (relying again on the finite model property of GNFO). Note also that GNFO views V can check integrity constraints (e.g. inclusion dependencies) as well as return results. Using the above, we can get:

Theorem 9. *Suppose a set of answer-guarded GNFO views $\{\phi_v : v \in V\}$ determine an answer-guarded GNFO ϕ_Q on finite structures satisfying a set of GNFO sentences Σ. Then there is a GNFO rewriting of ϕ_Q using $\{\phi_v : v \in V\}$ that is valid over structures satisfying Σ. Furthermore, there is an algorithm that, given ϕ_i's and ϕ_Q and Σ satisfying the hypothesis, effectively finds such a formula ρ.*

In particular, this holds if the view definitions ϕ_v are answer-guarded UCQs, ϕ_Q is an answer-guarded UCQ, and Σ consists of inclusion dependencies and LAV constraints.

Note also that "$\{\phi_v : v \in V\}$ determine ϕ_Q" (when the ϕ_v and ϕ_Q are answer-guarded and in GNFO) can be checked in 2ExpTime, since the property can again be expressed as a GNFO sentence, after which Theorem 1 can be applied.

Negative Results for the Guarded and Packed Fragments. We now prove that PBDP fails for the guarded fragment. This shows, intuitively, that if we want to express explicit definitions even for GFO implicitly-definable relations, we will need to use all of GNFO.

Theorem 10. *The PBDP fails for GFO.*

Proof. Consider the GF sentence ϕ that is the conjunction of the following:
$$\forall x \, C(x) \rightarrow \exists yzu(G(x,y,z,u) \wedge E(x,y) \wedge E(y,z) \wedge E(z,u) \wedge E(u,x))$$
$$\forall xy \, E(x,y) \wedge \neg C(x) \rightarrow P_0(x) \wedge \neg P_1(x) \wedge \neg P_2(x)$$
$$\forall xy \, P_i(x) \wedge E(x,y) \rightarrow P_{(i+1 \bmod 3)}(y) \quad \text{for all } 0 \leq i < 3$$

The first sentence forces that if $C(x)$ holds, then x lies on a directed E-cycle of length 4. The remaining two sentences force that if $\neg C(x)$ holds, then x only lies on directed E-cycles whose length is a multiple of 3. Clearly, the relation C is implicitly defined in terms of E. However, there is no explicit definition in GFO in terms of E, because no formula of GFO can distinguish the directed E-cycle of length k from the directed E-cycle of length ℓ for $3 \leq k < \ell$ [AvBN98]. □

It follows from Theorem 10 that GFO lacks CIP as well, which was already known [HMO99]. Furthermore, the above argument can be adapted to show that determinacy does not imply rewritability for views and queries defined in GFO: consider the set of views $\{\phi_{v_1}, \phi_{v_2}\}$, where $\phi_{v_1} = \phi$ and $\phi_{v_2}(x,y) = E(x,y)$. Clearly, $\{\phi_{v_1}, \phi_{v_2}\}$ determine the query $Q(x) = \phi \wedge C(x)$. On the other hand, any rewriting would constitute an explicit definition in GFO of C in terms of E, relative to ϕ, which we know does not exist.

In [Mar07, Lemma 4.4] it was asserted that PBDP holds for an extension of the Guarded Fragment, called the *Packed Fragment*, in which a guard $R(\mathbf{x})$ may be a conjunction of atomic formulas, as long as every pair of variables from $\mathbf{x}$ co-occurs in one of these conjuncts. The proof of Theorem 10, however, shows that PBDP fails for the Packed Fragment, because known results (cf. [Mar07]) imply that no formula of the Packed Fragment can distinguish the cycle of length k from the cycle of length ℓ for $4 \leq k < \ell$. Indeed, it turns out that there is a flaw in the proof of Lemma 4.4 in [Mar07].

5   Rewriting GNFO Dependencies

Given a finite structure I, a set of integrity constraints Σ, and a query $Q(x_1 \ldots x_k)$, the *certain answers to Q on I (under Σ)* are the set of tuples $c_1 \ldots c_k \in I$ such that $\mathbf{c} \in Q(M)$ for every M containing I and satisfying Σ. Calculating the certain answers is a central problem in information integration and ontologies (in the former case one restricts to M finite, but for our constraints there will be no distinction). One of the key advantages of GNFO is that one can compute the certain answers for every Q and Σ in GNFO, and thus in particular for every Σ in GNFO and conjunctive query Q [BtCO12]. Baget et al. [BLM10] proved that for every set of frontier-guarded dependencies Σ and conjunctive query Q, the certain answers can be computed in polynomial time in I. However, it is known that there are guarded TGDs and conjunctive queries such that the certain answers can not be computed by a first-order query. We say that conjunctive query Q is *first-order rewritable* under constraints Σ if there is a first-order formula ϕ such that on any finite structure I $\phi(I)$ is exactly the certain answer to Q on I under Σ. Our next goal will be to show that we can decide, given a set Σ of frontier-guarded TGDs and a conjunctive query Q, whether or not Q is first-order rewritable. We will proceed by first capturing the certain answers in a fragment of Datalog. In proving this, we will follow (and correct) the approach of Baget et al. [BMRT11b], who argued that the certain answers of conjunctive queries under frontier-guarded TGDs are rewritable in Datalog. For guarded TGDs, this result had been announced by Marnette [Mar11]. The proof of Baget et al. [BMRT11a] revolves around a "bounded base lemma" showing that whenever a set of facts is not closed under "chasing" with FGTGDs, there is a small subset that is not closed (Lemma 4 of [BMRT11a]). However both the exact statement of that lemma and its proof are flawed. Our proof corrects the argument, making use of model-theoretic techniques (including Lemma 1) to prove the bounded base lemma. It then follows the rest of the argument in [BMRT11a] to show not only Datalog-rewritability, but rewritability into a Datalog program comprised of frontier-guarded rules. A conjunctive query is *answer-guarded* if it includes an atom that guards all free variables. In particular all Boolean conjunctive queries are answer-guarded.

Theorem 11. *For every set Σ of frontier-guarded TGDs, and for every answer-guarded conjunctive query $Q(x)$, one can effectively find a frontier-guarded Datalog program that computes the certain answers to Q.*

Note that entailment can be interpreted either in the classical sense or in the finite sense, since we have the finite model property. Indeed, in our proofs, we use constructions that make use of infinite structures, but the conclusion hold in the finite.

In [BtCO12], a fragment of Datalog, denoted *GN-Datalog* was defined, and it was shown that for this fragment one can decide whether a query is equivalent to a first-order query (equivalently, as shown in [BtCO12], to some query obtained by unfolding the Datalog rules finitely many times). Since GN-Datalog contains frontier-guarded Datalog, we can combine the decision procedure from [BtCO12] with Theorem 11 to obtain:

Corollary 1. *FO-rewritability of conjunctive queries under sets of frontier-guarded TGDs is decidable.*

Acknowledgements. The authors gratefully acknowledge their debt to Martin Otto and to Maarten Marx for enlightening discussions on the subject.

References

[AHV95] Abiteboul, S., Hull, R., Vianu, V.: Foundations of Databases. Add.-Wesley (1995)

[AvBN98] Andréka, H., van Benthem, J., Németi, I.: Modal languages and bounded fragments of predicate logic. J. Phil. Logic 27, 217–274 (1998)

[BCM⁺03] Baader, F., Calvanese, D., McGuinness, D.L., Nardi, D., Patel-Schneider, P.F. (eds.): The description logic handbook. Cambridge University Press (2003)

[BGO10] Bárány, V., Gottlob, G., Otto, M.: Querying the guarded fragment. In: LICS (2010)

[BLM10] Baget, J.-F., Leclère, M., Mugnier, M.-L.: Walking the decidability line for rules with existential variables. In: KR (2010)

[BMRT11a] Baget, J.-F., Mugnier, M.-L., Rudolph, S., Thomazo, M.: Complexity boundaries for generalized guarded existential rules (2011) Research Report LIRMM 11006

[BMRT11b] Baget, J.-F., Mugnier, M.-L., Rudolph, S., Thomazo, M.: Walking the complexity lines for generalized guarded existential rules. In: IJCAI (2011)

[BtCO12] Bárány, V., ten Cate, B., Otto, M.: Queries with guarded negation. In: VLDB (2012)

[BtCS11] Bárány, V., ten Cate, B., Segoufin, L.: Guarded negation. In: Aceto, L., Henzinger, M., Sgall, J. (eds.) ICALP 2011, Part II. LNCS, vol. 6756, pp. 356–367. Springer, Heidelberg (2011)

[CGL09] Calì, A., Gottlob, G., Lukasiewicz, T.: A general datalog-based framework for tractable query answering over ontologies. In: PODS (2009)

[CK90] Chang, C.C., Keisler, J.: Model Theory. North-Holland (1990)

[EF99] Ebbinghaus, H.-D., Flum, J.: Finite Model Theory. Springer (1999)

[Fag82] Fagin, R.: Horn clauses and database dependencies. J. ACM 29(4), 952–985 (1982)

[FKMP05] Fagin, R., Kolaitis, P.G., Miller, R.J., Popa, L.: Data exchange: Semantics and query answering. TCS 336(1), 89–124 (2005)

[Fri76] Friedman, H.: The complexity of explicit definitions. AIM 20(1), 18–29 (1976)

[GLS03] Gottlob, G., Leone, N., Scarcello, F.: Robbers, marshals, and guards: game theoretic and logical characterizations of hypertree width. J. Comput. Syst. Sci. 66(4), 775–808 (2003)

[HMO99] Hoogland, E., Marx, M., Otto, M.: Beth definability for the guarded fragment. In: Ganzinger, H., McAllester, D., Voronkov, A. (eds.) LPAR 1999. LNCS, vol. 1705, pp. 273–285. Springer, Heidelberg (1999)

[Hoo00] Hoogland, E.: Definability and interpolation: model-theoretic investigations. PhD thesis, University of Amsterdam (2000)

[Len02] Lenzerini, M.: Data integration: A theoretical perspective. In: PODS (2002)

[Mar07] Marx, M.: Queries determined by views: pack your views. In: PODS (2007)

[Mar11] Marnette, B.: Resolution and datalog rewriting under value invention and equality constraints. Technical report (2011), http://arxiv.org/abs/1212.0254

[NSV10] Nash, A., Segoufin, L., Vianu, V.: Views and queries: Determinacy and rewriting. ACM Trans. Database Syst. 35(3) (2010)

[Ott04] Otto, M.: Modal and guarded characterisation theorems over finite transition systems. APAL 130, 173–205 (2004)

[Ros97] Rosen, E.: Modal logic over finite structures. JLLI 6(4), 427–439 (1997)

[Ros08] Rossman, B.: Homomorphism preservation theorems. J. ACM 55(3) (2008)

[tC05] ten Cate, B.: Interpolation for extended modal languages. JSL 70(1), 223–234 (2005)

[vB83] van Benthem, J.F.A.K.: Modal Logic and Classical Logic. Humanities Pr. (1983)

[Yan81] Yannakakis, M.: Algorithms for acyclic database schemes. In: VLDB (1981)

Parity Games and Propositional Proofs

Arnold Beckmann[1,*], Pavel Pudlák[2,**], and Neil Thapen[2,**]

[1] Department of Computer Science, College of Science,
Swansea University, Swansea SA2 8PP, UK
a.beckmann@swansea.ac.uk

[2] Institute of Mathematics, Academy of Sciences of the Czech Republic, Žitná 25,
115 67 Praha 1, Czech Republic
{pudlak,thapen}@math.cas.cz

Abstract. A propositional proof system is *weakly automatizable* if there is a polynomial time algorithm which separates satisfiable formulas from formulas which have a short refutation in the system, with respect to a given length bound. We show that if the resolution proof system is weakly automatizable, then parity games can be decided in polynomial time. We also define a combinatorial game and prove that resolution is weakly automatizable if and only if one can separate, by a set decidable in polynomial time, the games in which the first player has a positional winning strategy from the games in which the second player has a positional winning strategy.

1   Introduction

Parity games, mean payoff games and simple stochastic games are three classes of two player games, played by moving a token around a finite graph. In particular parity games have important applications in automata theory, logic, and verification [11]. The main computational problem for all of these games is to decide, given an instance of a game, which player has a positional winning strategy. From this point of view, parity games are reducible to mean payoff games, and mean payoff games are reducible to simple stochastic games [19, 23]. It is known that the decision problem for simple stochastic games is reducible to a search problem in the intersection of the classes PLS and PPAD [6, 13] (which are believed to be incomparable [4]). None of the decision problems is known to be in P, despite intensive research work on developing algorithms for them. For several of the existing algorithms, exponential lower bounds on their runtime have been given recently [9, 10].

Automatizability is an important concept for automated theorem proving. Call a propositional proof system *automatizable* if there is an algorithm which, given a tautology, produces a proof in time polynomial in the size of its smallest

---

* This research was partially done while the authors were visiting fellows at the Isaac Newton Institute for the Mathematical Sciences in the programme "Semantics & Syntax".
** Partially supported by grant IAA100190902 of GA AV ČR.

K. Chatterjee and J. Sgall (Eds.): MFCS 2013, LNCS 8087, pp. 111–122, 2013.
© Springer-Verlag Berlin Heidelberg 2013

proof—this time condition is the best we can hope for, assuming NP $\neq$ coNP. Automatizability is a very strict notion. For example, Alekhnovich and Razborov [1] have shown that resolution is not automatizable under a reasonable assumption in parameterised complexity theory. *Weak automatizability* is a relaxation of automatizability, where proofs of tautologies can be given in an arbitrary proof system, and only the time of finding such proofs is restricted to polynomial in the size of the smallest proof in a given proof system. This characterisation of weak automatizability is equivalent to the existence of a polynomial time algorithm which separates satisfiable formulas from formulas which have a short refutation in the system with respect to a given length bound.

Two recent papers have shown a connection between weak automatizability and the above mentioned games. Atserias and Maneva showed that if a certain proof system (called PK_1 in our notation) is weakly automatizable, then the decision problem for mean payoff games is in P [3]. Huang and Pitassi strengthened this to the decision problem for simple stochastic games [12]. In this paper we extend these results to resolution and parity games. In Sect. 2 below we show that if resolution is weakly automatizable, then parity games can be decided in polynomial time.

In order to obtain a kind of reverse direction of this result, in Sect. 3 we define a new game, the *point-line game*, also about moving a token around a finite graph. We show that its complexity is equivalent to that of resolution, in a certain sense. In particular, resolution is weakly automatizable if and only if one can separate, by a set in P, the games in which the first player has a positional winning strategy from the games in which the second player has a positional winning strategy.

The essential part of the argument in Sect. 2, together with one direction of Sect. 3, is to show that there is a polynomial-size propositional proof that winning strategies cannot exist simultaneously for both players in a game. Propositional proofs are complicated combinatorial objects, and constructing them by hand can be difficult. Instead, we work with weak first-order bounded arithmetic theories which capture the logical content of these proof systems, and rely on known translations of these to do the hard work of actually constructing the propositional proofs for us. These translations go back to Paris and Wilkie [17]. Later work has given finer results about the logical depth of the propositional proofs. The main result we need, a first order-theory which translates into polynomial-size resolution, is essentially due to Krajíček [14–16].

The full version of this paper, in preparation, will extend our methods to give simplified proofs of the results mentioned above relating weak automatizability of the proof system PK_1 to the decision problem for mean payoff and simple stochastic games [3, 12]. Furthermore, it will include a detailed proof of the translation of first order-theories into polynomial-size propositional proofs, with extra information about the fan-in k of connectives located at the maximum depth of propositional formulas, namely that k can be bounded by a constant that we can read directly from the formulas appearing in the first-order proof.

Finding a polynomial time algorithm to solve parity games is a long-standing open problem, so it is tempting to interpret our main result about parity games and resolution as evidence either that resolution is not weakly automatizable, or at least that if it is, then this will be hard to prove. On the other hand, modern SAT solvers typically use algorithms which, given a formula, generate either a satisfying assignment or what is essentially a resolution proof that the formula is unsatisfiable. Thus it seems that a necessary condition for a formula to be tractable by these SAT solvers is that the formula is either satisfiable, or has a short resolution refutation. Our reduction can be used to translate a parity game into a formula that satisfies at least this necessary condition. Hence, a possible application is to try to combine our reduction with a SAT solver, to obtain a new algorithm for solving parity games.

1.1   Resolution Proof Systems

For $k \geq 1$, the propositional proof system $\mathrm{Res}(k)$ is defined as follows. Propositional formulas are formed from propositional variables $p_0, p_1, p_2, \ldots$, negation $\neg$, and unbounded fan-in conjunctions and disjunctions $\bigwedge$ and $\bigvee$. Variables are called *atoms*, and atoms and negated atoms are together called *literals*. Formulas are then defined inductively: each literal is a formula, and if Φ is a finite non-empty set of formulas then $\bigwedge \Phi$ and $\bigvee \Phi$ are formulas. For a formula φ, we use $\neg \varphi$ as an abbreviation for the formula formed from φ by interchanging $\bigwedge$ and $\bigvee$ and interchanging atoms and their negations. We treat the binary connectives $\wedge$ and $\vee$ as the obvious set operations, for example $\bigvee \Phi \vee \bigvee \Psi = \bigvee (\Phi \cup \Psi)$. If a formula is a conjunction, we will sometimes treat it as the set of its conjuncts, and vice versa.

A k-DNF is a disjunction of conjunctions of literals, where each conjunction is of size at most k. Each line in a $\mathrm{Res}(k)$-proof is a k-DNF, usually written as the list of disjuncts separated by commas. The rules of $\mathrm{Res}(k)$ are as follows, where Γ, Δ stand for sets of formulas, possibly empty, A, B for formulas, and a_i for literals:

$$\wedge\text{-introduction} \quad \frac{\Gamma, A \qquad \Gamma, B}{\Gamma, A \wedge B}$$

$$\text{weakening} \ \frac{\Gamma}{\Gamma, \Delta} \qquad \text{cut} \ \frac{\Gamma, a_1 \wedge \ldots \wedge a_m \qquad \Gamma, \neg a_1, \ldots, \neg a_m}{\Gamma}$$

We also allow introduction of logical axioms $\overline{a, \neg a}$ for atoms a.

A $\mathrm{Res}(k)$ *refutation* of a set of disjunctions Γ is a sequence of disjunctions ending with the empty disjunction, such that each line in the proof is either in Γ, or a logical axiom, or follows from earlier disjunctions in the sequence by a rule. The system $\mathrm{Res}(1)$ is called *resolution* and is denoted by Res.

We will also consider the proof system PK_1, which is defined in the same way as $\mathrm{Res}(k)$ but now dropping the bound on the number of literals in conjunctions. That is, lines in PK_1 proofs are unrestricted DNFs, instead of k-DNFs in case of $\mathrm{Res}(k)$.

1.2   Bounded Arithmetic

We could obtain the results of this paper by a careful use of the conventional Buss-style bounded arithmetic theory T_1^2 [5]. However, these would introduce unnecessary complications to deal with sharply bounded quantification, so instead we will work with simpler systems.

For $r \in \mathbb{N}$, we will say that a function $f: \mathbb{N}^r \to \mathbb{N}$ is *polynomially bounded* if there is some polynomial p such that $f(\bar{x}) \leq p(\bar{x})$ for all $\bar{x}$. Let L be the language consisting of the constant symbols 0 and 1, and, for every $r \in \mathbb{N}$, a function symbol for every polynomially bounded function $\mathbb{N}^r \to \mathbb{N}$ and a relation symbol for every relation on $\mathbb{N}^r$. If the reader is uncomfortable with such a large language, it can be replaced by any reasonably rich language extending $\{0, 1, +, \cdot, <\}$ as long as all functions in the language are polynomially bounded. Let BASE be the set of true universal L-sentences. We will use this as our base theory.

We extend L to a language $L^+ = L \cup \bar{R}$ by adding a tuple $\bar{R}$ of finitely many new relation symbols. We will use these to stand for edges in a graph, or strategies in a game, or whatever other objects we need to reason about.

Adapting notation from Wilmers [22], we define a *strict U_d formula* to be one consisting of d alternating blocks of bounded quantifiers, beginning with a universal block, followed by a quantifier-free L^+ formula. To obtain optimal results about the depth of the propositional translations of these formulas, we add a technical requirement: the quantifier-free part should have the form of a CNF if d is odd, or a DNF if d is even. Any quantifier-free formula is logically equivalent to one in either form, so in the first-order proofs we construct in this paper we can ignore this requirement. A U_d *formula* is a subformula of a strict U_d formula. The strict E_d formulas and the E_d formulas are defined dually.

We remark that we will almost always work with bounded rather than unbounded quantifiers, and we will often not write the bounds if they are obvious, for example if we are quantifying over the vertices of a given finite graph.

For $d \geq 0$, we define U_d-IND to be BASE together with the usual induction scheme

$$\forall a, \; \phi(0) \wedge \forall x < a[\phi(x) \to \phi(x+1)] \to \phi(a)$$

for each U_d formula $\phi(x)$, which may also contain other parameters. The theory E_d-IND is defined similarly.

Similarly we define U_d-MIN to be the usual scheme asserting that any nonempty U_d (with parameters) subset of an interval $[0, a)$ has a least element. The schemes E_d-MIN, U_d-MAX and E_d-MAX are the obvious variants of this.

Lemma 1. *For $d \geq 0$, the following hold over* BASE:
1. E_d-IND *is equivalent to* U_d-IND
2. E_d-MAX *is equivalent to* E_d-MIN
3. U_d-MAX *is equivalent to* U_d-MIN
4. U_{d+1}-IND *proves* U_d-MAX *and* E_{d+1}-MAX.   □

We now define a version of the Paris-Wilkie translation of first-order proofs in bounded arithmetic into small propositional proofs [17]. We will use this as a tool for constructing resolution refutations out of U_2-IND proofs. For each relation symbol in $\bar{R}$ of arity s, we fix a propositional variable $r_{i_1,\ldots,i_s}$ for each tuple of numbers $i_1,\ldots,i_s$. We assume that all these propositional variables, for all relation symbols in $\bar{R}$, are pairwise distinct.

Let $\top$ and $\bot$ denote the truth values *true* and *false*, respectively. An assignment α is a total map from first-order variables to numbers, in which at most finitely many variables are assigned non-zero values. For an assignment α, a variable x and a number n, we write $\alpha[x \mapsto n]$ for the assignment which maps x to n and leaves the mapping of all other variables unchanged. We write $[x \mapsto n]$ for the assignment which maps x to n and all other variables to 0.

Definition 2. *We compute propositional translations as follows.*
1. *Any L formula ϕ has a definite truth value under α. If ϕ evaluates to true we let $\langle \phi \rangle_\alpha$ be $\top$, and if it evaluates to false we let $\langle \phi \rangle_\alpha$ be $\bot$.*
2. *For t an L-term, we let $\langle t \rangle_\alpha$ be the evaluation of t under α.*
3. *For R an s-ary relation symbol in $\bar{R}$, and $\bar{t}$ an s-tuple of L-terms, we let $\langle R(\bar{t}) \rangle_\alpha$ be the propositional variable $r_{i_1,\ldots,i_s}$ where each $i_j = \langle t_j \rangle_\alpha$, and let $\langle \neg R(\bar{t}) \rangle_\alpha$ be the negated variable $\neg r_{i_1,\ldots,i_s}$.*
4. *We let $\langle \phi \wedge \psi \rangle_\alpha$ be $\langle \phi \rangle_\alpha \wedge \langle \psi \rangle_\alpha$ and let $\langle \phi \vee \psi \rangle_\alpha$ be $\langle \phi \rangle_\alpha \vee \langle \psi \rangle_\alpha$.*
5. *We let $\langle \forall x < t\ \phi(x) \rangle_\alpha$ be $\bigwedge \{ \langle \phi \rangle_{\alpha[x \mapsto m]} : m < \langle t \rangle_\alpha \}$. Bounded existential quantifiers are similarly translated into disjunctions.*

Finally we simplify by inductively removing $\top$ from conjunctions, removing $\bot$ from disjunctions, replacing conjunctions containing $\bot$ with $\bot$, and replacing disjunctions containing $\top$ with $\top$.

Theorem 3. *Suppose that $\phi_1(x), \ldots, \phi_\ell(x)$ are U_2 formulas, with x the only free variable, such that U_2-IND proves $\forall x\, \neg(\phi_1(x) \wedge \ldots \wedge \phi_\ell(x))$. Then for some $k \in \mathbb{N}$ the family*

$$\Phi_n := \langle \phi_1(x) \rangle_{[x \mapsto n]} \cup \cdots \cup \langle \phi_\ell(x) \rangle_{[x \mapsto n]}$$

has polynomial size $\mathrm{Res}(k)$ refutations. $\qquad\qquad\square$

1.3   Disjoint NP Pairs

A *disjoint NP pair* is simply a pair of disjoint NP sets. In the context of proof complexity, these were first studied by Razborov in [20]. Our presentation follows [18]. A pair (A, B) is *polynomially reducible* to a pair (C, D) if there is a polynomial time function f, defined on all strings, such that $f[A] \subseteq C$ and $f[B] \subseteq D$. A pair (A, B) is *polynomially equivalent* to a pair (C, D) if polynomial reducibility holds in both directions. A pair (A, B) is *polynomially separable* if there is a polynomial time function which takes the value 0 on strings in A and the value 1 on strings in B.

If $\mathcal{P}$ is a propositional proof system, the *canonical pair* $\mathbf{C}_\mathcal{P}$ of $\mathcal{P}$ is the pair (A, B) where

$$A = \{(\phi, 1^m) : \phi \text{ is satisfiable}\}$$
$$B = \{(\phi, 1^m) : \phi \text{ has a } \mathcal{P}\text{-refutation of size at most } m\}.$$

We say that $\mathcal{P}$ is *weakly automatizable* if $\mathbf{C}_{\mathcal{P}}$ is polynomially separable. In other words, $\mathcal{P}$ is weakly automatizable if there is a polynomial time algorithm which separates satisfiable formulas from formulas which have a short refutation in the system with respect to a given length bound. This definition of weakly automatizability is equivalent to others in the literature (see [2]).

To define the *interpolation pair* $\mathbf{I}_{\mathcal{P}}$ of $\mathcal{P}$, let $\Delta_{\mathcal{P}}$ be the set of triples (ϕ, θ, π) where ϕ and θ are propositional formulas in disjoint variables and π is a $\mathcal{P}$-refutation of $\phi \wedge \theta$. Then $\mathbf{I}_{\mathcal{P}}$ is the pair (A, B) where

$$A = \{(\phi, \theta, \pi) \in \Delta_{\mathcal{P}} : \phi \text{ is satisfiable}\}$$
$$B = \{(\phi, \theta, \pi) \in \Delta_{\mathcal{P}} : \theta \text{ is satisfiable}\}.$$

Given a triple $(\phi, \theta, \pi) \in \Delta_{\mathcal{P}}$, at least one of ϕ and θ must be unsatisfiable. We say that $\mathcal{P}$ has *feasible interpolation* if there is a polynomial time function which, given such a triple as input, outputs 0 if ϕ is unsatisfiable and 1 if θ is unsatisfiable. It is easy to show that $\mathcal{P}$ has feasible interpolation if and only if $\mathbf{I}_{\mathcal{P}}$ is polynomially separable.

Proposition 4 ([2])

1. *Resolution has feasible interpolation.*
2. *The following list of* NP *pairs are pairwise equivalent: The canonical pairs of* Res, Res(2), Res(3), $\ldots$, *and the interpolation pairs of* Res(2), Res(3), Res(4), $\ldots$, *and of* PK_1. $\square$

Finally, we define the *canonical pair* of a class of two-player games to be the pair (A_0, A_1) where A_i is the set of games in which player i has a positional winning strategy. Naturally, for this to make sense we need there to be a definition of what a positional strategy is, and for it to be possible to recognise a positional winning strategy in NP.

2   Parity Games

Following Stirling [21] we will describe parity games in a simplified form, which is linear-time equivalent to the usual definition. A *parity game* G is given by a finite directed graph with vertices V and edges E satisfying the following properties. The set V is the disjoint union of two sets V_0 and V_1 which we think of as the vertices belonging respectively to player 0 and to player 1. The graph has a designated *start vertex* s, and every vertex has at least one outgoing edge. We identify V with the interval $[n] = \{0, \ldots, n-1\}$ where $n = |V|$. Below when we talk about the "least" vertex we mean the least with respect to the usual order on $[n]$. Without loss of generality, $s = 0$.

The game begins with a pebble placed on the start vertex s. On each turn, the pebble is moved from its current vertex v along an edge in the graph. If $v \in V_0$ then player 0 chooses which edge to move it along. If $v \in V_1$ then player 1 chooses. A *play* of the game is the infinite sequence $v_1, v_2, \ldots$ of vertices visited by the pebble. To decide the winner of a play, let v be the least vertex which occurs infinitely often. If $v \in V_0$ then player 0 wins and if $v \in V_1$ then player 1 wins.

A *positional strategy* σ for player 0 is a map $\sigma \colon V_0 \to V$ such that $(x, \sigma(x))$ is an edge in E for each $x \in V_0$. Similarly, a positional strategy τ for player 1 is a map $\tau \colon V_1 \to V$ such that $(x, \tau(x)) \in E$ for each $x \in V_1$.

The following theorem has been proven by Emerson [8] independently of a similar result for mean payoff games by Ehrenfeucht and Mycielski [7]; the reduction from parity to mean payoff games was found later by Puri [19].

Theorem 5 (Emerson [8]). *In each parity game, one of the players has a positional winning strategy.* $\qquad\square$

From now on we will only discuss positional strategies, so we will usually omit the word "positional". Given a strategy σ for player 0, we will use E^σ to mean the edge relation obtained from E by, for each vertex $v \in V_0$, removing all outgoing edges except for the one chosen in σ. We will similarly use E^τ to mean E restricted by a strategy τ for player 1.

It is straightforward to show that the strategy σ is winning for player 0 if and only if for every vertex t reachable from s in E^σ, for every path from t to t in E^σ, the least vertex on the path is in V_0. To prove our main result in this section, we formalise this characterisation in such a way that we can prove in U_2-IND that player 0 and player 1 cannot simultaneously have winning strategies. In our formalisation below, all quantifiers are implicitly bounded by n.

Expand the language L to include relation symbols E, V_0, V_1, E^σ, $R_{\min}^\sigma$, E^τ, $R_{\min}^\tau$ and a constant symbol n. We will write G to stand for the tuple E, V_0, V_1, n representing the structure of the game. The intended meaning of E^σ is as described above. The intended meaning of the ternary relation $R_{\min}^\sigma(x, y, z)$ is that there is a non-trivial path in E^σ from x to y on which the least vertex visited is z. The relations E^τ and $R_{\min}^\tau$ are similar.

Let $\mathrm{Game}(G)$ be a formula asserting that G is a suitable graph for a parity game, that is, that V_0 and V_1 partition the vertices, and that every vertex has at least one outgoing edge. Let $\mathrm{Strategy}_0(G, E^\sigma)$ be a formula asserting that E^σ represents a strategy for player 0, that is, that every vertex in V_0 has an outgoing edge in E^σ. Let $\mathrm{Strategy}_1(G, E^\tau)$ be a similar formula for player 1. It is clear that these can all be written as U_2 formulas.

Let $\mathrm{Win}_0(G, E^\sigma, R_{\min}^\sigma)$ be the conjunction of the universal closures of
1. $\mathrm{Strategy}_0(G, E^\sigma)$
2. $E^\sigma(x, y) \wedge z = \min(x, y) \to R_{\min}^\sigma(x, y, z)$
3. $R_{\min}^\sigma(x, y, u) \wedge R_{\min}^\sigma(y, z, v) \wedge w = \min(u, v) \to R_{\min}^\sigma(x, z, w)$
4. $R_{\min}^\sigma(s, x, u) \wedge R_{\min}^\sigma(x, x, v) \to v \in V_0$.

Let $\mathrm{Win}_1(G, E^\tau, R_{\min}^\tau)$ be a similar formula for player 1.

Lemma 6. *If player 0 has a winning strategy in G, then there exist E^σ and $R_{\min}^\sigma$ satisfying $\mathrm{Win}_0(G, E^\sigma, R_{\min}^\sigma)$. Similarly for player 1 and $\mathrm{Win}_1(G, E^\tau, R_{\min}^\tau)$.* $\square$

Theorem 7. *Provably in U_2-IND, it is impossible to satisfy formulas $\mathrm{Game}(G)$, $\mathrm{Win}_0(G, E^\sigma, R_{\min}^\sigma)$ and $\mathrm{Win}_1(G, E^\tau, R_{\min}^\tau)$ simultaneously.*

Proof. Let $R^*(x, y)$ be the formula $\exists v,\, R^\sigma_{\min}(x, y, v) \wedge R^\tau_{\min}(x, y, v)$. By condition 3 of Win_0 and Win_1, the relation $R^*(x, y)$ is transitive. Moreover for every x there is at least one y such that $R^*(x, y)$, since we can take y to be the unique successor of x in $E^\sigma \cap E^\tau$ and take v to be $\min(x, y)$.

Let $A(x)$ be the formula $R^*(s, x) \wedge \forall y > x\, \neg R^*(x, y)$. Using E_1-MAX, let x be maximum such that $R^*(s, x)$ holds. It follows that $A(x)$ holds. Hence using E_2-MIN, we let t be minimum such that $A(t)$. Now using E_1-MAX, let t' be maximum such that $R^*(t, t')$. By the transitivity of R^*, we know that $R^*(s, t')$ and also that for all $y > t'$ we have $\neg R^*(t', y)$. Hence $A(t')$ holds, and therefore $t' \geq t$ by minimality of t. On the other hand, since $A(t)$ and $R^*(t, t')$, we know $t' \leq t$. We conclude that $t' = t$.

We now have that $R^*(s, t)$ and $R^*(t, t)$. Hence there are vertices u and v such that both $R^\sigma_{\min}(s, t, u) \wedge R^\sigma_{\min}(t, t, v)$ and $R^\tau_{\min}(s, t, u) \wedge R^\tau_{\min}(t, t, v)$ hold. Therefore condition 4 must be false in either Win_0 or Win_1, since either $v \in V_0$ or $v \in V_1$. $\square$

The formula $\mathrm{Win}_0(G, E^\sigma, R^\sigma_{\min})$ is a conjunction of U_2 formulas. Suppose we are given a parity game G, with n vertices. Let α map the constant symbol n of our language (which we treat here as a free variable) to the number n. Then for some $k \in \mathbb{N}$ we can translate each such formula ϕ into a conjunction $\langle \phi \rangle_\alpha$ of k-DNFs, with propositional variables for the relations E^σ, $R^\sigma_{\min}$ and for the structure of the game G. We abuse notation and write $\langle \mathrm{Win}_0(E^\sigma, R^\sigma_{\min}) \rangle_G$ for the propositional formula obtained by taking the set of all the formulas $\langle \phi \rangle_\alpha$ and substituting in, for the propositional variables describing the structure of G, the values given by the actual game G.

In other words, $\langle \mathrm{Win}_0(E^\sigma, R^\sigma_{\min}) \rangle_G$ is the propositional formula obtained by translating Win_0 and substituting in the real values of G. It is satisfiable if and only if player 0 has a winning strategy in G. The formula $\langle \mathrm{Win}_1(E^\tau, R^\tau_{\min}) \rangle_G$ is similar.

Corollary 8. *There is a number $k \in \mathbb{N}$ and a polynomial p such that for every game G, the formula $\langle \mathrm{Win}_0(E^\sigma, R^\sigma_{\min}) \rangle_G \cup \langle \mathrm{Win}_1(E^\tau, R^\tau_{\min}) \rangle_G$ has a $\mathrm{Res}(k)$ refutation of size $p(n)$.*

Proof. Take the proof given by Theorem 3, and substitute in the real values of G. Observe that G satisfies $\mathrm{Game}(G)$, so all the initial formulas coming from $\mathrm{Game}(G)$ vanish. $\square$

Corollary 9. *The canonical pair for parity games is reducible to the canonical pair for resolution.*

Proof. Let p and k be as in Corollary 8. By Proposition 4, it is enough to show reducibility to the canonical pair for $\mathrm{Res}(k)$. The reduction function is given by

$$G \mapsto (\,\langle \mathrm{Win}_0(E^\sigma, R^\sigma_{\min}) \rangle_G,\, 1^{p(n)}\,).$$

If player 0 has a winning strategy for G then $\langle \mathrm{Win}_0(E^\sigma, R^\sigma_{\min}) \rangle_G$ is satisfiable. On the other hand, if player 1 has a winning strategy for G then $\langle \mathrm{Win}_1(E^\tau, R^\tau_{\min}) \rangle_G$

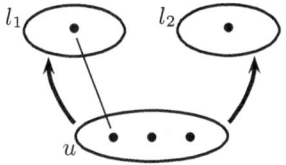

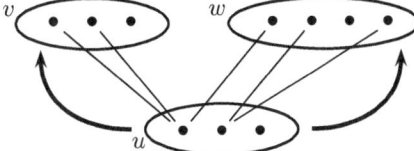

Vertex u connected to leaves l_1 Non-leaf vertices with points and lines
and l_2 with points and lines

Fig. 1. Components of point-line game graphs

is satisfiable, and substituting the satisfying assignment into the $\mathrm{Res}(k)$ refutation from Corollary 8 yields the required refutation of $\langle \mathrm{Win}_0(E^\sigma, R^\sigma_{\min})\rangle_G$ of size $p(n)$.   □

Corollary 10. *If resolution is weakly automatizable, then parity games can be decided in polynomial time.*   □

3   A Game Equivalent to Resolution

In this section we will define the *point-line game* and prove the following:

Theorem 11. *The canonical pair for the point-line game is equivalent to the canonical pair for resolution.*

An instance of the point-line game is given by a finite directed acyclic graph (V, E) with some extra structure. Namely, the set V is the disjoint union of sets V_0, V_1 and F, where vertices in V_0 and V_1 belong respectively to player 0 and player 1, and F contains exactly the leaf vertices, that is, those of out-degree 0. There is a designated start vertex s of in-degree 0. Each vertex v contains a set S_v of *points*. The start vertex is empty (contains no points) and every leaf contains exactly one point. Vertices do not share points. If there is an edge (u, v) in E, then some points in u may be connected to some points in v by *lines*. A point in u may have lines out to many points in v, but each point in v has a line in from at most one point in u, as in Fig. 1. During the game some points will be assigned *colours*, either black, for player 0, or white, for player 1.

The game starts with a pebble on s. At the beginning of a general turn, the pebble is on some vertex u and every point in u has a colour. As before, the player who owns vertex u moves the pebble along an outgoing edge to a new vertex v. Every point p in v that is connected by a line to some point q in u is then coloured with q's colour. Every other point in v is coloured with the colour of the player who did not move. The game ends when the pebble reaches a leaf w. The winner is the player whose colour is on the single point in w.

As before, a *positional strategy* is a function $\sigma \colon V_0 \to V$ or $\tau \colon V_1 \to V$ assigning a choice of outgoing edge to each of a player's vertices, regardless of the history of the game or the colouring of the current vertex. However in this case, it is not in general true that a winning strategy exists if and only if a positional winning strategy exists. One can give an example of such a game in which neither player has a positional winning strategy, while at the same time one of the players must, as in any finite game, have a (non-positional) winning strategy.

Lemma 12. *Given such a game G and a positional strategy σ for player 0, it is decidable in polynomial time whether σ is a winning strategy. Hence the canonical pair for point-line games is a disjoint NP pair.*

Proof. We describe a polynomial time algorithm which, working backwards from the leaves, labels each vertex u with either a set $B_u \subseteq S_u$ of points or a symbol "$Losing_0$". This labelling will have the property that if u is labelled "$Losing_0$" then, regardless of the colouring of u, if the pebble reaches u then player 1, playing optimally, will win the game if player 0 plays according to σ. If u is not labelled "$Losing_0$" then if player 0 plays according to σ and player 1 plays optimally, player 0 will win the game from u if and only if all points in B_u are coloured black. Thus σ is a winning strategy for player 0 if and only if the start vertex s is not labelled "$Losing_0$".

The algorithm labels a vertex u using the following rules.

1. If u is a leaf, set B_u to be the (unique) point in u.
2. If $u \in V_1$, suppose that u has children $v_1, \ldots, v_k$ and that these have all been labelled. If any child v_i is labelled "$Losing_0$", then label u as "$Losing_0$". Otherwise, let B_u contain every point in u which is connected by a line to some point in B_{v_i} for some child v_i (in other words, let B_u be the union of the pre-images of the sets B_{v_i}).
3. If $u \in V_0$, let $v = \sigma(u)$. Suppose that v has been labelled. If v is labelled "$Losing_0$" then label u as "$Losing_0$". If not, there are two possibilities. If there is a point in B_v that is not connected by a line to any point in u, label u as "$Losing_0$". Otherwise, let B_u be the set of points of u which are connected by a line to some point in B_v.   □

Theorem 13. *The canonical pair for the point-line game is reducible to the canonical pair for $Res(k)$ for some $k \in \mathbb{N}$, and hence to the canonical pair for resolution by Proposition 4.*

Proof. (Sketch.) We can write a formula Win_0 which is satisfiable if and only if there is a strategy σ for player 0 and a corresponding labelling of the graph, as in the previous lemma, in which no leaf reachable from s under σ is labelled "$Losing_0$". We can write a similar formula Win_1 wrt. a strategy τ for player 1 and a corresponding labelling. The proof that Win_0 and Win_1 cannot be satisfied simultaneously is then essentially a proof that the labelling algorithm works. We prove, working from the leaves of the graph down to s, that if any node v is reachable from s under both σ and τ, then $B_v^\sigma \cap W_v^\tau$ is non-empty, where W^τ is player 1's version of the relation B^σ and represents points that must be coloured white for player 1 to win using strategy τ. This gives a contradiction when we reach s, which contains no points.

This argument formalises as a U_2 induction (we also need to add relations R^σ and R^τ, for reachability under σ and τ, respectively to Win_0 and Win_1, as in the previous section). Thus, it translates into a $\text{Res}(k)$ refutation, which gives us our result, as in Corollaries 8 and 9. □

The other direction of Theorem 11 can be proven by showing that the interpolation pair for PK_1, which is equivalent to the canonical pair for resolution by Proposition 4, is reducible to the canonical pair for the game.

Theorem 14. *The interpolation pair of* PK_1 *is reducible to the canonical pair for the point-line game.*

Proof. (Sketch.) Starting from a PK_1-refutation of two sets of clauses Φ and Ψ in disjoint sets of variables X and Y, we can construct in polynomial time a game G such that if Φ is satisfiable then player 0 has a positional winning strategy in G, and if Ψ is satisfiable then player 1 has such a strategy.

The game has one vertex for each DNF that forms a line in the proof, and that vertex contains one point for each conjunction in the DNF. Additionally it has one leaf vertex for each literal z arising from a variable in $X \cup Y$. Each such leaf vertex contains a single point.

The structure of the game is similar to that of the proof. The edges reflect the structure of the proof, and two points are connected by a line if the corresponding conjunctions stand in a natural direct ancestor relation. The vertices corresponding to cut and $\wedge$-introduction rules belong to player 0 if an X variable is involved in the rule, and to player 1 if it is a Y variable. Vertices corresponding to clauses from Φ belong to player 0, similarly for Ψ and player 1.

The game is constructed so that the following is true. Suppose player 0 knows an assignment A to the X variables that satisfies Φ. Then he can use A to make choices in the game guaranteeing that, whenever the pebble moves to a non-leaf vertex u, then for every point p in u which corresponds to a conjunction whose X-literals are all satisfied by A, p gets coloured black. This means that when the game reaches a node corresponding to an initial clause of the proof, then if the clause is from Φ at least one point will be black, and if it is from Ψ then all the points will be black. Either way, player 0 will win. We have the symmetrical property for player 1. □

A question motivated by our results is to find a direct reduction of parity games to point-line games with positional strategies. Using such a reduction one may be able to define a subclass of point-line games that always have positional strategies, for which one could try to find a polynomial time algorithm instead of working directly with parity games.

References

1. Alekhnovich, M., Razborov, A.A.: Resolution is not automatizable unless W[P] is tractable. SIAM J. Comput. 38(4), 1347–1363 (2008)
2. Atserias, A., Bonet, M.L.: On the automatizability of resolution and related propositional proof systems. Inform. and Comput. 189(2), 182–201 (2004)

3. Atserias, A., Maneva, E.: Mean-payoff games and propositional proofs. Inform. and Comput. 209(4), 664–691 (2011)
4. Buresh-Oppenheim, J., Morioka, T.: Relativized NP search problems and propositional proof systems. In: Proceedings of the 19th IEEE Annual Conference on Computational Complexity, pp. 54–67. IEEE (2004)
5. Buss, S.R.: Bounded arithmetic, Studies in Proof Theory. Lecture Notes, vol. 3. Bibliopolis, Naples (1986)
6. Condon, A.: On algorithms for simple stochastic games. In: Advances in Computational Complexity Theory (New Brunswick, NJ, 1990). DIMACS Ser. Discrete Math. Theoret. Comput. Sci, vol. 13, pp. 51–71. Amer. Math. Soc., Providence (1993)
7. Ehrenfeucht, A., Mycielski, J.: Positional strategies for mean payoff games. Internat. J. Game Theory 8(2), 109–113 (1979)
8. Emerson, E.A.: Automata, tableaux, and temporal logics. In: Parikh, R. (ed.) Logic of Programs 1985. LNCS, vol. 193, pp. 79–88. Springer, Heidelberg (1985)
9. Friedmann, O.: An exponential lower bound for the latest deterministic strategy iteration algorithms. Log. Methods Comput. Sci. 7(3), 3:19, 42 (2011)
10. Friedmann, O.: Recursive algorithm for parity games requires exponential time. RAIRO Theor. Inform. Appl. 45(4), 449–457 (2011)
11. Grädel, E., Thomas, W., Wilke, T. (eds.): Automata, Logics, and Infinite Games. LNCS, vol. 2500. Springer, Heidelberg (2002)
12. Huang, L., Pitassi, T.: Automatizability and simple stochastic games. In: Aceto, L., Henzinger, M., Sgall, J. (eds.) ICALP 2011, Part I. LNCS, vol. 6755, pp. 605–617. Springer, Heidelberg (2011)
13. Juba, B.: On the Hardness of Simple Stochastic Games. Master's thesis, Carnegie Mellon University (2005)
14. Krajíček, J.: Lower bounds to the size of constant-depth Frege proofs. J. Symbolic Logic 59, 73–86 (1994)
15. Krajíček, J.: Interpolation theorems, lower bounds for proof systems, and independence results for bounded arithmetic. J. Symbolic Logic 62, 457–486 (1997)
16. Krajíček, J.: On the weak pigeonhole principle. Fund. Math. 170(1-2), 123–140 (2001)
17. Paris, J., Wilkie, A.: Counting problems in bounded arithmetic. In: Di Prisco, C.A. (ed.) Methods in Mathematical Logic. LNCS, vol. 1130, pp. 317–340. Springer, Heidelberg (1996)
18. Pudlák, P.: On reducibility and symmetry of disjoint NP pairs. Theoret. Comput. Sci. 295(1-3), 323–339 (2003)
19. Puri, A.: Theory of hybrid systems and discrete event structures. Ph.D. thesis, University of California, Berkeley (1995)
20. Razborov, A.A.: On provably disjoint NP-pairs. Tech. Rep. RS-94-36, Basic Research in Computer Science Center, Aarhus, Denmark (November 1994)
21. Stirling, C.: Modal and Temporal Properties of Processes. Texts in Computer Science. Springer (2001)
22. Wilmers, G.: Bounded existential induction. J. Symbolic Logic 50(1), 72–90 (1985)
23. Zwick, U., Paterson, M.: The complexity of mean payoff games on graphs. Theoret. Comput. Sci. 158(1-2), 343–359 (1996)

Logic and Branching Automata

Nicolas Bedon

LITIS (EA CNRS 4108) – Université de Rouen – France
Nicolas.Bedon@univ-rouen.fr

Abstract. The first result presented in this paper is the closure under complementation of the class of languages of finite N-free posets recognized by branching automata. Relying on this, we propose a logic, named *Presburger-MSO* or *P-MSO* for short, precisely as expressive as branching automata. The P-MSO theory of the class of all finite N-free posets is decidable.

Keywords: N-free posets, series-parallel posets, sp-rational languages, automata, commutative monoids, monadic second-order logic, Presburger logic.

1 Introduction

In computer science, if Kleene automata, or equivalently, rational expressions or finite monoids, are thought of as models of sequential programs, then introducing commutativity allows access to models of programs with permutation of instructions, or to concurrent programming. Among the formal tools for the study of commutativity in programs, let us mention for example Mazurkiewicz's traces, integer vector automata or commutative monoids.

In this paper, we are interested in another approach: the *branching automata*, introduced by Lodaya and Weil [13–16]. Branching automata are a generalisation of Kleene automata for languages of words to languages of finite N-free posets. This class of automata takes into account both sequentiality and the fork-join notion of parallelism, in which an execution flow f that splits into $f_1, \ldots, f_n$ concurrent execution flows, joins $f_1, \ldots, f_n$ before it continues. Divide-and-conquer concurrent programming naturally uses this fork-join principle. Lodaya and Weil generalized several important results of the theory of Kleene automata to branching automata, for example, a notion of rational expression with the same expressivity as branching automata. They also investigated the question of the algebraic counterpart of branching automata: the sp-algebras are sets equipped with two different associative products, one of them being also commutative. Contrary to the theory of Kleene automata, branching automata do not coincide any more with finite sp-algebras, and it is not known if the class of languages recognized by branching automata is closed under complementation.

An interesting particular case is the bounded-width rational languages [15], where the cardinality of the antichains of the posets of languages are bounded by an integer n. They correspond to fork-join models of concurrent programs with n

K. Chatterjee and J. Sgall (Eds.): MFCS 2013, LNCS 8087, pp. 123–134, 2013.
© Springer-Verlag Berlin Heidelberg 2013

as the upper bound of the number of execution flows (n is the number of physical processors). Bounded-width rational languages have a natural characterisation in rational expressions, branching automata, and sp-algebras. Taking into account those characterisations, the expressiveness of branching automata corresponds exactly to the finite sp-algebras. Furthermore, Kuske [12] proved that in this case, branching automata coincide also with monadic second-order logic, as it is the case for the rational languages of finite words. As in the general case monadic-second order logic is less expressive than branching automata, the question of an equivalent logic was left open.

This paper contains two new results:

1. first, the closure under complementation of the class of rational languages (Theorem 3);
2. second, we define a logic, named *P-MSO logic*, which basically is monadic second-order logic enriched with Presburger arithmetic, that is exactly as expressive as branching automata (Theorem 6).

The paper is organized as follows. Section 2 recalls basic definitions on posets. Section 3 is devoted to branching automata and rational expressions. Finally P-MSO is presented in Section 4.

All the proofs of the results of this paper are effective. As a consequence, the P-MSO theory of the class of finite N-free posets is decidable.

2   Notation and Basic Definitions

Let E be a set. We denote by $\mathcal{P}(E)$, $\mathcal{P}^+(E)$ and $\mathcal{M}^{>1}(E)$ respectively the set of subsets of E, the set of non-empty subsets of E and the set of multi-subsets of E with at least two elements. For any integer n, the group of permutations of $\{1, \ldots, n\}$ is denoted by S_n. The cardinality of E is denoted by $|E|$.

A *poset* $(P, <_P)$ is composed of a set P equipped with a partial ordering $<_P$. In this paper we consider only finite posets. For simplicity, by *poset* we always mean *finite* poset. A *chain* of length n in P is a sequence $p_1 <_P \cdots <_P p_n$ of elements of P. An *antichain* E in P is a set of elements of P mutually incomparable for $<_P$. The *width* of P is the size of a maximal antichain of P. An *alphabet* is a finite set whose elements are called *letters*. A poset $(P, <_P, \rho)$ *labelled* by A is composed of a poset $(P, <_P)$ and a map $\rho : P \to A$ which associates a letter A with any element of P. Observe that the posets of width 1 labelled by A correspond precisely to the usual finite words: finite totally ordered sequences of letters. Throughout this paper, we use labelled posets as a generalisation of words. In order to lighten the notation we write P for $(P, <_P, \rho)$ when no confusion is possible. The unique empty poset is denoted by ϵ.

Let $(P, <_P, \rho_P)$ and $(Q, <_Q, \rho_Q)$ be two disjoint posets labelled respectively by the alphabets A and A'. The *parallel product* of P and Q, denoted $P \parallel Q$, is the set $P \cup Q$ equipped with the orderings on P and Q such that the elements of P and Q are incomparable, and labelled by $A \cup A'$ by preservation of the labels from P and Q. It is defined as $(P \cup Q, <, \rho)$ where $\rho(x) = \rho_P(x)$ if $x \in P$,

$\rho(x) = \rho_Q(x)$ if $x \in Q$, and $x < y$ if and only if $(x, y \in P$ and $x <_P y)$ or $(x, y \in Q$ and $x <_Q y)$.

The *sequential product* of P and Q, denoted by $P \cdot Q$ or PQ for simplicity, is the poset $(P \cup Q, <, \rho)$ labelled by $A \cup A'$, such that $\rho(x) = \rho_P(x)$ if $x \in P$, $\rho(x) = \rho_Q(x)$ if $x \in Q$, and $x < y$ if and only if one of the following conditions is true:

- $x \in P$, $y \in P$ and $x <_P y$;   - $x \in P$ and $y \in Q$
- $x \in Q$, $y \in Q$ and $x <_Q y$;

Observe that the parallel product is an associative and commutative operation on posets, whereas the sequential product does not commute (but is associative). The parallel and sequential products can be generalized to finite sequences of posets. Let $(P_i)_{i \leq n}$ be a sequence of posets. We denote by $\prod_{i \leq n} P_i = P_0 \cdots \cdot P_n$ and $\|_{i \leq n} P_i = P_0 \| \cdots \| P_n$.

The class of *series-parallel* posets, denoted SP, is defined as the smallest set containing the posets with zero and one element and closed under finite parallel and sequential product. It is well known that this class corresponds precisely to the class of N-free posets [22, 23], in which the exact ordering relation between any four elements x_1, x_2, x_3, x_4 cannot be $x_1 < x_2$, $x_3 < x_2$ and $x_3 < x_4$. The class of series-parallel posets over an alphabet A is denoted $SP(A)$ (or $SP^+(A)$ when the empty poset is not considered).

A *block* B of a poset $(P, <)$ is a nonempty subset of P such that, if $b, b' \in B$ such that $b < b'$, then for all elements of $p \in P$, if $b \leq p \leq b'$ then $p \in B$. We say that B is *connected* if, for any different and incomparable $b, b' \in B$ there exists $b'' \in B$ such that $b, b' \leq b''$ or $b'' \leq b, b'$. A subset G of P is *good* if, for all $p \in P$, if p is comparable to an element of G and incomparable to another, then $p \in G$.

3   Rational Languages and Automata

A *language* over an alphabet A is a subset of $SP(A)$. The sequential and parallel product of labelled posets can naturally be extended to languages. If $L_1, L_2 \subseteq SP(A)$, then $L_1 \cdot L_2 = \{P_1 \cdot P_2 \mid P_1 \in L_1, P_2 \in L_2\}$ and $L_1 \| L_2 = \{P_1 \| P_2 \mid P_1 \in L_1, P_2 \in L_2\}$.

3.1   Rational Languages

Let A and B be two alphabets and let $P \in SP(A)$, $L \subseteq SP(B)$ and $\xi \in A$. We define the language $L \circ_\xi P$ of posets labelled by $A \cup B$ by substituting non-uniformly in P each element labelled by ξ by a labelled poset of L. This substitution $L\circ_\xi$ is the homomorphism from $(SP(A), \|, \cdot)$ into the powerset algebra $(\mathcal{P}(SP(A \cup B)), \|, \cdot)$ with $a \mapsto \{a\}$ for all $a \in A$, $a \neq \xi$, and $\xi \mapsto L$. It can be easily extended from labelled posets to languages of posets. Using this, we define the substitution and the iterated substitution on languages. By the way the usual Kleene rational operations [11] are recalled. Let L and L' be languages of $SP(A)$:

$$L \circ_\xi L' = \bigcup_{P \in L'} L \circ_\xi P$$

$$L^{*\xi} = \bigcup_{i \in \mathbb{N}} L^{i\xi} \text{ with } L^{0\xi} = \{\xi\} \text{ and } L^{(i+1)\xi} = \left(\bigcup_{j \leq i} L^{j\xi}\right) \circ_\xi L$$

$$L^* = \left\{\prod_{i<n} P_i : n \in \mathbb{N}, P_i \in L\right\}$$

A language $L \subseteq SP(A)$ is *rational* if it is empty, or obtained from the letters of the alphabet A using usual rational operators : finite union $\cup$, finite concatenation $\cdot$, and finite iteration $*$, and using also the finite parallel product $\|$, substitution $\circ_\xi$ and iterated substitution $*\xi$, provided that in $L^{*\xi}$ any element labelled by ξ in a labelled poset $P \in L$ is incomparable with another element of P. This latter condition excludes from the rational languages those of the form $(a\xi b)^{*\xi} = \{a^n \xi b^n : n \in \mathbb{N}\}$, for example, which are known to be not Kleene rational. Observe also that the usual Kleene rational languages are a particular case of the rational languages defined above, in which the operators $\|$, $\circ_\xi$ and $*\xi$ are not used.

Example 1. Let $A = \{a, b, c\}$ and $L = c \circ_\xi (a \| (b\xi))^{*\xi}$. Then L is the smallest language containing c and such that if $p \in L$, then $a \| (bx) \in L$.

$$L = \{c, a \| (bc), a \| (b(a \| (bc))), \dots\}$$

Let L be a language where the letter ξ is not used. In order to lighten the notation we use the following abreviation:

$$L^\circledast = \{\epsilon\} \circ_\xi (L \| \xi)^{*\xi} = \{\|_{i<n} P_i : n \in \mathbb{N}, P_i \in L\}$$

L^* is the sequential iteration of L whereas $L^\circledast$ is its parallel iteration.

A language L is $\|$-*rational* if it is rational without using the operators $\cdot$, $\circ_\xi$, $*$ and $*\xi$ (but $\circledast$ is allowed).

Remark 1. Any rational language L which does not make use of sequentiality (i.e. $PP' \notin L$ for all $P, P' \in SP^+(A)$) is $\|$-rational.

A subset L of $A^\circledast$ is *linear* if it has the form

$$L = a_1 \| \cdots \| a_k \| \left(\cup_{i \in I}(a_{i,1} \| \cdots \| a_{i,k_i})\right)^\circledast$$

where the a_i and $a_{i,j}$ are elements of A and I is a finite set. It is *semi-linear* if it is a finite union of linear sets. We refer to [5] for a proof of the following result:

Theorem 1. *Let A be an alphabet and $L \subseteq A^\circledast$. Then L is $\|$-rational if and only if it is semi-linear.*

3.2   Branching Automata

Branching automata are a generalisation of usual Kleene automata. They were introduced by Lodaya and Weil [13–15].

A *branching automaton* (or just *automaton* for short) over an alphabet A is a tuple $\mathcal{A} = (Q, A, E, I, F)$ where Q is a finite set of states, $I \subseteq Q$ is the set of *initial states*, $F \subseteq Q$ the set of *final states*, and E is the set of *transitions* of $\mathcal{A}$. The set of transitions of E is partitioned into $E = (E_{seq}, E_{fork}, E_{join})$, according to the different kinds of transitions:

- $E_{seq} \subseteq (Q \times A \times Q)$ contains the *sequential* transitions, which are usual transitions of Kleene automata;
- $E_{fork} \subseteq Q \times \mathcal{M}^{>1}(Q)$ and $E_{join} \subseteq \mathcal{M}^{>1}(Q) \times Q$ are respectively the sets of *fork* and *join* transitions.

Sequential transitions $(p, a, q) \in Q \times A \times Q$ are sometimes denoted by $p \xrightarrow{a} q$.

We now turn to the definition of paths in automata. The definition we use in this paper is different, but equivalent to, the one of Lodaya and Weil [13–16]. Paths in automata are posets labelled by transitions. A *path* γ from a state p to a state q is either the empty poset (in this case $p = q$), or a non-empty poset labelled by transitions, with a unique minimum and a unique maximum element. The minimum element of γ is mapped either to a sequential transition of the form (p, a, r) for some $a \in A$ and $r \in Q$ or to a fork transition of the form (p, R) for some $R \in \mathcal{M}^{>1}(Q)$. Symmetrically, the maximum element of γ is mapped either to a sequential transition of the form (r', a, q) for some $a \in A$ and $r' \in Q$ or to a join transition of the form (R', q) for some $R' \in \mathcal{M}^{>1}(Q)$. The states p and q are respectively called *source* (or *origin*) and *destination* of γ. Two paths γ and γ' are *consecutive* if the destination of γ is also the source of γ'. Formally, the paths γ labelled by $P \in SP(A)$ in $\mathcal{A}$ are defined by induction on the structure of P:

- the empty poset ϵ is a path from p to p, labelled by $\epsilon \in SP(A)$, for all $p \in Q$;
- for any transition $t = (p, a, q)$, then t is a path from p to q, labelled by a;
- for any finite set of paths $\{\gamma_0, \ldots, \gamma_k\}$ (with $k > 1$) respectively labelled by $P_0, \ldots, P_k$, from $p_0, \ldots, p_k$ to $q_0, \ldots, q_k$, if $t = (p, \{p_0, \ldots, p_k\})$ is a fork transition and $t' = (\{q_0, \ldots, q_k\}, q)$ a join transition, then $\gamma = t(\|_{j \leq k} \gamma_j)t'$ is a path from p to q and labelled by $\|_{j \leq k} P_j$;
- for any non-empty finite sequence $\gamma_0, \ldots, \gamma_k$ of consecutive paths respectively labelled by $P_0, \ldots, P_k$, then $\prod_{j<k+1} \gamma_j$ is a path labelled by $\prod_{j<k+1} P_j$ from the source of γ_0 to the destination of γ_k;

Observe that paths are labelled posets of three different forms: ϵ, t or tPt' for some transitions t, t' and some labelled poset P. In an automaton $\mathcal{A}$, a path γ from p to q labelled by $P \in SP(A)$ is denoted by $\gamma : p \xRightarrow[\mathcal{A}]{P} q$. A state s is a *sink* if s is the destination of any path originating in s.

A labelled poset is *accepted* by an automaton if it is the label of a path, called *successful*, leading from an initial state to a final state. The language $L(\mathcal{A})$ is the set of labelled posets accepted by the automaton $\mathcal{A}$. A language L is *regular* if there exists an automaton $\mathcal{A}$ such that $L = L(\mathcal{A})$.

Theorem 2 (Lodaya and Weil [13]). *Let A be an alphabet, and $L \subseteq SP(A)$. Then L is regular if and only if it is rational.*

Example 2. Figure 1 represents the automaton $\mathcal{A} = (\{1,2,3,4,5,6\}, \{a,b\}, E, \{1\}, \{1,6\})$, with $E_{seq} = \{(2,a,4),(3,b,5)\}$, $E_{fork} = \{(1,\{1,1\}),(1,\{2,3\})\}$ and $E_{join} = \{(\{6,6\},6),(\{4,5\},6)\}$, and an accepting path labelled by $a \parallel b \parallel a \parallel b$. Actually, $L(\mathcal{A}) = (a \parallel b)^{\circledast}$.

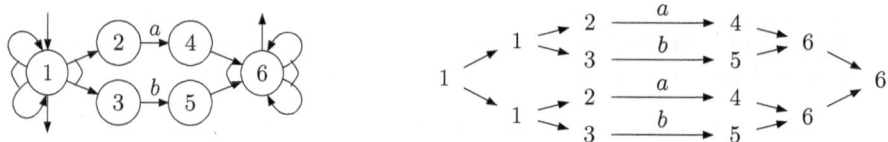

Fig. 1. An automaton $\mathcal{A}$ with $L(\mathcal{A}) = (a \parallel b)^{\circledast}$ and an accepting path labelled by $a \parallel b \parallel a \parallel b$

It is known from Lodaya and Weil [15] that the regular languages of $SP(A)$ are closed under finite union and finite intersection, but the closure under complementation was still unexplored.

The first result of this paper is stated by the following Theorem which implies that the class of regular languages of N-free posets is closed under boolean operations.

Theorem 3. *Let A be an alphabet. The class of regular languages of $SP(A)$ is effectively closed under complement.*

The proof relies on an algebraic approach of regular languages, which was first introduced by Lodaya and Weil [13–15]. Algebras considered here are of the form $(S, \cdot, \parallel)$ (or just S for short) such that $(S, \cdot)$ and $(S, \parallel)$ are respectively a semigroup and a commutative semigroup, which may be infinites. The first step consists in the construction of a morphism $\varphi : SP(A) \to S$, where S is build from an automaton $\mathcal{A}$ and $L(\mathcal{A}) = \varphi^{-1}(X)$ for some $X \subseteq S$. Then we show that $\varphi^{-1}(S - X)$ is regular by a reduction of the problem to the finitely generated commutative semigroup case, and we conclude by the use of the following result

Theorem 4 (Eilenberg and Schützenberger [5]). *If X and Y are rational subsets of a commutative monoid M, then $Y - X$ is also a rational subset of M.*

As emphasized in [18], if M is finitely generated then Theorem 4 is effective.

4 P-MSO

In this section we define a logical formalism called P-MSO, which is a mix between Presburger [17] and monadic second-order logic, and that has exactly the same expressivity as branching automata. As all the constructions involved in the proof are effective, then the P-MSO theory of the class of finite N-free posets is decidable.

Let us recall useful elements of monadic second-order logic, and settle some notation. For more details about MSO logic we refer e.g. to Thomas' survey paper [4, 20]. The monadic second-order (MSO) logic is classical in set theory, and was first set up by Büchi-Elgot-Trakhtenbrot for words [2, 6, 21]. In our case, the domain of interpretation is the class of finite N-free posets.

Monadic second-order logic is an extension of first-order logic that allows to quantify over elements as well as subsets of the domain of the structure. A MSO-formula is given by the following grammar

$$\psi ::= R_a(x) \mid x \in X \mid x < y \mid \psi_1 \vee \psi_2 \mid \psi_1 \wedge \psi_2 \mid \neg\psi$$
$$\mid \exists x \psi \mid \exists X \psi \mid \forall x \psi \mid \forall X \psi$$

where $a \in A$, x, y and X are respectively first- and second-order variables, $R_a(x)$ is interpreted as "x is labelled by a" (also denoted $a(x)$ for readability), and all other symbols have their usual meaning. The language L_ψ of ψ is the class of posets $(P, <, \rho)$ labelled over A that satisfy ψ. Logical equivalence of formulæ corresponds to the equality of their languages. In order to enhance readability of formulæ we use several notations and abbreviations for properties expressible in MSO. The following are usual and self-understanding: $\phi \to \psi$, $X \subseteq Y$, $x = y$. We also write $\exists^X x \psi$ for $\exists x \ x \in X \wedge \psi$, and extend this notion of *relative quantification* to universal quantification and second-order variables. MSO logic is strictly less expressive than automata. There is no MSO-formula that defines the language $(a \parallel b)^\circledast$. On the contrary, MSO-definability implies rationality.

In order to capture the expressiveness of automata with logic we need to add Presburger expressivity to MSO. Presburger logic is the first-order logic over the structure $(\mathbb{N}, +)$ where $+ = \{(a, b, c) : a + b = c\}$. A language $L \subseteq \mathbb{N}^n$ is a *Presburger set* of $\mathbb{N}^n$ if $L = \{(x_1, \ldots, x_n) : \varphi(x_1, \ldots, x_n) \text{ is true }\}$ for some Presburger formula $\varphi(x_1, \ldots, x_n)$. If $\varphi(x_1, \ldots, x_n)$ is given then L is called the *Presburger set* of $\varphi(x_1, \ldots, x_n)$ (or of φ for short). Presburger logic provides tools to manipulate semi-linear sets of $A^\circledast$ with formulæ. Indeed, let $A = \{a_1, \ldots, a_n\}$ be an alphabet ($n > 0$). As a word u of $A^\circledast$ can be thought of as a n-tuple $(|u|_{a_1}, \ldots, |u|_{a_n})$ of non-negative integers, where $|u|_a$ denotes the number of occurences of letter a in u, then $A^\circledast$ is isomorphic to $\mathbb{N}^n$.

Example 3. Let $A = \{a, b, c\}$ and $L = \{u \in A^\circledast : |u|_a \leq |u|_b \leq |u|_c\}$. Then L is isomorphic to $\{(n_a, n_b, n_c) \in \mathbb{N}^3 : n_a \leq n_b \leq n_c\}$, and thus the Presburger set of

$$\varphi(n_a, n_b, n_c) \equiv (\exists x \ n_b = n_a + x) \wedge (\exists y \ n_c = n_b + y)$$

Semi-linear sets and Presburger sets are connected by the following Theorem:

Theorem 5 (Ginsburg and Spanier [8], Theorem 1.3). *Let A be an alphabet and $L \subseteq A^\circledast$. Then L is semi-linear if and only if it is a Presburger set. Furthermore, the construction of one description from the other is effective.*

The *P-MSO logic* is a melt of Presburger and MSO logics. From the syntactic point of view, P-MSO logic contains MSO logic, and in addition formulæ of the form

$$Q(Z, (\psi_1(R_1), x_1), \ldots, (\psi_n(R_n), x_n), \varphi(x_1, \ldots, x_n))$$

where Z is the name of a (free) second-order variable, $\psi_i(R_i)$ (for each $i \in 1 \ldots n$) a P-MSO formula having no free first-order variables, and only quantifications relative to R_i, and $\varphi(x_1, \ldots, x_n)$ a Presburger formula with n free variables $x_1, \ldots, x_n$. Considering the formula $\psi(Z) = Q(Z, (\psi_1(R_1), x_1), \ldots, (\psi_n(R_n), x_n), \varphi(x_1, \ldots, x_n))$ the only variable that counts as *free* in $\psi(Z)$ is Z. Note that as n can be any positive integer then P-MSO does not really fit into the framework of usual formal propositional logic (where the arity of connectors are usually fixed).

As in monadic second-order logic, the class of syntactically correct P-MSO formulæ is closed under boolean operations, and existential and universal quantification over first and second-order variables of a P-MSO formula that are interpreted over elements or sets of elements of the domain of the structure. Semantics of P-MSO formulæ is defined below by extension of semantics of Presburger and MSO logics. The notions of a language and definability naturally extend from MSO to P-MSO.

Before continuing with formal definitions, let us give some intuition on the meaning of $\psi(Z) = Q(Z, (\psi_1(R_1), x_1), \ldots, (\psi_n(R_n), x_n), \varphi(x_1, \ldots, x_n))$. Let X be an interpretation of a second-order variable Z in P, such that X is a good block of P. That means, X is the poset associated with a sub-term of a term on A (a full binary tree whose leaves are elements of A, and nodes are a sequential or a parallel product) describing P, and is the parallel composition of $m \geq 1$ connected blocks: $X = X_1 \parallel \cdots \parallel X_m$. Take n different colors $c_1, \ldots, c_n$. To each X_i we associate a color c_j with the condition that X_i satisfies $\psi_j(X_i)$. Observe that this coloring may not be unique, and may not exist. Denote by x_j the number of uses of c_j in the coloring of X. Then $P, X \models \psi(Z)$ if there exists such a coloring with $x_1, \ldots, x_n$ satisfying the Presburger condition $\varphi(x_1, \ldots, x_n)$.

More formally, let $P \in SP(A)$, $\psi(Z) = Q(Z, (\psi_1(R_1), x_1), \ldots, (\psi_n(R_n), x_n), \varphi(x_1, \ldots, x_n))$ be a P-MSO formula, $X \subseteq P$ be an interpretation of Z in P such that X is a good block of P. Then $P, X \models \psi(Z)$ if there exist non negative integers $v_1, \ldots, v_n$ and a partition $(Z_{1,1}, \ldots, Z_{1,v_1}, \ldots, Z_{n,1}, \ldots, Z_{n,v_n})$ of X into connected blocks $Z_{i,j}$ such that

- $(v_1, \ldots, v_n)$ belongs to the Presburger set of $\varphi(x_1, \ldots, x_n)$,
- $z \in Z_{i,j}$, $z' \in Z_{i',j'}$ implies that z and z' are incomparable, for all possible (i, j) and (i', j') with $(i, j) \neq (i', j')$,
- $P, Z_{i,j} \models \psi_i(Z_{i,j})$ for all $i \in 1 \ldots n$ and $j \in 1 \ldots v_i$.

Example 4. Let L be the language of Example 3, and $\varphi(n_a, n_b, n_c)$ be the Presburger formula of Example 3. For all $\alpha \in A$, set $\psi_\alpha(X) \equiv \mathtt{Card}_1(X) \wedge \forall^X x\, \alpha(x)$, where $\mathtt{Card}_1(X)$ is a MSO formula (thus a P-MSO formula) which is true if and only if the interpretation of X has cardinality 1. Then L is the language of the following P-MSO sentence:

$$\forall P \; (\forall p \; p \in P) \to Q(P, (\psi_a(X), n_a), (\psi_b(X), n_b), (\psi_c(X), n_c), \varphi(n_a, n_b, n_c))$$

Theorem 6. *Let A be an alphabet, and $L \subseteq SP(A)$. Then L is rational if and only if is P-MSO definable.*

The proof uses usual arguments adapted to the case of N-free posets.

The inclusion from left to right relies on the ideas of Büchi on words: the encoding of accepting paths of a branching automaton $\mathcal{A}$ into a P-MSO formula. Each letter of the poset is mapped to a sequential transition of $\mathcal{A}$, and each part of the poset of the form $P = P_1 \parallel \cdots \parallel P_n$ ($n > 1$), as great as possible relatively to inclusion and such that each P_i is a connected block of P, is mapped to a pair (p, q) of states; informally speaking, p and q are the states that are supposed to respectively begin and finish the part of the path labelled by P. The formula guarantees that pairs of states and sequential transitions are chosen consistently with the transitions of $\mathcal{A}$, and that, if $P = P_1 \parallel \cdots \parallel P_n$ as above and $p_i \overset{P_i}{\underset{\mathcal{A}}{\Longrightarrow}} q_i$ for all $i \in 1 \ldots n$, then there exists a combination of fork transitions that connects p to $p_1, \ldots, p_n$, a sequence of join transitions that connects $q_1, \ldots, q_n$ to q, such that a path $p \overset{P}{\underset{\mathcal{A}}{\Longrightarrow}} q$ in $\mathcal{A}$ is formed.

The inclusion from right to left relies on well-known techniques from words adapted to posets. In this part of the proof posets are not just labelled by elements of the alphabet A, but by elements of $A \times \mathcal{P}(V_1) \times \mathcal{P}(V_2)$, where V_1 and V_2 are sets that contain respectively the names of the free first and second-order variables of the formula (we do not consider here the variables that are interpreted over nonnegative integers). When formulæ are sentences, then the posets are labelled by $A \times \emptyset \times \emptyset$, which is similar to A. Observe that an interpretation of the variables $\{x_1, \ldots, x_n\} = V_1$, $\{X_1, \ldots, X_m\} = V_2$ in P induces a unique poset labelled by elements of $A \times \mathcal{P}(V_1) \times \mathcal{P}(V_2)$, and reciprocally. This allows us to use indifferently one representation or the other in order to lighten the notation. This labelling of posets by elements of $A \times \mathcal{P}(V_1) \times \mathcal{P}(V_2)$ has a unique restriction: the name of a free first-order variable x must appear at most once in the labels of elements of the poset. An automaton $\mathcal{A}_r$ that accepts a poset if and only if this condition is verified on its label can easily be constructed. We may assume, up to an intersection with $\mathcal{A}_r$ (the regular languages are closed under intersection), that all the constructions of automata below have posets in $L(\mathcal{A}_r)$ as inputs.

We build, by induction on the structure of $\varphi(x_1, \ldots, x_n, X_1, \ldots, X_m)$, an automaton $\mathcal{A}_\varphi$ such that $P, x_1, \ldots, x_n, X_1, \ldots, X_m \models \varphi(x_1, \ldots, x_n, X_1, \ldots, X_m)$ if and only if $P, x_1, \ldots, x_n, X_1, \ldots, X_m \in L(\mathcal{A}_\varphi)$. The case $n = m = 0$ gives the inclusion from right to left of Theorem 6. For formulæ of the form $x < y$ it suffices to build an automaton that checks if the poset has two elements p_1 and p_2 respectively labelled by (a_1, X_1, X_2) and (a_2, Y_1, Y_2) such that $p_1 < p_2$, $x \in X_1$ and $y \in y_1$. An automaton that checks if the poset contains an element labelled by (a, X_1, X_2) with $x \in X_1$ can easily be constructed for formulæ of the form $a(x)$. The case of formulæ of the form $x \in X$ is similar. Constructions of automata for the boolean connectors $\vee, \wedge$ and $\neg$ are a consequence of Theorem 3 and the closure under finite union and intersection of regular languages. For formulæ of the form $\exists x \phi$ or $\exists X \phi$, constructions are a consequence of the closure under projection of regular languages. We finally turn to the last case where the formula ψ has the form $Q(Z, (\psi_1(R_1), x_1), \ldots, (\psi_n(R_n), x_n), \varphi(x_1, \ldots, x_n))$. Recall here that $x_1, \ldots, x_n$ are variables that are interpreted over nonnegative integers, and that each ψ_i, $i \in 1 \ldots n$, has only one free variable R_i, which is second-order. By induction hypothesis, there is an automaton $\mathcal{A}_{\psi_i}$ such that $P, R \models \psi_i(R)$ if and only if $P, R \in L(\mathcal{A}_{\psi_i})$. According to the semantics of $Q(Z, (\psi_1(R_1), x_1), \ldots, (\psi_n(R_n), x_n), \varphi(x_1, \ldots, x_n))$, the only interpretations of R in P verify (1) $R = P$ and (2) P is a connected block. The conjunction of (1) and (2) is a MSO-definable property of R, and thus it can be checked by an automaton $\mathcal{B}$. As a consequence of the closure under intersection of regular languages there exists an automaton $\mathcal{A}'_{\psi_i}$ such that $L_i = L(\mathcal{A}'_{\psi_i}) = L(\mathcal{A}_{\psi_i}) \cap L(\mathcal{B})$. Now, let $B = \{b_1, \ldots, b_n\}$ be a new alphabet disjoint from A. As a consequence of Theorems 5, 1 and 2 there is an automaton $\mathcal{C}$ over the alphabet B such that $L(\mathcal{C})$ is the Presburger set of $\varphi(x_1, \ldots, x_n)$ over B. Then $L_\psi = L_1 \circ_{b_1} (\ldots (L_n \circ_{b_n} L(\mathcal{C})))$ thus L_ψ is regular according to Theorem 2.

Example 5. Let L be the language over the alphabet $A = \{a, b\}$ composed of the sequential products of posets of the form $P = P_1 \parallel \cdots \parallel P_n$ such that each P_i is a nonempty totally ordered poset (i.e., a word), and that the number of P_i that starts with an a is $\frac{2}{3}n$. Set $L_1 = aA^*$ and $L_2 = bA^*$. Then L is the language of the rational expression $((L_1 \parallel L_1 \parallel L_2)^{\circledast})^*$. We define L by a P-MSO sentence as follows. Given two elements of the poset denoted by first order variables x and y, one can easily write a MSO formula $\mathtt{Succ}(x, y)$ (resp. $\mathtt{Pred}(x, y)$) that is true if and only if x is a successor (resp. predecessor) of y. Set

$$\mathtt{Lin}(X) \equiv \forall^X x \forall^X y \forall^X z \, ((\mathtt{Succ}(y, x) \wedge \mathtt{Succ}(z, x)) \to y = z)$$
$$\wedge \, ((\mathtt{Pred}(y, x) \wedge \mathtt{Pred}(z, x)) \to y = z)$$
$$\psi_1(X) \equiv \mathtt{Lin}(X) \wedge \exists^X x \, a(x) \wedge \forall^X y \, x = y \vee x < y$$
$$\psi_2(X) \equiv \mathtt{Lin}(X) \wedge \exists^X x \, b(x) \wedge \forall^X y \, x = y \vee x < y$$
$$\varphi(n_a, n_b) \equiv n_a = 2n_b$$

Then L is the language of the following P-MSO sentence

$$\psi \equiv \forall P (\forall p \; p \in P) \to \exists X_1 \exists X_2 \; P = X_1 \oplus X_2$$
$$\wedge \, \forall U ((\mathtt{MaxBlock}(U, X_1) \vee \mathtt{MaxBlock}(U, X_2)) \to$$
$$Q(U, (\psi_1(R_1), n_a), (\psi_2(R_2), n_b), \varphi(n_a, n_b))$$

with $X = U \oplus V \equiv \mathtt{Partition}(U, V, X) \wedge (\forall u \forall v \; u \in U \wedge v \in V \to \neg u \parallel v)$. In the formula above, $\mathtt{Partition}(U, V, X)$ and $u \parallel v$ respectively express with MSO formulæ that (U, V) partitions X, and that u and v are different and not comparable. The MSO formula $\mathtt{MaxBlock}(U, X)$ express that U is a block of X, maximal relatively to inclusion.

5   Conclusion

As all the constructions involved in the proof of Theorem 6 are effective, and emptiness is decidable for languages of branching automata, P-MSO is decidable:

Theorem 7. *Let A be an alphabet. The P-MSO theory of $SP(A)$ is decidable.*

In [15], Lodaya and Weil asked for logical characterizations of several classes of rational languages. As it is equivalent to branching automata, P-MSO is the natural logic to investigate such questions, that are still open.

Among the works connected to ours, let us mention Esik and Németh [7], which itself has been influenced by the work of Hoogeboom and ten Pas [9, 10] on text languages. They study languages of biposets from an algebraic, automata and regular expressions based point of view, and the connections with MSO. A biposet is a set equipped with two partial orderings; thus, N-free posets are a generalisation of N-free biposets, where commutation is allowed in the parallel composition.

MSO and Presburger logic were also mixed in other works, but for languages of trees instead of N-free posets. Motivated by reasoning about XML documents, Dal Zilio and Lugiez [3], and independently Seidl, Schwentick and Muscholl [19], defined a notion of tree automata which combines regularity and Presburger arithmetic. In particular in [19], MSO is enriched with Presburger conditions on the children of nodes in order to select XML documents, and proved equivalent to unranked tree automata. Observe that unranked trees are a particular case of N-free posets. The logic named *Unordered Presburger MSO logic* in [19] is contained in our P-MSO logic.

The quality of this paper has been enhanced by the comments of the anonymous referees. One of them noticed that Theorem 3 might also be retrieved using the notion of Commutative Hedge automata (see e.g. [1]), as N-free posets can be assimilated to terms over the operations of parallel and sequential products.

References

1. Bouajjani, A., Touili, T.: On computing reachability sets of process rewrite systems. In: Giesl, J. (ed.) RTA 2005. LNCS, vol. 3467, pp. 484–499. Springer, Heidelberg (2005)
2. Richard Büchi, J.: Weak second-order arithmetic and finite automata. Zeit. Math. Logik. Grund. Math. 6, 66–92 (1960)
3. Dal-Zilio, S., Lugiez, D.: XML Schema, Tree Logic and Sheaves Automata. In: Nieuwenhuis, R. (ed.) RTA 2003. LNCS, vol. 2706, pp. 246–263. Springer, Heidelberg (2003)
4. Ebbinghaus, H.-D., Flum, J.: Finite model theory, 2nd edn. Springer monographs in mathematics. Springer (1999)
5. Eilenberg, S., Schützenberger, M.-P.: Rational sets in commutative monoids. Journal of Algebra 13(2), 173–191 (1969)
6. Elgot, C.C.: Decision problems of finite automata design and related arithmetics. Trans. Amer. Math. Soc. 98, 21–51 (1961)

7. Ésik, Z., Németh, Z.L.: Automata on series-parallel biposets. In: Kuich, W., Rozenberg, G., Salomaa, A. (eds.) DLT 2001. LNCS, vol. 2295, pp. 217–227. Springer, Heidelberg (2002)
8. Ginsburg, S., Spanier, E.H.: Semigroups, Presburger formulas, and languages. Pacific Journal of Mathematics 16(2), 285–296 (1966)
9. Hoogeboom, H.J., ten Pas, P.: Text languages in an algebraic framework. Fund. Inform. 25, 353–380 (1996)
10. Hoogeboom, H.J., ten Pas, P.: Monadic second-order definable languages. Theory Comput. Syst. 30, 335–354 (1997)
11. Kleene, S.C.: Representation of events in nerve nets and finite automata. In: Shannon, McCarthy (eds.) Automata Studies, pp. 3–42. Princeton University Press, Princeton (1956)
12. Kuske, D.: Infinite series-parallel posets: Logic and languages. In: Welzl, E., Montanari, U., Rolim, J.D.P. (eds.) ICALP 2000. LNCS, vol. 1853, pp. 648–662. Springer, Heidelberg (2000)
13. Lodaya, K., Weil, P.: A Kleene iteration for parallelism. In: Arvind, V., Sarukkai, S. (eds.) FST TCS 1998. LNCS, vol. 1530, pp. 355–367. Springer, Heidelberg (1998)
14. Lodaya, K., Weil, P.: Series-parallel posets: algebra, automata and languages. In: Morvan, M., Meinel, C., Krob, D. (eds.) STACS 1998. LNCS, vol. 1373, pp. 555–565. Springer, Heidelberg (1998)
15. Lodaya, K., Weil, P.: Series-parallel languages and the bounded-width property. Theoret. Comput. Sci. 237(1-2), 347–380 (2000)
16. Lodaya, K., Weil, P.: Rationality in algebras with a series operation. Inform. Comput. 171, 269–293 (2001)
17. Presburger, M.: Über die vollstandigkeit eines gewissen systems der arithmetic ganzer zahlen, in welchem die addition als einzige operation hervortritt. In: Proc. Sprawozdaniez I Kongresu Matematykow Krajow Slowianskich, Warsaw, pp. 92–101 (1930); English translation: On the completeness of certain system of arithmetic of whole numbers in which addition occurs as the only operation. Hist. Philos. Logic 12, 92–101 (1991)
18. Sakarovitch, J.: Éléments de théorie des automates. Vuibert (2003); English (and revised) version: Elements of automata theory. Cambridge University Press (2009)
19. Seidl, H., Schwentick, T., Muscholl, A.: Counting in trees. In: Flum, J., Grädel, E., Wilke, T. (eds.) Logic and Automata. Texts in Logic and Games, vol. 2, pp. 575–612. Amsterdam University Press (2008)
20. Thomas, W.: Languages, automata, and logic. In: Rozenberg, G., Salomaa, A. (eds.) Handbook of Formal Languages, vol. III, pp. 389–455. Springer (1997)
21. Trakhtenbrot, B.A.: Finite automata and monadic second order logic. Siberian Math. 3, 101–131 (1962) (Russian) English translation in AMS Transl. 59, 23–55 (1966)
22. Valdes, J.: Parsing flowcharts and series-parallel graphs. Technical Report STAN-CS-78-682, Computer science departement of the Stanford University, Standford, Ca (1978)
23. Valdes, J., Tarjan, R.E., Lawler, E.L.: The recognition of series parallel digraphs. SIAM J. Comput. 11, 298–313 (1982)

A Constant Factor Approximation
for the Generalized Assignment Problem
with Minimum Quantities and Unit Size Items

Marco Bender[1,*], Clemens Thielen[2], and Stephan Westphal[1]

[1] Institute for Numerical and Applied Mathematics, University of Goettingen,
Lotzestr. 16-18, D-37083 Goettingen, Germany
{m.bender,s.westphal}@math.uni-goettingen.de
[2] Department of Mathematics, University of Kaiserslautern,
Paul-Ehrlich-Str. 14, D-67663 Kaiserslautern, Germany
thielen@mathematik.uni-kl.de

Abstract. We consider a variant of the generalized assignment problem (GAP) where the items have unit size and the amount of space used in each bin is restricted to be either zero (if the bin is not opened) or above a given lower bound (a *minimum quantity*). This problem is known to be strongly NP-complete and does not admit a polynomial time approximation scheme (PTAS).

By using randomized rounding, we obtain a randomized 3.93-approximation algorithm, thereby providing the first nontrivial approximation result for this problem.

Keywords: generalized assignment problem, combinatorial optimization, approximation algorithms, randomized rounding.

1 Introduction

The generalized assignment problem (GAP) is a classical generalization of both the (multiple) knapsack problem and the bin packing problem. In the classical version of GAP (cf., for example, [1, 2]), one is given m bins, a capacity B_j for each bin j, and n items such that each item i has size $s_{i,j}$ and yields profit $p_{i,j}$ when packed into bin j. The goal is to find a feasible packing of the items into the bins that maximizes the total profit. The problem has many practical applications, for which we refer to [2] and the references therein.

Recently, Krumke and Thielen [3] introduced the generalized assignment problem with minimum quantities (GAP-MQ), which is a variation of the generalized assignment problem where the amount of space used in each bin is restricted to be either zero (if the bin is not opened) or above a given lower bound (a *minimum quantity*). This additional restriction is motivated from many practical

---

* Partially supported by DFG RTG 1703 "Resource Efficiency in Interorganizational Networks".

K. Chatterjee and J. Sgall (Eds.): MFCS 2013, LNCS 8087, pp. 135–145, 2013.

packing problems where it does often not make sense to open an additional container (bin) if not at least a certain amount of space in it will be used. While it is not hard to see that it is NP-hard to compute any feasible solution with positive profit for the general version of GAP-MQ (and, hence, no polynomial time approximation algorithm exists for the problem unless $P = NP$), computing nontrivial feasible solutions is easy when all items have unit size. Due to its application in assigning students (unit size items) to seminars (bins) at a university such that the total satisfaction (profit) of the students is maximized, this special case of GAP-MQ where all items have unit size was termed *seminar assignment problem* (SAP) in [3] and is formally defined as follows:

Definition 1 (Seminar Assignment Problem (SAP))
INSTANCE: The number n of items, m bins with capacities $B_1, \ldots, B_m \in \mathbb{N}$
  and minimum quantities $q_1, \ldots, q_m \in \mathbb{N}$ (where $q_j \leq B_j \leq n$ for
  all $j = 1, \ldots, m$), and a profit $p_{i,j} \in \mathbb{N}$ resulting from assigning
  item i to bin j for $i = 1, \ldots, n$ and $j = 1, \ldots, m$.
TASK: *Find an assignment of a subset of the items to the bins such that the*
  number of items in each bin j is either zero (if bin j is not opened)
  or at least q_j and at most B_j and the total profit is maximized.

Note that, in the above definition and throughout the paper, we always assume $\mathbb{N}$ to contain zero and denote the positive integers by $\mathbb{N}_+$.

Even though computing nontrivial feasible solutions for SAP is easy, a gap-preserving reduction from the 3-bounded 3-dimensional matching problem (3DM-3) given in [3] shows the existence of a constant $\epsilon_0 > 0$ such that it is strongly NP-hard to approximate SAP within a factor smaller than $(1 + \epsilon_0)$ even if all profits $p_{i,j}$ are in $\{0, 1\}$ and the minimum quantities and bin capacities of all bins are fixed to three. In particular, the problem does not admit a polynomial time approximation scheme (PTAS). Apart from these negative results, however, the approximability of SAP (and, in particular, the existence of a constant factor approximation) remained open. As most standard techniques for designing deterministic approximation algorithms fail for this problem due to the minimum quantity restrictions (cf. [3]), it natural to consider randomization and to ask whether a constant approximation ratio can be obtained by a randomized algorithm.

In this paper, we answer this question by presenting a randomized 3.93-approximation algorithm for SAP, which is the first nontrivial approximation result for this problem. Our randomized rounding algorithm uses a packing-based integer programming formulation, for which we show that the linear relaxation can be solved in polynomial time by using column generation. In particular, by using the probabilistic method (cf., for example, [4]), our result implies that the integrality gap of this formulation is no larger than 3.93.

1.1   Previous Work

The classical GAP is well-studied in literature. A comprehensive introduction to the problem can be found in [1]. A survey of algorithms for GAP is given in [2].

For a survey on different variants of assignment problems studied in literature, we refer to [5].

GAP is known to be APX-hard [6], but there exists a 2-approximation algorithm [7, 6]. Cohen et al. [8] showed how any polynomial time α-approximation algorithm for the knapsack problem can be translated into a polynomial time $(1 + \alpha)$-approximation algorithm for GAP. A $(1, 2)$-approximation algorithm for the equivalent minimization version of GAP, in which assigning item i to bin j causes a cost $c_{i,j}$, was provided by Shmoys and Tardos [7]: For every feasible instance of GAP, their algorithm computes a solution that violates the bin capacities by at most a factor of 2 and whose cost is at most as large as the cost of the best solution that satisfies the bin capacities strictly.

GAP is a generalization of both the (multiple) knapsack problem (cf. [1, 6, 9]) and the bin packing problem (cf. [10–12]). The multiple knapsack problem is the special case of GAP where the size and profit of an item are independent of the bin (knapsack) it is packed into. The bin packing problem can be seen as the special case of the decision version of GAP in which all bins have the same capacity and all profits are one. The question of deciding whether a packing of total profit equal to the number of items exists is then equivalent to asking whether all items can be packed into the given number of bins.

A dual version of bin packing (often called *bin covering*) in which minimum quantities are involved was introduced in [13, 14]. Here, the problem is to pack a given set of items with sizes that do not depend on the bins so as to maximize the number of bins used, subject to the constraint that each bin contains items of total size at least a given threshold T (upper bin capacities are not considered due to the nature of the objective function). Hence, the bin covering problem can be seen as a variant of GAP-MQ in which the minimum quantity is the same for each bin and the objective is to maximize the number of bins used. Since any approximation algorithm with approximation ratio strictly smaller than 2 would have to solve the NP-complete partition problem when applied to instances in which the sizes of the items sum up to two, it follows that (unless P = NP) no polynomial time $(2 - \epsilon)$-approximation for bin covering exists for any $\epsilon > 0$. In contrast, the main result of Assmann et al. [14] is an $\mathcal{O}(n \log^2 n)$ time algorithm that yields an *asymptotic* approximation ratio of 4/3 for bin covering, while easier algorithms based on next fit and first fit decreasing are shown to yield asymptotic approximation ratios of 2 and 3/2, respectively. Later, an asymptotic PTAS [15] and an asymptotic FPTAS [16] for bin covering were developed.

Minimum quantities have recently been studied for minimum cost network flow problems [17–19]. In this setting, minimum quantities for the flow on each arc are considered, which results in the minimum cost flow problem becoming strongly NP-complete [18]. Moreover, it was shown in [18] that (unless P = NP) no polynomial time $g(|I|)$-approximation for the problem exists for any polynomially computable function $g : \mathbb{N}_+ \to \mathbb{N}_+$, where $|I|$ denotes the encoding length of the given instance. The special case of the maximum flow problem with minimum quantities has recently been studied in [20], where it was shown that the problem is strongly NP-hard to approximate in general, but admits a $(2 - \frac{1}{\lambda})$-approximation in the case of an identical minimum quantity λ on all arcs.

The generalized assignment problem with minimum quantities (GAP-MQ) and the seminar assignment problem (SAP) were introduced in [3], where it was shown that the general version of GAP-MQ does not admit any polynomial time approximation algorithm unless P = NP. For SAP, it was shown by a gap-preserving reduction from the 3-bounded 3-dimensional matching problem (3DM-3) that there exists a constant $\epsilon_0 > 0$ such that it is strongly NP-hard to approximate SAP within a factor smaller than $(1 + \epsilon_0)$ even if all profits $p_{i,j}$ are in $\{0, 1\}$ and the minimum quantities and bin capacities of all bins are fixed to three. In particular, the problem does not admit a polynomial time approximation scheme (PTAS). Apart from these negative results, however, the approximability of SAP (and, in particular, the existence of a constant factor approximation) remained open.

2   Overview of the Algorithm

Before we present our randomized rounding algorithm for SAP in detail, we give a brief overview of the different steps of our procedure and its analysis.

The algorithm is based on an integer programming formulation of SAP that is introduced in Section 3. For each bin j, the integer program contains a binary variable x_t for every feasible packing of j (i.e., for every assignment of $q_j \le l \le B_j$ items to bin j), where $x_t = 1$ means that packing t is selected for bin j. As we show in Theorem 1, the linear relaxation of this integer program can be solved in polynomial time by column generation even though it contains an exponential number of variables x_t.

After solving the linear relaxation of the integer program, our algorithm independently selects a packing for each bin by using the value of variable x_t in the optimal solution (scaled by a suitably chosen factor $\alpha \in [0, 1]$) as the probability of using packing t for the corresponding bin j. The expected profit of the set of packings obtained in this way is exactly α times the objective value of the optimal solution of the linear relaxation used for the rounding, but the set of packings will, in general, not correspond to a feasible integral solution as items may be packed several times into different bins. Hence, in order to obtain a feasible integral solution, we apply a clean-up procedure that works in two steps: In the first step, we discard a subset of the bins opened in order to ensure that the total number of places used in the bins is at most n. In the second step, we can then replace all remaining multiply assigned items in the solution by unassigned items in order to obtain a feasible integral solution. Overall, we show that, in expectation, the profit decreases by at most a factor 3.93 during the clean-up procedure, which yields the desired approximation guarantee.

3   An Integer Programming Formulation

We start by introducing the IP-formulation on which our randomized rounding algorithm is based.

Definition 2. *A (feasible) packing of bin j is an incidence vector of a subset of the items with cardinality at least q_j and at most B_j, i.e., a vector $t = (t_1, \ldots, t_n) \in \{0,1\}^n$ such that $q_j \leq \sum_{i=1}^{n} t_i \leq B_j$. The profit of t is $p_t := \sum_{i=1}^{n} p_{ij} \cdot t_i$. The set of all feasible packings of bin j will be denoted by $T(j)$ and we write $T := \dot{\bigcup}_{j=1}^{m} T(j)$.*

Using this definition, we can formulate SAP as the following integer program:

$$\max \quad \sum_{j=1}^{m} \sum_{t \in T(j)} x_t p_t \tag{1a}$$

$$\text{s.t.} \quad \sum_{t \in T(j)} x_t \leq 1 \qquad \forall\, j \in \{1, \ldots, m\} \tag{1b}$$

$$\sum_{j=1}^{m} \sum_{t \in T(j)} x_t t_i \leq 1 \qquad \forall\, i \in \{1, \ldots, n\} \tag{1c}$$

$$x_t \in \{0, 1\} \qquad \forall\, t \in T \tag{1d}$$

Here, variable x_t for $t \in T(j)$ is one if and only if packing t is selected for bin j. Constraint (1b) ensures that at most one packing is selected for each bin while constraint (1c) ensures that each item is packed into at most one bin.

We now show that, even though the number of variables in IP (1) exponential in the encoding length of the given instance of SAP, we can solve its linear relaxation in polynomial time by using column generation. To this end, it suffices to show that we can find a column (packing) of minimum reduced cost in polynomial time, i.e., solve the pricing problem in polynomial time (cf. [21, 22]). Denoting the dual variables corresponding to the constraints (1b) by $y_j, j = 1, \ldots, m$, and the dual variables corresponding to the constraints (1c) by $z_i, i = 1, \ldots, n$, the reduced cost of a packing $t \in T(j)$ of bin j is

$$\bar{c}_t = p_t - y_j - \sum_{i=1}^{n} t_i z_i = -y_j + \sum_{i=1}^{n} t_i (p_{ij} - z_i).$$

Hence, the pricing problem is

$$\min_{j=1,\ldots,m} \min_{t \in T(j)} -y_j + \sum_{i=1}^{n} t_i (p_{ij} - z_i).$$

This problem can be solved in polynomial time as follows: For each bin j, finding a packing $t \in T(j)$ of minimum reduced cost means solving a 0-1-knapsack problem with n unit size items, profit $-(p_{ij} - z_i)$ for item i, and the additional constraint that at least q_j items have to be packed into the knapsack. This problem can be solved by greedily selecting the item i with minimum value $p_{ij} - z_i$ until we have either selected B_j items, or the next item i satisfies $p_{ij} - z_i \geq 0$. If this procedure returns an infeasible packing with less than q_j items, we continue selecting the item i with minimum value $p_{ij} - z_i$ (which is now always nonnegative) until we have selected exactly q_j items. Afterwards, we can solve the pricing problem by simply comparing the best packings obtained for all bins in order to find a packing of globally minimum reduced cost. Hence, we obtain:

Theorem 1. *The linear relaxation of IP (1) can be solved in time polynomial in the encoding length of the given instance of SAP.*

4   The Randomized Rounding Procedure

We now present our randomized 3.93-approximation algorithm for SAP. In the algorithm, we first solve the linear relaxation of of IP (1) obtaining an optimal fractional solution $x \in [0,1]^{|T|}$. We then multiply all values x_t by a factor $\alpha \in [0,1]$ (which will be chosen later) and consider the resulting value $\bar{x}_t := \alpha x_t \in [0,1]$ as the probability of using packing $t \in T(j)$ for bin j. More precisely, we independently select a packing for each bin j at random, where packing $t \in T(j)$ is selected with probability $\bar{x}_t = \alpha x_t$, and with probability $1 - \sum_{t \in T(j)} \bar{x}_t$, bin j is not opened. Since we select at most one packing for each bin, the resulting vector $x^{\mathrm{IP}} \in \{0,1\}^{|T|}$ (where $x_t^{\mathrm{IP}} = 1$ if and only if packing t was selected) then satisfies constraint (1b), but is, in general, not a feasible solution to IP (1) since it may violate constraint (1c) (an item may be packed several times into different bins). In particular, the total number of items assigned to bins in x^{IP} may be larger than n (when counted with multiplicities). The expected profit $\mathbb{E}(\mathrm{PROFIT}(x^{\mathrm{IP}}))$, however, is exactly equal to $\alpha \cdot \mathrm{PROFIT}(x)$, i.e., exactly α times the profit $\mathrm{PROFIT}(x) =: \mathrm{OPT}_{\mathrm{LP}}$ of the optimal fractional solution x obtained for the linear relaxation. We note this fact for later reference:

Observation 1. The vector $x^{\mathrm{IP}} \in \{0,1\}^{|T|}$ obtained from the randomized rounding process satisfies $\mathbb{E}(\mathrm{PROFIT}(x^{\mathrm{IP}})) = \alpha \cdot \mathrm{OPT}_{\mathrm{LP}}$.

We now show how we can turn x^{IP} into a feasible solution of IP (1) while only decreasing the expected profit by a constant factor. Our procedure works in two steps: In the first step, we discard a subset of the bins opened in x^{IP} in order to ensure that the total number of places used in the bins is at most n. In the second step, we can then replace all remaining multiply assigned items in the solution by unassigned items in order to obtain a feasible integral solution.

We start by describing the first step of the procedure. Given the vector $x^{\mathrm{IP}} \in \{0,1\}^{|T|}$ obtained from the randomized rounding process, we consider the following instance of the 0-1-knapsack problem (0-1-KP): The objects are the packings $t \in T$ with $x_t^{\mathrm{IP}} = 1$, i.e., the packings selected by x^{IP}. The size of object t is the number of items contained in packing t and its profit is the profit p_t of the packing. The knapsack capacity is set to n.

Assuming that the total number of places used in the bins in x^{IP} is in $[kn, (k+1)n)$ for some $k \in \mathbb{N}$, it is easy to compute an integral solution to this knapsack instance with profit at least $\frac{1}{2k+1} \cdot \mathrm{PROFIT}(x^{\mathrm{IP}})$: We can assign all objects fractionally to at most $(k+1)$ knapsacks of size n each such that at most k objects are fractionally assigned (cf. Figure 1). Since the size of each object is at most n, we can then remove the fractionally assigned objects from the knapsacks and put each of them into its own (additional) knapsack, which yields an integral assignment of all objects to at most $2k+1$ knapsacks. Since all objects together have total profit $\mathrm{PROFIT}(x^{\mathrm{IP}})$, this implies that the objects in the most profitable one among these $2k+1$ knapsacks correspond to an integral solution of the knapsack instance with profit at least $\frac{1}{2k+1} \cdot \mathrm{PROFIT}(x^{\mathrm{IP}})$ as desired[1] and, by choosing only the corresponding packings, we lose profit at most

$$\left(1 - \frac{1}{2k+1}\right) \cdot \mathrm{PROFIT}(x^{\mathrm{IP}}) = \frac{2k}{2k+1} \cdot \mathrm{PROFIT}(x^{\mathrm{IP}}). \qquad (2)$$

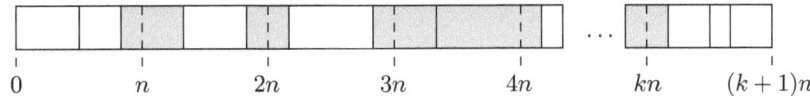

$$0 \qquad n \qquad 2n \qquad 3n \qquad 4n \qquad kn \qquad (k+1)n$$

Fig. 1. Fractional assigment of the objects to $(k+1)$ knapsacks of size n each. Fractionally assigned objects are shown in grey.

In order to bound the expected loss in profit resulting from using only the packings in our solution to the knapsack instance, we now consider the probability $\mathrm{Pr}(k)$ that the total number of places used in the bins in x^{IP} is at least kn for each $k \in \{1, 2, \ldots\}$ (if at most n places are used, we can use all packings selected by x^{IP}, so we do not lose any profit in this step). To this end, note that, by constraint (1c), the total number of places used in the optimal fractional solution x of IP (1) is at most n. Hence, since we used each packing $t \in T(j)$ with probability $\bar{x}_t = \alpha x_t$, the expected number of places used in x^{IP} is at most αn. Thus, Markov's inequality yields that

$$\mathrm{Pr}(k) = \mathrm{Pr}\left(\#(\text{places used in } x^{\mathrm{IP}}) \geq kn\right) \leq \frac{\alpha n}{kn} = \frac{\alpha}{k} \text{ for } k \in \{1, 2, \ldots\}. \qquad (3)$$

In the following, $\mathrm{Pr}\left([kn, (k+1)n)\right)$ will denote the probability that the total number of places used in the bins in x^{IP} is in $[kn, (k+1)n)$ and $\ln(\cdot)$ will denote

---

[1] Note that this bound on the profit of an integral solution of the knapsack instance is tight as long as $k < \frac{n}{4} - \frac{1}{2}$ as the example of $2k+1$ objects of size $\lfloor \frac{n}{2} \rfloor + 1 > \frac{n}{2}$ with unit profits shows. Hence, also computing an optimal solution for the knapsack instance (which is possible in polynomial time as the knapsack capacity n is polynomial) would not yield a better bound in general.

the natural logarithm. By (2) and (3), we then obtain that, in expectation, we lose at most the following factor times the profit of x^{IP} in the first step:

$$\sum_{k=1}^{\infty} \Pr\left([kn,(k+1)n)\right) \cdot \frac{2k}{2k+1}$$

$$= \sum_{k=1}^{\infty} \left(\sum_{l=k}^{\infty} \Pr\left([ln,(l+1)n)\right) - \sum_{l=k+1}^{\infty} \Pr\left([ln,(l+1)n)\right) \right) \cdot \frac{2k}{2k+1}$$

$$= \sum_{k=1}^{\infty} \sum_{l=k}^{\infty} \Pr\left([ln,(l+1)n)\right) \cdot \frac{2k}{2k+1} - \sum_{k=2}^{\infty} \sum_{l=k}^{\infty} \Pr\left([ln,(l+1)n)\right) \cdot \frac{2(k-1)}{2(k-1)+1}$$

$$= \sum_{l=1}^{\infty} \Pr\left([ln,(l+1)n)\right) \cdot \frac{2}{3} + \sum_{k=2}^{\infty} \sum_{l=k}^{\infty} \Pr\left([ln,(l+1)n)\right) \cdot \left(\frac{2k}{2k+1} - \frac{2(k-1)}{2(k-1)+1} \right)$$

$$= \Pr(1) \cdot \frac{2}{3} + \sum_{k=2}^{\infty} \Pr(k) \cdot \left(\frac{2k}{2k+1} - \frac{2(k-1)}{2(k-1)+1} \right)$$

$$= \Pr(1) \cdot \frac{2}{3} + \sum_{k=2}^{\infty} \Pr(k) \cdot \frac{2}{4k^2-1}$$

$$\leq \alpha \cdot \frac{2}{3} + \sum_{k=2}^{\infty} \frac{\alpha}{k} \cdot \frac{2}{4k^2-1}$$

$$= \alpha \cdot \sum_{k=1}^{\infty} \frac{2}{k(4k^2-1)}$$

$$= \alpha \cdot (4\ln(2) - 2)$$

$$= 2\alpha \cdot (2\ln(2) - 1)$$

Using Observation 1, this proves the following result:

Proposition 1. *The packings obtained after the first step contain at most n items in total and have expected profit at least $\alpha \cdot (1 - 2\alpha\,(2\ln(2) - 1)) \cdot \mathrm{OPT_{LP}}$.*

In the second step of our procedure, we now have to get rid of all multiply assigned items in the solution obtained after the first step. Denoting by $j_1(i), \ldots, j_{k(i)}(i)$ the bins a multiply assigned item i is currently assigned to, we simply delete item i from all bins but the one among $j_1(i), \ldots, j_{k(i)}(i)$ in which it yields the highest profit. Doing so for all multiply assigned items yields a solution in which no item is packed more than once. The minimum quantities of the bins, however, may not be satisfied any more after deleting the multiply assigned items. But since the total number of places used in the bins after the first step was no more than the total number n of items available, we know that, for each item i that was assigned to $l \geq 2$ bins, there must be $l - 1$ items that were not assigned to any bin after the first step. Hence, we can refill the $l - 1$ places vacated by deleting item i from all but one bin with items that were previously unassigned, and doing so for all multiply assigned items yields a feasible integral solution to the given instance of SAP.

In order to bound the expected loss in profit resulting from the second step of our procedure, we want to bound the loss in profit resulting from deleting a single item i from all but the most profitable bin it was previously assigned to. To do so, we use that, by constraint (1c) and the scaling of the probabilities x_t given by the optimal fractional solution of IP (1) by α, the expected number of bins item i was assigned to *before* the first step of the procedure is at most α. Hence, by Markov's inequality, the probability that item i was assigned to at least k bins before the first step of the procedure can be upper bounded as

$$\Pr(i \text{ in} \geq k \text{ bins}) \leq \frac{\alpha}{k}. \tag{4}$$

Clearly, discarding a subset of the bins opened cannot increase the number of bins item i is assigned to, so inequality (4) is still valid after the first step of our procedure. Hence, denoting the probability that item i was assigned to *exactly* k bins after the first step by $\Pr(i \text{ in } k \text{ bins})$, we lose at most the following factor times the total profit obtained from all copies of item i in the solution from the first step:

$$\sum_{k=2}^{m} \Pr(i \text{ in } k \text{ bins}) \cdot \frac{k-1}{k}$$

$$= \sum_{k=2}^{m} \left(\sum_{l=k}^{m} \Pr(i \text{ in } l \text{ bins}) - \sum_{l=k+1}^{m} \Pr(i \text{ in } l \text{ bins}) \right) \cdot \frac{k-1}{k}$$

$$= \sum_{k=2}^{m} \sum_{l=k}^{m} \Pr(i \text{ in } l \text{ bins}) \cdot \frac{k-1}{k} - \sum_{k=3}^{m} \sum_{l=k}^{m} \Pr(i \text{ in } l \text{ bins}) \cdot \frac{k-2}{k-1}$$

$$= \sum_{l=2}^{m} \Pr(i \text{ in } l \text{ bins}) \cdot \frac{1}{2} + \sum_{k=3}^{m} \sum_{l=k}^{m} \Pr(i \text{ in } l \text{ bins}) \cdot \left(\frac{k-1}{k} - \frac{k-2}{k-1} \right)$$

$$= \Pr(i \text{ in} \geq 2 \text{ bins}) \cdot \frac{1}{2} + \sum_{k=3}^{m} \Pr(i \text{ in} \geq k \text{ bins}) \cdot \left(\frac{k-1}{k} - \frac{k-2}{k-1} \right)$$

$$= \Pr(i \text{ in} \geq 2 \text{ bins}) \cdot \frac{1}{2} + \sum_{k=3}^{m} \Pr(i \text{ in} \geq k \text{ bins}) \cdot \frac{1}{k(k-1)}$$

$$\leq \frac{\alpha}{2} \cdot \frac{1}{2} + \sum_{k=3}^{m} \frac{\alpha}{k} \cdot \frac{1}{k(k-1)}$$

$$\leq \alpha \cdot \sum_{k=2}^{\infty} \frac{1}{k^2(k-1)}$$

$$= \alpha \cdot \left(2 - \frac{\pi^2}{6} \right)$$

Together with the bound on the profit of the packings obtained after the first step given in Proposition 1, this shows the following result:

Proposition 2. *The second step of the procedure yields a feasible integral solution to the given instance of SAP with expected profit at least*

$$\left(1 + \left(\frac{\pi^2}{6} - 2\right)\alpha\right) \cdot \left(1 - 2\alpha(2\ln(2) - 1)\right)\alpha \cdot \text{OPT}_{\text{LP}}.$$

Choosing the value α^* maximizing the expected profit in Proposition 2 (which is approximately 0.556339) yields an expected profit of at least $0.254551 \cdot \text{OPT}_{\text{LP}}$. As OPT_{LP} is an upper bound on the profit of the optimal integral solution of the given instance, taking the inverse of this factor and rounding up yields the following theorem:

Theorem 2. *With the right choice of α, the randomized rounding procedure yields a randomized 3.93-approximation algorithm for SAP.*

Using the probabilistic method (cf., for example, [4]), Theorem 2 yields an upper bound of 3.93 on the integrality gap of IP (1): Since the expected profit of the solution returned by the randomized rounding algorithm is at least $0.254551 \cdot \text{OPT}_{\text{LP}}$, it follows that we obtain a feasible integral solution with profit at least $0.254551 \cdot \text{OPT}_{\text{LP}}$ with positive probability. In particular, there always exists a feasible integral solution with profit at least $0.254551 \cdot \text{OPT}_{\text{LP}}$, which (by again taking the inverse of this factor and rounding up) proves the following result:

Corollary 1. *The integrality gap of IP (1) is at most 3.93.*

5   Conclusion and Open Problems

In this paper, we obtained the first nontrivial approximation result for SAP by providing a randomized 3.93-approximation algorithm. We believe that the approximation factor of 3.93 obtained for our algorithm is not tight and can be slightly improved by using stronger probability bounds in some places in the analysis. A natural open question is whether a constant factor approximation for SAP can also be obtained by a deterministic algorithm. We believe that such deterministic approximation algorithms exist, but will likely require techniques different from the ones commonly used in approximation algorithms for the generalized assignment problem without minimum quantities.

References

1. Martello, S., Toth, P.: Knapsack Problems: Algorithms and Computer Implementations. John Wiley & Sons, Chichester (1990)
2. Cattrysse, D.G., Van Wassenhove, L.N.: A survey of algorithms for the generalized assignment problem. European Journal of Operational Research 60(3), 260–272 (1992)
3. Krumke, S.O., Thielen, C.: The generalized assignment problem with minimum quantities. European Journal of Operational Research 228(1), 46–55 (2013)
4. Alon, N., Spencer, J.H.: The Probabilistic Method. John Wiley & Sons (1992)

5. Pentico, D.W.: Assignment problems: A golden anniversary survey. European Journal of Operational Research 176(2), 774–793 (2007)
6. Chekuri, C., Khanna, S.: A PTAS for the multiple knapsack problem. SIAM Journal on Computing 35(3), 713–728 (2006)
7. Shmoys, D.B., Tardos, É.: An approximation algorithm for the generalized assignment problem. Mathematical Programming 62, 461–474 (1993)
8. Cohen, R., Katzir, L., Raz, D.: An efficient approximation algorithm for the generalized assignment problem. Information Processing Letters 100(4), 162–166 (2006)
9. Kellerer, H., Pferschy, U., Pisinger, D.: Knapsack Problems. Springer (2004)
10. Nemhauser, G.L., Wolsey, L.A.: Integer and Combinatorial Optimization. Wiley-Interscience, New York (1988)
11. Vazirani, V.V.: Approximation Algorithms. Springer (2001)
12. Coffman Jr., E.G., Garey, M.R., Johnson, D.S.: Approximation algorithms for bin packing: A survey. In: [23], pp. 46–93
13. Assmann, S.F.: Problems in Discrete Applied Mathematics. PhD thesis, Massachusetts Institute of Technology (1983)
14. Assmann, S.F., Johnson, D.S., Kleinman, D.J., Leung, J.Y.T.: On a dual version of the one-dimensional bin packing problem. Journal of Algorithms 5(4), 502–525 (1984)
15. Csirik, J., Johnson, D.S., Kenyon, C.: Better approximation algorithms for bin covering. In: Proceedings of the 12th ACM-SIAM Symposium on Discrete Algorithms (SODA), pp. 557–566 (2001)
16. Jansen, K., Solis-Oba, R.: An asymptotic fully polynomial time approximation scheme for bin covering. Theoretical Computer Science 306, 543–551 (2003)
17. Seedig, H.G.: Network flow optimization with minimum quantities. In: Operations Research Proceedings 2010: Selected Papers of the Annual International Conference of the German Operations Research Society, pp. 295–300. Springer (2010)
18. Krumke, S.O., Thielen, C.: Minimum cost flows with minimum quantities. Information Processing Letters 111(11), 533–537 (2011)
19. Zhu, X., Yuan, Q., Garcia-Diaz, A., Dong, L.: Minimal-cost network flow problems with variable lower bounds on arc flows. Computers and Operations Research 38(8), 1210–1218 (2011)
20. Thielen, C., Westphal, S.: Complexity and approximability of the maximum flow problem with minimum quantities. Networks (2013)
21. Mehlhorn, K., Ziegelmann, M.: Resource constrained shortest paths. In: Paterson, M. (ed.) ESA 2000. LNCS, vol. 1879, pp. 326–337. Springer, Heidelberg (2000)
22. Minoux, M.: A class of combinatorial problems with polynomially solvable large scale set covering/partitioning relaxations. RAIRO Recherche Opérationelle 21(2), 105–136 (1987)
23. Hochbaum, D.S. (ed.): Approximation Algorithms for NP-Hard Problems. PWS Publishing, Boston (1997)

Determinacy and Rewriting
of Top-Down and MSO Tree Transformations[*]

Michael Benedikt[1], Joost Engelfriet[2], and Sebastian Maneth[1]

[1] University of Oxford
first.last@cs.ox.ac.uk
[2] LIACS, Leiden University
engelfr@liacs.nl

Abstract. A query is determined by a view, if the result to the query can be re-constructed from the result of the view. We consider the problem of deciding for two given tree transformations, whether one is determined by the other. If the view transformation is induced by a tree transducer that may copy, then determinacy is undecidable, even for identity queries. For a large class of non-copying views, namely compositions of functional extended linear top-down tree transducers with regular look-ahead, we show that determinacy is decidable, where queries are given by deterministic top-down tree transducers with regular look-ahead or by MSO tree transducers. We also show that if a query is determined, then it can be rewritten into a query that works directly over the view and is in the same class as the given query. The proof relies on the decidability of equivalence for the two considered classes of queries, and on their closure under composition.

1 Introduction

Given a transformation between data structures, a basic question is what sort of information it preserves. In some contexts, one desires a transformation that is "fully information-preserving" – one can recover the input from the output. In other cases it may be acceptable, or even important, to hide certain pieces of information in the input; but necessarily there is *some* important information in the input that must be recoverable from the output. This notion has been studied in the database community [29,26]: a query q is determined by another query v if there exists a function f such that $f(v(s)) = q(s)$ for every input s. The query v is referred to as "view". Note that nothing is said about how efficiently f can be computed (or if it can be computed at all). We can then strengthen determinacy by requiring the function f to lie within a certain class $\mathcal{C}$; then f is a "rewriting in $\mathcal{C}$". These notions have received considerable attention in the database setting [29,26,27,1].

In this paper we study determinacy and rewriting for classes of tree transformations (or, tree translations). Injectivity is undecidable for deterministic top-down tree transducers [15,17]; hence, one cannot decide if the identity query is determined by such a transducer. This holds for transducers that only copy once. We therefore restrict our attention to views induced by *linear* tree transducers. For the same reason we

[*] Benedikt and Maneth were supported by the Engineering and Physical Sciences Research Council project "Enforcement of Constraints on XML streams" (EPSRC EP/G004021/1).

K. Chatterjee and J. Sgall (Eds.): MFCS 2013, LNCS 8087, pp. 146–158, 2013.

restrict to a single view (while in database research, normally multiple views are considered). Our main result is that determinacy is decidable for views that are compositions of functional extended linear top-down tree transducers (with regular look-ahead) and for queries that are either deterministic top-down tree transducers (with regular look-ahead) or deterministic MSO definable tree transducers (where MSO stands for Monadic Second-Order logic). Extended transducers generalize the left-hand sides of conventional finite-state tree transducers (from one input symbol to an arbitrary "pattern tree"). They were invented by Arnold and Dauchet [3] and have recently been studied in [24,25,9]. Extended linear transducers are convenient because (1) they are more powerful than ordinary linear top-down or bottom-up transducers and (2) they allow to elegantly capture the inverses of translations.

As an example, consider the transformation v taking binary trees as input, with internal nodes labeled a, b, c, and leaves labeled l. It relabels the b nodes as a nodes, and otherwise copies the tree as is. A linear top-down transducer implementing this translation v has a single state p and these translation rules:

$$p(a(x,y)) \rightarrow a(p(x), p(y)) \qquad p(b(x,y)) \rightarrow a(p(x), p(y))$$
$$p(c(x,y)) \rightarrow c(p(x), p(y)) \qquad p(l()) \quad\;\; \rightarrow l()$$

Information about the (labels of) b nodes and a nodes is lost in the translation – e.g., from the output of v we cannot determine the answer to the identity query q_0. In contrast, information about the l nodes and their relationship to c nodes is maintained. For example, the query q_1 that removes a and b nodes but keeps c and l nodes is determined by v. Our algorithm can decide that q_0 is not determined and q_1 is.

Our decision procedure for determinacy establishes several results that are interesting on their own. For a view v realized by an extended linear top-down tree transducer, its inverse v^{-1} is a binary relation on trees. Our approach converts v^{-1} into a composition of two nondeterministic translations, a translation τ_1 of a very simple form and a translation τ_2 in the same class as v. We then construct uniformizers u_1, u_2 of τ_1, τ_2 and compose them to form a uniformizer u of v^{-1}. A *uniformizer* of a binary relation R is a function u such that $u \subseteq R$ and u has the same domain as R; thus u "selects" one of the possibly several elements that R associates with an element of its domain. It is easy to see that a query q is determined by v if and only if $v \circ u \circ q = q$ (where $\circ$ denotes sequential composition, see the Preliminaries). We show that if q is a deterministic top-down or MSO definable tree translation, then so is $v \circ u \circ q$. This is achieved by proving that u_1, u_2, and v are deterministic top-down *and* MSO definable tree translations. Since our two query classes are closed under composition and $u = u_1 \circ u_2$, this shows that $v \circ u \circ q$ is in the same class as q. We then decide $v \circ u \circ q = q$, and hence determinacy, making use of the decidability of equivalence for deterministic top-down or MSO definable tree translations ([16,13] or [12]). The same proof also shows that if q is determined by v, then $u \circ q$ is a rewriting belonging to the same class as q.

Related Work. The notion of view-query determinacy was introduced by Segoufin and Vianu in [29]. They focus on relational queries definable in first-order logic and show that if such queries are determined over arbitrary structures, then they can be rewritten in first-order, but that if they are determined over finite structures, they may require a much more powerful relational query to be rewritten. Nash, Segoufin, and Vianu [26] summarize a number of other results on the relational case. Due to the differing data models and notions of equality used in relational queries and tree structures, results on determinacy for queries in the relational case do not (directly) apply to transducers, and vice versa. In the context of unranked trees, determinacy is considered in Groz's thesis [20] for XML views and queries, see also [21]. Two notions of determinacy are considered, depending on whether or not the output trees preserve provenance information (i.e., node identities) from the input document. It is shown that both notions of determinacy are undecidable for views and queries defined using a transformation language that can select subtrees using regular XPath filters. On the positive side, it is shown that if the views are "interval-bounded" – there is a bound on the number of consecutive nodes skipped along a path – then determinacy can be tested effectively. The most related work is [23], which considers the determinacy problem (and rewriting) explicitly for tree transducers, and solves it for functional extended linear bottom-up views and deterministic bottom-up queries. Their approach is to decide determinacy by testing functionality of the inverse of the view composed with the query. To this end they generalize the functionality test for bottom-up transducers in [31] to extended bottom-up transducers with "grafting" (needed for the inverse of the view). Our main result generalizes the one of [23], and provides an alternative proof of it.

2   Preliminaries

For $k \in \mathbb{N} = \{0, 1, \ldots\}$ let $[k]$ denote the set $\{1, \ldots, k\}$. For a binary relation R and a set A we denote by $R(A)$ the set $\{y \mid \exists x \in A : (x, y) \in R\}$, and by $R(x)$ the set $R(\{x\})$. If $R \subseteq B \times C$ for sets B and C, then $\mathrm{ran}(R) = R(B)$ and $\mathrm{dom}(R) = R^{-1}(C)$. For two relations R and S we denote the sequential composition "R followed by S" by $R \circ S$, i.e., for an element x, $(R \circ S)(x) = S(R(x))$. Note this is in contrast to the conventional use of $\circ$. If $\mathcal{R}, \mathcal{S}$ are classes of binary relations, then $\mathcal{R} \circ \mathcal{S} = \{R \circ S \mid R \in \mathcal{R}, S \in \mathcal{S}\}$, $\mathcal{R}^* = \{R_1 \circ \cdots \circ R_n \mid n \geq 1, R_i \in \mathcal{R}\}$, and $\mathcal{R}^{-1} = \{R^{-1} \mid R \in \mathcal{R}\}$.

We define determinacy and rewritability, following [26]. Let $\mathcal{Q}, \mathcal{V}$ be classes of partial functions and let $q \in \mathcal{Q}$ and $v \in \mathcal{V}$. We say that q *is determined by* v, if there exists a function f such that $v \circ f = q$. Note that the latter means that the domains of $v \circ f$ and q coincide, and that $f(v(s)) = q(s)$ for each s in that domain. *Determinacy for $\mathcal{Q}$ under $\mathcal{V}$* is the problem that takes as input $q \in \mathcal{Q}$ and $v \in \mathcal{V}$ and outputs "yes" if q is determined by v, and "no" otherwise. Determinacy says that there is a functional dependency of q on v, with no limit on how complex it is to reconstruct the answer to q from the answer to v. A finer notion requires that the reconstruction be in a given class: a class $\mathcal{Q}'$ of partial functions is *complete for $\mathcal{V}$-to-$\mathcal{Q}$ rewritings*, if for every $q \in \mathcal{Q}$ and $v \in \mathcal{V}$ such that q is determined by v, there is an $f \in \mathcal{Q}'$ with $v \circ f = q$.

Trees and Tree Automata. A ranked alphabet consists of a finite set Σ together with a mapping $\mathrm{rank}_\Sigma : \Sigma \to \mathbb{N}$. We write $a^{(k)}$ to denote that $\mathrm{rank}_\Sigma(a) = k$ and define $\Sigma^{(k)}$ as the set $\{a \in \Sigma \mid \mathrm{rank}_\Sigma(a) = k\}$. The set of (ranked, ordered, node-labeled, finite) trees over Σ, denoted by T_Σ, is the set of words defined recursively as the smallest set T such that $a(s_1, \ldots, s_k) \in T$ if $a \in \Sigma^{(k)}$, $k \geq 0$, and $s_1, \ldots, s_k \in T$. For a tree $a()$ we simply write a. For a set T of trees, we denote by $T_\Sigma(T)$ the set of trees obtained from trees in T_Σ by replacing arbitrary leaves by trees in T. We fix a countably infinite set $X = \{x_1, x_2, \dots\}$ of variables. For $k \in \mathbb{N}$, let X_k be the ranked alphabet $\{x_1^{(0)}, \dots, x_k^{(0)}\}$. For $k \in \mathbb{N}$, an X_k-context (over Σ) is a tree C in $T_\Sigma(X_k)$ such that each variable in X_k occurs exactly once in C. For such a context C and trees $s_1, \dots, s_k$, $C[s_1, \dots, s_k]$ denotes the tree obtained from C by replacing each $x_i \in X_k$ by s_i. Let $T_1, \dots, T_n$ be sets of trees. For trees $s_1, \dots, s_n$ that are not subtrees of another, we denote by $s[s_i \leftarrow T_i \mid i \in [n]]$ the set of trees obtained from s by replacing each occurrence of a subtree s_i of s by a tree from T_i (where different occurrences of s_i need not be replaced by the same tree). For a ranked alphabet Q with $Q^{(1)} = Q$ we denote by $Q(X_k)$ the set of trees $\{q(x_i) \mid q \in Q, i \in [k]\}$. A *deterministic bottom-up tree automaton* (dbta) over Σ is a tuple $A = (P, \Sigma, F, \delta)$ where P is a finite set of states, Σ is a ranked alphabet, $F \subseteq P$ is the set of final states, and δ is the transition function. For every $a \in \Sigma^{(k)}$, $k \geq 0$, and $p_1, \dots, p_k \in P$, $\delta(a, p_1, \dots, p_k)$ is an element of P. The function δ is extended to trees s in T_Σ in the usual way; the resulting function from T_Σ to P is denoted δ as well. Thus $\delta(s)$ is the state reached by A at the root of s. The language accepted by A is $L(A) = \{s \in T_\Sigma \mid \delta(s) \in F\}$.

Convention: All lemmas, theorems, etc., stated in this paper (except in Section 4) are *effective*.

3   Extended Top-Down and Bottom-Up Tree Transducers

An *extended top-down tree transducer with regular look-ahead* (ET^{R} transducer) is a tuple $M = (Q, \Sigma, \Delta, I, R, A)$ where Q is a ranked alphabet of states all of rank 1, Σ and Δ are ranked alphabets of input and output symbols, respectively, $I \subseteq Q$ is a set of initial states, $A = (P, \Sigma, F, \delta)$ is a dbta called the look-ahead automaton, and R is a finite set of rules of the form $q(C) \to \zeta \ \langle p_1, \dots, p_k \rangle$, where $q \in Q$, $C \neq x_1$ is an X_k-context over Σ, $k \geq 0$, $\zeta \in T_\Delta(Q(X_k))$, and $p_1, \dots, p_k \in P$. For an input tree $s \in T_\Sigma$, the q-translation $[\![q]\!]_M(s)$ is the smallest set of trees $T \subseteq T_\Delta$ such that for every rule $q(C) \to \zeta \ \langle p_1, \dots, p_k \rangle$ and all $s_1, \dots, s_k \in T_\Sigma$, if $s = C[s_1, \dots, s_k]$ and $\delta(s_i) = p_i$ for every $i \in [k]$, then T contains the set of trees $\zeta[q'(x_i) \leftarrow [\![q']\!]_M(s_i) \mid q' \in Q, i \in [k]]$. The translation $[\![M]\!]$ realized by M is the binary relation $\{(s, t) \in T_\Sigma \times T_\Delta \mid s \in L(A), t \in \cup_{q \in I} [\![q]\!]_M(s)\}$. The class of all translations realized by ET^{R} transducers is denoted ET^{R} (and similarly for other transducers). The transducer M is *linear*, if the right-hand side ζ of each rule is linear in the set of variables X, i.e., each variable x_i occurs at most once in ζ. We use "L" to abbreviate "linear", i.e., $\mathrm{ELT}^{\mathrm{R}}$ is the class of $[\![M]\!]$ where M is a linear ET^{R} transducer. Transducers *without* look-ahead are defined by transducers with a trivial one-state look-ahead automaton (accepting T_Σ); this is indicated by omitting the superscript "R" for transducers and classes. Extended top-down transducers are studied in, e.g., [3,24,25,9]. [1]

An *extended linear bottom-up tree transducer* (ELB transducer) is a tuple $B = (Q, \Sigma, \Delta, F, R)$ where Q is a ranked alphabet of states all of rank 1, Σ and Δ are ranked alphabets of input and output symbols, respectively, $F \subseteq Q$ is a set of final states, and R is a finite set of rules of the form $C[q_1(x_1), \dots, q_k(x_k)] \to q(\zeta)$, where

---

[1] The class ELT is denoted l-XTOP$_{\mathrm{ef}}$ in [24], where "ef" denotes epsilon-freeness, meaning the left-hand side of rules are not of the form $q(x_i)$.

$k \geq 0$, $C \neq x_1$ is an X_k-context over Σ, $q_1, \ldots, q_k, q \in Q$, and $\zeta \in T_\Delta(X_k)$ is linear in X_k. If $t_i \in [\![q_i]\!]_B(s_i)$ then $[\![q]\!]_B(C[s_1, \ldots, s_k])$ contains the tree $\zeta[x_i \leftarrow \{t_i\} \mid i \in [k]]$. The translation realized by B is $[\![B]\!] = \cup_{q \in F} [\![q]\!]_B$. Extended linear bottom-up transducers are studied in, e.g., [3,9,23]. In [9,23], the left-hand side of a rule is allowed to be of the form $q(x_i)$, and the corresponding (larger) class of translations is denoted l-XBOT. It is easy to show that such rules can effectively be removed from a transducer B when it is known that $[\![B]\!]$ is a function.

As we show in Section 4, determinacy is undecidable if the view transducers copy. We therefore define views using linear transducers. We first show that for linear extended transducers, top-down (with look-ahead) gives the same translations as bottom-up, just as for non-extended transducers (see Theorem 2.8 of [7]). The following result was already pointed out below Proposition 5 in [9] (see also Theorem 3.1 of [19]).

Theorem 1. $\mathrm{ELT}^R = \mathrm{ELB}$.

Proof. $\subseteq$: Let $M = (Q, \Sigma, \Delta, I, R, (P, \Sigma, F, \delta))$ be an ELT^R transducer. We construct the ELB transducer B with the set $P \cup (Q \times P)$ of states and the set $I \times F$ of final states. Its rules are defined as follows. For the first set of rules let d_0 be a fixed element of $\Delta^{(0)}$. Let $p_1, \ldots, p_k, p \in P$ and $a \in \Sigma^{(k)}$ such that $\delta(a, p_1, \ldots, p_k) = p$. (1) We add $a(p_1(x_1), \ldots, p_k(x_k)) \to p(d_0)$ as a rule of B. This rule outputs d_0 and changes the state to p, but recursive calls to it will only make use of the computed state, not the output. (2) If M has the rule $q(C) \to \zeta$ $\langle p_1, \ldots, p_k \rangle$, then we add $C[\mathsf{state}(x_1), \ldots, \mathsf{state}(x_k)] \to \langle q, p \rangle(\mathsf{erase}(\zeta))$ as a rule of B, where $\mathsf{erase}(\zeta)$ is obtained from ζ by replacing every $q'(x_i)$ by x_i, while $\mathsf{state}(x_i) = \langle q', p_i \rangle(x_i)$ if q' is the unique state such that $q'(x_i)$ occurs in ζ and $\mathsf{state}(x_i) = p_i(x_i)$ if no such q' exists. The correctness of the construction follows from the following claim (for $q \in I$ and $p \in F$), which can be proved by a straightforward induction on the structure of s. Let $s \in T_\Sigma$, $q \in Q$, and $p \in P$.

Claim. (1) $\delta(s) = p$ if and only if $[\![p]\!]_B(s) \neq \varnothing$. (2) If $s \in \mathrm{dom}([\![\langle q, p \rangle]\!]_B)$, then $p = \delta(s)$. (3) $[\![\langle q, \delta(s) \rangle]\!]_B(s) = [\![q]\!]_M(s)$.

$\supseteq$: By the proof of Lemma 6 of [9], ELB is included in the class of all tree translations $\{(f(s), g(s)) \mid s \in L\}$ where f is a linear non-deleting non-erasing tree homomorphism, g is a linear tree homomorphism, and L is a regular tree language. By Theorem 17 of [24], ELT^R is equal to this class. $\qquad\square$

We now give a useful property of the composition closure of ELT^R.

Lemma 2. If $\tau \in (\mathrm{ELT}^R)^*$ and R is a regular tree language, then $\mathrm{dom}(\tau)$, $\mathrm{ran}(\tau)$, $\tau(R)$, and $\tau^{-1}(R)$ are regular tree languages.

Proof. We consider ELB translations, which suffices by Theorem 1. By Lemma 6 of [9], every ELB translation is of the form $\{(f(s), g(s)) \mid s \in L\}$ where f, g are linear tree homomorphisms and L is a regular tree language. From this it follows (as stated in Corollary 7 of [9]) that ELB translations preserve regularity. This implies that $\mathrm{ran}(\tau)$ is regular. The above form means that inverse ELB translations are also of that form and hence preserve regularity. This implies that $\mathrm{dom}(\tau)$ is regular. $\qquad\square$

Functionality Test. Later when we prove determinacy results, we restrict our views to classes of transducers that realize *functions*. In particular, we use the class $(\text{fu-ELT}^R)^*$ of compositions of functional translations in ELT^R, which properly contains fu-ELT^R by the proof of Theorem 5.2 of [25]. It is therefore important to know the next proposition.

Proposition 3. For an ELT^R transducer M it is decidable whether $\llbracket M \rrbracket$ is functional.

Proof. By Theorem 4.8 of [25], $\text{ET}^R = \text{T}^R$. The result follows because functionality is decidable for T^R transducers by [16] (see the sentence after Theorem 8 of [16]).   □

We note that it can be shown, using a variation of the Lemma of [8], that our class $(\text{fu-ELT}^R)^*$ is equal to the class $\text{fu-}(\text{ELT}^R)^*$ of functional compositions of ELT^R translations. However, we do not know whether functionality is decidable for such compositions. Note also that it was recently shown in [18] that $\text{ELT}^R \circ \text{ELT}^R \circ \text{ELT}^R = (\text{ELT}^R)^*$.

Ordinary Top-Down Tree Transducers. The ET^R transducer M is an (ordinary, not extended) *top-down tree transducer with regular look-ahead* (T^R transducer) if the left-hand side C of each of its rules contains exactly one symbol in Σ, i.e., each rule is of the form $q(a(x_1, \ldots, x_k)) \to \zeta \;\; \langle p_1, \ldots, p_k \rangle$ with $a \in \Sigma^{(k)}$ and $k \geq 0$. A T^R transducer is *deterministic* if it has exactly one initial state and for each q, a, and $\langle p_1, ..., p_k \rangle$ it has at most one rule as above. Determinism is denoted by the letter "D", thus we have DT^R and DLT^R transducers. A T^R transducer M is *finite-copying* (a T^R_{fc} transducer) if each input node is translated only a bounded number of times. Formally this means there exists a number K such that for every $p \in P$, $s \in T_\Sigma(\{\Box\})$, and $t \in \llbracket M_p \rrbracket(s)$, if $\Box$ occurs exactly once in s, then $\Box$ occurs $\leq K$ times in t; here we assume that $\Box$ is a new input and output symbol of rank 0, and that M_p is M extended with the look-ahead transition $\delta(\Box) = p$ and the rules $q(\Box) \to \Box$ for every state q. A DT^R_{fc} transducer is a deterministic T^R_{fc} transducer. Note that $\text{LT}^R \subseteq \text{T}^R_{\text{fc}}$ and that translations τ in T^R_{fc} are of linear size increase [11], i.e., there is a number N such that the size of t is at most N times the size of s for every $(s, t) \in \tau$.
    We later need the following four results. Let DMSOTT be the class of deterministic (or, parameterless) MSO definable tree translations (see, e.g., Chapter 8 of [5]).

Proposition 4. (1) $\text{DT}^R_{\text{fc}} \circ \text{DT}^R_{\text{fc}} \subseteq \text{DT}^R_{\text{fc}}$, (2) $\text{DT}^R_{\text{fc}} \subseteq \text{DMSOTT}$, (3) $\text{DT}^R \circ \text{DT}^R \subseteq \text{DT}^R$, and (4) $\text{DMSOTT} \circ \text{DMSOTT} \subseteq \text{DMSOTT}$.

For result (4) see, e.g., [5]. Results (1) and (2) follow from Proposition 2 of [4] and Theorem 7.4 of [10]. Result (1) is already mentioned in Theorem 5.4 of [14]. Result (3) is in Theorem 2.11 of [7]. It is not difficult to prove (1) and (3) directly via straightforward product constructions.

4   Undecidability Results

Let HOM denote the class of tree homomorphisms, i.e., translations realized by total (see next paragraph) one-state DT transducers. As observed in [23], a function v is injective if and only if q is determined by v, where q is the identity on $\text{dom}(v)$. Since the injectivity problem for HOM is undecidable by [17], one obtains (as stated in Theorem 17 of [23]) undecidability of the determinacy problem for ID under HOM, where ID is the class of identity translations on T_Σ, for any ranked alphabet Σ.

We show that determinacy is undecidable for ID under total copy-once DT transducers (tot-DT$_{co}$ transducers). A DT transducer is *total* if for each state q and input symbol a, it has a rule with left-hand side $q(a(x_1, \ldots, x_k))$. It is *copy-once* if for every rule $q(a(x_1, \ldots, x_k)) \to \zeta$ the initial state q_0 does not occur in ζ, and ζ is linear in X if $q \neq q_0$. Thus the transducer copies at most once, at the root node of the input tree. The undecidability of injectivity for non-total DT$_{co}$ transducers was proved by Ésik in [15], and in his PhD thesis (in Hungarian). Our proof for total DT$_{co}$ transducers (deferred to the full version) is a slight variation of Ésik's proof.

Theorem 5. Determinacy for ID under tot-DT$_{co}$ is undecidable.

Since, obviously, every DT$_{co}$ transducer is a DT$_{fc}$ transducer, and DT$_{fc}$ is (effectively) included in DMSOTT by Proposition 4(2), this immediately gives undecidability of determinacy for ID under DMSOTT (which slightly strengthens Theorem 19 of [23]).

One often considers determinacy for a query q under a *set of views* $\mathcal{V}$. The extended definition states that if two inputs give the same output for each view in $\mathcal{V}$, then they give the same output for q. In this case one has undecidability even when the views are deterministic finite-state word transformations. Thus, in what follows we consider only a single non-copying view.

5   Inverses and Uniformizers of Linear Extended Transducers

As Theorem 5 shows, determinacy cannot be decided under view transducers that copy, not even for a single initial copy at the input root node. Let us therefore restrict our attention to classes induced by linear view transducers. The results in this section hold for arbitrary linear extended transducers. When we want to decide determinacy in Section 6, we restrict the views to *functional* linear translations.

5.1   Inverses of Extended Linear Bottom-Up Transducers

Given an ELB transducer B, we would like to construct a transducer realizing its *inverse* $[\![B]\!]^{-1}$. Since B can translate the set of all input trees in T_Σ to a single output tree, a transducer realizing $[\![B]\!]^{-1}$ may need to translate a tree back to any tree in T_Σ. This is not possible by our extended top-down or bottom-up tree transducers because the height of an output tree is linearly bounded by the height of the input tree. The next, easy lemma "factors out" this problem by decomposing an ELB transducer into a component that can be inverted as an extended top-down transducer, and a component of a very simple form – a "projection mapping". Let n-ELB denote the class of *non-deleting non-erasing* ELB transducers: those in which every rule is of the form $C[q_1(x_1), \ldots, q_k(x_k)] \to q(\zeta)$, such that each variable in X_k occurs in ζ and $\zeta \neq x_1$. The phrase "non-deleting" indicates that we do not drop an input x_i, thus removing an entire subtree from the input. Non-erasing indicates that we do not have a rule such as $q(a(x_1, b)) \to q'(x_1)$, which "erases" the symbols a and b. Let Δ be a ranked alphabet and H a set of symbols disjoint from Δ each of rank at least 1. The *projection mapping from* $\Delta \cup H$ *to* Δ is the tree homomorphism $\pi_{\Delta, H} = \pi : T_{\Delta \cup H} \to T_\Delta$ defined as: $\pi(h(s_1, \ldots, s_k)) = \pi(s_1)$ for $h \in H^{(k)}$ and $\pi(d(s_1, \ldots, s_k)) = d(\pi(s_1), \ldots, \pi(s_k))$ for $d \in \Delta^{(k)}$, for all $s_1, \ldots, s_k \in T_{\Delta \cup H}$. We denote by PROJ the class of all projection mappings.

Lemma 6. ELB $\subseteq$ n-ELB $\circ$ PROJ.

Proof. Let $B = (Q, \Sigma, \Delta, F, R)$ be an ELB transducer and let m be the maximal number of variables that occur in the left-hand side of any rule of B. We define the ranked alphabet $H = \{\#_n^{(n+1)} \mid 0 \leq n \leq m\}$ and the n-ELB transducer $B' = (Q, \Sigma, \Delta \cup H, F, R')$. For every rule $C[q_1(x_1), \ldots, q_k(x_k)] \to q(\zeta)$ in R we let the rule $C[q_1(x_1), \ldots, q_k(x_k)] \to q(\#_n(\zeta, x_{i_1}, \ldots, x_{i_n}))$ be in R', where $x_{i_1}, \ldots, x_{i_n}$ are all the variables from X_k that do not occur in ζ. Clearly, $[\![B']\!] \circ \pi_{\Delta, H} = [\![B]\!]$. $\square$

As shown in [3] (TIA$^{-1} \subseteq$ TID), the inverse of an n-ELB can be converted to an ELT by just "inverting the rules".

Lemma 7. n-ELB$^{-1} \subseteq$ ELT.

Proof. Let $B = (Q, \Sigma, \Delta, F, R)$ be an n-ELB transducer. We construct the ELT transducer $M = (Q, \Delta, \Sigma, F, R')$ realizing B's inverse. For every rule $C[q_1(x_1), \ldots, q_k(x_k)] \to q(\zeta)$ in R let the rule $q(\zeta) \to C[q_1(x_1), \ldots, q_k(x_k)]$ be in R'. It should be clear that $[\![M]\!] = [\![B]\!]^{-1}$. $\square$

These two lemmas imply that ELB$^{-1} \subseteq$ PROJ$^{-1} \circ$ ELT.

5.2 Uniformizers

Let $\tau \subseteq A \times B$ be a translation and u a function from A to B. We say that u is a *uniformizer of* τ if $u \subseteq \tau$ and dom$(u) =$ dom(τ). For classes $\mathcal{T}, \mathcal{U}$ of translations we say that $\mathcal{T}$ *has uniformizers in* $\mathcal{U}$ if for every $\tau \in \mathcal{T}$ we can construct a uniformizer u of τ such that $u \in \mathcal{U}$. We say that the sequence $\tau_1, \ldots, \tau_n$ of translations is *compatible*, if for $i \in [n-1]$, ran$(\tau_i) \subseteq$ dom$(\tau_{i+1} \circ \cdots \circ \tau_n)$. It is easy to see that if $u_1, \ldots, u_n$ are uniformizers of $\tau_1, \ldots, \tau_n$, respectively, and $\tau_1, \ldots, \tau_n$ is compatible, then $u_1 \circ \cdots \circ u_n$ is a uniformizer of $\tau_1 \circ \cdots \circ \tau_n$. Our goal is to show that $((\text{ELT}^R)^*)^{-1}$ and $(\text{ELT}^R)^*$ have uniformizers in DT$_{\text{fc}}^R$. We do this by decomposing into compatible translations, constructing uniformizers in DT$_{\text{fc}}^R$ for them, and then obtaining a uniformizer in DT$_{\text{fc}}^R$ through Proposition 4(1). A similar idea was used in [8] to obtain uniformizers for compositions of top-down and bottom-up tree translations in DT^R.

Lemma 8. ELT^R has uniformizers in DT$_{\text{fc}}^R$.

Proof. By Theorem 4.8 of [25], ET$^R =$ T^R. For a T^R transducer M with $[\![M]\!] \in$ ELT^R, we construct a dbta A recognizing its domain (cf. Corollary 2.7 of [7]). We now change M so that the look-ahead automaton checks M's domain (by building a product automaton with A and changing the rules of the transducer appropriately). The resulting transducer can be decomposed (by an obvious variant of Theorem 2.6 of [7]) into a finite state relabeling B with the same domain as M, followed by a top-down tree transducer

T. Note that $[\![B]\!]$ is a function; for the notion of finite-state relabeling see Definition 3.14 in [6]. It follows that $\operatorname{ran}([\![B]\!]) \subseteq \operatorname{dom}([\![T]\!])$. Thus, $[\![B]\!], [\![T]\!]$ are compatible. A finite state relabeling can be seen as a top-down tree transducer, so by the Lemma in [8] we obtain uniformizers for $[\![B]\!]$ and $[\![T]\!]$, both in DT^R. Since DT^R is closed under composition by Proposition 4(3), the composition of these uniformizers is a uniformizer of $[\![M]\!]$, in DT^R. Obviously, ELT^R translations are of linear size increase, and so this uniformizer is of linear size increase. We obtain the desired result because DT^R translations of linear size increase are in $\mathrm{DT}^R_{\mathrm{fc}}$ by Section 7.1 of [11] (in fact, by the obvious generalization of the latter result to partial transducers: introduce output dummies for undefined rules and remove them later). $\qquad\square$

Note that there is an alternative proof to Lemma 8 which avoids the last step (of applying linear size increase): First, it follows from the construction in the proof of Theorem 4.8 of [25] that $\mathrm{ELT}^R \subseteq \mathrm{T}^R_{\mathrm{fc}}$. Second, the proof of the Lemma in [8] can easily be modified into a proof that every T^R transducer has a uniformizer in DT^R, and the proof preserves the finite-copying property.

An FTA *transducer* is a dbta A, seen as a tree transducer realizing the translation $[\![A]\!]$, which is the identity function on $L(A)$; composing a tree translation τ with $[\![A]\!]$ amounts to restricting the range of τ to $L(A)$: $\tau \circ [\![A]\!] = \{(s,t) \in \tau \mid t \in L(A)\}$.

Lemma 9. $\mathrm{PROJ}^{-1} \circ \mathrm{FTA}$ has uniformizers in DLT^R.

Proof. Let $\tau = \pi^{-1} \circ [\![A]\!]$ where $\pi \in \mathrm{PROJ}$ and A is an FTA transducer. Thus, $\pi = \pi_{\Sigma,H}$ for disjoint ranked alphabets Σ and H such that $H^{(0)} = \varnothing$, and A is a dbta $(Q, \Sigma \cup H, F, \delta)$. Let $\mathcal{C}$ be the set of all X_1-contexts C over $\Sigma \cup H$ such that the left-most leaf of C has label x_1 and all the ancestors of this leaf have labels in H. For every $a \in \Sigma^{(k)}$ and $q, q_1, \ldots, q_k \in Q$, let $\mathcal{C}(a, q, q_1, \ldots, q_k)$ be the set of $C \in \mathcal{C}$ such that $\delta(C[a(t_1, \ldots, t_k)]) = q$ for all $t_1, \ldots, t_k \in T_{\Sigma \cup H}$ with $\delta(t_i) = q_i$ for every $i \in [k]$. Let $C_0(a, q, q_1, \ldots, q_k)$ be one (fixed) such C – since the set $\mathcal{C}(a, q, q_1, \ldots, q_k)$ is effectively regular, one can always compute such an element C if the set is nonempty. If there does not exist such a C then $C_0(a, q, q_1, \ldots, q_k)$ is undefined.

Since the construction in the proof of the Lemma of [8] preserves linearity, LT has uniformizers in DLT^R. Hence, it suffices to construct an LT transducer M with $[\![M]\!] \subseteq \tau$ and $\operatorname{dom}([\![M]\!]) = \operatorname{dom}(\tau)$. We define $M = (Q, \Sigma, \Sigma \cup H, F, R')$ where R' consists of all rules $q(a(x_1, \ldots, x_k)) \to C_0(a, q, q_1, \ldots, q_k)[a(q_1(x_1), \ldots, q_k(x_k))]$ such that $C_0(a, q, q_1, \ldots, q_k)$ is defined. Intuitively, for $s \in T_\Sigma$, M simulates top-down the state behavior of A on some tree t in $\pi^{-1}(s)$ and, at each node of s, outputs a context in $\mathcal{C}$ on which A has the same state behavior as on the context in $\mathcal{C}$ that is "above" the corresponding node in t. Formally, the correctness of the construction follows from the following claim (for $q \in F$), which can easily be proved by structural induction on s and induction on the size of t, respectively. Let $q \in Q$, $s \in T_\Sigma$, and $t \in T_{\Sigma \cup H}$.

Claim. (1) If $t \in [\![q]\!]_M(s)$, then $\pi(t) = s$ and $\delta(t) = q$. (2) If $\delta(t) = q$, then $\pi(t) \in \operatorname{dom}([\![q]\!]_M)$.

In both proofs one uses that $\pi(C[t]) = \pi(t)$ for every $C \in \mathcal{C}$. In the proof of (2) one uses that t is of the form $C[a(t_1, \ldots, t_k)]$ with $C \in \mathcal{C}$, $k \geq 0$, $a \in \Sigma^{(k)}$ and $t_1, \ldots, t_k \in T_{\Sigma \cup H}$, and one applies the induction hypothesis to $t_1, \ldots, t_k$. $\qquad\square$

Lemma 10. ELB^{-1} has uniformizers in DT$_{\text{fc}}^{\text{R}}$.

Proof. Let $\tau \in$ ELB^{-1}. By Lemmas 6 and 7, $\tau \in$ PROJ$^{-1} \circ$ ELT. The domains of translations in ELT are effectively regular by Lemma 2, thus we obtain $\tau = \tau_1 \circ \tau_2$ such that the translations $\tau_1 \in$ PROJ$^{-1} \circ$ FTA, $\tau_2 \in$ ELT are compatible (by definition of the FTA transducer). For τ_1, τ_2 we obtain, by Lemmas 9 and 8, uniformizers $u_1, u_2 \in$ DT$_{\text{fc}}^{\text{R}}$. Then $u_1 \circ u_2$ is a uniformizer for τ; it is in DT$_{\text{fc}}^{\text{R}}$ by Proposition 4(1). □

Theorem 11. $((\text{ELT}^{\text{R}})^*)^{-1}$ has uniformizers in DT$_{\text{fc}}^{\text{R}}$.

Proof. Let $T_1, \ldots, T_n$ be ELT^{R} transducers. We change the T_i so that the sequence of translations $[\![T_1]\!]^{-1}, \ldots, [\![T_n]\!]^{-1}$ is compatible, i.e., ran($[\![T_i]\!]^{-1}$) $\subseteq$ dom($[\![T_{i+1}]\!]^{-1} \circ \cdots \circ [\![T_n]\!]^{-1}$): We change the domain of T_i to be included in the range of $[\![T_n]\!] \circ \cdots \circ [\![T_{i+1}]\!]$. This range is regular by Lemma 2. The domain of T_i is changed using look-ahead, as in the proof of Lemma 8. Using Theorem 1 and Lemma 10 we obtain uniformizers in DT$_{\text{fc}}^{\text{R}}$ for the $[\![T_i]\!]^{-1}$. This proves the theorem, by Proposition 4(1). □

Theorem 12. $(\text{ELT}^{\text{R}})^*$ has uniformizers in DT$_{\text{fc}}^{\text{R}}$.

Proof. Let $T_1, \ldots, T_n$ be ELT^{R} transducers. We change the T_i so that $[\![T_1]\!], \ldots, [\![T_n]\!]$ is compatible, i.e., restrict T_i's range to $D = $ dom($[\![T_{i+1}]\!] \circ \cdots \circ [\![T_n]\!]$), which is regular by Lemma 2. The range of T_i can be restricted to D as follows. As mentioned in the proof of Theorem 1, ELT^{R} is the class of all translations of the form $\tau = \{(f(s), g(s)) \mid s \in L\}$ where f is a linear non-deleting non-erasing tree homomorphism, g is a linear tree homomorphism, and L is a regular tree language. The restriction of the range of τ to D is $\{(f(s), g(s)) \mid s \in L, g(s) \in D\} = \{(f(s), g(s)) \mid s \in L \cap g^{-1}(D)\}$. Since $L \cap g^{-1}(D)$ is regular, this translation is again of the above form and hence in ELT^{R}. We obtain uniformizers in DT$_{\text{fc}}^{\text{R}}$ for the $[\![T_i]\!]$ by Lemma 8, and a uniformizer for $[\![T_1]\!] \circ \cdots \circ [\![T_n]\!]$ by Proposition 4(1). □

6 Decidability of Determinacy and Rewriting

Consider a query q, a view v, and a uniformizer u of v^{-1}, each of them a partial function. Clearly, q is determined by v if and only if $v \circ u \circ q = q$. For queries in DT^{R} or DMSOTT, equivalence is decidable [13,12], and they are closed under left composition with DT$_{\text{fc}}^{\text{R}}$ by Proposition 4. Thus, if v and u are in DT$_{\text{fc}}^{\text{R}}$, then we can decide determinacy. We will show that this holds for the views in the class (fu-ELT^{R})* of compositions of functions in ELT^{R}. If v is in this class, then v^{-1} has a uniformizer u in DT$_{\text{fc}}^{\text{R}}$ by Theorem 11. As the next corollary states, v itself is also in DT$_{\text{fc}}^{\text{R}}$. The inclusion is a direct consequence of Theorem 12; it is proper by Lemma 2 (because DT$_{\text{fc}}^{\text{R}}$ contains translations that do not preserve regularity).

Corollary 13. (fu-ELT^{R})$^* \subsetneq$ DT$_{\text{fc}}^{\text{R}}$.

The main results of this paper are presented in the next two theorems.

Theorem 14. Determinacy is decidable for DT^{R} and DMSOTT under (fu-ELT^{R})*.

Proof. Let $v \in (\text{fu-ELT}^R)^*$. According to Corollary 13 and Theorem 11 we construct DT^R_{fc} transducers M_1, M_2 such that $[\![M_1]\!] = v$ and $[\![M_2]\!] = u$ is a uniformizer of v^{-1}. If a query is given as a DT^R (DMSOTT) transducer N, then DT^R (DMSOTT) transducers N' and N'' can be constructed with $[\![N']\!] = u \circ [\![N]\!]$ and $[\![N'']\!] = v \circ [\![N']\!] = v \circ u \circ [\![N]\!]$. This follows from Proposition 4. We can decide if $[\![N'']\!] = [\![N]\!]$ because equivalence is decidable for DT^R and DMSOTT transducers ([16,13] and [12]).   □

The proof of Theorem 14 also proves Theorem 15.

Theorem 15. Let $\mathcal{V} = (\text{fu-ELT}^R)^*$, $v \in \mathcal{V}$, and let N be a DT^R (DMSOTT) transducer such that $[\![N]\!]$ is determined by v. A DT^R (DMSOTT) transducer N' can be constructed such that $v \circ [\![N']\!] = [\![N]\!]$. That is, DT^R (DMSOTT) is complete for $\mathcal{V}$-to-DT^R ($\mathcal{V}$-to-DMSOTT) rewritings.

Since the class fu-B of functional bottom-up translations is included in DT^R by [8], it is immediate from Theorems 14 and 1 that determinacy is decidable for fu-B under fu-ELB, as proved in Theorem 16 of [23]. In Theorem 21 of [23] it is shown to be decidable for $q \in$ fu-B and $v \in$ fu-ELB whether there exists $f \in$ fu-B such that $q = v \circ f$. Theorem 15 shows that such an f can always be found in DT^R.

Weakly Determined Queries. A query q is determined by a view v if there exists a function f such that (1) $\text{dom}(v \circ f) = \text{dom}(q)$ and (2) $f(v(s)) = q(s)$ for every $s \in \text{dom}(q)$. For practical purposes, condition (1) could be weakened to $\text{dom}(v \circ f) \supseteq \text{dom}(q)$. For a given element s, one first checks if $s \in \text{dom}(q)$, and if so, obtains $q(s)$ as $f(v(s))$. We say that q *is weakly determined by* v if there exists f with $f(v(s)) = q(s)$ for every $s \in \text{dom}(q)$. As an example consider $q = \{(1,1)\}$ and $v = \{(1,1),(2,1)\}$. Then q is *not* determined by v, but is weakly determined. Let $\mathcal{Q}, \mathcal{V}, \mathcal{Q}'$ be classes of partial functions. We say that $\mathcal{Q}'$ is *complete for weak $\mathcal{V}$-to-$\mathcal{Q}$ rewritings*, if for every $q \in \mathcal{Q}$ and $v \in \mathcal{V}$ such that q is weakly determined by v, there is an $f \in \mathcal{Q}'$ with $f(v(s)) = q(s)$ for every $s \in \text{dom}(q)$. For a function $\tau : A \to B$ and a set L let $\tau \restriction L$ denote the restriction of τ to inputs in L. Then q is weakly determined by v if and only if q is determined by $v \restriction \text{dom}(q)$, with the same functions f. For $q \in \text{DT}^R$ or $q \in$ DMSOTT, $\text{dom}(q)$ is effectively regular. And if $v \in (\text{fu-ELT}^R)^*$ then $v \restriction L$ is in $(\text{fu-ELT}^R)^*$ for every regular tree language L (simply by adding it to the look-ahead of the first transducer). Hence Theorems 14 and 15 also hold for weak determinacy.

Corollary 16. Let $\mathcal{V} = (\text{fu-ELT}^R)^*$. Weak determinacy is decidable for DT^R and for DMSOTT under $\mathcal{V}$. The classes DT^R and DMSOTT are complete for weak $\mathcal{V}$-to-DT^R and weak $\mathcal{V}$-to-DMSOTT rewritings, respectively.

Future Work. We would like to know the complexity of deciding determinacy. The complexity of our algorithm is dominated by that of the equivalence tests in [13,12]: double exponential time for DT^R, non-elementary for DMSOTT (and nondeterministic exponential time for streaming tree transducers [2]). Can we find subclasses of tree transducers for which determinacy is polynomial-time testable (cf. [13,22,30])? Can our results be extended to larger classes of tree transducers, such as deterministic macro tree transducers (see, e.g., [10,11])? For those transducers, decidability of equivalence is a long-standing open problem. It is interesting and practically important (for XML) to study determinacy for unranked tree transducers, e.g., those of [28].

References

1. Afrati, F.N.: Determinacy and query rewriting for conjunctive queries and views. Theor. Comput. Sci. 412(11), 1005–1021 (2011)
2. Alur, R., D'Antoni, L.: Streaming tree transducers. In: Czumaj, A., Mehlhorn, K., Pitts, A., Wattenhofer, R. (eds.) ICALP 2012, Part II. LNCS, vol. 7392, pp. 42–53. Springer, Heidelberg (2012)
3. Arnold, A., Dauchet, M.: Bi-transductions de forêts. In: ICALP (1976)
4. Bloem, R., Engelfriet, J.: A comparison of tree transductions defined by monadic second order logic and by attribute grammars. J. Comput. Syst. Sci. 61(1), 1–50 (2000)
5. Courcelle, B., Engelfriet, J.: Graph Structure and Monadic Second-Order Logic, a Language-Theoretic Approach. Cambridge University Press (2012)
6. Engelfriet, J.: Bottom-up and top-down tree transformations - a comparison. Math. Systems Theory 9(3), 198–231 (1975)
7. Engelfriet, J.: Top-down tree transducers with regular look-ahead. Math. Systems Theory 10, 289–303 (1977)
8. Engelfriet, J.: On tree transducers for partial functions. Inf. Proc. Lett. 7(4), 170–172 (1978)
9. Engelfriet, J., Lilin, E., Maletti, A.: Extended multi bottom-up tree transducers. Acta Inf. 46(8), 561–590 (2009)
10. Engelfriet, J., Maneth, S.: Macro tree transducers, attribute grammars, and MSO definable tree translations. Inf. Comput. 154(1), 34–91 (1999)
11. Engelfriet, J., Maneth, S.: Macro tree translations of linear size increase are MSO definable. SIAM J. Comput. 32(4), 950–1006 (2003)
12. Engelfriet, J., Maneth, S.: The equivalence problem for deterministic MSO tree transducers is decidable. Inf. Proc. Lett. 100(5), 206–212 (2006)
13. Engelfriet, J., Maneth, S., Seidl, H.: Deciding equivalence of top-down XML transformations in polynomial time. J. Comput. Syst. Sci. 75(5), 271–286 (2009)
14. Engelfriet, J., Rozenberg, G., Slutzki, G.: Tree transducers, L systems, and two-way machines. J. Comput. Syst. Sci. 20(2), 150–202 (1980)
15. Ésik, Z.: On decidability of injectivity of tree transformations. In: Les Arbres en Algèbre et en Programmation, Lille, pp. 107–133 (1978)
16. Ésik, Z.: Decidability results concerning tree transducers I. Acta Cybern. 5, 1–20 (1981)
17. Fülöp, Z., Gyenizse, P.: On injectivity of deterministic top-down tree transducers. Inf. Proc. Lett. 48(4), 183–188 (1993)
18. Fülöp, Z., Maletti, A.: Composition closure of ϵ-free linear extended top-down tree transducers. In: Béal, M.-P., Carton, O. (eds.) DLT 2013. LNCS, vol. 7907, pp. 239–251. Springer, Heidelberg (2013)
19. Fülöp, Z., Maletti, A., Vogler, H.: Weighted extended tree transducers. Fundam. Inform. 111(2), 163–202 (2011)
20. Groz, B.: XML Security Views: Queries, Updates, and Schemas. PhD thesis, Université Lille 1 (2012)
21. Groz, B., Staworko, S., Caron, A.-C., Roos, Y., Tison, S.: Static analysis of XML security views and query rewriting. Inf. Comput. (to appear, 2013)
22. Gurari, E.M., Ibarra, O.H.: A note on finite-valued and finitely ambiguous transducers. Math. Systems Theory 16(1), 61–66 (1983)
23. Hashimoto, K., Sawada, R., Ishihara, Y., Seki, H., Fujiwara, T.: Determinacy and subsumption for single-valued bottom-up tree transducers. In: Dediu, A.-H., Martín-Vide, C., Truthe, B. (eds.) LATA 2013. LNCS, vol. 7810, pp. 335–346. Springer, Heidelberg (2013)
24. Maletti, A.: Compositions of extended top-down tree transducers. Inf. Comput. 206(9-10), 1187–1196 (2008)

25. Maletti, A., Graehl, J., Hopkins, M., Knight, K.: The power of extended top-down tree trans-ducers. SIAM J. Comput. 39(2), 410–430 (2009)
26. Nash, A., Segoufin, L., Vianu, V.: Views and queries: Determinacy and rewriting. ACM Trans. Database Syst. 35(3) (2010)
27. Pasailă, D.: Conjunctive queries determinacy and rewriting. In: ICDT (2011)
28. Perst, T., Seidl, H.: Macro forest transducers. Inf. Proc. Lett. 89(3), 141–149 (2004)
29. Segoufin, L., Vianu, V.: Views and queries: determinacy and rewriting. In: PODS (2005)
30. Seidl, H.: Single-valuedness of tree transducers is decidable in polynomial time. Theor. Comput. Sci. 106(1), 135–181 (1992)
31. Seidl, H.: Equivalence of finite-valued tree transducers is decidable. Math. Systems Theory 27(4), 285–346 (1994)

On the Speed of Constraint Propagation and the Time Complexity of Arc Consistency Testing

Christoph Berkholz[1] and Oleg Verbitsky[2,*]

[1] RWTH Aachen University, Aachen, Germany
[2] Humboldt-University of Berlin, Berlin, Germany

Abstract. Establishing arc consistency on two relational structures is one of the most popular heuristics for the constraint satisfaction problem. We aim at determining the time complexity of arc consistency testing. The input structures G and H can be supposed to be connected colored graphs, as the general problem reduces to this particular case. We first observe the upper bound $O(e(G)v(H) + v(G)e(H))$, which implies the bound $O(e(G)e(H))$ in terms of the number of edges and the bound $O((v(G) + v(H))^3)$ in terms of the number of vertices. We then show that both bounds are tight up to a constant factor as long as an arc consistency algorithm is based on constraint propagation (as all current algorithms are).

Our argument for the lower bounds is based on examples of slow constraint propagation. We measure the speed of constraint propagation observed on a pair G, H by the size of a proof, in a natural combinatorial proof system, that Spoiler wins the existential 2-pebble game on G, H. The proof size is bounded from below by the game length $D(G, H)$, and a crucial ingredient of our analysis is the existence of G, H with $D(G, H) = \Omega(v(G)v(H))$. We find one such example among old benchmark instances for the arc consistency problem and also suggest a new, different construction.

1 Introduction

According to the framework of [10], the *constraint satisfaction problem* (CSP) takes two finite relational structures as input and asks whether there is a homomorphism between these structures. In this paper we consider structures with unary and binary relations and refer to unary relations as colors and to binary relations as directed edges. In fact, most of the time we deal with structures where the only binary relation E is symmetric and irreflexive relation, i.e., with vertex-colored graphs. This is justified by a linear time reduction from the CSP on binary structures to its restriction on colored graphs.

Let G and H be an input of the CSP. It is customary to call the vertices of G *variables* and the vertices of H *values*. A mapping from $V(G)$ to $V(H)$ then corresponds to an assignment of values to the variables, and the assignment is

* Supported by DFG grant VE 652/1–1. On leave from the Institute for Applied Problems of Mechanics and Mathematics, Lviv, Ukraine.

K. Chatterjee and J. Sgall (Eds.): MFCS 2013, LNCS 8087, pp. 159–170, 2013.

satisfying if the mapping defines a homomorphism. Let a *domain* $D_x \subseteq V(H)$ of a variable $x \in V(G)$ be a set of values such that for every homomorphism $h : G \to H$ it holds that $h(x) \in D_x$. The aim of the arc consistency heuristic is to find small domains in order to shrink the search space. The first step of the arc consistency approach is to ensure *node consistency*, that is, D_x is initialized to the set of vertices in H that are colored with the same color as x. The second step is to iteratively shrink the domains according to the following rule:

If there exists an $a \in D_x$ and a variable $y \in V(G)$ such that $\{x, y\} \in E(G)$ and $\{a, b\} \notin E(H)$ for all $b \in D_y$, then delete a from D_x.

A pair of graphs augmented with a set of domains is *arc consistent* if the above rule cannot be applied and all domains are nonempty. We say that arc consistency *can be established* for G and H, if there exists a set of domains such that G and H augmented with these domains is arc consistent. Our aim is to estimate the complexity of the following decision problem.

AC-PROBLEM

Input: Two colored graphs G and H.
Question: Can arc consistency be established on G and H?

We observe that the AC-PROBLEM can be solved in time $O(v(G)e(H) + e(G)v(H))$, where $v(G)$ and $e(G)$ denote the number of vertices and the number of edges respectively. This upper bound has never been stated explicitly although it can be obtained using known techniques. In terms of the overall input size $e(G) + e(H)$ this gives us only a quadratic upper bound, so there could be a chance for improvement: Is it possible to solve the AC-PROBLEM in sub-quadratic or even linear time? In fact, we cannot rule out this possibility completely. The first author [4] recently obtained lower bounds for higher levels of k-consistency (note that arc consistency is equivalent to 2-consistency). In particular, 15-consistency cannot be established in linear time and establishing 27-consistency requires more than quadratic time on multi-tape Turing machines. The lower bounds are obtained in [4] via the time hierarchy theorem and, unfortunately, these methods are not applicable to arc consistency because of the blow-up in the reduction.

However, we show lower bounds for every algorithm that is based on constraint propagation. A *propagation-based arc consistency algorithm* is an algorithm that solves the AC-PROBLEM by iteratively shrinking the domains via the arc consistency rule above. Note that all currently known arc consistency algorithms (e.g. AC-1, AC-3 [15]; AC-3.1/AC-2001 [7]; AC-3.2, AC-3.3; AC-3_d; AC-4 [17]; AC-5; AC-6; AC-7; AC-8; AC-* [18]) are propagation-based in this sense. Different AC algorithms differ in the principle of ordering propagation steps; for a general overview we refer the reader to [7]. The upper bound $O(v(G)e(H) + e(G)v(H))$ implies $O(e(G)e(H))$ in terms of the number of edges and $O(n^3)$ in terms of the number of vertices $n = v(G) + v(H)$. Our main result, Theorem 10 in Section 5, states that both bounds are tight up to a constant factor for any propagation-based algorithm.

We obtain the lower bounds by exploring a connection between the *existential 2-pebble game* and propagation-based arc consistency algorithms. In its general form, the existential k-pebble game is an Ehrenfeucht–Fraïssé like game that determines whether two finite structures can be distinguished in the existential-positive k-variable fragment of first order logic. It has found applications also outside finite model theory: to study the complexity and expressive power of Datalog [13], k-consistency tests [14,12,1,4] and bounded-width resolution [2,5]. It turns out that the existential 2-pebble game exactly characterizes the power of arc consistency [14], i.e., Spoiler wins the existential 2-pebble game on two colored graphs G and H iff arc consistency cannot be established.

The connection between the existential 2-pebble game and arc consistency algorithms is deeper than just a reformulation of the AC-PROBLEM. We show that every constraint propagation-based arc consistency algorithm computes in passing a proof of Spoiler's win on instances where arc consistency cannot be established. On the one hand these proofs of Spoiler's win naturally correspond to a winning strategy for Spoiler in the game. On the other hand they reflect the propagation steps performed by an algorithm. We consider three parameters to estimate the complexity of such proofs: length, size and depth. The length corresponds to the number of propagation steps, whereas size also takes the cost of propagation into account. The depth corresponds to the number of "nested" propagation steps and precisely matches the number of rounds $D(G, H)$ Spoiler needs to win the game. We observe that the minimum size of a proof of Spoiler's win on G and H bounds from below the running time of sequential propagation-based algorithms, whereas the minimal depth matches the running time of parallel algorithms.

We exhibit pairs of colored graphs G, H where $D(G, H) = \Omega(v(G)v(H))$ and hence many nested propagation steps are required to detect arc inconsistency. Because these graphs have a linear number of edges this implies that there is no sub-quadratic propagation-based arc consistency algorithm. It should be noted that CSP instances that are hard for sequential and parallel arc consistency algorithms, in the sense that they require many propagation steps, were explored very early in the AI-community [9,19]. Such examples were also proposed to serve as benchmark instances to compare different arc consistency algorithms [8]. Graphs G and H with large $D(G, H)$ can be derived from the old DOMINO example [8], consisting of structures with two binary relations. We also provide a new example, which we call CO-WHEELS, that shows the same phenomenon of slow constraint propagation for a more restricted class of rooted loopless digraphs.

2   Preliminaries

The *existential 2-pebble game on binary structures A and B* [13] is played by two players, Spoiler and Duplicator, to whom we will refer as he and she respectively.

Each player has a pair of distinct pebbles p and q. A *round* consists of a move of Spoiler followed by a move of Duplicator. Spoiler takes a pebble, p or q, and puts it on a vertex in A. Then Duplicator has to put her copy of this pebble on a vertex of B. Duplicator's objective is to keep the following condition true after each round: the pebbling should determine a partial homomorphism from A to B.

For each positive integer r, the r-round existential 2-pebble game on A and B is a two-person game of perfect information with a finite number of positions. Therefore, either Spoiler or Duplicator has a *winning strategy* in this game, that is, a strategy winning against every strategy of the opponent. Let $D(A, B)$ denote the minimum r for which Spoiler has a winning strategy. If such r does not exist, we will write $D(A, B) = \infty$. As it is well known [13, Theorem 4.8], $D(A, B) \leq r$ if and only if A can be distinguished from B by a sentence of quantifier rank r in the existential-positive two-variable logic.

Suppose that $D(A, B) < \infty$. We say that Spoiler plays *optimally* if he never loses an opportunity to win as soon as possible. More specifically, after a round is ended in a position P (determined by the pebbled vertices), Spoiler makes the next move according to a strategy that allows him to win from the position P in the smallest possible number of rounds.

Lemma 1. *If Spoiler plays optimally, then the following conditions are true.*

1. *Spoiler uses the pebbles alternately, say, p in odd and q in even rounds.*
2. *Whenever Spoiler moves a pebble, he moves it to a new position. That is, if $x_i \in V(A)$ denotes the vertex choosen in the i-th round, then $x_{i+2} \neq x_i$. Moreover, if $x_{i+1} = x_i$, then $x_{i+2} \neq x_{i-1}$.*
3. *For all i, (x_i, x_{i+1}) or (x_{i+1}, x_i) satisfies at least one binary relation.*

Lemma 1 has several useful consequences. The first of them is that, without loss of generality, we can restrict our attention to connected structures. Two distinct vertices of a binary structure A are adjacent in its *Gaifman graph* G_A if they satisfy at least one binary relation of A. *Connected components* of A are considered with respect to G_A. Let A consist of connected components $A_1, \ldots, A_k$ and B consist of connected components $B_1, \ldots, B_l$. Then it easily follows from part 3 of Lemma 1 that $D(A, B) = \min_i \max_j D(A_i, B_j)$. Another consequence follows from parts 2 and 3.

Corollary 2. *Suppose that the Gaifman graph G_A of A is a tree. If $D(A, B) < \infty$, then $D(A, B) < 2\, v(A)$.*

Furthermore, we now can state a general upper bound for $D(A, B)$.

Corollary 3. *If $D(A, B) < \infty$, then $D(A, B) \leq v(A)v(B) + 1$.*

Proof. Assume that Spoiler plays optimally. Let $x_i \in V(A)$ and $u_i \in V(B)$ denote the vertices pebbled in the i-th round. By part 1 of Lemma 1, we can further assume that Spoiler's move in the $(i + 1)$-th round depends only on the (x_i, u_i). It readily follows that, if the game lasts r rounds, then the pairs $(x_1, u_1), \ldots, (x_{r-1}, u_{r-1})$ are pairwise different, and hence $r - 1 \leq v(A)v(B)$. $\square$

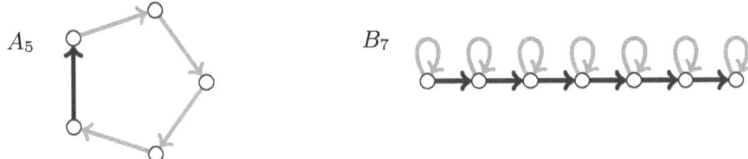

Fig. 1. The DOMINO example

The bound of Corollary 3 is tight, at least, up to a factor of $1/2$. A suitable lower bound can be obtained from the CSP instances that appeared in [9,8] under the name of *DOMINO problem* and were used for benchmarking arc consistency algorithms. A DOMINO instance consists of two digraphs A_m and B_n whose arrows are colored red and blue; see Fig. 1. A_m is a directed cycle of length m with one arrow colored blue and the others red. B_n is a blue directed path where red loops are attached to all its n vertices. Spoiler can win the existential 2-pebble game on A_m and B_n by moving the pebbles along the cycle A_m, always in the same direction. By Lemma 1, this is the only way for him to win in the minimum number of rounds. When Spoiler passes red edges, Duplicator stays with both pebbles at the same vertex of B_n. Only when Spoiler passes the blue edge, Duplicator passes a blue edge in B_n. Thus, if Duplicator starts playing in the middle of B_n, she survives for at least $\frac{1}{2} m(n-1)$ rounds.

3   More Examples of Slow Constraint Propagation

The DOMINO pairs are remarkable examples of binary structures on which constraint propagation is as slow as possible, up to a constant factor of $1/2$. An important role in the DOMINO example is played by the fact that we have two different edge colors. We now show that essentially the same lower bound holds true over a rather restricted class of structures, namely *rooted loopless digraphs*, where edges are uncolored, there is a single color for vertices, and only a single *root* vertex is colored in it. It is also supposed that every vertex of a rooted digraph is reachable from the root along a directed path.

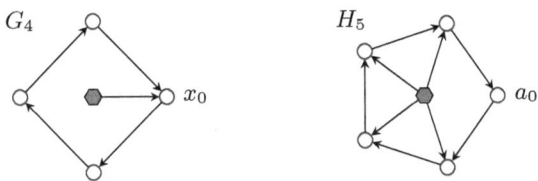

Fig. 2. An example of CO-WHEELS

By the *wheel* W_n we mean the rooted digraph with $n+1$ vertices where there are arrows from the root to all the other n vertices and these vertices form a directed cycle. We call a pair of rooted digraphs G_m and H_n *co-wheels* if G_m is obtained from W_m by removal of all but one of the arrows from the root and H_n is obtained from W_n by removal of one arrow from the root; see an example in Fig. 2.

Lemma 4. *Let G_m and H_n be co-wheels. If m and n are coprime, then $D(G_m, H_n) < \infty$ and $D(G_m, H_n) > \frac{1}{2} m(n-3)$.*

Proof. Let $V(G_m) = \{x_{\text{root}}, x_0, \ldots, x_{m-1}\}$ and $V(H_n) = \{a_{\text{root}}, a_0, \ldots, a_{n-1}\}$. Assume that x_0 is adjacent to the root x_{root} of G_m, a_0 is non-adjacent to the root a_{root} of H_n, and the indices increase in the direction of arrows. We first argue that Spoiler has a winning strategy in the existential 2-pebble game on G_m and H_n. Let Spoiler pebble x_0 in the first round. If Duplicator pebbles the root, then Spoiler wins by pebbling x_{n-1}. Assume that Duplicator responds with a_t. If $t = 0$, Spoiler wins by putting the other pebble on the root. If $t > 0$, Spoiler is able to force pebbling the pair (x_0, a_0) in a number of rounds. Indeed, if Spoiler moves the pebbles alternately along the cycle so that the pebbled vertices are always adjacent, then after ℓm rounds Spoiler passes the cycle ℓ times and arrives again at x_0, while Duplicator is forced to come to $a_{t+\ell m}$, where the index is computed modulo n. Since m and n are coprime, $m \bmod n$ is a generator of the cyclic group $\mathbb{Z}_n$. It follows that the parameter ℓ can be chosen so that $t + \ell m = 0$ (mod n), and then $a_{t+\ell m} = a_0$.

We now have to show that Duplicator is able to survive at least $\frac{1}{2} m(n-3)$ rounds. When considering the length of the game, we can assume that Spoiler plays according to an optimal strategy. It follows by Lemma 1 that Spoiler begins playing in a non-root vertex x_s and forces pebbling the pair (x_0, a_0) as explained above, by moving along the cycle always in the same direction. Let $D(x_s, a_t)$ denote the minimum number of moves needed for Spoiler to reach this configuration if Duplicator's move in the first round is a_t.

Suppose first that $s = 0$ and also that Spoiler moves in the direction of arrows. Then he can force pebbling (x_0, a_0) only in ℓm moves with ℓ satisfying $t + \ell m = 0$ (mod n). Denote $r = \lfloor n/2 \rfloor$ and let Duplicator choose $t = (-rm) \bmod n$. Then the smallest possible positive value of ℓ is equal to r. If Spoiler decides to move in the opposite direction, we have the relation $t - \ell m = 0$ (mod n), which gives us $\ell \geq \lceil n/2 \rceil$. In both cases $D(x_0, a_t) \geq \frac{1}{2} m(n-1)$.

Suppose now that $s > 0$. Let Duplicator pebble $a_{t'}$ in the first round with $t' = (t + s) \bmod n$, where t is fixed as above. Note that, from the position (x_0, a_t), Spoiler can force the position $(x_s, a_{t'})$ in s moves. Therefore, $D(x_0, a_t) \leq s + D(x_s, a_{t'})$, which implies that $D(x_s, a_{t'}) \geq D(x_0, a_t) - (m-1) > \frac{1}{2} m(n-3)$, as claimed. $\qquad\square$

The parameter $D(G_m, H_n)$ does not change if we attach extra arrows to the roots of the digraphs. In this way, Lemma 4 leads to the following result.

Theorem 5. *For every pair of numbers $M \geq 5$ and $N \geq 5$, there is a pair of rooted loopless digraphs G and H with $v(G) = M$ and $v(H) = N$ such that $D(G, H) < \infty$ and $D(G, H) \geq (\frac{1}{2} - o(1))MN$. Here the $o(1)$-term is a function of $\min(M, N)$.*

Fig. 3. Co-Wheels as colored graphs

Using a simple gadget, in the Co-Wheels pattern we can make edges undirected simulating directions by vertex colors. In this way, we can construct examples of pairs with large $D(G, H)$ also for colored graphs; see Fig. 3.

Corollary 6. *Theorem 5 holds true also for colored undirected graphs with bound* $D(G, H) \geq (\frac{1}{6} - o(1))MN$.

Corollary 6 can be obtained also from the Domino pattern, though with a smaller factor $\frac{1}{8} - o(1)$; see Fig. 4. It is worth noting that G will be a unicyclic graph while H will be a tree. Note that this result is best possible in the sense that, by Corollary 2, G cannot be acyclic.

Corollary 7. *For every* $M \geq 7$ *there is a unicyclic colored graph* G_M *with* M *vertices and for every* $N \geq 1$ *there is a tree* H_N *with* N *vertices such that* $D(G_M, H_N) < \infty$ *and* $D(G_M, H_N) > \frac{1}{8}(M - 1)(N - 5)$.

Fig. 4. The colored graphs obtained from the Domino example in Fig. 1.

4   Winner Proof Systems

Inspired by [3], we now introduce a notion that allows us to define a few useful parameters measuring the speed of constraint propagation. In the next section it will serve as a link between the length of the existential 2-pebble game on (G, H) and the running time of an AC algorithm on input (G, H).

Let G and H be connected colored graphs, both with at least 2 vertices, and $N(x)$ denote the set of vertices adjacent to x. A *proof system of Spoiler's win* on (G, H) consists of *axioms*, that are pairs $(y, b) \in V(G) \times V(H)$ with y and b colored differently, and derivations of pairs $(x, a) \in V(G) \times V(H)$ and a special symbol $\perp$ by the following *rules*:

- (x, a) is derivable from any set $\{y\} \times N(a)$ such that $y \in N(x)$;
- $\perp$ is derivable from a set $\{y\} \times V(H)$.

A *proof* is a sequence $P = p_1, \ldots, p_{\ell+1}$ such that if $i \leq \ell$, then $p_i \in V(G) \times V(H)$ and it is either an axiom or is derived from a set $\{p_{i_1}, \ldots, p_{i_s}\}$ of preceding pairs p_{i_j}; also, $p_{\ell+1} = \perp$ is derived from a set of preceding elements of P. More precisely, we regard P as a dag on $\ell + 1$ nodes where a derived p_i sends arrows to each p_{i_j} used in its derivation. Moreover, we always assume that P contains a directed path from $\perp$ to each node, that is, every element of P is used while deriving $\perp$.

We define the *length* and the *size* of the proof P as $length(P) = v(P) - 1$ and $size(P) = e(P)$ respectively. Note that $length(P)$ is equal to ℓ, the total number of axioms and intermediate derivations in the proof. Since it is supposed that the underlying graph of P is connected, we have $length(P) \leq size(P)$, with equality exactly when P is a tree. The *depth* of P will be denoted by $depth(P)$ and defined to be the length of a longest directed path in P. Obviously, $depth(P) \leq length(P)$.

It is easy to show that a proof P exists iff $D(G, H) < \infty$ (see part 1 of Theorem 8 below). Given such G and H, define the *(proof) depth* of (G, H) to be the minimum depth of a proof for Spoiler's win on (G, H). The *(proof) length* and the *(proof) size* of (G, H) are defined similarly. We denote the three parameters by $depth(G, H)$, $length(G, H)$, and $size(G, H)$, respectively. Note that $depth(G, H) \leq length(G, H) \leq size(G, H)$.

Theorem 8. *Let G and H be connected colored graphs, both with at least 2 vertices, such that $D(G, H) < \infty$.*

1. *$depth(G, H) = D(G, H)$.*
2. *$depth(G, H) \leq length(G, H) \leq v(G)v(H)$ and this is tight up to a constant factor: for every pair of integers $M, N \geq 2$ there is a pair of colored graphs G, H with $v(G) = M$ and $v(H) = N$ such that $depth(G, H) \geq (\frac{1}{6} - o(1)) MN$.*
3. *$size(G, H) < 2\,v(G)e(H) + v(H)$.*
4. *For every N there is a pair of colored graphs G_N and H_N both with N vertices such that $size(G_N, H_N) > \frac{1}{128} N^3$ for all large enough N.*

Note that part 3 implies that $size(G, H) < N^3$ if both both G and H have N vertices. Therefore, part 4 shows that the upper bound of part 3 is tight up to a constant factor.

Proof. **1.** It suffices to prove that, for every $r \geq 0$, Spoiler has a strategy allowing him to win in at most r rounds starting from a position (x, a) if and only if the pair (x, a) is derivable with depth r. This equivalence follows by a simple inductive argument on r.

2. The upper bound on $depth(G, H)$ follows from a simple observation that any proof can be rewritten so that every axiom used and every derived pair appears in it exactly once. The lower bound follows by part 1 from Corollary 6.

3. Consider a proof P where each pair (x, a) appears at most once. Since the derivation of (x, a) contributes $\deg a$ arrows in P, and the derivation of $\perp$ contributes $v(H)$ arrows, we have

$$size(P) < \sum_{(x,a)} \deg a + v(H) = v(G) \sum_{a} \deg a + v(H) = 2\,v(G)e(H) + v(H).$$

The inequality is strict because there must be at least one axiom node, which has out-degree 0.

4. Note that $size(G,H) \geq depth(G,H)\delta(H)$, where $\delta(H)$ denotes the minimum vertex degree of H. Therefore, we can take graphs G and H with almost the same number of vertices and with quadratic $depth(G,H)$, and make $\delta(H)$ large by adding linearly many universal vertices of a new color to each of the graphs. A *universal* vertex is adjacent to all other vertices in the graph. If each of the graphs receives at least two new vertices, they have no influence on the duration of the existential 2-pebble game.

More specifically, we use the co-wheels from Lemma 4 with coprime parameters $m = n-1$ converted to colored graphs as in Corollary 6; see Fig. 3. Thus, we have colored graphs G and H with $v(G) = 3n - 2$ and $v(H) = 3n + 1$ such that $D(G,H) > \frac{1}{2}(n-1)(n-3)$. Add green universal vertices so that the number of vertices in each graph becomes $N = \lfloor \frac{9}{2}n \rfloor$. For the new graphs G_N and H_N we still have $depth(G_N, H_N) > \frac{1}{2}(n-1)(n-3)$ while now $\delta(H_N) \geq \frac{3}{2}n$. $\qquad\square$

5   Time Complexity of Arc Consistency

5.1   An Upper Bound

We now establish an upper bound of $O(v(G)e(H) + e(G)v(H))$ for the time complexity of the AC-PROBLEM. One way to obtain this result is to use the linear-time reduction from arc consistency to the satisfiability problem for propositional Horn clauses (HORN-SAT) presented by Kasif [11]. The reduction transforms the input graphs G and H into a propositional Horn formula of size $v(G)e(H) + e(G)v(H)$ that is satisfiable iff arc consistency can be established on G and H. The upper bound then follows by applying any linear time HORN-SAT algorithm. Going a different way, we here show that the same bound can be achieved by a propagation-based algorithm, that we call AC'13. On the one hand, AC'13 does much the same of what a linear time HORN-SAT solver would do (after applying Kasif's reduction). On the other hand, it can be seen as a slightly accelerated version of the algorithm AC-4 in [17].

Theorem 9. *AC'13 solves the* AC-PROBLEM *in time* $O(v(G)e(H)+e(G)v(H))$.

Proof. A detailed proof of the algorithm's correctness can be found in a full version of the paper [6]. Let us analyze the running time. The initialization phase requires $O(v(G)v(H))$ steps. The propagation phase takes $\deg a$ steps for every $(x,a) \in Q$ and $\deg x$ steps for every (x,b) such that `counter[x,b]` gets 0. Since every pair is only put once on the queue and every counter voids out only once, the total running time of the propagation phase is bounded by $\sum_{(x,a)\in V(G)\times V(H)} (\deg x + \deg a) = v(G)e(H) + e(G)v(H).$ $\qquad\square$

Algorithm 1. AC'13

Input: Two colored connected graphs G and H.
/*INITIALIZATION*/
Let Q be an empty queue.
for all $x \in V(G)$ **do**
    $D_x \leftarrow \{a \in V(H) \mid a$ has the same color as $x\}$;
    if $D_x = \emptyset$ **then return** reject;
for all $x \in V(G),\, a \in V(H)$ **do**
    counter$[x,a] \leftarrow \deg a$;
    if $a \notin D_x$ **then** add (x,a) to Q;
/*PROPAGATION*/
while Q not empty **do**
    Select and remove (x,a) from Q;
    for all $b \in N(a)$ **do**
        counter$[x,b] \leftarrow$ counter$[x,b]-1$;
        if counter$[x,b]= 0$ **then**
            for all $y \in N(x)$ **do**
                if $b \in D_y$ **then**
                    Delete b from D_y;
                    Add (y,b) to Q;
                    if $D_y = \emptyset$ **then return** reject;
  return accept;

5.2 Lower Bounds

Recall that by a *propagation-based arc consistency algorithm* we mean an algorithm that solves the AC-PROBLEM by iteratively deleting possible assignments a to a variable x from the domain D_x according to the arc consistency rule and rejects iff one domain gets empty. Let us maintain a list L of deleted variable–value pairs by putting a pair (x,a) there once a is deleted from D_x. If the algorithm detects arc inconsistency, then it is evident that L, prepended with axioms and appended with $\perp$, forms a proof of Spoiler's win. Thus, a propagation-based arc consistency algorithm can be viewed as a proof search algorithm that produces (in passing) a proof P of Spoiler's win. This situation is related to the concept of a *certifying algorithm* [16]: Propagation-based algorithms not only detect Spoiler's win but also produce its certificate. For every derived element of P, an algorithm has to recognize its already derived parents. This allows us to relate the running time to the proof size. Specifically, given an arbitrary propagation-based algorithm for the AC-PROBLEM, let $T(G, H)$ denote the time it takes on input (G, H). If the input (G, H) is arc inconsistent, then

$$T(G, H) \geq size(G, H). \tag{1}$$

Theorem 10. *Fix an arbitrary propagation-based algorithm.*

1. *Let $T_1(k, l)$ denote the worst-case running time of this algorithm over colored graphs G and H with $e(G) = k$ and $e(H) = l$. Then $T_1(k, l) > \frac{1}{8}(k-1)(l-4)$ for all k and l.*

2. Let $T_2(n)$ denote the worst-case running time of the algorithm on inputs (G, H) with $v(G) + v(H) = n$. Then $T_2(n) > \frac{1}{16} n^3$ for all large enough n.

Proof. By Corollary 7, there are colored graphs G_k with $e(G_k) = v(G_k) = k$ and H_l with $e(H_l) = v(H_l) - 1 = l$ for which $\frac{1}{8}(k-1)(l-4) < D(G_k, H_l) < \infty$. By the relation (1), on input (G_k, H_l) the algorithm takes time at least $size(G_k, H_l)$, for which we have $size(G_k, H_l) \geq depth(G_k, H_l) = D(G_k, H_l)$ by part 1 of Theorem 8.

Part 2 follows from part 4 of Theorem 8.   □

Corollary 11. *In terms of the parameters $e(G)$ and $e(H)$, the time bound $O(e(G) \cdot e(H))$ is optimal up to a constant factor among propagation-based algorithms.*

Note that $O(e(G)v(H) + v(G)e(H)) = O((v(G) + v(H))^3)$.

Corollary 12. *In terms of the parameter $n = v(G) + v(H)$, the time bound $O(n^3)$ is best possible for a propagation-based algorithm.*

5.3   Parallel Complexity

It is known that the AC-PROBLEM is PTIME-complete under logspace reductions [11,12]. Under the assumption that PTIME $\neq$ NC, it follows that the AC-PROBLEM cannot be parallelized. However, several parallel algorithms with a polynomial number of processors appear in the literature (e.g., [19]). We are able to show a tight connection between the running time of a parallel algorithm and the round complexity of the existential 2-pebble game. The following result is worth noting since $D(G, H) = depth(G, H)$ can be much smaller than $size(G, H)$, and then a parallel propagation-based algorithm can be much faster than any sequential propagation-based algorithm.

Theorem 13.   *1. AC-PROBLEM can be solved in time $O(D(G, H))$ on a CRCW-PRAM with polynomially many processors.*
   2. Any parallel propagation-based arc consistency algorithm needs time $D(G, H)$ on arc inconsistent instances (G, H).

6   Conclusion and Further Questions

We investigated the round complexity $D(G, H) = D^2(G, H)$ of the existential 2-pebble game on colored graphs and established lower bounds of the form $\Omega(v(G)v(H))$, which translate to lower bounds on the nested propagation steps in arc consistency algorithms. The next step in this line of research is to investigate the number of rounds $D^3(G, H)$ in the existential 3-pebble game that interacts with *path consistency* (i.e., 3-consistency) algorithms in the same way as the 2-pebble game with arc consistency. Note that $D^3(G, H) = O(v(G)^2 v(H)^2)$ and we conjecture that this bound is tight.

Finally, we want to stress that our lower bounds for the time complexity of arc consistency hold only for constraint propagation-based algorithms. Is there a faster way to solve the AC-PROBLEM using a different approach?

References

1. Atserias, A., Bulatov, A.A., Dalmau, V.: On the power of k-consistency. In: Arge, L., Cachin, C., Jurdziński, T., Tarlecki, A. (eds.) ICALP 2007. LNCS, vol. 4596, pp. 279–290. Springer, Heidelberg (2007)
2. Atserias, A., Dalmau, V.: A combinatorial characterization of resolution width. J. Comput. Syst. Sci. 74(3), 323–334 (2008)
3. Atserias, A., Kolaitis, P.G., Vardi, M.Y.: Constraint propagation as a proof system. In: Wallace, M. (ed.) CP 2004. LNCS, vol. 3258, pp. 77–91. Springer, Heidelberg (2004)
4. Berkholz, C.: Lower bounds for existential pebble games and k-consistency tests. In: LICS 2012, pp. 25–34. IEEE Computer Society, Los Alamitos (2012)
5. Berkholz, C.: On the complexity of finding narrow proofs. In: FOCS 2012, pp. 351–360. IEEE Computer Society, Los Alamitos (2012)
6. Berkholz, C., Verbitsky, O.: On the speed of constraint propagation and the time complexity of arc consistency testing. E-print: arxiv.org/abs/1303.7077 (2013)
7. Bessière, C.: Constraint Propagation. In: Handbook of Constraint Programming. Elsevier, Amsterdam (2006)
8. Bessière, C., Régin, J.C., Yap, R.H.C., Zhang, Y.: An optimal coarse-grained arc consistency algorithm. Artificial Intelligence 165(2), 165–185 (2005)
9. Dechter, R., Pearl, J.: A problem simplification approach that generates heuristics for constraint-satisfaction problems. Tech. rep., Cognitive Systems Laboratory, Computer Science Department, University of California, Los Angeles (1985)
10. Feder, T., Vardi, M.Y.: The computational structure of monotone monadic SNP and constraint satisfaction: A study through datalog and group theory. SIAM Journal on Computing 28(1), 57–104 (1998)
11. Kasif, S.: On the parallel complexity of discrete relaxation in constraint satisfaction networks. Artificial Intelligence 45(3), 275–286 (1990)
12. Kolaitis, P.G., Panttaja, J.: On the complexity of existential pebble games. In: Baaz, M., Makowsky, J.A. (eds.) CSL 2003. LNCS, vol. 2803, pp. 314–329. Springer, Heidelberg (2003)
13. Kolaitis, P.G., Vardi, M.Y.: On the expressive power of datalog: Tools and a case study. J. Comput. Syst. Sci. 51(1), 110–134 (1995)
14. Kolaitis, P.G., Vardi, M.Y.: A game-theoretic approach to constraint satisfaction. In: Kautz, H.A., Porter, B.W. (eds.) AAAI/IAAI 2000, pp. 175–181. AAAI Press/The MIT Press, California (2000)
15. Mackworth, A.K.: Consistency in networks of relations. Artificial Intelligence 8(1), 99–118 (1977)
16. McConnell, R.M., Mehlhorn, K., Näher, S., Schweitzer, P.: Certifying algorithms. Computer Science Review 5(2), 119–161 (2011)
17. Mohr, R., Henderson, T.C.: Arc and path consistency revisited. Artificial Intelligence 28(2), 225–233 (1986)
18. Régin, J.-C.: AC-*: A configurable, generic and adaptive arc consistency algorithm. In: van Beek, P. (ed.) CP 2005. LNCS, vol. 3709, pp. 505–519. Springer, Heidelberg (2005)
19. Samal, A., Henderson, T.: Parallel consistent labeling algorithms. International Journal of Parallel Programming 16, 341–364 (1987)

Validity of Tree Pattern Queries with Respect to Schema Information

Henrik Björklund[1,*], Wim Martens[2,**], and Thomas Schwentick[3]

[1] Umeå University, Sweden
[2] University of Bayreuth, Germany
[3] TU Dortmund University, Germany

Abstract. We prove that various containment and validity problems for tree pattern queries with respect to a schema are EXPTIME-complete. When one does not require the root of a tree pattern query to match the root of a tree, validity of a non-branching tree pattern query with respect to a Relax NG schema or W3C XML Schema is already EXPTIME-hard when the query does not branch and uses only child axes. These hardness results already hold when the alphabet size is fixed. Validity with respect to a DTD is proved to be EXPTIME-hard already when the query only uses child axes and is allowed to branch only once.

1 Introduction

Tree pattern queries are omnipresent in query and schema languages for XML. They form a logical core of the query languages XPath, XQuery, and XSLT, and they are needed to define key constraints in XML Schema. Static analysis problems such as containment, satisfiability, validity, and minimization for tree pattern queries have been studied for over a decade [15,10,17,3] since their understanding helps us, for example, in the development of query optimization procedures. Since queries can usually be optimized more if schema information is taken into account, these static analysis problems are also relevant in settings with schema information [17,3]. This is the setting that we consider.

The literature uses the term "tree pattern query" for a variety of query languages. In this paper, we use the tree pattern queries as in [15], which can use labels, wildcards (*), the child relation (/), the descendant relation (//), and filtering ([·]) which allows them to branch. In the following, we us the terms *path query* for tree pattern queries without [·] and *child-only query* for tree pattern queries without //. Containment, satisfiability, and validity of tree pattern queries are closely related to each other in the usual way, i.e., satisfiability and validity are special cases of containment. Since tree pattern queries are not closed under the Boolean operations, satisfiability and validity often have a lower complexity than containment. Taking schema information into account usually

* Supported by the Swedish Research Council grant 621-2011-6080.
** Supported by grant number MA 4938/2–1 from the Deutsche Forschungsgemein-schaft (Emmy Noether Nachwuchsgruppe).

K. Chatterjee and J. Sgall (Eds.): MFCS 2013, LNCS 8087, pp. 171–182, 2013.
© Springer-Verlag Berlin Heidelberg 2013

increases the computational complexity. For example, containment of tree pattern queries is coNP-complete [15] but becomes EXPTIME-complete if schema information, even in its weakest form (a DTD) is provided [17].[1]

We investigate the complexity of the validity problem (with schema information) and obtain complexity lower bounds that contrast rather sharply with known upper bounds. Hashimoto et al. [12] showed that validity of path queries with respect to DTDs is in PTIME. We prove:

- Validity of path queries with respect to tree automata is EXPTIME-hard, even if the tree automata are XSDs with a constant-size alphabet (Theorems 10 and 11).
- Validity of child-only tree pattern queries with respect to DTDs is already EXPTIME-hard even if the tree pattern queries branch only once and the branch has only one node (Theorem 12).
- As a simpler application of our techniques we prove as a warm-up: inclusion of a DFA in a regular expression of the form $\Sigma^* a \Sigma^n b \Sigma^*$ is PSPACE-complete over $\Sigma = \{a, b, c\}$ (Theorem 9). This means that validity of very simple child-only path queries is PSPACE-hard, even if trees don't branch.[2]

Each case is only a very slight extension of the above mentioned PTIME scenario of Hashimoto et al. [12]. Our semantics of path and tree pattern queries is such that the root of the query does not need to be matched by the root of the tree. For our EXPTIME-hardness results to hold when using the more restricted semantics of [12], we would need queries to have one additional descendant axis, placed at the root. On the other hand, the PTIME upper bound of [12] also holds in our setting.

Our lower bounds are also relevant in terms of conjunctive queries over trees. For example, Benedikt et al. ([2], Corollary 3) proved a matching EXPTIME upper bound for validity of UCQs (Unions of Conjunctive Queries) with respect to a tree automaton. Here, UCQs form a class of queries that do not use the descendant axis but are strictly more general than child-only tree pattern queries since their syntactic structure is not required to be tree-shaped. Recently, static analysis for such queries (with schema information) has also been investigated in [5,16], with complexity results ranging from tractable to 2EXPTIME-complete.

In our proofs we use restricted variants of tiling games (Section 3) that may be interesting in their own right.

2   Preliminaries

We use standard definitions and notation for regular expressions and deterministic finite automata (DFAs). DFAs are denoted as tuples $(Q, \Sigma, \delta, \{q_0\}, F)$ where Q is the (finite) set of states, Σ is the (finite) alphabet, δ the transition function, q_0 the initial state and $F \subseteq Q$ the set of accepting states.

---

[1] Schemas can be given as DTDs (the weakest form), XSDs (in the middle), or tree automata (the strongest form; defining regular tree languages), see [14].

[2] This result has already been used in the context of XML key inference [1].

Trees and Tree Pattern Queries. Schema languages for XML recognize trees which are rooted, ordered, finite, labeled, unranked, and directed from the root downwards. For this reason, we consider finite trees in which nodes can have arbitrarily many children, ordered from left to right. However, we note that the results in this paper hold equally well for automata and DTDs recognizing unordered trees, that is, trees in which the children can occur in any order. More formally, we view a tree t as a relational structure over a finite number of unary labeling relations $a(\cdot)$, for $a \in \Sigma$, and binary relations $Child(\cdot, \cdot)$ and $NextSibling(\cdot, \cdot)$. Here, $a(u)$ expresses that u is a node with label a, and $Child(u, v)$ (respectively, $NextSibling(u, v)$) expresses that v is a child (respectively, the right sibling) of u. We denote the set of nodes of a tree t by $\mathrm{Nodes}(t)$. We assume that trees are non-empty, i.e., $\mathrm{Nodes}(t) \neq \emptyset$. By $\mathrm{Edges}(t)$ we denote the set of child edges of t. For a node u, we denote by $\mathrm{lab}^t(u)$ the unique symbol a such that $a(u)$ holds in t. We often omit t from this notation when t is clear from the context. By $\mathrm{root}(t)$ we denote the root node of t. For a node u of t, we denote by $\text{anc-str}^t(u)$ the string obtained by concatenating all labels on the path from the root of t to u. That is, $\text{anc-str}^t(u) = \mathrm{lab}^t(u_1) \cdots \mathrm{lab}^t(u_k)$ where $u_1 = \mathrm{root}(t)$, $u_k = u$, and $u_1 \cdots u_k$ is the path from u_1 to u_k. Similarly, $\text{ch-str}^t(u)$ is the concatenation of the labels of all children of u, from left to right.

Definition 1. [Tree Pattern Query, Path Query] A *tree pattern query, (TPQ)*, over Σ is a tuple $T = (p, \mathrm{Anc})$, where p is a tree that uses the labeling alphabet $\Sigma \uplus \{*\}$ and $\mathrm{Anc} \subseteq \mathrm{Edges}(p)$ is the set of *ancestor edges*. A *path query* is a TPQ in which each node in p has at most one child.

Here, we use $*$ as a wildcard label. More formally, the semantics of TPQs is defined as follows. Let $T = (p, \mathrm{Anc})$ be a TPQ and let s be a tree. Let $v_p \in \mathrm{Nodes}(p)$ and $v_s \in \mathrm{Nodes}(s)$. We say that v_s *matches* v_p if either $\mathrm{lab}(v_p) = *$ or $\mathrm{lab}(v_s) = \mathrm{lab}(v_p)$. An *embedding* of $T = (p, \mathrm{Anc})$ on a tree s is a mapping m from $\mathrm{Nodes}(p)$ to $\mathrm{Nodes}(s)$ such that,

- for every node $v \in \mathrm{Nodes}(p)$, $m(v)$ matches v, and
- for every two nodes $v_1, v_2 \in \mathrm{Nodes}(p)$,
  - if $(v_1, v_2) \in \mathrm{Edges}(p) \setminus \mathrm{Anc}$, then $(m(v_1), m(v_2)) \in \mathrm{Edges}(s)$;
  - if $(v_1, v_2) \in \mathrm{Anc}$, then $(m(v_1), m(v_2))$ is in the transitive closure of $\mathrm{Edges}(s)$.

Notice that the root of p does not need to be mapped to the root of s, which is important when comparing our results to related work. The *language* defined by T is denoted $L(T)$ and consists of all trees s for which there is an embedding of T into s. Notice that our semantics defines tree pattern queries as Boolean queries. Tree pattern queries form a natural fragment of the XPath query language [8]. We assume familiarity with the standard XPath notation of tree pattern queries (see, e.g., [15]). Figure 1 contains an example of a tree pattern query and its corresponding XPath notation.

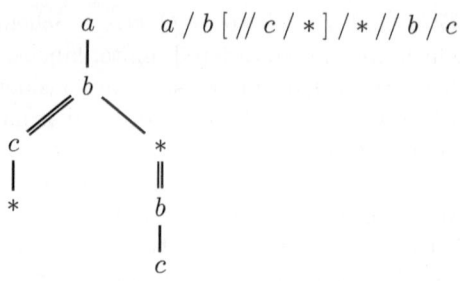

Fig. 1. A tree pattern query $T = (p, \text{Anc})$ depicted as a tree (on the left) and in XPath notation (on the right). On the left, edges in Anc are drawn as double lines. On the right, the bracketed part corresponds to the left branch in the tree. Slashes ('/') represent edges and double slashes ('//') represent edges in Anc.

Schemas. We introduce our abstractions of Document Type Definition (DTD) [6], XML Schema [19], and Relax NG schemas [9].

Definition 2. A *Document Type Definition (DTD)* over Σ is a triple $D = (\Sigma, d, S)$ where $S \subseteq \Sigma$ is the set[3] of start symbols and d is a set of rules of the form $a \rightarrow R$, where $a \in \Sigma$ and R is a regular expression over Σ. No two rules have the same left-hand side.

A tree t *satisfies* D if *(i)* $\text{lab}^t(\text{root}(t)) \in S$ and, *(ii)* for every $u \in \text{Nodes}(t)$ with label a and n children $u_1, \ldots, u_n$ from left to right, there is a rule $a \rightarrow R$ in d such that $\text{lab}^t(u_1) \cdots \text{lab}^t(u_n) \in L(R)$. By $L(D)$ we denote the set of trees satisfying D.

We abstract XML Schema Definitions as *DFA-based XSDs*. DFA-based XSDs were introduced by Martens, Neven, Schwentick, and Bex [14,13] as formal model for XML Schema convenient in proofs.[4]

Definition 3. A *DFA-based XSD* is a pair (A, λ), where $A = (Q, \Sigma, \delta, \{q_{\text{init}}\}, \emptyset)$ is a DFA with initial state q_{init} and λ is a function mapping each state in $Q \setminus \{q_{\text{init}}\}$ to a regular expression over Σ.

An tree t *satisfies* (A, λ) if, for every node u, $A(\text{anc-str}^t(u)) = \{q\}$ implies that $\text{ch-str}^t(u)$ is in the language defined by $\lambda(q)$.

We abstract from Relax NG schemas [9] by unranked tree automata.

Definition 4. A *nondeterministic (unranked) tree automaton (NTA)* over Σ is a quadruple $A = (Q, \Sigma, \delta, F)$, where Q is a finite set of states, $F \subseteq Q$ is the set of accepting states, and δ is a set of transition rules of the form $(q, a) \rightarrow L$,

---

[3] DTDs usually have a single start symbol in the literature. Our abstraction is slightly closer to reality; it has no influence on our complexity results.

[4] XML Schema Definitions are sometimes also abstracted as single-type EDTDs, but it is well-known that DFA-based XSDs and single-type EDTDs can be converted back and forth in polynomial time [11]. DFA-based XSDs [13] are called DFA-based *DTDs* in [11] but are the same thing. Since they are a formal model for XSDs, we choose to reflect this in their name.

where $q \in Q$, $a \in \Sigma$, and L is a regular string language over Q, represented by a regular expression.[5]

A *run* of A on a tree t is a labeling $r : \text{Nodes}(t) \to Q$ such that, for every $u \in \text{Nodes}(t)$ with label a and children $u_1, \ldots, u_n$ from left to right, there exists a rule $(q, a) \to L$ such that $r(u) = q$ and $r(u_1) \cdots r(u_n) \in L$. Note that when u has no children, the criterion reduces to $\varepsilon \in L$, where ε denotes the empty string. A run on t is *accepting* if the root of t is labeled with an accepting state, that is, $r(\text{root}(t)) \in F$. A tree t is *accepted* if there is an accepting run of A on t. The set of all accepted trees is denoted by $L(A)$ and is called a *regular tree language*. From now on, we use the word "schema" to refer to DTDs, DFA-based XSDs, or NTAs.

It is well-known that DTDs are less expressive than DFA-based XSDs, which in turn are less expressive than NTAs [14]. Likewise, DTDs can be polynomial-time converted into DFA-based XSDs, which can be polynomial-time converted into NTAs.

We are concerned with the following decision problem:

Definition 5. *Validity w.r.t. a schema*: Given a TPQ T and a schema S, is $L(S) \subseteq L(T)$?

3   Tiling Problems and Games

We recall definitions and properties of tiling systems, corridor tilings, and their associated games. We define a restricted form of corridor tiling games that remains EXPTIME-complete and may be of interest in its own right.

A *tiling system* $S = (T, V, H, t_{\text{fin}})$ consists of a finite set T of *tiles*, two sets $V, H \subseteq T \times T$ of *vertical* and *horizontal constraints*, respectively, and a *final tile* $t_{\text{fin}} \in T$. A *solution* for a tiling system S is a mapping $\tau : \{1, \ldots, n\} \times \{1, \ldots, m\} \to T$ for some $n, m \geq 2$ such that (i) the horizontal constraints are fulfilled, that is, for every $i \in \{1, \ldots, n-1\}, j \in \{1, \ldots, m\}$: $(\tau(i, j), \tau(i+1, j)) \in H$; (ii) the vertical constraints are fulfilled, that is, for every $i \in \{1, \ldots, n\}, j \in \{1, \ldots, m-1\}$: $(\tau(i, j), \tau(i, j+1)) \in V$; and (iii) the final tile is correct, that is, $\tau(n, m) = t_{\text{fin}}$.

In the *corridor tiling* problem one is given a tiling system S and a word $w = w_1 \cdots w_n \in T^*$ of tiles, called the *initial row*. The problem asks whether there exists a solution to S with bottom row w, that is a mapping $\tau : \{1, \ldots, n\} \times \{1, \ldots, m\} \to T$ as above with $n = |w|$ such that $\tau(i, 1) = w_i$ for every $i \in \{1, \ldots, n\}$.

It is well-known that the corridor tiling problem is PSPACE-complete [7]. However, this result even holds for some fixed tiling systems S. For a tiling system S, we write $\text{Tiling}(S)$ for the set of strings w such that S has a solution with initial row w.

---

[5] For our complexity results, it does not matter whether the languages L are represented by regular expressions, nondeterministic string automata, deterministic string automata, or even as a finite set of strings.

Theorem 6 ([7], Section 4). *There is a tiling system S such that Tiling(S) is PSPACE-hard.*

This strengthening can be obtained by applying the argument of Section 4 in [7] to some fixed PSPACE-complete language L and a TM M for L.

Tiling systems can also be used to define two-player games. The input for a tiling game is the same as for the corridor tiling problem but the underlying idea is different: two players, CONSTRUCTOR and SPOILER, alternatingly choose tiles. CONSTRUCTOR's goal is to build a solution for the tiling system and SPOILER's goal is to prevent that.

More formally, we associate with a tiling system S and an initial row w for S a 2-player game as follows. The word w induces a mapping $\tau : \{1, \ldots, n\} \times \{1\} \to T$, where $n = |w|$. The two players alternatingly choose tiles $t \in T$, implicitly defining $\tau(1,2), \tau(2,2), \ldots, \tau(n,2), \tau(1,3)$, etc. A move is *legal* if it satisfies the constraints. More precisely, a tile t is a legal move as $\tau(i,j)$ if $(\tau(i-1,j), \tau(i,j)) \in H$ and $(\tau(i,j-1), \tau(i,j)) \in V$. Players are not allowed to play a non-legal move. CONSTRUCTOR loses the game if, at any point, one of the players cannot make a legal move. On the other hand, CONSTRUCTOR wins if at some point a correct corridor tiling for S is constructed (for some m).

For a tiling system S, we denote by TilingWinner(S) the set of all strings w such that CONSTRUCTOR has a winning strategy for the game induced by S and w. From [7] the following theorem immediately follows.[6]

Theorem 7 ([7], Theorem 5.1).
(a) For every tiling system S, TilingWinner$(S) \in$ EXPTIME, and
(b) there is a tiling system S, for which TilingWinner(S) is EXPTIME-hard.

For our reductions we need to work with suitably restricted tiling systems which we define next. Given a system S and an initial row w, a *valid rectangle* for S and w is a tiling that is a solution except that the last tile need not be t_{fin}, i.e., it is a mapping $R : \{1, \ldots, n\} \times \{1, \ldots, m\} \to T$ with initial row w, where $n = |w|$ and $m \geq 1$ that respects V and H. A *tiling prefix* for S and w is a valid rectangle plus the beginning of a next row, that is, a mapping P from $\{1, \ldots, n\} \times \{1, \ldots, m\} \cup \{1, \ldots, i\} \times \{m+1\}$ to T, for some $i \in \{1, \ldots, n\}$ and $m \geq 1$, with bottom row w that respects V and H. A tiling prefix for S and w is *valid* if the partial row can be completed to form a valid rectangle. We define the *length* of P to be $nm + i$. In particular, every valid rectangle is also a valid prefix. Given a tiling prefix P and a tile t, we write $P.t$ for the extension of P by t.

We call a tiling system $S = (T, V, H, t_{\text{fin}})$ *restricted* if the following holds for every initial row w.

(1) If $|w|$ is odd, then $w \notin$ Tiling(S).
(2) For every valid prefix P there are exactly two tiles t_1 and t_2 such that $P.t_1$ and $P.t_2$ are valid prefixes.

---

[6] Similarly as for Tiling(S), it suffices to fix an ATM for some EXPTIME-complete language in the proof to infer Theorem 7 (b) from Theorem 5.1 in [7].

(3) For every odd length valid prefix P, there are exactly two tiles t_1 and t_2 such that $P.t_1$ and $P.t_2$ are tiling prefixes.

The restriction guarantees that, if CONSTRUCTOR has a winning strategy, she has one in which SPOILER always has exactly two legal moves.

Proposition 8. *There is a restricted tiling system S, for which Tiling Winner(S) is EXPTIME-hard.*

4   String Languages

Theorem 9. *Validity of a regular expression r w.r.t. a DFA A, i.e., whether $L(A) \subseteq L(r)$, is PSPACE-complete even for regular expressions[7] of the form $\Sigma^* a \Sigma^n b \Sigma^*$ and DFAs over the alphabet $\Sigma = \{a, b, c\}$.*

Proof. Obviously, the problem is in PSPACE since it can be reduced in logarithmic space to the containment problem for NFAs, which is known to be PSPACE-complete [18]. We show the lower bound by reduction from Tiling(S), i.e., corridor tiling with a fixed tiling system. Let $S = (T, V, H, t_{\text{fin}})$ with $T = \{t_1, \ldots, t_k\}$ be a tiling system such that Tiling(S) is PSPACE-hard. Notice that S exists by Theorem 6. Given S and an initial row w, we will construct a regular expression q_w and a DFA $A(S, w)$ such that $L(A(S, w)) \subseteq L(q_w)$ if and only if $w \notin$ Tiling(S). Since PSPACE is closed under complement, this yields the desired result.

We associate with every tiling function $\tau : \{1, \ldots, n\} \times \{1, \ldots, m\} \to T$ a string w_τ by simply concatenating all tiles in row-major order. More precisely, w_τ is the string from T^* of length mn that carries at position $(j - 1)n + i$ tile $\tau(i, j)$, for every $i \in \{1, \ldots, n\}$ and $j \in \{1, \ldots, m\}$.

From S and an initial row w a DFA $A(S, w)$ can be constructed in polynomial time that tests whether a word $v \in T^*$ has the following properties: (i) v has prefix w; (ii) the length of v is a multiple of $n \stackrel{\text{def}}{=} |w|$; (iii) the tiling function $\tau : \{1, \ldots, n\} \times \{1, \ldots, m\} \to T$ corresponding to v fulfills the horizontal constraints; and (iv) $\tau(n, m) = t_{\text{fin}}$.

The task of q_w will be to accept all words with a violation of some vertical constraint. To achieve this with the limited form that we allow for q_w we use a more elaborate encoding of tiles for the actual reduction and define the DFA accordingly. We simultaneously encode tiles and their relevant vertical constraints as strings of length $2k$. For each $i, j \in \{1, \ldots, k\}$ we let e_{ij} be a symbol that encodes whether $(t_i, t_j) \in V$ as follows.

$$e_{ij} \stackrel{\text{def}}{=} \begin{cases} a & \text{if } (t_i, t_j) \notin V \\ c & \text{otherwise.} \end{cases}$$

Then, for each $i \in \{1, \ldots, k\}$, we encode tile t_i as the string
$$\text{enc}(t_i) \stackrel{\text{def}}{=} cc \cdots cbc \cdots ce_{i1} \cdots e_{ik}$$

---

[7] Σ^n abbreviates concatenations of n symbols from Σ.

of length $2k$ in which the entry labeled b is at position i. For a string $v \in T^*$, we write $\mathrm{enc}(v)$ for the symbol-wise encoding of v. It is straightforward to construct from $A(S, w)$ an automaton $A'(S, w)$ that accepts all encodings $\mathrm{enc}(v)$ of strings $v \in L(A(S, w))$. Finally, the regular expression q_w is just $(a + b + c)^* a (a + b + c)^{(2n-1)k-1} b (a + b + c)^*$. □

5   Hardness Results on Trees

In this section we are going to prove the following three results.

Theorem 10. *Validity of tree pattern queries w.r.t. an NTA is EXPTIME-complete even for path queries of the form $a/*/*/\cdots/*/b$ over schemas with three symbols.*

Theorem 11. *Validity of tree pattern queries w.r.t. a DFA-based XSD is EXPTIME-complete even for path queries of the form $a/*/*/\cdots/*/b$ over schemas with four symbols.*

Theorem 12. *Validity of tree pattern queries w.r.t. a DTD is EXPTIME-complete even for tree pattern queries of the form $*[/a]/*/\cdots/*/b$ over DTDs.*

Notice the subtle differences between the three cases: In the NTA case it suffices to have path queries and three alphabet symbols. For DFA-based XSDs we use one more alphabet symbol due to their limited expressiveness when compared to NTAs. If we limit the expressiveness even more to DTDs, then validity of path queries is not hard anymore, as was shown by Hashimoto et al. [12].

Theorem 13 ([12], Theorem 3). *Validity of path queries w.r.t. a DTD is in PTIME.*

However, even allowing the path to have one additional leaf branching off makes the validity problem EXPTIME-hard, even w.r.t. DTDs. Thus, compared to Theorem 13, allowing a single branching as opposed to a pure path query or using DFA-based XSDs as opposed to DTDs results in a provably exponential blow-up in the time complexity for Validity.

All the problems considered in this section are in EXPTIME because of the following result.

Theorem 14. *[[4], Theorem 3.1] Validity of tree pattern queries w.r.t. an NTA is in EXPTIME.*

The lower bounds in Theorems 11 and 12 are shown by reductions from the complement of the TilingWinner problem. For the purpose of these reductions we use the restricted form of tiling games from Section 3.

Strategies for CONSTRUCTOR for some tiling system S and initial row w can be represented by *strategy trees* as usual. The nodes of such a tree carry the tiles chosen in the game. Each node that corresponds to a tile chosen by SPOILER has a child labelled with the symbol that is chosen according to CONSTRUCTOR's strategy. Each node v that corresponds to a tile chosen by CONSTRUCTOR has one child for every possible legal move by SPOILER. Every internal node in the tree corresponds to a tiling prefix, induced by the path from the root to that node.

Proof (of Theorem 11). Let $S = (T, V, H, t_{\text{fin}})$ be a restricted tiling system for which TilingWinner(S) is EXPTIME-hard and $w \in T^*$ an initial row. By definition of restricted tiling systems, the following holds for every tree s representing a winning strategy of CONSTRUCTOR.

(i) Each path represents a solution for S (with initial row w), and
(ii) Each node corresponding to a tile chosen by CONSTRUCTOR has exactly two children labelled by different tiles.

Thus, the following two statements are equivalent.

(a) $w \in$ TilingWinner(S).
(b) There is a strategy tree s for CONSTRUCTOR with the properties (i) and (ii).

In the following we define an encoding function enc that maps strategy trees fulfilling property (ii) to trees over alphabet $\{a, b, c, c'\}$. Furthermore, we construct from S and w a DFA-based XSD (B, λ), and a path query P such that the following are equivalent.

(c) There is a tree s' of the form $s' = \text{enc}(s)$ for some strategy tree s for CONSTRUCTOR with properties (i) and (ii).
(d) P is *not* valid w.r.t. (B, λ).

By combining the two above equivalences with the obvious equivalence between (b) and (c) we get that $w \in$ TilingWinner(S) if and only if P is *not* valid w.r.t. (B, λ). The theorem then follows because we have a reduction from the complement of TilingWinner(S) to the validity problem and the former is EXPTIME-complete because EXPTIME is closed under complementation. The encoding of strategy trees is similar to the encoding of strings in the proof of Theorem 9. We basically replace nodes of the tree by paths of length $2k$, where $k = |T|$.

Let s be a strategy tree. We describe how the encoded tree enc(s) is obtained from s. We use the definitions of e_{ij} and enc(t_i) from Section 4.

For technical reasons that will become apparent below, we use the alphabet $\Sigma' = \{a, b, c, c'\}$ and allow additional encodings of tiles as follows. For every $i, j \in \{1, \ldots, k\}$ with $i < j$ we let

$$\text{enc}_i(t_j) \stackrel{\text{def}}{=} cc \cdots cc'c \cdots cbc \cdots ce_{j1} \cdots e_{jk},$$

be the string obtained from enc(t_j) by replacing the symbol c at position i by c'. In the following, we identify strings enc(t_i) and enc$_i(t_j)$ with paths consisting of $2k$ nodes that are labelled according to enc(t_i) and enc$_i(t_j)$, respectively.

We associate with each strategy tree s for CONSTRUCTOR an *encoded tree* enc(s) in two stages. The first stage proceeds in a top-down fashion. We replace the root with tile t by the path enc(t). We replace every node u that is the only child of its parent and is labelled with some t_i by enc(t_i). For all siblings u and

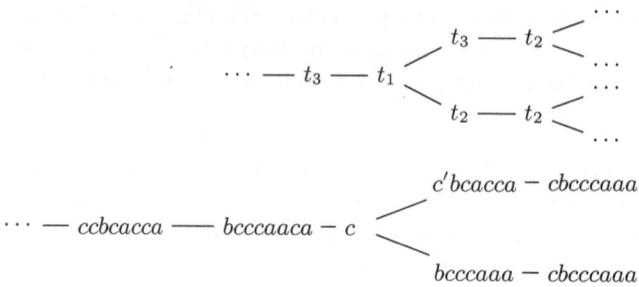

Fig. 2. At the top we see part of a strategy tree for CONSTRUCTOR in a game with four tile types. Below is the encoding of the same part of the tree.

v in s labelled by tiles t_i and t_j, respectively, with $i < j$, we replace u by $\mathrm{enc}(t_i)$ and v by $\mathrm{enc}_i(t_j)$.

In the second stage we combine the two paths $\mathrm{enc}(t_i)$ and $\mathrm{enc}_i(t_j)$ of a pair of siblings u, v by a *prefix tree* that is obtained by identifying their prefixes of length $i - 1$ and put the resulting tree (or forest of two paths, if $i = 1$) below the lowest node of the encoding of the parent of u and v. After this, we no longer have siblings that carry the same label. The resulting tree is $\mathrm{enc}(s)$. We illustrate the encoding with an example.

Example 15. Figure 2 shows an example of how the encoding in the proof of Theorem 11 works. Let $T = \{t_1, t_2, t_3, t_4\}$, $\{(t_1, t_2), (t_1, t_3), (t_2, t_2), (t_3, t_1), (t_3, t_2)\} \subseteq H$, and $V = \{(t_1, t_3), (t_2, t_1), (t_3, t_2), (t_3, t_3), (t_4, t_1)\}$. Thus, $\mathrm{enc}(t_1) = bcccaaca$, $\mathrm{enc}(t_2) = cbcccaaa$, $\mathrm{enc}(t_3) = ccbcacca$, and $\mathrm{enc}(t_4) = cccbcaaa$. On the top, we see a possible part of a strategy tree for CONSTRUCTOR, where the tile t_1 corresponds to a move of CONSTRUCTOR and t_2 and t_3 are the two possible next legal moves of SPOILER. In the lower part we illustrate the encoding of this tree fragment. We use strings to represent unary tree fragments to simplify the picture. The encoding of the siblings t_2 and t_3 share a node because they have the same prefix c.

The DFA-based XSD (B, λ) is constructed from S and w as follows. The DFA B is a slight extension of the DFA $A'(S, w)$ constructed in the proof of Theorem 9. It tests, for a path in the given tree whether its label sequence is an encoding of a string $x \in T^*$ such that

- x has prefix w;
- the length of x is a multiple of $n \overset{\mathrm{def}}{=} |w|$;
- the tiling function $\tau : \{1, \ldots, n\} \times \{1, \ldots, m\} \to T$ corresponding to x fulfills the horizontal constraints; and
- $\tau(n, m) = t_{\mathrm{fin}}$.

Here, the encoding is over Σ' and allows substrings of the form $\mathrm{enc}_i(t_j)$ beyond $\mathrm{enc}(t_i)$ in even columns (chosen by SPOILER). The DFA $A'(S, w)$ basically has states[8] of the form (t, t', i), where t is the previous tile (or $\#$ in column 1), t' the current tile (or $\#$ if this has not yet been determined) and i a counter modulo $2kn$. In B, $\#$ also indicates that the current tile is not yet determined, but if c' is read, t' changes into "?" to indicate that the prefix tree has already branched (but the tile on the current branch is still unknown). For the sake of clarity later in the proof, we briefly assume that B has accepting states (that, by definition, can only be reached after reading t_{fin}). These states are only important to define below where λ allows a node to be a leaf. We do not require accepting states in B in our DFA-based XSD. The function λ is defined as follows.

- For states of the form $q = (t, \#, i)$ with $(i \mod 2k) < k$, for which $(t, t_{i \mod 2k}) \notin H$, $\lambda(q) = c$.
- For states of the form $q = (t, \#, i)$ with $(i \mod 2k) < k$, for which i indicates an odd column (where CONSTRUCTOR is about to move) and $(t, t_{i \mod 2k}) \in H$ or $t = \#$, $\lambda(q) = c + b$.
- For states of the form $q = (t, \#, i)$ with $(i \mod 2k) < k$, for which i indicates an even column (where SPOILER is about to move) and $(t, t_{i \mod 2k}) \in H$, $\lambda(q) = c + bc'$.
- For states of the form $q = (t, t', i)$ with $(i \mod 2k) < k$ and $t' \in T$, $\lambda(q) = c$.
- For states of the form $q = (t, ?, i)$ with $(i \mod 2k) < k$, $\lambda(q) = c + b$.
- For states of the form $q = (t, t', i)$ with $(i \mod 2k) \geq k$ and $t' \in T$, $\lambda(q) = e_{\ell j}$ where $t_\ell = t'$ and $j = i \mod 2k$.
- For states q corresponding to the first row, $\lambda(q)$ is just the next symbol from the encoding of w.
- For every "accepting state" q of B, $\lambda(q) = \varepsilon + b + c$.

Finally, the path query P has the form

$$\underbrace{a/*/*/\cdots/*/*/b}_{2nk-k+1 \text{ labels}}.$$

To complete the proof it only remains to show that (c) and (d) are indeed equivalent.

To this end, let us first assume that (c) holds, that is, there is a tree s' of the form $s' = \text{enc}(s)$ for some strategy tree s for CONSTRUCTOR with properties (i) and (ii). Since s is a strategy tree, each of its paths represents a solution for S with initial row w. As such, each path from root to leaf satisfies the horizontal and vertical constraints. Since it has the correct length and satisfies the horizontal constraints, B has a run over each path that ends in a state q such that $\lambda(q) = \varepsilon + b + c$. Since each node that corresponds to a SPOILER tile has exactly two children labelled by different tiles, we also have that, by construction, the conditions on λ are fulfilled. Thus, $\text{enc}(s)$ is satisfied by (B, λ). Finally, since s does not have any violations against the vertical constraints, we also have that P does not match s' by construction. Therefore s' is a witness for the fact that P is *not* valid w.r.t. (B, λ).

---

[8] It has further states for the prefix w and a rejecting sink state.

For the other direction, let us assume that P is *not* valid w.r.t. (B, λ). Let s' be a tree that conforms to (B, λ) and in which P does not match. By construction $s' = \mathrm{enc}(s)$, for some tree s such that (ii) holds. Furthermore, as s' conforms to (B, λ) it follows that every path fulfills the horizontal constraints of S and ends with the final tile t_{fin}. Finally, as P does not match in s' there is no violation of any vertical constraint in s. Therefore, s also fulfills (i) and thus (c) holds.   □

The proofs of Theorems 10 and 12 can be obtained by adapting the above proof.

References

1. Arenas, M., Daenen, J., Neven, F., Van den Bussche, J., Ugarte, M., Vansummeren, S.: Discovering XSD keys from XML data. In: SIGMOD (to appear, 2013)
2. Benedikt, M., Bourhis, P., Senellart, P.: Monadic datalog containment. In: Czumaj, A., Mehlhorn, K., Pitts, A., Wattenhofer, R. (eds.) ICALP 2012, Part II. LNCS, vol. 7392, pp. 79–91. Springer, Heidelberg (2012)
3. Benedikt, M., Fan, W., Geerts, F.: XPath satisfiability in the presence of DTDs. J. ACM 55(2) (2007)
4. Björklund, H., Gelade, W., Martens, W.: Incremental XPath evaluation. ACM Trans. Database Syst. 35(4), 29 (2010)
5. Björklund, H., Martens, W., Schwentick, T.: Optimizing conjunctive queries over trees using schema information. In: Ochmański, E., Tyszkiewicz, J. (eds.) MFCS 2008. LNCS, vol. 5162, pp. 132–143. Springer, Heidelberg (2008)
6. Bray, T., Paoli, J., Sperberg-McQueen, C.M., Maler, E., Yergeau, F.: Extensible Markup Language XML 1.0 (5th edn.). World Wide Web Consortium (2008)
7. Chlebus, B.S.: Domino-tiling games. JCSS 32(3), 374–392 (1986)
8. Clark, J., DeRose, S.: XML Path Language (XPath) version 1.0. Technical report, World Wide Web Consortium (1999), http://www.w3.org/TR/xpath/
9. Clark, J., Murata, M.: Relax NG specification (2001), http://www.relaxng.org
10. Flesca, S., Furfaro, F., Masciari, E.: On the minimization of XPath queries. J. ACM 55(1) (2008)
11. Gelade, W., Idziaszek, T., Martens, W., Neven, F.: Simplifying XML Schema: Single-type approximations of regular tree languages. In: PODS (2010)
12. Hashimoto, K., Kusunoki, Y., Ishihara, Y., Fujiwara, T.: Validity of positive XPath queries with wildcard in the presence of DTDs. In: DBPL (2011)
13. Martens, W., Neven, F., Schwentick, T.: Simple off the shelf abstractions for XML Schema. SIGMOD Record 36(3), 15–22 (2007)
14. Martens, W., Neven, F., Schwentick, T., Bex, G.J.: Expressiveness and complexity of XML Schema. ACM Trans. Database Syst. 31(3), 770–813 (2006)
15. Miklau, G., Suciu, D.: Containment and equivalence for a fragment of XPath. J. ACM 51(1), 2–45 (2004)
16. Murlak, F., Ogiński, M., Przybyłko, M.: Between tree patterns and conjunctive queries: Is there tractability beyond acyclicity? In: Rovan, B., Sassone, V., Widmayer, P. (eds.) MFCS 2012. LNCS, vol. 7464, pp. 705–717. Springer, Heidelberg (2012)
17. Neven, F., Schwentick, T.: On the complexity of XPath containment in the presence of disjunction, DTDs, and variables. LMCS 2(3) (2006)
18. Stockmeyer, L., Meyer, A.: Word problems requiring exponential time: Preliminary report. In: STOC, pp. 1–9 (1973)
19. Thompson, H.S., Mendelsohn, N., Beech, D., Maloney, M.: XML Schema Definition Language (XSD) 1.1, http://www.w3.org/TR/xmlschema11-1/

Auctions for Partial Heterogeneous Preferences

Piero A. Bonatti, Marco Faella, Clemente Galdi, and Luigi Sauro

Università degli Studi di Napoli Federico II

Abstract. Online privacy provides fresh motivations to generalized auctions where: (i) preferences may be partial, because of lack of knowledge and formalization difficulties; (ii) the preferences of auctioneers and bidders may be heterogeneous and unrelated. We tackle these generalized scenarios by introducing a few natural generalizations of first-price and second-price auctions, and by investigating which of their classical properties are preserved under which conditions.

1 Introduction

In traditional economic models, partial preferences—that reflect lack of knowledge [1]—are turned into total orders by means of utility functions [5]. This approach does not readily apply to new markets where the "currency" is more complex than money and may involve information. Consider online privacy. Some previous works used bargaining [4] and a generalization of procurement auctions [3] in order to reduce personal information disclosure and improve privacy. In this context bids are contracts involving privacy policies and/or information requests (e.g. login information, credit card data, etc.). The risks associated to information disclosure contribute to the costs incurred by the user, and the value of user information becomes part of the utility of service providers. The complexity of these domains makes the adopted mechanisms depart from their classical counterparts in several respects. For instance:

1) Preferences on bids are partial. Users do not have enough information or cognitive/social capabilities to carry out a detailed risk analysis [12]. For instance, this happens when preferences result from complex tradeoffs between the different parameters of a contract (additional functionalities, cost, Quality of Service (QoS), information disclosure risks, etc.) or when each user actually consist of a group of persons where internal debates do not easily end up with a complete preference. Additionally, it is almost impossible to formalize preferences unambiguously. Human choices are not necessarily consistent along time. One might toss a coin, but the way random choices are made affects the properties of the game, as discussed later. So it is important to investigate partially ordered bids in order to accurately analyze the possible consequences of uncertainty.

2) The preferences of bidders and auctioneers may be largely unrelated (while in the classical setting they are based on a single totally ordered domain, i.e. money, and the auctioneer's preference is exactly the opposite of the bidders'). So, in general, an agent does not know which offers are preferred by the others. As an example of heterogeneous preferences consider a user with three mail addresses: a private "anonymous" address a_p, an institutional address a_i related

K. Chatterjee and J. Sgall (Eds.): MFCS 2013, LNCS 8087, pp. 183–194, 2013.
© Springer-Verlag Berlin Heidelberg 2013

to her current role at work, and a personal office address a_o. The user in general prefers to disclose a_p for privacy reasons, while a vendor prefers a_i, instead, which is more stable and independent from who is currently playing that role. Spammers find the three addresses equally preferable.

We contribute to the analysis of this kind of games by introducing a few natural generalizations of second-price auctions, and investigating their properties, with particular attention to truthfulness. Section 2 introduces the generalized framework. Section 3 discusses the main drawbacks of linearizing preferences by means of utility functions. Then, in Sec. 4 and 5, we introduce two mechanisms with different properties, such as the amount of disclosed private information. A final discussion concludes the paper. We assume the reader to be familiar with the basics of mechanism design. Our notation follows [9].

2   The Playground

We start by introducing the common features of our generalized auctions. The agents of such games consist of an auctioneer a and a set of bidders $\mathscr{B} = \{1, \dots, n\}$. The bids contain elements of an abstract set of possible values $\mathbf{V}$. To help intuition, $\mathbf{V}$ can be viewed as a set of possible contracts specifying both the functional features of a service provided by a bidder, and its nonfunctional properties, such as cost, QoS, and the information the user (the auctioneer) has to disclose in order to access the service. Here, we assume that $\mathbf{V}$ is finite.[1]

An agent might not be able—or willing—to accept some outcomes $v \in \mathbf{V}$ (because of budget limitations, unsatisfactory privacy policies, etc.); this is modelled by associating each agent i to a set of admissible values $V_i^* \subseteq \mathbf{V}$. The idea is that if a bidder i wins the auction with outcome $v \notin V_i^*$, then i incurs a loss due to penalties, bad reputation, costs higher than the value of the transaction, etc.

In general, a bidder $b \in \mathscr{B}$ does not know which values are preferred by a (cf. point 2 in the introduction); furthermore, b may be allowed to bid a *set* of equally preferable values (from b's point of view), $V_b \subseteq \mathbf{V}$, in order to offer a wider choice to the auctioneer and improve the chance of winning the auction. The auctioneer a may restrict acceptable values to a set V_a (e.g. in classical settings a may set a maximum price).

Based on the bids, a mechanism determines which agent is to provide the service and under which conditions, by selecting a winner $b \in \mathscr{B}$ and a set of possible outcomes $po \subseteq V_b \cap V_a$, from which the winner can freely choose its preferred options. We are interested in stochastic mechanisms, so the outputs of our auctions, from a bidder's viewpoint, can be modelled by probability distributions $f \in \Delta(\mathscr{B} \times \mathcal{P}(\mathbf{V}))$,[2] such that $f(i, po)$ is the probability that i wins and can choose the result of the auction from po.

Clearly, the goal of each agent $x \in \mathscr{B} \cup \{a\}$ is to get a maximally preferred value, according to a private preference relation $\leq_x$ over $\mathbf{V}$. Sometimes, agents

---

[1] Most of our results do not strictly require this assumption, but it remarkably simplifies formal details without jeopardizing the applicability of our framework, since options, rankings, etc. typically range over finite domains, in the real world.

[2] $\Delta(X)$ denotes the set of all probability distributions over the set X.

have not enough information to compare a given pair of values (e.g. it may be difficult to evaluate the risks associated to two different information disclosures, or the optimal tradeoff between quality, cost, and risks). In other cases, agents may consider two values equally good (e.g. when two services are delivered under the same preconditions and their functional differences are irrelevant to the user). As usual, this is modelled by assuming $\leq_x$ to be an arbitrary preorder. As usual, $w <_x v$ means that $w \leq_x v$ and $v \not\leq_x w$. In the following, given a set of values $V \subseteq \mathbf{V}$, $\min_x[V]$ denotes the elements of V that are minimal w.r.t. the agent x:

$$\min_x[V] = \{v \in V \mid \forall w \in V \; w \not<_x v\}.$$

Similarly, $\max_x[V]$ denotes V's maximal values (w.r.t. x).

In strategic reasoning, the relations $\leq_i$ must be lifted to preference orderings $\preceq_i$ over stochastic auction outputs $(\Delta(\mathcal{B} \times \mathcal{P}(\mathbf{V})))$. At this stage we do not specify how; we only assume that $\preceq_i$ is any relation satisfying:

PA $f \preceq_i g$ if (sufficient condition, not necessary)
1. $\sum_{X \cap V_i^* = \emptyset} f(i, X) \geq \sum_{X \cap V_i^* = \emptyset} g(i, X)$, and
2. for all V such that $V \cap V_i^* \neq \emptyset$ and $f(i, V) > 0$, there exists $W \supseteq V \cap V_i^*$ such that: $g(i, W) \geq \sum_{X \cap V_i^* \subseteq W} f(i, X)$.

Condition 1 reflects the meaning of V_i^*: it says that it is better to reduce the probability of a loss. Condition 2 says that g is preferable if it simultaneously raises the probability of winning and enlarges the space of admissible choices from which i can pick the outcome. Axiom **PA** shall be refined in Sec. 5.

3   Re-using Classical Second Price Auctions

A natural approach to truthful auctions in the new framework consists in: (i) setting $\leq_a$ to one of the linearizations of the true (partial) preference relation $\leq_a^*$ of the auctioneer, and (ii) applying the classical second-price mechanism using $\leq_a$. We will see that this plan has several drawbacks.

Utility theory defines linearizations in terms of a real-valued utility function u such that $v \leq_a^* w$ iff $u(v) \leq u(w)$ and $v <_a^* w$ iff $u(v) < u(w)$. This tight correspondence is possible under the assumption that $\leq_a^*$ is a *weak order*, i.e. asymmetric and *negatively transitive* (if $x \not<_a^* y$ and $y \not<_a^* z$ then $x \not<_a^* z$) [5]. Currently, there is no evidence that preferences are weak orders in our scenarios.

The next issue (robustness) is that non-best offers might not belong to the admissible values V_b^* for the winner b. Then b incurs a loss, and it might be unable to complete the transaction, which damages the auctioneer, too. For example, the user might send her Mastercard instead of a Paypal receipt, but the provider might only be equipped for Paypal payments.

The third problem is that two or more bidders may submit a maximal offer. The naive solution consists in choosing the winner among them according to some criterion (say, an arbitrary, a priori ordering of bidders, a random choice, etc.). Unfortunately, the resulting game is not truthful: it is easy to find cases in which an agent finds it profitable to bid something higher than the truthful bid, in order to raise the probability of winning to 1. This problem affects also classical second-price auctions; indeed, Vickrey assumes that the maximal offer is unique.[3] While this might be realistic in some standard auctions based on purely monetary payments, it is not realistic to assume it in our reference scenarios, where many different offers may be equally good or incomparable.

The fourth problem is that a priori linearizations of $\leq_a$ introduce a bias over bidders that may affect truthfulness, damage some bidders, and cause lock-in effects. Even if linearizations were randomly generated anew for each auction, the mechanism would not be truthful. Consider an auction with 3 bidders and assume that a linearization and a winner (among those who make a $<_a$-maximal bid) are selected randomly with uniform probability. Assume also that the winner can choose a preferred item from those that are ranked second-best or higher by $<_a$, like in [6]. Suppose that $\mathbf{V} = \{v_1, v_2, v_3\}$, $v_1 <_a^* v_2$, and v_3 is not comparable (in $\leq_a^*$) with v_1 and v_2. There are three strict linearizations ($v_1 <_a v_2 <_a v_3$, $v_1 <_a v_3 <_a v_2$, and $v_3 <_a v_1 <_a v_2$) plus two linearizations where v_3 is given the same rank as v_1 and v_2, respectively. Let the bids of agents 1 and 2 be v_1 and v_2, respectively. Bidder 3 is such that $V_3^* = \{v_1, v_3\}$ and $v_2 <_3 v_3 <_3 v_1$. The reader may easily verify that if bidder 3 offers its true preferred value v_1 then it loses in all cases. Bidding the (non-admissible) value v_2 is the best strategy in 4 out of five linearizations. If the outcome had to be taken exactly from the second-best offers, then v_2 would be a dominant strategy.

Finally, from a practical perspective, preference linearizations exclude a priori some of the possible outcomes; this precludes any chance of re-introducing the user in the game and have her make the final choice among a set of incomparable options, possibly using situation-specific knowledge and preferences that have not been formalized nor encoded in the user agent.

4   Relaxing the Second Price Notion

The notion of second price can be relaxed according to two simple principles: (a) the outcome v should belong to $V_a \cap V_b$, where b is the winner; (b) v should not be worse (for a) than any other bid $w \in V_a \cap V_i$ ($i \neq b$). We introduce mechanisms based on these principles that address the problems raised in Sec. 3.

Mechanism 1's inputs are vectors $\sigma = \langle (V_a, \leq_a), V_1, \ldots, V_n \rangle$ that represent the auctioneer's restrictions and preference $(V_a, \leq_a)$ and the bids. For all such σ, we will denote by $\sigma_{-i} = \langle (V_a, \leq_a), V_1, \ldots, V_{i-1}, \emptyset, V_{i+1}, \ldots, V_n \rangle$. The set of *admissible outcomes* for σ is:

$$adm(\sigma) = V_a \cap \bigcup_{i \in \mathscr{B}} V_i \,. \tag{1}$$

The *candidate winners* are those that submit an optimal offer:

$$cw(\sigma) = \{i \in \mathscr{B} \mid V_i \cap \max_a[adm(\sigma)] \neq \emptyset\} \,. \tag{2}$$

---

[3] There exist auctions for selling multiple instances of a same good that yield multiple winners, but in our reference scenarios users eventually choose only one provider.

The actual winner can be selected in many possible ways, with or without bias. Formally, let a *selection function* be a function $sel : \mathcal{P}(\mathcal{B}) \setminus \emptyset \to \Delta(\mathcal{B})$ such that, for each non-empty $P \subseteq \mathcal{B}$, $sel(P)(u) > 0$ iff $u \in P$. As the set of candidate winners gets larger, the probability of winning should not increase, so we require sel to be *inverse monotonic*, i.e., for all P, Q such that $\emptyset \subset P \subseteq Q \subseteq \mathcal{B}$, $\forall u \in P$, $sel(P)(u) \geq sel(Q)(u)$.

Finally, the *possible outcomes* for i, formalizing principles (a) and (b), are:

$$po(\sigma, i) = \{v \in V_a \cap V_i \mid \forall w \in adm(\sigma_{-i}), v \not\prec_a w\}. \tag{3}$$

The mechanism, for all given σ, consists of three phases:

- Phase one: The mechanism computes an *extended* set of candidate winners $cw^+(\sigma)$. If $|cw(\sigma)| > 1$ then $cw^+(\sigma) = cw(\sigma)$; else, if $cw(\sigma) = \{j\}$, then $cw^+(\sigma) = cw(\sigma) \cup cw(\sigma_{-j})$. The winner is selected from $cw^+(\sigma)$ with probability distribution $sel(cw^+(\sigma))$.
- Phase two: The selected winner i receives from the mechanism an *extended* set of possible outcomes $po^+(\sigma, i)$. If $cw(\sigma) = \{j\}$ and $i \neq j$, then $po^+(\sigma, i) = po(\sigma_{-j}, i)$; else $po^+(\sigma, i) = po(\sigma, i)$. The winner then selects a non-empty set $value(\sigma, i) \subseteq po^+(\sigma, i)$ and returns it to the auctioneer.
- Phase three: The auctioneer chooses the outcome of the auction from the set $\max_a[value(\sigma, i)]$.

Example 1. Let $\mathcal{B} = \{1, 2, 3\}$ and $\mathbf{V} = P \times \{cc, pp\}$, where P is the set of all prices with two decimal digits, cc represents credit card information and pp Paypal payments. The intended meaning of $(p, m) \in \mathbf{V}$ is that the requested service costs p and should be payed with method m. Let $(p_1, m_1) \leq_a (p_2, m_2)$ iff $p_1 \geq p_2$ and either $m_1 = m_2$ or $m_1 = pp$ and $m_2 = cc$ (since Paypal payments do not reveal credit card information). For $i \in \mathcal{B}$, $v_1 \leq_i v_2$ iff $v_2 \leq_a v_1$. Let $V_a = \{v \mid v \leq_a (22, cc) \lor v \leq_a (25, pp)\}$, $V_1 = \{(p, cc) \mid p \geq 20\}$, $V_2 = \{(p, m) \mid p \geq 20 \land m \in \{cc, pp\}\}$, and $V_3 = \{(p, pp) \mid p \geq 21\}$. The best offer is $(20, pp) \in V_2$ and hence $cw(\sigma) = \{2\}$. The other optimal bids are not comparable, so $cw^+(\sigma) = \mathcal{B}$. Bidder 1 can choose the results of the auction from $po^+(\sigma, 1) = \{(p, cc) \mid 20 \leq p \leq 20.99\}$, as higher prices are dominated by the offers of bidder 3, and method pp is not admissible for 1. Bidder 2's space of choices is $po^+(\sigma, 2) = \{(20, cc)\} \cup \{(p, pp) \mid 20 \leq p \leq 21\}$. The values (p, pp) with $p > 21$ are excluded because they are dominated by the offer of bidder 3. Finally, $po^+(\sigma, 3) = \{(p, pp) \mid 21 \leq p \leq 25\}$; the upper bound is set by V_a. □

Strategic Reasoning. The relevant result of Mechanism 1 for a winner i is $(i, po^+(\sigma, i))$, because i can freely select its preferred outcomes from $po^+(\sigma, i)$. Accordingly, the stochastic outputs of Mechanism 1 are the distributions:

$$f_\sigma(i, V) = \begin{cases} sel(cw^+(\sigma))(i) & \text{if } V = po^+(\sigma, i), \\ 0 & \text{otherwise.} \end{cases} \tag{4}$$

As usual, let (V, σ_{-i}) denote the input obtained from σ by replacing V_i with V. Mechanism 1 is *truthful* if bidding the true admissible set is a *dominant strategy*, that is, for all inputs σ and $\sigma' = (V_i^*, \sigma_{-i})$, $f_\sigma \preceq_i f_{\sigma'}$. In general, Mechanism 1 is not truthful, as shown by the following example.

Example 2. Let $\mathscr{B} = \{1, 2, 3\}$ and $\mathbf{V} = \{a, b, c\} = V_a$. Assume that $V_1^* = \{a\}$ and $V_2^* = V_3^* = \{b\}$. The only preference expressed by the auctioneer is $b \leq_a c$ and the selection function is the uniform distribution. Let $V_3 = \{b, c\}$, $\sigma = \langle (V_a, \leq_a), V_1^*, V_2^*, V_3 \rangle$ and σ' be the truthful strategy profile. By adding the value c to her bid, bidder 3 manages to exclude bidder 2 from the auction, so that $cw^+(\sigma) = \{1, 3\}$, whereas $cw^+(\sigma') = \{1, 2, 3\}$. Moreover, $po^+(\sigma, 3) = \{b, c\}$, so bidder 3 can safely choose value b in Phase two of the auction. It is possible (i.e., consistent with **PA**) that bidder 3 strictly prefers σ over σ', thus violating truthfulness.   □

It can be proved that whenever i bids non-admissible values, there exists a context such that the outcome is not admissible and i incurs a loss. Formally, for all $V_i \not\subseteq V_i^*$, there exist $\sigma = \langle (V_a, \leq_a), V_1, \ldots, V_i, \ldots, V_n \rangle$ and W such that $f_\sigma(i, W) = 1$ and $W \cap V_i^* = \emptyset$. If the expected loss is high enough, then i is induced to bid only admissible values, i.e. $V_i \subseteq V_i^*$. In this case, the truthful strategy is dominant:

Theorem 1. *For all $\sigma = \langle (V_a, \leq_a), V_1, \ldots, V_n \rangle$ such that $V_i \subseteq V_i^*$, $f_\sigma \preceq_i f_{(V_i^*, \sigma_{-i})}$.*

Proof. (Sketch) $V_i \subseteq V_i^*$ implies that: (i) if $i \in cw^+(\sigma)$ then $i \in cw^+(\sigma')$, where $\sigma' = (V_i^*, \sigma_{-i})$, (ii) $cw^+(\sigma) \supseteq cw^+(\sigma')$, and (iii) $po^+(\sigma, i) \subseteq po^+(\sigma', i)$. By (i), (ii), and the inverse monotonicity of sel, $f_\sigma(i, po^+(\sigma, i)) \leq f_{\sigma'}(i, po^+(\sigma', i))$. Then, by (iii) and axiom **PA**, we obtain $f_\sigma \preceq_i f_{\sigma'}$.   □

Truthfulness for unrestricted bids can be proved if $\leq_a$ consists of a totally ordered sequence of layers such that (i) the members of each layer are strictly better than all the members of the preceding layers; (ii) within each layer, values are either equivalent or incomparable. Formally, $\leq_a$ should be reflexive, transitive and *quasi negatively transitive*, i.e., if $y \not\leq_a x$, $x \not\leq_a y$, and $y \not\leq_a z$, then $x \not\leq_a z$. We call such a preorder *superweak*, as its strict version $<_a$ generalizes weak orders. Notice that the preference relation $\leq_a$ of Example 2 is not superweak.

Theorem 2. *If $\leq_a$ is superweak then Mechanism 1 is truthful.*

Proof. Let $\sigma = (V_i, \sigma_{-i})$ and $\sigma' = (V_i^*, \sigma_{-i})$. Assume that $V_i \not\subseteq V_i^*$ (the other cases are covered by Theorem 1). First suppose that $i \notin cw^+(\sigma)$; then for all W, $f_\sigma(i, W) = 0$; moreover, by definition, $po^+(\sigma', i) \subseteq V_i^*$, so the first condition of **PA** is satisfied by $f = f_\sigma$ and $g = f_{\sigma'}$; the second condition is vacuously true, so $f_\sigma \preceq_i f_{\sigma'}$. Next suppose that $i \in cw^+(\sigma)$ and $i \notin cw^+(\sigma')$. Then the offers of the other bidders dominate those in $V_i \cap V_i^*$ and hence $po^+(\sigma, i) \cap V_i^* = \emptyset$. From the properties of sel and **PA** it follows that $f_\sigma \preceq_i f_{\sigma'}$. Finally suppose that $i \in cw^+(\sigma) \cap cw^+(\sigma')$. By quasi negative transitivity, the best offers of i, that is $V_i \cap \max_a[adm(\sigma)]$, are either in the same layer as the best offers of the other bidders in $cw^+(\sigma)$, or strictly preferred to all the offers of the bidders in $cw(\sigma_{-i})$. Similarly for V_i^* and σ'. In all cases, $cw^+(\sigma) = cw(\sigma_{-i}) \cup \{i\} = cw(\sigma'^{-i}) \cup \{i\} = cw^+(\sigma')$ and hence $f_\sigma(i, po^+(\sigma, i)) = f_{\sigma'}(i, po^+(\sigma', i))$. Moreover, $po^+(\sigma, i) \cap V_i^* = po^+((V_i \cap V_i^*, \sigma_{-i}), i) \subseteq po^+((V_i^*, \sigma_{-i}), i) = po^+(\sigma', i)$ so, by **PA**, $f_\sigma \preceq_i f_{\sigma'}$.   □

Note that if $\mathbf{V}$ is the classic, totally ordered set of monetary values and the winner chooses the minimum price from po^+, then Mechanism 1 says: (i) if the maximal offer v_1 is submitted by a single bidder b, then the price paid is the second best offer, v_2, and the winner may be either b or any of the bidders that offered v_2; (ii) if two or more bidders offer v_1, then the winner is one of those agents and the price paid is v_1. This variation of second price auctions affects the probability of winning, not the outcome. It is truthful for any number of maximal offers, so Vickrey's uniqueness assumption can be dropped. This improvement is due to the extended sets cw^+ and po^+ that are insensitive to overbidding.

5   Second Family of Mechanisms

A drawback of Mechanism 1 is that the winner has no incentive to maximize its choice $value(\sigma, i)$; on the contrary, i may prefer to return a single value to hide its preferences from the auctioneer, which makes Phase 3 useless. In the following, we tackle this limitation by having the mechanism act like a trusted third party that optimizes the outcome using the preferences of all agents (of course agents may lie).

   Mechanism 2's input space Σ consists of all $\sigma = \langle\langle V_a, \leq_a\rangle, \langle V_1, \leq_1\rangle, \ldots, \langle V_n, \leq_n\rangle\rangle$ where, for each agent $x \in \mathscr{B} \cup \{a\}$, $V_x \subseteq \mathbf{V}$ is the set of x's offers and $\leq_x$ is a preference relation over $\mathbf{V}$. Each V_i is assumed to be upward-closed w.r.t. the corresponding $\leq_i$, that is, if $v \leq_i w$ and $v \in V_i$ then $w \in V_i$.

   We decompose the mechanism into a deterministic module $res : \Sigma \to \mathcal{P}(\mathscr{B} \times \mathbf{V})$ and a stochastic module $smod : \mathcal{P}(\mathscr{B} \times \mathbf{V}) \to \Delta(\mathscr{B} \times \mathbf{V})$. Intuitively, $res(\sigma)$ returns the set of all eligible adjudgements (i, v) where i awards the auction by providing the value v. Thus, the set of candidate winners is formally defined as $cw(\sigma) = \{i \in \mathscr{B} \mid \exists v \in \mathbf{V}\ (i, v) \in res(\sigma)\}$. Once the deterministic module has returned a set of all the possible results $res(\sigma) = P$, the mechanism draws a final result $(i, v) \in P$ according with the distribution $smod(P)$.[4] In particular, $smod$ does not enable to draw a rejected pair; hence, it returns a distribution that assigns positive values to the elements of $res(\sigma)$ only. Finally, a mechanism is defined as the composition $mech = smod \circ res : \Sigma \to \Delta(\mathscr{B} \times \mathbf{V})$. Notice that, differently from Mechanism 1, once an input σ is provided to $mech$, a unique winner and value are returned without any further interaction with the parties.

   Let $po(\sigma, i)$ be defined as in Sec. 4; $value(\sigma, i)$ is directly calculated by the mechanism by means of the preference relation $\leq_i$ provided in the input σ, $value(\sigma, i) = \max_i[po(\sigma, i)]$. Then, as in the 3rd phase of Mechanism 1, the $\leq_a$-maximal values $v \in value(\sigma, i)$ are selected and paired with i, hence $res(\sigma) = \{(i, v) \in \mathscr{B} \times \mathbf{V} \mid v \in \max_a[value(\sigma, i)]\}$.

---

[4] This can be regarded as an instance of the stochastic outputs $f \in \Delta(\mathscr{B}, \mathcal{P}(\mathbf{V}))$ of Sec. 2 by identifying each (i, v) with $(i, \{v\})$.

It can be seen that *(i) res* is robust in the sense that whenever there exists an admissible result $adm(\sigma) \neq \emptyset$, then *res* does not fail (i.e. $res(\sigma) \neq \emptyset$); *(ii) res* generalizes the standard second-price mechanism: When $\leq_x$ is the same total order for all $x \in \mathscr{B}$ and $\leq_a$ is the opposite of $\leq_x$, we obtain $res(\sigma) = \{(i, v)\}$, where V_i contains the maximal bid, and v is the second-best bid.

In what follows, we are interested in two particular stochastic modules that we call *totally random* module, $smod_t$, and *candidate-value* random module, $smod_{cv}$. The former makes each element in $res(\sigma)$ equiprobable. Formally, given $res(\sigma) = P$, $smod_t$ returns the distribution $f = smod_t(P)$ over $\mathscr{B} \times \mathbf{V}$ defined as follows: $f((i, v)) = \frac{1}{|P|}$ if $(i, v) \in P$, where $|P|$ is as usual the cardinality of P, 0 otherwise. Then, the probability of winning is defined as

$$pw_t(\sigma, i) = \sum_{(i,v) \in \mathscr{B} \times \mathbf{V}} f((i, v)) = \frac{|P^i|}{|P|} ,$$

where P^i restricts P to the pairs of type (i, v), for some $v \in \mathbf{V}$. Notice that $smod_t$ does not equally distribute the probability of winning over the set $cw(\sigma)$ of candidate winners. For example, if $res(\sigma) = \{(1, v), (1, v'), (2, w)\}$, then $pw_t(1) = \frac{2}{3}$ whereas $pw_t(2) = \frac{1}{3}$.

The candidate-value random module, instead, returns a distribution over $\mathscr{B} \times \mathbf{V}$ that makes the probability of winning equal for all the candidate winners. Formally, given $res(\sigma) = P$ and $g = smod_{cv}(P)$, we have

$$g((i, v)) = \begin{cases} \frac{1}{|cw(\sigma)| \cdot |P^i|} & \text{if } (i, v) \in P \\ 0 & \text{otherwise.} \end{cases}$$

Clearly, we have that the probability of winning is: $pw_{cv}(\sigma, i) = \frac{1}{|cw(\sigma)|}$.

Strategic Reasoning. An auction can be seen as a game where each agent has its own private set of admissible values and preference relation $\langle V_x^*, \leq_x^* \rangle$, but it can possibly lie if this provides a better result. In this perspective, the input $\sigma \in \Sigma$ is a strategy profile and $mech(\sigma)$ is the outcome. Here, we consider the viewpoint of the bidders where a particular $\langle V_a, \leq_a \rangle$ is fixed in σ and each bidder $i \in \mathscr{B}$ plays $\langle V_i, \leq_i \rangle$.

In order to act strategically, a bidder has to compare different outcomes, which means "lifting" a preference relation $\leq_i^*$ over $\mathbf{V}$ to a preference relation $\preceq_i$ over output distributions, i.e., elements of $\Delta(\mathscr{B} \times \mathbf{V})$. This will be done in two steps. In the first step, we define the preference relation over pairs in $\mathscr{B} \times \mathbf{V}$.

Definition 1. *Given the pair $\langle V_i^*, \leq_i^* \rangle$ for the bidder i, the extension of $\leq_i^*$ over the set of outputs $\mathscr{B} \times \mathbf{V}$ is defined as follows, for all $j, h \neq i$ and $v, w \in \mathbf{V}$:*
(i) $(i, v) \leq_i^* (i, w)$ iff $v \leq_i^* w$, (ii) $(j, v) \leq_i^* (h, w)$ and $(h, w) \leq_i^* (j, v)$, (iii) $(j, w) <_i^* (i, v)$ if $v \in V_i^*$, and $(i, v) <_i^* (j, w)$ otherwise.

Intuitively, each bidder i cares for its own outputs (i, v), according to its preference $\leq_i^*$. Moreover, admissible values are better than values assigned to the other bidders, whereas non-admissible values are judged as a loss — hence, the bidder would prefer to abandon the auction. Once we have defined a preference over outputs $(j, v) \in \mathscr{B} \times \mathbf{V}$, we do not need to look inside them anymore and use y, z, possibly with subscripts, as typical letters to denote them.

In the second step we lift the preference relation to distributions in $\Delta(\mathscr{B} \times \mathbf{V})$. In the following, given two probability distributions $f, g \in \Delta(\mathscr{B} \times \mathbf{V})$ and $\alpha \in [0, 1]$, $\alpha f + (1-\alpha)g$ denotes the convex combination such that $(\alpha f + (1-\alpha)g)(z) = \alpha f(z) + (1 - \alpha)g(z)$. Moreover, for a preference relation $\preceq_i$, $f \sim_i g$ means that $f \preceq_i g$ and $g \preceq_i f$ and $f \prec_i g$ means that $f \preceq_i g$ and $g \not\preceq_i f$. For $y \in \mathscr{B} \times \mathbf{V}$, we denote by $[y]$ the distribution that assigns 1 to y.

Definition 2. *Given $\langle V_i^*, \leq_i^* \rangle$, a partial preorder $\preceq_i$ over $\Delta(\mathscr{B} \times \mathbf{V})$ is a preference relation for i iff*

1. *for all $j, h \in \mathscr{B}$ and $v, w \in \mathbf{V}$, $[(j, v)] \preceq_i [(h, w)]$ iff $(j, v) \leq_i^* (h, w)$ (proper lifting);*
2. *if $f \prec_i g$ and $0 \leq \alpha < \beta \leq 1$, then $\alpha g + (1 - \alpha)f \prec_i \beta g + (1 - \beta)f$;*
3. *if $f_1 \preceq_i g_1$, $f_2 \preceq_i g_2$, and $0 \leq \alpha \leq 1$, then $\alpha f_1 + (1-\alpha)f_2 \preceq_i \alpha g_1 + (1-\alpha)g_2$;*
4. *if $f_1 \prec_i g_1$, $f_2 \preceq_i g_2$, and $0 \leq \alpha \leq 1$, then $\alpha f_1 + (1-\alpha)f_2 \prec_i \alpha g_1 + (1-\alpha)g_2$;*
5. *if $0 \leq \alpha \leq 1$, and $\alpha f_1 + (1 - \alpha)f_2 \prec_i \alpha g_1 + (1 - \alpha)g_2$, then there exist $j, k \in \{1, 2\}$ such that $f_j \prec_i g_k$.*

The first of the above axioms connects $\preceq_i$ with the preference over pairs (i, v) introduced in Definition 1. Axioms 2-4 are borrowed from classic decision theory [8]. Finally, Axiom 5 states a necessary condition for strictly preferring a mixed distribution over another one. When preference relations are assumed to be linear orders, as in classical decision theory, Axiom 5 is a consequence of Axioms 2-4. Distributions f of the type $\alpha f_1 + (1-\alpha)f_2$ can be seen as a random choice which picks f_1 with probability α and f_2 with probability $1 - \alpha$. Comparing such distribution with another one g of the same type $\alpha g_1 + (1 - \alpha)g_2$ encompasses four possible draws: (f_1, g_1), (f_1, g_2), (f_2, g_1), and (f_2, g_2). If there is no draw in which the second component is better than the second, Axiom 5 requires that g is not strictly preferred to f.

For a distribution $f \in \Delta(\mathscr{B} \times \mathbf{V})$, we denote by $supp(f)$ the *support* of f, i.e., the set of items to which f assigns positive probability. The following result can be proven by induction on the total size of the two supports of f and g.

Lemma 1. *Let $f, g \in \Delta(\mathscr{B} \times \mathbf{V})$ be such that, for all $x \in supp(f)$ and $y \in supp(g)$, $x \not\lesssim_i y$ (resp., $x \not\lesssim_i^* y$). Then, $f \not\lesssim_i g$ (resp., $f \not\prec_i g$).*

Truthfulness. As seen in Sec. 3, if we assume that two or more bidders can provide the same bid, even standard second-price mechanisms are not truthful. Since standard second-price mechanisms is a particular case of *mech*, in general *mech* is not truthful, either. However, some aspects deserve attention: First, by not playing truthfully a bidder may increase its probability of winning but it cannot obtain a better value. Second, bluffing is profitable only if bids include some value which is not admissible for the bidder. The following theorems show that the first aspect holds in general (Theorem 3), whereas the second one depends on which stochastic module is adopted. In particular, by adopting the candidate-value random module, an analogue of Theorem 1 holds (Theorem 4). On the contrary, Theorem 5 shows that with the totally random module a bidder i can safely profit from lying.

In this section we obtain a weaker form of truthfulness than in Section 4: whereas Theorem 1 states that being truthful is *as good as* lying ($mech(\sigma) \preceq_i mech(\sigma^*)$, where σ^* is a truthful profile), the forthcoming Theorem 4 states that lying is *not strictly better* than telling the truth ($mech(\sigma^*) \nprec_i mech(\sigma)$), i.e. truthful and non-truthful strategies may lead to *incomparable* outcomes. In the following, let $\sigma_i^* = \langle V_i^*, \leq_i^* \rangle$. First we prove that bidders cannot obtain better values by deviating from truthful strategies.

Theorem 3. *Let $\sigma = (\sigma_i, \sigma_{-i}) \in \Sigma$ be an input and $\sigma^* = (\sigma_i^*, \sigma_{-i})$. There do not exist $(i, v) \in res(\sigma)$ and $(i, w) \in res(\sigma^*)$, such that $w <_i^* v$.*

Proof. Let $S = \{v \in V_a \mid \forall w \in adm(\sigma_{-i}), v \nleq_a w\}$, $A = \max_{\leq_i} S$ and $A^* = \max_{\leq_i^*} S$. Clearly, there do not exist $v \in A$ and $w \in A^*$, such that $w <_i^* v$. By definition and upward-closure of V_i and V_i^*, $value(\sigma, i) \subseteq A$ and $value(\sigma^*, i) \subseteq A^*$. Then, thesis immediately follows from the fact that $res(\sigma, i) \subseteq \{i\} \times value(\sigma, i)$ and $res(\sigma^*, i) \subseteq \{i\} \times value(\sigma^*, i)$. $\qquad\square$

Theorem 4 proves that using the candidate-value random module, a bidder i can improve σ_i^* (which means excluding some candidate winner, by Theorem 3) only by declaring a value not in V_i^* (which introduces the possibility of a loss). On the contrary, Theorem 5 shows that with the total random module, i might profitably adopt a non-truthful strategy σ_i with no risks (as $V_i \subseteq V_i^*$). Lemma 2 is used in the proof of Theorem 4.

Lemma 2. *Let $\sigma = (\sigma_i, \sigma_{-i}) \in \Sigma$ be an input and $\sigma^* = (\sigma_i^*, \sigma_{-i})$. If $V_i \subseteq V_i^*$, then $pw_{cv}(\sigma, i) \leq pw_{cv}(\sigma^*, i)$.*

Theorem 4. *Let $mech = smod_{cv} \circ res$, $\sigma = (\sigma_i, \sigma_{-i}) \in \Sigma$ be a strategy profile and $\sigma^* = (\sigma_i^*, \sigma_{-i})$. If $V_i \subseteq V_i^*$, then $mech(\sigma^*) \nprec_i mech(\sigma)$.*

Proof. Let $f^* = mech(\sigma^*)$ and $f = mech(\sigma)$. Split f^* (resp., f) into the sub-distribution f_i^* that assigns probability to the pairs (i, v) and the sub-distribution f_{-i}^* that assigns probability to the pairs (j, v), for $j \neq i$, so that $f^* = \alpha \cdot f_{-i}^* + (1 - \alpha) \cdot f_i^*$ and $f = \beta \cdot f_{-i} + (1 - \beta) \cdot f_i$. Since $1 - \alpha = pw_{cv}(\sigma^*, i)$ and $1 - \beta = pw_{cv}(\sigma, i)$, by Lemma 2 it holds that $\alpha \leq \beta$. Then

$$f = \alpha \cdot f_{-i} + (\beta - \alpha) \cdot f_{-i} + (1 - \beta) \cdot f_i = \alpha \cdot f_{-i} + (1 - \alpha)\Big(\frac{\beta - \alpha}{1 - \alpha} \cdot f_{-i} + \frac{1 - \beta}{1 - \alpha} \cdot f_i\Big)$$

$$= \alpha \cdot f_{-i} + (1 - \alpha)f'.$$

Now, by Def. 1, $f_{-i} \sim_i f_{-i}^*$ and $f_{-i} \prec_i f_i^*$. Assume by contradiction that $f^* \prec_i f$. By applying Axiom 3 twice, we have $f_i^* \prec_i f'$. By Axiom 5, $f_i^* \prec_i f_{-i}$ or $f_i^* \prec_i f_i$. The former is an immediate contradiction. The latter is a contradiction, too, because by Theorem 3 we can apply Lemma 1 to f_i^* and f_i, and obtain $f_i^* \nprec_i f_i$. $\qquad\square$

Theorem 5. *Let $mech = smod_t \circ res$, there exists a strategy profile $\sigma = (\sigma_i, \sigma_{-i}) \in \Sigma$ such that $V_i \subseteq V_i^*$ and $mech(\sigma^*) \prec_i mech(\sigma)$, where $\sigma^* = (\sigma_i^*, \sigma_{-i})$.*

6   Related Work

Traditionally, neither partial nor heterogeneous preferences over auction bids are considered [10,9]. In economic models, preferences are typically linearized through utility functions (e.g. [2]); moreover, it is typically assumed that preferences are already *almost* total orders (i.e. weak orders) [5]. Partial preferences are first class citizens in mechanisms without money; usually results on incentive compatibility are negative, e.g. the Gibbard-Satterthwaite theorem in social choice theory ([9,11]). Second-price auctions have been generalized to partial preferences in [3] according to principles (a) and (b) (cf. Sec. 4). That approach is based on a bid domain that represents only information disclosures, and two ad hoc preference relations that are not suitable for modelling service cost, QoS, functional differences, etc, as discussed in [3]. Our mechanisms, instead, are general enough to cover all these features and a much wider range of preferences.

In [7] a generic model for matching with contracts (using a doctor-hospital metaphor) is introduced. If offers are made by doctors, hospital preferences satisfy the law of aggregate demand, and doctors are substitutes, then revealing the doctors' true preferences is a dominant strategy. However definitions and results rely on the assumption that such preferences are total orders, and if any of the above hypotheses is dropped, then truthfulness does not hold.

Finally, [6] introduces Vickrey auctions without payments, using qualitative preference relations. Auctioneers and bidders have independent preferences, as in our framework; moreover, infinite bid domains are considered. There are three major differences, though: First, the preference relation of the auctioneer is restricted to *total* preorders (actually, plain linear orders in the finite case), and the auction's definition (based on a pretty standard notion of *second best offer*) is not well-defined for the unrestricted partial preferences we deal with. Second, in [6] the auctioneer's preferences *must* be published in advance, while we admit games—such as Mechanism 1—where preferences can be kept private (the key is defining bids as *sets* of alternative values). Third, in [6] the tie-breaking over multiple maximal offers is dealt with by assuming the auctioneer's preferences to be *equipeaked* (all local maxima are also global maxima). Using cw^+ and po^+, instead, we can prove truthfulness for all superweak auctioneer preference orders, that cover all the preference orders of [6] as a special case. Moreover, we prove weaker forms of truthfulness for unrestricted partial preferences.

7   Conclusions

When the preferences on bids are partial and heterogeneous, generalizing second price auctions in a way that preserves truthfulness is a nontrivial task. The approach based on linearizing preferences leaves a number of problems open (cf. Sec. 3). The generalization of the second price notion based on principles (a) and (b) (cf. Sec. 4), in general, does not yield truthful mechanisms, since overbidding may be profitable in some contexts. Still, overbidding yields potential losses, and if risks are higher than potential gain, then the truthful strategy is dominant.

These results hold for two different kinds of mechanism: Mechanism 1 does not require agents to publish their preferences, but the only guarantee for the auctioneer is that the outcome is not worse than any of the offers of non-winners (by principle (b)). The instances of Mechanism 2, instead, need preferences to be disclosed; however, they yield a $\leq_a$-maximal value from the $\leq_i$-maximal options of winner i. Mechanism 1 is truthful in a stronger sense: telling the truth is at least as good as any other strategy, while in Mechanism 2 telling the truth is one of possibly many, incomparable optimal strategies; moreover, the winner must be chosen with some care to achieve truthfulness. Both mechanisms are robust, that is, they introduce no additional transaction failures. In Mechanism 1 we experimented with extended sets of candidate winners and possible outcomes (cw^+ and po^+) that induce unconditional truthfulness when the auctioneer's preference relation $\leq_a$ is "layered" (superweak order). This approach applies also to the classical scenarios based on monetary payments where multiple optimal offers may occur (they are explicitly excluded in classical truthfulness proofs); the auctioneer's outcome is at least as good as in standard second-price auctions. We expect similar techniques to yield analogous results in Mechanism 2.

References

1. Aumann, R.J.: Subjective programming. Econometric Res. Prog. Research memorandum n. 22. Princeton University (February 10, 1961); Also in Shelly, M.W., Bryan, G.L. (eds.) Human Judgments and Optimality. Wiley, New York (1964)
2. Aumann, R.J.: Utility theory without the completeness axiom. Econometric Res. Prog. Research memorandum n. 26. Princeton University (April 3, 1961)
3. Bonatti, P.A., Faella, M., Galdi, C., Sauro, L.: Towards a mechanism for incentivating privacy. In: Atluri, V., Diaz, C. (eds.) ESORICS 2011. LNCS, vol. 6879, pp. 472–488. Springer, Heidelberg (2011)
4. Cranor, L.F., Resnick, P., Resnick, P.: Protocols for automated negotiations with buyer anonymity and seller reputations. Netnomics 2(1), 1–23 (2000)
5. Fishburn, P.: Utility Theory for Decision Making. Wiley, New York (1970)
6. Harrenstein, P., de Weerdt, M., Conitzer, V.: A qualitative vickrey auction. In: Chuang, J., Fortnow, L., Pu, P. (eds.) ACM Conference on Electronic Commerce, pp. 197–206. ACM (2009)
7. Hatfield, J.W., Milgrom, P.R.: Matching with contracts. American Economic Review, 913–935 (2005)
8. Myerson, R.: Game Theory: Analysis of Conflict. Harvard University Press (1997)
9. Nisan, N., Roughgarden, T., Tardos, E., Vazirani, V.V.: Algorithmic Game Theory. Cambridge University Press, New York (2007)
10. Osborne, M., Rubinstein, A.: A Course in Game Theory. MIT Press (1994)
11. Reffgen, A.: Generalizing the Gibbard-Satterthwaite theorem: partial preferences, the degree of manipulation, and multi-valuedness. Social Choice and Welfare 37(1), 39–59 (2011)
12. Simon, H.: A behavioral model of rational choice. The Quarterly Journal of Economics 69(1), 99–118 (1955)

New Polynomial Cases of the Weighted Efficient Domination Problem[*]

Andreas Brandstädt[1], Martin Milanič[2], and Ragnar Nevries[1]

[1] Institut für Informatik, Universität Rostock, D-18051 Rostock, Germany
{andreas.brandstaedt,ragnar.nevries}@uni-rostock.de
[2] UP IAM and UP FAMNIT, University of Primorska, SI6000 Koper, Slovenia
martin.milanic@upr.si

Abstract. Let G be a finite undirected graph. A vertex *dominates* itself and its neighbors in G. A vertex set D is an *efficient dominating set* (*e.d.* for short) of G if every vertex of G is dominated by exactly one vertex of D. The *Efficient Domination* (ED) problem, which asks for the existence of an e.d. in G, is known to be $\mathbb{NP}$-complete even for very restricted graph classes.

In particular, the ED problem remains $\mathbb{NP}$-complete for $2P_3$-free graphs and thus for P_7-free graphs. We show that the weighted version of the problem (abbreviated WED) is solvable in polynomial time on various subclasses of P_7-free graphs, including $(P_2 + P_4)$-free graphs, P_5-free graphs and other classes.

Furthermore, we show that a minimum weight e.d. consisting only of vertices of degree at most 2 (if one exists) can be found in polynomial time. This contrasts with our $\mathbb{NP}$-completeness result for the ED problem on planar bipartite graphs with maximum degree 3.

Keywords: efficient domination, P_k-free graphs, polynomial-time algorithm, robust algorithm.

1 Introduction

Packing and covering problems in graphs and hypergraphs and their relationships belong to the most fundamental topics in combinatorics and graph algorithms and have a wide spectrum of applications in computer science, operations research and many other fields. Packing problems ask for a maximum collection of objects which are not "in conflict", while covering problems ask for a minimum collection of objects which "cover" some or all others. A good example is the Exact Cover Problem (X3C [SP2] in the monograph by Garey and Johnson [19]) asking for a subset $\mathcal{F}'$ of a set family $\mathcal{F}$ over a ground set, say V, covering every

---

[*] This work is supported in part by the Slovenian Research Agency (research program P1–0285 and research projects J1–4010, J1–4021, MU-PROM/2012–022 and N1–0011: GReGAS, supported in part by the European Science Foundation).

K. Chatterjee and J. Sgall (Eds.): MFCS 2013, LNCS 8087, pp. 195–206, 2013.

vertex in V exactly once. It is well known that this problem is NP-complete even for set families containing only 3-element sets (see [19]) as shown by Karp [21].

The following variant of the domination problem is closely related to the Exact Cover Problem: Let $G = (V, E)$ be a finite undirected graph. A vertex v *dominates* itself and its neighbors. A vertex subset $D \subseteq V$ is an *efficient dominating set* (*e.d.* for short) of G if every vertex of G is dominated by exactly one vertex in D. Obviously, D is an e.d. of G if and only if the subfamily of all closed neighborhoods of vertices in D is an exact cover of the vertex set of G. Note that not every graph has an e.d.; the EFFICIENT DOMINATING SET (ED) problem asks for the existence of an e.d. in a given graph G.

The notion of efficient domination was introduced by Biggs [3] under the name *perfect code*. In [1, 2], among other results, it was shown that the ED problem is NP-complete. It is known that ED is NP-complete even for bipartite graphs [36], chordal graphs [36], planar bipartite graphs [29], chordal bipartite graphs [29], and planar graphs with maximum degree 3 [16, 22]. Efficient dominating sets are also called *independent perfect dominating sets* in various papers, and a lot of work has been done on the ED problem which is motivated by various applications, among them coding theory and resource allocation in parallel computer networks; see, e.g., [1–3, 13, 24–26, 29, 32, 35, 36].

In this paper, we will also consider the weighted version of the ED problem:

WEIGHTED EFFICIENT DOMINATION (WED)

Instance: A connected graph $G = (V, E)$ with vertex weights $\omega : V \to \mathbb{N}$.
Task: Find an e.d. of minimum total weight,
         or determine that G contains no e.d.

The WED (and consequently the ED) problem is solvable in polynomial time for trees [35], cocomparability graphs [10, 13], split graphs [11], interval graphs [12, 13], circular-arc graphs [12], permutation graphs [24], trapezoid graphs [24, 25], bipartite permutation graphs [29], distance-hereditary graphs [29], block graphs [36] and hereditary efficiently dominatable graphs [32].

For a set $\mathcal{F}$ of graphs, a graph G is called $\mathcal{F}$-*free* if G contains no induced subgraph from $\mathcal{F}$. For two graphs F and G, we say that G is F-free if it is $\{F\}$-free. Let P_k denote a chordless path with k vertices, and let $P_i + P_j$ denote the disjoint union of P_i and P_j. We write $2P_i$ for $P_i + P_i$. From the proof of the NP-completeness result for chordal graphs in [36] it follows that for $2P_3$-free graphs, the ED problem remains NP-complete and thus, it is also NP-complete for P_7-free graphs.

A set M of edges in a graph G is an *efficient edge dominating set* of G if and only if it is an e.d. in the line graph $L(G)$ of G (these sets are also called *dominating induced matchings* in some papers). It is known that deciding if a given graph has an efficient edge dominating set is NP-complete, see e.g. [4, 8, 9, 20, 28, 30]. Hence, we have:

Corollary 1. *For line graphs, the ED problem is NP-complete.*

Since line graphs are claw-free, ED is $\mathbb{NP}$-complete for claw-free graphs. Moreover, we mentioned already that ED is $\mathbb{NP}$-complete for bipartite graphs (and thus for triangle-free graphs) and for chordal graphs (and thus for C_k-free graphs where C_k is a cycle of order $k \geq 4$). Therefore, if F contains an induced cycle or claw then ED is $\mathbb{NP}$-complete on F-free graphs. This is why we subsequently consider F-free graphs where F is cycle- and claw-free, i.e., a disjoint union of paths.

In this paper, we present polynomial-time algorithms for the WED problem for various subclasses of $2P_3$-free graphs as well as of P_7-free graphs and also sharpen one of the $\mathbb{NP}$-completeness results by showing that the ED problem remains $\mathbb{NP}$-complete for planar bipartite graphs of maximum degree 3. Most of our algorithms are robust in the sense of [34]: For the algorithm working on a given graph class $\mathcal{C}$, it is not necessary to recognize whether the input graph is in $\mathcal{C}$; the algorithm either solves the problem or finds out that the input graph is not in $\mathcal{C}$. For the class of P_5-free graphs, we give two different polynomial-time algorithms for ED which lead to incomparable time bounds. The algorithms are formulated as search algorithms for WED, but with minor modifications, all e.d.s also can be enumerated.

Contrary to the above $\mathbb{NP}$-completeness result on planar bipartite graphs of maximum degree 3, we show that it can be decided in polynomial-time whether an input graph G contains an e.d. D containing only vertices of degree at most 2 in G, and if this is the case, such an e.d. of minimum weight can also be found efficiently.

Due to space limitation, we omit most of the proofs and some procedures; see [7] for a full version of this paper.

2   Basic Notions and Results

All graphs considered in this paper will be finite, undirected and simple (i.e., without loops and multiple edges). For a graph G, let $V(G)$ denote its vertex set and $E(G)$ its edge set; for short, let $V = V(G)$ and $E = E(G)$. Let $|V| = n$ and $|E| = m$. A graph is *nontrivial* if it has at least two vertices. For a vertex $v \in V$, $N(v) = \{u \in V \mid uv \in E\}$ denotes its *open neighborhood*, and $N[v] := \{v\} \cup N(v)$ denotes its *closed neighborhood*. The *degree* of a vertex x in a graph G is $d(x) := |N(x)|$. A vertex v *sees* the vertices in $N(v)$ and *misses* all the others. A vertex u is *universal* for $G = (V, E)$ if $N[u] = V$. For standard graph notions such as independent sets, complement graph, and connected components we refer to [5].

Let $\delta_G(v, w)$ ($\delta(v, w)$ for short if G is clear from the context) denote the distance between v and w in G. The *square* of a graph $G = (V, E)$ is the graph $G^2 = (V, E^2)$ such that $uv \in E^2$ if and only if $\delta_G(u, v) \in \{1, 2\}$. In [6,23,32], the following relationship between the ED problem on a graph G and the maximum weight independent set (MWIS) problem on G^2 is used:

Lemma 1. *Let $G = (V, E)$ be a graph and $\omega(v) := |N[v]|$ a vertex weight function for G. Then the following are equivalent for any subset $D \subseteq V$:*

(i) D is an efficient dominating set in G.

(ii) D is a (maximum weight) independent set in G^2 with $\omega(D) = |V|$.

Thus, the ED problem on a graph class $\mathcal{C}$ can be reduced to the MWIS problem on the squares of graphs in $\mathcal{C}$. In Section 5, we will give an example for this reduction which leads to a polynomial-time solution for ED on P_5-free graphs; for most classes, however, the direct way for ED is more efficient.

Given a graph $G = (V, E)$ and a vertex $v \in V$, we define the *distance levels* $N_i(v) = \{w \in V \mid \delta(v, w) = i\}$ for all $i \in \mathbb{N}$. If v is fixed, we denote $N_i(v)$ by N_i.

3   The WED Problem for $2P_2$-Free Graphs

A graph $G = (V, E)$ is a *split graph* if V can be partitioned into a clique C and an independent set I with $C \cap I = \emptyset$. In [11], the WED problem (with vertex and edge weights) was solved in linear time for split graphs.

Since a graph is a split graph if and only if it is $\{2P_2, C_4, C_5\}$-free [17], $2P_2$-free graphs generalize split graphs.

Theorem 1. *For $2P_2$-free graphs, the WED problem can be solved in linear time* $O(n + m)$.

For showing Theorem 1, we need some definitions and preparing steps; see [18] for the following basic notions on modular decomposition. A set H of at least two vertices of a graph G is called *homogeneous* if $H \neq V(G)$ and every vertex outside H is either adjacent to all vertices in H, or to no vertex in H. Obviously, H is homogeneous in G if and only if H is homogeneous in the complement graph $\overline{G}$. A graph is *prime* if it contains no homogeneous set. A homogeneous set H is *maximal* if no other homogeneous set properly contains H. It is well known that in a connected graph G with connected complement $\overline{G}$, the maximal homogeneous sets are pairwise disjoint and can be determined in linear time (see, e.g., [31]). The *characteristic graph* G^* of G results from G by contracting each of the maximal homogeneous sets H of G to a single representative vertex $h \in H$, and connecting two such vertices by an edge if and only if they are adjacent in G. It is well known that G^* is a prime graph.

Let G be a connected $2P_2$-free graph. If $\overline{G}$ is not connected, then the existence of an e.d. D in G implies $|D| = 1$. Thus, in this case, we have to test whether G has a universal vertex. Hence, from now on assume that $\overline{G}$ is connected. The characteristic graph G^* of G is well-defined and prime.

If G has an e.d. D, then for every homogeneous set H of G: $|H \cap D| \leq 1$.   (1)

Proof of (1): Assume that there is a homogeneous set H of G and $d, d' \in D$ with $d \neq d'$ and $d, d' \in H$. Since G is connected, there is a vertex $x \in V \setminus H$ with $dx \in E$ and $d'x \in E$ – a contradiction to the e.d. property. $\square$

If G has an e.d. D, then no $d \in D$ is in a homogeneous set of G.   (2)

Proof of (2): Assume that there is $d \in D$ in a homogeneous set H. Let $x \in H$ be another vertex in H. If $dx \notin E$, there must be $d' \in D$ with $d' \neq d$ and $xd' \in E$. By (1), $d' \notin H$. Since H is a homogeneous set, $dd' \in E$ – a contradiction. Hence, $dx \in E$.

Since $\overline{G}$ is connected, G has no universal vertex and thus $|D| > 1$. Let $d' \in D$ with $d' \neq d$. By (1), $d' \notin H$. Since G is connected, d' has at least one neighbor, say x'. Since D is an e.d., $dd' \notin E$ and hence $x' \notin H$. By the e.d. property and since H is a homogeneous set, $xx' \notin E$ and $dx' \notin E$. Hence, d, x, d', x' induce a $2P_2$ in G – a contradiction.   □

Next we claim:

For every $d \in D$ with $|N(d)| \geq 2$, $N(d)$ is a homogeneous set in G.   (3)

Proof of (3): Assume that for $d \in D$ with $|N(d)| \geq 2$, $N(d)$ is not homogeneous. Then there are $x, y \in N(d)$, and $z \notin N(d)$ such that $xz \in E$ and $yz \notin E$. Since $z \notin N(d)$ and, by the e.d. property, $z \notin D$, there is a vertex $d' \in D$ with $d' \neq d$ and $d'z \in E$, but now, d, y, d', z induce a $2P_2$, a contradiction.   □

Furthermore,

if D is an e.d. of G, then D is an e.d. of G^*.   (4)

Proof of (4): Let D be an e.d. of G. By (2), no $d \in D$ is in a homogeneous set of G. Therefore, all vertices of D are contained in G^*. By construction of G^*, set D is an e.d. in it.   □

Hence, to find an e.d. of a $2P_2$-free graph G, by (2) and (4), it suffices to check if G^* admits an e.d. D^* such that no vertex of D^* is in a homogeneous set of G. To do so, we need the following notion:

A *thin spider* is a split graph $G = (V, E)$ with partition $V = C \cup I$ into a clique C and an independent set I such that every vertex of C has exactly one neighbor in I and vice versa. We claim:

A nontrivial prime $2P_2$-free graph G has an e.d. $\Leftrightarrow$ G is a thin spider.   (5)

Proof of (5): Obviously, in a thin spider the independent set I is an e.d. Conversely, let D be an e.d. of G. By the e.d. property, D is an independent set.

We claim that $|N(d)| = 1$ for every $d \in D$: Since G is connected, $|N(d)| \geq 1$ holds for all $d \in D$. Assume that $|N(d)| > 1$ for some $d \in D$. Then by (3), $N(d)$ is a homogeneous set – a contradiction.

We claim that $G[V \setminus D]$ is a clique: If there are $x, x' \in V \setminus D$ with $xx' \notin E$, there are $d, d' \in D$ with $xd \in E$ and $x'd' \in E$. Then by the e.d. property d, x, d', x' induce a $2P_2$ in G – a contradiction.

Since every vertex of $V \setminus D$ has exactly one neighbor in D, G is a thin spider.   □

Thus, an algorithm for solving the WED problem on $2P_2$-free graphs does the following: For a given nontrivial connected $2P_2$-free graph G:

1. Check whether $\overline{G}$ is connected. If not, then check whether G has a universal vertex. If not, then G has no e.d. Otherwise, minimize $\omega(u)$ over all universal vertices u of G.
2. (*Now G and $\overline{G}$ are connected.*) Construct the characteristic graph G^* of G and check whether G^* is a thin spider. If not, then G has no e.d. If G^* is a thin spider, then let $V(G^*) = C \cup I$ be its split partition. Check if any vertex of I is in a homogeneous set of G. If so, then G has no e.d., otherwise, I is the minimum weight e.d. for G.

Since modular decomposition can be computed in linear time [31], Theorem 1 follows. Note that it is not known whether $2P_2$-free graphs can be recognized in linear time; Theorem 1 assumes that the input graph is known to be $2P_2$-free.

4   A Direct Solution for the WED Problem on P_5-Free Graphs and Related Classes

Since the ED problem is $\mathbb{NP}$-complete for P_7-free graphs, it is interesting to study the complexity of the WED problem for subclasses of P_7-free graphs. We start with P_5-free graphs. Note that the complexity of the closely related MWIS problem on P_5-free graphs is a long standing open problem [27, 33].

Theorem 2. *The WED problem is solvable in time $O(nm)$ on P_5-free graphs in a robust way.*

To prove Theorem 2, we need some preparing steps: Assume that G admits an e.d. D. Let $v \in D$ and let $N_1, N_2, \ldots$ be its distance levels. If G is P_5-free, clearly $N_i = \emptyset$ for all $i > 3$. Moreover, clearly

$$N_1 \cap D = N_2 \cap D = \emptyset. \tag{6}$$

Furthermore,

$$\text{for every edge } yz \in E \text{ in } G[N_3] : N(y) \cap N_2 = N(z) \cap N_2. \tag{7}$$

Proof of (7): Assume without loss of generality that there is $x \in (N(y) \setminus N(z)) \cap N_2$. Let $w \in N(x) \cap N_1$. Then v, w, x, y, z is a P_5 in G—a contradiction.   □

Let H be a component of $G[N_3]$. By (6) and because $N_i = \emptyset$ for all $i > 3$, all vertices of N_3 must be dominated by vertices in $D \cap N_3$ and by (7), all vertices of H have at least one common neighbor in N_2. Hence,

$$H \text{ contains a nonempty set of vertices } U_H \text{ which are universal in } H, \tag{8}$$

and since the choice of a universal vertex of H for D is independent from the choice in the other components of $G[N_3]$, we may assume that

$$\text{for every component } H, D \text{ contains a vertex of } U_H \text{ with minimum weight.} \tag{9}$$

By (6), the vertices of N_2 must be dominated by vertices of N_3, hence every vertex of N_2 has at least one neighbor in N_3. Together with (7) and (9) this implies that

$$\text{for all } w \in N_2, N(w) \cap N_3 \text{ is a component of } G[N_3], \qquad (10)$$

because otherwise a vertex of N_2 would have two neighbors in D.

Conversely:

Claim 1 *If $D \subseteq V(G)$ such that for every $w \in N_2$, $N(w) \cap N_3$ is a connected component of $G[N_3]$, and D contains a universal vertex u of every component of $G[N_3]$, then $D \cup \{v\}$ is an e.d. of G.*

Proof. Clearly, the assumptions imply that D is an independent set. Moreover, D contains no vertices with common neighbors, because v has distance 3 to all other vertices of D and if there are two vertices in $D \cap N_3$ with a common neighbor w, then $w \in N_2$ by construction, contradicting the assumption that $N(w) \cap N_3$ is a connected component of $G[N_3]$. All vertices of N_1 are connected to v, all vertices in N_2 have a neighbor in $D \cap N_3$ and all vertices in $N_3 \setminus D$ have a neighbor in D. Hence, D is dominating, and thus an e.d. □

Claim 1 enables us to give the following algorithm:

Algorithm: Robust-P_5-Free-WED
Input: A connected graph $G = (V, E)$ with vertex weights $\omega : V \to \mathbb{N}$.
Output: One of the following: An e.d. D of G of minimum weight, or a proof that G admits no e.d., or a P_5 in G.

1. Set $\mathcal{D} := \emptyset$.
2. For every vertex $v \in V$, do
   2.1. Determine the distance levels $N_1, N_2, \ldots$ of v.
   2.2. If $N_4 \neq \emptyset$ then Stop—G is not P_5-free.
   2.3. Find the components $H_1, \ldots, H_k$ of $G[N_3]$, and for every H_i let U_i be the set of universal vertices of H_i.
   2.4. Check for every $w \in N_2$ and every H_i if w sees either every or no vertex of H_i. If not then Stop—G is not P_5-free.
   2.5. Check for every $w \in N_2$ if there is an H_i such that w sees exactly the vertices of H_i in N_3. If not, then v is an unsuccessful choice—Continue with next loop iteration.
   2.6. Check if every U_i is nonempty. If not, then v is an unsuccessful choice—Continue with next loop iteration.
   2.7. Let $u_i \in U_i$ of min. weight for every U_i. Set $\mathcal{D} := \mathcal{D} \cup \{\{v, u_1, \ldots, u_k\}\}$.
3. For every $D \in \mathcal{D}$, check if D is an e.d. of G and calculate its weight.
4. If $\mathcal{D}$ contains no e.d. of G, Stop, otherwise, Return a set $D \in \mathcal{D}$ that is an e.d. of G of minimum weight.

A proof that the algorithm is correct and runs in linear time can be found in [7]. This completes the proof of Theorem 2.

The technique used in algorithm `Robust-`P_5`-Free-WED`, that is, testing for every vertex v if there is an e.d. D with $v \in D$ by analyzing the distance levels, can be extended to some superclasses of P_5-free graphs and related classes. The algorithms and their analysis are much more involved than in the P_5-free case; the details can be found in [7]. The graph $S_{1,2,2}$ consists of a chordless path (a, b, c, d, e) and an additional vertex f adjacent to c (i.e., $S_{1,2,2}$ results from a claw by subdividing two of the edges by one vertex each). Since $S_{1,2,2}$ contains the claw as induced subgraph, the ED problem is NP-complete on $S_{1,2,2}$-free graphs.

Theorem 3. *For $\{P_6, S_{1,2,2}\}$-free graphs, the WED problem can be solved in time $O(n^2 m)$ by a robust algorithm.*

Theorem 3 is mostly motivated by the fact that the complexity of ED is open for P_6-free graphs.

Theorem 4. *For $\{2P_3, S_{1,2,2}\}$-free graphs, the WED problem can be solved in time $O(n^5)$ by a robust algorithm.*

Theorem 4 is mostly motivated by the fact that ED is NP-complete for $2P_3$-free graphs as well as for $S_{1,2,2}$-free graphs.

Theorem 5. *The WED problem can be solved on $(P_2 + P_4)$-free graphs in time $O(nm)$ by a robust algorithm.*

5   Solving the ED Problem via Squares of Graphs

As already mentioned, Lemma 1 gives a close relationship between the ED problem on G and the MWIS problem on G^2. This relationship was used in [32] to show that the ED problem is polynomially solvable in the class of $\{S_{1,2,2}, net\}$-free graphs, by reducing it to the MWIS problem in claw-free graphs. The *net* is the triangle with three pendant edges, that is, the graph (V, E) with $V = \{a, b, c, d, e, f\}$ and $E = \{ab, bc, ca, ad, be, cf\}$.

This approach can also be used to solve the ED problem in the class of P_5-free graphs. Let P denote an induced copy of P_4 in G with vertices a, b, c, d and edges ab, bc, cd. Then a and d are the *endpoints* of P, and b and c are the *midpoints* of P.

Proposition 1. *In a P_5-free graph G, midpoints of an induced P_4 are not in any e.d. of G.*

Proof. Let G be a P_5-free graph having an e.d. D, and let a, b, c, d induce a P_4 in G with edges ab, bc, cd. Assume to the contrary that $b \in D$. Then, since $d \notin D$, there is some $d' \in D$ with $dd' \in E$. Now, by the e.d. property, a, b, c, d, d' induce a P_5, a contradiction. $\square$

Theorem 6. *If graph G is P_5-free and has an e.d. then G^2 is P_4-free.*

Proof (sketch). Let $G = (V, E)$ be a P_5-free graph having an e.d. D, and assume to the contrary that G^2 contains an induced P_4 (a, b, c, d). Then $\delta_G(a, b) \leq 2$, $\delta_G(b, c) \leq 2$, and $\delta_G(c, d) \leq 2$ while $\delta_G(a, c) \geq 3$, $\delta_G(a, d) \geq 3$, and $\delta_G(b, d) \geq 3$. Since (a, b, c, d) is a P_4 in G^2, $\delta_G(a, b) = \delta_G(b, c) = \delta_G(c, d) = 1$ is impossible. Thus, there are additional vertices of G in the subgraph $G[P]$ which lead to the P_4 (a, b, c, d) in G^2. If there is only one additional vertex $x \in G$ being adjacent to b and c then $P = (a, b, x, c, d)$ is an induced P_5 in G, a contradiction. Thus, there are at least two additional vertices x, y. If there are only two, say x, y, such that x sees a and b and y sees b and c then, since G is P_5-free, $xy \in E$ but now a, x, y, c, d induce a P_5, a contradiction. Thus, the only remaining cases are the following two:

(1) There are two vertices $x, y \in G$ such that $P = (a, x, b, c, y, d)$ is a path in G with $xy \in E$.

(2) There are three vertices $x, y, z \in G$ such that $P = (a, x, b, y, c, z, d)$ is a path in G with $xy, xz, yz \in E$.

Case (1): We first claim that none of the vertices a, x, b, c, y, d are in D: By Proposition 1, $b, c, x, y \notin D$. Then there is $c' \in D$ with $cc' \in E$. Suppose that $a \in D$. Then $c'x \notin E$ by the e.d. property. Since a, x, b, c, c' do not induce a P_5, $c'b \in E$ follows. Since by Proposition 1, c' is not a midpoint of a P_4 (d, c', b, x), it follows that $c'd \notin E$. Since c', b, x, y, d do not induce a P_5, $c'y \in E$ follows but now c' is a midpoint of P_4 (b, c', y, d), a contradiction. Thus, $a \notin D$ and by symmetry, also $d \notin D$.

Now $a, x, b, c, y, d \notin D$. Thus, there is $a' \in D$ with $aa' \in E$. By the distances in G^2, a' misses c and d, and thus, there is $c' \in D$ with $c'c \in E$ and $c' \neq a'$. Since $a' \in D$ is not a midpoint of a P_4, a, a', b, c do not induce a P_4 and thus $a'b \notin E$. Since a', a, x, b, c do not induce a P_5, $a'x \in E$ and thus by the e.d. property, $c'x \notin E$. Since $a' \in D$ is not a midpoint of a P_4, a, a', y, c do not induce a P_4 and thus $a'y \notin E$. Since a', x, y, c, c' do not induce a P_5, $c'y \in E$ holds. Since a', x, b, c, c' do not induce a P_5, $c'b \in E$ but now (b, c', y, d) is a P_4 with midpoint c', a contradiction.

Case (2) works with similar arguments as Case (1), but we omit the details due to space constraints. See [7] for the complete proof.   □

Let $T(n, m)$ be the best time bound for constructing G^2 from given graph G. Using the fact that the MWIS and recognition problems are solvable in linear time for P_4-free graphs [14, 15], we have, by Lemma 1:

Corollary 2. *For a given P_5-free graph G, the ED problem can be solved in time $T + O(|E(G^2)|)$.*

Since G^2 can be computed from G using matrix multiplication, this time bound is incomparable with the $O(nm)$ bound obtained in Theorem 2.

6   The Bounded-Degree WED Problem

Various NP-completeness results from the literature can be sharpened in the following way:

Theorem 7. *The ED problem is* NP*-complete on planar bipartite graphs of maximum degree 3.*

This raises the question about e.d.'s consisting of vertices with bounded degree. For a non-negative integer k, an e.d. D in a graph G is said to be k-*bounded* if every vertex in D has degree at most k in G. For short, a k-bounded e.d. will also be referred to as a k-*b.e.d.*. The task of the k-Bounded Weighted Efficient Domination (k-BWED) problem is to determine whether a given vertex-weighted graph G admits a k–b.e.d., and if so, to compute one of minimum weight. Clearly, a graph G admits a 0-b.e.d. if and only if it is edgeless. It is also straightforward to see that G admits a 1-b.e.d. if and only if each connected component of G is either K_1, K_2, or the vertices of degree 1 in it form an ED set. Therefore, the k-BWED problem is solvable in linear time for $k \in \{0, 1\}$. On the other hand, since the ED problem is NP-complete for graphs of maximum degree 3 by Theorem 7, the k-BWED problem is NP-complete for every $k \geq 3$.

Theorem 8. *The 2-BWED problem is solvable in polynomial time.*

Proofs of Theorems 7 and 8 can be found in [7].

7   Conclusion

In this paper, we studied the Weighted Efficient Domination on F-free graphs in a systematic way. As described in the introduction, it follows from known results that ED is NP-complete for F-free graphs whenever F contains a cycle or claw. Thus, we focus on graphs F that are the disjoint union of paths. Furthermore, it follows from the proof of the NP-completeness result for chordal graphs in [36] that ED is NP-complete for $2P_3$-free graphs and thus for P_7-free graphs. We obtained new polynomial-time results for various subclasses of P_7-free graphs as shown in Figure 1 below. The results for $\{2P_3, S_{1,2,2}\}$-free and $\{P_6, S_{1,2,2}\}$-free graphs complement the polynomial-time result for $\{S_{1,2,2}, net\}$-free graphs from [32].

By previous results, the results of this paper, and some simple observations (among them the observation that for $3P_2$-free graphs, ED is solvable in polynomial time), it follows that the complexity of ED for the class of F-free graphs is known for all graphs F with at most 6 vertices, except for $F = P_6$. Thus, two of the most challenging related questions seem to be:

(i) What is the complexity of ED for P_6-free graphs?
(ii) Can WED be solved in linear time for P_5-free graphs?

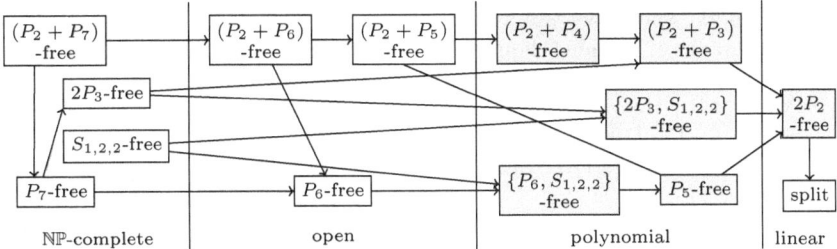

Fig. 1. The complexity of the Efficient Dominating Set Problem on several graph classes. The arrows denote graph class inclusions. The results for the grey highlighted classes are introduced in this paper, and hold for the weighted case of the problem.

References

1. Bange, D.W., Barkauskas, A.E., Slater, P.J.: Efficient dominating sets in graphs. In: Ringeisen, R.D., Roberts, F.S. (eds.) Applications of Discrete Math., pp. 189–199. SIAM, Philadelphia (1988)
2. Bange, D.W., Barkauskas, A.E., Host, L.H., Slater, P.J.: Generalized domination and efficient domination in graphs. Discrete Math. 159, 1–11 (1996)
3. Biggs, N.: Perfect codes in graphs. J. of Combinatorial Theory (B) 15, 289–296 (1973)
4. Brandstädt, A., Hundt, C., Nevries, R.: Efficient Edge Domination on Hole-Free graphs in Polynomial Time. In: López-Ortiz, A. (ed.) LATIN 2010. LNCS, vol. 6034, pp. 650–661. Springer, Heidelberg (2010)
5. Brandstädt, A., Le, V.B., Spinrad, J.P.: Graph Classes: A Survey. SIAM Monographs on Discrete Math. Appl., vol. 3. SIAM, Philadelphia (1999)
6. Brandstädt, A., Leitert, A., Rautenbach, D.: Efficient Dominating and Edge Dominating Sets for Graphs and Hypergraphs. In: Chao, K.-M., Hsu, T.-S., Lee, D.-T. (eds.) ISAAC 2012. LNCS, vol. 7676, pp. 267–277. Springer, Heidelberg (2012)
7. Brandstädt, A., Milanič, M., Nevries, R.: New polynomial cases of the weighted efficient domination problem, Technical report arXiv:1304.6255 (2013)
8. Brandstädt, A., Mosca, R.: Dominating induced matchings for P_7-free graphs in linear time. In: Asano, T., Nakano, S.-i., Okamoto, Y., Watanabe, O. (eds.) ISAAC 2011. LNCS, vol. 7074, pp. 100–109. Springer, Heidelberg (2011)
9. Cardoso, D.M., Korpelainen, N., Lozin, V.V.: On the complexity of the dominating induced matching problem in hereditary classes of graphs. Discrete Applied Math. 159, 521–531 (2011)
10. Chang, M.-S.: Weighted domination of cocomparability graphs. Discrete Appl. Math. 80, 135–148 (1997)
11. Chang, M.-S., Liu, Y.-C.: Polynomial algorithms for the weighted perfect domination problems on chordal graphs and split graphs. Information Processing Letters 48, 205–210 (1993)
12. Chang, M.-S., Liu, Y.-C.: Polynomial algorithms for weighted perfect domination problems on interval and circular-arc graphs. J. Inf. Sci. Eng. 11, 549–568 (1994)
13. Chang, G.J., Pandu Rangan, C., Coorg, S.R.: Weighted independent perfect domination on co-comparability graphs. Discrete Applied Math. 63, 215–222 (1995)
14. Corneil, D.G., Perl, Y., Stewart, L.K.: A linear recognition algorithm for cographs. SIAM J. Computing 14, 926–934 (1985)

15. Courcelle, B., Makowsky, J.A., Rotics, U.: Linear time solvable optimization problems on graphs of bounded clique width. Theory of Computing Systems 33, 125–150 (2000)
16. Fellows, M.R., Hoover, M.N.: Perfect Domination. Australasian J. of Combinatorics 3, 141–150 (1991)
17. Földes, S., Hammer, P.L.: Split graphs. Congressus Numerantium 19, 311–315 (1977)
18. Gallai, T.: Transitiv orientierbare Graphen. Acta Math. Acad. Sci. Hung. 18, 25–66 (1967)
19. Garey, M.R., Johnson, D.S.: Computers and Intractability – A Guide to the Theory of NP-completeness. Freeman, San Francisco (1979)
20. Grinstead, D.L., Slater, P.L., Sherwani, N.A., Holmes, N.D.: Efficient edge domination problems in graphs. Information Processing Letters 48, 221–228 (1993)
21. Karp, R.M.: Reducibility among combinatorial problems. In: Complexity of Computer Computations, pp. 85–103. Plenum Press, New York (1972)
22. Kratochvíl, J.: Perfect codes in general graphs, Rozpravy Československé Akad. Věd Řada Mat. Přírod Vǔ 7, Akademia, Praha (1991)
23. Leitert, A.: Das Dominating Induced Matching Problem für azyklische Hypergraphen, Diploma Thesis, University of Rostock, Germany (2012)
24. Liang, Y.D., Lu, C.L., Tang, C.Y.: Efficient domination on permutation graphs and trapezoid graphs. In: Jiang, T., Lee, D.T. (eds.) COCOON 1997. LNCS, vol. 1276, pp. 232–241. Springer, Heidelberg (1997)
25. Lin, Y.-L.: Fast algorithms for independent domination and efficient domination in trapezoid graphs. In: Chwa, K.-Y., Ibarra, O.H. (eds.) ISAAC 1998. LNCS, vol. 1533, pp. 267–275. Springer, Heidelberg (1998)
26. Livingston, M., Stout, Q.: Distributing resources in hypercube computers. In: Proc. 3rd Conf. on Hypercube Concurrent Computers and Applications, pp. 222–231 (1988)
27. Lozin, V.V., Mosca, R.: Maximum independent sets in subclasses of P_5-free graphs. Information Processing Letters 109, 319–324 (2009)
28. Lu, C.L., Tang, C.Y.: Solving the weighted efficient edge domination problem on bipartite permutation graphs. Discrete Applied Math. 87, 203–211 (1998)
29. Lu, C.L., Tang, C.Y.: Weighted efficient domination problem on some perfect graphs. Discrete Applied Math. 117, 163–182 (2002)
30. Lu, C.L., Ko, M.-T., Tang, C.Y.: Perfect edge domination and efficient edge domination in graphs. Discrete Applied Math. 119, 227–250 (2002)
31. McConnell, R.M., Spinrad, J.P.: Modular decomposition and transitive orientation. Discrete Math. 201, 189–241 (1999)
32. Milanič, M.: Hereditary Efficiently Dominatable Graphs. Available online in: Journal of Graph Theory (2012)
33. Randerath, B., Schiermeyer, I.: On maximum independent sets in P_5-free graphs. Discrete Applied Mathematics 158, 1041–1044 (2010)
34. Spinrad, J.P.: Efficient Graph Representations, Fields Institute Monographs. American Math. Society (2003)
35. Yen, C.-C.: Algorithmic aspects of perfect domination, Ph.D. Thesis, Institute of Information Science, National Tsing Hua University, Taiwan (1992)
36. Yen, C.-C., Lee, R.C.T.: The weighted perfect domination problem and its variants. Discrete Applied Math. 66, 147–160 (1996)

Bringing Order to Special Cases
of Klee's Measure Problem

Karl Bringmann*

Max Planck Institute for Informatics

Abstract. Klee's Measure Problem (KMP) asks for the volume of the union of n axis-aligned boxes in $\mathbb{R}^d$. Omitting logarithmic factors, the best algorithm has runtime $\mathcal{O}^*(n^{d/2})$ [Overmars,Yap'91]. There are faster algorithms known for several special cases: CUBE-KMP (where all boxes are cubes), UNITCUBE-KMP (where all boxes are cubes of equal side length), HYPERVOLUME (where all boxes share a vertex), and k-GROUNDED (where the projection onto the first k dimensions is a HYPERVOLUME instance).

In this paper we bring some order to these special cases by providing reductions among them. In addition to the trivial inclusions, we establish HYPERVOLUME as the easiest of these special cases, and show that the runtimes of UNITCUBE-KMP and CUBE-KMP are polynomially related. More importantly, we show that any algorithm for one of the special cases with runtime $T(n, d)$ implies an algorithm for the general case with runtime $T(n, 2d)$, yielding the first non-trivial relation between KMP and its special cases. This allows to transfer W[1]-hardness of KMP to all special cases, proving that no $n^{o(d)}$ algorithm exists for any of the special cases assuming the Exponential Time Hypothesis. Furthermore, assuming that there is no *improved* algorithm for the general case of KMP (no algorithm with runtime $\mathcal{O}(n^{d/2-\varepsilon})$) this reduction shows that there is no algorithm with runtime $\mathcal{O}(n^{\lfloor d/2 \rfloor/2-\varepsilon})$ for any of the special cases. Under the same assumption we show a tight lower bound for a recent algorithm for 2-GROUNDED [Yıldız,Suri'12].

1   Introduction

Klee's measure problem (KMP) asks for the volume of the union of n axis-aligned boxes in $\mathbb{R}^d$, where d is considered to be a constant. This is a classic problem with a long history [2, 10, 11, 15, 17, 18]. The fastest algorithm has runtime $\mathcal{O}(n^{d/2} \log n)$ for $d \geqslant 2$, given by Overmars and Yap [17], which was slightly improved to $n^{d/2} 2^{\mathcal{O}(\log^* n)}$ by Chan [10]. Thus, for over twenty years there has been no improvement over the runtime bound $n^{d/2}$. As already expressed in [10], one might conjecture that no *improved* algorithm for KMP exists, i.e., no algorithm with runtime $\mathcal{O}(n^{d/2-\varepsilon})$ for some $\varepsilon > 0$.

* Karl Bringmann is a recipient of the *Google Europe Fellowship in Randomized Algorithms*, and this research is supported in part by this Google Fellowship.

K. Chatterjee and J. Sgall (Eds.): MFCS 2013, LNCS 8087, pp. 207–218, 2013.

However, no matching lower bound is known, not even under reasonable complexity theoretic assumptions. The best unconditional lower bound is $\Omega(n \log n)$ for any dimension d [11]. Chan [10] proved that KMP is W[1]-hard by giving a reduction to the k-Clique problem. Since his reduction has $k = d/2$, we can transfer runtime lower bounds from k-Clique to KMP, implying that there is no $n^{o(d)}$ algorithm for KMP assuming the Exponential Time Hypothesis (see [16]). However, this does not determine the correct constant in the exponent. Moreover, Chan argues that since no "purely combinatorial" algorithm with runtime $\mathcal{O}(n^{k-\varepsilon})$ is known for Clique, it might be that there is no such algorithm with runtime $\mathcal{O}(n^{d/2-\varepsilon})$ for KMP, but this does not rule out faster algorithms using, e.g., fast matrix multiplication techniques.

Since no further progress was made for KMP for a long time, research turned to the study of the following special cases. For each one we list the asymptotically fastest results.

- **Cube-KMP:** Here the given boxes are cubes, not necessarily all with the same side length. This case can be solved in time $\mathcal{O}(n^{(d+2)/3})$ for $d \geqslant 2$ [6]. In dimension $d = 3$ this has been improved to $\mathcal{O}(n \log^4 n)$ by Agarwal [1]. In dimensions $d \leqslant 2$ even the general case can be solved in time $\mathcal{O}(n \log n)$. As described in [6], there are simple reductions showing that the case of cubes is roughly the same as the case of "α-fat boxes", where all side lengths of a box differ by at most a constant factor α.

- **Unitcube-KMP:** Here the given boxes are cubes, all of the same side length. This is a specialization of Cube-KMP, so all algorithms from above apply. The combinatorial complexity of a union of unit cubes is $\mathcal{O}(n^{\lfloor d/2 \rfloor})$ [5]. Using this, there are algorithms with runtime $\mathcal{O}(n^{\lfloor d/2 \rfloor} \text{polylog} \, n)$ [14] and $\mathcal{O}(n^{\lceil d/2 \rceil - 1 + \frac{1}{\lceil d/2 \rceil}} \text{polylog} \, n)$ [9]. Again, there is a generalization to "α-fat boxes of roughly equal size" with the same computational complextiy [6].

- **Hypervolume:** Here all boxes have a common vertex. Without loss of generality, we can assume that they share the vertex $(0, \ldots, 0) \in \mathbb{R}^d$ and lie in the positive orthant $\mathbb{R}^d_{\geqslant 0}$. This special case is of particular interest for practice, as it is used as an indicator of the quality of a set of points in the field of Evolutionary Multi-Objective Optimization [3, 13, 20, 21]. Improving upon the general case of KMP, there is an algorithm with runtime $\mathcal{O}(n \log n)$ for $d = 3$ [4]. The same paper also shows an unconditional lower bound of $\Omega(n \log n)$ for $d > 1$, while #P-hardness in the number of dimensions was shown in [8]. Recently, an algorithm with runtime $\mathcal{O}(n^{(d-1)/2} \log n)$ for $d \geqslant 3$ was presented in [19].

- **k-Grounded:** Here the projection of the input boxes to the first k dimensions is a Hypervolume instance, where $0 \leqslant k \leqslant d$, the other coordinates are arbitrary. This rather novel special case appeared in [19], where an algorithm with runtime $\mathcal{O}(n^{(d-1)/2} \log^2 n)$ for $d \geqslant 3$ was given for 2-Grounded.

Note that for none of these special cases W[1]-hardness is known, so there is no larger lower bound than $\Omega(n \log n)$ (for constant or slowly growing d), not even under reasonable complexity theoretic assumptions. Also note that there are trivial inclusions of some of these special cases: Each special case can be seen as a subset of all instances of the general case. As such subsets, the following inclusions hold.

- UNITCUBE-KMP $\subseteq$ CUBE-KMP $\subseteq$ KMP.
- $(k+1)$-GROUNDED $\subseteq$ k-GROUNDED for all k.
- d-GROUNDED = HYPERVOLUME and 0-GROUNDED = KMP.

This allows to transfer some results listed above to other special cases.

1.1 Our results

We present several reductions among the above four special cases and the general case of KMP. They provide bounds on the runtimes needed for these variants and, thus, yield some order among the special cases.

Our first reduction relates HYPERVOLUME and UNITCUBE-KMP.

Theorem 1. *If there is an algorithm for* UNITCUBE-KMP *with runtime* $T_{\text{UNITCUBE-KMP}}(n, d)$, *then there is an algorithm for* HYPERVOLUME *with runtime*

$$T_{\text{HYPERVOLUME}}(n, d) \leqslant \mathcal{O}(T_{\text{UNITCUBE-KMP}}(n, d)).$$

Note that if HYPERVOLUME were a subset of UNITCUBE-KMP, then the same statement would hold, with the constant hidden by the $\mathcal{O}$-notation being 1. Hence, this reduction can nearly be seen as an inclusion. Moreover, together with the trivial inclusions this reduction establishes HYPERVOLUME as the easiest of all studied special cases.

Corollary 1. *For all studied special cases,* HYPERVOLUME, UNITCUBE-KMP, CUBE-KMP, *and* k-GROUNDED *(for any* $0 \leqslant k \leqslant d$*), we have the unconditional lower bound* $\Omega(n \log n)$ *for any* $d > 1$.

One can find contradicting statements regarding the feasibility of a reduction as in Theorem 1 in the literature. On the one hand, existence of such a reduction has been mentioned in [19]. On the other hand, a newer paper [12] contains this sentence: "Better bounds have been obtained for the KMP on unit cubes ..., but reducing the hypervolume indicator to such problems is not possible in general." In any case, to the best of our knowledge a proof of such a statement cannot be found anywhere in the literature.

Our second reduction substantiates the intuition that the special cases CUBE-KMP and UNITCUBE-KMP are very similar, by showing that their runtimes differ by at most a factor of $\mathcal{O}(n)$. Recall that UNITCUBE-KMP $\subseteq$ CUBE-KMP was one of the trivial inclusions. We prove an inequality in the other direction in the following theorem.

Theorem 2. *If there is an algorithm for* UNITCUBE-KMP *with runtime* $T_{\text{UNITCUBE-KMP}}(n, d)$, *then there is an algorithm for* CUBE-KMP *with runtime*

$$T_{\text{CUBE-KMP}}(n, d) \leqslant \mathcal{O}(n \cdot T_{\text{UNITCUBE-KMP}}(n, d)).$$

Our third and last reduction finally allows to show lower bounds for all special cases. We show an inequality between the general case of KMP and $2k$-GROUNDED, in the opposite direction than the trivial inclusions. For this, we have to increase the dimension in which we consider $2k$-GROUNDED.

Theorem 3. *If there is an algorithm for $2k$-GROUNDED in dimension $d + k$ with runtime $T_{2k\text{-}GROUNDED}(n, d + k)$, then there is an algorithm for KMP in dimension d with runtime*

$$T_{KMP}(n, d) \leqslant \mathcal{O}(T_{2k\text{-}GROUNDED}(n, d + k)).$$

Note that, if we set $k = d$, the special case $2k$-GROUNDED in $d + k$ dimensions becomes HYPERVOLUME in $2d$ dimensions. Since we established HYPERVOLUME as the easiest variant, the above reduction allows to transfer W[1]-hardness from the general case to all special cases. Since the dimension is increased only by a constant factor, even the tight lower bound on the runtime can be transferred to all special cases.

Corollary 2. *There is no $n^{o(d)}$ algorithm for any of the special cases HYPER-VOLUME, UNITCUBE-KMP, CUBE-KMP, and k-GROUNDED, assuming the Exponential Time Hypothesis.*

We immediately get more precise lower bounds if we assume that no improved algorithm exists for KMP (no algorithm with runtime $\mathcal{O}(n^{d/2-\varepsilon})$).

Corollary 3. *If there is no improved algorithm for KMP, then there is no algorithm with runtime $\mathcal{O}(n^{\lfloor d/2 \rfloor/2-\varepsilon})$ for any of HYPERVOLUME, UNITCUBE-KMP, CUBE-KMP, and k-GROUNDED, for any $\varepsilon > 0$.*

This shows the first lower bound for all studied special cases that is larger than $\Omega(n \log n)$. Note that there is, however, a wide gap to the best known upper bound of $\mathcal{O}(n^{(d+2)/3})$ for HYPERVOLUME, UNITCUBE-KMP, and CUBE-KMP.

Furthermore, setting $k = 1$, Theorem 3 immediately implies that the recent algorithm for 2-GROUNDED with runtime $\mathcal{O}(n^{(d-1)/2} \log^2 n)$ [19] is optimal (apart from logarithmic factors and if there is no improved algorithm for KMP).

Corollary 4. *If there is no improved algorithm for KMP, then there is no algorithm for 2-GROUNDED with runtime $\mathcal{O}(n^{(d-1)/2-\varepsilon})$ for any $\varepsilon > 0$.*

To simplify our runtime bounds, in some proofs we use the following technical lemma. Informally, it states that for any k-GROUNDED algorithm with runtime $T(n, d)$ we have $T(\mathcal{O}(n), d) \leqslant \mathcal{O}(T(n, d))$. Note that in this paper we hide by the $\mathcal{O}$-notation any functions depending solely on d.

Lemma 1. *Fix $0 \leqslant k \leqslant d$ and $c > 1$. If there is an algorithm for k-GROUNDED with runtime $T_{k\text{-}GROUNDED}(n, d)$ then there is another algorithm for k-GROUNDED with runtime $T'_{k\text{-}GROUNDED}(n, d)$ satisfying*

$$T'_{k\text{-}GROUNDED}(cn, d) \leqslant \mathcal{O}(T_{k\text{-}GROUNDED}(n, d)).$$

Due to space constraints, the proofs of this and other statements can be found in the full version of this paper [7].

1.2   Notation and Organization

A *box* is a set of the form $B = [a_1, b_1] \times \ldots \times [a_d, b_d] \subset \mathbb{R}^d$, $a_i, b_i \in \mathbb{R}$, $a_i \leqslant b_i$. A *cube* is a box with all side lengths equal, i.e., $|b_1 - a_1| = \ldots = |b_d - a_d|$. Moreover, a KMP *instance* is simply a set M of n boxes. In CUBE-KMP all these boxes are cubes, and in UNITCUBE-KMP all these boxes are cubes of common side length. In HYPERVOLUME, all input boxes share the vertex $(0, \ldots, 0) \in \mathbb{R}^d$, i.e., each input box is of the form $B = [0, b_1] \times \ldots \times [0, b_d]$. In k-GROUNDED, the projection of each input box to the first k dimensions is a HYPERVOLUME box, meaning that each input box is of the form $B = [a_1, b_1] \times \ldots \times [a_d, b_d]$ with $a_1 = \ldots = a_k = 0$.

We write the usual Lebesgue measure of a set $A \subseteq \mathbb{R}^d$ as $\text{VOL}(A)$. For sets $R, A \subseteq \mathbb{R}^d$ we write $\text{VOL}_R(A) := \text{VOL}(R \cap A)$, the volume of A restricted to R. For a KMP instance M we let $\mathcal{U}(M) := \bigcup_{B \in M} B$. To shorten notation we write $\text{VOL}(M) := \text{VOL}(\mathcal{U}(M))$ and $\text{VOL}_R(M) := \text{VOL}(R \cap \mathcal{U}(M))$.

In the next section we present the proof of Theorem 1. In Section 3 we prove Theorem 2. The proof of Theorem 3 is split into Section 4 and Section 5: We first give the reduction for 2-GROUNDED (again split into the case $d = 1$ and a generalization to larger dimensions) and then generalize this result to $2k$-GROUNDED, $k > 1$. We close with an extensive list of open problems.

2   Hypervolume $\leqslant$ Unitcube-KMP

In this section we prove Theorem 1 by giving a reduction from HYPERVOLUME to UNITCUBE-KMP.

Given an instance of HYPERVOLUME, let Δ be the largest coordinate of any box. We extend all boxes to cubes of side length Δ, yielding a UNITCUBE-KMP instance. In this process, we make sure that the new parts of each box will not lie in the positive orthant $\mathbb{R}^d_{\geqslant 0}$, but in the other orthants, as depicted in Figure 1. This means that the volume of the newly constructed cubes - restricted to $\mathbb{R}^d_{\geqslant 0}$ - is the same as the volume of the input boxes. To compute this restricted volume, we compute the volume of the constructed UNITCUBE-KMP instance once with and once without an additional cube $C = [0, \Delta]^d$. From this we can infer the volume of the input HYPERVOLUME instance.

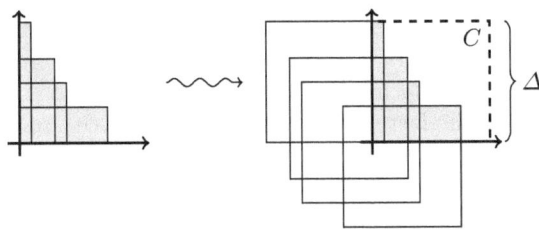

Fig. 1. Construction in the proof of Theorem 1

3   Unitcube-KMP $\geqslant$ Cube-KMP

In this section we prove Theorem 2 by giving a reduction from CUBE-KMP to UNITCUBE-KMP.

Given a CUBE-KMP instance, let C be the cube with smallest side length. We will compute the *contribution* v of C, i.e., the volume of space that is contained in C but no other cube. Having this, we can delete C and recurse on the remaining boxes. Adding up yields the total volume of the input instance.

To compute v, we modify each cube such that it becomes a cube of C's side length and its restriction to C stays the same, as depicted in Figure 2. Applying this construction to all input boxes, we get a UNITCUBE-KMP instance that, inside C, looks the same as the input CUBE-KMP instance. Computing the volume of this new instance once with and once without C allows to infer v.

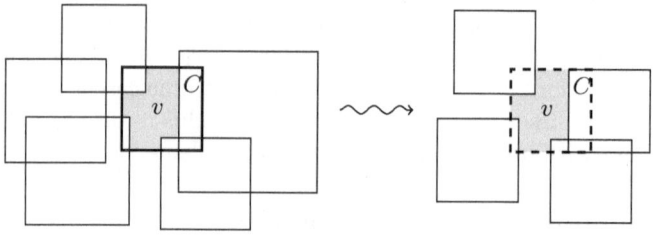

Fig. 2. Construction in the proof of Theorem 2

4   2-Grounded $\geqslant$ KMP

We first show the reduction of Theorem 3 for 2-GROUNDED, i.e., we show $T_{\mathrm{KMP}}(n,d) \leqslant \mathcal{O}(T_{\text{2-GROUNDED}}(n, d+1))$ by giving a reduction from KMP to 2-GROUNDED. This already implies Corollary 4 and lays the foundations for the complete reduction given in the next section.

We begin by showing the reduction for $d = 1$. As a second step we show how to generalize this to larger dimensions.

4.1   Dimension $d = 1$

We want to give a reduction from KMP in 1 dimension to 2-GROUNDED in 2 dimensions. Note that the latter is the same as HYPERVOLUME in 2 dimensions. Let M be an instance of KMP in 1 dimension, i.e., a set of n intervals in $\mathbb{R}$. We will reduce the computation of $\mathrm{VOL}(M)$ to two instances of 2-GROUNDED.

Denote by $x_1 < \ldots < x_m$ the endpoints of all intervals in M (if all endpoints are distinct then $m = 2n$). We can assume that $x_1 = 0$ after translation. Consider the boxes

$$A_i := [m - i - 1, m - i] \times [x_i, x_{i+1}]$$

in $\mathbb{R}^2$ for $1 \leqslant i \leqslant m-1$, as depicted in Figure 3. Denote the union of these boxes by A. Note that the volume of box A_i is the same as the length of the interval $[x_i, x_{i+1}]$. This means that we took the chain of intervals $\{[x_i, x_{i+1}]\}$ and made it into a staircase of boxes $\{A_i\}$, where each box has the same volume as the corresponding interval.

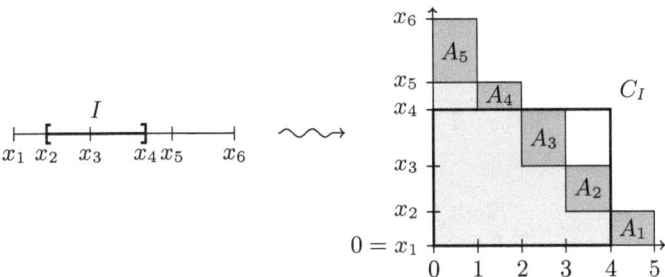

Fig. 3. The left hand side depicts all endpoints $0 = x_1 \leqslant \ldots \leqslant x_6$ of a 1-dimensional KMP instance. An input interval I is indicated. The right hand side shows the result of our transformation. Each interval $[x_i, x_{i+1}]$ to the left corresponds to a box A_i to the right. The interval I gets mapped to the box C_I. The shaded regions depict the set A (■, the union of all A_i) and the set T_0 ( ).

Now consider an interval $I = [x_j, x_k] \in M$. We construct the box

$$C_I := [0, m - j] \times [0, x_k],$$

also shown in Figure 3. Then C_I contains the boxes A_i with $j \leqslant i < k$ and (its interior) has no common intersection with any other box A_i. This is easily seen as $A_i \subseteq C_I$ iff $m - i \leqslant m - j$ and $x_{i+1} \leqslant x_k$. Hence, for any interval $I \in M$ we constructed a box C_I that contains exactly those boxes A_i whose corresponding interval $[x_i, x_{i+1}]$ is contained in I, or in other words

$$[x_i, x_{i+1}] \subseteq I \Leftrightarrow A_i \subseteq C_I,$$
$$\mathrm{VOL}([x_i, x_{i+1}] \cap I) = \mathrm{VOL}(A_i \cap C_I).$$

From these properties it follows that the volume of C_I restricted to A is the same as the length of I, i.e.,

$$\mathrm{VOL}_A(C_I) = \mathrm{VOL}(I).$$

Furthermore, considering the whole set M of intervals, the interval $[x_i, x_{i+1}]$ is contained in some interval in M iff the box A_i is contained in some box in $C_M := \{C_I \mid I \in M\}$. This yields

$$\mathrm{VOL}(M) = \mathrm{VOL}_A(C_M).$$

It remains to reduce the computation of $\text{VOL}_A(C_M)$ to two 2-GROUNDED instances. For this we consider

$$T_0 := \bigcup_{1 \leqslant i \leqslant m} C_{[x_i, x_i]}.$$

Informally speaking, T_0 consists of all points "below" A, as depicted in Figure 3. Note that no set A_j is contained in T_0. Moreover, we consider the set $T_1 := T_0 \cup A$. Observe that we can write

$$T_1 = \bigcup_{1 \leqslant i \leqslant m-1} C_{[x_i, x_{i+1}]},$$

since $A_i \subseteq C_{[x_i, x_{i+1}]}$. Note that both sets T_0 and T_1 are unions of $\mathcal{O}(n)$ 2-GROUNDED boxes. Informally, T_0 is the maximum 2-GROUNDED instance that has $\text{VOL}_A(T_0) = 0$, and T_1 is the minimum 2-GROUNDED instance with $\text{VOL}_A(T_1) = \text{VOL}(A)$. Now, we can compute $\text{VOL}_A(C_M)$ as follows.

Lemma 2. *In the above situation we have*

$$\text{VOL}_A(C_M) = \text{VOL}(A) + \text{VOL}(T_0 \cup \mathcal{U}(C_M)) - \text{VOL}(T_1 \cup \mathcal{U}(C_M)).$$

Proof. Set $U := \mathcal{U}(C_M)$. Using $T_0 \subseteq T_1$ and $A = T_1 \setminus T_0$ in a sequence of simple transformations, we get

$$\begin{aligned}
\text{VOL}(T_1 \cup U) - \text{VOL}(T_0 \cup U) &= \text{VOL}((T_1 \cup U) \setminus (T_0 \cup U)) \\
&= \text{VOL}((T_1 \setminus T_0) \setminus U) \\
&= \text{VOL}(A \setminus U) \\
&= \text{VOL}(A) - \text{VOL}(A \cap U) \\
&= \text{VOL}(A) - \text{VOL}_A(C_M),
\end{aligned}$$

which proves the claim.   □

Note that $\text{VOL}(A) = \sum_i \text{VOL}(A_i) = \sum_i |x_{i+1} - x_i| = |x_m - x_1|$ is trivial. Also note that both sets T_0 and T_1 are the union of $\mathcal{O}(n)$ 2-GROUNDED boxes, so that $\text{VOL}(T_b \cup \mathcal{U}(C_M))$ can be seen as a 2-GROUNDED instance of size $\mathcal{O}(n)$, for both $b \in \{0, 1\}$. Hence, we reduced the computation of the input instance's volume $\text{VOL}(M)$ to $\text{VOL}_A(C_M)$ and further to the 2-GROUNDED instances $\text{VOL}(T_0 \cup \mathcal{U}(C_M))$ and $\text{VOL}(T_1 \cup \mathcal{U}(C_M))$.

As we have to sort the given intervals first, we get

$$T_{\text{KMP}}(n, 1) \leqslant \mathcal{O}(T_{\text{2-GROUNDED}}(\mathcal{O}(n), 2) + n \log n).$$

Note that this inequality alone gives no new information, as already Klee [15] showed that $T_{\text{KMP}}(n, 1) \leqslant \mathcal{O}(n \log n)$. However, we get interesting results when we generalize this reduction to higher dimensions in the next section.

4.2   Larger Dimensions

In this section we show how the reduction from the last section carries over to larger dimensions, yielding a reduction from KMP in d dimensions to 2-GROUNDED in $d+1$ dimensions. This implies $T_{\text{KMP}}(n, d) \leqslant \mathcal{O}(T_{\text{2-GROUNDED}}(n, d+1))$.

Assume we are given a KMP instance M in dimension d. The idea is that we use the dimension doubling reduction from the last section on the first dimension and leave all other dimensions untouched. More precisely, for a box $B \in M$ let $\pi_1(B)$ be its projection onto the first dimension and let $\pi_*(B)$ be its projection onto the last $d-1$ dimensions, so that $B = \pi_1(B) \times \pi_*(B)$. Now follow the reduction from the last section on the instance $M' := \{\pi_1(B) \mid B \in M\}$. This yields sets A, T_0, T_1, and a box C_I for each $I \in M'$.

We set $C_B := C_{\pi_1(B)} \times \pi_*(B)$ and $C_M = \{C_B \mid B \in M\}$. A possible way of generalizing A would be to set $A'' := A \times \mathbb{R}^{d-1}$. Then we would be interested in $\text{VOL}_{A''}(C_M)$, which can be seen to be exactly $\text{VOL}(M)$. This definition of A'' is, however, not simple enough, as it is not a difference of 2-GROUNDED instances (unlike $A = T_1 \setminus T_0$). To give a different definition, assume (after translation) that all coordinates of the input instance are non-negative and let Δ be the maximal coordinate in any dimension. We set $A' := A \times [0, \Delta]^{d-1}$ and still get the same volume $\text{VOL}_{A'}(C_M) = \text{VOL}(M)$. This allows to generalize T_0 and T_1 to $T_0' := T_0 \times [0, \Delta]^{d-1}$ and $T_1' := T_1 \times [0, \Delta]^{d-1}$, while still having

$$\text{VOL}_{A'}(C_M) = \text{VOL}(A') + \text{VOL}(T_0' \cup \mathcal{U}(C_M)) - \text{VOL}(T_1' \cup \mathcal{U}(C_M)).$$

Note that T_0' and T_1' are also a union of $\mathcal{O}(n)$ 2-GROUNDED boxes, so a volume such as $\text{VOL}(T_0' \cup \mathcal{U}(C_M))$ can be seen as a 2-GROUNDED instance. This completes the reduction and yields the time bound

$$T_{\text{KMP}}(n, d) \leqslant \mathcal{O}(T_{\text{2-GROUNDED}}(\mathcal{O}(n), d+1) + n \log n).$$

Using the lower bound $\Omega(n \log n)$ of Corollary 1 we can hide the additional $n \log n$ in the first summand. Moreover, first using the technical Lemma 1 we can finally simplify this to the statement of Corollary 4,

$$T_{\text{KMP}}(n, d) \leqslant \mathcal{O}(T_{\text{2-GROUNDED}}(n, d+1)).$$

5   2k-Grounded $\geqslant$ KMP

It is left to show the full version of Theorem 3, i.e., to give a reduction from KMP in dimension d to 2k-GROUNDED in dimension $d+k$. A full proof of this can be found in the full version of this paper, here we only present an outline.

The first steps of generalizing the reduction from the last section to the case $k > 1$ are straightforward. We want to use the dimension doubling reduction from Section 4.1 on each one of the first k dimensions. For any box $B \in \mathbb{R}^d$ denote its projection onto the i-th dimension by $\pi_i(B)$, $1 \leqslant i \leqslant k$, and its projection onto dimensions $k + 1, \ldots, d$ by $\pi_*(B)$. We use the reduction from Section 4.1 on each dimension $1 \leqslant i \leqslant k$, i.e., on each instance $M^{(i)} := \{\pi_i(B) \mid B \in M\}$, yielding sets $A^{(i)}, T_0^{(i)}, T_1^{(i)}$, and a box $C_I^{(i)}$ for each $I \in M^{(i)}$.

For a box $B \in M$ we now define $C_B := C_{\pi_1(B)}^{(1)} \times \ldots \times C_{\pi_k(B)}^{(k)} \times \pi_*(B)$. This is a box in $\mathbb{R}^{d+k}$, it is even a $2k$-GROUNDED box, as its projection onto the first $2k$ coordinates has the vertex $(0, \ldots, 0)$. Let $C_M := \{C_B \mid B \in M\}$. Setting $\Omega := [0, \Delta]$ and $A := A^{(1)} \times \ldots \times A^{(k)} \times \Omega^{d-k}$, we now can show the following.

Lemma 3. *We have* $\text{VOL}(M) = \text{VOL}_A(C_M)$.

The hard part of the reduction that remains to show is that the right hand side of this can indeed be computed using $2k$-GROUNDED calls, although A is a non-trivial region.

For $1 \leqslant i \leqslant k$ and $b \in \{0, 1\}$ we set $\tilde{T}_b^{(i)} := \Omega^{2(i-1)} \times T_b^{(i)} \times \Omega^{d+k-2i}$. This set in Ω^{d+k} consists of all points x whose projection to dimensions $2i - 1$ and $2i$ is contained in $T_b^{(i)}$. Note that each set $\tilde{T}_b^{(i)}$ can be written as the union of $\mathcal{O}(n)$ $2k$-GROUNDED boxes, since $T_b^{(i)}$ is the union of $\ell = \mathcal{O}(n)$ 2-GROUNDED boxes in $\mathbb{R}^2$. Thus, we can use an algorithm for $2k$-GROUNDED to compute any volume of the form $\text{VOL}(\tilde{T}_b^{(i)} \cup V)$, where V is a union of $\mathcal{O}(n)$ $2k$-GROUNDED boxes.

Furthermore, define for $S \subseteq [k]$

$$D_S := \left(\bigcup_{i \in S} \tilde{T}_1^{(i)} \right) \cup \bigcup_{i \in [k] \setminus S} \tilde{T}_0^{(i)}.$$

Note that $D_S \subseteq D_{S'}$ holds for $S \subseteq S'$. We can express A using the sets D_S as shown by the following lemma.

Lemma 4. *We have* $A = \bigcap_{1 \leqslant i \leqslant k} D_{\{i\}} \setminus D_\emptyset$.

Moreover, each D_S can be written as the union of $\mathcal{O}(n)$ $2k$-GROUNDED instances, since the same was true for the sets $\tilde{T}_b^{(i)}$. Hence, we can use an algorithm for $2k$-GROUNDED to compute the volume

$$H_S := \text{VOL}(D_S \cup \mathcal{U}(C_M)).$$

Finally, we show that we can compute $\text{VOL}_A(C_M)$ from the H_S by an interesting usage of the inclusion-exclusion principle, finishing the reduction.

Lemma 5. *We have* $\text{VOL}_A(C_M) = \text{VOL}(A) + \sum_{S \subseteq [k]} (-1)^{|S|} H_S$.

Proof (Sketch). In this proof we write for short $U := \mathcal{U}(C_M)$. Using Lemma 4 and the inclusion-exclusion principle we arrive at

$$\text{VOL}(A) - \text{VOL}_A(U) = \text{VOL}(A \setminus U) = \text{VOL}\left(\bigcap_{1 \leqslant i \leqslant k} D_{\{i\}} \setminus (D_\emptyset \cup U) \right)$$

$$= \sum_{\emptyset \neq S \subseteq [k]} (-1)^{|S|+1} \text{VOL}\left(\bigcup_{i \in S} D_{\{i\}} \setminus (D_\emptyset \cup U) \right).$$

Together with $H_S - H_\emptyset = \text{VOL}(D_S \setminus (D_\emptyset \cup U)) = \text{VOL}\left(\bigcup_{i \in S} D_{\{i\}} \setminus (D_\emptyset \cup U) \right)$ and some simplifications, this proves the claim. $\qquad \square$

6   Conclusion

We presented reductions between the special cases Cube-KMP, Unitcube-KMP, Hypervolume, and k-Grounded of Klee's measure problem. These reductions imply statements about the runtime needed for these problem variants. We established Hypervolume as the easiest among all studied special cases, and showed that the variants Cube-KMP and Unitcube-KMP have polynomially related runtimes. Moreover, we presented a reduction from the general case of KMP to $2k$-Grounded. This allows to transfer W[1]-hardness from KMP to all special cases, proving that no $n^{o(d)}$ algorithm exists for any of the special cases assuming the Exponential Time Hypothesis. Moreover, assuming that no improved algorithm exists for KMP, we get a tight lower bound for a recent algorithm for 2-Grounded, and a lower bound of roughly $n^{(d-1)/4}$ for all other special cases. Thus, we established some order among the special cases of Klee's measure problem.

Our results lead to a number of open problems, both asking for new upper and lower bounds:

- Is there a polynomial relation between Hypervolume and Unitcube-KMP, similar to Cube-KMP and Unitcube-KMP, or do both problems have significantly different runtimes?
- Show that no improved algorithm exists for KMP, e.g., assuming the Strong Exponential Time Hypothesis, as has been done for the Dominating Set problem, see [16]. Or give an improved algorithm.
- Assuming that no improved algorithm for KMP exists, we know that the optimal runtimes of Hypervolume and Cube-KMP/Unitcube-KMP are of the form $n^{c_d \cdot d \pm \mathcal{O}(1)}$, with $c_d \in [1/4, 1/3]$. Determine the correct value of c_d.
- Generalize the $\mathcal{O}(n^{(d-1)/2} \log^2 n)$ algorithm for 2-Grounded [19] to an $\mathcal{O}(n^{(d-k)/2+o(1)})$ algorithm for $2k$-Grounded. This would again be optimal by Theorem 3.
- We showed the relation $T_{\mathrm{KMP}}(n,d) \leqslant \mathcal{O}(T_{2k\text{-Grounded}}(n, d+k))$. Show an inequality in the opposite direction, i.e., a statement of the form $T_{k\text{-Grounded}}(n,d) \leqslant \mathcal{O}(T_{\mathrm{KMP}}(n,d'))$ with $d' < d$.

References

[1] Agarwal, P.K.: An improved algorithm for computing the volume of the union of cubes. In: Proc. 2010 Annual Symposium on Computational Geometry (SoCG 2010), pp. 230–239. ACM, New York (2010)

[2] Bentley, J.L.: Algorithms for Klee's rectangle problems. Department of Computer Science, Carnegie Mellon University (1977) (unpublished notes)

[3] Beume, N., Naujoks, B., Emmerich, M.T.M.: SMS-EMOA: Multiobjective selection based on dominated hypervolume. European Journal of Operational Research 181, 1653–1669 (2007)

[4] Beume, N., Fonseca, C.M., López-Ibáñez, M., Paquete, L., Vahrenhold, J.: On the complexity of computing the hypervolume indicator. IEEE Trans. Evolutionary Computation 13, 1075–1082 (2009)

[5] Boissonnat, J.-D., Sharir, M., Tagansky, B., Yvinec, M.: Voronoi diagrams in higher dimensions under certain polyhedral distance functions. Discrete & Computational Geometry 19, 485–519 (1998)

[6] Bringmann, K.: An improved algorithm for Klee's measure problem on fat boxes. Computational Geometry: Theory and Applications 45, 225–233 (2012)

[7] Bringmann, K.: Bringing order to special cases of Klee's measure problem. CoRR, abs/1301.7154 (2013)

[8] Bringmann, K., Friedrich, T.: Approximating the least hypervolume contributor: NP-hard in general, but fast in practice. Theoretical Computer Science 425, 104–116 (2012)

[9] Chan, T.M.: Semi-online maintenance of geometric optima and measures. SIAM J. Comput. 32, 700–716 (2003)

[10] Chan, T.M.: A (slightly) faster algorithm for Klee's measure problem. Computational Geometry: Theory and Applications 43, 243–250 (2010)

[11] Fredman, M.L., Weide, B.W.: On the complexity of computing the measure of $\bigcup[a_i, b_i]$. Commun. ACM 21, 540–544 (1978)

[12] Guerreiro, A.P., Fonseca, C.M., Emmerich, M.T.M.: A fast dimension-sweep algorithm for the hypervolume indicator in four dimensions. In: Proc. 24th Canadian Conference on Computational Geometry (CCCG 2012), pp. 77–82 (2012)

[13] Igel, C., Hansen, N., Roth, S.: Covariance matrix adaptation for multi-objective optimization. Evolutionary Computation 15, 1–28 (2007)

[14] Kaplan, H., Rubin, N., Sharir, M., Verbin, E.: Counting colors in boxes. In: Proc. 18th Annual ACM-SIAM Symposium on Discrete Algorithms (SODA 2007), pp. 785–794 (2007)

[15] Klee, V.: Can the measure of $\bigcup[a_i, b_i]$ be computed in less than $O(n \log n)$ steps? American Mathematical Monthly 84, 284–285 (1977)

[16] Lokshtanov, D., Marx, D., Saurabh, S.: Lower bounds based on the exponential time hypothesis, The Complexity Column by V. Arvind. Bulletin of the EATCS 105, 41–72 (2011)

[17] Overmars, M.H., Yap, C.-K.: New upper bounds in Klee's measure problem. SIAM J. Comput. 20, 1034–1045 (1991)

[18] van Leeuwen, J., Wood, D.: The measure problem for rectangular ranges in d-space. J. Algorithms 2, 282–300 (1981)

[19] Yıldız, H., Suri, S.: On Klee's measure problem for grounded boxes. In: Proc. 2012 Annual Symposuim on Computational Geometry (SoCG 2012), pp. 111–120. ACM, New York (2012)

[20] Zitzler, E., Künzli, S.: Indicator-based selection in multiobjective search. In: Yao, X., et al. (eds.) PPSN 2004. LNCS, vol. 3242, pp. 832–842. Springer, Heidelberg (2004)

[21] Zitzler, E., Thiele, L.: Multiobjective evolutionary algorithms: A comparative case study and the strength Pareto approach. IEEE Trans. Evolutionary Computation 3, 257–271 (1999)

Random Shortest Paths: Non-euclidean Instances for Metric Optimization Problems⋆

Karl Bringmann[1],[⋆⋆], Christian Engels[2],
Bodo Manthey[3], and B.V. Raghavendra Rao[4]

[1] Max Planck Institute for Informatics
kbringma@mpi-inf.mpg.de
[2] Saarland University
engels@cs.uni-saarland.de
[3] University of Twente
b.manthey@utwente.nl
[4] Indian Institute of Technology Madras
bvrr@cse.iitm.ac.in

Abstract. Probabilistic analysis for metric optimization problems has mostly been conducted on random Euclidean instances, but little is known about metric instances drawn from distributions other than the Euclidean.

This motivates our study of random metric instances for optimization problems obtained as follows: Every edge of a complete graph gets a weight drawn independently at random. The length of an edge is then the length of a shortest path (with respect to the weights drawn) that connects its two endpoints.

We prove structural properties of the random shortest path metrics generated in this way. Our main structural contribution is the construction of a good clustering. Then we apply these findings to analyze the approximation ratios of heuristics for matching, the traveling salesman problem (TSP), and the k-center problem, as well as the running-time of the 2-opt heuristic for the TSP. The bounds that we obtain are considerably better than the respective worst-case bounds. This suggests that random shortest path metrics are easy instances, similar to random Euclidean instances, albeit for completely different structural reasons.

1 Introduction

For large-scale optimization problems, finding optimal solutions within reasonable time is often impossible, because many such problems, like the traveling salesman problem (TSP), are NP-hard. Nevertheless, we often observe that simple heuristics succeed surprisingly quickly in finding close-to-optimal solutions. Many such heuristics perform well in practice but have a poor worst-case performance. In order to explain the performance of such heuristics, probabilistic

---

⋆ A full version with all proofs is available at http://arxiv.org/abs/1306.3030.

⋆⋆ Karl Bringmann is a recipient of the *Google Europe Fellowship in Randomized Algorithms*, and this research is supported in part by this Google Fellowship.

K. Chatterjee and J. Sgall (Eds.): MFCS 2013, LNCS 8087, pp. 219–230, 2013.
© Springer-Verlag Berlin Heidelberg 2013

analysis has proved to be a useful alternative to worst-case analysis. Probabilistic analysis of optimization problems has been conducted with respect to arbitrary instances (without the triangle inequality) [15,23] or instances embedded in Euclidean space. In particular, the limiting behavior of various heuristics for many of the Euclidean optimization problems is known precisely [35].

However, the average-case performance of heuristics for general metric instances is not well understood. This lack of understanding can be explained by two reasons: First, independent random edge lengths (without the triangle inequality) and random geometric instances are relatively easy to handle from a technical point of view – the former because of the independence of the lengths, the latter because Euclidean space provides a structure that can be exploited. Second, analyzing heuristics on random metric spaces requires an understanding of random metric spaces in the first place. While Vershik [33] gave an analysis of a process for obtaining random metric spaces, using this directly to analyze algorithms seems difficult.

In order to initiate systematic research of heuristics on general metric spaces, we use the following model, proposed by Karp and Steele [24, Section 3.4]: Given an undirected complete graph, we draw edge weights independently at random. Then the length of an edge is the length of a shortest path connecting its endpoints. We call such instances *random shortest path metrics*.

This model is also known as *first-passage percolation*, and has been introduced by Broadbent and Hemmersley as a model for passage of fluid in a porous medium [6, 7]. More recently, it has also been used to model shortest paths in networks such as the internet [12]. The appealing feature of random shortest path metrics is their simplicity, which enables us to use them for the analysis of heuristics.

1.1 Known and Related Results

There has been significant study of random shortest path metrics or first-passage percolation. The expected length of an edge is known to be $\Theta(\log n/n)$ [9,21], and the same asymptotic bound holds also for the longest edge almost surely [18, 21]. This model has been used to analyze algorithms for computing shortest paths [16, 18, 28]. Kulkarni and Adlakha have developed algorithmic methods to compute distribution and moments of several optimization problems [25–27]. Beyond shortest path algorithms, random shortest path metrics have been applied only rarely to analyze algorithms. Dyer and Frieze, answering a question raised by Karp and Steele [24, Section 3.4], analyzed the patching heuristic for the asymmetric TSP (ATSP) in this model. They showed that it comes within a factor of $1 + o(1)$ of the optimal solution with high probability. Hassin and Zemel [18] applied their findings to the 1-center problem.

From a more structural point of view, first-passage percolation has been analyzed in the area of complex networks, where the hop-count (the number of edges on a shortest path) and the length of shortest path trees have been analyzed [20]. These properties have also been studied on random graphs with random edge weights [5, 19]. More recently, Addario-Berry et. al. [1] showed that the number of edges in the longest of the shortest paths is $O(\log n)$ with high probability, and hence the shortest path trees have depth $O(\log n)$.

1.2   Our Results

As far as we are aware, simple heuristics such as greedy heuristics have not been studied in this model yet. Understanding the performance of such algorithms is particularly important as they are easy to implement and used in many applications.

We provide a probabilistic analysis of simple heuristics for optimization under random shortest path metrics. First, we provide structural properties of random shortest path metrics (Section 3). Our most important structural contribution is proving the existence of a good clustering (Lemma 3.8). Then we use these structural insights to analyze simple algorithms for minimum weight matching and the TSP to obtain better expected approximation ratios compared to the worst-case bounds. In particular, we show that the greedy algorithm for minimum-weight perfect matching (Theorem 4.2), the nearest-neighbor heuristic for the TSP (Theorem 4.3), and every insertion heuristic for the TSP (Theorem 4.4) achieve constant expected approximation ratios. We also analyze the 2-opt heuristic for the TSP and show that the expected number of 2-exchanges required before the termination of the algorithm is bounded by $O(n^8 \log^3 n)$ (Theorem 4.5). Investigating further the structural properties of random shortest path metrics, we then consider the k-center problem (Section 5), and show that the most trivial procedure of choosing k arbitrary vertices as k-centers yields a $1 + o(1)$ approximation in expectation, provided $k = O(n^{1-\varepsilon})$ for some $\varepsilon > 0$ (Theorem 5.2).

2   Model and Notation

We consider undirected complete graphs $G = (V, E)$ without loops. First, we draw *edge weights* $w(e)$ independently at random according to the exponential distribution with parameter 1. (Exponential distributions are technically the easiest to handle because they are memoryless. However, our results hold also for other distributions, in particular for the uniform distribution on $[0, 1]$. We briefly discuss this in Section 6.) Second, let the *distances* or *lengths* $d : V \times V \to [0, \infty)$ be given by the lengths of the shortest paths between the vertices with respect to the weights thus drawn. In particular, we have $d(v, v) = 0$ for all $v \in V$, we have $d(u, v) = d(v, u)$ because G is undirected, and we have the triangle inequality: $d(u, v) \leq d(u, x) + d(x, v)$ for all $u, x, v \in V$. We call the complete graph with edge lengths d obtained from random weights w a *random shortest path metric*.

We use the following notation: Let $\Delta_{\max} = \max_{e \in E} d(e)$ denote the longest edge in the random shortest path metric. Let $N_\Delta^v = \{u \in V \mid d(u, v) \leq \Delta\}$ be the set of all nodes in a Δ-environment of v, and let $k_\Delta^v = |N_\Delta^v|$ the number of nodes around v in a Δ-environment. We denote the minimal Δ such that there are at least k nodes within a distance of Δ of v by Δ_k^v. Formally, we define

$\Delta_k^v = \min\{\Delta \mid k_\Delta^v \geq k\}$. Note that $v \in N_\Delta^v$ for any $\Delta \geq 0$ because the distance of v to itself is 0. Consequently, we have $\Delta_1^v = 0$ and $k_0^v \geq 1$.

By $\mathrm{Exp}(\lambda)$, we denote the exponential distribution with parameter λ. By exp, we denote the exponential function. For $n \in \mathbb{N}$, let $[n] = \{1, \ldots, n\}$, and let $H_n = \sum_{i=1}^n 1/i$ be the n-th harmonic number.

3   Structural Properties of Shortest Path Metrics

3.1   Random Process

To understand random shortest path metrics, it is convenient to fix a starting vertex v and see how the lengths from v to the other vertices develop. In this way, we analyze the distribution of Δ_k^v.

The values Δ_k^v are generated by a simple birth process as follows. (The same process has been analyzed by Davis and Prieditis [9], Janson [21], and also in subsequent papers.) For $k = 1$, we have $\Delta_k^v = 0$. For $k \geq 1$, we are looking for the closest vertex to any vertex in $N_{\Delta_k^v}^v$ in order to obtain Δ_{k+1}^v. This conditions all edges (u, x) with $u \in N_{\Delta_k^v}^v$ and $x \notin N_{\Delta_k^v}^v$ to be of length at least $\Delta_k^v - d(v, u)$. Otherwise, x would already be in $N_{\Delta_k^v}^v$. The set $N_{\Delta_k^v}^v$ contains k vertices. Thus, there are $k \cdot (n - k)$ connections from $N_{\Delta_k^v}^v$ to the rest of the graph. Consequently, the difference $\delta_k = \Delta_k^v - \Delta_{k-1}^v$ is distributed as the minimum of $k(n - k)$ exponential random variables (with parameter 1), or, equivalently, as an exponential random variable with parameter $k \cdot (n - k)$. We obtain that $\Delta_{k+1}^v = \sum_{i=1}^k \mathrm{Exp}(i \cdot (n - i))$. (Note that the exponential distributions and the random variables $\delta_1, \ldots, \delta_n$ are independent.)

Exploiting linearity of expectation and that the expected value of $\mathrm{Exp}(\lambda)$ is $1/\lambda$ yields the following theorem.

Theorem 3.1. *For any $k \in [n]$ and any $v \in V$, we have $\mathbb{E}(\Delta_k^v) = \frac{1}{n} \cdot (H_{k-1} + H_{n-1} - H_{n-k})$ and Δ_k^v is distributed as $\sum_{i=1}^{k-1} \mathrm{Exp}(i \cdot (n - i))$.*

From this result, we can easily deduce two known results: averaging over k yields that the expected length of an edge is $\frac{H_{n-1}}{n-1} \approx \ln n/n$ [9, 21]. By considering Δ_n^v, we obtain that the longest edge incident to a fixed vertex has an expected length of $2H_{n-1}/n \approx 2 \cdot \ln n/n$ [21]. For completeness, the length of the longest edge in the whole graph is roughly $3 \cdot \ln n/n$ [21].

3.2   Distribution of Δ_k^v

Let us now have a closer look at the distribution of Δ_k^v for fixed $v \in V$ and $k \in [n]$. Let F_k^v denote the cumulative distribution function (CDF) of Δ_k^v, i.e., $F_k^v(x) = \mathbb{P}(\Delta_k^v \leq x)$. A careful analysis of the distribution of a sum of exponential random variables yields the following two lemmas.

Lemma 3.2. *For every $\Delta \geq 0$, $v \in V$, and $k \in [n]$, we have*

$$\left(1 - \exp(-(n - k)\Delta)\right)^{k-1} \leq F_k^v(\Delta) \leq \left(1 - \exp(-n\Delta)\right)^{k-1}.$$

Proof. We have already seen that Δ_k^v is a sum of exponentially distributed random variables with parameters $\lambda_i = i(n - i) \in [(n - k)i, ni]$ for $i \in [k - 1]$. We approximate the parameters by ci for $c \in \{n - k, n\}$. The distribution with $c = n$ is stochastically dominated by the true distribution, which is in turn dominated by the distribution obtained for $c = n - k$.

We keep c as a parameter and obtain the following density function for the sum of exponentially distributed random variables with parameters $c, \ldots, (k-1) \cdot c$ [31, p. 308ff]:

$$\sum_{i=1}^{k-1} \left(\prod_{j \in [k-1] \setminus \{i\}} \frac{j}{j - i} \right) \cdot ci \cdot \exp(-cix) = \sum_{i=1}^{k-1} \frac{\frac{(k-1)!}{i} \cdot (-1)^{i-1}}{(i - 1)!(k - 1 - i)!} \cdot ci \cdot \exp(-cix)$$

$$= \sum_{i=1}^{k-1} \binom{k - 1}{i}(-1)^{i-1} \cdot ci \cdot \exp(-cix).$$

Integrating plus the binomial theorem yields

$$\sum_{i=1}^{k-1} \binom{k - 1}{i} \left(-\exp(-cix) \right)(-1)^{i-1} \cdot ci \cdot \exp(-cix) = \left(-\exp(-cx) + 1 \right)^{k-1} - 1.$$

Taking the difference of the function values at Δ and 0 yields $\left(1 - \exp(-c\Delta) \right)^{k-1}$, which yields the bounds claimed by choosing $c = n - k$ and $c = n$. $\qquad\square$

Lemma 3.3. *Fix $\Delta \geq 0$ and a vertex $v \in V$. Then*

$$\left(1 - \exp(-(n - k)\Delta) \right)^{k-1} \leq \mathbb{P}\left(k_\Delta^v \geq k \right) \leq \left(1 - \exp(-n\Delta) \right)^{k-1}.$$

We can improve Lemma 3.2 slightly in order to obtain even closer lower and upper bounds. For $n, k \geq 2$, combining Lemmas 3.2 and 3.4 yields tight lower and upper bounds if we disregard the constants in the exponent, namely $F_k^v(\Delta) = \left(1 - \exp(-\Theta(n\Delta)) \right)^{\Theta(k)}$.

Lemma 3.4. *For all $v \in V$, $k \in [n]$, and $\Delta \geq 0$, we have $F_k^v(\Delta) \geq (1 - \exp(-(n - 1)\Delta/4))^{n-1}$ and $F_k^v(\Delta) \geq (1 - \exp(-(n - 1)\Delta/4))^{\frac{4}{3}(k-1)}$.*

3.3   Tail Bounds for k_Δ^v and $\Delta_{\max}$

Our first tail bound for k_Δ^v, which is the number of vertices within a distance of Δ of a given vertex v, follows directly from Lemma 3.2.

From this lemma we derive the following corollary, which is a crucial ingredient for the existence of good clusterings and, thus, for the analysis of the heuristics in the remainder of this paper.

Corollary 3.5. *Let $n \geq 5$ and fix $\Delta \geq 0$ and a vertex $v \in V$. Then we have*

$$\mathbb{P}\left(k_\Delta^v < \min \left\{ \exp\left(\Delta n/5 \right), \frac{n + 1}{2} \right\} \right) \leq \exp\left(-\Delta n/5 \right).$$

Corollary 3.5 is almost tight according to the following result.

Corollary 3.6. *Fix $\Delta \geq 0$, a vertex $v \in V$, and any $c > 1$. Then*

$$\mathbb{P}\big(k_\Delta^v \geq \exp(c\Delta n)\big) < \exp\big(-(c-1)\Delta n\big).$$

Janson [21] derived the following tail bound for the length $\Delta_{\max}$ of the longest edge. A qualitatively similar bound can be proved using Lemma 3.3 and can also be derived from Hassin and Zemel's analysis [18]. However, Janson's bound is stronger with respect to the constants in the exponent.

Lemma 3.7 (Janson [21, p. 352]). *For any fixed $c > 3$, we have $\mathbb{P}(\Delta_{\max} > c\ln(n)/n) \leq O(n^{3-c}\log^2 n)$.*

3.4   Stars and Clusters

In this section, we show our main structural contribution, which is a more global property of random shortest path metrics. We show that such instances can be divided into a small number of clusters of any given diameter.

From now on, let $\#(n, \Delta) = \min\{\exp(\Delta n/5), (n+1)/2\}$, as in Corollary 3.5. If the number k_Δ^v of vertices within a distance of Δ of v is at least $\#(n, \Delta)$, then we call the vertex v a *dense Δ-center*, and we call the set N_Δ^v of vertices within a distance of at most Δ of v (including v itself) the *Δ-star of v*. Otherwise, if $k_\Delta^v < \#(n, \Delta)$, and v is not part of any Δ-star, we call the vertex v a *sparse Δ-center*. Any two vertices in the same Δ-star have a distance of at most 2Δ because of the triangle inequality. If Δ is clear from the context, then we also speak about centers and stars without parameter. We can bound, by Corollary 3.5, the expected number of sparse Δ-centers to be at most $O(n/\#(n, \Delta))$.

We want to partition the graph into a small number of clusters, each of diameter at most 6Δ. For this purpose, we put each sparse Δ-center in its own cluster (of size 1). Then the diameter of each such cluster is $0 \leq 6\Delta$ and the number of these clusters is expected to be at most $O(n/\#(n, \Delta))$.

We are left with the dense Δ-centers, which we cluster using the following algorithm: Consider an auxiliary graph whose vertices are all Δ-centers. We draw an edge between two dense Δ-centers u and v if $N_\Delta^u \cap N_\Delta^v \neq \emptyset$. Now consider any maximal independent set of this auxiliary graph (for instance, a greedy independent set), and let t be the number of its vertices. Then we form initial clusters $C_1', \ldots, C_t'$, each containing one of the Δ-stars corresponding to the vertices in the independent set. By the independence, all t Δ-stars are disjoint, which implies $t \leq n/\#(n, \Delta)$. The star of every remaining center v has at least one vertex (maybe v itself) in one of the C_i'. We add all remaining vertices of N_Δ^v to such a C_i' to form the final clusters $C_1, \ldots, C_t$. Now, the maximum distance within each C_i is at most 6Δ: Consider any two vertices $u, v \in C_i$. The distance of u towards its closest neighbor in the initial star C_i' is at most 2Δ. The same holds for v. Finally, the diameter of the initial star C_i' is also at most 2Δ.

With this partitioning, we have obtained the following structure: We have an expected number of $O(n/\#(n, \Delta))$ clusters of size 1 and diameter 0, and a number of $O(n/\#(n, \Delta))$ clusters, each of size at least $\#(n, \Delta)$ and diameter at most 6Δ. Thus, we have $O(n/\#(n,\Delta)) = O(1 + n/\exp(\Delta n/5))$ clusters in total. We summarize these findings in the following lemma. It will be the crucial ingredient for bounding the expected approximation ratios of the greedy, nearest-neighbor, and insertion heuristics.

Lemma 3.8. *Consider a random shortest path metric and let $\Delta \geq 0$. If we partition the instance into clusters, each of diameter at most 6Δ, then the expected number of clusters needed is $O(1 + n/\exp(\Delta n/5))$.*

4   Analysis of Heuristics

In order to bound approximation ratios, we will exploit a simple upper bound on the probability that an optimal TSP tour or matching has a length of at most c for some small constant c. Note that the expected lengths of the minimum-length perfect matching and the optimal TSP are $\Theta(1)$ even without taking shortest paths [15, 34]. Thus, both the optimal TSP and the optimal matching have an expected length of $O(1)$ for random shortest path metrics.

4.1   Greedy Heuristic for Minimum-Length Perfect Matching

Finding minimum-length perfect matchings in metric instances is the first problem that we consider. This problem has been widely considered in the past and has applications in, e.g., optimizing the speed of mechanical plotters [29,32]. The worst-case running-time of $O(n^3)$ for finding an optimal matching is prohibitive if the number n of points is large. Thus, simple heuristics are often used, with the greedy heuristic being probably the simplest one: at every step, choose an edge of minimum length incident to the unmatched vertices and add it to the partial matching. Let GREEDY denote the cost of the matching output by this greedy matching heuristic, and let MM denote the optimum value of the minimum weight matching. The worst-case approximation ratio for greedy matching on metric instances is $\Theta(n^{\log_2(3/2)})$ [29], where $\log_2(3/2) \approx 0.58$. In the case of Euclidean instances, the greedy algorithm has an approximation ratio of $O(1)$ with high probability on random instances [3]. For independent random edge weights (without the triangle inequality), the expected weight of the matching computed by the greedy algorithm is $\Theta(\log n)$ [10] whereas the optimal matching has a weight of $\Theta(1)$ with high probability, which gives an $O(\log n)$ approximation ratio.

We show that greedy matching finds a matching of constant expected length on random shortest path metrics. The proof is similar to the ones of Theorems 4.3 and 4.4, and we include it as an example.

Theorem 4.1. $\mathbb{E}[\mathsf{GREEDY}] = O(1)$.

Proof. Set $\Delta_i = i/n$ for $i \in \{0, 1, \ldots, \log n\}$. We divide the run of GREEDY in phases as follows: We say that GREEDY is in phase i if the lengths of the edges it inserts are in the interval $(6\Delta_{i-1}, 6\Delta_i]$. Lemma 3.7 allows to show that the expected sum of all lengths of edges longer than $6\Delta_{O(\log n)}$ is $o(1)$, so we can ignore them.

Since the lengths of the edges that GREEDY adds increases monotonically, GREEDY goes through phases i with increasing i (while a phase can be empty). We now estimate the contribution of phase i to the matching computed by GREEDY. Using Lemma 3.8, after phase $i - 1$, we can find a clustering into clusters of diameter at most $6\Delta_{i-1}$ using an expected number of $O(1+n/e^{(i-1)/5})$ clusters. Each such cluster can have at most one unmatched vertex. Thus, we have to add at most $O(1 + n/e^{(i-1)/5})$ edges in phase i, each of length at most $6\Delta_i$. Thus, the contribution of phase i is $O(\Delta_i(1 + n/e^{(i-1)/5}))$ in expectation. Summing over all phases yields the desired bound:

$$\mathbb{E}[\mathsf{GREEDY}] = o(1) + \sum_{i=1}^{\log n} O\left(\frac{i}{e^{(i-1)/5}} + \frac{i}{n}\right) = O(1).$$

$\square$

Careful analysis allows us to even bound the expected approximation ratio.

Theorem 4.2. *The greedy algorithm for minimum-length matching has constant approximation ratio on random shortest path metrics, i.e.,*

$$\mathbb{E}\left[\frac{\mathsf{GREEDY}}{\mathsf{MM}}\right] \in O(1).$$

4.2   Nearest-Neighbor Algorithm for the TSP

A greedy analogue for the traveling salesman problem (TSP) is the *nearest neighbor* heuristic: Start with a vertex v as the current vertex, and at every iteration choose the nearest yet unvisited neighbor u of the current vertex as the next vertex in the tour and move to the next iteration with the new vertex u as the current vertex. Let NN denote both the nearest-neighbor heuristic itself and the cost of the tour computed by it. Let TSP denote the cost of an optimal tour. The nearest-neighbor heuristic NN achieves a worst-case ratio of $O(\log n)$ for metric instances and also an average-case ratio (for independent, non-metric edge lengths) of $O(\log n)$ [2]. We show that NN achieves a constant approximation ratio on random shortest path instances. The proof is similar to the ones of Theorems 4.1 and 4.2.

Theorem 4.3. $\mathbb{E}[\mathsf{NN}] = O(1)$ *and* $\mathbb{E}\left[\frac{\mathsf{NN}}{\mathsf{TSP}}\right] \in O(1)$.

4.3   Insertion Heuristics

An insertion heuristic for the TSP is an algorithm that starts with an initial tour on a few vertices and extends this tour iteratively by adding the remaining vertices. In every iteration, a vertex is chosen according to some rule, and this vertex is inserted at the place in the current tour where it increases the total tour length the least. Certain insertion heuristics such as nearest neighbor insertion (which is different from the nearest neighbor algorithm from the previous section) are known to achieve constant approximation ratios [30]. The random insertion algorithm, where the next vertex is chosen uniformly at random from the remaining vertices, has a worst-case approximation ratio of $\Omega(\log \log n / \log \log \log n)$, and there are insertion heuristics with a worst-case approximation ratio of $\Omega(\log n / \log \log n)$ [4].

For random shortest path metrics, we show that any insertion heuristic produces a tour whose length is expected to be within a constant factor of the optimal tour. This result holds irrespective of which insertion strategy we actually use. It holds even in the (admittedly a bit unrealistic) scenario, where an adversary specifies the order in which the vertices have to be inserted after the random instance is drawn.

Theorem 4.4. *The expected cost of the TSP tour obtained with any insertion heuristics is bounded from above by $O(1)$. This holds even against an adaptive adversary, i.e., if an adversary chooses the order in which the vertices are inserted after the edge weights are drawn.*

Furthermore, the expected approximation ratio of any insertion heuristic is also $O(1)$.

4.4   Running-Time of 2-Opt for the TSP

The 2-opt heuristic for the TSP starts with an initial tour and successively improves the tour by so-called 2-exchanges until no further refinement is possible. In a 2-exchange, a pair of edges $e_1 = \{u, v\}$ and $e_2 = \{x, y\}$ are replaced by a pair of edges $f_1 = \{u, y\}$ and $f_2 = \{x, v\}$ to get a shorter tour. The 2-opt heuristic is easy to implement and widely used. In practice, it usually converges quite quickly to close-to-optimal solutions [22]. However, its worst-case running-time is exponential [14]. To explain 2-opt's performance on geometric instances, Englert et al. [14] have proved that the number of iterations that 2-opt needs is bounded by a polynomial in a smoothed input model for geometric instances. Also for random shortest path metrics, the expected number of iterations that 2-opt needs is bounded by a polynomial. The proof is similar to Englert et al.'s analysis [14].

Theorem 4.5. *The expected number of iterations that 2-opt needs to find a local optimum is bounded by $O(n^8 \log^3 n)$.*

5 k-Center

In the (metric) k-center problem, we are given a finite metric space (V, d) and should pick k points $U \subseteq V$ such that $\sum_{v \in V} \min_{u \in U} d(v, u)$ is minimized. We call the set U a k-center. Gonzalez [17] gave a simple 2-approximation for this problem and showed that finding a $(2 - \varepsilon)$-approximation is NP-hard.

In this section, we consider the k-center problem in the setting of random shortest path metrics. In particular we examine the approximation ratio of the algorithm TRIVIAL, which picks k points independent of the metric space, e.g., $U = \{1, \ldots, k\}$, or k random points in V. We show that TRIVIAL yields a $(1 + o(1))$-approximation for $k = O(n^{1-\varepsilon})$. This can be seen as an algorithmic result since it improves upon the worst-case approximation factor of 2, but it is essentially a structural result on random shortest path metrics. It means that any set of k points is, with high probability, a very good k-center, which gives some knowledge about the topology of random shortest path metrics. For larger, but not too large k, i.e., $k \leq (1 - \varepsilon)n$, TRIVIAL still yields an $O(1)$-approximation.

The main insight comes from generalizing the growth process described in Section 3.2 Fixing $U = \{v_1, \ldots, v_k\} \subseteq V$ we sort the vertices $V \setminus U$ by their distance to U in ascending order, calling the resulting order $v_{k+1}, \ldots, v_n$. Now we consider $\delta_i = d(v_{i+1}, U) - d(v_i, U)$ for $k \leq i < n$. These random variables are generated by an easy growth process analogously to Section 3.2, which shows that the δ_i are independent and $\delta_i \sim \mathrm{Exp}(i(n - i))$. Since the cost of U as a k-center can be expressed using the δ_i's and since $a\,\mathrm{Exp}(1) \sim \mathrm{Exp}(1/a)$, we have $\mathsf{cost}(U) = \sum_{i=k}^{n-1}(n-i) \cdot \delta_i \sim \sum_{i=k}^{n-1}(n-i) \cdot \mathrm{Exp}(i(n-i)) \sim \sum_{i=k}^{n-1} \mathrm{Exp}(i)$. From this, we can read off the expected cost of U immediately, and thus the expected cost of TRIVIAL.

Lemma 5.1. *Fix $U \subseteq V$ of size k. We have $\mathbb{E}[\mathsf{TRIVIAL}] = \mathbb{E}[\mathsf{cost}(U)] = H_{n-1} - H_{k-1} = \ln(n/k) + \Theta(1)$.*

By closely examining the random variable $\sum_{i=k}^{n-1} \mathrm{Exp}(i)$, we can show good tail bounds for the probability that the cost of U is lower than expected. Together with the union bound this yields tail bounds for the optimal k-center CENTER, which implies the following theorem. In this theorem, the approximation ratio becomes $1 + O\big(\frac{\ln \ln(n)}{\ln(n)}\big)$ for $k = O(n^{1-\varepsilon})$.

Theorem 5.2. *Let $k \leq (1 - \varepsilon)n$ for some constant $\varepsilon > 0$. Then $\mathbb{E}\left[\frac{\mathsf{TRIVIAL}}{\mathsf{CENTER}}\right] = O(1)$. If we even have $k \leq cn$ for some sufficiently small constant $c \in (0, 1)$, then $\mathbb{E}\left[\frac{\mathsf{TRIVIAL}}{\mathsf{CENTER}}\right] = 1 + O\big(\frac{\ln \ln(n/k)}{\ln(n/k)}\big)$.*

6 Remarks and Open Problems

The results of this paper carry over to the case of edge weights drawn according to the uniform distribution on the interval $[0, 1]$. The analysis remains basically identical, but gets technically a bit more difficult because we lose the memorylessness of the exponential distribution. The intuition is that, because the longest edge has a length of $O(\log n/n) = o(1)$, only the behavior of the distribution in a small, shrinking interval $[0, o(1)]$ is relevant. Essentially, if the probability that an edge weight is smaller than t is $t + o(t)$, then our results carry over. We refer to Janson's coupling argument [21] for more details.

To conclude the paper, let us list the open problems that we consider most interesting:

1. While the distribution of edge lengths in asymmetric instances does not differ much from the symmetric case, an obstacle in the application of asymmetric random shortest path metrics seems to be the lack of clusters of small diameter (see Section 3). Is there an asymmetric counterpart for this?
2. Is it possible to prove even an $1 + o(1)$ (like Dyer and Frieze [11] for the patching algorithm) approximation ratio for any of the simple heuristics that we analyzed?
3. What is the approximation ratio of 2-opt in random shortest path metrics? In the worst case on metric instances, it is $O(\sqrt{n})$ [8]. For edge lengths drawn uniformly at random from the interval $[0, 1]$ without taking shortest paths, the expected approximation ratio is $O(\sqrt{n} \cdot \log^{3/2} n)$ [13]. For d-dimensional geometric instances, the smoothed approximation ratio is $O(\phi^{1/d})$ [14], where ϕ is the perturbation parameter.

   We easily get an approximation ratio of $O(\log n)$ based on the two facts that the length of the optimal tour is $\Theta(1)$ with high probability and that $\Delta_{\max} = O(\log n / n)$ with high probability. Can we prove that the expected ratio of 2-opt is $o(\log n)$?

References

1. Addario-Berry, L., Broutin, N., Lugosi, G.: The longest minimum-weight path in a complete graph. Combin. Probab. Comput. 19(1), 1–19 (2010)
2. Ausiello, G., Crescenzi, P., Gambosi, G., Kann, V., Marchetti-Spaccamela, A., Protasi, M.: Complexity and Approximation. Springer (1999)
3. Avis, D., Davis, B., Steele, J.M.: Probabilistic analysis of a greedy heuristic for Euclidean matching. Probab. Engrg. Inform. Sci. 2, 143–156 (1988)
4. Azar, Y.: Lower bounds for insertion methods for TSP. Combin. Probab. Comput. 3, 285–292 (1994)
5. Bhamidi, S., van der Hofstad, R., Hooghiemstra, G.: First passage percolation on the Erdős-Rényi random graph. Combin. Probab. Comput. 20(5), 683–707 (2011)
6. Blair-Stahn, N.D.: First passage percolation and competition models. arXiv:1005.0649v1 [math.PR] (2010)
7. Broadbent, S.R., Hemmersley, J.M.: Percolation processes. I. Crystals and mazes. Proceedings of the Cambridge Philosophical Society 53(3), 629–641 (1957)
8. Chandra, B., Karloff, H.J., Tovey, C.A.: New results on the old k-opt algorithm for the traveling salesman problem. SIAM J. Comput. 28(6), 1998–2029 (1999)
9. Davis, R., Prieditis, A.: The expected length of a shortest path. Inform. Process. Lett. 46(3), 135–141 (1993)
10. Dyer, M., Frieze, A.M., Pittel, B.: The average performance of the greedy matching algorithm. Ann. Appl. Probab. 3(2), 526–552 (1993)
11. Dyer, M.E., Frieze, A.M.: On patching algorithms for random asymmetric travelling salesman problems. Math. Program. 46, 361–378 (1990)
12. Eckhoff, M., Goodman, J., van der Hofstad, R., Nardi, F.R.: Short paths for first passage percolation on complete graphs. arXiv:1211.4569v1 [math.PR] (2012)
13. Engels, C., Manthey, B.: Average-case approximation ratio of the 2-opt algorithm for the TSP. Oper. Res. Lett. 37(2), 83–84 (2009)

14. Englert, M., Röglin, H., Vöcking, B.: Worst case and probabilistic analysis of the 2-opt algorithm for the TSP. In: Proc. of the 18th Ann. ACM-SIAM Symp. on Discrete Algorithms (SODA), pp. 1295–1304. SIAM (2007)
15. Frieze, A.M.: On random symmetric travelling salesman problems. Math. Oper. Res. 29(4), 878–890 (2004)
16. Frieze, A.M., Grimmett, G.R.: The shortest-path problem for graphs with random arc-lengths. Discrete Appl. Math. 10, 57–77 (1985)
17. Gonzalez, T.F.: Clustering to minimize the maximum intercluster distance. Theoret. Comput. Sci. 38, 293–306 (1985)
18. Hassin, R., Zemel, E.: On shortest paths in graphs with random weights. Math. Oper. Res. 10(4), 557–564 (1985)
19. van der Hofstad, R., Hooghiemstra, G., van Mieghem, P.: First passage percolation on the random graph. Probab. Engrg. Inform. Sci. 15(2), 225–237 (2001)
20. van der Hofstad, R., Hooghiemstra, G., van Mieghem, P.: Size and weight of shortest path trees with exponential link weights. Combin. Probab. Comput. 15(6), 903–926 (2006)
21. Janson, S.: One, two, three times $\log n/n$ for paths in a complete graph with edge weights. Combin. Probab. Comput. 8(4), 347–361 (1999)
22. Johnson, D.S., McGeoch, L.A.: Experimental analysis of heuristics for the STSP. In: Gutin, G., Punnen, A.P. (eds.) The Traveling Salesman Problem and its Variations. Kluwer (2002)
23. Karp, R.M.: Probabilistic analysis of partitioning algorithms for the traveling-salesman problem in the plane. Math. Oper. Res. 2(3), 209–224 (1977)
24. Karp, R.M., Steele, J.M.: Probabilistic analysis of heuristics. In: Lawler, E.L., et al. (eds.) The Traveling Salesman Problem, pp. 181–205. Wiley (1985)
25. Kulkarni, V.G., Adlakha, V.G.: Maximum flow in planar networks in exponentially distributed arc capacities. Comm. Statist. Stochastic Models 1(3), 263–289 (1985)
26. Kulkarni, V.G.: Shortest paths in networks with exponentially distributed arc lengths. Networks 16(3), 255–274 (1986)
27. Kulkarni, V.G.: Minimal spanning trees in undirected networks with exponentially distributed arc weights. Networks 18(2), 111–124 (1988)
28. Peres, Y., Sotnikov, D., Sudakov, B., Zwick, U.: All-pairs shortest paths in $o(n^2)$ time with high probability. In: Proc. of the 51st Ann. IEEE Symp. on Foundations of Computer Science (FOCS), pp. 663–672. IEEE (2010)
29. Reingold, E.M., Tarjan, R.E.: On a greedy heuristic for complete matching. SIAM J. Comput. 10(4), 676–681 (1981)
30. Rosenkrantz, D.J., Stearns, R.E., Lewis II, P.M.: An analysis of several heuristics for the traveling salesman problem. SIAM J. Comput. 6(3), 563–581 (1977)
31. Ross, S.M.: Introduction to Probability Models. Academic Press (2010)
32. Supowit, K.J., Plaisted, D.A., Reingold, E.M.: Heuristics for weighted perfect matching. In: Proc. of the 12th Ann. ACM Symp. on Theory of Computing (STOC), pp. 398–419. ACM (1980)
33. Vershik, A.M.: Random metric spaces and universality. Russian Math. Surveys 59(2), 259–295 (2004)
34. Walkup, D.W.: On the expected value of a random assignment problem. SIAM J. Comput. 8(3), 440–442 (1979)
35. Yukich, J.E.: Probability Theory of Classical Euclidean Optimization Problems. Springer (1998)

Semilinearity and Context-Freeness of Languages Accepted by Valence Automata*

P. Buckheister and Georg Zetzsche

Fachbereich Informatik, Technische Universität Kaiserslautern

Abstract. Valence automata are a generalization of various models of automata with storage. Here, each edge carries, in addition to an input word, an element of a monoid. A computation is considered valid if multiplying the monoid elements on the visited edges yields the identity element. By choosing suitable monoids, a variety of automata models can be obtained as special valence automata. This work is concerned with the accepting power of valence automata. Specifically, we ask for which monoids valence automata can accept only context-free languages or only languages with semilinear Parikh image, respectively. First, we present a characterization of those graph products (of monoids) for which valence automata accept only context-free languages. Second, we provide a necessary and sufficient condition for a graph product of copies of the bicyclic monoid and the integers to yield only languages with semilinear Parikh image when used as a storage mechanism in valence automata. Third, we show that all languages accepted by valence automata over torsion groups have a semilinear Parikh image.

1 Introduction

A valence automaton is a finite automaton in which each edge carries, in addition to an input word, an element of a monoid. A computation is considered valid if multiplying the monoid elements on the visited edges yields the identity element. By choosing suitable monoids, one can obtain a wide range of automata with storage mechanisms as special valence automata. Thus, they offer a framework for generalizing insights about automata with storage. For examples of automata as valence automata, see [4,19].

In this work, we are concerned with the accepting power of valence automata. That is, we are interested in relationships between the structure of the monoid representing the storage mechanism and the class of languages accepted by the corresponding valence automata. On the one hand, we address the question for which monoids valence automata accept only *context-free* languages. Since the context-free languages constitute a very well-understood class, insights in this direction promise to shed light on the acceptability of languages by transferring results about context-free languages.

A very well-known result on context-free languages is Parikh's Theorem, which states that the Parikh image (that is, the image under the canonical morphism

---

* The full version of this work is available at `http://arxiv.org/abs/1306.3260`.

K. Chatterjee and J. Sgall (Eds.): MFCS 2013, LNCS 8087, pp. 231–242, 2013.
© Springer-Verlag Berlin Heidelberg 2013

onto the free commutative monoid) of each context-free language is semilinear (in this case, the language itself is also called semilinear). It has various applications in proving that certain languages are not context-free and its effective nature (one can actually compute the semilinear representation) allows it to be used in decision procedures for numerous problems (see [15] for an example from group theory and [10] for others). It is therefore our second goal to gain understanding about which monoids cause the corresponding valence automata to accept only languages with a *semilinear Parikh image*.

Our contribution is threefold. First, we obtain a characterization of those graph products (of monoids) whose corresponding valence automata accept only context-free languages. Graph products are a generalization of the free and the direct product in the sense that for each pair of participating factors, it can be specified whether they should commute in the product. Since valence automata over a group accept only context-free languages if and only if the group's word problem (and hence the group itself) can be described by a context-free grammar, such a characterization had already been available for groups in a result by Lohrey and Sénizergues [13]. Therefore, our characterization is in some sense an extension of Lohrey and Sénizergues' to monoids.

Second, we present a necessary and sufficient condition for a graph product of copies of the bicyclic monoid and the integers to yield, when used in valence automata, only languages with semilinear Parikh image. Although this is a smaller class of monoids than arbitrary graph products, it still covers a number of storage mechanisms found in the literature, such as pushdown automata, blind multicounter automata, and partially blind multicounter automata (see [19] for more information). Hence, our result is a generalization of various semilinearity results about these types of automata.

Third, we show that every language accepted by a valence automaton over a torsion group has a semilinear Parikh image. On the one hand, this is particularly interesting because of a result by Render [16], which states that for every monoid M, the languages accepted by valence automata over M either (1) coincide with the regular languages, (2) contain the blind one-counter languages, (3) contain the partially blind one-counter languages, or (4) are those accepted by valence automata over an infinite torsion group (which is not locally finite). Hence, our result establishes a strong language theoretic property in the fourth case and thus contributes to completing the picture of language classes that can arise from valence automata.

On the other hand, Lohrey and Steinberg [15] have used the fact that for certain groups, valence automata accept only semilinear languages (in different terms, however) to obtain decidability of the rational subset membership problem. However, their procedures require that the semilinear representation can be obtained effectively. Since there are torsion groups where even the word problem is undecidable [1], our result yields examples of groups that have the semilinearity property but which do not permit the computation of a corresponding representation. Our proof is based on well-quasi-orderings (see, e.g., [11]).

2   Basic Notions

In this section, we will fix some notation and introduce basic concepts.

A *monoid* is a set M together with an associative operation and a neutral element. Unless defined otherwise, we will denote the neutral element of a monoid by 1 and its operation by juxtaposition. That is, for a monoid M and elements $a, b \in M$, $ab \in M$ is their product. In each monoid M, we have the submonoids $\mathsf{R}(M) = \{a \in M \mid \exists b \in M : ab = 1\}$ and, $\mathsf{L}(M) = \{a \in M \mid \exists b \in M : ba = 1\}$. When using a monoid M as part of a control mechanism, the subset $\mathsf{J}(M) = \{a \in M \mid \exists b, c \in M : bac = 1\}$ plays an important role[1]. A *subgroup* of a monoid is a subset that is closed under the operation and is a group.

For an alphabet X, we will write X^* for the set of words over X. The empty word is denoted by $\lambda \in X^*$. Given alphabets X and Y, subsets of X^* and $X^* \times Y^*$ are called *languages* and *transductions*, respectively. A *family* is a set of languages that is closed under isomorphism and contains at least one non-trivial member. For a transduction $T \subseteq X^* \times Y^*$ and a language $L \subseteq X^*$, we write $T(L) = \{v \in Y^* \mid \exists u \in L : (u, v) \in T\}$. For any finite subset $S \subseteq M$ of a monoid, let X_S be an alphabet in bijection with S. Let $\varphi_S : X_S^* \to M$ be the morphism extending this bijection. Then the set $\{w \in X_S^* \mid \varphi_S(w) = 1\}$ is called the *identity language of M with respect to S*.

Let $\mathcal{F}$ be a family of languages. An *$\mathcal{F}$-grammar* is a quadruple $G = (N, T, P, S)$ where N and T are disjoint alphabets and $S \in N$. P is a finite set of pairs (A, M) with $A \in N$ and $M \subseteq (N \cup T)^*$, $M \in \mathcal{F}$. A pair $(A, M) \in P$ will also be denoted by $A \to M$. We write $x \Rightarrow_G y$ if $x = uAv$ and $y = uwv$ for some $u, v, w \in (N \cup T)^*$ and $(A, M) \in P$ with $w \in M$. The *language generated by G* is $L(G) = \{w \in T^* \mid S \Rightarrow_G^* w\}$. Languages generated by $\mathcal{F}$-grammars are called *algebraic over $\mathcal{F}$*. The family of all languages that are algebraic over $\mathcal{F}$ is called the *algebraic extension of $\mathcal{F}$*. The algebraic extension of the family of finite languages is denoted CF, its members are called *context-free*.

Given an alphabet X, we write $X^\oplus$ for the set of maps $\alpha : X \to \mathbb{N}$. Elements of $X^\oplus$ are called *multisets*. By way of pointwise addition, written $\alpha + \beta$, $X^\oplus$ is a monoid. The *Parikh mapping* is the mapping $\Psi : X^* \to X^\oplus$ such that $\Psi(w)(x)$ is the number of occurrences of x in w for every $w \in X^*$ and $x \in X$.

Let A be a (not necessarily finite) set of symbols and $R \subseteq A^* \times A^*$. The pair (A, R) is called a *(monoid) presentation*. The smallest congruence of A^* containing R is denoted by $\equiv_R$ and we will write $[w]_R$ for the congruence class of $w \in A^*$. The *monoid presented by (A, R)* is defined as $A^*/\equiv_R$. Note that since we did not impose a finiteness restriction on A, every monoid has a presentation. By $\mathbb{B}$, we denote the monoid presented by (A, R) with $A = \{x, \bar{x}\}$ and $R = (x\bar{x}, \lambda)$ and called *bicyclic monoid*. The elements $[x]_R$ and $[\bar{x}]_R$ are called its *positive* and *negative generator*, respectively. The set D_1 of all $w \in \{x, \bar{x}\}^*$ with $[w]_R = [\lambda]_R$ is called the *Dyck language*. The group of integers is denoted $\mathbb{Z}$. We call $1 \in \mathbb{Z}$ its *positive* and $-1 \in \mathbb{Z}$ its *negative generator*.

---

[1] Note that $\mathsf{R}(M)$, $\mathsf{L}(M)$, and $\mathsf{J}(M)$ are the $\mathcal{R}$-, $\mathcal{L}$-, and $\mathcal{J}$-class, respectively, of the identity and hence are important concepts in the theory of semigroups [7].

Let M be a monoid. An *automaton over M* is a tuple $A = (Q, M, E, q_0, F)$, in which Q is a finite set of *states*, E is a finite subset of $Q \times M \times Q$ called the set of *edges*, $q_0 \in Q$ is the *initial state*, and $F \subseteq Q$ is the set of *final states*. The *step relation* $\Rightarrow_A$ of A is a binary relation on $Q \times M$, for which $(p, a) \Rightarrow_A (q, b)$ if and only if there is an edge (p, c, q) such that $b = ac$. The set *generated by A* is then $S(A) = \{a \in M \mid \exists q \in F : (q_0, 1) \Rightarrow_A^* (q, a)\}$. A set $R \subseteq M$ is called *rational* if it can be written as $R = S(A)$ for some automaton A over M. Rational languages are also called *regular*, the corresponding class is denoted REG. A class $\mathcal{C}$ for which $L \in \mathcal{C}$ implies $T(L) \in \mathcal{C}$ for rational transductions T is called a *full trio*.

For $n \in \mathbb{N}$ and $\alpha \in X^{\oplus}$, we use $n\alpha$ to denote $\alpha + \cdots + \alpha$ (n summands). A subset $S \subseteq X^{\oplus}$ is *linear* if there are elements $\alpha_0, \ldots, \alpha_n$ such that $S = \{\alpha_0 + \sum_{i=1}^n m_i \alpha_i \mid m_i \in \mathbb{N}, 1 \leq i \leq n\}$. A set $S \subseteq C$ is called *semilinear* if it is a finite union of linear sets. In slight abuse of terminology, we will sometimes call a language L semilinear if the set $\Psi(L)$ is semilinear.

A *valence automaton over M* is an automaton A over $X^* \times M$, where X is an alphabet. Instead of $A = (Q, X^* \times M, E, q_0, F)$, we then also write $A = (Q, X, M, E, q_0, F)$ and for an edge $(p, (w, m), q) \in E$, we also write (p, w, m, q). The *language accepted by A* is defined as $L(A) = \{w \in X^* \mid (w, 1) \in S(A)\}$. The class of languages accepted by valence automata over M is denoted by $\mathsf{VA}(M)$. It is well-known that $\mathsf{VA}(M)$ is the smallest full trio containing every identity language of M (see, for example, [9]).

A *graph* is a pair $\Gamma = (V, E)$ where V is a finite set and $E \subseteq \{S \subseteq V \mid 1 \leq |S| \leq 2\}$. The elements of V are called *vertices* and those of E are called *edges*. If $\{v\} \in E$ for some $v \in V$, then v is called a *looped* vertex, otherwise it is *unlooped*. A *subgraph* of Γ is a graph (V', E') with $V' \subseteq V$ and $E' \subseteq E$. Such a subgraph is called *induced (by V')* if $E' = \{S \in E \mid S \subseteq V'\}$, i.e. E' contains all edges from E incident to vertices in V'. By $\Gamma \setminus \{v\}$, for $v \in V$, we denote the subgraph of Γ induced by $V \setminus \{v\}$. Given a graph $\Gamma = (V, E)$, its *underlying loop-free graph* is $\Gamma' = (V, E')$ with $E' = E \cap \{S \subseteq V \mid |S| = 2\}$. For a vertex $v \in V$, the elements of $N(v) = \{w \in V \mid \{v, w\} \in E\}$ are called *neighbors* of v. Moreover, a *clique* is a graph in which any two distinct vertices are adjacent. A *simple path of length n* is a sequence $x_1, \ldots, x_n$ of pairwise distinct vertices such that $\{x_i, x_{i+1}\} \in E$ for $1 \leq i < n$. If, in addition, we have $\{x_n, x_1\} \in E$, it is called a *cycle*. Such a cycle is called *induced* if $\{x_i, x_j\} \in E$ implies $|i - j| = 1$ or $\{i, j\} = \{1, n\}$. A loop-free graph $\Gamma = (V, E)$ is *chordal* if it does not contain an induced cycle of length ≥ 4. It is well-known that every chordal graph contains a vertex whose neighborhood is a clique [3]. By C_4 and P_4, we denote the cycle of length 4 and the simple path of length 4, respectively. A loop-free graph is called a *transitive forest* if it is the disjoint union of comparability graphs of rooted trees. A result by Wolk [17] states that a loop-free graph is a transitive forest if and only if it contains neither C_4 nor P_4 as an induced subgraph.

Let $\Gamma = (V, E)$ be a loop-free graph and M_v a monoid for each $v \in V$ with a presentation (A_v, R_v) such that the A_v are pairwise disjoint. Then the *graph product* $M = \mathbb{M}(\Gamma, (M_v)_{v \in V})$ is the monoid given by the presentation (A, R), where $A = \bigcup_{v \in V} A_v$ and $R = \{(ab, ba) \mid a \in A_v, b \in A_w, \{v, w\} \in E\} \cup \bigcup_{v \in V} R_v$.

Note that for each $v \in V$, there is a map $\varphi_v : M \to M_v$ such that φ_v is the identity map on M_v. When $V = \{0, 1\}$ and $E = \emptyset$, we also write $M_0 * M_1$ for M and call this the *free product* of M_0 and M_1. Given a subset $U \subseteq V$, we write $M \!\upharpoonright_U$ for the product $\mathbb{M}(\Gamma', (M_v)_{v \in U})$, where Γ' is the subgraph induced by U.

Let $\Gamma = (V, E)$ be a (not necessarily loop-free) graph. Furthermore, for each $v \in V$, let M_v be a copy of $\mathbb{B}$ if v is an unlooped vertex and a copy of $\mathbb{Z}$ if v is looped. If Γ^- is obtained from Γ by removing all loops, we write $\mathbb{M}\Gamma$ for the graph product $\mathbb{M}(\Gamma^-, (M_v)_{v \in V})$. For information on valence automata over monoids $\mathbb{M}\Gamma$, see [19]. For $i \in \{0, 1\}$, let M_i be a monoid and let $\varphi_i : N \to M_i$ be an injective morphism. Let $\equiv$ be the smallest congruence in $M_0 * M_1$ such that $\varphi_0(a) \equiv \varphi_1(a)$ for every $a \in N$. Then the monoid $(M_0 * M_1)/\equiv$ is denoted by $M_0 *_N M_1$ and called a *free product with amalgamation*.

3   Auxiliary Results

In this section, we present auxiliary results that are used in later sections. In the following, we will call a monoid M an *FRI-monoid* (or say that M has the FRI-property) if for every finitely generated submonoid N of M, the set $\mathsf{R}(N)$ is finite. In [16] and independently in [18], the following was shown.

Theorem 1. $\mathsf{VA}(M) = \mathsf{REG}$ *if and only if M is an FRI-monoid.*

The first lemma states a well-known fact from semigroup theory.

Lemma 1. *For each monoid M, exactly one of the following holds: Either $\mathsf{J}(M)$ is a group or M contains a copy of $\mathbb{B}$ as a submonoid.*

We will employ a result by van Leeuwen [12] stating that semilinearity of all languages is preserved by building the algebraic extension of a language family.

Theorem 2. *Let $\mathcal{F}$ be a family of semilinear languages. Then every language that is algebraic over $\mathcal{F}$ is also semilinear.*

In light of the previous theorem, the following implies that the class of monoids M for which $\mathsf{VA}(M)$ contains only semilinear languages is closed under taking free products with amalgamation over a finite identified subgroup that contains the identity of each factor. In the case where the factors are residually finite groups, this was already shown in [15, Lemma 8] (however, for a more general operation than free products with amalgamation). The following also implies that if $\mathsf{VA}(M_i)$ contains only context-free languages for $i \in \{0, 1\}$, then this is also true for $\mathsf{VA}(M_0 *_F M_1)$.

Theorem 3. *For each $i \in \{0, 1\}$, let M_i be a finitely generated monoid and F be a subgroup that contains M_i's identity. Every language in $\mathsf{VA}(M_0 *_F M_1)$ is algebraic over $\mathsf{VA}(M_0) \cup \mathsf{VA}(M_1)$.*

Proof. Since the algebraic extension of a full trio is again a full trio, it suffices to show that with respect to some generating set $S \subseteq M_0 *_F M_1$, the identity language of $M_0 *_F M_1$ is algebraic over $\mathsf{VA}(M_0) \cup \mathsf{VA}(M_1)$.

For $i \in \{0, 1\}$, let $S_i \subseteq M_i$ be a finite generating set for M_i such that $F \subseteq S_i$. Furthermore, let X_i be an alphabet in bijection with S_i and let $\varphi_i : X_i^* \to M_i$ be the morphism extending this bijection. Moreover, let $Y_i \subseteq X_i$ be the subset with $\varphi_i(Y_i) = F$. Let $\psi_i : M_i \to M_0 *_F M_1$ be the canonical morphism. Since F is a subgroup of M_0 and M_1, ψ_0 and ψ_1 are injective (see e.g. [7, Theorem 8.6.1]). Let $X = X_0 \cup X_1$ and let $\varphi : X^* \to M_0 *_F M_1$ be the morphism extending $\psi_0 \varphi_0$ and $\psi_1 \varphi_1$. Then the identity language of $M_0 *_F M_1$ is $\varphi^{-1}(1)$ and we shall prove the theorem by showing that $\varphi^{-1}(1)$ is algebraic over $\mathsf{VA}(M_0) \cup \mathsf{VA}(M_1)$. We will make use of the following fact about free products with amalgamation of monoids with a finite identified subgroup. Let $s_1, \ldots, s_n, s_1', \ldots, s_m' \in (X_0^* \setminus \varphi_0^{-1}(F)) \cup (X_1^* \setminus \varphi_1^{-1}(F))$, such that $s_j \in X_i^*$ if and only if $s_{j+1} \in X_{1-i}^*$ for $1 \le j < n$, $i \in \{0, 1\}$ and $s_j' \in X_i^*$ if and only if $s_{j+1}' \in X_{1-i}^*$ for $1 \le j < m$, $i \in \{0, 1\}$. Then the equality $\varphi(s_1 \cdots s_n) = \varphi(s_1' \cdots s_m')$ implies $n = m$. A stronger statement was shown in [14, Lemma 10]. We will refer to this as the *syllable property*.

For each $i \in \{0, 1\}$ and $f \in F$, we define $L_{i,f} = \varphi_i^{-1}(f)$ and write y_f for the symbol in Y_i with $\varphi_i(y_f) = f^{-1}$. Then clearly $L_{i,1} \in \mathsf{VA}(M_i)$. Furthermore, since $L_{i,f} = \{w \in X_i^* \mid y_f w \in L_{i,1}\}$, (here we again use that F is a group) we can obtain $L_{i,f}$ from $L_{i,1}$ by a rational transduction and hence $L_{i,f} \in \mathsf{VA}(M_i)$.

Let $\mathcal{F} = \mathsf{VA}(M_0) \cup \mathsf{VA}(M_1)$. Since for each $\mathcal{F}$-grammar G, it is clearly possible to construct an $\mathcal{F}$-grammar G' such that $L(G')$ consists of all sentential forms of G, it suffices to construct an $\mathcal{F}$-grammar $G = (N, T, P, S)$ with $N \cup T = X$ and $S \Rightarrow_G^* w$ if and only if $\varphi(w) = 1$ for $w \in X^*$. We construct $G = (N, T, P, S)$ as follows. Let $N = Y_0 \cup Y_1$ and $T = (X_0 \cup X_1) \setminus (Y_0 \cup Y_1)$. As productions, we have $y \to L_{1-i,f}$ for each $y \in Y_i$ where $f = \varphi_i(y)$. Since $1 \in F$, we have an $e_i \in Y_i$ with $\varphi_i(e_i) = 1$. As the start symbol, we choose $S = e_0$. We claim that for $w \in X^*$, we have $S \Rightarrow_G^* w$ if and only if $\varphi(w) = 1$.

The "only if" is clear. Let $w \in X^*$ with $\varphi(w) = 1$. Write $w = w_1 \cdots w_n$ such that $w_j \in X_0^* \cup X_1^*$ for all $1 \le j \le n$ such that $w_j \in X_i^*$ if and only if $w_{j+1} \in X_{1-i}^*$ for $i \in \{0, 1\}$ and $1 \le j < n$. We show by induction on n that $S \Rightarrow_G^* w$. For $n \le 1$, we have $w \in X_i^*$ for some $i \in \{0, 1\}$. Since $1 = \varphi(w) = \psi_i(\varphi_i(w))$ and ψ_i is injective, we have $\varphi_i(w) = 1$ and hence $w \in L_{i,1}$. This means $S = e_0 \Rightarrow_G w$ or $S = e_0 \Rightarrow_G e_1 \Rightarrow_G w$, depending on whether $i = 1$ or $i = 0$.

Now let $n \ge 2$. We claim that there is a $1 \le j \le n$ with $\varphi(w_j) \in F$. Indeed, if $\varphi(w_j) \notin F$ for all $1 \le j \le n$ and since $\varphi(w_1 \cdots w_n) = 1 = \varphi(\lambda)$, the syllable property implies $n = 0$, against our assumption. Hence, let $f = \varphi(w_j) \in F$. Furthermore, let $w_j \in X_i^*$ and choose $y \in Y_{1-i}$ so that $\varphi_{1-i}(y) = f$. Then $\psi_i(\varphi_i(w_j)) = \varphi(w_j) = f$ and the injectivity of ψ_i yields $\varphi_i(w_j) = f$. Hence, $w_j \in L_{i,f}$ and thus $w' = w_1 \cdots w_{j-1} y w_{j+1} \cdots w_n \Rightarrow_G w$. For w' the induction hypothesis holds, meaning $S \Rightarrow_G^* w'$ and thus $S \Rightarrow_G^* w$. $\square$

4   Context-Freeness

In this section, we are concerned with the context-freeness of languages accepted by valence automata over graph products. The first lemma is a simple observation and we will not provide a proof. In the case of groups, it appeared in [5].

Lemma 2. *Let $\Gamma = (V, E)$ and $M = \mathbb{M}(\Gamma, (M_v)_{v \in V})$ be a graph product. Then for each $v \in V$, we have $M \cong (M \upharpoonright_{V \setminus \{v\}}) *_{M \upharpoonright_{N(v)}} (M \upharpoonright_{N(v)} \times M_v)$.*

The following is a result by Lohrey and Sénizergues [13].

Theorem 4. *Let G_v be a non-trivial group for each $v \in V$. Then $\mathbb{M}(\Gamma, (G_v)_{v \in V})$ is virtually free if and only if (1) for each $v \in V$, G_v is virtually free, (2) if G_v and G_w are infinite and $v \neq w$, then $\{v, w\} \notin E$, (3) if G_v is infinite, G_u and G_w are finite and $\{v, u\}, \{v, w\} \in E$, then $\{u, w\} \in E$, and (4) Γ is chordal.*

Aside from Theorem 3, the following is the key tool to prove our result on context-freeness. We call a monoid M *context-free* if $\mathsf{VA}(M) \subseteq \mathsf{CF}$.

Lemma 3. *The direct product of monoids M_0 and M_1 is context-free if and only if for some $i \in \{0, 1\}$, M_i is context-free and M_{1-i} is an FRI-monoid.*

Proof. Suppose M_i is context-free and M_{1-i} is an FRI-monoid. Then each language $L \in \mathsf{VA}(M_i \times M_{1-i})$ is contained in $\mathsf{VA}(M_i \times N)$ for some finitely generated submonoid N of M_{1-i}. Since M_{1-i} is an FRI-monoid, N has finitely many right-invertible elements and hence $\mathsf{J}(N)$ is a finite group. Since no element outside of $\mathsf{J}(N)$ can appear in a product yielding the identity, we may assume that $L \in \mathsf{VA}(M_i \times \mathsf{J}(N))$. This means, however, that L can be accepted by a valence automaton over M_i by keeping the right component of the storage monoid in the state of the automaton. Hence, $L \in \mathsf{VA}(M_i)$ is context-free.

Suppose $\mathsf{VA}(M_0 \times M_1) \subseteq \mathsf{CF}$. Then certainly $\mathsf{VA}(M_i) \subseteq \mathsf{CF}$ for each $i \in \{0, 1\}$. This means we have to show that at least one of the monoids M_0 and M_1 is an FRI-monoid and thus, toward a contradiction, assume that none of them is.[2]

By Lemma 1, for each i, either $\mathsf{J}(M_i)$ is a subgroup of M_i or M_i contains a copy of $\mathbb{B}$ as a submonoid. Since every infinite virtually free group contains an element of infinite order, we have that for each i, either (1) $\mathsf{J}(M_i)$ is an infinite group and hence contains a copy of $\mathbb{Z}$ or (2) M_i contains a copy of $\mathbb{B}$. In any case, $\mathsf{VA}(M_0 \times M_1)$ contains $\{a^n b^m c^n d^m \mid n, m \geq 0\}$, which is not context-free. □

We are now ready to prove our main result on context-freeness. Since for a graph product $M = \mathbb{M}(\Gamma, (M_v)_{v \in V})$, there is a morphism $\varphi_v : M \to M_v$ for each $v \in V$ that restricts to the identity on M_v, we have $\mathsf{J}(M) \cap M_v = \mathsf{J}(M_v)$: While the inclusion "$\supseteq$" is true for any submonoid, given $b \in \mathsf{J}(M) \cap M_v$ with $abc = 1$, $a, c \in M$, we also have $\varphi_v(a) b \varphi_v(c) = \varphi_v(abc) = 1$ and hence $b \in \mathsf{J}(M_v)$. This means no element of $M_v \setminus \mathsf{J}(M_v)$ can appear in a product yielding the identity. In particular, removing a vertex v with $\mathsf{J}(M_v) = \{1\}$ will not change $\mathsf{VA}(M)$. Hence, our requirement that $\mathsf{J}(M_v) \neq \{1\}$ is not a serious restriction.

Theorem 5. *Let $\Gamma = (V, E)$ and let $\mathsf{J}(M_v) \neq \{1\}$ for any $v \in V$. $M = \mathbb{M}(\Gamma, (M_v)_{v \in V})$ is context-free if and only if*
(1) for each $v \in V$, M_v is context-free,
(2) if M_v and M_w are not FRI-monoids and $v \neq w$, then $\{v, w\} \notin E$,

---

[2] In the full version, we have a second proof for the fact that $\mathsf{VA}(M_0 \times M_1)$ contains non-context-free languages in this case. It is elementary in the sense that it does not invoke the fact that context-free groups are virtually free.

(3) if M_v is not an FRI-monoid, M_u and M_w are FRI-monoids and $\{v, u\}$, $\{v, w\} \in E$, then $\{u, w\} \in E$, and

(4) the graph Γ is chordal.

Proof. First, we show that conditions (1)–(4) are necessary. For (1), this is immediate and for (2), this follows from Lemma 3. If (3) is violated then for some $u, v, w \in V$, $M_v \times (M_u * M_w)$ is a submonoid of M such that M_u and M_w are FRI-monoids and M_v is not. Since M_u and M_w contain non-trivial (finite) subgroups, $M_u * M_w$ contains an infinite group and is thus not an FRI-monoid, meaning $M_v \times (M_u * M_w)$ is not context-free by Lemma 3.

Suppose (4) is violated for context-free M. By (2) and (3), any induced cycle of length at least four involves only vertices with FRI-monoids. Each of these, however, contains a non-trivial finite subgroup. This means M contains an induced cycle graph product of non-trivial finite groups, which is not virtually free by Theorem 4 and hence has a non-context-free identity language.

In order to prove the other direction, we note that $\mathsf{VA}(M) \subseteq \mathsf{CF}$ follows if $\mathsf{VA}(M') \subseteq \mathsf{CF}$ for every finitely generated submonoid $M' \subseteq M$. Since every such submonoid is contained in a graph product $N = \mathbb{M}(\Gamma, (N_v)_{v \in V})$ where each N_v is a finitely generated submonoid of M_v, it suffices to show that for such graph products, we have $\mathsf{VA}(N) \subseteq \mathsf{CF}$. This means whenever M_v is an FRI-monoid, N_v has finitely many right-invertible elements. Moreover, since $N_v \cap \mathsf{J}(N) = \mathsf{J}(N_v)$, no element of $N_v \setminus \mathsf{J}(N_v)$ can appear in a product yielding the identity. Hence, if N_v is generated by $S \subseteq N_v$, replacing N_v by the submonoid generated by $S \cap \mathsf{J}(N_v)$ does not change the identity languages of the graph product. Thus, we assume that each N_v is generated by a finite subset of $\mathsf{J}(N_v)$. Therefore, whenever M_v is an FRI-monoid, N_v is a finite group.

We first establish sufficiency in the case that M_v is an FRI-monoid for every $v \in V$ and proceed by induction on $|V|$. This means that N_v is a finite group for every $v \in V$. Since Γ is chordal, there is a $v \in V$ whose neighborhood is a clique. This means $N\!\restriction_{N(v)}$ is a finite group and hence $N\!\restriction_{N(v)} \times N_v$ context-free by Lemma 3. Since $N\!\restriction_{V \setminus \{v\}}$ is context-free by induction, Theorem 3 and Lemma 2 imply that N is context-free.

To complete the proof, suppose there are n vertices $v \in V$ for which M_v is not an FRI-monoid. We proceed by induction on n. The case $n = 0$ is treated above. Choose $v \in V$ such that M_v is not an FRI-monoid. For each $u \in N(v)$, M_u is an FRI-monoid by condition (2), and hence N_u a finite group. Furthermore, condition (3) guarantees that $N(v)$ is a clique and hence $N\!\restriction_{N(v)}$ is a finite group. As above, Theorem 3 and Lemma 2 imply that N is context-free.   □

5   Semilinearity

A well-known theorem by Chomsky and Schützenberger [2] was re-proved and phrased in terms of valence automata in the following way by Kambites [9].

Theorem 6. $\mathsf{VA}(\mathbb{Z} * \mathbb{Z}) = \mathsf{CF}$.

Using standard methods of formal language theory, one can show:

Lemma 4 ([15,19]). *If every language in* $\mathsf{VA}(M)$ *is semilinear, then so is every language in* $\mathsf{VA}(M \times \mathbb{Z})$.

The following is a consequence of the results of Greibach [6] and Jantzen [8].

Lemma 5. $\mathsf{VA}(\mathbb{B} \times \mathbb{B})$ *contains a non-semilinear language.*

The next result also appears in [19], where, however, it was not made explicit that the language is unary. A proof can be found in the full version.

Lemma 6. *If* Γ*'s underlying loop-free graph contains* P_4 *as an induced subgraph, then* $\mathsf{VA}(\mathbb{M}\Gamma)$ *contains an undecidable unary language.*

We are now ready to show the first main result of this section. Note that the first condition of the following theorem is similar to conditions (2) and (3) in Theorem 5 (and 4): we have looped vertices instead of FRI-monoids (finite groups) and unlooped vertices instead of non-FRI-monoids (infinite groups).

Theorem 7. *All languages in* $\mathsf{VA}(\mathbb{M}\Gamma)$ *are semilinear if and only if (1)* Γ *contains neither* •——• *nor* ⊶——⊶ *as an induced subgraph and (2)* Γ*'s underlying loop-free graph contains neither* C_4 *nor* P_4 *as an induced subgraph.*

Proof. First, observe that if $\mathsf{VA}(N_i) \subseteq \mathsf{VA}(M_i)$ for $i = 0, 1$ then $\mathsf{VA}(N_0 \times N_1) \subseteq \mathsf{VA}(M_0 \times M_1)$. Let $\Gamma = (V, E)$. Suppose conditions 1 and 2 hold. We proceed by induction on $|V|$. 2 implies that Γ's underlying loop-free graph is a transitive forest. If Γ is not connected, then $\mathbb{M}\Gamma$ is a free product of graph products $\mathbb{M}\Gamma_1$ and $\mathbb{M}\Gamma_2$, for which $\mathsf{VA}(\mathbb{M}\Gamma_i)$ contains only semilinear languages by induction. Hence, by Theorems 2 and 3, every language in $\mathsf{VA}(\mathbb{M}\Gamma)$ is semilinear. If Γ is connected, there is a vertex $v \in V$ that is adjacent to every vertex other than itself. We distinguish two cases.

If v is a looped vertex, then $\mathsf{VA}(\mathbb{M}\Gamma) = \mathsf{VA}(\mathbb{Z} \times \mathbb{M}(\Gamma \setminus \{v\}))$, which contains only semilinear languages by induction and Lemma 4. If v is an unlooped vertex, then by 1, $V \setminus \{v\}$ induces a clique of looped vertices. Thus, $\mathbb{M}\Gamma \cong \mathbb{B} \times \mathbb{Z}^{|V|-1}$, meaning $\mathsf{VA}(\mathbb{M}\Gamma)$ contains only semilinear languages by Lemma 4.

We shall now prove the other direction. If Γ contains •——• as an induced subgraph, then $\mathsf{VA}(\mathbb{B} \times \mathbb{B})$ is included in $\mathsf{VA}(\mathbb{M}\Gamma)$ and the former contains a non-semilinear language by Lemma 5. If Γ contains ⊶——⊶, then $\mathbb{M}\Gamma$ contains a copy of $\mathbb{B} \times (\mathbb{Z} * \mathbb{Z})$ as a submonoid. By Theorem 6, we have $\mathsf{VA}(\mathbb{B}) \subseteq \mathsf{VA}(\mathbb{Z} * \mathbb{Z})$ and hence the observation above implies $\mathsf{VA}(\mathbb{B} \times \mathbb{B}) \subseteq \mathsf{VA}(\mathbb{B} \times (\mathbb{Z} * \mathbb{Z}))$.

Suppose Γ's underlying loop-free graph contains C_4 as an induced subgraph. Since we have already shown that the presence of •——• or ⊶——⊶ as an induced subgraph guarantees a non-semilinear language in $\mathsf{VA}(\mathbb{M}\Gamma)$, we may assume that all four participating vertices are looped. Hence, $\mathbb{M}\Gamma$ contains a copy of $(\mathbb{Z} * \mathbb{Z}) \times (\mathbb{Z} * \mathbb{Z})$. By Theorem 6 and the observation above, this means $\mathsf{VA}(\mathbb{B} \times \mathbb{B}) \subseteq \mathsf{VA}(\mathbb{M}\Gamma)$. Thus, $\mathsf{VA}(\mathbb{M}\Gamma)$ contains non-semilinear languages. Finally, if Γ's underlying loop-free graph contains P_4 as an induced subgraph, Lemma 6 provides the existence of an undecidable unary language in $\mathsf{VA}(\mathbb{M}\Gamma)$. Since such a language cannot be semilinear, the lemma is proven.   □

Torsion Groups. A *torsion group* is a group G in which for each $g \in G$, there is a $k \in \mathbb{N} \setminus \{0\}$ with $g^k = 1$. In the following, we show that for torsion groups G, all languages in $\mathsf{VA}(G)$ are semilinear. The key ingredient in our proof is showing that a certain set of multisets is upward closed with respect to a well-quasi-ordering. A *well-quasi-ordering on A* is a reflexive transitive relation $\leq$ on A such that for every infinite sequence $(a_n)_{n \in \mathbb{N}}$, $a_n \in A$, there are indices $i < j$ with $a_i \leq a_j$. We call a subset $B \subseteq A$ *upward closed* if $a \in B$ and $a \leq b$ imply $b \in B$. A basic observation about well-quasi-ordered sets states that for each upward closed set $B \subseteq A$, the set of its minimal elements is finite and B is the set of those $a \in A$ with $m \leq a$ for some minimal $m \in B$ (see [11]).

Given multisets $\alpha, \beta \in X^{\oplus}$ and $k \in \mathbb{N}$, we write $\alpha \equiv_k \beta$ if $\alpha(x) \equiv \beta(x)$ (mod k) for each $x \in X$. We write $\alpha \leq_k \beta$ if $\alpha \leq \beta$ and $\alpha \equiv_k \beta$. Clearly, $\leq_k$ is a well-quasi-ordering on $X^{\oplus}$: Since $\equiv_k$ has finite index in $X^{\oplus}$, we find in any infinite sequence $\alpha_1, \alpha_2, \ldots \in X^{\oplus}$ an infinite subsequence $\alpha'_1, \alpha'_2, \ldots \in X^{\oplus}$ of $\equiv_k$-equivalent multisets. Furthermore, $\leq$ is well-known to be a well-quasi-ordering and yields indices $i < j$ with $\alpha'_i \leq \alpha'_j$ and hence $\alpha'_i \leq_k \alpha'_j$. If $S \subseteq X^{\oplus}$ is upward closed with respect to $\leq_k$, we also say S is *k-upward-closed*. The observation above means in particular that every k-upward-closed set is semilinear.

Theorem 8. *For every torsion group G, the languages in $\mathsf{VA}(G)$ are semilinear.*

Proof. Let G be a torsion group and K be accepted by the valence automaton $A = (Q, X, G, E, q_0, F)$. We regard the finite set E as an alphabet and define the automaton $\hat{A} = (Q, E, G, \hat{E}, q_0, F)$ such that $\hat{E} = \{(p, (p, w, g, q), g, q) \mid (p, w, g, q) \in E\}$. Let $\hat{K} = L(\hat{A})$. Clearly, in order to prove Theorem 8, it suffices to show that $\hat{K}$ is semilinear.

For a word $w \in E^*$, $w = (p_1, x_1, g_1, q_1) \cdots (p_n, x_n, g_n, q_n)$, we write $\sigma(w)$ for the set $\{p_i, q_i \mid 1 \leq i \leq n\}$. w is called a *p, q-computation* if $p_1 = p$, $q_n = q$, and $q_i = p_{i+1}$ for $1 \leq i < n$. A q, q-computation is also called a *q-loop*. Moreover, a q-loop w is called *simple* if $q_i \neq q_j$ for $i \neq j$.

For each $S \subseteq Q$, let F_S be the set of all words $w \in E^*$ with $\sigma(w) = S$ and for which there is a $q \in F$ such that w is a q_0, q-computation and $|w| \leq |Q| \cdot (2^{|Q|} + 1)$. Let $L_S \subseteq E^*$ consist of all $w \in E^*$ such that w is a simple q-loop for some $q \in S$ and $\sigma(w) \subseteq S$. Note that L_S is finite, which allows us to define the alphabet Y_S so as to be in bijection with L_S. Let $\varphi : Y_S \to L_S$ be this bijection and let $\tilde{\varphi} : Y_S^{\oplus} \to E^{\oplus}$ be the morphism with $\tilde{\varphi}(y) = \Psi(\varphi(y))$ for $y \in Y_S$.

For p, q-computations $v, w \in E^*$, we write $v \vdash w$ if $\sigma(v) = \sigma(w)$ and $w = rst$ such that r is a p, q'-computation, s is a simple q'-loop, t is a q', q-computation, and $v = rt$. Moreover, let $\preceq$ be the reflexive transitive closure of $\vdash$. In other words, $v \preceq w$ means that w can be obtained from v by inserting simple q-loops for states $q \in Q$ without increasing the set of visited states. For each $v \in F_S$, we define

$$U_v = \{\mu \in Y_S^{\oplus} \mid \exists w \in \hat{K} : v \preceq w, \ \Psi(w) = \Psi(v) + \tilde{\varphi}(\mu)\}$$

(note that there is only one $S \subseteq Q$ with $v \in F_S$). We claim that

$$\Psi(\hat{K}) = \bigcup_{S \subseteq Q} \bigcup_{v \in F_S} \Psi(v) + \tilde{\varphi}(U_v). \tag{$*$}$$

The inclusion "$\supseteq$" holds by definition. For the other direction, we show by induction on n that for any $q_f \in F$ and any q_0, q_f-computation $w \in E^*$, $|w| = n$, there is a $v \in F_S$ for $S = \sigma(w)$ and a $\mu \in Y_S^{\oplus}$ with $v \preceq w$ and $\Psi(w) = \Psi(v) + \tilde{\varphi}(\mu)$. If $|w| \leq |Q| \cdot (2^{|Q|} + 1)$, this is satisfied by $v = w$ and $\mu = 0$. Therefore, assume $|w| > |Q| \cdot (2^{|Q|} + 1)$ and write $w = (p_1, x_1, g_1, q_1) \cdots (p_n, x_n, g_n, q_n)$. Since $n = |w| > |Q| \cdot (2^{|Q|} + 1)$, there is a $q \in Q$ that appears more than $2^{|Q|} + 1$ times in the sequence $q_1, \ldots, q_n$. Therefore, we can write

$$w = w_0(p_1', x_1', g_1', q)w_1 \cdots (p_m', x_m', g_m', q)w_m$$

with $m > 2^{|Q|} + 1$. Observe that for each $1 \leq i < m$, the word $w_i(p_{i+1}', x_{i+1}', g_{i+1}', q)$ is a q-loop. Since $m - 1 > 2^{|Q|}$, there are indices $1 \leq i < j < m$ with $\sigma(w_i(p_{i+1}', x_{i+1}', g_{i+1}', q)) = \sigma(w_j(p_{j+1}', x_{j+1}', g_{j+1}', q))$. Furthermore, we can find a simple q-loop ℓ as a subword of $w_i(p_{i+1}', x_{i+1}', g_{i+1}', q)$. This means for the word $w' \in E^*$, which is obtained from w by removing ℓ, we have $\sigma(w') = \sigma(w)$ and thus $w' \vdash w$. Moreover, with $S = \sigma(w)$ and $\varphi(y) = \ell$, $y \in Y_S$, we have $\Psi(w) = \Psi(w') + \tilde{\varphi}(y)$. Finally, since $|w'| < |w|$, the induction hypothesis guarantees a $v \in F_S$ and a $\mu \in Y_S^{\oplus}$ with $v \preceq w'$ and $\Psi(w') = \Psi(v) + \tilde{\varphi}(\mu)$. We have $v \preceq w$ and $\Psi(w) = \Psi(v) + \tilde{\varphi}(\mu + y)$ and the induction is complete. In order to prove "$\subseteq$" of $(*)$, suppose $w \in \hat{K}$. Since w is a q_0, q_f-computation for some $q_f \in F$, we can find the above $v \in F_S$, $S = \sigma(w)$, and $\mu \in Y_S^{\oplus}$ with $v \preceq w$ and $\Psi(w) = \Psi(v) + \tilde{\varphi}(\mu)$. This means $\mu \in U_v$ and hence $\Psi(w)$ is contained in the right hand side of $(*)$. This proves $(*)$.

By $(*)$ and since F_S is finite for each $S \subseteq Q$, it suffices to show that U_v is semilinear for each $v \in F_S$ and $S \subseteq Q$. Let $\gamma : E^* \to G$ be the morphism with $\gamma((p, x, g, q)) = g$ for $(p, x, g, q) \in E$. Since G is a torsion group, the finiteness of L_S permits us to choose a $k \in \mathbb{N}$ such that $\gamma(\ell)^k = 1$ for any $\ell \in L_S$. We claim that U_v is k-upward-closed. It suffices to show that for $\mu \in U_v$, we also have $\mu + k \cdot y \in U_v$ for any $y \in Y_S$. Hence, let $\mu \in U_v$ with $w \in \hat{K}$ such that $v \preceq w$ and $\Psi(w) = \Psi(v) + \tilde{\varphi}(\mu)$ and let $\mu' = \mu + k \cdot y$. Let $\ell = \varphi(y) \in L_S$ be a simple q-loop. Then $q \in S$ and since $\sigma(w) = \sigma(v) = S$, we can write $w = r(q_1, x_1, g_1, q)s$, $r, s \in E^*$. The fact that $w \in \hat{K}$ means in particular $\gamma(w) = 1$. Thus, the word $w' = r(q_1, x_1, g_1, q)\ell^k s$ is a q_0, q_f-computation for some $q_f \in F$ with $\gamma(w') = 1$ since $\gamma(\ell)^k = 1$. This means $w' \in \hat{K}$ and $\Psi(w') = \Psi(w) + k \cdot \Psi(\ell) = \Psi(v) + \tilde{\varphi}(\mu + k \cdot y)$. We also have $\sigma(\ell) \subseteq S$ and hence $v \preceq w \preceq w'$. Thus, $\mu' = \mu + k \cdot y \in U_v$. This proves U_v to be k-upward-closed and thus semilinear. $\qquad\square$

Render [16] proved that for every monoid M, the class $\mathsf{VA}(M)$ either (1) coincides with the regular languages, (2) contains the blind one-counter languages, (3) contains the partially blind one-counter languages, or (4) consists of those accepted by valence automata over an infinite torsion group. Hence, we obtain:

Corollary 1. *For each monoid M, at least one of the following holds: (1) $\mathsf{VA}(M)$ contains only semilinear languages. (2) $\mathsf{VA}(M)$ contains the languages of blind one-counter automata. (3) $\mathsf{VA}(M)$ contains the languages of partially blind one-counter automata.*

There are torsion groups with an undecidable word problem [1], hence:

Corollary 2. *There is a group G with an undecidable word problem such that all languages in $\mathsf{VA}(G)$ are semilinear.*

As another application, we can show that the one-sided Dyck language is not accepted by any valence automaton over $G \times \mathbb{Z}^n$ where G is a torsion group.

Corollary 3. *For torsion groups G and $n \in \mathbb{N}$, we have $D_1 \notin \mathsf{VA}(G \times \mathbb{Z}^n)$.*

Acknowledgements. We are indebted to one of the anonymous referees, who pointed out a misuse of terminology in a previous version of Theorem 3.

References

1. Adian, S.I.: The Burnside problem and related topics. Russian Mathematical Surveys 65(5), 805–855 (2010)
2. Chomsky, N., Schützenberger, M.P.: The algebraic theory of context-free languages. In: Computer Programming and Formal Systems, pp. 118–161. North-Holland, Amsterdam (1963)
3. Dirac, G.: On rigid circuit graphs. Abhandlungen aus dem Mathematischen Seminar der Universität Hamburg 25(1-2), 71–76 (1961)
4. Gilman, R.H.: Formal Languages and Infinite Groups. DIMACS Series in Discrete Mathematics and Theoretical Computer Science, vol. 25, pp. 27–51. American Mathematical Society (1996)
5. Green, R.E.: Graph Products of Groups. Ph.D. thesis, University of Leeds (1990)
6. Greibach, S.A.: Remarks on blind and partially blind one-way multicounter machines. Theoretical Computer Science 7(3), 311–324 (1978)
7. Howie, J.M.: Fundamentals of Semigroup Theory. Clarendon Press, Oxford (1995)
8. Jantzen, M.: Eigenschaften von Petrinetzsprachen. Ph.D. thesis, Universität Hamburg (1979)
9. Kambites, M.: Formal languages and groups as memory. Communications in Algebra 37, 193–208 (2009)
10. Kopczynski, E., To, A.: Parikh images of grammars: Complexity and applications. In: Proceedings of LICS 2010, pp. 80–89 (2010)
11. Kruskal, J.B.: The theory of well-quasi-ordering: A frequently discovered concept. Journal of Combinatorial Theory, Series A 13(3), 297–305 (1972)
12. van Leeuwen, J.: A generalisation of Parikh's theorem in formal language theory. In: Loeckx, J. (ed.) ICALP 1974. LNCS, vol. 14, pp. 17–26. Springer, Heidelberg (1974)
13. Lohrey, M., Sénizergues, G.: When is a graph product of groups virtually-free? Communications in Algebra 35(2), 617–621 (2007)
14. Lohrey, M., Sénizergues, G.: Rational subsets in HNN-extensions and amalgamated products. Internat. J. Algebra Comput. 18(01), 111–163 (2008)
15. Lohrey, M., Steinberg, B.: The submonoid and rational subset membership problems for graph groups. J. Algebra 320(2), 728–755 (2008)
16. Render, E.: Rational Monoid and Semigroup Automata. Ph.D. thesis, University of Manchester (2010)
17. Wolk, E.S.: A note on "the comparability graph of a tree". Proceedings of the American Mathematical Society 16(1), 17–20 (1965)
18. Zetzsche, G.: On the capabilities of grammars, automata, and transducers controlled by monoids. In: Aceto, L., Henzinger, M., Sgall, J. (eds.) ICALP 2011, Part II. LNCS, vol. 6756, pp. 222–233. Springer, Heidelberg (2011)
19. Zetzsche, G.: Silent transitions in automata with storage. In: Fomin, F.V., Freivalds, R., Kwiatkowska, M., Peleg, D. (eds.) ICALP 2013, Part II. LNCS, vol. 7966, pp. 434–445. Springer, Heidelberg (2013), http://arxiv.org/abs/1302.3798

Learning Reductions to Sparse Sets

Harry Buhrman[1,*], Lance Fortnow[2,**],
John M. Hitchcock[3,***], and Bruno Loff[4,†]

[1] CWI and University of Amsterdam
buhrman@cwi.nl
[2] Northwestern University
fortnow@eecs.northwestern.edu
[3] University of Wyoming
jhitchco@cs.uwyo.edu
[4] CWI
bruno.loff@cwi.nl

Abstract. We study the consequences of NP having non-uniform polynomial size circuits of various types. We continue the work of Agrawal
and Arvind [1] who study the consequences of SAT being many-one reducible to functions computable by non-uniform circuits consisting of a
single weighted threshold gate. (SAT $\leq_m^p$ LT$_1$). They claim that P = NP
follows as a consequence, but unfortunately their proof was incorrect.

We take up this question and use results from computational learning
theory to show that if SAT $\leq_m^p$ LT$_1$ then PH = P^{NP}.

We furthermore show that if SAT disjunctive truth-table (or majority truth-table) reduces to a sparse set then SAT $\leq_m^p$ LT$_1$ and hence a
collapse of PH to P^{NP} also follows. Lastly we show several interesting
consequences of SAT $\leq_{dtt}^p$ SPARSE.

1 Introduction

In this paper we study consequences of NP having non-uniform polynomial size
circuits of various types. This question is intimately related to the existence of
sparse hard sets for SAT under different types of reductions, and has played a
central role in complexity theory starting with the work of Berman, Hartmanis,
Karp, Lipton and Mahaney [13, 23, 26].

Karp and Lipton showed that if NP is Turing reducible to a sparse set then the
polynomial time hierarchy collapses to its second level. This was later improved
to a collapse of PH = ZPP^{NP} [24, 14], and finally PH = S$_2^p$ [15]. Improvement
of this result to a deeper collapse is a challenging open question whose positive
solution would imply new unconditional circuit lower bounds.

Mahaney [26] showed that if SAT reduces many-one to a sparse set then in
fact P = NP. This implication was subsequently improved by Ogiwara and

---

* Supported by a Vici grant from NWO, and EU-grant QCS.
** Supported in part by NSF grants CCF-0829754 and DMS-0652521.
*** Supported in part by an NWO visiting scholar grant and by NSF grants 0652601
and 0917417. Research done while visiting CWI.
† Supported by FCT grant SFRH/BD/43169/2008.

K. Chatterjee and J. Sgall (Eds.): MFCS 2013, LNCS 8087, pp. 243–253, 2013.
© Springer-Verlag Berlin Heidelberg 2013

Watanabe [29] to bounded truth-table reductions, and later work extended this result to other weak reductions [6, 7, 8, 30, 5, 9, 10]. Notoriously open is to show a similar result for disjunctive truth-table reductions. The best known consequence of this is a collapse of PH to P^{NP} [11].

Agarwal and Arvind [1] took a geometric view of this question and studied the consequences of SAT many-one reducing to LT_1, the class of languages accepted by non-uniform circuits consisting of a single weighted linear-threshold gate. They claimed that SAT $\leq_m^p LT_1$ implies $P = NP$ — unfortunately, the proof in that paper was flawed, as it relied essentially on their incorrect *Splitting Lemma* (p. 203).[1]

We take a fresh look at this approach and connect it with results in learning theory. We use an efficient deterministic algorithm from Maass and Turán [25] for learning half spaces, to obtain a collapse of the polynomial-time hierarchy to P^{NP} from the assumption that SAT $\leq_m^p LT_1$. Interestingly the main ingredient in the learning algorithm is the use of linear programming, which also featured prominently in the work of Agrawal and Arvind.

The use of learning theory in this area of complexity theory is not new and was used before by [24, 14, 18, 19], however the use of *deterministic* learning algorithms in relationship with the polynomial time hierarchy is new.

Next we examine the consequences of SAT $\leq_{dtt}^p$ SPARSE and make a link with the geometric approach above. Using the leftset technique from [29] it is easy to show for conjunctive truth-table reductions that if SAT $\leq_{ctt}^p$ SPARSE then $P = NP$. Frustratingly, for disjunctive truth table reductions the best known consequence is PH = P^{NP}, a result due to Arvind et al.[11], who use a complicated argument. We use error-correcting codes to show that SAT $\leq_{dtt}^p$ SPARSE implies that SAT $\leq_m^p LT_1$, which with our previous result gives a new and more modular proof of the collapse to P^{NP}. Our new approach enables us to obtain the same collapse for majority reductions.

We finish with a handful of new consequences of SAT $\leq_{dtt}^p$ SPARSE and SAT $\leq_{maj}^p$ SPARSE. Interestingly it turns out that in the case of disjunctive reductions, improvement of the above results to PH = $P_{\parallel}^{NP}$ is sufficient to obtain the full collapse to $P = NP$.

2   Preliminaries

We assume that the reader is familiar with computational complexity, as expounded, for instance, in [4]. In particular, we make use of

$$A \in P/poly \iff A \leq_T^p SPARSE,$$

so a reduction to a sparse set can be seen as a polynomial-time circuit. The weaker the reduction, the weaker the access to non-uniformity.

---

[1] The mistake in this *Splitting Lemma* was not seen by any of the paper's referees, but instead was accidentally discovered years later. For an anecdotal account of the episode, please consult `http://blog.computationalcomplexity.org/2009/10/thanks-for-fuzzy-memories.html` .

The least common notation we use, is: $P_{\parallel}^{NP}$ and $FP_{\parallel}^{NP}$, which are the classes of sets and functions, respectively, that are polynomial-time computable with non-adaptive queries to an NP oracle; $P^{NP[q]}$, and $FP^{NP[q]}$, the classes of sets and functions that are polynomial-time computable by asking no more than $q(n)$ (possibly adaptive) queries to an NP oracle.

A linear threshold function $L : \{0,1\}^m \to \{0,1\}$ is defined by a vector of m real numbers $w \in \mathbb{R}^m$, called weights, a threshold $\theta \in \mathbb{R}$, and the equation

$$L(z) = \begin{cases} 1 & \text{if } z \cdot w > \theta, \text{ and} \\ 0 & \text{if } z \cdot w \leq \theta. \end{cases}$$

Here $z \cdot w$ denotes the inner product $\sum_{i=1}^{m} z_i w_i$.

We let $LT_1(m)$ denote the class of linear-threshold functions with m-bit binary inputs. We may freely assume, for functions in $LT_1(m)$, that the weights and thresholds are integers of bit-length $m \log m$ [27, Thm. 16].

In this paper we are concerned with three kinds of reductions:

Definition 1. (*dtt reductions*) *A set A disjunctive truth-table reduces to a set S, written $A \leq_{dtt}^{p} S$, if there exists a polytime computable function Q, outputting a set of queries, such that*

$$x \in A \iff Q(x) \cap S \neq \varnothing.$$

(majority reductions) *A set A majority truth-table reduces to a set S, written $A \leq_{maj}^{p} S$, if there exists a function Q, as above, such that*

$$x \in A \iff |Q(x) \cap S| > \frac{|Q(x)|}{2}$$

(LT$_1$ reductions) *A set A reduces to linear-threshold functions, written $A \leq_m^p LT_1$, if there exists a polytime computable function f, and a family $\{L_n\}_{n \in \mathbb{N}}$ of linear threshold functions, such that[2]*

$$x \in A^{=n} \iff L_n(f(x)) = 1.$$

3   If SAT $\leq_m^p$ LT$_1$ ...

Attempting to derive $P = NP$ should prove difficult, since by the results in the next section this would imply the same collapse for *dtt* and majority reductions to sparse sets. Since $A \leq_m^p LT_1$ implies $A \in P/poly$, then from SAT $\leq_m^p LT_1$ and [15] we get $PH = S_2^p$. This collapse can be improved in the following way:

Theorem 1. *If SAT $\leq_m^p$ LT$_1$, then $PH = P^{NP}$.*

---

[2] Notice that the length of $f(x)$ must be a function of the length of x.

We take a similar approach as [14]: the existence of a suitable learning algorithm will, under the assumption that $\text{SAT} \leq_m^p \text{LT}_1$, collapse the polynomial-time hierarchy. The difference being that we have a *deterministic* learning algorithm for linear threshold functions, but only (zero-error) probabilistic algorithms with access to an NP oracle are known that can learn general circuits.

Our learning model is the online learning model of Angluin [3] for learning with counter-examples. In our case, the learner wishes to identify an unknown linear threshold function, say $L \in \text{LT}_1(m)$. At each learning step, the algorithm proposes some hypothesis $H \in \text{LT}_1(m)$. If $H \neq L$, then the algorithm is given a counter-example x such that $H(x) \neq L(x)$. The algorithm is not allowed to make any assumptions on the choice of counter-example, which could very well be adversarially chosen. Based on the previous counter-examples and hypotheses, the algorithm suggests a new hypothesis which is correct on the inputs seen so far, and the process is repeated until $H = L$. The learning complexity of such an algorithm is the maximum number of these steps that it will need in order to learn any function in $\text{LT}_1(m)$.

Theorem 2 ([25]). *There is a deterministic polynomial-time algorithm for learning $\text{LT}_1(m)$ functions in $O(m^3 \log m)$ steps.*

As a corollary, will be able to prove Theorem 1, and the forthcomming Theorem 3. It should be noted that both of these theorems hold for polynomial-time many-one reductions to any class of functions which, like LT_1, have a polynomial-time algorithm for learning with counter-examples.

Proof (of Theorem 1). Suppose $\text{SAT} \leq_m^p \text{LT}_1$, and let L_n be a family of linear threshold functions, and f a polytime reduction, such that

$$\psi \in \text{SAT}^{=n} \iff L_n(f(\psi)) = 1. \tag{1}$$

For a given formula of length n, we use the algorithm of Theorem 2 in order to uncover a linear threshold function H with the same property (1) as L_n, in polynomial time with the help of an NP oracle.

Let $m = |f(\psi)|$ on inputs ψ of length n. We proceed as follows: we start with an initial hypothesis H for L_n, given by the learning algorithm for $\text{LT}_1(m)$. Then at each step in the learning process we ask the NP oracle if there exists some formula ψ of length n such that:

1. ψ has no variables and evaluates to true, but $H(f(\psi)) = 0$, or
2. ψ has no variables and evaluates to false, but $H(f(\psi)) = 1$, or
3. $H(f(\psi)) = 1$ but both $H(f(\psi_0)) = 0$ and $H(f(\psi_1)) = 0$, or
4. $H(f(\psi)) = 0$, but $H(f(\psi_0)) = 1$ or $H(f(\psi_1)) = 1$.

Above, ψ_0 and ψ_1 are obtained by replacing the first variable of ψ respectively with 0 or 1. Essentially, we are asking whether the set $\text{SAT}(H) = \{\psi | H(f(\psi)) = 1\}$ violates the self-reducibility of SAT. If this is not the case, then necessarily $\text{SAT}(H) = \text{SAT}^{=n}$, and we are done.

But if the self-reducibility is violated, then for at least one $\phi \in \{\psi, \psi_0, \psi_1\}$, we must have $H(f(\phi)) \neq L_n(f(\phi))$, and so $f(\phi)$ gives us a counter-example to update the hypothesis H. We use prefix-search to obtain such a formula ϕ, and from equation (1) this will provide us with a counter-example, i.e., $H(f(\phi)) \neq L_n(f(\phi))$.

After $O(m^3 \log m) = poly(n)$ many iterations, we will either have learnt L_n, or otherwise obtained an hypothesis H suitable for the purpose of querying $\mathrm{SAT}^{=n}$.

By feeding the NP oracle the suitable linear-threshold functions, it now becomes possible to simulate a Σ_2^p computation. So Σ_2^p, and consequently all of PH, collapses to P^{NP}.   □

The algorithm above is non-adaptive, and in order to solve $\mathrm{SAT}^{=n}$, it potentially asks $\Omega(nm^3 \log m)$-many queries to SAT. We can be a bit more clever, and actually reduce this number to n. This will essentially give us the following:

Theorem 3. *If* $\mathrm{SAT} \leq_m^p \mathrm{LT}_1$, *then* $\mathrm{NP}^{\mathrm{SAT}^{=n}} \subseteq \mathrm{P}^{\mathrm{SAT}[n]} \cap \mathrm{NP}/lin$.

Proof. The idea is to use the self-reducibility of SAT once again, in order to learn $\mathrm{SAT}^{=n}$ first for formulas with no variables (formulas over constants, which evaluate to true or false), then for formulas with 1 variable, then 2 variables, and so on. Let $\mathrm{SAT}_k^{=n}$ be the set of satisfiable formulas having exactly k variables. Starting with the initial hypothesis H, we set out to learn $\mathrm{SAT}_0^{=n}$. What is the largest number of mistakes that we can make, i.e., how many times might we need to change our hypothesis H until we have properly learned $\mathrm{SAT}_0^{=n}$?

Using a SAT oracle, we can ask: *is there a sequence* $\psi_1, \ldots, \psi_\ell$ *of* ℓ *formulas, having 0 vars, such that* ψ_{i+1} *is always a counter-example to the hypothesis constructed by our learning algorithm after seeing* $\psi_1, \ldots, \psi_i$?[3]

We know that such a sequence will have at most $poly(n)$ formulas, and so using binary search, then by making $O(\log n)$ such queries, we can find the length of the largest sequence of counter-examples which can be given to our learning algorithm before it necessarily learns $\mathrm{SAT}_0^{=n}$. Let this length be ℓ_0.

Then because ℓ_0 is maximal, at this point we know that if the learning algorithm is given *any* sequence of ℓ_0-many counter-examples having no variables, the constructed hypothesis H will be correct on $\mathrm{SAT}_0^{=n}$, in the sense that $\psi \in \mathrm{SAT}_0^{=n} \iff H(f(\psi)) = 1$.

Now that we know ℓ_0, we set out to learn $\mathrm{SAT}_1^{=n}$. Using SAT as an oracle, we may ask: *Is there a sequence of* ℓ_0 *counter-examples with 0 vars, followed by* ℓ *counter-examples with 1 var?* Thus we may obtain ℓ_1, the length of the largest sequence of counter-examples with 1 var, that can be given to the learning algorithm *after it has already learned every possible formula with 0 vars.*

---

[3] Formalizing the question as an NP-set gives us:

$$A = \{\langle 0^n, 0^\ell \rangle \mid \exists \boldsymbol{\psi}, \boldsymbol{H} \forall i \; H_i = \mathrm{Learner}(\psi_1, \ldots, \psi_i) \wedge H_{i-1}(f(\psi_i)) \neq \mathrm{SAT}(\psi_i)\},$$

where $\boldsymbol{\psi}$ is a sequence of ℓ-many formulas with 0 vars, $\boldsymbol{H}$ is a sequence of ℓ-many threshold functions, and $i \in \{1, \ldots, \ell\}$. Notice that $H_{i-1}(f(\psi_i)) \neq \mathrm{SAT}(\psi_i)$ is decidable in polynomial time because the formulas ψ_i have no variables.

In general we know $\ell_0, \ldots, \ell_{k-1}$, and we set out to learn $\text{SAT}_k^{=n}$. Using SAT as an oracle, we ask: *Is there a sequence of ℓ_0 counter-examples with 0-vars, followed by ℓ_1 counter-examples with 1-var, ..., followed by ℓ_{k-1} counter-examples with $k-1$ vars, followed by ℓ counter-examples with k vars?*

The key observation is that in order for the SAT oracle to be able to tell whether a formula ψ with k variables is a counter-example to hypothesis H, i.e., whether $H(f(\psi)) \neq \text{SAT}(\psi)$, it will need to know whether ψ is or is not satisfiable. In order to know this, the SAT oracle uses H itself, which at this point is known to be correct for formulas with $k-1$ variables, and thus $\psi \in \text{SAT} \iff H(f(\psi_0)) = 1$ or $H(f(\psi_1)) = 1$.

In the end we have $n+1$ numbers $\ell_0, \ldots, \ell_n$, and we know that if the learning algorithm is given *any* sequence of ℓ_0-many counter-examples having no variables, followed by ℓ_1 counter-examples having 1 variable, ..., followed by ℓ_n counter-examples having n variables, then the constructed hypothesis H will be correct on all of $\text{SAT}^{=n}$. Furthermore, such a sequence must exist by construction.

These numbers take up at most $O(n \log n)$ many bits, and each bit is the outcome of one (much larger, adaptive) query to SAT. Having access to $\ell_0, \ldots, \ell_n$, an NP machine can guess a proper sequence of counter-examples, and it will thus obtain an hypothesis H which it can use to answer any query to $\text{SAT}^{=n}$. Thus $\text{NP}^{\text{SAT}^{=n}} \subseteq \text{P}^{\text{SAT}[n \log n]}$, and $\text{NP}^{\text{SAT}^{=n}} \subseteq \text{NP}/n \log n$.

In order to improve $n \log n$ into n bits, or even $\frac{n}{c \log n}$ bits, the proof is similar, but instead of learning how to decide $\text{SAT}^{=n}$ for one extra variable at a time, we learn $O(\log n)$ many extra variables at a time — this requires us to unfold the self-reduction tree $O(\log n)$-deep. $\square$

Under the assumption that SAT has polynomial-size circuits, we may decide, in CONP, whether a given string $\alpha(n)$ encodes a circuit correct for $\text{SAT}^{=n}$. However, there will possibly be many strings with this property — the following theorem gives us a way to single out, in CONP, a *unique* advice string $\alpha(n)$ suitable to decide $\text{SAT}^{=n}$.

Theorem 4. *If $\text{NP} \subseteq \text{P}/poly$, and $\text{PH} \subseteq \text{P}^{\text{NP}}$, then $\text{PH} \subseteq \text{P}/\alpha$ for some polynomial advice function $0^n \mapsto \alpha(n)$ whose graph $G_\alpha = \{\langle 0^n, \alpha(n)\rangle | n \in \mathbb{N}\} \in$ CONP.*

Proof. Let A be Δ_2-complete. Then there is a polytime machine $\mathcal{M}$ that decides $A^{=n}$ with polynomially-long advice $\gamma(n)$, where $\gamma(n)$ codes a circuit solving $\text{SAT}^{=m}$, for some $m = poly(n)$. The machine $\mathcal{M}$ uses $\gamma(n)$ to answer the queries needed in the Δ_2 computation of A. Furthermore, the function $0^n \mapsto \tilde{\alpha}(n)$, given by

$$\tilde{\alpha}(n) \text{ is the lexicographically smallest string}$$
$$\text{such that } x \in A^{=n} \iff \mathcal{M}(x)/\tilde{\alpha}(n) = 1,$$

is in PH and thus in FP^{SAT}. Then let $\mathcal{N}$ be a polytime machine computing $\tilde{\alpha}$ with a SAT oracle, and let's say it makes k queries to compute $\tilde{\alpha}(n)$. Let $S \in$ CONP be the set of strings $\langle 0^n, \tilde{\alpha}, a_1, \ldots, a_k, y_1, \ldots, y_k\rangle$ such that

1. $\mathcal{N}^a(0^n) = \tilde{\alpha}$ (i.e., when $a_1, \ldots, a_k$ are given as answers to the queries of $\mathcal{N}$),
2. if $a_i = 1$ then y_i is the lexicographically smallest satisfying assignment of the i-th formula queried by $\mathcal{N}^a$, and
3. if $a_i = 0$ then $y_i = \lambda$ (the empty string) and the i-th formula queried by $\mathcal{N}^a$ is not satisfiable.

Notice that for a given n, the string $\langle 0^n, \tilde{\alpha}, a_1, \ldots, a_k, y_1, \ldots, y_k \rangle \in S$ is uniquely defined, so S is the graph of $\alpha(n) = \langle \tilde{\alpha}, a_1, \ldots, a_k, y_1, \ldots, y_k \rangle$. When given $\alpha(n)$, an algorithm for A can simply check if $\mathcal{M}(x)/\tilde{\alpha} = 1$. $\qquad \square$

Corollary 5. *If* SAT $\leq^p_m$ LT$_1$, *then* PH $\subseteq$ P$/\alpha$ *for some polynomial advice function* $0^n \mapsto \alpha(n)$ *whose graph* $G_\alpha \in$ coNP.

4 LT$_1$ versus *dtt* and *maj* Reductions

In this section we show that LT$_1$ reductions can simulate *dtt* and majority reductions to sparse sets. Thus, effectively, the collapses we have proven for LT$_1$ reductions imply similar collapses for *dtt* and majority reductions.

Theorem 6. *If* $A \leq^p_{dtt}$ SPARSE *or* $A \leq^p_{maj}$ SPARSE, *then* $A \leq^p_m$ LT$_1$.

Proof. We will use a Reed-Solomon code to construct the LT$_1$ reduction. Suppose $A \leq^p_{dtt} S \in$ SPARSE, and assume w.l.o.g. that the *dtt* reduction is given by a polytime computable function Q, such that

$$x \in A^{=n} \iff S^{=m} \cap Q(x) \neq \varnothing, \tag{2}$$

$$|S^{=m}| = m, \text{ and}$$

$$|Q(x)| = d.$$

That is, for every input x of length n, $Q(x) = \{y_1, \ldots, y_d\}$ always queries the same number of $d = d(n)$ strings of the same length $m = m(n)$, and that there will be exactly m many such strings in $S^{=m}$. Such an assumption can always be made by tweaking the reduction and changing S accordingly.

We will be working over the field $\mathbb{F}_{2^\ell}$, for $\ell \geq \lceil \log dm^2 \rceil$. For any given binary string s of length m, we define the polynomial $p_s(z) = \sum_{i=1}^m s_i z^{i-1}$. Now let $C(s)$ be the encoding of s as a $2^\ell \times 2^\ell$-long binary string: this string is the concatenation of $p_s(a)$, as a goes through all the 2^ℓ elements of $\mathbb{F}_{2^\ell}$; each $p_s(a)$ is in turn encoded by a binary string of length 2^ℓ, having a 1 at position $p_s(a)$ (for some fixed enumeration of $\mathbb{F}_{2^\ell}$), and 0s elsewhere.

Note that $|C(s)| = O(d^2 m^4) = poly(n)$. Then vitally note that by encoding strings this way, the number of bit positions where $C(s)$ and $C(y)$ are both 1, given by the inner product $C(s) \cdot C(y)$,[4] is exactly the number of elements

---

[4] Note that the binary strings $C(s)$ and $C(y)$ are seen as 0-1 vectors, and that the inner product is a natural number $\sum_{j=1}^{2^\ell \times 2^\ell} C(s)_j C(y)_j$.

$a \in \mathbb{F}_{2^\ell}$ where $p_s(a) = p_y(a)$. So for any two words $s, y \in \{0,1\}^m$, using the fact that $p_s - p_y$ is either identically zero, or has at most $m - 1$ roots,

$$\begin{cases} C(y) \cdot C(s) \leq m - 1 & \text{if } y \neq s, \text{ and} \\ C(y) \cdot C(s) \geq dm^2 & \text{if } y = s. \end{cases}$$

Define $g(x) = \bigvee_{i=1}^{d} C(y_i)$, where $Q(x) = \{y_1, \ldots, y_d\}$, and by $\bigvee$ we mean bitwise-OR. Then

$$\begin{cases} g(x) \cdot C(s) \leq \sum_{i=1}^{d} C(y_i) \cdot C(s) \leq d(m-1) & \text{if } s \notin Q(x), \text{ and} \\ g(x) \cdot C(s) \geq dm^2 & \text{if } s \in Q(x). \end{cases}$$

Finally, let $w_n = \bigoplus_{s \in S^{=m}} C(s)$, and $f(x) = (g(x))^{\oplus m}$, where by $\oplus$ we mean the direct sum of vectors / concatenation of strings. Then $f(x) \cdot w_n = \sum_{s \in S^{=m}} g(x) \cdot C(s)$, and we come to

$$\begin{cases} f(x) \cdot w_n \leq md(m-1) & \text{if } S^{=m} \cap Q(x) = \varnothing, \text{ and} \\ f(x) \cdot w_n \geq dm^2 & \text{if } S^{=m} \cap Q(x) \neq \varnothing. \end{cases} \tag{3}$$

So $x \in A \iff f(x) \cdot w_n > dm(m-1)$, showing that $A \leq_m^p \text{LT}_1$.

The transformation for maj reductions is similar. We begin with a dtt reduction function Q, which is like before, except that now Equation (2) is replaced with

$$x \in A^{=n} \iff |S^{=m} \cap Q(x)| > \frac{d}{2}.$$

Then both the LT_1 reduction function f, and the set of weights w_n are constructed exactly in the same way, but over a slightly larger field. Working through the proof, if 2^ℓ is the size of our chosen field, and $K = |S^{=m} \cap Q(x)|$, then Equation (3) becomes:

$$2^\ell K \leq f(x) \cdot w_n \leq 2^\ell K + d(m-1)(m-K).$$

Now choose $\ell \geq \lceil \log 4dm^2 \rceil$ as the size of our field. Using the defining property of the maj reduction, a small computation will show us that

$$x \in A^{=n} \iff K > \frac{d}{2} \iff f(x) \cdot w_n > 2^\ell \left(\frac{d}{2} + \frac{1}{4} \right)$$

— this defines our LT_1 reduction. $\square$

5   If SAT $\leq_{dtt}^p$ SPARSE ...

Disjunctive truth-table reductions to sparse sets are powerful enough to simulate bounded truth-table reductions to sparse sets [2]. But the collapses that are known, under the assumption that SAT $\leq_{dtt}^p$ SPARSE, are not as strong as those for btt reductions. We can summarize what was known about SAT $\leq_{dtt}^p$ SPARSE, in the following two theorems:

Consequense 1 ([16, 12]). ... *then* $\mathrm{FP}_{\parallel}^{\mathrm{NP}} = \mathrm{FP}^{\mathrm{NP}[\log]}$, $\mathrm{UP} \subseteq \mathrm{P}$, *and* $\mathrm{NP} = \mathrm{RP}$. □

Consequense 2 ([11]). ... *then* $\mathrm{PH} = \mathrm{P}^{\mathrm{NP}} = \mathrm{P}^{\mathrm{RP}} = \mathrm{BPP}$. □

To these consequences, we append our own observations, which follow from the results in the previous sections.

Consequense 3. ... *then* $\mathrm{NP}^{\mathrm{SAT}^{=n}} \subseteq \mathrm{P}^{\mathrm{SAT}[n]}$, $\mathrm{NP}^{\mathrm{SAT}^{=n}} \subseteq \mathrm{NP}/lin$. □

Consequense 4. ... *then* $\mathrm{PH} \subseteq \mathrm{P}/\alpha$ *for some function* $0^n \mapsto \alpha(n)$ *whose graph* $G_\alpha \in \mathrm{coNP}$. □

Finally, we note that we are not far away from obtaining the final consequence $\mathrm{P} = \mathrm{NP}$.

Consequense 5. ... *then* $\mathrm{E} \not\subseteq \mathrm{NP}/\log$.

Consequense 6. ... *then* $\mathrm{E}^{\mathrm{NP}} \not\subseteq \mathrm{SIZE}(2^{\varepsilon n})$ *for some* $\varepsilon > 0$.

Consequense 7. ... *then the following statements are all equivalent:*

1. $\mathrm{P} = \mathrm{NP}$.
2. $\mathrm{P}^{\mathrm{NP}} = \mathrm{P}_{\parallel}^{\mathrm{NP}}$.
3. $\mathrm{coNP} \cap \mathrm{SPARSE} \subseteq \mathrm{NP}$.
4. $\mathrm{E}^{\mathrm{NP}} = \mathrm{E}_{\parallel}^{\mathrm{NP}}$.

Proof (of Consequence 5). [17] show that

$$\mathrm{EXP} \subseteq \mathrm{P}_{\parallel}^{\mathrm{NP}} \iff \mathrm{EXP} \subseteq \mathrm{NP}/\log.$$

But if we had $\mathrm{EXP} \subseteq \mathrm{P}_{\parallel}^{\mathrm{NP}}$, then we could compute the lexicographically least satisfying assignment of a given formula in $\mathrm{FP}_{\parallel}^{\mathrm{NP}}$, and thus in $\mathrm{FP}^{\mathrm{NP}[\log]}$, by Consequence 1. But then we could also do it in FP alone, simply by trying every possible answer to the queries made by the $\mathrm{FP}^{\mathrm{NP}[\log]}$ computation. But then $\mathrm{P} = \mathrm{NP}$, and the necessary conclusion $\mathrm{EXP} \subseteq \mathrm{PH} \subseteq \mathrm{P}$ would contradict the time-hierarchy theorem. □

Proof (of Consequence 6). By counting there is a function $f : \{0,1\}^{\log n} \to \{0,1\} \notin \mathrm{SIZE}(n^\varepsilon)$ which can be found in P^{Σ_2} [cf. 22], and thus, by Consequence 2, in P^{NP}. Translating this upwards we get a set in E^{NP} with no circuits of size $2^{\varepsilon n}$. □

Proof (of Consequence 7). As in the proof of Consequence 5, $\mathrm{P} = \mathrm{NP}$ follows if we are able to compute the least satisfying assignment of a given formula in $\mathrm{FP}_{\parallel}^{\mathrm{NP}}$. This is trivially the case when $\mathrm{P}^{\mathrm{NP}} = \mathrm{P}_{\parallel}^{\mathrm{NP}}$.

Now if $\mathrm{SPARSE} \cap \mathrm{coNP} \subseteq \mathrm{NP}$, then, from Consequence 4, we get $\mathrm{PH} \subseteq \mathrm{NP}^{G_\alpha} \subseteq \mathrm{NP}^{\mathrm{NP} \cap \mathrm{SPARSE}}$: the non-deterministic machine just guesses the advice α and checks it using the oracle. But $\mathrm{NP}^{\mathrm{NP} \cap \mathrm{SPARSE}} \subseteq \mathrm{NP}$ [cf 21], and thus the least satisfying assignment of a given formula can be obtained in $\mathrm{FP}_{\parallel}^{\mathrm{NP}}$.

To see the third equivalence, notice that $\mathrm{E}^{\mathrm{NP}} = \mathrm{E}_{\parallel}^{\mathrm{NP}}$, then Consequence 6 implies we can derandomise BPP in $\mathrm{P}_{\parallel}^{\mathrm{NP}}$ [cf. 28, 20]; since $\mathrm{PH} \subseteq \mathrm{BPP}$, this implies that the least satisfying assignment can be found in $\mathrm{FP}_{\parallel}^{\mathrm{NP}}$. □

6 Final Remarks

We remark that our paper also draws new conclusions from $\text{SAT} \leq^p_{maj} \text{SPARSE}$. It was previously known that, under this hypothesis, NP = RP, but it remains open to show that $\text{FP}^{\text{NP}}_{\|} = \text{FP}^{\text{NP}[\log]}$ [cf. 12]. However, the results in this paper imply that Consequences 2, 3 and 4 of the previous section also apply to the $\text{SAT} \leq^p_{maj} \text{SPARSE}$ case, which was previously unknown.

Clearly, the most important open question is to prove that $\text{SAT} \leq^p_m \text{LT}_1$ implies that P = NP, or otherwise show a relativized world where only the former holds.

References

[1] Agrawal, M., Arvind, V.: Geometric sets of low information content. Theor. Comput. Sci. 158(1-2), 193–219 (1996)
[2] Allender, E., Hemachandra, L.A., Ogiwara, M., Watanabe, O.: Relating equivalence and reducibility to sparse sets. SIAM J. Comput. 21(3), 521–539 (1992)
[3] Angluin, D.: Queries and concept learning. Mach. Learn. 2(4), 319–342 (1987)
[4] Arora, S., Barak, B.: Computational Complexity: A Modern Approach. Cambridge University Press (2009)
[5] Arvind, V., Han, Y., Hemachandra, L., Köbler, J., Lozano, A., Mundhenk, M., Ogiwara, M., Schöning, U., Silvestri, R., Thierauf, T.: Reductions to sets of low information content. In: Ambos-Spies, K., Homer, S., Schöning, U. (eds.) Complexity Theory: Current Research, pp. 1–45. Cambridge University Press (1993)
[6] Arvind, V., Köbler, J., Mundhenk, M.: Bounded truth-table and conjunctive reductions to sparse and tally sets. Technical report, University of Ulm (1992)
[7] Arvind, V., Köbler, J., Mundhenk, M.: Lowness and the complexity of sparse and tally descriptions. In: Ibaraki, T., Iwama, K., Yamashita, M., Inagaki, Y., Nishizeki, T. (eds.) ISAAC 1992. LNCS, vol. 650, pp. 249–258. Springer, Heidelberg (1992)
[8] Arvind, V., Köbler, J., Mundhenk, M.: On bounded truth-table, conjunctive, and randomized reductions to sparse sets. In: Proc. 12th CFSTTCS, pp. 140–151. Springer (1992)
[9] Arvind, V., Köbler, J., Mundhenk, M.: Hausdorff reductions to sparse sets and to sets of high information content. In: Borzyszkowski, A.M., Sokolowski, S. (eds.) MFCS 1993. LNCS, vol. 711, pp. 232–241. Springer, Heidelberg (1993)
[10] Arvind, V., Köbler, J., Mundhenk, M.: Monotonous and randomized reductions to sparse sets. Theo. Inform. and Appl. 30(2), 155–179 (1996)
[11] Arvind, V., Köbler, J., Mundhenk, M.: Upper bounds for the complexity of sparse and tally descriptions. Theor. Comput. Syst. 29, 63–94 (1996)
[12] Arvind, V., Torán, J.: Sparse sets, approximable sets, and parallel queries to NP. In: Meinel, C., Tison, S. (eds.) STACS 1999. LNCS, vol. 1563, pp. 281–290. Springer, Heidelberg (1999)
[13] Berman, L., Hartmanis, J.: On isomorphisms and density of NP and other complete sets. In: Proc. 8th STOC, pp. 30–40 (1976)
[14] Bshouty, N.H., Cleve, R., Gavaldà, R., Kannan, S., Tamon, C.: Oracles and queries that are sufficient for exact learning. J. Comput. Syst. Sci. 52(3), 421–433 (1996)
[15] Cai, J.-Y.: $S^p_2 \subseteq \text{ZPP}^{\text{NP}}$. J. Comput. Syst. Sci. 73(1), 25–35 (2002)

[16] Cai, J.-Y., Naik, A.V., Sivakumar, D.: On the existence of hard sparse sets under weak reductions. In: Puech, C., Reischuk, R. (eds.) STACS 1996. LNCS, vol. 1046, pp. 307–318. Springer, Heidelberg (1996)

[17] Fortnow, L., Klivans, A.: NP with small advice. In: Proc. 20th CCC, pp. 228–234 (2005)

[18] Harkins, R., Hitchcock, J.M.: Dimension, halfspaces, and the density of hard sets. Theor. Comput. Syst. 49(3), 601–614 (2011)

[19] Hitchcock, J.M.: Online learning and resource-bounded dimension: Winnow yields new lower bounds for hard sets. SIAM J. Comput. 36(6), 1696–1708 (2007)

[20] Impagliazzo, R., Wigderson, A.: P = BPP if E requires exponential circuits. In: Proc. 29th STOCS, pp. 220–229 (1997)

[21] Kadin, J.: $P^{NP[O(\log n)]}$ and sparse Turing-complete sets for NP. J. Comput. Syst. Sci. 39(3), 282–298 (1989)

[22] Kannan, R.: Circuit-size lower bounds and non-reducibility to sparse sets. Inform. Comput. 55(1-3), 40–56 (1982)

[23] Karp, R., Lipton, R.: Some connections between nonuniform and uniform complexity classes. In: Proc. 12th STOC, pp. 302–309 (1980)

[24] Köbler, J., Watanabe, O.: New collapse consequences of NP having small circuits. In: Fülöp, Z., Gécseg, F. (eds.) ICALP 1995. LNCS, vol. 944, pp. 196–207. Springer, Heidelberg (1995)

[25] Maass, W., Turán, G.: How fast can a threshold gate learn? In: Worksh. Comput. Learn. Theor. & Natur. Learn. Syst., vol. 1, pp. 381–414. MIT Press, Cambridge (1994)

[26] Mahaney, S.: Sparse complete sets for NP: Solution of a conjecture of Berman and Hartmanis. J. Comput. Syst. Sci. 25(2), 130–143 (1982)

[27] Muroga, S., Toda, I., Takasu, S.: Theory of majority decision elements. J. Franklin. I. 271(5), 376–418 (1961)

[28] Nisan, N., Wigderson, A.: Hardness vs randomness. J. Comput. Syst. Sci. 49(2), 149–167 (1994)

[29] Ogiwara, M., Watanabe, O.: Polynomial-time bounded truth-table reducibility of NP sets to sparse sets. SIAM J. Comput. 20(3), 471–483 (1991)

[30] Ranjan, D., Rohatgi, P.: On randomized reductions to sparse sets. In: Proc. 7th STOC, pp. 239–242 (1992)

Probabilistic Automata with Isolated Cut-Points

Rohit Chadha[1], A. Prasad Sistla[2], and Mahesh Viswanathan[3],[*]

[1] University of Missouri
[2] University of Illinois, Chicago
[3] University of Illinois, Urbana-Champaign

Abstract. We consider various decision problems for probabilistic finite automata (PFA)s with isolated cut-points. Recall that a cut-point x is said to be isolated for a PFA if the acceptance probability of all finite strings is bounded away from x. First we establish the exact level of undecidability of the problem of determining if a cut-point is isolated; we show this problem to be Σ_2^0-complete. Next we introduce a new class of PFAs called eventually weakly ergodic PFAs that generalize ergodic and weakly ergodic PFAs. We show that the emptiness and universality problem for these PFAs is decidable provided the cut-point is isolated.

1   Introduction

A probabilistic finite automaton (PFA) [21,20] is like a deterministic finite automaton except that after reading an input symbol the automaton rolls a dice to determine the next state. Thus the transition function of a PFA associates a probability distribution on next states with each state and input symbol. Given an acceptance threshold or cut-point x and an initial distribution μ, the language recognized by a PFA $\mathcal{B}$ (denoted as $\mathsf{L}_{>x}(\mathcal{B}, \mu)$) is the collection of all finite words u that reach a final state with probability $> x$ when $\mathcal{B}$ is started with initial distribution μ. Surprisingly, even though PFAs have only finitely many states, they are known to recognize non-regular languages [21].

One semantic restriction that has been extensively studied is that of cut-points being isolated [21,4,5,15] — a cut-point x is isolated for PFA $\mathcal{B}$ and initial distribution μ, if there is an $\epsilon > 0$ such that any input word u is either accepted with probability at most $x - \epsilon$ or with probability at least $x + \epsilon$. Thus, the acceptance probability of any input word is bounded away from the cut-point x. Isolated cut-points are important because algorithms described by PFAs are useful mainly when there is a separation between the probability of accepting the good inputs from the probability of accepting the bad inputs. Isolation allows one to use standard algorithmic techniques like amplification by running multiple copies of the algorithm to drive down the probability of error. PFAs with isolated cut-points are the constant space analogues of probabilistic polynomial time complexity classes like **BPP** and **RP**.

In this paper, we consider various decision problems for PFAs with isolated cut-points. The first problem we consider is that of determining if a cut-point x is isolated for a PFA $\mathcal{B}$ and initial distribution μ. The problem was shown to be undecidable (in fact, **r.e.**-hard) by Bertoni [4,5] when $x \in (0, 1)$. Recently, the problem was shown to

---

[*] Supported by NSF CCF 1016989 and NSF CNS 1016791.

K. Chatterjee and J. Sgall (Eds.): MFCS 2013, LNCS 8087, pp. 254–265, 2013.
© Springer-Verlag Berlin Heidelberg 2013

be undecidable (in fact, **co-r.e.**-hard) even when x is either 0 or 1 [15]. Determining the exact level of undecidability was posed as a problem in Bertoni's original paper and has remained open until now. We show that this problem is Σ_2^0-complete (Theorem 1).

Next, we consider the emptiness and universality problems for PFAs with isolated cut-points. The decidability of these problems is still open. We conjecture that these problems are undecidable. Our belief in this result stems from the undecidability of the problem of determining if a cut-point is isolated. Also, Condon-Lipton's [13] undecidability proof for the emptiness problem for PFAs (without the cut-point being necessarily isolated) can be modified to establish the undecidability of the emptiness problem for PFAs with semi-isolated cut-points — x is a semi-isolated cut-point for PFA $\mathcal{B}$ and initial distribution μ if there is an $\epsilon > 0$ such that every input is either accepted with probability at most x or with probability at least $x + \epsilon$. Thus, if x is semi-isolated then $x + \epsilon/2$ is an isolated cut-point (for some unknown ϵ).

Given our belief in the undecidability of the emptiness and universality problems for general PFAs with isolated cut-points, we consider restricted classes of PFAs. Ergodicity and weak ergodicity have played an important role in the study of Markov Chains and non-homogeneous Markov Chains, and have been considered in the context of PFAs in the past [22,19,17,7]. Recall that a Markov Chain is ergodic if its transition graph forms an aperiodic, strongly connected component. Weak ergodicity for non-homogeneous Markov Chains means that any sequence of input symbols has only one terminal strongly connected component and this component is aperiodic. In this paper we generalize both ergodic and weakly ergodic PFAs to define a new class that we call *eventually weakly ergodic* (see Definition 4). Informally, these are PFAs such that the states can be partitioned into sets $Q_T, Q_1, \ldots Q_r$ and there is an ℓ such that in the transition graph on any word of length ℓ, $Q_1, \ldots Q_r$ are the terminal strongly connected components, and these are aperiodic. Any state in Q_T has a non-zero probability of reaching some state in $\cup_i Q_i$ on any word of length ℓ. Note that any Markov chain is eventually weakly ergodic (see Proposition 1).

There are several natural classes of systems that can be modeled as eventually weakly ergodic PFAs. One such class of protocols is randomized leadership election protocols in which a leader is elected amongst a set of "equally" likely candidates. Such a protocol usually proceeds in rounds until a leader is elected. Once a leader is elected the protocol stops and each "elected" choice forms a closed communicating class. Furthermore, leadership election protocols normally ensure that there is a constant number k, such that in every k rounds the probability that a leader is elected is > 0.

Another class of systems relates to "Dolev-Yao" modeling of probabilistic security protocols such as probabilistic anonymity protocols. In this setting, protocol participants are modeled as processes that can send and receive messages, and the communication is mediated through an attacker than can intercept messages, inject and modify messages. The attacker keeps track of the messages exchanged and *nondeterministically* chooses to send new messages to protocol participants. An "attack" is a particular resolution of the nondeterministic choices of the attacker, and protocol analysis checks for security under *every* possible attack. For a faithful analysis [14,12,6,16,11,10], we have to consider *view-consistent* [10] attack strategies in which, at any instance, the attacker must do the same actions in all computations in which its view is the same upto that instance. Under suitable bounds (memory of the attacker, message size and number of sessions), we can model the resulting system as a PFA: an input letter being a "view-consistent" function from the (bounded) view of the attacker to the set of its possible choices. The resulting PFA is also likely to satisfy eventual weak ergodicity because there will often be a constant number k such that each session finishes within k-steps with probability > 0.

We establish the following two results for eventually weakly ergodic PFAs: (a) the problem of determining if x is isolated is **r.e.**-complete (as opposed to Σ_2^0-complete for general PFAs) (Theorem 3), and (b) $\mathsf{L}_{>x}(\mathcal{B}, \mu)$ is regular and can be computed, if x is isolated (Theorem 2). The second observation allows us to conclude that the emptiness and universality problems for eventually weakly ergodic PFAs with isolated cut-points is decidable. These results are useful when we know that x is an isolated cut point and we want to know whether at least one string is accepted with probability $> x$. Note that if the cut-point is not isolated, the emptiness and universality problems for such special PFAs is undecidable (Proposition 3).

Related Work. As already mentioned above, the problem of checking emptiness of PFAs is undecidable [13]. For weakly ergodic PFAs, [7] shows that the problem of checking emptiness/universality is decidable under the assumption that the cut-point is isolated. The class of eventually weakly ergodic PFAs is a strict superset of weakly ergodic PFAs (See Example 2). Hence that result does not apply to our setting. Furthermore, the proof of that result relies on the existence of a unique compact non-empty set of distributions W which is invariant on the set of inputs (that is $W = \{\mu\delta_a \mid \mu \in W, \delta_a$ is the transition matrix on the input $a\}$). Eventually weakly ergodic matrices do not enjoy these properties and we have to appeal to different proof methods.

A decidability result under the assumption of isolation is also obtained in [18]. However, our results are incomparable to the results in [18]. They consider contracting PFAs which are different from eventually weakly ergodic matrices (see Remark 1 on Page 260). Furthermore, they only consider emptiness/universality problem relative to restricted sets of inputs (and not over the whole language).

2   Preliminaries

We assume that the reader is familiar with regular languages and basic measure theory. We will also assume that the reader is familiar with the basic theory of Markov Chains. The set of natural numbers will be denoted by $\mathbb{N}$. The powerset of any set A will be denoted by 2^A. Given any set Σ, $\Sigma^*(\Sigma^+$ respectively) will denote the set of finite words (nonempty finite words respectively) over Σ. A set $\mathsf{L} \subseteq \Sigma^*$ is said to be a language over Σ. Given $\rho \in \Sigma^*$, $|\rho|$ will denote the length of ρ. Given $\ell \in \mathbb{N}$, Σ^ℓ will denote the set $\{u \in \Sigma^* \mid |u| = \ell\}$ and $\Sigma^{<\ell}$ will denote the set $\{u \in \Sigma^* \mid |u| < \ell\}$.

2.1   Arithmetical Hierarchy

Let Δ be a finite alphabet. A language L over Δ is a set of finite strings over Δ. Arithmetical hierarchy consists of classes of languages Σ_n^0, Π_n^0 for each integer $n > 0$.

Fix an $n > 0$. A language $L \in \Sigma_n^0$ iff there exists a recursive predicate $\phi(u, \boldsymbol{x}_1, ..., \boldsymbol{x}_n)$ where u is a variable ranging over Δ^*, and for each i, $0 < i \leq n$, $\boldsymbol{x}_i$ is a finite sequence of variables ranging over integers such that

$$L \; = \; \{u \in \Delta^* \mid \exists \boldsymbol{x}_1, \forall \boldsymbol{x}_2, \ldots, Q_n \boldsymbol{x}_n \; \phi(u, \boldsymbol{x}_1, ..., \boldsymbol{x}_n)\}$$

where Q_n is an existential quantifier if n is odd, else it is a universal quantifier. Note that the quantifiers in the above equation are alternating starting with an existential quantifier. The class $\boldsymbol{\Pi}_n^0$ is exactly the class of languages that are complements of languages in Σ_n^0. Σ_1^0, $\boldsymbol{\Pi}_1^0$ are exactly the class of **R.E.**-sets and **co-R.E.**-sets. Let $\mathcal{C}$ be a class in the arithmetic hierarchy. $L \in \mathcal{C}$ is said to be $\mathcal{C}$-complete if for every $L' \in \mathcal{C}$ there is a computable function f such that $x \in L'$ iff $f(x) \in L$. A well known Σ_2^0-complete language is the set of deterministic Turing machine encodings that halt on finitely many inputs.

2.2 Distributions and Stochastic Matrices

Distributions. A *probability distribution* over a *finite* set Q is a map $\mu : Q \to [0, 1]$ s.t. $\sum_{q \in Q} \mu(q) = 1$. For $Q' \subseteq Q$, we shall write $\mu(Q')$ for $\sum_{q \in Q'} \mu(q)$. $\mathrm{Dist}(Q)$ will denote the set of all distributions over Q. The map $d : \mathrm{Dist}(Q) \times \mathrm{Dist}(Q) \to [0, 1]$ defined as

$$d(\mu, \nu) = \frac{\sum_{q \in Q} |\mu(q) - \nu(q)|}{2} = \max_{Q' \subseteq Q} |\mu(Q') - \nu(Q')|$$

defines a metric on the set $\mathrm{Dist}(Q)$. Note that $d(\mu, \nu) \leq 1$. Unless otherwise stated, we assume that $\mu(q)$ is a rational number.

Stochastic Matrices. A stochastic matrix over a *finite* set Q is a matrix $\delta : Q \times Q \to [0, 1]$ s.t. $\forall q \in Q$. $\sum_{q' \in Q} \delta(q, q') = 1$. $\mathrm{Mat}_{=1}(Q)$ will denote the set of all stochastic matrices over the set Q. For $\delta \in \mathrm{Mat}_{=1}(Q)$ and $\mu \in \mathrm{Dist}(Q)$, $\mu\delta$ denotes the distribution, given by $\mu\delta(q) = \sum_{q' \in Q} \mu(q')\delta(q', q)$. Unless otherwise stated, we assume that $\delta(q, q')$ is a rational number. Given $\delta_1, \delta_2 \in \mathrm{Mat}_{=1}(Q)$, we write $\delta_1\delta_2$ to denote the matrix product of δ_1 and δ_2 and we write $(\delta_1)^\ell$ to denote the ℓ-times product of δ_1.

Given a state $q \in Q$, and a matrix $\delta \in \mathrm{Mat}_{=1}(Q)$, we write $\mathrm{post}(q, \delta) = \{q' \mid \delta(q, q') > 0\}$. Given $Q' \subseteq Q$, we write $\mathrm{post}(Q', \delta) = \cup_{q \in Q'} \mathrm{post}(q, \delta)$. $Q' \subseteq Q$ is said to be *closed for* δ if $\mathrm{post}(Q', \delta) \subseteq Q'$. It is easy to see that if Q' is closed for δ, then the matrix $\delta|_{Q'}$ obtained by restricting δ to $Q' \times Q'$ is a stochastic matrix over Q'. Given $\Delta \subseteq \mathrm{Mat}_{=1}(Q)$, $Q' \subseteq Q$ is said to be *closed for* Δ if Q' is closed for each $\delta \in \Delta$. If Q' is closed for Δ, we let $\Delta|_{Q'} = \{\delta|_{Q'} \mid \delta \in \Delta\}$.

$\delta \in \mathrm{Mat}_{=1}(Q)$ is said to be *irreducible* if for each $q, q' \in Q$, there is an $\ell > 0$ s.t. $\delta^\ell(q, q') > 0$. The *period* of q, written $period_\delta(q)$, is defined to be the the greatest common divisor of $\{j \mid \delta^j(q, q) > 0\}$. δ is said to be *aperiodic* if for every $q \in Q$, $period_\delta(q) = 1$. δ is said to be *ergodic* if it is aperiodic and irreducible.

Markov chains. A Markov chain $\mathcal{M}$ is a tuple (Q, δ, μ) s.t. Q is a finite set of *states*, $\delta \in \mathrm{Mat}_{=1}(Q)$ and an *initial distribution* $\mu \in \mathrm{Dist}(Q)$. A Markov chain defines a sequence of distributions $\mu_0, \mu_1, \cdots$ where $\mu_i = \mu\delta^i$.

2.3 Probabilistic Finite Automata

A probabilistic finite automaton [20,21] is like a deterministic automaton except that the transition function from a state on a given input is described as a probability distribution that determines the probability of transitioning to the next state.

Definition 1. *A Probabilistic Automaton (PFA) is a tuple $\mathcal{B} = (\Sigma, Q, Q_f, \Delta = \{\delta_a\}_{a\in\Sigma})$ where Σ is a finite nonempty set of input symbols and is called the input alphabet, Q is a finite set of states, $Q_f \subseteq Q$ is the set of accepting/final states and $\Delta = \{\delta_a\}_{a\in\Sigma}$ is a collection of stochastic matrices, one each for each input letter a.*

Notation: Given a PFA $\mathcal{B} = (\Sigma, Q, Q_f, \Delta = \{\delta_a\}_{a\in\Sigma})$ and a word $u = a_0 \cdots a_m \in \Sigma^*$, we denote the matrix $\delta_{a_0} \cdots \delta_{a_m}$ by δ_u. If u is the empty word then δ_u shall denote the identity matrix. Given a nonempty set $\Sigma_1 \subseteq \Sigma$, we denote the set of matrices $\{\delta_a\}_{a\in\Sigma_1}$ by Δ_{Σ_1}.

Language of a PFA. Language of a PFA is defined relative to an initial distribution and a cut-point. Formally, given a PFA $\mathcal{B} = (\Sigma, Q, Q_f, \Delta = \{\delta_a\}_{a\in\Sigma})$, an *initial distribution* $\mu \in \text{Dist}(Q)$ and $v \in \Sigma^*$, the quantity $\mu\delta_v(Q_f)$ is called the *probability of $\mathcal{B}$ accepting v when started in μ* and shall be denoted by $\text{Pr}^{acc}_{\mathcal{B},\mu}(v)$. For $x \in [0,1]$, the set of words

$$\mathsf{L}_{>x}(\mathcal{B}, \mu) = \{v \in \Sigma^* \mid \text{Pr}^{acc}_{\mathcal{B},\mu}(v) > x\}$$

is said to be the *language accepted by $\mathcal{B}$ with initial distribution μ and cut-point x.*

PFAs can recognize non-regular languages [21]. Furthermore, the problem of deciding emptiness and universality respectively for PFAs is undecidable [20,13].

2.4 Isolated Cut-Points

A much celebrated result of PFAs concerns isolated cut-points. A cut-point x is said to be isolated if there is an ϵ such that every word is either accepted with probability at least $x + \epsilon$ or accepted with probability at most $x - \epsilon$. Formally,

Definition 2. *Given a PFA $\mathcal{B} = (\Sigma, Q, Q_f, \Delta = \{\delta_a\}_{a\in\Sigma})$ and an initial distribution μ, x is said to be an isolated cut-point for $(\mathcal{B}, \mu)$ with a degree of isolation $\epsilon > 0$ if for each $v \in \Sigma^*$, $|\text{Pr}^{acc}_{\mathcal{B},\mu}(v) - x| > \epsilon$. x is said to be an isolated cut-point for $(\mathcal{B}, \mu)$ if there is an $\epsilon > 0$ s.t. x is an isolated cut-point for $(\mathcal{B}, \mu)$ with a degree of isolation ϵ.*

A famous result of Rabin [21] says that if x is an isolated cut-point for $(\mathcal{B}, \mu)$ then $\mathsf{L}_{>x}(\mathcal{B}, \mu)$ is a regular language. This fact raises two interesting questions.

The first one asks if there is an algorithm that decides given a PFA $\mathcal{B}$, an initial distribution μ and a cut-point $x \in [0, 1]$, whether x is an isolated cut-point for $(\mathcal{B}, \mu)$ or not. Bertoni [4,5] showed that the problem is undecidable when $x \in (0, 1)$. A close examination of the proof reveals that this problem is **r.e.**-hard. Recently, Gimbert and Oualhadj [15] showed that the problem remains undecidable even when x is 0 or 1. A close examination of their proof reveals that this problem is **co-r.e.**-hard also. However, the exact level of undecidability of this problem remained open.

The second question asks if there is a decision procedure that given a PFA $\mathcal{B}$, an initial distribution μ and a cut-point $x \in [0, 1]$ isolated for $(\mathcal{B}, \mu)$ decides whether the language $L_{>x}(\mathcal{B}, \mu)$ is empty or not. This problem seems to be less studied in literature. A close examination of Rabin's proof shows that if a degree of isolation ϵ is known, then $L_{>x}(\mathcal{B}, \mu)$ can be computed as the proof computes an upper bound on the number of states of the deterministic automaton recognizing $L_{>x}(\mathcal{B}, \mu)$ in terms of ϵ and the number of states of $\mathcal{B}$. However, the status of the problem when the degree of isolation is not known, remains open.

3 Checking Isolation in PFAs

The following theorem states that the problem of checking whether a given x is isolated for a given PFA A is Σ_2^0-complete, thus settling the open problem posed in [4,5] (please note that the results also apply when x is 0 or 1).

Theorem 1. *Given a PFA $\mathcal{B} = (\Sigma, Q, Q_f, \Delta = \{\delta_a\}_{a \in \Sigma})$, an initial distribution $\mu \in \mathrm{Dist}(Q)$ and rational $x \in [0, 1]$, the problem of checking if x is an isolated cut-point for $(\mathcal{B}, \mu)$ is Σ_2^0-complete.*

Proof. Please note that x is isolated for $(\mathcal{B}, \mu)$ iff

$$\exists n \in \mathbb{N}, n > 0. \, \forall u \in \Sigma^*. |\mathrm{Pr}_{\mathcal{B},\mu}^{acc}(u) - x| > \frac{1}{n}.$$

This demonstrates that the problem of checking if x is isolated for $(\mathcal{B}, \mu)$ is in Σ_2^0.

For the lower bound, please observe that it suffices to show that the problem of checking if 1 is isolated for $(\mathcal{B}, \mu)$ is Σ_2^0-hard. We will demonstrate this by a reduction from emptiness checking problem of Probabilistic Büchi Automata (PBA)s [2]. A PBA $\mathcal{B}' = (\Sigma', Q', Q'_f, \Delta' = \{\delta'_a\}_{a \in \Sigma'})$ is like a PFA except that it is used to define languages over infinite words. Given a PBA $\mathcal{B}'$ and an initial distribution μ', $\mathcal{L}_{>0}(\mathcal{B}', \mu') \subseteq \Sigma^\omega$ denotes the set of infinite words accepted by $\mathcal{B}'$ with probability > 0. Intuitively, $\mathcal{B}$ accepts an infinite word α with probability > 0 if on input α, the measure of all (infinite) paths that visit Q'_f infinitely often is > 0. The exact definition of what it means for a PBA to accept an infinite word α with probability > 0 is beyond the scope of the paper and the interested reader is referred to [1]. We recall the necessary results.

We had shown the following problem to be Σ_2^0-complete in [9]: Given a PBA $\mathcal{B}'$ and an initial distribution μ', check whether $\mathcal{L}_{>0}(\mathcal{B}', \mu') = \emptyset$. We will use this decision problem to establish the lower bound result.

Given a PBA, $\mathcal{B}' = (\Sigma', Q', Q'_f, \Delta' = \{\delta'_a\}_{a \in \Sigma'})$, an initial distribution μ' and $Q'' \subseteq Q'$, let $reachable(Q'')$ be the predicate $(\exists u \in \Sigma^*. (\mu' \delta_u)(Q'') > 0)$. Given $q \in Q$ and $v \in \Sigma^+$, let $\mathrm{Pr}_{q,v}^{Q'_f}(Q'')$ be the probability that the PFA $\mathcal{B}'$, on input v, when started in q reaches Q'' after passing through a state in Q'_f.

We had shown in [9] that $\mathcal{L}_{>0}(\mathcal{B}', \mu') \neq \emptyset$ iff

$\exists Q'' \subseteq Q'. (reachable(Q'')$ and

$\quad (\forall n \in \mathbb{N}, n > 0. \, \exists v \in \Sigma^+. \, \forall q \in Q''. \, \mathrm{Pr}_{q,v}^{Q'_f}(Q'') > 1 - \frac{1}{2^n}))$.

Observe first that the predicate $reachable(\cdot)$ is a recursive predicate. Now, pick a new element $\dagger$ not in Q' and for all subsets $Q'' \subseteq Q'$ s.t. $reachable(Q'')$ is true, construct a PFA $\mathcal{B}_{Q''}$ and an initial distribution $\mu_{Q''}$ as follows. The input alphabet is Σ'. The states of $\mathcal{B}_{Q''}$ are $Q' \cup (Q' \times \{\dagger\})$. The set of final states of $\mathcal{B}_{Q''}$ are $Q'' \times \{\dagger\}$. The set of transitions $\Delta'' = \{\delta_a''\}_{a \in \Sigma'}$ is as follows. For each $a \in \Sigma'$:

- $\delta_a''(q_1, q_2) = \delta_a'(q_1, q_2)$ if $q_1 \in Q'$ and $q_2 \in Q' \setminus Q_f'$.
- $\delta_a''(q_1, (q_2, \dagger)) = \delta_a'(q_1, q_2)$ if $q_1 \in Q'$ and $q_2 \in Q_f'$.
- $\delta_a''((q_1, \dagger), (q_2, \dagger)) = \delta_a'(q_1, q_2)$ if $q_1, q_2 \in Q'$.

Let $\mu_{Q''}$ be the distribution assigns probability $\frac{1}{|Q''|}$ to each $q \in Q''$ where $|Q''|$ is the number of elements of Q''. It can be easily shown that $\mathcal{L}_{>0}(\mathcal{B}', \mu') = \emptyset$ iff $\forall Q'' \subseteq Q$, $reachable(Q'')$ implies that 1 is an isolated cut-point for $(\mathcal{B}_{Q''}, \mu_{Q''})$.

While the reduction above is a truth-table reduction, note that by taking "disjoint" union of the PFAs $\mathcal{B}_{Q''}$, adding a new initial state, a new reject state and new input symbols, we can easily construct a many-to-one reduction. The result now follows.   □

Remark 1. We can conclude from the proof of Theorem 1 that the problem of checking whether 1 is an isolated cut-point for a PFA A is equivalent to the problem of checking whether a PBA $\mathcal{B}$ accepts an infinite word with probability > 0. The proof of Theorem 1 establishes one side of this equivalence and the converse is established in [3] (see Remark 5.7 on Page 41).

4   Weak Ergodicity and Eventually Weak Ergodicity

Ergodicity is an important concept that is useful in the study of stochastic matrices. We will recall this notion shortly and its extension to sets of stochastic matrices. We shall need one notation: Given a nonempty finite set $\Delta \subseteq \mathsf{Mat}_{=1}(Q)$ of stochastic matrices and $\ell > 0$, let $\Delta^\ell = \{\delta_1 \delta_2 \cdots \delta_\ell \mid \delta_i \in \Delta\}$.

Recall that a stochastic matrix is *ergodic* if it is irreducible and aperiodic. A Markov chain $\mathcal{M} = (Q, \delta, \mu)$ is ergodic if the matrix δ is ergodic. Ergodic chains have a special property that they converge to a *unique stationary distribution* in the limit irrespective of the starting distribution. More generally, this fact generalizes to Markov chains that a) have a single closed communicating class and b) this class is aperiodic. The notion of ergodicity has been extended to sets of stochastic matrices [22,19,17] and such sets are called *weakly ergodic sets*. Analogous to the convergence to the stationary distribution, if Δ is weakly ergodic then for any "long enough sequence" $\delta_1, \cdots, \delta_\ell$, any two distributions $\mu_1 \delta_1 \cdots \delta_\ell, \mu_2 \delta_1 \cdots \delta_\ell$ are "very close." We recall the formal definition of weakly ergodic matrices introduced in [7].

Definition 3. *A finite set of stochastic matrices Δ over a finite state space Q is said to be strongly semi-regular if for each $\delta \in \Delta$ there is a state q_δ s.t. for each $q \in Q$ $\delta(q, q_\delta) > 0$. Δ is weakly ergodic if there is an $\ell > 0$ s.t. Δ^ℓ is strongly semi-regular. A PFA $\mathcal{B} = (\Sigma, Q, Q_f, \Delta = \{\delta_a\}_{a \in \Sigma})$ is weakly ergodic if Δ is weakly ergodic.*

Example 1. Consider the 2-element set Δ shown in Figure 1.a). Δ can be seen to be strongly semi-regular and hence weakly ergodic as follows: the transition represented by solid edges always "hits" the state H and the transition corresponding to the dashed edges always "hits" the state G.

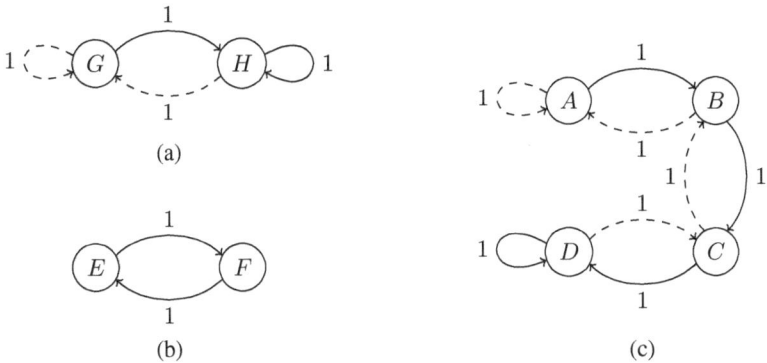

(a)

(b)

(c)

Fig. 1. a) Δ shown is weakly ergodic. There are two matrices in Δ: transitions of first one are shown as solid edges, while transitions of second one are shown as dashed edges. b) Δ is singleton and hence is eventually weakly ergodic. Δ is not weakly ergodic. c) Δ shown is not eventually weakly ergodic. There are two matrices in Δ: transitions of first one are shown as solid edges, while transitions of second one are shown as dashed edges.

Remark 2. There are several other equivalent definitions of weakly ergodic set of matrices. For example, one formulation [22] says that a finite set of matrices is weakly ergodic if every finite product of matrices has only one closed communicating class that is irreducible and aperiodic. There is an algorithm [22,19] that given a nonempty, finite set $\Delta \subseteq \mathsf{Mat}_{=1}(Q)$ checks if Δ is weakly ergodic or not.

Eventually weakly ergodic sets. Even when stochastic matrices are not ergodic, the notion of ergodicity proves useful for analysis of Markov chains. This is because for any stochastic matrix δ, there is an $\ell > 0$ such that Q can be written as a disjoint sum $Q = Q_T \cup Q'_1 \cdots Q'_m$ where Q'_j is an aperiodic, closed communicating class for δ^ℓ and Q_T is the set of transient states for δ^ℓ. This observation motivates the following:

Definition 4. $\Delta \subseteq \mathsf{Mat}_{=1}(Q)$ *is said to be eventually weakly ergodic if there is a partition* $Q_T, Q_1, \ldots Q_r$ *of Q and a natural number $\ell > 0$ s.t. for each $1 \leq i \leq r$ the following conditions hold–*

- *Q_i is closed for Δ^ℓ.*
- *$\Delta^\ell|_{Q_i}$ is strongly semi-regular.*
- *For each $q \in Q_T$ and each $\delta \in \Delta^\ell$, $\mathsf{post}(q, \delta) \cap (\cup_{1 \leq j \leq r} Q_j) \neq \emptyset$.*

The tuple $(\ell, Q_T, (Q_0, \ldots, Q_r))$ is said to be a witness of eventual weak ergodicity. A PFA $\mathcal{B} = (\Sigma, Q, Q_f, \Delta = \{\delta_a\}_{a \in \Sigma})$ is eventually weakly ergodic if Δ is eventually weakly ergodic.

Remark 3. Please note that we are requiring that each Q_i be closed for Δ^ℓ and not for Δ. This means that Q_i is closed for all $\Delta^{\ell'}$ s.t. ℓ' is a multiple of ℓ. For ℓ' which is not a multiple of ℓ, a $\delta \in \Delta^{\ell'}$ may take a state in Q_i to a state not in Q_i. For example, consider the singleton Δ in Figure 1.b). Note that Δ is eventually weakly ergodic (but not weakly ergodic) with witness $(2, \emptyset, (\{E\}, \{F\}))$. Note that the sets $\{E\}$ and $\{F\}$ are closed for $\Delta^{\ell'}$ iff ℓ' is even.

As expected, each singleton turns out to be eventually weakly ergodic.

Proposition 1. *If $\Delta \subseteq \mathsf{Mat}_{=1}(Q)$ and $|\Delta| = 1$ then Δ is eventually weakly ergodic.*

Example 2. Observe that the Δ shown in Figure 1.c) is not eventually weakly ergodic. This can be seen as follows. For each ℓ, the only closed class of Δ^ℓ is the set of all states. Let δ_{solid} and δ_{dashed} be the matrices shown with solid lines and dashed lines respectively. For any $\delta \in \delta^\ell$ of the form $\delta_{solid}\delta_{dashed}\delta_{solid}\delta_{dashed} \cdots$, $\mathrm{post}(\delta, \{B\}) \cap \mathrm{post}(\delta, \{C\}) = \emptyset$. On the other hand, Δ shown in Figure 1.b) is eventually weakly ergodic with $(2, \emptyset, (\{E\}, \{F\}))$ as a witness of eventual weak ergodicity.

Remark 4. The *contracting* PFAs considered in [18] are different from eventually weakly ergodic matrices. Contracting PFAs are PFAs in which each transition matrix has only one closed communicating class which is aperiodic. The set Δ in Figure (1.c) is not eventually weakly ergodic but is contracting and the set Δ in Figure (1.b) is eventually weakly ergodic but not contracting.

The algorithm for checking whether a finite set of matrices are weakly ergodic can be extended to checking whether a finite set of matrices is eventually weakly ergodic.

Proposition 2. *The problem of checking given nonempty, finite set $\Delta \subseteq \mathsf{Mat}_{=1}(Q)$, whether Δ is eventually weakly ergodic is decidable. Furthermore, if Δ is eventually weakly ergodic then a witness of eventual weak ergodicity can be computed.*

5   Decision Problems for Eventually Weakly Ergodic PFAs

We shall now study the problems of deciding emptiness and isolation for eventually weakly ergodic PFAs. We start by discussing the emptiness problem.

5.1   Emptiness/Universality Checking for Eventually Weakly Ergodic PFAs

The problem of checking emptiness of a PFA is undecidable [20,13]. The problem continues to remain undecidable for eventually weakly ergodic PFAs. The proof of undecidability is similar to the one used in [8] to show that the problem of checking emptiness of *finite probabilistic monitors* is undecidable.

Proposition 3. *The following problems are undecidable: Given an eventually weakly ergodic PFA $\mathcal{B} = (\Sigma, Q, Q_f, \Delta = \{\delta_a\}_{a \in \Sigma})$, an initial distribution $\mu \in \mathrm{Dist}(Q)$ and rational $x \in (0, 1)$ check whether a) $\mathsf{L}_{>x}(\mathcal{B}, \mu) = \emptyset$ and whether b) $\mathsf{L}_{>x}(\mathcal{B}, \mu) = \Sigma^*$.*

Therefore, it follows that the *syntactic restriction* of eventually weak ergodicity is not enough for deciding emptiness of probabilistic automata. However, we will show that the problem of checking emptiness becomes decidable under the promise that the cut-point is isolated. In order to establish this result, we shall first establish a useful lemma (Lemma 1). In order to state this lemma, we need one auxiliary definition. Note that by the notation $m|n$ we mean that m divides n.

Definition 5. *Given an alphabet Σ and natural numbers $\ell, \ell' > 0$ such that $\ell'|\ell$, let $c_{(\ell,\ell')} : \Sigma^* \to \Sigma^*$ be defined as follows.*

$$c_{(\ell,\ell')}(u) = \begin{cases} u & \text{if } |u| < \ell' + 2\ell; \\ u_0 u_1 v_1 & \text{if } u = u_0 u_1 w v_1, |u_0| < \ell', |u_1| = \ell, w \in (\Sigma^{\ell'})^+ \text{ and } |v_1| = \ell \end{cases}$$

Informally, given ℓ, ℓ' and Σ such that $\ell'|\ell$, the function $c_{(\ell,\ell')}(\cdot)$ works as follows. If u is a string whose length is a multiple of ℓ', then $c_{(\ell,\ell')}(u)$ keeps the prefix of length ℓ of u and the suffix of the length ℓ of u and "cuts" away the rest of the string. If the length of the u is not a multiple of ℓ' then it selects the largest suffix whose length is a multiple of ℓ' and applies $c_{(\ell,\ell')}(\cdot)$ to it.

The following lemma states that if $\mathcal{B}$ is eventually weakly ergodic then the distribution obtained by inputting a word in Σ^* is determined up to a given ϵ by the initial distribution and an appropriate "cut" (which depends only on $\mathcal{B}$ and ϵ).

Lemma 1. *Given an eventually weakly ergodic PFA $\mathcal{B} = (\Sigma, Q, Q_f, \Delta = \{\delta_a\}_{a\in\Sigma})$ and $\epsilon > 0$, there are $\ell > 0$ and $\ell' > 0$ s.t. $\ell'|\ell$ and*

$$\forall \mu_0 \in \text{Dist}(Q) . \forall u \in \Sigma^* . d(\mu_0 \delta_u, \mu_0 \delta_{c_{(\ell,\ell')}(u)}) < \epsilon.$$

Furthermore, if ϵ is rational then ℓ, ℓ' can be computed from $\mathcal{B}$ and ϵ.

We shall now show that if $\mathcal{B}$ is eventually weakly ergodic and x is an isolated cut point for $(\mathcal{B}, \mu)$ then there is an algorithm that computes the regular language $L_{>x}(\mathcal{B}, \mu)$. Recall that regularity is a consequence of Rabin's theorem on isolated cut-points (see Section 2.4). Indeed, as observed in Section 2.4, Rabin's proof can also be used to compute the regular language, provided we can compute a degree of isolation. The eventually weak ergodicity condition allows us to compute this.

Lemma 2. *There is a procedure that given an eventually weakly ergodic PFA $\mathcal{B} = (\Sigma, Q, Q_f, \Delta = \{\delta_a\}_{a\in\Sigma})$, a distribution $\mu \in \text{Dist}(Q)$ and a rational $x \in [0,1]$ such that x is an isolated cut-point for $(\mathcal{B}, \mu)$ terminates and outputs $\epsilon > 0$ such that ϵ is a degree of isolation.*

Proof. Consider the procedure in Figure 2. Thanks to Lemma 1, if the procedure terminates then the ϵ returned is a degree of isolation. Hence, it suffices to show that the procedure terminates if x is an isolated cut-point for $(\mathcal{B}, \mu)$. Let ϵ_0 be a degree of isolation and fix it. Thus, for all $u \in \Sigma^*$, $|\text{Pr}^{acc}_{\mathcal{B},\mu}(u) - x| > \epsilon_0$. Let $\epsilon^{(n)}$ be the value of variable ϵ at the beginning of the nth unrolling of the while loop. As long as $isolation_{found}$ is false, $\epsilon^{(n)} = \frac{1}{2^n}$. Let $N_0 = \lfloor \log_2 \epsilon_0 \rfloor + 1$. As $\epsilon^{(N_0)} < \epsilon_0$ and $\forall u \in \Sigma^*$, $|\text{Pr}^{acc}_{\mathcal{B},\mu}(u) - x| > \epsilon_0$, $isolation_{found}$ must become true at the end of N_0th unrolling, if not before. $\square$

Lemma 2 along with the proof of Rabin's theorem on isolated cut-points can now be used to establish our main result.

Theorem 2. *There is a procedure that given an eventually weakly ergodic PFA $\mathcal{B} = (\Sigma, Q, Q_f, \Delta = \{\delta_a\}_{a\in\Sigma})$, a distribution $\mu \in \text{Dist}(Q)$ and a rational $x \in [0,1]$*

Input: $\mathcal{B}, \mu, x$ where
$\quad\quad \mathcal{B} = (\Sigma, Q, Q_f, \Delta = \{\delta_a\}_{a \in \Sigma})$ is an eventually weakly ergodic PFA,
$\quad\quad \mu \in \mathrm{Dist}(Q)$ is the initial distribution and $x \in [0, 1]$ is a rational number
{
$isolation_{found} := false;$
$\epsilon := \frac{1}{2};$
while $not(isolation_{found})$
$\quad$ do
$\quad\quad$ Compute $\ell, \ell' > 0$ such that
$\quad\quad\quad \forall \mu_0 \in \mathrm{Dist}(Q). \forall u \in \Sigma^*. \ d(\mu_0 \delta_u, \mu_0 \delta_{c_{(\ell, \ell')}(u)}) < \epsilon;$
$\quad\quad curr_{isolation} := \min_{v \in (\Sigma) < 2\ell + \ell'} |\mu \delta_v(Q_f) - x|;$
$\quad\quad$ If $curr_{isolation} \le \epsilon$ then $\epsilon := \frac{\epsilon}{2}$
$\quad\quad\quad$ else $\{\epsilon := curr_{isolation} - \epsilon; \ isolation_{found} = true; \}$
$\quad$ od;
$\quad$ return(ϵ);
}

Fig. 2. Procedure for computing the degree of isolation

such that x is an isolated cut-point for $(\mathcal{B}, \mu)$ terminates and outputs the regular language $\mathsf{L}_{>x}(\mathcal{B}, \mu)$. Therefore, the following problems are decidable: Given an eventually
weakly ergodic PFA $\mathcal{B} = (\Sigma, Q, Q_f, \Delta = \{\delta_a\}_{a \in \Sigma})$, a distribution $\mu \in \mathrm{Dist}(Q)$ and
rational $x \in [0, 1]$ s.t. x is an isolated cut-point, a) check whether $\mathsf{L}_{>x}(\mathcal{B}, \mu) = \emptyset$ and
b) check whether $\mathsf{L}_{>x}(\mathcal{B}, \mu) = \Sigma^*$.

5.2 Checking Isolation for Weakly Ergodic PFAs

A close examination of the proof of undecidability of checking isolation in PFAs given
in Bertoni [4,5] reveals that the problem of checking isolation is **r.e.**-hard even for
eventually weakly ergodic automata. Furthermore, if we run the procedure in Lemma 2
on an arbitrary (i.e., not necessarily isolated) cut-point x then the procedure terminates
if and only if x is an isolated cut-point, implying that checking isolation is in **r.e.**.

Theorem 3. *The following problem is **r.e.**-complete: Given an eventually weakly er-
godic PFA $\mathcal{B} = (\Sigma, Q, Q_f, \Delta = \{\delta_a\}_{a \in \Sigma})$, a distribution $\mu \in \mathrm{Dist}(Q)$ and rational
x, check if x is isolated for $(\mathcal{B}, \mu)$.*

6 Conclusions

We have established the exact level of undecidability of checking if a given threshold x
is an isolated cut point for a given PFA A, showing it to be Σ_2^0-complete. We have also
proved decidability of non-emptiness (and universality) for eventually weakly ergodic
automata, given that the automaton has an isolated cut-point. The problem of decid-
ability/undecidability of checking non-emptiness for arbitrary PFAs with isolated cut
points is still an open problem.

References

1. Baier, C., Bertrand, N., Größer, M.: On decision problems for probabilistic Büchi automata. In: Amadio, R.M. (ed.) FOSSACS 2008. LNCS, vol. 4962, pp. 287–301. Springer, Heidelberg (2008)
2. Baier, C., Größer, M.: Recognizing ω-regular languages with probabilistic automata. In: Proceedings of LICS, pp. 137–146 (2005)
3. Baier, C., Größer, M., Bertrand, N.: Probabilistic ω-automata. J. ACM 59(1), 1 (2012)
4. Bertoni, A.: The solution of problems relative to probabilistic automata in the frame of the formal languages theory. In: Siefkes, D. (ed.) GI 1974. LNCS, vol. 26, pp. 107–112. Springer, Heidelberg (1975)
5. Bertoni, A., Mauri, G., Torelli, M.: Some recursive unsolvable problems relating to isolated cutpoints in probabilistic automata. In: Salomaa, A., Steinby, M. (eds.) ICALP 1977. LNCS, vol. 52, pp. 87–94. Springer, Heidelberg (1977)
6. Canetti, R., Cheung, L., Kaynar, D., Liskov, M., Lynch, N., Pereira, P., Segala, R.: Task-Structured Probabilistic I/O Automata. In: Workshop on Discrete Event Systems (2006)
7. Chadha, R., Korthikranthi, V.A., Viswanathan, M., Agha, G., Kwon, Y.: Model checking mdps with a unique compact invariant set of distributions. In: Proceedings of QEST, pp. 121–130 (2011)
8. Chadha, R., Sistla, A.P., Viswanathan, M.: On the expressiveness and complexity of randomization in finite state monitors. Journal of the ACM 56(5) (2009)
9. Chadha, R., Sistla, A.P., Viswanathan, M.: Power of randomization in automata on infinite strings. In: Bravetti, M., Zavattaro, G. (eds.) CONCUR 2009. LNCS, vol. 5710, pp. 229–243. Springer, Heidelberg (2009)
10. Chadha, R., Sistla, A.P., Viswanathan, M.: Model checking concurrent programs with nondeterminism and randomization. In: Proceedings of FSTTCS, pp. 364–375 (2010)
11. Chatzikokolakis, K., Palamidessi, C.: Making Random Choices Invisible to the Scheduler. Information and Computation 208(6), 694–715 (2010)
12. Cheung, L.: Reconciling Nondeterministic and Probabilistic Choices. PhD thesis, Radboud University of Nijmegen (2006)
13. Condon, A., Lipton, R.J.: On the complexity of space bounded interactive proofs (extended abstract). In: Proceedings of FOCS, pp. 462–467 (1989)
14. de Alfaro, L.: The Verification of Probabilistic Systems under Memoryless Partial Information Policies is Hard. In: PROBMIV (1999)
15. Gimbert, H., Oualhadj, Y.: Probabilistic automata on finite words: Decidable and undecidable problems. In: Abramsky, S., Gavoille, C., Kirchner, C., Meyer auf der Heide, F., Spirakis, P.G. (eds.) ICALP 2010. LNCS, vol. 6199, pp. 527–538. Springer, Heidelberg (2010)
16. Giro, S., D'Argenio, P.R.: Quantitative model checking revisited: Neither decidable nor approximable. In: Raskin, J.-F., Thiagarajan, P.S. (eds.) FORMATS 2007. LNCS, vol. 4763, pp. 179–194. Springer, Heidelberg (2007)
17. Hajnal, J., Bartlett, M.S.: Weak ergodicity in non-homogeneous markov chains. Mathematical Proceedings of the Cambridge Philosophical Society 54(02), 233–246 (1958)
18. Korthikanti, V.A., Viswanathan, M., Agha, G., Kwon, Y.: Reasoning about mdps as transformers of probability distributions. In: Proceedings of QEST, pp. 199–208 (2010)
19. Paz, A.: Definite and quasidefinite sets of stochastic matrices. Proceedings of the American Mathematical Society 16(4), 634–641 (1965)
20. Paz, A.: Introduction to Probabilistic Automata. Academic Press (1971)
21. Rabin, M.O.: Probabilitic automata. Information and Control 6(3), 230–245 (1963)
22. Wolfowitz, J.: Products of indecomposable, aperiodic, stochastic matrices. Proceedings of the American Mathematical Society 14(5), 733–737 (1963)

On Stochastic Games with Multiple Objectives

Taolue Chen, Vojtěch Forejt, Marta Kwiatkowska,
Aistis Simaitis, and Clemens Wiltsche

Department of Computer Science, University of Oxford, United Kingdom

Abstract. We study two-player stochastic games, where the goal of one player is to satisfy a formula given as a positive boolean combination of expected total reward objectives and the behaviour of the second player is adversarial. Such games are important for modelling, synthesis and verification of open systems with stochastic behaviour. We show that finding a winning strategy is PSPACE-hard in general and undecidable for deterministic strategies. We also prove that optimal strategies, if they exists, may require infinite memory and randomisation. However, when restricted to disjunctions of objectives only, memoryless deterministic strategies suffice, and the problem of deciding whether a winning strategy exists is NP-complete. We also present algorithms to approximate the Pareto sets of achievable objectives for the class of stopping games.

1 Introduction

Stochastic games [20] have many applications in semantics and formal verification, and have been used as abstractions for probabilistic systems [15], and more recently for quantitative verification and synthesis of competitive stochastic systems [8]. Two-player games, in particular, provide a natural representation of open systems, where one player represents the system and the other its environment, in this paper referred to as Player 1 and Player 2, respectively. Stochasticity models uncertainty or randomisation, and leads to a game where each player can select an outgoing edge in states he controls, while in stochastic states the choice is made according to a state-dependent probability distribution. A *strategy* describes which actions a player picks. A fixed pair of strategies and an initial state determines a probability space on the runs of a game, and yields expected values of given objective (payoff) functions. The problem is then to determine if Player 1 has a strategy to ensure that the expected values of the objective functions meet a given set of criteria for all strategies that Player 2 may choose.

Various objective functions have been studied, for example reachability, ω-regular, or parity [4]. We focus here on *reward functions*, which are determined by a reward structure, annotating states with rewards. A prominent example is the reward function evaluating *total reward*, which is obtained by summing up rewards for all states visited along a path. Total rewards can be conveniently used to model consumption of resources along the execution of the system, but (with a straightforward modification of the game) they can also be used to encode other objective functions, such as reachability.

K. Chatterjee and J. Sgall (Eds.): MFCS 2013, LNCS 8087, pp. 266–277, 2013.

Although objective functions can express various useful properties, many situations demand considering not just the value of a single objective function, but rather values of several such functions simultaneously. For example, we may wish to maximise the number of successfully provided services and, at the same time, ensure minimising resource usage. More generally, given multiple objective functions, one may ask whether an arbitrary boolean combination of upper or lower bounds on the expected values of these functions can be ensured (in this paper we restrict only to positive boolean combinations, i.e. we do not allow negations). Alternatively, one might ask to compute or approximate the *Pareto set*, i.e. the set of all bounds that can be assured by exploring trade-offs. The simultaneous optimisation of a conjunction of objectives (also known as multi-objective, multi-criteria or multi-dimensional optimisation) is actively studied in operations research [21] and used in engineering [17]. In verification it has been considered for Markov decision processes (MDPs), which can be seen as one-player stochastic games, for discounted objectives [5] and general ω-regular objectives [10]. Multiple objectives for non-stochastic games have been studied by a number of authors, including in the context of energy games [22] and strategy synthesis [6].

In this paper, we study *stochastic games* with multi-objective queries, which are expressed as positive boolean combinations of total reward functions with upper or lower bounds on the expected reward to be achieved. In that way we can, for example, give several alternatives for a valid system behaviour, such as "the expected consumption of the system is at most 10 units of energy and the probability of successfully finishing the operation is at least 70%, or the expected consumption is at most 50 units, but the probability of success is at least 99%". Another motivation for our work is assume-guarantee compositional verification [19], where the system satisfies a set of guarantees φ whenever a set of assumptions ψ is true. This can be formulated using multi-objective queries of the form $\bigwedge \psi \Rightarrow \bigwedge \varphi$. For MDPs it has been shown how to formulate assume-guarantee rules using multi-objective queries [10]. The results obtained in this paper would enable us to explore the extension to stochastic games.

Contributions. We first obtain nondeterminacy by a straightforward modification of earlier results. Then we prove the following novel results for multi-objective stochastic games:

- We prove that, even in a pure conjunction of objectives, infinite memory and randomisation are required for the winning strategy of Player 1, and that the problem of finding a *deterministic* winning strategy is undecidable.
- For the case of a pure disjunction of objectives, we show that memoryless deterministic strategies are sufficient for Player 1 to win, and we prove that determining the existence of such strategies is an NP-complete problem.
- For the general case, we show that the problem of deciding whether Player 1 has a winning strategy in a game is PSPACE-hard.
- We provide Pareto set approximation algorithms for stopping games. This result directly applies to the important class of *discounted rewards* for non-stopping games, due to an off-the-shelf reduction [9].

Related Work. Multi-objective optimisation has been studied for various subclasses of stochastic games. For non-stochastic games, multi-dimensional objectives have been considered in [6,22]. For MDPs, multiple discounted objectives [5], long-run objectives [2], ω-regular objectives [10] and total rewards [12] have been analysed. The objectives that we study in this paper are a special case of branching time temporal logics for stochastic games [3,1]. However, already for MDPs, such logics are so powerful that it is not decidable whether there is an optimal controller [3]. A special case of the problem studied in this paper is the case where the goal of Player 1 is to achieve a *precise value* of the expectation of an objective function [9]. As regards applications, stochastic games with a single objective function have been employed and implemented for quantitative abstraction refinement for MDP models in [15]. The usefulness of techniques for verification and strategy synthesis for stochastic games with a single objective is demonstrated, e.g., for smart grid protocols [8]. Applications of multi-objective verification include assume-guarantee verification [16] and controller synthesis [13] for MDPs.

2   Preliminaries

We begin this section by introducing notations used throughout the paper. We then provide the definition of stochastic two-player games together with the concepts of strategies and paths of the game. Finally, we introduce the objectives that are studied in this paper.

2.1   Notation

Given a vector $\boldsymbol{x} \in \mathbb{R}^n$, we use x_i to refer to its i-th component, where $1 \leq i \leq n$, and define the norm $\|\boldsymbol{x}\| \stackrel{\text{def}}{=} \sum_{i=1}^{n} |x_i|$. Given a number $y \in \mathbb{R}$, we use $\boldsymbol{x} \pm y$ to denote the vector $(x_1 \pm y, x_2 \pm y, \ldots, x_n \pm y)$. Given two vectors $\boldsymbol{x}, \boldsymbol{y} \in \mathbb{R}^n$, the *dot product* of $\boldsymbol{x}$ and $\boldsymbol{y}$ is defined by $\boldsymbol{x} \cdot \boldsymbol{y} = \sum_{i=1}^{n} x_i \cdot y_i$, and the comparison operator $\leq$ on vectors is defined to be the componentwise ordering. The sum of two sets of vectors $X, Y \subseteq \mathbb{R}^n$ is defined by $X + Y = \{\boldsymbol{x} + \boldsymbol{y} \mid \boldsymbol{x} \in X, \boldsymbol{y} \in Y\}$. Given a set X, we define the *downward closure* of X as $\mathsf{dwc}(X) \stackrel{\text{def}}{=} \{\boldsymbol{y} \mid \exists \boldsymbol{x} \in X . \boldsymbol{y} \leq \boldsymbol{x}\}$ and the *upward closure* as $\mathsf{up}(X) \stackrel{\text{def}}{=} \{\boldsymbol{y} \mid \exists \boldsymbol{x} \in X . \boldsymbol{x} \leq \boldsymbol{y}\}$. We denote by $\mathbb{R}_{\pm\infty}$ the set $\mathbb{R} \cup \{+\infty, -\infty\}$, and we define the operations $\cdot$ and $+$ in the expected way, defining $0 \cdot x = 0$ for all $x \in \mathbb{R}_{\pm\infty}$ and leaving $-\infty + \infty$ undefined. We also define function $\mathrm{sgn}(x) : \mathbb{R}_{\pm\infty} \to \mathbb{N}$ to be 1 if $x > 0$, -1 if $x < 0$ and 0 if $x = 0$.

A *discrete probability distribution* (or just *distribution*) over a (countable) set S is a function $\mu : S \to [0, 1]$ such that $\sum_{s \in S} \mu(s) = 1$. We write $\mathcal{D}(S)$ for the set of all distributions over S. Let $\mathsf{supp}(\mu) = \{s \in S \mid \mu(s) > 0\}$ be the *support set* of $\mu \in \mathcal{D}(S)$. We say that a distribution $\mu \in \mathcal{D}(S)$ is a *Dirac distribution* if $\mu(s) = 1$ for some $s \in S$. We represent a distribution $\mu \in \mathcal{D}(S)$ on a set $S = \{s_1, \ldots, s_n\}$ as a map $[s_1 \mapsto \mu(s_1), \ldots, s_n \mapsto \mu(s_n)]$ and omit the elements of S outside $\mathsf{supp}(\mu)$ to simplify the presentation. If the context is clear we sometimes identify a Dirac distribution μ with the unique element in $\mathsf{supp}(\mu)$.

2.2 Stochastic Games

In this section we introduce turn-based stochastic two-player games.

Stochastic Two-player Games. A *stochastic two-player game* is a tuple $\mathcal{G} = \langle S, (S_\Box, S_\Diamond, S_\bigcirc), \Delta \rangle$ where S is a finite set of states partitioned into sets $S_\Box$, $S_\Diamond$, and $S_\bigcirc$; $\Delta : S \times S \to [0,1]$ is a probabilistic transition function such that $\Delta(\langle s,t\rangle) \in \{0,1\}$ if $s \in S_\Box \cup S_\Diamond$ and $\sum_{t \in S} \Delta(\langle s,t\rangle) = 1$ if $s \in S_\bigcirc$.

$S_\Box$ and $S_\Diamond$ represent the sets of states controlled by players Player 1 and Player 2, respectively, while $S_\bigcirc$ is the set of stochastic states. For a state $s \in S$, the set of successor states is denoted by $\Delta(s) \stackrel{\text{def}}{=} \{t \in S \mid \Delta(\langle s,t\rangle) > 0\}$. We assume that $\Delta(s) \neq \emptyset$ for all $s \in S$. A state from which no other states except for itself are reachable is called *terminal*, and the set of terminal states is denoted by $\mathsf{Term} \stackrel{\text{def}}{=} \{s \in S \mid \Delta(\langle s,t\rangle)=1 \text{ iff } s = t\}$.

Paths. An infinite *path* λ of a stochastic game $\mathcal{G}$ is an infinite sequence $s_0 s_1 \ldots$ of states such that $s_{i+1} \in \Delta(s_i)$ for all $i \geq 0$. A finite path is a finite such sequence. For a finite or infinite path λ we write $\mathsf{len}(\lambda)$ for the number of states in the path. For $i < \mathsf{len}(\lambda)$ we write λ_i to refer to the i-th state s_i of λ. For a finite path λ we write $\mathsf{last}(\lambda)$ for the last state of the path. For a game $\mathcal{G}$ we write $\Omega_{\mathcal{G}}^+$ for the set of all finite paths, and $\Omega_{\mathcal{G}}$ for the set of all infinite paths, and $\Omega_{\mathcal{G},s}$ for the set of infinite paths starting in state s. We denote the set of paths that reach a state in $T \subseteq S$ by $\Diamond T \stackrel{\text{def}}{=} \{\omega \in \Omega_{\mathcal{G}} \mid \exists i . \omega_i \in T\}$.

Strategies. A *strategy* of Player 1 is a (partial) function $\pi : \Omega_{\mathcal{G}}^+ \to \mathcal{D}(S)$, which is defined for $\lambda \in \Omega_{\mathcal{G}}^+$ only if $\mathsf{last}(\lambda) \in S_\Box$, such that $s \in \mathsf{supp}(\pi(\lambda))$ only if $\Delta(\langle \mathsf{last}(\lambda), s\rangle) = 1$. A strategy π is a *finite-memory* strategy if there is a finite automaton $\mathcal{A}$ over the alphabet S such that $\pi(\lambda)$ is determined by $\mathsf{last}(\lambda)$ and the state of $\mathcal{A}$ in which it ends after reading the word λ. We say that π is *memoryless* if $\mathsf{last}(\lambda)=\mathsf{last}(\lambda')$ implies $\pi(\lambda)=\pi(\lambda')$, and *deterministic* if $\pi(\lambda)$ is Dirac for all $\lambda \in \Omega_{\mathcal{G}}^+$. If π is a memoryless strategy for Player 1 then we identify it with the mapping $\pi \colon S_\Box \to \mathcal{D}(S)$. A strategy σ for Player 2 is defined similarly. We denote by Π and Σ the sets of all strategies for Player 1 and Player 2, respectively.

Probability Measures. A stochastic game $\mathcal{G}$, together with a strategy pair $(\pi, \sigma) \in \Pi \times \Sigma$ and a starting state s, induces an infinite Markov chain on the game (see e.g. [9]). We define the probability measure of this Markov chain by $\mathrm{Pr}_{\mathcal{G},s}^{\pi,\sigma}$. The expected value of a measurable function $f \colon S^\omega \to \mathbb{R}_{\pm\infty}$ is defined as $\mathbb{E}_{\mathcal{G},s}^{\pi,\sigma}[f] \stackrel{\text{def}}{=} \int_{\Omega_{\mathcal{G},s}} f \, d\mathrm{Pr}_{\mathcal{G},s}^{\pi,\sigma}$. We say that a game $\mathcal{G}$ is a *stopping game* if, for every pair of strategies π and σ, a terminal state is reached with probability 1.

Rewards. A reward function $r : S \to \mathbb{Q}^n$ assigns a reward vector $r(s) \in \mathbb{Q}^n$ to each state s of the game $\mathcal{G}$. We use r_i for the function defined by $r_i(t) = r(t)_i$ for all t. We assume that for each i the reward assigned by r_i is either non-negative or non-positive for all states (we adopt this approach in order to express minimisation problems via maximisation, as explained in the next subsection). The analysis of more general reward functions is left for future work. We define the vector of *total reward* random variables $rew(\boldsymbol{r})$ such that, given a path λ, $rew(\boldsymbol{r})(\lambda) = \sum_{j\geq 0} \boldsymbol{r}(\lambda_j)$.

2.3   Multi-objective Queries

A *multi-objective query* (MQ) φ is a positive boolean combination (i.e. disjunctions and conjunctions) of predicates (or *objectives*) of the form $r \bowtie v$, where r is a reward function, $v \in \mathbb{Q}$ is a bound and $\bowtie \in \{\geq, \leq\}$ is a comparison operator. The validity of an MQ is defined inductively on the structure of the query: an objective $r \bowtie v$ is true in a state s of $\mathcal{G}$ under a pair of strategies (π, σ) if and only if $\mathbb{E}_{\mathcal{G},s}^{\pi,\sigma}[rew(r)] \bowtie v$, and the truth value of disjunctions and conjunctions of queries is defined straightforwardly. Using the definition of the reward function above, we can express the operator $\leq$ by using $\geq$, applying the equivalence $r \leq v \equiv (-r \geq -v)$. Thus, throughout the paper we often assume that MQs only contain the operator $\geq$.

We say that Player 1 *achieves* the MQ φ (i.e., *wins* the game) in a state s if it has a strategy π such that for all strategies σ of Player 2 the query φ evaluates to true under (π, σ). An MQ φ is a *conjunctive query* (CQ) if it is a conjunction of objectives, and a *disjunctive query* (DQ) if it is a disjunction of objectives.

For a MQ φ containing n objectives $r_i \bowtie_i v_i$ for $1 \leq i \leq n$ and for $\boldsymbol{x} \in \mathbb{R}^n$ we use $\varphi[\boldsymbol{x}]$ to denote φ in which each $r_i \bowtie_i v_i$ is replaced with $r_i \bowtie_i x_i$.

Reachability. We can enrich multi-objective queries with *reachability objectives*, i.e. objectives $\Diamond T \geq p$ for a set of target states $T \subseteq S$, where $p \in [0,1]$ is a bound. The objective $\Diamond T \geq p$ is true under a pair of strategies (π, σ) if $\Pr_{\mathcal{G},s}^{\pi,\sigma}(\Diamond T) \geq p$, and notions such as achieving a query are defined straightforwardly. Note that queries containing reachability objectives can be reduced to queries with total expected reward only (see [7] for a reduction). It also follows from the construction that if all target sets contain only terminal states, the reduction works in polynomial time.

Pareto Sets. Let φ be an MQ containing n objectives. The vector $\boldsymbol{v} \in \mathbb{R}^n$ is a *Pareto vector* if and only if (a) $\varphi[\boldsymbol{v} - \varepsilon]$ is achievable for all $\varepsilon > 0$, and (b) $\varphi[\boldsymbol{v} + \varepsilon]$ is not achievable for any $\varepsilon > 0$. The set P of all such vectors is called a *Pareto set*. Given $\varepsilon > 0$, an *ε-approximation of a Pareto set* is a set of vectors Q satisfying that, for any $\boldsymbol{w} \in Q$, there is a vector $\boldsymbol{v}$ in the Pareto set such that $\|\boldsymbol{v} - \boldsymbol{w}\| \leq \varepsilon$, and for every $\boldsymbol{v}$ in the Pareto set there is a vector $\boldsymbol{w} \in Q$ such that $\|\boldsymbol{v} - \boldsymbol{w}\| \leq \varepsilon$.

Example. Consider the game $\mathcal{G}$ from Figure 1 (left). It consists of one Player 1 state s_0, one Player 2 state s_1, six stochastic states s_2, s_3, s_4, s_5, t_1 and t_2, as well as two terminal states t_1' and t_2'. Outgoing edges of stochastic states are assigned uniform distributions by convention. For the MQ $\varphi_1 = r_1 \geq \frac{2}{3} \wedge r_2 \geq \frac{1}{6}$, where the reward functions are defined by $r_1(t_1) = r_2(t_2) = 1$ and all other values are zero, the Pareto set for the initial state s_0 is shown in Figure 1 (centre). Hence, φ_1 is satisfied at s_0, as $(\frac{2}{3}, \frac{1}{6})$ is in the Pareto set. For the MQ $\varphi_2 = r_1 \geq \frac{2}{3} \wedge -r_2 \geq -\frac{1}{6}$, Figure 1 (right) illustrates the Pareto set for s_0, showing that φ_2 is not satisfied

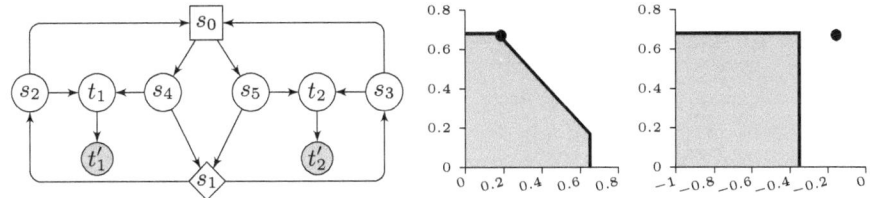

Fig. 1. An example game (left), Pareto set for φ_1 at s_0 (centre), and Pareto set for φ_2 at s_0 (right), with bounds indicated by a dot. Note that the sets are unbounded towards $-\infty$.

at s_0. Note that φ_1 and φ_2 correspond to the combination of reachability and safety objectives, i.e., $\Diamond\{t'_1\} \geq \frac{2}{3} \wedge \Diamond\{t'_2\} \geq \frac{1}{6}$ and $\Diamond\{t'_1\} \geq \frac{2}{3} \wedge \Diamond\{t'_2\} \leq \frac{1}{6}$.

3 Conjunctions of Objectives

In this section we present the results for CQs. We first recall that the games are not determined, and then show that Player 1 may require an infinite-memory randomised strategy to win, while it is not decidable whether deterministic winning strategies exist. We also provide fixpoint equations characterising the Pareto sets of achievable vectors and their successive approximations.

Theorem 1 (Non-determinacy, optimal strategies [9]). *Stochastic games with multiple objectives are, in general, not determined, and optimal strategies might not exist, already for CQs with two objectives.*

Theorem 1 carries over from the results for precise value games, because the problem of reaching a set of terminal states $T \subseteq \mathsf{Term}$ with probability precisely p is a special case of multi-objective stochastic games and can be expressed as a CQ $\varphi = \Diamond T \geq p \wedge \Diamond T \leq p$.

Theorem 2 (Infinite memory). *An infinite-memory randomised strategy may be required for Player 1 to win a multi-objective stochastic game with a CQ even for stopping games with reachability objectives.*

Proof. To prove the theorem we will use the example game from Figure 2. We only explain the intuition behind the need of infinite memory here; the formal proof is presented in [7]. First, we note that it is sufficient to consider deterministic counter-strategies for Player 2, since, after Player 1 has proposed his strategy, the resulting model is an MDP with finite branching [18]. Consider the game starting in the initial state s_0 and a CQ $\varphi = \bigwedge_{i=1}^{3} \Diamond T_i \geq \frac{1}{3}$, where the target sets T_1, T_2 and T_3 contain states labelled 1, 2 and 3, respectively. We note that target sets are terminal and disjoint, and for any π and σ we have that $\sum_{i=1}^{3} \mathrm{Pr}_{\mathcal{G},s_0}^{\pi,\sigma}(\Diamond T_i) = 1$, and hence for any winning Player 1 strategy π it must be the case that, for any σ, $\mathrm{Pr}_{\mathcal{G},s_0}^{\pi,\sigma}(\Diamond T_i) = \frac{1}{3}$ for $1 \leq i \leq 3$.

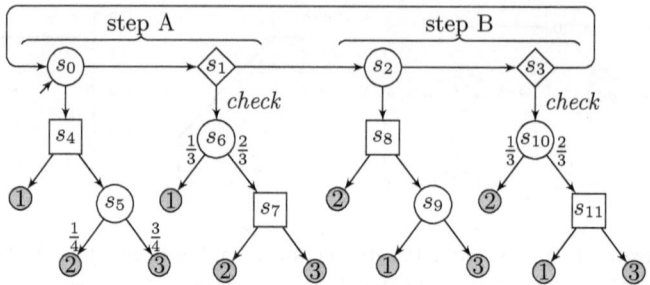

Fig. 2. Game where Player 1 requires infinite memory to win.

Let E be the set of runs which never take any transition *check*. The game proceeds by alternating between the two steps A and B as indicated in Figure 2. In step A, Player 1 chooses a probability to go to T_1 from state s_4, and then Player 2 gets an opportunity to "verify" that the probability $\mathrm{Pr}_{\mathcal{G},s_0}^{\pi,\sigma}(\lozenge T_1|E)$ of runs reaching T_1 conditional on the event that no *check* action was taken is $\frac{1}{3}$. She can do this by taking the action *check* and so ensuring that $\mathrm{Pr}_{\mathcal{G},s_0}^{\pi,\sigma}(\lozenge T_1|\Omega_{\mathcal{G}}\backslash E) = \frac{1}{3}$. If Player 2 again does not choose to take *check*, the game continues in step B, where the same happens for T_2, and so on.

When first performing step A, Player 1 has to pick probability $\frac{1}{3}$ to go to T_1. But since the probability of going from s_4 to T_2 is $< \frac{1}{3}$, when step B is performed for the first time, Player 1 must go to T_2 with probability $y_0 > \frac{1}{3}$ to compensate for the "loss" of the probability in step A. However, this decreases the probability of reaching T_1 at step B, and so Player 1 must compensate for it in the subsequent step A by taking probability $> \frac{1}{3}$ of going to T_1. This decreases the probability of reaching T_2 in the second step B even more (compared to first execution of step A), for which Player 1 must compensate by picking $y_1 > y_0 > \frac{1}{3}$ in the second execution of step B, and so on. So, in order to win, Player 1 has to play infinitely many different probability distributions in states s_4 and s_8. Note that, if Player 2 takes action "check", Player 1 can always randomise in states s_7 and s_{11} to achieve expectations exactly $\frac{1}{3}$ for all objectives.   □

In fact, the above idea allows us to encode natural numbers together with operations of increment and decrement, and obtain a reduction of the location reachability problem in the two-counter machine (which is known to be undecidable [14]) to the problem of deciding whether there exists a *deterministic* winning strategy for Player 1 in a multi-objective stochastic game.

Theorem 3 (Undecidability). *The problem whether there exists a* determin- *istic winning strategy for* Player 1 *in a multi-objective stochastic game is unde- cidable already for stopping games and conjunctions of reachability objectives.*

Our proof is inspired by the proof of [3] which shows that the problem of existence of a winning strategy in an MDP for a PCTL formula is undecidable. However, the proof of [3] relies on branching time features of PCTL to ensure the counter

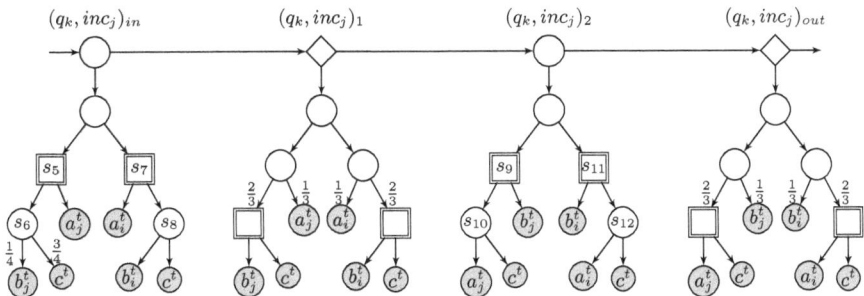

Fig. 3. Increment gadget for counter j

values of the two-counter machine are encoded correctly. Since MQs only allow us to express combinations of linear-time properties, we need to take a different approach, utilising ideas of Theorem 2. We present the proof idea here; for the full proof see [7]. We encode the counter machine instructions in gadgets similar to the ones used for the proof of Theorem 2, where Player 1 has to change the probabilities with which he goes to the target states based on the current value of the counter. For example, the gadget in Figure 3 encodes the instruction to *increment* the counter j. The basic idea is that, if the counter value is c_j when entering the increment gadget, then in state s_5 Player 1 has to assign probability exactly $\frac{2}{3 \cdot 2^{c_j}}$ to the edge $\langle s_5, s_6 \rangle$, and then probability $\frac{2}{3 \cdot 2^{c_j+1}}$ to the edge $\langle s_9, s_{10} \rangle$ in s_9, resulting in the counter being incremented. The gadgets for counter decrement and zero-check can be found in [7]. The resulting query contains six target sets. In particular, there is a conjunct $\Diamond T_t \geq 1$, where the set T_t is not reached with probability 1 only if the gadget representing the target counter machine location is reached. The remaining five objectives ensure that Player 1 updates the counter values correctly (by picking corresponding probability distributions) and so the strategy encodes a valid computation of the two-counter machine. Hence, the counter machine terminates if and only if there does not exist a winning strategy for Player 1.

We note that the problem of deciding whether there is a *randomised* winning strategy for Player 1 remains open, since the gadgets modelling decrement instructions in our construction rely on the strategy being deterministic. Nevertheless, for stopping games, in Theorem 4 below we provide a functional that, given a CQ φ, computes $\varepsilon-$approximations of the Pareto sets, i.e. the sets containing the bounds $\boldsymbol{x}$ so that Player 1 has a winning strategy for $\varphi[\boldsymbol{x} - \varepsilon]$. As a corollary of the theorem, using a simple reduction (see e.g. [9]) we get an approximation algorithm for the Pareto sets in non-stopping games with (multiple) *discounted reward objectives*.

Theorem 4 (Pareto set approximation). *For a stopping game $\mathcal{G}$ and a CQ $\varphi = \bigwedge_{i=1}^{n} r_i \geq v_i$, an $\varepsilon-$approximation of the Pareto sets for all states can be* computed in $k = |S| + \lceil |S| \cdot \frac{\ln(\varepsilon \cdot (n \cdot M)^{-1})}{\ln(1-\delta)} \rceil$ iterations of the operator $F : (S \to \mathcal{P}(\mathbb{R}^n)) \to (S \to \mathcal{P}(\mathbb{R}^n))$ defined by

$$
F(X)(s) \stackrel{\text{def}}{=} \begin{cases} \mathsf{dwc}(\mathsf{conv}(\bigcup_{t \in \Delta(s)} X_t) + r(s)) & \text{if } s \in S_\square \\ \mathsf{dwc}(\bigcap_{t \in \Delta(s)} X_t + r(s)) & \text{if } s \in S_\diamond \\ \mathsf{dwc}(\sum_{t \in \Delta(s)} \Delta(\langle s, t \rangle) \cdot X_t + r(s)) & \text{if } s \in S_\bigcirc, \end{cases}
$$

where the initial sets are $X_s^0 \stackrel{\text{def}}{=} \{\boldsymbol{x} \in \mathbb{R}^n \mid \boldsymbol{x} \le r(s)\}$ for all $s \in S$, and $M = |S| \cdot \frac{\max_{s \in S, i} |r_i(s)|}{\delta}$ for $\delta = p_{\min}^{|S|}$ and $p_{\min}$ being the smallest positive probability in $\mathcal{G}$.

We first explain the intuition behind the operations when $r(s) = \boldsymbol{0}$. For $s \in S_\square$, Player 1 can randomise between successor states, so any convex combination of achievable points in X_t^{k-1} for the successors $t \in \Delta(s)$ is achievable in X_s^k, and so we take the convex closure of the union. For $s \in S_\diamond$, a value in X_s^k is achievable if it is achievable in X_s^{k-1} for all successors $t \in \Delta(s)$, and hence we take the intersection. Finally, stochastic states $s \in S_\bigcirc$ are like Player 1 states with a fixed probability distribution, and hence the operation performed is the weighted Minkowski sum. When $r(s) \ne \boldsymbol{0}$, the reward is added as a contribution to what is achievable at s.

Proof (Outline). The proof, presented in [7], consists of two parts. First, we prove that the result of the k-th iteration of F contains exactly the points achievable by some strategy in k steps; this is done by applying induction on k. As the next step, we observe that, since the game is *stopping*, after $|S|$ steps the game has terminated with probability at least $\delta = p_{\min}^{|S|}$. Hence, the maximum change to any dimension to any vector in X_s^k after k steps of the iteration is less than $M \cdot (1-\delta)^{\lfloor \frac{k}{|S|} \rfloor}$. It follows that $k = |S| + \lceil |S| \cdot \frac{\ln(\varepsilon \cdot (n \cdot M)^{-1})}{\ln(1-\delta)} \rceil$ iterations of F suffice to yield all points which are within ε from the Pareto points for r.

4   General Multi-objective Queries

In this section we consider the general case where the objective is expressed as an arbitrary MQ. The nondeterminacy result from Theorem 1 carries over to the more general MQs, and, even if we restrict to DQs, the games stay nondetermined (see [7] for a proof). The following theorem establishes lower complexity bounds for the problem of deciding the existence of the winning strategy for Player 1.

Theorem 5. *The problem of deciding whether there is a winning strategy for Player 1 for an MQ φ is PSPACE-hard in general, and NP-hard if φ is a DQ.*

The above theorem is proved by reductions from QBF and 3SAT, respectively (see [7] for the proofs). The reduction from QBF is similar to the one in [11], the major differences being that our results apply even when the target states are terminal, and that we need to deal with possible randomisation of the strategies.

We now establish conditions under which a winning strategy for Player 1 exists. Before we proceed, we note that it suffices to consider MQs in conjunctive normal form (CNF) that contain no negations, since any MQ can be converted to CNF using standard methods of propositional logic. Before presenting the proof of Theorem 6, we give the following reformulation of the separating hyperplane theorem, proved in [7].

Lemma 1. *Let $W \subseteq \mathbb{R}^m_{\pm\infty}$ be a convex set satisfying the following. For all j, whenever there is $\boldsymbol{x} \in W$ such that $sgn(x_j) \geq 0$ (resp. $sgn(x_j) \leq 0$), then $sgn(y_j) \geq 0$ (resp. $sgn(y_j) \leq 0$) for all $\boldsymbol{y} \in W$. Let $\boldsymbol{z} \in \mathbb{R}^m$ be a point which does not lie in the closure of $\mathsf{up}(W)$. Then there is a non-zero vector $\boldsymbol{x} \in \mathbb{R}^m$ such that the following conditions hold:*

1. *for all $1 \leq j \leq m$ we have $x_j \geq 0$;*
2. *for all $1 \leq j \leq m$, if there is $\boldsymbol{w} \in W$ satisfying $w_j = -\infty$, then $x_j = 0$; and*
3. *for all $\boldsymbol{w} \in W$, the product $\boldsymbol{w} \cdot \boldsymbol{x}$ is defined and satisfies $\boldsymbol{w} \cdot \boldsymbol{x} \geq \boldsymbol{z} \cdot \boldsymbol{x}$.*

Theorem 6. *Let $\psi = \bigwedge_{i=1}^{n} \bigvee_{j=1}^{m} q_{i,j} \geq u_{i,j}$ be an MQ in CNF, and let π be a strategy of Player 1. The following two conditions are equivalent.*

- *The strategy π achieves ψ.*
- *For all $\varepsilon > 0$ there are nonzero vectors $\boldsymbol{x}_1, \ldots \boldsymbol{x}_n \in \mathbb{R}^m_{\geq 0}$, such that π achieves the conjunctive query $\varphi = \bigwedge_{i=1}^{n} r_i \geq v_i$, where $r_i(s) = \boldsymbol{x}_i \cdot (q_{i,1}(s), \ldots, q_{i,m}(s))$ and $v_i = \boldsymbol{x}_i \cdot (u_{i,1} - \varepsilon, \ldots, u_{i,m} - \varepsilon)$ for all $1 \leq i \leq n$.*

Proof (Sketch). We only present high-level intuition here, see [7] for the full proof. Using the separating hyperplane theorem we show that if there exists a winning strategy for Player 1, then there exist separating hyperplanes, one per conjunct, separating the objective vectors within each conjunct from the set of points that Player 2 can enforce, and vice versa. This allows us to reduce the MQ expressed in CNF into a CQ, by obtaining one reward function per conjuct, which is constructed by weigthing the original reward function by the characteristic vector of the hyperplane.

When we restrict to DQs only, it follows from Theorem 6 that there exists a strategy achieving a DQ if and only if there is a strategy achieving a certain single-objective expected total reward, and hence we obtain the following theorem.

Theorem 7 (Memoryless deterministic strategies). *Memoryless deterministic strategies are sufficient for Player 1 to achieve a DQ.*

Since memoryless deterministic strategies suffice for optimising single total reward, to determine whether a DQ is achievable we can guess such a strategy for Player 1, which uniquely determines an MDP. We can then use the polynomial time algorithm of [10] to verify that there exists no winning Player 2 strategy. This NP algorithm, together with Theorem 5, gives us the following corollary.

Corollary 1. *The problem whether a DQ is achievable is NP-complete.*

Using Theorem 6 we can construct an approximation algorithm computing Pareto sets for disjunctive objectives for stopping games, which performs multiple calls to the algorithm for computing optimal value for the single-objective reward.

Theorem 8 (Pareto sets). *For stopping games, given a vector $r = (r_1, \ldots, r_m)$ of reward functions, an ε-approximation of the Pareto sets for disjunction of objectives for r can be computed by $(\frac{2 \cdot m^2 \cdot (M+1)}{\varepsilon})^{m-1}$ calls to a NP$\cap$coNP algorithm computing single-objective total reward, where M is as in Theorem 4.*

Proof (Sketch). By Theorem 6 and a generalisation of Lemma 1 (see [7]), we have that a DQ $\varphi = \bigvee_{j=1}^{m} r_j \geq v_j$ is achievable if and only if there exists π and $x \in \mathbb{R}_{\geq 0}^m$ such that $\forall \sigma \in \Sigma. \mathbb{E}_{\mathcal{G},s}^{\pi,\sigma}[x \cdot rew(r)] \geq x \cdot v$, which is a single-objective query decidable by an NP$\cap$coNP oracle. Given a finite set $X \subseteq \mathbb{R}^m$, we can compute values $d_x = \sup_\pi \inf_\sigma \mathbb{E}_{\mathcal{G},s}^{\pi,\sigma}[x \cdot rew(r)]$ for all $x \in X$, and define $U_X = \bigcup_{x \in X} \{p \mid x \cdot p \leq d_x\}$. It is not difficult to see that U_X yields an under-approximation of achievable points. Let $\tau = \frac{\varepsilon}{2 \cdot m^2 \cdot (M+1)}$. We argue that when we let X be the set of all non-zero vectors x such that $\|x\| = 1$, and where all x_i are of the form $\tau \cdot k_i$ for some $k_i \in \mathbb{N}$, we obtain an ε-approximation of the Pareto set by taking all Pareto points on U_X (see [7] for a proof).

The above approach, together with the algorithm for Pareto set approximations for CQs from Theorem 4, can be used to compute ε-approximations of the Pareto sets for MQs expressed in CNF. The set U_X would then contain tuples of vectors, one per conjunct.

5 Conclusions

We studied stochastic games with multiple expected total reward objectives, and analysed the complexity of the related algorithmic problems. There are several interesting directions for future research. Probably the most obvious is settling the question whether the problem of existence of a strategy achieving a MQ is decidable. Further, it is natural to extend the algorithms to handle long-run objectives containing mean-payoff or ω-regular goals, or to lift the restriction on reward functions to allow both negative and positive rewards at the same time. Another direction is to investigate practical algorithms for the solution for the problems studied here, such as more sophisticated methods for the approximation of Pareto sets.

Acknowledgements. The authors would like to thank Klaus Draeger, Ashutosh Trivedi and Michael Ummels for the discussions about the problem. The authors are partially supported by ERC Advanced Grant VERIWARE, the Institute for the Future of Computing at the Oxford Martin School, EPSRC grant EP/F001096, and the German Academic Exchange Service (DAAD). V. Forejt was suppoted by the Newton Fellowship of Royal Society and is also affiliated with the Faculty of Informatics, Masaryk University, Czech Republic.

References

1. Baier, C., Brázdil, T., Größer, M., Kucera, A.: Stochastic game logic. Acta Inf. 49(4), 203–224 (2012)
2. Brázdil, T., Brožek, V., Chatterjee, K., Forejt, V., Kučera, A.: Two views on multiple mean-payoff objectives in Markov decision processes. In: LICS (2011)
3. Brázdil, T., Brožek, V., Forejt, V., Kučera, A.: Stochastic games with branching-time winning objectives. In: LICS, pp. 349–358 (2006)
4. Chatterjee, K.: Stochastic Omega-Regular Games. PhD thesis, EECS Department, University of California, Berkeley (October 2007)
5. Chatterjee, K., Majumdar, R., Henzinger, T.A.: Markov decision processes with multiple objectives. In: Durand, B., Thomas, W. (eds.) STACS 2006. LNCS, vol. 3884, pp. 325–336. Springer, Heidelberg (2006)
6. Chatterjee, K., Randour, M., Raskin, J.-F.: Strategy synthesis for multi-dimensional quantitative objectives. In: Koutny, M., Ulidowski, I. (eds.) CONCUR 2012. LNCS, vol. 7454, pp. 115–131. Springer, Heidelberg (2012)
7. Chen, T., Forejt, V., Kwiatkowska, M., Simaitis, A., Wiltsche, C.: On stochastic games with multiple objectives. Technical Report RR-13-06, Oxford U. DCS (2013)
8. Chen, T., Forejt, V., Kwiatkowska, M., Parker, D., Simaitis, A.: Automatic verification of competitive stochastic systems. In: Flanagan, C., König, B. (eds.) TACAS 2012. LNCS, vol. 7214, pp. 315–330. Springer, Heidelberg (2012)
9. Chen, T., Forejt, V., Kwiatkowska, M., Simaitis, A., Trivedi, A., Ummels, M.: Playing stochastic games precisely. In: Koutny, M., Ulidowski, I. (eds.) CONCUR 2012. LNCS, vol. 7454, pp. 348–363. Springer, Heidelberg (2012)
10. Etessami, K., Kwiatkowska, M., Vardi, M., Yannakakis, M.: Multi-objective model checking of Markov decision processes. LMCS 4(4), 1–21 (2008)
11. Fijalkow, N., Horn, F.: The surprizing complexity of reachability games. CoRR, abs/1010.2420 (2010)
12. Forejt, V., Kwiatkowska, M., Norman, G., Parker, D., Qu, H.: Quantitative multi-objective verification for probabilistic systems. In: Abdulla, P.A., Leino, K.R.M. (eds.) TACAS 2011. LNCS, vol. 6605, pp. 112–127. Springer, Heidelberg (2011)
13. Forejt, V., Kwiatkowska, M., Parker, D.: Pareto curves for probabilistic model checking. In: Chakraborty, S., Mukund, M. (eds.) ATVA 2012. LNCS, vol. 7561, pp. 317–332. Springer, Heidelberg (2012)
14. Harel, D.: Effective transformations on infinite trees, with applications to high undecidability, dominoes, and fairness. J. ACM 33(1), 224–248 (1986)
15. Kattenbelt, M., Kwiatkowska, M.Z., Norman, G., Parker, D.: A game-based abstraction-refinement framework for markov decision processes. In: FMSD (2010)
16. Kwiatkowska, M., Norman, G., Parker, D., Qu, H.: Assume-guarantee verification for probabilistic systems. In: Esparza, J., Majumdar, R. (eds.) TACAS 2010. LNCS, vol. 6015, pp. 23–37. Springer, Heidelberg (2010)
17. Marler, R.T., Arora, J.S.: Survey of multi-objective optimization methods for engineering. Structural and Multidisciplinary Optimization 26(6), 369–395 (2004)
18. Martin, D.: The determinacy of Blackwell games. JSL 63(4), 1565–1581 (1998)
19. Pnueli, A.: Logics and models of concurrent systems. Springer (1985)
20. Shapley, L.S.: Stochastic games. PNAS 39(10), 1095 (1953)
21. Suman, B., Kumar, P.: A survey of simulated annealing as a tool for single and multiobjective optimization. J. Oper. Res. Soc. 57(10), 1143–1160 (2005)
22. Velner, Y., Chatterjee, K., Doyen, L., Henzinger, T.A., Rabinovich, A., Raskin, J.-F.: The complexity of multi-mean-payoff and multi-energy games. In: CoRR 2012 (2012)

Minimal Indices for Successor Search

(Extended Abstract)

Sarel Cohen, Amos Fiat, Moshik Hershcovitch, and Haim Kaplan

School of Computer Science, Tel Aviv University
{sarelcoh,fiat,haimk}@post.tau.ac.il,
moshik1@gmail.com

Abstract. We give a new successor data structure which improves upon the index size of the Pătraşcu-Thorup data structures, reducing the index size from $O(nw^{4/5})$ bits to $O(n \log w)$ bits, with optimal probe complexity. Alternatively, our new data structure can be viewed as matching the space complexity of the (probe-suboptimal) z-fast trie of Belazzougui et al. Thus, we get the best of both approaches with respect to both probe count and index size. The penalty we pay is an extra $O(\log w)$ inter-register operations. Our data structure can also be used to solve the weak prefix search problem, the index size of $O(n \log w)$ bits is known to be optimal for any such data structure.

The technical contributions include highly efficient single word indices, with out-degree $w/\log w$ (compared to the $w^{1/5}$ out-degree of fusion tree based indices). To construct such high efficiency single word indices we device highly efficient bit selectors which, we believe, are of independent interest.

Keywords: Predecessor Search, Succinct Data Structures, Cell Probe Model, Fusion Trees, Tries, Word RAM model.

1 Introduction

A fundamental problem in data structures is the successor problem: given a RAM with w bit word operations, and n keys (each w bits long), give a data structure that answers successor queries efficiently. We distinguish between the space occupied by the n input keys themselves, which is $O(nw)$ bits, and the additional space requires by the data structure which we call the *index*. The two other performance measures of the data structure which are of main interest are how many accesses to memory (called *probes*) it performs per query, and the query time or the total number of machine operations performed per query, which could be larger than the number of probes. We can further distinguish between probes to the index and probes to the input keys themselves. The motivation is that if the index is small and fits in cache probes to the index would be cheaper. We focus on constructing a data structure for the successor problem that requires sublinear $o(nw)$ extra bits.

The simplest successor data structure is a sorted list, this requires no index, and performs $O(\log n)$ probes and $O(\log n)$ operations per binary search.

K. Chatterjee and J. Sgall (Eds.): MFCS 2013, LNCS 8087, pp. 278–289, 2013.
© Springer-Verlag Berlin Heidelberg 2013

This high number of probes that are widely dispersed can makes this solution inefficient for large data sets.

Fusion trees of Fredman and Willard [10] (see also [11]) reduce the number of probes and time to $O(\log_w n)$. A fusion tree node has outdegree $B = w^{1/5}$ and therefore fusion trees require only $O(nw/B) = O(nw^{4/5})$ extra bits.

Another famous data structure is the y-fast trie of Willard [18]. It requires linear space ($O(nw)$ extra bits) and $O(\log w)$ probes and time per query.

Pătraşcu and Thorup [14] solve the successor problem optimally (to within an $O(1)$ factor) for any possible point along the probe count/space tradeoff, and for any value of n and w. However, they do not distinguish between the space required to store the input and the extra space required for the index. They consider only the total space which cannot be sublinear.

Pătraşcu and Thorup's linear space data structure for successor search is an improvement of three previous data-structures and achieves the following bounds.

1. For values of n such that $\log n \in [0, \frac{\log^2 w}{\log \log w}]$ their data structure is a fusion tree and therefore the query time is $O(\log_w n)$. This bound increases monotonically with n.

2. For n such that $\log n \in [\frac{\log^2 w}{\log \log w}, \sqrt{w}]$ their data structure is a generalization of the data structure of Beame & Fich [3] that is suitable for linear space, and has the bound $O(\frac{\log w}{\log \log w - \log \log \log n})$. This bound increases from $O(\frac{\log w}{\log \log w})$ at the beginning of this range to $O(\log w)$ at the end of the range.

3. For values of n such that $\log n \in [\sqrt{w}, w]$ their data structure is a slight improvement of the van Emde Boas (vEB) data structure [17] and has the bound of $O(\max\{1, \log(\frac{w - \log n}{\log w})\})$. This bound decreases with n from $O(\log w)$ to $O(1)$.

A recent data structure of Belazzougui et al. [5] called the probabilistic z-fast trie, reduces the extra space requirement to $O(n \log w)$ bits, but requires a (sub-optimal) expected $O(\log w)$ probes (and $O(\log n)$ probes in the worst case). See Table 1 for a detailed comparison between various data structures for the successor porblem with respect to the space and probe parameters under consideration.

Consider the following multilevel scheme to reduce index size: (a) partition the keys into consecutive sets of $w^{1/5}$ keys, (b) build a Fusion tree index structure for each such set (one w bit word), and (c) index the smallest key in every such group using any linear space data structure. The number of fusion tree nodes that we need $n/w^{1/5}$ and the total space required for these nodes and the data structure that is indexing them is $O(nw^{4/5})$.

This standard bucketing trick shows that we can get indices of smaller size by constructing a "fusion tree node" of larger outdegree. That is we seek a data structure, which we refer to as a *word-index*, that by using $O(1)$ words can answer successor queries with respect to as many keys as possible.

Our main contribution is such a word index that can handle $w/\log w$ keys (rather than $w^{1/5}$ for fusion trees).[1] However, this new highly compact index

---

[1] The $w/\log w$ keys take more than $O(1)$ words but are not considered part of the word index.

Table 1. Requirements of various data structures for the successor problem. The word length is w and the number of keys is n. Indexing groups of $w/\log w$ consecutive keys with our new word indices we can reduce the space of any of the linear space data structures above to $O(n \log w)$ bits while keeping the number of probes the same and increasing the query time by $O(\log w)$.

Data Structure	Ref.	Index size (in bits)		# Non-index Probes	Total # Probes	# operations
Binary Search		–		$O(\log n)$	$O(\log n)$	$O(\#probes)$
van Emde Boas	[17]	$O(2^w)$		$O(1)$	$O(\log w)$	$O(\#probes)$
x-fast trie	[18]	$O(nw^2)$		$O(1)$	$O(\log w)$	$O(\#probes)$
y-fast trie	[18]	$O(nw)$		$O(1)$	$O(\log w)$	$O(\#probes)$
x-fast trie on "splitters" poly(w) apart	Folklore	$O(n/\text{poly}(w))$		$O(\log w)$	$O(\log w)$	$O(\#probes)$
Beame & Fich	[3]	$\Theta(n^{1+\epsilon}w)$		$O(1)$	$O\left(\frac{\log w}{\log\log w}\right)$	$O(\#probes)$
Fusion Trees	[10]	$O(nw^{4/5})$		$O(1)$	$O\left(\frac{\log n}{\log w}\right)$	$O(\#probes)$
z-fast trie	[5,4,6]	$O(n\log w)$	exp.	$O(1)$	$O(\log w)$	$O(\#probes)$
			w.c.	$O(\log n)$	$O(\log n)$	
Pătraşcu & Thorup	[14]	$O(nw)$ or $O(nw^{4/5})$		$O(1)$	Optimal given linear space	$O(\#probes)$
Pătraşcu & Thorup + γ-nodes	This Paper	$O(n\log w)$		$O(1)$	Optimal given linear space	$O(\#probes + \log w)$

requires $\Theta(\log w)$ operations per search (versus the $O(1)$ operations required by Fusion trees).

Using these word indices we obtain, as described above, a (deterministic) data structure that, for any n, w, answers a successor query with an optimal number of probes (within an $O(1)$ factor), and requires only $O(n \log w)$ extra bits. We remark that we only probe $O(1)$ non-index words (which is true of Pătraşcu-Thorup data structures as well, with minor modifications). The penalty we pay is an additional $O(\log w)$ in the time complexity.

Indices of small size are particularly motivated today by the multicore shared memory architectures abundant today [8,15]. When multiple cores access shared cache/memory, contention arises. Such contention is deviously problematic because it may cause serialization of memory accesses, making a mockery of multicore parallelism. Multiple memory banks and other hardware are attempts to deal with such problems, to various degrees. Thus, the goals of reducing the index size, so it fits better in fast caches, reducing the number of probes extraneous to the index, and the number of probes within the index, become critical.

2   High Level Overview of Our Results and Their Implications

Computation Model: We assume a RAM model of computation with w bits per word. A key (or query) is one word (w bits long). We can operate on the registers using at least a basic instruction set consisting of (as defined in [7]): Direct and indirect addressing, conditional jump, and a number of *inter-register operations*, including addition, subtraction, bitwise Boolean operations and left and right shifts. All operations are unit cost. One of our construction does not require multiplication.

We give three variants of high outdegree single word indices which we call α nodes, β nodes, and γ nodes. Each of these structures index $w/\log w$ keys and answer successor queries using only $O(1)$ w-bit words, $O(\log w)$ time, and $O(1)$ extra-index probes (in expectation for α and β nodes, worst case for γ nodes) to get at most two of the $w/\log w$ keys.

The α node is simply a z-fast trie ([5]) applied to $w/\log w$ keys. Given the small number of keys, the z-fast trie can be simplified. A major component of the z-fast trie involves translating a prefix match (between a query and a trie node) to the query rank. As there are only $w/\log w$ keys involved, we can discard this part of the z-fast trie and store ranks explicitly in $O(1)$ words.

Based on a different set of ideas, β nodes are arguably simpler than the z-fast trie, and have the same performance as the α nodes. As β-nodes are not our penultimate construction, the full description of β-nodes is in appendix, in the full version of this article [1].

Our penultimate variant, γ nodes, has the advantages that it is deterministic and gives worst case $O(1)$ non-index probes, and, furthermore, requires no multiplication.

To get the γ nodes we introduce highly efficient bit-selectors (see section 2.2) that may be of independent interest. Essentially, a bit-selector selects a multiset of bits from a binary input string and outputs a rearrangement of these bits within a shorter output string.

Thorup [16] proved that it is impossible to have $O(1)$ time successor search in a "standard AC(0) model", for any non-constant number of keys, unless one uses enormous space, $2^{\Omega(w)}$, where w is the number of bits per word. This means that it would be impossible to derive an improved γ-node (or Fusion tree node) with $O(1)$ time successor search in the "standard AC(0) model".

2.1   Succinct Successor Data Structure

As mentioned in the introduction we obtain using our word indices a successor data structure that requires $O(n \log w)$ bits in addition to the input keys. The idea is standard and simple: We divide the keys into consecutive chunks of size $w/\log w$ keys each. We index each chunk with one of our word indices and index the chunks (that is the first key in each chunk) using another linear space data structure. This has the following consequences depending upon the linear space data structure which we use to index the chunks. (We henceforth refer to our γ-nodes, but similar results can be obtained using either α or β nodes in expectation.)

Fusion Trees + γ-nodes: This data structure answers successor queries with $O(\log_w n)$ probes, and $O(\log_w n + \log w)$ time.

The Optimal Structure of Pătraşcu & Thorup + γ-nodes: Here the number of probes to answer a query is optimal, the time is $O(\#probes + \log w)$.

y-fast-trie + γ-nodes: This gives an improvement upon the recently introduced [probabilistic] z-fast-trie, [5,4] (we omit the "probabilistic" prefix hereinafter). The worst-case probes and query time improves from $O(\log n)$ to $O(1)$ probes and $O(\log w)$ query time, and the data structure is deterministic.

The Weak Prefix Search Problem: In this problem the query is as follows. Given a bit-string p, such that p is the prefix of at least one key among the n input keys, return the range of ranks of those input keys having p as a prefix.

It is easy to modify the index of our successor data structures to a new data structure for "weak prefix search". We construct a word x containing the query p padded to the right with trailing zeros, and a word y containing the query p padded to the right with trailing ones. Searching for the rank of the successor of x in S and the rank of the predecessor of y in S gives the required range.

We note that we can carry out the search of the successor of x and the predecessor of y without accessing the keys indexed by the γ nodes. As we will see, our γ nodes implement a succinct blind tree. Searching a blind trie for the right rank of the successor typically requires accessing one of the indexed keys. But, as implicitly used in [6], this access can be avoided if the query is a padded prefix of an indexed key such as x and y above. This implies that the keys indexed by the γ nodes can in fact be discarded and not stored at all. We get a data structure of overall size $O(n \log w)$ bits for weak prefix search.

Belazzougui et al., [6], show that any data structure supporting "weak prefix search" must have size $\Omega(n \log w)$ bits. Hence, our index size is optimal for this related problem.

2.2   Introducing Bit-Selectors and Building a (k, k)-Bit Selector

To construct the γ-nodes we define and construct bit selectors as follows. A (k, L) bit-selector, $1 \le k \le L \le w$, consists of a preprocessing phase and a query phase, (see figure in the appendix of the full version of this article [1]):

- The preprocessing phase: The input is a sequence of length k (with repetitions),

$$I = I[1], I[2], \ldots, I[k],$$

where $0 \le I[j] \le w - 1$ for all $j = 1, \ldots, k$. Given I, we compute the following:
  - A sequence of k strictly increasing indices,
    $0 \le j_1 < j_2 < \cdots < j_k \le L - 1$, and,
  - An $O(1)$ word data structure, $D(I)$.
- The query phase: given an input word x, and using $D(I)$, produces an output word y such that

$$y_{j_\ell} = x_{I[\ell]}, \qquad 1 \le \ell \le k,$$
$$y_m = 0, \qquad m \in \{0, \ldots w - 1\} - \{j_\ell\}_{\ell=1}^{k}.$$

One main technically difficult result is a construct for (k, k) bit-selectors for all $1 \leq k \leq w/\log w$ (Section 3). The bit selector query time is $O(\log w)$, while the probe complexity and space are constant.

With respect to upper bounds, Brodnik, Miltersen, and Munro [7], give several bit manipulation primitives, similar to some of the components we use for bit-selection, but the setting of [7] is somewhat different, and there is no attempt to optimize criteria such as memory probes and index size. The use of Benes networks to represent permutations also appears in Munro et. al [13].

Note that, for (k, k)-bit-selectors, it must be that $j_\ell = \ell - 1$, $1 \leq \ell \leq k$, independently of I. For a sequence of indices I, we define $x[I]$ to be the bits of x in these positions (ordered as in I), if I has multiplicities then $x[I]$ also has multiplicities. With this notation a (k, k) bit selector $D(I)$ computes $x[I]$ for a query x in $O(\log w)$ time.

A $(w^{1/5}, w^{4/5})$ bit-selector is implicit in fusion trees and lie at the core of the data structure. Figure 1 compares the fusion tree bit-selector with our construction.

	k	L	$|D(I)|$ in words	# Operations	Multiplication?
Fusion tree bit-selector	$1 \leq k \leq w^{1/5}$	k^4	$O(1)$	$O(1)$	Yes
Our bit-selector	$1 \leq k \leq w/\log w$	k	$O(1)$	$O(\log w)$	No

Fig. 1. The bit-selector used for Fusion Trees in [10,11] *vs.* our bit-selector

We remark that Andersson, Miltersen, and Thorup [2] give an AC(0) implementation of fusion trees, *i.e.*, they use special purpose hardware to implement a (k, k) bit-selector (that produces a sketch of length k containing k bits of the key). Ignoring other difficulties, computing a [perfect] sketch in AC(0) is easy: just lead wires connecting the source bits to the target bits. With this interpretation, our bit-selector is a software implementation in $O(\log w)$ time that implements the special purpose hardware implementation of [2].

Our bit-selectors are optimal with respect to query time, when considering implementation on a "practical RAM" (no multiplication is allowed) as defined by Miltersen [12]. This follows from Brodnik et al. [7] (Theorem 17) who prove that in the "practical RAM" model, any (k, k)-bit-selector, with $k \geq \log^{10} w$, requires at least $\Omega(\log k)$ time per bit-selector query. (Observe that the bit-reversal of Theorem 17 in [7] is a special case of bit-selection).

3   Bit Selectors

In this section we describe both the preprocessing and selection operations for our bit-selectors. We sketch the selection process, which makes use of $D(I)$, the output of the preprocessing. A more extensive description and figures can be found in the appendix, in the full version of this article [1].

$D(I)$ consists of $O(1)$ words and includes precomputed constants used during the selection process. As $D(I)$ is $O(1)$ words, we assume that $D(I)$ is loaded into registers at the start of the selection process. Also, the total working memory required throughout the selection is $O(1)$ words, all of whom we assume to reside within the set of registers.

Partition the sequence $\sigma = 0, 1, \ldots, w - 1$ into $w/\log w$ blocks (consecutive, disjoint, subsequences of σ), each of length $\log w$. Let B_j denote the jth block of a word, i.e., $B_j = j \log w, j \log w + 1, \ldots, (j+1) \log w - 1, 0 \leq j \leq w/\log w - 1$.

Given an input word x and the precomputed $D(I)$, the selection process goes through the seven phases sketched below.

In this high level explanation we give an example input using the following parameters: The word length $w = 16$ bits, a bit index requires $\log w = 4$ bits, I consists of $w/\log w = 4$ indices (with repetitions). A "block" consists of $\log w = 4$ bits, and there are $w/\log w = 4$ blocks.

As a running example let the input word be $x = 1000\ 1101\ 1110\ \mathbf{0}011$ and let $I =< 0, 15, 12, 15 >$, the required output is $x[I] = 1101$.

Phase 0: Zero irrelevant bits. We take the mask M with ones at positions in I, and set $x = x$ AND M. For our example this gives
Input   : $M = 1000\ 0000\ 0000\ 1001$, $x = 1000\ 1101\ 1110\ \mathbf{0}011$;
Phase 0: $M = 1000\ 0000\ 0000\ 1001$, $x = 1000\ 0000\ 0000\ \mathbf{0}001$.

Phase 1: Packing blocks to the Left: All bits of x whose index belongs to some block are shifted to the left within the block. We modify the mask M accordingly. Let the number of such bits in block j be b_j. This phase transforms M and x as follows:
Phase 0: $M = 1000\ 0000\ 0000\ 1001$, $x = 1000\ 0000\ 0000\ \mathbf{0}001$;
Phase 1: $M = 1000\ 0000\ 0000\ 1100$, $x = 1000\ 0000\ 0000\ \mathbf{0}100$;
Note that $b_0 = 1$, $b_1 = b_2 = 0$, and $b_3 = 2$. Phase 1 requires $O(\log w)$ operations on a constant number of words (or registers).

Phase 2: Sorting Blocks in descending order of b_j (defined in Phase 1 above). This phase transforms M and x as follows:
Phase 1: $M = 1000\ 0000\ 0000\ 1100$, $x = 1000\ 0000\ 0000\ \mathbf{0}100$;
Phase 2: $M = 1100\ 1000\ 0000\ 0000$, $x = \mathbf{0}100\ 1000\ 0000\ 0000$;
Technically, phase 2 uses a Benes network to sort the blocks in descending order of b_j, in our running example this means block 3 should come first, then block 0, then blocks 2 and 3 in arbitrary order. Brodnik, Miltersen, and Munro [7] show how to simulate a Benes network on bits of a word, we extend this so as to sort entire blocks of $\log w$ bits.
The precomputed $D(I)$ includes $O(1)$ words to encode this Benes network. Phase 2 requires $O(\log w)$ bit operations on $O(1)$ words.

Phase 3: Dispersing bits: reorganize the word produced in Phase 2 so that each of the different bits whose index is in I will occupy the leftmost bit of a unique block. As there may be less distinct indices in I than blocks, some of the blocks may be empty, and these will be the rightmost blocks. This process requires $O(\log w)$ word operations to reposition the bits. This phase transforms M and x as follows:

Phase 2: $M = 1100\ 1000\ 0000\ 0000$, $x = 0100\ 1000\ 0000\ 0000$;
Phase 3: $M = 1000\ 1000\ 1000\ 0000$, $x = 0000\ 1000\ 1000\ 0000$;

Phase 4: Packing bits. The goal now is to move the bits positioned by Phase 3 at the leftmost bits of the leftmost r blocks (r being the number of indices in I without repetitions). Again, by appropriate bit manipulation, this can be done with $O(\log w)$ word operations (see appendix in [1]). This phase transforms M and x as follows:

Phase 3: $M = 1000\ 1000\ 1000\ 0000$, $x = 0000\ 1000\ 1000\ 0000$;
Phase 4: $M = 1110\ 0000\ 0000\ 0000$, $x = 0110\ 0000\ 0000\ 0000$;

We remark that if $r = k$, i.e., if I contains no duplicate indices, then we can skip Phases 5 and 6 whose purpose is to duplicate those bits required several times in I.

Phase 5: Spacing the bits. Once again, we simulate a Benes network on the k leftmost bits. The purpose of this permutation is to space out and rearrange the bits so that bits who appear multiple times in I are placed so that multiple copies can be made.

In our running example, phase 5 changes neither M nor x, but this is coincidental – for other inputs ($I' \neq I$) phase 5 would not be the identity function. Phase 5 is yet another application of a Benes network and requires $O(\log w)$ word operations.

Phase 6: Duplicating bits - we duplicate the bits for which space was prepared during Phase 5. This phase transforms M and x as follows:

Phase 5: $M = 1110\ 0000\ 0000\ 0000$, $x = 0110\ 0000\ 0000\ 0000$;
Phase 6: $M = 1111\ 0000\ 0000\ 0000$, $x = 0111\ 0000\ 0000\ 0000$;

Technically, phase 6 makes use of shift and OR operations, where the shifts are decreasing powers of two.

Phase 7: Final positioning: The bits are all now in the k leftmost positions of a word, every bit appears the same number of times it's index appears in I, and we need to run one last Benes network simulation so as to permute these k bits. This permutation gives the final outcome. This phase transforms M and x as follows:

Phase 6: $M = 1111\ 0000\ 0000\ 0000$, $x = 0111\ 0000\ 0000\ 0000$;
Phase 7: $M = 1111\ 0000\ 0000\ 0000$, $x = 1101\ 0000\ 0000\ 0000$;

Note the leftmost $|I| = w/\log w = 4$ bits of x contain the required output of the bit selector.

4 γ-Nodes

In this section we use the $(w/\log w, w/\log w)$−bit-selector, described above, to build a γ-node defined as follows.

Definition 1. *A γ-node answers successor queries over a static set S of at most $w/\log w$ w-bit keys. The γ-node uses a compact index of $O(1)$ w-bit words, in addition to the input S. Successor queries perform $O(1)$ word probes, and $O(\log w)$ operations.*

We describe the γ-node data structure in stages, beginning with a *slow γ-node* below. A slow γ-node is defined as a γ-node but performs $O(w/\log w)$ operations rather than $O(\log w)$.

4.1   Construction of Slow γ-Nodes

We build a blind trie over the set of keys $S = y_1 < y_2, \ldots, < y_k$, $k \leq w/\log w$. We denote this trie by $T(S)$. The trie $T(S)$ is a full binary tree with k leaves, each corresponds to a key, and $k-1$ internal nodes. (We do not think of the keys as part of the trie.) We store $T(S)$ in $O(1)$ w-bit words. (The keys, of course require $|S|w$ bits.) $T(S)$ has the following structure:

1. Each internal node of $T(S)$ has pointers to its left and right children.
2. An internal node u includes a bit index, i_u, in the range $0, \ldots, w - 1$, i_u is the length of the longest common prefix of the keys associated with the leaves in the subtree rooted at u.
3. Key y_i corresponds to the ith leaf from left to right. We store i in this leaf and denote this leaf by $\ell(y_i)$.
4. Keys associated with descendants of the left-child of u have bit i_u equals to zero. Analogously, keys associated with descendants of the right-child of u have bit i_u equals to one.

In addition to $T(S)$, we assume that the keys in S are stored in memory, consecutively in sorted order.

Indices both in internal nodes and leaves are in the range $0, \ldots, w - 1$ and thereby require $O(\log w)$ bits. Since $T(S)$ has $O(w/\log w)$ nodes, a pointer to a node also requires $O(\log w)$ bits. Thus, in total, each node in $T(S)$ requires only $O(\log w)$ bits. It follows that $T(S)$ (internal nodes and leaves) requires only $O(w)$ bits (or, equivalently, can be packed into $O(1)$ words).

Fundamentally, a *blind-search* follows a root to leaf path in blind trie $T(S)$, ignoring intermediate bits. Searching $T(S)$ for a query x always ends at leaf of the trie (which contains the index of some key). Let $\mathrm{bs}(x, S)$ denote the index stored at this leaf, and let $\mathrm{bkey}(x)$ be $y_{\mathrm{bs}(x,S)}$. I.e., blind search for query x in $T(S)$ leads to a leaf that points to $\mathrm{bkey}(x)$. In general, $\mathrm{bkey}(x)$ is *not* the answer to the successor query, but it does have the longest common prefix of x amongst all keys in S. (See [9].)

To arrive at the successor of x, we retrieve $\mathrm{bkey}(x)$ and compute its longest common prefix with x. Let b be the next bit of x, after $\mathrm{LCP}(x, \mathrm{bkey}(x))$. We use b to pad the remaining bits, let $\|$ denote concatenation, and let

$$z = \mathrm{LCP}(x, \mathrm{bkey}(x)) \| b^{w - |\mathrm{LCP}(x, \mathrm{bkey}(x))|}.$$

Finally, we perform a second blind-search on z. The result of this second search gives us the index of the successor to x to within ± 1.

Overall, the number of probes required for such a search is $O(1)$. However, the computation time is equal to the length of the longest root to leaf path in $T(S)$, which is $O(w/\log w)$.

4.2   Improving the Running Time

Using our $(w/\log w, w/\log w)$-bit-selector we can reduce the search time in the blind trie from $O(w/\log w)$ to $O(\log w)$ operations while still representing the trie in $O(1)$ words. For that we change the first part of the query, that is the *blind-search* for $\mathrm{bs}(x, S)$ (the index of $\mathrm{bkey}(x)$). Rather than walking top down along a path in the trie we use a binary search as follows.

We need the following notation. Any node $u \in T(S)$, internal node or leaf, defines a unique root to u path in $T(S)$. Denote this path by $\pi_u = v_0, v_1, \ldots, v_{|\pi_u|}$ where v_0 is the root, $v_{|\pi_u|} = u$, and v_i is the parent of v_{i+1}. For any node $u \in T(S)$ let I_u be the sequence of indices i_v for all internal nodes v along π_u. Also, let ζ_u be a sequence of zeros and ones, one entry per edge in π_u, zero for an edge pointing left, one otherwise. For all $1 \le q \le |S|$ we define $\pi_q = \pi_{\ell(y_q)}$, $I_q = I_{\ell(y_q)}$, and $\zeta_q = \zeta_{\ell(y_q)}$. The following lemma is straightforward.

Lemma 1. *For any index $1 \le q \le |S|$, query x, we have that*

$$\zeta_q \text{ is lexicographically smaller than } x[I_q] \Rightarrow y_q < \mathrm{bkey}(x)$$
$$\zeta_q = x[I_q] \Rightarrow y_q = \mathrm{bkey}(x),$$
$$\zeta_q \text{ is lexicographically larger than } x[I_q] \Rightarrow y_q > \mathrm{bkey}(x).$$

Based on Lemma 1, given query x, we can do binary search to find $\mathrm{bs}(S, x)$:

$L \leftarrow 1, R \leftarrow |S|, q \leftarrow \lfloor (L+R)/2 \rfloor$
while $\zeta_q \ne x[I_q]$ **do**
    if $\zeta_q < x[I_q]$ **then** $R \leftarrow q$
    else $L \leftarrow q$
    end if
    $q \leftarrow \lfloor (L+R)/2 \rfloor$
end while
return q

Lemma 2. *The above binary search algorithm returns $\mathrm{bs}(x, S)$ and has $O(\log w)$ iterations.*

Next we show how to implement each iteration of this binary search and compare $x[I_q]$ and ζ_q in $O(1)$ time while keeping the trie stored in $O(1)$ words.

For this we devise a sequence I of bit indices, of length $w/\log w$. Prior to running the binary search we use the bit-selector of Section 3 to compute $x[I]$ and later we use $x[I]$ to construct $x[I_q]$ in every iteration in $O(1)$ time. We extract $x[I_q]$ from $x[I]$ and retrieve ζ_q using $O(1)$ additional words. The details are as follows.

The $O(1)$ Words Which Form the γ Node: For each $1 \le q \le |S|$ there is a unique interval $[L_q, R_q]$ of which q may be the splitting point (i.e. $q = \lfloor (L_q + R_q)/2 \rfloor$) during the binary search. Let $\pi_q = u_1, u_2, \ldots, u_t = \ell(y_q)$ be the path to q as defined above. Define $j_{L_q} \in 1, \ldots, t$ to be the length of the longest common prefix of π_q and π_{L_q}. That is u_{j_L} is the lowest common ancestor of the leaves $\ell(y_q)$ and $\ell(y_{L_q})$. Define j_{R_q} analogously, and let $j = \max(j_{L_q}, j_{R_q})$.

Let $\tilde{\pi}_q$ be the suffix of π_q starting at node u_{j+1}, and let $\tilde{I}_q$ be the suffix of I_q starting at $I_q[j+1]$. (These are the indices stored in $u_{j+1}, u_{j+2}, \ldots, u_{t-1}$). Similarly, let $\tilde{\zeta}_q$ be the suffix of ζ_q, starting at the jth element.

Given S, for every $1 \leq q \leq |S|$ we precompute and store the following data: j_{L_q}, $j_{R_q}, \tilde{I}_q, \tilde{\zeta}_q$. It is easy to verify that $O(1)$ words suffice to store the $4|S|$ values above. Indeed, j_{L_q} and j_{R_q} are indices in $1, \ldots, |S|$, $O(\log w)$ bits each. As the number of keys $|S| \leq w/\log w$, all the j_{L_q}'s, and j_{R_q}'s fit in $O(1)$ words. Since $\tilde{\pi}_q$ paths are pairwise disjoint, the sum of their path lengths is $O(|S|) = O(w/\log w)$. Hence, storing all the sequences $\tilde{\zeta}_q$, $1 \leq q \leq w/\log w$, requires no more than $O(w/\log w)$ bits. We store the $\tilde{\zeta}_q$'s concatenated in increasing order of q in a single word Z.

The sequence I for which we construct the bit selector is the concatenation of the $\tilde{I}_q$ sequences, in order of q. As above, it follows that I is a sequence of $O(w/\log w)$ $\log w$-bit indices. The bit selector $D(I)$ is also stored as part of the γ node.

For each q we also compute and store the index s_q of the starting position of $\tilde{\zeta}_q$ in Z. This is the same as the index of the starting position of I_q in I. Clearly all these indices s_q can be stored in a single word.

Implementing the Blind Search: As we mentioned, given x as a query to the γ-node, we compute $x[I]$ (once) from x and $D(I)$, which requires $O(\log w)$ operations and no more than $O(1)$ probes.

At the start of an iteration of the binary search, we have a new value of q, and access to the following values, all of whom are in $O(1)$ registers from previous iterations:
$$x[I], \quad j_{L_q}, \quad j_{R_q} \quad L_q, \quad R_q, \quad \zeta_{L_q}, \quad \zeta_{R_q}, \quad x[I_{L_q}], \quad x[I_{R_q}].$$
For the rest of this section let $L = L_q$ and $R = R_q$. We now compute ζ_q and $x[I_q]$. We retrieve j_L, j_R from the data-structure, and we also retrieve $x[\tilde{I}_q]$ from $x[I]$ and $\tilde{\zeta}_q$ from Z (note that $x[\tilde{I}_q]$ is stored consecutively in $x[I]$ and $\tilde{\zeta}_q$ is stored consecutively in Z, and we use s_q to know where they start).

If $j_L \geq j_R$, we compute $x[I_q] \leftarrow (x[I_L][1, \ldots, j_L]) \| (x[\tilde{I}_q])$ and $\zeta_q \leftarrow (\zeta_L[1, \ldots, j_L]) \| (\tilde{\zeta}_q)$.

Analogously, if $j_L < j_R$, and we compute
$$x[I_q] \leftarrow (x[I_R][1, \ldots, j_R]) \| (x[\tilde{I}_q])$$
$$\zeta_q \leftarrow (\zeta_R[1, \ldots, j_R]) \| (\tilde{\zeta}_q).$$

All these operations are easily computed using $O(1)$ SHIFT, AND, OR operations.

5   Open Issues

1. Our (k, k)-bit selector takes $O(\log w)$ operations, which are optimal when $k \geq w^\epsilon$ for any constant $\epsilon > 0$. What can be done for smaller values of k? (E.g., for $k = O(1)$ one can definitely do better).
2. It follows from Thorup ([16]) that, in the practical-RAM model, a search node with fan-out $\frac{w}{\log w}$ requires $\Omega(\log \log w)$ operations. Our γ nodes have fan out $w/\log w$ and require $O(\log w)$ operations. Can this gap be bridged?
3. A natural open question is if the additive $O(\log w)$ in time complexity is required or not.

Acknowledgments. We wish to thank Nir Shavit for introducing us to the problems of contention in multicore environments, for posing the question of multicore efficient data structures, and for many useful discussions. We also wish to thank Mikkel Thorup for his kindness and useful comments.

References

1. Cohen, S., Fiat, A., Hershcovitch, M., Kaplan, H.: Minimal Indices for Successor Search [Full Vestion], in arXiv.org.
2. Andersson, A., Miltersen, P.B., Thorup, M.: Fusion trees can be implemented with AC(0) instructions only. Theor. Comput. Sci. 215(1-2), 337–344 (1999)
3. Beame, P., Fich, F.E.: Optimal bounds for the predecessor problem and related problems. Journal of Computer and System Sciences 65(1), 38–72 (2002)
4. Belazzougui, D., Boldi, P., Pagh, R., Vigna, S.: Theory and practice of monotone minimal perfect hashing. J. Exp. Algorithmics 16, 3.2 (2011)
5. Belazzougui, D., Boldi, P., Pagh, R., Vigna, S.: Monotone minimal perfect hashing: searching a sorted table with o(1) accesses. In: SODA, pp. 785–794 (2009)
6. Belazzougui, D., Boldi, P., Pagh, R., Vigna, S.: Fast prefix search in little space, with applications. In: de Berg, M., Meyer, U. (eds.) ESA 2010, Part I. LNCS, vol. 6346, pp. 427–438. Springer, Heidelberg (2010)
7. Brodnik, A., Miltersen, P.B., Munro, J.I.: Trans-dichotomous algorithms without multiplication - some upper and lower bounds. In: Rau-Chaplin, A., Dehne, F., Sack, J.-R., Tamassia, R. (eds.) WADS 1997. LNCS, vol. 1272, pp. 426–439. Springer, Heidelberg (1997)
8. Drepper, U.: What every programmer should know about memory (2007), http://lwn.net/Articles/250967/
9. Ferragina, P., Grossi, R.: The string B-tree: a new data structure for string search in external memory and its applications. J. ACM 46, 236–280 (1999)
10. Fredman, M.L., Willard, D.E.: Surpassing the information theoretic bound with fusion trees. Journal of Computer and System Sciences 47(3), 424–436 (1993)
11. Fredman, M.L., Willard, D.E.: Trans-dichotomous Algorithms for Minimum Spanning Trees and Shortest Paths. In: FOCS, pp. 719–725 (1990)
12. Miltersen, P.B.: Lower bounds for static dictionaries on rams with bit operations but no multiplication. In: Meyer auf der Heide, F., Monien, B. (eds.) ICALP 1996. LNCS, vol. 1099, pp. 442–453. Springer, Heidelberg (1996)
13. Munro, J.I., Raman, R., Raman, V., Rao, S.S.: Succinct representations of permutations. In: Baeten, J.C.M., Lenstra, J.K., Parrow, J., Woeginger, G.J. (eds.) ICALP 2003. LNCS, vol. 2719, pp. 345–356. Springer, Heidelberg (2003)
14. Pătraşcu, M., Thorup, M.: Time-space trade-offs for predecessor search. In: STOC, pp. 232–240 (2006)
15. Shavit, N.: Data structures in the multicore age. Commun. ACM 54(3), 76–84 (2011)
16. Thorup, M.: On AC0 implementations of fusion trees and atomic heaps. In: SODA, pp. 699–707 (2003)
17. van Emde Boas, P.: Preserving order in a forest in less than logarithmic time and linear space. Inf. Process. Lett. 6(3), 80–82 (1977)
18. Willard, D.E.: Log-logarithmic worst-case range queries are possible in space t(n). Information Processing Letters 17(2), 81–84 (1983)

Paradigms for Parameterized Enumeration[*]

Nadia Creignou[1], Arne Meier[2], Julian-Steffen Müller[2],
Johannes Schmidt[3], and Heribert Vollmer[2]

[1] Aix-Marseille Université
nadia.creignou@lif.univ-mrs.fr
[2] Leibniz Universität Hannover
{meier,mueller,vollmer}@thi.uni-hannover.de
[3] Linköping University
johannes.schmidt@liu.se

Abstract. The aim of the paper is to examine the computational complexity and algorithmics of enumeration, the task to output all solutions of a given problem, from the point of view of parameterized complexity. First we define formally different notions of efficient enumeration in the context of parameterized complexity. Second we show how different algorithmic paradigms can be used in order to get parameter-efficient enumeration algorithms in a number of examples. These paradigms use well-known principles from the design of parameterized decision as well as enumeration techniques, like for instance kernelization and self-reducibility. The concept of kernelization, in particular, leads to a characterization of fixed-parameter tractable enumeration problems.

1   Introduction

This paper is concerned with algorithms for and complexity studies of enumeration problems, the task of generating all solutions of a given computational problem. The area of enumeration algorithms has experienced tremendous growth over the last decade. Prime applications are query answering in databases and web search engines, data mining, web mining, bioinformatics and computational linguistics.

Parameterized complexity theory provides a framework for a refined analysis of hard algorithmic problems. It measures complexity not only in terms of the input size, but in addition in terms of a parameter. Problem instances that exhibit structural similarities will have the same or similar parameter(s). Efficiency now means that for fixed parameter, the problem is solvable with reasonable time resources. A parameterized problem is fixed-parameter tractable (in FPT) if it can be solved in polynomial time for each fixed value of the parameter, where the degree of the polynomial does not depend on the parameter. Much like in the classical setting, to give evidence that certain algorithmic problems are not in

[*] Supported by a Campus France/DAAD Procope grant, Campus France Projet No 28292TE, DAAD Projekt-ID 55892324.

K. Chatterjee and J. Sgall (Eds.): MFCS 2013, LNCS 8087, pp. 290–301, 2013.

FPT one shows that they are complete for superclasses of FPT, like the classes in what is known as the W-hierarchy.

Our main goal is to initiate a study of enumeration from a parameterized complexity point of view and in particular to develop parameter-efficient enumeration algorithms. Preliminary steps in this direction have been undertaken by H. Fernau [5]. He considers algorithms that output *all* solutions of a problem to a given instance in polynomial time for each fixed value of the parameter, where, as above, the degree of the polynomial does not depend on the parameter (let us briefly call this fpt-time). We subsume problems that exhibit such an algorithm in the class Total-FPT. (A similar notion was studied by Damaschke [4]). Algorithms like these can of course only exists for algorithmic problems that possess only relatively few solutions for an input instance. We therefore consider algorithms that exhibit a delay between the output of two different solutions of fpt-time, and we argue that this is the "right way" to define tractable parameterized enumeration. The corresponding complexity class is called Delay-FPT.

We then study the techniques of kernelization (stemming from parameterized complexity) and self-reducibility (well-known in the design of enumeration algorithms) under the question if they can be used to obtain parameter-efficient enumeration algorithms. We study these techniques in the context of different algorithmic problems from the context of propositional satisfiability (and vertex cover, which can, of course, also be seen as a form of weighted 2-CNF satisfiability question). We obtain a number of upper and lower bounds on the enumerability of these problems.

In the next section we introduce parameterized enumeration problems and suggest four hopefully reasonable complexity classes for their study. In the following two sections we study in turn kernelization and self-reducibility, and apply them to the problems VERTEX-COVER, MAXONES-SAT and detection of strong Horn-backdoor sets. We conclude with some open questions about related algorithmic problems.

2   Complexity Classes for Parameterized Enumeration

Because of the amount of solutions that enumeration algorithms possibly produce, the size of their output is often much larger (e.g., exponentially larger) than the size of their input. Therefore, polynomial time complexity is not a suitable yardstick of efficiency when analyzing their performance. As it is now agreed, one is more interested in the regularity of these algorithms rather than in their total running time. For this reason, the efficiency of an enumeration algorithm is better measured by the delay between two successive outputs, see e.g., [7]. The same observation holds within the context of parametrized complexity and we can define parameterized complexity classes for enumeration based on this time elapsed between two successive outputs. Let us start with the formal definition of a parameterized enumeration problem.

Definition 1. *A parameterized enumeration problem (over a finite alphabet Σ) is a triple $E = (Q, \kappa, \text{Sol})$ such that*

- $Q \subseteq \Sigma^*$,
- κ *is a parameterization of* Σ^*, *that is* $\kappa \colon \Sigma^* \to \mathbb{N}$ *is a polynomial time computable function.*
- Sol $\colon \Sigma^* \to \mathcal{P}(\Sigma^*)$ *is a function such that for all* $x \in \Sigma^*$, Sol(x) *is a finite set and* Sol$(x) \neq \emptyset$ *if and only if* $x \in Q$.

If $E = (Q, \kappa, \mathrm{Sol})$ is a parameterized enumeration problem over the alphabet Σ, then we call strings $x \in \Sigma^*$ instances of E, the number $\kappa(x)$ the corresponding parameter, and Sol(x) the set of solutions of x. As an example we consider the problem of enumerating all vertex covers with bounded size of a graph.

Problem:	ALL-VERTEX-COVER
Input:	An undirected graph G and a positive integer k
Parameter:	k
Output:	The set of all vertex covers of G of size $\leq k$

An *enumeration algorithm* $\mathcal{A}$ for the enumeration problem $E = (Q, \kappa, \mathrm{Sol})$ is an algorithm, which on the input x of E, outputs exactly the elements of Sol(x) without duplicates, and which terminates after a finite number of steps on every input.

At first we need to fix the notion of delay for algorithms.

Definition 2 (Delay). *Let* $E = (Q, \kappa, \mathrm{Sol})$ *be a parameterized enumeration problem and* $\mathcal{A}$ *an enumeration algorithm for* E. *Let* $x \in Q$, *then we say that the* i-*th delay of* $\mathcal{A}$ *is the time between outputting the* i-*th and* $(i+1)$-*st solutions in* Sol(x). *Further, we define the* 0-*th delay as the* precalculation time *as the time from the start of the computation to the first output statement. Analogously, the* n-*th delay, for* $n = |\mathrm{Sol}(x)|$, *is the* postcalculation time *which is the time needed after the last output statement until* $\mathcal{A}$ *terminates.*

We are now ready to define different notions of fixed-parameter tractability for enumeration problems.

Definition 3. *Let* $E = (Q, \kappa, \mathrm{Sol})$ *be a parameterized enumeration problem and* $\mathcal{A}$ *an enumeration algorithm for* E.

1. *The algorithm* $\mathcal{A}$ *is a* Total-FPT *algorithm if there exist a computable function* $t \colon \mathbb{N} \to \mathbb{N}$ *and a polynomial* p *such that for every instance* $x \in \Sigma^*$, $\mathcal{A}$ *outputs all solutions of* Sol(x) *in time at most* $t(\kappa(x)) \cdot p(|x|)$.
2. *The algorithm* $\mathcal{A}$ *is a* Delay-FPT *algorithm if there exist a computable function* $t \colon \mathbb{N} \to \mathbb{N}$ *and a polynomial* p *such that for every* $x \in \Sigma^*$, $\mathcal{A}$ *outputs all solutions of* Sol(x) *with delay of at most* $t(\kappa(x)) \cdot p(|x|)$.

Though this will not be in the focus of the present paper, we remark that, in analogy to the non-parameterized case (see [3,15]), one can easily adopt the definition for Inc-FPT algorithms whose ith delay is at most $t(\kappa(x)) \cdot p(|x| + i)$. Similarly, one gets the notion of Output-FPT algorithms which is defined by a runtime of at most $t(\kappa(x)) \cdot p(|x| + |\mathrm{Sol}(x)|)$.

Definition 4. *The class* Total-FPT *(resp.,* Delay-FPT*) is the class of all parameterized enumeration problems that admit a* Total-FPT *(resp.,* Delay-FPT*) enumeration algorithm.*

Observe that Fernau's notion of fixed parameter enumerable [5] is equivalent to our term of Total-FPT. Obviously the existence of a Total-FPT enumeration algorithm requires that for every instance x the number of solution is bounded by $f(\kappa(x)) \cdot p(|x|)$, which is quite restrictive. Nevertheless, Fernau was able to show that the problem MINIMUM-VERTEX-COVER (where we are only interested in vertex covers of minimum cardinality) is in Total-FPT, but by the just given cardinality constraint, ALL-VERTEX-COVER is not in Total-FPT. In the upcoming section we will prove that ALL-VERTEX-COVER is in Delay-FPT; hence we conclude:

Corollary 5. Total-FPT $\subsetneq$ Delay-FPT.

We consider that Delay-FPT should be regarded as the good notion of tractability for parameterized enumeration complexity.

3   Enumeration by Kernelization

Kernelization is one of the most successful techniques in order to design parameter-efficient algorithms, and actually characterizes parameter-tractable problems. Remember that kernelization consists in a pre-processing, which is a polynomial time many-one reduction of a problem to itself with the additional property that the (size of the) image is bounded in terms of the parameter of the argument (see e.g., [6]).

In the following we propose a definition of an *enum-kernelization*, which should be seen as a pre-processing step suitable for an efficient enumeration.

Definition 6. *Let* $(Q, \kappa, \mathrm{Sol})$ *be a parameterized enumeration problem over* Σ. *A polynomial time computable function* $K \colon \Sigma^* \to \Sigma^*$ *is an* enum-kernelization *of* $(Q, \kappa, \mathrm{Sol})$ *if there exist:*

1. *a computable function* $h \colon \mathbb{N} \to \mathbb{N}$ *such that for all* $x \in \Sigma^*$ *we have*
$$(x \in Q \Leftrightarrow K(x) \in Q) \text{ and } |K(x)| \leq h(\kappa(x)),$$
2. *a computable function* $f \colon \Sigma^{*2} \to \mathcal{P}(\Sigma^*)$, *which from a pair* (x, w) *where* $x \in Q$ *and* $w \in \mathrm{Sol}(K(x))$, *computes a subset of* $\mathrm{Sol}(x)$, *such that*
   (a) *for all* $w_1, w_2 \in \mathrm{Sol}(K(x))$, $w_1 \neq w_2 \Rightarrow f(x, w_1) \cap f(x, w_2) = \emptyset$,
   (b) $\displaystyle\bigcup_{w \in \mathrm{Sol}(K(x))} f(x, w) = \mathrm{Sol}(x)$
   (c) *there exists an enumeration algorithm* $\mathcal{A}_f$, *which on input* (x, w), *where* $x \in Q$ *and* $w \in \mathrm{Sol}(K(x))$, *enumerates all solutions of* $f(x, w)$ *with delay* $p(|x|) \cdot t(\kappa(x))$, *where* p *is a polynomial and* t *is a computable function.*

If K *is an enum-kernelization of* $(Q, \kappa, \mathrm{Sol})$, *then for every instance* x *of* Q *the image* $K(x)$ *is called an* enum-kernel *of* x *(under* K*).*

An enum-kernelization is a reduction K from a parameterized enumeration problem to itself. As in the decision setting it has the property that the image is bounded in terms of the parameter argument. For a problem instance x, $K(x)$ is the kernel of x. Observe that if K is an enum-kernelization of the enumeration problem $(Q, \kappa, \mathrm{Sol})$, then it is also a kernelization for the associated decision problem. In order to fit for enumeration problems, enum-kernelizations have the additional property that the set of solutions of the original instance x can be rebuilt from the set of solutions of the image $K(x)$ with Delay-FPT. This can be seen as a generalization of the notion of *full kernel* from [4], appearing in the context of what is called *subset minimization problems*. A full kernel is a kernel that contains all minimal solutions, since they represent in a certain way all solutions. In the context of backdoor sets (see the next section), what is known as a loss-free kernel [13] is a similar notion. In our definition, an enum-kernel is a kernel that represents all solutions in the sense that they can be obtained with FPT delay from the solutions for the kernel.

Vertex cover is a very famous problem whose parameterized complexity has been extensively studied. It is a standard example when it comes to kernelization. Let us examine it in the light of the notion of enum-kernelization.

Proposition 7. ALL-VERTEX-COVER *has an enum-kernelization.*

Proof. Given a graph $G = (V, E)$ and a positive integer k, we are interested in enumerating all vertex covers of G of size at most k. We prove that the famous Buss' kernelization [6, pp. 208ff] provides an enum-kernelization. Let us remember that Buss' algorithm consists in applying repeatedly the following rules until no more reduction can be made:

1. If v is a vertex of degree greater than k, remove v from the graph and decrease k by one.
2. If v is an isolated vertex, remove it.

The algorithm terminates and the kernel $K(G)$ is the reduced graph (V_K, E_K) so obtained if it has less than k^2 edges, and the complete graph $\mathcal{K}_{k+1}$ otherwise.

One verifies that whenever in a certain step of the removing process rule (1) is applicable to a vertex v, and v is not removed immediately, then rule (1) remains applicable to v also in any further step, until it is removed. Therefore, whenever we have a choice during the removal process, our choice does not influence the finally obtained graph: the kernel is unique.

Suppose that $K(G) = (V_K, E_K)$. Let V_D be the set of vertices (of large degree) that are removed by the rule (1) and V_I the set of vertices (isolated) that are removed by the rule (2). On the one hand every vertex cover of size $\leq k$ of G has to contain V_D. On the other hand, no vertex from V_I is part of a minimal vertex cover. Thus, all vertex covers of G are obtained in considering all the vertex covers of $K(G)$, completing them by V_D and by some vertices of V_I up to the cardinality k. Therefore, given W a vertex cover of $K(G)$, then we define $f(G, W) = \{W \cup V_D \cup V' \mid V' \subseteq V_I, |V'| \leq k - |W| - |V_D|\}$. It is then clear that for $W_1 \neq W_2$, $W_1, W_2 \in \mathrm{Sol}(K(G))$, we have that $f(G, W_1) \cap f(G, W_2) = \emptyset$. From the discussion above we have that $\bigcup_{W \in \mathrm{Sol}(K(G))} f(G,W)$ is the set of all $\leq k$-vertex covers of G. Finally, given W a vertex cover of $K(G)$, after a polynomial time pre-processing of G by Buss's kernelization in order to compute V_D and V_I, the enumeration of $f(G,W)$ comes down to an enumeration of all subsets of V_I of size at most $k - |W| - |V_D|$. Such an enumeration can be done with polynomial delay by standard algorithms. Therefore, the set $f(G,W)$ can be enumerated with polynomial delay and, *a fortiori*, with Delay-FPT.   □

As in the context of decision problems, enum-kernelization actually characterizes the class of enumeration problems having Delay-FPT-algorithm, as shown in the following theorem.

Theorem 8. *For every parameterized enumeration problem $(Q, \kappa, \mathrm{Sol})$ over Σ, the following are equivalent:*

1. *$(Q, \kappa, \mathrm{Sol})$ is in* Delay-FPT
2. *For all $x \in \Sigma^*$ the set $\mathrm{Sol}(x)$ is computable and $(Q, \kappa, \mathrm{Sol})$ has an enum-kernelization.*

Proof. (2) ⇒ (1): Let K be an enum-kernelization of $(Q, \kappa, \mathrm{Sol})$. Given an instance $x \in \Sigma^*$ the following algorithm enumerates all solution in $\mathrm{Sol}(x)$ with Delay-FPT: compute $K(x)$ in polynomial time, say $p'(|x|)$. Compute $\mathrm{Sol}(K(x))$, this requires a time $g(\kappa(x))$ for some function g since the size of $K(x)$ is bounded in terms of the parameter argument. Apply successively the enumeration algorithm $\mathcal{A}_f$ to the input (x, w) for each $w \in \mathrm{Sol}(K(x))$. Since $\mathcal{A}_f$ requires a delay $p(|x|) \cdot t(\kappa(x))$, the delay of this enumeration algorithm is bounded from above by $(p'(|x|) + p(|x|)) \cdot (g(\kappa(x)) + t(\kappa(x)))$. The correctness of the algorithm follows from the definition of an enum-kernelization (Item 2.(a) ensures that there is no repetition, Item 2.(b) that all solutions are output).

(1) ⇒ (2): Let $\mathcal{A}$ be an enumeration algorithm for $(Q, \kappa, \mathrm{Sol})$ that requires delay $p(n) \cdot t(k)$ where p is a polynomial and t some computable function. Without loss of generality we assume that $p(n) \geq n$ for all positive integer n. If $Q = \emptyset$ or $Q = \Sigma^*$ then $(Q, \kappa, \mathrm{Sol})$ has a trivial kernelization that maps every $x \in \Sigma^*$ to the empty string ϵ. If $Q = \emptyset$ we are done. If $Q = \Sigma^*$, then fix $w_\epsilon \in \mathrm{Sol}(\epsilon)$ and set for all x, $f(x, w_\epsilon) = \mathrm{Sol}(x)$ and $f(x, w) = \emptyset$ for $w \in \mathrm{Sol}(\epsilon) \setminus \{w_\epsilon\}$. Otherwise, we fix $x_0 \in \Sigma^* \setminus Q$, and $x_1 \in Q$ with $w_1 \in \mathrm{Sol}(x_1)$.

The following algorithm $\mathcal{A}'$ computes an enum-kernelization for $(Q, \kappa, \mathrm{Sol})$: Given $x \in \Sigma^*$ with $n := |x|$ and $k = \kappa(x)$,

1. the algorithm simulates $p(n) \cdot p(n)$ steps of $\mathcal{A}$.
2. If it stops with the answer "no solution", then set $K(x) = x_0$ (since $x_0 \notin Q$, the function f does not need to be defined).
3. If a solution is output within this time, then set $K(x) = x_1$, $f(x, w_1) = \mathrm{Sol}(x)$ and $f(x, w) = \emptyset$ for all $w \in \mathrm{Sol}(x_1) \setminus \{w_1\}$.
4. If it does not output a solution within this time, then it holds $n \leq p(n) \leq t(k)$ and then we set $K(x) = x$, and $f(x, w) = \{w\}$ for all $w \in \mathrm{Sol}(x)$.

Clearly $K(x)$ can thus be computed in time $p(n)^2$, $|K(x)| \leq |x_0| + |x_1| + t(k)$, $(x \in Q \Leftrightarrow K(x) \in Q)$, and the function f we have obtained satisfies all the requirements of Theorem 6, in particular the enumeration algorithm $\mathcal{A}$ can be used to enumerate $f(x, w)$ when applicable. Therefore K provides indeed an enum-kernelization for (Q, κ, Sol). □

Corollary 9. ALL-VERTEX-COVER *is in* Delay-FPT.

Remark 10. Observe that in the proof of Theorem 7, the enumeration of the sets of solutions obtained from a solution W of $K(G)$ is enumerable even with polynomial-delay, we do not need fpt delay. We will show in the full paper that this is a general property: Enum-kernelization can be equivalently defined as FPT-preprocessing followed by enumeration with polynomial delay.

4   Enumeration by Self-reducibility

In this section we would like to exemplify the use of the algorithmic paradigm of self-reducibility ([16,8,15]), on which various enumeration algorithms are based in the literature. The self-reducibility property of a problem allows a "search-reduces-to-decision" algorithm to enumerate the solutions. This technique seems quite appropriate for satisfiability related problems. We will first investigate the enumeration of models of a formula having weight at least k, and then turn to strong HORN-backdoor sets of size k. In the first example the underlying decision problem can be solved in using kernelization (see [9]), while in the second it is solved in using the bounded-search-tree technique.

4.1   Enumeration Classification for MaxOnes-SAT

The self-reducibility technique was in particular applied in order to enumerate all satisfying assignments of a generalized CNF-formula [1], thus allowing to identify classes of formulas which admit efficient enumeration algorithms. In the context of parameterized complexity a natural problem is MAXONES-SAT, in which the question is to decide whether there exists a satisfying assignment of weight at least k, the integer k being the parameter. We are here interested in the corresponding enumeration problem, and we will study it for generalized CNF formulas, namely in Schaefer's framework. In order to state the problem we are interested in more formally, we need some notation.

A *logical relation* of arity k is a relation $R \subseteq \{0, 1\}^k$. By abuse of notation we do not make a difference between a relation and its predicate symbol. A *constraint*, C, is a formula $C = R(x_1, \ldots, x_k)$, where R is a logical relation of arity k and the x_i's are (not necessarily distinct) variables. If u and v are two variables, then $C[u/v]$ denotes the constraint obtained from C in replacing each occurrence of v by u. An assignment m of truth values to the variables *satisfies* the constraint C if $(m(x_1), \ldots, m(x_k)) \in R$. A *constraint language* Γ is a finite set of logical relations. A Γ-*formula* ϕ, is a conjunction of constraints using only logical relations from Γ and is hence a quantifier-free first order formula.

With $\mathrm{Var}(\phi)$ we denote the set of variables appearing in ϕ. A Γ-formula ϕ is satisfied by an assignment $m : \mathrm{Var}(\phi) \rightarrow \{0,1\}$ if m satisfies all constraints in ϕ simultaneously (such a satisfying assignment is also called a *model* of ϕ). The *weight of a model* is given by the number of variables set to true. Assuming a canonical order on the variables we can regard models as tuples in the obvious way and we do not distinguish between a formula ϕ and the logical relation R_ϕ it defines, i.e., the relation consisting of all models of ϕ. In the following we will consider two particular constraints, namely $\mathrm{Imp}(x,y) = (x \rightarrow y)$ and $\mathrm{T}(x) = (x)$.

We are interested in the following parameterized enumeration problem.

Problem:	ENUM-MAXONES-SAT(Γ)
Input:	A Γ-formula φ and a positive integer k
Parameter:	k
Output:	All assignments satisfying φ of weight $\geq k$

The corresponding decision problem, denoted by MAXONES-SAT(Γ), i.e., the problem to decide if a given formula has a satisfying assignment of a given weight, has been studied by Kratsch et al. [9]. They completely settle the question of its parameterized complexity in Schaefer's framework. To state their result we need some terminology concerning types of Boolean relations.

Well known already from Schaefer's original paper [14] are the following seven classes: We say that a Boolean relation R is *a-valid* (for $a \in \{0,1\}$) if $R(a,\ldots,a) = 1$. A relation R is *Horn* (resp., *dual Horn*) if R can be defined by a CNF formula which is Horn (resp., dual Horn), i.e., every clause contains at most one positive (resp., negative) literal. A relation R is *bijunctive* if R can be defined by a 2-CNF formula. A relation R is *affine* if it can be defined by an *affine* formula, i.e., conjunctions of XOR-clauses (consisting of an XOR of some variables plus maybe the constant 1)—such a formula may also be seen as a system of linear equations over GF[2]. A relation R is *complementive* if for all $m \in R$ we have also $\mathbf{1} \oplus m \in R$.

Kratsch et al. [9] introduce a new restriction of the class of bijunctive relations as follows. For this they use the notion of *frozen implementation*, stemming from [12]. Let φ be a formula and $x \in \mathrm{Var}(\varphi)$, then x is said to be *frozen* in φ if it is assigned the same truth value in all its models. Further, we say that Γ *freezingly implements* a given relation R if there is a Γ-formula φ such that $R(x_1,\ldots x_n) \equiv \exists X \varphi$, where φ uses variables from $X \cup \{x_1,\ldots x_n\}$ only, and all variables in X are frozen in φ. For sake of readability, we denote by $\langle \Gamma \rangle_{fr}$ the set of all relations that can be freezingly implemented by Γ. A relation R is *strongly bijunctive* if it is in $\langle \{(x \vee y), (x \neq y), (x \rightarrow y)\} \rangle_{fr}$.

Finally, we say that a constraint language Γ has one of the just defined properties if every relation in Γ has the property.

Theorem 11. *[9, Thm. 7] If Γ is 1-valid, dual-Horn, affine, or strongly bijunctive, then* MAXONES-SAT(Γ) *is in* FPT. *Otherwise* MAXONES-SAT(Γ) *is* W[1]-*hard.*

Interestingly we can get a complete classification for enumeration as well. The fixed-parameter efficient enumeration algorithms are obtained through the algorithmic paradigm of self-reducibility.

We would like to mention that an analogously defined decision problem MINONES-SAT(Γ) is in FPT (by a bounded search-tree algorithm) and the enumeration problem has FPT-delay for all constraint langauges Γ. The decision problem EXACTONES-SAT(Γ) has been studied by Marx [10] and shown to be in FPT iff Γ has a property called "weakly separable". We remark that it can be shown, again by making use of self-reducibility, that under the same conditions, the corresponding enumeration algorithm has FPT-delay. This will be presented in the full paper. In the present submission we concentrate on the, as we think, more interesting maximization problem, since here, the classification of the complexity of the enumeration problem differs from the one for the decision problem, as we state in the following theorem.

Theorem 12. *If Γ is dual-Horn, affine, or strongly bijunctive, then there is a Delay-FPT algorithm for* ENUM-MAXONES-SAT(Γ). *Otherwise such an algorithm does not exist unless* W[1] = FPT.

It would be interesting for those cases of Γ that do not admit a Delay-FPT algorithm to determine an upper bound besides the trivial exponential time bound to enumerate all solutions. In particular, are there such sets Γ for which ENUM-MAXONES-SAT(Γ) is in Output-FPT?

Proof. (of Theorem 12) We first propose a canonical algorithm for enumerating all satisfying assignments of weight at least k. The function HasMaxOnes(ϕ, k) tests if the formula ϕ has a model of weight at least k.

Algorithm 2. Enumerate the models of weight at least k

Input: A formula ϕ with Var(ϕ) = $\{x_1, \ldots, x_n\}$, an integer k
Output: All sat. assignments (given as sets of variables) of ϕ of weight $\geq k$.
1 **if** HasMaxOnes(ϕ, k) **then** Generate($\phi, \emptyset, k, n$)

Procedure Generate(ϕ, M, w, p) :

1 **if** $w = 0$ *or* $p = 0$ **then return** M
2 **else**
3 $\quad$ **if** HasMaxOnes($\phi[x_p = 1], w - 1$) **then**
4 $\quad\quad$ Generate($\phi[x_p = 1], M \cup \{x_p\}, w - 1, p - 1$)
5 $\quad$ **if** HasMaxOnes($\phi[x_p = 0], w$) **then** Generate($\phi[x_p = 0], M, w, p - 1$)

Observe that if Γ is dual-Horn, affine, or strongly bijunctive, then according to Theorem 12 the procedure HasMaxOnes(ϕ, k) can be performed in FPT. Moreover essentially if ϕ is dual-Horn (resp., affine, strongly bijunctive) then so are $\phi[x_p = 0]$ and $\phi[x_p = 1]$ for any variable x_p. Therefore, in all these cases the proposed enumeration algorithm has clearly Delay-FPT.   □

A full version of the proof can be found in the arXiv [2].

4.2   Enumeration of Strong HORN-Backdoor Sets

We consider here the enumeration of strong backdoor sets. Let us introduce some relevant terminology [18]. Consider a formula ϕ, a set V of variables of ϕ, $V \subseteq \mathrm{Var}(\phi)$. For a truth assignment τ, $\phi(\tau)$ denotes the result of removing all clauses from ϕ which contain a literal x with $\tau(x) = 1$ and removing literals y with $\tau(y) = 0$ from the remaining clauses.

The set V is a *strong HORN-backdoor set* of ϕ if for all truth assignment $\tau\colon V \to \{0,1\}$ we have $\phi(\tau) \in \mathrm{HORN}$. Observe that equivalently V is a strong HORN-backdoor set of ϕ if $\phi|_V$ is HORN, where $\phi|_V$ denotes the formula obtained from ϕ in deleting in ϕ all occurrences of variables from V.

Now let us consider the following enumeration problem.

> *Problem:*   EXACT-STRONG-BACKDOORSET[HORN]
> *Input:*   A formula ϕ in CNF
> *Parameter:* k
> *Output:*   The set of all strong HORN-backdoor sets of ϕ of size exactly k

From [11] we know that detection of strong HORN-backdoor sets is in FPT. In using a variant of bounded-search tree the authors use in their FPT-algorithm, together with self-reducibility we get an efficient enumeration algorithm for all strong HORN-backdoor sets of size k.

Theorem 13. EXACT-STRONG-BACKDOORSET[HORN] *is in* Delay-FPT.

Proof. The procedure GenerateSBDS(ϕ, B, k, V) depicted in Algorithm 1 enumerates all sets $S \subseteq V$ of size k such that $B \cup S$ is a strong HORN-backdoor set for ϕ, while the function Exists-SBDS(ϕ, k, V) tests if ϕ has a strong HORN-backdoor set of size exactly k made of variables from V.

The point that this algorithm is indeed in Delay-FPT relies on the fact that the function Exists-SBDS depicted in Algorithm 2 is in FPT. This function is an adaptation of the one proposed in [11]. There Nishimura et al. use an important fact holding for non-HORN clauses (i.e., clauses contains at least two positive literals): if p_1, p_2 are two positive literals then either one of them must belong to any strong backdoor set of the complete formula.

In their algorithm they just go through all clauses for these occurrences. However for our task, the enumeration of the backdoor sets, it is very important to take care of the ordering of variables. The reason for this is the following. Using the algorithm without changes makes it impossible to enumerate the backdoor sets because wrong sets would be considered: e.g., for some formula ϕ and variables $x_1, \dots, x_n$ let $B = \{x_2, x_4, x_5\}$ be the only strong backdoor set. Then, during the enumeration process, one would come to the point where the sets with x_2 have been investigated (our algorithm just enumerates from the smallest variable index to the highest). When we start investigating the sets containing x_4, the procedure would then wrongly say "yes there is a backdoor set containing x_4" which is not desired in this situation because we finished considering x_2 (and only want to investigate backdoor sets that do not contain x_2).

Algorithm 3. Enumerate all strong HORN-backdoor sets of size k

---

Input: A formula ϕ, an integer k
Output: All strong HORN-backdoor sets of size k.
1 **if** Exists-SBDS$(\phi, k, \text{Var}(\phi))$ **then** GenerateSBDS$(\phi, \emptyset, k, \text{Var}(\phi))$

   Procedure GenerateSBDS(ϕ, B, k, V) :

1 **if** $k = 0$ *or* $V = \emptyset$ **then return** B
2 **else**
3   **if** Exists-SBDS$(\phi|_{B \cup \{\min(V)\}}, k-1, V \setminus \{\min(V)\})$ **then**
4   GenerateSBDS$(\phi, B \cup \{\min(V)\}, k-1, V \setminus \{\min(V)\})$
5   **if** Exists-SBDS$(\phi|_B, k, V \setminus \{\min(V)\})$ **then**
6   GenerateSBDS$(\phi, B, k, V \setminus \{\min(V)\})$

   Function Exists-SBDS(ϕ, k, V) :

1 **if** $k = 0$ *or* $V = \emptyset$ **then**
2   **if** $\phi|_V \in \text{HORN}$ **then return** true **else return** false
3 **if** *there is a clause* C *with two positive literals* p_1, p_2 **then**
4   **if** *exactly one of* p_1 *and* p_2 *is in* V, *say* $p_1 \in V, p_2 \notin V$ **then**
5   **if** Exists-SBDS$(\phi|_{\{p_1\}}, k-1, V \setminus \{p_1\})$ **then return** true
6   **else**
7   **if** $p_1 \in V$ *and* $p_2 \in V$ **then**
8   **if** Exists-SBDS$(\phi|_{\{p_1\}}, k-1, V \setminus \{p_1\})$ **then return** true
9   **if** Exists-SBDS$(\phi|_{\{p_2\}}, k-1, V \setminus \{p_2\})$ **then return** true
10   **return** false
11 **else return** true

Therefore the algorithm needs to consider only the variables in the set V where in each recursive call the minimum variable (i.e., the one with smallest index) is removed from the set V of considered variables. $\square$

5 Conclusion

We made a first step to develop a computational complexity theory for parameterized enumeration problems by defining a number of, as we hope, useful complexity classes. We examined two design paradigms for parameterized algorithms from the point of view of enumeration. Thus we obtained a number of upper bounds and also some lower bounds for important algorithmic problems, mainly from the area of propositional satisfiability.

As further promising problems we consider the cluster editing problem [4] and the k-flip-SAT problem [17].

Of course it will be very interesting to examine further algorithmic paradigms for their suitability to obtain enumeration algorithms. Here, we think of the technique of bounded search trees and the use of structural graph properties like treewidth.

Acknowledgements. We are very thankful to Frédéric Olive for helpful discussions. We also acknowledge many helpful comments from the reviewers.

References

1. Creignou, N., Hébrard, J.-J.: On generating all solutions of generalized satisfiability problems. Theoretical Informatics and Applications 31(6), 499–511 (1997)
2. Creignou, N., Meier, A., Müller, J.-S., Schmidt, J., Vollmer, H.: Paradigms for parameterized enumeration. CoRR, arXiv:1306.2171 (2013)
3. Creignou, N., Olive, F., Schmidt, J.: Enumerating all solutions of a Boolean CSP by non-decreasing weight. In: Sakallah, K.A., Simon, L. (eds.) SAT 2011. LNCS, vol. 6695, pp. 120–133. Springer, Heidelberg (2011)
4. Damaschke, P.: Parameterized enumeration, transversals, and imperfect phylogeny reconstruction. TCS 351(3), 337–350 (2006)
5. Fernau, H.: On parameterized enumeration. Computing and Combinatorics (2002)
6. Flum, J., Grohe, M.: Parameterized Complexity Theory. Springer (2006)
7. Johnson, D.S., Papadimitriou, C.H., Yannakakis, M.: On generating all maximal independent sets. IPL 27(3), 119–123 (1988)
8. Khuller, S., Vazirani, V.V.: Planar graph coloring is not self-reducible, assuming P $\neq$ NP. TCS 88(1), 183–189 (1991)
9. Kratsch, S., Marx, D., Wahlström, M.: Parameterized complexity and kernelizability of max ones and exact ones problems. In: Hliněný, P., Kučera, A. (eds.) MFCS 2010. LNCS, vol. 6281, pp. 489–500. Springer, Heidelberg (2010)
10. Marx, D.: Parameterized complexity of constraint satisfaction problems. Computational Complexity (14), 153–183 (2005)
11. Nishimura, N., Ragde, P., Szeider, S.: Detecting backdoor sets with respect to horn and binary clauses. In: Proc. SAT (2004)
12. Nordh, G., Zanuttini, B.: Frozen boolean partial co-clones. In: Proc. ISMVL, pp. 120–125 (2009)
13. Samer, M., Szeider, S.: Backdoor trees. In: Proc. AAAI, pp. 363–368. AAAI Press (2008)
14. Schaefer, T.J.: The complexity of satisfiability problems. In: Proc. STOC, pp. 216–226. ACM Press (1978)
15. Schmidt, J.: Enumeration: Algorithms and complexity. Master's thesis, Leibniz Universität Hannover (2009)
16. Schnorr, C.P.: Optimal algorithms for self-reducible problems. In: Proc. ICALP, pp. 322–337 (1976)
17. Szeider, S.: The parameterized complexity of k-flip local search for SAT and MAX SAT. Discrete Optimization 8(1), 139–145 (2011)
18. Williams, R., Gomes, C., Selman, B.: Backdoors to typical case complexity. In: Proc. IJCAI, pp. 1173–1178 (2003)

Complexity of Checking Bisimilarity between Sequential and Parallel Processes

Wojciech Czerwiński[1], Petr Jančar[2], Martin Kot[2], and Zdeněk Sawa[2,*]

[1] Institute of Computer Science, University of Bayreuth
[2] Dept. of Computer Science, FEI, Technical University of Ostrava
wczerwin@mimuw.edu.pl, {petr.jancar,martin.kot,zdenek.sawa}@vsb.cz

Abstract. Decidability of bisimilarity for Process Algebra (PA) processes, arising by mixing sequential and parallel composition, is a long-standing open problem. The known results for subclasses contain the decidability of bisimilarity between basic sequential (i.e. BPA) processes and basic parallel processes (BPP). Here we revisit this subcase and derive an exponential-time upper bound. Moreover, we show that the problem if a given basic parallel process is inherently sequential, i.e. bisimilar with an unspecified BPA process, is PSPACE-complete. We also introduce a model of one-counter automata, with no zero tests but with counter resets, that capture the behaviour of processes in the intersection of BPA and BPP.

1 Introduction

Bisimilarity (i.e. bisimulation equivalence) is a fundamental behavioral equivalence in concurrency and process theory. Related decidability and complexity questions on various classes of infinite-state processes are an established research topic; see e.g. [2,17] for surveys. One of long-standing open problems in this area is the decidability question for process algebra (PA) processes where sequential and parallel compositions are mixed. An involved procedure working in double-exponential nondeterministic time is known for the normed subclass of PA [7].

More is known for the subclasses of PA where only one type of composition is allowed. The class Basic Process Algebra (BPA) is the "sequential" subclass, while Basic Parallel Processes (BPP) is the "parallel" subclass. Bisimilarity of BPA processes is in 2-EXPTIME [3,10], and EXPTIME-hard [14]. On BPP, bisimilarity is PSPACE-complete [12,16]. For normed subclasses of BPA and BPP, the problem is polynomial [9,8]. A unified polynomial algorithm [5] decides bisimilarity on a superclass of both normed BPP and normed BPA.

The most difficult part of the algorithm for normed PA [7] deals with the case when (a process expressed as) sequential composition is bisimilar with (a process expressed as) parallel composition. A proper analysis when a BPA process is bisimilar with a BPP seems to be a natural prerequisite for understanding this

---

* P. Jančar, M. Kot and Z. Sawa are supported by the Grant Agency of the Czech Rep. (project GAČR: P202/11/0340).

K. Chatterjee and J. Sgall (Eds.): MFCS 2013, LNCS 8087, pp. 302–313, 2013.
© Springer-Verlag Berlin Heidelberg 2013

difficult part. Comparing normed BPA and normed BPP was shown decidable in exponential time [4], and later in polynomial time [11].

For comparing general (unnormed) BPA processes with BPP processes only decidability has been known [13]. The algorithm in [13] checks if a BPP process can be modelled by a (special) pushdown automaton. In the negative case this BPP process cannot be bisimilar to any BPA process; in the positive case, a special one-counter automaton with resets, bisimilar to the BPP process, can be constructed. The BPA-BPP decidability then follows from the decidability of bisimilarity for pushdown processes, which is an involved result by Sénizergues [15]; the latter problem has been recently shown to be non-elementary [1].

Here we revisit the bisimilarity problem comparing BPA and BPP processes and improve the decidability result [13] by showing an exponential-time upper bound; the known lower bound is PTIME-hardness, inherited already from finite-state processes. We also get a completeness result: we show that deciding if a given BPP process is BPA-equivalent, i.e. equivalent to some (unspecified) BPA process, is PSPACE-complete. PSPACE-hardness of this problem follows by a straightforward use of the results in [16], more difficult has been to show the upper bound; this is done in Sect. 3. (We have no upper bound for the opposite problem, asking if a given BPA process is equivalent to some BPP process.) When a BPP process is found to be BPA-equivalent then we can construct a concrete equivalent BPA process, as is also shown in Sect. 3; the construction yields a double exponential bound on its size. To achieve a single exponential upper bound (in Sect. 4) when comparing a given BPP process with a given BPA process, we need to go in more details, and substantially improve the previous constructions. If a given BPP process is BPA-equivalent then we construct a special exponentially bounded *one-counter net with resets* (OCNR) bisimilar with this BPP process. The last step is deciding bisimilarity between the OCNR and a given BPA process. The idea of the algorithm guaranteeing the overall exponential upper bound is sketched in Sect. 4.

2 Notation, Definitions, and Results

Sect. 2.1 provides the definitions, and Sect. 2.2 summarizes the results. Sect. 2.3 recalls the notion of dd-functions and their properties, to be used in the proofs.

2.1 Basic Definitions and Notation

For a set A, by A^* we denote the set of finite sequences of elements of A, i.e., of *words* over A; ε denotes the empty word, and $|w|$ denotes the length of $w \in A^*$. We use $\mathbb{N}$ to denote the set of nonnegative integers $\{0, 1, 2, \ldots\}$.

LTS. A *labelled transition system (LTS)* is a tuple $\mathcal{L} = (S, \mathcal{A}, (\xrightarrow{a})_{a \in \mathcal{A}})$ where S is a set of *states*, $\mathcal{A}$ is a set of *actions*, and $\xrightarrow{a} \subseteq S \times S$ is a set of transitions labelled with a; we put $\longrightarrow = \bigcup_{a \in \mathcal{A}} \xrightarrow{a}$. We write $s \xrightarrow{a} s'$ instead of $(s, s') \in \xrightarrow{a}$, and $s \longrightarrow s'$ instead of $(s, s') \in \longrightarrow$. For $w \in \mathcal{A}^*$, we define $s \xrightarrow{w} s'$ inductively: $s \xrightarrow{\varepsilon} s$; if $s \xrightarrow{a} s'$ and $s' \xrightarrow{u} s''$ then $s \xrightarrow{au} s''$. By $s \longrightarrow^* s'$ we denote that s' is *reachable* from s, i.e., $s \xrightarrow{w} s'$ for some $w \in \mathcal{A}^*$.

Bisimilarity. Given an LTS $\mathcal{L} = (S, \mathcal{A}, (\xrightarrow{a})_{a \in \mathcal{A}})$, a *symmetric* relation $\mathcal{B} \subseteq S \times S$ is a *bisimulation* if for any $(s,t) \in \mathcal{B}$ and $s \xrightarrow{a} s'$ there is t' such that $t \xrightarrow{a} t'$ and $(s', t') \in \mathcal{B}$. Two states s, t are *bisimilar*, i.e., *bisimulation equivalent*, if there is a bisimulation containing (s,t); we write $s \sim t$ to denote that s, t are bisimilar. The relation $\sim$ is indeed an equivalence on S; it is the maximal bisimulation, i.e., the union of all bisimulations. When comparing the states from different LTSs $\mathcal{L}_1$, $\mathcal{L}_2$, we implicitly refer to the disjoint union of $\mathcal{L}_1$ and $\mathcal{L}_2$.

BPA (Basic Process Algebra, or basic sequential processes). A *BPA system* is a tuple $\Sigma = (V, \mathcal{A}, \mathcal{R})$, where V is a finite set of *variables*, $\mathcal{A}$ is a finite set of *actions*, and $\mathcal{R}$ is a finite set of *rules* of the form $A \xrightarrow{a} \alpha$ where $A \in V$, $a \in \mathcal{A}$, and $\alpha \in V^*$. A BPA system $\Sigma = (V, \mathcal{A}, \mathcal{R})$ gives rise to the LTS $\mathcal{L}_\Sigma = (V^*, \mathcal{A}, (\xrightarrow{a})_{a \in \mathcal{A}})$ where the relations $\xrightarrow{a}$ are induced by the following (deduction) rule: if $X \xrightarrow{a} \alpha$ is in $\mathcal{R}$ then $X\beta \xrightarrow{a} \alpha\beta$ for any $\beta \in V^*$. A *BPA process* is a pair (Σ, α) where $\Sigma = (V, \mathcal{A}, \mathcal{R})$ is a BPA system and $\alpha \in V^*$; we often write just α when Σ is clear from context.

BPP (Basic Parallel Processes). A BPP system can be defined as arising from a BPA system when the concatenation is viewed as commutative, thus standing for a parallel composition instead of a sequential one. For later technical reasons we present BPP systems as *communication-free Petri nets*, called *BPP-nets* here; these are classical place/transition nets with labelled transitions where each transition has exactly one input place. A *BPP net* is thus a tuple $\Delta = (P, Tr, \text{PRE}, \text{POST}, \mathcal{A}, \lambda)$ where P is a finite set of *places*, Tr is a finite set of *transitions*, $\text{PRE} : Tr \to P$ is a function assigning an input place to each transition, $\text{POST} : Tr \times P \to \mathbb{N}$ is (equivalent to) a function assigning a multiset of output places to each transition, $\mathcal{A}$ is a finite set of *actions*, and $\lambda : Tr \to \mathcal{A}$ is a function labelling each transition with an action. A *marking* $M : P \to \mathbb{N}$ is a multiset of places, also viewed as a function assigning a nonnegative number of *tokens* to each place. (We could also view P as variables and Tr as rules.)

A BPP net $\Delta = (P, Tr, \text{PRE}, \text{POST}, \mathcal{A}, \lambda)$ gives rise to the *transition-based* LTS $\mathcal{L}_\Delta^{Tr} = (\mathbb{N}^P, Tr, (\xrightarrow{t})_{t \in Tr})$ where $M \xrightarrow{t} M'$ iff $M(\text{PRE}(t)) \geq 1$, $M'(\text{PRE}(t)) = M(\text{PRE}(t)) - 1 + \text{POST}(t, \text{PRE}(t))$, and $M'(p) = M(p) + \text{POST}(t, p)$ for each $p \neq \text{PRE}(t)$. The *action-based* LTS $\mathcal{L}_\Delta = (\mathbb{N}^P, \mathcal{A}, (\xrightarrow{a})_{a \in \mathcal{A}})$ arises from $\mathcal{L}_\Delta^{Tr}$ by putting $M \xrightarrow{a} M'$ iff $M \xrightarrow{t} M'$ for some t where $\lambda(t) = a$.

A *BPP process* is a pair (Δ, M) where Δ is a BPP net and M is a state in $\mathcal{L}_\Delta$ (i.e., a marking); we write just M when Δ is clear from context.

2.2 Results

We assume some standard presentation of the inputs; it does not matter if the numbers $\text{POST}(t, p)$ in the BPP definitions are presented in unary or in binary. The first result clarifies the complexity question of deciding if a basic parallel process is inherently sequential. The second result gives an upper bound on the complexity of deciding bisimulation equivalence of a given pair of one sequential and one parallel process. The known lower bound is PTIME-hardness in this case. For the counterpart of the question in Theorem 1 we get only a lower bound. The lower bounds in Theorem 1 and Proposition 3 can be derived routinely by using the PSPACE-hardness of regularity shown in [16]. The result of clarifying the intersection of BPA and BPP by using OCNR (one-counter nets with resets) is not stated explicitly here.

Theorem 1. *It is* PSPACE-*complete to decide for a given BPP process (Δ, M) if there is a BPA process (Σ, α) such that $\alpha \sim M$.*

Theorem 2. *The problem to decide, given a BPA process (Σ, α) and a BPP process (Δ, M), if $\alpha \sim M$ is in* EXPTIME.

Proposition 3. *It is* PSPACE-*hard to decide for a given BPA process (Σ, α) if there is a BPP process (Δ, M) such that $\alpha \sim M$.*

2.3   Distance-to-Disabling Functions (dd-functions)

We add further notation and recall the notion of dd-functions introduced in [12].

Let $\mathbb{N}_\omega = \mathbb{N} \cup \{\omega\}$ where ω stands for an infinite number satisfying $n < \omega$, $n + \omega = \omega + n = \omega - n = \omega + \omega = \omega - \omega = \omega$ for all $n \in \mathbb{N}$.

Distance. Let $\mathcal{L} = (S, \mathcal{A}, (\xrightarrow{a})_{a \in \mathcal{A}})$ be an LTS. We capture the *(reachability) distance* of a state $s \in S$ to a set of states $U \subseteq S$ by the function DIST $: S \times 2^S \to \mathbb{N}_\omega$ given by the following definition, where we put $\min \emptyset = \omega$:

$$\text{DIST}(s, U) = \min\{\ell \in \mathbb{N} \mid \text{there are } w \in \mathcal{A}^*, \, s' \in U \text{ where } |w| = \ell, \, s \xrightarrow{w} s'\}.$$

We note that $s \longrightarrow s'$ implies $\text{DIST}(s', U) \geq \text{DIST}(s, U) - 1$, i.e., the distance can drop by at most 1 in one step; moreover, if $\text{DIST}(s, U) = \omega$ then $\text{DIST}(s', U) = \omega$. On the other hand, a finite distance can increase even to ω in one step. A one-step change thus belongs to $\mathbb{N}_{\omega, -1} = \mathbb{N}_\omega \cup \{-1\}$. By our definitions, if $\text{DIST}(s, U) = \text{DIST}(s', U) = \omega$ then $\text{DIST}(s, U) + x = \text{DIST}(s', U)$ for any $x \in \mathbb{N}_{\omega, -1}$; formally any $x \in \mathbb{N}_{\omega, -1}$ can be viewed as a respective change in this case.

DD-Functions. *Distance-to-disabling functions* (related to the LTS $\mathcal{L}$), or *dd-functions* for short, are defined inductively. By $s \xrightarrow{a}$ we denote that $a \in \mathcal{A}$ is *enabled* in s, i.e., $s \xrightarrow{a} s'$ for some s'. By $s \xcancel{\xrightarrow{a}}$ we denote that a is *disabled* in s, i.e., $\neg(s \xrightarrow{a})$. We put $\text{DISABLED}_a = \{s \in S \mid s \xcancel{\xrightarrow{a}}\}$. For each $a \in \mathcal{A}$, the function $dd_a : S \to \mathbb{N}_\omega$ defined by $dd_a(s) = \text{DIST}(s, \text{DISABLED}_a)$ is a (basic) dd-function.

If $\mathcal{F} = (d_1, d_2, \ldots, d_k)$ is a tuple of dd-functions and $\delta = (x_1, x_2, \ldots, x_k) \in (\mathbb{N}_{\omega, -1})^k$ then $\text{DISABLED}_{a, \mathcal{F}, \delta} = \{s \in S \mid \text{for any } s' \in S, \text{ if } s \xrightarrow{a} s' \text{ then there is } i \in \{1, 2, \ldots, k\} \text{ such that } d_i(s) + x_i \neq d_i(s')\}$. (Hence $s \in \text{DISABLED}_{a, \mathcal{F}, \delta}$ has no outgoing a-transition which would cause the change δ of the values of dd-functions in $\mathcal{F}$.) The function $dd_{a, \mathcal{F}, \delta} : S \to \mathbb{N}_\omega$ defined by $dd_{a, \mathcal{F}, \delta}(s) = \text{DIST}(s, \text{DISABLED}_{a, \mathcal{F}, \delta})$ is also a dd-function.

A *path* $s_1 \xrightarrow{a_1} s_2 \xrightarrow{a_2} \cdots s_m \xrightarrow{a_m} s_{m+1}$ in $\mathcal{L}$ is *d-reducing*, for a dd-function d, if $d(s_{i+1}) - d(s_i) = -1$ for all $i \in \{1, 2, \ldots, m\}$.

It is easy to verify (inductively) that $s \sim s'$ implies $d(s) = d(s')$ for every dd-function d. If the *LTS* $\mathcal{L} = (S, \mathcal{A}, (\xrightarrow{a})_{a \in \mathcal{A}})$ is *image-finite*, i.e., the set $\{s' \mid s \xrightarrow{a} s'\}$ is finite for any $s \in S$ and $a \in \mathcal{A}$ (which is the case of our $\mathcal{L}_\Sigma$, $\mathcal{L}_\Delta$) then we get a full characterization of bisimilarity on S:

Proposition 4. *For any image-finite LTS* $\mathcal{L} = (S, \mathcal{A}, (\xrightarrow{a})_{a \in \mathcal{A}})$, *the set* $\{(s, s') \mid d(s) = d(s')$ *for every dd-function* $d\}$ *is the maximal bisimulation (i.e., the relation* $\sim$ *on* S).

DD-Functions on BPP. Let $\Delta = (P, Tr, \text{PRE}, \text{POST}, \mathcal{A}, \lambda)$ be a BPP net; $\mathcal{L}_\Delta = (\mathbb{N}^P, \mathcal{A}, (\xrightarrow{a})_{a \in \mathcal{A}})$ is the respective LTS. For $Q \subseteq P$ we put $\text{UNMARK}(Q) = \{M \in \mathbb{N}^P \mid M(p) = 0$ for each $p \in Q\}$, and $\text{NORM}_Q(M) = \text{DIST}(M, \text{UNMARK}(Q))$. The next proposition is standard (by a use of dynamic programming); we stipulate $0 \cdot \omega = \omega \cdot 0 = 0$ and $n \cdot \omega = \omega \cdot n = \omega$ when $n \geq 1$.

Proposition 5. *There is a polynomial-time algorithm that, given a BPP net* $\Delta = (P, Tr, \text{PRE}, \text{POST}, \mathcal{A}, \lambda)$ *and* $Q \subseteq P$, *computes a function* $c : Q \to \mathbb{N}_\omega$ *such that for any* $M \in \mathbb{N}^P$ *we have* $\text{NORM}_Q(M) = \sum_{p \in Q} c(p) \cdot M(p)$.

We note that the coefficient $c(p)$ attached to $p \in Q$ either is ω or is at most exponential (in the size of Δ). The places $p \in Q$ with $c_p = \omega$ constitute a trap, in fact the maximal trap in Q; we call $R \subseteq P$ a *trap* if each $t \in Tr$ with $\text{PRE}(t) \in R$ satisfies $\text{POST}(t, p) \geq 1$ for at least one $p \in R$. We also note that each transition $t \in Tr$ has an associated $\delta_Q^t \in \mathbb{N}_{\omega, -1}$ such that $M \xrightarrow{t} M'$ implies $\text{NORM}_Q(M') = \text{NORM}_Q(M) + \delta_Q^t$ (which is trivial when $\text{NORM}_Q(M) = \omega$); we have $\delta_Q^t = \omega$ if t puts a token in a trap in Q. The next lemma follows from [12].

Lemma 6.
1. *Given a BPP net* $\Delta = (P, Tr, \text{PRE}, \text{POST}, \mathcal{A}, \lambda)$, *any dd-function* d *in* $\mathcal{L}_\Delta$ *has the associated set* $Q_d \subseteq P$ *such that* $d(M) = \text{NORM}_{Q_d}(M)$.
2. *The problem to decide if a given set* $Q \subseteq P$ *is* important, *i.e., associated with a dd-function, is* **PSPACE**-*complete*.

Propositions 4, 5 and Lemma 6 imply that the question whether $M \nsim M'$ can be decided by a nondeterministic polynomial-space algorithm, guessing a set Q and verifying that Q is important and $\text{NORM}_Q(M) \neq \text{NORM}_Q(M')$. Bisimilarity of BPP processes is thus in **PSPACE**.

DD-Functions on BPA. We now assume a BPA system $\Sigma = (V, \mathcal{A}, \mathcal{R})$ and the respective LTS $\mathcal{L}_\Sigma$. For any $\alpha \in V^*$ we define the norm of α as $\|\alpha\| = \text{DIST}(\alpha, \{\varepsilon\})$. If $\|\alpha\| = \omega$ then obviously $\alpha \sim \alpha\beta$ for any β. For any considered α we can thus assume that either α is *normed*, i.e., $\|\alpha\| < \omega$, or $\alpha = \beta U$ where $\|\beta\| < \omega$ and $U \in V$ is an *unnormed* variable, i.e., $\|U\| = \omega$; the *pseudo-norm* $pn(\alpha)$ is equal to $\|\alpha\|$ in the first case, and to $\|\beta\|$ in the second case. A transition $X\beta \xrightarrow{a} \gamma\beta$ is *pn-reducing* if $\|\gamma\| = \|X\| - 1 < \omega$.

A dd-function d is *prefix-encoded above* $C \in \mathbb{N}$ if for any $\alpha \in V^*$ satisfying $C < d(\alpha) < \omega$ we have that each transition $\alpha \xrightarrow{a} \alpha'$ is d-reducing iff it is *pn-reducing*; d is *prefix-encoded* if it is prefix-encoded above some $C \in \mathbb{N}$.

The next lemma is shown in [13]; it is intuitively clear: a BPA process can "remember" large values only by long strings.

Lemma 7. *For any BPA system, every dd-function is prefix-encoded.*

3   Sequentiality of Basic Parallel Processes is in PSPACE

In this section we prove the PSPACE upper bound stated in Theorem 1; this will follow from Proposition 9 and Lemmas 10 and 11.

Given an LTS $\mathcal{L} = (S, \mathcal{A}, (\xrightarrow{a})_{a \in \mathcal{A}})$, by REACH($s$) we denote the set $\{s' \mid s \longrightarrow^* s'\}$ of the states reachable from s. A *state* $s \in S$ is *BPA-equivalent* if there is some BPA process (Σ, α) such that $s \sim \alpha$; in this case all $s' \in$ REACH(s) are BPA-equivalent.

We say that a path $s_1 \xrightarrow{a_1} s_2 \xrightarrow{a_2} \cdots s_m \xrightarrow{a_m} s_{m+1}$ in $\mathcal{L}$ is a *d-down path*, for a dd-function d, if $d(s_{m+1}) < d(s_i)$ for all $i \in \{1, 2, \dots, m\}$. (Note that a d-down path might contain steps which are not d-reducing.) The difference $d(s_1) - d(s_{m+1})$ is called the *d-drop* of the path.

We now formulate a crucial condition that is necessary for a state to be BPA-equivalent. It is motivated by this observation based on Lemma 7: If $d(X\alpha)$ is finite and large, for a dd-function d and a BPA process $X\alpha$, then any *d-down path* from $X\alpha$ with the *d-drop* $\|X\|$ finishes in α. (By "large" we also mean larger than $d(\gamma)$ for all unnormed right-hand sides γ in the BPA rules.) In the next definition it might be useful to imagine $s \sim X\alpha$ and $k = \|X\|$.

Definition 8. Given an LTS, a *state* s_0 is *down-joining* if for any dd-functions d_1, d_2 (not necessarily different) there are $B, C \in \mathbb{N}$ such that for every $s \in$ REACH(s_0) where $\omega > d_1(s) > C$ and $\omega > d_2(s) > C$ we have the following: there is k such that $1 \leq k \leq B$ and for any d_1-down path $s \xrightarrow{w_1} s_1$ with the d_1-drop k and any d_2-down path $s \xrightarrow{w_2} s_2$ with the d_2-drop k we have $s_1 \sim s_2$.

Proposition 9. *If s_0 in an LTS is BPA-equivalent then s_0 is down-joining.*

Proof. Let (Σ, α_0), where $\Sigma = (V, \mathcal{A}, \mathcal{R})$, be a BPA process such that $s_0 \sim \alpha_0$. We put $B = \max\{\|X\|; X \in V, \|X\| < \omega\}$ (where $\max \emptyset = 0$). For dd-functions d_1, d_2 we choose some sufficiently large C so that we can apply the observation before Def. 8 to both d_1 and d_2. The claim can be thus verified easily.   □

In the case of BPP processes, the down-joining property will turn out to be also sufficient for BPA-equivalence, and to be verifiable in polynomial space. The next lemma is a crucial step to show this. It also says that if a BPP process M_0 is down-joining then there is an exponential constant C such that for the LTS restricted to REACH(M_0) we have: the values of dd-functions for $M \in$ REACH(M_0) that are finite and large, i.e. larger than C, are all equal; if a dd-function becomes large (by performing a transition) then all previously large dd-function have been already set to ω; if a large dd-function is sufficiently decreased (by a sequence of transitions) then the values of small dd-functions are determined, independently of the particular way and value of this decreasing.

Lemma 10. *There is a polynomial-space algorithm deciding if a given BPP process* (Δ, M_0) *is down-joining. Moreover, in the positive case the algorithm returns exponentially bounded* $C \in \mathbb{N}$ *such that for any* $M \in \text{REACH}(M_0)$ *and any dd-functions* d_1, d_2, d_3, d, d' *we have:*

1. *If* $C < d_1(M) < \omega$ *and* $C < d_2(M) < \omega$ *then* $d_1(M') = d_2(M')$ *for all* $M' \in \text{REACH}(M)$; *moreover, if* $d_3(M) \neq d_1(M)$ *and* $M \longrightarrow^* M' \longrightarrow M''$ *where* $C < d_3(M'') < \omega$ *then* $d_1(M') = d_2(M') = \omega$.
2. *If* $M \xrightarrow{w_1} M_1$ *is a d-down path with the d-drop* $C_1 \geq C$ *and* $M \xrightarrow{w_2} M_2$ *is a d-down path with the d-drop* $C_2 \geq C$, *and* $d'(M) \neq d(M)$, *then* $d'(M_1) = d'(M_2)$.

Proof. (Sketch of the idea.) Let $\Delta = (P, Tr, \text{PRE}, \text{POST}, \mathcal{A}, \lambda)$ be a BPP net. We recall that each dd-function d coincides with NORM_Q for some important set $Q \subseteq P$ (and there thus exist at most exponentially many pairwise different dd-functions). Each $t \in Tr$ has an associated change δ_Q^t as we have already discussed; recall that t also has the associated label $\lambda(t) \in \mathcal{A}$. We also recall that it is PSPACE-complete to decide if a given Q is important.

We now assume a given M_0 and restrict ourselves to $\text{REACH}(M_0)$. Our claimed algorithm will be using a subprocedure for deciding if some sets are important, and we can allow ourselves even the luxurious NPSPACE-upper bound for questions in our analysis (since PSPACE = NPSPACE).

The reachability relation on $\mathcal{L}_\Delta$ was studied in detail by Esparza [6], and we could use deciding various questions which are reducible to Integer Linear Programming by [6]. A crucial point is simple: In a BPP net, each token can move freely between connected places, possibly generating other tokens; travelling along a cycle can "pump" some places above any bound. We can decide, e.g., if a concrete place $p \in P$ can get arbitrarily large values $M(p)$ for $M \in \text{REACH}(M_0)$ where we might also have some specified constraints, like that some traps are not marked by M (have no tokens in M) and that some specific transitions are enabled in M (or in some $M' \in \text{REACH}(M)$).

We can thus check (in nondeterministic polynomial space) if there are two important sets Q_1, Q_2 such that for any $b \in \mathbb{N}$ there is $M \in \text{REACH}(M_0)$ such that $\text{NORM}_{Q_1}(M)$, $\text{NORM}_{Q_2}(M)$ are finite, bigger than b, and different. If this is the case (i.e., we have found some appropriate "pumping" cycles) then M_0 is surely not down-joining, as can be verified by a straightforward analysis.

A full technical proof would require a complete analysis of all possible violations of the down-joining property. In principle, it is a routine (omitted here due to the limited space); some exponential C claimed for the case with no violations can be also derived by a straightforward technical analysis. $\square$

Lemma 11. *Any down-joining BPP process* (Δ, M_0) *is BPA-equivalent.*

Proof. Let $\Delta = (P, Tr, \text{PRE}, \text{POST}, \mathcal{A}, \lambda)$ be a BPP net, and let M_0 be down-joining. We will construct a BPA process (Σ, α) such that $M_0 \sim \alpha$; the size of (Σ, α) will be double exponential in the size of (Δ, M_0). We note that in this proof the size plays no role, since just the existence of some such (Σ, α) is sufficient; in Sect. 4 we will discuss the details of the one-counter net (OCNR) that is single exponential.

Let $d_1 = \text{NORM}_{Q_1}, \ldots, d_m = \text{NORM}_{Q_m}$ be all pairwise different dd-functions, given by all important sets $Q_i \subseteq P$. We put $\mathbb{D}(M) = (d_1(M), \ldots, d_m(M)) \in (\mathbb{N}_\omega)^m$ and recall that $M \sim M'$ iff $\mathbb{D}(M) = \mathbb{D}(M')$. We also note that $m \leq 2^{|P|}$.

Let $\mathcal{L}_\Delta^D = (\{\mathbb{D}(M) \mid M \in \mathbb{N}^P\}, \mathcal{A}, (\xrightarrow{a})_{a \in \mathcal{A}})$ be the LTS where $M \xrightarrow{a} M'$ in $\mathcal{L}_\Delta$ induces $\mathbb{D}(M) \xrightarrow{a} \mathbb{D}(M')$ in $\mathcal{L}_\Delta^D$. It is straightforward to verify that $M \sim \mathbb{D}(M)$. We also note that for deciding if a $label\text{-}change$ $(a, \delta) \in \mathcal{A} \times (\mathbb{N}_{\omega,-1})^m$ is $enabled$ in D, i.e., if $D \xrightarrow{a} (D + \delta)$, it suffices to know $\text{TYPE}(D) \in \{0, +, \omega\}^m$ where $\text{TYPE}(D)(i) = 0, +, \omega$ if $D(i) = 0, 0 < D(i) < \omega, D(i) = \omega$, respectively.

We define $\mathcal{L}$ as the restriction of $\mathcal{L}_\Delta^D$ to the state set $S = \{\mathbb{D}(M) \mid M \in \text{REACH}(M_0)\}$; we note that $D_0 = \mathbb{D}(M_0)$ is down-joining in $\mathcal{L}$. Let $C \in \mathbb{N}$ be the constant guaranteed by Lemma 10; we assume, moreover, that $D_0(i) \leq C$ for all $i \in \{1, 2, \ldots, m\}$ such that $D_0(i) < \omega$, and that C is bigger than any possible finite increase of any d_i in one step. For any $D \in S$ we say that $D(i)$ is $small$ if $D(i) \leq C$ or $D(i) = \omega$; otherwise $D(i)$ is big.

We build a BPA system $\Sigma = (V, \mathcal{A}, \mathcal{R})$ where variables in V are tuples of the form $(\text{VEC}, \text{BIG}, \perp)$ or $(\text{VEC}, \text{BIG}, \text{DET}, \not\perp)$ where $\text{VEC} \in (\{0, 1, \ldots, C\} \cup \{\omega\})^m$, $\text{BIG} \subseteq \{1, 2, \ldots, m\}$, and $\text{DET} : (\{1, 2, \ldots, m\} \smallsetminus \text{BIG}) \to (\{0, 1, \ldots, C\} \cup \{\omega\})$. We aim to achieve $D_0 \sim (D_0, \emptyset, \perp)$ (in the disjoint union of $\mathcal{L}$ and $\mathcal{L}_\Sigma$). In fact, we will stepwise construct a bijection between the paths $D_0 \xrightarrow{a_1} D_1 \xrightarrow{a_2} \cdots \xrightarrow{a_r} D_r$ in $\mathcal{L}$ and $\alpha_0 \xrightarrow{a_1} \alpha_1 \xrightarrow{a_2} \cdots \xrightarrow{a_r} \alpha_r$ in $\mathcal{L}_\Sigma$, where $\alpha_0 = (D_0, \emptyset, \perp)$; we will have $D_x \sim \alpha_x$. In general, $\alpha_x \in V^*$ corresponding to D_x in two paths related by the bijection will be either a variable $(\text{VEC}, \emptyset, \perp)$, in which case $D_x = \text{VEC}$, or of the form

$$(\text{VEC}_1, \text{BIG}, \text{DET}, \not\perp), (\text{VEC}_2, \text{BIG}, \text{DET}, \not\perp) \ldots (\text{VEC}_{\ell-1}, \text{BIG}, \text{DET}, \not\perp), (\text{VEC}_\ell, \text{BIG}, \perp), \tag{1}$$

for $\ell \geq 1$ and $\text{BIG} \neq \emptyset$, where the following will hold:

1. for any $i_1, i_2 \in \text{BIG}$ we have $\text{VEC}_j(i_1) = \text{VEC}_j(i_2)$ for all $j \in \{1, 2, \ldots, \ell\}$;
2. for any $i \in \text{BIG}$, $\text{SUM}(i) = \sum_{j=1}^\ell \text{VEC}_j(i)$ is finite, and equal to $D_x(i)$;
3. for any $i \in \text{BIG}$, $\text{VEC}_j(i)$ is positive for each $j \in \{1, 2, \ldots, \ell\}$, with the possible exception in the case $\ell = 1$ where we might have $\text{VEC}_1(i) = 0$;
4. for any $i \notin \text{BIG}$, $\text{VEC}_1(i) = D_x(i)$;
5. for any $i \notin \text{BIG}$ and $j \in \{2, 3, \ldots, \ell\}$ we have $\text{VEC}_j(i) = \text{DET}(i)$.

We note that $i \in \text{BIG}$ does not necessarily imply that $\text{SUM}(i)$ is big; this just signals that $D_y(i)$ was big for some $y \leq x$. By Lemma 10(2), the values $\text{VEC}_j(i)$ in 5. are thus determined; this will be clarified below.

We now inductively define the sets V and $\mathcal{R}$ in Σ; we start with putting $(D_0, \emptyset, \perp)$ in V. We leave implicit a verification of the soundness of our construction and of the above claimed conditions. Each $(\text{VEC}, \text{BIG}, \perp)$ will be unnormed, and such a variable always finishes our considered strings α_x.

Suppose $(\text{VEC}, \text{BIG}, \text{DET}, \text{BOT}) \in V$ is the first variable in some α_x, corresponding to some D_x, as given around (1); here $\text{BOT} \in \{\perp, \not\perp\}$ and DET is assumed to be missing if $\text{BOT} = \bot$. Suppose also some concrete (a, δ) which is enabled by $\text{TYPE}(\text{VEC})$ (i.e., $D_x \xrightarrow{a} (D_x + \delta)$ in $\mathcal{L}_\Delta^D$; note that $\text{TYPE}(D_x) = \text{TYPE}(\text{VEC})$). In this case we proceed as follows (using Lemma 10 implicitly):

1. If $\text{BOT} = \cancel{\bot}$ and $\text{VEC}(i) + \delta(i) = 0$ for some $i \in \text{BIG}$ (which implies $\text{VEC}(i) + \delta(i) = 0$ for each $i \in \text{BIG}$), then we add the rule $(\text{VEC}, \text{BIG}, \cancel{\bot}) \xrightarrow{a} \varepsilon$.

2. If $\text{VEC}(i) + \delta(i) = \omega$ for some $i \in \text{BIG}$ (which implies $\text{VEC}(i) + \delta(i) = \omega$ for each $i \in \text{BIG}$) then we add $(\text{VEC}, \text{BIG}, \text{DET}, \text{BOT}) \xrightarrow{a} ((\text{VEC} + \delta), \emptyset, \bot)$.

3. If none of 1.,2. applies and $\text{VEC}(i) + \delta(i) \in \{0, 1, \ldots, C\} \cup \{\omega\}$ for all i then we add $(\text{VEC}, \text{BIG}, \text{DET}, \text{BOT}) \xrightarrow{a} ((\text{VEC} + \delta), \text{BIG}, \text{DET}, \text{BOT})$.

4. If $C < \text{VEC}(i) + \delta(i) < \omega$ for some i (in which case none of 1.,2.,3. applies): Denote $\text{BIG}' = \{i \mid C < \text{VEC}(i) + \delta(i) < \omega\}$; our assumptions imply that there is $k, 1 \le k < C$, such that $\text{VEC}(i) + \delta(i) = C + k$ for each $i \in \text{BIG}'$, and, moreover, $\text{BIG}' = \text{BIG}$ if $\text{BIG} \ne \emptyset$. If $\text{BOT} = \cancel{\bot}$ then we add
   $(\text{VEC}, \text{BIG}, \text{DET}, \cancel{\bot}) \xrightarrow{a} (\text{VEC}', \text{BIG}', \text{DET}, \cancel{\bot})(\text{VEC}'', \text{BIG}', \text{DET}, \cancel{\bot})$
   where we put $\text{VEC}'(i) = C$ and $\text{VEC}''(i) = k$ for each $i \in \text{BIG}'$, and $\text{VEC}'(i) = \text{VEC}(i) + \delta(i)$ and $\text{VEC}''(i) = \text{DET}(i)$ for each $i \notin \text{BIG}'$.
   If $\text{BOT} = \bot$ then we add
   $(\text{VEC}, \text{BIG}, \bot) \xrightarrow{a} (\text{VEC}', \text{BIG}', \text{DET}, \cancel{\bot})(\text{VEC}'', \text{BIG}', \bot)$
   where we put $\text{VEC}'(i) = C$ and $\text{VEC}''(i) = k$ for each $i \in \text{BIG}'$, and $\text{VEC}'(i) = \text{VEC}(i) + \delta(i)$ for each $i \notin \text{BIG}'$; DET is defined by using Lemma 10(2): for some $i' \in \text{BIG}'$ we take a $d_{i'}$-down path $(D_x + \delta) \xrightarrow{w} D'$ with the $d_{i'}$-drop C and put $\text{DET}(i) = \text{VEC}''(i) = D'(i)$ for each $i \notin \text{BIG}'$.   □

4 Bisimilarity between BPA and BPP in EXPTIME

In this section we give the main ideas of the proof of Theorem 2. We assume a fixed instance of the problem — a fixed BPA $\Sigma = (V, \mathcal{A}, \mathcal{R})$ with the initial configuration α_0 and a fixed BPP $\Delta = (P, Tr, \text{PRE}, \text{POST}, \mathcal{A}, \lambda)$ with the initial marking M_0, for which we have already checked (in polynomial space) that M_0 is down-joining (otherwise obviously $\alpha_0 \not\sim M_0$).

We recall the exponential constant C discussed in and before Lemma 10. The discussion and the construction of the BPA in Lemma 11 suggests that (Δ, M_0) can be represented by a certain kind of one-counter process, called a *one-counter net with resets (OCNR)*. It stores the values of "small" dd-functions (that are either ω or less than C) in the control unit and the value of big dd-functions in the counter. The transitions that set the big dd-functions to ω will be represented by special *reset* transitions that reset the value of the counter to some fixed value, independent of the previous value of the counter.

On the high level, the algorithm works as follows. For a given BPP process (Δ, M_0) it constructs a bisimilar OCNR Γ with an initial configuration c_0 such that $M_0 \sim c_0$. The size of Γ is at most exponential w.r.t. the size of (Δ, M_0) and Γ can be constructed in exponential time. The algorithm then decides in exponential time if $\alpha_0 \sim c_0$.

OCNR. A *one-counter net with resets* is a tuple $\Gamma = (\mathcal{F}, \mathcal{A}, R_{=0}, R_{>0})$, where $\mathcal{F}$ is a finite set of *control states*, $\mathcal{A}$ is a (finite) set of *actions*, and $R_{=0}, R_{>0} \subseteq$

$\mathcal{F} \times \mathcal{A} \times \mathsf{RuleTypes} \times (\mathbb{N} \cup \{-1\}) \times \mathcal{F}$ are finite sets of *rules*, where $\mathsf{RuleTypes} = \{change, reset\}$. Informally, $R_{=0}$ are the rules, which are enabled when the value of the counter is zero, and $R_{>0}$ are the rules, which are enabled when the counter is non-zero. We require that $(g, a, \xi, d, g') \in R_{=0}$ implies $d \geq 0$, and that $R_{=0} \subseteq R_{>0}$, as there is no test for zero.

Configurations of an OCNR $\Gamma = (\mathcal{F}, \mathcal{A}, R_{=0}, R_{>0})$ are pairs (g, k), where $g \in \mathcal{F}$ and $k \in \mathbb{N}$ is the value of the counter. To denote configurations, we will write $g(k)$ instead of (g, k). We also use $c_1, c_2, \ldots$ to denote configurations of Γ. The OCNR Γ generates the LTS $(S, \mathcal{A}, \longrightarrow)$ where $S = \mathcal{F} \times \mathbb{N}$ and where the transitions are defined as follows:

- $g(k) \xrightarrow{a} g'(k+d)$ iff $(g, a, change, d, g') \in R'$
- $g(k) \xrightarrow{a} g'(d)$ iff $(g, a, reset, d, g') \in R'$,

where $R' = R_{=0}$ for $k = 0$, and $R' = R_{>0}$ for $k > 0$.

A transition performed due to some rule $(g, a, t, reset, g) \in R_{=0} \cup R_{>0}$ is called a *reset*, and a transition performed due to some rule $(g, a, t, change, g) \in R_{=0} \cup R_{>0}$ is called a *change*.

Note that OCNR can be easily encoded into a pushdown automaton, but not in BPA, as, intuitively, we need states.

Construction of an OCNR Bisimilar to (Δ, M_0). Let us start with some technical definitions. A marking M is *big*, if there is some dd-function d such that $C \leq d(M) < \omega$. A marking, which is not big, is *small*.

Let REACH(M_0) be the set of markings reachable from M_0, and let $\mathcal{M}_{big}$ be the set of the big markings in REACH(M_0). We define a function $cnt : \mathcal{M}_{big} \to \mathbb{N}$, where $cnt(M)$ is the value $d(M)$ for the dd-functions d that are big in M.

Let $\simeq_C \subseteq$ REACH$(M_0) \times$ REACH(M_0) be the equivalence where $M \simeq_C M'$ iff M, M' differ only on values of big dd-functions (i.e., $d(M) \neq d(M')$ implies $C \leq d(M) < \omega$ and $C \leq d(M') < \omega$). Let $\mathcal{B}$ be the partition of REACH(M_0) according to $\simeq_C$, i.e., the elements of $\mathcal{B}$ are sets of markings, where M, M' are in the same set $B \in \mathcal{B}$ iff $M \simeq_C M'$. We will show later that the number of classes in $\mathcal{B}$ is at most exponential.

A class $B \in \mathcal{B}$ is *small* if it contains only small markings, and *big* otherwise. (Note that in a big class, all markings are big.)

For each class $B \in \mathcal{B}$, Γ contains a corresponding control state f_B. The control states corresponding to small classes are called *fs-states*, and the control states corresponding to big classes are called *oc-states*. The sets of fs-states and oc-states are denoted $\mathcal{F}_{fs}$ and $\mathcal{F}_{oc}$, respectively.

The OCNR Γ is constructed in such a way that each configuration $f_B(0)$, where B is small, is bisimilar to any marking $M \in B$, and each configuration $f_B(k)$, where B is big and $k \geq C$, is bisimilar to any marking $M \in B$ with $cnt(M) = k$. In each configuration $f_B(k)$ where B is big and $k \geq 0$, the values of dd-functions will be the same as the values of these functions in markings in B, except the functions, which are big in markings in B, which will have value k.

The transitions of Γ are constructed in an obvious way to meet the above requirement. In particular, the only resets in Γ are transitions in states from

$\mathcal{F}_{oc}$ that correspond to setting big dd-functions to ω. The initial configuration c_0 is the configuration corresponding to M_0.

By $\mathcal{C}_\Gamma$ we denote the set of configurations $\{f(0) \mid f \in \mathcal{F}_{fs}\} \cup \{g(k) \mid g \in \mathcal{F}_{oc}, k \geq 0\}$. Note that $\text{REACH}(c_0) \subseteq \mathcal{C}_\Gamma$.

Bounding the Size of Γ. Because the number of (different) dd-functions on Δ is exponential and each small dd-function has at most exponential value, we can naively estimate the number of control states of Γ as double exponential. A closer analysis reveals that this number is single exponential.

For this purpose, it is useful to introduce so called *symbolic markings*. A symbolic marking $\overline{M}$ is obtained from a marking M by replacing the values $M(p)$, where $M(p) \geq C$, with some special symbol $*$. Let $symb_C$ be the function that assigns to each marking the corresponding symbolic marking, and let $\mathcal{S}_C = \{symb_C(M) \mid M \in \text{REACH}(M_0)\}$. It is clear that for given a symbolic marking $\overline{M}$ we can check in polynomial space whether $\overline{M} \in \mathcal{S}_C$. Moreover, from $\overline{M}$ we can easily determine, which transitions (and so, which actions and changes on values of dd-functions) are enabled in any marking M such that $symb_C(M) = \overline{M}$. It is also clear that $\mathcal{S}_C$ contains at most $K = (C+1)^{|P|}$ symbolic markings.

Observation 12 *For each $M, M' \in \text{REACH}(M_0)$, $symb_C(M) = symb_C(M')$ implies $M \simeq_C M'$.*

From Observation 12 we see that $\simeq_C$ has at most K equivalence classes, which means that Γ has at most exponential number of control states. By using sets of symbolic markings as a succinct representation of control states of Γ, Γ can be constructed in exponential time.

The constructed OCNR Γ has some additional special properties that allow us to decide bisimilarity between BPA processes (Σ, α_0) and the OCNR process (Γ, c_0) in exponential time, w.r.t. the original BPA-BPP instance. The OCNR with these additional properties is called a *special OCNR* (sOCNR). Due to lack of space, the description of these properties together with the description of the rest of the algorithm are omitted here.

Lemma 13. *There is an exponential time algorithm that for a given BPP process (Δ, M_0) constructs an sOCNR process (Γ, c_0) such that $M_0 \sim c_0$.*

Lemma 14. *There is an algorithm deciding for a given BPA process (Σ, α_0) and the constructed sOCNR process (Γ, c_0), whether $\alpha_0 \sim c_0$. The running time of the algorithm is exponential wrt the size of the original instance of the problem.*

Intuitively, the basic idea, on which the algorithm from Lemma 14 is based, is the following. When $A\beta \sim c$, where $A \in V$ is normed, $\beta \in V^*$ and $c \in \mathcal{C}_\Gamma$, then there must exist some $c' \in \mathcal{C}_\Gamma$ such that $\beta \sim c'$. This means that β can be replaced with c' in $A\beta$, by which we obtain the configuration Ac' in a transition system that can be viewed as a sequential composition of BPA Σ and sOCNR Γ. We can then characterize the bisimulation equivalence in this combined system by a bisimulation base consisting of pairs of configurations of the form (Ac', c) where $Ac' \sim c$, resp. (A, c) where $A \sim c$.

This bisimulation base is still infinite but it can be represented succinctly due to fact that there is some computable exponential constant B such that if $Af(k) \sim g(\ell)$, where A is normed, then if k or ℓ is greater than B, then $\|A\|+k = \ell$ and it holds for each $k \geq B$ that $Af(k) \sim g(\ell)$ iff $Af(k+1) \sim g(\ell+1)$.

References

1. Benedikt, M., Göller, S., Kiefer, S., Murawski, A.: Bisimilarity of pushdown systems is nonelementary. In: Proc. 28th LiCS. IEEE Computer Society (to appear, 2013)
2. Burkart, O., Caucal, D., Moller, F., Steffen, B.: Verification on infinite structures. In: Handbook of Process Algebra, pp. 545–623. Elsevier (2001)
3. Burkart, O., Caucal, D., Steffen, B.: An elementary decision procedure for arbitrary context-free processes. In: Hájek, P., Wiedermann, J. (eds.) MFCS 1995. LNCS, vol. 969, pp. 423–433. Springer, Heidelberg (1995)
4. Černá, I., Křetínský, M., Kučera, A.: Comparing expressibility of normed BPA and normed BPP processes. Acta Informatica 36, 233–256 (1999)
5. Czerwinski, W., Fröschle, S.B., Lasota, S.: Partially-commutative context-free processes: Expressibility and tractability. Information and Computation 209(5), 782–798 (2011)
6. Esparza, J.: Petri nets, commutative context-free grammars, and basic parallel processes. Fundamenta Informaticae 31(1), 13–25 (1997)
7. Hirshfeld, Y., Jerrum, M.: Bisimulation equivalence is decidable for normed process algebra. In: Wiedermann, J., Van Emde Boas, P., Nielsen, M. (eds.) ICALP 1999. LNCS, vol. 1644, pp. 412–421. Springer, Heidelberg (1999)
8. Hirshfeld, Y., Jerrum, M., Moller, F.: A polynomial-time algorithm for deciding bisimulation equivalence of normed basic parallel processes. Mathematical Structures in Computer Science 6, 251–259 (1996)
9. Hirshfeld, Y., Jerrum, M., Moller, F.: A polynomial algorithm for deciding bisimilarity of normed context-free processes. Theor. Comput. Sci. 158, 143–159 (1996)
10. Jančar, P.: Bisimilarity on basic process algebra is in 2-ExpTime (an explicit proof). Logical Methods in Computer Science 9(1) (2013)
11. Jančar, P., Kot, M., Sawa, Z.: Complexity of deciding bisimilarity between normed BPA and normed BPP. Information and Computation 208(10), 1193–1205 (2010)
12. Jančar, P.: Strong bisimilarity on basic parallel processes is PSPACE-complete. In: Proc. 18th LiCS, pp. 218–227. IEEE Computer Society (2003)
13. Jančar, P., Kučera, A., Moller, F.: Deciding bisimilarity between BPA and BPP processes. In: Amadio, R.M., Lugiez, D. (eds.) CONCUR 2003. LNCS, vol. 2761, pp. 159–173. Springer, Heidelberg (2003)
14. Kiefer, S.: BPA bisimilarity is EXPTIME-hard. Information Processing Letters 113(4), 101–106 (2013)
15. Sénizergues, G.: The bisimulation problem for equational graphs of finite outdegree. SIAM J. Comput. 34(5), 1025–1106 (2005)
16. Srba, J.: Strong bisimilarity of simple process algebras: Complexity lower bounds. Acta Informatica 39, 469–499 (2003)
17. Srba, J.: Roadmap of infinite results. In: Current Trends In Theoretical Computer Science, The Challenge of the New Century, vol. 2, pp. 337–350. World Scientific Publishing Co. (2004), for an updated version see
http://users-cs.au.dk/srba/roadmap/

Guarding Orthogonal Art Galleries Using Sliding Cameras: Algorithmic and Hardness Results

Stephane Durocher* and Saeed Mehrabi**

Department of Computer Science,
University of Manitoba, Winnipeg, Canada
{durocher,mehrabi}@cs.umanitoba.ca

Abstract. Let P be an orthogonal polygon. Consider a sliding camera that travels back and forth along an orthogonal line segment $s \subseteq P$ as its *trajectory*. The camera can see a point $p \in P$ if there exists a point $q \in s$ such that pq is a line segment normal to s that is completely contained in P. In the *minimum-cardinality sliding cameras problem*, the objective is to find a set S of sliding cameras of minimum cardinality to guard P (i.e., every point in P can be seen by some sliding camera in S) while in the *minimum-length sliding cameras problem* the goal is to find such a set S so as to minimize the total length of trajectories along which the cameras in S travel.

In this paper, we first settle the complexity of the minimum-length sliding cameras problem by showing that it is polynomial tractable even for orthogonal polygons with holes, answering a question posed by Katz and Morgenstern [9]. Next we show that the minimum-cardinality sliding cameras problem is NP-hard when P is allowed to have holes, which partially answers another question posed by Katz and Morgenstern [9].

1 Introduction

The art gallery problem is well known in computational geometry, where the objective is to cover a geometric shape (e.g., a polygon) with the union of the visibility regions of a set of point guards while minimizing the number of guards. The problem's multiple variants have been examined extensively (e.g., see [1,15,17]) and can be classified based on the type of guards (e.g., points or line segments), the type of visibility model, and the geometric shape (e.g., simple polygons, orthogonal polygons [6], or polyominoes [2]).

In this paper, we consider a variant of the orthogonal art gallery problem introduced by Katz and Morgenstern [9], in which *sliding cameras* are used to guard the gallery. Let P be an orthogonal polygon with n vertices. A sliding camera travels back and forth along an orthogonal line segment s inside P. The camera (i.e., the guarding line segment s) can *see* a point $p \in P$ (equivalently, p

* Work of the author is supported in part by the Natural Sciences and Engineering Research Council of Canada (NSERC).
** Work of the author is supported in part by a University of Manitoba Graduate Fellowship (UMGF).

K. Chatterjee and J. Sgall (Eds.): MFCS 2013, LNCS 8087, pp. 314–324, 2013.
© Springer-Verlag Berlin Heidelberg 2013

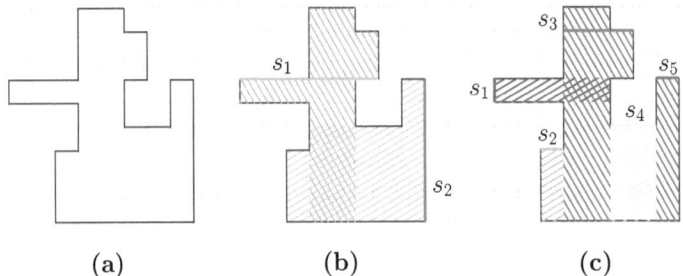

(a)   (b)   (c)

Fig. 1. An illustration of the MCSC and MLSC problems. Each grid cell has size 1×1. (a) A simple orthogonal polygon P. (b) The trajectories of two sliding cameras s_1 and s_2 are shown in pink and green, respectively; each shaded region indicates the visibility region of the corresponding camera. This set of two cameras is an optimal solution to the MCSC problem on P. (c) A set of five sliding cameras whose total length is 8, which is an optimal solution for the MLSC problem on P.

is *orthogonally visible* to s) if and only if there exists a point q on s such that pq is normal to s and is completely contained in P. We study two variants of this problem: in the *minimum-cardinality sliding cameras (MCSC) problem*, we wish to minimize the number of sliding cameras so as to guard P entirely, while in the *minimum-length sliding cameras (MLSC) problem* the objective is to minimize the total length of trajectories along which the cameras travel; we assume that in both variants of the problem, polygon P and sliding cameras are constrained to be orthogonal. In both problems, every point in P must be visible to some camera. See Figure 1.

Throughout the paper, we denote an orthogonal polygon with n vertices by P. Moreover, we denote the set of vertices and the set of edges of P by $V(P)$ and $E(P)$, respectively. We consider P to be a closed set; therefore, a camera's trajectory may include an edge of P. We also assume that a camera can see any point on its trajectory. We say that a set T of orthogonal line segments contained in P is a *cover of* P, if the corresponding cameras can collectively see any point in P; equivalently, we say that the line segments in T *guard* P entirely.

Related Work. The art gallery problem was first introduced by Klee in 1973. Two years later, Chvátal [3] gave an upper bound proving that $\lfloor n/3 \rfloor$ point guards are always sufficient and sometimes necessary to guard a simple polygon with n vertices. The orthogonal art gallery problem was first studied by Kahn et al. [7] who proved that $\lfloor n/4 \rfloor$ guards are always sufficient and sometimes necessary to guard the interior of a simple orthogonal polygon. Lee and Lin [12] showed that the problem of guarding a simple polygon using the minimum number of guards is NP-hard. Moreover, the problem was also shown to be NP-hard for orthogonal polygons [16]. Even the problem of guarding the vertices of an orthogonal polygon using the minimum number of guards is NP-hard [10].

Limiting visibility allows some versions of the problem to be solved in polynomial time. Motwani et al. [14] studied the art gallery problem under s-visibility, where a guard point $p \in P$ can see all points in P that can be connected to p by an orthogonal staircase path contained in P. They use a perfect graph approach to solve the problem in polynomial time. Worman and Keil [18] defined r-visibility, in which a guard point $p \in P$ can see all points $q \in P$ such that the bounding rectangle of p and q (i.e., the axis-parallel rectangle with diagonal $\overline{pq}$) is contained in P. Given that P has n vertices, they use a similar approach to Motwani et al. [14] to solve this problem in $\tilde{O}(n^{17})$ time, where $\tilde{O}()$ hides poly-logarithmic factors. Moreover, Lingas et al. [13] presented a linear-time 3-approximation algorithm for this problem.

Recently, Katz and Morgenstern [9] introduced sliding cameras as another model of visibility to guard a simple orthogonal polygon P; they study the MCSC problem. They first consider a restricted version of the problem, where cameras are constrained to travel only vertically inside the polygon. Using a similar approach to Motwani et al. [14] they construct a graph G corresponding to P and then show that (i) solving this problem on P is equivalent to solving the minimum clique cover problem on G, and that (ii) G is chordal. Since the minimum clique cover problem is polynomial-time solvable on chordal graphs, they solve the vertical-camera MCSC problem in polynomial time. They also generalize the problem such that both vertical and horizontal cameras are allowed (i.e., the MCSC problem); they present a 2-approximation algorithm for this problem under the assumption that the given input is an x-monotone orthogonal polygon. They leave open the complexity of the problem and mention studying the minimum-length sliding cameras problem as future work.

A *histogram* H is a simple orthogonal polygon that has an edge, called the *base*, whose length is equal to the sum of the lengths of the edges of H that are parallel to the base. Moreover, a *double-sided histogram* is the union of two histograms that share the same base edge and that are located on opposite sides of the base. It is easy to observe that the MCSC problem is equivalent to the problem of covering P with minimum number of double-sided histograms. Fekete and Mitchell [4] proved that partitioning an orthogonal polygon (possibly with holes) into a minimum number of histograms is NP-hard. However, their proof does not directly imply that the MCSC problem is also NP-hard for orthogonal polygons with holes.

Our Results. In this paper, we first answer a question posed by Katz and Morgenstern [9] by proving that the MLSC problem is solvable in polynomial time even for orthogonal polygons with holes (see Section 2). We next show that the MCSC problem is NP-hard for orthogonal polygons with holes (see Section 3) that partially answers another question posed by Katz and Morgenstern [9]. We conclude the paper in Section 4.

2   The MLSC Problem: An Exact Algorithm

In this section, we give an algorithm that solves the MLSC problem exactly in polynomial time even when P has holes. Let T be a cover of P. In this section, we say that T is an *optimal cover for P* if the total length of trajectories along which the cameras in T travel is minimum over that of all covers of P. Our algorithm relies on reducing the MLSC problem to the *minimum-weight vertex cover problem* in bipartite graphs. We remind the reader of the definition of the minimum-weight vertex cover problem:

Definition 1. *Given a graph $G = (V, E)$ with positive vertex weights, the minimum-weight vertex cover problem is to find a subset $V' \subseteq V$ that is a vertex cover of G (i.e., every edge in E has at least one endpoint in V') such that the sum of the weights of vertices in V' is minimized.*

The minimum-weight vertex cover problem is NP-hard in general [8]. However, König's theorem [11] that describes the equivalence between maximum matching and vertex cover in bipartite graphs implies that the minimum-weight vertex cover problem in bipartite graphs is solvable in polynomial time. Given P, we first construct a vertex-weighted graph G_P and then we show (i) that the MLSC problem on P is equivalent to the minimum-weight vertex cover problem on G_P, and (ii) that graph G_P is bipartite.

Similar to Katz and Morgenstern [9], we define a partition of an orthogonal polygon P into rectangles as follows. Extend the two edges of P incident to every reflex vertex in $V(P)$ inward until they hit the boundary of P. Let $S(P)$ be the set of the extended edges and the edges of P whose endpoints are both non-reflex vertices of P. We refer to elements of $S(P)$ simply as *edges*. The edges in $S(P)$ partition P into a set of rectangles; let $R(P)$ denote the set of resulting rectangles. We observe that in order to guard P entirely, it suffices to guard all rectangles in $R(P)$. The following observations are straightforward:

Observation 1. *Let T be a cover of P and let s be an orthogonal line segment in T. Then, for any partition of s into line segments $s_1, s_2, \ldots, s_k$ the set $T' = (T \setminus \{s\}) \cup \{s_1, \ldots, s_k\}$ is also a cover of P and the respective sums of the lengths of segments in T and T' are equal.*

Observation 2. *Let T be a cover of P. Moreover, let T' be the set of line segments obtained from T by translating every vertical line segment in T horizontally to the nearest boundary of P to its right and every horizontal line segment in T vertically to the nearest boundary of P below it. Then, T' is also a cover of P and the respective sums of the lengths of line segments in T and T' are equal. We call T' a regular cover of P.*

We first need the following result.

Lemma 1. *Let $R \in R(P)$ be a rectangle and let T be a cover of P. Then, there exists a set $T' \subseteq T$ such that all line segments in T' have the same orientation (i.e., they are all vertical or they are all horizontal) and they collectively guard R entirely.*

Proof. Suppose no such set T' exists. Let R_v (resp., R_h) be the subregion of R that is guarded by the union of the vertical (resp., horizontal) line segments

Fig. 2. An illustration of the reduction; each grid cell has size 1×1. (a) An orthogonal polygon P along with the elements of $B(P)$ labelled as $a, b, c, \ldots, i$. (b) The graph G_P associated with P; the integer value besides each vertex indicates the weight of the vertex. The vertices of a vertex cover on G_P and their corresponding guarding line segments for P are shown in red.

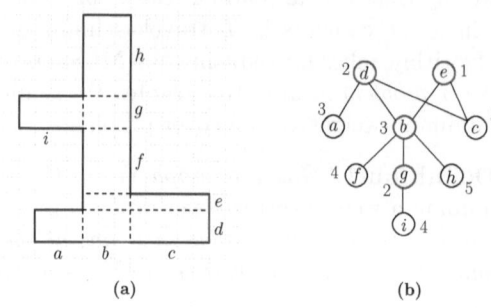

(a)   (b)

in T and let $R_v^c = R \setminus R_v$ (resp., $R_h^c = R \setminus R_h$). Since R cannot be guarded exclusively by vertical line segments (resp., horizontal line segments), we have $R_v^c \neq \emptyset$ (resp., $R_h^c \neq \emptyset$). Choose any point $p \in R_v^c$ and let L_h be the maximal horizontal line segment inside R that crosses p. Since no vertical line segment in T can guard p, we conclude that no point on L_h is guarded by a vertical line segment in T. Similarly, choose any point $q \in R_h^c$ and let L_v be the maximal vertical line segment inside R that contains q. By an analogous argument, we conclude that no point on L_v is guarded by a horizontal line segment. Since L_h and L_v are maximal and have perpendicular orientations, L_h and L_v intersect inside R. Therefore, no orthogonal line segment in T can guard the intersection point of L_h and L_v, which is a contradiction.   □

Given P, let $H(P)$ denote the subset of the boundary of P consisting of line segments that are immediately to the right of or below P; in other words, for each edge $e \in H(P)$, the region of the plane immediately to the right of or below e does not belong to the interior of P. Let $B(P)$ denote the partition of $H(P)$ into line segments induced by the edges in $S(P)$. The following lemma follows by Lemma 1 and Observations 1 and 2:

Lemma 2. *Every orthogonal polygon P has an optimal cover $T \subseteq B(P)$.*

Observation 3. *Let P be an orthogonal polygon and consider its corresponding set $R(P)$ of rectangles induced by edges in $S(P)$. Every rectangle $R \in R(P)$ is seen by exactly one vertical line segment in $B(P)$ and exactly one horizontal line segment in $B(P)$. Furthermore, if $T \subseteq B(P)$ is a cover of P, then every rectangle in $R(P)$ must be seen by at least one horizontal or one vertical line segment in T.*

We denote the horizontal and vertical line segments in $B(P)$ that can see a rectangle $R \in R(P)$ by R_V and R_H, respectively. Using Observation 3, we now describe a reduction of the MLSC problem to the minimum-weight vertex cover problem. We construct an undirected weighted graph $G_P = (V, E)$ associated with P as follows: each line segment $s \in B(P)$ corresponds to a vertex $v_s \in V$ such that the weight of v_s is the length of s. We denote the vertex in V that corresponds to the line segment $s \in B(P)$ by v_s. Two vertices $v_s, v_{s'} \in V$ are adjacent in G_P if and only if the line segments s and s' can both see a common rectangle $R \in R(P)$. See Figure 2. By Observation 3 the following result is straightforward:

Observation 4. *There is a bijection between rectangles in $R(P)$ and edges in G_P.*

Next we show equivalency between the two problems and then prove that graph G_P is bipartite.

Theorem 1. *The MLSC problem on P reduces to the minimum-weight vertex cover problem on G_P.*

Proof. Let S_0 be a vertex cover of G_P and let C_0 be a cover of P defined in terms of S_0; the mapping from S_0 to C_0 will be defined later. Moreover, for each vertex v of G_P let $w(v)$ denote the weight of v and for each line segment $s \in C_0$ let $len(s)$ denote the length of s. We need to prove that S_0 is a minimum-weight vertex cover of G_P if and only if C_0 is an optimal cover of P. We show the following stronger statements: (i) for any vertex cover S of G_P, there exists a cover C of P such that

$$\sum_{s \in C} len(s) = \sum_{v \in S} w(v),$$

and (ii) for any cover C of P, there exists a vertex cover S of G_P such that

$$\sum_{v \in S} w(v) = \sum_{s \in C} len(s).$$

Part 1. Choose any vertex cover S of G_P. We find a cover C for P as follows: for each edge $(v_s, v_{s'}) \in E$, if $v_s \in S$ we locate a guarding line segment on the boundary of P that is aligned with the line segment $s \in B(P)$. Otherwise, we locate a guarding line segment on the boundary of P that is aligned with the line segment $s' \in B(P)$. Since at least one of v_s and $v_{s'}$ is in S, we conclude by Observation 4 that every rectangle in $R(P)$ is guarded by at least one line segment located on the boundary of P and so C is a cover of P. Moreover, for each vertex in S we locate exactly one guarding line segment on the boundary of P whose length is the same as the weight of the vertex. Therefore,

$$\sum_{s \in C} len(s) = \sum_{v \in S} w(v).$$

Part 2. Choose any cover C of P. We construct a vertex cover S for G_P as follows. By Observation 2, let T' be the regular cover obtained from C. Moreover, let M be the partition of T' into line segments induced by the edges in $S(P)$. By Lemma 1, for any rectangle $R \in R(P)$, there exists a set $C'_R \subseteq C$ such that all line segments in C'_R have the same orientation and collectively guard R. Therefore, M is also a cover of P. Now, let S be the subset of the vertices of G_P such that $v_s \in S$ if and only if $s \in M$. Since M is a cover of G_P we conclude, by Observation 4, that S is a vertex cover of G_P. Moreover, we observe that

$$\sum_{v \in S} w(v) = \sum_{s \in M} len(s) = \sum_{s \in C} len(s). \qquad \square$$

Lemma 3. *Graph G_P is bipartite.*

Proof. The proof follows from the facts that (i) we have two types of vertices in G_P; those that correspond to the vertical line segments in $B(P)$ and those that correspond to the horizontal line segments in $B(P)$, and that (ii) no two vertical line segments in $B(P)$ nor any two horizontal line segments in $B(P)$ can see a fixed rectangle in $R(P)$.   □

It is easy to see that the construction in the proof of Theorem 1 can be completed in polynomial time. Therefore, by Theorem 1, Lemma 3 and the fact that minimum-weight vertex cover is solvable in polynomial time on bipartite graphs [11], we have the main result of this section:

Theorem 2. *Given an orthogonal polygon P with n vertices, there exists an algorithm that finds an optimal cover of P in time polynomial in n.*

3   The MCSC Problem

In this section, we show that the following problem is NP-hard:

MCSC With Holes
Input: An orthogonal polygon P, possibly with holes and an integer k.
Output: Yes, if there exists k orthogonal line segments inside P that guard P entirely; No, otherwise.

Fig. 3. An L-hole gadget; each grid cell has size $\frac{1}{12} \times \frac{1}{12}$

We show NP-hardness by a reduction from the *minimum hitting of horizontal unit segments* problem, which we call the Min Segment Hitting problem. The Min Segment Hitting problem is defined as follows [5]:

Min Segment Hitting
Input: n pairs (a_i, b_i), $i = 1, \ldots, n$, of integers and an integer k
Output: Yes, if there exist k orthogonal lines $l_1, \ldots, l_k$ in the plane, i.e., for each i, l_i is horizontal or vertical, such that each line segment $[(a_i, b_i), (a_i + 1, b_i)]$ is hit by at least one of the lines; No, otherwise.

Hassin and Megiddo [5] prove that the Min Segment Hitting problem is NP-complete. Let I be an instance of the Min Segment Hitting problem, where I is a set of n horizontal unit-length segments with integer coordinates. We construct an orthogonal polygon P (with holes) such that there exists a set of k orthogonal lines that hit the segments in I if and only if there exists a set C of $k + 1$ orthogonal line segments inside P that collectively guard P. Throughout this section, we refer to the segments in I as *unit segments* and to the segments in C as *line segments*.

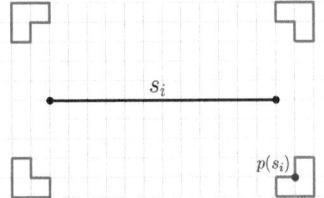

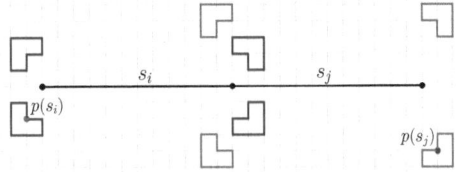

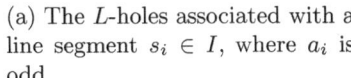

(a) The L-holes associated with a line segment $s_i \in I$, where a_i is odd.

(b) An illustration of the L-holes associated with two line segments in I that share a common endpoint.

Fig. 5. An illustration of the gadgets used in the reduction

Gadgets. We first observe that any two unit segments in I can share at most one point, which must be a common endpoint of the two unit segments. For each unit segment $s_i \in I$, $1 \le i \le n$, we denote the left endpoint of s_i by (a_i, b_i) and, therefore, the right endpoint of s_i is $(a_i + 1, b_i)$. Moreover, let $N(s_i)$ denote the set of unit segments in I that have at least one endpoint with x-coordinate equal to a_i or $a_i + 1$. Our reduction refers to an L-hole,

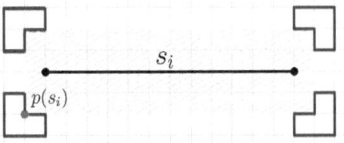

Fig. 4. The L-holes associated with a line segment $s_i \in I$, where a_i is even

which we define as a minimum-area orthogonal polygon with six vertices at grid coordinates such that exactly one is a reflex vertex. Figure 3 shows an L-hole. We constrain each grid cell to have size $\frac{1}{12} \times \frac{1}{12}$. An L-hole may be rotated by $\pi/2$, π or $3\pi/2$. For each unit segment $s_i \in I$, we associate exactly four L-holes with s_i depending on the parity of a_i: if a_i is even, then Figure 4 shows the L-holes associated with s_i. If a_i is odd, then Figure 5a shows the L-holes associated with s_i. Note that, in this case, the L-holes are located such that the vertical distance between any point on an L-hole and s_i is at least $3/12$. Note the red vertex on the bottom left L-hole of s_i in Figure 4 and the blue vertex on the bottom right L-hole of s_i in Figure 5a; we call this vertex the *visibility vertex* of s_i, which we denote $p(s_i)$.

Observe that the L-holes associated with s_i do not interfere with the L-holes associated with the line segments in $N(s_i)$ because for any unit segment $s_j \in N(s_i)$ the vertical distance d between s_i and s_j is either zero or at least one. If $d \ge 1$, then it is trivial that the L-holes of s_i do not interfere with those of s_j. Now, suppose that s_i and s_j share a common endpoint; that is $d = 0$. Since s_i and s_j have unit lengths a_i and a_j have different parities and, therefore, the L-holes associated with s_i and s_j do not interfere with each other. Figure 5b shows an example of such two unit segments s_i and s_j and their corresponding L-holes. We now describe the reduction.

Reduction. Given an instance I of the MIN SEGMENT HITTING problem, we first associate each unit segment in $s_i \in I$ with four L-holes depending on

Fig. 6. A complete example of the reduction, where $I = \{s_1, s_2, \ldots, s_9\}$, with the assumption that a_1 is even. Each line segment that has a bend represents an L-hole associated with a unit segment. The visibility vertices of the unit segments in I are shown red or blue appropriately. Note the green vertex on the lower left corner of the smaller rectangle; this vertex is only visible to the line segments that pass through the interior of the smaller rectangle, which in turn cannot intersect any unit segment in I.

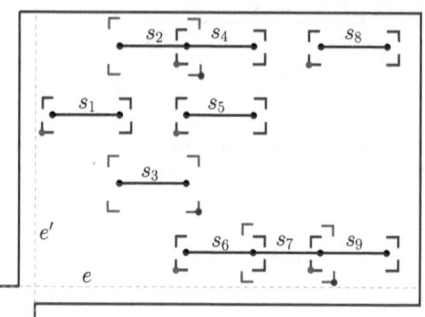

whether a_i is even or odd. After adding the corresponding L-holes, we enclose I in a rectangle such that all unit segments and the L-holes associated with them lie in its interior. Finally, we create a small rectangle on the bottom left corner of the bigger rectangle (see Figure 6) such that any orthogonal line that passes through the smaller rectangle cannot intersect any of the unit segments in I. See Figure 6 for a complete example of the reduction. Let P be the resulting orthogonal polygon. Observe from Figure 4 (see also Figure 5a) that the left endpoint (resp., the right endpoint) of every unit segment $s \in I$ is vertically aligned with the rightmost edges (resp., leftmost edges) of the two left L-holes (resp., right L-holes) associated with s. This provides the following observation.

Observation 5. *Let s be a unit segment in I and let l be a vertical line segment contained in P that can see $p(s)$. Moreover, let l' be the maximal vertical line segment that is aligned with l. If l' does not intersect s, then $p(s')$ is not orthogonally visible to l' for all $s' \in I \setminus \{s\}$.*

We now show the following lemma.

Lemma 4. *There exist k orthogonal lines such that each unit segment in I is hit by one of the lines if and only if there exists $k + 1$ orthogonal line segments contained in P that collectively guard P.*

Proof. ($\Rightarrow$) Suppose there exists a set S of k lines such that each unit segment in I is hit by at least one line in S. Let $L \in S$ and let $L_P = L \cap P$. If L is horizontal, then it is easy to see that L, and therefore L_P, does not cross any L-hole inside P. Similarly, if L is vertical and passes through an endpoint of some unit segment(s) in I, then neither L nor L_P passes through the interior of any L-hole in P.[1] Now, suppose that L is vertical and passes through the interior of some unit segment $s \in I$. Translate L_P horizontally such that it passes through the midpoint of s. Since unit segments have endpoints on adjacent integer grid point, L_P still crosses the same set of unit segments of I as it did before this move.

---

[1] Note that it is possible for L to pass through the boundary of some L-hole.

Moreover, this ensures that L_P does not cross any L-hole inside P. Consider the set $S' = \{L_P \mid L \in S\}$.

We observe that the line segments in S' cannot guard the interior of the smaller rectangle. Moreover, if all line segments in S' are vertical or all are horizontal, then they cannot collectively guard the outer rectangle entirely.[2] In order to guard P entirely, we add one more orthogonal line segment C as follows: if all line segments in S' are vertical (resp., horizontal), then C is the maximal horizontal (resp., the maximal vertical) line segment inside P that aligns the upper edge (resp., the right edge) of the smaller rectangle of P; see the line segment e (resp., e') in Figure 6. If the line segments in S' are a combination of vertical and horizontal line segments, then C can be either e or e'. It is easy to observe that now the line segments in S' along with C collectively guard P entirely. Therefore, we have established that the entire polygon P is guarded by $k + 1$ orthogonal line segments inside P in total.

($\Leftarrow$) Now, suppose that there exists a set M of $k+1$ orthogonal line segments contained in P that collectively guard P. Let $c \in M$ and let L_c denote the line induced by c. We now describe how to find k lines that form a solution to instance I by moving the line segments in M accordingly such that each unit segment in I is hit by at least one of the corresponding lines. Let $c_0 \in M$ be the line segment that guards the bottom left vertex of the smaller rectangle of P. We know that L_{c_0} cannot guard $p(s)$ for any unit segment $s \in I$. For each unit segment $s \in I$ in order, consider a line segment $l \in M \setminus \{c_0\}$ that guards $p(s)$; let l' be the maximal line segment inside P that is aligned with l. We observe that l' must intersect the rectangle whose endpoints are the reflex vertices of the L-holes associated with unit segment s (see the pink rectangle in Figure 4 for an example). If l' is horizontal and $L_{l'}$ does not align s, then move l' accordingly up or down until it aligns with s. Thus, $L_{l'}$ is a line that hits s. Now, suppose that l' is vertical. If l' intersects s, then $L_{l'}$ also intersects s. It might be possible that l' is vertical and guards $p(s)$, but $L_{l'}$ does not intersect s; in this case, by Observation 5, $p(s)$ is the only visibility vertex that is visible to l'. So, move l' horizontally to the left or to the right until it hits s. Therefore, $L_{l'}$ is a line that hits s after this move.

We observe that we obtained exactly one line from each line segment in $M \setminus \{c_0\}$. Therefore, we have found k lines such that each unit segment in I is hit by at least one of the lines. This completes the proof of the lemma.   □

By Lemma 4 we obtain the main result of this section:

Theorem 3. *The* MCSC WITH HOLES *is* NP-*hard.*

4   Conclusion

In this paper, we studied the problem of guarding an orthogonal polygon P using sliding cameras that was introduced by Katz and Morgenstern [9]. We considered

---

[2] Specifically, in either cases, there are regions between two L-holes associated with different unit segments that cannot be guarded by any line segment.

two variants of this problem: the MCSC problem (in which the objective is to minimize the number of sliding cameras used to guard P) and the MLSC problem (in which the objective is to minimize the total length of trajectories along which the cameras travel).

We gave a polynomial-time algorithm that solves the MLSC problem exactly even for orthogonal polygons with holes, answering a question posed by Katz and Morgenstern [9]. We also showed that the MCSC problem is NP-hard when P contains holes, which partially answers another question posed by Katz and Morgenstern [9]. Although we settled the complexity of the MLSC problem, the complexity of the MCSC problem for any simple orthogonal polygon remains open. Giving an approximation algorithm for the MCSC problem on any simple orthogonal polygon is also another direction for future work.

References

1. Amit, Y., Mitchell, J.S.B., Packer, E.: Locating guards for visibility coverage of polygons. Int. J. Comput. Geometry Appl. 20(5), 601–630 (2010)
2. Biedl, T.C., Irfan, M.T., Iwerks, J., Kim, J., Mitchell, J.S.B.: The art gallery theorem for polyominoes. Disc. & Comp. Geom. 48(3), 711–720 (2012)
3. Chvátal, V.: A combinatorial theorem in plane geometry. J. Comb. Theory, Ser. B 18, 39–41 (1975)
4. Fekete, S.P., Mitchell, J.S.B.: Terrain decomposition and layered manufacturing. Int. J. of Comp. Geom. & App. 11(6), 647–668 (2001)
5. Hassin, R., Megiddo, N.: Approximation algorithms for hitting objects with straight lines. Disc. App. Math. 30(1), 29–42 (1991)
6. Hoffmann, F.: On the rectilinear art gallery problem. In: Paterson, M. (ed.) ICALP 1990. LNCS, vol. 443, pp. 717–728. Springer, Heidelberg (1990)
7. Kahn, J., Klawe, M.M., Kleitman, D.J.: Traditional galleries require fewer watchmen. SIAM J. on Algebraic Disc. Methods 4(2), 194–206 (1983)
8. Karp, R.M.: Reducibility among combinatorial problems. In: Complexity of Computer Computations, pp. 85–103 (1972)
9. Katz, M.J., Morgenstern, G.: Guarding orthogonal art galleries with sliding cameras. Int. J. of Comp. Geom. & App. 21(2), 241–250 (2011)
10. Katz, M.J., Roisman, G.S.: On guarding the vertices of rectilinear domains. Comput. Geom. 39(3), 219–228 (2008)
11. König, D.: Gráfok és mátrixok. Matematikai és Fizikai Lapok 38, 116–119 (1931)
12. Lee, D.T., Lin, A.K.: Computational complexity of art gallery problems. IEEE Trans. on Inf. Theory 32(2), 276–282 (1986)
13. Lingas, A., Wasylewicz, A., Żyliński, P.: Linear-time 3-approximation algorithm for the r-star covering problem. In: Nakano, S.-i., Rahman, M. S. (eds.) WALCOM 2008. LNCS, vol. 4921, pp. 157–168. Springer, Heidelberg (2008)
14. Motwani, R., Raghunathan, A., Saran, H.: Covering orthogonal polygons with star polygons: the perfect graph approach. In: Proc. ACM SoCG, pp. 211–223 (1988)
15. O'Rourke, J.: Art gallery theorems and algorithms. Oxford University Press (1987)
16. Schuchardt, D., Hecker, H.-D.: Two NP-hard art-gallery problems for orthopolygons. Math. Logic Quarterly 41(2), 261–267 (1995)
17. Urrutia, J.: Art gallery and illumination problems. In: Handbook of Comp. Geom., pp. 973–1027. North-Holland (2000)
18. Worman, C., Keil, J.M.: Polygon decomposition and the orthogonal art gallery problem. Int. J. of Comp. Geom. & App. 17(2), 105–138 (2007)

Linear-Space Data Structures for Range Frequency Queries on Arrays and Trees*

Stephane Durocher[1], Rahul Shah[2],
Matthew Skala[1], and Sharma V. Thankachan[2]

[1] University of Manitoba, Winnipeg, Canada
{durocher,mskala}@cs.umanitoba.ca
[2] Louisiana State University, Baton Rouge, USA
{rahul,thanks}@csc.lsu.edu

Abstract. We present $O(n)$-space data structures to support various range frequency queries on a given array $A[0 : n - 1]$ or tree T with n nodes. Given a query consisting of an arbitrary pair of pre-order rank indices (i, j), our data structures return a least frequent element, mode, or α-minority of the multiset of elements in the unique path with endpoints at indices i and j in A or T. We describe a data structure that supports range least frequent element queries on arrays in $O(\sqrt{n/w})$ time, improving the $\Theta(\sqrt{n})$ worst-case time required by the data structure of Chan et al. (SWAT 2012), where $w \in \Omega(\log n)$ is the word size in bits. We describe a data structure that supports range mode queries on trees in $O(\log \log n \sqrt{n/w})$ time, improving the $\Theta(\sqrt{n} \log n)$ worst-case time required by the data structure of Krizanc et al. (ISAAC 2003). Finally, we describe a data structure that supports range α-minority queries on trees in $O(\alpha^{-1} \log \log n)$ time, where $\alpha \in [0, 1]$ is specified at query time.

1 Introduction

The *frequency*, denoted $\mathrm{freq}_{A[i:j]}(x)$, of an element x in a multiset stored as an array $A[i : j]$ is the number of occurrences of x in $A[i : j]$. Elements a and b in $A[i : j]$ are respectively a *mode* and a *least frequent* element of $A[i : j]$ if for all $c \in A[i : j]$, $\mathrm{freq}_{A[i:j]}(a) \geq \mathrm{freq}_{A[i:j]}(c) \geq \mathrm{freq}_{A[i:j]}(b)$. Finally, given $\alpha \in [0, 1]$, an α-*minority* of $A[i : j]$ is an element $d \in A[i : j]$ such that $1 \leq \mathrm{freq}_{A[i:j]}(d) \leq \alpha|j - i + 1|$. Conversely, d is an α-*majority* of $A[i : j]$ if $\mathrm{freq}_{A[i:j]}(d) > \alpha|j - i + 1|$.

We study the problem of indexing a given array $A[0 : n - 1]$ to construct data structures that can be stored using $O(n)$ words of space and support efficient range frequency queries. Each query consists of a pair of input indices (i, j) (along with a value $\alpha \in [0, 1]$ for α-minority queries), for which a mode, least frequent element, or α-minority of $A[i : j]$ must be returned. Range queries generalize to trees, where they are called *path queries*: given a tree T and a pair of indices

* Work supported in part by the Natural Sciences and Engineering Research Council of Canada (NSERC) and National Science Foundation (NSF) Grants CCF–1017623 (R. Shah and J. S. Vitter) and CCF–1218904 (R. Shah).

K. Chatterjee and J. Sgall (Eds.): MFCS 2013, LNCS 8087, pp. 325–336, 2013.

(i, j), a query is applied to the multiset of elements stored at nodes along the unique path in T whose endpoints are the two nodes with pre-order traversal ranks i and j.

Krizanc et al. [12] presented $O(n)$-space data structures that support range mode queries in $O(\sqrt{n} \log \log n)$ time on arrays and $O(\sqrt{n} \log n)$ time on trees. Chan et al. [3,4] achieved $o(\sqrt{n})$ query time with an $O(n)$-space data structure that supports queries in $O(\sqrt{n/w}) \subseteq O(\sqrt{n/\log n})$ time on arrays, where $w \in \Omega(\log n)$ is the word size in bits.

For range least frequent elements, Chan et al. [5] presented an $O(n)$-space data structure that supports queries in $O(\sqrt{n})$ time on arrays. Range mode and range least frequent queries on arrays appear to require significantly longer times than either range minimum or range selection queries; respective reductions from boolean matrix multiplication show that query times significantly lower than $\sqrt{n}$ are unlikely for either problem with linear space [3,5]. Whereas an $O(n)$-space data structure that supports range mode queries on arrays in $o(\sqrt{n})$ time is known [3], the space reduction techniques applied to achieve the time improvement are not directly applicable to the setting of least frequent elements. Chan et al. [5] ask whether $o(\sqrt{n})$ query time is possible in an $O(n)$-space linear data structure, observing that "unlike the frequency of the mode, the frequency of the least frequent element does not vary monotonically over a sequence of elements. Furthermore, unlike the mode, when the least frequent element changes [in a sequence], the new element of minimum frequency is not necessarily located in the block in which the change occurs" [5, p. 11]. By applying different techniques, this paper presents the first $O(n)$-space data structure that supports range least frequent element queries on arrays in $o(\sqrt{n})$ time; specifically, we achieve $O(\sqrt{n/w}) \subseteq O(\sqrt{n/\log n})$ query time.

Finally, the range α-majority query problem was introduced by Durocher et al. [7,8], who presented an $O(n \log(\alpha^{-1}))$-space data structure that supports queries in $O(\alpha^{-1})$ time for any $\alpha \in (0, 1)$ fixed during preprocessing. When α is specified at query time, Gagie et al. [9] and Chan et al. [5] presented $O(n \log n)$-space data structures that support queries in $O(\alpha^{-1})$ time, and Belazzougui et al. [2] presented an $O(n)$-space data structure that supports queries in $O(\alpha^{-1} \log \log(\alpha^{-1}))$ time. For range α-minority queries, Chan et al. [5] described an $O(n)$-space data structure that supports queries in $O(\alpha^{-1})$ time, where α is specified at query time.

After revisiting some necessary previous work in Section 2, in Section 3 we describe the first $O(n)$-space data structure that achieves $o(\sqrt{n})$ time for range least frequent queries on arrays, supporting queries in $O(\sqrt{n/w})$ time. We then extend this data structure to the setting of trees. In Section 4 we present an $O(n)$-space data structure that supports path mode queries on trees in $O(\log \log n \sqrt{n/w})$ time. To do so, we construct $O(n)$-space data structures that support colored nearest ancestor queries on trees in $O(\log \log n)$ time (find the nearest ancestor with value k of node i, where i and k are given at query time); path frequency queries on trees in $O(\log \log n)$ time (count the number of instances of k on the path between nodes i and j, where i, j, and k are given at query time); and

Table 1. worst-case query times of previous best and new $O(n)$-space data structures

range query	input	previous best	new (this paper)
least frequent element	array	$O(\sqrt{n})$ [5]	$O(\sqrt{n/w})$
	tree	no previous result	$O(\log\log n\sqrt{n/w})$
mode	array	$O(\sqrt{n/w})$ [3,4]	
	tree	$O(\sqrt{n}\log n)$ [11,12]	$O(\log\log n\sqrt{n/w})$
α-minority	array	$O(\alpha^{-1})$ [5]	
	tree	no previous result	$O(\alpha^{-1}\log\log n)$

k-nearest distinct ancestor queries on trees in $O(k)$ time (return k ancestors of node i such that each ancestor stores a distinct value and the distance to the furthest ancestor from i is minimized, where i and k are given at query time). Finally, in Section 5 we present an $O(n)$-space data structure that supports path α-minority query on trees in $O(\alpha^{-1}\log\log n)$ time, where α is given at query time. Our contributions are summarized in Table 1.

We assume the Word RAM model of computation using words of size $w \in \Omega(\log n)$ bits, where n denotes the number of elements stored in the input array/tree. Unless explicitly specified otherwise, space requirements are expressed in multiples of words. We use the notation $\log^{(k)}$ to represent logarithm iterated k times; that is, $\log^{(1)} n = \log n$ and $\log^{(k)} n = \log\log^{(k-1)} n$ for any integer $k > 1$. To avoid ambiguity, we use the notation $(\log n)^2$ instead of $\log^2 n$.

2 Chan et al.'s Framework for Range Least Frequent Element Query on Arrays

Our data structure for range least frequent element queries on an arbitrary given input array $A[0 : n - 1]$ uses a technique introduced by Chan et al. [5]. Upon applying a rank space reduction to A, all elements in A are in the range $\{0,\ldots,\Delta - 1\}$, where Δ denotes the number of distinct elements in the original array A. Before returning the result of a range query computation, the corresponding element in the rank-reduced array is mapped to its original value in constant time by a table lookup [3,5]. Chan et al. [5] prove the following result.

Theorem 1 (Chan et al. [5]). *Given any array $A[0 : n - 1]$ and any fixed $s \in [1, n]$, there exists an $O(n + s^2)$-word space data structure that supports range least frequent element query on A in $O(n/s)$ time and requires $O(n \cdot s)$ preprocessing time.*

The data structure of Chan et al. includes index data that occupy a linear number of words, and two tables D_t and E_t whose sizes ($O(s^2)$ words each) depend on the parameter s. Let t be an integer blocking factor. Partition $A[0 : n - 1]$ into $s = \lceil n/t \rceil$ blocks of size t (except possibly the last block which has size $1 + [(n - 1) \bmod t]$). For every pair (i, j), where $0 \le i < j \le s - 1$, the contents of the tables D_t and E_t are as follows:

- $D_t(i,j)$ stores a least frequent element in $A[i \cdot t : j \cdot t - 1]$, and
- $E_t(i,j)$ stores an element which is least frequent in the multiset of elements that are in $A[i \cdot t : j \cdot t - 1]$ but not in $A[i \cdot t : (i+1)t - 1] \cup A[(j-1)t : j \cdot t - 1]$.

In the data structure of Chan et al. [5], the tables D_t and E_t are the only components whose space bound depends on s. The cost of storing and accessing the tables can be computed separately from the costs incurred by the rest of the data structure. The proof for Theorem 1 given by Chan et al. implies the following result.

Lemma 1 (Chan et al. [5]). *If the tables D_t and E_t can be stored using $S(t)$ bits of space to support lookup queries in $T(t)$ time, then, for any $\{i,j\} \subseteq \{0,\ldots,n-1\}$, a least frequent element in $A[i:j]$ can be computed in $O(T(t)+t)$ time using an $O(S(t) + n \log n)$-bit data structure.*

When $t \in \Theta(\sqrt{n})$, the tables D_t and E_t can be stored explicitly in linear space. In that case, $S(t) \in O((n/\sqrt{n})^2 \log n) = O(n \log n)$ bits and $T(t) \in O(1)$, resulting in an $O(n \log n)$-bit ($O(n)$-word) space data structure that supports $O(\sqrt{n})$-time queries [5]. In the present work, we describe how to encode the tables using fewer bits per entry, allowing them to contain more entries, and therefore allowing a smaller value for t and lower query time.

We also refer to the following lemma by Chan et al. [3]:

Lemma 2 (Chan et al. [3]). *Given an array $A[0 : n - 1]$, there exists an $O(n)$-space data structure that returns the index of the q-th instance of $A[i]$ in $A[i : n-1]$ in $O(1)$ time for any $0 \leq i \leq n - 1$ and any q.*

3   Faster Range Least Frequent Element Query on Arrays

We first describe how to calculate the table entries for a smaller block size using lookups on a similar pair of tables for a larger block size and some index data that fits in linear space. Then, starting from the $t = \sqrt{n}$ tables which we can store explicitly, we apply that block-shrinking operation $\log^* n$ times, ending with blocks of size $O(\sqrt{n/w})$, which gives the desired lookup time.

At each level of the construction, we partition the array into three levels of blocks whose sizes are t (*big blocks*), t' (*small blocks*), and t'' (*micro blocks*), where $1 \leq t'' \leq t' \leq t \leq n$. We will compute table entries for the small blocks, $D_{t'}$ and $E_{t'}$, assuming access to table entries for the big blocks, D_t and E_t. The micro block size t'' is a parameter of the construction but does not directly determine which queries the data structure can answer. Lemma 3 follows from Lemmas 4 and 5 (see Section 3.1). The bounds in Lemma 3 express only the cost of computing small block table entries $D_{t'}$ and $E_{t'}$, not for answering a range least frequent element query at the level of individual elements.

Lemma 3. *Given block sizes $1 \leq t'' \leq t' \leq t \leq n$, if the tables D_t and E_t can be stored using $S(t)$ bits of space to support lookup queries in $T(t)$ time, then*

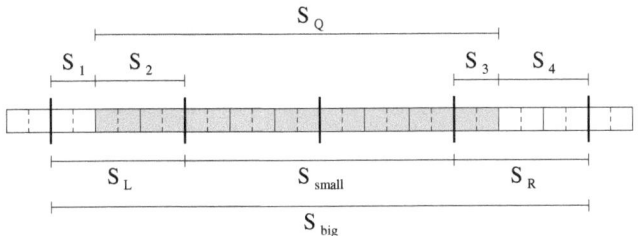

Fig. 1. illustration in support of Lemma 3

the tables $D_{t'}$ and $E_{t'}$ can be stored using $S(t')$ bits of space to support lookup queries in $T(t')$ time, where

$$S(t') = S(t) + O(n + (n/t')^2 \log(t/t''))\,, \quad and \tag{1}$$
$$T(t') = T(t) + O(t'')\,. \tag{2}$$

Following Chan et al. [3,5], we call a consecutive sequence of blocks in A a *span*. For any span S_Q among the $\Theta((n/t')^2)$ possible spans of small blocks, we define S_{big}, S_{small}, S_L, and S_R, as follows (see Figure 1):

- S_{big}: the unique minimal span of big blocks containing S_Q,
- S_L: the leftmost big block in S_{big},
- S_R: the rightmost big block in S_{big}, and
- S_{small}: the span of big blocks obtained by removing S_L and S_R from S_{big}.
- S_L is divided into S_1 (outside S_Q) and S_2 (inside S_Q).
- S_R is divided into S_3 (inside S_Q) and S_4 (outside S_Q).

Let $S_{big} = A[i : j]$, hence $S_{small} = A[i + t : j - t]$ and $S_Q = A[i_Q : j_Q]$. In Sections 3.1 and 3.2 we show how to encode the entries in $D_{t'}(\cdot, \cdot)$ and $E_{t'}(\cdot, \cdot)$ in $O(\log(t/t''))$ bits. In brief, we store an *approximate index* and *approximate frequency* for each entry and decode the exact values at query time.

3.1   Encoding and Decoding of $D_{t'}(\cdot, \cdot)$

We denote the least frequent element in S_Q by π and its frequency in S_Q by f_π. We consider three cases based on the indices at which π occurs in S_{big} as follows. The case that applies to any particular span can be indicated by 2 bits, hence $O(2(n/t')^2)$ bits in total. We use the same notation for representing a span as for the set of distinct elements within it.

Case 1: π is present in $S_L \cup S_R$ but not in S_{small} As explicit storage of π is costly, we store the approximate index at which π occurs in $S_L \cup S_R$, and the approximate value of f_π, in $O(\log(t/t''))$ bits. Later we show how to decode π and f_π in $O(t'')$ time using the stored values.

The approximate value of f_π can be encoded using the following observations. We have $|S_L \cup S_R| \le 2t$. Therefore $f_\pi \in [1, 2t]$. Explicitly storing f_π requires

$\log(2t)$ bits. However, an approximate value of f_π (with an additive error at most of t'') can be encoded in fewer bits. Observe that $t''\lfloor f_\pi/t''\rfloor \le f_\pi < t''\lfloor f_\pi/t''\rfloor+t''$. Therefore the value $\lfloor f_\pi/t''\rfloor \in [0, 2t/t'')$ can be stored using $O(\log(t/t''))$ bits and accessed in $O(1)$ time. The approximate location of π is a reference to a micro block within $S_L \cup S_R$ (among $2t/t''$ micro blocks) which contains π and whose index can be encoded in $O(\log(t/t''))$ bits. There can be many such micro blocks, but we choose one carefully from among the following possibilities:

- the rightmost micro block in S_1 which contains π,
- the leftmost micro block in S_2 which contains π,
- the rightmost micro block in S_3 which contains π, and
- the leftmost micro block in S_4 which contains π.

Next we show how to decode the exact values of π and f_π. Consider the case when the micro block (say B_m) containing π is in S_1. First initialize π' to any arbitrary element and f'_π to τ (an approximate value of f_π), such that $\tau - t'' \le f_\pi < \tau$. Upon terminating the following algorithm, we obtain the exact values of π and f_π as π' and f'_π respectively. Scan the elements in B_m from left to right and let k denote the current index. While k is an index in B_m, do:

1. If the second occurrence of $A[k]$ in $A[k : n-1]$ is in S_1, then go to Step 1 with $k \leftarrow k+1$.
2. If the $(f'_\pi+1)$st occurrence of $A[k]$ in $A[k : n-1]$ is in S_Q, then go to Step 1 with $k \leftarrow k+1$.
3. Set $f'_\pi \leftarrow f'_\pi - 1$, $\pi' \leftarrow A[k]$, and go Step 2.

This algorithm finds the rightmost occurrence of π within B_m, i.e., the rightmost occurrence of π before the index i_Q. Correctness can be proved via induction as follows: after initializing π' and f'_π, at each step we check whether the element $A[k]$ is a least frequent element in S_Q among all the elements in B_m which we have seen so far. Step 1 discards the position k if the rightmost occurrence of $A[k]$ in B_m is not at k, because we will see the same element eventually. Note that if the rightmost occurrence of $A[k]$ in B_m is at the position k, then the frequency of the element $A[k]$ in $S_Q = A[i_Q : j_Q]$ is exactly one less than its frequency in $A[k : j_Q]$. Using this property, we can check in $O(1)$ time whether the frequency of $A[k]$ in S_Q is less than f'_π (Step 2). If so, we update the current best answer π' by $A[k]$ and compute the exact frequency of $A[k]$ in S_Q in Step 3. We scan all elements in B_m and on completion the value stored at π' represents the least frequent element in S_Q among all elements present in B_m. Since π is present in B_m, π is the same as π', and $f_\pi = f'_\pi$. By Lemma 2, each step takes constant time. Since $\tau - f_\pi \le t''$, the total time is proportional to $|B_m| = t''$, i.e., $O(t'')$ time.

The remaining three cases, in which B_m is within S_2, S_3, and S_4, respectively, can be analyzed similarly.

Case 2: π is present in $S_L \cup S_R$ and in S_{small} The approximate position of π is encoded as in Case 1. In this case, however, f_π can be much larger than $2t$. Observe that $\alpha \le f_\pi \le \alpha + 2t$, where α is the frequency of the least frequent element in

S_{small}, which is already stored and can be retrieved in $T(t)$ time. Therefore, an approximate value $f_\pi - \alpha$ (with an additive error of at most t'') can be stored using $O(\log(t/t''))$ bits and decoded in $T(t) + O(1)$ time. The approximate location of π among the four possibilities as described in Case 1 is also maintained. By the algorithm above we can decode π and f_π in $T(t) + O(t'')$ time.

Case 3: π is present in S_{small} but in neither S_L nor S_R Since π is the least frequent element in S_Q, and does not appear in $S_L \cup S_R$, it is the least frequent element in S_{small} that does not appear in $S_L \cup S_R$. This implies π is the least frequent element in S_{big} that does not appear in $S_L \cup S_R$ (which is precomputed as stored). Therefore the time required for decoding the values of π and f_π is $T(t) + O(1)$.]

Lemma 4. *The table $D_{t'}(\cdot, \cdot)$ can be stored using $O((n/t')^2 \log(t/t''))$ bits in addition to $S(t)$ and any value within it can be decoded in $T(t) + O(t'')$ time.*

3.2   Encoding and Decoding of $E_{t'}(\cdot, \cdot)$

Let ϕ denote the least frequent element in S_Q that does not appear in the leftmost and rightmost small blocks in S_Q and let f_ϕ denote its frequency in S_Q. As before, we consider three cases for the indices at which ϕ occurs in S_{big}. The case that applies to any particular span can be indicated by 2 bits, hence $O((n/t')^2 \times 2)$ bits in total for any single given value of t'.

For each small block (of size t') we maintain a hash table that can answer whether a given element is present within the small block in $O(1)$ time. We can maintain each hash table in $O(t')$ bits for an overall space requirement of $O(n)$ bits for any single given value of t', using perfect hash techniques such as those of Schmidt and Siegel [14], Hagerup and Tholey [10], or Belazzougui et al. [1].

Case 1: ϕ is present in $S_L \cup S_R$ but not in S_{small} In this case, $f_\phi \in [1, 2t]$, and its approximate value and approximate position (i.e., the relative position of a small block) can be encoded in $O(\log(t/t''))$ bits. Encoding is the same as the encoding of π in Case 1 of $D_{t'}(\cdot, \cdot)$. For decoding we modify the algorithm for $D_{t'}(\cdot, \cdot)$ to use the hash table for checking that $A[k]$ is not present in the first and last small blocks of S_Q. The decoding time can be bounded by $O(t'')$.

Case 2: ϕ is present in $S_L \cup S_R$ and in S_{small} The approximate position of ϕ is stored as in Case 1. The encoding of f_ϕ is more challenging. Let α denote the frequency of the least frequent element in S_{small}, which is already stored and can be retrieved in $T(t)$ time. If $f_\phi > \alpha + 2t$, the element ϕ cannot be the least frequent element of any span S, where S contains S_{small} and is within S_{big}. In other words, ϕ is useful if and only if $f_\phi \leq \alpha + 2t$. Moreover, $f_\phi \geq \alpha$. Therefore we store the approximate value of f_ϕ if and only if it is useful information, and in such cases we can do it using only $O(\log(t/t''))$ bits. Using similar arguments to those used before, the decoding time can be bounded by $T(t) + O(t'')$.

Case 3: ϕ is present in S_{small} but in neither S_L nor S_R Since ϕ is the least frequent element in S_Q that does not appear in the leftmost and rightmost small blocks in

S_Q, and does not appear in $S_L \cup S_R$, it is the least frequent element in S_Q that does not appear in $S_L \cup S_R$. Therefore, π it is the least frequent element in S_{small} (as well as S_{big}) that does not appear in $S_L \cup S_R$ (which is precomputed as stored). Hence ϕ and f_ϕ can be retrieved in $T(t) + O(1)$ time.

Lemma 5. *The table $E_{t'}(\cdot, \cdot)$ can be encoded in $O(n + (n/t')^2 \log(t/t''))$ bits in addition to $S(t)$ and any value within it can be decoded in $T(t) + O(t'')$ time.*

By applying Lemma 3 with carefully chosen block sizes, followed by Lemma 1 for the final query on a range of individual elements, we show the following result.

Theorem 2. *Given any array $A[0 : n - 1]$, there exists an $O(n)$-word space data structure that supports range least frequent element queries on A in $O(\sqrt{n/w})$ time, where $w = \Omega(\log n)$ is the word size.*

Proof. Let $t_h = \log^{(h)} n \sqrt{n/w}$ and $t''_h = \sqrt{n/w}/\log^{(h+1)} n$, where $h \geq 1$. Then by applying Lemma 3 with $t = t_h$, $t' = t_{h+1}$, and $t'' = t''_h$, we obtain the following:

$$S(t_{h+1}) = S(t_h) + O\big(n + (n/t_{h+1})^2 \log(t_h/t''_h)\big) \in S(t_h) + O(nw/\log^{(h+1)} n)$$
$$T(t_{h+1}) = T(t_h) + O(t''_h) \qquad\qquad \in T(t_h) + O(\sqrt{n/w}/\log^{(h+1)} n) \, .$$

By storing D_{t_1} and E_{t_1} explicitly, we have $S(t_1) \in O(n)$ bits and $T(t_1) \in O(1)$. Applying Lemma 1 to $\log^* n$ levels of the recursion gives $t_{\log^* n} = \sqrt{n/w}$ and

$$S(\sqrt{n/w}) \in O\left(nw \sum_{h=1}^{\log^* n} \frac{1}{\log^{(h)} n} \right) = O(nw)$$

$$T(\sqrt{n/w}) \in O\left(\sqrt{n/w} \sum_{h=1}^{\log^* n} \frac{1}{\log^{(h)} n} \right) = O(\sqrt{n/w}) \, . \qquad \square$$

4   Path Frequency Queries on Trees

In this section, we generalize the range frequency query data structures to apply to trees (path mode query). The linear time bound of Chan et al. [5] for range mode queries on arrays depends on the ability to answer a query of the form "is the frequency of element x in the range $A[i : j]$ greater than k?" in constant time. There is no obvious way to generalize the data structure for such queries on arrays to apply to trees. Instead, we use the following lemma for an exact calculation of path frequency (not just whether it is greater than k). The proof is omitted due to space constraints.

Lemma 6. *Given any tree T of n nodes, there exists an $O(n)$-word data structure that can compute the number of occurrences of x on the path from i to j in $O(\log \log n)$ time for any nodes i and j in T and any element x.*

The following lemma describes a scheme for selecting some nodes in T as marked nodes, which split the tree into blocks over which we can apply the same kinds of block-based techniques that were effective in the array versions of the problems. The proof is omitted due to space constraints.

Lemma 7. *Given a tree T with n nodes and an integer $t < n$ which we call the blocking factor, we can choose a subset of the nodes, called the* marked *nodes, such that:*

- *at most $O(n/t)$ nodes are marked;*
- *the lowest common ancestor of any two marked nodes is marked; and*
- *the path between any two nodes contains $\leq t$ consecutive unmarked nodes.*

4.1   A Simple Data Structure for Path Mode Query

A simple path mode data structure follows naturally: we store the answers explicitly for all pairs of marked nodes, then use the data structure of Lemma 6 to compute exact frequencies for a short list of candidate modes. We let the blocking factor be a parameter, to support later use of this as part of a more efficient data structure.

Lemma 8. *For any blocking factor t, if we can answer path mode queries between marked nodes in time $T(t)$ with a data structure of $S(t)$ bits, then we can answer path mode queries between any nodes in time $T(t)+O(t \log \log n)$ with a data structure of $S(t) + O(n \log n)$ bits.*

Proof. As in the array case considered by Chan et al. [5], we can split the query path into a prefix of size $O(t)$, a span with both endpoints marked, and a suffix of size $O(t)$ using Lemma 7. The mode of the query must either be the mode of the span, or it must occur within the prefix or the suffix. We find the mode of the span in $T(t)$ time by assumption, and compute its frequency in $O(\log \log n)$ time using the data structure of Lemma 6. Then we also compute the frequencies of all elements in the prefix and suffix, for a further time cost of $O(t \log \log n)$. The result follows.   □

Setting $t = \sqrt{n}$ and using a simple lookup table for the marked-node queries gives $O(\sqrt{n} \log \log n)$ query time with $O(n)$ words of space.

4.2   A Faster Data Structure for Path Mode Query

To improve the time bound by an additional factor of $\sqrt{w}$, we derive the following lemma and apply it recursively.

Lemma 9. *For any blocking factor t, given a data structure that can answer path mode queries between marked nodes in time $T(t)$ with a space requirement of $S(t)$ bits, there exists a data structure answering path mode queries between marked nodes for blocking factor t' in time $T(t') = T(t) + O(t'' \log \log n)$ with a space requirement $S(t') = S(t) + O(n + (n/t')^2 \log(t/t''))$ bits, where $t > t' > t''$.*

Proof. (Sketch) Assume the nodes in T are marked based on a blocking factor t using Lemma 7, and the mode between any two marked nodes can be retrieved in $T(t)$ time using an $S(t)$-bit structure. Now we are interested in encoding the mode corresponding to the path between any two nodes i' and j', which are marked based on a smaller blocking factor t'. Note that there are $O((n/t')^2)$ such pairs. The tree structure along with this new marking information can be maintained in $O(n)$ bits using succinct data structures [13]. Where i and j are the first and last nodes in the path from i' to j', marked using t as the blocking factor, the path between i' and j' can be partitioned as follows: the path from i' to i, which we call the *path prefix*; the path from i to j; and the path from j to j', which we call the *path suffix*. The mode in the path from i' to j' must be either (i) the mode of i to j path or (ii) an element in the path prefix or path suffix.

In case (i), the answer is already stored using $S(t)$ bits and can be retrieved in $T(t)$ time. Case (ii) is more time-consuming. Note that the number of nodes in the path prefix and path suffix is $O(t)$. In case (ii) our answer must be stored in a node in the path prefix which is $k < t$ nodes away from i', or in a node in the path suffix which is $k < t$ nodes away from j'. Hence an approximate value of k (call it k', with $k < k' \leq k + t''$) can be maintained in $O(\log(t/t''))$ bits. In order to obtain a candidate list, we first retrieve the node corresponding to k' using a constant number of level ancestor queries (each taking $O(1)$ time [13]) and its $O(t'')$ neighboring nodes in the i' to j' path. The final answer can be computed by evaluating the frequencies of these $O(t'')$ candidates using Lemma 6 in $O(t'' \log \log n)$ overall time.   □

The following theorem is our main result on path mode query.

Theorem 3. *There exists a linear-space (in words; that is, $O(n \log n)$ bits) data structure that answers path mode queries on trees in $O(\log \log n \sqrt{n/w})$ time.*

Proof. Let $t_h = \log^{(h)} n \sqrt{n/w}$ and $t''_h = \sqrt{n/w}/\log^{(h+1)} n$, where $h \geq 1$. Then by applying Lemma 9 with $t = t_h$, $t' = t_{h+1}$, and $t'' = t''_h$, we obtain the following:

$$S(t_{h+1}) = S(t_h) + O\big(n + (n/t_{h+1})^2 \log(t_h/t''_h)\big) \in S(t_h) + O(nw/\log^{(h+1)} n)$$

$$T(t_{h+1}) = T(t_h) + O(t''_h \log \log n) \in T(t_h) + O(\log \log n \sqrt{n/w}/\log^{(h+1)} n).$$

By storing D_{t_1} and E_{t_1} explicitly, we have $S(t_1) \in O(n)$ bits and $T(t_1) \in O(1)$. Applying Lemma 8 to $\log^* n$ levels of the recursion gives $t_{\log^* n} = \sqrt{n/w}$ and

$$S(\sqrt{n/w}) \in O\left(nw \sum_{h=1}^{\log^* n} \frac{1}{\log^{(h)} n} \right) \qquad = O(nw)$$

$$T(\sqrt{n/w}) \in O\left(\log \log n \sqrt{n/w} \sum_{h=1}^{\log^* n} \frac{1}{\log^{(h)} n} \right) = O(\log \log n \sqrt{n/w}).   □$$

Similar techniques lead to a data structure for tree path least frequent element queries; we defer the proof to the full version due to space constraints.

Theorem 4. *There exists a linear-space data structure that answers path least frequent element queries on trees in $O(\log \log n \sqrt{n/w})$ time.*

5   Path α-Minority Query on Trees

An α-*minority* in a multiset A, for some $\alpha \in [0, 1]$, is an element that occurs at least once and as no more than α proportion of A. If there are n elements in A, then the number of occurrences of the α-minority in A can be at most αn. Elements in A that are not α-minorities are called α-*majorities*. Chan et al. studied α-minority range queries in arrays [5]; here, we generalize the problem to path queries on trees. In general, an α-minority is not necessarily unique; given a query consisting of a pair of tree node indices and a value $\alpha \in [0, 1]$ (specified at query time), our data structure returns one α-minority, if at least one exists, regardless of the number of distinct α-minorities. As in the previous section, we can compute path frequencies in $O(\log \log n)$ time (Lemma 6); then a data structure similar to the one for arrays gives us distinct elements within a path in constant time per distinct element. Combining the two gives a bound of $O(\alpha^{-1} \log \log n)$ time for α-minority queries.

As discussed by Chan et al. for the case of arrays [5], examining α^{-1} distinct elements in a query range allows us to guarantee either that we have examined an α-minority, or that no α-minority exists. So we construct a data structure based on the hive graph of Chazelle [6] for the k-nearest distinct ancestor problem: given a node i, find a sequence $a_1, a_2, \ldots$ of ancestors of i such that $a_1 = i$, a_2 is the nearest ancestor of i distinct from a_1, a_3 is the nearest ancestor of i distinct from a_1 and a_2, and so on. Queries on the data structure return the distinct ancestors in order and in constant time each. The proof is omitted due to space constraints.

Lemma 10. *There exists a linear-space data structure that answers k-nearest distinct ancestor queries on trees in $O(k)$ time, returning them in nearest-to-furthest order in $O(1)$ time each, so that k can be chosen interactively.*

Lemmas 6 and 10 give the following theorem.

Theorem 5. *There exists a linear-space data structure that answers path α-minority queries on trees in $O(\alpha^{-1} \log \log n)$ time (where α and the path's endpoints are specified at query time).*

Proof. We construct the data structures of Lemma 6 and Lemma 10, both of which use linear space. To answer a path α-minority query between two nodes i and j, we find the α^{-1} nearest distinct ancestors (or as many as exist, if that is fewer) above each of i and j. That takes α^{-1} time. If an α-minority exists between i and j, then one of these candidates must be an α-minority. We can test each one in $O(\log \log n)$ using the path frequency data structure, and the result follows.   □

6   Discussion and Directions for Future Research

Our data structures for path queries refer to Lemma 6. Consequently, each has query time $O(\log \log n)$ times greater than the corresponding time on arrays. For arrays, Chan et al. [3] use $O(1)$-time range frequency queries for the case in which the element whose frequency is being measured is at an endpoint of query range. Generalizing this technique to path queries on trees should allow each data structure's query time to be decreased accordingly.

Acknowledgements. The authors thank the anonymous reviewers for their helpful suggestions. Part of this work was done while the fourth author was visiting the University of Manitoba in July 2012. The fourth author thanks the members of the Computational Geometry Lab at University of Manitoba and N. M. Prabhakaran.

References

1. Belazzougui, D., Botelho, F.C., Dietzfelbinger, M.: Hash, displace, and compress. In: Fiat, A., Sanders, P. (eds.) ESA 2009. LNCS, vol. 5757, pp. 682–693. Springer, Heidelberg (2009)
2. Belazzougui, D., Gagie, T., Navarro, G.: Better space bounds for parameterized range majority and minority. In: Dehne, F., Solis-Oba, R., Sack, J.-R. (eds.) WADS 2013. LNCS, vol. 8037, pp. 121–132. Springer, Heidelberg (2013)
3. Chan, T.M., Durocher, S., Larsen, K.G., Morrison, J., Wilkinson, B.T.: Linear-space data structures for range mode query in arrays. In: Proc. STACS, vol. 14, pp. 291–301 (2012)
4. Chan, T.M., Durocher, S., Larsen, K.G., Morrison, J., Wilkinson, B.T.: Linear-space data structures for range mode query in arrays. Theory Comp. Sys. (2013)
5. Chan, T.M., Durocher, S., Skala, M., Wilkinson, B.T.: Linear-space data structures for range minority query in arrays. In: Fomin, F.V., Kaski, P. (eds.) SWAT 2012. LNCS, vol. 7357, pp. 295–306. Springer, Heidelberg (2012)
6. Chazelle, B.: Filtering search: A new approach to query-answering. SIAM J. Comp. 15(3), 703–724 (1986)
7. Durocher, S., He, M., Munro, J.I., Nicholson, P.K., Skala, M.: Range majority in constant time and linear space. In: Aceto, L., Henzinger, M., Sgall, J. (eds.) ICALP 2011, Part I. LNCS, vol. 6755, pp. 244–255. Springer, Heidelberg (2011)
8. Durocher, S., He, M., Munro, J.I., Nicholson, P.K., Skala, M.: Range majority in constant time and linear space. Inf. & Comp. 222, 169–179 (2013)
9. Gagie, T., He, M., Munro, J.I., Nicholson, P.K.: Finding frequent elements in compressed 2D arrays and strings. In: Grossi, R., Sebastiani, F., Silvestri, F. (eds.) SPIRE 2011. LNCS, vol. 7024, pp. 295–300. Springer, Heidelberg (2011)
10. Hagerup, T., Tholey, T.: Efficient minimal perfect hashing in nearly minimal space. In: Ferreira, A., Reichel, H. (eds.) STACS 2001. LNCS, vol. 2010, pp. 317–326. Springer, Heidelberg (2001)
11. Krizanc, D., Morin, P., Smid, M.: Range mode and range median queries on lists and trees. In: Ibaraki, T., Katoh, N., Ono, H. (eds.) ISAAC 2003. LNCS, vol. 2906, pp. 517–526. Springer, Heidelberg (2003)
12. Krizanc, D., Morin, P., Smid, M.: Range mode and range median queries on lists and trees. Nordic J. Computing 12, 1–17 (2005)
13. Sadakane, K., Navarro, G.: Fully-functional succinct trees. In: Proc. ACM-SIAM SODA, pp. 134–149 (2010)
14. Schmidt, J.P., Siegel, A.: The spatial complexity of oblivious k-probe hash functions. SIAM J. Comput. 19(5), 775–786 (1990)

Noninterference with Local Policies

Sebastian Eggert, Henning Schnoor, and Thomas Wilke

Institut für Informatik, Christian-Albrechts-Universität zu Kiel, 24098 Kiel, Germany
{sebastian.eggert,henning.schnoor,thomas.wilke}@email.uni-kiel.de

Abstract. We develop a theory for state-based noninterference in a setting where different security policies—we call them local policies—apply in different parts of a given system. Our theory comprises appropriate security definitions, characterizations of these definitions, for instance in terms of unwindings, algorithms for analyzing the security of systems with local policies, and corresponding complexity results.

1 Introduction

Research in formal security aims to provide rigorous definitions for different notions of security as well as methods to analyse a given system with regard to the security goals. Restricting the information that may be available to a user of the system (often called an agent) is an important topic in security. Noninterference [GM82, GM84] is a notion that formalizes this. Noninterference uses a security policy that specifies, for each pair of agents, whether information is allowed to flow from one agent to the other. To capture different aspects of information flow, a wide range of definitions of noninterference has been proposed, see, e.g., [YB94, Mil90, vO04, WJ90].

In this paper, we study systems where in different parts different policies apply. This is motivated by the fact that different security requirements may be desired in different situations, for instance, a user may want to forbid interference between his web browser and an instant messenger program while visiting banking sites but when reading a news page, the user may find interaction between these programs useful.

As an illustrating example, consider the system depicted in Fig. 1, where three agents are involved: an administrator A and two users H and L. The rounded boxes represent system states, the arrows represent transitions. The labels of the states indicate what agent L observes in the respective state; the labels of the arrows denote the action, either action a performed by A or action h performed by H, inducing the respective transition. Every action can be performed in every state; if it does not change the state (i. e., if it induces a loop), the corresponding transition is omitted in the picture.

The lower part of the system constitutes a secure subsystem with respect to the bottom policy: when agent H performs the action h in the initial state, the observation of agent L changes from 0 to 1, but this is allowed according to the policy, as agent H may interfere with agent L—there is an edge from H to L.

K. Chatterjee and J. Sgall (Eds.): MFCS 2013, LNCS 8087, pp. 337–348, 2013.

Similarly, the upper part of the system constitutes a secure subsystem with respect to the top policy: interference between H and L is not allowed—no edge from H to L—and, in fact, there is no such interference, because L's observation does not change when h performs an action.

However, the entire system is clearly insecure: agent A must not interfere with anyone—there is no edge starting from A in either policy—but when L observes "1" in the lower right state, L can conclude that A did *not* perform the a action depicted.

Note that interference between H and L is allowed, unless A performs action a. But L must not get to know whether a was performed. To achieve

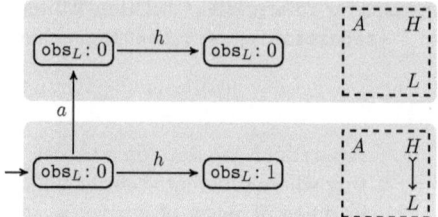

Fig. 1. System with local policies

this, interference between H and L must never be allowed. Otherwise, as we have just argued, L can—by observing H's actions—conclude that in the current part of the system, interference between H and L is still legal and thus A did not perform a. In other words, in the policy of the lower part, the edge connecting H and L can never be "used" for an actual information flow. We call such edges *useless*.—Useless edges are a key issue arising in systems with local policies.

Our results. We develop a theory of noninterference with local policies which takes the aforementioned issues into account. Our contributions are as follows:

1. We provide new and natural definitions for noninterference with local policies, both for the transitive [GM82, GM84] (agent L may only be influenced by agent H if there is an edge from H to L in the policy) and for the intransitive setting [HY87] (interference between H and L via "intermediate steps" is also allowed).
2. We show that policies can always be rewritten into a normal form which does not contain any "useless" edges (see above).
3. We provide characterizations of our definitions based on unwindings, which demonstrate the robustness of our definitions and from which we derive efficient verification algorithms.
4. We provide results on the complexity of verifying noninterference. In the transitive setting, noninterference can be verified in nondeterministic logarithmic space (NL). In the intransitive setting, the problem is NP-complete, but fixed-parameter tractable with respect to the number of agents.

Our results show significant differences between the transitive and the intransitive setting. In the transitive setting, one can, without loss of generality, always assume a policy is what we call uniform, which means that each agent may "know" (in a precise epistemic sense) the set of agents that currently may interfere with him. Assuming uniformity greatly simplifies the study of noninterference with local policies in the transitive setting. Moreover, transitive noninterference with local policies can be characterized by a simple unwinding, which yields very efficient algorithms.

In the intransitive setting, the situation is more complicated. Policies cannot be assumed to be uniform, verification is NP-complete, and, consequently, we only give an unwinding condition that requires computing exponentially many relations. However, for *uniform* policies, the situation is very similar to the transitive setting: we obtain simple unwindings and efficient algorithms.

As a consequence of our results for uniform policies, we obtain an unwinding characterization of IP-security [HY87] (which uses a single policy for the entire system). Prior to our results, only an unwinding characterization that was *sound*, but not *complete* for IP-security was known [Rus92]. Our new unwinding characterization immediately implies that IP-security can be verified in nondeterministic logarithmic space, which improves the polynomial-time result obtained in [E+11]. Proofs and additional details can be found in [ESW13].

Related Work. Our definition for intransitive security is a generalization of IP-security [HY87] mentioned above. The issues raised against IP-security in [vdM07] are orthogonal to the issues arising from local policies. We therefore study local policies in the framework of IP-security, which is technically simpler than, e.g., TA-security as defined in [vdM07].

Several extensions of intransitive noninterference have been discussed, for instance, in [RG99, MSZ06]. In [Les06], a definition of intransitive noninterference with local policies is given, however, the definition in [Les06] does not take into account the aforementioned effects, and that work does not provide complete unwinding characterizations nor complexity results.

2   State-Based Systems with Local Policies

We work with the standard state-observed system model, that is, a system is a deterministic finite-state automaton where each action belongs to a dedicated agent and each agent has an observation in each state. More formally, a *system* is a tuple $M = (S, s_0, A, \text{step}, \text{obs}, \text{dom})$, where S is a finite set of *states*, $s_0 \in S$ is the *initial state*, A is a finite set of *actions*, $\text{step}: S \times A \to S$ is a *transition function*, $\text{obs}: S \times D \to O$ is an *observation function*, where O is an arbitrary set of observations, and $\text{dom}: A \to D$ associates with each action an agent, where D is an arbitrary finite set of agents (or security domains).

For a state s and an agent u, we write $\text{obs}_u(s)$ instead of $\text{obs}(s, u)$. For a sequence $\alpha \in A^*$ of actions and a state $s \in S$, we denote by $s \cdot \alpha$ the state obtained when performing α starting in s, i.e., $s \cdot \epsilon = s$ and $s \cdot \alpha a = \text{step}(s \cdot \alpha, a)$.

A *local policy* is a reflexive relation $\rightarrowtail \, \subseteq D \times D$. To keep our notation simple, we do not define subsystems nor policies for subsystems explicitly. Instead, we assign a local policy to every state and denote the policy in state s by $\rightarrowtail_s$. We call the collection of all local policies $(\rightarrowtail_s)_{s \in S}$ the *policy* of the system. If $(u, v) \in \, \rightarrowtail_s$ for some $u, v \in D$, $s \in S$, we say $u \rightarrowtail_s v$ is an *edge* in $(\rightarrowtail_s)_{s \in S}$. A system has a *global policy* if all local policies $\rightarrowtail_s$ are the same in all states, i.e., if $u \rightarrowtail_s v$ does not depend on s. In this case, we denote the single policy by $\rightarrowtail$ and only write $u \rightarrowtail v$. We define the set $u_s^{\leftharpoonup}$ as the set of agents that *may interfere* with u in s, i.e., the set $\{v \mid v \rightarrowtail_s u\}$.

In the following, we fix an arbitrary system M and a policy $(\rightarrowtail_s)_{s \in S}$.

In our examples, we often identify a state with an action sequence leading to it from the initial state s_0, that is, we write α for $s_0 \cdot \alpha$, which is well-defined, because we consider deterministic systems. For example, in the system from Fig. 1, we denote the initial state by ϵ and the upper right state by ah. In each state, we write the local policy in that state as a graph. In the system from Fig. 1, we have $H \rightarrowtail_\epsilon L$, but $H \not\rightarrowtail_a L$. In general, we only specify the agents' observations as far as relevant for the example, which usually is only the observation of the agent L. We adapt the notation from Fig. 1 to our definition of local policies, which assigns a local policy to every state: we depict the graph of the local policy inside the rounded box for the state, see Fig. 2.

3   The Transitive Setting

In this section, we define noninterference for systems with local policies in the transitive setting, give several characterizations, introduce the notion of useless edge, and discuss it. The basic idea of our security definition is that an occurrence of an action which, according to a local policy, should not be observable by an agent u must not have any influence on u's future observations.

Definition 3.1 (t-security). *The system M is t-secure iff for all $u \in D$, $s \in S$, $a \in A$ and $\alpha \in A^*$ the following implication holds:*

$$\text{If } \text{dom}(a) \not\rightarrowtail_s u, \text{ then } \text{obs}_u(s \cdot \alpha) = \text{obs}_u(s \cdot a\alpha) \ .$$

Fig. 2 shows a t-secure system. In contrast, the system in Fig. 1 is not t-secure, since $A \not\rightarrowtail_\epsilon L$, but $\text{obs}_L(ah) \neq \text{obs}_L(h)$.

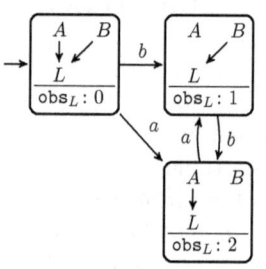

Fig. 2. A t-secure system

3.1   Characterizations of t-Security

In Theorem 3.4, we give two characterizations of t-security, underlining that our definition is quite robust. The first characterization is based on an operator which removes all actions that must not be observed. It is essentially the definition from Goguen and Meseguer [GM82, GM84] of the **purge** operator generalized to systems with local policies.

Definition 3.2 (purge for local policies). *For all $u \in D$ and $s \in S$ let* $\text{purge}(\epsilon, u, s) = \epsilon$ *and for all $a \in A$ and $\alpha \in A^*$ let*

$$\text{purge}(a\alpha, u, s) = \begin{cases} a \ \text{purge}(\alpha, u, s \cdot a) & \text{if } \text{dom}(a) \rightarrowtail_s u \\ \text{purge}(\alpha, u, s) & \text{otherwise} \ . \end{cases}$$

The other characterization is in terms of unwindings, which we define for local policies in the following, generalizing the definition of Haigh and Young [HY87].

Definition 3.3 (transitive unwinding with local policies). *A transitive unwinding for M with a policy $(\rightarrowtail_s)_{s\in S}$ is a family of equivalence relations $(\sim_u)_{u\in D}$ such that for every agent $u \in D$, all states $s, t \in S$ and all $a \in A$, the following holds:*

- *If $\mathsf{dom}(a) \not\rightarrowtail_s u$, then $s \sim_u s \cdot a$.* $\qquad$ (LR$_t$)—local respect
- *If $s \sim_u t$, then $s \cdot a \sim_u t \cdot a$.* $\qquad$ (SC$_t$)—step consistency
- *If $s \sim_u t$, then $\mathsf{obs}_u(s) = \mathsf{obs}_u(t)$.* $\qquad$ (OC$_t$)—output consistency

Our characterizations of t-security are spelled out in the following theorem.

Theorem 3.4 (characterizations of t-security). *The following are equivalent:*

1. *The system M is t-secure.*
2. *For all $u \in D$, $s \in S$, and $\alpha, \beta \in A^*$ with $\mathsf{purge}(\alpha, u, s) = \mathsf{purge}(\beta, u, s)$, we have $\mathsf{obs}_u(s \cdot \alpha) = \mathsf{obs}_u(s \cdot \beta)$.*
3. *There exists a transitive unwinding for M with the policy $(\rightarrowtail_s)_{s\in S}$.*

Unwinding relations yield efficient verification procedure. For verifying t-security, it is sufficient to compute for every $u \in D$ the smallest equivalence relation satisfying (LR$_t$) and (SC$_t$) and check that the function obs_u is constant on every equivalence class. This can be done with nearly the same algorithm as is used for global policies, described in [E+11]. The above theorem directly implies that t-security can be verified in nondeterministic logarithmic space.

3.2   Useless Edges

An "allowed" interference $v \rightarrowtail_s u$ may contradict a "forbidden" interference $v \not\rightarrowtail_{s'} u$ in a state s' that should be indistinguishable to s for u. In this case, the edge $v \rightarrowtail_s u$ is useless. What this means is that an edge $v \rightarrowtail_s u$ in the policy may be deceiving and should not be interpreted as "it is allowed that v interferes with u", rather, it should be interpreted as "it is not explicitly forbidden that v interferes with u". To formalize this, we introduce the following notion:

Definition 3.5 (t-similarity). *States s, s' are t-similar for an agent $u \in D$, denoted $s \approx_u s'$, if there exist $t \in S$, $a \in A$, and $\alpha \in A^*$ such that $\mathsf{dom}(a) \not\rightarrowtail_t u$, $s = t \cdot a\alpha$, and $s' = t \cdot \alpha$.*

Observe that t-similarity is identical with the smallest equivalence relation satisfying (LR$_t$) and (SC$_t$). Also observe that the system M is t-secure if and only if for every agent u, if $s \approx_u s'$, then $\mathsf{obs}_u(s) = \mathsf{obs}_u(s')$.

The notion of t-similarity allows us to formalize the notion of a useless edge:

Definition 3.6 (useless edge). *An edge $v \rightarrowtail_s u$ is useless if there is a state s' with $s \approx_u s'$ and $v \not\rightarrowtail_{s'} u$.*

For example, consider again the system in Fig. 1. Here, the local policy in the initial state allows information flow from H to L. However, if L is allowed to observe H's action in the initial state, then L would know that the system is in the initial state, and would also know that A has not performed an action. This is an information flow from A to L, which is prohibited by the policy.

Useless edges can be removed without any harm:

Theorem 3.7 (removal of useless edges). *Let* $(\rightarrowtail'_s)_{s \in S}$ *be defined by*

$$\rightarrowtail'_s = \rightarrowtail_s \setminus \{v \rightarrowtail_s u \mid v \rightarrowtail_s u \text{ is useless}\} \qquad \text{for all } s \in S.$$

Then M is t-secure w. r. t. $(\rightarrowtail_s)_{s \in S}$ *iff M is t-secure w. r. t.* $(\rightarrowtail'_s)_{s \in S}$.

The policy $(\rightarrowtail'_s)_{s \in S}$ in Theorem 3.7 has no useless edges, hence every edge in one of its local policies represents an *allowed* information flow—no edge contradicts an edge in another local policy. Another interpretation is that any information flow that is *forbidden* is *directly* forbidden via the absence of the corresponding edge. In that sense, the policy is closed under logical deduction.

We call a policy $(\rightarrowtail_s)_{s \in S}$ *uniform* if $u_s^\leftharpoonup = u_{s'}^\leftharpoonup$ holds for all states s and s' with $s \approx_u s'$. In other words, in states that u should not be able to distinguish, the exact same set of agents may interfere with u. Hence u may "know" the set of agents that currently may interfere with him. Note that a policy is uniform if and only if it does not contain useless edges. (This is not true in the intransitive setting, hence the seemingly complicated definition of uniformity.) Uniform policies have several interesting properties, for example, with a uniform policy the function **purge** behaves very similarly to the setting with a global policy: it suffices to verify action sequences that start in the initial state of the system and **purge** satisfies a natural associativity condition on a uniform policy.

4   The Intransitive Setting

In this section, we consider the intransitive setting, where, whenever an agent performs an action, this event may transmit information about the actions the agent has performed himself as well as information about actions by other agents that was previously transmitted to him. The definition follows a similar pattern as that of t-security: if performing an action sequence $a\alpha$ starting in a state s should not transmit the action a (possibly via several intermediate steps) to the agent u, then u should be unable to deduce from his observations whether a was performed. To formalize this, we use Leslie's extension [Les06] of Rushby's definition [Rus92] of **sources**.

Definition 4.1 (sources). *For an agent u let* $\mathrm{src}(\epsilon, u, s) = \{u\}$ *and for $a \in A$, $\alpha \in A^*$, if* $\mathrm{dom}(a) \rightarrowtail_s v$ *for some* $v \in \mathrm{src}(\alpha, u, s \cdot a)$, *then let* $\mathrm{src}(a\alpha, u, s) = \mathrm{src}(\alpha, u, s \cdot a) \cup \{\mathrm{dom}(a)\}$, *and else let* $\mathrm{src}(a\alpha, u, s) = \mathrm{src}(\alpha, u, s \cdot a)$.

The set $\mathrm{src}(a\alpha, u, s)$ contains the agents that "may know" whether the action a has been performed in state s after the run $a\alpha$ is performed: initially, this is only the set of agents v with $\mathrm{dom}(a) \rightarrowtail_s v$. The knowledge may be spread by every action performed by an agent "in the know:" if an action b is performed in a later state t, and $\mathrm{dom}(b)$ already may know that the action a was performed, then all agents v with $\mathrm{dom}(b) \rightarrowtail_t v$ may obtain this information when b is performed. Following the discussion above, we obtain a natural definition of security:

Definition 4.2 (i-security). *The system M is i-secure iff for all $s \in S$, $a \in A$, and $\alpha \in A^*$, the following implication holds.*

$$\text{If } \text{dom}(a) \notin \text{src}(a\alpha, u, s), \text{ then } \text{obs}_u(s \cdot a\alpha) = \text{obs}_u(s \cdot \alpha).$$

The definition formalizes the above: if, on the path $a\alpha$, the action a is not transmitted to u, then u's observation must not depend on whether a was performed; the runs $a\alpha$ and α must be indistinguishable for u.

Consider the example in Fig. 1. The system remains insecure in the intransitive setting: as A must not interfere with any agent in any state, we have $\text{dom}(a) \notin \text{src}(ah, L, \epsilon)$, where again, according to our convention, ϵ denotes the initial state. So, the system is insecure, since $\text{obs}_L(ah) \neq \text{obs}_L(h)$.

4.1   Characterizations and Complexity of i-Security

We now establish two characterizations of intransitive noninterference with local policies and study the complexity of verifying i-security. Our characterizations are analogous to the ones obtained for the transitive setting in Theorem 3.4. The first one is based on a purge function, the second one uses an unwinding condition. This demonstrates the robustness of our definition and strengthens our belief that i-security is indeed a natural notion.

We first extend Rushby's definition of ipurge to systems with local policies.

Definition 4.3 (intransitive purge for local policies). *For all $u \in D$ and all $s \in S$, let* $\text{ipurge}(\epsilon, u, s) = \epsilon$ *and, for all $a \in A$ and $\alpha \in A^*$, let*

$$\text{ipurge}(a\alpha, u, s) = \begin{cases} a\ \text{ipurge}(\alpha, u, s \cdot a) & \text{if } \text{dom}(a) \in \text{src}(a\alpha, u, s), \\ \text{ipurge}(\alpha, u, s) & \text{otherwise.} \end{cases}$$

The crucial point is that in the case where a must remain hidden from agent u, we define $\text{ipurge}(a\alpha, u, s)$ as $\text{ipurge}(\alpha, u, s)$ instead of the possibly more intuitive choice $\text{ipurge}(\alpha, u, s \cdot a)$, on which the security definition in [Les06] is based.

We briefly explain the reasoning behind this choice. To this end, let ipurge$'$ denote the alternative definition of ipurge outlined above. Consider the sequence ah, performed from the initial state in the system in Fig. 1. Clearly, the action a is purged from the trace, thus the result of ipurge$'$ is the same as applying ipurge$'$ to the sequence h starting in the upper left state. However, in this state, the action h is invisible for L, hence ipurge$'$ removes it, and thus purging ah results in the empty sequence. On the other hand, if we consider the sequence h also starting in the initial state, then h is not removed by ipurge$'$, since H may interfere with L. Hence ah and h do not lead to the same purged trace— a security definition based on ipurge$'$ does not require ah and h to lead to states with the same observation. Therefore, the system is considered secure in the ipurge$'$-based security definition from [Les06]. However, a natural definition must require ah and h to lead to the same observation for agent L, as the action a must always be hidden from L.

We next define unwindings for i-security and then give a characterization of i-security based on them.

Definition 4.4 (intransitive unwinding). *An* intransitive unwinding *for the system M with a policy $(\rightarrowtail_s)_{s \in S}$ is a family of relations $(\precsim_{D'})_{D' \subseteq D}$ such that $\precsim_{D'} \subseteq S \times S$ and for all $D' \subseteq D$, all $s, t \in S$ and all $a \in A$, the following hold:*

- $s \precsim_{\{u \in D \mid \text{dom}(a) \not\rightarrowtail_s u\}} s \cdot a$. (LR$_i$)
- *If $s \precsim_{D''} t$, then $s \cdot b \precsim_{D''} t \cdot b$, where $D'' = D'$ if $\text{dom}(b) \in D'$, and else $D'' = D' \cap \{u \mid \text{dom}(b) \not\rightarrowtail_s u\}$.* (SC$_i$)
- *If $s \precsim_{D'} t$ and $u \in D'$, then $\text{obs}_u(s) = \text{obs}_u(t)$,* (OC$_i$)

Intuitively, $s \precsim_{D'} t$ expresses that there is a common reason for all agents in D' to have the same observations in s as in t, i.e., if there is a state $\tilde{s}$, an action a and a sequence α such that $s = \tilde{s} \cdot a\alpha$, $t = \tilde{s} \cdot \alpha$, and $\text{dom}(a) \notin \text{src}(a\alpha, u, \tilde{s})$ for all agents $u \in D'$.

Theorem 4.5 (characterization of i-security). *The following are equivalent:*

1. *The system M is i-secure.*
2. *For all agents u, all states s, and all action sequences α and β with $\text{ipurge}(\alpha, u, s) = \text{ipurge}(\beta, u, s)$, we have $\text{obs}_u(s \cdot \alpha) = \text{obs}_u(s \cdot \beta)$.*
3. *There exists an intransitive unwinding for M and $(\rightarrowtail_s)_{s \in S}$.*

In contrast to the transitive setting, the unwinding characterization of i-security does not lead to a polynomial-time algorithm to verify security of a system, because the number of relations needed to consider is exponential in the number of agents in the system. Unless P = NP, we cannot do significantly better, because the verification problem is NP-complete; our unwinding characterization, however, yields an FPT-algorithm.

Theorem 4.6 (complexity of i-security). *Deciding whether a given system is i-secure with respect to a policy is NP-complete and fixed-parameter tractable with the number of agents as parameter.*

4.2 Intransitively Useless Edges

In our discussion of t-security we observed that local policies may contain edges that can never be used. This issue also occurs in the intransitive setting, but the situation is more involved. In the transitive setting, it is sufficient to "remove any incoming edge for u that u must not know about" (see Theorem 3.7). In the intransitive setting it is not: when the system in Fig. 3 is in state h_1, then agent L must not know that the edge $D \rightarrowtail L$ is present, since states ϵ and

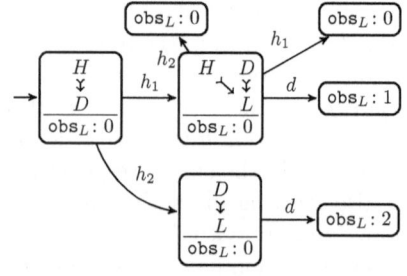

Fig. 3. Intransitively useless edge

h_1 should be indistinguishable for L, but clearly, the edge cannot be removed without affecting security. However, useless edges still exist in the intransitive setting, even in the system from Figure 3, as we will show below.

To formally define useless edges, we adapt t-similarity to the intransitive setting in the natural way.

Definition 4.7 (i-similarity). *For an agent u, let $\approx_u^i$ be the smallest equivalence relation on the states of M such that for all $s \in S$, $a \in A$, $\alpha \in A^*$, if $\mathrm{dom}(a) \notin \mathrm{src}(a\alpha, u, s)$, then $s \cdot a\alpha \approx_u^i s \cdot \alpha$. We call states s and s' with $s \approx_u^i s'$ i-similar for u.*

Using this, we can now define intransitively useless edges:

Definition 4.8 (intransitively useless edge). *Let e be an edge in a local policy of $(\rightarrowtail_s)_{s \in S}$ and let $(\hat{\rightarrowtail}_s)_{s \in S}$ be the policy obtained from $(\rightarrowtail_s)_{s \in S}$ by removing e. Let $\approx_u^i$ and $\hat{\approx}_u^i$ be the respective i-similarity relations. Then e is intransitively useless if $s \approx_u^i s'$ if and only if $s \hat{\approx}_u^i s'$ for all states s and s' and all agents u.*

An edge is intransitively useless if removing it does not forbid any information flow that was previously allowed. In particular, such an edge itself cannot be used directly. Whether an edge is useless does not depend on the observation function of the system, but only on the policy and the transition function, whereas a definition of security compares observations in different states.

If the policy does not contain any intransitively useless edges, then there is no edge in any of its local policies that is contradicted by other aspects of the policy. In other words, the set of information flows *forbidden* by such a policy is closed under logical deduction—every edge that can be shown to represent a forbidden information flow is absent in the policy.

Fig. 3 shows a secure system with an intransitively useless edge. The system is secure (agent L knows whether in the initial state, h_1 or h_2 was performed, as soon as this information is transmitted by agent D). The edge $H \rightarrowtail_{h_1} L$ is intransitively useless, as explained in what follows.

The edge allows L to distinguish between the states h_1, h_1h_1, h_1h_2. However, one can verify that $h_2h_1 \approx_L^i h_1$, $h_2h_1h_1 \approx_L^i h_2h_1$, $h_2h_1h_1 \approx_L^i h_1h_1$, $h_2h_1h_2 \approx_L^i h_2h_1$, and $h_2h_1h_2 \approx_L^i h_1h_2$ all hold. Symmetry and transitivity of $\approx_L^i$ imply that all the three states h_1, h_1h_1, h_1h_2 are $\approx_L^i$-equivalent. Hence the edge $H \rightarrowtail_{h_1} L$ is indeed intransitively useless (and the system would be insecure if h_1, h_1h_1, and h_1h_2 would not have the same observations).

Intransitively useless edges can be removed without affecting security:

Theorem 4.9 (removal of intransitively useless edges). *Let $(\rightarrowtail_s')_{s \in S}$ be obtained from $(\rightarrowtail_s)_{s \in S}$ by removing a set of edges which are intransitively useless. Then M is i-secure with respect to $(\rightarrowtail_s)_{s \in S}$ if and only if M is i-secure with respect to $(\rightarrowtail_s')_{s \in S}$.*

This theorem implies that for every policy $(\rightarrowtail_s)_{s \in S}$, a policy $(\rightarrowtail_s')_{s \in S}$ without intransitively useless edges that is equivalent to $(\rightarrowtail_s)_{s \in S}$ can be obtained from $(\rightarrowtail_s)_{s \in S}$ by removing all intransitively useless edges.

4.3   Sound Unwindings and Uniform Intransitive Policies

The exponential size unwinding of i-security given in Section 4.1 does not yield a polynomial-time algorithm for security verification. Since the problem is NP-complete, such an algorithm—and hence an unwinding that is both small and easy to compute—does not exist, unless P = NP. In this section, we define unwinding conditions that lead to a polynomial-size unwinding and are *sound* for i-security, and are *sound and complete* for i-secure in the case of uniform policies. Uniform policies are (as in the transitive case) policies in which every agent "may know" the set of agents who may currently interfere with him, that is, if an agent u must not distinguish two states by the security definition, then the set of agents that may interfere with u must be identical in these two states. Formally, we define this property as follows.

Definition 4.10 (intransitive uniform). *A policy* $(\rightarrowtail_s)_{s \in S}$ *is intransitively uniform, if for all agents u and states s, s' with $s \approx_u^i s'$, we have that $u_s^\leftarrow = u_{s'}^\leftarrow$.*

Note that this definition is very similar to the uniformity condition for the transitive setting, but while in the transitive setting, uniform policies and policies without useless edges coincide, this is not true for intransitive noninterference (in fact, neither implication holds).

   Uniformity, on an abstract level, is a natural requirement and often met in concrete systems, since an agent usually knows the sources of information available to him. In the uniform setting, many of the subtle issues with local policies do not occur anymore; as an example, i-security and the security definition from [Les06] coincide for uniform policies. Uniformity also has nice algorithmic properties, as both, checking whether a system has a uniform policy and checking whether a system with a uniform policy satisfies i-security, can be performed in polynomial time. This follows from the characterizations of i-security in terms of the unwindings we define next.

Definition 4.11 (uniform intransitive unwinding). *A* uniform intransitive unwinding *for M with a policy* $(\rightarrowtail_s)_{s \in S}$ *is a family of equivalence relations* $\sim_u^{\tilde{s},v}$ *for each choice of states $\tilde{s}$ and agents v and u, such that for all $s, t \in S$, and all $a \in A$, the following holds:*

- *If $s \sim_u^{\tilde{s},v} t$, then $\mathrm{obs}_u(s) = \mathrm{obs}_u(t)$.* $\qquad$ (OC$_i^u$)
- *If $s \sim_u^{\tilde{s},v} t$, then $u_s^\leftarrow = u_t^\leftarrow$.* $\qquad$ (PC$_i^u$)
- *If $s \sim_u^{\tilde{s},v} t$ and $a \in A$ with $v \not\rightarrowtail_{\tilde{s}} \mathrm{dom}(a)$, then $s \cdot a \sim_u^{\tilde{s},v} t \cdot a$.* $\qquad$ (SC$_i^u$)
- *If $\mathrm{dom}(a) \not\rightarrowtail_{\tilde{s}} u$, then $\tilde{s} \sim_u^{\tilde{s},\mathrm{dom}(a)} \tilde{s} \cdot a$.* $\qquad$ (LR$_i^u$)

In the following theorem intransitive uniformity and i-security (for uniform policies) are characterized by almost exactly the same unwinding. The only difference is that for uniformity we require policy consistency (PC$_i^u$), since we are concerned with having the same *local policies* in certain states, while for security, we require (OC$_i^u$), since we are interested in *observations*.

Theorem 4.12 (uniform unwinding characterizations)

1. *The policy $(\rightarrowtail_s)_{s \in S}$ is intransitively uniform if and only if there is a uniform intransitive unwinding for M and $(\rightarrowtail_s)_{s \in S}$ that satisfies* $(\mathrm{PC_i^u})$, $(\mathrm{SC_i^u})$, *and* $(\mathrm{LR_i^u})$.
2. *If $(\rightarrowtail_s)_{s \in S}$ is intransitively uniform, then M is i-secure if and only if there is a uniform intransitive unwinding that satisfies* $(\mathrm{OC_i^u})$, $(\mathrm{SC_i^u})$ *and* $(\mathrm{LR_i^u})$.

In particular, if an unwinding satisfying all four conditions exists, then a system is secure. Due to Theorem 4.6, we cannot hope that the above unwindings completely characterize i-security, and indeed the system in Fig. 3 is i-secure but not intransitively uniform. However, for uniform policies, Theorem 4.12 immediately yields efficient algorithms to verify the respective conditions via a standard dynamic programming approach:

Corollary 4.13 (uniform unwinding verification)

1. *Verifying whether a policy is intransitively uniform can be performed in nondeterministic logarithmic space.*
2. *For systems with intransitively uniform policies, verifying whether a system is i-secure can be performed in nondeterministic logarithmic space.*

The above shows that the complexity of intransitive noninterference with local policies comes from the *combination* of local policies that do not allow agents to "see" their allowed sources of information with an intransitive security definition. In the transitive setting, this interplay does not arise, since there a system always can allow agents to "see" their incoming edges (see Theorem 3.7).

4.4   Unwinding for IP-Security

In the setting with a global policy, i-security is equivalent to IP-security as defined in [HY87]. For IP-security, Rushby gave unwinding conditions that are sufficient, but not necessary. This left open the question whether there is an unwinding condition that *exactly* characterizes IP-security, which we can now answer positively as follows. Clearly, a policy that assigns the same local policy to every state is intransitively uniform. Hence our results immediately yield a characterization of IP-security with the above unwinding conditions, and from these, an algorithm verifying IP-security in nondeterministic logarithmic space can be obtained in the straight-forward manner.

Corollary 4.14 (unwinding for IP-security)

1. *A system is IP-secure if and only if it has an intransitive unwinding satisfying* $(\mathrm{OC_i^u})$, $(\mathrm{SC_i^u})$, *and* $(\mathrm{LR_i^u})$.
2. *IP-security can be verified in nondeterministic logarithmic space.*

5   Conclusion

We have shown that noninterference with local policies is considerably different from noninterference with a global policy: an allowed interference in one state may contradict a forbidden interference in another state. Our new definitions address this issue. Our purge- and unwinding-based characterizations show that our definitions are natural, and directly lead to our complexity results.

We have studied generalizations of Rusby's IP-security [Rus92]. An interesting question is to study van der Meyden's TA-security [vdM07] in a setting with local policies. Preliminary results indicate that such a generalization needs to use a very different approach from the one used in this paper.

References

[E+11]   Eggert, S., van der Meyden, R., Schnoor, H., Wilke, T.: The complexity of intransitive noninterference. In: IEEE Symposium on Security and Privacy, pp. 196–211. IEEE Computer Society (2011)

[ESW13]   Eggert, S., Schnoor, H., Wilke, T.: Noninterference with local policies. CoRR, abs/1208.5580 (2013)

[GM82]   Goguen, J.A., Meseguer, J.: Security policies and security models. In: Proc. IEEE Symp. on Security and Privacy, Oakland, pp. 11–20 (1982)

[GM84]   Goguen, J.A., Meseguer, J.: Unwinding and inference control. In: IEEE Symp. on Security and Privacy (1984)

[HY87]   Haigh, J.T., Young, W.D.: Extending the noninterference version of MLS for SAT. IEEE Trans. on Software Engineering SE-13(2), 141–150 (1987)

[Les06]   Leslie, R.: Dynamic intransitive noninterference. In: Proc. IEEE International Symposium on Secure Software Engineering (2006)

[Mil90]   Millen, J.K.: Hookup security for synchronous machines. In: CSFW, pp. 84–90 (1990)

[MSZ06]   Myers, A.C., Sabelfeld, A., Zdancewic, S.: Enforcing robust declassification and qualified robustness. Journal of Computer Security 14(2), 157–196 (2006)

[RG99]   Roscoe, A.W., Goldsmith, M.H.: What is intransitive noninterference? In: IEEE Computer Security Foundations Workshop, pp. 228–238 (1999)

[Rus92]   Rushby, J.: Noninterference, transitivity, and channel-control security policies. Technical Report CSL-92-02, SRI International (December 1992)

[vdM07]   van der Meyden, R.: What, indeed, is intransitive noninterference? In: Biskup, J., López, J. (eds.) ESORICS 2007. LNCS, vol. 4734, pp. 235–250. Springer, Heidelberg (2007)

[vO04]   von Oheimb, D.: Information flow control revisited: Noninfluence = noninterference + nonleakage. In: Samarati, P., Ryan, P.Y.A., Gollmann, D., Molva, R. (eds.) ESORICS 2004. LNCS, vol. 3193, pp. 225–243. Springer, Heidelberg (2004)

[WJ90]   Todd Wittbold, J., Johnson, D.M.: Information flow in nondeterministic systems. In: IEEE Symposium on Security and Privacy, pp. 144–161 (1990)

[YB94]   Young, W.D., Bevier, W.R.: A state-based approach to non-interference. In: CSFW, pp. 11–21 (1994)

In-Place Binary Counters

Amr Elmasry[1,2] and Jyrki Katajainen[2]

[1] Department of Computer Engineering and Systems, Alexandria University
Alexandria 21544, Egypt
[2] Department of Computer Science, University of Copenhagen
Universitetsparken 5, 2100 Copenhagen East, Denmark

Abstract. We introduce a binary counter that supports increments and decrements in $O(1)$ worst-case time per operation. (We assume that arithmetic operations on an index variable that is stored in one computer word can be performed in $O(1)$ time each.) To represent any integer in the range from 0 to $2^n - 1$, our counter uses an array of at most n bits plus few words of $\lceil \lg(1 + n) \rceil$ bits each. Extended-regular and strictly-regular counters are known to also support increments and decrements in $O(1)$ worst-case time per operation, but the implementation of these counters would require $O(n)$ words of extra space, whereas our counter only needs $O(1)$ words of extra space. Compared to other space-efficient counters, which rely on Gray codes, our counter utilizes codes with binary weights allowing for its usage in the construction of efficient data structures.

1 Introduction

A *numeral system* provides a specification for how to represent integers. In a positional numeral system, a string $\boldsymbol{d} = \langle d_{n-1}, \ldots, d_1, d_0 \rangle$ of *digits* is used to represent an integer, n being the length of the representation. As in the decimal system, d_0 denotes the least-significant digit, d_{n-1} the most-significant digit, and $d_{n-1} \neq 0$. If w_i is the *weight* of d_i, the string represents the decimal number $value(\boldsymbol{d}) = \sum_{i=0}^{n-1} d_i w_i$. (An empty string can be used to represent zero.)

A numeral system comprises four components:

1. The *digit set* specifies the values that the digits can take. For example, in a *redundant binary system* the digit set is $\{0, 1, 2\}$.
2. The *weight set* specifies the weights that the digits represent when determining the decimal value of the underlying integer. For example, a binary system uses the *binary weights* $w_i = 2^i$ where $i \in \{0, 1, \ldots, n-1\}$.
3. The *rule set* specifies the rules that the representation of each integer must obey. For example, the *regular system* [3] is a redundant binary system where every two 2's have at least one 0 in between.
4. The *operation set* specifies the operations that are to be supported. In this paper we only consider in details *increment* (increase the value by one) and *decrement* (decrease the value by one) operations. However, it is also relevant to support other arithmetic operations like additions and subtractions.

K. Chatterjee and J. Sgall (Eds.): MFCS 2013, LNCS 8087, pp. 349–360, 2013.

A representation of an integer that is subject to increments and decrements is called a *counter*. To represent an integer in the range $[0 \ldots 2^n - 1]$, the ordinary binary counter is space-efficient requiring n bits, but an *increment* or a *decrement* requires $\Theta(n)$ bit flips in the worst case. A *regular counter*—a counter using the regular system [3]—supports increments (of arbitrary digits) with a constant number of digit changes per operation. An *extended-regular counter* [3,10,13] uses the digit set $\{0, 1, 2, 3\}$ imposing a rule set that between any two 3's there is a digit other than 2 and between any two 0's there is a digit other than 1. Such a counter supports both increments and decrements (of arbitrary digits) with a constant number of digit changes per operation. A *strictly-regular counter* [9] provides the same guarantee, but it uses the digit set $\{0, 1, 2\}$. Unfortunately, the implementation of the aforementioned regular counters would require up to n indices in addition to the space needed by the digits themselves.

Recently, efficient and more space-economical counters were proposed by Bose et al. [1], Brodal et al. [2], and Rahman and Munro [16]. In these papers the complexity of the operations was analysed in the bit-probe model. In [2], using a representation of $n + 1$ bits, each of the *increment* and *decrement* operations could be accomplished by reading $\lg n + O(1)$ bits and writing $O(1)$ bits.

We use the word-RAM model [12] as our model of computation. If a counter requires n bits, these bits are kept compactly in an array of $\lceil n/w \rceil$ words, where w is the size of the machine word in bits. In addition to these words, we only allow the usage of $O(1)$ other words. Also we assume that, for a problem of size n, $w \geq \lceil \lg(1 + n) \rceil$, i.e. a variable counting the number of bits and a variable referring to a position in the array of bits can each fit in one computer word.

In this paper we introduce an in-place binary counter; it uses $n + O(\lg n)$ bits, and supports *increment* and *decrement* operations in $O(1)$ worst-case time. This solves an open problem stated, for example, in Demaine's lecture notes [4]. In the bit-probe model, both our *increment* and *decrement* operations involve $O(\lg n)$ bit accesses and modifications. However, the bits accessed and modified are stored in a constant number of words, and we only perform $O(1)$ word operations per *increment* and *decrement*. We can also test whether the value of the counter is zero in $O(1)$ worst-case time. Conceptually, our counter is a modification of a regular counter; instead of giving preference to handling the carries (2's) at the least-significant end of the representation, we handle the carries at the most-significant end first. Although increments are easy and direct, incorporating decrements is more involved and tricky. A simple consequence of our construction is a new representation of positive integers, using binary weights, in which a positive integer with value K is coded using $\lg K + O(\lg \lg K)$ bits and the encoding differs from that of $K + 1$ only in $O(\lg \lg K)$ bits.

Compared to other space-efficient counters, our counter has a significant advantage: It can be applied in the construction of efficient data structures [3,17]. For a survey on numeral systems and their applications to data structures, we refer to [14, Chapter 9]. The idea is to relate the number of objects of a specific type in a data structure to the value of a digit. Often two objects of the same size can be combined efficiently, and one object can be split into two objects of the same size. These operations are the exact counterparts for the carry and borrow operations that are employed by binary counters. On the other hand, the known space-economical counters [1,2,16] rely on some variant of a Gray code [11], and more involved operations, like bit flips, required by a Gray code can seldom be simulated at the level of the data structure. Because the upper bound for the sum of the digits of our counter is optimized (the sum of the digits representing a positive integer K is at most $\lceil \lg(1 + K) \rceil$), the number of objects in the corresponding data structure is bounded from above as well.

The drawback of the ordinary binary counter is that *increment* costs $O(1)$ only in the amortized (not wost-case) sense, and that it cannot support both *increment* and *decrement* operations efficiently at the same time. Since our in-place counter supports both *increment* and *decrement* operations efficiently, it can be used as a replacement for the ordinary binary counter even in applications where space-saving is not the main goal. As for the data structure, the cost of the *insert* operation, which resembles *increment* and appends a new element to the data structure, is $O(1)$ in the worst case. Additionally, the cost of the *borrow* operation, which resembles *decrement* and removes an arbitrary element from the data structure, is $O(1)$ in the worst case. The importance of fast *borrow* has been demonstrated in several earlier papers; see for example [5,6,7,13].

2   The Data Structure

Our objective is to implement a counter that represents an integer in the range $[0 .. 2^n - 1]$ with at most $n + O(\lg n)$ bits. Assuming $\lceil \lg(1 + n) \rceil$ bits fit in one computer word, our counter uses a constant number of words in addition to the n bits. To represent a positive integer K, the counter has the following characteristics:

$\mathbf{C_1}$. The sum of the digits is at most $\lceil \lg(1 + K) \rceil$.

$\mathbf{C_2}$. Other than the least-significant digit, at most one digit has value 2, and all the other digits are 0's and 1's.

$\mathbf{C_3}$. The most-significant digit is always non-zero. Due to the binary weights, we have $K \geq 2^{n-1}$ implying $n \leq \lceil \lg(1 + K) \rceil$.

Let ℓ denote the current length of the number representation and assume that the string of digits in the representation is $\boldsymbol{d} = \langle d_{\ell-1}, d_{n-2}, \ldots, d_0 \rangle$. A straightforward implementation of our in-place counter is to store the value of d_0, which is at most ℓ, in one word that we call x. In addition, we store the index of the digit with value 2 (if any such digit other than the least-significant one exists); this also consumes one word that we call α. Each of the other digits is either 0 or 1. To store these bits, we assume the availability of an (infinite) array $\boldsymbol{b}$, the first ℓ bits (or $\lceil \ell/w \rceil$ words) of which are in use. To be able to efficiently incorporate the *decrement* operations, we shall also use two more variables: β and ζ. The meaning of β will be explained later; ζ counts the number of 0 bits in $\boldsymbol{d}$. The representation of the integer 50 is given in Fig. 1.

To distinguish whether there is a 2 in the representation or not, while still using as few bits as indicated, we use an extra bit called *carry* (which could actually be the unused bit b_0) and adopt the following convention:

$$\ell = 6$$

$\langle\ 1,\ 0,\ 1,\ 1,\ 0,\ -\ \rangle$ $x = 2$

$b_5\ b_4\ b_3\ b_2\ b_1\ b_0$ $carry = 1$ (indicates that there is a 2)

$$\alpha = 2$$

$$\zeta = 2$$

Fig. 1. Representation of the integer 50 when only *increment* is supported

- If $carry = 1$, then the actual value of d_α is 2. Otherwise, d_α equals b_α.

We still maintain the stored bit in b_α to be 1 when $carry = 1$. Accordingly,

$$value(\boldsymbol{d}) = \begin{cases} x + \sum_{i=1}^{\ell-1} b_i \cdot 2^i & \text{if } carry = 0, \\ x + 2^\alpha + \sum_{i=1}^{\ell-1} b_i \cdot 2^i & \text{if } carry = 1. \end{cases}$$

The following procedure is used to initialize our counter to zero.

Algorithm *initialize*$(\boldsymbol{b}, \ell, x, carry, \alpha, \beta, \zeta)$

1: $\ell \leftarrow 1$
2: $x \leftarrow 0$
3: $carry \leftarrow 0$
4: $\alpha \leftarrow 1$
5: $\beta \leftarrow 1$
6: $\zeta \leftarrow 0$

2.1 Increments

To maintain $\mathbf{C}_1$ we need a mechanism to reduce the sum of the digits within the *increment* operations. Instead of monitoring this sum until it reaches the threshold, we reduce the sum of the digits by one with every *increment* operation whenever possible. We use a procedure called *fix-carry* that works as follows: If there exists an index $\alpha \neq 0$ where d_α is a 2 (i.e. $carry = 1$), we set d_α to 0 and increase $d_{\alpha+1}$ by one. Otherwise, if $d_0 \geq 2$, we decrease d_0 by two and increase d_1 by one. Note that a *fix-carry* does not change the value of the number, and in addition it maintains all the aforementioned characteristics.

To increment a number, we perform a *fix-carry* operation and add one to x.

Algorithm *increment*$(\boldsymbol{b}, \ell, x, carry, \alpha, \zeta)$

1: *fix-carry*$(\boldsymbol{b}, \ell, x, carry, \alpha, \zeta)$
2: $x \leftarrow x + 1$

Algorithm *fix-carry*($\boldsymbol{b}$, ℓ, x, *carry*, α, ζ)

1: **if** *carry* $= 1$
2:      $b_\alpha \leftarrow 0$
3:      $\zeta \leftarrow \zeta + 1$
4:      $\alpha \leftarrow \alpha + 1$
5: **else if** $x \geq 2$
6:      $x \leftarrow x - 2$
7:      $\alpha \leftarrow 1$
8: **else**
9:      **return**
10: **if** $\alpha = \ell$
11:     $\ell \leftarrow \ell + 1$
12:     $b_{\ell-1} \leftarrow 1$
13:     *carry* $\leftarrow 0$
14: **else if** $b_\alpha = 0$
15:     $b_\alpha \leftarrow 1$
16:     $\zeta \leftarrow \zeta - 1$
17:     *carry* $\leftarrow 0$
18: **else**
19:     *carry* $\leftarrow 1$

For more illustration, here are the first 50 positive integers obtained by applying *increment* repeatedly: 1, 2, 11, 12, 21, 102, 111, 112, 121, 202, 1003, 1012, 1021, 1102, 1111, 1112, 1121, 1202, 2003, 10004, 10013, 10022, 10103, 10112, 10121, 10202, 11003, 11012, 11021, 11102, 11111, 11112, 11121, 11202, 12003, 20004, 100005, 100014, 100023, 100104, 100113, 100122, 100203, 101004, 101013, 101022, 101103, 101112, 101121, 101202.

We would further characterize any representation of our counter, resulting from a sequence of *increment* operations, with the following properties:

P$_1$. The value of the least-significant digit is greater than 0.
P$_2$. If there is a digit with value 2 other than the least-significant digit, all the digits between this 2 and the least-significant digit are 0's.

In consequence, the number $\boldsymbol{d}$ is kept as a string of the form $(1(0|1)^*)^*x$ or $(1(0|1)^*)^*20^*x$, where x is a positive integer and $*$ denotes zero or more repetitions of the preceding digit or string of digits.

P$_3$. If $\boldsymbol{d}$ is of the form 1^*2, there is obviously no 0 digits in $\boldsymbol{d}$. Otherwise, if $\boldsymbol{d}$ is of the form 1^*20^*x or if there is no digit in $\boldsymbol{d}$ with value 2 other than (possibly) the least-significant digit, the number of 0 digits in $\boldsymbol{d}$ equals $x-1$. Otherwise, the number of 0 digits in $\boldsymbol{d}$ equals x.

We can thus express this property using the following trichotomy:

$$\begin{cases} \zeta = 0 & \text{if } \boldsymbol{d} \in 1^*2, \\ \zeta = x - 1 & \text{if } \boldsymbol{d} \in 1^*20^*x \cup (1(0|1)^*)^*x, \\ \zeta = x & \text{otherwise.} \end{cases} \tag{1}$$

As a consequence of **P$_3$**, the following property also holds:

P$_4$. If no digit is larger than 1, then all the digits are 1's.

This property results from **P$_3$** because $carry = 0$ and $x = 1$ imply $\zeta = 0$.

Lemma 1. *The increment operation sustains the characteristics and properties.*

Proof. Trivially the characteristics are valid for a counter whose value is one. Increasing the least-significant digit by one may only break **C$_1$** if the sum of the digits exceeds the threshold by one. The *fix-carry* operation decreases the sum of the digits by one returning it back below the threshold, unless no digit with a value greater than one exists. In this latter case, assuming that the sum of the digits after the *increment* operation is s, then the resulting integer is at least 2^{s-1}; this ensures the validity of **C$_1$**. If before this operation there exists $\alpha \neq 0$ where d_α has value 2, then after the *increment* operation d_α is 0 and at most one digit other than d_0 with value 2 may exist; this digit is $d_{\alpha+1}$. If before the *increment* operation there was no digit other than d_0 with value 2, at most one digit may have value 2 after the *increment* operation; this digit is d_1. The validity of **C$_2$** follows.

It is easy to verify that the *increment* operation maintains **P$_1$** and **P$_2$**. We show next, using induction on the counter values, that **P$_3$** and accordingly **P$_4$** are true. Initially, for $d = 12$ the first case of (1) holds. Later on, an *increment* applied to $d \in 11^*2$ results in $d \in 1^*21$ and the second case of (1) holds. Starting with $d \in 1^*20^*x$, an *increment* will increase both x and ζ by one, and the second equation of (1) will still be valid. Starting with $d \in 1(0|1)^*0x$ when $x \geq 2$, an *increment* will decrease both x and ζ by one, and the second equation of (1) will still be valid. Alternatively, starting with $d \in 1(0|1)^*0(0|1)^*1x$ when $x \geq 2$, an *increment* will decrease x by one, while ζ will not change, and the third case of (1) will then be fulfilled. A complementing state for these last two is when $x = 1$; we may assume then, using **P$_4$**, that $d \in 1^*1$. In such a case, an *increment* will result in $d \in 1^*2$, and we are back to the first case of (1). Lastly, consider the third case where $d \in 1(0|1)^*0(0|1)^*20^*x$. Starting with $d \in 1(0|1)^*0(0|1)^*120^*x$, an *increment* will increase both x and ζ by one, and the third equation of (1) will still be valid. Staring with $d \in 1(0|1)^*020^*x$, an *increment* will increase x by one, while ζ will not change, and the second case of (1) will then be fulfilled. □

Note that, for any sequence of *increment* operations, **C$_1$** can even be shown to hold with equality. To advocate for this, we point out that every *fix-carry* operation decreases the sum of the digits by one, except when all the digits are 1's. In other words, the only case the sum of the digits increases is when the value of the counter becomes a power of two after the *increment* operation.

2.2   Decrements

Our objective is to implement the *decrement* operations as the reverse of the corresponding *increment* operations. To efficiently implement *fix-borrow*, we need to figure out the changes that were made to the counter when it was last increased beyond its current value. More precisely, we need to know the index of the digit that the corresponding *fix-carry* operation changed back in history. The *fix-carry* operation used the index, say γ, of the currently second-least-significant non-zero digit. Assuming that γ is available, the *fix-borrow* operation decreases the corresponding digit d_γ by one and increases its preceding digit by two. This may result in losing track of γ!

Algorithm *fix-borrow*(b, ℓ, x, *carry*, γ, ζ)

1: **if** *carry* $= 1$
2:    $b_\gamma \leftarrow 1$
3: **else if** $\gamma = \ell - 1$
4:    $\ell \leftarrow \ell - 1$
5: **else**
6:    $b_\gamma \leftarrow 0$
7:    $\zeta \leftarrow \zeta + 1$
8:    **if** $\gamma > 1$
9:       $\gamma \leftarrow \gamma - 1$
10:      $b_\gamma \leftarrow 1$
11:      *carry* $\leftarrow 1$
12:      $\zeta \leftarrow \zeta - 1$
13:   **else**   // we lose track of the correct value of γ if *carry* $= 0$
14:      $x \leftarrow x + 2$
15:      *carry* $\leftarrow 0$

If it happens that $d_\alpha \neq 0$, then we are lucky as γ is equal to α that we are already maintaining. However, d_α may become 0 following the *decrement* operation. To see the problem, consider for example the case when *decrement* is to be applied to a number $10000001x$. Before this operation, γ is equal to 1. But the preceding number that resulted in this number via the corresponding *increment* operation is $10000000(x+1)$ with $\gamma = 8$. How would we find the new value of γ in constant time within the *decrement* operation?

The critical property that gets us to a worst-case constant-time implementation of *decrement* is $\mathbf{P}_3$. Our idea is to decrease the value of the least-significant digit with every *decrement*, while possibly not performing a *fix-borrow*. More precisely, within some *decrement* operations we incrementally walk through the zero digits, two digits at a time, until we reach the digit d_γ. We then perform the postponed *fix-borrow* operations within the upcoming *decrement* operations working with double the speed (two *fix-borrow* operations at a time). To efficiently implement the *decrement* operations, we maintain the index α that we are up to so far in our search for γ, leaving behind its old (before starting the search) value stored as β. In accordance, the *decrement* operations work in three modes: In the *normal mode*, when $d_\beta \neq 0$ (which implies $\alpha = \beta$), each *decrement* operation performs one *fix-borrow* operation using the index α as the argument. In the *search mode*, each *decrement* operation sequentially traverses the next two digits and increases α by two as long as both digits are 0's. Once a *decrement* operation reaches a non-zero digit, i.e. $\alpha = \gamma$, we switch to the *rapid mode* as there are postponed *fix-borrow* operations; in this mode each *decrement* operation performs two *fix-borrow* operations using the index α as the argument.

Another subtle issue to be considered is when the corresponding *increment* operation has not performed a *fix-carry*. This only happens if the number we want to decrease is $d \in 1^*2$. To be able to distinguish this case from the case $d \in 11^*01^*2$, we check if there are no 0's in the number, i.e. $x = 2$ and $\zeta = 0$. In such a case, we skip the *fix-borrow* operation altogether. We also skip the *fix-borrow* operation if the number that is to be decreased is 1.

Algorithm *decrement*$(b, \ell, x, carry, \alpha, \beta, \zeta)$

1:   **if** $\ell = 1$ **or** $(x = 2$ **and** $\zeta = 0)$
2:      $x \leftarrow x - 1$
3:      **return**
4:   **if** $b_\beta \neq 0$ // work in normal mode
5:      *fix-borrow*$(b, \ell, x, carry, \alpha, \zeta)$
6:      $\beta \leftarrow \alpha$
7:   **else if** $b_\alpha = 0$ // work in search mode
8:      $\alpha \leftarrow \alpha + 1$
9:      **if** $b_\alpha = 0$
10:         $\alpha \leftarrow \alpha + 1$
11:    **else**   // switch to rapid mode
12:      *fix-borrow*$(b, \ell, x, carry, \alpha, \zeta)$
13: **else**   // work in rapid mode
14:      *fix-borrow*$(b, \ell, x, carry, \alpha, \zeta)$
15:      *fix-borrow*$(b, \ell, x, carry, \alpha, \zeta)$
16: $x \leftarrow x - 1$

The correctness of the construction is a consequence of property $\mathbf{P}_3$ that implies that x will always be non-zero, and hence we can decrease its value while the *decrement* operations are working in any of the three modes. However, when incorporating the *decrement* operations, $\mathbf{P}_3$ would not hold as tight as before and must be relaxed as follows:

$\mathbf{P}_3'$. If there is a digit in d with value 2 other than the least-significant digit, the number of zero digits in d is bounded by *twice* the value of the least-significant digit. Otherwise, the number of zero digits is bounded by *twice* the value of the least-significant digit minus two.

$$\begin{cases} \zeta \leq 2x - 2 & \text{if } carry = 0, \\ \zeta \leq 2x & \text{if } carry = 1. \end{cases}$$

In the revised implementation of *increment* we have to consider the case when there are postponed *fix-borrow* operations. To be able to detect this, we make use of the index β. Whenever $\alpha > \beta$, it means that there have been more *decrement* operations than *increment* operations since the time we switched to the search mode. Here the *increment* operation only needs to undo what a preceding *decrement* has done. More precisely, in this case the *increment* operation increases the least-significant digit and instead of performing a *fix-carry* operation it moves β two steps forward. Once β and α meet, i.e. $\alpha = \beta$, this means that there is no postponed *fix-borrow* operations, and the *increment* operation should work in the normal mode by calling *fix-carry*.

$$\ell = 6$$
$$x = 2$$

$b_5 \ b_4 \ b_3 \ b_2 \ b_1 \ b_0$

$$carry = 1$$

$\langle \ 1, \ 0, \ 1, \ 0, \ 0, \ - \ \rangle$

$$\alpha = 3$$
$$\beta = 1$$
$$\zeta = 3$$

Fig. 2. A representation of the integer 50 when both *increment* and *decrement* are supported; this representation was obtained by starting from the representation of Fig. 1, performing three *increment* operations followed by three *decrement* operations

---

Algorithm *increment*(b, ℓ, x, *carry*, α, β, ζ)

1:   **if** $\alpha = \beta$ // work in normal mode
2:      *fix-carry*(b, ℓ, x, *carry*, α, ζ)
3:      $\beta \leftarrow \alpha$
4:   **else**   // work in search mode
5:      $\beta \leftarrow \beta + 2$
6:   $x \leftarrow x + 1$

In Fig. 2 one representation of the integer 50 is given when the *decrement* operation and the revised *increment* operation are in use. This example shows that the representation of our integers is not unique, since a number can have several representations depending on the sequence of operations applied.

Note that the *increment* and *decrement* operations keep either $\alpha = \beta$ or $\alpha > \beta$ with $\alpha - \beta$ being an even positive integer.

Lemma 2. *The increment and decrement operations sustain the characteristics and the relaxed properties.*

Proof. The *fix-borrow* operation increases the sum of the digits by one. When two *fix-borrow* operations are performed per *decrement* in the rapid mode, the sum of the digits increases as a result. However, a second *fix-borrow* is executed only if it was postponed in a previous *decrement*, indicating that $\mathbf{C}_1$ was satisfied with a strict inequality; this ensures the validity of $\mathbf{C}_1$ following any *decrement* operation. On the other hand, the *increment* operation skips calling *fix-carry* only when $\alpha > \beta$. We start the search mode when $\alpha = \beta = 1$, a *decrement* moves α two steps forward, and an *increment* moves β two steps forward. The condition $\alpha > \beta$ then implies that there are postponed *fix-borrow* operations. Hence, there is no need to execute a *fix-carry* within such *increment* operation; this ensures the validity of $\mathbf{C}_1$ following any *increment* operation.

If there is a digit d_α whose value is 2, the *fix-borrow* operation decreases d_α to 1 and then adds two to the preceding digit. In accordance, at most one 2 may exist and is preceded by 0's up to (and not including) the least-significant digit. The same fact holds as a result of the *fix-carry* operation. The validity of $\mathbf{C}_2$ and that of $\mathbf{P}_2$ are thus sustained.

To prove $\mathbf{P_3}$', we use the relation $\alpha - \beta \leq \zeta$ that is true because $d_j = 0$ for all j satisfying $\alpha - 1 \geq j \geq \beta$. It follows that we only need to show the dichotomy:

$$\begin{cases} \zeta \leq x + (\alpha - \beta)/2 - 1 & \text{if } carry = 0, \\ \zeta \leq x + (\alpha - \beta)/2 & \text{if } carry = 1. \end{cases} \tag{2}$$

The proof is by induction on the operations sequence. The base case follows by noting that an initial sequence of *increment* operations maintains (1) and $\alpha = \beta$. When the *decrement* operations work in the search mode, $carry = 0$ and the first inequality of (2) holds. With each of those *decrement* operations, x decreases by one and α increases by two; the first inequality is still guaranteed. As an exception, the last *decrement* operation working in the search mode may execute a *fix-borrow* operation. As a result, the value of α does not change; but then we have $carry = 1$, guaranteeing the second inequality of (2). For each of the following *decrement* operations working in the rapid mode, x decreases by one and α decreases by two, but ζ decreases by two (via two *fix-borrow* operations); the second inequality of (2) is still valid. For a *decrement* operation to reset the *carry* bit back to 0, the 2 must vanish; this only happens if $\alpha = 1$. In such a case, as a result of the *decrement* operation, x increases by one and both α and ζ do not change; so now the first inequality of (2) is guaranteed. Each of the other *decrement* operations working in the normal mode keeps $\alpha = \beta$ and either decreases both x and ζ by one or increases both by one. An exception is when the *decrement* operation is applied to a number where $x = 2$ and $\zeta = 0$. In this case, the resulting number has all 1's, and the first inequality of (2) is still valid. An *increment* operation working in the search mode increases x by one and β by two. We also need to mention that the *increment* operations working in the normal mode maintain the induction hypothesis. This follows using arguments similar to those of Lemma 1 and by noting that these operations keep $\alpha = \beta$. In conclusion, the two operations in all modes maintain (2), and $\mathbf{P_3}$' is satisfied. It directly follows from the first inequality of $\mathbf{P_3}$' that $\mathbf{P_4}$ is also satisfied.

Using $\mathbf{P_3}$', since $\zeta \geq 0$, the only case where x could have possibly been 0 is when $carry = 1$. Contradictorily, it follows that in this case $\zeta = 0$. The value of the least-significant digit must then be greater than 0, and $\mathbf{P_1}$ holds.   □

In fact, we can prove a tighter version of $\mathbf{C_1}$. We namely argue that $\sum_{i=0}^{\ell-1} d_i = \lceil \lg(1 + K) \rceil - (\alpha - \beta)/2$. The equation is true when $\alpha = \beta$ (as we have shown for the *increment* operations); that is when the operations work in the normal mode. A *decrement* operation working in the search mode increases α by two, and hence decreases the right-hand side by one, but it decreases the left-hand side by one as well. A *decrement* operation working in the rapid mode calls *fix-borrow* twice and decreases x by one. As a result, the left-hand side increases by one and the right-hand side also increases by one (as α decreases by two). An *increment* operation working in the search mode increases the left-hand side by one, but then it increases the right-hand side by one (as β increases by two).

3   Remarks

Let $\langle d_{n-1}, \ldots, d_1, d_0 \rangle$ be a string of bits representing a positive integer, where $d_{n-1} \neq 0$. To distinguish all the possible representations, at least $n - O(1)$ bits are needed by any counter. A binary counter stores the bits and the length of the representation. Thus, the total space usage is $n + O(\lg n)$ bits. Our in-place counter achieves the same space bound. In addition, our counter supports *increment* and *decrement* operations in $O(1)$ worst-case time.

The main ingredients of our counter are: 1) a binary encoding of an integer in the leading $n - 1$ bits of the representation, 2) at most one carry whose position is recalled, 3) the least-significant digit that can take any value up to n, and 4) at most one delayed query "find the next non-zero digit" that is in progress.

We can efficiently support the addition of two of our counters by summing the individual bit-array representations, the two carries, and the values of the two least significant digits for both, and then converting the resulting binary number back to the required form. It is possible to do this conversion such that the properties are maintained. Most importantly, the least-significant digit must be made larger than the number of 0 bits. When implemented this way, an addition requires $O(n)$ bit operations, n being the number of bits of the longer counter. Moreover, for the addition of binary numbers we can rely on word-wise operations so that the addition takes $O(n/w)$ worst-case time on the word RAM, w being the size of the machine word in bits.

A binary counter that supports increments, decrements, and additions is often used in the implementation of mergeable priority queues. For example, a *binomial queue* [17], which is a sequence of heap-ordered binomial trees (for the definition of a binomial tree, we refer to any well-equipped textbook on data structures, e.g. [15, Section 11.4]), relies on a binary counter. A 1 bit at position r in the numeral representation of N, where N is the number of elements stored, means that the data structure contains a binomial tree of 2^r nodes. When a carry is propagated, two binomial trees are joined, and when a borrow is propagated, a binomial tree is split into two; both of these operations can be carried out in $O(1)$ worst-case time on binomial trees. By replacing the ordinary binary counter with our counter, the data structure can support both *insert* and *borrow* operations in $O(1)$ worst-case time, whereas Vuillemin's original implementation supports both of these operations in $\Theta(\lg N)$ worst-case time.

A *run-relaxed heap* [5], which is a sequence of almost heap-ordered binomial trees, uses a counter that bounds the sum of the digits to at most $\lceil \lg(1 + N) \rceil$, where N is the number of elements. A *two-tier relaxed heap* [8] uses the zeroless version of the extended-regular counter [3,10,13] to keep track of the trees in the structure (for the description of zeroless counters, see [14, Chapter 9]). In both of these data structures the counters used could be replaced by our counter. This replacement has an interesting trade-off: *borrow* and *delete* operations would become more efficient, but *decrease* and *union* operations would become less efficient. The main reason why we cannot support the last two operations efficiently is that our counter does not support increments of arbitrary digits; it just efficiently supports the increment and decrement of the value by one.

We leave it as an open problem to extend our in-place counter to support a larger operation set efficiently.

References

1. Bose, P., Carmi, P., Jansens, D., Maheshwari, A., Morin, P., Smid, M.: Improved methods for generating quasi-gray codes. In: Kaplan, H. (ed.) SWAT 2010. LNCS, vol. 6139, pp. 224–235. Springer, Heidelberg (2010)
2. Brodal, G.S., Greve, M., Pandey, V., Rao, S.S.: Integer representations towards efficient counting in the bit probe model. In: Ogihara, M., Tarui, J. (eds.) TAMC 2011. LNCS, vol. 6648, pp. 206–217. Springer, Heidelberg (2011)
3. Clancy, M.J., Knuth, D.E.: A programming and problem-solving seminar. Technical Report STAN-CS-77-606, Computer Science Department, Stanford University, Stanford (1977)
4. Demaine, E., et al.: Advanced data structures: Lecture 17 (2012), http://courses.csail.mit.edu/6.851/spring12/scribe/L17.pdf
5. Driscoll, J.R., Gabow, H.N., Shrairman, R., Tarjan, R.E.: Relaxed heaps: An alternative to Fibonacci heaps with applications to parallel computation. Commun. ACM 31(11), 1343–1354 (1988)
6. Elmasry, A., Jensen, C., Katajainen, J.: Multipartite priority queues. ACM Trans. Algorithms 5(1), Acticle 14 (2008)
7. Elmasry, A., Jensen, C., Katajainen, J.: Two new methods for constructing double-ended priority queues from priority queues. Computing 83(4), 193–204 (2008)
8. Elmasry, A., Jensen, C., Katajainen, J.: Two-tier relaxed heaps. Acta Inform. 45(3), 193–210 (2008)
9. Elmasry, A., Jensen, C., Katajainen, J.: Strictly-regular number system and data structures. In: Kaplan, H. (ed.) SWAT 2010. LNCS, vol. 6139, pp. 26–37. Springer, Heidelberg (2010)
10. Elmasry, A., Katajainen, J.: Worst-case optimal priority queues via extended regular counters. In: Hirsch, E.A., Karhumäki, J., Lepistö, A., Prilutskii, M. (eds.) CSR 2012. LNCS, vol. 7353, pp. 125–137. Springer, Heidelberg (2012)
11. Gray, F.: Pulse code communications. U.S. Patent 2632058 (1953)
12. Hagerup, T.: Sorting and searching on the word RAM. In: Morvan, M., Meinel, C., Krob, D. (eds.) STACS 1998. LNCS, vol. 1373, pp. 366–398. Springer, Heidelberg (1998)
13. Kaplan, H., Shafrir, N., Tarjan, R.E.: Meldable heaps and Boolean union-find. In: STOC 2002, pp. 573–582. ACM, New York (2002)
14. Okasaki, C.: Purely Functional Data Structures. Cambridge University Press, Cambridge (1998)
15. Preiss, B.R.: Data Structures and Algorithms with Object-Oriented Design Patterns in C++. John Wiley & Sons, Inc., New York (1999)
16. Rahman, M.Z., Munro, J.I.: Integer representation and counting in the bit probe model. Algorithmica 56(1), 105–127 (2010)
17. Vuillemin, J.: A data structure for manipulating priority queues. Commun. ACM 21(4), 309–315 (1978)

Rent or Buy Problems
with a Fixed Time Horizon

Leah Epstein and Hanan Zebedat-Haider

Department of Mathematics, University of Haifa, 31905 Haifa, Israel
lea@math.haifa.ac.il, hnan_haider@hotmail.com

Abstract. We study several variants of a fixed length ski rental problem and related scheduling problems with rejection. A ski season consists of m days, and an equipment of cost 1 is to be used during these days. The equipment can be bought on any day, in which case it can be used without any additional cost starting that day and until the vacation ends. On each day, the algorithm is informed with the current non-negative cost of renting the equipment. As long as the algorithm did not buy the equipment, it must rent it every day of the vacation, paying the rental cost of each day of rental. We consider the case of arbitrary, non-increasing, and non-decreasing rental costs. We consider the case where the season cannot end before the mth day, and the case that it can end without prior notice. We propose optimal online algorithms for all values of m for all variants. The optimal competitive ratios are either defined by solutions of equations (closed formulas or finite recurrences) or sets of mathematical programs, and tend to 2 as m grows.

1   Introduction

We consider deterministic ski rental problems with a finite time horizon, and related scheduling problems. In such ski rental problems, a positive integer $m \geq 2$ is given, and the input corresponds to a process that either consists of exactly m steps, or of at most m steps, as follows. Every step consists of a request and dealing with it subsequently, and it is called a *day*. The goal of the algorithm is to use a certain equipment incurring a minimum total cost. One option for each step is buying the equipment for a cost of 1. Once the equipment was bought, further requests (of future steps) do not incur any cost. For $1 \leq j \leq m$, request j has a rental cost $q_j \geq 0$, and the alternative way of dealing with it is renting it for the cost q_j. The value q_j is unknown prior to the jth day. Thus, since buying is possible at all times, given a new request such that the equipment was not bought previously, it is possible to deal with it either by buying the equipment or by renting it. The cost of an algorithm is the total rental cost paid in all steps, plus the buying cost, if the equipment was ever bought. The classic (deterministic) ski rental problem [16,20,15,14,13] is the case where all rental costs are equal, and the number of days is not known in advance (an upper bound on the number of days is not known either). This problem is attributed to Rudolph (see [16]). In this paper we study problems where one has to decide (every day) whether to buy or rent skis when it is known that the ski season

K. Chatterjee and J. Sgall (Eds.): MFCS 2013, LNCS 8087, pp. 361–372, 2013.

lasts m days. In some of the variants it could end earlier without a prior notice (but it cannot end later). We study the general scenario, where the rental costs can be very different on different days. In the case of arbitrary costs, they can go up and down, due to supply and demand. In the case of non-increasing costs, which is probably the most realistic case, the cost cannot go up, since as the season proceeds, the demand is unlikely to increase. We also consider the case of non-decreasing costs, where the merchants try to pressure the skiers to buy as the season is closer to its end and increase the requested rental costs.

Formally, we consider three variants with exactly m days; the general case of arbitrary rental costs, denoted by $SR(m)$, and the cases with non-increasing and non-decreasing rental costs, denoted by $SRD(m)$ and $SRIF(m)$, respectively. We are also interested in the variants with *at most* m days. In the cases of $SR(m)$ and $SRD(m)$, the variant with at most m days is the same as the one with exactly m days, as requests of zero rental cost can be added at the end of the input. For the case of non-decreasing rental costs, this is an additional variant that we call $SRIT(m)$. We show that this case is indeed different from $SRIF(m)$. We study related problems of online scheduling with rejection of jobs with equal job processing times (also called unit size jobs or unit jobs) on m identical machines. The input consists of jobs arriving one by one. An arriving job j is presented together with a non-negative rejection penalty w_j. An algorithm has to make an immediate decision for each arriving job, that is, the job needs to be either rejected, in which case the algorithm pays its penalty, or otherwise it must be scheduled to run on a machine, in which case it contributes its processing time to the completion time of that machine. The objective is to minimize the sum of the makespan of the schedule for all accepted jobs and the total penalty of all the rejected jobs. The makespan of a schedule is defined as the maximum completion time of any machine. If the number of input jobs is at most m, then the problem is equivalent to $SR(m)$ for the online variant (with arbitrary rejection penalties), and it is equivalent to $SRD(m)$ and $SRIT(m)$, for the semi-online variants with non-increasing and non-decreasing rejection penalties, respectively, (and if it is known that the rejection penalties are non-decreasing and the number of jobs is exactly m, then the problem is equivalent to $SRIF(m)$). In the full version of this paper we also consider each of these problems with an arbitrary number of jobs n (known or unknown in advance), and their relation to the ski rental problems. We give more details regarding the relation between the problems in the full version of this paper. Throughout the paper, we let OPT denote an optimal offline algorithm as well as its cost. The competitive ratio of an algorithm is the worst-case ratio (over all inputs) between the cost of the algorithm and OPT. The optimal competitive ratio is the supremum value such that no algorithm can have a smaller competitive ratio, and an algorithm that has this competitive ratio is called an optimal online algorithm. An algorithm is called r-competitive if its competitive ratio does not exceed r. Note that our optimal online algorithms use their (optimal) competitive ratios as a parameter. If instead of the exact value of the optimal competitive ratio only an approximation is known, then in all cases the algorithm can be applied instead with an upper bound on this value.

Previous Work. In standard ski rental, it is assumed that both the rental cost and the buying cost are fixed (but the length of the skiing season is unknown). For this problem, the optimal competitive ratio is 2 [16]. Many variants of ski rental were studied. In one type of variants, the influence of economic effects on prices, such as various interest rates, was studied [7,5,4,22,24]. Another approach was to define versions of ski rental where the algorithm can choose not necessarily between buying and renting, but possibly between other options, some of which are some mixture between buying and renting [11,17,8,23,18]. For example, an option can be a situation where it is possible to buy a member card that gives a discount on rental. Additional variants were studied in [6,1,19]. Bienkowski [3] studied a problem where costs can change over time, but the ratio between the buying cost and rental cost is fixed. He considered variants with known game end and with unknown game end. Irani and Ramanathan [12] studied a version where the buying cost varies, but the rental cost stays fixed. The problem of scheduling unit jobs without rejection is trivial (jobs are scheduled in a round-robin manner), while the scheduling problem of unit jobs with rejection is a non-trivial problem (see [2,10]). Multiprocessor scheduling with rejection was first introduced by Bartal at al. [2] and many variants have been studied since then (see for example [21,9]). Bartal at al. [2] designed an optimal online algorithm of competitive ratio $1 + \phi$ for arbitrary values of m, where $\phi = \frac{\sqrt{5}+1}{2} \approx 1.618$ is the golden ratio. They improved this result for $m = 2$ by giving an optimal online algorithm with a competitive ratio of ϕ. All the lower bounds in that paper consist of unit jobs (while the algorithms can act on inputs of jobs with arbitrary sizes). They also presented a sequence of lower bounds on the optimal competitive ratio for fixed values of m, where this sequence of lower bounds on the competitive ratios tends to 2 for large values of m. The lower bounds are related to our work and we will discuss them in detail.

Our Results and the Structure of the Paper. We provide optimal online algorithms for all four variants. These competitive ratios are not always given by a closed formula, but in all cases, they can be computed. We present the competitive ratios for $2 \le m \le 20$ in Table 1. All variants except for $\mathrm{SRD}(m)$ are studied in Section 2. The optimal competitive ratio in each one of these cases can be computed for any $m \ge 2$, and it is either a closed expression, or it can be found by solving an equation that is based on a finite recurrence. The variant $\mathrm{SRD}(m)$ requires a more delicate treatment and it is studied in Section 3. For this case the optimal competitive ratio is obtained by solving $O(m)$ non-trivial mathematical programs. All optimal competitive ratios are monotonically non-decreasing functions of m, tending to 2 for $m \to \infty$. We give a direct proof of this property for each one of the cases. Note that the overall upper bound of 2 follows from previous work, and the overall lower bound of 2 follows from previous work in all cases except for $\mathrm{SRIF}(m)$. In the full version of the paper we discuss the relation to scheduling problems (with arbitrary numbers of unit jobs) and show that in all cases except for the case of arbitrary rejection penalties the best competitive ratio is the same as for the corresponding ski rental variant.

2   Threshold Based Algorithms for Three Variants

In this section we consider the cases of unrestricted inputs and inputs with non-decreasing rental costs. For these problems we use algorithms with relatively simple structures that we now define. We define a sequence $\Theta_i \geq 0$ (for $0 \leq i \leq m - 1$), where Θ_i is called the ith threshold. The requirements for thresholds are $\Theta_0 = 0$, and that Θ_i is a monotonically non-decreasing function of i. Let $\theta_i = \Theta_i - \Theta_{i-1} \geq 0$ for $1 \leq i \leq m - 1$. A threshold based algorithm (or a threshold algorithm) is defined as follows. Given the ith request (for $1 \leq i \leq m$), if the equipment was already bought by the algorithm, then the algorithm does nothing. Otherwise, for $i < m$, if $\sum_{j=1}^{i} q_j \leq \Theta_i$, then the algorithm rents the equipment, and otherwise buys it, and for the mth request, it rents the equipment if $q_m \leq 1$, and otherwise buys it.

The simple algorithm for the classic ski rental [16] problem can be seen as a threshold algorithm with $\Theta_i = 1$ for $i = 1, 2, \ldots, m - 1$. It is 2-competitive; if OPT < 1, the algorithm is optimal since it always rents. Otherwise, OPT $= 1$, and the total rental cost of the algorithm (excluding the rental cost of the last request, if it is rented) never exceeds 1, while the cost of buying (or the cost of

Table 1. A comparison of the competitive ratios for the different variants

c.r. for:	SR(m)	SRD(m)	SRIT(m)	SRIF(m)
denoted by:	$\rho(m)$	$\mathcal{R}(m)$	$\lambda(m)$	$\omega(m)$
$m = 2$	1.618034	$\sqrt{2} = 1.414213$	$\frac{\sqrt{5}+1}{2} \approx 1.618034$	$\frac{\sqrt{3}+1}{2} \approx 1.366025$
$m = 3$	1.839287	1.618034	$\frac{\sqrt{17}+3}{4} \approx 1.7807764$	1.513312
$m = 4$	1.927562	1.686141	1.847127	1.595093
$m = 5$	1.965948	1.816496	1.882782	1.648127
$m = 6$	1.983583	1.839287	1.904988	1.685787
$m = 7$	1.991964	1.866025	1.920133	1.714170
$m = 8$	1.996031	1.894427	1.931119	1.736479
$m = 9$	1.998029	1.912868	1.939451	1.754571
$m = 10$	1.999019	1.925819	1.945986	1.769601
$m = 11$	1.999510	1.927562	1.951249	1.782329
$m = 12$	1.999756	1.935414	1.955578	1.793278
$m = 13$	1.999878	1.942809	1.959201	1.802818
$m = 14$	1.999939	1.948683	1.962278	1.811223
$m = 15$	1.999969	1.953462	1.964923	1.818697
$m = 16$	1.999985	1.957426	1.967222	1.825397
$m = 17$	1.999992	1.960769	1.969238	1.831445
$m = 18$	1.999996	1.963624	1.971021	1.836939
$m = 19$	1.999998	1.965948	1.972608	1.841957
$m = 20$	1.999999	1.966089	1.974030	1.846562
$m = 37$	almost 2	1.983738	1.986208	1.891795
$m = 70$	almost 2	1.992031	1.992780	1.923967
$m = 135$	almost 2	1.996055	1.996276	1.946731

renting the last request if the algorithm never buys) if at most 1, for a total cost
of at most 2. This algorithm can be used for the variants studied here. Thus, all
the competitive ratios that we will find are at most 2, and in order to show that
a sequence of competitive ratios tends to 2 for large m, it is sufficient to show
that the limit (for m tending to infinity) is no smaller than 2. The known tight
result for classic ski rental also implies that the optimal competitive ratios for
$SR(m)$, $SRIT(m)$, and $SRD(m)$, tend to 2 as m grows. Specifically, a worst-
case instance consists of requests of rental costs $\frac{1}{M}$ for an integer M, stopping
the input after the algorithm buys, or after there were $2M - 1$ requests, if the
algorithm did not buy by then. The optimal competitive ratio for this input is
$2 - \frac{1}{M}$. Letting $m = 2M - 1$ results in lower bounds for these three variants.

Threshold algorithms are a generalization of this method. Thresholds are not
necessarily universal in the sense that the algorithm still buys once its total rental
cost is about to reach a certain number, but this target number can depend on
the index of the request. Such algorithms are useful for inputs of known length.
In the next theorem we analyze such algorithms, and prove lower bounds using
instances where the rental costs are based on differences between consecutive
thresholds (i.e., the values θ_i are used as rental costs in the difficult inputs). The
upper bounds the are proved for threshold algorithms are called R_t and R'_t.

Theorem 1. *Consider a sequence $\Theta_{m-1} \geq \Theta_{m-2} \geq \cdots \geq \Theta_1 \geq \Theta_0 = 0$, and
let $\theta_i = \Theta_i - \Theta_{i-1}$ for $1 \leq i \leq m - 1$.*

*A threshold based algorithm with the thresholds Θ_i ($0 \leq i \leq m - 1$) has a
competitive ratio of at most $R_t = \max\left\{\Theta_{m-1} + 1, \max_{1 \leq i \leq m-1}\left\{\frac{\Theta_{i-1}+1}{\Theta_i}\right\}\right\}$ for
$SR(m)$ (and for $SRIT(m)$), and at most*

$$R'_t = \max\left\{\Theta_{m-1} + 1, \max_{1 \leq i \leq m-1}\left\{\frac{\Theta_{i-1} + 1}{\Theta_{i-1} + (m - i + 1)\theta_i}\right\}\right\}$$

*for $SRIF(m)$. The competitive ratio of any algorithm for $SR(m)$ is at least
$\min\left\{\Theta_{m-1} + 1, \min_{1 \leq i \leq m-1}\left\{\frac{\Theta_{i-1}+1}{\Theta_i}\right\}\right\}$. If the sequence θ_i is monotonically
non-decreasing, and $\theta_{m-1} \leq 1$, then the last expression is a lower bound on
the competitive ratio of any algorithm for $SRIT(m)$ as well, and under the same
conditions on the sequence, the competitive ratio of any algorithm for $SRIF(m)$
is at least $\min\left\{\Theta_{m-1} + 1, \min_{1 \leq i \leq m-1}\left\{\frac{\Theta_{i-1}+1}{\Theta_{i-1}+(m-i+1)\theta_i}\right\}\right\}$.*

A lower bound for $SR(m)$ was given by [2]. They present it for the more general
problem of scheduling with rejection on identical machines, but it is not difficult
to see that it holds for $SR(m)$, and for completeness, we present it using our terminology. Let $g_m(\rho) = \rho^m - \sum_{j=1}^{m} \rho^{j-1}$, and let $\rho(m)$ be a solution of the equation
$g_m(\rho) = 0$ in $(1, 2)$. As $g_m(2) = 1$, and $g_m(1) = -(m - 1) < 0$ for $m \geq 2$, such a
solution must exist (since g_m is continuous). We have $\rho(2) = \frac{\sqrt{5}+1}{2} = \phi \approx 1.618$.
Let $\theta_j = \frac{1}{(\rho(m))^j}$ for $1 \leq j \leq m - 1$, and $\Theta_i = \sum_{j=1}^{i} \theta_j$. By Theorem 1, the
competitive ratio of the threshold algorithm that uses these thresholds is at most

$\max\{\Theta_{m-1}+1, \max_{1\le i\le m-1}\{\frac{\Theta_{i-1}+1}{\Theta_i}\}\}$. We have $\Theta_{m-1}+1 = \sum_{j=1}^{m-1}\theta_j + 1 = \sum_{j=0}^{m-1}\frac{1}{(\rho(m))^j} = \rho(m)$, by the definition of $\rho(m)$. Additionally,

$$\frac{\Theta_{i-1}+1}{\Theta_i} = \frac{\sum_{j=1}^{i-1}\theta_j + 1}{\sum_{j=1}^{i}\theta_j} = \frac{\sum_{j=0}^{i-1}\frac{1}{(\rho(m))^j}}{\sum_{j=1}^{i}\frac{1}{(\rho(m))^j}} = \rho(m) \ .$$

Since this shows that $\rho(m)$ is the optimal competitive ratio for $SR(m)$, the value $\rho(m)$ is unique. Note that $\rho(m+1) > \rho(m)$ since by using $(\rho(m))^m = \sum_{j=1}^{m}(\rho(m))^{j-1}$ we find

$$g_{m+1}(\rho(m)) = (\rho(m))^{m+1} - \sum_{j=1}^{m+1}(\rho(m))^{j-1} = \sum_{j=1}^{m}(\rho(m))^j - \sum_{j=0}^{m}(\rho(m))^j < 0 \ ,$$

while $g_{m+1}(2) > 0$ for $m \ge 1$. Therefore, $\rho(m+1)$ that satisfies $g_{m+1}(\rho(m+1)) = 0$ also satisfies $\rho(m) < \rho(m+1) < 2$. We can show that $\rho(m)$ tends to 2 as m grows by showing $\rho(m) \ge 2 - \frac{1}{2^{m-1}}$. To show this, we prove $g_m(2 - \frac{1}{2^{m-1}}) \le 0$ (and therefore the solution of $g_m(\rho) = 0$ is in $[2 - \frac{1}{2^{m-1}}, 2)$). We have $g_m(\rho) = \rho^m - \frac{\rho^m-1}{\rho-1} = \frac{\rho^{m+1}-2\rho^m+1}{\rho-1}$. Thus, showing $g_m(\rho) \le 0$ is equivalent to showing $\rho^m(2-\rho) \ge 1$. Letting $\rho = 2-\frac{1}{2^{m-1}}$, we would like to show $(2-\frac{1}{2^{m-1}})^m/2^{m-1} \ge 1$, or alternatively, $(1 - \frac{1}{2^m})^m \ge \frac{1}{2}$. We prove by induction that $(1 - \frac{1}{2^m})^c \ge 1 - \frac{1}{2^{m-c+1}}$ holds for any integer $1 \le c \le m$. For $c = 1$ this inequality holds with equality. Assume that it holds for $c \ge 1$, that is, $(1 - \frac{1}{2^m})^c \ge 1 - \frac{1}{2^{m-c+1}}$. To prove $(1 - \frac{1}{2^m})^{c+1} \ge 1 - \frac{1}{2^{m-c}}$ we write $(1 - \frac{1}{2^m})^{c+1} = (1 - \frac{1}{2^m})^c \cdot (1 - \frac{1}{2^m}) \ge (1 - \frac{1}{2^{m-c+1}}) \cdot (1 - \frac{1}{2^m}) > 1 - \frac{1}{2^{m-c+1}} - \frac{1}{2^m} > 1 - \frac{1}{2^{m-c+1}} - \frac{1}{2^{m-c+1}} = 1 - \frac{1}{2^{m-c}}$. The case $c = m$ proves the above condition. We have proved the following proposition (the lower bound is implied also by the results of [2]).

Proposition 1. *The optimal competitive ratio for* $SR(m)$ *is* $\rho(m)$, *a monotonically increasing function of* m, *tending to 2 as* m *grows to infinity.*

Next, we consider $SRIF(m)$. For $1 \le r \le 2$, define the following thresholds. Let $\Theta_0(r) = 0$, $\Theta_i(r) = \Theta_{i-1}(r) + \frac{1-(r-1)\Theta_{i-1}(r)}{r(m-i+1)}$ (and $\theta_i(r) = \Theta_i(r) - \Theta_{i-1}(r)$ for $1 \le i \le m - 1$). Informally, since the process ends when the algorithm buys, an adversary who tries to force a large competitive ratio would define thresholds such that the algorithm is forced to pay as much as possible for renting. The thresholds are chosen such that $\Theta_i(r)$ is the maximum amount that an algorithm can be forced to pay for renting the first i requests without buying the equipment, if it does not want to reach or exceed the competitive ratio r (and buying the equipment would cause this). Once it already paid $\Theta_{i-1}(r)$, and the ith request rental cost $\theta_i(r)$, if it buys the equipment, then its competitive ratio will be r. For example, $\theta_1(r) = \Theta_1(r) = \frac{1}{rm}$. If it buys the equipment, its cost is 1, while if all further rental costs are equal to $\theta_1(r)$, then the cost of rejecting all of them is $\frac{1}{r}$. For $i > 1$, we have $\theta_i(r) = \frac{1-(r-1)\Theta_{i-1}(r)}{r(m-i+1)}$. If the algorithm buys the equipment when the ith request is given, then its cost is $\Theta_{i-1}(r) + 1$, and if

all further rental costs are equal to $\theta_i(r)$, then the cost of renting all requests is
$\Theta_{i-1}(r) + (m - i + 1)\theta_i(r) = \frac{1+\Theta_{i-1}(r)}{r}$ (by definition).

Theorem 2. *The optimal competitive ratio for* SRIF(m) *is a monotonically non-decreasing function of* m, *tending to 2 as* m *grows to infinity.*

The function $\omega(m)$ denotes the optimal competitive ratio for SRIF(m).

Finally, consider SRIT(m). In this case we will use different thresholds for the lower bound and the upper bound, and apply the two parts of Theorem 1. Let $\alpha(m)$ be the positive solution of $\alpha^2(m-1)^2 + \alpha - 1 = 0$, that is, $\frac{1}{m} <$ $\alpha(m) = \frac{\sqrt{4(m-1)^2+1}-1}{2(m-1)^2} < \frac{1}{m-1}$ (the bounds on $\alpha(m)$ can be proved using simple algebra, by substituting into the quadratic equation it can be seen that there is a root between these values, while the other root is negative). Let $\lambda(m) =$ $(m-1)\alpha(m)+1 = \frac{(m-2)\alpha(m)+1}{(m-1)\alpha(m)} = \frac{\sqrt{4(m-1)^2+1}+2m-3}{2(m-1)}$ (this is the greater solution of $(m-1)\lambda^2 + (3 - 2m)\lambda - 1 = 0$). Using simple calculus it can be seen that $\lambda(m)$ is a monotonically increasing function of m that tends to 2 for sufficiently large values of m.

Letting $\theta_i = \alpha(m)$ for $1 \leq i \leq m - 1$, and using the second part of Theorem 1 we find a lower bound that is the minimum between $1 + (m - 1)\alpha(m) = \lambda(m)$ and $\min_{1 \leq i \leq m-1}\{\frac{(i-1)\alpha(m)+1}{i\alpha(m)}\}$. The minimum of the last expression is obtained for $i = m - 1$, giving $\frac{(m-2)\alpha(m)+1}{(m-1)\alpha(m)} = \lambda(m)$.

For the algorithm, we use $\Theta_i = (m-1)\alpha(m)$ for $1 \leq i \leq m-1$. This algorithm requires a different analysis. If neither the algorithm nor OPT buy the equipment, then the algorithm is optimal. If OPT buys the equipment but the algorithm does not, or the algorithm buys the equipment upon the arrival of the mth request, then OPT $= 1$, while the cost of the algorithm never exceeds $(m - 1)\alpha(m) + 1$, giving a competitive ratio of at most $\lambda(m)$. We are left with the case that the algorithm buys the equipment on the ith day, where $i < m$. Since the algorithm has no additional costs afterwards, we can assume that this is the last request, and $\sum_{j=1}^{i-1} q_j \leq (m - 1)\alpha(m) < \sum_{j=1}^{i} q_j$. If OPT $= 1$, then since $\sum_{j=1}^{i-1} q_i \leq$ $(m - 1)\alpha(m)$, while the cost of the algorithm is $\sum_{j=1}^{i-1} q_j + 1 \leq (m - 1)\alpha(m) + 1$, the competitive ratio is at most $\lambda(m)$. Finally, if OPT < 1, then OPT $= \sum_{j=1}^{i} q_j$. Since $q_1 \leq q_2 \leq \ldots \leq q_i$, we have $\sum_{j=1}^{i-1} q_j \leq \frac{i-1}{i}\sum_{j=1}^{i} q_j$. We find a competitive ratio of at most $\frac{\frac{i-1}{i}\sum_{j=1}^{i} q_j+1}{\sum_{j=1}^{i} q_j} = \frac{i-1}{i} + \frac{1}{\sum_{j=1}^{i} q_j} \leq 1 - \frac{1}{i} + \frac{1}{(m-1)\alpha(m)}$. Using $i \leq m - 1$, this is at most $1 - \frac{1}{m-1} + \frac{1}{(m-1)\alpha(m)} = \frac{(m-2)\alpha(m)+1}{(m-1)\alpha(m)} = \lambda(m)$.

We have proved the following theorem.

Theorem 3. *The optimal competitive ratio for* SRIT(m) *is* $\lambda(m)$, *a monotonically non-decreasing function of* m, *tending to 2 as* m *grows to infinity.*

3   The Variant SRD(m)

Threshold algorithms perform well when the worst-case scenario is of the following form. The algorithm is forced to keep renting, by ensuring that at each step the rental cost paid so far plus the buying cost is too high compared to the cost of an optimal solution (so if the algorithm buys, then it immediately has a high competitive ratio). Finally, just before the input ends, the mth request has a high rental cost and the algorithm is forced to buy (or to rent for the cost of buying). However, this situation cannot be enforced for $\mathrm{SRD}(m)$, since if the rental cost for the first day it at least 1, then the algorithm should obviously buy the equipment immediately, and otherwise, the rental cost of the last request cannot be 1. Thus, buying (after paying a large amount for renting previous requests) cannot be enforced by a single request but it is enforced by a subsequence of the requests that appear last in the sequence, where renting all of them is too expensive (for example, if the last j requests have each a rental cost of $\frac{1}{j}$, then their total rental cost is the same as the cost of buying). Moreover, in the previous cases, the adversary could choose rental costs greedily, i.e., the worst-case scenario occurred when each request had a maximum rental cost under the condition that it still forces the algorithm to rent it. Since large rental costs increase the cost of the algorithm but also of optimal solutions, it is not necessarily the worst-case scenario for all variants. In this section we encounter a variant where the adversary whose strategy is to present at each time a request of maximum rental cost that will still be rented by the algorithm does not lead to the worst-case inputs. In this section, we study the remaining variant $\mathrm{SRD}(m)$ and show that the optimal competitive ratio can be expressed as the maximum solution of a class of mathematical programs. The algorithm uses this competitive ratio as a parameter, and instead of using thresholds on the total rental cost, it makes its decisions based both on the total rental cost so far and the rental cost of the current request.

3.1   The Mathematical Program

We present a class of mathematical programs, denoting one program by $\Pi_k(m)$. We let $\mathcal{R}_k(m)$ denote the objective value of $\Pi_k(m)$ (we later show that this value is well-defined). The program corresponds to a specific value of m, and additionally, to an integer parameter $1 \le k \le m$. Let $\mathcal{R}(m) = \max_{1 \le k \le m} \mathcal{R}_k(m)$. Later, we will show that the optimal competitive ratio for $\mathrm{SRD}(m)$ is exactly $\mathcal{R}(m)$. Given a solution of $\Pi_k(m)$, we will show that the variables $p_1, \ldots, p_k$ provide a worst-case input with the competitive ratio $\mathcal{R}_k(m)$. This proof will also provide some intuition for the constraints of $\Pi_k(m)$.

The mathematical program $\Pi_k(m)$

$$\text{maximize} \quad R \quad \text{s.t.}$$

$$R \ge 1 \tag{1}$$

$$p_i \ge p_{i+1} \qquad \text{for} \quad 1 \le i \le k-1 \tag{2}$$

$$p_k \ge 0 \tag{3}$$

$$\textstyle\sum_{i=1}^{k} p_i \le 1 \tag{4}$$

$$\sum_{i=1}^{k} p_i \geq R - 1 \tag{5}$$

$$\sum_{i=1}^{k} p_i + (m - k) \cdot p_k \geq R \tag{6}$$

$$\sum_{i=1}^{j-1} p_i \cdot (R - 1) + p_j \cdot R \leq 1 \qquad \text{for } 1 \leq j \leq k \tag{7}$$

Proposition 2. *The value $\mathcal{R}_k(m)$ is well-defined, and $1 \leq \mathcal{R}_k(m) \leq 2$ for all m. The optimal competitive ratio for $\mathrm{SRD}(m)$ is at least $\mathcal{R}_k(m)$ for any $1 \leq k \leq m$.*

Table 2 contains the competitive ratios for $2 \leq m \leq 20$, as well as the values of k and the rental costs of an optimal solution for $\Pi_k(m)$.

Table 2. The values $\mathcal{R}(m)$ are given together with values of k such that $\mathcal{R}(m) = \mathcal{R}_k(m)$, and the list of rental costs of the first k requests of a lower bound instance (the remaining costs should add a cost of 1, so their total rental cost must be at least 1, no matter if the algorithm buys the equipment or not, and if their costs are equal, then it take any value in $[\frac{1}{m-k}, q_k]$). Note that there is no clear structure for the worst-case inputs, and in most cases the first few values cannot be obtained using a greedy process (of selecting the rental costs to be as large as possible or as small as possible).

m	competitive ratio	k	rental costs for the worst-case instance $q_1, q_2, \ldots$
2	1.414213	1	0.707107
3	1.618034	1	0.618034
4	1.686141	2	0.421535, 0.421535
5	1.816496	2	0.483163, $1/3 \approx 0.333333$
6	1.839287	2	0.543689, 0.295598
7	1.866025	3	0.349563, 0.266463, $1/4 = 0.25$
8	1.894427	3	0.438646, 0.255781, $1/5 = 0.2$
9	1.912868	3	0.487951, 0.258253, $1/6 \approx 0.166667$
10	1.925819	3	0.519148, 0.263816, $1/7 \approx 0.142857$
11	1.927561	3	0.518790, 0.269143, 0.139628
12	1.935414	4	0.352506, 0.264382, 0.193525, $1/8 = 0.125$
13	1.942809	4	0.366610, 0.274818, 0.190270, $1/9 \approx 0.111111$
14	1.948683	4	0.389994, 0.279508, 0.179181, $1/10 = 0.1$
15	1.953462	4	0.418022, 0.280420, 0.164111, $1/11 \approx 0.090909$
16	1.957426	4	0.439401, 0.284652, 0.150039, $1/12 \approx 0.083333$
17	1.960769	4	0.461537, 0.282966, 0.139343, $1/13 \approx 0.076923$
18	1.963624	4	0.488031, 0.268670, 0.135494, $1/14 \approx 0.071429$
19	1.965948	4	0.508660, 0.258735, 0.131608, 0.066944
20	1.966089	5	0.385565, 0.248135, 0.169317, 0.096408, $1/15 \approx 0.066667$

We present some additional properties of the function $\mathcal{R}(m)$ in the following proposition. These properties are useful for calculating the value of $\mathcal{R}(m)$ for fixed values of m (and we used them in the process of computing Table 1), since they reduce the number of values of k that have to be checked.

Proposition 3. *For any $m \geq 2$, $\mathcal{R}(m+1) \geq \mathcal{R}(m)$. The function $\mathcal{R}(m)$ tends to 2 when m grows to infinity. For $k \geq \lfloor \frac{m}{2} \rfloor$, $\mathcal{R}_k(m) \geq \mathcal{R}_{k+1}(m)$ holds. Let ℓ be such that $\ell + 2^\ell \leq m < \ell + 2^{\ell+1}$. We have $\mathcal{R}_\ell(m) > \mathcal{R}_i(m)$, for $i < \ell$.*

3.2 The Algorithm

We define an algorithm for $\mathrm{SRD}(m)$. At each step, given a new request (if the algorithm did not buy the equipment yet), it examines the current request and decides whether to buy the equipment. If the algorithm does not buy the equipment, it can decide to rent all further requests, or to rent just the current request, in which case it continues and considers the next request. Thus, the algorithm is applied on the first request, and then on each further request as long as the algorithm does not buy and it does not decide to rent all the remaining requests. The algorithm calls the procedure Examine that has three parameters. The parameter $1 \leq i \leq m$ is the index of the current request, the parameter q is its rental cost, and Q is the total rental cost of requests $1, \ldots, i$. If the procedure returns RENT_ALL, then from now on, all further requests are rented, that is, requests $i, \ldots, m$ are rented. If the algorithm returns BUY, then the algorithm buys now, and if it returns RENT, then the current request is rented, and the next request is examined using the same procedure. We define the main procedure. The variable k is used for the index of the current request. The main procedure for m is defined as follows. It initializes $k = 1$ and $Q = 0$. For request k, Q is increased by q_k and Examine(k, q, Q) is applied. If it returns BUY, then the equipment is bought and the algorithm halts. If it returns RENT_ALL, then the equipment is rented for requests $k, k+1, \ldots, m$, and halts after the mth request. If it returns RENT, then request k is rented; in this case if $k = m$, then the algorithm halts, and otherwise it lets $k = k + 1$, and applies examine again with the new value of k.

Examine(i, q, Q)
1. If $Q + (m - i)q \leq \mathcal{R}(m)$, then return RENT_ALL.
2. If $Q > 1$, then return BUY.
3. If $Q - q + 1 \leq \mathcal{R}(m) \cdot Q$, then return BUY.
4. Return RENT.

Theorem 4. *The competitive ratio of the algorithm is $\mathcal{R}(m)$.*

Proof. We split the analysis of the competitive ratio into several cases, according to the last call of Examine (Examine is applied at least once, for $i = 1$). Let $1 \leq k \leq m$ denote the last index of any request that Examine was applied for, and Q and q denote the last values of the corresponding variables. The cases correspond to the step of Examine that was applied in the last call for it (since exactly one of the four steps is applied in each call).

Case 1. Examine(k, q, Q) applied step 1, and returned RENT_ALL. We have $Q + (m - k)q \leq \mathcal{R}(m)$, and the algorithm rents all m requests, so its cost is $\sum_{j=1}^m q_j$. If OPT does not buy the equipment, then its cost is $\sum_{j=1}^m q_j$ as well, and the algorithm is optimal. Otherwise OPT buys the equipment and OPT $= 1$. The cost of the algorithm is $\sum_{j=1}^{k} q_j + \sum_{j=k+1}^{m} q_j \leq Q + (m-k)q$, since $q_j \leq q_k = q$ for $k < j \leq m$, and since OPT $= 1$, the competitive ratio does not exceed $\mathcal{R}(m)$.

Case 2. Examine(k, q, Q) applied step 4 and returned RENT. Since the algorithm must examine the next request, if it exists, using Examine, the only case that this is the last application of Examine is $k = m$. Since the algorithm reached this step, the condition of step 2 does not hold, and $\sum_{j=1}^{m} q_j = Q \leq 1$. Moreover, since the algorithm reached this step, it rented all requests, and its cost is $\sum_{j=1}^{m} q_j$, which is the optimal cost as well. Thus, in this case the algorithm is optimal.

Case 3. Examine(k, q, Q) applied step 3 and returned BUY. Since the algorithm reached this step, it must hold that $\sum_{j=1}^{k} q_j \leq 1$, as step 2 was not applied. We find OPT $= \min\{1, \sum_{j=1}^{m} q_j\} \geq \sum_{j=1}^{k} q_j = Q$. The algorithm buys the equipment, so its cost is $\sum_{j=1}^{k-1} q_j + 1 = Q - q + 1$. By the condition of this case, this is at most $\mathcal{R}(m) \cdot Q$, that is, the competitive ratio does not exceed $\mathcal{R}(m)$.

Case 4. Examine(k, q, Q) applied step 2 and returned BUY. Since the condition of step 2 holds for k, $\sum_{i=1}^{k} q_i > 1$, and therefore OPT $= 1$.

If $k = 1$, then the algorithm buys the equipment already for the first request, so its cost is $1 = $ OPT, and thus the algorithm is optimal. In what follows we assume $k > 1$. As the condition of step 1 does not hold during the first application of Examine, we have $q_1 > \frac{\mathcal{R}(m)}{m} > 0$, and we get $\sum_{i=1}^{j} q_i > 0$ for $1 \leq j \leq m$. Let $k' = k - 1 \geq 1$. The cost of the algorithm is $\sum_{i=1}^{k'} q_i + 1$. Assume by contradiction that the competitive ratio exceeds $\mathcal{R}(m)$, and denote the competitive ratio by $\tilde{R}$. We have $\sum_{i=1}^{k'} q_i + 1 = \tilde{R}$, since OPT $= 1$. Since step 4 was applied in all the previous iterations, the condition of step 3 did not hold, so for all $j \leq k'$, $\sum_{i=1}^{j-1} q_i + 1 > \mathcal{R}(m) \cdot (\sum_{i=1}^{j} q_i)$. Let $\tilde{r} = \min_{1 \leq j \leq k'} \frac{\sum_{i=1}^{j-1} q_i + 1}{\sum_{i=1}^{j} q_i}$. By the previous properties, $\tilde{r} > \mathcal{R}(m)$. Additionally, when Examine was applied for request k', the conditions of steps 1 and 2 did not hold, and thus $\sum_{i=1}^{k'} q_i + (m - k')q_{k'} > \mathcal{R}(m)$ and $\sum_{i=1}^{k'} q_i \leq 1$. Let $\hat{R} = \min\{\tilde{R}, \sum_{i=1}^{k'} q_i + (m - k')q_{k'}, \tilde{r}\} > \mathcal{R}(m)$.

Consider the mathematical program $\Pi_{k'}(m)$. We show that the set of variables $p_i = q_i$ for $1 \leq i \leq k'$, and $R = \hat{R}$, give a feasible solution. Since $\mathcal{R}(m) \geq 1$, $\hat{R} > 1$, and constraint (1) holds. Constraints (2) and (3) hold by the properties of the input (having non-negative non-increasing rental costs). The properties mentioned above show that constraints (6) and (4) must hold, and additionally, the family of constraints (7) holds for $1 \leq j \leq k'$. Since $\sum_{i=1}^{k'} q_i + 1 = \tilde{R} \geq \hat{R}$, constraint (5) holds as well. Thus, there is a feasible solution whose objective value is $\hat{R} > \mathcal{R}(m) \geq \mathcal{R}_{k'}(m)$, contradicting the definition of $\mathcal{R}(m)$. $\square$

References

1. Azar, Y., Bartal, Y., Feuerstein, E., Fiat, A., Leonardi, S., Rosén, A.: On capital investment. Algorithmica 25(1), 22–36 (1999)
2. Bartal, Y., Leonardi, S., Marchetti-Spaccamela, A., Sgall, J., Stougie, L.: Multiprocessor scheduling with rejection. SIAM Journal on Discrete Mathematics 13(1), 64–78 (2000)

3. Bienkowski, M.: Price fluctuations: To buy or to rent. In: Bampis, E., Jansen, K. (eds.) WAOA 2009. LNCS, vol. 5893, pp. 25–36. Springer, Heidelberg (2010)

4. Dong, Y., Xu, Y., Xu, W.: The on-line rental problem with risk and probabilistic forecast. In: Preparata, F.P., Fang, Q. (eds.) FAW 2007. LNCS, vol. 4613, pp. 117–123. Springer, Heidelberg (2007)

5. El-Yaniv, R., Kaniel, R., Linial, N.: Competitive optimal on-line leasing. Algorithmica 25(1), 116–140 (1999)

6. El-Yaniv, R., Karp, R.: Nearly optimal competitive online replacement policies. Mathematics of Operations Research 22(4), 814–839 (1997)

7. El-Yaniv, R., Karp, R.M.: The mortgage problem. In: Dolev, D., Rodeh, M., Galil, Z. (eds.) ISTCS 1992. LNCS, vol. 601, pp. 304–312. Springer, Heidelberg (1992)

8. Epstein, L., Fiat, A., Levy, M.: Caching content under digital rights management. In: Bampis, E., Skutella, M. (eds.) WAOA 2008. LNCS, vol. 5426, pp. 188–200. Springer, Heidelberg (2009)

9. Epstein, L., Zebedat-Haider, H.: Online scheduling with rejection and withdrawal. Theo. Comp. Sci. 412(48), 6666–6674 (2011)

10. Epstein, L., Zebedat-Haider, H.: Online scheduling with rejection and reordering: exact algorithms for unit size jobs. Journal of Combinatorial Optimization (2013)

11. Fleischer, R.: On the bahncard problem. Theo. Comp. Sci. 268(1), 161–174 (2001)

12. Irani, S., Ramanathan, D.: The problem of renting versus buying (1994) (manuscript)

13. Karlin, A.R., Kenyon, C., Randall, D.: Dynamic TCP acknowledgment and other stories about $e/(e-1)$. Algorithmica 36(3), 209–224 (2003)

14. Karlin, A.R., Manasse, M.S., McGeoch, L.A., Owicki, S.S.: Competitive randomized algorithms for nonuniform problems. Algorithmica 11(6), 542–571 (1994)

15. Karlin, A.R., Manasse, M.S., Rudolph, L., Sleator, D.D.: Competitive snoopy caching. Algorithmica 3, 77–119 (1988)

16. Karp, R.M.: On-line algorithms versus off-line algorithms: How much is it worth to know the future? In: Proc. of the IFIP 12th World Computer Congress (IFIP1992), Algorithms, Software, Architecture - Information Processing. IFIP Transactions, vol. A-12, pp. 416–429 (1992)

17. Lotker, Z., Patt-Shamir, B., Rawitz, D.: Ski rental with two general options. Information Processing Letters 108(6), 365–368 (2008)

18. Lotker, Z., Patt-Shamir, B., Rawitz, D.: Rent, lease, or buy: Randomized algorithms for multislope ski rental. SIAM Journal on Discrete Mathematics 26(2), 718–736 (2012)

19. Meyerson, A.: The parking permit problem. In: Proc. of the 46th Annual IEEE Symposium on Foundations of Computer Science (FOCS 2005), pp. 274–284 (2005)

20. Rudolph, L., Segall, Z.: Dynamic paging schemes for MIMD parallel processors. Technical report, CS Department, Carnegie-Mellon University (1986)

21. Seiden, S.S.: Preemptive multiprocessor scheduling with rejection. Theo. Comp. Sci. 262(1), 437–458 (2001)

22. Yang, X., Zhang, W., Zhang, Y., Xu, W.: Optimal randomized algorithm for a generalized ski-rental with interest rate. Information Processing Letters 112(13), 548–551 (2012)

23. Zhang, G., Poon, C.K., Xu, Y.: The ski-rental problem with multiple discount options. Information Processing Letters 111(18), 903–906 (2011)

24. Zhang, Y., Zhang, W., Xu, W., Li, H.: A risk-reward model for the on-line leasing of depreciable equipment. Information Processing Letters 111(6), 256–261 (2011)

On the Recognition of Four-Directional Orthogonal Ray Graphs*

Stefan Felsner[1], George B. Mertzios[2], and Irina Mustață[1]

[1] Institut für Mathematik, Technische Universität Berlin, Germany
{felsner,mustata}@math.tu-berlin.de
[2] School of Engineering and Computing Sciences, Durham University, UK
george.mertzios@durham.ac.uk

Abstract. Orthogonal ray graphs are the intersection graphs of horizontal and vertical rays (i.e. half-lines) in the plane. If the rays can have any possible orientation (left/right/up/down) then the graph is a *4-directional orthogonal ray graph (4-DORG)*. Otherwise, if all rays are only pointing into the positive x and y directions, the intersection graph is a *2-DORG*. Similarly, for *3-DORGs*, the horizontal rays can have any direction but the vertical ones can only have the positive direction. The recognition problem of 2-DORGs, which are a nice subclass of bipartite comparability graphs, is known to be polynomial, while the recognition problems for 3-DORGs and 4-DORGs are open. Recently it has been shown that the recognition of unit grid intersection graphs, a superclass of 4-DORGs, is NP-complete. In this paper we prove that the recognition problem of 4-DORGs is polynomial, given a partition $\{L, R, U, D\}$ of the vertices of G (which corresponds to the four possible ray directions). For the proof, given the graph G, we first construct two cliques G_1, G_2 with both directed and undirected edges. Then we successively augment these two graphs, constructing eventually a graph $\widetilde{G}$ with both directed and undirected edges, such that G has a 4-DORG representation if and only if $\widetilde{G}$ has a transitive orientation respecting its directed edges. As a crucial tool for our analysis we introduce the notion of an *S-orientation* of a graph, which extends the notion of a transitive orientation. We expect that our proof ideas will be useful also in other situations. Using an independent approach we show that, given a permutation π of the vertices of U (π is the order of y-coordinates of ray endpoints for U), while the partition $\{L, R\}$ of $V \setminus U$ is not given, we can still efficiently check whether G has a 3-DORG representation.

1 Introduction

Segment graphs, i.e. the intersection graphs of segments in the plane, have been the subject of wide spread research activities (see e.g. [2, 12]). More tractable subclasses of segment graphs are obtained by restricting the number of directions

---

* This work was partially supported by (i) the DFG ESF EuroGIGA projects COMPOSE and GraDR, (ii) the EPSRC Grant EP/K022660/1, (iii) the EPSRC Grant EP/G043434/1, and (iv) the Berlin Mathematical School.

K. Chatterjee and J. Sgall (Eds.): MFCS 2013, LNCS 8087, pp. 373–384, 2013.
© Springer-Verlag Berlin Heidelberg 2013

for the segments to some fixed positive integer k [4, 11]. These graphs are called *k-directional segment graphs*. For the easiest case of $k = 2$ directions, segments can be assumed to be parallel to the x- and y-axis. If intersections of parallel segments are forbidden, then 2-directional segment graphs are bipartite and the corresponding class of graphs is also known as *grid intersection graphs* (GIG), see [9]. The recognition of GIGs is NP-complete [10].

Since segment graphs are a fairly complex class, it is natural to study the subclass of *ray intersection graphs* [1]. Again, the number of directions can be restricted by an integer k, which yields the class of *k-directional ray intersection graphs*. Particularly interesting is the case where all rays are parallel to the x- or y-axis. The resulting class is the class of *orthogonal ray graphs*, which the subject of this paper. A *k-directional orthogonal ray graph*, for short a k-DORG ($k \in \{2, 3, 4\}$), is an orthogonal ray graph with rays in k directions. If $k = 2$ we assume that all rays point in the positive x- and the positive y-direction, if $k = 3$ we additionally allow the negative x-direction.

The class of 2-DORGs was introduced in [19], where it is shown that the class of 2-DORGs coincides with the class of bipartite graphs whose complements are circular arc graphs, i.e. intersection graphs of arcs on a circle. This characterization implies the existence of a polynomial recognition algorithm (see [13]), as well as a characterization based on forbidden subgraphs [5]. Alternatively, 2-DORGs can also be characterized as the comparability graphs of ordered sets of height two and interval dimension two. This yields another polynomial recognition algorithm (see e.g. [7]), and due to the classification of 3-interval irreducible posets ([6], [21, sec 3.7]) a complete description of minimally forbidden subgraphs. In a very nice recent contribution on 2-DORGs [20], a clever solution has been presented for the jump number problem for the corresponding class of posets and shows a close connection between this problem and a hitting set problem for axis aligned rectangles in the plane.

4-DORGs in VLSI Design. In [18] 4-DORGs were introduced as a mathematical model for defective nano-crossbars in PLA (programmable logic arrays) design. A nano-crossbar is a rectangular circuit board with $m \times n$ orthogonally crossing wires. Fabrication defects may lead to disconnected wires. The bipartite intersection graph that models the surviving crossbar is an orthogonal ray graph.

We briefly mention two problems for 4-DORGs that are tackled in [18]. One of them is that of finding, in a nano-crossbar with disconnected wire defects, a maximal surviving square (perfect) crossbar, which translates into finding a maximal k such that the balanced complete bipartite graph $K_{k,k}$ is a subgraph of the orthogonal ray graph modeling the crossbar. This *balanced biclique problem* is NP-complete for general bipartite graphs but turns out to be polynomially solvable on 4-DORGs [18]. The other problem, posed in [16], asks how difficult it is to find a subgraph that would model a given logic mapping and is shown in [18] to be NP-hard.

4-DORGs and UGIGs. A *unit grid intersection graph* (UGIG) is a GIG that admits an orthogonal segment representation with all segments of equal (unit) length. Every 4-DORG is a GIG. This can be seen by intersecting the ray representation

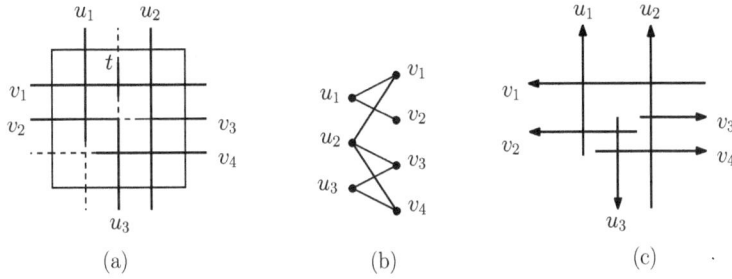

Fig. 1. (a) A nano-wire crossbar with disconnected wire defects, (b) the bipartite graph modeling this crossbar, and (c) a 4-DORG representation of this graph. Note that vertex t is not present, since the corresponding wire is not connected to the crossbar boundary, hence with the remaining circuit.

with a rectangle R, that contains all intersections between the rays in the interior. To see that every 4-DORG is a UGIG, we first fix an appropriate length for the segments, e.g. the length d of the diagonal of R. If we only keep the initial part of length d from each ray we get a UGIG representation. Essentially this construction was already used in [18].

Unit grid intersection graphs were considered in [15]. There it is shown that UGIG contains P_6-free bipartite graphs, interval bigraphs and bipartite permutation graphs. Actually, these classes are already contained in 2-DORG. Another contribution of [15] is to provide an example showing that the inclusion of UGIG in GIG is proper. In [17] it is shown that interval bigraphs belong to UGIG. Hardness of Hamiltonian cycle for inputs from UGIG and hardness of graph isomorphism for inputs from GIG have been shown in [22]. Very recently it was shown that the recognition of UGIGs is NP-complete [14]. With this last result we find 4-DORG nested between 2-DORG and UGIG with easy and hard recognition, respectively. This fact was central for our motivation to attack the open recognition problem for 4-DORGs [19].

Our Contribution. In this paper we prove that, given a graph G along with a partition $\{L, R, U, D\}$ of its vertices, it can be efficiently checked whether G has a 4-DORG representation such that the vertices of L (resp. the vertices of R, U, D) correspond to the rays pointing leftwards (resp. rightwards, upwards, downwards). To obtain our result, we first construct two cliques G_1, G_2 that have both directed and undirected edges. Then we iteratively augment G_1 and G_2, constructing eventually a graph $\widetilde{G}$ with both directed and undirected edges. As we prove, the input graph G has a 4-DORG representation if and only if $\widetilde{G}$ has a transitive orientation respecting its directed edges. As a crucial tool for our results, we introduce the notion of an S-orientation of an arbitrary graph, which extends the notion of a transitive orientation. By setting $D = \emptyset$, our results trivially imply that, given a partition $\{L, R, U\}$ of the vertices of G, it can be efficiently checked whether G has a 3-DORG representation according to this partition. With an independent approach, we show that if we are given a permutation π of the vertices of U (which represents the order of y-coordinates of ray-endpoints for the set U) but the partition $\{L, R\}$ of $V \setminus U$ is unknown, then we can still efficiently check whether G has a 3-DORG representation. The method we use to prove this result can be viewed as a particular *partition refinement technique*. Such techniques have various applications in string sorting, automaton minimization, and graph algorithms (see [8] for an overview).

Notation. We consider in this article simple undirected and directed graphs. For a graph G, we denote its vertex and edge set by $V(G)$ and $E(G)$, respectively. In an undirected graph G, the edge between vertices u and v is denoted by uv, and in this case u and v are said to be *adjacent* in G. The set $N(v) = \{u \in V : uv \in E\}$ is called the *neighborhood* of the vertex v of G. If the graph G is directed, we denote by $\langle uv \rangle$ the oriented arc from u to v. If G is the complete graph (i.e. a clique), we call an orientation λ of all (resp. of some) edges of G a *(partial) tournament* of G. If in addition λ is transitive, then we call it a (partial) transitive tournament. Given two matrices A and B of size $n \times n$ each, we call by $O(\text{MM}(n))$ the time needed by the fastest known algorithm for multiplying A and B; currently this can be done in $O(n^{2.376})$ time [3].

Let G be a 4-DORG. Then, in a 4-DORG representation of G, every ray is completely determined by one point on the plane and the direction of the ray. We call this point the *endpoint* of this ray. Given a 4-DORG G along with a 4-DORG representation of it, we may not distinguish in the following between a vertex of G and the corresponding ray in the representation, whenever it is clear from the context. Furthermore, for any vertex u of G we will denote by u_x and u_y the x-coordinate and the y-coordinate of the endpoint of the ray of u in the representation, respectively.

2   4-Directional Orthogonal Ray Graphs

In this section we investigate some fundamental properties of 4-DORGs and their representations, which will then be used for our recognition algorithm. The next observation on a 4-DORG representation is crucial for the rest of the section.

Observation 1 *Let $G = (V, E)$ be a graph that admits a 4-DORG representation, in which L (resp. R, U, D) is the set of leftwards (resp. rightwards, upwards, downwards) oriented rays. If $u \in U$ and $v \in R$ (resp. $v \in L$), then $uv \in E$ if and only if $u_x > v_x$ (resp. $u_x < v_x$) and $u_y < v_y$. Similarly, if $u \in D$ and $v \in R$ (resp. $v \in L$), then $uv \in E$ if and only if $u_x > v_x$ (resp. $u_x < v_x$) and $u_y > v_y$.*

For the remainder of the section, let $G = (V, E)$ be an arbitrary input graph with vertex partition $V = L \cup R \cup U \cup D$, such that $E \subseteq (L \cup R) \times (U \cup D)$.

The Oriented Cliques G_1 and G_2. In order to decide whether the input graph $G = (V, E)$ admits a 4-DORG representation, in which L (resp. R, U, D) is the set of leftwards (resp. rightwards, upwards, downwards) oriented rays, we first construct two auxiliary cliques G_1 and G_2 with $|V|$ vertices each. We partition the vertices of G_1 (resp. G_2) into the sets L_x, R_x, U_x, D_x (resp. L_y, R_y, U_y, D_y). The intuition behind this notation for the vertices of G_1 and G_2 is that, if G

has a 4-DORG representation with respect to the partition $\{L, R, U, D\}$, then each of these vertices of G_1 (resp. G_2) corresponds to the x-coordinate (resp. y-coordinate) of the endpoint of a ray of G in this representation.

We can now define some orientation of the edges of G_1 and G_2. The intuition behind these orientations comes from Observation 1: if the input graph G is a 4-DORG, then it admits a 4-DORG representation such that, for every $u \in U \cup D$ and $v \in L \cup R$, we have that $u_x > v_x$ (resp. $u_y > v_y$) in this representation if and only if $\langle u_x v_x \rangle$ (resp. $\langle u_y v_y \rangle$) is an oriented edge of the clique G_1 (resp. G_2). That is, since all x-coordinates (resp. y-coordinates) of the endpoints of the rays in a 4-DORG representation can be linearly ordered, these orientations of the edges of G_1 (resp. G_2) build a transitive tournament.

Therefore, the input graph G admits a 4-DORG representation if and only if some edges of G_1, G_2 are forced to have specific orientations in these transitive tournaments of G_1 and G_2, while some pairs of edges of G_1, G_2 are not allowed to have a specific *pair* of orientations in these tournaments. Motivated by this, we introduce in the next two definitions the notions of *type-1-mandatory* orientations and of *forbidden pairs* of orientations, which will be crucial for our analysis in the remainder of Section 2.

Definition 1 (type-1-mandatory orientations). *Let $u \in U \cup D$ and $v \in L \cup R$, such that $uv \in E$. If $u \in U$ and $v \in R$ (resp. $v \in L$) then the orientations $\langle u_x v_x \rangle$ (resp. $\langle v_x u_x \rangle$) and $\langle u_y u_y \rangle$ of G_1 and G_2 are called* type-1-mandatory. *If $u \in D$ and $v \in R$ (resp. $v \in L$) then the orientations $\langle u_x v_x \rangle$ (resp. $\langle v_x u_x \rangle$) and $\langle u_y v_y \rangle$ of G_1 and G_2 are called* type-1-mandatory. *The set of all type-1-mandatory orientations of G_1 and G_2 is denoted by M_1.*

Definition 2 (forbidden pairs of orientations). *Let $u \in U \cup D$ and $v \in R \cup L$, such that $uv \notin E$. If $u \in U$ and $v \in R$ (resp. $v \in L$) then the pair $\{\langle u_x v_x \rangle, \langle v_y u_y \rangle\}$ (resp. the pair $\{\langle v_x u_x \rangle, \langle v_y u_y \rangle\}$) of orientations of G_1 and G_2 is called* forbidden. *If $u \in D$ and $v \in R$ (resp. $v \in L$) then the pair $\{\langle u_x v_x \rangle, \langle u_y v_y \rangle\}$ (resp. the pair $\{\langle v_x u_x \rangle, \langle u_y v_y \rangle\}$) of orientations of G_1 and G_2 is called* forbidden.

For simplicity of notation in the remainder of the paper, we introduce in the next definition the notion of *optional edges*.

Definition 3 (optional edges). *Let $\{\langle pq \rangle, \langle ab \rangle\}$ be a pair of forbidden orientations of G_1 and G_2. Then each of the (undirected) edges pq and ab is called optional edges.*

The Augmented Oriented Cliques G_1^* and G_2^*. We iteratively augment the cliques G_1 and G_2 into the two larger cliques G_1^* and G_2^*, respectively, as follows. For every *optional* edge pq of G_1 (resp. of G_2), where $p \in U_x \cup D_x$ and $q \in L_x \cup R_x$ (resp. $p \in U_y \cup D_y$ and $q \in L_y \cup R_y$), we add two vertices $r_{p,q}$ and $r_{q,p}$ and we add all needed edges to make the resulting graph G_1^* (resp. G_2^*) a clique. Note that, if the initial graph G has n vertices and m non-edges (i.e. $\binom{n}{2} - m$ edges), then G_1^* and G_2^* are cliques with $n + 2m$ vertices each. We now introduce the notion of *type-2-mandatory* orientations of G_1^* and G_2^*.

Definition 4 (type-2-mandatory orientations). *For every optional edge pq of* G_1^*, *the orientations* $\langle pr_{p,q}\rangle$ *and* $\langle qr_{q,p}\rangle$ *of* G_1^* *are called* type-2-mandatory *orientations of* G_1^*. *For every optional edge pq of* G_2^*, *the orientations* $\langle r_{p,q}p\rangle$ *and* $\langle r_{q,p}q\rangle$ *of* G_2^* *are called* type-2-mandatory *orientations of* G_2^*. *The set of all type-2-mandatory orientations of* G_1^* *and* G_2^* *is denoted by* M_2.

The Coupling of G_1^* **and** G_2^* **into the Oriented Clique** G^*. Now we iteratively construct the clique G^* from the cliques G_1^* and G_2^*, as follows. Initially G^* is the union of G_1^* and G_2^*, together with all needed edges such that G^* is a clique. Then, for every pair $\{\langle pq\rangle, \langle ab\rangle\}$ of *forbidden orientations* of G_1^* and G_2^* (where $pq \in E(G_1)$ and $ab \in E(G_2)$, cf. Definition 2), we *merge* in G^* the vertices $r_{b,a}$ and $r_{p,q}$, i.e. we have $r_{b,a} = r_{p,q}$ in G^*. Recall that each of the cliques G_1^* and G_2^* has $n + 2m$ vertices. Therefore, since G_1^* and G_2^* have m pairs $\{\langle pq\rangle, \langle ab\rangle\}$ of forbidden orientations, the resulting clique G^* has $2n + 3m$ vertices. We now introduce the notion of *type-3-mandatory* orientations of G^*.

Definition 5 (type-3-mandatory orientations). *For every pair* $\{\langle pq\rangle, \langle ab\rangle\}$ *of forbidden orientations of* G_1^* *and* G_2^*, *the orientation* $\langle r_{q,p}r_{a,b}\rangle$ *is called a* type-3-mandatory *orientation of* G^*. *The set of all type-3-mandatory orientations of* G^* *is denoted by* M_3.

Whenever the orientation of an edge uv of G^* is type-1 (resp. type-2, type-3)-mandatory, we may say for simplicity that the *edge* uv (instead of its *orientation*) is type-1 (resp. type-2, type-3)-mandatory. An example for the construction of G^* from G_1^* and G_2^* is illustrated in Figure 2, where it is shown how two optional edges $pq \in E(G_1^*)$ and $ab \in E(G_2^*)$ are joined together in G^*, where $\{\langle pq\rangle, \langle ab\rangle\}$ is a pair of forbidden orientations of G_1^* and G_2^*. For simplicity of the presentation, only the optional edges pq and ab, the type-2-mandatory edges $pr_{p,q}$, $qr_{q,p}$, $ar_{a,b}$, $br_{b,a}$, and the edges $r_{p,q}r_{q,p}$ and $r_{a,b}r_{b,a}$ are shown in Figure 2. Furthermore, the type-2-mandatory orientations $\langle pr_{p,q}\rangle$, $\langle qr_{q,p}\rangle$, $\langle r_{a,b}a\rangle$, and $\langle r_{b,a}b\rangle$, as well as the type-3-mandatory orientation $\langle r_{q,p}r_{a,b}\rangle$, are drawn with double arrows in Figure 2 for better visibility.

In the next theorem we provide a characterization of 4-DORGs in terms of a transitive tournament λ^* of the clique G^*. The main novelty of the characterization of Theorem 1 is that it does not rely on the *forbidden pairs* of orientations.

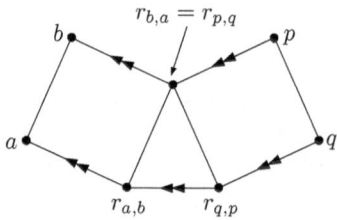

Fig. 2. An example of joining in G^* the pair of optional edges $\{pq, ab\}$, where $pq \in E(G_1)$ and $ab \in E(G_2)$.

This characterization will be used in Section 4, in order to provide our main result of the paper, namely the recognition of 4-DORGs with respect to the vertex partition $\{L, R, U, D\}$.

Theorem 1. *The next two conditions are equivalent:*

1. *The graph $G = (V, E)$ with n vertices has a 4-DORG representation with respect to the vertex partition $\{L, R, U, D\}$.*
2. *There exists a transitive tournament λ^* of G^*, such that $M_1 \cup M_2 \cup M_3 \subseteq \lambda^*$, and in addition:*
   (a) *let pq be an optional edge of G_1^* and $pw \notin M_2$ be an incident edge of pq in G_1^*; then $\langle wr_{p,q} \rangle \in \lambda^*$ implies that $\langle wp \rangle \in \lambda^*$,*
   (b) *let pq be an optional edge of G_2^* and $pw \notin M_2$ be an incident edge of pq in G_2^*; then $\langle r_{p,q}w \rangle \in \lambda^*$ implies that $\langle pw \rangle \in \lambda^*$,*
   (c) *let pq be an optional edge of G_1^* (resp. G_2^*), where $p \in U_x \cup D_x$ (resp. $p \in U_y \cup D_y$); then we have:*
      (i) *either $\langle pq \rangle, \langle r_{p,q}q \rangle, \langle r_{p,q}r_{q,p} \rangle \in \lambda^*$ or $\langle qp \rangle, \langle qr_{p,q} \rangle, \langle r_{q,p}r_{p,q} \rangle \in \lambda^*$,*
      (ii) *for any incident optional edge pq' of G_1^* (resp. G_2^*), either $\langle pq \rangle, \langle r_{p,q'}q \rangle \in \lambda^*$ or $\langle qp \rangle, \langle qr_{p,q'} \rangle \in \lambda^*$,*
      (iii) *for any incident optional edge $p'q$ of G_1^* (resp. G_2^*), either $\langle r_{p,q}q \rangle, \langle r_{p,q}r_{q,p'} \rangle \in \lambda^*$ or $\langle qr_{p,q} \rangle, \langle r_{q,p'}r_{p,q} \rangle \in \lambda^*$.*

Furthermore, as we can prove, given a transitive tournament λ^* of G^* as in Theorem 1, a 4-DORG representation of G can be computed in $O(n^2)$ time. An example of the orientations of condition 2(c) in Theorem 1 (for the case of G_1^*) is shown in Figure 3. For simplicity of the presentation, although G_1^* is a clique, we show in Figure 3 only the edges that are needed to illustrate Theorem 1.

3 S-Orientations of Graphs

In this section we introduce a new way of augmenting an *arbitrary* graph G by adding a new vertex and some new edges to G. This type of augmentation process is done with respect to a particular edge $e_i = x_i y_i$ of the graph G, and is called the *deactivation* of e_i in G. In order to do so, we first introduce the crucial notion of an *S-orientation* of a graph G (cf. Definition 7), which extends the classical notion of a transitive orientation. For the remainder of this section, G denotes an arbitrary graph, and not the input graph discussed in Section 2.

Definition 6. *Let $G = (V, E)$ be a graph and let (x_i, y_i), $1 \leq i \leq k$, be k ordered pairs of vertices of G, where $x_i y_i \in E$. Let V_{out}, V_{in} be two disjoint vertex subsets of G, where $\{x_i : 1 \leq i \leq k\} \subseteq V_{out} \cup V_{in}$. For every $i = 1, 2, \ldots, k$:*

- *a special neighborhood of x_i is a vertex subset $S(x_i) \subseteq \left(N(x_i) \cap \left(\bigcap_{x_j = x_i} N(y_j)\right)\right) \setminus \{x_j : 1 \leq j \leq k\}$,*
- *the forced neighborhood orientation of x_i is:*
   - *the set $F(x_i) = \{\langle x_i z \rangle : z \in S(x_i)\}$ of oriented edges of G, if $x_i \in V_{out}$,*
   - *the set $F(x_i) = \{\langle z x_i \rangle : z \in S(x_i)\}$ of oriented edges of G, if $x_i \in V_{in}$.*

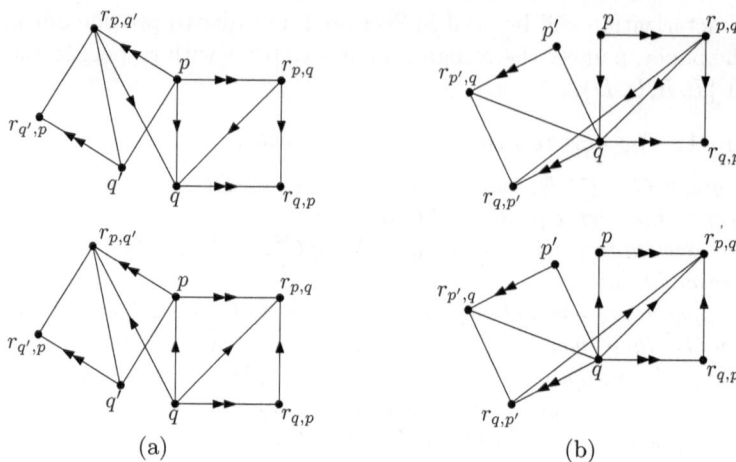

Fig. 3. An example of the orientations of the clique G_1^* in the transitive tournament λ^*, where $p \in U_x \cup D_x$ (cf. condition 2(c) in Theorem 1): (a) both possible orientations where the optional edges pq and pq' are incident and (b) both possible orientations where the optional edges pq and $p'q$ are incident. In both (a) and (b), the orientations of the type-2-mandatory edges are drawn with double arrows. The case for G_2 is the same, except that the orientation of the type-2-mandatory edges is the opposite.

Definition 7. *Let $G = (V, E)$ be a graph. For every $i = 1, 2, \ldots, k$ let $S(x_i)$ be a special neighborhood in G. Let T be a transitive orientation of G. Then T is an S-orientation of G on the special neighborhoods $S(x_i)$, $1 \leq i \leq k$, if for every $i = 1, 2, \ldots, k$:*

1. *$F(x_i) \subseteq T$ and*
2. *for every $z \in S(x_i)$, $\langle x_i y_i \rangle \in T$ if and only if $\langle z y_i \rangle \in T$.*

Definition 8. *Let $G = (V, E)$ be a graph. For every $i = 1, 2, \ldots, k$ let $S(x_i)$ be a special neighborhood in G. Let T be an S-orientation of G on the sets $S(x_i)$, $1 \leq i \leq k$. Then T is consistent if, for every $i = 1, 2, \ldots, k$, it satisfies the following conditions, whenever $zw \in E$, where $z \in S(x_i)$ and $w \in (N(x_i) \cap N(y_i)) \setminus S(x_i)$:*

- *if $x_i \in V_{out}$, then $\langle wz \rangle \in T$ implies that $\langle wx_i \rangle \in T$,*
- *if $x_i \in V_{in}$, then $\langle zw \rangle \in T$ implies that $\langle x_i w \rangle \in T$.*

In the next definition we introduce the notion of *deactivating* an edge $e_i = x_i y_i$ of a graph G, where $S(x_i)$ is a special neighborhood in G. In order to deactivate edge e_i of G, we augment appropriately the graph G, obtaining a new graph $\widetilde{G}(e_i)$ that has one new vertex.

Definition 9. *Let $G = (V, E)$ be a graph and let $S(x_i)$ be a special neighborhood in G. The graph $\widetilde{G}(e_i)$ obtained by deactivating the edge $e_i = x_i y_i$ (with respect to S_i) is defined as follows:*

1. *$V(\widetilde{G}(e_i)) = V \cup \{a_i\}$ (i.e. add a new vertex a_i to G),*
2. *$E(\widetilde{G}(e_i)) = E \cup \{za_i : z \in N(x_i) \setminus S(x_i)\}$.*

Algorithm 1. Recognition of 4-DORGs

Input: An undirected graph $G = (V, E)$ with a vertex partition $V = L \cup R \cup U \cup D$
Output: A 4-DORG representation for G, or the announcement that G is not a 4-
DORG graph

1: $n \leftarrow |V|$; $m \leftarrow \binom{n}{2} - |E|$ {m is the number of non-edges in G}
2: Construct from G the clique G_1 with vertex set $L_x \cup R_x \cup U_x \cup D_x$ and the clique G_2
   with vertex set $L_y \cup R_y \cup U_y \cup D_y$
3: Construct the set M_1 of type-1-mandatory orientations in G_1 and G_2
4: Construct the m forbidden pairs of orientations of G_1 and G_2
5: Construct from G_1, G_2 the augmented cliques G_1^*, G_2^* and the set M_2 of type-2-
   mandatory orientations
6: Construct from G_1^*, G_2^* the clique G^* and the set M_3 of type-3-mandatory orientations
7: **for** $i = 1$ to m **do**
8:    Let $p_i q_i \in E(G_1), a_i b_i \in E(G_2)$ be the optional edges in the ith pair of forbidden
      orientations, where $p_i \in U_x \cup D_x$, $q_i \in L_x \cup R_x$, $a_i \in U_y \cup D_y$, $b_i \in L_y \cup R_y$
9:    $(x_{2i-1}, y_{2i-1}) \leftarrow (p_i, q_i)$; $(x_{2i}, y_{2i}) \leftarrow (q_i, r_{p_i, q_i})$
10:   $(x_{2m+2i-1}, y_{2m+2i-1}) \leftarrow (a_i, b_i)$; $(x_{2m+2i}, y_{2m+2i}) \leftarrow (b_i, r_{a_i, b_i})$
11:   $S(x_i) \leftarrow \{r_{x_j, y_i} : x_j = x_i\}$
12: Construct the graph $\widetilde{G}^*$ by iteratively deactivating all edges $x_i y_i$, $1 \le i \le 4m$
13: **if** $\widetilde{G}^*$ has a transitive orientation $\widetilde{T}$ such that $M_1 \cup M_2 \cup M_3 \subseteq \widetilde{T}$ **then**
14:    **return** the 4-DORG representation of G computed by Theorem 1
15: **else**
16:    **return** "G is not a 4-DORG graph with respect to the partition $\{L, R, U, D\}$"

After deactivating the edge e_k of G, obtaining the graph $\widetilde{G}(e_k)$, we can continue
by sequentially deactivating the edges $e_{k-1}, e_{k-2}, \ldots, e_1$, obtaining eventually
the graph $\widetilde{G}$.

Theorem 2. *Let $G = (V, E)$ be a graph and $S(x_i)$, $1 \le i \le k$, be a set of k
special neighborhoods in G. Let M_0 be an arbitrary set of edge orientations of
G, and let $\widetilde{G}$ be the graph obtained after deactivating all edges $e_i = x_i y_i$, where
$1 \le i \le k$.*

- *If G has a consistent S-orientation T on $S(x_1), S(x_2), \ldots, S(x_k)$ such that
  $M_0 \subseteq T$, then $\widetilde{G}$ has a transitive orientation $\widetilde{T}$ such that $M_0 \cup F(x_i) \subseteq \widetilde{T}$
  for every $i = 1, 2, \ldots, k$.*
- *If $\widetilde{G}$ has a transitive orientation $\widetilde{T}$ such that $M_0 \cup F(x_i) \subseteq \widetilde{T}$ for every
  $i = 1, 2, \ldots, k$, then G has an S-orientation T on $S(x_1), S(x_2), \ldots, S(x_k)$
  such that $M_0 \subseteq T$.*

4   Efficient Recognition of 4-DORGs

In this section we complete our analysis in Sections 2 and 3 and we present our 4-
DORG recognition algorithm (cf. Algorithm 1). Let $G = (V, E)$ be an arbitrary

input graph that is given along with a vertex partition $V = L \cup R \cup U \cup D$, such that $E \subseteq (L \cup R) \times (U \cup D)$. Assume that G has n vertices and m non-edges (i.e. $\binom{n}{2} - m$ edges). First we construct from G the cliques G_1, G_2, then we construct the augmented cliques G_1^*, G_2^*, and finally we combine G_1^* and G_2^* to produce the clique G^* (cf. Section 2). Then, for a specific choice of $4m$ ordered pairs (x_i, y_i) of vertices, where $1 \leq i \leq 4m$ (cf. Algorithm 1), and for particular sets $S(x_i)$ and neighborhood orientations $F(x_i)$, $1 \leq i \leq 4m$ (cf. Definitions 6 and 7), we iteratively deactivate the edges $x_i y_i$, $1 \leq i \leq 4m$ (cf. Section 3), constructing thus the graph $\widetilde{G}^*$. Then, we can prove that for a specific partial orientation of the graph $\widetilde{G}^*$, $\widetilde{G}^*$ has a transitive orientation that extends this partial orientation if and only if the input graph G has a 4-DORG representation with respect to the vertex partition $\{L, R, U, D\}$. The proof of correctness of Algorithm 1 and the timing analysis are given in the next theorem.

Theorem 3. *Let $G = (V, E)$ be a graph with n vertices, given along with a vertex partition $V = L \cup R \cup U \cup D$, such that $E \subseteq (L \cup R) \times (U \cup D)$. Then Algorithm 1 constructs in $O(\mathrm{MM}(n^2))$ time a 4-DORG representation for G with respect to this vertex partition, or correctly announces that G does not have a 4-DORG representation.*

5   Recognizing 3-DORGs with Partial Representation Restrictions

In this section we consider a bipartite graph $G = (A, B, E)$, where $|A| = m$ and $|B| = n$, given along with an ordering $\pi = (v_1, v_2, \ldots, v_m)$ of the vertices of A. The question we address is the following: "Does G admit a 3-DORG representation where A (resp. B) is the set of rays oriented upwards (resp. horizontal, i.e. either leftwards or rightwards), such that, whenever $1 \leq i < j \leq m$, the y-coordinate of the endpoint of $v_i \in A$ is greater than that of $v_j \in A$?" Our approach uses the adjacency relations in G to recursively construct an x-coordinate ordering of the endpoints of the rays in the set A. If during the process we do not reach a contradiction, we eventually construct a 3-DORG representation for G, otherwise we conclude that such a representation does not exist.

Definition 10. *Let P_1, P_2 be two ordered partitions of the same base set S. Then P_1 and P_2 are compatible if there exists an ordered partition R of S which is refining and order preserving for both P_1 and P_2. A linear order L respects an ordered partition P of S, if L and P are compatible.*

Here we provide the main ideas and an overview of our algorithm. We start with the trivial partition of the set A (consisting of a single set including all elements of A). During the algorithm we process each vertex of $V = A \cup B$ once, and each time we process a new vertex we refine the current partition of the vertices of A, where the final partition of A implies an x-coordinate ordering of the rays of A. In particular, the algorithm proceeds in $|A| = m$ phases, where during phase i we process vertex $v_i \in A$ (the sequence of the vertices in A is according to

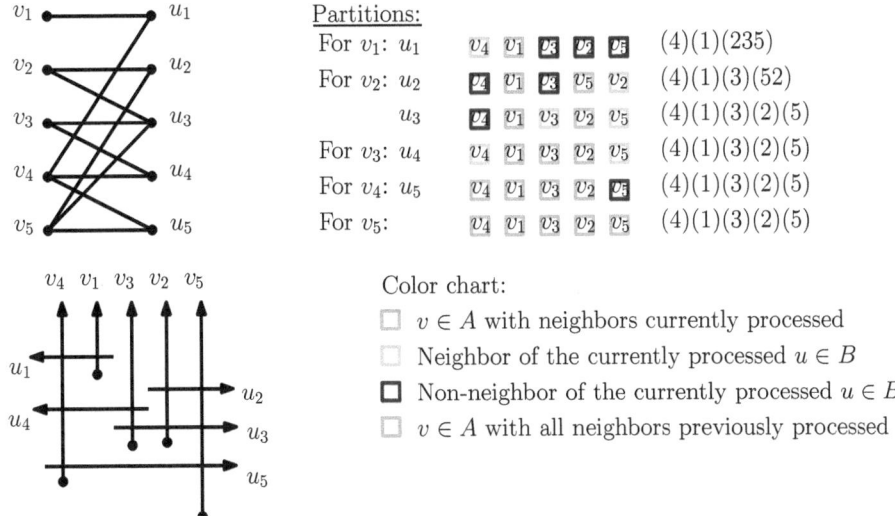

Fig. 4. Construction of a 3-DORG representation. Top left of the figure: the bipartite graph G with the given vertex ordering $\pi = (v_1, v_2, v_3, v_4, v_5)$. Top-right: the chain of partition refinements. Bottom left: The 3-DORG representation of G as read from the partition chain.

the given ordering π). During phase i, we process sequentially every neighbor $u \in N(v_i) \subseteq B$ that has not been processed in any previous phase $j < i$.

For every $i = 1, 2, \ldots, m$ let $A_i = \{v_i, v_{i+1}, \ldots, v_m\}$ be the set of vertices of A that have not been processed before phase i. At the end of every phase i, we fix the position of vertex $v_i \in A$ in the final partition of A, and we ignore v_i in the subsequent phases (i.e. during the phases $j > i$ we consider only the restriction of the current partition to the vertices of A_{i+1}). Phase i starts with the partition of A_i that results at the end of phase $i-1$. For any vertex $u \in N(v_i)$ that we process during phase i, we check whether the current partition P of A_i is compatible with at least one of the ordered partitions $Q_1 = (N(u), A_i \backslash N(u))$ and $Q_2 = (A_i \backslash N(u), N(u))$. If not, then we conclude that G is not a 3-DORG with respect to the given ordering π of A. Otherwise we refine the current partition P into an ordered partition that is also a refinement of Q_1 (resp. Q_2). In the case where P is compatible with both Q_1 and Q_2, it does not matter if we compute a common refinement of P with Q_1 or Q_2. If we can execute all m phases of this algorithm without returning that a 3-DORG representation does not exist, then we can compute a 3-DORG representation of G in which the y-coordinates of the endpoints of the rays of A respect the ordering π. In this extended abstract this construction is illustrated in the example of Figure 4.

Theorem 4. *Given a bipartite graph $G = (V, E)$ with color classes A, B and an ordering π of A, we can decide in $O(|V|^2)$ time whether G admits a 3-DORG representation where A are the vertical rays and the y-coordinates of their endpoints respect the ordering π.*

References

1. Cabello, S., Cardinal, J., Langerman, S.: The clique problem in ray intersection graphs. In: Epstein, L., Ferragina, P. (eds.) ESA 2012. LNCS, vol. 7501, pp. 241–252. Springer, Heidelberg (2012)
2. Chalopin, J., Gonçalves, D.: Every planar graph is the intersection graph of segments in the plane: extended abstract. In: Proc STOC 2009, pp. 631–638 (2009)
3. Coppersmith, D., Winograd, S.: Matrix multiplication via arithmetic progressions. Journal of Symbolic Computation 9(3), 251–280 (1990)
4. Estivill-Castro, V., Noy, M., Urrutia, J.: On the chromatic number of tree graphs. Discrete Mathematics 223(1-3), 363–366 (2000)
5. Feder, T., Hell, P., Huang, J.: List homomorphisms and circular arc graphs. Combinatorica 19, 487–505 (1999)
6. Felsner, S.: 3-interval irreducible partially ordered sets. Order 11, 12–15 (1994)
7. Felsner, S., Habib, M., Möhring, R.H.: On the interplay of interval dimension and dimension. SIAM Journal on Discrete Mathematics 7, 32–40 (1994)
8. Habib, M., Paul, C., Viennot, L.: Partition refinement techniques: An interesting algorithmic tool kit. Int. J. Found. Comput. Sci. 10(2), 147–170 (1999)
9. Hartman, I.-A., Newman, I., Ziv, R.: On grid intersection graphs. Discrete Mathematics 87(1), 41–52 (1991)
10. Kratochvíl, J.: A special planar satisfiability problem and a consequence of its NP-completeness. Discrete Applied Mathematics 52(3), 233–252 (1994)
11. Kratochvíl, J., Matoušek, J.: NP-hardness results for intersection graphs. Commentationes Mathematicae Universitatis Carolinae 30(4), 761–773 (1989)
12. Kratochvíl, J., Matoušek, J.: Intersection graphs of segments. J. Comb. Theory, Ser. B 62(2), 289–315 (1994)
13. McConnell, R.: Linear-time recognition of circular-arc graphs. In: Proceedings of the 42nd IEEE Symposium on Foundations of Computer Science, pp. 386–394 (2001)
14. Mustață, I., Pergel, M.: Unit grid intersection graphs: recognition and properties (in preparation, 2013)
15. Otachi, Y., Okamoto, Y., Yamazaki, K.: Relationships between the class of unit grid intersection graphs and other classes of bipartite graphs. Discrete Applied Mathematics 155(17), 2383–2390 (2007)
16. Rao, W., Orailoglu, A., Karri, R.: Logic mapping in crossbar-based nanoarchitectures. IEEE Design & Test of Computers 26(1), 68–77 (2009)
17. Richerby, D.: Interval bigraphs are unit grid intersection graphs. Discrete Mathematics, 1718–1719 (2009)
18. Shrestha, A., Tayu, S., Ueno, S.: Orthogonal ray graphs and nano-PLA design. In: Proc. IEEE ISCAS 2009, pp. 2930–2933 (2009)
19. Shrestha, A.M.S., Tayu, S., Ueno, S.: On orthogonal ray graphs. Discrete Appl. Math. 158(15), 1650–1659 (2010)
20. Soto, J.A., Telha, C.: Jump number of two-directional orthogonal ray graphs. In: Günlük, O., Woeginger, G.J. (eds.) IPCO 2011. LNCS, vol. 6655, pp. 389–403. Springer, Heidelberg (2011)
21. Trotter, W.: Combinatorics and Partially Ordered Sets. The Johns Hopkins University Press (1992)
22. Uehara, R.: Simple geometrical intersection graphs. In: Nakano, S.-I., Rahman, M. S. (eds.) WALCOM 2008. LNCS, vol. 4921, pp. 25–33. Springer, Heidelberg (2008)

Reachability Analysis of Recursive Quantum Markov Chains

Yuan Feng, Nengkun Yu, and Mingsheng Ying

University of Technology, Sydney, Australia
Tsinghua University, China

Abstract. We introduce the notion of *recursive quantum Markov chain*
(RQMC) for analysing recursive quantum programs with procedure calls.
RQMCs are natural extension of Etessami and Yannakakis's recursive
Markov chains where the probabilities along transitions are replaced by
completely positive and trace-nonincreasing super-operators on a state
Hilbert space of a quantum system.

We study the *reachability* problem for RQMCs and establish a reduction from it to computing the least solution of a system of polynomial
equations in the semiring of super-operators. It is shown that for an
important subclass of RQMCs, namely *linear* RQMCs, the reachability
problem can be solved in polynomial time. For general case, technique of
Newtonian program analysis recently developed by Esparza, Kiefer and
Luttenberger is employed to approximate reachability super-operators.
A polynomial time algorithm that computes the support subspaces of
the reachability super-operators in general case is also proposed.

1 Introduction

The model of recursive Markov chains (RMCs), defined in [9], is an extension
of both Markov chains and recursive state machines [2,3]. A RMC consists of a
collection of finite-state Markov chains with the ability to invoke each other in
a potentially recursive manner; it is especially suitable to describe probabilistic
programs with procedures. RMCs have been proven to be linear-time equivalent to probabilistic Pushdown Automata presented in [8], and they both have
been widely used as the mathematical models for software model checking and
program analysis.

Quantum Markov chains (QMCs) have been widely used in quantum optics
and quantum information theory. Recently, they have also been employed in the
studies of quantum programs. In the literature, there were quite a few different notions of QMC, introduced by authors from different research communities. These
models roughly fall into two categories: 1) the *fully quantum* ones [1,6,10,19,20]
where the Hilbert space of the quantum system is regarded as the state space of
the Markov model (thus infinite), while all possible quantum operations constitute
transitions between the states; 2) the *semi-quantum* ones [12,11] where the state
space of the model is still taken classical (and usually can be finite), while the transitions between them are quantum - transitions are labelled with super-operators

K. Chatterjee and J. Sgall (Eds.): MFCS 2013, LNCS 8087, pp. 385–396, 2013.

on the Hilbert space associated with the quantum system. This treatment is actually stemmed from the well-known slogan *quantum data, classical control* for quantum computers architecture [16]: the (classical) states are used to label the nodes in the control flow of a program, while super-operators along the transitions encode the manipulation applied on the quantum data.

In this paper, we introduce the notion of *recursive quantum Markov chain (RQMC)* based on the second type of QMCs. It can also be regarded as a quantum extension of RMC where the probabilities along the transitions are replaced with super-operators on a given Hilbert space. Similar to RMC, this notion of RQMC is especially suitable for defining semantics for recursive quantum programs which include procedure calls.

The distinct advantage of the second type of QMCs, and thus our model of RQMC as well, for model checking purpose, is twofold: (1) They provide a way to check *once for all* in that once a property is checked to be valid, it holds for all input quantum states. For example, for reachability problem considered in this paper we indeed calculate the accumulated *super-operator*, say $\mathcal{E}$, along all possible paths. As a result, the reachability *probability* when the program is executed on the input quantum state ρ is simply $\text{tr}(\mathcal{E}(\rho))$. (2) In these models, as all quantum effects are encoded in the super-operators labelling the transitions, and the nodes are kept purely classical, techniques from classical model checking can be adapted to verification of quantum systems. This has been shown in [11] for QMCs, and will be further extended to RQMCs defined in this paper.

A fundamental problem for RMCs is the following: given two vertices u and v, what is the probability of reaching v, starting from u, both with the empty context? This is called reachability problem, and has wide applications for many other analyses of RMCs. In this paper, we are going to investigate reachability problem for RQMCs, but instead of computing reachability *probability* from u to v, we now need to compute the accumulated *super-operator* along all possible paths leading from u to v. Similar to [9], this problem can be reduced to the *termination problem* where both the initial and the destination vertices are from the same component, and the destination vertex is an exit node. The contribution of the paper is three-fold:

(1) For an important subclass of RQMCs, namely linear RQMCs, we provide a polynomial time algorithm to compute all termination super-operators. Even for this simplest case, computing the termination super-operators is highly non-trivial, and some techniques beyond those introduced in previous works such as [11] and [9] must be introduced.

(2) For general RQMCs, we adapt the Newtonian program analysis recently developed in [7] to approximate termination super-operators. Note that the technique proposed in [7] cannot be applied directly because there the coefficient domain is required to be an ω-continuous semiring, while the set of super-operators, although being a semiring, is not ω-continuous.

(3) For general RQMCs, we propose a polynomial time algorithm to compute the support subspaces of the termination super-operators, which are of special interest for the purpose of safety analysis.

2 Preliminaries

2.1 Semiring of Super-Operators

We fix a finite dimensional Hilbert space $\mathcal{H}$ throughout this paper. Let $\mathcal{L}(\mathcal{H})$, $\mathcal{D}(\mathcal{H})$, and $\mathcal{S}(\mathcal{H})$ be the sets of linear operators, density operators, and super-operators on $\mathcal{H}$, respectively. In this paper, by super-operators we mean completely positive super-operators. Let $\mathcal{I}_{\mathcal{H}}$ and $0_{\mathcal{H}}$ be the identity and null super-operators on $\mathcal{H}$, respectively. Then obviously, $(\mathcal{S}(\mathcal{H}), +, \circ)$ forms a semiring where $\circ$ is the composition of super-operators defined by $(\mathcal{E} \circ \mathcal{F})(\rho) = \mathcal{E}(\mathcal{F}(\rho))$ for any $\rho \in \mathcal{D}(\mathcal{H})$. We always omit the symbol $\circ$ and write $\mathcal{E}\mathcal{F}$ directly for $\mathcal{E} \circ \mathcal{F}$. We will use two different orders over $\mathcal{S}(\mathcal{H})$.

Definition 1. *Let $\mathcal{E}, \mathcal{F} \in \mathcal{S}(\mathcal{H})$.*

(1) $\mathcal{E} \sqsubseteq \mathcal{F}$ if there exists $\mathcal{G} \in \mathcal{S}(\mathcal{H})$ such that $\mathcal{G} + \mathcal{E} = \mathcal{F}$;
(2) $\mathcal{E} \lesssim \mathcal{F}$ if for any $\rho \in \mathcal{D}(\mathcal{H})$, $\mathrm{tr}(\mathcal{E}(\rho)) \leq \mathrm{tr}(\mathcal{F}(\rho))$.

It is easy to check that if $\mathcal{E} \sqsubseteq \mathcal{F}$, then there exits a *unique* $\mathcal{G}$ such that $\mathcal{G} + \mathcal{E} = \mathcal{F}$. Note that the trace of a (unnormalised) quantum state is exactly the probability that the (normalised) state is reached [16]. Intuitively, $\mathcal{E} \lesssim \mathcal{F}$ if and only if the success probability of performing $\mathcal{E}$ is always not greater than that of performing $\mathcal{F}$, whatever the initial state is. Let $\approx$ be the kernel of $\lesssim$; that is, $\approx = \lesssim \cap \gtrsim$. Then $\mathcal{E} \approx \mathcal{I}_{\mathcal{H}}$ if and only if $\mathcal{E}$ is trace-preserving while $\mathcal{E} \approx 0_{\mathcal{H}}$ if and only if $\mathcal{E} = 0_{\mathcal{H}}$.

Let $\mathcal{S}^1(\mathcal{H}) = \{\mathcal{E} \in \mathcal{S}(\mathcal{H}) : \mathcal{E} \lesssim \mathcal{I}_{\mathcal{H}}\}$ be the set of trace-nonincreasing super-operators on $\mathcal{H}$. Let $\mathcal{S}^1(\mathcal{H})^n$ be the set of n-size vectors over $\mathcal{S}^1(\mathcal{H})$, and extend the partial order $\sqsubseteq$ componentwise to it. Then for any $n \geq 1$, $(\mathcal{S}^1(\mathcal{H})^n, \sqsubseteq)$ is a complete partially ordered set [16] with the least element $0_{\mathcal{H}} = (0_{\mathcal{H}}, \ldots, 0_{\mathcal{H}})$.

The next lemma, which is very useful for our purpose, shows that $\lesssim$ is preserved by the right-composition of super-operators, while $\sqsubseteq$ is preserved by both the left and the right compositions of super-operators.

Lemma 1. *Let $\mathcal{E}, \mathcal{F}, \mathcal{G} \in \mathcal{S}(\mathcal{H})$.*

(1) If $\mathcal{E} \lesssim \mathcal{F}$, then $\mathcal{E}\mathcal{G} \lesssim \mathcal{F}\mathcal{G}$. Especially, $\mathcal{E}\mathcal{F} \lesssim \mathcal{F}$ provided that $\mathcal{E} \in \mathcal{S}^1(\mathcal{H})$.
(2) If $\mathcal{E} \sqsubseteq \mathcal{F}$, then both $\mathcal{G}\mathcal{E} \sqsubseteq \mathcal{G}\mathcal{F}$ and $\mathcal{E}\mathcal{G} \sqsubseteq \mathcal{F}\mathcal{G}$.

2.2 Polynomials over Super-Operators

Let $\widetilde{X}$ be a finite set of variables. A *valuation* is a mapping $\mathbf{v} : \widetilde{X} \to \mathcal{S}^1(\mathcal{H})$. We write $\mathbf{v}_X$ for $\mathbf{v}(X)$ and denote the set of valuations by $\mathbf{V}$. Then $\mathbf{V}$ is isomorphic to the set $\mathcal{S}^1(\mathcal{H})^{|\widetilde{X}|}$ when an arbitrary total order is defined in $\widetilde{X}$. Consequently, we will not distinguish these two notations in this paper. A *monomial* is a finite expression $\mathcal{E}_0 X_0 \mathcal{E}_1 X_1 \cdots \mathcal{E}_{k-1} X_{k-1} \mathcal{E}_k$ where $k \geq 0$, each $\mathcal{E}_i \in \mathcal{S}^1(\mathcal{H})$, and each $X_i \in \widetilde{X}$. Given a monomial $m = \mathcal{E}_0 X_0 \mathcal{E}_1 X_1 \cdots \mathcal{E}_{k-1} X_{k-1} \mathcal{E}_k$ and a valuation $\mathbf{v}$, we define $m(\mathbf{v})$, the value of m at $\mathbf{v}$, as $\mathcal{E}_0 \mathbf{v}_{X_0} \mathcal{E}_1 \mathbf{v}_{X_1} \cdots \mathcal{E}_{k-1} \mathbf{v}_{X_{k-1}} \mathcal{E}_k$. We denote by $last(m) = \mathcal{E}_k$ the rightmost super-operator in m. A *polynomial* is an expression of the form $\sum_{1 \leq i \leq k} m_i$ where $k \geq 1$ and for each i, m_i is a monomial. The value of $f = \sum_i m_i$ at $\mathbf{v}$ is defined by $f(\mathbf{v}) = \sum_i m_i(\mathbf{v})$.

Definition 2. *A polynomial* $f = \sum_{1 \leq i \leq k} m_i$ *is called* trace-nonincreasing *if* $\sum_{1 \leq i \leq k} last(m_i) \lesssim \mathcal{I}_{\mathcal{H}}$.

By Lemma 1, we can easily check that a trace-nonincreasing polynomial is actually a mapping from $\mathbf{V}$ to $\mathcal{S}^1(\mathcal{H})$. In the rest of this paper, if not otherwise stated, any polynomial is assumed to be trace-nonincreasing.

A vector $\mathbf{f}$ of polynomials is a mapping that assigns to each variable $X \in \widetilde{X}$ a polynomial. Naturally, $\mathbf{f}$ induces a mapping, denoted again by $\mathbf{f}$ for simplicity, from $\mathbf{V}$ to $\mathbf{V}$ as follows. For each $\mathbf{v} \in \mathbf{V}$, let $\mathbf{f}(\mathbf{v})$ be the valuation such that for every $X \in \widetilde{X}$, $(\mathbf{f}(\mathbf{v}))_X = \mathbf{f}_X(\mathbf{v})$ where we write $\mathbf{f}_X$ for $\mathbf{f}(X)$. It is easy to observe that $\mathbf{f}$ is a Scott continuous function over the complete partially ordered set $\mathcal{S}^1(\mathcal{H})^{|\widetilde{X}|}$. Thus we have the following theorem, which is a special case of Kleene's well-known fixed point theorem.

Theorem 1. *Let $\mathbf{f}$ be a vector of polynomials. Then $\mathbf{f}$ has a unique least fixed point $\mu\mathbf{f}$ in $\mathbf{V}$. Furthermore, $\mu\mathbf{f}$ is the supremum (with respect to $\sqsubseteq$) of the Kleene sequence given by $\mathbf{v}^{(0)} = \mathbf{f}(\mathbf{0}_{\mathcal{H}})$, and $\mathbf{v}^{(i+1)} = \mathbf{f}(\mathbf{v}^{(i)})$ for $i \geq 0$.*

2.3   Quantum Markov Chains

To conclude this section, we recall the definition of quantum Markov chains which was first used to model check quantum programs as well as quantum cryptographic protocols in [11].

Definition 3. *A super-operator weighted Markov chain, or quantum Markov chain (QMC), over $\mathcal{H}$ is a tuple (S, δ), where*

- *S is a finite or countably infinite set of states;*
- *$\delta : S \times S \to \mathcal{S}^1(\mathcal{H})$ is called the transition relation such that for each $s \in S$, $\sum_{t \in S} \delta(s, t) \approx \mathcal{I}_{\mathcal{H}}$.*

The only difference between QMC and classical Markov chain (MC) is that the probabilities along the transitions in MCs are replaced by super-operators on a given Hilbert space in QMCs (and accordingly, the condition that all probabilities from a given state sum up to 1 is replaced by that all super-operators from it sum up to a trace-preserving one). This similarity, together with some nice properties of the set of super-operators, makes it possible to adapt techniques from classical model checking to the quantum case, as shown in [11].

3   Recursive Quantum Markov Chains

Now we present the main definition of this paper, which can be regarded as an extension of both the QMC in [11] and the RMC defined in [9]. The definition follows closely from [3] and [9].

Definition 4. *A recursive super-operator weighted Markov chain, or recursive quantum Markov chain (RQMC), over $\mathcal{H}$ is a tuple $A = (A_1, \ldots, A_k)$ where for each $1 \leq i \leq k$, the component A_i is again a tuple $(N_i, B_i, \Gamma_i, En_i, Ex_i, \delta_i)$ consisting of*

- A finite set N_i of nodes, with two special subsets En_i of entry nodes and Ex_i of exit nodes;
- A finite set B_i of boxes, and a mapping $\Gamma_i : B_i \to \{1, \ldots, k\}$ that assigns to each box (the index of) one of the components $A_1, \ldots, A_k$;
- To each box $b \in B_i$ we associate a set of call ports $Call_b = \{(b, en) : en \in En_{\Gamma_i(b)}\}$, and a set of return ports $Return_b = \{(b, ex) : ex \in Ex_{\Gamma_i(b)}\}$;
- Let

$$V_i = (N_i \backslash Ex_i) \cup \bigcup_{b \in B_i} Return_b \quad \text{and} \quad W_i = (N_i \backslash En_i) \cup \bigcup_{b \in B_i} Call_b.$$

A transition relation δ_i is a mapping $\delta_i : V_i \times W_i \to \mathcal{S}^1(\mathcal{H})$ such that for each $u \in V_i$ we have $\sum_{v \in W_i} \delta_i(u, v) \approx \mathcal{I}_{\mathcal{H}}$.

Let $Q_i = N_i \cup \bigcup_{b \in B_i}(Return_b \cup Call_b)$, and $Q = \cup_{i=1}^k Q_i$ be the set of all vertices. Let $B = \cup_{i=1}^k B_i$. Similar to [9], we can construct a (countably infinite state) QMC based on the RQMC A as $M_A = (S_A, \delta_A)$ where $S_A \subseteq B^* \times Q$ and $\delta_A : S_A \times S_A \to \mathcal{S}^1(\mathcal{H})$ as follows

(1) $\langle \epsilon, u \rangle \in S_A$ for all $u \in Q$, and $\delta_A(\langle \epsilon, u \rangle, \langle \epsilon, u \rangle) = \mathcal{I}_{\mathcal{H}}$ if u is an exit node. Here ϵ is the empty string;
(2) if $\langle \beta, u \rangle \in S_A$ and $\delta_i(u, v) \neq 0_{\mathcal{H}}$ for some i, then $\langle \beta, v \rangle \in S_A$ and

$$\delta_A(\langle \beta, u \rangle, \langle \beta, v \rangle) = \delta_i(u, v);$$

(3) if $\langle \beta, (b, en) \rangle \in S_A$ for some $(b, en) \in Call_b$, then $\langle \beta b, en \rangle \in S_A$ and

$$\delta_A(\langle \beta, (b, en) \rangle, \langle \beta b, en \rangle) = \mathcal{I}_{\mathcal{H}};$$

(4) if $\langle \beta b, ex \rangle \in S_A$ for some $(b, ex) \in Return_b$, then $\langle \beta, (b, ex) \rangle \in S_A$ and

$$\delta_A(\langle \beta b, ex \rangle, \langle \beta, (b, ex) \rangle) = \mathcal{I}_{\mathcal{H}};$$

(5) S_A is the smallest set satisfying (1) to (4), and $\delta_A(\langle \beta, u \rangle, \langle \alpha, v \rangle) = 0_{\mathcal{H}}$ if it is not defined in (1) to (4).

Intuitively, given a pair $\langle \beta, u \rangle$ in S_A, the sequence β denotes the stack of pending recursive calls and u the current control state.

Lemma 2. *For any RQMC A, the M_A defined above is indeed a QMC.*

Example 1. Suppose two players, Alice and Bob, want to randomly choose a winner among them, by taking a qubit system q as the coin. The protocol of Alice goes as follows. She first measures the system q according to the observable $M_A = 0|\psi\rangle\langle\psi| + 1|\psi^\perp\rangle\langle\psi^\perp|$ where $\{|\psi\rangle, |\psi^\perp\rangle\}$ is an orthonormal basis of $\mathcal{H}_q$. If the outcome 0 is observed, then she is the winner. Otherwise, she gives the quantum system to Bob and lets him decide. After Bob's manipulation, Alice performs a super-operator $\mathcal{F}$ on q if Bob is the winner. Bob's protocol goes similarly, except that his measurement operator is $M_B = 0|\phi\rangle\langle\phi| + 1|\phi^\perp\rangle\langle\phi^\perp|$ for another orthonormal basis $\{|\phi\rangle, |\phi^\perp\rangle\}$ of $\mathcal{H}_q$, and his post-Alice super-operator, if the winner is Alice, is $\mathcal{G}$.

We can formally describe such a protocol as a quantum program with procedure calls as follows.

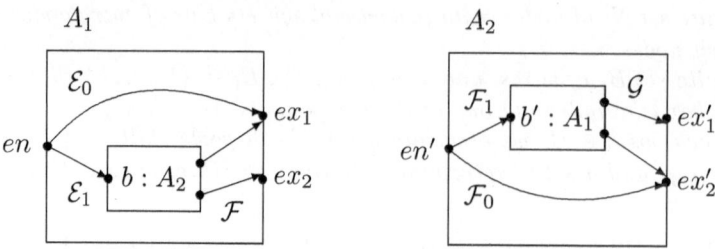

Fig. 1. The RQMC for a quantum program with recursive procedure calls

Global variables *winner* : String, q : **qubit**	
Program Alice	**Program** Bob
switch $M_A[q]$ **do**	**switch** $M_B[q]$ **do**
case 0	**case** 0
winner := 'A';	*winner* := 'B';
case 1	**case** 1
Call Bob;	Call Alice;
if *winner* = 'B' **then**	**if** *winner* = 'A' **then**
$q := \mathcal{F}[q]$;	$q := \mathcal{G}[q]$;
od	**od**

The semantics of this program can be described by the RQMC depicted in Figure 1 where the exits ex_1 and ex_1' represent that Alice wins, while ex_2 and ex_2' represent Bob wins. The super-operators $\mathcal{E}_0 = \{|\psi\rangle\langle\psi|\}$ and $\mathcal{E}_1 = \{|\psi^\perp\rangle\langle\psi^\perp|\}$ correspond respectively to the measurement outcomes 0 and 1 when M_A is applied. Similarly, $\mathcal{F}_0 = \{|\phi\rangle\langle\phi|\}$ and $\mathcal{F}_1 = \{|\phi^\perp\rangle\langle\phi^\perp|\}$. For simplicity, we omit all $\mathcal{I}_\mathcal{H}$ labels.

4   Reachability Problems

In this paper, we will focus on some fundamental reachability problems for RQMCs. Note that given a RQMC A over $\mathcal{H}$, there are two different notions of states: *classical* states in S_A of the underlying QMC M_A, and the *quantum* states in the associated Hilbert space $\mathcal{H}$. Thus naturally, we have three versions of reachability problems:

- *Classical state reachability.* Given a pair of vertices $u, v \in Q$, let $\mathcal{E}_{u,v}$ be the accumulated super-operator along all the possible paths in M_A from the initial state $\langle \epsilon, u \rangle$ to the destination state $\langle \epsilon, v \rangle$. We call $\mathcal{E}_{u,v}$ the reachability super-operator from u to v, and our goal is to compute it for any u and v in Q.
- *Quantum state reachability.* We take an initial classical state $\langle \epsilon, u \rangle$ in M_A and an initial quantum state $\rho \in \mathcal{D}(\mathcal{H})$. Then a quantum state $\sigma \in \mathcal{D}(\mathcal{H})$ is reachable if along some finite path in M_A which starts from $\langle \epsilon, u \rangle$, the associated quantum system evolves from ρ into σ.

– *Quantum subspace reachability.* Note that for the purpose of safety analysis, sometimes we are only concerned with the subspaces of $\mathcal{H}$ spanned by the reachable states, while the exact forms of them are irrelevant. In this setting, we would like to compute the support subspace of $\mathcal{E}_{u,v}$, called reachable subspace, for all $u, v \in Q$.

Obviously, the last two problems can be regarded as special cases of the first one: given $u \in Q$ and $\rho \in \mathcal{D}(\mathcal{H})$, if we are able to compute the reachability super-operators $\mathcal{E}_{u,v}$ for all $v \in Q$, then a state σ is reachable if and only if $\sigma \in \{\mathcal{E}_{u,v}(\rho) : v \in Q\}$, and the reachable subspace from u to u' is simply the subspace spanned by the Kraus operators of $\mathcal{E}_{u,u'}$. Furthermore, similar to [9], we can reduce classical state reachability problem to the *termination problem* where both the initial and the destination vertices are from the same component A_i, and the destination vertex is an exit node. To be more specific, the reachability super-operator for two given vertices of a RQMC is exactly a certain termination super-operator in another efficiently constructible RQMC. For this reason, we are going to focus on termination problem in the rest of the paper.

To conclude this section, we reduce the calculation of termination super-operators to the problem of computing the least solution to a system of polynomial equations.

Definition 5. *Let $A = (A_1, \ldots, A_k)$ be a RQMC where $A_i = (N_i, B_i, \Gamma_i, En_i, Ex_i, \delta_i)$. We construct a system of polynomial equations, with super-operator coefficients, over the variables $\widetilde{X} = \{X_{u,ex} : u \in Q_i, ex \in Ex_i, 1 \leq i \leq k\}$ as follows. The equations are indexed by u and ex, and have the form $X_{u,ex} = f_{u,ex}$ where $f_{u,ex}$ is defined, for any $u \in Q_i$ and $ex \in Ex_i$, by*

(1) $f_{u,ex} = \mathcal{I}_{\mathcal{H}}$ if $u = ex$, and $f_{u,ex} = 0_{\mathcal{H}}$ if $u \in Ex_i \backslash \{ex\}$;
(2) If $u \in N_i \backslash Ex_i$ or u is a return port, then $f_{u,ex} = \sum_{v \in W_i} X_{v,ex} \delta_i(u, v)$;
(3) If $u = (b, en)$ is a call point, then $f_{u,ex} = \sum_{ex' \in Ex_{\Gamma_i(b)}} X_{(b,ex'),ex} X_{en,ex'}$.

Given a RQMC $A = (A_1, \cdots, A_k)$, we let $\xi = \max_{i \in \{1, \ldots, k\}} |Ex_i|$ be the maximum number of exit nodes among all components. Then the number of all possible vertex-exit pairs is $|\widetilde{X}| \leq \sum_{i=1}^{k} |Q_i| \cdot |Ex_i| \leq |Q|\xi$. That is, the number of polynomials $\{f_{u,ex}\}$ defined in Definition 5 is bounded by $|Q|\xi$.

Theorem 2. *Let A be a RQMC. Then the system of polynomials $\{f_{u,ex}\}$ defined in Definition 5 is a continuous function from $\mathcal{S}^1(\mathcal{H})^{|\widetilde{X}|}$ to $\mathcal{S}^1(\mathcal{H})^{|\widetilde{X}|}$. Furthermore, for any u and ex from the same component of A, $\mathcal{E}_{u,ex}$ is exactly the (u, ex)-component of the least solution to the system of equations $\{X_{u,ex} = f_{u,ex}\}$.*

5   Computing Reachability Super-Operators for Linear RQMCs

This section is devoted to the calculation of reachability super-operators for an important class of RQMCs, namely *linear* RQMCs where there is no path in any component from a return port to a call port. The solution here also serves as a key ingredient of the Newtonian method for approximating the termination super-operators for general RQMCs in Sec 6.

Note that in the worst case, $O(d^4)$ complex numbers are needed to represent a super-operator on a d-dimensional Hilbert space (maximally d^2 Kraus operators, each being a $d \times d$ matrix). For a RQMC $A = (A_1, \cdots, A_k)$ over $\mathcal{H}$ with $\dim(\mathcal{H}) = d$, we let $|A| = |Q|d^4$ be the size of A. Then we have the following theorem.

Theorem 3. *Given a linear RQMC A, we can calculate in time complexity $O(|A|^4 \xi^4)$ the reachability super-operators $\mathcal{E}_{u,ex}$ for all vertex-exit pairs (u, ex) from the same component of A.*

6   Newtonian Method for Approximating Termination Super-Operators of General RQMCs

As our RQMC model includes classical RMCs as a subset, we cannot hope to have a finite method to solve termination problem for general RQMCs, and an approximate algorithm is the only possibility. Kleene's method (Theorem 1 of this paper) has already provided a natural way for the approximation, but as pointed out in [9] and [7], it normally has a very slow convergence rate, and an extension of Newtonian method can be employed to give a faster approximation, both for RMCs and for equations over ω-*continuous semirings*. In this section, we are going to explore the possibility of extending this method to approximate termination super-operators for general RQMCs. Note that the technique proposed in [7] cannot be applied directly because there the coefficient domain is required to be an ω-continuous semiring in which every ω-chain has a supremum. The set of super-operators, although being a semiring, is not ω-continuous, and it seems impossible to simply add an extra element to make it ω-continuous, just like what we did for the semiring of nonnegative real numbers.

However, as $\mathcal{S}^1(\mathcal{H})^{|\tilde{X}|}$ is a complete partially ordered set, and the system of equations which concerns us is a continuous map from $\mathcal{S}^1(\mathcal{H})^{|\tilde{X}|}$ to itself, it is possible to adapt the Newtonian program analysis for our purpose.

Recall that for a polynomial f over $\mathcal{S}^1(\mathcal{H})$ and $X \in \tilde{X}$, the differential of f with respect to X at the point $\mathbf{v}$ is the mapping $D_X f|_{\mathbf{v}} : \mathbf{V} \to \mathcal{S}^1(\mathcal{H})$ inductively defined as [7]:

$$D_X f|_{\mathbf{v}}(\mathbf{w}) = \begin{cases} 0_{\mathcal{H}}, & \text{if } f \in \mathcal{S}^1(\mathcal{H}) \text{ or } f \in \tilde{X} \backslash \{X\}; \\ \mathbf{w}_X, & \text{if } f = X; \\ D_X g|_{\mathbf{v}}(\mathbf{w}) h(\mathbf{v}) + g(\mathbf{v}) D_X h|_{\mathbf{v}}(\mathbf{w}), & \text{if } f = gh; \\ \sum_i D_X f_i|_{\mathbf{v}}(\mathbf{w}), & \text{if } f = \sum_i f_i. \end{cases}$$

The differential of f at $\mathbf{v} \in \mathbf{V}$ is the function $Df|_{\mathbf{v}} = \sum_{X \in \tilde{X}} D_X f|_{\mathbf{v}}$, and the differential of a vector $\mathbf{f}$ of polynomials is defined as the function $D\mathbf{f}|_{\mathbf{v}} : \mathbf{V} \to \mathbf{V}$ such that $(D\mathbf{f}|_{\mathbf{v}}(\mathbf{w}))_X = Df_X|_{\mathbf{v}}(\mathbf{w})$.

Definition 6. *Let* $\mathbf{f} : \mathbf{V} \to \mathbf{V}$ *be a vector of polynomials. For* $i \geq 0$*, the* i*th Newton approximant* $\mathbf{v}^{(i)}$ *of* $\mu\mathbf{f}$ *is inductively defined by* $\mathbf{v}^{(0)} = \mathbf{f}(\mathbf{0}_{\mathcal{H}})$ *and* $\mathbf{v}^{(i+1)} = \mathbf{v}^{(i)} + \mathbf{\Delta}^{(i)}$*, where* $\mathbf{\Delta}^{(i)}$ *is the least solution of*

$$D\mathbf{f}|_{\mathbf{v}^{(i)}}(\mathbf{X}) + \boldsymbol{\delta}^{(i)} = \mathbf{X} \tag{1}$$

and $\boldsymbol{\delta}^{(i)}$ *is the (unique) valuation satisfying* $\mathbf{f}(\mathbf{v}^{(i)}) = \mathbf{v}^{(i)} + \boldsymbol{\delta}^{(i)}$.

The sequence $(\mathbf{v}^{(i)})_{i \in \mathbb{N}}$ in the previous definition is called the Newton sequence of $\mathbf{f}$, and its well-definedness is guaranteed by the following theorem.

Theorem 4. *For any vector of polynomial over* $\mathcal{S}^1(\mathcal{H})$*, Newton sequence exists and is unique.*

Finally, we can show that compared with Kleene's method, the Newton sequence performs better in approximating the least fixed point for a given vector of polynomials.

Theorem 5. *Let* $\mathbf{f}$ *be a vector of polynomials with the Newton sequence* $(\mathbf{v}^{(i)})_{i \in \mathbb{N}}$*. Then*

(1) for any i*,* $\mathbf{k}^{(i)} \sqsubseteq \mathbf{v}^{(i)} \sqsubseteq \mu\mathbf{f} = \sup_j \mathbf{k}^{(i)}$ *where* $(\mathbf{k}^{(i)})_{i \in \mathbb{N}}$ *is the Kleene sequence for* $\mathbf{f}$*, defined in Theorem 1.*
(2) $\mu\mathbf{f} = \lim_{i \to \infty} \mathbf{v}^{(i)}$*.*

Note that for each $i \geq 0$, the left hand side of Eq.(1) is a vector of linear polynomials. The result shown in Sec 5 can be employed to compute its least fixed point $\mathbf{\Delta}^{(i)}$. Thus Theorem 5 indeed provides a way to approximate the termination super-operators for any RQMCs.

7   Computing Reachable Subspaces for General RQMCs

The problem of computing reachability super-operators is very hard for general RQMCs, and the last section is devoted to a Newtonian method for approximating them. In this section, we turn to a simplified version of this problem which is of special interest in safety analysis: instead of calculating the exact form of reachability super-operators, we are concerned with the support subspaces of them. Here the support subspace of a super-operator $\mathcal{E} \in \mathcal{S}(\mathcal{H})$ with the Kraus operators $\{E_i : i \in I\}$ is defined to be $\mathrm{supp}(\mathcal{E}) = \mathrm{span}\{E_i : i \in I\}$. It turns out that this simplified problem can be solved in polynomial time.

We first show that the support subspaces of the least fixed-point for a vector of polynomials can be calculated with at most $|\widetilde{X}|d^2$ Kleene iterations. The idea is similar to [21].

Lemma 3. *Let* $\mathbf{f}$ *be a vector of polynomials over* $\mathcal{S}^1(\mathcal{H})$*, and* $\mathbf{k}^{(0)}, \mathbf{k}^{(1)}, \cdots$ *the Kleene sequence of* $\mathbf{f}$ *defined in Theorem 1. Then for any* $X \in \widetilde{X}$*,* $\mathrm{supp}[(\mu\mathbf{f})_X] = \mathrm{supp}[(\mathbf{k}^{(n)})_X]$ *for some* $n \leq |\widetilde{X}|d^2$*, where* $d = \dim(\mathcal{H})$ *is the dimension of* $\mathcal{H}$*.*

Algorithm 1. Computing the support subspaces of the least fixed-point for a vector of polynomials

---

Input : A vector of polynomials $\mathbf{f} = \{\mathbf{f}_X = \sum_{i \in I_X} m_i^X : X \in \widetilde{X}\}$ with $m_i^X = \mathcal{E}_0^{X,i} Y_0^{X,i} \cdots \mathcal{E}_{k_i-1}^{X,i} Y_{k_i-1}^{X,i} \mathcal{E}_{k_i}^{X,i}$.

Output: An orthonormal basis $\mathcal{B}_X$ of $\text{supp}[(\mu\mathbf{f})_X]$ for each $X \in \widetilde{X}$.

for $X \in \widetilde{X}$ **do**
     set of matrices $\mathcal{B}_X \leftarrow \emptyset$;
     integer $n_X \leftarrow 0$;
end

for $j = 1 : |\widetilde{X}|d^2$ **do**
     for $X \in \widetilde{X}$ **do**
        $n'_X \leftarrow n_X$;
        for $i \in I_X$ **do**
           for $E_0 \in \mathcal{E}_0^{X,i}, B_0 \in \mathcal{B}_{Y_0^{X,i}}, \cdots, E_{k_i} \in \mathcal{E}_{k_i}^{X,i}$ **do**
              $C \leftarrow E_0 B_0 \cdots E_{k_i-1} B_{k_i-1} E_{k_i}$;   (**)
              (*Gram-Schmidt orthonormalization*)
              $C \leftarrow C - \sum_{B \in \mathcal{B}_X} \text{tr}(B^\dagger C)B$;
              if $C \neq 0$ **then**
                 $n_X \leftarrow n_X + 1$;
                 $\mathcal{B}_X \leftarrow \mathcal{B}_X \cup \{C/\text{tr}(C^\dagger C)\}$;
              end
           end
        end
     end
     if $n_X = n'_X$ *for all* X **then**
        return $\{\mathcal{B}_X : X \in \widetilde{X}\}$
     end
end

Algorithm 1 implements Lemma 3 where for each X, i, and j, the super-operator $\mathcal{E}_j^{X,i}$ is given by the set of its Kraus operators. The termination of Algorithm 1 follows directly from Lemma 3, and the correctness is guaranteed by the following Lemma.

Lemma 4. *For any* $\mathcal{E}, \mathcal{F} \in \mathcal{S}(\mathcal{H})$, *if* $\mathcal{E} = \{E_i : i \in I\}$ *and* $\mathcal{B}$ *is an orthonormal basis of* $\text{supp}(\mathcal{F})$, *then* $\text{supp}(\mathcal{E}\mathcal{F}) = span\{E_i B : i \in I, B \in \mathcal{B}\}$ *and* $\text{supp}(\mathcal{F}\mathcal{E}) = span\{BE_i : i \in I, B \in \mathcal{B}\}$.

As the number of Kraus operators for a super-operator on $\mathcal{H}$ can be assumed to be no more than d^2, we have $|I_X| \leq d^2$ for each X. Then the time complexity of Algorithm 1 is calculated as

$$|\widetilde{X}|d^2 \cdot \sum_{X \in \widetilde{X}} \sum_{i \in I_X} \sum_{E_0 \in \mathcal{E}_0^{X,i}, \cdots, E_{k_i} \in \mathcal{E}_{k_i}^{X,i}} d^3(2k_i + 1) = O(|\widetilde{X}|^2 k d^{4k+9}) \qquad (2)$$

where $k = \max\{k_i\}$ is the maximum degree of the monomials in $\mathbf{f}$.

Observe that from Definition 5, in the system of equations corresponding to a given RQMC A, each nontrivial polynomial $f_{u,ex}$ has the simple form $\sum_i X_i \mathcal{E}_i$ or $\sum_i X_i Y_i$. That means at the line (**) in Algorithm 1, only two matrix multiplications are needed; thus the innermost for-loop takes time $O(d^2 \cdot d^2 \cdot d^3)$. Consequently, for RQMC A the complexity in Eq.(2) can be refined as $O(|\widetilde{X}|^2 d^{11})$. Noting again that $|A| = |Q| d^4$ and $|\widetilde{X}| \leq |Q| \xi$, we immediately obtain the following theorem.

Theorem 6. *Given a RQMC A, we can compute in time $O(|A|^2 \xi^2 d^3)$ an orthonormal basis of $\mathrm{supp}(\mathcal{E}_{u,ex})$ for all vertex-exit pairs (u, ex) from the same component of A, where $d = \dim(\mathcal{H})$ is the dimension of $\mathcal{H}$.*

To conclude this section, we would like to point out the implication of our result here for classical RMCs, which correspond to the case of $d = 1$. Note that a 1-dimensional Hilbert space $\mathcal{H}$ has only two subspaces: 0-dimensional subspace $\{0\}$ and $\mathcal{H}$ itself. Computing the support subspace of a reachability super-operator $\mathcal{E}_{u,ex}$ is equivalent to determining if ex is reachable from u. Thus our algorithm indeed solves the (qualitative) reachability problem for classical RMCs in time $O(|A|^2 \xi^2)$. Note that the same problem has been proven in [2] to be decidable in time $O(|A| \theta \xi)$ where $\theta = \max_{i \in \{1,\dots,k\}} \min\{|En_i|, |Ex_i|\}$ is the maximum, over all components, of the minimum of the number of entry nodes and the number of exit nodes.

8   Conclusion and Future Work

In this paper, the notion of quantum recursive Markov chain is defined, and fundamental problems such as reachability and termination, characterised by the accumulated super-operators along certain paths of the underlying QMCs, are identified. For linear QRMCs, we show that the reachability problem can be solved in polynomial time. For general case, we are able to approximate the reachability super-operators for all vertex-exit pairs by adapting the technique of Newtonian program analysis [7], or compute the support subspaces of them in polynomial time.

We note that our algorithm in Theorem 3 for linear RQMCs is based on the Jordan decomposition for matrices. Although Jordan decomposition is not very expensive in theory - it takes $O(n^4)$ time to decompose an $n \times n$ matrix, in practice it is not numerically stable thus usually is avoided. For future work we are going to exploit other more stable ways to deal with this problem.

Acknowledgement. This work was partially supported by Australian Research Council (grant numbers DP110103473, DP130102764, and FT100100218). The authors are also supported by the Overseas Team Program of Academy of Mathematics and Systems Science, Chinese Academy of Sciences.

References

1. Accardi, L.: Nonrelativistic quantum mechanics as a noncommutative Markov process. Advances in Mathematics 20(3), 329–366 (1976)
2. Alur, R., Benedikt, M., Etessami, K., Godefroid, P., Reps, T., Yannakakis, M.: Analysis of recursive state machines. ACM Transactions on Programming Languages and Systems 27(4), 786–818 (2005)
3. Alur, R., Etessami, K., Yannakakis, M.: Analysis of recursive state machines. In: Berry, G., Comon, H., Finkel, A. (eds.) CAV 2001. LNCS, vol. 2102, pp. 207–220. Springer, Heidelberg (2001)
4. Bennett, C.H., Brassard, G., Crepeau, C., Jozsa, R., Peres, A., Wootters, W.: Teleporting an unknown quantum state via dual classical and EPR channels. Physical Review Letters 70, 1895–1899 (1993)
5. Bennett, C.H., Wiesner, S.J.: Communication via one- and two-particle operators on Einstein-Podolsky-Rosen states. Physical Review Letters 69(20), 2881–2884 (1992)
6. Breuer, H., Petruccione, F.: The theory of open quantum systems. Oxford University Press, New York (2002)
7. Esparza, J., Kiefer, S., Luttenberger, M.: Newtonian program analysis. Journal of the ACM 57(6), 33 (2010)
8. Esparza, J., Kucera, A., Mayr, R.: Model checking probabilistic pushdown automata. In: IEEE Symposium on Logic in Computer Science (LICS 2004), pp. 12–21 (July 2004)
9. Etessami, K., Yannakakis, M.: Recursive Markov chains, stochastic grammars, and monotone systems of nonlinear equations. Journal of the ACM 56(1), 1 (2009)
10. Faigle, U., Schönhuth, A.: Discrete Quantum Markov Chains. Arxiv.org/abs/1011.1295 (2010)
11. Feng, Y., Yu, N., Ying, M.: Model checking quantum Markov chains. Journal of Computer and System Sciences 79, 1181–1198 (2013)
12. Gudder, S.: Quantum Markov chains. Journal of Mathematical Physics 49(7), 072105, 14 (2008)
13. Kraus, K.: States, Effects and Operations: Fundamental Notions of Quantum Theory. Springer, Berlin (1983)
14. von Neumann, J.: Mathematical Foundations of Quantum Mechanics. Princeton University Press, Princeton (1955)
15. Nielsen, M., Chuang, I.: Quantum computation and quantum information. Cambridge University Press (2000)
16. Selinger, P.: Towards a quantum programming language. Mathematical Structures in Computer Science 14(4), 527–586 (2004)
17. Steel, A.: A new algorithm for the computation of canonical forms of matrices over fields. Journal of Symbolic Computation 24(3-4), 409–432 (1997)
18. Watrous, J.: Lecture Notes on Theory of Quantum Information (2011), https://cs.uwaterloo.ca/~watrous/CS766/
19. Ying, M., Li, Y., Yu, N., Feng, Y.: Model-Checking Linear-Time Properties of Quantum Systems. Arxiv.org/abs/1101.0303. Submitted to ACM Transactions on Computational Logic (revised)
20. Ying, M., Yu, N., Feng, Y., Duan, R.: Verification of Quantum Programs. Science of Computer Programming 78, 1679–1700 (2013)
21. Yu, N., Ying, M.: Reachability and Termination Analysis of Concurrent Quantum Programs. In: Koutny, M., Ulidowski, I. (eds.) CONCUR 2012. LNCS, vol. 7454, pp. 69–83. Springer, Heidelberg (2012)

Ordering Metro Lines by Block Crossings

Martin Fink[1] and Sergey Pupyrev[2,3]

[1] Lehrstuhl für Informatik I, Universität Würzburg, Germany
[2] Department of Computer Science, University of Arizona, USA
[3] Ural Federal University, Ekaterinburg, Russia

Abstract. A problem that arises in drawings of transportation networks is to minimize the number of crossings between different transportation lines. While this can be done efficiently under specific constraints, not all solutions are visually equivalent. We suggest merging crossings into *block crossings*, that is, crossings of two neighboring groups of consecutive lines. Unfortunately, minimizing the total number of block crossings is NP-hard even for very simple graphs. We give approximation algorithms for special classes of graphs and an asymptotically worst-case optimal algorithm for block crossings on general graphs.

1 Introduction

In many metro maps and transportation networks some edges, that is, railway track or road segments, are used by several lines. To visualize such networks, lines that share an edge are drawn individually along the edge in distinct colors. Often, some lines must cross, and it is desirable to draw the lines with few crossings. The *metro-line crossing minimization* problem has recently been introduced [4]. The goal is to order the lines along each edge such that the number of crossings is minimized. So far, the focus has been on the number of crossings and not on their visualization, although two line orders with the same crossing number may look quite differently; see Fig. 1.

Our aim is to improve the readability of metro maps by computing line orders that are aesthetically more pleasing. To this end, we merge *pairwise* crossings into crossings of blocks of lines minimizing the number of *block crossings* in the map. Informally, a block crossing is an intersection of two neighboring groups of consecutive lines sharing the same edge; see Fig. 1(b). We consider two variants of the problem. In the first variant, we want to find a line ordering with the minimum number of block crossings. In the second variant, we want to minimize both pairwise and block crossings.

Problem Definition. The input consists of an embedded graph $G = (V, E)$, and a set $L = \{l_1, \ldots, l_{|L|}\}$ of simple paths in G. We call G the *underlying network* and the paths

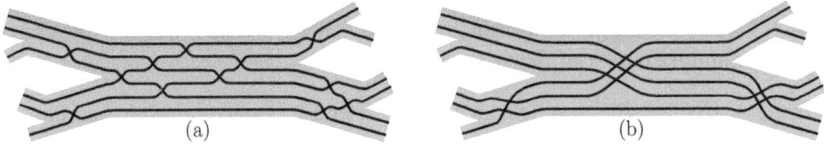

Fig. 1. Optimal orderings of a metro network: (a) 12 pairwise crossings; (b) 3 block crossings

K. Chatterjee and J. Sgall (Eds.): MFCS 2013, LNCS 8087, pp. 397–408, 2013.
© Springer-Verlag Berlin Heidelberg 2013

lines. The nodes of G are *stations* and the endpoints v_0, v_k of a line $(v_0, \ldots, v_k) \in L$ are *terminals.* For each edge $e = (u, v) \in E$, let L_e be the set of lines passing through e. For $i \leq j < k$, a *block move* (i, j, k) on the sequence $\pi = [\pi_1, \ldots, \pi_n]$ of lines on e is the exchange of two consecutive blocks $\pi_i, \ldots, \pi_j$ and $\pi_{j+1}, \ldots, \pi_k$. We are interested in *line orders* $\pi^0(e), \ldots, \pi^{t(e)}(e)$ on e, so that $\pi^0(e)$ is the order of lines L_e on e close to u, $\pi^{t(e)}(e)$ is the order close to v, and each $\pi^i(e)$ is an ordering of L_e so that $\pi^{i+1}(e)$ is constructed from $\pi^i(e)$ by a block move. We say that there are t *block crossings* on e.

Following previous work [1,13], we use the *edge crossings* model, that is, we do not hide crossings under station symbols if possible. Two lines sharing at least one common edge either do not cross or cross each other on an edge but never in a node; see Fig. 2(a). For pairs of lines sharing a vertex but no edges, crossings at the vertex are allowed and not counted as they exist in any solution. We call them *unavoidable vertex crossings*; see Fig. 2(b). If the line orders on the edges incident to a vertex v produce only edge crossings and unavoidable vertex crossings, we call them *consistent* in v. Line orders for all edges are consistent if they are consistent in all nodes. More formally, we can check consistency of line orders in a vertex v by looking at each incident edge e. Close to v the order of lines L_e on e is fixed. The other edges $e_1, \ldots, e_k$ incident to v contain lines of L_e. The combined order of L_e on the edges $e_1, \ldots, e_k$ must be the same as the order on e; otherwise, lines of L_e would cross

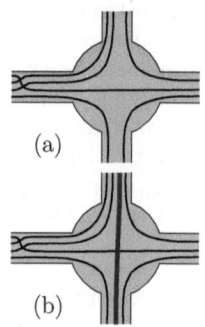

Fig. 2. Consistent line orders (a) without, (b) with an unavoidable vertex crossing

in v. The *block crossing minimization* problem (BCM) is defined as follows.

*Problem 1 (**BCM**).* Let $G = (V, E)$ be an embedded graph and let L be a set of lines on G. For each edge $e \in E$, find line orders $\pi^0(e), \ldots, \pi^{t(e)}(e)$ such that the total number of block crossings, $\sum_{e \in E} t(e)$, is minimum and the line orders are consistent.

In this paper, we restrict our attention to instances with two additional properties. First, any line terminates at nodes of degree one and no two lines terminate at the same node (**path terminal property**). Second, the intersection of two lines, that is, the edges and vertices they have in common, forms a path (**path intersection property**). This includes the cases that the intersection is empty or a single node. If both properties hold, a pair of lines either has to cross, that is, a crossing is *unavoidable,* or it can be kept crossing-free, that is, a crossing is *avoidable.* The orderings that are optimal with respect to pairwise crossings are exactly the orderings that contain just unavoidable crossings (Lemma 2 in [13]); that is, any pair of lines crosses at most once, in an equivalent formulation. As this is a very reasonable condition also for block crossings, we use it to define the *monotone block crossing minimization* problem (MBCM) whose feasible solutions must have the minimum number of pairwise crossings.

*Problem 2 (**MBCM**).* Given an instance of BCM, find a feasible solution that minimizes the number of block crossings subject to the constraint that no two lines cross twice.

On some instances BCM does allow fewer crossings than MBCM does; see Fig. 3.

Table 1. Overview of our results for BCM and MBCM

graph class	BCM		MBCM	
single edge	11/8-approximation	[7]	3-approximation	Sec. 2
path	3-approximation	Sec. 3	3-approximation	[9]
tree	$\leq 2|L| - 3$ crossings	Sec. 4	$\leq 2|L| - 3$ crossings	Sec. 4
upward tree	—		6-approximation	Sec. 4
general graph	$O(|L|\sqrt{|E|})$ crossings	Sec. 5	$O(|L|\sqrt{|E|})$ crossings	Sec. 5

Our Contribution. We introduce the new problems BCM and MBCM. To the best of our knowledge, ordering lines by block crossings is a new direction in graph drawing. So far BCM has been investigated only for the case that the *skeleton*, that is, the graph without terminals, is a single edge [2], while MBCM is a completely new problem.

We first analyze MBCM on a single edge (Sec. 2), exploiting, to some extent, the similarities to *sorting by transpositions* [2]. Then, we use the notion of *good pairs* of lines, that is, lines that should be neighbors, for developing an approximation algorithm for BCM on graphs whose skeleton is a path (Sec. 3); we properly define good pairs so that changes between adjacent edges are taken into account. Yet, good pairs can not always be kept close; we introduce a good strategy for breaking pairs when needed.

Unfortunately, the approximation algorithm does not generalize to trees. We do, however, develop a worst-case optimal algorithm for trees (Sec. 4). It needs $2|L| - 3$ block crossings and there are instances in which this number of block crossings is necessary in any solution. We then use our algorithm for obtaining approximate solutions for MBCM on the special class of *upward trees*.

As our main result, we develop an algorithm for obtaining a solution for (M)BCM on general graphs (Sec. 5). We show that it uses only monotone block moves and analyze the upper bound on the number of block crossings. While the algorithm itself is simple and easy to implement, proving the upper bound is non-trivial. We also show that our algorithm is asymptotically worst-case optimal. Table 1 summarizes our results.

Related Work. Line crossing problems in transportation networks were initiated by Benkert et al. [4], who considered the problem of *metro-line crossing minimization* (MLCM) on a single edge. MLCM in its general model is challenging; its complexity is open and no efficient algorithms are known for the case of two or more edges. Bekos et al. [3] addressed the problem on paths and trees. They also proved that a variant in which all lines must be placed outermost in their terminals is NP-hard. Subsequently, Argyriou et al. [1] and Nöllenburg [13] devised polynomial-time algorithms for general graphs with the path terminal property. A lot of recent research, both in graph drawing and information visualization, is devoted to *edge bundling* where some edges are drawn close together—like metro lines—which emphasizes the structure of the graph [14]. Pupyrev et al. [14] studied MLCM in this context and suggested a linear-time algorithm for MLCM on instances with the path terminal property. All these works are dedicated to pairwise crossings; the optimization criterion being the number of crossing pairs of lines.

A closely related problem arises in VLSI design, where the goal is to minimize intersections between nets (physical wires) [10,12]. Net patterns with fewer crossings most likely have better electrical characteristics and require less wiring area; hence, it is an important optimization criterion in circuit board design. Marek-Sadowska and Sarrafzadeh [12] considered not only minimizing the number of crossings, but also suggested distributing the crossings among circuit regions in order to simplify net routing.

BCM on a *single* edge is equivalent to the problem of sorting a permutation by block moves, which is well studied in computational biology for DNA sequences; it is known as *sorting by transpositions* [2,6]. The task is to find the shortest sequence of block moves transforming a given permutation into the identity permutation. The complexity of the problem was open for a long time; only recently it has been shown to be NP-hard [5]. The currently best known algorithm has an approximation ratio of $11/8$ [7]; no tight upper bound is known. There are several variants of sorting by transpositions; see the survey of Fertin et al. [8]. For instance, Vergara et al. [11] used *correcting short block moves* to sort a permutation. In our terminology, these are monotone moves such that the combined length of exchanged blocks does not exceed three. Hence, their problem is a restricted variant of MBCM on a single edge; its complexity is unknown.

2   Block Crossings on a Single Edge

First, we restrict our attention to networks consisting of a single edge with multiple lines passing through it. BCM then can be reformulated as follows. Given two permutations π and τ (determined by the order of terminals on both sides of the edge), find the shortest sequence of block moves transforming π into τ. By relabeling we can assume that τ is the identity permutation, and the goal is to sort π. This problem is known as *sorting by transpositions* [2]. We concentrate on the new problem of sorting with monotone block moves; that means that the relative order of any pair of elements changes at most once. The problems are not equivalent; see Fig.3 for an example where non-monotonicity allows fewer crossings. In what follows, we give lower and upper bounds on the number of block crossings for MBCM on a single edge. Additionally, we present a simple 3-approximation algorithm for the problem.

We first introduce some terminology following the one from previous works where possible. Let $\pi = [\pi_1, \ldots, \pi_n]$ be a permutation of n elements. For convenience, we assume there are extra elements $\pi_0 = 0$ and $\pi_{n+1} = n + 1$ at the beginning of the permutation and at the end, respectively. A *block* in π is a sequence of consecutive elements $\pi_i, \ldots, \pi_j$ with $i \leq j$. A *block move* (i, j, k) with $i \leq j < k$ on π maps $[\ldots \pi_{i-1} \pi_i \ldots \pi_j \pi_{j+1} \ldots \pi_k \pi_{k+1} \ldots]$ to $[\ldots \pi_{i-1} \pi_{j+1} \ldots \pi_k \pi_i \ldots \pi_j \pi_{k+1} \ldots]$. We say that a block move (i, j, k) is *monotone* if $\pi_q > \pi_r$ for all $i \leq q \leq j < r \leq k$. We denote the minimum number

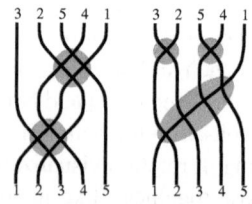

Fig. 3. Permutation [3 2 5 4 1] is sorted with 2 block moves and 3 monotone block moves

of monotone block moves needed to sort π by $\mathrm{bc}(\pi)$. An ordered pair (π_i, π_{i+1}) is a *good pair* if $\pi_{i+1} = \pi_i + 1$, and a *breakpoint* otherwise. Intuitively, sorting π is a process of creating good pairs (or destroying breakpoints) by block moves. A permutation is *simple* if it has no good pairs. Any permutation can be uniquely simplified—by gluing good pairs together and relabeling—without affecting its distance to the identity permutation [6]. A breakpoint (π_i, π_{i+1}) is a *descent* if $\pi_i > \pi_{i+1}$, and a *gap* otherwise. We use $\mathrm{bp}(\pi)$, $\mathrm{des}(\pi)$, and $\mathrm{gap}(\pi)$ to denote the number of breakpoints, descents, and gaps in π. The *inverse* of π is the permutation π^{-1} in which each element and the index of its position are exchanged, that is, $\pi^{-1}_{\pi_i} = i$ for $1 \leq i \leq n$. A descent in π^{-1}, that is, a pair of elements $\pi_i = \pi_j + 1$ with $i < j$, is called an *inverse descent* in π. Analogously, an *inverse gap* is a pair of elements $\pi_i = \pi_j + 1$ with $i > j + 1$. Now, we give lower and upper bounds for MBCM.

A lower bound. It is easy to see that a block move affects three pairs of adjacent elements. Therefore the number of breakpoints can be reduced by at most three in a move. As only the identity permutation has no breakpoints, this implies $\mathrm{bc}(\pi) \geq \mathrm{bp}(\pi)/3$ for a simple permutation [2]. The following observations yield better lower bounds.

Lemma 1. *In a monotone block move, the number of descents in a permutation decreases by at most one, and the number of gaps decreases by at most two.*

Proof. Consider a monotone move $[\ldots ab \ldots cd \ldots ef \ldots] \Rightarrow [\ldots ad \ldots eb \ldots cf \ldots]$; it affects three adjacencies. Suppose a descent is destroyed between a and b, that is, $a > b$ and $a < d$. Then, $b < d$ contradicting monotonicity. Similarly, no descent can be destroyed between e and f. Since $c > d$, no gap can be destroyed between c and d. $\qquad\square$

Lemma 2. *In a monotone block move, the number of inverse descents decreases by at most one, and the number of inverse gaps decreases by at most two.*

Proof. Consider a monotone exchange of blocks $\pi_i, \ldots, \pi_j$ and $\pi_{j+1}, \ldots, \pi_k$. Note that inverse descents can only be destroyed between elements π_q $(i \leq q \leq j)$ and π_r $(j + 1 \leq r \leq k)$. Suppose that the move destroys two inverse descents such that the first block contains elements $x + 1$ and $y + 1$, and the second block contains x and y. Since the block move is monotone, $y + 1 > x$ and $x + 1 > y$, which means that $x = y$.

On the other hand, there cannot be inverse gaps between elements π_q $(i \leq q \leq j)$ and $\pi_r (j + 1 \leq r \leq k)$. Therefore, there are only two possible inverse gaps between π_{i-1} and $\pi_r (j < r \leq k)$, and between $\pi_q (i \leq q \leq j)$ and π_{k+1}. $\qquad\square$

Theorem 1. *A lower bound on the number of monotone block moves needed to sort a permutation is $\mathrm{bc}(\pi) \geq \max(\mathrm{des}(\pi), \mathrm{gap}(\pi)/2, \mathrm{des}(\pi^{-1}), \mathrm{gap}(\pi^{-1})/2)$.*

An upper bound. We suggest a simple algorithm for sorting a simple permutation π: In each step find the smallest i such that $\pi_i \neq i$ and move element i to position i, that is, exchange block $\pi_i, \ldots, \pi_{k-1}$ and π_k, where $\pi_k = i$. Clearly, the step destroys at least one breakpoint. Therefore $\mathrm{bc}(\pi) \leq \mathrm{bp}(\pi)$ and the algorithm yields a 3-approximation.

Theorem 2. *There exists an $O(n^2)$-time 3-approximation algorithm for MBCM on a single edge.*

3   Block Crossings on a Path

Now we consider an embedded graph $G = (V, E)$ consisting of a path $P = (V_P, E_P)$ with attached terminals. In every node $v \in V_P$ the clockwise order of terminals adjacent to v is given, and we assume the path is oriented from left to right. We say that a line l starts at v if v is the leftmost vertex on P that lies on l and ends at its rightmost vertex of the path. As we consider only crossings of lines sharing an edge, we assume that the terminals connected to any path node v are in such an order that first lines end at v and then lines start at v; see Fig. 5.

We suggest a 3-approximation algorithm for BCM. Similar to the single edge case, the basic idea of the algorithms is to consider good pairs of lines. A *good pair* is, in-tuitively, an ordered pair of lines that will be adjacent—in this order—in any feasible solution when one of the lines ends. We argue that our algorithm creates at least one additional good pair per block crossing, while even the optimum creates at most three new good pairs per crossing. To describe our algorithm we first define good pairs.

Definition 1 (Good pair).
  (i) *If two lines a and b end on the same node, and a and b are consecutive in clockwise order, then (a, b) is a good pair (as it is in the case of a single edge in Sec. 2).*
  (ii) *Let v be a node with edges (u, v) and (v, w) on P, let a_1 be the first line starting on v above P, and let a_2 be the last line ending on v above P as in Fig. 4. If (a_1, b) is a good pair, then (a_2, b) also is a good pair. We say that (a_2, b) is inherited from (a_1, b), and identify (a_1, b) with (a_2, b), which is possible as a_1 and a_2 do not share an edge. Analogously, there is inheritance for lines starting/ending below P.*

As a preprocessing step, we add a virtual line t_e (b_e) for each edge $e \in E_P$. The line t_e (b_e) is the last line starting before e, and the first line ending after e to the top (bottom). Although virtual lines are never moved, t_e (b_e) does partic-ipate in good pairs, which model the fact that the first lines ending after an edge should be brought to the top (bottom).

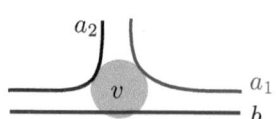

Fig. 4. Inheritance of a good pair above node v

There are important properties of good pairs (see full ver-sion for proofs [9]). On an edge $e \in E_P$ there is, for each line l, at most one good pair (l', l) and at most one good pair (l, l''). If $e \in E_P$ is the last edge before line l ends to the top (bottom), then there exists a good pair (l', l) $((l, l''))$ on e.

In what follows, we say that a solution (or algorithm) creates a good pair in a block crossing if the two lines of the good pair are brought together in the right order by that block crossing; analogously, we speak of breaking good pairs. It is easy to see that any solution, especially an optimal one, has to create all good pairs, and a block crossing can create at most three new pairs. There are only two possible ways for creating a good pair (a, b): (i) a and b start at the same node consecutively in the right order, that is, they form an *initial good pair*, or (ii) a block crossing brings a and b together. Similarly, good pairs can only be destroyed by crossings before both lines end.

Using good pairs, we formulate our algorithm as follows; see Fig. 5 for an example.

We follow P from left to right. On an edge $e = (u, v)$ there are red lines that end at v to the top, green lines that end at v to the bottom, and black lines that continue on the

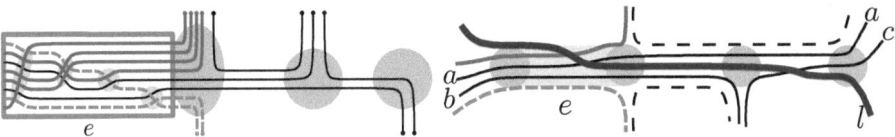

Fig. 5. Ordering the lines on edge e in a step of the algorithm

Fig. 6. The (necessary) insertion of line l forces breaking the good pair (a, b) ($\equiv (a, c)$) on edge e

next edge. We bring the red lines in the right order to the top by moving them upwards. Doing so, we keep existing good pairs together. If a line is to be moved, we consider the lines below it consecutively. As long as the current line forms a good pair with the next line, we extend the block that will be moved. We stop at the first line that does not form a good pair with its successor. Finally we move the whole block of lines linked by good pairs in one block move to the top. Next, we bring the green lines in the right order to the bottom, again keeping existing good pairs together. There is an exception, where one good pair on e cannot be kept together. If the moved block is a sequence of lines containing both red and green lines, and possibly some—but not all—black lines, then it has to be broken; see Fig. 6. Note that this can only happen in the last move on an edge. There are two cases:

(i) A good pair in the sequence contains a black line and has been created by the algorithm previously. We break the sequence at this good pair.

(ii) All pairs with a black line are initial good pairs, that is, were not created by a crossing. We break at the pair that ends last of these. Inheritance is also considered, that is, a good pair ends only when the last of the pairs that are linked by inheritance ends.

After an edge has been processed, the lines ending to the top and to the bottom are on their respective side in the right relative order. Hence, our algorithm produces a feasible solution. We show that it produces a 3-approximation for the number of block crossings. A key property is that our strategy for case (ii) is optimal (see full version for the proof [9]).

Theorem 3. *Let* $\mathrm{bc_{alg}}$ *and* OPT *be the number of block crossings created by the algorithm and an optimal solution, respectively. Then,* $\mathrm{bc_{alg}} \leq 3\,\mathrm{OPT}$.

The algorithm needs $O(|L|(|L| + |E_P|))$ time. Note that it does normally not produce orderings with monotone block crossings. It can be turned into a 3-approximation algorithm for MBCM. To this end, the definition of inheritance of good pairs, as well as the step of destroying good pairs is adjusted (see full version [9]).

Theorem 4. *There exist* $O(|L|(|L|+|E_P|))$*-time 3-approximation algorithms for BCM and MBCM on a path.*

4 Block Crossings on Trees

In what follows we focus on instances of (M)BCM that are trees. We first give an algorithm that bounds the number of block crossings. Then, we consider trees with an additional constraint on the lines; for these we develop a 6-approximation for MBCM.

Theorem 5. *For any tree T and lines L on T, we can order the lines with at most* $2|L| - 3$ *monotone block crossings in* $O(|L|(|L| + |E|))$ *time.*

Proof. We give an algorithm in which paths are inserted one by one into the current order; for each newly inserted path we create at most 2 monotone block crossings. The first line cannot create a crossing, and the second line crosses the first one at most once.

We start at an edge (u, v) incident to a terminal. When processing the edge the paths L_{uv} are already in the correct order;
they do not need to cross on yet unprocessed edges of T. We
consider all unprocessed edges $(v, a), (v, b), \dots$ incident to
v and build the correct order for them. The relative order of
lines also passing through (u, v) is kept unchanged. For all
lines passing through v that were not treated before, we apply
an insertion procedure; see Fig. 7. Consider, e.g., the insertion of a line passing through (v, a) and (v, b). Close to v we
add l on both edges at the innermost position such that we do
not get vertex crossings with lines that pass through (v, a) or
(v, b). We find its correct position in the current order of lines
L_{va} close to a, and insert it using one block crossing. This

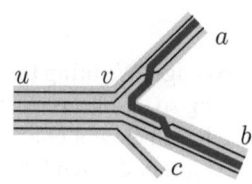

Fig. 7. Insertion of a new line (red, fat) into the current order on edges (v, a) and (v, b)

crossing will be the last one on (v, a) going from v to a. Similarly, l is inserted into
L_{vb}. We have to make sure that lines that do not have to cross are inserted in the right
order. As we know the right relative order for a pair of such lines we can make sure that
the one that has to be innermost at node v is inserted first. Similarly, by looking at the
clockwise order of edges around v, we know the right order of line insertions such that
there are no avoidable vertex crossings. When all new paths are inserted the orders on
$(v, a), (v, b), \dots$ are correct; we proceed by recursively processing these edges.

When inserting a line, we create at most 2 block crossings, one per edge of l incident
to v. After inserting the first two lines into the drawing there is at most one crossing.
Hence, we get at most $2|L| - 3$ block crossings in total. Suppose monotonicity would be
violated, that is, there is a pair of lines that crosses twice. The crossings then have been
introduced when inserting the second of those lines on two edges incident to a node v.
This can, however, not happen, as at node v the two edges are inserted in the right order.
Hence, the block crossings of the solution are monotone. □

While the upper bound that our algorithm yields is tight (see full version [9]), it does
not guarantee an approximation. Next, we introduce an additional constraint on the
lines, which helps us to approximate the minimum number of block crossings.

Upward Trees. We consider MBCM on an *upward* tree T, that is, a tree that has a
planar upward drawing in which all paths are monotone in vertical direction, and all
path sources are on the same height as well as all path sinks; see Fig. 8. Note that a
graph whose skeleton is a path is not necessarily an upward tree. Our algorithm consists
of three steps. First, we perform a simplification step removing some lines. Second, we
use the algorithm for trees given in Sec. 4 on a simplified instance. Finally, we reinsert
the removed lines into the constructed order. We first analyze the upward embedding.

Given an upward drawing of T, we read a permutation π produced by the terminals
on the top; we assume that the terminals produce the identity permutation on the bottom.

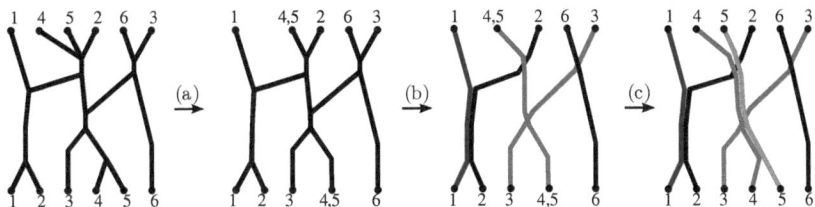

Fig. 8. Algorithm for upward trees: (a) simplification, (b) line ordering, (c) reinsertion

Similar to the single edge case the goal is to sort π by a shortest sequence of block moves. Edges of T restrict some block moves on π; e.g., the blocks 1 4 and 5 in Fig. 8 cannot be exchanged as there is no suitable edge. However, we can use the lower bound for block crossings on a single edge, see Sec. 2: For sorting a simple permutation π, at least $\mathrm{bp}(\pi)/3$ block moves are necessary. We stress that simplicity of π is crucial here. To get an approximation, we show how to simplify a tree.

Consider two non-intersecting paths a and b that are adjacent in both permutations and share a common edge. We prove that one of these paths can be removed without changing the optimal number of block crossings. First, if any other line c crosses a then it also crosses b (i). This is implied by planarity and y-monotonicity of the drawing. Second, if c crosses both a and b then all three paths share a common edge (ii); otherwise, there would be a cycle due to planarity. Hence, for any solution for the paths $L - \{b\}$, we can construct a solution for L by inserting b without any new block crossing. To insert b, we must first move all block crossings on a to the common subpath with b. This is possible due to observation (ii). Finally, we can place b parallel to a.

To get a 6-approximation for an upward tree T, we first remove lines until the tree is simple. Then we apply the insertion algorithm presented in Sec. 4, and finally reinsert the lines removed in the first step. The number of block crossings is at most $2|L'|$, where L' is the set of lines of the simplified instance. As an optimal solution has at least $|L'|/3$ block crossings for this simple instance, and reinserting lines does not create new block crossings, we get the following theorem.

Theorem 6. *The algorithm yields a 6-approximation for MBCM on upward trees.*

5   Block Crossings on General Graphs

Finally, we consider general graphs. We suggest an algorithm that achieves an upper bound on the number of block crossings and show that it is asymptotically worst-case optimal. Our algorithm uses monotone block moves, that is, each pair of lines crosses at most once. The algorithm works on any embedded graph; it does not even need to be planar, we just need to know the circular order of incident edges around each vertex.

The idea of our algorithm is simple. We go through the edges in some arbitrary order, similar to previous work on standard metro-line crossing minimization [1,13]. When we treat an edge, we completely sort the lines that traverse it. A crossing between a pair of lines can be created on the edge only if this edge is the first one treated by the algorithm

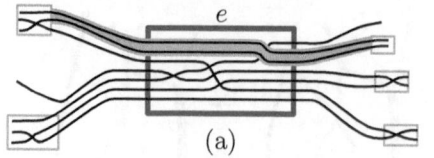 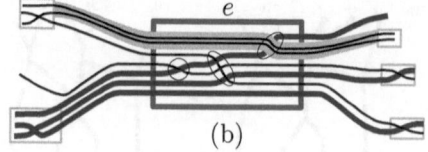

Fig. 9. Sorting the lines on edge e. (a) Cutting edges (marked) define groups. The lines marked in gray are merged as they are in the same group on both sides. (b) Sorting by insertion into the largest group (red, fat); the merged lines always stay together, especially in their block crossing.

that is used by both lines of the pair; see Algorithm 1. The crucial part is sorting the lines on an edge. Suppose we currently deal with edge e and want to sort L_e. Due to the path intersection property, the edge set used by the lines in L_e forms a tree on each side of e; see Fig. 9. We cut these trees at those edges that have already been processed by our algorithm. Now, each line on e starts at a leaf on one side and ends at a leaf on the other side. Note that multiple lines can start or end at the same leaf.

From the tree structure and the orderings on the edges processed previously, we get two orders of the lines, one on each side of e. We consider *groups of lines* that start or end at a common leaf of the tree (like the red lines in Fig. 9). All lines of a group have been seen on a common edge, and, hence, have been sorted. Therefore lines of the same group form a consecutive subsequence on one side of e, and have the same relative order on the other side of e.

Let g and g' be a group of lines on the left and on the right side of e, respectively. Suppose the set L' of lines starting in g and ending in g' consists of multiple lines. As the lines of g as well as the lines of g' stay parallel on e, L' must form a consecutive subsequence (in the same order) on both sides of e. Now we *merge* L' into one representative, that is, we remove all lines of L' and replace them by a single line that is in the position of the lines of L' on the sequences on both sides of e. Once we find a solution, we replace the representative by the sequence without changing the number of block crossings. Consider a crossing that involves the representative of L', that is, it is part of one of the moved blocks. After replacing it, the sequence L' of parallel lines is completely contained in the same block. Hence, we do not need additional block crossings.

We apply this merging for all pairs of groups on the left and right end of E. Then, we identify a group with the largest number of lines after merging, and insert all remaining lines into it one by one. Clearly, each insertion requires at most one block crossing; in Fig. 9 we need three block crossings to insert the lines into the largest (red) group. After computing the crossings, we undo the merging step and get a solution for edge e.

---

foreach edge e with $|L_e| > 1$ **do**
    Build order of lines on both sides of e
    Merge lines that are in the same group on both sides
    Find the largest group of consecutive lines that stay parallel on e
    Insert all other lines into this group and undo merging

Algorithm 1. Ordering the lines on a graph

Theorem 7. *Algorithm 1 sorts all lines in $O(|E|^2|L|)$ time by monotone block moves. The resulting number of block crossings is $O(|L|\sqrt{|E'|})$, where E' is the set of edges with at least two lines on them.*

Proof. First, it is easy to see that no avoidable crossings are created, due to the path intersection property. Additionally, we care about all edges with at least two lines, which ensures that all unavoidable crossings will be placed. Hence, we get a feasible solution using monotone crossings. Our algorithm sorts the lines on an edge in $O(|L||E|)$ time. We can build the tree structure and find the orders and groups by following all lines until we find a terminal or an edge that was processed before in $O(|L||E|)$ time. Merging lines and finding the largest group need $O(|L|)$ time; sorting by insertion into this group and undoing the merging can be done in $O(|L|^2)$ time. Note that $|L| \leq |E|$ due to the path terminal property.

For analyzing the total number of block crossings, we maintain an *information table* T with $|L|^2$ entries. Initially, all the entries are empty. After processing an edge e in our algorithm, we fill entries $T[l, l']$ of the table for each pair (l, l') of lines that we see together for the first time. The main idea is that with b_e block crossings on edge e we fill at least b_e^2 new entries of T. The upper bound then can be concluded.

More precisely, let the *information gain* $I(e)$ be the number of pairs of (not necessarily distinct) lines l, l' that we see together on a common edge e for the first time. Clearly, $\sum_{e \in E} I(e) \leq |L|^2$. Suppose that $b_e^2 \leq I(e)$ for each edge e. Then, $\sum_{e \in E} b_e^2 \leq \sum_{e \in E} I(e) \leq |L|^2$. Using the Cauchy-Schwarz inequality $|\langle x, y \rangle| \leq \sqrt{\langle x, x \rangle \cdot \langle y, y \rangle}$ with x as the vector of the b_e and y as a vector of 1-entries, we see that the total number of block crossings is $\sum_{e \in E'} b_e \leq |L|\sqrt{|E'|}$.

Let us show that $b_e^2 \leq I(e)$ for an edge e. We analyze the lines after the merging step. Consider the groups on both sides of e; we number the groups on the left side $\mathfrak{L}_1, \ldots, \mathfrak{L}_n$ and the groups on the right side $\mathfrak{R}_1, \ldots, \mathfrak{R}_m$ with $l_i = |\mathfrak{L}_i|, r_j = |\mathfrak{R}_j|$ for $1 \leq i \leq n, 1 \leq j \leq m$. Without loss of generality, we can assume that $\mathfrak{L}_1$ is the largest group into which all remaining lines are inserted. Then, $b_e \leq |L_e| - l_1$. Let s_{ij} be the number of lines that are in group $\mathfrak{L}_i$ on the left side and in group $\mathfrak{R}_j$ on the right side of e. Note that $s_{ij} \in \{0, 1\}$, otherwise we could still merge lines. Then $l_i = \sum_j s_{ij}, r_j = \sum_i s_{ij}, s := |L_e| = \sum_{ij} s_{ij}$, and $b_e = s - l_1$. The information gain is $I(e) = s^2 - \sum_i l_i^2 - \sum_j r_j^2 + \sum_{ij} s_{ij}^2$. Using $s_{ij} \in \{0, 1\}$, it is easy to see that $b_e^2 \leq I(e)$. To complete the proof, note that the unmerging step cannot decrease $I(e)$. $\square$

In the full version [9] we show that the upper bound that our algorithm achieves is tight by using the existence of special Steiner systems for building (non-planar) worst-case examples of arbitrary size in which many block crossings are necessary.

6   Conclusion and Open Problems

We introduced a new variant of the metro-line crossing minimization problem, and presented algorithms for single edges, paths, trees, and general graphs. The following is a list of some interesting open problems and future work.

1. What is the complexity status of MBCM on a single edge?
2. Can we derive an approximation algorithm for (M)BCM on trees and general graphs?
3. For making a metro line easy to follow the important criterion is the number of its bends. Hence, an interesting question is how to sort metro lines using the minimum total number of bends.

Acknowledgments. We are grateful to Sergey Bereg, Alexander E. Holroyd, and Lev Nachmanson for the initial discussion of the block crossing minimization problem, and for pointing out a connection with sorting by transpositions. We thank Jan-Henrik Haunert, Joachim Spoerhase, and Alexander Wolff for fruitful suggestions.

References

1. Argyriou, E.N., Bekos, M.A., Kaufmann, M., Symvonis, A.: On metro-line crossing minimization. J. Graph Algorithms Appl. 14(1), 75–96 (2010)
2. Bafna, V., Pevzner, P.A.: Sorting by transpositions. SIAM J. Discrete Math. 11(2), 224–240 (1998)
3. Bekos, M.A., Kaufmann, M., Potika, K., Symvonis, A.: Line crossing minimization on metro maps. In: Hong, S.-H., Nishizeki, T., Quan, W. (eds.) GD 2007. LNCS, vol. 4875, pp. 231–242. Springer, Heidelberg (2008)
4. Benkert, M., Nöllenburg, M., Uno, T., Wolff, A.: Minimizing intra-edge crossings in wiring diagrams and public transportation maps. In: Kaufmann, M., Wagner, D. (eds.) GD 2006. LNCS, vol. 4372, pp. 270–281. Springer, Heidelberg (2007)
5. Bulteau, L., Fertin, G., Rusu, I.: Sorting by transpositions is difficult. SIAM J. Discr. Math. 26(3), 1148–1180 (2012)
6. Christie, D.A., Irving, R.W.: Sorting strings by reversals and by transpositions. SIAM J. Discr. Math. 14(2), 193–206 (2001)
7. Elias, I., Hartman, T.: A 1.375-approximation algorithm for sorting by transpositions. IEEE/ACM Trans. Comput. Biol. Bioinformatics 3(4), 369–379 (2006)
8. Fertin, G., Labarre, A., Rusu, I., Tannier, E., Vialette, S.: Combinatorics of Genome Rearrangements. The MIT Press (2009)
9. Fink, M., Pupyrev, S.: Ordering metro lines by block crossings. ArXiv report (2013), http://arxiv.org/abs/1305.0069
10. Groeneveld, P.: Wire ordering for detailed routing. IEEE Des. Test 6(6), 6–17 (1989)
11. Heath, L.S., Vergara, J.P.C.: Sorting by bounded block-moves. Discrete Applied Mathematics 88(1-3), 181–206 (1998)
12. Marek-Sadowska, M., Sarrafzadeh, M.: The crossing distribution problem. IEEE Trans. CAD Integrated Circuits Syst. 14(4), 423–433 (1995)
13. Nöllenburg, M.: An improved algorithm for the metro-line crossing minimization problem. In: Eppstein, D., Gansner, E.R. (eds.) GD 2009. LNCS, vol. 5849, pp. 381–392. Springer, Heidelberg (2010)
14. Pupyrev, S., Nachmanson, L., Bereg, S., Holroyd, A.E.: Edge routing with ordered bundles. In: van Kreveld, M., Speckmann, B. (eds.) GD 2011. LNCS, vol. 7034, pp. 136–147. Springer, Heidelberg (2012)

Reachability in Register Machines
with Polynomial Updates

Alain Finkel[1,*], Stefan Göller[2], and Christoph Haase[1,*]

[1] LSV - CNRS & ENS Cachan, France
{finkel,haase}@lsv.ens-cachan.fr
[2] Institut für Informatik, Universität Bremen, Germany
goeller@informatik.uni-bremen.de

Abstract. This paper introduces a class of register machines whose registers can be updated by polynomial functions when a transition is taken, and the domain of the registers can be constrained by linear constraints. This model strictly generalises a variety of known formalisms such as various classes of Vector Addition Systems with States. Our main result is that reachability in our class is PSPACE-complete when restricted to one register. We moreover give a classification of the complexity of reachability according to the type of polynomials allowed and the geometry induced by the range-constraining formula.

1 Introduction

Register machines are a class of abstract machines comprising a finite-state controller with a finite number of integer-valued registers that can be manipulated or tested when a transition is taken. A prominent instance are *counter machines* due to Minsky [18], which are obtained by restricting registers to range over the naturals, allowing for addition of integers to the registers along transitions, and testing registers for zero. A seminal result by Minsky states that counter machines are Turing powerful in the presence of at least two registers. Decidability can be obtained by further restricting counter machines and disallowing zero tests, which yields a class of register machines known as *Vector Addition Systems with States (VASS)* or *Petri nets*. Their reachability problem is known to be decidable and EXPSPACE-hard [17,16].

A number of extensions, generalisations and restrictions of VASS can be found in the literature. For instance, various extensions that increase the power of transitions have been studied, including Reset/Transfer (Petri) nets [6], Petri nets with inhibitory arcs [3], or Affine nets [8] which extend VASS such that transitions can be any non-decreasing affine function; any of these extensions lead to undecidability of reachability in the presence of more than one register. On the other hand, relaxing the domain of the registers of a VASS to the integers, or restricting VASS to just one register renders reachability NP-complete [12].

---

* The authors are supported by the French Agence Nationale de la Recherche, REACHARD (grant ANR-11-BS02-001).

K. Chatterjee and J. Sgall (Eds.): MFCS 2013, LNCS 8087, pp. 409–420, 2013.

In summary, we can identify three parameters in which the aforementioned classes of register machines differ and which impact their expressiveness and the complexity of reachability: (1) the number of registers available, (2) the shape of the domain of the registers, and (3) the class of the transition functions used. In this paper, we generalise (3) and study the decidability and complexity of reachability when allowing for *polynomial functions with integer coefficients* to update register values. To this end, we introduce *polynomial register machines (PRMs)*, a class of register machines in which the previously mentioned classes of register machines embed smoothly. Of course, their undecidability results carry over, but on the positive side we are able to identify a decidable class of PRMs that is not contained in any of them.

The main result of this paper is to show that reachability in PRMs is PSPACE-complete when restricted to one register. As a motivating example, consider the question whether the following loop involving a single register variable x terminates:

```
int x := 0
while (x < 5):
    x := x**3 - 2x**2 - x + 2
```

This example is inspired by an example given in [1], and in this example x alternates between 0 and 2, and thus the loop never terminates. In fact, it is not difficult to see that the loop never terminates for all values $x < 3$. However for polynomials of higher degree and loops with a richer control structure, deciding termination becomes non-obvious. Even in dimension one, problems of this nature can become intriguingly difficult, see *e.g.* [2] for a discussion on open problems of this kind. Reachability for non-deterministically applied affine transformations from a finite set in dimension one has been shown to be decidable in 2-EXPTIME by Fremont [9].

There are a number of obstacles making it challenging to show decidability and complexity results for reachability in PRMs. In some classes of register machines, semi-linearity of the reachability set can be exploited in order to show decidability. However, taking a single-state PRM with one self-loop that updates the only register x with the polynomial $p(x) = x^2$, we see that the reachability set is not semi-linear. Moreover, the representation of the values that the register x can take grows exponentially with the number of times the self-loop is taken, which makes it not obvious how to decide reachability in polynomial space only.

The property that the reachability set is not semi-linear separates languages generated by PRMs from classes of machines that have semi-linear reachability sets, such as VASS in dimension one. More interestingly, PRMs can generate languages that cannot be generated by general VASS, which do also not have semi-linear reachability sets: the language $L = \{a^{n^2} : n > 0\}$ over the singleton alphabet $\{a\}$ can easily be generated by a PRM with two control locations, but not by any VASS [15].

Besides the aforementioned related work, as indicated by the example above, work related to ours can be found in the area of program verification. In [1], Babić *et al.* describe a semi decision procedure for proving termination of loops involving polynomial updates, similar to the one above. Another example is the work by Bradley *et al.* [4] which provides a semi decision procedure for so-called multipath polynomial programs. However, to the best of our knowledge, no sound and complete algorithm for problems of this kind exists.

Due to space constraints, we had to omit some proof details. An extended version of this paper containing the omitted proofs in an appendix can be obtained from the authors.

2   Preliminaries

Before we formally introduce PRMs, we provide some technical definitions and known results on elementary algebra and number theory.

2.1   Technical Definitions and Known Results

By $\mathbb{N}, \mathbb{Z}, \mathbb{R}$ and $\mathbb{C}$ we denote the naturals, integers, reals and complex numbers, respectively. All integers in this paper are assumed to be encoded in binary unless stated otherwise. For $z \in \mathbb{Z}$, we denote by $\operatorname{sgn} z$ the *sign of z*, and by $|z|$ its absolute value. For $r_1 \leq r_2 \in \mathbb{R}$, we denote by $[r_1, r_2]$ the *closed interval* $\{r \in \mathbb{R} : r_1 \leq r \leq r_2\}$.

By $\mathbb{Z}[\overline{x}]$ we denote the ring of polynomials with integer coefficients over variables $\overline{x} = (x_1, \ldots, x_n)$. A polynomial $p(x) \in \mathbb{Z}[x]$ will be written as $p(x) = a_n x^n + \cdots + a_1 x + a_0$, and represented in sparse encoding by a sequence of pairs $(i, a_i)_{i \in I}$, where $I \subseteq \{0, \ldots, n\}$ contains those indexes for which $a_i \neq 0$. Given $z \in \mathbb{Z}$ and $p(x)$ in our representation, deciding $p(z) > 0$ is known to be computable in polynomial time [5]. Given a root $c \in \mathbb{C}$ of $p(x)$, we will make use of the following bound from [14] on the magnitude of c:

$$|c| \leq 1 + \sum_{0 \leq i < n} |a_i/a_n|. \tag{1}$$

Recall that for all $m > 0$, $p(a) \equiv p(b) \bmod m$ whenever $a \equiv b \bmod m$ for all $m > 0$, *i.e.* all $p(x) \in \mathbb{Z}[x]$ are invariant w.r.t. residual classes. Given pairwise co-prime $m_1, \ldots, m_k > 0$ and $b_1, \ldots, b_k \in \mathbb{Z}$, the *Chinese remainder theorem* states that a system of k linear congruences $x \equiv b_i \bmod m_i$, $1 \leq i \leq k$ has a unique solution modulo $m_1 m_2 \cdots m_k$. Moreover, recall that the prime number theorem states that the number $\pi(n)$ of primes below n grows as $\pi(n) \sim n/\ln n$. In particular, this implies that $O(\log n)$ bits are sufficient to represent the n-th prime number.

A *linear constraint* $\phi(\overline{x})$ is a conjunction of atoms of the form $\top$ and $p(\overline{x}) \sim z$, where $p \in \mathbb{Z}[\overline{x}]$ is linear, $z \in \mathbb{Z}$ and $\sim \in \{<, \leq, =, \geq, >\}$. The set of *solutions* of $\phi(\overline{x})$ is $\{\overline{z} \in \mathbb{Z}^d : \phi[\overline{z}/\overline{x}]$ is true$\}$ that we also denote by $[\![\phi(\overline{x})]\!]$. We say that $[\![\phi(\overline{x})]\!]$ is *upward closed* if whenever $\overline{z} \in [\![\phi(\overline{x})]\!]$ then $\overline{z'} \in [\![\phi(\overline{x})]\!]$ for all $\overline{z'}$ such that $\overline{z} \preceq \overline{z'}$. Here, $\preceq$ denotes the natural component-wise extension of the order $\leq$ on $\mathbb{Z}$ to tuples over $\mathbb{Z}$.

2.2   Polynomial Register Machines

This section introduces polynomial register machines. We only give full definitions for dimension one, since the major part of this paper in Section 3 focuses on this class. From the definitions below, it is easy to generalise to higher dimensions, which we are only going to discuss briefly in Section 4.

A *polynomial register machine (PRM)* is a tuple $\mathcal{R} = (Q, \Delta, \lambda, \phi)$, where Q is a finite set of *states* or *control locations*, $\Delta \subseteq Q \times Q$ is the *transition relation*, $\lambda : \Delta \to \mathbb{Z}[x]$ is the *transition labelling function*, labelling each transition with an *update polynomial*, and $\phi(x)$ is a *global invariant*, which is a linear constraint. As a convention, we assume $0 \in [\![\phi(x)]\!]$, though all results in this paper hold without this assumption. We write $q \xrightarrow{p(x)} q'$ whenever $(q, q') \in \Delta$ and $\lambda(q, q') = p(x)$. The set $C(\mathcal{R})$ of *configurations of* $\mathcal{R}$ is $C(\mathcal{R}) \stackrel{\text{def}}{=} Q \times [\![\phi(x)]\!] \subseteq Q \times \mathbb{Z}$, and we write configurations in $C(\mathcal{R})$ as $q(z)$. The size $|\mathcal{R}|$ of $\mathcal{R}$ is the number of bits required to write down $\mathcal{R}$ and $P(\mathcal{R})$ denotes *the set of polynomials that occur in* $\mathcal{R}$.

The semantics of $\mathcal{R}$ is given by a transition system $T(\mathcal{R}) = (C(\mathcal{R}), \to_{\mathcal{R}})$, where $q(z) \to_{\mathcal{R}} q'(z')$ if $q \xrightarrow{p(x)} q'$ and $z' = p(z)$. *Reachability* is to decide, given $q, q' \in Q$ and $z, z' \in \mathbb{Z}$, does $q(z) \to_{\mathcal{R}}^* q'(z')$ hold? Clearly, this problem can be reduced in logarithmic space to deciding $q(0) \to_{\mathcal{R}'}^* q'(0)$ for some PRM $\mathcal{R}'$ linear in the size of $\mathcal{R}$, z and z'.

For dimensions $d > 1$, a d-PRM is obtained by amending the above definitions such that $\phi(\overline{x})$ is free in $\overline{x} = (x_1, \ldots, x_d)$ and transitions are labelled with vectors of polynomials $(p_1(x_1), \ldots, p_d(x_d))$ that are applied componentwise. The next example shows how some of the classes of register machines mentioned in the introduction can be embedded into PRMs.

Example 1. A dimension d-VASS is a d-PRM with global invariant $\phi(\overline{x}) = \bigwedge_{1 \leq i \leq d} x_i \geq 0$ and transition polynomials of the form $p_i(x_i) = x_i + a_i$; a bounded d-counter automaton [13] with bounds $\overline{b} = (b_1, \ldots, b_d) \in \mathbb{N}^d$ is a d-PRM with the same transition polynomials and $\phi(\overline{x}) = \bigwedge_{1 \leq i \leq d}(x_i \geq 0 \wedge x_i \leq b_i)$. A reset d-VASS [6] can be simulated by employing polynomials of the form $p_i(x_i) = 0$ for resets.

The previous examples lead us to a classification of update polynomials. We call a polynomial of the form $p(x) = a_1 x + a_0$ a *counter polynomial* if $a_1 = 1$, *counter-like polynomial* if $a_1 \in \{-1, 1\}$, and if the degree of $p(x)$ is one then $p(x)$ is called an *affine polynomial*.

3   Reachability for One Register

This section proves the main theorem of this paper and shows that reachability in PRMs is decidable and PSPACE-complete. For the lower bound, we show that reachability becomes PSPACE-hard for update polynomials of degree two, even if the global invariant is unconstrained and thus upward closed. Subsequently, we show a matching upper bound which involves a thorough analysis of paths in the transition systems generated by PRMs.

3.1 Hardness for PSPACE

We reduce from the reachability problem for *linear-bounded automata (LBA)*, which is a well-known PSPACE-complete problem. An LBA (without input alphabet) is a tuple $\mathcal{M} = (Q_\mathcal{M}, \Gamma, \Delta_\mathcal{M})$, where $Q_\mathcal{M}$ is a finite set of *states* and Γ is a finite *tape alphabet* implicitly containing two distinguished symbols $\triangleright$ and $\triangleleft$ acting as left delimiter ($\triangleright$) and right delimiter ($\triangleleft$). The *transition relation* is a relation $\Delta_\mathcal{M} \subseteq Q_\mathcal{M} \times \Gamma \times Q_\mathcal{M} \times \Gamma \times \{\leftarrow, \rightarrow\}$ such that $(q, \gamma, q', \gamma', d) \in \Delta_\mathcal{M}$ implies that whenever $\mathcal{M}$ is in state q reading γ at the current head position on the tape then $\mathcal{M}$ switches to the state q' writing γ' onto the tape and moving the head in direction $d \in \{\leftarrow, \rightarrow\}$. We assume $\Delta_\mathcal{M}$ to be constrained such that it respects the delimiters, *i.e.*, it fulfils the conditions

(i) $(q, \triangleright, q', \gamma, d) \in \Delta_\mathcal{M}$ implies $\gamma = \triangleright$ and $d = \rightarrow$; and
(ii) $(q, \triangleleft, q', \gamma, d) \in \Delta_\mathcal{M}$ implies $\gamma = \triangleleft$ and $d = \leftarrow$.

A *configuration of* $\mathcal{M}$ is a tuple $(q, \triangleright w \triangleleft, i)$, where $q \in Q$ is the current state, $w \in (\Gamma \setminus \{\triangleright, \triangleleft\})^*$ is the *tape content* and $i \in \{0, |w| + 1\}$ is the position of the read-write head. Hence, at head position 0 the tape content is $\triangleright$ and at position $|w| + 1$ it is $\triangleleft$. The *successor relation* $\rightarrow_\mathcal{M}$ between two configurations is defined in the standard way.

Deciding whether $(q_0, \triangleright 0^n \triangleleft, 0) \rightarrow_\mathcal{M}^* (q_f, \triangleright 0^n \triangleleft, 0)$ for given $n \in \mathbb{N}$ (in unary) and given states $q_0, q_f \in Q_\mathcal{M}$ of a given LBA $\mathcal{M}$ working on the alphabet $\{\triangleright, 0, 1, \triangleleft\}$ is well-known to be PSPACE-complete. For our reduction, let us fix such an LBA $\mathcal{M}$ and $n \in \mathbb{N}$. The goal of the remainder of this section is to show how we can compute in polynomial time from $\mathcal{M}$ and n a PRM $\mathcal{R} = (Q, \Delta, \lambda, \top)$ with particular control locations $q_\mathcal{R}, q'_\mathcal{R}$ such that $(q_0, \triangleright 0^n \triangleleft, 0) \rightarrow_\mathcal{M}^* (q_f, \triangleright 0^n \triangleleft, 0)$ if, and only if, $q_\mathcal{R}(0) \rightarrow_\mathcal{R}^* q'_\mathcal{R}(0)$, which gives PSPACE-hardness of reachability in PRMs.

To begin with, let us discuss an encoding of configurations of $\mathcal{M}$. In the following, let p_i denote the $(i+3)$-th prime number, *i.e.*, $p_1 = 7$, $p_2 = 11$, $p_3 = 13$, etc. Recall that by the prime number theorem p_i can be represented using $O(\log i)$ bits. Set $P \overset{\text{def}}{=} \prod_{1 \leq i \leq n} p_i$, we call a residue class r modulo P *valid* if for each $1 \leq i \leq n$ there is some $b_i \in \{0, 1\}$ such that $r \equiv b_i \bmod p_i$. Otherwise, r is called *invalid*. Our idea is to encode a tape configuration $\triangleright w \triangleleft$ of $\mathcal{M}$ with $w = w_1 \cdots w_n \in \{0, 1\}^n$ via the unique valid residue class r modulo P satisfying $r \equiv w_i \bmod p_i$ for all $1 \leq i \leq n$. Consequently, we can establish a one-to-one correspondence between valid residue classes modulo P and tape contents of $\mathcal{M}$. Thus, modulo each prime p_i, we naturally view the residue classes 0 and 1 to encode the Boolean values 0 and 1, respectively. During the simulation of $\mathcal{M}$ by $\mathcal{R}$, we will need a way to remember that an error has occurred. For that reason, we extend the set of valid residue classes to the set S of *sane residue classes modulo* P. Let $0 \leq r < P$, we call r *sane* if for every $1 \leq i \leq n$ there is some $b_i \in \{0, 1, 2\}$ such that $r \equiv b_i \bmod p_i$. We regard the residue class 2 as *erroneous*. Finally, let us introduce some additional notation that allows us to

$$flip_i(r) \stackrel{\text{def}}{=} \begin{cases} r[1 \bmod p_i] & \text{if } r \equiv 0 \bmod p_i \\ r[0 \bmod p_i] & \text{if } r \equiv 1 \bmod p_i \\ r[2 \bmod p_i] & \text{if } r \equiv 2 \bmod p_i \end{cases} \qquad eqzero_i(r) \stackrel{\text{def}}{=} \begin{cases} r[0 \bmod p_i] & \text{if } r \equiv 0 \bmod p_i \\ r[2 \bmod p_i] & \text{if } r \equiv 1 \bmod p_i \\ r[2 \bmod p_i] & \text{if } r \equiv 2 \bmod p_i \end{cases}$$

Fig. 1. The mappings $flip_i$ and $eqzero_i$

alter a residue class r *locally*. Let $0 \le r < P$, $1 \le i \le n$ and $0 \le a < p_i$, we denote by $r[a \bmod p_i]$ the unique residue class r' modulo P satisfying

$$r' \equiv a \bmod p_i; \text{ and}$$
$$r' \equiv r \bmod p_j \text{ for all } 1 \le j \le n \text{ such that } j \ne i.$$

The existence of r' is guaranteed by the Chinese remainder theorem.

For each $1 \le i \le n$, we define mappings $flip_i, eqzero_i, eqone_i : S \to S$ that allow us to perform tests and operations on sane residue classes. The definitions of $flip_i$ and $eqzero_i$ are given in Figure 1. Given $0 \le r < P$, $flip_i(r)$ flips the bit encoded in the residue class modulo p_i, provided it is not erroneous. If it is erroneous, it remains so after an application of $flip_i$. Similarly, $eqzero_i$ allows for "guess-testing" of the bit encoded in the residue class modulo p_i: if $r \equiv 0 \bmod p_i$ then this value is preserved by the application of $eqzero_i$. Otherwise, $eqzero_i$ maps r to 2 so that it informally speaking "remembers" the wrong guess by mapping to a value r' such that $r' \equiv 2 \bmod p_i$. The mapping $eqone_i$ is defined analogously to $eqzero_i$ and allows for "guess-testing" whether $r \equiv 1 \bmod p_i$. The crucial point of our reduction is that $flip_i$, $eqzero_i$ and $eqone_i$ can be implemented via quadratic polynomials with coefficients of polynomial bit size.

Lemma 2. *For any $1 \le i \le n$ and any of $flip_i, eqzero_i, eqone_i : S \to S$, there is a quadratic polynomial with coefficients from $\{0, \ldots, P-1\}$ that realises the respective function.*

Proof. Let us first give polynomials for each of the mappings that work in $\mathbb{Z}/p_i\mathbb{Z}$. One easily verifies that the polynomials

$$p_{eqzero}(x) \stackrel{\text{def}}{=} -x^2 + 3x \qquad p_{flip}(x) \stackrel{\text{def}}{=} 3 \cdot 2^{-1} \cdot x^2 - 5 \cdot 2^{-1} \cdot x + 1$$
$$p_{eqone}(x) \stackrel{\text{def}}{=} x^2 - 2x + 2$$

realise the respective mappings. Here, it is important to recall that $p_i \ge 7$ and that 2 has a multiplicative inverse. However, the polynomials above are generally *not* realising the identity in $\mathbb{Z}/p_j\mathbb{Z}$ for $j \ne i$, which is required by the definition of $flip_i$, $eqzero_i$ and $eqone_i$. For instance, in $\mathbb{Z}/7\mathbb{Z}$ we do not have $x^2 - 2x + 2 \equiv x$. Thus, for each of the three polynomials $p_{flip}(x)$, $p_{eqzero}(x)$, $p_{eqone}(x)$, written as $a_2x^2 + a_1x + a_0$, in order to obtain corresponding polynomials $p_{flip,i}(x)$, $p_{eqzero,i}(x)$, $p_{eqone,i}(x)$, we apply the Chinese remainder theorem and for every $k \in \{0, 1, 2\}$ replace a_k with a'_k, where a'_k is the unique solution in $\mathbb{Z}/P\mathbb{Z}$ to the system of congruences $x \equiv a_k \bmod p_i$ and $x \equiv b_k \bmod p_j$ for each $1 \le j \ne i \le n$ with $b_1 \stackrel{\text{def}}{=} 1$ and $b_0 = b_2 \stackrel{\text{def}}{=} 0$. $\qquad \square$

$(q, i, b) \xrightarrow{eqzero_{i+1}} (q, i+1, 0)$ if $b = b', d = \rightarrow$ $\qquad$ $(q, i, b) \xrightarrow{eqzero_{i+1} \circ flip_i} (q, i+1, 0)$ if $b \neq b', d = \rightarrow$

$(q, i, b) \xrightarrow{eqone_{i+1}} (q, i+1, 1)$ if $b = b', d = \rightarrow$ $\qquad$ $(q, i, b) \xrightarrow{eqone_{i+1} \circ flip_i} (q, i+1, 1)$ if $b \neq b', d = \rightarrow$

$(q, i, b) \xrightarrow{eqzero_{i-1}} (q, i-1, 0)$ if $b = b', d = \leftarrow$ $\qquad$ $(q, i, b) \xrightarrow{eqzero_{i-1} \circ flip_i} (q, i-1, 0)$ if $b \neq b', d = \leftarrow$

$(q, i, b) \xrightarrow{eqone_{i-1}} (q, i-1, 1)$ if $b = b', d = \leftarrow$ $\qquad$ $(q, i, b) \xrightarrow{eqone_{i-1} \circ flip_i} (q, i-1, 1)$ if $b \neq b', d = \leftarrow$

Fig. 2. Transitions of $\mathcal{R}$ for simulating a transition (q, b, q', b', d) of $\mathcal{M}$

We have now accumulated all ingredients that enable us to simulate $\mathcal{M}$ with a PRM $\mathcal{R}$. Subsequently, we will identify each mapping $flip_i$, $eqzero_i$ and $eqone_i$ with the corresponding polynomial from Lemma 2. We now define the control locations of Q, the transitions Δ and the labelling function λ of $\mathcal{R}$. The control locations of $\mathcal{R}$ contain those of $\mathcal{M}$ paired with the head position and a guess of the contents of the tape cell at the current head position, i.e., $Q \stackrel{\text{def}}{=} Q_{\mathcal{M}} \times \{0, \ldots, n+1\} \times \{0, 1\}$.

For every control location (q, i, b) of $\mathcal{R}$ such that $1 \leq i \leq n$ and every transition $(q, b, q', b', d) \in \Delta_{\mathcal{M}}$ of $\mathcal{M}$, Δ contains the transitions shown in Figure 2, and an additional transition $(q_f, i, 0) \xrightarrow{x-P} (q_f, i, 0)$ for each $b \in \{0, 1\}$. The degree of the polynomials in Figure 2 is actually four, but quadratic polynomials can be regained by replacing a single transition with two consecutive transitions. Also, for brevity we have omitted the cases when the head moves to position 0 or $n+1$, whose behaviour can easily be hard-wired into $\mathcal{R}$.

The transitions of $\mathcal{R}$ are chosen such that every time we simulate a move of the head of $\mathcal{M}$, we guess the contents of the next tape cell. The guess is instantaneously verified through the application of the polynomials $eqzero_{i-1}$, $eqzero_{i+1}$, $eqone_{i-1}$ and $eqone_{i+1}$ along the transition: if the guess was wrong, the value of the register x becomes 2 modulo some prime p_i and will remain 2 modulo this prime forever. Simulating writing to a cell is done via the $flip_i$ polynomials, which are only applied if the currently read bit differs from the bit that is ought to be written. Finally, there is a self-loop at the control locations (q_f, i, b) subtracting P allows for checking that we end with a register value z such that $z \equiv 0 \mod P$. Setting $q_{\mathcal{R}} = (q_0, 0, 0)$ and $q'_{\mathcal{R}} = (q_f, 0, 0)$, by induction on the length of the run of $\mathcal{M}$ and $\mathcal{R}$ respectively, it is easily verified that $(q_0, \triangleright 0^n \triangleleft, 0) \rightarrow_{\mathcal{M}}^* (q_f, \triangleright 0^n \triangleleft, 0)$ if, and only if, $q_{\mathcal{R}}(0) \rightarrow_{\mathcal{R}}^* q'_{\mathcal{R}}(0)$.

3.2   Membership in PSPACE

We now show the existence of a PSPACE algorithm that decides reachability in PRMs in the most unconstrained case where register values come from $\mathbb{Z}$. We will generalise this to the case of general formulas in the end of this section. Due to space constraints, it is not possible to give all technical details and formal proofs, we rather prefer presenting our algorithm on a high level and only state the most important technical results that give the PSPACE upper bound. All formal details can be found in the appendix of the extended version of this paper.

For the remainder of this section, let us fix a PRM $\mathcal{R} = (Q, \Delta, \lambda, \phi)$ with one register x and control locations $q, q' \in Q$ for which we wish to decide $q(0) \to^*_{\mathcal{R}} q'(0)$. Denote by a and d the largest absolute value of all coefficients of the update polynomials in $\mathcal{R}$ and their maximum degree, respectively. We assume that every $p(x) \in P(\mathcal{R})$ is a non-constant polynomial. Otherwise, reachability can be reduced to a bounded number of reachability queries in PRMs with no constant update polynomials by guessing the order in which these transitions are traversed. The same approach would also enable us to additionally equip PRMs with zero tests.

Note that in the following, when referring to the size of a number, we refer to the number of bits required for its representation. On a high level, we can identify three key observations and ideas that lead us to our upper bound:

(i) There exists a bound b of size polynomial in $|\mathcal{R}|$ such that once the absolute register value of x goes above b, only counter-like polynomials can *decrement* the absolute value of x due to monotonicity properties of non-counter-like polynomials. A similar observation is part of the argument in [9] to show decidability of reachability for non-deterministic applications of affine polynomials.

(ii) The previous observation suggests that we should extract a 1-VASS $\mathcal{C}$ from the transitions from $\mathcal{R}$ labelled with counter-like polynomials that can simulate $\mathcal{R}$ acting on those transitions. This in turn enables us to make use of the property that reachability relations for 1-VASS are ultimately periodic with some period m of size polynomially bounded in $|\mathcal{C}|$ [10,11] and hence $|\mathcal{R}|$, and that reachability in 1-VASS can be decided in NP [12] and hence in PSPACE. In particular, this makes it possible to witness the existence of paths in $T(\mathcal{R})$ decrementing the register value from arbitrarily large absolute register values x, provided we know the residue class of x modulo m.

(iii) Observation (i) additionally enables us to show that paths in $T(\mathcal{R})$ whose absolute register value stays above b allow for deriving paths with special properties such that in particular residue classes modulo m of the register values occurring on the derived path are preserved. More precisely, we can derive paths for which a bound on the length of sequences that strictly decrease the absolute values of the register x exists. This in turn enables us to witness in PSPACE the *existence* of paths that end with a register value in a certain residue class modulo m by simulating $\mathcal{R}$ on residue classes modulo m without explicitly constructing those paths.

By gluing (ii) and (iii) together, we can then show that the PSPACE upper bound for reachability in PRMs follows. In the following, set $b \overset{\text{def}}{=} d(a+2)$. Observation (i) above is a consequence of the following lemma.

Lemma 3. *Let $p(x) \in P(\mathcal{R})$ be non-counter-like. Then $p(x)$ is monotonically increasing or decreasing in $\mathbb{Z} \setminus [-b, b]$, and $|p(z)| \geq 2|z|$ for all $z \in \mathbb{Z} \setminus [-b, b]$.*

The proof of the lemma is a straight-forward application of the inequality (1) in Section 2.1. It allows us to conclude that non-counter-like polynomials behave monotonically outside $[-b, b]$.

Before we start with formally discussing Observations (ii) and (iii), we need to introduce some auxiliary technical notation. A $q(z)$-$q'(z')$ *path* π *in* $T(\mathcal{R})$ *of length* n is a finite sequence of configurations $\pi : q_1(z_1)q_2(z_2) \cdots q_{n+1}(z_{n+1})$ such that $q(z) = q_1(z_1)$, $q'(z') = q_{n+1}(z_{n+1})$ and $q_i(z_i) \to_{\mathcal{R}} q_{i+1}(z_{i+1})$ for all $1 \leq i \leq n$. We write $\pi : q(z) \to_{\mathcal{R}}^* q'(z')$ if π is a $q(z)$-$q'(z')$ path and denote the length of π by $|\pi|$. Let $I \subseteq \mathbb{Z}$, we say that π *stays in* I if $z_i \in I$ for all $1 \leq i \leq n$. A path is *counter-like* if for all $q_i \xrightarrow{p(x)} q_{i+1}$, $p(x)$ is counter-like.

Now turning towards Observation (ii), the 1-VASS $\mathcal{C} \stackrel{\text{def}}{=} (Q_{\mathcal{C}}, \Delta_{\mathcal{C}}, \lambda_{\mathcal{C}})$ discussed above is obtained from the counter-like transitions of $\mathcal{R}$ as follows, where $\Delta_{\mathcal{C}} \stackrel{\text{def}}{=} \Delta_1 \cup \Delta_2$:

$$Q_{\mathcal{C}} \stackrel{\text{def}}{=} \{q^{\sim} : q \in Q, \sim \in \{+, -\}\};$$

$$\Delta_1 \stackrel{\text{def}}{=} \{(q_1^{\sim}, q_2^{\sim}) : q_1, q_2 \in Q, q_1 \xrightarrow{p(x) = x + a_0} q_2 \in \Delta\};$$

$$\Delta_2 \stackrel{\text{def}}{=} \{(q_1^{\sim_1}, q_2^{\sim_2}) : q_1, q_2 \in Q, q_1 \xrightarrow{p(x) = -x + a_0} q_2 \in \Delta, \sim_1 \neq \sim_2\}$$

$$\lambda_{\mathcal{C}} \stackrel{\text{def}}{=} (q_1^{\sim_1}, q_2^{\sim_2}) \mapsto x + \sim_2 a_0 \text{ if } q_1 \xrightarrow{a_1 x + a_0} q_2 \in \Delta.$$

The idea behind this construction is as follows. The counter of $\mathcal{C}$ stores the *absolute value* of the register x of $\mathcal{R}$. The control locations of $\mathcal{C}$ are control locations from $\mathcal{R}$ with an indicator of the sign of the register x, e.g. q^- indicates that the control location is q and the value of the register x is negative. The transitions in Δ_1 and Δ_2 are defined such that they obey a flip of the sign. The following lemma, which can easily be shown by induction, enables us to relate paths in $T(\mathcal{R})$ and $T(\mathcal{C})$.

Lemma 4. *Let $q_1(z_1), q_2(z_2) \in C(\mathcal{R})$ and let $z = \min\{|z_1|, |z_2|\}$ such that $z > a$. There exists a counter-like path $\pi : q_1(z_1) \to_{\mathcal{R}}^* q_2(z_2)$ staying in $\mathbb{Z} \setminus (-z, z)$ if, and only if, there exists a path $\pi' : q_1^{\sim_1}(|z_1| - z) \to_{\mathcal{C}}^* q_2^{\sim_2}(|z_2| - z)$ for $\sim_i = \text{sgn}(z_i)$.*

The benefit we get from extracting a 1-VASS from the counter-like transitions of $\mathcal{R}$ is that we can employ known periodicity properties for counter automata. The following proposition is a consequence of Lemma 5.1.9, pp. 139 in [11]. It allows us to conclude that reachability in 1-VASS is ultimately periodic with a small period of polynomial size.

Proposition 5 ([10,11]). *Let $\mathcal{C} = (Q_{\mathcal{C}}, \Delta_{\mathcal{C}}, \lambda_{\mathcal{C}})$ be a 1-VASS with maximum absolute increment a. There exists a fixed polynomial p and a period $m \leq (|Q_{\mathcal{C}}|a)^{|Q_{\mathcal{C}}|}$ such that for any $q, q' \in Q_{\mathcal{C}}$ and $n' \in \mathbb{N}$ there exists a set of residue classes $R \subseteq \{0, \ldots, m - 1\}$ such that for all $n > 2^{p(|\mathcal{C}|)} + n'$,*

$$q(n) \to_{\mathcal{C}}^* q'(n') \text{ if, and only if, } n \equiv r \bmod m \text{ for some } r \in R.$$

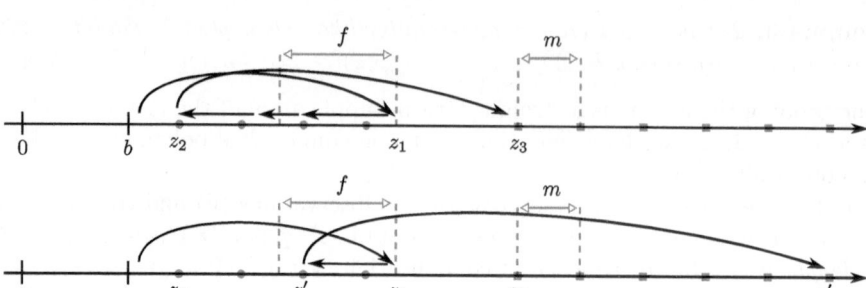

Fig. 3. Illustration of making a path non-dropping

For the remainder of this section, fix m to be the period from Proposition 5. We will now elaborate on Observation (iii) and turn towards normalising paths of $\mathcal{R}$ in $T(\mathcal{R})$ whose register values stay in $\mathbb{Z} \setminus [-b, b]$. To this end, we define a partial order $\preceq_m \subseteq \mathbb{Z} \times \mathbb{Z}$ such that

$$z_1 \preceq_m z_2 \text{ if } \operatorname{sgn}(z_1) = \operatorname{sgn}(z_2), |z_1| \leq |z_2| \text{ and } z_1 \equiv z_2 \bmod m.$$

Informally speaking, we show that if the register values encountered along a path fluctuate too much then we can decrease the magnitude of fluctuation while staying invariant w.r.t. $\preceq_m$. Formally, set $f \stackrel{\text{def}}{=} (2|Q|m)^2 + (2|Q|m)$ and let $\pi : q(z) \to_\mathcal{R}^* q'(z')$ be a path. We say that π is *dropping* if there are $1 \leq i < j < |\pi|$ such that $\pi(i) = q_i(z_i)$, $\pi(j) = q_j(z_j)$ and $|z_i| - |z_j| > f$. Observe that for any non-dropping path $\pi : q(z) \to q'(z')$, we have $|z'| \geq |z| - f$.

Lemma 6. *Let $\pi : q(z) \to_\mathcal{R}^* q'(z')$ be a path staying in $\mathbb{Z} \setminus [-b, b]$. If π is dropping then there is a path π' staying in $\mathbb{Z} \setminus [b, b]$ such that $\pi' : q(z) \to_\mathcal{R}^* q'(z'')$ for some $z'' \in \mathbb{Z}$, $z' \preceq_m z''$ and $|\pi'| < |\pi|$.*

Figure 3 illustrates the main idea. There, the illustrated path on top is dropping between register values z_1 and z_2. A counting argument shows that we can then find some z'_2 in the interval $[z_1 - f, z_1]$ such that $z_2 \preceq_m z'_2$, which allows us to chop the path. We can then mimic the remainder of the path and end with some register value z'_3 such that $z_3 \preceq_m z'_3$, illustrated at the bottom of Figure 3. A repeated application of the lemma allows us to make any path non-dropping, and it is not difficult to see that witnessing the existence of a non-dropping path reaching a certain residue class modulo m can be done in space polynomial in $|\mathcal{R}|$. This brings us to the main theorem of this paper.

Theorem 7. *Reachability in PRMs is* PSPACE-*complete.*

Proof (sketch). The main idea is that we can simulate $\mathcal{R}$ as long as its register values stay inside $[-B, B]$ for some sufficiently large $B \in \mathbb{N}$ of polynomial bit-size in $|\mathcal{R}|$. Here, it is important that checking whether an application of an update polynomial $p(x)$ to the current register value $z \in \mathbb{Z}$ leaves the interval $[-B, B]$

Table 1. Complexity landscape of reachability in d-PRMs

	counter polynomials			arbitrary polynomials		
	finite	upward closed	$\mathbb{Z}^d$	finite	upward closed	$\mathbb{Z}^d$
$d = 1$	PSPACE-c.[7]	NP-complete [12]		PSPACE-complete		
$d > 1$	PSPACE-c.[13]	EXPSPACE-h.,dec.[16,17]	NP-c.	PSPACE-c.	undec. [19]	

can be decided in polynomial time, and that $p(z)$ can be computed in polynomial time if $p(z) \in [-B, B]$ [5]. If the interval $[-B, B]$ were to be left, we compute $p(z)$ mod m, guess a residue class r modulo m, and then check using Lemma 6 in polynomial space for the existence of a non-dropping path starting with register value $p(z)$ mod m reaching some register value in the residue class r. Moreover, we can use $\mathcal{C}$ together with Proposition 5 to check in polynomial space that from the residue class r there is a counter-like path back into $[-B, B]$.   □

The proof of PSPACE-completeness can straight-forwardly be adapted to the case where the global invariant $\phi(x)$ imposes an upward-closed domain on the register x. The main difference is that $\mathcal{C}$ constructed above must not allow for flipping of signs, and when simulating $\mathcal{R}$ on the residue classes modulo m in the proof of Theorem 7 no transitions can be taken that result in a flip of the sign of the register.

Remark 8. A possible generalisation of PRMs could be to allow the global invariant to be a Presburger formula open in one variable x. Since the sets defined by such formulas are ultimately periodic below and above zero, it is not difficult to adapt the techniques used for showing the PSPACE upper bound in order to show that reachability is decidable. However, unsurprisingly the complexity of reachability may potentially increase by several exponents.

4   Concluding Remarks

This paper introduced polynomial register machines, a class of infinite-state systems comprising a finite number of integer-valued registers, whose domain is constrained by a linear constraint, with a finite-state controller which can update the registers along transitions by an application of a polynomial function. Our main result is that reachability with one register is PSPACE-complete. For higher dimensions, as discussed in the introduction, reachability becomes quickly undecidable, in particular already in the presence of two integer-valued registers and affine polynomials with *integer* coefficients [19].

A detailed complexity landscape classifying the complexity of reachability according to the number of registers, the type of update polynomials and the domain constraint is given in Table 1 together with bibliographic references. The results of this paper are emphasised by grey background colour.

References

1. Babić, D., Cook, B., Hu, A.J., Rakamarić, Z.: Proving termination of nonlinear command sequences. Formal Aspects of Computing, 1–15 (2012)
2. Bell, P., Potapov, I.: On undecidability bounds for matrix decision problems. Theoretical Computer Science 391(1-2), 3–13 (2008)
3. Bonnet, R.: The reachability problem for vector addition system with one zero-test. In: Murlak, F., Sankowski, P. (eds.) MFCS 2011. LNCS, vol. 6907, pp. 145–157. Springer, Heidelberg (2011)
4. Bradley, A.R., Manna, Z., Sipma, H.B.: Termination of polynomial programs. In: Cousot, R. (ed.) VMCAI 2005. LNCS, vol. 3385, pp. 113–129. Springer, Heidelberg (2005)
5. Cucker, F., Koiran, P., Smale, S.: A polynomial time algorithm for Diophantine equations in one variable. Journal of Symbolic Computation 27(1), 21–29 (1999)
6. Dufourd, C., Finkel, A., Schnoebelen, P.: Reset nets between decidability and undecidability. In: Larsen, K.G., Skyum, S., Winskel, G. (eds.) ICALP 1998. LNCS, vol. 1443, pp. 103–115. Springer, Heidelberg (1998)
7. Fearnley, J., Jurdziński, M.: Reachability in two-clock timed automata is PSPACE-complete. In: Fomin, F.V., Freivalds, R., Kwiatkowska, M., Peleg, D. (eds.) ICALP 2013, Part II. LNCS, vol. 7966, pp. 212–223. Springer, Heidelberg (2013)
8. Finkel, A., McKenzie, P., Picaronny, C.: A well-structured framework for analysing Petri net extensions. Information and Computation 195(1-2), 1–29 (2004)
9. Fremont, D.: The reachability problem for affine functions on the integers (2012), http://web.mit.edu/~dfremont/www/reachability.pdf
10. Göller, S., Haase, C., Ouaknine, J., Worrell, J.: Branching-time model checking of parametric one-counter automata. In: Birkedal, L. (ed.) FOSSACS 2012. LNCS, vol. 7213, pp. 406–420. Springer, Heidelberg (2012)
11. Haase, C.: On the Complexity of Model Checking Counter Automata. PhD thesis, University of Oxford, UK (2012)
12. Haase, C., Kreutzer, S., Ouaknine, J., Worrell, J.: Reachability in succinct and parametric one-counter automata. In: Bravetti, M., Zavattaro, G. (eds.) CONCUR 2009. LNCS, vol. 5710, pp. 369–383. Springer, Heidelberg (2009)
13. Haase, C., Ouaknine, J., Worrell, J.: On the relationship between reachability problems in timed and counter automata. In: Finkel, A., Leroux, J., Potapov, I. (eds.) RP 2012. LNCS, vol. 7550, pp. 54–65. Springer, Heidelberg (2012)
14. Hirst, H.P., Macey, W.T.: Bounding the roots of polynomials. The College Mathematics Journal 28(4), 292–295 (1997)
15. Lambert, J.-L.: A structure to decide reachability in petri nets. Theoretical Computer Science 99(1), 79–104 (1992)
16. Lipton, R.: The reachability problem is exponential-space-hard. Technical report, Yale University, New Haven, CT (1976)
17. Mayr, E.W.: An algorithm for the general Petri net reachability problem. In: Proc. STOC, pp. 238–246. ACM, New York (1981)
18. Minsky, M.L.: Recursive Unsolvability of Post's Problem of "Tag" and other Topics in Theory of Turing Machines. The Annals of Mathematics 74(3), 437–455 (1961)
19. Reichert, J.: Personal communication (2013)

On the Parameterized Complexity
of Cutting a Few Vertices from a Graph

Fedor V. Fomin[1], Petr A. Golovach[1], and Janne H. Korhonen[2]

[1] Department of Informatics, University of Bergen
Norway
[2] Helsinki Institute for Information Technology HIIT
& Department of Computer Science, University of Helsinki
Finland

Abstract. We study the parameterized complexity of separating a small set of vertices from a graph by a small vertex-separator. That is, given a graph G and integers k, t, the task is to find a vertex set X with $|X| \leq k$ and $|N(X)| \leq t$. We show that

- the problem is fixed-parameter tractable (FPT) when parameterized by t but W[1]-hard when parameterized by k, and
- a terminal variant of the problem, where X must contain a given vertex s, is W[1]-hard when parameterized either by k or by t alone, but is FPT when parameterized by $k + t$.

We also show that if we consider edge cuts instead of vertex cuts, the terminal variant is NP-hard.

1 Introduction

We investigate two related problems that concern separating a small vertex set from a graph $G = (V, E)$. Specifically, we consider finding a vertex set X of size at most k such that

1. X is separated from the rest of V by a small cut (e.g. *finding communities in a social network*, cf. [14]), or
2. X is separated from the rest of V by a small cut and contains a specified terminal vertex s (e.g. *isolating a dangerous node*, cf. [11,13]).

We focus on *parameterized complexity* of the *vertex-cut versions* of these problems.

Parameterized Vertex Cuts. Our interest in the vertex-cut version stems from the following parameterized separation problem, studied by Marx [16]. Let $N(X)$ denote the vertex-neighborhood of X.

Cutting k Vertices
Input: Graph $G = (V, E)$, integers $k \geq 1$, $t \geq 0$
Parameter 1: k
Parameter 2: t
Question: Is there a set $X \subseteq V$ such that $|X| = k$ and $|N(X)| \leq t$?

K. Chatterjee and J. Sgall (Eds.): MFCS 2013, LNCS 8087, pp. 421–432, 2013.
© Springer-Verlag Berlin Heidelberg 2013

In particular, Marx showed that CUTTING k VERTICES is W[1]-hard even when parameterized by both k and t. We contrast this result by investigating the parameterized complexity of the two related separation problems with relaxed requirement on the size of the separated set X.

CUTTING AT MOST k VERTICES
Input: Graph $G = (V, E)$, integers $k \geq 1$, $t \geq 0$
Parameter 1: k
Parameter 2: t
Question: Is there a non-empty set $X \subseteq V$ such that $|X| \leq k$ and $|N(X)| \leq t$?

CUTTING AT MOST k VERTICES WITH TERMINAL
Input: Graph $G = (V, E)$, terminal vertex s, integers $k \geq 1$, $t \geq 0$
Parameter 1: k
Parameter 2: t
Question: Is there a non-empty set $X \subseteq V$ such that $s \in X$, $|X| \leq k$ and $|N(X)| \leq t$?

We show that these closely related problems exhibit quite different complexity behaviors. In particular, we show that CUTTING AT MOST k VERTICES is fixed-parameter tractable (FPT) when parameterized by the size of the separator t, while we need both k and t as parameters to obtain an FPT algorithm for CUTTING AT MOST k VERTICES WITH TERMINAL. A full summary of the parameterized complexity of these problems and our results is given in Table 1.

The main algorithmic contribution of our paper is the proof that CUTTING AT MOST k VERTICES is FPT when parameterized by t (Theorem 2). To obtain this result, we utilize the concept of *important separators* introduced by Marx [16]. However, a direct application of important separators—guess a vertex contained in the separated set, and find a minimal set containing this vertex that can be separated from the remaining graph by at most t vertices—does not work. Indeed, pursuing this approach would bring us to essentially solving CUTTING AT MOST k VERTICES WITH TERMINAL, which is W[1]-hard when parameterized by t. Our FPT algorithm is based on new structural results about unique important

Table 1. Parameterized complexity of CUTTING k VERTICES, CUTTING AT MOST k VERTICES, and CUTTING AT MOST k VERTICES WITH TERMINAL

Parameter	CUTTING k VERTICES	CUTTING $\leq k$ VERTICES	CUTTING $\leq k$ VERTICES WITH TERMINAL
k	W[1]-hard, [16]	W[1]-hard, Thm 3	W[1]-hard, Thm 3
t	W[1]-hard, [16]	FPT, Thm 2	W[1]-hard, Thm 5
k and t	W[1]-hard, [16]	FPT, Thm 1	FPT, Thm 1

separators of minimum size separating pairs of vertices. We also observe that it is unlikely that CUTTING AT MOST k VERTICES has a polynomial kernel.

Edge Cuts. Although our main focus is on vertex cuts, we will also make some remarks on the edge-cut versions of the problems. In particular, the edge-cut versions again exhibit a different kind of complexity behavior. Let $\partial(X)$ denote the edge-boundary of X.

CUTTING AT MOST k VERTICES BY EDGE-CUT
Input: Graph $G = (V, E)$, integers $k \geq 1$, $t \geq 0$
Parameter 1: k
Parameter 2: t
Question: Is there a non-empty set $X \subseteq V$ such that $|X| \leq k$ and $|\partial(X)| \leq t$?

CUTTING k VERTICES BY EDGE-CUT WITH TERMINAL
Input: Graph $G = (V, E)$, terminal vertex s, integers $k \geq 1$, $t \geq 0$
Parameter 1: k
Parameter 2: t
Question: Is there a set $X \subseteq V$ such that $s \in X$, $|X| \leq k$ and $|\partial(X)| \leq t$?

Results by Watanabe and Nakamura [19] imply that CUTTING AT MOST k VERTICES BY EDGE-CUT can be done in polynomial time even when k and t are part of the input; more recently, Armon and Zwick [2] have shown that this also holds in the edge-weighted case. Lokshtanov and Marx [15] have proven that CUTTING AT MOST k VERTICES BY EDGE-CUT WITH TERMINAL is fixed-parameter tractable when parameterized by k or by t; see also [5]. We complete the picture by showing that CUTTING AT MOST k VERTICES BY EDGE-CUT WITH TERMINAL is NP-hard (Theorem 6). The color-coding techniques we employ in Theorem 1 also give a simple algorithm with running time $2^{k+t+o(k+t)} \cdot n^{O(1)}$.

Related edge-cut problems have received attention in the context of approximation algorithms. In contrast to CUTTING AT MOST k VERTICES BY EDGE-CUT, finding a minimum-weight edge-cut that separates exactly k vertices is NP-hard. Feige et al. [9] give a PTAS for $k = O(\log n)$ and an $O(k/\log n)$-approximation for $k = \Omega(\log n)$; Li and Zhang [14] give an $O(\log n)$-approximation. Approximation algorithms have also been given for unbalanced s-t-cuts, where s and t are specified terminal vertices and the task is to find an edge cut $(X, V \setminus X)$ with $s \in X$ and $t \in V \setminus S$ such that (a) $|X| \leq k$ and weight of the cut is minimized [11,14], or (b) weight of the cut is at most w and $|X|$ is minimized [13].

2   Basic Definitions and Preliminaries

Graph Theory. We follow the conventions of Diestel [7] with graph-theoretic notations. We only consider finite, undirected graphs that do not contain loops or multiple edges. The vertex set of a graph G is denoted by $V(G)$ and the edge set is denoted by $E(G)$, or simply by V and E, respectively. Typically we use n to denote the number of vertices of G and m the number of edges.

For a set of vertices $U \subseteq V(G)$, we write $G[U]$ for the subgraph of G induced by U, and $G-U$ for the graph obtained form G by the removal of all the vertices of U, i.e., the subgraph of G induced by $V(G) \setminus U$. Similarly, for a set of edges A, the graph obtained from G by the removal of all the edges in A is denoted by $G - A$.

For a vertex v, we denote by $N_G(v)$ its (open) neighborhood, that is, the set of vertices which are adjacent to v. The degree of a vertex v is $d_G(v) = |N_G(v)|$. For a set of vertices $U \subseteq V(G)$, we write $N_G(U) = \cup_{v \in U} N_G(v) \setminus U$ and $\partial_G(U) = \{uv \in E(G) \mid u \in U, v \in V(G) \setminus U\}$. We may omit subscripts in these notations if there is no danger of ambiguity.

Submodularity. We will make use of the well-known fact that given a graph G, the mapping $2^V \to \mathbb{Z}$ defined by $U \mapsto |N(U)|$ is submodular. That is, for $A, B \subseteq V$ we have

$$|N(A \cap B)| + |N(A \cup B)| \le |N(A)| + |N(B)| . \tag{1}$$

Important Separators. Let G be a graph. For disjoint sets $X, Y \subseteq V$, a vertex set $S \subseteq V \setminus (X \cup Y)$ is a (vertex) (X,Y)-separator if there is no path from X to Y in $G - S$. An edge (X,Y)-separator $A \subseteq E$ is defined analogously. Note that we do not allow deletion of vertices in X and Y, and thus there are no vertex (X,Y)-separators if X and Y are adjacent. As our main focus is on the vertex-cut problems, all separators are henceforth vertex-separators unless otherwise specified.

We will make use of the concept of important separators, introduced by Marx [16]. A vertex v is reachable from a set $X \subseteq V$ if G has a path that joins a vertex of X and v. For any sets S and $X \subseteq V \setminus S$, we denote the set of vertices reachable from X in $G - S$ by $R(X, S)$. An (X,Y)-separator S is minimal if no proper subset of S is an (X,Y)-separator. For (X,Y)-separators S and T, we say that T dominates S if $|T| \le |S|$ and $R(X, S)$ is a proper subset of $R(X, T)$. For singleton sets, we will write x instead of $\{x\}$ in the notations defined above.

Definition 1 ([16]). An (X,Y)-separator S is important if it is minimal and there is no other (X,Y)-separator dominating S.

In particular, this definition implies that for any (X,Y)-separator S there exists an important (X,Y)-separator T with $|T| \le |S|$ and $R(X, T) \supseteq R(X, S)$. If S is not important, then at least one of the aforementioned relations is proper.

The algorithmic usefulness of important separators follows from the fact that the number of important separators of size at most t is bounded by t alone, and furthermore, these separators can be listed efficiently. Moreover, minimum-size important separators are unique and can be found in polynomial time. That is, we will make use of the following lemmas.

Lemma 1 ([6]). *For any disjoint sets $X, Y \subseteq V$, the number of important (X, Y)-separators of size at most t is at most 4^t, and all important (X, Y)-separators of size at most t can be listed in time $4^t \cdot n^{O(1)}$.*

Lemma 2 ([16]). *For any sets $X, Y \subseteq V$, if there exists an (X, Y)-separator, then there is exactly one important (X, Y)-separator of minimum size. This separator can be found in polynomial time.*

Parameterized Complexity. We will briefly review the basic notions of parameterized complexity, though we refer to the books of Downey and Fellows [8], Flum and Grohe [10], and Niedermeier [18] for a detailed introduction. Parameterized complexity is a two-dimensional framework for studying the computational complexity of a problem; one dimension is the input size n and another one is a parameter k. A parameterized problem is *fixed-parameter tractable* (or FPT) if it can be solved in time $f(k) \cdot n^{O(1)}$ for some function f, and in the class XP if it can be solved in time $O\left(n^{f(k)}\right)$ for some function f.

Between FPT and XP lies the class W[1]. One of basic assumptions of the parameterized complexity theory is the conjecture that W[1] $\neq$ FPT, and it is thus held to be unlikely that a W[1]-hard problem would be in FPT. For exact definition of W[1], we refer to the books mentioned above. We mention only that INDPENDENT SET and CLIQUE parameterized by solution size are two fundamental problems that are known to be W[1]-complete.

The basic way of showing that a parameterized problem is unlikely to be fixed-parameter tractable is to prove W[1]-hardness. To show that a problem is W[1]-hard, it is enough to give a *parameterized reduction* from a known W[1]-hard problem. That is, let A, B be parameterized problems. We say that A is (uniformly many-one) *FPT-reducible* to B if there exist functions $f, g : \mathbb{N} \to \mathbb{N}$, a constant $c \in \mathbb{N}$ and an algorithm $\mathcal{A}$ that transforms an instance (x, k) of A into an instance $(x', g(k))$ of B in time $f(k)|x|^c$ so that $(x, k) \in A$ if and only if $(x', g(k)) \in B$.

Cutting Problems with Parameters k and t. In the remainder of this section, we consider CUTTING AT MOST k VERTICES, CUTTING AT MOST k VERTICES WITH TERMINAL, and CUTTING AT MOST k VERTICES BY EDGE-CUT WITH TERMINAL with parameters k and t. We first note that if there exists a solution for one of the problems, then there is also a solution in which X is connected; indeed, it suffices to take any maximal connected $Y \subseteq X$. Furthermore, we note finding a connected set X with $|X| = k$ and $|N(X)| \leq t$ is fixed-parameter tractable with parameters k and t due to a result by Marx [16, Theorem 13], and thus CUTTING AT MOST k VERTICES is also fixed-parameter tractable with parameters k and t.

We now give a simple color-coding algorithm [1,4] for the three problems with parameters k and t, in particular improving upon the running time of the aforementioned algorithm for CUTTING AT MOST k VERTICES.

Theorem 1. CUTTING AT MOST k VERTICES, CUTTING AT MOST k VERTICES WITH TERMINAL, *and* CUTTING AT MOST k VERTICES BY EDGE-CUT WITH TERMINAL *can be solved in time* $2^{k+t} \cdot (k+t)^{O(\log(k+t))} \cdot n^{O(1)}$.

Proof. We first consider a 2-colored version of CUTTING AT MOST k VERTICES. That is, we are given a graph G where each vertex is either colored red or blue (this is not required to be a proper coloring), and the task is to find a connected red set X with $|X| \leq k$ such that $N(X)$ is blue and $|N(X)| \leq t$. If such a set exists, it can be found in polynomial time by trying all maximal connected red sets.

Now let $G = (V, E)$ be a graph. Assume that there is a set X with $|X| \leq k$ and $|N(X)| \leq t$; we may assume that X is connected. It suffices to find a coloring of V such that X is colored red and $N(X)$ is colored blue. This can be done by coloring each vertex v either red or blue independently and uniformly at random. Indeed, this gives a desired coloring with probability at least $2^{-(k+t)}$, which immediately yields a $2^{k+t} \cdot n^{O(1)}$ time randomized algorithm for CUTTING AT MOST k VERTICES.

This algorithm can be derandomized in standard fashion using universal sets (compare with Cai et al. [4]). Recall that a (n, ℓ)-*universal set* is a collection of binary vectors of length n such that for each index subset of size ℓ, each of the 2^ℓ possible combinations of values appears in some vector of the set. A construction of Naor et al. [17] gives a (n, ℓ)-universal set of size $2^\ell \cdot \ell^{O(\log \ell)} \log n$ that can be listed in linear time. It suffices to try all colorings induced by a $(n, k+t)$-universal set obtained trough this construction.

The given algorithm works for CUTTING AT MOST k VERTICES WITH TERMI-NAL with obvious modifications. That is, given a coloring, we simply check if the terminal s is red and its connected red component is a solution. This also works for CUTTING AT MOST k VERTICES BY EDGE-CUT WITH TERMINAL, as we have $|N(X)| \leq |\partial(X)|$. $\qquad\square$

3 Cutting at Most k Vertices Parameterized by t

In this section we show that CUTTING AT MOST k VERTICES is fixed-parameter tractable when parameterized by the size of the separator t only. Specifically, we will prove the following theorem.

Theorem 2. CUTTING AT MOST k VERTICES *can be solved in time* $4^t \cdot n^{O(1)}$.

The remainder of this section consists of the proof of Theorem 2. Note that we may assume $\frac{3}{4}t < k < n - t$. Indeed, if $k \leq ct$ for a fixed constant $c < 1$, then we can apply the algorithm of Theorem 1 to solve CUTTING AT MOST k VERTICES in time $4^t n^{O(1)}$. On the other hand, if $k \geq n - t$, then any vertex set X of size k is a solution, as $|N(X)| \leq n - k \leq t$.

We start by guessing a vertex $u \in V$ that belongs to a solution set X if one exists; specifically, we can try all choices of u. We cannot expect to necessarily find a solution X that contains the chosen vertex u, even if the guess is correct, as the terminal variant is W[1]-hard. We will nonetheless try; turns out that the only thing that can prevent us from finding a solution containing u is that we find a solution *not* containing u.

With u fixed, we compute for each $v \in V \setminus (\{u\} \cup N(u))$ the unique minimum important (u, v)-separator S_v. This can be done in polynomial time by Lemma 2. Let V_0 be set of those v with $|S_v| \leq t$, and denote $R(v) = R(v, S_v)$. Finally, let X be a set family consisting of those $R(v)$ for $v \in V_0$ that are inclusion-minimal, i.e., if $R(v) \in X$, then there is no $w \in V_0$ such that $R(w) \subsetneq R(v)$. Note that we can compute the sets V_0, $R(v)$ and X in polynomial time.

There are now three possible cases that may occur.

1. If $V_0 = \emptyset$, we conclude that we have no solution containing u.
2. If there is $v \in V_0$ such that $|R(v)| \leq k$, then $X = R(v)$ gives a solution, and we stop and return a YES-answer.
3. Otherwise, X is non-empty and for all sets $A \in X$ we have $|A| > k$.

We only have to consider the last case, as otherwise we are done. We will show that in that case, the sets $A \in X$ can be used to find a solution X containing u if one exists. For this, we need the following structural results about the sets $R(v)$.

Lemma 3. *For any $v, w \in V_0$, if $w \in R(v)$ then $R(w) \subseteq R(v)$.*

Proof. Let $A = R(v)$ and $B = R(w)$. Since $S_v = N(A)$ is a (u, v)-separator of minimum size, we must have $|N(A \cup B)| \geq |N(A)|$. By (1), we have

$$|N(A \cap B)| \leq |N(A)| + |N(B)| - |N(A \cup B)| \leq |N(B)| \ .$$

Because $w \in A$, the set $N(A \cap B)$ is a (u, w)-separator. Thus, if $B \neq A \cap B$, then $N(A \cap B)$ is a (u, w)-separator that witnesses that S_w is not an important separator. But this is not possible by the definition of S_w, so we have $B = A \cap B \subseteq A$. $\square$

Lemma 4. *Any distinct $A, B \in X$ are disjoint.*

Proof. Assume that $A, B \in X$ are distinct and intersect. Then there is $v \in A \cap B$. Since $v \in A$, the set $N(A)$ is a (u, v)-separator of size at most t, and $v \in V_0$. Recall that X contains inclusion-minimal sets $R(w)$ for $w \in V_0$. But by Lemma 3, $R(v)$ is a proper subset of both A and B, which is not possible by the definition of X. $\square$

Now assume that the input graph G has a solution for CUTTING AT MOST k VERTICES containing u. In particular, then there is an inclusion-minimal set $X \subseteq V$ with $u \in X$ satisfying $|X| \leq k$ and $|N(X)| \leq t$. Let us fix one such set X.

Lemma 5. *For all $A \in X$, the set A is either contained in $X \cup N(X)$ or does not intersect it.*

Proof. Suppose that there is a set $A \in \mathcal{X}$ that intersects both $X \cup N(X)$ and its complement. Let $Y = V \setminus (X \cup N(X))$.

Now let $v \in A \cap Y$. By Lemma 3, we have $R(v) = A$. If $|N(A \cup Y)| > |N(Y)|$ then it follows from (1) that

$$|N(A \cap Y)| \leq |N(A)| + |N(Y)| - |N(A \cup Y)| < |N(A)| \, .$$

However, this would imply that $N(A \cap Y)$ is a (u, v)-separator smaller than $S_v = N(A)$.

Thus, we have $|N(A \cup Y)| \leq |N(Y)|$. But $X' = X \setminus (A \cup Y \cup N(A \cup Y))$ is a proper subset of X; furthermore, any vertex of $N(X')$ that is not in $N(A \cup Y)$ is also in $N(X) \setminus N(Y)$, so we have $|N(X')| \leq |N(X)| \leq t$. This is in contradiction with the minimality of X. $\square$

Lemma 6. *Let Z be the union of all $A \in \mathcal{X}$ that do not intersect $X \cup N(X)$. Then $Z \neq \emptyset$ and there is an important (Z, u)-separator S of size at most t such that $|R(u, S)| + |S| \leq k + t$.*

Proof. Let $S = N(X)$. Consider an arbitrary $v \in V \setminus (X \cup S)$; such vertex exists, since $k + t < n$. Since S separates v from u, the set $R(v)$ is well-defined.

Suppose now that $R(v)$ is not contained in $R(v, S)$. Let $B = R(u, S_v)$. Since S_v is a minimum-size (u, v)-separator we have $|N(B)| = |N(R(v))|$. But $N(X \cup B)$ also separates u and v, so we have $|N(X \cup B)| \geq |N(R(v))| = |N(B)|$. By (1), we have

$$|N(X \cap B)| \leq |N(X)| + |N(B)| - |N(X \cup B)| \leq |N(X)| \leq t \, .$$

But since $R(v)$ is not contained $R(v, S)$, it follows that $X \cap B$ is a proper subset of X, which contradicts the minimality of X.

Thus we have $R(v) \subseteq R(v, S)$. It follows that $R(v, S)$ contains a set $A \in \mathcal{X}$, which implies that $Z \neq \emptyset$ and $v \in R(A, S) \subseteq R(Z, S)$. Furthermore, since $v \in V \setminus (X \cup S)$ was chosen arbitrarily, we have that $R(Z, S) = V \setminus (X \cup S)$.

If S is an important (Z, u)-separator, we are done. Otherwise, there is an important (Z, u)-separator T with $|T| \leq |S|$ and $R(Z, S) \subseteq R(Z, T)$. But then we have $|T| \leq t$, and $R(u, T) \cup T \subseteq X \cup S$, that is, $|R(u, T) \cup T| \leq k + t$. $\square$

Recall now that we may assume $|A| > k$ for all $A \in \mathcal{X}$. Furthermore, we have $|X \cup N(X)| \leq k + t < \left(2 + \frac{1}{3}\right) k$ and the sets $A \in \mathcal{X}$ are disjoint by Lemma 4. Thus, at most two sets $A \in \mathcal{X}$ fit inside $X \cup N(X)$ by Lemma 5. This means that if we let Z be the union of all $A \in \mathcal{X}$ that do not intersect $X \cup N(X)$, then as we have already computed $\mathcal{X}$, we can guess Z by trying all $O(n^2)$ possible choices.

Assume now that X is a minimal solution containing u and our guess for Z is correct. We enumerate all important (Z, u)-separators of size at most t. We will find by Lemma 6 an important (Z, u)-separator S such that $|S| \leq t$ and $|R(u, S)| + |S| \leq k + t$. If $|R(u, S)| \leq k$, we have found a solution. Otherwise, we delete a set S' of $|R(u, S)| - k$ elements from $R(u, S)$ to obtain a solution X'. To see that this suffices, observe that $N(X') \subseteq S' \cup S$. Therefore, $|N(X')| \leq |S'| + |S| = |R(u, S)| - k + |S| \leq t$. As all important (Z, u)-separators can be listed in time $4^t \cdot n^{O(1)}$ by Lemma 1, the proof of Theorem 2 is complete.

4   Hardness Results

We start this section by complementing Theorem 2, as we show that CUTTING AT MOST k VERTICES is NP-complete and W[1]-hard when parameterized by k. We also show that same holds for CUTTING AT MOST k VERTICES WITH TERMINAL. Note that both of these problems are in XP when parameterized by k, as they can be solved by checking all vertex subsets of size at most k.

Theorem 3. CUTTING AT MOST k VERTICES *and* CUTTING AT MOST k VERTICES WITH TERMINAL *are* NP-*complete and* W[1]-*hard with the parameter* k.

Proof. We prove the W[1]-hardness claim for CUTTING AT MOST k VERTICES by a reduction from CLIQUE. Recall that this W[1]-complete (see [8]) problem asks for a graph G and a positive integer k where k is a parameter, whether G contains a clique of size k. Let (G, k) be an instance of CLIQUE, $n = |V(G)|$ and $m = |E(G)|$; we construct an instance (G', k', t) of CUTTING AT MOST k VERTICES as follows. Let H_V be a clique of size n^3 and identify n vertices of H_V with the vertices of G. Let H_E be a clique of size m and identify the vertices of H_E with the edges of G. Finally, add an edge between vertex v of H_V and vertex e of H_E whenever v is incident to e in G. Set $k' = \binom{k}{2}$ and $t = k + m - \binom{k}{2}$. The construction is shown in Fig. 1 a).

If G has a k-clique K, then for the set X that consists of the vertices e of H_E corresponding to edges of K we have $|X| = \binom{k}{2}$ and $|N_{G'}(X)| = k + m - \binom{k}{2}$. On the other hand, suppose that there is a set of vertices X of G' such that $|X| \leq k'$ and $|N_{G'}(X)| \leq t$. First, we note that X cannot contain any vertices of H_V, as then $N_{G'}(X)$ would be too large. Thus, the set X consists of vertices of H_E. Furthermore, we have that $|X| = \binom{k}{2}$. Indeed, assume that this is not the case. If $|X| \leq \binom{k-1}{2} = \binom{k}{2} - k$, then, since X has at least one neighbor in H_V, we have

$$|N_{G'}(X)| \geq m - |X| + 1 \geq m - \binom{k}{2} + k + 1\,,$$

and if $\binom{k-1}{2} < |X| < \binom{k}{2}$, then X has at least k neighbors in H_V, and thus

$$|N_{G'}(X)| \geq m - |X| + k > m - \binom{k}{2} + k\,.$$

Thus, we have that X only consist of vertices of H_E and $|X| = \binom{k}{2}$. But then the vertices of H_V that are in $N_{G'}(X)$ form a k-clique in G.

The W[1]-hardness proof for CUTTING AT MOST k VERTICES WITH TERMINAL uses the same arguments. The only difference is that we add the terminal s in the clique H_E and let $k' = \binom{k}{2} + 1$ (see Fig. 1 b).

Because CLIQUE is well known to be NP-complete [12] and our parameterized reductions are polynomial in k, it immediately follows that CUTTING AT MOST k VERTICES and CUTTING AT MOST k VERTICES WITH TERMINAL are NP-complete. $\qquad\square$

While we have an FPT-algorithm for CUTTING AT MOST k VERTICES when parameterized by k and t or by t only, it is unlikely that the problem has a polynomial kernel (we refer to [8,10,18] for the formal definitions of kernels). Let G be a graph with s connected components $G_1, \ldots, G_s$, and let $k \geq 1$, $t \geq 0$ be integers. Now (G, k, t) is a YES-instance of CUTTING AT MOST k VERTICES if and only if (G_i, k, t) is a YES-instance for some $i \in \{1, \ldots, s\}$, because it can always be assumed that a solution is connected. By the results of Bodlaender et al. [3], this together with Theorem 3 implies the following.

Theorem 4. CUTTING AT MOST k VERTICES *has no polynomial kernel when parameterized either by k and t or by t only, unless* NP $\subseteq$ coNP/poly.

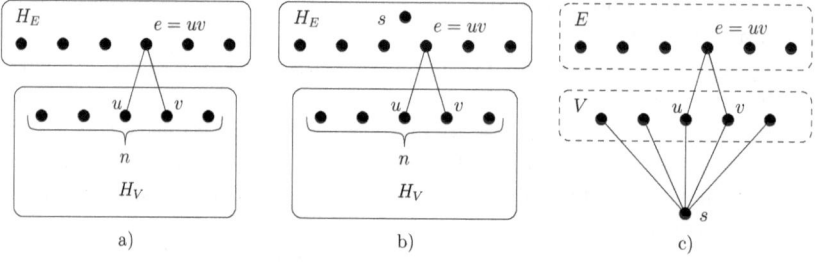

Fig. 1. Constructions of G' in the proofs of Theorems 3 and 5

We will next show that when we consider the size of the separator t as the sole parameter, adding a terminal makes the problem harder. Indeed, while CUTTING AT MOST k VERTICES WITH TERMINAL with parameter t is trivially in XP, we next show that it is also W[1]-hard, in contrast to Theorem 2.

Theorem 5. CUTTING AT MOST k VERTICES WITH TERMINAL *is* W[1]-*hard with parameter t.*

Proof. Again, we prove the claim by a reduction from CLIQUE. Let (G, k) be a clique instance, $n = |V(G)|$ and $m = |E(G)|$; we create an instance (G', k', t, s) of CUTTING AT MOST k VERTICES WITH TERMINAL. The graph G' is constructed as follows. Create a new vertex s as the terminal. For each vertex and edge of G, add a corresponding vertex to G', and add an edge between vertices v and e in G' when e is incident to v in G. Finally, connect all vertices of G' corresponding to vertices of G to the terminal s, and set $k' = n - k + m - \binom{k}{2} + 1$ and $t = k$. The construction is shown in Fig. 1 c).

If G has a k-clique K, then cutting away the k vertices of G' corresponding to K leaves exactly $n - k + m - \binom{k}{2} + 1$ vertices in the connected component of $G' - K$ containing s. Now suppose that $X \subseteq V(G')$ is a set with $s \in X$ such that $|X| \leq k'$ and $|N_{G'}(X)| \leq t$, and let $S = N_{G'}(X)$. Note that the elements of $V(G')$ that do not belong to X are exactly the elements $v \in S$ and the vertices corresponding to $e = uv$ such that $u, v \in S$ and $e \notin S$; denote this latter set of elements by E_0. Since X is a solution, we have $|S| + |E_0| \geq \binom{k}{2} + k$, and thus $|E_0| \geq \binom{k}{2}$. But this is only possible if S is a k-clique in G.   □

Finally, we show that CUTTING AT MOST k VERTICES BY EDGE-CUT WITH TERMINAL is also NP-hard.

Theorem 6. CUTTING AT MOST k VERTICES BY EDGE-CUT WITH TERMINAL *is* NP-*complete.*

Proof. We give a reduction from the CLIQUE problem. It is known that this problem is NP-complete for regular graphs [12]. Let (G, k) be an instance of CLIQUE, with G being a d-regular n-vertex graph. We create an instance (G', k', t, s) of CUTTING k VERTICES BY EDGE-CUT WITH TERMINAL as follows. The graph G' is constructed by starting from a base clique of size dn. One vertex in this base clique is selected as the terminal s, and we additionally distinguish d special vertices. For each $v \in V(G)$, we add a new vertex to G', and add an edge between this vertex and all of the d distinguished vertices of the base clique. For each edge $e = uv$ in G, we also add a new vertex to G', and add edges between this vertex and vertices corresponding to u and v. The construction is shown in Fig. 2. We set $k' = dn + k + \binom{k}{2}$ and $t = dn - 2\binom{k}{2}$.

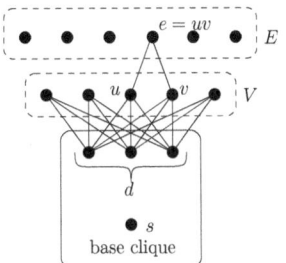

Fig. 2. Construction of G' in the proof of Theorem 6

If G has a k-clique K, then selecting as X the base clique and all vertices of G' corresponding to vertices and edges of K gives a solution to (G', k', t, s), as we have $|X| = dn + k + \binom{k}{2}$ and $|\partial(X)| = (dn - dk) + \left(dk - 2\binom{k}{2}\right) = dn - 2\binom{k}{2}$. For the other direction, consider any solution X to instance (G', k', t, s). The set X must contain the whole base clique, as otherwise there are at least $dn - 1$ edges inside the base clique that belong to $\partial(X)$. Let $V_0 \subseteq V$ and $E_0 \subseteq E$ be the subsets of X corresponding to vertices and edges of G, respectively. If $E_0 = \emptyset$, then $|\partial(X)| = dn$. Assume now that V_0 is fixed, and consider how adding vertices to E_0 changes $|\partial(X)|$. For each edge $e \in E(G)$, if neither of the endpoints of e is in V_0, then adding e to E_0 adds 2 to $|\partial(X)|$. If exactly one of the endpoints of e is in V_0, then adding e to E_0 does not change $|\partial(X)|$. Finally, if both of the endpoints of e are in V_0, then adding e to E_0 reduces $|\partial(X)|$ by 2. Thus, in order to have $|\partial(X)| \leq dn - 2\binom{k}{2}$, we must have that $|E_0| \geq \binom{k}{2}$ and the endpoints of all edges in E_0 are in V_0. But due to the requirement that $|X| \leq dn + k + \binom{k}{2}$, this is only possible if V_0 induces a clique in G. $\square$

Acknowledgments. We thank the anonymous reviewers for pointing out the related work of Lokshtanov and Marx. This work is supported by the European Research Council (ERC) via grant Rigorous Theory of Preprocessing, reference 267959 (F.F., P.G.) and by the Helsinki Doctoral Programme in Computer Science – Advanced Computing and Intelligent Systems (J.K.).

References

1. Alon, N., Yuster, R., Zwick, U.: Color-coding. Journal of the ACM 42(4), 844–856 (1995)
2. Armon, A., Zwick, U.: Multicriteria global minimum cuts. Algorithmica 46(1), 15–26 (2006)
3. Bodlaender, H.L., Downey, R.G., Fellows, M.R., Hermelin, D.: On problems without polynomial kernels. Journal of Computer and System Sciences 75(8), 423–434 (2009)
4. Cai, L., Chan, S.M., Chan, S.O.: Random separation: A new method for solving fixed-cardinality optimization problems. In: Bodlaender, H.L., Langston, M.A. (eds.) IWPEC 2006. LNCS, vol. 4169, pp. 239–250. Springer, Heidelberg (2006)
5. Cao, Y.: A note on small cuts for a terminal (2013), arXiv:1306.2578 [cs.DS]
6. Chen, J., Liu, Y., Lu, S.: An improved parameterized algorithm for the minimum node multiway cut problem. Algorithmica 55(1), 1–13 (2009)
7. Diestel, R.: Graph theory, 4th edn. Springer (2010)
8. Downey, R.G., Fellows, M.R.: Parameterized complexity. Springer (1999)
9. Feige, U., Krauthgamer, R., Nissim, K.: On cutting a few vertices from a graph. Discrete Applied Mathematics 127(3), 643–649 (2003)
10. Flum, J., Grohe, M.: Parameterized complexity theory. Springer (2006)
11. Galbiati, G.: Approximating minimum cut with bounded size. In: Pahl, J., Reiners, T., Voß, S. (eds.) INOC 2011. LNCS, vol. 6701, pp. 210–215. Springer, Heidelberg (2011)
12. Garey, M.R., Johnson, D.S.: Computers and intractability: a guide to the theory of NP-completeness. W. H. Freeman and Co., San Francisco (1979)
13. Hayrapetyan, A., Kempe, D., Pál, M., Svitkina, Z.: Unbalanced graph cuts. In: Brodal, G.S., Leonardi, S. (eds.) ESA 2005. LNCS, vol. 3669, pp. 191–202. Springer, Heidelberg (2005)
14. Li, A., Zhang, P.: Unbalanced graph partitioning. Theory of Computing Systems 53(3), 454–466 (2013)
15. Lokshtanov, D., Marx, D.: Clustering with local restrictions. Information and Computation 222, 278–292 (2013)
16. Marx, D.: Parameterized graph separation problems. Theoretical Computer Science 351(3), 394–406 (2006)
17. Naor, M., Schulman, L., Srinivasan, A.: Splitters and near-optimal derandomization. In: 36th Annual Symposium on Foundations of Computer Science (FOCS 1995), pp. 182–191. IEEE (1995)
18. Niedermeier, R.: Invitation to fixed-parameter algorithms. Oxford University Press (2006)
19. Watanabe, T., Nakamura, A.: Edge-connectivity augmentation problems. Journal of Computer and System Sciences 35(1), 96–144 (1987)

On Fixed-Polynomial Size Circuit Lower Bounds for Uniform Polynomials in the Sense of Valiant

Hervé Fournier[1,*], Sylvain Perifel[1,**], and Rémi de Verclos[2,***]

[1] Univ Paris Diderot, Sorbonne Paris Cité
[2] ENS Lyon
fournier@math.univ-paris-diderot.fr,
sylvain.perifel@liafa.univ-paris-diderot.fr,
remi.de_joannis_de_verclos@ens-lyon.fr

Abstract. We consider the problem of fixed-polynomial lower bounds on the size of arithmetic circuits computing uniform families of polynomials. Assuming the Generalised Riemann Hypothesis (GRH), we show that for all k, there exist polynomials with coefficients in MA having no arithmetic circuits of size $O(n^k)$ over $\mathbb{C}$ (allowing any complex constant). We also build a family of polynomials that can be evaluated in AM having no arithmetic circuits of size $O(n^k)$. Then we investigate the link between fixed-polynomial size circuit bounds in the Boolean and arithmetic settings. In characteristic zero, it is proved that $\mathsf{NP} \not\subset \mathsf{size}(n^k)$, or $\mathsf{MA} \subset \mathsf{size}(n^k)$, or $\mathsf{NP} = \mathsf{MA}$ imply lower bounds on the circuit size of uniform polynomials in n variables from the class VNP over $\mathbb{C}$, assuming GRH. In positive characteristic p, uniform polynomials in VNP have circuits of fixed-polynomial size if and only if both $\mathsf{VP} = \mathsf{VNP}$ over $\mathbb{F}_p$ and $\mathsf{Mod}_p\mathsf{P}$ has circuits of fixed-polynomial size.

Keywords: Arithmetic circuits, Circuit lower bounds, Valiant's classes, Complex field, Arthur-Merlin.

1 Introduction

Baur and Strassen [1] proved in 1983 that the number of arithmetic operations needed to compute the polynomials $x_1^n + \ldots + x_n^n$ is $\Omega(n \log n)$. This is still the best lower bound on uniform polynomials on n variables and of degree $n^{O(1)}$, if uniformity means having circuits computed in polynomial time.

If no uniformity condition is required, lower bounds for polynomials have been known since Lipton [10]. For example, Schnorr [14], improving on [10] and Strassen [15], showed for any k a lower bound $\Omega(n^k)$ on the complexity of a family (P_n) of univariate polynomials of degree polynomial in n – even allowing arbitrary complex constants in the circuits. The starting point of Schnorr's

---

[*] Institut de Mathématiques de Jussieu, UMR 7586 CNRS, F-75205 Paris, France.
[**] LIAFA, UMR 7089 CNRS, F-75205 Paris, France.
[***] ENS Lyon. F-69342 Lyon, France.

K. Chatterjee and J. Sgall (Eds.): MFCS 2013, LNCS 8087, pp. 433–444, 2013.
© Springer-Verlag Berlin Heidelberg 2013

method is to remark that the coefficients of a polynomial computed by a circuit using constants $\alpha = (\alpha_1, \ldots, \alpha_p)$ is given by a polynomial mapping in α. Hence, finding hard polynomials reduces to finding a point outside the image of the mapping associated to some circuit which is universal for a given size. This method has been studied and extended by Raz [12].

In the Boolean setting, this kind of fixed-polynomial lower bounds has already drawn a lot of attention, from Kannan's result [7] proving that for all k, Σ_2^p does not have circuits of size n^k, to [3], delineating the frontier of Boolean classes which are known to have fixed-polynomial size circuits lower bounds. It might seem easy to prove similar lower bounds in the algebraic world, but the fact that arbitrary constants from the underlying field (e.g. $\mathbb{C}$) are allowed prevents from readily adapting Boolean techniques.

Different notions of uniformity can be thought of, either in terms of the circuits computing the polynomials, or in terms of the complexity of computing the coefficients. For instance, an inspection of the proof of Schnorr's result mentioned above shows that the coefficients of the polynomials can be computed in exponential time. But this complexity is generally considered too high to qualify these polynomials as uniform.

The first problem we tackle is the existence of hard polynomials (i.e. without small circuits over $\mathbb{C}$) but with coefficients that are "easy to compute". The search for a uniform family of polynomials with no circuits of size n^k was pursued recently by Jansen and Santhanam [6]. They show in particular that there exist polynomials with coefficients in MA (thus, uniform in some sense) but not computable by arithmetic circuits of size n^k over $\mathbb{Z}$.[1] Assuming the Generalised Riemann Hypothesis (GRH), we extend their result to the case of circuits over the complex field. GRH is used to eliminate the complex constants in the circuits, by considering solutions over $\mathbb{F}_p$ of systems of polynomial equations, for a small prime p, instead of solutions over $\mathbb{C}$. In fact, the family of polynomials built by Jansen and Santhanam is also uniform in the following way: it can be evaluated at integer points in MA. Along this line, we obtain families of polynomials without arithmetic circuits of size n^k over $\mathbb{C}$ and that can be evaluated in AM. The arbitrary complex constants prevents us to readily adapt Jansen and Santhanam's method and we need to use in addition the AM protocol of Koiran [8] in order to decide whether a system of polynomial equations has a solution over $\mathbb{C}$.

Another interesting and robust notion of uniformity is provided by Valiant's algebraic class VNP, capturing the complexity of the permanent. The usual definition is non-uniform, but a natural uniformity condition can be required and gives two equivalent characterisations: in terms of the uniformity of circuits and in terms of the complexity of the coefficients. This is one of the notions we shall study in this paper and which is also used by Raz [12] (where the term *explicit* is used to denote uniform families of VNP polynomials). The second problem we study is therefore to give an $\Omega(n^k)$ lower bound on the complexity of an n-variate polynomial in the uniform version of the class VNP. Note that from Valiant's criterion, it corresponds

---

[1] Even though this result is not stated explicitly in their paper, it is immediate to adapt their proof to our context.

to the coefficients being in GapP, so it is a special case of coefficients that are easy to compute. Even though MA may seem a small class in comparison with GapP (in particular due to Toda's theorem $PH \subseteq P^{\#P}$), the result obtained above does not yield lower bounds for the uniform version of VNP.

We show how fixed-polynomial circuit size lower bound on uniform VNP is connected to various questions in Boolean complexity. For instance, the hypothesis that NP does not have circuits of size n^k for all k, or the hypothesis that MA has circuits of size n^k for some k, both imply the lower bound on the uniform version of VNP assuming GRH. Concerning the question on finite fields, we show an equivalence between lower bounds on uniform VNP and standard problems in Boolean and algebraic complexity.

The paper is organised as follows. Definitions, in particular of the uniform versions of Valiant's classes, are given in Section 2. Hard families of polynomials with easy to compute coefficients, or that are easy to evaluate, are built in Section 3. Finally, conditional lower bounds on uniform VNP are presented in the last section.

2 Preliminaries

Arithmetic Circuits. An arithmetic circuit over a field K is a directed acyclic graph whose vertices have indegree 0 or 2 and where a single vertex (called the output) has outdegree 0. Vertices of indegree 0 are called inputs and are labelled either by a variable x_i or by a constant $\alpha \in K$. Vertices of indegree 2 are called gates and are labelled by $+$ or $\times$.

The polynomial computed by a vertex is defined recursively as follows: the polynomial computed by an input is its label; a $+$ gate (resp. $\times$ gate), having incoming edges from vertices computing the polynomials f and g, computes the polynomial $f + g$ (resp. fg). The polynomial computed by a circuit is the polynomial computed by its output gate.

A circuit is called *constant-free* if the only constant appearing at the inputs is -1. The *formal degree* of a circuit is defined by induction in the following way: the formal degree of a leaf is 1, and the formal degree of a sum (resp. product) is the maximum (resp. sum) of the formal degree of the incoming subtrees (thus constants "count as variables" and there is no possibility of cancellation).

We are interested in sequences of arithmetic circuits $(C_n)_{n \in \mathbb{N}}$, computing sequences of polynomials $(P_n)_{n \in \mathbb{N}}$ (we shall usually drop the subscript "$n \in \mathbb{N}$").

Definition 1. *Let K be a field. If $s : \mathbb{N} \to \mathbb{N}$ is a function, a family (P_n) of polynomials over K is in $\mathsf{asize}_K(s(n))$ if it is computed by a family of arithmetic circuits of size $O(s(n))$ over K.*

Similarly, $\mathsf{size}(s(n))$ denotes the set of (Boolean) languages decided by Boolean circuits of size $O(s(n))$.

Counting Classes. A function $f : \{0,1\}^\star \to \mathbb{N}$ is in $\#P$ if there exists a polynomial $p(n)$ and a language $A \in P$ such that for all $x \in \{0,1\}^\star$

$$f(x) = |\{y \in \{0,1\}^{p(|x|)}, \ (x,y) \in A\}|.$$

A function $g : \{0,1\}^\star \to \mathbb{Z}$ is in GapP if there exist two functions $f, f' \in \#P$ such that $g = f - f'$. The class $C_=P$ is the set of languages $A = \{x, g(x) = 0\}$ for some function $g \in$ GapP. The class $\oplus P$ is the set of languages $A = \{x, f(x) \text{ is odd}\}$ for some function $f \in \#P$. We refer the reader to [4] for more details on counting classes.

Valiant's Classes and Their Uniform Counterpart. Let us first recall the usual definition of Valiant's classes.

Definition 2 (Valiant's classes). *Let K be a field. A family (P_n) of polynomials over K is in the class VP_K if the degree of P_n is polynomial in n and (P_n) is computed by a family (C_n) of polynomial-size arithmetic circuits over K.*

A family $(Q_n(x))$ of polynomials over K is in the class VNP_K if there exists a family $(P_n(x,y)) \in \mathsf{VP}_K$ such that

$$Q_n(x) = \sum_{y \in \{0,1\}^{\ell_n}} P_n(x,y)$$

where ℓ_n denotes the length of y in P_n.

The size of x and y is limited by the circuits for P_n and is therefore polynomial. Note that the only difference between VP_K and $\mathsf{asize}_K(\text{poly})$ is the constraint on the degree of P_n. If the underlying field K is clear, we shall drop the subscript "K" and speak only of VP and VNP. Based on these usual definitions, we now define uniform versions of Valiant's classes.

Definition 3 (Uniform Valiant's classes). *Let K be a field. A family of circuits (C_n) is called uniform if the (usual, Boolean) encoding of C_n can be computed in time $n^{O(1)}$. A family of polynomials (P_n) over K is in the class $\mathsf{unif\text{-}VP}_K$ if it is computed by a uniform family of constant-free arithmetic circuits of polynomial formal degree.*

A family of polynomials $(Q_n(x))$ over K is in the class $\mathsf{unif\text{-}VNP}_K$ if Q_n has n variables $x = x_1, \ldots, x_n$ and there exists a family $(P_n(x,y)) \in \mathsf{unif\text{-}VP}_K$ such that

$$Q_n(x) = \sum_{y \in \{0,1\}^{\ell_n}} P_n(x,y)$$

where ℓ_n denotes the length of y in P_n.

The uniformity condition implies that the size of the circuit C_n in the definition of $\mathsf{unif\text{-}VP}$ is polynomial in n. Note that $\mathsf{unif\text{-}VP}_K$ and $\mathsf{unif\text{-}VNP}_K$ only depend on the characteristic of the field K (indeed, since no constant from K is allowed in the circuits, these classes are equal to the ones defined over the prime subfield of K).

In the definition of $\mathsf{unif\text{-}VNP}$, we have chosen to impose that Q_n has n variables because this enables us to give a very succinct and clear statement of our questions. This is *not* what is done in the usual non-uniform definition where the number of variables is only limited by the (polynomial) size of the circuit.

The well-known "Valiant's criterion" is easily adapted to the uniform case in order to obtain the following alternative characterisation of unif-VNP.

Proposition 1 (Valiant's criterion). *In characteristic zero, a family (P_n) is in* unif-VNP *iff P_n has n variables, a polynomial degree and its coefficients are computable in* GapP*; that is, the function mapping $(c_1, \ldots, c_n)$ to the coefficient of $X_1^{c_1} \cdots X_n^{c_n}$ in P_n is in* GapP*.*
*The same holds in characteristic $p > 0$ with coefficients in "*GapP $\bmod p$*"[2].*

Over a field K, a polynomial $P(x_1, \ldots, x_n)$ is said to be a projection of a polynomial $Q(y_1, \ldots, y_m)$ if $P(x_1, \ldots, x_n) = Q(a_1, \ldots, a_m)$ for some choice of $a_1, \ldots, a_m \in \{x_1, \ldots, x_n\} \cup K$. A family (P_n) reduces to (Q_n) (via projections) if P_n is a projection of $Q_{q(n)}$ for some polynomially bounded function q.

The Hamiltonian Circuit polynomials are defined by

$$\mathrm{HC}_n(x_{1,1}, \ldots, x_{n,n}) = \sum_{\sigma} \prod_{i=1}^{n} x_{i,\sigma(i)},$$

where the sum is on all cycles $\sigma \in S_n$ (i.e. on all the Hamiltonian cycles of the complete graph over $\{1, \ldots, n\}$). The family (HC_n) is known to be VNP-complete over any field [16] (for projections).

Elimination of Complex Constants in Circuits. The weight of a polynomial $P \in \mathbb{C}[X_1, \ldots, X_n]$ is the sum of the absolute values of its coefficients. We denote it by $\omega(P)$. It is well known that for $P, Q \in \mathbb{C}[X_1, \ldots, X_n]$ and $\alpha \in \mathbb{C}$, $\omega(PQ) \leqslant \omega(P)\omega(Q)$, $\omega(P + Q) \leqslant \omega(P) + \omega(Q)$ and $\omega(\alpha P) = |\alpha|\omega(P)$.

The following result gives a bound on the weight of a polynomial computed by a circuit.

Lemma 1. *Let P be a polynomial computed by an arithmetic circuit of size s and formal degree d with constants of absolute value bounded by $M \geqslant 2$, then $\omega(P) \leqslant M^{s \cdot d}$.*

Proof. We prove it by induction on the structure of the circuit C which computes P. The inequality is clear if the output of C is a constant or a variable since $\omega(P) \leqslant M$, $s \geqslant 1$ and $d \geqslant 1$ in this case. If the output of P is a $+$ gate then P is the sum of the value of two polynomials P_1 and P_2 calculated by subcircuits of C of formal degree at most d and size at most $s-1$. By induction hypothesis, we have $\omega(P_1) \leqslant M^{d(s-1)}$ and $\omega(P_1) \leqslant M^{d(s-1)}$. We have $\omega(P) \leqslant \omega(P_1) + \omega(P_2)$ so $\omega(P) \leqslant 2 \cdot M^{d(s-1)} \leqslant M^{d(s-1)+1} \leqslant M^{ds}$. If the output of C in a $\times$ gate, P is the product some polynomials P_1 and P_2 each calculated by circuits of size at most $s-1$ and degrees d_1 and d_2 respectively such that $d_1 + d_2 = d$. Then $\omega(P) \leqslant \omega(P_1)\omega(P_2) \leqslant M^{(s-1)d_1} M^{(s-1)d_2} = M^{(s-1)d} \leqslant M^{sd}$. $\square$

---

[2] This is equivalent to the fact that for all $v \in \mathbb{F}_p$, the set of monomials having coefficient v is in $\mathrm{Mod}_p\mathrm{P}$.

For $a \in \mathbb{N}$, we denote by $\pi(a)$ the number of prime numbers smaller than or equal to a. For a system S of polynomial equations with integer coefficients, we denote by $\pi_S(a)$ the number of prime numbers $p \leqslant a$ such that S has a solution over $\mathbb{F}_p$. The following lemma will be useful for eliminating constants from $\mathbb{C}$. (Note that the similar but weaker statement first shown by Koiran [8] as a step in his proof of Theorem 1 would be enough for our purpose.)

Lemma 2 (Bürgisser [2, p. 64]). *Let S be a system of polynomial equations*

$$P_1(x) = 0, \ldots, P_m(x) = 0$$

with coefficients in $\mathbb{Z}$ and with the following parameters : n unknowns, and for all i, degree of P_i at most d and $\omega(P_i) \leqslant w$. If the system S has a solution over $\mathbb{C}$ then under GRH,

$$\pi_S(a) \geqslant \frac{\pi(a)}{d^{O(n)}} - \sqrt{a} \log(wa).$$

At last, we need a consequence of VNP having small arithmetic circuits over the complex field.

Lemma 3. *Assume GRH. If* VP = VNP *over* $\mathbb{C}$, *then* CH = MA.

Proof. Assume VP = VNP over $\mathbb{C}$. From the work on Boolean parts of Valiant's classes [2, Chapter 4], this implies P/poly = PP/poly = CH/poly, therefore MA = CH [11]. □

3   Hard Polynomials with Coefficients in MA

We begin with lower bounds on polynomials with coefficients in PH before bringing them down to MA.

Hard Polynomials with Coefficients in PH. We first need to recall a couple of results. The first one is an upper bound on the complexity of the following problem called HN (named after Hilbert's Nullstellensatz):

Input A system $S = \{P_1 = 0, \ldots, P_m = 0\}$ of n-variate polynomial equations with integer coefficients, each polynomial $P_i \in \mathbb{Z}[x_1, \ldots, x_n]$ being given as a constant-free arithmetic circuit.
Question Does the system S have a solution over $\mathbb{C}^n$?

Theorem 1 (Koiran [8]). *Assuming GRH is true,* $HN \in$ PH.

Koiran's result is stated here for polynomials given by arithmetic circuits, instead of the list of their coefficients. Adapting the result of the original paper in terms of arithmetic circuits is not difficult: it is enough to add one equation per gate expressing the operation made by the gate, thus simulating the whole circuit.

   The second result is used in the proof of Schnorr's result mentioned in the introduction.

Lemma 4 (Schnorr [14]). *Let (U_n) be the family of polynomials defined inductively as follows:*

$$\begin{cases} U_1 = a_0^{(1)} + b_0^{(1)} x & \text{where } a_0^{(1)}, b_0^{(1)} \text{ and } x \text{ are new variables} \\ U_n = \left(\sum_{i=1}^{n-1} a_i^{(n)} U_i \right) \left(\sum_{i=1}^{n-1} b_i^{(n)} U_i \right) & \text{where } a_i^{(n)}, b_i^{(n)} \text{ are new variables.} \end{cases}$$

Thus U_n has variables x, $a_i^{(j)}$ and $b_i^{(j)}$ (for $1 \leqslant j \leqslant n$ and $0 \leqslant i < j$). For simplicity, we will write $U_n(a,b,x)$, where the total number of variables in the tuples a, b is $n(n+1)$.

For every univariate polynomial $P(x)$ over $\mathbb{C}$ computed by an arithmetic circuit of size s, there are constants $a, b \in \mathbb{C}^{s(s+1)}$ such that $P(x) = U_s(a, b, x)$.

The polynomials U_s in this lemma are universal in the sense that they can simulate any circuit of size s; the definition of such a polynomial indeed reproduces the structure of an arbitrary circuit by letting at each gate the choice of the inputs and of the operation, thanks to new variables.

The third result we'll need is due to Hrubeš and Yehudayoff [5] and relies on Bézout's Theorem. Showing Theorem 2 could also be done without using algebraic geometry, but this would complicate the overall proof.

Lemma 5 (Hrubeš and Yehudayoff [5]). *Let $F : \mathbb{C}^n \to \mathbb{C}^m$ be a polynomial map of degree $d > 0$, that is, $F = (F_1, \ldots, F_m)$ where each F_i is a polynomial of degree at most d. Then $|F(\mathbb{C}^n) \cap \{0,1\}^m| \leqslant (2d)^n$.*

We are now ready to give our theorem.

Theorem 2. *Assume GRH is true. For any constant k, there is a family (P_n) of univariate polynomials with coefficients in $\{0,1\}$ satisfying:*

- *$\deg(P_n) = n^{O(1)}$ (polynomial degree);*
- *the coefficients of P_n are computable in PH, that is, on input $(1^n, i)$ we can decide in PH if the coefficient of x^i is 1;*
- *(P_n) is not computed by arithmetic circuits over $\mathbb{C}$ of size n^k.*

Proof. Fix $s = n^k$. Consider the universal polynomial $U_s(a, b, x)$ of Lemma 4 simulating circuits of size s. If $\alpha_i^{(s)}$ denotes the coefficient of x^i in U_s, then we have the relation

$$\alpha_i^{(s)} = \sum_{\substack{i_1 + i_2 = i \\ s_1, s_2 < s}} a_{s_1}^{(s)} b_{s_2}^{(s)} \alpha_{i_1}^{(s_1)} \alpha_{i_2}^{(s_2)}.$$

By induction, the coefficient $\alpha_i^{(s)}$ is therefore a polynomial in a, b of degree $\leqslant (i+1)2^{2s}$.

Now, we would like to find a polynomial whose coefficients are different from the $\alpha_i^{(s)}$ for any value of a, b. This will be done thanks to Lemma 5, but we have to use it in a clever way because our method requires to use interpolation on $d+1$ points to identify two polynomials of degree d: hence we need to "truncate" the polynomial U_s to degree d.

Fix $d = s^4$. It follows from the beginning of the proof that the map computing the first $(d+1)$ coefficients of U_s

$$F : \mathbb{C}^{s(s+1)} \to \mathbb{C}^{d+1}$$
$$(a, b) \quad \mapsto (\alpha_0^{(s)}, \dots, \alpha_d^{(s)})$$

is a polynomial map of degree at most $(d+1)2^{2s}$. Since $((d+1)2^{2s})^{s(s+1)} < 2^{d+1}$, by Lemma 5 there exist coefficients $(\beta_0, \dots, \beta_d) \in \{0,1\}^{d+1}$ not in $F(\mathbb{C}^{s(s+1)})$. In other words, for any values of a, b in $\mathbb{C}$, the first $(d+1)$ coefficients of U_s differ from $(\beta_0, \dots, \beta_d)$.

Let $P_\beta(x)$ be the polynomial $\sum_{i=0}^{d} \beta_i x^i$ and let us call $U_{s|d}$ the truncation of U_s up to degree d, that is, the sum of all the monomials of degree $\leqslant d$ in x. For any instantiation of a, b in $\mathbb{C}$, we have $U_{s|d}(a, b, x) \neq P_\beta(x)$. Since both polynomials are of degree smaller than or equal to d, this means that there exists an integer $m \in \{0, \dots, d\}$ such that $U_{s|d}(a, b, m) \neq P_\beta(m)$. Therefore the following system of polynomial equations with unknowns a, b:

$$S_\beta = \{U_{s|d}(a, b, m) = P_\beta(m) \ : \ m \in \{0, \dots, d\}\}$$

has no solution over $\mathbb{C}$.

Conversely, consider now this system for other coefficients than β, that is, S_γ for $\gamma_0, \dots, \gamma_d \in \{0, 1\}$. If S_γ does not have a solution over $\mathbb{C}$, this means that for any instantiation of $a, b \in \mathbb{C}$ we have $U_{s|d}(a, b, x) \neq P_\gamma(x)$, hence P_γ is not computable by a circuit of size s by Lemma 4.

The goal now is then to find values of $\gamma \in \{0, 1\}^{d+1}$ such that S_γ does not have a solution over $\mathbb{C}$.

Remark first that on input $\gamma_0, \dots, \gamma_d \in \{0, 1\}$ and $m \in \{0, \dots, d\}$, we can describe in polynomial time a circuit $C_{\gamma, m}(a, b)$ computing the polynomial $U_{s|d}(a, b, m) - P_\gamma(m)$. Indeed, U_s is computable by an easily described circuit following its definition, hence its truncation to degree d also is (by computing the homogeneous components up to degree d), and a circuit for P_γ is also immediate if we are given γ. Therefore, we can describe in polynomial time the system S_γ to be used in Theorem 1.

The algorithm in PH to compute the coefficients of a polynomial P_β without circuits of size s is then the following on input $(1^n, i)$:

- Find the lexicographically first $\gamma_0, \dots, \gamma_d \in \{0, 1\}$ such that $S_\gamma \notin \mathrm{HN}$;
- accept iff $\gamma_i = 1$.

This algorithm is in $\mathrm{PH}^{\mathrm{HN}}$. By Theorem 1, if we assume GRH then the problem HN is in PH. We deduce that computing the coefficients of P_γ can be done in PH. $\qquad \square$

Hard Polynomials with Coefficients in MA. Allowing n variables instead of only one, we can even obtain lower bounds for polynomials with coefficients in MA.

Corollary 1. *Assume GRH is true. For any constant k, there is a family (P_n) of polynomials on n variables, with coefficients in $\{0,1\}$, of degree $n^{O(1)}$, with coefficients computable in* MA, *and such that $(P_n) \notin$ asize$_\mathbb{C}(n^k)$.*

Proof. If the Hamiltonian family (HC_n) does not have circuits of polynomial size over $\mathbb{C}$, consider the following variant of a family with n variables: $\mathrm{HC}'_n(x_1, \ldots, x_n) = \mathrm{HC}_{\lfloor\sqrt{n}\rfloor}(x_1, \ldots, x_{\lfloor\sqrt{n}\rfloor^2})$. This is a family whose coefficients are in P (hence in MA) and without circuits of size n^k.

On the other hand, if the Hamiltonian family (HC_n) has circuits of polynomial size over $\mathbb{C}$, then PH $=$ MA by Lemma 3. Therefore the family of polynomials of Theorem 2 has its coefficients in MA. □

Hard Polynomials That Can Be Evaluated in AM. A family of polynomials $(P_n(x_1, \ldots, x_n))$ is said to be evaluable in AM if the language

$$\{(x_1, \ldots, x_n, i, b) \mid \text{the } i\text{-th bit of } P_n(x_1, \ldots, x_n) \text{ is } b\}$$

is in AM, where $x_1, \ldots, x_n, i$ are integers given in binary and $b \in \{0,1\}$. In the next proposition, we show how to obtain polynomials which can be evaluated in AM. The method is based on Santhanam [13] and Koiran [9] (proof omitted due to space constraints).

Proposition 2. *Assume GRH is true. For any constant k, there is a family (P_n) of polynomials on n variables, with coefficients in $\{0,1\}$, of degree $n^{O(1)}$, evaluable in* AM *and such that $(P_n) \notin$ asize$_\mathbb{C}(n^k)$.*

4   Conditional Lower Bounds for Uniform VNP

In Characteristic Zero. In this whole section we assume GRH is true. Our main result in this section is that if for all k, C$_=$P has no circuits of size n^k, then the same holds for unif-VNP (in characteristic zero). For the clarity of exposition, we first prove the weaker result where the assumption is on the class NP instead.

Lemma 6. *If there exists k such that unif-VNP $\subseteq$ asize$_\mathbb{C}(n^k)$, then there exists ℓ such that NP $\subseteq$ size(n^ℓ).*

Proof. Let us assume that unif-VNP $\subseteq$ asize$_\mathbb{C}(n^k)$. Let $L \in$ NP. There is a polynomial q and a polynomial time computable relation $\phi : \{0,1\}^* \times \{0,1\}^* \to \{0,1\}$ such that for all $x \in \{0,1\}^n$, $x \in L$ if and only if $\exists y \in \{0,1\}^{q(n)}\ \phi(x,y) = 1$.

We define the polynomial P_n by

$$P_n(X_1, \ldots, X_n) = \sum_{x \in \{0,1\}^n} \left(\sum_{y \in \{0,1\}^{q(n)}} \phi(x,y) \right) \prod_{i=1}^n X_i^{x_i}(1 - X_i)^{1-x_i}.$$

Note that for $x \in \{0,1\}^n$, $P_n(x)$ is the number of elements y in relation with x via ϕ. By Valiant's criterion (Proposition 1), the family (P_n) belongs to unif-VNP in characteristic 0. By hypothesis, there exists a family of arithmetic circuits (C_n) over $\mathbb{C}$ computing (P_n), with C_n of size $t = O(n^k)$.

Let $\alpha = (\alpha_1, \ldots, \alpha_t)$ be the complex constants used by the circuit. We have $P_n(X_1, \ldots, X_n) = C_n(X_1, \ldots, X_n, \alpha)$. Take one unknown Y_i for each α_i and one additional unknown Z, and consider the following system S:

$$\begin{cases} \left(\prod_{x \in L \cap \{0,1\}^n} C_n(x, Y)\right) \cdot Z = 1 \\ C_n(x, Y) = 0 \text{ for all } x \in \{0,1\}^n \setminus L. \end{cases}$$

Note that introducing one equation for each $x \in L \cap \{0,1\}^n$ (as we did for each $x \in \{0,1\}^n \setminus L$) would not work since it would require to introduce an exponential number of new variables.

Let $\beta = \left(\prod_{x \in L \cap \{0,1\}^n} C_n(x, \alpha)\right)^{-1}$. Then (α, β) is a solution of S over $\mathbb{C}$.

The system S has $t + 1 = O(n^k)$ unknowns. The degree of $C_n(x, Y)$ is bounded by 2^t; hence the degree of S is at most $2^{O(n^k)}$. Moreover, the weight of the polynomials in S is bounded by $2^{2^{O(n^k)}}$ using Lemma 1.

Since the system S has the solution (α, β) over $\mathbb{C}$, by Lemma 2 it has a solution over $\mathbb{F}_p$ for some p small enough. We recall that $\pi(p) \sim p/\log p$; hence the system S has a solution over $\mathbb{F}_p$ for $p = 2^{O(n^{2k})}$.

Consider p as above and (α', β') a solution of the system S over $\mathbb{F}_p$. By definition of S, when the circuit C_n is evaluated over $\mathbb{F}_p$, the following is satisfied:

$$\begin{cases} \forall x \in L \cap \{0,1\}^n, & C_n(x, \alpha') \neq 0, \\ \forall x \in \{0,1\}^n \setminus L, & C_n(x, \alpha') = 0. \end{cases}$$

Computations over $\mathbb{F}_p$ can be simulated by Boolean circuits, using $\log_2 p$ bits to represent an element of $\mathbb{F}_p$, and $O(\log^2 p)$ gates to simulate an arithmetic operation. This yields Boolean circuits of size n^ℓ for $\ell = O(k)$ to decide the language L.   □

Theorem 3. *Assume GRH is true. Suppose one of the following conditions holds:*

1. NP $\not\subset$ size(n^k) *for all k;*
2. C$_=$P $\not\subset$ size(n^k) *for all k;*
3. MA $\subset$ size(n^k) *for some k;*
4. NP = MA.

Then unif-VNP $\not\subset$ asize$_{\mathbb{C}}(n^k)$ *for all k.*

Proof. The first point is proved in Lemma 6.

The second point subsumes the first since coNP $\subseteq$ C$_=$P. It can be proved in a very similar way. Indeed consider $L \in$ C$_=$P and $f \in$ GapP such that $x \in L \iff f(x) = 0$, and its associated family of polynomials

$$P_n(X_1, \ldots, X_n) = \sum_{x \in \{0,1\}^n} f(x) \prod_{i=1}^n X_i^{x_i} (1 - X_i)^{1-x_i}$$

as in the proof of Lemma 6. Then for all $x \in \{0,1\}^n$, $P_n(x) = 0$ iff $x \in L$. The family (P_n) belongs to unif-VNP and thus, assuming unif-VNP $\subset$ asize$_{\mathbb{C}}(n^k)$, has arithmetic circuits (C_n) over $\mathbb{C}$ of size $t = O(n^k)$. Constants of $\mathbb{C}$ are replaced with elements of a small finite field by considering the system:

$$\begin{cases} C_n(x, Y) = 0 \text{ for all } x \in L \cap \{0,1\}^n \\ \left(\prod_{x \in \{0,1\}^n \setminus L} C_n(x, Y) \right) \cdot Z = 1. \end{cases}$$

The end of the proof is similar.

For the third point, let us assume unif-VNP $\subset$ asize$_{\mathbb{C}}$(poly). It implies VP = VNP thanks to the VNP-completeness of the uniform family (HC_n), then MA = PP by Lemma 3. This implies MA $\not\subset$ size(n^k) for all k since PP $\not\subset$ size(n^k) for all k [17].

For the last point, assume NP = MA. If NP is without n^k circuits for all k, then the conclusion comes from the first point. Otherwise MA has n^k-size circuits and the conclusion follows from the previous point. □

For any constant c, the class $\mathsf{P}^{\mathsf{NP}[n^c]}$ is the set of languages decided by a polynomial time machine making $O(n^c)$ calls to an NP oracle. It is proven in [3] that NP $\subset$ size(n^k) implies $\mathsf{P}^{\mathsf{NP}[n^c]} \subset$ size(n^{ck^2}). Hence, it is enough to assume fixed-polynomial lower bounds on this larger class $\mathsf{P}^{\mathsf{NP}[n^c]}$ for some c to get fixed-polynomial lower bounds on unif-VNP$_{\mathbb{C}}$.

An Unconditional Lower Bound in Characteristic Zero. In this part we do not allow arbitrary constants in circuits. We consider instead circuits with -1 as the only scalar that can label the leaves. For $s : \mathbb{N} \to \mathbb{N}$, let asize$_0(s)$ be the family of polynomials computed by families of unbounded degree constant-free circuits of size $O(s)$ (in characteristic zero). Note that the formal degree of these circuits are not polynomially bounded: hence, large constants produced by small arithmetic circuits can be used. (The proof of the next theorem has been omitted due to space constraints).

Theorem 4. unif-VNP $\not\subset$ asize$_0(n^k)$ for all k.

In Positive Characteristic. This subsection deals with fixed-polynomial lower bounds in positive characteristic. The results are presented in characteristic 2 but they hold in any positive characteristic p (replacing $\oplus$P with Mod$_p$P). (The proof has been omitted due to space constraints.)

Theorem 5. *The following are equivalent:*

- unif-VNP$_{\mathbb{F}_2} \subset$ asize$_{\mathbb{F}_2}(n^k)$ *for some* k;
- VP$_{\mathbb{F}_2}$ = VNP$_{\mathbb{F}_2}$ *and* $\oplus$P $\subset$ size(n^k) *for some* k.

Acknowledgements. We thank Guillaume Malod for useful discussions and Thomas Colcombet for some advice on the presentation.

References

1. Baur, W., Strassen, V.: The complexity of partial derivatives. Theor. Comput. Sci. 22, 317–330 (1983)
2. Bürgisser, P.: Completeness and reduction in algebraic complexity theory. Algorithms and Computation in Mathematics, vol. 7. Springer, Berlin (2000)
3. Fortnow, L., Santhanam, R., Williams, R.: Fixed-polynomial size circuit bounds. In: IEEE Conference on Computational Complexity, pp. 19–26 (2009)
4. Hemaspaandra, L.A., Ogihara, M.: The complexity theory companion. Texts in Theoretical Computer Science. An EATCS Series. Springer, Berlin (2002)
5. Hrubes, P., Yehudayoff, A.: Arithmetic complexity in ring extensions. Theory of Computing 7(1), 119–129 (2011)
6. Jansen, M.J., Santhanam, R.: Stronger lower bounds and randomness-hardness trade-offs using associated algebraic complexity classes. In: STACS, pp. 519–530 (2012)
7. Kannan, R.: Circuit-size lower bounds and non-reducibility to sparse sets. Information and Control 55(1-3), 40–56 (1982)
8. Koiran, P.: Hilbert's Nullstellensatz is in the polynomial hierarchy. J. Complexity 12(4), 273–286 (1996)
9. Koiran, P.: Hilbert's Nullstellensatz is in the polynomial hierarchy. Technical Report 96-27, DIMACS (July 1996)
10. Lipton, R.J.: Polynomials with 0-1 coefficients that are hard to evaluate. In: FOCS, pp. 6–10 (1975)
11. Lund, C., Fortnow, L., Karloff, H.J., Nisan, N.: Algebraic methods for interactive proof systems. In: FOCS, pp. 2–10 (1990)
12. Raz, R.: Elusive functions and lower bounds for arithmetic circuits. Theory of Computing 6(1), 135–177 (2010)
13. Santhanam, R.: Circuit lower bounds for merlin–arthur classes. SIAM J. Comput. 39(3), 1038–1061 (2009)
14. Schnorr, C.-P.: Improved lower bounds on the number of multiplications/divisions which are necessary of evaluate polynomials. Theor. Comput. Sci. 7, 251–261 (1978)
15. Strassen, V.: Polynomials with rational coefficients which are hard to compute. SIAM J. Comput. 3(2), 128–149 (1974)
16. Valiant, L.G.: Completeness classes in algebra. In: STOC, pp. 249–261 (1979)
17. Vinodchandran, N.V.: A note on the circuit complexity of PP. Theor. Comput. Sci. 347(1-2), 415–418 (2005)

A Parameterized Complexity Analysis
of Combinatorial Feature Selection Problems*

Vincent Froese, René van Bevern, Rolf Niedermeier, and Manuel Sorge

Institut für Softwaretechnik und Theoretische Informatik, TU Berlin, Germany
{vincent.froese,rene.vanbevern,rolf.niedermeier,
manuel.sorge}@tu-berlin.de

Abstract. We examine the algorithmic tractability of NP-hard combinatorial feature selection problems in terms of parameterized complexity
theory. In combinatorial feature selection, one seeks to discard dimensions from high-dimensional data such that the resulting instances fulfill
a desired property. In parameterized complexity analysis, one seeks to
identify relevant problem-specific quantities and tries to determine their
influence on the computational complexity of the considered problem.
In this paper, for various combinatorial feature selection problems, we
identify parameterizations and reveal to what extent these govern computational complexity. We provide tractability as well as intractability
results; for example, we show that the DISTINCT VECTORS problem on
binary points is polynomial-time solvable if each pair of points differs in
at most three dimensions, whereas it is NP-hard otherwise.

1 Introduction

Feature selection in a high-dimensional data space means to choose a subset of
features (that is, dimensions) such that some desirable data properties are preserved or achieved. *Combinatorial* feature selection [14, 5] is a well-motivated
alternative to the more frequently studied affine feature selection: While affine
feature selection combines features to reduce dimensionality, combinatorial feature selection chooses a subspace by discarding some dimensions. The advantage
of the latter is that the resulting reduced feature space is easier to interpret. See
Charikar et al. [5] for a more extensive discussion in favor of combinatorial feature
selection. Unfortunately, combinatorial feature selection problems are typically
computationally very hard to solve (NP-hard and also hard to approximate [5]),
resulting in the use of heuristic approaches in practice [2, 8, 12, 13].

In this work, mainly following Charikar et al. [5], who provided classical
computational hardness results (NP-hardness and inapproximability), we adopt
the fresh perspective of parameterized complexity analysis. We thus refine the
known picture of the computational complexity landscape of combinatorial feature selection problems. Intuitively speaking, our guiding principle is to identify

* Vincent Froese was supported by DFG, project DAMM (NI 369/13). René van Bevern and Manuel Sorge were supported by DFG, project DAPA (NI369/12).

problem-specific parameters (quantities such as number of dimensions to discard or number of dimensions to keep) and to analyze how these quantities influence the problem complexity. The point here is that in relevant applications these parameters can be small. Hence, the central question is whether the considered problems become computationally tractable in the case of small parameters.

We revisit two categories of combinatorial feature selection problems (namely dimension reduction and clustering problems) as introduced by Charikar et al. [5]. Within their framework they defined (amongst others) two problems called DISTINCT VECTORS and HIDDEN CLUSTERS. In this work, we consider DISTINCT VECTORS and introduce a new problem called L_p-HIDDEN CLUSTER GRAPH which is based on HIDDEN CLUSTERS. For both problems, we shed new light on the (non-)existence of provably tractable special cases.

DISTINCT VECTORS is a dimension reduction problem defined as follows:

DISTINCT VECTORS
Input: A multiset $S = \{x_1, \ldots, x_n\} \subseteq \Sigma^d$ of n distinct points in d dimensions and $k \in \mathbb{N}$.
Question: Is there a subset $K \subseteq \{1, \ldots, d\}$ of dimensions with $|K| \leq k$ such that all points in $S_{|K}$ are still distinct?

Throughout this work, $S_{|K} := \{x_{1|K}, \ldots, x_{n|K}\}$ denotes the multiset of projections $x_{i|K}$ of the points in S into the dimensions in K, that is, dimensions not in K are set to zero. DISTINCT VECTORS is NP-hard to approximate within a logarithmic factor [5]. It is also known as the MINIMAL REDUCT problem in *rough set theory* [17] and was already earlier proven to be NP-hard [18].

In the clustering category, we assume that the input data would cluster well once some noise is removed. The representative problem for this category is HIDDEN CLUSTERS [5]. The goal is to maximize the number of dimensions that allow for a clustering of the data into a predefined number of cluster centers of a given radius. Notably, the number of sought clusters has to be known in advance. This is not always realistic. Hence, we would like also to reveal clusterings in our data without knowing the number of clusters beforehand. To this end, we employ a clustering notion from graph-based data clustering: Instead of formulating a cluster as a point set within a given radius r from some center as in HIDDEN CLUSTERS, we now formulate a cluster as a set of points of pairwise distance at most r. Such sets of points form cliques in a "threshold graph" that contains an edge between two points whenever their distance is at most r. The search for a clustering now essentially becomes the search for a graph whose connected components are cliques. In contrast to HIDDEN CLUSTERS, this also expresses the need of points in different clusters to be dissimilar to each other.

L_p-HIDDEN CLUSTER GRAPH
Input: A set $S = \{x_1, \ldots, x_n\} \subseteq \Sigma^d$ with $\Sigma \subseteq \mathbb{Q}$, $r \in \mathbb{Q}_0^+$, $k \in \mathbb{N}$.
Question: Is there a subset $K \subseteq \{1, \ldots, d\}$ of dimensions with $|K| \geq k$ such that the graph $G_K = (V, E_K)$ with $V := S$, $E_K := \{\{x_i, x_j\} \mid x_i \neq x_j \in V, \text{dist}_{|K}^{(p)}(x_i, x_j) \leq r\}$ is a cluster graph (that is, a union of disjoint cliques)?

Herein, $\text{dist}_{|K}^{(p)}$ is a metric computing the distance between two points from Σ^d projected to the dimensions in K. We explicitly consider the distance functions induced by the L_p-norm: $\text{dist}^{(p)}(x, y) := \sum_{j=1}^{d} |(x - y)_j|^p$ for $p \in \mathbb{N}$ and $\text{dist}^{(\infty)}(x, y) := \max_{j \in \{1, \ldots, d\}} |(x - y)_j|$. By $(x)_j$ we denote the value of $x \in \Sigma^d$ in the j-th dimension. Note that G_K is a so-called unit ball graph.

Parameterized complexity preliminaries. The computational complexity of a parameterized problem is measured in terms of two quantities: one is the input size, the other is the *parameter* (usually a positive integer). A parameterized problem $L \subseteq \Sigma^* \times \mathbb{N}$ is called *fixed-parameter tractable* with respect to a parameter k if it can be solved in $f(k) \cdot |x|^{O(1)}$ time, where f is a computable function only depending on k, and $|x|$ is the size of the input instance x. A *problem kernel* for a parameterized problem is a many-one self-reduction that runs in polynomial time such that the produced instances have size upper-bounded by some function exclusively depending on the parameter. Existence of a problem kernel is equivalent to fixed-parameter tractability [10, 11, 16].

A *parameterized reduction* from a parameterized problem P to another parameterized problem P' is a function that, given an instance (x, k), computes in $f(k) \cdot |x|^{O(1)}$ time an instance (x', k') (with k' only depending on k) such that (x, k) is a "yes"-instance of P if and only if (x', k') is a "yes"-instance of P'. The two basic complexity classes for showing (presumable) fixed-parameter intractability are called W[1] and W[2]; the standard assumption is that W[1]-hard and W[2]-hard problems are not fixed-parameter tractable [10, 11, 16].

Throughout this work we assume that arithmetic operations such as additions and comparisons of numbers can be done in $O(1)$ time.

Our contributions. For DISTINCT VECTORS we prove W[2]-hardness with respect to the solution size k. In addition, we observe that it cannot be solved in $d^{o(k)} \cdot |x|^{O(1)}$ time unless W[1] = FPT (which is strongly believed not to be the case). Moreover, for DISTINCT VECTORS restricted to a binary input alphabet, we give the following complexity dichotomy: if the maximum pairwise Hamming distance h between input points is at most three, then DISTINCT VECTORS is polynomial-time solvable, and it is NP-complete for $h \geq 4$. The latter NP-completeness proof also implies W[1]-hardness with respect to the parameter $d - k$ ("number of dimensions to discard"). In contrast, we provide some problem kernels with respect to the combined parameters "alphabet size combined with k" and "h combined with k".

For L_p-HIDDEN CLUSTER GRAPH, we show that it is W[2]-hard with respect to the number t of discarded dimensions for all $p \in \mathbb{N}$, whereas it is fixed-parameter tractable with respect to t combined with the radius r. L_∞-HIDDEN CLUSTER GRAPH even is polynomial-time solvable in general.

Due to the lack of space, several technical details are deferred to a full version.

2   Distinct Vectors

Skowron and Rauszer [18] first proved NP-hardness for MINIMAL REDUCT (which is equivalent to DISTINCT VECTORS) by a reduction from HITTING SET.

Charikar et al. [5] additionally showed that there is some constant c such that DISTINCT VECTORS is not polynomial-time approximable within a factor of $c \log d$ unless P = NP. We analyze various restricted scenarios for the DISTINCT VECTORS problem and conduct a more fine-grained computational complexity analysis which, unfortunately, yields further hardness results in most cases. More specifically, we consider the cases of (i) retaining *few* dimensions, (ii) deleting *few* dimensions, and (iii) *small* pairwise differences between points.

We first present results for a binary input alphabet in Section 2.1 and then proceed with results for larger and unbounded alphabet size in Section 2.2.

2.1   Bounded Pairwise Hamming Distance: A Complexity Dichotomy

Throughout this subsection we focus on instances with a binary input alphabet $\Sigma = \{0, 1\}$. We further restrict our considerations to instances with points of bounded "*degree of distinctiveness*". Herein, we refer to instances where each pair of points differs in at most h dimensions. In other words, the Hamming distance of any pair of points is bounded from above by h. For example, this situation can arise for *sparse* data sets where the points mainly contain 0's. Intuitively, if the data set consists of points that are all "similar" to each other, one could hope to be able to solve the instance efficiently since there are at most h dimensions to choose from in order to distinguish two points. The following theorem, however, shows that this intuition is deceptive: when crossing a certain threshold of dissimilarity, the complexity suddenly changes.

Theorem 1. *For a binary input alphabet* $\Sigma = \{0, 1\}$, DISTINCT VECTORS *is*

  i) *solvable in* $O(n^3 d)$ *time if the maximum pairwise Hamming distance h of the input vectors is at most three, and*
  ii) *NP-hard for* $h \geq 4$.

In order to prove (i), we use the following combinatorial lemma.

Lemma 2. *Let* $m, n \in \mathbb{N}$ *with* $m > n + 1$ *and let* $\mathcal{A} = \{A_1, \ldots, A_m\}$ *be a family of pairwise different sets of size n each with* $\forall A_i \neq A_j : |A_i \cap A_j| = n - 1$. *Then,* $\forall A_i \neq A_j : A_i \cap A_j = \bigcap_{k=1}^{m} A_k$.

Now, we can sketch a proof of Theorem 1(i).

Proof (Sketch, Theorem 1(i)). We give a search tree algorithm that solves a given DISTINCT VECTORS instance (S, k). The restriction $h = 3$ guarantees that there are not "too many" branches in the search tree to consider and, hence, that the search tree has polynomial size. For $x \in S$ and $i \in \mathbb{N}$ we define $D_x := \{j \in \{1, \ldots, d\} \mid (x)_j = 1\}$ and $S_i := \{x \in S \mid i = |D_x|\}$. Without loss of generality, we can assume that $\mathbf{0} := (0, \ldots, 0) \in S$. If this is not the case, then we can simply fix an arbitrary point $x_0 \in S$ and exchange 1's and 0's in all points in S in all dimensions where x_0 equals 1. This yields an equivalent instance with $x_0 = \mathbf{0} \in S$ in linear time.

x_1	1	1	1			
x_2	1	1		1		
x_3	1	1			1	
x_4	1	1				1
x_5	1	1					1

Fig. 1. The points in $S_{3|D^3} \subseteq \{0,1\}^7$ represented as rows of a matrix with columns corresponding to the dimensions in D^3. Empty cells represent zero entries. Each pair of points shares a 1 in two dimensions. For more than four points there exist two dimensions in which all points equal 1. At most one of the other dimensions is not contained in a solution.

Let (S, k), $S \subseteq \{0,1\}^d$, be an instance of DISTINCT VECTORS with $|S| = n$. The bound $h = 3$ implies that each point in S contains at most three 1's since otherwise it differs in more than three dimensions from $\mathbf{0}$. Thus, we can partition the data set $S = \{\mathbf{0}\} \uplus S_1 \uplus S_2 \uplus S_3$. Moreover, the restriction $h = 3$ also implies the following two conditions, which constitute the crucial aspects for our proof.

$$\forall x, y \in S_3 : |D_x \cap D_y| = 2, \tag{1}$$
$$\forall x, y \in S_2 : |D_x \cap D_y| = 1. \tag{2}$$

Both conditions have to be met since otherwise there exists a pair of points differing in at least four dimensions. The algorithm starts with considering the subset S_3. The points in S_3 can only be distinguished from each other by a subset of the dimensions $D^3 := \bigcup_{x \in S_3} D_x$. If $|S_3| \leq 4$, then we simply branch over all possible subsets of D^3. With a constant number of at most four distinct points in S_3, the size of D^3 is also bounded by a constant and so there are only constantly many subsets to try. If $|S_3| > 4$, then statement (1) together with Lemma 2 implies that $C^3 := \bigcap_{x \in S_3} D_x$ contains two dimensions. It follows that for each dimension $j \in D^3 \setminus C^3$ there exists exactly one point $x \in S_3$ with $(x)_j = 1$. This situation is depicted in Figure 1. In order to distinguish all points in S_3 from each other, any solution contains at least all but one dimension from $D^3 \setminus C^3$. Hence, we can try out all subsets of $D^3 \setminus C^3$ of size at least $|S_3| - 1$. Together with the four possible subsets of C^3 we end up with at most $4(n+1)$ subsets of D^3 to branch over. Similarly, we obtain that we have to branch over at most $2(n+1)$ subsets of dimensions to distinguish all points in S_2. Thus, we end up with $O(n^2)$ possible subset selections. For the set S_1 no branching is necessary. For each selection we check whether it is a solution or not. This can be done in $O(nd)$ time by sorting the data set lexicographically with radix sort and comparing successive points. Overall, we obtain a search tree algorithm with running time of $O(n^3 d)$.   □

When the pairwise Hamming distance h of the input vectors is at least four, the conditions (1) and (2) from the proof of Theorem 1(i) do not hold. Therefore, we cannot apply Lemma 2, which is crucial in that it guarantees a regular structure of the data set that makes the instance easy to solve. Instead, we can observe that, if a pair of points is allowed to take on different values in at least four dimensions, then the data set can "encode" arbitrary graphs. We exploit this to prove Theorem 1(ii), that is, that DISTINCT VECTORS is NP-complete for $h \geq 4$. To this end, we describe a polynomial-time many-one reduction from a special variant of the INDEPENDENT SET problem in graphs, which is defined as follows.

DISTANCE-3 INDEPENDENT SET

Input: An undirected graph $G = (V, E)$ and $k \in \mathbb{N}$.

Question: Is there a subset of vertices $I \subseteq V$ of size at least k such that any pair of vertices from I has distance at least three?

Here, the distance of two vertices is the number of edges contained in a shortest path between them. DISTANCE-3 INDEPENDENT SET can easily be shown to be NP-hard by a reduction from INDUCED MATCHING [3].

We are now ready to prove that DISTINCT VECTORS is NP-complete for $h \geq 4$, even if the input alphabet Σ is binary.

Proof (Theorem 1(ii)). It is easy to check that DISTINCT VECTORS is in NP. To show NP-hardness, let $(G = (V, E), k)$ with $|V| = n$ and $|E| = m$ be an instance of DISTANCE-3 INDEPENDENT SET and let Z be the $m \times n$ transposed incidence matrix of G with rows corresponding to edges and columns to vertices. The data set S of our DISTINCT VECTORS instance (S, k') is defined to contain all m row vectors of Z and the null point $\mathbf{0} = (0, \ldots, 0) \in \{0, 1\}^n$. The sought solution size is set to $k' := n - k$. Notice that each point in S contains exactly two 1's (except for $\mathbf{0}$). Thus, each pair of points differs in at most $h = 4$ dimensions. The instance (S, k') can be computed in $O(nm)$ time.

Correctness of the reduction follows by the following argument: The subset $I \subseteq V$ is a solution of (G, k) if and only if it is of size k and every edge in G has at least one endpoint in $V \setminus I$ and no vertex in $V \setminus I$ has two neighbors in I. In other words, the latter condition says that no two edges with an endpoint in I share the same endpoint in $V \setminus I$. Equivalently, for the subset K of dimensions corresponding to the vertices in $V \setminus I$, it holds that all row vectors of Z in $S_{|K}$ contain at least one 1 and no two vectors contain only a single 1 in the same dimension. This holds if and only if K is a solution for (S, k'), because S contains the null point and thus two points can only be identical in $S_{|K}$ if either they consist of 0's only or contain a single 1 in the same dimension.   □

We remark that from a W[1]-hardness result for INDUCED MATCHING [15] we can infer W[1]-hardess for DISTANCE-3 INDEPENDENT SET with respect to k. Since the proof of Theorem 1(i) yields a parameterized reduction from DISTANCE-3 INDEPENDENT SET parameterized by k to DISTINCT VECTORS parameterized by the number $n - k' = k$ of dimensions to discard, we have the following:

Corollary 3. DISTINCT VECTORS *is* W[1]-*hard with respect to the number of dimensions to delete.*

2.2 Distinct Vectors with an Arbitrary Alphabet

As we have seen in Section 2.1, DISTINCT VECTORS is NP-complete and W[1]-hard with respect to the number of dimensions to be deleted even in the case of a binary alphabet when the pairwise Hamming distance of the vectors is bounded by four. Nevertheless, we note later in this section that some tractability results are achievable even for larger alphabets. First, however, we mention that HITTING SET parameterized by the sought solution size (which is W[2]-hard, as shown by Downey and Fellows [10]) is parameterized reducible to DISTINCT VECTORS in the case of an arbitrary alphabet size, which yields the following:

Theorem 4. *Allowing an arbitrary alphabet size,* DISTINCT VECTORS *is $W[2]$-hard with respect to the parameter k.*

Proof. We give a parameterized reduction from HITTING SET:

HITTING SET
Input: A finite universe U, a collection $\mathcal{C}$ of subsets of U and a nonnegative integer k.
Question: Is there a subset $K \subseteq U$ with $|K| \leq k$ such that K contains at least one element from each subset in $\mathcal{C}$?

Given an instance $(U, \mathcal{C}, k)$ of HITTING SET with $U = \{u_1, \ldots, u_m\}$ and $\mathcal{C} = \{C_1, \ldots, C_n\}$, we construct a DISTINCT VECTORS instance (S, k') with $S := \{x_1, \ldots, x_n, \mathbf{0}\} \subseteq \mathbb{N}^m$ and $k' := k$, where $\mathbf{0} = (0, \ldots, 0)$ and

$$(x_i)_j := \begin{cases} i, & u_j \in C_i \\ 0, & u_j \notin C_i \end{cases} \text{ for all } i \in \{1, \ldots, n\}, j \in \{1, \ldots, m\}.$$

The above instance is polynomial-time computable. If $K \subseteq U$ is a solution of $(U, \mathcal{C}, k)$, then $K \cap C_i \neq \emptyset$ for all $C_i \in \mathcal{C}$ and thus for each $x_i \in S$ there is a dimension corresponding to some element in K, such that x_i equals i in this dimension and is thus different from all other points in S. Conversely, in order to distinguish any $x_i \in S$ from $\mathbf{0}$, any solution K' of (S, k') has to contain a dimension where x_i is different from 0. This implies that the subset of U corresponding to K' contains at least one element of each C_i and is thus a solution of the original instance. Finally, note that this is a parameterized reduction since $k' = k$. $\square$

It was shown by Chen et al. [6] that, unless FPT = W[1], HITTING SET cannot be solved in $|U|^{o(k)} \cdot |x|^{O(1)}$ time. Since the reduction from HITTING SET yields an instance with $d = |U|$ dimensions and solution size k in polynomial time, it follows that DISTINCT VECTORS cannot be solved in $d^{o(k)} \cdot |x|^{O(1)}$ time unless FPT = W[1]. On the positive side, DISTINCT VECTORS can trivially be solved by trying out all subsets of dimensions of size k within $d^k \cdot |x|^{O(1)}$ time. Consequently, we obtain the following corollary.

Corollary 5. *If* FPT $\neq$ W[1], *then the fastest algorithm solving* DISTINCT VECTORS *has a running time of $d^{\Theta(k)} \cdot |x|^{O(1)}$.*

Although Theorem 4 shows that DISTINCT VECTORS is W[2]-hard with respect to the parameter k, we can provide a problem kernel for DISTINCT VECTORS if we additionally consider the input alphabet size $|\Sigma|$ as parameter. The size of the problem kernel is superexponential in the parameter $(k, |\Sigma|)$. Clearly, a problem kernel of polynomial size would be desirable. However, based on the complexity-theoretic assumption that the polynomial hierarchy does not collapse, polynomial-size kernels do not exist even with the additional parameter n of input points:

Theorem 6.

i) There exists an $O(|\Sigma|^{|\Sigma|^k+k}/|\Sigma|! \cdot \log|\Sigma|)$-size problem kernel computable in $O(d^2 n^2)$ time for DISTINCT VECTORS.

ii) Unless $\mathrm{NP} \subseteq \mathrm{coNP}/\mathrm{poly}$, DISTINCT VECTORS does not admit a polynomial-size kernel with respect to the combined parameter $(n, |\Sigma|, k)$.

Proof (Sketch). (i) The idea is that k dimensions can distinguish at most $|\Sigma|^k$ points. Observe that every dimension partitions the data set into at most $|\Sigma|$ non-empty subsets. If any two dimensions yield the same partitioning, we can simply delete one of them. Thus, any "yes"-instance has at most $|\Sigma|^{|\Sigma|^k}/|\Sigma|!$ essentially different dimensions. Any larger instance can be discarded as "no"-instance.

(ii) The reduction from HITTING SET in the proof of Theorem 4 can easily be turned into a reduction from the closely related SET COVER. For SET COVER, Dom et al. [9] showed that there is no polynomial-size kernel, which in combination with the reduction also excludes polynomial-size kernels for DISTINCT VECTORS. □

Besides parameterizing by the alphabet size, the maximum Hamming distance h of all pairs of points also yields tractability results. It is possible to reduce DISTINCT VECTORS to h-HITTING SET for which problem kernels with respect to (h, k) are known [1]. These can be used to obtain problem kernels for DISTINCT VECTORS in turn. We omit the details here and refer to a full version.

In this subsection we have seen that DISTINCT VECTORS can basically be regarded as a special HITTING SET problem. Interestingly, HITTING SET with respect to the solution size is W[2]-hard in general, but for constant-size alphabets, DISTINCT VECTORS is fixed-parameter tractable (Theorem 6). Thus, the set systems induced by instances of DISTINCT VECTORS involve a certain structure that makes them easier to solve.

3 Hidden Cluster Graph

This section investigates the complexity of HIDDEN CLUSTER GRAPH. It turns out that, in contrast to the HIDDEN CLUSTERS problem—which is NP-hard for the radius $r = 0$ and, hence, for arbitrary metrics—the choice of the distance function has a considerable influence on the tractability of HIDDEN CLUSTER GRAPH.

Theorem 7.

i) L_∞-HIDDEN CLUSTER GRAPH is solvable in $O(d(n^2 d + n^3))$ time.

ii) For $p \in \mathbb{N}$, L_p-HIDDEN CLUSTER GRAPH is NP-complete and even W[2]-hard with respect to the parameter "maximum number t of allowed dimension deletions".

Proof (Sketch). The proof of (i) is deferred to a full version of the paper. The basic idea is to insert missing edges by deleting all dimensions in which the corresponding endpoints differ more than r.

To prove (ii), first observe that L_p-HIDDEN CLUSTER GRAPH is contained in NP: given a solution set K, we can build the corresponding graph G_K and check whether it is a cluster graph in polynomial time. To show NP- and W[2]-hardness, we give a polynomial-time executable parameterized many-one reduction from the NP-hard and W[2]-hard LOBBYING problem [7, 4] occurring in computational social choice.

LOBBYING

Input: A matrix $A \in \{0,1\}^{m \times n}$ with an odd number n of columns and an integer $k > 0$.

Question: Can one modify (set to zero) at most k columns in A such that in the resulting matrix each row contains at least as many zeros as ones?

Compared with the problem definition of Bredereck et al. [4], we exchanged the roles of ones and zeros and of rows and columns. This clearly does not change the complexity. Moreover, we ask for "at least as many" instead of "more" zeros than ones per row. Since the problem is W[2]-hard with respect to k if the number of columns n is odd [7], these conditions are equivalent and our variant is also W[2]-hard. We assume that every row of A contains more ones than zeros because otherwise we could delete it from the input without changing the answer to the question.

Our reduction works as follows: Let (A, k) be an instance of LOBBYING with $A \in \{0,1\}^{m \times n}$ containing m rows $a_1, \ldots, a_m \in \{0,1\}^n$. We define an L_p-HIDDEN CLUSTER GRAPH instance (S, r, k') with

$$S := \bigcup_{1 \le i \le m} \{u_i, v_i, w_i\} \subseteq \Sigma^n, \quad r := 2^{p-1} n, \quad k' := n - k.$$

The idea is to let S contain three data points u_i, v_i, and w_i for every row a_i in A such that their induced subgraph $H_i := G_{\{1,\ldots,n\}}[\{u_i, v_i, w_i\}]$ is a P_3, that is, a path with three vertices. To this end, let

$$u_1 := \mathbf{0}, \qquad w_1 := 2a_1, \qquad v_1 := \frac{u_1 + w_1}{2},$$

$$u_i := w_{i-1} + 2\mathbf{n}, \qquad w_i := u_i + 2a_i, \qquad v_i := \frac{u_i + w_i}{2},$$

for $i \in \{2, \ldots, m\}$, where $\mathbf{x} := (x, \ldots, x) \in \Sigma^n$ for $x \in \Sigma$. The above construction requires $\mathbb{N} \subseteq \Sigma$ in order to be well-defined. It is computable in $O(mn)$ time. Note that this is a parameterized reduction with respect to t since $t = n - k' = k$. Figure 2 illustrates the constructed data set. Now, for all $i = 1, \ldots, m$,

$$\text{dist}^{(p)}(u_i, w_i) = \sum_{j=1}^{n} 2^p \cdot |(a_i)_j|^p \ge 2^p \cdot \left(\frac{n+1}{2}\right) > r$$

and $\text{dist}^{(p)}(u_i, v_i) = \text{dist}^{(p)}(v_i, w_i) \le n \le r$. Since $G_{\{1,\ldots,n\}}$ is defined to contain an edge between two vertices if and only if the distance of their corresponding points in S is at most r, it follows indeed that H_i is a P_3. By construction, the

Fig. 2. A two-dimensional illustration of the constructed L_p-HIDDEN CLUSTER GRAPH instance: For each row a_i in the lobbying matrix A there are three points u_i, v_i, w_i in the data set S such that, for every non-empty subset of dimensions K, they induce a P_3 in G_K. This is achieved by recursively setting $v_i = u_i + a_i$, $w_i = v_i + a_i$ and choosing an appropriate radius $\|a_i\|_p^p \leq r < \|2a_i\|_p^p$. Note that the point u_{i+1} is defined such that its distance to w_i is greater than r in every dimension, which ensures that there is no edge between vertices from different P_3's for any K.

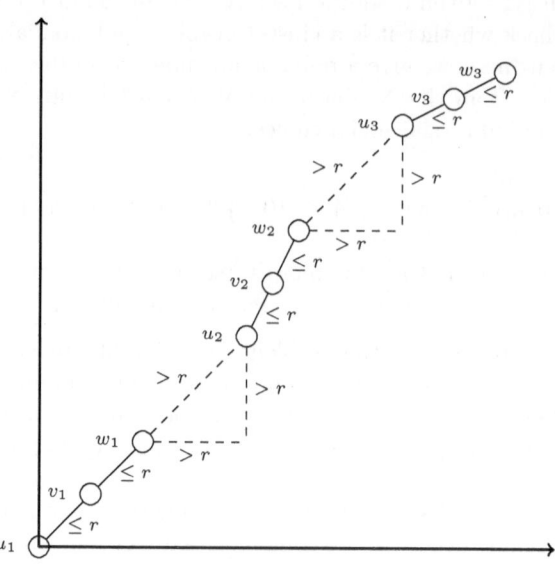

subgraphs H_i are independent of each other in the sense that, for every non-empty subset $K \subseteq \{1, \ldots, n\}$ of dimensions, G_K never contains an edge between any vertices from H_i and H_j for $i \neq j$. To verify this, let $1 \leq i < j \leq m$ and note that, by construction, the smallest distance between any vertices from H_i and H_j is the distance of w_i and u_j. For every non-empty subset K of dimensions, $\mathrm{dist}_{|K}^{(p)}(u_j, w_i)$ is

$$\sum_{l \in K} \left| \left(w_i + (j - i) \cdot 2\boldsymbol{n} + \sum_{k=1}^{j-i-1} 2a_{i+k} \right)_l - (w_i)_l \right|^p$$

$$\geq \sum_{l \in K} 2^p |(\boldsymbol{n})_l|^p = 2^p |K| \cdot n \geq 2^p n > r.$$

Thus, there cannot be an edge in G_K between vertices from H_i and H_j for any K. It follows that the only solution of this instance is the cluster graph consisting of the m disjoint triangles obtained by inserting the missing edge in each H_i. In order to insert the missing edge between u_i and w_i in every H_i, we have to find a subset of dimensions K such that

$$\mathrm{dist}_{|K}^{(p)}(u_i, w_i) = 2^p \sum_{j \in K} |(a_i)_j|^p \leq r = 2^{p-1} n$$

holds for all $i = 1, \ldots, m$. In other words, we have to delete at most t dimensions (that is, setting entries in a_i to zero) such that for the remaining dimensions K it holds that $\sum_{j \in K} |(a_i)_j|^p \leq n/2$. Since a_i is a binary vector, this upper bound states that the modified a_i contains at least as many zeros as ones, which is exactly our LOBBYING problem. So, the L_p-HIDDEN CLUSTER GRAPH instance is a "yes"-instance if and only if the initial LOBBYING instance is a "yes"-instance.   □

The reduction in the proof of Theorem 7(ii) is not only running in polynomial time but also is a polynomial parameter transformation in the sense that the number of data points n equals three times the number of rows of A, the number t of dimensions to discard equals k and the number d of dimensions equals the number of columns of A. Hence, we can transfer some problem kernel lower bound results for LOBBYING [4, Theorems 3 & 4] to L_p-HIDDEN CLUSTER GRAPH.

Corollary 8. *Unless* $NP \subseteq coNP/poly$, L_p-HIDDEN CLUSTER GRAPH *does neither admit a polynomial-size kernel with respect to* (n, t) *nor with respect to* d.

One easily observes that the proof of Theorem 7(ii) generates instances of L_p-HIDDEN CLUSTER GRAPH of unbounded diameter δ, which is defined as the maximum distance between any two vectors in S. This scenario seems not always realistic in practice since features often take on values around some expected value. And indeed, we can show that if δ and the number t of dimensions to be deleted are constant, then L_p-HIDDEN CLUSTER GRAPH is solvable in cubic time. To this end, observe that if $r > \delta$ in an input instance, we can immediately answer "yes", since the graph $G_{\{1,...,d\}}$ is then a clique and thus a cluster graph. For $r \le \delta$, we can prove the following theorem using a search tree algorithm. For bounding the search tree size, we need the additional condition that the data set only contains integers.

Theorem 9. L_p-HIDDEN CLUSTER GRAPH *is* $O((2^p r)^t \cdot (n^2 d + n^3))$-*time solvable for* $p \in \mathbb{N}$ *and an alphabet* $\Sigma \subseteq \mathbb{Z}$.

Obviously, Theorem 9 does not yield an algorithm that is applicable to large data sets. Yet it shows that, despite the hardness of the problem in the general case, the development of efficient algorithms on realistic data might be possible.

4   Outlook

We conclude with some directions for future research. As to DISTINCT VECTORS, our kernelization results in Theorem 6 (lower and upper bounds) are still far apart and ask for closing this gap. Further, it would be interesting to find improved kernels for the parameterization by Hamming distance h and number of retained dimensions k. Here, exploiting structural restrictions in context with connections to HITTING SET seems promising. Finally, we left open to generalize the polynomial-time algorithm for pairwise Hamming distance at most three from binary alphabets (see Theorem 1) to general alphabets.

As to HIDDEN CLUSTER GRAPH, spotting further natural and useful parameterizations is desirable.

Acknowledgements. We are grateful to anonymous MFCS referees for extensive and constructive feedback.

References

[1] van Bevern, R.: Towards optimal and expressive kernelization for d-hitting set. Algorithmica (2013)

[2] Blum, A., Langley, P.: Selection of relevant features and examples in machine learning. Artificial Intelligence 97(1-2), 245–271 (1997)

[3] Brandstädt, A., Mosca, R.: On distance-3 matchings and induced matchings. Discrete Appl. Math. 159(7), 509–520 (2011)

[4] Bredereck, R., Chen, J., Hartung, S., Kratsch, S., Niedermeier, R., Suchý, O.: A multivariate complexity analysis of lobbying in multiple referenda. In: Proc. 26th AAAI, pp. 1292–1298 (2012)

[5] Charikar, M., Guruswami, V., Kumar, R., Rajagopalan, S., Sahai, A.: Combinatorial feature selection problems. In: Proc. 41st FOCS, pp. 631–640 (2000)

[6] Chen, J., Chor, B., Fellows, M., Huang, X., Juedes, D., Kanj, I.A., Xia, G.: Tight lower bounds for certain parameterized NP-hard problems. Information and Computation 201(2), 216–231 (2005)

[7] Christian, R., Fellows, M.R., Rosamond, F., Slinko, A.: On complexity of lobbying in multiple referenda. Review of Economic Design 11(3), 217–224 (2007)

[8] Dasgupta, A., Drineas, P., Harb, B., Josifovski, V., Mahoney, M.W.: Feature selection methods for text classification. In: Proc. 13th ACM SIGKDD, pp. 230–239 (2007)

[9] Dom, M., Lokshtanov, D., Saurabh, S.: Incompressibility through colors and IDs. In: Albers, S., Marchetti-Spaccamela, A., Matias, Y., Nikoletseas, S., Thomas, W. (eds.) ICALP 2009, Part I. LNCS, vol. 5555, pp. 378–389. Springer, Heidelberg (2009)

[10] Downey, R.G., Fellows, M.R.: Parameterized Complexity. Springer (1999)

[11] Flum, J., Grohe, M.: Parameterized Complexity Theory. Springer (2006)

[12] Forman, G.: An extensive empirical study of feature selection metrics for text classification. J. Mach. Learn. Res. 3, 1289–1305 (2003)

[13] Guyon, I., Elisseeff, A.: An introduction to variable and feature selection. J. Mach. Learn. Res. 3, 1157–1182 (2003)

[14] Koller, D., Sahami, M.: Towards optimal feature selection. In: Proc. 13th ICML, pp. 284–292 (1996)

[15] Moser, H., Thilikos, D.M.: Parameterized complexity of finding regular induced subgraphs. J. Discrete Algorithms 7(2), 181–190 (2009)

[16] Niedermeier, R.: Invitation to Fixed-Parameter Algorithms. Oxford University Press (2006)

[17] Pawlak, Z.: Rough Sets: Theoretical Aspects of Reasoning about Data. Kluwer Academic (1991)

[18] Skowron, A., Rauszer, C.: The discernibility matrices and functions in information systems. In: Slowinski, R. (ed.) Intelligent Decision Support—Handbook of Applications and Advances of the Rough Sets Theory, pp. 331–362. Kluwer Academic (1992)

Meta-kernelization with Structural Parameters*

Robert Ganian, Friedrich Slivovsky, and Stefan Szeider

Institute of Information Systems, Vienna University of Technology, Vienna, Austria
{rganian,fslivovsky}@gmail.com, stefan@szeider.net

Abstract. Meta-kernelization theorems are general results that provide polynomial kernels for large classes of parameterized problems. The known meta-kernelization theorems, in particular the results of Bodlaender et al. (FOCS'09) and of Fomin et al. (FOCS'10), apply to optimization problems parameterized by *solution size*. We present meta-kernelization theorems that use *structural parameters* of the input and not the solution size. Let C be a graph class. We define the *C-cover number* of a graph to be the smallest number of modules the vertex set can be partitioned into such that each module induces a subgraph that belongs to the class C.

We show that each graph problem that can be expressed in Monadic Second Order (MSO) logic has a polynomial kernel with a linear number of vertices when parameterized by the C-cover number for any fixed class C of bounded rank-width (or equivalently, of bounded clique-width, or bounded Boolean-width). Many graph problems such as c-COLORING, c-DOMATIC NUMBER and c-CLIQUE COVER are covered by this meta-kernelization result.

Our second result applies to MSO expressible optimization problems, such as MINIMUM VERTEX COVER, MINIMUM DOMINATING SET, and MAXIMUM CLIQUE. We show that these problems admit a polynomial annotated kernel with a linear number of vertices.

1 Introduction

Kernelization is an algorithmic technique that has become the subject of a very active field in parameterized complexity, see, e.g., the references in [14,21,27]. Kernelization can be considered as a *preprocessing with performance guarantee* that reduces an instance of a parameterized problem in polynomial time to a decision-equivalent instance, the *kernel*, whose size is bounded by a function of the parameter alone [14,21,17]; if the reduced instance is an instance of a different problem, then it is called a *bikernel*. Once a kernel or bikernel is obtained, the time required to solve the original instance is bounded by a function of the parameter and therefore independent of the input size. Consequently one aims at (bi)kernels that are as small as possible.

Every fixed-parameter tractable problem admits a kernel, but the size of the kernel can have an exponential or even non-elementary dependence on the parameter [16]. Thus research on kernelization is typically concerned with the question of whether a fixed-parameter tractable problem under consideration admits a small, and in particular a *polynomial*, kernel. For instance, the parameterized MINIMUM VERTEX COVER

---

* Research supported by the European Research Council (ERC), project COMPLEX REASON 239962.

K. Chatterjee and J. Sgall (Eds.): MFCS 2013, LNCS 8087, pp. 457–468, 2013.
© Springer-Verlag Berlin Heidelberg 2013

problem (does a given graph have a vertex cover consisting of k vertices?) admits a polynomial kernel containing at most $2k$ vertices.

There are many fixed-parameter tractable problems for which no polynomial kernels are known. Recently, theoretical tools have been developed to provide strong theoretical evidence that certain fixed-parameter tractable problems do not admit polynomial kernels [3]. In particular, these techniques can be applied to a wide range of graph problems parameterized by treewidth and other width parameters such as clique-width, or rank-width (see e.g., [3,5]). Thus, in order to get polynomial kernels, structural parameters have been suggested that are somewhat weaker than treewidth, including the vertex cover number, max-leaf number, and neighborhood diversity [15,23]. While these parameters do allow polynomial kernels for some problems, no meta-kernelization theorems are known. The general aim here is to find a parameter that admits a polynomial kernel for the given problem while being as general as possible.

We extend this line of research by using results from modular decompositions and rank-width to introduce new structural parameters for which large classes of problems have polynomial kernels. Specifically, we study the *rank-width-d cover number*, which is a special case of a *$\mathcal{C}$-cover number* (see Section 3 for definitions). We establish the following result which is an important prerequisite for our kernelization results.

Theorem 1. *For every constant d, a smallest rank-width-d cover of a graph can be computed in polynomial time.*

Hence, for graph problems parameterized by rank-width-d cover number, we can always compute the parameter in polynomial time. The proof of Theorem 1 relies on a combinatorial property of modules of bounded rank-width that amounts to a variant of partitivity [9].

Our kernelization results take the shape of *algorithmic meta-theorems*, stated in terms of the evaluation of formulas of monadic second order logic (MSO) on graphs. Monadic second order logic over graphs extends first order logic by variables that may range over sets of vertices (sometimes referred to as MSO_1 logic). Specifically, for an MSO formula φ, our first meta-theorem applies to all problems of the following shape, which we simply call *MSO model checking* problems.

MSO-MC$_\varphi$
Instance: A graph G.
Question: Does $G \models \varphi$ hold?

Many NP-hard graph problems can be naturally expressed as MSO model checking problems, for instance c-COLORING, c-DOMATIC NUMBER and c-CLIQUE COVER.

Theorem 2. *Let $\mathcal{C}$ be a graph class of bounded rank-width. Every MSO model checking problem, parameterized by the $\mathcal{C}$-cover number of the input graph, has a polynomial kernel with a linear number of vertices.*

While MSO model checking problems already capture many important graph problems, there are some well-known optimization problems on graphs that cannot be captured in this way, such as MINIMUM VERTEX COVER, MINIMUM DOMINATING SET, and MAXIMUM CLIQUE. Many such optimization graph problems can be equivalently stated as decision problems, in the following way. Let $\varphi = \varphi(X)$ be an MSO formula with one free set variable X and $\Diamond \in \{\leq, \geq\}$.

MSO-OPT$_{\varphi}^{\Diamond}$
Instance: A graph G and an integer $r \in \mathbb{N}$.
Question: Is there a set $S \subseteq V(G)$ such that $G \models \varphi(S)$ and $|S| \Diamond r$?

We call problems of this form *MSO optimization problems*. MSO optimization problems form a large fragment of the so-called *LinEMSO* problems [2]. There are dozens of well-known graph problems that can be expressed as MSO optimization problems.

We establish the following result (cf. Section 2 for the definition of a *bikernel*)

Theorem 3. *Let $\mathcal{C}$ be a graph class of bounded rank-width. Every MSO optimization problem, parameterized by the $\mathcal{C}$-cover number of the input graph, has a polynomial bikernel with a linear number of vertices.*

In fact, the obtained bikernel is an instance of an annotated variant of the original MSO optimization problem [1]. Hence, Theorem 3 provides a polynomial kernel for an annotated version of the original MSO optimization problem.

We would like to point out that a class of graphs has bounded rank-width iff it has bounded clique-width iff it has bounded Boolean-width [7]. Hence, we could have equivalently stated the theorems in terms of clique-width or Boolean-width. Furthermore we would like to point out that the theorems hold also for some classes $\mathcal{C}$ where we do not know whether $\mathcal{C}$ can be recognized in polynomial time, and where we do not know how to compute the partition in polynomial time. For instance, the theorems hold if $\mathcal{C}$ is a graph class of bounded clique-width (it is not known whether graphs of clique-width at most 4 can be recognized in polynomial time).

Note: Some proofs were omitted due to space constraints. A full version of this paper is available on arxiv.org (arXiv:1303.1786).

2 Preliminaries

The set of natural numbers (that is, positive integers) will be denoted by $\mathbb{N}$. For $i \in \mathbb{N}$ we write $[i]$ to denote the set $\{1, \ldots, i\}$.

Graphs. We will use standard graph theoretic terminology and notation (cf. [12]). A *module* of a graph $G = (V, E)$ is a nonempty set $X \subseteq V$ such that for each vertex $v \in V \setminus X$ it holds that either no element of X is a neighbor of v or every element of X is a neighbor of v. We say two modules $X, Y \subseteq V$ are *adjacent* if there are vertices $x \in X$ and $y \in Y$ such that x and y are adjacent. A *modular partition* of a graph G is a partition $\{U_1, \ldots, U_k\}$ of its vertex set such that U_i is a module of G for each $i \in [k]$.

Monadic Second-Order Logic on Graphs. We assume that we have an infinite supply of individual variables, denoted by lowercase letters x, y, z, and an infinite supply of set variables, denoted by uppercase letters X, Y, Z. *Formulas of monadic second-order logic* (MSO) are constructed from atomic formulas $E(x, y)$, $X(x)$, and $x = y$ using the connectives ¬ (negation), ∧ (conjunction) and existential quantification $\exists x$ over individual variables as well as existential quantification $\exists X$ over set variables. Individual variables range over vertices, and set variables range over sets of vertices. The atomic formula $E(x, y)$ expresses adjacency, $x = y$ expresses equality, and $X(x)$ expresses that vertex x in the set X. From this, we define the semantics of monadic second-order logic in the standard way (this logic is sometimes called MSO_1).

Free and bound variables of a formula are defined in the usual way. A *sentence* is a formula without free variables. We write $\varphi(X_1, \ldots, X_n)$ to indicate that the set of free variables of formula φ is $\{X_1, \ldots, X_n\}$. If $G = (V, E)$ is a graph and $S_1, \ldots, S_n \subseteq V$ we write $G \models \varphi(S_1, \ldots, S_n)$ to denote that φ holds in G if the variables X_i are interpreted by the sets S_i, for $i \in [n]$.

We review MSO *types* roughly following the presentation in [24]. The *quantifier rank* of an MSO formula φ is defined as the nesting depth of quantifiers in φ. For non-negative integers q and l, let $MSO_{q,l}$ consist of all MSO formulas of quantifier rank at most q with free set variables in $\{X_1, \ldots, X_l\}$.

Let $\varphi = \varphi(X_1, \ldots, X_l)$ and $\psi = \psi(X_1, \ldots, X_l)$ be MSO formulas. We say φ and ψ are equivalent, written $\varphi \equiv \psi$, if for all graphs G and $U_1, \ldots, U_l \subseteq V(G)$, $G \models \varphi(U_1, \ldots, U_l)$ if and only if $G \models \psi(U_1, \ldots, U_l)$. Given a set F of formulas, let $F/\equiv$ denote the set of equivalence classes of F with respect to $\equiv$. A system of representatives of $F/\equiv$ is a set $R \subseteq F$ such that $R \cap C \neq \emptyset$ for each equivalence class $C \in F/\equiv$. The following statement has a straightforward proof using normal forms (see Proposition 7.5 in [24] for details).

Fact 1. *Let q and l be fixed non-negative integers. The set $MSO_{q,l}/\equiv$ is finite, and one can compute a system of representatives of $MSO_{q,l}/\equiv$.*

We will assume that for any pair of non-negative integers q and l the system of representatives of $MSO_{q,l}/\equiv$ given by Fact 1 is fixed.

Definition 4 (MSO Type). *Let q, l be a non-negative integers. For a graph G and an l-tuple U of sets of vertices of G, we define $type_q(G, U)$ as the set of formulas $\varphi \in MSO_{q,l}$ such that $G \models \varphi(U)$. We call $type_q(G, U)$ the MSO rank-q type of U in G.*

It follows from Fact 1 that up to logical equivalence, every type contains only finitely many formulas. This allows us to represent types using MSO formulas as follows.

Lemma 5. *Let q and l be non-negative integer constants, let G be a graph, and let U be an l-tuple of sets of vertices of G. One can compute a formula $\Phi \in MSO_{q,l}$ such that for any graph G' and any l-tuple U' of sets of vertices of G' we have $G' \models \Phi(U')$ if and only if $type_q(G, U) = type_q(G', U')$. Moreover, if $G \models \varphi(U)$ can be decided in polynomial time for any fixed $\varphi \in MSO_{q,l}$ then Φ can be computed in time polynomial in $|V(G)|$.*

Proof. Let R be a system of representatives of $MSO_{q,l}/\equiv$ given by Fact 1. Because q and l are constant, we can consider both the cardinality of R and the time required to compute it as constants. Let $\Phi \in MSO_{q,l}$ be the formula defined as $\Phi = \bigwedge_{\varphi \in S} \varphi \wedge \bigwedge_{\varphi \in R \setminus S} \neg \varphi$, where $S = \{\varphi \in R : G \models \varphi(U)\}$. We can compute Φ by deciding $G \models \varphi(U)$ for each $\varphi \in R$. Since the number of formulas in R is a constant, this can be done in polynomial time if $G \models \varphi(U)$ can be decided in polynomial time for any fixed $\varphi \in \mathrm{MSO}_{q,l}$.

Let G' be an arbitrary graph and U' an l-tuple of subsets of $V(G')$. We claim that $type_q(G, U) = type_q(G', U')$ if and only if $G' \models \Phi(U')$. Since $\Phi \in \mathrm{MSO}_{q,l}$ the forward direction is trivial. For the converse, assume $type_q(G, U) \neq type_q(G', U')$. First suppose $\varphi \in type_q(G, U) \setminus type_q(G', U')$. The set R is a system of representatives of $\mathrm{MSO}_{q,l}/\equiv$, so there has to be a $\psi \in R$ such that $\psi \equiv \varphi$. But $G' \models \Phi(U')$ implies $G' \models \psi(U')$ by construction of Φ and thus $G' \models \varphi(U')$, a contradiction. Now suppose $\varphi \in type_q(G', U') \setminus type_q(G, U)$. An analogous argument proves that there has to be a $\psi \in R$ such that $\psi \equiv \varphi$ and $G' \models \neg\psi(U')$. It follows that $G' \not\models \varphi(U')$, which again yields a contradiction. $\qquad\square$

Fixed-Parameter Tractability and Kernels. A *parameterized problem* P is a subset of $\Sigma^* \times \mathbb{N}$ for some finite alphabet Σ. For a problem instance $(x, k) \in \Sigma^* \times \mathbb{N}$ we call x the main part and k the parameter. A parameterized problem P is *fixed-parameter tractable* (FPT in short) if a given instance (x, k) can be solved in time $O(f(k) \cdot p(|x|))$ where f is an arbitrary computable function of k and p is a polynomial function.

A *bikernelization* for a parameterized problem $P \subseteq \Sigma^* \times \mathbb{N}$ into a parameterized problem $Q \subseteq \Sigma^* \times \mathbb{N}$ is an algorithm that, given $(x, k) \in \Sigma^* \times \mathbb{N}$, outputs in time polynomial in $|x| + k$ a pair $(x', k') \in \Sigma^* \times \mathbb{N}$ such that (i) $(x, k) \in P$ if and only if $(x', k') \in Q$ and (ii) $|x'| + k' \leq g(k)$, where g is an arbitrary computable function. The reduced instance (x', k') is the *bikernel*. If $P = Q$, the reduction is called a *kernelization* and (x', k') a *kernel*. The function g is called the *size* of the (bi)kernel, and if g is a polynomial then we say that P admits a *polynomial (bi)kernel*.

It is well known that every fixed-parameter tractable problem admits a generic kernel, but the size of this kernel can have an exponential or even non-elementary dependence on the parameter [13]. Since recently there have been workable tools available for providing strong theoretical evidence that certain parameterized problems do not admit a polynomial kernel [3,25].

Rank-width. The graph invariant rank-width was introduced by Oum and Seymour [26] with the original intent of investigating the graph invariant clique-width. It later turned out that rank-width itself is a useful parameter, with several advantages over clique-width.

For a graph G and $U, W \subseteq V(G)$, let $A_G[U, W]$ denote the $U \times W$-submatrix of the adjacency matrix over the two-element field $\mathrm{GF}(2)$, i.e., the entry $a_{u,w}$, $u \in U$ and $w \in W$, of $A_G[U, W]$ is 1 if and only if $\{u, w\}$ is an edge of G. The *cut-rank* function ρ_G of a graph G is defined as follows: For a bipartition (U, W) of the vertex set $V(G)$, $\rho_G(U) = \rho_G(W)$ equals the rank of $A_G[U, W]$ over $\mathrm{GF}(2)$.

A *rank-decomposition* of a graph G is a pair (T, μ) where T is a tree of maximum degree 3 and $\mu : V(G) \to \{t : t \text{ is a leaf of } T\}$ is a bijective function. For an edge e of T, the connected components of $T - e$ induce a bipartition (X, Y) of the set of leaves of T. The *width* of an edge e of a rank-decomposition (T, μ) is $\rho_G(\mu^{-1}(X))$. The *width* of (T, μ) is the maximum width over all edges of T. The *rank-width* of G is the minimum width over all rank-decompositions of G.

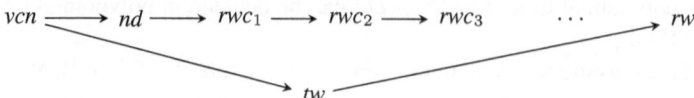

Fig. 1. Relationship between graph invariants: the vertex cover number (vcn), the neighborhood diversity (nd), the rank-width-d cover number (rwc_d), the rank-width (rw), and the treewidth (tw). An arrow from A to B indicates that for any graph class for which B is bounded also A is bounded. See Proposition 8 and [26] for references.

Theorem 6 ([22]). *Let $k \in \mathbb{N}$ be a constant and $n \geq 2$. For an n-vertex graph G, we can output a rank-decomposition of width at most k or confirm that the rank-width of G is larger than k in time $O(n^3)$.*

Theorem 7 ([20]). *Let $d \in \mathbb{N}$ be a constant and let φ and $\psi = \psi(X)$ be fixed MSO formulas. Given a graph G with $rw(G) \leq d$, we can decide whether $G \models \varphi$ in polynomial time. Moreover, a set $S \subseteq V(G)$ of minimum (maximum) cardinality such that $G \models \psi(S)$ can be found in polynomial time, if one exists.*

3   Rank-Width Covers

Let $\mathcal{C}$ be a graph class containing all trivial graphs, i.e., all graphs consisting of only a single vertex. We define a *$\mathcal{C}$-cover of G* as a modular partition $\{U_1, \ldots, U_k\}$ of $V(G)$ such that the induced subgraph $G[U_i]$ belongs to the class $\mathcal{C}$ for each $i \in [k]$. Accordingly, the *$\mathcal{C}$-cover number* of G is the size of a smallest $\mathcal{C}$-cover of G.

Of special interest to us are the classes $\mathcal{R}_d$ of graphs of *rank-width* at most d. We call the $\mathcal{R}_d$-cover number also the *rank-width-d cover number*. If $\mathcal{C}$ is the class of all complete graphs and all edgeless graphs, then the $\mathcal{C}$-cover number equals the neighborhood diversity [23], and clearly $\mathcal{C} \subsetneq \mathcal{R}_1$. Figure 1 shows the relationship between the rank-width-d cover number and some other graph invariants.

We state some further properties of rank-width-d covers.

Proposition 8. *Let vcn, nd, and rw denote the vertex cover number, the neighborhood diversity, and the rank-width of a graph G, respectively. Then the following (in)equalities hold for any $d \in \mathbb{N}$:*

1. *$rwc_d(G) \leq nd(G) \leq 2^{vcn(G)}$,*
2. *if $d \geq rw(G)$, then $|rwc_d(G)| = 1$.*

Proof. (1) The neighborhood diversity of a graph is also a rank-width-1 cover. The neighborhood diversity is known to be upper-bounded by $2^{vcn(G)}$ [23].

(2) This follows immediately from the definition of rank-width-d covers.   □

3.1   Finding the Cover

Next we state several properties of modules of graphs. These will be used to obtain a polynomial algorithm for finding smallest rank-width-d covers.

The *symmetric difference* of sets A, B is $A \triangle B = (A \setminus B) \cup (B \setminus A)$. Sets A and B *overlap* if $A \cap B \neq \emptyset$ but neither $A \subseteq B$ nor $B \subseteq A$.

Definition 9. *Let $\mathcal{S} \subseteq 2^S$ be a family of subsets of a set S. We call $\mathcal{S}$ partitive if it satisfies the following properties:*

1. *$S \in \mathcal{S}$, $\emptyset \notin \mathcal{S}$, and $\{x\} \in \mathcal{S}$ for each $x \in S$.*
2. *For every pair of overlapping subsets $A, B \in \mathcal{S}$, the sets $A \cup B, A \cap B, A \triangle B, A \setminus B$, and $B \setminus A$ are contained in $\mathcal{S}$.*

Theorem 10 ([9]). *The family of modules of a graph G is partitive.*

Lemma 11 ([6]). *Let G be a graph and $x, y \in V(G)$. There is a unique minimal (with respect to set inclusion) module M of G such that $x, y \in M$, and M can be computed in time $O(|V(G)|^2)$.*

Definition 12. *Let G be a graph and $d \in \mathbb{N}$. We define a relation $\sim_d^G$ on $V(G)$ by letting $v \sim_d^G w$ if and only if there is a module M of G with $v, w \in M$ and $rw(G[M]) \leq d$. We drop the superscript from $\sim_d^G$ if the graph G is clear from context.*

Proposition 13. *For every graph G and $d \in \mathbb{N}$ the relation $\sim_d$ is an equivalence relation, and each equivalence class U of $\sim_d$ is a module of G with $rw(G[U]) \leq d$.*

Corollary 14. *Let G be a graph and $d \in \mathbb{N}$. The equivalence classes of $\sim_d$ form a smallest rank-width-d cover of G.*

Proposition 15. *Let $d \in \mathbb{N}$ be a constant. Given a graph G and two vertices $v, w \in V(G)$, we can decide whether $v \sim_d w$ in polynomial time.*

Proof (of Theorem 1). Let $d \in \mathbb{N}$ be a constant. Given a graph G, we can compute the set of equivalence classes of $\sim_d$ by testing whether $v \sim_d w$ for each pair of vertices $v, w \in V(G)$. By Proposition 15, this can be done in polynomial time, and by Corollary 14, $V(G)/\sim_d$ is a smallest rank-width-d cover of G. $\qquad\square$

4   Kernels for MSO Model Checking

In this section, we show that every MSO model checking problem admits a polynomial kernel when parameterized by the $\mathcal{C}$-cover number of the input graph, where $\mathcal{C}$ is some recursively enumerable class of graphs satisfying the following properties:

(I) $\mathcal{C}$ contains all trivial graphs, and a $\mathcal{C}$-cover of a graph G with minimum cardinality can be computed in polynomial time.
(II) There is an algorithm $\mathbb{A}$ that decides whether $G \models \varphi$ in time polynomial in $|V(G)|$ for any fixed MSO sentence φ and any graph $G \in \mathcal{C}$.

For obtaining the kernel for MSO model checking problems, we proceed as follows. First, we compute a smallest rank-width-d cover of the input graph G in polynomial time. Second, we compute for each module a small representative of constant size. Third, we replace each module with a constant size module, which results in the kernel. We show how to carry out the second and third steps below.

Let G be a graph and $U \subseteq V(G)$. Let $\boldsymbol{V}$ be an l-tuple of sets of vertices of G. We write $\boldsymbol{V}|_U = (V_1 \cap U, \dots, V_l \cap U)$ to refer to the elementwise intersection of $\boldsymbol{V}$ with U. If $\{U_1, \dots, U_k\}$ is a modular partition of G and $i \in [k]$ we will abuse notation and write $\boldsymbol{V}|_i = \boldsymbol{V}_{U_i}$ if there is no ambiguity about what partition the index refers to.

Definition 16 (Congruent). *Let q and l be non-negative integers and let G and G' be graphs with modular partitions $\boldsymbol{M} = \{M_1, \dots, M_k\}$ and $\boldsymbol{M'} = \{M'_1, \dots, M'_k\}$, respectively. Let $\boldsymbol{V_0}$ be an l-tuple of subsets of $V(G)$ and let $\boldsymbol{U_0}$ be an l-tuple of subsets of $V(G')$. We say $(G, \boldsymbol{M}, \boldsymbol{V_0})$ and $(G', \boldsymbol{M'}, \boldsymbol{U_0})$ are q-congruent if the following conditions are met:*

1. *For every $i, j \in [k]$ with $i \neq j$, M_i and M_j are adjacent in G if and only if M'_i and M'_j are adjacent in G'.*
2. *For each $i \in [k]$, $type_q(G[M_i], \boldsymbol{V_0}|_i) = type_q(G'[M'_i], \boldsymbol{U_0}|_i)$.*

We begin by showing how congruents are related to the previously introduced notion of types.

Lemma 17. *Let q and l be non-negative integers and let G and G' be graphs with modular partitions $\boldsymbol{M} = \{M_1, \dots, M_k\}$ and $\boldsymbol{M'} = \{M'_1, \dots, M'_k\}$. Let $\boldsymbol{V_0}$ be an l-tuple of subsets of $V(G)$ and let $\boldsymbol{U_0}$ be an l-tuple of subsets of $V(G')$. If $(G, \boldsymbol{M}, \boldsymbol{V_0})$ and $(G', \boldsymbol{M'}, \boldsymbol{U_0})$ are q-congruent, then $type_q(G, \boldsymbol{V_0}) = type_q(G', \boldsymbol{U_0})$.*

Next, we showcase the tool we use to replace a graph G by a small representative.

Lemma 18. *Let $\mathcal{C}$ be a recursively enumerable graph class and let q be a non-negative integer constant. Let $G \in \mathcal{C}$ be a graph. If $G \models \varphi$ can be decided in time polynomial in $|V(G)|$ for any fixed $\varphi \in MSO_{q,0}$ then one can in polynomial time compute a graph $G' \in \mathcal{C}$ such that $|V(G')|$ is bounded by a constant and $type_q(G) = type_q(G')$.*

Finally, in Lemma 19 below we use Lemma 18 to obtain our polynomial kernels.

Lemma 19. *Let q be a non-negative integer constant, and let $\mathcal{C}$ be a recursively enumerable graph class satisfying (II). Then given a graph G and a $\mathcal{C}$-cover $\{U_1, \dots, U_k\}$, one can in polynomial time compute a graph G' with modular partition $\{U'_1, \dots, U'_k\}$ such that $(G, \boldsymbol{U})$ and $(G', \boldsymbol{U'})$ are q-congruent and for each $i \in [k]$, $G'[U'_i] \in \mathcal{C}$ and the number of vertices in U'_i is bounded by a constant.*

Proposition 20. *Let φ be a fixed MSO sentence. Let $\mathcal{C}$ be a recursively enumerable graph class satisfying (I) and (II). Then MSO-MC$_\varphi$ has a polynomial kernel parameterized by the $\mathcal{C}$-cover number of the input graph.*

Proof (of Theorem 2). Immediate from Theorems 1, 6, and 7 in combination with Proposition 20. □

Corollary 21. *The following problems have polynomial kernels when parameterized by the rank-width-d cover number of the input graph: c-COLORING, c-DOMATIC NUMBER, c-PARTITION INTO TREES, c-CLIQUE COVER, c-PARTITION INTO PERFECT MATCHINGS, c-COVERING BY COMPLETE BIPARTITE SUBGRAPHS.*

5   Kernels for MSO Optimization

By definition, MSO formulas can only directly capture decision problems such as 3-coloring, but many problems of interest are formulated as optimization problems. The usual way of transforming decision problems into optimization problems does not work here, since the MSO language cannot handle arbitrary numbers.

Nevertheless, there is a known solution. Arnborg, Lagergren, and Seese [2] (while studying graphs of bounded tree-width), and later Courcelle, Makowsky, and Rotics [10] (for graphs of bounded clique-width), specifically extended the expressive power of MSO logic to define so-called LINEMS optimization problems, and consequently also showed the existence of efficient (parameterized) algorithms for such problems in the respective cases.

The class of so-called MSO optimization problems (problems which may be stated as MSO-$\text{OPT}_\varphi^\Diamond$) considered here are a streamlined and simplified version of the formalism introduced in [10]. Specifically, we consider only a single free variable X, and ask for a satisfying assignment of X with minimum or maximum cardinality. To achieve our results, we need a recursively enumerable graph class $\mathcal{C}$ that satisfies (I) and (II) along with the following property:

(III) Let $\varphi = \varphi(X)$ be a fixed MSO formula. Given a graph $G \in \mathcal{C}$, a set $S \subseteq V(G)$ of minimum (maximum) cardinality such that $G \models \varphi(S)$ can be found in polynomial time, if one exists.

Our approach will be similar to the MSO kernelization algorithm, with one key difference: when replacing the subgraph induced by a module, the cardinalities of subsets of a given q-type may change, so we need to keep track of their cardinalities in the original subgraph.

To do this, we introduce an annotated version of MSO-$\text{OPT}_\varphi^\Diamond$. Given a graph $G = (V, E)$, an *annotation* $\mathcal{W}$ is a set of triples (X, Y, w) with $X \subseteq V, Y \subseteq V, w \in \mathbb{N}$. For every set $Z \subseteq V$ we define

$$\mathcal{W}(Z) = \sum_{(X,Y,w)\in\mathcal{W}, X\subseteq Z, Y\cap Z=\emptyset} w.$$

We call the pair $(G, \mathcal{W})$ an *annotated graph*. If the integer w is represented in binary, we can represent a triple (X, Y, w) in space $|X| + |Y| + \log_2(w)$. Consequently, we may assume that the size of the encoding of an annotated graph $(G, \mathcal{W})$ is polynomial in $|V(G)| + |\mathcal{W}| + \max_{(X,Y,w)\in\mathcal{W}} \log_2 w$.

Each MSO formula $\varphi(X)$ and $\Diamond \in \{\leq, \geq\}$ gives rise to an *annotated MSO-optimization problem*.

aMSO-$\text{OPT}_\varphi^\Diamond$
Instance: A graph G with an annotation $\mathcal{W}$ and an integer $r \in \mathbb{N}$.
Question: Is there a set $Z \subseteq V(G)$ such that $G \models \varphi(Z)$ and $\mathcal{W}(Z) \Diamond r$?

Notice that any instance of MSO-$\text{OPT}_\varphi^\Diamond$ is also an instance of aMSO-$\text{OPT}_\varphi^\Diamond$ with the trivial annotation $\mathcal{W} = \{ (\{v\}, \emptyset, 1) : v \in V(G) \}$. The main result of this section is a bikernelization algorithm which transforms any instance of MSO-OPT$_\varphi^\diamondsuit$ into an instance of aMSO-OPT$_\varphi^\diamondsuit$; this kind of bikernel is called an *annotated kernel* [1].

The results below are stated and proved for minimization problems aMSO-OPT$_\varphi^\leq$ only. This is without loss of generality—the proofs for maximization problems are symmetric.

Lemma 22. *Let q and l be non-negative integers and let G and G' be a graphs such that G and G' have the same $q + l$ MSO type. Then for any l-tuple $\boldsymbol{V}$ of sets of vertices of G, there exists an l-tuple $\boldsymbol{U}$ of sets of vertices of G' such that $type_q(G, \boldsymbol{V}) = type_q(G', \boldsymbol{U})$.*

Proof. Suppose there exists an l-tuple $\boldsymbol{V}$ of sets of vertices of G, and a formula $\varphi = \varphi(X_1, \ldots, X_l) \in \mathrm{MSO}_{q,l}$ such that $G \models \varphi(V_1, \ldots, V_l)$ but for every l-tuple $\boldsymbol{U}$ of sets of vertices of G' we have $G' \not\models \varphi(U_1, \ldots, U_l)$. Let $\psi = \exists X_1 \ldots \exists X_l\, \varphi$. Clearly, $\psi \in \mathrm{MSO}_{q+l,0}$ and $G \models \psi$ but $G' \not\models \psi$, a contradiction. $\square$

Using Lemma 22 and the results of Section 4, we may proceed directly to the construction of our annotated kernel.

Lemma 23. *Let $\varphi = \varphi(X)$ be a fixed MSO formula and $\mathcal{C}$ be a recursively enumerable graph class satisfying (II) and (III). Then given an instance (G, r) of MSO-OPT$_\varphi^\leq$ and a $\mathcal{C}$-cover $\{U_1, \ldots, U_k\}$ of G, an annotated graph $(G', \mathcal{W})$ satisfying the following properties can be computed in polynomial time.*

1. *$(G, r) \in$ MSO-OPT$_\varphi^\leq$ if and only if $(G', \mathcal{W}, r) \in a$MSO-OPT$_\varphi^\leq$.*
2. *$|V(G')| \in O(k)$.*
3. *The encoding size of $(G', \mathcal{W})$ is $O(k \log(|V(G)|))$.*

The last obstacle we face is that the annotation itself may be "too large" for the kernel. Here we use the following simple folklore result, which allows us to prove that either our annotated kernel is "small enough", or we can solve our problem in polynomial time (and subsequently output a trivial yes/no instance).

Fact 2 (Folklore). *Given an MSO sentence φ and a graph G, one can decide whether $G \models \varphi$ in time $O(2^{nl})$, where $n = |V(G)|$ and $l = |\varphi|$.*

Proposition 24. *Let $\varphi = \varphi(X)$ be a fixed MSO formula, and let $\mathcal{C}$ be a recursively enumerable graph class satisfying (I), (II), and (III). Then MSO-OPT$_\varphi^\leq$ has a polynomial bikernel parameterized by the $\mathcal{C}$-cover number of the input graph.*

Proof (of Theorem 3). Immediate from Theorems 1, 6, and 7 when combined with Proposition 24. $\square$

Corollary 25. *The following problems have polynomial bikernels when parameterized by the rank-width-d cover number of the input graph:* MINIMUM DOMINATING SET, MINIMUM VERTEX COVER, MINIMUM FEEDBACK VERTEX SET, MAXIMUM INDEPENDENT SET, MAXIMUM CLIQUE, LONGEST INDUCED PATH, MAXIMUM BIPARTITE SUBGRAPH, MINIMUM CONNECTED DOMINATING SET.

6 Conclusion

Recently Bodlaender et al. [4] and Fomin et al. [18] established *meta-kernelization theorems* that provide polynomial kernels for large classes of parameterized problems. The known meta-kernelization theorems apply to optimization problems parameterized by *solution size*. Our results are, along with very recent results parameterized by the modulator to constant-treedepth [19], the first meta-kernelization theorems that use a *structural parameter* of the input and not the solution size. In particular, we would like to emphasize that our Theorem 3 applies to a large class of optimization problems where the solution size can be arbitrarily large.

It is also worth noting that our structural parameter, the rank-width-d cover number, provides a trade-off between the maximum rank-width of modules (the constant d) and the maximum number of modules (the parameter k). Different problem inputs might be better suited for smaller d and larger k, others for larger d and smaller k. This two-dimensional setting could be seen as a contribution to *multivariate complexity analysis* as advocated by Fellows et al. [15].

We conclude by mentioning possible directions for future research. We believe that some of our results can be extended from modular partitions to partitions into splits [8,11][1]. This would indeed result in more general parameters, however the precise details require further work (one problem is that while all modules are partitive, only strong splits have this property). Another direction is to focus on polynomial kernels for problems which cannot be described by MSO logic, such as HAMILTONIAN PATH or CHROMATIC NUMBER.

References

1. Abu-Khzam, F.N., Fernau, H.: Kernels: Annotated, proper and induced. In: Bodlaender, H.L., Langston, M.A. (eds.) IWPEC 2006. LNCS, vol. 4169, pp. 264–275. Springer, Heidelberg (2006)
2. Arnborg, S., Lagergren, J., Seese, D.: Easy problems for tree-decomposable graphs. J. Algorithms 12(2), 308–340 (1991)
3. Bodlaender, H.L., Downey, R.G., Fellows, M.R., Hermelin, D.: On problems without polynomial kernels. J. of Computer and System Sciences 75(8), 423–434 (2009)
4. Bodlaender, H.L., Fomin, F.V., Lokshtanov, D., Penninkx, E., Saurabh, S., Thilikos, D.M.: (meta) kernelization. In: FOCS 2009, pp. 629–638. IEEE Computer Society (2009)
5. Bodlaender, H.L., Jansen, B.M.P., Kratsch, S.: Preprocessing for treewidth: A combinatorial analysis through kernelization. In: Aceto, L., Henzinger, M., Sgall, J. (eds.) ICALP 2011, Part I. LNCS, vol. 6755, pp. 437–448. Springer, Heidelberg (2011)
6. Bui-Xuan, B.-M., Habib, M., Limouzy, V., de Montgolfier, F.: Algorithmic aspects of a general modular decomposition theory. Discr. Appl. Math. 157(9), 1993–2009 (2009)
7. Bui-Xuan, B.-M., Telle, J.A., Vatshelle, M.: Boolean-width of graphs. Theoretical Computer Science 412(39), 5187–5204 (2011)
8. Charbit, P., de Montgolfier, F., Raffinot, M.: Linear time split decomposition revisited. SIAM J. Discrete Math. 26(2), 499–514 (2012)
9. Chein, M., Habib, M., Maurer, M.: Partitive hypergraphs. Discrete Math. 37(1), 35–50 (1981)

---

[1] We thank Sang-il Oum for pointing this out to us.

10. Courcelle, B., Makowsky, J.A., Rotics, U.: Linear time solvable optimization problems on graphs of bounded clique-width. Theory Comput. Syst. 33(2), 125–150 (2000)
11. Cunningham, W.H.: Decomposition of directed graphs. SIAM J. Algebraic Discrete Methods 3(2), 214–228 (1982)
12. Diestel, R.: Graph Theory, 2nd edn. Graduate Texts in Mathematics, vol. 173. Springer, New York (2000)
13. Downey, R., Fellows, M.R., Stege, U.: Parameterized complexity: A framework for systematically confronting computational intractability. In: Contemporary Trends in Discrete Mathematics: From DIMACS and DIMATIA to the Future. AMS-DIMACS, vol. 49, pp. 49–99. American Mathematical Society (1999)
14. Fellows, M.R.: The lost continent of polynomial time: Preprocessing and kernelization. In: Bodlaender, H.L., Langston, M.A. (eds.) IWPEC 2006. LNCS, vol. 4169, pp. 276–277. Springer, Heidelberg (2006)
15. Fellows, M.R., Jansen, B.M., Rosamond, F.: Towards fully multivariate algorithmics: Parameter ecology and the deconstruction of computational complexity. European J. Combin. 34(3), 541–566 (2013)
16. Flum, J., Grohe, M.: Parameterized Complexity Theory. Texts in Theoretical Computer Science. An EATCS Series, vol. XIV. Springer, Berlin (2006)
17. Fomin, F.V.: Kernelization. In: Ablayev, F., Mayr, E.W. (eds.) CSR 2010. LNCS, vol. 6072, pp. 107–108. Springer, Heidelberg (2010)
18. Fomin, F.V., Lokshtanov, D., Misra, N., Saurabh, S.: Planar f-deletion: Approximation, kernelization and optimal fpt algorithms. In: FOCS 2012, pp. 470–479. IEEE Computer Society (2012)
19. Gajarský, J., Hliněný, P., Obdržálek, J., Ordyniak, S., Reidl, F., Rossmanith, P., Villaamil, F.S., Sikdar, S.: Kernelization using structural parameters on sparse graph classes. CoRR, abs/1302.6863 (2013)
20. Ganian, R., Hliněný, P.: On parse trees and Myhill-Nerode-type tools for handling graphs of bounded rank-width. Discr. Appl. Math. 158(7), 851–867 (2010)
21. Guo, J., Niedermeier, R.: Invitation to data reduction and problem kernelization. ACM SIGACT News 38(2), 31–45 (2007)
22. Hliněný, P., Oum, S.: Finding branch-decompositions and rank-decompositions. SIAM J. Comput. 38(3), 1012–1032 (2008)
23. Lampis, M.: Algorithmic meta-theorems for restrictions of treewidth. Algorithmica 64(1), 19–37 (2012)
24. Libkin, L.: Elements of Finite Model Theory. Springer (2004)
25. Misra, N., Raman, V., Saurabh, S.: Lower bounds on kernelization. Discrete Optimization 8(1), 110–128 (2011)
26. Oum, S., Seymour, P.: Approximating clique-width and branch-width. J. Combin. Theory Ser. B 96(4), 514–528 (2006)
27. Rosamond, F.: Table of races. In: Parameterized Complexity Newsletter, pp. 4–5 (2010), http://fpt.wikidot.com/

Separating Hierarchical and General Hub Labelings

Andrew V. Goldberg[1], Ilya Razenshteyn[2,*], and Ruslan Savchenko[3,*]

[1] Microsoft Research Silicon Valley
[2] CSAIL, MIT
[3] Department of Mech. and Math., MSU

Abstract. In the context of distance oracles, a labeling algorithm computes vertex labels during preprocessing. An s, t query computes the corresponding distance using the labels of s and t only, without looking at the input graph. Hub labels is a class of labels that has been extensively studied. Performance of the hub label query depends on the label size. Hierarchical labels are a natural special kind of hub labels. These labels are related to other problems and can be computed more efficiently. This brings up a natural question of the quality of hierarchical labels. We show that there is a gap: optimal hierarchical labels can be polynomially bigger than the general hub labels. To prove this result, we give tight upper and lower bounds on the size of hierarchical and general labels for hypercubes.

1 Introduction

The point-to-point shortest path problem is a fundamental optimization problem with many applications. Dijkstra's algorithm [6] solves this problem in near-linear time [10] on directed and in linear time on undirected graphs [13], but some applications require sublinear distance queries. This is possible for some graph classes if preprocessing is allowed (e.g., [5,8]). Peleg introduced a *distance labeling* algorithm [12] that precomputes a *label* for each vertex such that the distance between any two vertices s and t can be computed using only their labels. A special case is *hub labeling* (HL) [8]: the label of u consists of a collection of vertices (the *hubs* of u) with their distances from u. Hub labels satisfy the *cover property*: for any two vertices s and t, there exists a vertex w on the shortest s–t path that belongs to both the label of s and the label of t.

Cohen et al. [4] give a polynomial-time $O(\log n)$-approximation algorithm for the smallest labeling (here n denotes the number of vertices). (See [3] for a generalization.) The complexity of the algorithm, however, is fairly high, making it impractical for large graphs. Abraham et al. [1] introduce a class of *hierarchical labelings* (HHL) and show that HHL can be computed in $O^*(nm)$ time, where m is the number of arcs. This makes preprocessing feasible for moderately large graphs, and for some problem classes produces labels that are sufficiently

---

* Part of this work done while at Microsoft.

small for practical use. In particular, this leads to the fastest distance oracles for continental-size road networks [2]. However, the algorithm of [1] does not have theoretical guarantees on the size of the labels.

HHL is a natural algorithm that is closely related to other widely studied problems, such as vertex orderings for contraction hierarchies [9] and elimination sequences for chordal graphs (e.g., [11]). This provides additional motivation for studying HHL. This motivation is orthogonal the relationship of HHL to HL, which is not directly related to the above-mentioned problems.

HHL is a special case of HL, so a natural question is how the label size is affected by restricting the labels to be hierarchical. In this paper we show that HHL labels can be substantially bigger than the general labels. Note that it is enough to show this result for a special class of graphs. We study hypercubes, which have a very simple structure. However, proving tight bounds for them is non-trivial: Some of our upper bound constructions and lower bound proofs are fairly involved.

We obtain upper and lower bounds on the optimal size for both kinds of labels in hypercubes. In particular, for a hypercube of dimension d (with 2^d vertices), we give both upper and lower bounds of 3^d on the HHL size. For HL, we also give a simple construction producing labels of size 2.83^d, establishing a polynomial separation between the two label classes. A more sophisticated argument based on the primal-dual method yields $(2.5 + o(1))^d$ upper and lower bounds on the HL size. Although the upper bound proof is non-constructive, it implies that the Cohen et al. approximation algorithm computes the labels of size $(2.5 + o(1))^d$, making the bound constructive.

The paper is organized as follows. After introducing basic definitions in Section 2, we prove matching upper and lower bounds on the HHL size in Section 3. Section 4 gives a simple upper bound on the size of HL that is polynomially better than the lower bound on the size of HHL. Section 5 strengthens these results by proving a better lower bound and a near-matching upper bound on the HL size. Section 6 contains the conclusions.

2   Preliminaries

In this paper we consider shortest paths in an undirected graph $G = (V, E)$, with $|V| = n$, $|E| = m$, and length $\ell(a) > 0$ for each arc a. The length of a path P in G is the sum of its arc lengths. The *distance query* is as follows: given a source s and a target t, to find the distance $\text{dist}(s, t)$ between them, i.e., the length of the shortest path P_{st} between s and t in G. Often we will consider unweighted graphs ($\ell \equiv 1$).

Dijkstra's algorithm [6] solves the problem in $O(m + n \log n)$ [7] time in the comparison model and in linear time in weaker models [13]. However, for some applications, even linear time is too slow. For faster queries, *labeling* algorithms preprocess the graph and store a *label* with each vertex; the s–t distance can be computed from the labels of s and t. We study *hub labelings* (HL), a special case of the labeling method. For each vertex $v \in V$, HL precomputes a label

$L(v)$, which contains a subset of vertices (*hubs*) and, for every hub u the distance $\text{dist}(v, u)$. Furthermore, the labels obey the *cover property*: for any two vertices s and t, $L(s) \cap L(t)$ must contain at least one vertex on the shortest s–t path.

For an s–t query, among all vertices $u \in L(s) \cap L(t)$ we pick the one minimizing $\text{dist}(s, u) + \text{dist}(u, t)$ and return the corresponding sum. If the entries in each label are sorted by hub vertex ID, this can be done with a sweep over the two labels, as in mergesort. The *label size of v*, $|L(v)|$, is the number of hubs in $L(v)$. The time for an s–t query is $O(|L(s)| + |L(t)|)$.

The *labeling* L is the set of all labels. We define its *size* as $\sum_v (|L(v)|)$. Cohen et al. [4] show how to generate in $O(n^4)$ time a labeling whose size is within a factor $O(\log n)$ of the optimum.

Given two distinct vertices v, w, we say that $v \preceq w$ if $L(v)$ contains w. A labeling is *hierarchical* if $\preceq$ is a partial order. We say that this order is *implied* by the labeling. Labelings computed by the algorithm of Cohen et al. are not necessarily hierarchical. Given a total order on vertices, the *rank function* $r : V \to [1 \ldots n]$ ranks the vertices according to the order. We will call the corresponding order r.

We define a d-dimensional hypercube $H = (V, E)$ graph as follows. Let $n = 2^d$ denote the number of vertices. Every vertex v has an d-bit binary ID that we will also denote by v. The bits are numbered from the most to the least significant one. Two vertices v, w are connected iff their IDs differ in exactly one bit. If i is the index of that bit, we say that (v, w) *flips i*. We identify vertices with their IDs, and $v \oplus w$ denotes exclusive or. We also sometimes view vertices as subsets of $\{1 \ldots d\}$, with bits indicating if the corresponding element is in or out of the set. Then $v \oplus w$ is the symmetric difference. The graph is undirected and unweighted.

3   Tight Bounds for HHL on Hypercubes

In this section we show that a d-dimensional hypercube has a labeling of size 3^d, and this labeling is optimal.

Consider the following labeling: treat vertex IDs as sets. $L(v)$ contains all vertices whose IDs are subsets of that of v. It is easy to see that this is a valid hierarchical labeling. The size of the labeling is

$$\sum_{i=0}^{d} 2^i \binom{d}{i} = 3^d.$$

Lemma 1. *A d-dimensional hypercube has an HHL of size 3^d.*

Next we show that 3^d is a tight bound. Given two vertices v and w of the hypercube, the *induced hypercube* H_{vw} is the subgraph induced by the vertices that have the same bits in the positions where the bits of v and w are the same, and arbitrary bits in other positions. H_{vw} contains all shortest paths between v and w. For a fixed order of vertices $v_1, v_2, \ldots, v_n$ (from least to most important), we define a *canonical labeling* as follows: w is in the label of v iff w is the maximum vertex of H_{vw} with respect to the vertex order. The labeling is valid because for any s, t, the maximum vertex of H_{st} is in $L(s)$ and $L(t)$, and is on the s–t shortest path. The labeling is HHL because all hubs of a vertex v have ranks greater or equal to the rank of v. The labeling is minimal because if w is the maximum vertex in H_{vw}, then w is the only vertex of $H_{vw} \cap L(w)$, so $L(v)$ must contain w.

Lemma 2. *The size of a canonical labeling is independent of the vertex ordering.*

Proof. It is sufficient to show that any transposition of neighbors does not affect the size. Suppose we transpose v_i and v_{i+1}. Consider a vertex w. Since only the order of v_i and v_{i+1} changed, $L(w)$ can change only if either $v_i \in H_{v_{i+1}w}$ or $v_{i+1} \in H_{v_i w}$, and v_{i+1} is the most important vertex in the corresponding induced hypercubes. In the former case v_{i+1} is removed from $L(w)$ after the transposition, and in the latter case v_i is added. There are no other changes to the labels.

Consider a bijection $b : H \Rightarrow H$, obtained by flipping all bits of w in the positions in which v_i and v_{i+1} differ. We show that v_{i+1} is removed from $L(w)$ iff v_i is added to $L(b(w))$. This fact implies the lemma.

Suppose v_{i+1} is removed from $L(w)$, i.e., $v_i \in H_{v_{i+1}w}$ and before the transposition v_{i+1} is the maximum vertex in $H_{v_{i+1}w}$. From $v_i \in H_{v_{i+1}w}$ it follows that v_i coincides with v_{i+1} in the positions in which v_{i+1} and w coincide. Thus b doesn't flip bits in the positions in which v_{i+1} and w coincide. So positions in which v_{i+1} and w coincide are exactly the same in which $b(v_{i+1})$ and $b(w)$ coincide. Moreover, in these positions all four v_{i+1}, w, $b(v_{i+1})$ and $b(w)$ coincide. So each vertex from $H_{v_{i+1},w}$ contains in $H_{b(v_{i+1}),b(w)}$ and vice versa, thus implying $H_{v_{i+1}w} = H_{b(v_{i+1})b(w)}$. Note that $b(v_{i+1}) = v_i$, and therefore $H_{v_{i+1}w} = H_{v_i b(w)}$. Before the transposition, v_{i+1} is the maximum vertex of $H_{v_i b(w)}$ and therefore $L(b(w))$ does not contain v_i. After the transposition, v_i becomes the maximum vertex, so $L(b(w))$ contains v_i.

This proves the if part of the claim. The proof of the only if part is similar. □

The hierarchical labeling of size 3^d defined above is canonical if the vertices are ordered in the reverse order of their IDs. Therefore we have the following theorem.

Theorem 1. *Any hierarchical labeling of a hypercube has size of at least 3^d.*

4   An $O(2.83^d)$ HL for Hypercubes

Next we show an HL for the hypercube of size $O(2.83^d)$. Combined with the results of Section 3, this implies that there is a polynomial (in $n = 2^d$) gap between hierarchical and non-hierarchical label sizes.

Consider the following HL L: For every v, $L(v)$ contains all vertices with the first $\lfloor d/2 \rfloor$ bits of ID identical to those of v and the rest arbitrary, and all vertices with the last $\lceil d/2 \rceil$ bits of ID identical to those of v and the rest arbitrary. It is easy to see that this labeling is non-hierarchical. For example, consider two distinct vertices v, w with the same $\lfloor d/2 \rfloor$ first ID bits. Then $v \in L(w)$ and $w \in L(v)$.

To see that the labeling is valid, fix s, t and consider a vertex u with the first $\lfloor d/2 \rfloor$ bits equal to t and the last $\lceil d/2 \rceil$ bits equal to s. Clearly u is in $L(s) \cap L(t)$. The shortest path that first changes bits of the first half of s to those of t and then the last bits passes through u.

The size of the labeling is $2^d \cdot (2^{\lfloor d/2 \rfloor} + 2^{\lceil d/2 \rceil}) = O(2^{\frac{3}{2}d}) = O(2.83^d)$. We have the following result.

Theorem 2. *A d-dimensional hypercube has an* HL *of size $O(2.83^d)$.*

5   Better HL Bounds

The bound of Theorem 2 can be improved. Let OPT be the optimal hub labeling size for a d-dimensional hypercube. In this section we prove the following result.

Theorem 3. OPT $= (2.5 + o(1))^d$

The proof uses the primal-dual method. Following [4], we view the labeling problem as a special case of **SET-COVER**. We state the problem of finding an optimal hub labeling of a hypercube as an integer linear program (ILP) which is a special case of a standard ILP formulation of **SET-COVER** (see e.g. [14]), with the sets corresponding to the shortest paths in the hypercube. For every vertex $v \in \{0,1\}^d$ and every subset $S \subseteq \{0,1\}^d$ we introduce a binary variable $x_{v,S}$. In the optimal solution $x_{v,S} = 1$ iff S is the set of vertices whose labels contain v. For every *unordered* pair of vertices $\{i, j\} \subseteq \{0,1\}^d$ we introduce the following constraint: there must be a vertex $v \in \{0,1\}^d$ and a subset $S \subseteq \{0,1\}^d$ such that $v \in H_{ij}$ (recall that the subcube H_{ij} consists of vertices that lie on the shortest paths from i to j), $\{i, j\} \subseteq S$, and $x_{v,S} = 1$. Thus, OPT is the optimal value of the following integer linear program:

$$\min \sum |S| \cdot x_{v,S} \quad \text{subject to}$$
$$\begin{cases} x_{v,S} \in \{0,1\}^{v,S} & \forall\, v \in \{0,1\}^d, S \subseteq \{0,1\}^d \\ \sum_{\substack{S \supseteq \{i,j\} \\ v \in H_{ij}}} x_{v,S} \geq 1 & \forall\, \{i,j\} \subseteq \{0,1\}^d \end{cases} \quad (1)$$

We consider the following LP-relaxation of (1):

$$\min \sum_{v,S} |S| \cdot x_{v,S} \quad \text{subject to}$$
$$\begin{cases} x_{v,S} \geq 0 & \forall\, v \in \{0,1\}^d, S \subseteq \{0,1\}^d \\ \sum_{\substack{S \supseteq \{i,j\} \\ v \in H_{ij}}} x_{v,S} \geq 1 & \forall\, \{i,j\} \subseteq \{0,1\}^d \end{cases} \quad (2)$$

We denote the optimal value of (2) by LOPT, and bound OPT as follows:

Lemma 3. LOPT $\leq$ OPT $\leq O(d) \cdot$ LOPT

Proof. The first inequality follows from the fact that (2) is a relaxation of (1).

As (1) corresponds to the standard ILP-formulation of **SET-COVER**, and (2) is the standard LP-relaxation for it, we can use the well-known (e.g., [14], Theorem 13.3) result: The integrality gap of LP-relaxation for **SET-COVER** is logarithmic in the number of elements we want to cover, which in our case is $O(n^2) = O(2^{2d})$. This implies the second inequality.   □

Now consider the dual program to (2).

$$\max \sum_{\{i,j\}} y_{\{i,j\}} \quad \text{subject to}$$
$$\begin{cases} y_{\{i,j\}} \geq 0 & \forall \{i,j\} \subseteq \{0,1\}^d \\ \sum_{\substack{\{i,j\} \subseteq S \\ H_{ij} \ni v}} y_{\{i,j\}} \leq |S| & \forall v \in \{0,1\}^d, S \subseteq \{0,1\}^d \end{cases} \quad (3)$$

The dual problem is a path packing problem. The strong duality theorem implies that LOPT is also the optimal solution value for (3).

We strengthen (3) by requiring that the values $y_{\{i,j\}}$ depend only on the distance between i and j. Thus, we have variables $\tilde{y}_0, \tilde{y}_1, \ldots, \tilde{y}_d$. Let N_k denote the number of vertex pairs at distance k from each other. Note that since $\tilde{y}$'s depend only on the distance and the hypercube is symmetric, it is enough to add constraints only for one vertex (e.g., 0^d); other constraints are redundant. We have the following linear program, which we call *regular*, and denote its optimal value by ROPT.

$$\max \sum_k N_k \cdot \tilde{y}_k \quad \text{subject to}$$
$$\begin{cases} \tilde{y}_k \geq 0 & \forall \, 0 \leq k \leq d \\ \sum_{\substack{\{i,j\} \subseteq S \\ H_{ij} \ni 0^d}} \tilde{y}_{\mathrm{dist}(i,j)} \leq |S| & \forall S \subseteq \{0,1\}^d \end{cases} \quad (4)$$

Clearly ROPT ≤ LOPT. The following lemma shows that in fact the two values are the same.

Lemma 4. ROPT ≥ LOPT

Proof. Intuitively, the proof shows that by averaging a solution for (3), we obtain a feasible solution for (4) with the same objective function value.

Given a feasible solution $y_{\{i,j\}}$ for (3), define

$$\tilde{y}_k = \frac{\sum_{\{i,j\}:\mathrm{dist}(i,j)=k} y_{\{i,j\}}}{N_k}.$$

From the definition,

$$\sum_{\{i,j\}} y_{\{i,j\}} = \sum_k N_k \cdot \tilde{y}_k.$$

We need to show that $\tilde{y}_k$ is a feasible solution for (4).

Consider a random mapping $\varphi \colon \{0,1\}^d \to \{0,1\}^d$ that is a composition of a mapping $i \mapsto i \oplus p$, where $p \in \{0,1\}^d$ is a uniformly random vertex, and a uniformly random permutation of coordinates. Then, clearly, we have the following properties:

- φ preserves distance;
- φ is a bijection;
- if the distance between i and j is k, then the pair $(\varphi(i), \varphi(j))$ is uniformly distributed among all pairs of vertices at distance k from each other.

Let $S \subseteq \{0,1\}^d$ be a fixed subset of vertices. As $y_{\{i,j\}}$ is a feasible solution of (3), we have

$$\sum_{\substack{\{i,j\} \subseteq S \\ H_{ij} \ni 0^d}} y_{\{i,j\}} \le |S|.$$

We define a random variable X as follows:

$$X = \sum_{\substack{\{i,j\} \subseteq \varphi(S) \\ H_{ij} \ni \varphi(0^d)}} y_{\{i,j\}}.$$

Since φ is a bijection and y is a feasible solution of (3), we have $\mathbf{E}_\varphi[X] \le |S|$. Furthermore, $\mathbf{E}_\varphi[X]$ is equal to

$$\mathbf{E}_\varphi\left[\sum_{\substack{\{i,j\} \subseteq \varphi(S) \\ H_{ij} \ni \varphi(0^d)}} y_{\{i,j\}}\right] = \mathbf{E}_\varphi\left[\sum_{\substack{\{i,j\} \subseteq S \\ H_{ij} \ni 0^d}} y_{\{\varphi(i),\varphi(j)\}}\right] = \sum_{\substack{\{i,j\} \subseteq S \\ H_{ij} \ni 0^d}} \mathbf{E}_\varphi\left[y_{\{\varphi(i),\varphi(j)\}}\right].$$

Since $(\varphi(i), \varphi(j))$ is uniformly distributed among all pairs of vertices at distance $\mathrm{dist}(i,j)$, the last expression is equal to $\sum_{\substack{\{i,j\} \subseteq S \\ H_{ij} \ni 0^d}} \tilde{y}_{\mathrm{dist}(i,j)}$. $\qquad\square$

Combining Lemmas 3 and 4, we get

$$\mathrm{ROPT} \le \mathrm{OPT} \le O(d) \cdot \mathrm{ROPT}.$$

It remains to prove that $\mathrm{ROPT} = (2.5 + o(1))^d$. For $0 \le k \le d$, let $\tilde{y}_k^*$ denote the maximum feasible value of $\tilde{y}_k$. It is easy to see that $\max_k N_k \tilde{y}_k^* \le \mathrm{ROPT} \le (d+1) \cdot \max_k N_k \tilde{y}_k^*$. Next we show that $\max_k N_k \tilde{y}_k^* = (2.5 + o(1))^d$.

To better understand (4), consider the graphs G_k for $0 \le k \le d$. Vertices of G_k are the same as those of the hypercube, interpreted as subsets of $\{1,\dots,d\}$. Two vertices are connected by an edge in G_k iff there is a shortest path of length k between them that passes through 0^d in the hypercube. This holds iff the corresponding subsets are disjoint and the cardinality of the union of the subsets is equal to k.

Consider connected components of G_k. By C_k^i ($0 \le i \le \lfloor k/2 \rfloor$) we denote the component that contains sets of cardinality i (and $k - i$).

If k is odd or $i \ne k/2$, C_k^i is a bipartite graph, with the right side vertices corresponding to sets of cardinality i, and the left side vertices – to sets of cardinality $k - i$. The number of these vertices is $\binom{d}{i}$ and $\binom{d}{k-i}$, respectively. C_k^i is a regular bipartite graph with vertex degree on the right side equal to $\binom{d-i}{k-i}$:

given a subset of i vertices, this is the number of ways to choose a disjoint subset of size $k - i$. The density of C_k^i is

$$\frac{\binom{d}{i} \cdot \binom{d-i}{k-i}}{\binom{d}{i} + \binom{d}{k-i}}.$$

If k is even and $i = k/2$, then G_k^i is a graph with $\binom{d}{i}$ vertices corresponding to the subsets of size i. The graph is regular, with the degree $\binom{d-i}{k-i}$. The density of C_k^i written to be consistent with the previous case is again

$$\frac{\binom{d}{i} \cdot \binom{d-i}{k-i}}{\binom{d}{i} + \binom{d}{k-i}}.$$

Next we prove a lemma about regular graphs, which may be of independent interest.

Lemma 5. *In a regular graph, density of any subgraph does not exceed the density of the graph. In a regular bipartite graph (i.e., degrees of each part are uniform), the density of any subgraph does not exceed the density of the graph.*

Proof. Let x be the degree of a regular graph. The density is a half of the average degree, and the average degree of any subgraph is at most x, so the lemma follows.

Now consider a bipartite graph with X vertices on the left side and Y vertices of the right side. Consider a subgraph with X' vertices on the left and Y' vertices on the right. Assume $X/X' \geq Y/Y'$; the other case is symmetric.

Let x be the degree of the vertices on the left size, then the graph density is $x \cdot X/(X + Y)$. For the subgraph, the number of edges adjacent to X' is at most $x \cdot X'$, so the subgraph density is at most

$$\frac{x \cdot X'}{X' + Y'} = \frac{x \cdot X}{X + Y'X/X'} \leq \frac{x \cdot X}{X + Y'Y/Y'} = \frac{x \cdot X}{X + Y}.$$

$\square$

By the lemma, each C_k^i is the densest subgraph of itself, and since C_k^i are connected components of G_k, the densest C_k^i is the densest subgraph of G_k.

Next we prove a lemma that gives (the inverse of) the value of maximum density of a subgraph of G_k.

Lemma 6. *For fixed d and k with $k \leq d$, the minimum of the expression*

$$\frac{\binom{d}{x} + \binom{d}{k-x}}{\binom{d}{x} \cdot \binom{d-x}{k-x}}$$

is achieved for $x = \lfloor k/2 \rfloor$ and $x = \lceil k/2 \rceil$ (with the two values being equal).

Proof. Using the standard identity

$$\binom{d}{x} \cdot \binom{d-x}{k-x} = \binom{d}{k-x} \cdot \binom{d-k+x}{x}$$

we write the expression in the lemma as

$$\frac{1}{\binom{d-x}{k-x}} + \frac{1}{\binom{d-k+x}{x}} = \frac{(d-k)!(k-x)!}{(d-x)!} + \frac{(d-k)!x!}{(d-k+x)!}.$$

Since $d-k$ is a constant, we need to minimize

$$\frac{1}{(k-x+1) \cdot \ldots \cdot (d-x)} + \frac{1}{(x+1) \cdot \ldots \cdot (d-k+x)}. \qquad (5)$$

Note that the expression is symmetric around $x = k/2$: for $y = k - x$, the expression becomes

$$\frac{1}{(y+1) \cdot \ldots \cdot (d-k+y)} + \frac{1}{(k-y+1) \cdot \ldots \cdot (d-y)}.$$

So it is enough to show that for $x \geq \lceil k/2 \rceil$, the minimum is achieved at $x = \lceil k/2 \rceil$. We will need the following auxiliary lemma.

Lemma 7. *If $0 \leq s \leq t$ and $\alpha \geq \beta \geq 1$, then $\alpha t + s/\beta \geq t + s$.*

Proof. Since $2 \leq \alpha + 1/\alpha \leq \alpha + 1/\beta$, we have $\alpha - 1 \geq 1 - 1/\beta$. Thus, $(\alpha - 1)t \geq s(1 - 1/\beta)$, and the lemma follows. □

It is clear that for every x the first term of (5) is not less than the second one. If we move from x to $x+1$, then the first term is multiplied by $(d-x)/(k-x)$, and the second term is divided by $(d-k+x+1)/(x+1)$. Since

$$\frac{d-x}{k-x} - \frac{d-k+x+1}{x+1} = \frac{(d-k)(2x+1-k)}{(x+1)(k-x)} \geq 0,$$

we can invoke Lemma 7 with t and s being equal to the first and the second term of (5), respectively, $\alpha = (d-x)/(k-x)$, $\beta = (d-k+x+1)/(x+1)$. □

Recall that $\tilde{y}_k^*$ denotes the maximum feasible value of $\tilde{y}_k$.

Lemma 8.

$$\tilde{y}_k^* = \begin{cases} 1 & k = 0 \\ 2/\binom{d-i}{i} & k = 2i, i > 0 \\ \left(\binom{d}{i} + \binom{d}{i+1}\right) / \left(\binom{d}{i} \cdot \binom{d-i}{i+1}\right) & k = 2i+1. \end{cases}$$

Proof. Fix k and consider the maximum density subgraph of G_k. Inverse of the subgraph density is an upper bound on a feasible value of $\tilde{y}_k$.

On the other hand, it is clear that we can set $\tilde{y}_k$ to the inverse density of the densest subgraph of G_k and other $\tilde{y}$'s to zero, and obtain the feasible solution of (4). By applying Lemma 6, we obtain the desired statement. □

Recall that N_k denotes the number of vertex pairs at distance k from each other. For each vertex v, we can choose a subset of k bit positions and flip bits in these positions, obtaining a vertex at distance k from v. This counts the ordered pairs, we need to divide by two to get the the number of unordered pairs:

$$N_k = 2^d \binom{d}{k}/2,$$

except for the case $k = 0$, where $N_0 = 2^d$.

Finally, we need to find the maximum value of

$$\psi(k) := N_k \cdot \tilde{y}_k^* = 2^d \cdot \begin{cases} \binom{d}{2i}/\binom{d-i}{i} & k = 2i \\ \binom{d}{2i+1} \cdot \left(\binom{d}{i} + \binom{d}{i+1}\right) / \left(2 \cdot \binom{d}{i} \cdot \binom{d-i}{i+1}\right) & k = 2i+1. \end{cases}$$

One can easily see that $\psi(2i+1)/\psi(2i) = (d+1)/(4i+2)$. So, if we restrict our attention to the case $k = 2i$, we could potentially lose only polynomial factors.

We have
$$\frac{\psi(2i+2)}{\psi(2i)} = \frac{d-i}{4i+2}.$$

This expression is greater than one if $i < (d-2)/5$. The optimal i has to be as close as possible to the bound. As $d \to \infty$, this is $\frac{d}{5} \cdot (1 + o(1))$.

We will use the standard fact: if for $n \to \infty, m/n \to \alpha$, then

$$\binom{n}{m} = (2^{H(\alpha)} + o(1))^n,$$

where H is the Shannon entropy function $H(\alpha) = -\alpha \log_2 \alpha - (1-\alpha) \log_2(1-\alpha)$.

Thus, if $d \to \infty, k/d \to 2/5$, then

$$\psi(k) = (2^{1+H(0.4)-0.8 \cdot H(0.25)} + o(1))^d.$$

One can verify that
$$2^{1+H(0.4)-0.8 \cdot H(0.25)} = 2.5,$$

so we have the desired result.

6   Concluding Remarks

We show a polynomial gap between the sizes of HL and HHL for hypercubes. Although our existence proof for $(2.5 + o(1))^d$-size HL is non-constructive, the approximation algorithm of [4] can build such labels in polynomial time. However, it is unclear how these labels look like. It would be interesting to have an explicit construction of such labels.

Little is known about the problem of computing the smallest HHL. We do not know if the problem is NP-hard, and we know no polynomial-time algorithm for it (exact or polylog-approximate). These are interesting open problems.

The HL vs. HHL separation we show does not mean that HHL labels are substantially bigger than the HL ones for any graphs. In particular, experiments suggest that HHL works well for road networks. It would be interesting to characterize the class of networks for which HHL works well.

Note that an arbitrary (non-hub) labelings for the hypercube can be small: we can compute the distances from the standard d-bit vertex IDs. It would be interesting to show the gap between HL and HHL for graph classes for which arbitrary labelings must be big.

We believe that one can prove an $\Theta^*(n^{1.5})$ bound for HL size on constant degree random graphs using the primal-dual method. However, for this graphs it is unclear how to prove tight bounds on the size of HHL.

References

1. Abraham, I., Delling, D., Goldberg, A.V., Werneck, R.F.: Hierarchical Hub Labelings for Shortest Paths. In: Epstein, L., Ferragina, P. (eds.) ESA 2012. LNCS, vol. 7501, pp. 24–35. Springer, Heidelberg (2012)
2. Abraham, I., Delling, D., Goldberg, A.V., Werneck, R.F.: A Hub-Based Labeling Algorithm for Shortest Paths on Road Networks. In: Pardalos, P.M., Rebennack, S. (eds.) SEA 2011. LNCS, vol. 6630, pp. 230–241. Springer, Heidelberg (2011)
3. Babenko, M., Goldberg, A.V., Gupta, A., Nagarajan, V.: Algorithms for hub label optimization. In: Fomin, F.V., Freivalds, R., Kwiatkowska, M., Peleg, D. (eds.) ICALP 2013, Part I. LNCS, vol. 7965, pp. 69–80. Springer, Heidelberg (2013)
4. Cohen, E., Halperin, E., Kaplan, H., Zwick, U.: Reachability and Distance Queries via 2-hop Labels. SIAM Journal on Computing 32 (2003)
5. Delling, D., Sanders, P., Schultes, D., Wagner, D.: Engineering Route Planning Algorithms. In: Lerner, J., Wagner, D., Zweig, K.A. (eds.) Algorithmics. LNCS, vol. 5515, pp. 117–139. Springer, Heidelberg (2009)
6. Dijkstra, E.W.: A Note on Two Problems in Connexion with Graphs. Numerische Mathematik 1, 269–271 (1959)
7. Fredman, M.L., Tarjan, R.E.: Fibonacci Heaps and Their Uses in Improved Network Optimization Algorithms. J. Assoc. Comput. Mach. 34, 596–615 (1987)
8. Gavoille, C., Peleg, D., Pérennes, S., Raz, R.: Distance Labeling in Graphs. Journal of Algorithms 53, 85–112 (2004)
9. Geisberger, R., Sanders, P., Schultes, D., Delling, D.: Contraction Hierarchies: Faster and Simpler Hierarchical Routing in Road Networks. In: McGeoch, C.C. (ed.) WEA 2008. LNCS, vol. 5038, pp. 319–333. Springer, Heidelberg (2008)
10. Goldberg, A.V.: A Practical Shortest Path Algorithm with Linear Expected Time. SIAM Journal on Computing 37, 1637–1655 (2008)
11. Golumbic, M.C.: Algorithmic Graph Theory and Perfect Graphs. Academic Press, New York (1980)
12. Peleg, D.: Proximity-preserving labeling schemes. Journal of Graph Theory 33(3), 167–176 (2000)
13. Thorup, M.: Undirected Single-Source Shortest Paths with Positive Integer Weights in Linear Time. J. Assoc. Comput. Mach. 46, 362–394 (1999)
14. Vazirani, V.V.: Approximation Algorithms. Springer (2001)

Solving 3-Superstring in $3^{n/3}$ Time

Alexander Golovnev[1], Alexander S. Kulikov[2,3], and Ivan Mihajlin[4]

[1] New York University
[2] St. Petersburg Department of Steklov Institute of Mathematics
[3] Algorithmic Biology Laboratory, St. Petersburg Academic University
[4] St. Petersburg Academic University

Abstract. In the shortest common superstring problem (SCS) one is given a set $s_1, \ldots, s_n$ of n strings and the goal is to find a shortest string containing each s_i as a substring. While many approximation algorithms for this problem have been developed, it is still not known whether it can be solved exactly in fewer than 2^n steps. In this paper we present an algorithm that solves the special case when all of the input strings have length 3 in time $3^{n/3}$ and polynomial space. The algorithm generates a combination of a de Bruijn graph and an overlap graph, such that a SCS is then a shortest directed rural postman path (DRPP) on this graph. We show that there exists at least one optimal DRPP satisfying some natural properties. The algorithm works basically by exhaustive search, but on the reduced search space of such paths of size $3^{n/3}$.

1 Introduction

The *shortest common superstring problem* (SCS) is: given a set $\{s_1, \ldots, s_n\}$ of n strings, find a shortest string containing each s_i as a substring (w.l.o.g., we assume that no input string is a subtstring of another). The problem is known to be NP-hard and has many practical applications including data storage, data compression, and genome assembly. For this reason, approximation algorithms for SCS are widely studied. For a long time the best known approximation ratio was 2.5 by Sweedyk [26] (the same bound also follows from 2/3-approximation for MAX-ATSP [15,24]). Very recently the bound was improved to $2\frac{11}{23}$ by Mucha [23]. The best known inapproximability ratio (under the P $\neq$ NP assumption) is $\frac{345}{344}$ by Karpinski and Schmied [17].

At the same time it is not known whether SCS can be solved in fewer than $O^*(2^n)$ steps ($O^*(\cdot)$ suppresses polynomial factors of input length). Note that SCS is a permutation problem: to find a string containing all s_i's in a given order one just overlaps the strings in this order. Thus, the trivial algorithm requires $O^*(n!)$ time. Now consider the following *suffix graph* of the given set of strings: the set of vertices is $\{s_1, \ldots, s_n\}$, vertices s_i and s_j are joined by an arc of weight $|\mathrm{suffix}(s_i, s_j)|$ where $\mathrm{suffix}(s_i, s_j)$ is the shortest string such that s_j is a suffix of $s_i \circ \mathrm{suffix}(s_i, s_j)$ (where $\circ$ denotes concatenation). SCS can be solved by finding a shortest traveling salesman path (TSP) in this graph. For TSP, the classical dynamic programming based $O^*(2^n)$ algorithm discovered by Bellman [2] and independently by Held and Karp [12] is well-known.

K. Chatterjee and J. Sgall (Eds.): MFCS 2013, LNCS 8087, pp. 480–491, 2013.
© Springer-Verlag Berlin Heidelberg 2013

There are two natural special cases of SCS: the case when the size of the alphabet is bounded by a constant and the case when all input strings have length r. The latter case is called r-*SCS*. Note that when both these parameters are bounded the problem degenerates as the number of possible input strings is then also bounded by a constant. It is known that both SCS over the binary alphabet and 3-SCS are NP-hard, while 2-SCS can be solved in linear time [10] and 2-SCS with multiplicities (where each input string is given together with the number of its occurrences in a superstring) can be solved in quadratic time [7]. Vassilevska [27] showed that SCS over the binary alphabet cannot be much easier than the general case. Namely, she provided a polynomial-time reduction of general SCS to SCS over the binary alphabet that preserves the number of input strings. It implies that α-approximation of SCS over the binary alphabet is not easier than α-approximation of general SCS. It also means that an $O^*(c^n)$-algorithm for SCS over the binary alphabet implies an $O^*(c^n)$-algorithm for the general case. Hence SCS for smaller size alphabet cannot be much easier. Our result suggests that SCS for shorter strings can actually be easier to solve.

In this paper we present an algorithm solving a special case when all of the input strings have length 3 in time $O^*(3^{n/3})$ and polynomial space. The approach is based on finding a shortest rural postman path in the de Bruijn graph of the given set of strings. The algorithm works basically by exhaustive search, but having reduced the search space to size $3^{n/3}$, and then inspecting each possibility in polynomial time. We show that for the case of 3-strings to find an optimal rural path it is enough to guess where such a path enters each weakly connected component formed by input strings. We then show that for a component on k arcs there are at most k such entry points. Since the total number of arcs is n, the running time is roughly $k^{n/k}$ and this does not exceed $3^{n/3}$ (for $k \in \mathbb{N}$).

The current situation with exact algorithms for SCS is similar to what is known for some other NP-hard problems — say, the satisfiability problem (SAT), the maximum satisfiability problem (MAX-SAT), and the traveling salesman problem (TSP). Namely, despite many efforts the best known algorithms for the general versions of these problems run in time $O^*(2^n)$ (n being the number of variables/vertices). At the same time better upper bounds are known for special cases of these problems: $O(1.308^n)$ for 3-SAT [13], $O(1.731^n)$ for MAX-2-SAT [28], $O(1.251^n)$ for TSP on cubic graphs [14], $O^*(1.109^n)$ for $(n,3)$-MAX-2-SAT [19], c^n (where $c < 2$) for SAT [5] and MAX-SAT [8,19] on formulas with constant clause density. Moreover, it is known that k-SAT can be solved in time $O((2 - 2/k)^n)$ [22] and TSP can be solved in time $O((2 - \epsilon)^n)$, where $\epsilon > 0$ depends only on the degree bound of a graph [4].

2   General Setting

Throughout the paper $\mathcal{S} = \{s_1, \ldots, s_n\}$ is an input set of strings over an alphabet Σ, and n is the number of strings. W.l.o.g. we assume that no s_i is a substring of s_j for any $i \neq j$.

For strings s and t, by $s \circ t$ we denote the concatenation of s and t. By overlap(s,t) we denote the longest suffix of s that is also a prefix of t.

By prefix(s, t) we denote the first $|s| - |\text{overlap}(s, t)|$ symbols of s and by suffix(s, t) we denote the last $|t| - |\text{overlap}(s, t)|$ symbols of t. Clearly,

$$\text{prefix}(s, t) \circ \text{overlap}(s, t) = s \text{ and overlap}(s, t) \circ \text{suffix}(s, t) = t.$$

2.1   Suffix Graphs and the Traveling Salesman Problem

Clearly, $s \circ \text{suffix}(s, t)$ is the shortest string containing s and t in this order. More generally, the shortest string containing strings $s_{i_1}, \ldots, s_{i_n}$ in this order is

$$s_{i_1} \circ \text{suffix}(s_{i_1}, s_{i_2}) \circ \cdots \circ \text{suffix}(s_{i_{n-1}}, s_{i_n}).$$

Thus, the goal of SCS is to find a permutation of n input strings minimizing the total length of the suffix function. As mentioned in the introduction, one can define a complete directed graph on the given set $\{s_1, \ldots, s_n\}$ of n strings as a set of vertices where vertices s_i and s_j are joined by an arc of weight $|\text{suffix}(s_i, s_j)|$ (suffix graph). Solving SCS then corresponds to solving TSP in this graph. This connection has been used in essentially all previous approximation algorithms for SCS. This graph however is asymmetric (i.e., directed) and the best known algorithm due to Bellman [2], Held and Karp [12] uses $O^*(2^n)$ time and space. There are also algorithms based on inclusion-exclusion with running time $O^*(2^n \cdot M)$ and space $O^*(M)$ [18,16,1] (here, M is the maximal arc weight). Lokshtanov and Nederlof [21] show how to solve TSP in $O^*(2^n \cdot M)$ time with only $O^*(1) = \text{poly}(n, \log M)$ space. For symmetric TSP, Björklund [3] recently came up with an $O^*(1.657^n \cdot M)$ time randomized algorithm. Note that for SCS, M does not exceed the size of the input, hence the mentioned inclusion-exclusion algorithm solves SCS in $O^*(2^n)$ time and polynomial space.

2.2   De Bruijn Graphs and the Rural Postman Problem

In this paper we deal with another useful concept, namely *de Bruijn graphs*. Such graphs are widely used in genome assembly, one of the most important practical applications of SCS [25]. At the same time they have only few applications in theoretical investigations of SCS. To simplify its definition from now on we stick to strings of length 3 only. So, let $\mathcal{S} = \{s_1, \ldots, s_n\}$ be a set of 3-strings over the alphabet Σ. The de Bruijn graph DG is a weighted complete directed graph (with loops, but without multiple arcs) with the set of vertices Σ^2. Distinct vertices s and t are joined by an arc of weight $|\text{suffix}(s, t)|$. Also, for each string from Σ^2 consisting of the same two symbols there is a loop of weight 1. Intuitively, the weight of an arc (s, t) is equal to the number of symbols we need to spell going from a string s to a string t. (Particularly, going from a string AA to itself we need to spell one more A, that is why loops are of weight 1.) Thus, all arcs in DG have weight either 1 or 2. Note that any 3-string s over Σ defines an arc of weight 1 in DG: the arc joins the prefix of s of length 2 and the suffix of s of length 2. Thus, what we are looking for in the SCS problem is a shortest path in DG going through all the arcs $E_{\mathcal{S}}$ given by $\mathcal{S}$.

This problem is known as *directed rural postman path problem* (DRPP). In DRPP one is given a weighted graph $G = (V, E)$ and a subset $E_R \subseteq E$ of its arcs and the goal is to find a shortest path in G going through all the arcs of E_R. The arcs from E_R are called *required*, all the remaining arcs are called *optional*. A path going through all the required arcs is called a *rural path* and a shortest such path is called an *optimal rural path*.[1]

DRPP has many practical applications (see, e.g., [9,11]). At the same time almost no non-trivial exact algorithms are known for DRPP. As with SCS and TSP, for DRPP there is a simple algorithm with the running time $O^*(n!)$ (for DRPP, by n we denote the number of required arcs) as well as dynamic programming based algorithm with running time $O^*(2^n)$. DRPP generalizes such problems as Chinese Postman Problem and Asymmetric TSP. If the set of required arcs forms a single weakly connected component (weakly connected components of a directed graph are just connected components in this graph with all directed arcs replaced by undirected edges), then the problem can be solved in polynomial time: all one needs to do is to add arcs of minimal total weight to imbalanced vertices. This can be done by finding a minimum weight perfect matching in an appropriate bipartite graph (details can be found, e.g., in [6]). If the set of required arcs forms more than just one weakly connected component then DRPP becomes NP-hard [20]. The reason is that now one not only needs to balance all the imbalanced vertices by adding arcs of minimal total weight but also to guarantee somehow that the resulting graph is connected. This turns out to be harder.

However 2-SCS can be solved in polynomial time even if the input strings form more than one weakly connected component. The important property of 2-SCS (as opposed to, say, 3-SCS) is that 2-strings from different weakly connected components always have zero overlap. This means that one can find an optimal rural path for each component separately.

3   Algorithm

The set $\mathcal{S}$ of 3-strings defines the required set of arcs $E_{\mathcal{S}}$ in the de Bruijn graph DG. What we are looking for is a shortest path in this graph going through all the required arcs (an optimal rural path). Optional arcs have weight 1 or 2 while all required arcs have weight 1. For each vertex of the graph we know the number of adjacent incoming and outgoing required arcs, but we do not know the number of adjacent incoming and outgoing optional arcs (in an optimal rural path).

Note the following two simple properties of an optimal rural path.

– An optimal rural path does not start and does not end with an optional arc (removing such an arc leaves a rural path of smaller weight).

---

[1] We use the term "path" to denote a path that may go through some vertices and arcs more than once (a term "walk" is also used in the literature for this). A simple path is a path without repeated vertices and arcs.

– There always exists an optimal rural path that does not contain an optional arc followed by another optional arc. Two such arcs can be replaced by a single arc. Since all arcs have weight 1 or 2 this does not increase the total weight of a path.

By $d_{in}^{re}(v)$ and $d_{out}^{re}(v)$ we denote the number of required incoming and outgoing arcs to v, respectively: $d_{in}^{re}(v) = |\{(u,v) \in E_S\}|$ and $d_{out}^{re}(v) = |\{(v,w) \in E_S\}|$. Similarly, for a path P in DG, by $d_{in}^{op}(P,v)$ and $d_{out}^{op}(P,v)$ we denote the number of optional incoming and outgoing arcs for the vertex v in the path P:

$$d_{in}^{op}(P,v) = |\{(u,v) \mid (u,v) \in P, (u,v) \notin E_S\}|,$$
$$d_{out}^{op}(P,v) = |\{(v,w) \mid (v,w) \in P, (v,w) \notin E_S\}|.$$

Recall that a path may go through a particular vertex more than once, so these degrees may be greater than 1. Also, the path P may go through a particular arc more than once hence the sets in the right hand side of the definition $d_{in}^{op}(P,v)$ and $d_{out}^{op}(P,v)$ are actually multisets. In other words, $d_{in}^{op}(P,v)$ ($d_{out}^{op}(P,v)$) is the number of times the path P enters (respectively, leaves) the vertex v by an optional arc.

Definition 1 (configuration). *A configuration is a pair $f = (f_{in}, f_{out})$ of functions from V to $\mathbb{N}$. A configuration tells for each vertex the number of incoming and outgoing optional arcs of a path. Consequently we say that a configuration is consistent with a path P iff $f_{in}(v) = d_{in}^{op}(P,v)$ and $f_{out}(v) = d_{out}^{op}(P,v)$ for each vertex v. A path in DG determines a configuration in a natural way.*

Definition 2 (normal configuration, special vertex). *We say that a configuration $f = (f_{in}, f_{out})$ is normal iff the following three conditions hold.*

– *It is consistent with at least one rural path. This, in particular, means that for all but two vertices v (the two exceptional vertices being the first and the last vertices of a path)*

$$f_{in}(v) + d_{in}^{re}(v) = f_{out}(v) + d_{out}^{re}(v). \tag{1}$$

– *For each weakly connected component C of E_S,*

$$\sum_{v \in C} \min\{f_{in}(v), f_{out}(v)\} \le 1. \tag{2}$$

I.e., each weakly connected component contains at most one vertex that has both incoming and outgoing optional arcs. Moreover if such a vertex exists then it has just one incoming or one outgoing arc. Such a vertex is called special.

– *For each vertex v, if v has only incoming (outgoing) required arcs then it has only outgoing (incoming) optional arcs:*

$$(d_{out}^{re}(v) = 0 \Rightarrow f_{in}(v) = 0) \text{ and } (d_{in}^{re}(v) = 0 \Rightarrow f_{out}(v) = 0). \tag{3}$$

In particular, a vertex v with $\min\{d_{in}^{re}(v), d_{out}^{re}(v)\} = 0$ cannot be special.

Definition 3 (normal path). *A normal path is a path with a normal configuration.*

The motivation for studying configurations is given by the following lemmas (which are proven below).

Lemma 1. *There exists an optimal rural path that is normal.*

Lemma 2. *Given a normal configuration f of an (unknown) optimal rural path we can find in polynomial time an optimal rural path consistent with f.*

It remains to show that the number of different normal configurations is not too large. This is guaranteed by the following lemmas.

Lemma 3. *A weakly connected component $\mathcal{C}$ of $E_{\mathcal{S}}$ consisting of k arcs has at most k different normal configurations.*

Lemma 4. *All normal configurations can be enumerated in time $O^*(3^{n/3})$ and polynomial space.*

Using these four lemmas the main result of the paper follows almost immediately.

Theorem 1. *The 3-SCS problem can be solved in time $O^*(3^{n/3})$ and polynomial space.*

Proof. Due to Lemma 4 we can enumerate all normal configurations in time $O^*(3^{n/3})$ and polynomial space. By Lemma 1 at least one of these configurations corresponds to an optimal rural path. Given such a configuration we can recover an optimal rural path by Lemma 2. □

3.1 Proofs

In this subsection, we complete the analysis of the algorithm by proving the lemmas given in the previous subsection. In the proofs, we often consider a path as a sequence of vertices. In this notation, lower case letters are used to denote vertices while upper case letters denote parts of a path, i.e., sequences of vertices (possibly empty). E.g., to specify that a path P starts with a vertex s, goes through a vertex v and ends in a vertex t we write $P = sAvBt$. In the pictures below, required arcs are shown in bold, optional arcs are thin and gray, snaked arcs denote just a part of a path.

Proof (of Lemma 1). Let P be an optimal rural path containing the minimal number of optional arcs. We show that if P is not a normal path, then the number of optional arcs in P can be decreased without increasing its weight. This is done by replacing two optional arcs with a new one. Since all arcs have weights 1 or 2, this replacement does not increase the weight of P.

Consider a weakly connected component $\mathcal{C}$ and let x be the last vertex of $\mathcal{C}$ in the path P. To guarantee that (2) holds we first transform P such that for all vertices v of $\mathcal{C}$ with the only possible exception of x we have

$$\min\{d_{\mathrm{in}}^{\mathrm{op}}(P, v), d_{\mathrm{out}}^{\mathrm{op}}(P, v)\} = 0\,.$$

Assume that a vertex $v \neq x$ not fulfilling this equality exists in $\mathcal{C}$. Denote incoming and outgoing optional arcs of v by (w_1, v) and (v, u_1), respectively. Let u_2 be a vertex such that the arc (u_2, v) precedes the arc (v, u_1) in P and w_2 be a vertex such that the arc (v, w_2) follows the arc (w_1, v) in P. Since the path does not contain two consecutive optional arcs, the arcs (u_2, v) and (v, w_2) differ from the arcs (w_1, v) and (v, u_1). We now consider the following two cases depending on whether the path first goes through (u_2, v) and (v, u_1) or through (w_1, v) and (v, w_2).

1. The path P has the form $sAu_2vu_1Bw_1vw_2CxDt$ (i.e., P first goes through (u_2, v) and (v, u_1) and only then through (w_1, v) and (v, w_2)). We transform it to $sAu_2vw_2CxDtu_1Bw_1$. Note that this transformation increases the number of optional arcs out of t, but it reduces the total number of optional arcs.

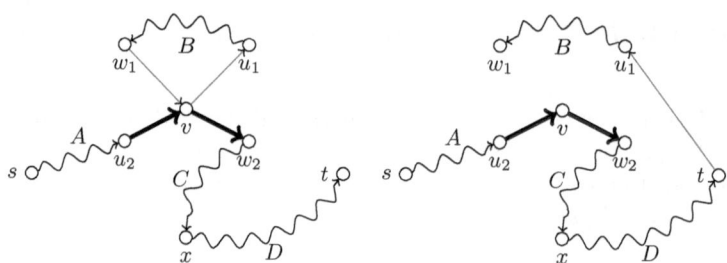

2. The path P has the form $sAw_1vw_2Bu_2vu_1CxDt$. We then replace the arcs (w_1, v) and (v, u_1) by a new arc (w_1, u_1). As a result we get the path sAw_1u_1CxDt and a cycle vw_2Bu_2v. Recall however that $\mathcal{C}$ is a weakly connected component. This means that the new path has at least one vertex in common with the cycle. Thus we can glue this cycle into this path.

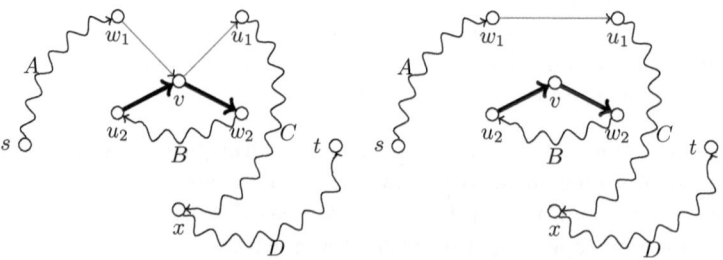

Clearly both transformations above do not break the path and decrease the total number of optional arcs.

We now show that P can be transformed so that $\min\{d_{\text{in}}^{\text{op}}(P,x), d_{\text{out}}^{\text{op}}(P,x)\} \leq$ 1. Assume for the sake of contradiction that x has in P at least two incoming and at least two outgoing optional arcs. Let (v,x) and (x,w) be the first optional incoming and outgoing arcs for x in P. Consider two subcases.

1. P first goes through (v,x) and then through (x,w). Since P has at least two optional arcs out of x the path P has the form $sAvxBxwCxDt$. We transform it to $sAvwCxBxDt$.
2. P first goes through (x,w). Then it has the form $sAxwBvxCt$ and can be transformed to $sAxCtwBv$.

Thus, P satisfies (2).

Finally, we show how to transform P so that (3) holds. Consider a vertex $v \in C$ and assume w.l.o.g. that it has no incoming required arcs (i.e., $d_{\text{in}}^{\text{re}}(v) = 0$). Assume that P also has an optional arc (v,w). Since P cannot start with an optional arc it has an arc (u,v) preceding (v,w) and this arc is also optional. But then two optional arcs (u,v) and (v,w) can be replaced with an arc (u,w). This again contradicts the assumption that P has the minimal possible number of optional arcs. The case $d_{\text{out}}^{\text{re}}(v) = 0$ is treated similarly. Thus, P satisfies (3).

We conclude that any rural path with the minimal number of optional arcs satisfies the properties (2) and (3). The property (1) holds for such a path for a trivial reason. Thus, any such path is normal. $\qquad\square$

Proof (of Lemma 2). In the following we assume that we know the first vertex s and the last vertex t of an optimal rural path that we are looking for. Since the first and the last arc of such a path are both required arcs, enumerating all such pairs (s,t) can be done in $O(n^2)$ time.

To find the required path we modify the graph DG and the set of required arcs E_S as follows:

- Introduce $|\Sigma|$ new vertices labeled by single symbols and join them to all other vertices by arcs of weight equal to the length of the suffix of the two corresponding strings. E.g., $w(\text{A},\text{AB}) = 1$, $w(\text{A},\text{BC}) = 2$, $w(\text{BC},\text{A}) = 1$, $w(\text{BA},\text{A}) = 0$, $w(\text{A},\text{B}) = 1$.
- For each vertex v of the initial graph DG labeled by AB add $f_{\text{in}}(v)$ copies of the arc (A,AB) and $f_{\text{out}}(v)$ copies of (AB,B) to the set of required arcs E_S.

Denote the resulting graph by DG' and the resulting set of required arcs by E_S'. It is worth to note that E_S' is a multiset, namely it might contain several copies of new required arcs (e.g., $f_{\text{in}}(\text{AB})$ copies of the arc (A,AB)).

Let $C_1, \ldots, C_p$ be the weakly connected components of E_S and $C_1', \ldots, C_q'$ be the weakly connected components of E_S'. Clearly $q \leq p$ and for each C_i there is C_j' such that $C_i \subseteq C_j'$.

First we show that the weight of an optimal rural path with configuraion f in DG is equal to the weight of an optimal rural path in DG'. Indeed, given an optimal rural path P consistent with f in DG one replaces each its optional arc (AB,CD) (of weight 2) with three arcs (AB,B), (B,C), (C,CD) (of total weight $0 + 1 + 1 = 2$) and each optional arc (AB,BC) (of weight 1) with two arcs (AB,B), (B,BC) (of total weight $0 + 1 = 1$). The resulting path P' is a rural path in DG':

we replaced exactly $f_{\text{out}}(\text{AB})$ optional arcs out of the vertex AB with new required arcs (AB, B). Moreover, this path clearly has exactly the same weight. Conversely, let P' be an optimal rural path in DG'. Just by removing all vertices labeled by single symbols we get a rural path P consistent with f whose weight is not greater than the weight of P'.

Now we show that an optimal rural path in DG' can be found in polynomial time. For this, we show that it is enough to solve the problem for each weakly connected component of E'_S separately.

Let P' be such an optimal rural path in DG'. Translate it back to a path P in DG by removing all vertices labeled by single symbols. Let $P = A_1 A_2 \ldots A_k$ where each sequence of vertices A_i lies inside the same weakly connected component C'_j of E'_S and A_i and A_{i+1} belong to different components. Denote by u_i, v_i the first and the last vertex of A_i (recall that the path does not contain an optional arc followed by another optional arc). A simple but crucial observation is that each arc (v_i, u_{i+1}) has weight 2. Indeed, if $w(v_i, u_{i+1}) = 1$ then $v_i = \text{AB}$ and $u_{i+1} = \text{BC}$. Note that (v_i, u_{i+1}) is an optional arc since v_i and u_{i+1} belong to different components of E'_S (and hence to different components of E_S). This means that $f_{\text{out}}(v_i) > 0$ and $f_{\text{in}}(u_{i+1}) > 0$. But then the arcs (AB, B) and (B, BC) are required in DG' and thus v_i and u_{i+1} lie in the same weakly connected component of E'_S.

We would like to show now that there exists an optimal rural path P' in DG' that goes through each component of E'_S separately. For this, we show that if P' enters the same component of E'_S more than once then we can reduce the number of optional arcs between the components by transforming a path (without increasing the total weight of the path). As before, translate the path P' back to P. Now assume that for some component C'_j, the path P enters C'_j at least two times, i.e., there are two optional arcs (a_1, b_1) and (a_2, b_2) in P such that $b_1, b_2 \in C'_j$ and $a_1, a_2 \notin C'_j$. Assume that C'_j is not the last component of the path P (the case when it is the last one is similar). This means that P must also leave the component C'_j two times. More formally, P contains two optional arcs (b_3, a_3) and (b_4, a_4) where $b_3, b_4 \in C'_j$ and $a_3, a_4 \notin C'_j$. Replace now the arcs (a_1, b_1) and (b_3, a_3) by (b_3, b_1) and (a_1, a_3). It is easy to see that such a transformation does not change the degrees of vertices. To guarantee that the resulting set of arcs is a single path but not a cycle and a path we note that b_1, b_2, b_3, b_4 lie in the same weakly connected component. Also, the weight of the path is not increased (since $w(a_1, b_1) = w(b_3, a_3) = 2$ while $w(b_3, b_1), w(a_1, a_3) \leq 2$).

Thus, to find an optimal rural path in DG' we can find an optimal path for each component of E'_S separately and then join the found paths arbitrarily (recall that solving DRPP for a weakly connected component is a polynomial problem). □

Proof (of Lemma 3). Let

$$\text{mindeg}^{\text{re}}(v) = \min\{d_{\text{in}}^{\text{re}}(v), d_{\text{out}}^{\text{re}}(v)\}, \ \text{mindeg}^{\text{op}}(v) = \min\{f_{\text{in}}(v), f_{\text{out}}(v)\}.$$

By definition of a normal configuration (see (3)) each component contains at most one special vertex, i.e., a vertex with $\text{mindeg}^{\text{op}} = 1$. Recall from the proof of Lemma 4 that we only need to know which vertex in a configuration is special (if any) to fully determine the configuration.

We now consider the following two cases.

$\mathcal{C}$ is Eulerian. Clearly $\mathcal{C}$ contains at most k vertices (and contains exactly k vertices when it is a simple cycle). Note that if E_S does not consist of $\mathcal{C}$ only then $\mathcal{C}$ must contain at least one special vertex in any rural path and hence the number of different normal configurations for $\mathcal{C}$ is k. At the same time, if E_S contains $\mathcal{C}$ only then an optimal rural path can be found in polynomial time.

$\mathcal{C}$ is not Eulerian. By (3), it is enough to show that $\mathcal{C}$ contains at most $(k-1)$ vertices with non-zero mindeg^{re}. Then either one of these $(k-1)$ vertices is special or there are no special vertices — thus, at most k different configurations.

To show that there are at most $(k-1)$ vertices in $\mathcal{C}$ with non-zero mindeg^{re} consider two subcases.

1. By removing directions of the arcs in $\mathcal{C}$ we get a simple path on k arcs. Then $\mathcal{C}$ contains $(k+1)$ vertices but both ends of this path have zero mindeg^{re}.
2. Otherwise $\mathcal{C}$ contains at most k vertices. If the number of vertices is strictly smaller than k then we are done. If the number of vertices is equal to k we find a vertex with zero mindeg^{re}. For this, take any vertex in $\mathcal{C}$ and start a path from it. As a result we either arrive to a vertex with zero out-degree (in this case we are done) or construct a cycle. Since $\mathcal{C}$ is weakly connected for at least one of the vertices of this cycle the sum of in-degree and out-degree is at least 3. But then $\mathcal{C}$ must contain a vertex with in-degree plus out-degree equal to 1 and we are done again. □

Proof (of Lemma 4). Let E_S consist of t weakly connected components $\mathcal{C}_1, \ldots, \mathcal{C}_t$, let also n_i be the number of required arcs in $\mathcal{C}_i$ (hence $n_1 + \cdots + n_t = n$). By Lemma 3 above, for $\mathcal{C}_i$ there are at most n_i different configurations. Thus, the total number of normal configurations for E_S is at most $\prod_{i=1}^{t} n_i$. We show that this is at most $3^{n/3}$ by induction on n. The base case $n = 1$ is clear. Induction step:

$$\prod_{i=1}^{t} n_i = n_t \cdot \prod_{i=1}^{t-1} n_i \leq 3^{\frac{n-n_t}{3}} n_t = 3^{\frac{n-n_t}{3} + \log_3 n_t} .$$

This does not exceed $3^{n/3}$ since $\log_3 n_t \leq n_t/3$ for any $n_t \in \mathbb{N}$.

Enumerating all normal configurations is easy: for each weakly connected component we just need to select a special vertex. Indeed, if a vertex $v \neq s, t$ is special then $\min\{f_{\text{in}}(v), f_{\text{out}}(v)\} = 1$, otherwise $\min\{f_{\text{in}}(v), f_{\text{out}}(v)\} = 0$. The exact values of $f_{\text{in}}(v)$ and $f_{\text{out}}(v)$ can be then derived from the equality (1). □

4 Further Directions

The natural open question is to solve SCS in less than 2^n steps. An apparently easier problem is to prove an upper bound $O^*(2^{\alpha(r)n})$ for r-SCS where $\alpha(r) < 1$ for all r.

Acknowledgments. Research is partially supported by Russian Foundation for Basic Research (12-01-31057-mol_a), RAS Program for Fundamental Research, Grant of the President of Russian Federation (NSh-3229.2012.1), the Ministry of Education and Science of the Russian Federation (8216) and Computer Science Club scholarship.

Also, we would like to thank the anonymous reviewers for many valuable comments that helped us to improve the readability of the paper.

References

1. Bax, E., Franklin, J.: A Finite-Difference Sieve to Count Paths and Cycles by Length. Inf. Process. Lett. 60, 171–176 (1996)
2. Bellman, R.: Dynamic Programming Treatment of the Travelling Salesman Problem. J. ACM 9, 61–63 (1962)
3. Björklund, A.: Determinant Sums for Undirected Hamiltonicity. In: Proceedings of the 2010 IEEE 51st Annual Symposium on Foundations of Computer Science, FOCS 2010, pp. 173–182. IEEE Computer Society, Washington, DC (2010)
4. Björklund, A., Husfeldt, T., Kaski, P., Koivisto, M.: The traveling salesman problem in bounded degree graphs. ACM Trans. Algorithms 8(2), 18:1–18:13 (2012)
5. Calabro, C., Impagliazzo, R., Paturi, R.: A Duality between Clause Width and Clause Density for SAT. In: Proceedings of the 21st Annual IEEE Conference on Computational Complexity, CCC 2006, pp. 252–260. IEEE Computer Society (2006)
6. Christofides, N., Campos, V., Corberan, A., Mota, E.: An algorithm for the Rural Postman problem on a directed graph. In: Netflow at Pisa. Mathematical Programming Studies, vol. 26, pp. 155–166. Springer, Heidelberg (1986)
7. Crochemore, M., Cygan, M., Iliopoulos, C., Kubica, M., Radoszewski, J., Rytter, W., Waleń, T.: Algorithms for three versions of the shortest common superstring problem. In: Amir, A., Parida, L. (eds.) CPM 2010. LNCS, vol. 6129, pp. 299–309. Springer, Heidelberg (2010)
8. Dantsin, E., Wolpert, A.: MAX-SAT for formulas with constant clause density can be solved faster than in $O(2^n)$ time. In: Biere, A., Gomes, C.P. (eds.) SAT 2006. LNCS, vol. 4121, pp. 266–276. Springer, Heidelberg (2006)
9. Eiselt, H.A., Gendreau, M., Laporte, G.: Arc Routing Problems, Part II: The Rural Postman Problem. Operations Research 43(3), 399–414 (1995)
10. Gallant, J., Maier, D., Storer, J.A.: On finding minimal length superstrings. Journal of Computer and System Sciences 20(1), 50–58 (1980)
11. Groves, G., van Vuuren, J.: Efficient heuristics for the Rural Postman Problem. ORiON 21(1), 33–51 (2005)
12. Held, M., Karp, R.M.: The Traveling-Salesman Problem and Minimum Spanning Trees. Mathematical Programming 1, 6–25 (1971)
13. Hertli, T.: 3-SAT Faster and Simpler - Unique-SAT Bounds for PPSZ Hold in General. In: Foundations of Computer Science, FOCS, pp. 277–284 (October 2011)
14. Iwama, K., Nakashima, T.: An Improved Exact Algorithm for Cubic Graph TSP. In: Lin, G. (ed.) COCOON 2007. LNCS, vol. 4598, pp. 108–117. Springer, Heidelberg (2007)
15. Kaplan, H., Lewenstein, M., Shafrir, N., Sviridenko, M.: Approximation Algorithms for Asymmetric TSP by Decomposing Directed Regular Multigraphs. J. ACM 52, 602–626 (2005)

16. Karp, R.M.: Dynamic Programming Meets the Principle of Inclusion and Exclusion. Operations Research Letters 1(2), 49–51 (1982)
17. Karpinski, M., Schmied, R.: Improved Lower Bounds for the Shortest Superstring and Related Problems. CoRR abs/1111.5442 (2011)
18. Kohn, S., Gottlieb, A., Kohn, M.: A Generating Function Approach to the Traveling Salesman Problem. In: ACN 1977: Proceedings of the 1977 Annual Conference, New York, NY, USA, pp. 294–300 (1977)
19. Kulikov, A., Kutzkov, K.: New upper bounds for the problem of maximal satisfiability. Discrete Mathematics and Applications 19, 155–172 (2009)
20. Lenstra, J.K., Kan, A.H.G.R.: Complexity of vehicle routing and scheduling problems. Networks 11(2), 221–227 (1981)
21. Lokshtanov, D., Nederlof, J.: Saving space by algebraization. In: Proceedings of the 42nd ACM Symposium on Theory of Computing, STOC 2010, pp. 321–330. ACM (2010)
22. Moser, R.A., Scheder, D.: A full derandomization of Schöning's k-SAT algorithm. In: Proceedings of the 43rd Annual ACM Symposium on Theory of Computing, STOC 2011, pp. 245–252. ACM (2011)
23. Mucha, M.: Lyndon Words and Short Superstrings. In: Proceedings of the Twenty-Fourth Annual ACM-SIAM Symposium on Discrete Algorithms, SODA 2013. Society for Industrial and Applied Mathematics (to appear, 2013)
24. Paluch, K., Elbassioni, K., van Zuylen, A.: Simpler Approximation of the Maximum Asymmetric Traveling Salesman Problem. In: STACS 2012. LIPIcs, vol. 14, pp. 501–506 (2012)
25. Pevzner, P.A., Tang, H., Waterman, M.S.: An Eulerian path approach to DNA fragment assembly. Proc. Natl. Acad. Sci. 98(17), 9748–9753 (2001)
26. Sweedyk, Z.: $2\frac{1}{2}$-Approximation Algorithm for Shortest Superstring. SIAM J. Comput. 29(3), 954–986 (1999)
27. Vassilevska, V.: Explicit Inapproximability Bounds for the Shortest Superstring Problem. In: Jedrzejowicz, J., Szepietowski, A. (eds.) MFCS 2005. LNCS, vol. 3618, pp. 793–800. Springer, Heidelberg (2005)
28. Williams, R.: A new algorithm for optimal 2-constraint satisfaction and its implications. Theoretical Computer Science 348(2-3), 357–365 (2005)

On the Parameterized Complexity
of the Maximum Edge 2-Coloring Problem

Prachi Goyal, Vikram Kamat, and Neeldhara Misra

Indian Institute of Science, Bangalore
{prachi.goyal,vkamat,neeldhara}@csa.iisc.ernet.in

Abstract. We investigate the parameterized complexity of the following edge coloring problem motivated by the problem of channel assignment in wireless networks. For an integer $q \geq 2$ and a graph G, the goal is to find a coloring of the edges of G with the maximum number of colors such that every vertex of the graph sees at most q colors. This problem is NP-hard for $q \geq 2$, and has been well-studied from the point of view of approximation. Our main focus is the case when $q = 2$, which is already theoretically intricate and practically relevant. We show fixed-parameter tractable algorithms for both the standard and the dual parameter, and for the latter problem, the result is based on a linear vertex kernel.

1   Introduction

Graph coloring problems are a broad and fundamental class of problems, involving an assignment of colors to the elements of a graph subject to certain constraints. They are often useful in modeling practical questions (map coloring, scheduling, register allocation, and pattern matching, to name a few), and have therefore been of central algorithmic interest. On the other hand, they have also been the subject of intensive structural study.

We are interested in the following edge coloring problem. For an integer $q \geq 2$ and a simple, undirected graph $G = (V, E)$, an assignment of colors to the edges of G is called an edge q-coloring if for every vertex $v \in V$, the edges incident on v are colored with at most q colors. An edge q-coloring that uses the maximum number of colors is called a maximum edge q-coloring. We note that the flavor of this question is quite different from the classical edge coloring question, which is a minimization problem, and the constraints require a vertex to be incident to completely distinct colors. This problem definition is motivated by the problem of channel assignment in wireless networks (as pointed out in [1,10], see also [18]). The interference between the frequency channels is understood to be a bottleneck for bandwidth in wireless networks. The goal is to minimize interference to optimize bandwidth. Some wireless LAN standards allow multiple non-overlapping frequency channels to be used simultaneously. In this scenario, a computer on the network equipped with multiple interface cards can use multiple channels. The goal is to maximize the number of channels used simultaneously, if all the nodes in the network have q interface cards. It turns out that the network can be

K. Chatterjee and J. Sgall (Eds.): MFCS 2013, LNCS 8087, pp. 492–503, 2013.
© Springer-Verlag Berlin Heidelberg 2013

modelled by a simple, undirected graph, while the channel assignment problem corresponds to a coloring of the edges where the edges incident to any given vertex are not colored with more than q colors. The maximum edge q-coloring is also considered in combinatorics, as a particular case of the anti-Ramsey number, see [1] for the details of this formulation.

The problem is already interesting for the special case when $q = 2$. It is known to be NP-complete and APX-hard [1] and also admits a 2-approximation algorithm [10]. In their work on this problem, Feng et al [10] show the problem to be polynomial time for trees and complete graphs for $q = 2$, and Adamaszek and Popa [1] demonstrate a 5/3-approximation algorithm for graphs which have a perfect matching. Given these developments, it is natural to pursue the parameterized complexity of the problem. Our main focus will be on the case when $q = 2$, and we note that this special case continues to be relevant in practice.

The goal of parameterized complexity is to find ways of solving NP-hard problems more efficiently than brute force. Here the aim is to restrict the combinatorial explosion to a parameter that is hopefully much smaller than the input size. It is a two-dimensional generalization of "P vs. NP" where, in addition to the overall input size n, one studies how a secondary measurement (called the *parameter*), that captures additional relevant information, affects the computational complexity of the problem in question. Parameterized decision problems are defined by specifying the input, the parameter, and the question to be answered. The two-dimensional analogue of the class P is decidability within a time bound of $f(k)n^c$, where n is the total input size, k is the parameter, f is some computable function and c is a constant that does not depend on k or n. A parameterized problem that can be decided in such a time-bound is termed *fixed-parameter tractable* (FPT). For general background on the theory of fixed-parameter tractability, see [9], [11], and [17].

A parameterized problem is said to admit a *polynomial kernel* if every instance (I, k) can be reduced in polynomial time to an equivalent instance with both size and parameter value bounded by a polynomial in k. The study of kernelization is a major research frontier of parameterized complexity and many important recent advances in the area are on kernelization. These include general results showing that certain classes of parameterized problems have polynomial kernels [2,5,12] or randomized kernels [16]. The recent development of a framework for ruling out polynomial kernels under certain complexity-theoretic assumptions [4,7,13] has added a new dimension to the field and strengthened its connections to classical complexity. For overviews of kernelization we refer to surveys [3,15] and to the corresponding chapters in books on parameterized complexity [11,17].

Our Contributions. We develop FPT algorithms and kernels for the maximum edge 2-coloring problem. The standard parameter is the solution size, or the number of colors used. On the other hand, it is known that the maximum number of colors used in an edge 2-coloring in a graph on n vertices is at most the number of vertices in the graph. This leads to a natural "dual" parameterization below an upper bound. Specifically, we ask if we can color the graph with at least $(n-k)$

colors, and we treat k as the parameter. As an aside, we also characterize the class of graphs that can be colored with n colors as being two-factors (this is implicit in several notions in the literature, and we state the proof for completeness).

Let us consider the problem with the standard parameter. A straightforward and well-understood observation [1,10] is that the maximum edge 2-coloring number is at least the size of the maximum matching of the graph. Therefore, if a graph G has a matching of size at least k, then G is a YES-instance. This is a simple polynomial time preprocessing step, and therefore we may assume throughout that the size of the maximum matching in the input graph is bounded by k. Consequently, the vertex cover of the input is bounded by $2k$ and the treewidth is bounded by $2k$. We do not consider these natural structural parameterizations separately, since they are implicitly bounded in terms of the solution size.

We note that the expressibility of the maximum edge 2-colorability question in MSO_2 is easily verified. Therefore, we may easily classify the the problem as being FPT (parameterized by the solution size), by an application of Courcelle's theorem [6]. However, the running time of the algorithm obtained from this meta theorem is impractical, and therefore, we explore the possibility of better algorithms specific to the problem. We first show an exponential kernel obtained by the application of some simple reduction rules, which also implies that the problem is FPT. We then present a concrete FPT algorithm that runs in time $\mathcal{O}^*(k^k)^1$ for the problem. Also, for the dual parameterization, we obtain a linear vertex kernel, with $\mathcal{O}(k)$ vertices and $\mathcal{O}(k^2)$ edges. This implies a FPT algorithm with running time $\mathcal{O}^*(k^{k^2})$.

This paper is organized as follows. In Section 2 we provide some basic definitions and facts. In section 3, we consider the standard parameter and present an exponential kernel and a FPT algorithm. In section 4, we consider the dual parameter and show a linear vertex kernel. Due to space constraints, some proofs have been omitted. Such statements are marked with a $\star$, and we refer the reader to a full version of the paper [14] for the complete details.

2   Preliminaries

In this section we state some basic definitions related to parameterized complexity and graph theory, and give an overview of the notation used in this paper. To describe running times of algorithms we sometimes use the O^* notation. Given $f : \mathbb{N} \to \mathbb{N}$, we define $O^*(f(n))$ to be $O(f(n) \cdot p(n))$, where $p(\cdot)$ is some polynomial function. That is, the O^* notation suppresses polynomial factors in the running-time expression. The set of natural numbers (that is, nonnegative integers) is denoted by $\mathbb{N}$. For a natural number n let $[n] := \{1, \dots, n\}$. By $\log n$ we mean $\lceil \log n \rceil$ if an integer is expected.

---

[1] The $\mathcal{O}^*$ notation is used to suppress polynomial factors in the running time (c.f. Section 2).

Graphs. In the following, let $G = (V, E)$ and $G' = (V', E')$ be graphs, and $U \subseteq V$ some subset of vertices of G. The union of graphs G and G' is defined as $G \cup G' = (V \cup V', E \cup E')$, and their intersection is defined as $G \cap G' = (V \cap V', E \cap E')$. A set U is said to be a *vertex cover* of G if every edge in G is incident to at least one vertex in U. U is said to be an *independent set* in G if no two elements of U are adjacent to each other. The *independence number* of G is the number of vertices in a largest independent set in G. U is said to be a *clique* in G if every pair of elements of U is adjacent to each other. A set U is said to be a *dominating set* in G if every vertex in $V \setminus U$ is adjacent to some vertex in U. A *two-factor* is a graph where every vertex has degree exactly two. We refer the reader to [8] for details on standard graph theoretic notation and terminology we use in the paper.

Parameterized Complexity. A parameterized problem Π is a subset of $\Gamma^* \times \mathbb{N}$, where Γ is a finite alphabet. An instance of a parameterized problem is a tuple (x, k), where k is called the parameter. A central notion in parameterized complexity is *fixed-parameter tractability (FPT)* which means, for a given instance (x, k), decidability in time $f(k) \cdot p(|x|)$, where f is an arbitrary function of k and p is a polynomial in the input size. The notion of *kernelization* is formally defined as follows.

Definition 1. [Kernelization] *[17,11] A kernelization algorithm for a parameterized problem $\Pi \subseteq \Gamma^* \times \mathbb{N}$ is an algorithm that, given $(x, k) \in \Gamma^* \times \mathbb{N}$, outputs, in time polynomial in $|x| + k$, a pair $(x', k') \in \Gamma^* \times \mathbb{N}$ such that (a) $(x, k) \in \Pi$ if and only if $(x', k') \in \Pi$ and (b) $|x'|, k' \leq g(k)$, where g is some computable function. The output instance x' is called the kernel, and the function g is referred to as the size of the kernel. If $g(k) = k^{O(1)}$ (resp. $g(k) = O(k)$) then we say that Π admits a polynomial (resp. linear) kernel.*

The Maximum Edge Coloring Problem Let $G = (V, E)$ be a graph, and let c be an assignment of k colors to the edges of G, that is, let c be a surjective function from E to $[k]$. We say that c is an edge coloring of the graph using k colors. For a subset F of the edge set E, let $c(F)$ denote the set of colors assigned to the edge set F, that is,

$$c(F) = \bigcup_{e \in F} c(e).$$

We say that c is *q-valid* if every vertex in the graph is incident to edges colored with at most q distinct colors. Formally, if F_v denotes the set of edges incident on a vertex v, then an edge coloring c is q-valid if $|c(F_v)| \leq q$ for all $v \in V$. We denote by $\sigma_q(G)$ the largest integer k for which there exists a q-valid edge coloring function with k colors. When considering the special case $q = 2$, we drop the subscript, and simply use $\sigma(G)$ to refer to the maximum number of colors with which G admits a 2-valid edge coloring. The first algorithmic question that arises is the following:

Max Edge 2-Coloring   Parameter: k
    Input: A graph G and an integer k
    Question: Is $\sigma(G) \geq k$, that is, is there a 2-valid edge coloring of G
        with at least k colors?

We first note that the MAX EDGE 2-COLORING problem is equivalent to its exact version:

Proposition 1 ($\star$). *For a graph G, $\sigma(G) \geq k$ if and only if there is a 2-valid edge coloring of G with exactly k colors.*

Therefore, when parameterizing by the standard parameter, we will address the question of whether there is a 2-valid edge coloring that uses exactly k colors, and we refer to this as the EXACT EDGE 2-COLORING problem. We now introduce the dual parameterization. We will need some terminology first. Let G be a graph and let $c : E \to [k]$ be a 2-valid edge coloring of G with k colors. For $1 \leq i \leq k$, let F_i denote the set of edges e for which $c(e) = i$, that is, $F_i = c^{-1}(i)$. Notice that each F_i is non-empty. Fix an arbitrary edge $e_i \in F_i$, and let H be the subgraph induced by $\{e_1, \ldots, e_k\}$. We call H the *character subgraph* of G. Notice that $\Delta(H) \leq 2$. It is also easy to argue that $\sigma(G) \leq |V|$ by examining the character subgraph and using the fact that it has at most $|V|$ edges (see [10]). Therefore, we may ask the following question:

$(n-k)$-**Edge 2-Coloring**   Parameter: k
    Input: A graph G and an integer k
    Question: Is $\sigma(G) \geq (n-k)$, that is, is there a 2-valid edge coloring of
        G with at least $(n-k)$ colors?

An useful notion is that of a *palette assignment* associated with an edge coloring c. Recall that for a vertex v, we use F_v to denote the set of edges incident on v. If $c : E \to [k]$ is an edge coloring, then the palette assignment associated with c is the function $c^\dagger$ defined as: $c^\dagger(v) = c(F_v)$. Note that in general, $c^\dagger$ is a function from V to $2^{[k]}$, however, if c is a 2-valid coloring, then $c^\dagger : V \to \binom{[k]}{2} \cup [k] \cup \{\emptyset\}$. We conclude this introduction to the maximum edge coloring problem with a straightforward characterization of graphs for which $\sigma(G) = |V|$.

Proposition 2 ($\star$). *A graph $G = (V, E)$ is a two factor if, and only if, $\sigma(G) = |V|$.*

3   A FPT Algorithm for Max Edge 2-Coloring

We begin by describing an exponential kernel for the EXACT EDGE 2-COLORING problem. We will subsequently describe a detailed FPT algorithm. We first observe that if G has a matching of k edges, then it is already a YES-instance of the problem.

Proposition 3 (⋆). *Let (G, k) be an instance of* EXACT EDGE 2-COLORING, *and let m denote the number of edges in G. If $m < k$, then G is a* NO-*instance. If the size of the maximum matching in G is at least $(k-1)$ and $m \geq k$, then G is a* YES-*instance.*

Let $(G = (V, E), k)$ be an instance of EXACT EDGE 2-COLORING. The first step towards an exponential kernel is to identify a matching of maximum size, say M, and return a trivial YES-instance if $|M| \geq k - 1$. If this is not the case, let $S \subseteq V$ be the set of both endpoints of every edge in M. We use I to denote $V \setminus S$. Note that $|S| \leq 2k - 4$ and I is an independent set.

For $T \subseteq S$, let $I_T \subseteq I$ denote the set of vertices v in I for which $N(v) = T$. Note that $\{I_T \mid T \subseteq S\}$ forms a partition of I into at most $2^{|S|}$ classes. We are now ready to suggest our first reduction rule.

(R1) For $T \subseteq S$, and let $r := \max\{10, |T| + 1\}$. If $|I_T| > r$, delete all but r vertices from I_T. The reduced instance has the same parameter as the original.

It is easy to see that this reduction rule may be applied in $O(|I|)$ time. We now prove the correctness of this rule.

Proposition 4 (⋆). *Let (G, k) be an instance of* EXACT EDGE 2-COLORING, *let S be a vertex cover of G and let $T \subseteq S$. Let (H, k) be the instance obtained by applying* **(R1)** *to G with respect to T. The instances (G, k) and (H, k) are equivalent.*

Lemma 1. EXACT EDGE 2-COLORING *has a kernel on $O(4^k \cdot (2k-4))$ vertices.*

Proof. Notice that once reduced with respect to **(R1)**, for every $T \subseteq S$, there are at most $\max\{10, |T| + 1\}$ vertices in G. Thus, a conservative upper bound on the number of vertices in a reduced instance would be $(|S| + 2^{|S|}|S|)$, and the lemma follows from the fact that $|S| \leq 2k - 4$. □

We now turn to a FPT algorithm for EXACT EDGE 2-COLORING. See Algorithm 1 in the full version [14] for a pseudocode-based description of the overall algorithm. Recall that the goal is to compute a 2-valid edge coloring that uses k colors. We begin by using Proposition 3 to accept instances with a maximum matching on at least $k - 1$ edges, and reject instances that have fewer than k edges. Otherwise, let S be the vertex cover obtained by choosing both endpoints of a maximum matching.

The algorithm begins by guessing a palette assignment τ to the vertices in S. First, some simple sanity checks are implemented. Note that if c is a 2-valid edge coloring of G that uses k colors, and S is a vertex cover of G, then $\bigcup_{v \in S} c^\dagger(v) = [k]$ (if not, the missing color cannot be attributed to any edge). Therefore, we ensure that $\bigcup_{v \in S} \tau(v) = [k]$. Also, for an edge in S, the palettes assigned to the endpoints clearly cannot be disjoint. Therefore, for $u, v \in S$, if $(u, v) \in E$, we ensure that $\tau(u) \cap \tau(v) \neq \emptyset$.

Let G be a Yes-instance of Exact Edge 2-Coloring, and suppose c is a 2-valid edge coloring of G that uses k colors. Let $X_c \subseteq [k]$ be the set of colors used by c on S. More formally, $X_c := \bigcup_{e \in G[S]} c(e)$. The second step of the algorithm involves guessing this subset of colors, that is, we consider all possible subsets of $[k]$ as candidates for being the exact set of colors that are realized by some 2-valid coloring when restricted to $G[S]$.

Let $X \subseteq [k]$ be the colors that are to be realized in S. All the colors in X are initially labelled *unused*. Note that for $u, v \in S$, if $(u, v) \in E$, $p_{uv} := \tau(u) \cap \tau(v)$ either has one or two colors. If the intersection has one color, say i, and $i \notin X$, then we reject the guess X. On the other hand, if $i \in X$, we assign i to the edge between u and v and update the label for i as *used*. Notice that this is a "forced" assignment, since this is the only way to extend c to the edge uv while respecting τ. On the other hand, suppose p_{uv} has two colors. If neither of these colors is in X, then we may reject this guess. If it has two colors and only one of them is in X, then we assign the color in X to (u, v) and update its label as used. Otherwise, we branch on the two possibilities of $c(u, v)$, which come from p_{uv}. Note that the count of colors labelled unused in $|X|$ drops by exactly one in both branches, so this is a two-way branching, where the corresponding search tree has depth bounded by $|X|$. This completes the description of the functionality of **CheckTop** (see also Function **CheckTop** in the full version [14]).

Finally, we need to realize the colors in $[k] \setminus X$ on the edges that have one endpoint each in S and $G \setminus S$. To this end, we compute the lists of feasible assignments of colors for each vertex in $G \setminus S$, based on τ. In particular, a pair of colors $\{i, j\}$ belongs to the feasibility list $\ell(u)$ of a vertex $u \in G \setminus S$ if there is a way of coloring the edges incident on u with the colors i and j while respecting the palette τ. In other words, one of the colors i or j appears in $\tau(v)$ for every $v \in N(u)$. If such a list is empty, then we know that no feasible extension of τ exists. On the other hand, if the list contains a unique set, then we may color the edges incident on u according to the unique possibility.

Other than the special cases above, we know, for the same reasons as in the proof of Proposition 4, that these lists either have constant size, or have one color in common. When the lists have one color in common, then this color can be removed from $[k] \setminus X$, as such a color will be used by any coloring c that respects τ.

For lists $\ell(u)$ of constant size, as long as at least two elements in the list contain a color from $[k] \setminus X$, we branch on such elements. Note that the depth of branching is bounded by $[k] \setminus X$ and the width is bounded by a constant (at most 10, see Proposition 4). If exactly one element in $\ell(u)$ contains a color from $[k] \setminus X$, then we color u according to that element. If no elements in $\ell(u)$ contain colors from $[k] \setminus X$, then color u according to any element in the list of its feasible assignments.

Finally, we are left with a situation where some colors from $[k] \setminus X$ still need to be assigned, and the only vertices from $G \setminus S$ that are left are those whose lists contain a common color. Now this is a question of whether every color that remains in $[k] \setminus X$ can be matched to a vertex from $G \setminus S$ whose feasibility list contains that color. To this end, we construct the bipartite graph $H = ((A \cup B), E)$ as follows. The vertex set A has one vertex for every color in $[k] \setminus X$. The vertex set B has one vertex for every $u \in G \setminus S$ for which the feasibility list of u has a common color. For $i \in A$ and $u \in B$, add the edge (i, u) if $\ell(u)$ has a set which contains i. Now we compute a maximum matching M in H, and it is easy to see that the remaining colors can be realized if and only if M saturates A (see also the pseudocode for function **CheckAcross** in the full version [14]).

This brings us to the main result of this section.

Theorem 1. *There is an algorithm with running time* $\mathcal{O}^*((20k)^k)$ EXACT EDGE 2-COLORING.

Proof. The correctness is accounted for in the description of the algorithm. Guessing the palette assignment requires time $\mathcal{O}^*((k + \binom{k}{2})^k)$ and guessing $X \subseteq [k]$ incurs an expense of 2^k. We note that the only branching steps happen in lines 22—28 of **CheckTop** and lines 47—52 in **CheckAcross**. The former is a two-way branching with a cost of $2^{|X|}$ and the latter is a 10-way branching with a cost of $10^{|[k] \setminus X|}$. Overall, therefore, the running time of these branching steps is bounded by 10^k. Therefore, the overall running time is bounded by $\mathcal{O}^*(20^k)$, as desired. We refer the reader to the full version of the paper [14] for the detailed pseudocode.   □

4   Parameterizing below an Upper Bound: A Linear Kernel

We now address the question of whether a given graph $G = (V, E)$ admits a 2-valid edge coloring using at least $(n - k)$ colors, where $n := |V|$. In this section, we show a polynomial kernel with parameter k. We note that the NP-hardness of the question is implicit in the NP-hardness of the MAX EDGE 2-COLORING PROBLEM shown in [1].

The kernel is essentially obtained by studying the structure of a YES-instances of the problem. We argue that if G is a YES-instance, c is a 2-valid edge coloring of G using at least $(n - k)$ colors, and H is a character subgraph of G with respect to c, then $|V(H)|$ must be at least $(n - k)$, or in other words, $G \setminus H$ is at most k. We then proceed to show that the components which are not cycles in H are also bounded. An easy but crucial observation is that any vertex cannot be adjacent to too many vertices whose palettes are disjoint. On the other hand, we are able to bound the number of vertices in H whose palettes are not disjoint. This leads to a bound on the maximum degree of G in terms of k. Finally, we show a reduction rule that applies to "adjacent degree two vertices", and this finally rounds off the analysis of the kernel size. We now formally describe the sequence of claims leading up to the kernel.

We begin by analyzing the structure of YES-instances of the problem. Let $G = (V, E)$ be a graph that admits a 2-valid edge coloring using at least $(n - k)$ colors. Let c be such a coloring, and let H be a character subgraph with respect to c. Since $\Delta(H) \leq 2$, the components of H comprise of paths and cycles.

Let $C_1, \ldots, C_r$ denote the components of H that are cycles and let $P_1, \ldots, P_s$ denote the components that are paths. Let the sizes of these components be $c_1, \ldots, c_r, p_1, \ldots, p_s$, respectively. We first claim that $s \leq k$.

Proposition 5 ($\star$). *Let c be a 2-valid edge coloring of G using at least $(n - k)$ colors, and let H be a character subgraph with respect to c. If H consists of s paths of lengths $p_1, \ldots, p_s$ and r cycles of lengths $c_1, \ldots, c_r$, then $s \leq k$.*

Next, we show that there are at most k vertices in G that are not in H.

Proposition 6. *Let c be a 2-valid edge coloring of G using at least $(n - k)$ colors, and let H be a character subgraph with respect to c. Then, $|G \setminus H| \leq k$.*

Proof. Suppose, for the sake of contradiction, that $|G \setminus H| > k$. This in turn implies that $|H| < n - k$. Recall, however, that $\Delta(H) \leq 2$, and therefore $|E(H)| \leq \frac{2|H|}{2} = |H| < n - k$. However, since H is character subgraph of G with respect to a coloring that uses at least $(n - k)$ colors, we have that $|E(H)| \geq n - k$. Therefore, the above amounts to a contradiction. □

Let $\mathcal{P}$ denote the set of endpoints of the paths $P_1, \ldots, P_s$. Notice that $|\mathcal{P}| \leq 2k$. Let $\mathcal{T}$ denote the remaining vertices in H, that is, $\mathcal{T} := H \setminus \mathcal{P}$. We now claim that the maximum degree of G in $\mathcal{T}$ is bounded:

Proposition 7. *For a graph G that admits a 2-valid edge coloring using at least $(n - k)$ colors, its character subgraph is such that, any vertex u in G is adjacent to at most six vertices in $\mathcal{T}$.*

Proof. Let c be a 2-valid edge coloring of G using at least $(n - k)$ colors, and let H, $\mathcal{P}$ and $\mathcal{T}$ be defined as above.

Suppose, for the sake of contradiction, that there is a vertex $u \in G$ that has more than six neighbors in $\mathcal{T}$. Since $\Delta(H[\mathcal{T}]) \leq 2$, in any subset of seven vertices of $\mathcal{T}$, there is at least one triplet of vertices, say x, y, and z that are mutually non-adjacent in H. By definition of H and $\mathcal{T}$, we know that the palettes of x, y and z with respect to c have two colors each and are mutually disjoint:

$$c^\dagger(x) \cap c^\dagger(y) = \emptyset;\ c^\dagger(x) \cap c^\dagger(z) = \emptyset;\ \text{and } c^\dagger(y) \cap c^\dagger(z) = \emptyset.$$

It follows that $|c^\dagger(x)| = |c^\dagger(y)| = |c^\dagger(z)| = 2$. Since u is adjacent to x, y and z, we conclude that there is no way to extend c to a 2-valid coloring of the edges (u, x), (u, y) and (u, z). Therefore, we contradict our assumption that c is a 2-valid edge coloring of G using at least $(n - k)$ colors, and conclude that all vertices in G have at most six neighbours in $\mathcal{T}$.

The following corollary is implied by the fact that there are at most $3k$ vertices in the graph other than $\mathcal{T}$.

Corollary 1. *Let G be a graph that admits a 2-valid edge coloring using at least $(n - k)$ colors. Then $\Delta(G) \leq 3k + 6$.*

We now state the reduction rules that define the kernelization.

(R1) If $\Delta(G) > 3k + 6$, then return a trivial No-instance.

(R2) Let u and v be adjacent vertices with $d(u) = d(v) = 2$, and let v' be the other neighbor of v. Delete v and add the edge (u, v'). Let the graph obtained thus be denoted by H. Then the reduced instance is $(H, n^* - k)$, where $n^* = |V(H)| = (n - 1)$. Notice that the parameter does not change.

It is easy to see that both the reduction rules above can be executed in linear time. The correctness of **(R1)** follows from Corollary 1. We now show the correctness of the second reduction rule.

Proposition 8. [⋆] *Let G be a graph where vertices u and v are adjacent, and $d(u) = d(v) = 2$. Let v' be the other neighbor of v. Let H be the graph obtained from G after an application of reduction rule **(R2)**. The graph G has a 2-valid edge coloring that uses at least $(n - k)$ colors if and only if the graph H has a 2-valid edge coloring that uses at least $(n - k - 1)$ colors.*

Observe that Proposition 8 implies the correctness of **(R2)**. We now turn to an analysis of the size of the kernel.

Lemma 2. *If $(G, n - k)$ is a Yes-instance of $(n - k)$-Edge 2-Coloring that is reduced with respect to **(R2)**, then $|V(G)| = O(k)$.*

Proof. Since G is a Yes-instance, it admits a 2-valid edge coloring c using at least $(n - k)$ colors. Let H be a character subgraph with respect to c. Let $C_1, \ldots, C_r$ denote the components of H that are cycles and let $P_1, \ldots, P_s$ denote the components that are paths.

Let $\mathcal{P}$ denote the set of endpoints of the paths $P_1, \ldots, P_s$ and let $\mathcal{T}$ denote the remaining vertices in H, that is, $\mathcal{T} := H \setminus \mathcal{P}$. Let $|\mathcal{P}_1| = |G \setminus H| + |\mathcal{P}|$. By Proposition 7, we know that every vertex in G, has at most six neighbors in $\mathcal{T}$. Since $|\mathcal{P}_1| \leq 3k$ (this follows from Proposition 5), the number of vertices in $\mathcal{T}$ that have neighbors in $\mathcal{P}_1$ is at most $3k \cdot 6 = 18k$. Notice that all other vertices in $\mathcal{T}$ have degree two in G. Therefore, we conclude that the number of vertices of G that have degree three or more is at most $3k + 18k = 21k$.

We now have that $|\mathcal{P}| \leq 2k$ and $|G \setminus H| \leq k$, hence it remains to bound the vertices in $\mathcal{T}$. Notice that the vertices of $\mathcal{T}$ have degree two or more in G. Among them, the vertices that have degree three or more in G are bounded by $21k$. The vertices left are the vertices in $\mathcal{T}$ that have degree two in G. Since the graph is reduced with respect to **(R2)**, the neighbors of these vertices have either degree one or degree three or more. Note that the number of degree one vertices is at most $|\mathcal{P}| + |G \setminus H| \leq 3k$. Hence the number of degree two vertices in $\mathcal{T}$ is at most $21k \cdot 6 + 3k = 129k$. Thus the total number of vertices in $\mathcal{T}$ is also $\mathcal{O}(k)$. This concludes our argument.

5 Concluding Remarks and Future Work

The most natural unresolved question is to settle the kernelization complexity of the maximum edge 2-coloring problem when parameterized by the solution size. The exponential kernel described in this work implies a polynomial kernel when the input is restricted to graphs where the maximum degree is a constant, and also if the input is restricted to graphs without cycles of length four. These observations are interesting because the problem continues to be NP-complete for both of these graph classes. The NP-hardness for graphs of bounded degree can be obtained by easy modifications to the reduction proposed in [1], and the NP-hardness on graphs without cycles of length four is given in the full version for completeness. Given these results, the question of whether the problem admits a polynomial kernel on general graphs is an interesting open problem.

Improved FPT algorithms for both the standard and the dual parameter, specifically with running time $O(c^k)$ for some constant c, will be of interest as well. It is also natural to pursue the above-guarantee version of the question, with the size of the maximum matching used as the guarantee. In particular, if γ is the size of a maximum matching in a graph G, we would like to study the question of checking if G can be colored with at least $(\gamma+k)$ colors, parameterized by k.

For the more general question of MAXIMUM EDGE q-COLORING, note that since the problem is NP-complete for fixed values of q, the question is para-NP-complete when parameterized by q alone. Generalizing some of the results that hold for $q = 2$ is also an interesting direction for future work.

References

1. Adamaszek, A., Popa, A.: Approximation and hardness results for the maximum edge q-coloring problem. In: Cheong, O., Chwa, K.-Y., Park, K. (eds.) ISAAC 2010, Part II. LNCS, vol. 6507, pp. 132–143. Springer, Heidelberg (2010)
2. Alon, N., Gutin, G., Kim, E.J., Szeider, S., Yeo, A.: Solving MAX-r-SAT above a tight lower bound. In: ACM-SIAM Symposium on Discrete Algorithms, SODA, pp. 511–517 (2010)
3. Bodlaender, H.L.: Kernelization: New upper and lower bound techniques. In: Chen, J., Fomin, F.V. (eds.) IWPEC 2009. LNCS, vol. 5917, pp. 17–37. Springer, Heidelberg (2009)
4. Bodlaender, H.L., Downey, R.G., Fellows, M.R., Hermelin, D.: On problems without polynomial kernels. J. Comput. Syst. Sci. 75(8), 423–434 (2009)
5. Bodlaender, H.L., Fomin, F.V., Lokshtanov, D., Penninkx, E., Saurabh, S., Thilikos, D.M.: (meta) kernelization. In: IEEE Annual Symposium on Foundations of Computer Science, FOCS, pp. 629–638 (2009)
6. Courcelle, B.: The monadic second-order theory of graphs. I. recognizable sets of finite graphs. Information and Computation 85(1), 12–75 (1990)
7. Dell, H., van Melkebeek, D.: Satisfiability allows no nontrivial sparsification unless the polynomial-time hierarchy collapses. In: ACM Symposium on the Theory of Computing, STOC, pp. 251–260 (2010)
8. Diestel, R.: Graph Theory, 3rd edn. Springer, Heidelberg (2005)
9. Downey, R.G., Fellows, M.R.: Parameterized Complexity, 530 p. Springer (1999)
10. Feng, W., Zhang, L., Wang, H.: Approximation algorithm for maximum edge coloring. Theor. Comput. Sci. 410(11), 1022–1029 (2009)
11. Flum, J., Grohe, M.: Parameterized Complexity Theory. Texts in Theoretical Computer Science. An EATCS Series. Springer-Verlag New York, Inc., Secaucus (2006)

12. Fomin, F.V., Lokshtanov, D., Saurabh, S., Thilikos, D.M.: Bidimensionality and kernels. In: ACM-SIAM Symposium on Discrete Algorithms, SODA, pp. 503–510 (2010)
13. Fortnow, L., Santhanam, R.: Infeasibility of instance compression and succinct PCPs for NP. In: ACM Symposium on the Theory of Computing, STOC, pp. 133–142 (2008)
14. Goyal, P., Kamat, V., Misra, N.: On the parameterized complexity of the maximum edge coloring problem (2013), http://arxiv.org/abs/1306.2931
15. Guo, J., Niedermeier, R.: Invitation to data reduction and problem kernelization. SIGACT News 38(1), 31–45 (2007)
16. Kratsch, S., Wahlström, M.: Compression via matroids: a randomized polynomial kernel for odd cycle transversal. In: ACM-SIAM Symposium on Discrete Algorithms, SODA, pp. 94–103 (2012)
17. Niedermeier, R.: Invitation to Fixed Parameter Algorithms. Oxford Lecture Series in Mathematics and Its Applications. Oxford University Press, USA (2006)
18. Raniwala, A., Gopalan, K., Cker Chiueh, T.: Centralized channel assignment and routing algorithms for multi-channel wireless mesh networks. Mobile Computing and Communications Review 8(2), 50–65 (2004)

A Note on Deterministic Poly-Time Algorithms for Partition Functions Associated with Boolean Matrices with Prescribed Row and Column Sums

Leonid Gurvits

The City College of New York, New York, NY 10031
gurvits@cs.ccny.cuny.edu

Abstract. We prove a new efficiently computable lower bound on the coefficients of stable homogeneous polynomials and present its algorthmic and combinatorial applications. Our main application is the first poly-time deterministic algorithm which approximates the partition functions associated with boolean matrices with prescribed row and column sums within simply exponential multiplicative factor. This new algorithm is a particular instance of new polynomial time deterministic algorithms related to the multiple partial differentiation of polynomials given by evaluation oracles.

1 Basic Definitions and Motivations

For given two integer vectors $\mathbf{r} = (r_1, ..., r_n)$ and $\mathbf{c} = (c_1, ..., c_m)$, we denote as $BM_{\mathbf{r},\mathbf{c}}$ the set of boolean $n \times m$ matrices with prescribed rows sums $\mathbf{r}$ and column sums $\mathbf{c}$.

Next, we introduce an analogue of the permanent (a partition function associated with $BM_{\mathbf{r},\mathbf{c}}$):

$$PE_{\mathbf{r},\mathbf{c}}(A) =: \sum_{B \in BM(\mathbf{r},\mathbf{c})} \prod_{1 \leq i \leq n; 1 \leq j \leq m} A(i,j)^{B(i,j)}, \tag{1}$$

where A is $n \times m$ complex matrix. Note that if A is a $n \times n$ matrix; $\mathbf{r} = \mathbf{c} = e_n$, where e_n is n-dimensional vector of all ones, then the definition (1) reduces to the permanent: $PE_{e_n,e_n}(A) = per(A)$.

The main focus of this note is on bounds and **deterministic** algorithms for $PE_{\mathbf{r},\mathbf{c}}(A)$ in the non-negative case $A \geq 0$. To avoid messy formulas, we will mainly focus below on the uniform square case, i.e. $n = m$ and $r_i = c_j = r, 1 \leq i, j \leq n$ and use simplified notations: $BM_{r e_n, r e_n} =: BM(r, n); PE_{r e_n, r e_n}(A) =: PE(r, A)$.

Boolean matrices with prescribed row and column sums is one of the most classical and intensely studied topics in analytic combimatorics, with applications to many areas from applied statistics to the representation theory. We, as many other researchers, are interested in the counting aspect, i.e. in computing/bounding/approximating the partition function $PE_{\mathbf{r},\mathbf{c}}(A)$. It was known

K. Chatterjee and J. Sgall (Eds.): MFCS 2013, LNCS 8087, pp. 504–515, 2013.

already to W.T.Tutte [17] that this partition function can be in poly-time reduced to the permanent. Therefore, if A is nonnegative the famous **FPRAS** [19] can be applied and this was already mentioned in [19] as one of the main applications. We are after deterministic poly-time algorithms. A. Barvinok initiated this, deterministic, line of algorithmic research in [14]. He also used the reduction to the permanent and the Van Der Waerden-Falikman-Egorychev (**VFE**) [10], [9] celebrated lower bound on the permanent of doubly-stochastic matrices. The techniques in [14] result in a deterministic poly-time algorithm approximating $PE(r, A)$ within multiplicative factor $(\Omega(\sqrt{n}))^n$ for any fixed r, even for $r = 1$. Such pure approximation is due the fact that the reduction to the permanent produces highly structured $n^2 \times n^2$ matrices. **VFE** bound is clearly a powerful algorithmic tool, as was recently effectively illustrated in [18]. Yet, neither **VFE** nor even more refined Schrijver's lower bound [2] are sharp enough for those structured matrices. This phenomenon was observed by A. Schrijver 30 years ago in [1]. The author introduced in [11] and [4] a new approach to lower bounds. We will give a brief description of the approach and refine it. The new lower bounds are asymptoticaly sharp and allow, for instance, to get a deterministic poly-time algorithm to approximate $PE(r, A)$ within multiplicative factor $f(r)^n$, where $f(r) = \frac{r!(n-r)!n^n}{r^r(n-r)^{n-r}n!} \approx \sqrt{2\pi \min(r, n-r)}$. Besides, we show that algorithm from [14] actually approximates within multiplicative factor $f(r)^{2n}$. So, for fixed r or $n-r$ the new bounds give simply exponential factor. But, say for $r = \frac{n}{2}$, the current factor is not simply exponential. *Is there a deterministic Non-Approximability result for $PE(\frac{n}{2}, A)$?*

We also study the sparse case, i.e. when, say, the columns of matrix A have relatively small number of non-zero entries. In this direction we generalize, reprove, sharpen the results of A. Schrijver [1] on how many k-regular subgraphs $2k$-regular bipartite graph can have.

The main moral of this paper is that when one needs to deal with the permanent of highly structured matrices the only (and often painless) way to get sharp lower bounds is to use **stable polynomials approach**.Prior to [11] and [4] **VFE** was, essentially, the only general pourpose non-trivial lower bound on the permanent. It is not true anymore.

1.1 Generating Polynomials

The goal of this subsection is to represent $PE_{\mathbf{r},\mathbf{c}}(A)$ as a coefficient of some effectively computable polynomial.

1. The following natural representation in the case of unit weights, i.e $A(i, j) \equiv 1$, was already in [16], the general case of it was used in [14].

$$PE_{\mathbf{r},\mathbf{c}}(A) = [\prod_{1 \leq i \leq n} y_i^{r_i} \prod_{1 \leq j \leq m} x_j^{c_j}] \prod_{1 \leq i \leq n, 1 \leq j \leq m} (1 + A(i, j)x_j y_i), \quad (2)$$

i.e. $PE_{\mathbf{r},\mathbf{c}}(A)$ is the coefficient of the monomial $\prod_{1 \leq i \leq n} y_i^{r_i} \prod_{1 \leq j \leq m} x_j^{c_j}$ in the non-homogeneous polynomial $\prod_{1 \leq i \leq n, 1 \leq j \leq m} (1 + A(i, j)x_j y_i)$.

It is easy to convert non-homogeneous formula (2) into a homogeneous one:

$$PE_{\mathbf{r},\mathbf{c}}(A) = [\prod_{1\leq j\leq m} x_j^{c_j} \prod_{1\leq i\leq n} z_i^{m-r_i}] \prod_{1\leq i\leq n, 1\leq j\leq m} (z_i + A(i,j)x_j). \qquad (3)$$

As the polynomial $\prod_{1\leq i\leq n, 1\leq j\leq m}(z_i + A(i,j)x_j)$ is a product of linear forms, the formula (3) allows to express $PE_{\mathbf{r},\mathbf{c}}(A)$ as the permanent of some $nm \times nm$ matrix, the fact essentially proved in a very different way in [17]. The permanent also showed up, in a similar context of Eulerian Orientations, in [1].

Indeed, associate with any $k \times l$ matrix B the product polynomial

$$Prod_B(x_1, ..., x_l) =: \prod_{1\leq i\leq k} \sum_{1\leq j\leq l} B(i,j)x_j. \qquad (4)$$

Then

$$[\prod_{1\leq j\leq l} x_j^{\omega_j}]Prod_B(x_1, ..., x_l) = per(B_{\omega_1,...,\omega_l}) \prod_{1\leq j\leq l} (\omega_j!)^{-1}, \qquad (5)$$

where $k \times k$ matrix $B_{\omega_1,...,\omega_l}$ consists of ω_j copies of the jth column of B, $1 \leq j \leq l$.

2. We will use below the following equally natural representation. Recall the definition of standard symmetric functions:

$$S_k(x_1, ..., x_m) = \sum_{1\leq i_1 < .. < i_k \leq m} \prod_{1\leq j\leq k} x_{i_j},$$

and define the following homogeneous polynomial

$$ES_{\mathbf{r},\mathbf{c};A}(x_1, ..., x_m) = \prod_{1\leq i\leq n} S_{r_i}(A(i,1)x_1, ..., A(i,m)x_m). \qquad (6)$$

Then

$$PE_{\mathbf{r},\mathbf{c}}(A) = [\prod_{1\leq j\leq n} x_j^{c_j}]ES_{\mathbf{r},\mathbf{c};A}(x_1, ..., x_m). \qquad (7)$$

Remark 1. Note that in the square case $n = m$, the polynomial $ES_{e_n,e_m;A} = Prod_A$. The polynomial $ES_{\mathbf{r},\mathbf{c};A}$ is, of course, related to the polynomial $TM(z_1, ..., z_n; x_1, ..., x_m) =: \prod_{1\leq i\leq n, 1\leq j\leq m}(z_i + A(i,j)x_j)$:

$$ES_{\mathbf{r},\mathbf{c};A}(x_1, ..., x_m) = const \prod_{1\leq i\leq n} \frac{\partial^{m-r_i}}{\partial z_i^{m-r_i}}TM(z_i = 0, 1 \leq i \leq n; x_1, ..., x_m).$$

$$(8)$$

∎

1.2   Exact Algorithms

It is well known that the coefficient $[\prod_{1\leq j\leq n} x_j^{c_j}]ES_{\mathbf{r},\mathbf{c};A}(x_1,...,x_m)$ can be computed by evaluating the polynomial $ES_{\mathbf{r},\mathbf{c};A}$ at $\prod_{1\leq j\leq n}(1+c_j)$ points. Which gives(see Remark (1)) an exact algorithm for $PE_{\mathbf{r},\mathbf{c}}(A)$ of complexity

$$O\left(\min(\prod_{1\leq j\leq n}(1+c_j)nm\log(m),\ \prod_{1\leq i\leq m}(1+r_i)nm\log(n))\right).$$

Thus if $n > m$ and m is fixed then the exists a polynomial in n exact deterministic algorithm to compute $PE_{\mathbf{r},\mathbf{c}}(A)$.

1.3   Previous Work

Estimation of the cardinality $|BM_{\mathbf{r},\mathbf{c}}| = PE_{\mathbf{r},\mathbf{c}}(A)$, where $A = J_{n,m} = e_n e_m^T$ is a matrix of all ones, is one of classical topics in analytic combinatorics. The reader may consult Barvinok's paper [14] for references to most major results on the topic.

To avoid messy formulas, we will mainly focus below on the uniform square case, i.e. $n = m$ and $r_i = c_j = r, 1 \leq i, j \leq n$ and use simplified notations:

$$BM_{\mathbf{re_n},\mathbf{re_n}} =: BM(r,n); PE_{\mathbf{re_n},\mathbf{re_n}} =: PE(r,A).$$

It is easy to see that $PE(r,A)$ is #P-Complete for all $1 \leq r < n$. The connection to the permanent implies that for non-negative matrices A there is **FPRAS** for $PE(r,A)$. We are interested in this paper in deterministic algorithms. Probably, the first published deterministic algorithm to approximate $PE(r,A)$ within a multiplicative factor appeared in Barvinok's paper [14]:
Define

$$\alpha(A) = \inf_{z_j,x_i>0} \frac{\prod_{1\leq i\leq n,1\leq j\leq m}(z_j + A(i,j)x_i)}{\prod_{1\leq i\leq n} x_i^{r_i} \prod_{1\leq j\leq m} z_j^{n-c_j}}.$$

Then

$$\alpha(A) \geq PE(r,A) \geq \frac{vdw(n^2)}{(vdw(n-r)vdw(r))^n}\alpha(A), \tag{9}$$

where $vdw(k) =: \frac{k!}{k^k}$. As the number $\log(\alpha(A))$ can computed(approximated within small additive error) via the convex minimization, the bounds (9) give a poly-time deterministic algorithm to approximate $PE(r,A)$ within multiplicative factor $\gamma_n =: (\frac{vdw(n^2)}{(vdw(n-r)vdw(r))^n})^{-1}$. The factor γ_n is not simply exponential even for $r = 1$, indeed $(\gamma_n)^{\frac{1}{n}} \approx const(\sqrt{n})$ for a fixed r. The proof of (9) in [14] is based on the **Sinkhorm's Scaling** and the Van Der Waerden-Falikman-Egorychev lower bound on the permanent of doubly-stochatic matrices.

2   Our Results

We prove and apply in this paper an optimized version of our lower bounds on the coefficients of **H-Stable** polynomials [8]. The lower bounds in [8] were obtained by a naive application of the lower on the mixed derivative of **H-Stable** polynomials.

When applied to the polynomial $\prod_{1 \leq i \leq n, 1 \leq j \leq m}(z_j + A(i,j)x_i)$, it implies the following bounds:

$$\alpha(A) \geq PE(r, A) \geq \left(\frac{vdw(n)}{vdw(n-r)vdw(r)}\right)^{2n} \alpha(A) \tag{10}$$

I.e. for the fixed r the Barvinok's approach gives a deterministic algorithm to approximate $PE(r, A)$ within simply exponential factor $(\frac{e^r}{vdw(r)})^{2n}$. We stress again that this result seems to be unprovable by using only Van Der Waerden-Falikman-Egorychev and alike purely permanental bounds, even the newest ones in [7].

When applied to the the the polynomial $ES_{\mathbf{r,c};A}(x_1, ..., x_m)$, our new bounds implies the following inequality

$$\mu(A) \geq PE(r, A) \geq \left(\frac{vdw(n)}{vdw(n-r)vdw(r)}\right)^n \prod_{2 \leq j \leq n}\left(\frac{vdw(n)}{vdw(n-c_j)vdw(c_j)}\right)\mu(A), \tag{11}$$

where

$$\mu(A) =: \inf_{x_j > 0} \frac{ES_{\mathbf{r,c};A}(x_1, ..., x_m)}{\prod_{1 \leq j \leq m} x_j^{c_j}}$$

Note that

$$\log(\mu) = \inf_{\sum_{1 \leq j \leq m} y_j = 0} \log(ES_{\mathbf{r,c};A}(exp(\frac{y_1}{c_1}), ..., exp(\frac{y_m}{c_m}))),$$

and the function $\log(ES_{\mathbf{r,c};A}(exp(\frac{y_1}{c_1}), ..., exp(\frac{y_m}{c_m})))$ is convex in ys. For the fixed r this gives a deterministic poly-time algorithm to approximate $PE(r, A)$ within simply exponential factor $(\frac{e^r}{vdw(r)})^n$. The detailed complexity analysis of the convex minimization will be described in the full version of the paper.

In the sparse case we get a much better lower bound(not fully optimized yet):

$$\mu(A) \geq PE(r, A) \geq \prod_{2 \leq j \leq n}\left(\frac{vdw(Col_j)}{vdw(Col_j - r)vdw(r)}\right)\mu(A), \tag{12}$$

where Col_j is the number of nonzer entries in the jth column of A.

Our final result is the following combinatorial lower bound: Let $A \in BM_{\mathbf{kr,kc}} \neq \emptyset$. Then

$$\inf_{x_j > 0} \frac{ES_{\mathbf{r,c};A}(x_1, ..., x_m)}{\prod_{1 \leq j \leq n} x_j^{c_j}} = \prod_{1 \leq i \leq n}\binom{kr_i}{r_i} \tag{13}$$

and

$$PE_{\mathbf{r},\mathbf{c}}(A) \geq \prod_{1 \leq i \leq n} \binom{kr_i}{r_i} \prod_{2 \leq j \leq m} \frac{vdw(kc_j)}{vdw(kc_j - c_j)vdw(c_j)} \tag{14}$$

The formula (14) can be sligthly, i.e. by $const(k,t) > 1$, improved in the regular case. In particular, $const(2,t) = \left(\binom{2t}{t}\right)^{-1} 2^{2t}$.

Let $A \in BM_{kte_n, kte_n}$, where k, t are positive integers. Then

$$PE_{te_n, te_n}(A) \geq \binom{kt}{t}^n \left(\frac{vdw(kt)}{vdw((k-1)t)vdw(t)}\right)^{n-k} \frac{vdw(kt)}{vdw(t)^k}. \tag{15}$$

The inequalities (14, 15) generalize and improve results from [1].

All the inequalities in this section are fairly direct corollaries of Theorem(3) (see the main inequality (25)).

3 Stable Homogeneous Polynomials

3.1 Definitions and Previous Results

The next definition introduces key notations and notions.

Definition 1. *1. The linear space of homogeneous polynomials with real (complex) coefficients of degree n and in m variables is denoted $Hom_R(m,n)$ ($Hom_C(m,n)$).*
We denote as $Hom_+(m,n)$ ($Hom_{++}(n,m)$) the closed convex cone of polynomials $p \in Hom_R(m,n)$ with nonnegative (positive) coefficients.
*2. For a polynomial $p \in Hom_+(n,n)$ we define its **Capacity** as*

$$Cap(p) = \inf_{x_i > 0, \prod_{1 \leq i \leq n} x_i = 1} p(x_1, \ldots, x_n) = \inf_{x_i > 0} \frac{p(x_1, \ldots, x_n)}{\prod_{1 \leq i \leq n} x_i}. \tag{16}$$

3. Consider a polynomial $p \in Hom_C(m,n)$,

$$p(x_1, \ldots, x_m) = \sum_{(r_1, \ldots, r_m)} a_{r_1, \ldots, r_m} \prod_{1 \leq i \leq m} x_i^{r_i}.$$

We define $Rank_p(S)$ as the maximal joint degree attained on the subset $S \subset \{1, \ldots, m\}$:

$$Rank_p(S) = \max_{a_{r_1, \ldots, r_m} \neq 0} \sum_{j \in S} r_j. \tag{17}$$

If $S = \{i\}$ is a singleton, we define $deg_p(i) = Rank_p(S)$.
*4. A polynomial $p \in Hom_C(m,n)$ is called **H-Stable** if $p(Z) \neq 0$ provided $Re(Z) > 0$; is called **H-SStable** if $p(Z) \neq 0$ provided $Re(Z) \geq 0$ and $\sum_{1 \leq i \leq m} Re(z_i) > 0$.*
*We coined the term "**H-Stable**" to stress two things: Homogeniety and Hurwitz' stability.*

5. *We define*

$$vdw(i) = \frac{i!}{i^i}; G(i) = \frac{vdw(i)}{vdw(i-1)} = \left(\frac{i-1}{i}\right)^{i-1}, i > 1; G(1) = 1. \qquad (18)$$

Note that $vdw(i)$ *and* $G(i)$ *are strictly decreasing sequences.* ∎

The main inequality in [4] was stated as the following theorem

Theorem 1. *Let* $p \in Hom_+(n, n)$ *be* **H-Stable** *polynomial. Then the following inequality holds*

$$\frac{\partial^n}{\partial x_1 \ldots \partial x_n} p(0, \ldots, 0) \geq \prod_{2 \leq i \leq n} G(\min(i, deg_p(i))) Cap(p). \qquad (19)$$

Associate with a polynomial $p \in Hom_+(n, n)$ the following sequence of polynomials $q_i \in Hom_+(i, i)$:

$$q_n = p, q_i(x_1, \ldots, x_i) = \frac{\partial^{n-i}}{\partial x_{i+1} \ldots \partial x_n} p(x_1, \ldots, x_i, 0, \ldots, 0); 1 \leq i \leq n - 1.$$

The inequality (19) is, actually, a corollary of the following inequality, which holds for **H-Stable** polynomials:

$$Cap(q_i) \geq Cap(q_{i-1}) \geq G(deg_{q_i}(i)) Cap(q_i), n \geq i \geq 2. \qquad (20)$$

As $Cap(q_1) = \frac{\partial^n}{\partial x_1 \ldots \partial x_n} p(0, \ldots, 0)$, one gets that

$$\frac{\partial^n}{\partial x_1 \ldots \partial x_n} p(0, \ldots, 0) \geq \prod_{2 \leq i \leq n} G(deg_{q_i}(i)) Cap(p). \qquad (21)$$

The inequality (19) follows from (21) because $G(i)$ is decreasing and $deg_{q_i}(i) \leq \min(i, deg_p(i))$.

3.2   New Observations

There were several reasons why the inequality (19) was stated as the main result:

1. It is simpler to understand than more general one (21). It was sufficient for the killer application: a short, transparent proof of the (improved) Schrijver's lower bound on the number of perfect matchings in k-regular bipartite graphs.
2. For the most of natural polynomials, the gegrees $deg_{q_i}(i)$ are straightforward to compute. Moreover, if a polynomial p with integer coefficients is given as an evaluation oracle then $Rank_p(S)$ can be computed in polynomial time via the univariate interpolation. On the other hand, if $i = n - [n^a], a > 0$ then even deciding whether $deg_{q_i}(i)$ is equal zero is **NP-HARD**. Indeed,

consider, for instance, the following family of polynomials, essentialy due to A. Barvinok:

$$p(x_1, ..., x_n) = Bar_A(x_n, ..., x_{n-[n^a]+1})(x_1 + ... + x_{n-[n^a]})^{n-[n^a]},$$

where $Bar_A(x_n, ..., x_{n-[n^a]}) = tr((Diag(x_n, ..., x_{n-[n^a]})A)^{[n^a]})$ and A is the adjacency matrix of an undirected graph. If the graph has a Hamiltonian cycle then $deg_{q_i}(i) = i$ and zero otherwise.

3.3   New Structural Results

The following simple bound was overlooked in [4]:
$deg_{q_i}(i) \leq \min(Rank_p(\{i, ..., n\}) - n + i, deg_p(i))$. So, if

$$Rank_p(\{j, ..., n\}) - n + j \leq k : k + 1 \leq j \leq n \tag{22}$$

then

$$\frac{\partial^n}{\partial x_1 ... \partial x_n} p(0, ..., 0) \geq Cap(p)G(k)^{n-k} vdwk. \tag{23}$$

Example 1. Let A be $n \times n$ doubly-stochastic matrix with the following pentagon shaped support: $A(i, j) = 0 : j - i \geq n - k$. Then the product polynomial $Prod_A(x_1, ..., x_n) = \prod_{1 \leq i \leq n} \sum_{[} 1 \leq j \leq nA(i, j)x_j$ satisfies the inequalities (22) and $cap(Prod_A) = 1$. Therefore $per(A) \geq G(k)^{n-k} vdwk$. This lower bound was proved by very different methods in [20], moreover it was shown there that it is sharp. Therefore, the more general bound (23) is sharp as well.   ∎

We remind the following result(combination of results in [6] and [11]).

Theorem 2. *Let* $p \in Hom_+(m, n)$, $p(x_1, ..., x_m) = \sum_{r_1 + ... + r_m = n} a_{r_1, ..., r_n} x_1^{r_1} ... x_m^{r_m}$ *be* **H-Stable**. *Then*

1.

$$a_{r_1, ..., r_m} > 0 \iff \sum_{j \in S} r_j \leq Rank_p(S) : S \subset \{1, ..., m\}. \tag{24}$$

2. *The set function* $Rank_p(S)$ *is submodular.*
3. *As* $a_{r_1, ..., r_m} > 0$ *iff* $\min_{S \subset \{1, ..., m\}} (Rank_p(S) - \sum_{j \in S} r_j) \geq 0$ *hence given the evaluation oracle for* p *there is poly-time detrministic algorithm to decide whether* $a_{r_1, ..., r_m} > 0$.

Lemma 1. *Let* $p \in Hom_+(n, n)$ *be* **H-Stable** *polynomial with integer coefficients given as an evaluation oracle. Then for any* $i \geq 1$ *there is a deterministic strongly polynomial algorithm to compute* $deg_{q_i}(i)$.

Proof: Associate with the number i and any polynomial $p \in Hom_+(n, n)$ the following polynomials
$P_l(y_1, ..., y_n) = p(z_1, ..., z_n)$ where$0 \leq l \leq n - i - 1$ and $z_j = y_1 + ... + y_l, 1 \leq j \leq i - 1; z_i = y_{l+1} + ... + y_{n-i}; z_{i+k} = y_{i+k}, 1 \leq k \leq n - i$.

Then $deg_{q_i}(i) \geq n - i - l$ iff $\frac{\partial^n}{\partial y_1 \ldots \partial y_n} P(0, \ldots, 0) > 0$. Now, if the original polynomial p is **H-Stable** then the polynomials P_l are **H-Stable**, have integer coefficients and evaluation oracles. Therefore we can apply the submodular minimization algorithm from Theorem (2) to decide whether the monomial $y_1 \ldots y_n$ is in the support of P_l. Running this algorithm at most $i \leq n$ times will give us $deg_{q_i}(i)$. ∎

Example 2. Consider the following **H-Stable** polynomial

$GR_{\mathbf{r},\mathbf{c}}(x_1, \ldots, x_m) = \prod_{1 \leq i \leq n} S_{r_i}(x_1, \ldots, x_m); \sum_j c_j = \sum_i r_i.$

Clearly, the monomial $\prod_{1 \leq j \leq n} x_j^{c_j}$ is in the support iff the set $BM_{\mathbf{r},\mathbf{c}}$ is not empty, i.e. there exists a boolean matrix with column sums $\mathbf{c}$ and row sums $\mathbf{r}$. It is easy to see that

$Rank_{GR_{\mathbf{r},\mathbf{c}}}(S) = \sum_{1 \leq i \leq n} \min(|S|, r_i)$. It follows from the characterization (24) that $BM_{\mathbf{r},\mathbf{c}}$ is not empty iff $\sum_{j \in S} c_j \leq \sum_{1 \leq i \leq n} \min(|S|, r_i)$ for all subsets $S \subset \{1, \ldots, m\}$. Equivalently, for the ordered column sums $c_{j_1} \geq c_{j_2} \geq \ldots \geq c_{j_m}$ the following inequalities hold:

$\sum_{1 \leq k \leq t} c_{j_k} \leq \sum_{1 \leq i \leq n} \min(t, r_i); 1 \leq t \leq m.$

These are the famous Gale-Ryser inequalities, albeit stated without Ferrers matrices. ∎

4   Main New Lower Bound

Let $p \in Hom_+(d, m)$ be a homogeneous polynomial in m variables, of degree d and with non-negative coefficients. We fix a monomial $\prod_{1 \leq j \leq m} x_j^{c_j}, \sum_{1 \leq j \leq m} c_j = d$ and assume WLOG that $c_j > 0, 1 \leq j \leq m$. Let $0 \leq a_{c_1, \ldots, c_m} = [\prod_{1 \leq j \leq m} x_j^{c_j}] p$ be a coefficient of the monomial. Define $Cap_{c_1, \ldots, c_m}(p) =: \inf_{x_j > 0} \frac{p(x_1, \ldots, x_m)}{\prod_{1 \leq j \leq m} x_j^{c_j}}$.

Clearly, $a_{c_1, \ldots, c_m} \leq Cap_{c_1, \ldots, c_m}(p)$.

Theorem 3. *Let $p \in Hom_+(d, m)$ be* **H-Stable.** *Define the following family of polynomials:*

$Q_m = p, Q_i \in Hom(d - (c_n + \ldots + c_{i+1}), i), m - 1 \geq i \geq 1$:

$$Q_i = \frac{\partial^{c_n + \ldots + c_{i+1}}}{\partial x_{i+1}^{c_{i+1}} \ldots \partial x_n^{c_n}} p(x_1, \ldots, x_i, 0, \ldots, 0); 1 \leq i \leq m - 1.$$

Denote $dg(i) =: deg_{Q_i}(i)$. Then the following inequality holds

$$a_{c_1, \ldots, c_m} \geq Cap_{c_1, \ldots, c_m}(p) \prod_{2 \leq j \leq m} \frac{vdw(dg(j))}{vdw(c_j)vdw(dg(j) - c_j)} \tag{25}$$

Corollary 1. *Let $p \in Hom_+(d, m)$ be* **H-Stable.** *Then the following (non-optimized but easy to use) lower bound holds:*

$$a_{c_1, \ldots, c_m} \geq Cap_{c_1, \ldots, c_m}(p) \prod_{1 \leq j \leq m} \frac{vdw(deg_p(j))}{vdw(c_j)vdw(deg_p(j) - c_j)} \tag{26}$$

Our proof is, similarly to [4], by induction, which is based on the following bivariate lemma.

Lemma 2. $p \in Hom_+(d, 2)$ be **H-Stable**, *i.e.* $p(x_1, x_2) = \sum_{0 \leq i \leq d} a_i x_1^{d-i} x_2^i$ and $1 \leq c_2 < d$. *Then*

$$a_{c_2} \geq Cap_{d-c_2,c_2}(p) \frac{vdw(d)}{vdw(c_2)vdw(d-c_2)}.$$

Proof: Define the following polynomial $P \in Hom_+(d, d)$:
$$P(y_1, ..., y_{d-c_2}; z_1, ..., z_{c_2}) = p(\frac{1}{d-c_2} \sum_{1 \leq k \leq d-c_2} y_k, \frac{1}{c_2} \sum_{1 \leq i \leq c_2} z_i.$$
It follows from the standard AG inequality that $Cap_{d-c_2,c_2}(p) = Cap(P)$ and it is easy to see that P is **H-Stable**. Consider the following polynomial $R(z_1, ..., z_{c_2}) =: \prod_{1 \leq k \leq d-c_2} \frac{\partial}{\partial y_k} P(y_k = 0, 1 \leq k \leq; z_1, ..., z_{c_2})$. First, it follows from (20) that $Cap(R) \geq G(d)...G(c_2 + 1)Cap(P)$. By the direct inspection, $R(z_1, ..., z_{c_2}) = a_{c_2} vdw(d-c_2)(\frac{1}{c_2} \sum_{1 \leq i \leq c_2} z_i)^{c_2}$. Therefore $Cap(R) = a_{c_2} vdw(d-c_2)$.

Putting things together gives that
$$a_{c_2} \geq \frac{G(d)...G(c_2+1)}{vdw(d-c_2)} Cap_{d-c_2,c_2}(p) = \frac{vdw(d)}{vdw(c_2)vdw(d-c_2)} Cap_{d-c_2,c_2}(p). \quad \blacksquare$$

Proof: [Sketch of a proof of Theorem (3)]. Let $p \in Hom_+(d, m)$ be **H-Stable**. Expand it in the last variable:
$$p(x_1, ..., x_m) = \sum_{0 \leq i \leq deg_p(m)} x_m^i T_i(x_1, ..., x_{m-1}).$$ Our goal is to prove that

$$Cap_{c_1,...,c_{m-1}}(T_{c_m}) \geq \frac{vdw(d)}{vdw(c_m)vdw(d - c_m)} Cap_{c_1,...,c_{m-1},c_m}(p). \quad (27)$$

Fix positive numbers $(y_1, ..., y_{m-1})$ and consider the following bivariate polynomial: $W(t, x_m) = p(ty_1, ..., ty_{m-1}, x_m)$. The polynomial W is of degree D and **H-Stable**. Note that $W(t, x_m) \geq Cap_{c_1,...,c_m}(p)t^{d-c_m} x_m^{c_m} \prod_{1 \leq i \leq m-1} y_i^{c_j}$. It follows from Lemma(2) that
$$T_i(y_1, ..., y_{m-1}) \geq \frac{vdw(d)}{vdw(c_m)vdw(d-c_m)} Cap_{c_1,...,c_m}(p) \prod_{1 \leq i \leq m-1} y_i^{c_j}, \quad \text{which}$$
proves the inequality (27). Now the polynomial $T_{c_m} \in Hom_+(d - c_m, m - 1)$ is also **H-Stable** [4]. Thus we can apply the same argument to the polynomial $T_{c_m}(x_1, ..., x_{m-1})$ and so on until only the first variable x_1 remains. $\quad \blacksquare$

Example 3. 1. The polynomial from [14] $TM(z_1, ..., z_n; x_1, ..., x_m) =: \prod_{1 \leq i \leq n, 1 \leq j \leq m}(z_i + A(i, j)x_j)$. Consider, just for the illustration, the square uniform case: $n = m$, $\mathbf{c} = (n - r, ..., n - r; r, ..., r)$. Note that the degrees of all variable are bounded by n. Using non-optimized lower bound (26) we get that the coefficient

$$a_{n-r,...,n-r;r,...,r} \geq Cap_{n-r,...,n-r;r,...,r}(TZ) \left(\frac{vdw(n)}{vdw(r)vdw(n - r)} \right)^{2n}$$

2. We give a lower bound on $|BM(r, n)|$. The polynomial is $Sym_{r,n}(\mathbf{x}) =: (S_r(x_1, ..., x_n))^n$. Degree of each variable is n. $Cap_{r,...,r}(Sym_{r,n}) = \binom{n}{r}^n$. The slightly optimized lower bound is

$$|BM(r,n)| \geq \binom{n}{r}^n \left(\frac{vdw(n)}{vdw(r)vdw(n-r)} \right)^{\frac{n(r-1)}{r}} vdw(n)(vdw(r)^{-\frac{n}{r}}). \quad (28)$$

One of the first asymptocally exact results was proved by Everrett in [15](his proof is rather involved):

$$|BM(r,n)| = \frac{(rn)!}{(r!)^{2n}} exp(-\frac{1}{2}(r-1)^2)\beta(r,n), \quad (29)$$

where $\lim_{n \to \infty} \beta(r,n) = 1$ for any fixed integer number r. We will compare our new lower bounds with (29).

Define $EVER(r,n) =: \frac{(rn)!}{(r!)^{2n}} exp(-\frac{1}{2}(r-1)^2)$ and

$HYP(r,n) = \binom{n}{r}^n \left(\frac{vdw(n)}{vdw(r)vdw(n-r)} \right)^{\frac{n(r-1)}{r}} vdw(n)(vdw(r)^{-\frac{n}{r}});$

i.e. $Ever(r,n)$ is the Everett-Stein asymptotically exact extimate and $Hyp(r,n)$ is our lower bound on $|BM(r,n)|$. Using the Stirling formula, one gets that $\lim_{n \to \infty} \frac{Ever(r,n)}{HYP(r,n)} = \sqrt{r}$ for a fixed r. Not bad at all, considering how computationally and conceptually simple is our derivation of (28)!

∎

Acknowledgements. The author acknowledges the support of NSF grant 116143.

References

1. Schrijver, A.: Bounds on permanents, and the number of 1-factors and 1-factorizations of bipartite graphs. In: Surveys in Combinatorics (Southampton, 1983). London Math. Soc. Lecture Note Ser., vol. 82, pp. 107–134. Cambridge Univ. Press, Cambridge (1983)
2. Schrijver, A.: Counting 1-factors in regular bipartite graphs. Journal of Combinatorial Theory, Series B 72, 122–135 (1998)
3. Laurent, M., Schrijver, A.: On Leonid Gurvits' proof for permanents. Amer. Math. Monthly 117(10), 903–911 (2010)
4. Gurvits, L.: Van der Waerden/Schrijver-Valiant like conjectures and stable (aka hyperbolic) homogeneous polynomials: one theorem for all. Electronic Journal of Combinatorics 15 (2008)
5. Gurvits, L.: A polynomial-time algorithm to approximate the mixed volume within a simply exponential factor. Discrete Comput. Geom. 41(4), 533–555 (2009)
6. Gurvits, L.: Combinatorial and algorithmic aspects of hyperbolic polynomials (2004), http://xxx.lanl.gov/abs/math.CO/0404474
7. Gurvits, L.: Unleashing the power of Schrijver's permanental inequality with the help of the Bethe Approximation (2011), http://arxiv.org/abs/1106.2844
8. Gurvits, L.: On multivariate Newton-like inequalities. In: Advances in Combinatorial Mathematics, pp. 61–78. Springer, Berlin (2009), http://arxiv.org/pdf/0812.3687v3.pdf
9. Egorychev, G.P.: The solution of van der Waerden's problem for permanents. Advances in Math. 42, 299–305 (1981)

10. Falikman, D.I.: Proof of the van der Waerden's conjecture on the permanent of a doubly stochastic matrix. Mat. Zametki 29(6), 931–938, 957 (1981) (in Russian)

11. Gurvits, L.: Hyperbolic polynomials approach to Van der Waerden/Schrijver-Valiant like conjectures: sharper bounds, simpler proofs and algorithmic applications. In: Proc. 38 ACM Symp. on Theory of Computing (StOC 2006), pp. 417–426. ACM, New York (2006)

12. Greenhill, C., McKay, B.D., Wang, X.: Asymptotic enumeration of sparse 0-1 matrices with irregular row and column sums. Journal of Combinatorial Theory. Series A 113, 291–324 (2006)

13. Greenhill, C., McKay, B.D.: Random dense bipartite graphs and directed graphs with specified degrees. Random Structures and Algorithms 35, 222–249 (2009)

14. Barvinok, A.: On the number of matrices and a random matrix with prescribed row and column sums and 0-1 entries. Adv. Math. 224(1), 316–339 (2010)

15. Everett, C.J., Stein, P.R.: The asymptotic number of integer stochastic matrices. Discrete Math. 1(1), 55–72 (1971/1972)

16. McKay, B.D.: Asymptotics for 0-1 matrices with prescribed line sums. In: Enumeration and Design, pp. 225–238. Academic Press, Canada (1984)

17. Tutte, W.T.: A short proof of the factor theorem for finite graphs. Canad. J. Math. 6, 347–352 (1954)

18. Vishnoi, N.K.: A Permanent Approach to the Traveling Salesman Problem. In: FOCS 2012, pp. 76–80 (2012)

19. Jerrum, M., Sinclair, A., Vigoda, E.: A polynomial-time approximation algorithm for the permanent of a matrix with nonnegative entries. Journal of the ACM 51, 671–697 (2004)

20. Hwang, S.G.: Matrix Polytope and Speech Security Systems. Korean J. CAM. 2(2), 3–12 (1995)

Polynomial Threshold Functions and Boolean Threshold Circuits

Kristoffer Arnsfelt Hansen[1] and Vladimir V. Podolskii[2]

[1] Aarhus University
arnsfelt@cs.au.dk
[2] Steklov Mathematical Institute
podolskii@mi.ras.ru

Abstract. We initiate a comprehensive study of the complexity of computing Boolean functions by polynomial threshold functions (PTFs) on general Boolean domains. A typical example of a general Boolean domain is $\{1,2\}^n$. We are mainly interested in the length (the number of monomials) of PTFs, with their degree and weight being of secondary interest.

First we motivate the study of PTFs over the $\{1,2\}^n$ domain by showing their close relation to depth two threshold circuits. In particular we show that PTFs of polynomial length and polynomial degree compute exactly the functions computed by polynomial size $\mathsf{THR} \circ \mathsf{MAJ}$ circuits. We note that known lower bounds for $\mathsf{THR} \circ \mathsf{MAJ}$ circuits extends to the likely strictly stronger model of PTFs. We also show that a "max-plus" version of PTFs are related to $\mathsf{AC}^0 \circ \mathsf{THR}$ circuits.

We exploit this connection to gain a better understanding of threshold circuits. In particular, we show that (super-logarithmic) lower bounds for 3-player randomized communication protocols with unbounded error would yield (super-polynomial) size lower bounds for $\mathsf{THR} \circ \mathsf{THR}$ circuits.

Finally, having thus motivated the model, we initiate structural studies of PTFs. These include relationships between weight and degree of PTFs, and a degree lower bound for PTFs of constant length.

1 Introduction

Let $f : X \to \{-1,1\}$ be a Boolean function on a domain $X \subseteq \mathbb{R}^n$. We say that a real n-variate polynomial P is a *polynomial threshold function* (PTF) computing f if for all $x \in X$ it holds that $f(x) = \mathrm{sgn}(P(x))$. Polynomial threshold function have been studied intensively for decades. Much of this work was motivated by questions in computer science [27], and PTFs are now an important object of study in areas such as Boolean circuit complexity [4,10,2,19], learning theory [16,17], and communication complexity [28]. The main motivation of this paper is Boolean circuit complexity. A major and long-standing open problem is to obtain an explicit super-polynomial lower bound for depth two threshold circuits. A long line of research have established lower bounds for several subclasses of depth two threshold circuits. The largest subclass for which super-polynomial lower bounds are known is the class $\mathsf{THR} \circ \mathsf{MAJ}$ of depth two threshold circuits, where all gates except the

K. Chatterjee and J. Sgall (Eds.): MFCS 2013, LNCS 8087, pp. 516–527, 2013.

output gate is required to compute threshold functions with polynomially bounded weights [8]. We shall see that PTFs on general Boolean domains are tightly connected to both these classes of circuits.

For a PTF P we will be interested in the several measures of complexity. The *length* of P, denoted by $\text{len}(P)$, is the number of monomials of P. The *degree* of P, denoted by $\deg(P)$, is the usual total degree of P. Finally, note that in the case that X is a finite domain, without loss of generality one may assume that the coefficients of P are integers, and can thus speak of the *weight* of P, meaning the largest magnitude of a coefficient of P.

We restrict our focus to the case of computing Boolean functions with Boolean inputs. More precisely we only consider the case when the domain X is a Boolean n-cube, $X = \{a, b\}^n$, for distinct $a, b \in \mathbb{R}$. Such representations of Boolean functions have been studied intensively due to their fundamental nature and vast number of applications. This research has almost exclusively focused on the two Boolean n-cubes, $\{0, 1\}^n$ and $\{-1, 1\}^n$, sometimes denoted as the "standard basis" and the "Fourier basis", respectively. Indeed, most often the notion of PTFs is defined specifically for the case of the domain $\{-1, 1\}^n$. This choice is, however, of little consequence when one disregards the length as a parameter and focuses on the degree, as is the case in many applications of PTFs. Note also that for these two domains any PTF can without loss of generality be assumed to be *multilinear*, meaning that all variables have individual degree at most 1.

Focusing on the length of a PTF rather than the degree, the choice of domain becomes crucial already for the case of the two domains $\{0, 1\}^n$ and $\{-1, 1\}^n$. This was studied in depth by Krause and Pudlák [19]. Minksy and Papert [20] has shown that the parity function requires exponential length over the domain $\{0, 1\}^n$ (cf. [9,1]), whereas it can be computed by a PTF of length 1 over the domain $\{-1, 1\}^n$. Conversely, Krause and Pudlák construct a PTF on domain $\{0, 1\}^n$ of length $\sqrt{n}$ that require length $2^{n^{\Omega(1)}}$ on domain $\{-1, 1\}^n$. For this construction, large weight is crucial. Indeed, Krause and Pudlák also show that any function computed by a polynomial length and polynomial weight PTF on the domain $\{0, 1\}^n$ can also be computed by a polynomial length and polynomial weight PTF on the domain $\{-1, 1\}^n$.

A notable exception to the focus on the domains $\{0, 1\}^n$ and $\{-1, 1\}^n$ is the work of Basu et al. [1] that consider representing the parity function (or rather, a natural generalization of the parity function) on domains of the form $X = A^n$, for a set $A \subseteq \mathbb{Z}$. They especially focus on the cases $A = \{0, 1, \ldots, m\}$ and $A = \{1, 2, \ldots, m\}$, where $m \geqslant 2$. It is important to note that on most Boolean domains $\{a, b\}^n$, it is not without loss of generality to assume that polynomials are multilinear. One may easily convert a given PTF into a multilinear PTF computing the same function, but such a conversion may change both the length as well as the weight significantly. Indeed, Basu et al. show that the parity function provides such an example. Namely they show that on the domain $\{1, 2\}^n$ there is a PTF of length $n + 1$ and degree n^2 computing the parity function, whereas any multilinear PTF computing the same function must have length 2^n. Thus evaluating PTFs on general Boolean domains has the effect that allowing high degree (meaning polynomial, exponential, or perhaps even higher), may help to greatly reduce the length needed to compute a given Boolean function.

In this paper our aim is to investigate in detail the computational power of PTFs of polynomial length over a general Boolean domain of the form $\{a, b\}^n$, $a \neq b$. Some of these domains essentially corresponds to the two usual domains $\{0, 1\}^n$ and $\{-1, 1\}^n$, namely those that are simple scalings $\{0, a\}^n$ and $\{a, -a\}^n$, and we shall hence not consider these further. In particular we shall by general Boolean domains, refer to any other Boolean domain $\{a, b\}$. For most of our results it turns out the precise choice of general domain does not matter (in fact all our results hold when $\text{sgn}(a) = \text{sgn}(b)$), and we shall henceforth develop our results in terms of the domain $\{1, 2\}^n$.

1.1   Our Results

Over the usual Boolean domains $\{0, 1\}^n$ and $\{-1, 1\}^n$ PTFs are basic extensions of linear threshold functions that are still very limited in expressive power. More specifically they correspond to the subclass of THR ∘ AND circuits with no negations of inputs in the case of domain $\{0, 1\}^n$ and THR∘XOR circuits in the case of domain $\{-1, 1\}^n$ (multiplication over $\{0, 1\}$ is AND, and over $\{-1, 1\}$ — XOR). In these cases PTFs require exponential length to compute simple functions such as symmetric Boolean functions [4,18]. Over a general Boolean domain the situation changes drastically. We show that in this case PTFs of just constant length can actually compute interesting classes of functions (see Proposition 1 and Proposition 2). More importantly, when moving to polynomial length PTFs obtain computational power right at the frontier of known circuit lower bounds for threshold circuits. Namely we show in Theorem 3 that PTFs of polynomial length and polynomial degree compute exactly the functions computed by polynomial size THR ∘ MAJ circuits. This tight connection is the main motivation to our studies of PTFs over general domains. The circuit class THR ∘ MAJ is the largest depth two threshold circuit class for which superpolynomial lower bounds are known (note, that the class MAJ ∘ THR is known to be strictly smaller [10]). These lower bounds were obtained by sign rank lower bounds of matrices, or equivalently lower bounds for unbounded error communication complexity [8], and this is still the only lower bound method known for this class of circuits. In Section 3.2 we show that this lower bound method applies to PTFs, even with no degree restriction. We tend to believe that allowing exponential or perhaps even larger degree allows for more Boolean functions to be computed by PTFs, and we relate this in Proposition 7 to a question about simulating large weights by small weights in threshold circuits in a very strong way. This in turn also gives an indication that the power of the sign rank lower bound method extends beyond THR ∘ MAJ circuits.

Our study of PTFs on general Boolean domains and its connection to the threshold circuits leads also to a possible way to approach the major open problem of proving lower bounds for THR ∘ THR circuits. Just as is the case of THR ∘ MAJ circuits, most lower bounds for classes of threshold circuits have been obtained using various models of communication complexity [11,15,10,23,8,29,26].

In Section 3.3 we generalize the notion of sign rank to higher order tensors, which captures a suitable generalization of unbounded error communication complexity to the multiparty number-on-the-forehead setting. We show that good lower bounds for order 3 tensors (or equivalently, 3-party communication protocols) would yield lower bounds for THR ∘ THR circuits. An important technical ingredient in this connection along with PTFs over general domain is a previous result showing that the threshold gates at the second level can be exchanged with *exact* threshold gates [13]. While we currently know *no* lower bounds for this communication model[1], we feel this relation is significant, given the previous successes of communication complexity for lower bounds for threshold circuit classes, and deserves further study. Multi-party communication complexity have been used earlier for threshold circuit classes, but in the bounded error setting. In particular, lower bounds have been obtained for depth 3 unweighted threshold circuits with small bottom fanin [15]. In the unbounded error setting we can additionally address depth 3 *weighted* threshold circuits with small bottom fanin.

In addition to PTFs on general domains we also consider a max-plus version of PTFs. The max-plus algebra works over the max-plus semiring, which is the set of integers with the max operation playing the role of addition and the usual addition playing the role of multiplication [30,6]. This setting arises as a "limit" case in several areas of mathematics and turns out to be helpful. In our case it turns out that max-plus PTFs are connected with PTFs over the general domains and are moreover connected to the hierarchy of $AC^0 \circ THR$ circuits.

The above relations between PTFs on general Boolean domains and threshold circuits further motivate an in-depth study of PTFs, besides them being a fundamental way to represent Boolean functions. For instance, it is tempting to conjecture that PTFs of polynomial length can only compute functions computable by polynomial size constant depth threshold circuits. It seems that before such questions can be addressed, one needs more insight into PTFs. We currently don't know how large PTF degree can be useful for computation. In Section 5 we show that the minimal degree of a PTF within a given length bound can be bounded in terms of its integer weights, and conversely the integer weights can be bounded in terms of its degree. These bounds are obtained by setting up suitable linear programs and integer linear programs, where the variables are exponents or weights respectively, and then using known bounds on feasible basic solutions and small integer feasible solutions.

Finally we study the relations between PTFs over different general domains. Though we are unable to completely resolve the questions arising here, we still can prove some nontrivial relations. In particular we can prove that if $|a| \neq |b|$ then PTFs of polynomial length over the domain $\{a,b\}^n$ are equivalent to those over domain $\{a^k, b^k\}^n$ for any k. In particular this shows the equivalence in expressive power of the domains $\{1,2\}^n$ and $\{1,-2\}^n$.

Due to space constraints, several proofs, remarks and even entire sections are omitted compared to the full version of the paper [14].

---

[1] Unlike the case of bounded error communication complexity, even obtaining non-explicit lower bounds pose a challenge, since counting arguments fail [24].

2   Preliminaries

Polynomial Threshold Functions. For given length bound $l(n)$ and degree bound $d(n)$, we let $\mathsf{PTF}_{a,b}(l(n), d(n))$ denote the class of Boolean functions on domain $\{a, b\}^n$ computed by polynomial threshold functions of length $l(n)$ and degree $d(n)$. That is $f \in \mathsf{PTF}_{a,b}(l(n), d(n))$ if and only if there is a polynomial $p(x) \in \mathbb{Z}[x]$ with $l(n)$ monomials of degree at most $d(n)$ and such that for all $x \in \{a, b\}^n$ we have $f(x) = 1$ if and only if $p(x) \geqslant 0$. Of particular interest is the case when $l(n)$ is a polynomial in n. For this reason we will abbreviate $\mathsf{PTF}_{a,b}(\mathrm{poly}(n), d(n))$ by $\mathsf{PTF}_{a,b}(d(n))$. If we do not wish to impose a degree bound we write this as $\mathsf{PTF}_{a,b}(l(n), \infty)$ and $\mathsf{PTF}_{a,b}(\infty)$, respectively. As mentioned in the introduction we state our results in terms of the specific domain $\{1, 2\}^n$. We remark that in most of our results one may replace $\{1, 2\}$ be any other domain $\{a, b\}$, where $|a| \neq |b|$, and $a, b \neq 0$. The exceptions to this are our results about PTFs of constant length[2], namely Propositions 1, 2, and 18 as well as Theorem 19. These results hold instead assuming $\mathrm{sgn}(a) = \mathrm{sgn}(b)$.

Exponential Form of PTFs. We shall find it convenient to switch back to the standard domain $\{0, 1\}^n$ even when considering PTFs over the domain $\{1, 2\}^n$. Given variables $y_1, \ldots, y_n \in \{1, 2\}$, define $x_1, \ldots, x_n \in \{0, 1\}$ by $x_i = \log_2(y_i)$. Correspondingly, $y_i = 2^{x_i}$. Under this change of variables monomials turn into exponential functions, $y_1^{a_1} \ldots y_n^{a_n} = 2^{a_1 x_1 + \cdots + a_n x_n}$ and more generally a polynomial $P(y) = \sum_{j=1}^{l} c_j \prod_{i=1}^{n} y_i^{a_{ij}}$, turns into a weighted sum of exponential functions: $P(y) = \sum_{j=1}^{l} c_j 2^{\sum_{i=1}^{n} a_{ij} x_i}$, where $a_{ij} \geqslant 0$ are the non-negative integer exponents of the polynomial. Rewriting a PTF in this way, we shall say it is in *exponential form*. We shall in general allow also for negative integer coefficients a_{ij} in the exponents. They can be easily made positive by simply multiplying the entire expression with the term $2^{\sum_{i=1}^{n} b_i x_i}$ for large enough b_i. This in turn requires us to redefine the degree of the polynomial in the natural way to suit this. Sometimes it may also be convenient to move the absolute value of the coefficients $|c_j|$ to the exponents as an additive term $\log_2(|c_j|)$ in order to make all coefficients of the exponential form be ± 1.

Boolean Functions and Circuit Classes. We use mostly standard definitions and notation of Boolean functions and circuits built from these, see e.g. [13].

3   PTFs and Threshold Circuits

3.1   Circuit Characterizations

We first note that PTFs already of length 2 on domain $\{1, 2\}^n$ can compute the class of linear threshold functions, and this is in fact an exact characterization. Also polynomial degree corresponds to polynomial weights.

Proposition 1. $\mathsf{PTF}_{1,2}(2, \infty) = \mathsf{THR}$ *and* $\mathsf{PTF}_{1,2}(2, \mathit{poly}(n)) = \mathsf{MAJ}$.

---

[2] Note that the XOR function can be computed by a length 1 PTF over the domain $\{1, -2\}$ but not over the domain $\{1, 2\}$.

With more work we can characterize the Boolean functions computed by constant length PTFs on domain $\{1,2\}^n$ as the class of constant size Boolean combinations of linear threshold functions. This class of functions was considered earlier in the setting of learning in [16].

Proposition 2. $\mathsf{PTF}_{1,2}(O(1),\infty) = \mathsf{ANY}_{O(1)} \circ \mathsf{THR}$, *and* $\mathsf{PTF}_{1,2}(O(1),poly(n)) = \mathsf{ANY}_{O(1)} \circ \mathsf{MAJ}$

In the case of polynomial degree we can characterize the Boolean functions computed by polynomial length PTFs on domain $\{1,2\}^n$ as a class of depth 2 threshold functions with polynomially bounded weights on the bottom level.

Theorem 3. $\mathsf{PTF}_{1,2}(poly(n)) = \mathsf{THR} \circ \mathsf{MAJ}$

Proof. We first construct a PTF given a $\mathsf{THR} \circ \mathsf{MAJ}$ circuit. Suppose that the output threshold gate is given by the inequality $(\sum_{k=1}^{s} w_k y_k) - t \geqslant 0$. Let $l_1(x), \ldots, l_s(x)$ be the linear expressions defining the s majority gates with integer coefficients. Let $p(n)$ be a polynomial such that $|l_k(x)| \leqslant p(n)$ for all $x \in \{0,1\}^n$ and all k. Let $m = 2p(n) + 1$ and define the $m \times m$ matrix $A = (a_{ij})$ by $a_{ij} = 2^{(i-p(n)-1)(j-1)}$ for $i, j = 1, \ldots, m$. Note that A is a Vandermonde matrix with distinct rows, and hence A is invertible. Define $u = (\underbrace{0, \ldots, 0}_{p(n)}, \underbrace{1, \ldots, 1}_{p(n)+1})^\mathsf{T}$, and let $v = A^{-1}u$. We now define PTFs by the exponential forms $E_1(x), \ldots, E_s(x)$ given by $E_k(x) = \sum_{j=1}^{m} v_j 2^{(j-1)l_k(x)}$. By construction we have $E_k(x) = u_{l_k(x)+p(n)+1}$. In other words, whenever $l_k(x) < 0$ we have $E_k(x) = 0$ and whenever $l_k(x) \geqslant 0$ we have $E_k(x) = 1$. We then obtain a PTF for the entire circuit by the exponential form $E(x) = (\sum_{k=1}^{s} w_k E_k(x)) - t$.

Conversely consider a PTF in its exponential form $E(x) = \sum_{k=1}^{s} c_k 2^{l_k(x)}$, where the coefficients of $l_k(x)$ are positive integers of polynomial magnitude. Thus there is a polynomial $p(n)$ such that $0 \leqslant l_k(x) \leqslant p(n)$ for all $x \in \{0,1\}^n$ and all k. We now construct a $\mathsf{THR} \circ \mathsf{EMAJ}$ circuit as follows. For every $k \in \{1, \ldots, s\}$ and for every $j \in \{0, \ldots, p(n)\}$, we take an EMAJ gate deciding whether $l(x) = j$, and then feed the output of this gate into the output THR gate with weight $c_k 2^j$. This $\mathsf{THR} \circ \mathsf{EMAJ}$ circuit is then easily converted into a $\mathsf{THR} \circ \mathsf{MAJ}$ circuit [13].

3.2   Lower Bounds for PTFs

The sign rank of a real matrix $A = (a_{ij})$ with nonzero entries is the minimum possible rank of a real matrix $B = (b_{ij})$ of same dimensions as A satisfying $\mathrm{sgn}(a_{ij}) = \mathrm{sgn}(b_{ij})$ for all i, j. We are interested in the sign rank of matrices defined from Boolean functions. Let $f : \{a,b\}^n \times \{a,b\}^n \to \{-1,1\}$ be a Boolean function of $2n$ bits partitioned in two blocks each of n bits. We associate with f a $2^n \times 2^n$ matrix M_f, the "communication matrix", indexed by $x, y \in \{a,b\}^n$ and defined by $(M_f)_{x,y} = f(x,y)$.

Lemma 4. *Assume* $f : \{1,2\}^n \times \{1,2\}^n \to \{-1,1\}$ *is computed by a PTF on* $\{1,2\}^n \times \{1,2\}^n$ *of length* s. *Then the matrix* M_f *has sign rank at most* s.

Proof. Consider the PTF for f in exponential form $E(x, y) = \sum_{j=1}^{s} c_j 2^{l_j(x,y)}$. For each j, define the $2^n \times 2^n$ matrix B_j, indexed by $x, y \in \{0, 1\}^n$, defined by $(B_j)_{x,y} = 2^{l_j(x,y)}$. From this definition we immediately have that the matrix $B = \sum_{j=1}^{s} c_j B_j$ is a sign representation of M_f. Now note that we can write $l_j(x, y) = l_j^{(1)}(x) + l_j^{(2)}(y)$. Hence B_j is an outer product, $B_j = b_j^{(1)} b_j^{(2)^\mathsf{T}}$, where $b_j^{(1)}, b_j^{(2)} \in \{0, 1\}^n$ are defined by $(b_j^{(1)})_x = 2^{l_j^{(1)}(x)}$ and $(b_j^{(2)})_y = 2^{l_j^{(2)}(y)}$. It follows that B_j is of rank at most 1, and hence $\mathrm{rank}(B) \leqslant \sum_{j=1}^{s} \mathrm{rank}(B_j) \leqslant s$.

Thus lower bounds on the sign rank of communication matrices of Boolean functions directly implies length lower bounds for PTFs on domain $\{1, 2\}^n$ not depending on the degree and weights. Strong lower bounds are now known for several of Boolean functions. We mention two of particular interest. Forster [7] proved that the sign rank of the $2^n \times 2^n$ matrix corresponding to the inner product mod 2 function, $\mathsf{IP}_2(x, y)$, has sign rank $2^{\frac{n}{2}}$. Razborov and Sherstov [26] proved that the sign rank of the $2^{m^3} \times 2^{m^3}$ matrix corresponding to the Boolean function $f_m(x, y) = \bigwedge_{i=1}^{m} \bigvee_{j=1}^{m^2} (x_{ij} \wedge y_{ij})$ is $2^{\Omega(m)}$. Combining these results with Lemma 4 we have the following.

Corollary 5. *Any PTF on domain $\{1, 2\}^n \times \{1, 2\}^n$ for IP_2 requires length $2^{\frac{n}{2}}$. Any PTF on domain $\{1, 2\}^{m^3} \times \{1, 2\}^{m^3}$ for f_m requires length $2^{\Omega(m)}$.*

Sign rank was previously used to give the first lower bounds for $\mathsf{THR} \circ \mathsf{MAJ}$ circuits and sign rank remains the only known method for obtaining such lower bounds. Since PTFs can compute all functions computed by $\mathsf{THR} \circ \mathsf{MAJ}$ circuits already with polynomial degree by Theorem 3, Corollary 5 indicates that the lower bound technique of sign rank is applicable to more general models of computation. Showing that these models are indeed stronger would require a different lower bound method for $\mathsf{THR} \circ \mathsf{MAJ}$ circuits. Instead we will relate the question whether PTFs with no degree restrictions are more expressive than PTFs of polynomial degree to a question about threshold circuits.

For this we will need the following lemma.

Lemma 6. *For any s we have $\mathsf{THR}_s \circ \mathsf{ETHR} \subseteq \mathsf{PTF}_{1,2}(s + 1, \infty) \circ \mathsf{AND}_2$. In particular, $\mathsf{THR} \circ \mathsf{THR} \subseteq \mathsf{PTF}_{1,2}(\infty) \circ \mathsf{AND}_2$.*

Proof. From [13] we have $\mathsf{THR} \circ \mathsf{THR} = \mathsf{THR} \circ \mathsf{ETHR}$, so the second statement follows from the first. Consider a $\mathsf{THR}_s \circ \mathsf{ETHR}$ circuit C with ETHR gates $g_1, \ldots, g_s$ defined by integer linear expressions $l_1(x), \ldots, l_s(x)$, and suppose the output THR gate is given by $\mathrm{sgn}(\sum_{j=1}^{s} w_j y_j - t)$, where $w_j \neq 0$ for all j. We define polynomials $p_1, \ldots, p_s$ by $p_j(x) = -C_j l_j(x)^2$ for large enough constants C_j. Then $2^{p_j(x)} = 1$ for $l_j(x) = 0$ and $2^{p_j(x)}$ close to zero for $l_j(x) \neq 0$, so $2^{p_j(x)}$ is a good approximation of ETHR gate corresponding to l_j. More specifically, let $0 < m \leqslant \min_{y \in \{0,1\}^s} |\sum_{j=1}^{s} w_j y_j - t|$ be such that $m/(2s|w_j|) < 1$ for all j, let $m_j = \min_{\{x \in \{0,1\}^n | l_j(x) \neq 0\}} l_j(x)^2$ and let $c_j = \lfloor \log_2((2s|w_j|)/m)/m_j \rfloor$. Let $x \in \{0, 1\}^n$ and define $y \in \{0, 1\}^s$ be $y_j = 1$ if and only if $l_j(x) = 0$. Then $|y_j - 2^{p_j(x)}| \leqslant m/2s|w_j|$ and thus $\left| \sum_{j=1}^{s} w_j y_j - \sum_{j=1}^{s} w_j 2^{p_j(x)} \right| \leqslant \frac{m}{2}$. It follows that the sign of $\sum_{j=1}^{s} w_j 2^{p_j(x)} - t$ corresponds to the output of the circuit. On the other hand, it is easy to see that after opening the brackets in the exponents this expression corresponds to a $\mathsf{PTF}_{1,2}(s+1, \infty) \circ \mathsf{AND}_2$ circuit.

Proposition 7. $\mathsf{PTF}_{1,2}(poly(n)) \subsetneq \mathsf{PTF}_{1,2}(\infty)$ *unless* $\mathsf{THR} \circ \mathsf{THR} \subseteq \mathsf{THR} \circ \mathsf{MAJ} \circ \mathsf{AND}_2$.

Proof. Assume $\mathsf{PTF}_{1,2}(poly(n)) = \mathsf{PTF}_{1,2}(\infty)$. Then $\mathsf{THR} \circ \mathsf{THR} \subseteq \mathsf{PTF}_{1,2}(\infty) \circ \mathsf{AND}_2 = \mathsf{PTF}_{1,2}(poly(n)) \circ \mathsf{AND}_2 = \mathsf{THR} \circ \mathsf{MAJ} \circ \mathsf{AND}_2$, where the first inclusion follows from Lemma 6 and the last equality follows from Theorem 3.

We tend to consider the inclusion $\mathsf{THR} \circ \mathsf{THR} \subseteq \mathsf{THR} \circ \mathsf{MAJ} \circ \mathsf{AND}_2$ as being unlikely to hold. Note that this would also mean $\mathsf{THR} \circ \mathsf{THR} \circ \mathsf{AND} = \mathsf{THR} \circ \mathsf{MAJ} \circ \mathsf{AND}$.

3.3   Sign Complexity of Tensors and Depth 2 Threshold Circuits

In this section we define the notion of sign complexity of an arbitrary order tensor, generalizing sign rank of matrices. The definition is made with the aim of capturing a notion of k-party unbounded error communication complexity. For simplicity we give the definition for the special case of order 3 tensors. The extension to tensors of any order k is direct.

Let $A = (a_{ijk})$ be an order 3 tensor. We say that A is a *cylinder tensor* if there is an order 2 tensor $A' = (a'_{ij})$ such either $a_{ijk} = a'_{jk}$, for all i, j, k, $a_{ijk} = a'_{ik}$ for all i, j, k, or $a_{ijk} = a'_{ij}$, for all i, j, k. In other words an order 3 tensor is a cylinder tensor if there are two indices such that every entry depends only on the value of these two indices. An order 3 tensor A is a *cylinder product* if it can be written as a Hadamard product $A_1 \odot A_2 \odot A_3$ where A_1, A_2, and A_3 are cylinder tensors. That is, $a_{ijk} = a_{jk}^{(1)} a_{ik}^{(2)} a_{ij}^{(3)}$, for all i, j, k, where $A_1 = (a_{jk}^{(1)}), A_2 = (a_{ik}^{(2)})$, $A_3 = (a_{ij}^{(3)})$. The *sign complexity* of an order 3 tensor $A = (a_{ijk})$ is the minimum r such that there exist cylinder product tensors $B_1, \ldots, B_r$, with $B_\ell = (b_{ijk}^{(\ell)})$, such that $\mathrm{sgn}(a_{ijk}) = \mathrm{sgn}\left(b_{ijk}^{(1)} + \cdots + b_{ijk}^{(r)}\right)$, for all i, j, k.

In the full version of this paper we generalize unbounded error communication complexity to the multi-party number-on-the-forehead (NOF) setting, and show that sign complexity of tensors essentially captures communication complexity in this setting.

For a Boolean function $f : \{0,1\}^n \times \{0,1\}^n \times \{0,1\}^n \to \{-1,1\}$ we associate with f a $2^n \times 2^n \times 2^n$ tensor T_f, the "communication tensor", indexed by $x, y, z \in \{0,1\}^n$ and defined by $(T_f)_{xyz} = f(x, y, z)$.

Proposition 8. *Assume that* $f : \{0,1\}^n \times \{0,1\}^n \times \{0,1\}^n \to \{-1,1\}$ *is computed by a* $\mathsf{THR}_s \circ \mathsf{ETHR}$ *circuit. Then the sign complexity of* T_f *is at most* $s + 1$.

Proof. From Lemma 6 we have that $\mathrm{sgn}\left(\sum_{j=1}^{s+1} w_j 2^{p_j(x,y,z)}\right) = f(x, y, z)$, for all $x, y, z \in \{0,1\}^n$, where $p_j(x, y, z)$ are degree 2 polynomials. Now notice that we can rewrite each p_j as $p_j(x, y, z) = p_j^{(1)}(y, z) + p_j^{(2)}(x, z) + p_j^{(3)}(x, y)$ and we can rewrite the above as $\text{sgn}\left(\sum_{j=1}^{s+1} w_j 2^{p_j^{(1)}(y,z)} 2^{p_j^{(2)}(x,z)} 2^{p_j^{(3)}(x,y)}\right) = f(x, y, z)$. Since each of the exponential expressions $w_j 2^{p_j^{(1)}(y,z)}$, $2^{p_j^{(2)}(x,z)}$, and $2^{p_j^{(3)}(x,y)}$ define $2^n \times 2^n \times 2^n$ cylinder tensors, this shows that the sign complexity of T_f is at most $s + 1$.

Using the result of [13] that $\mathsf{THR} \circ \mathsf{THR} = \mathsf{THR} \circ \mathsf{ETHR}$ this translates to a statement about $\mathsf{THR} \circ \mathsf{THR}$ circuits. Inspection of the proof of [13, Theorem 7] along with Proposition 8 above and the equivalence to unbounded error communication complexity gives us the following.

Corollary 9. *Assume that $f : \{0,1\}^n \times \{0,1\}^n \times \{0,1\}^n \to \{-1,1\}$ has unbounded error 3-player NOF communication complexity c. Then every $\mathsf{THR} \circ \mathsf{THR}$computing f must contain $2^c/poly(n)$ THR gates.*

Remark 10. The above result can be generalized to $\mathsf{THR} \circ \mathsf{THR} \circ \mathsf{AND}_k$ circuits by considering communication protocols with $2k + 1$ parties. If lower bounds for such circuits could be obtained for increasing k, they could using the switching lemma be generalized to $\mathsf{THR} \circ \mathsf{THR} \circ \mathsf{AND}$ circuits, or even $\mathsf{THR} \circ \mathsf{THR} \circ \mathsf{AC}^0$ circuits (cf. [25,12]).

4   Max-plus PTFs

Let $L_i(x)$ and $M_j(x)$ be integer linear forms, where $i = 1, \ldots, l_1$ and $j = 1, \ldots, l_2$. By max-plus PTFs we denote expressions of the form

$$\max_{i=1,\ldots,l_1} (L_i(x)) \geqslant \max_{j=1,\ldots,l_2} (M_j(x)). \tag{1}$$

The length of the PTF (1) is $l_1 + l_2$, the degree is the maximal sum of absolute values of all coefficients of $L_1, \ldots, L_{l_1}$ and $M_1, \ldots, M_{l_2}$ except the constant term. Note that max-plus PTFs are essentially just polynomial inequalities in the max-plus algebra.

Notations $\mathsf{mpPTF}(l(n), d(n))$, $\mathsf{mpPTF}(d(n))$, $\mathsf{mpPTF}(\infty)$, etc. are introduced analogously to usual PTFs. It turns out that the class $\mathsf{mpPTF}(\infty)$ is related to $\mathsf{AC}^0 \circ \mathsf{THR}$ circuits, but the other hand max-plus PTFs are not stronger than usual PTFs.

Lemma 11. $\mathsf{AND} \circ \mathsf{THR}, \mathsf{OR} \circ \mathsf{THR} \subseteq \mathsf{mpPTF}(\infty)$, *and* $\mathsf{mpPTF}(\infty) \subseteq \mathsf{AND} \circ \mathsf{OR} \circ \mathsf{THR}, \mathsf{OR} \circ \mathsf{AND} \circ \mathsf{THR}$

Lemma 12. *For all $b > 1$ there is a constant C such that $\mathsf{mpPTF}(l(n), d(n)) \subseteq \mathsf{PTF}_{1,b}(l(n), C \cdot d(n) \log l(n))$.*

From the lemma above and Corollary 5 we immediately obtain the following.

Corollary 13. *Any max-plus PTF computing IP_2 requires length $2^{\frac{n}{2}}$. Any max-plus PTF computing f_m requires length $2^{\Omega(m)}$.*

We note that for AND∘THR and OR∘THR circuits lower bounds are known (the standard proof of the lower bound on the size of DNF for the parity function works). For AND∘OR∘THR and OR∘AND∘THR circuits no superpolynomial lower bounds for explicit functions are known. Thus mpPTF(∞) is an intermediate class in the $\mathsf{AC}^0 \circ \mathsf{THR}$ hierarchy for which we do know a lower bound. On the other hand the class is still rather strong. We show that it contains some functions which are complicated for other complexity classes.

The ODD-MAX-BIT function (abbreviated here by OMB) was defined by Beigel [3] as $\mathrm{OMB}(x_1, \ldots, x_n) = 1$ if and only if $(\max\{i \mid x_i = 1\} \mod 2) = 1$.

Lemma 14. PARITY $\in$ mpPTF($poly(n)$), OMB $\circ$ THR $\subseteq$ mpPTF(∞).

Buhrman et al. [5] proved that the function OMB ∘AND$_2$ is not in the class MAJ ∘ MAJ, thus mpPTF(∞) $\not\subseteq$ MAJ ∘ MAJ. Besides max-plus PTFs which are just polynomial inequalities in the max-plus algebra, we can consider systems of max-plus polynomial inequalities. The results above shows that this class is equivalent reformulation of AND ∘ OR ∘ THR circuits.

Corollary 15. *The functions computed by systems of max-plus PTFs are exactly those computed by* AND ∘ OR ∘ THR *circuits.*

5   Weights and Degree

In this section we address the question of the minimal degree of a PTF computing a given Boolean function. Currently we are unable to give an upper bound on the degree required to compute a Boolean function given a bound on the length. That is, we don't know if $\mathsf{PTF}_{1,2}(poly(n), d(n)) = \mathsf{PTF}_{1,2}(\infty)$ for any function $d(n)$, and in particular we don't know if $\mathsf{PTF}_{1,2}(poly(n)) = \mathsf{PTF}_{1,2}(\infty)$, though as we have indicated we believe the latter to be false. We first show that the degree can be bounded in terms of the weight, and conversely the weight can be bounded in terms of the degree. We currently know of no method to bound the degree and weight simultaneously in terms of the length. The proof by Muroga, Toda and Takasu [22] (cf. [21]), showing that linear threshold functions needs integer weights of magnitude no more than $(n + 1)^{(n+1)/2}/2^n$ can readily be adapted to PTFs on domain $\{1, 2\}^n$ to give a bound on weight in terms of degree.

Proposition 16. *Suppose P is a PTF of degree d and length s. Then there is another PTF P' of degree d and length s having weight at most $s^{s/2}2^{ds}$ such that* $\mathrm{sgn}(P(x)) = \mathrm{sgn}(P'(x))$ *for all $x \in \{1,2\}^n$. Furthermore the set of monomials of P' is the same of P.*

A more complicated proof can give us a bound in another direction.

Proposition 17. *Suppose P is a PTF having integer coefficients, weight W and length s with n variables. Then there is another PTF P' of weight W and length s having degree at most $(sn+1)(sn)^{sn/2}\lceil \log_2 s + \log_2 W \rceil^{sn}$ such that $\mathrm{sgn}(P(x)) = \mathrm{sgn}(P'(x))$ for all $x \in \{1,2\}^n$. Furthermore the (multi)set of weights of P' is the same of P.*

Propositions 16 and 17 implies for example that for a PTF of length $\mathrm{poly}(n)$, if the degree is at most $2^{\mathrm{poly}(n)}$ then the weight can be assumed to be $2^{2^{\mathrm{poly}(n)}}$, and vise versa. Note that the proof of Proposition 2 together with the upper bound $n^{O(n)}$ on the weight of linear threshold functions [21] imply that any PTF P of constant length is equivalent to another PTF P' of constant length and degree $n^{O(n)}$. However the precise length of P' is exponential in the length of P. We are able to avoid the exponential increase only for length 3. The proof of this is rather complicated and we feel that this gives some indications about the difficulty of the general problem.

Proposition 18. *Any PTF of length 3 over the domain $\{1, 2\}^n$ is equivalent to a length 3 PTF with degree $n^{O(n)}$.*

Next we prove an exponential degree lower bound for PTFs of constant length. Namely, we prove a degree lower bound of the form $2^{\Omega(n^\epsilon)}$ for any PTF of constant length s computing OMB, where ϵ depends on s.

By Proposition 2 we have $\mathsf{PTF}_{1,2}(O(1), \infty) = \mathsf{ANY}_{O(1)} \circ \mathsf{THR}$. The proof along the same lines can give us that a PTF of individual degree at most d and length k can in fact be turned into a constant size DNF, i.e. we obtain an $\mathsf{OR}_{2^{2^{O(k^2)}}} \circ \mathsf{AND}_{2^{O(k^2)}} \circ \mathsf{THR}$ circuit where all threshold gates are computed with integer weights of magnitude at most d. We thus just give weight lower bounds for $\mathsf{OR}_{O(1)} \circ \mathsf{AND}_{O(1)} \circ \mathsf{THR}$ circuits.

Theorem 19. *Any circuit in the class $\mathsf{OR}_k \circ \mathsf{AND}_l \circ \mathsf{THR}$ computing OMB function on n variables require weights of size $2^{\Omega(n^{1/kl})}$.*

Finally we state our most notable result on relations between the domains.

Lemma 20. *For all $a, b \in \mathbb{R}$ such that $|a| \neq |b|$ and for any k we have $\mathsf{PTF}_{a,b}(\infty)$ $= \mathsf{PTF}_{a^k, b^k}(\infty)$.*

References

1. Basu, S., Bhatnagar, N., Gopalan, P., Lipton, R.J.: Polynomials that sign represent parity and Descartes' rule of signs. Computat. Complex. 17(3), 377–406 (2008)
2. Beigel, R.: The polynomial method in circuit complexity. In: CCC 1993, pp. 82–95. IEEE Computer Society Press (1993)
3. Beigel, R.: Perceptrons, PP, and the polynomial hierarchy. Comput. Complex. 4(4), 339–349 (1994)
4. Bruck, J., Smolensky, R.: Polynomial threshold functions, AC^0 functions, and spectral norms. SIAM J. Comput. 21(1), 33–42 (1992)
5. Buhrman, H., Vereshchagin, N.K., de Wolf, R.: On computation and communication with small bias. In: CCC 2007, pp. 24–32 (2007)
6. Butkovič, P.: Max-linear Systems: Theory and Algorithms. Springer (2010)
7. Forster, J.: A linear lower bound on the unbounded error probabilistic communication complexity. J. Comput. Syst. Sci. 65(4), 612–625 (2002)

8. Forster, J., Krause, M., Lokam, S.V., Mubarakzjanov, R., Schmitt, N., Simon, H.U.: Relations between communication complexity, linear arrangements, and computational complexity. In: Hariharan, R., Mukund, M., Vinay, V. (eds.) FSTTCS 2001. LNCS, vol. 2245, pp. 171–182. Springer, Heidelberg (2001)
9. Goldmann, M.: On the power of a threshold gate at the top. Inform. Process. Lett. 63(6), 287–293 (1997)
10. Goldmann, M., Håstad, J., Razborov, A.A.: Majority gates vs. general weighted threshold gates. Comput. Complex. 2(4), 277–300 (1992)
11. Hajnal, A., Maass, W., Pudlák, P., Szegedy, M., Turán, G.: Threshold circuits of bounded depth. J. Comput. Syst. Sci. 46(2), 129–154 (1993)
12. Hansen, K.A., Miltersen, P.B.: Some meet-in-the-middle circuit lower bounds. In: Fiala, J., Koubek, V., Kratochvíl, J. (eds.) MFCS 2004. LNCS, vol. 3153, pp. 334–345. Springer, Heidelberg (2004)
13. Hansen, K.A., Podolskii, V.V.: Exact threshold circuits. In: CCC 2010, pp. 270–279. IEEE Computer Society (2010)
14. Hansen, K.A., Podolskii, V.V.: Polynomial threshold functions and boolean threshold circuits. ECCC TR13-021 (2013)
15. Håstad, J., Goldmann, M.: On the power of small-depth threshold circuits. Comput. Complex. 1, 113–129 (1991)
16. Klivans, A.R., O'Donnell, R., Servedio, R.A.: Learning intersections and thresholds of halfspaces. J. Comput. Syst. Sci. 68(4), 808–840 (2004)
17. Klivans, A.R., Servedio, R.A.: Learning DNF in time $2^{\tilde{O}(n^{1/3})}$. J. Comput. Syst. Sci. 68(2), 303–318 (2004)
18. Krause, M., Pudlák, P.: On the computational power of depth-2 circuits with threshold and modulo gates. Theor. Comput. Sci. 174(1–2), 137–156 (1997)
19. Krause, M., Pudlák, P.: Computing boolean functions by polynomials and threshold circuits. Comput. Complex. 7(4), 346–370 (1998)
20. Minsky, M., Papert, S.: Perceptrons: An Introduction to Computational Geometry. MIT Press (1969)
21. Muroga, S.: Threshold Logic and its Applications. John Wiley & Sons, Inc. (1971)
22. Muroga, S., Toda, I., Takasu, S.: Theory of majority decision elements. Journal of the Franklin Institute 271, 376–418 (1961)
23. Nisan, N.: The communication complexity of threshold gates. In: Miklós, D., Szönyi, T., Sós, V.T. (ed.) Combinatorics, Paul Erdős is Eighty, Mathematical Studies, vol. 1, pp. 301–315. Bolyai Society (1993)
24. Paturi, R., Simon, J.: Probabilistic communication complexity. J. Comput. Syst. Sci. 33(1), 106–123 (1986)
25. Razborov, A., Wigderson, A.: $n^{\Omega(\log n)}$ lower bounds on the size of depth-3 threshold circuits with AND gates at the bottom. Inf. Proc. Lett. 45(6), 303–307 (1993)
26. Razborov, A.A., Sherstov, A.A.: The sign-rank of AC^0. SIAM J. Comput. 39(5), 1833–1855 (2010)
27. Saks, M.E.: Slicing the hypercube. In: Walker, K. (ed.) Surveys in Combinatorics. London Mathematical Society Lecture Note Series, vol. 187. Cambridge University Press (1993)
28. Sherstov, A.A.: Communication lower bounds using dual polynomials. Bulletin of the EATCS 95, 59–93 (2008)
29. Sherstov, A.A.: Separating AC^0 from depth-2 majority circuits. SIAM J. Comput. 38(6), 2113–2129 (2009)
30. Speyer, D., Sturmfels, B.: Tropical Mathematics. ArXiv math/0408099 (2004), http://arxiv.org/abs/math/0408099

Reachability in Higher-Order-Counters[*]

Alexander Heußner[1] and Alexander Kartzow[2]

[1] Otto-Friedrich-Universität Bamberg, Germany
[2] Universität Leipzig, Germany

Abstract. Higher-order counter automata (HOCA) can be either seen as a restriction of higher-order pushdown automata (HOPA) to a unary stack alphabet, or as an extension of counter automata to higher levels. We distinguish two principal kinds of HOCA: those that can test whether the topmost counter value is zero and those which cannot.

We show that control-state reachability for level k HOCA with 0-test is complete for $(k-2)$-fold exponential space; leaving out the 0-test leads to completeness for $(k-2)$-fold exponential time. Restricting HOCA (without 0-test) to level 2, we prove that global (forward or backward) reachability analysis is **P**-complete. This enhances the known result for pushdown systems which are subsumed by level 2 HOCA without 0-test.

We transfer our results to the formal language setting. Assuming that **P** $\subsetneq$ **PSPACE** $\subsetneq$ **EXPTIME**, we apply proof ideas of Engelfriet and conclude that the hierarchies of languages of HOPA and of HOCA form strictly interleaving hierarchies. Interestingly, Engelfriet's constructions also allow to conclude immediately that the hierarchy of collapsible pushdown languages is strict level-by-level due to the existing complexity results for reachability on collapsible pushdown graphs. This answers an open question independently asked by Parys and by Kobayashi.

1 From Higher-Order Pushdowns to Counters and Back

Higher-order pushdown automata (HOPA) — also known as iterated pushdown automata — were first introduced by Maslov in [15] and [16] as an extension of classical pushdown automata where the pushdown storage is replaced by a nested pushdown of pushdowns of ... of pushdowns. After being originally studied as acceptors of languages, these automata have nowadays obtained renewed interest as computational model due to their connection to safe higher-order recursion schemes. Recent results focus on algorithmic questions concerning the underlying configuration graphs, e.g., Carayol and Wöhrle [5] showed decidability of the monadic second-order theories of higher-order pushdown graphs due to the pushdown graph's connection to the Caucal-hierarchy [6], and Hague and Ong determined the precise complexity of the global backwards reachability problem for HOPA: for level k it is complete for **DTIME**$(\bigcup_{d \in \mathbb{N}} \exp_{k-1}(n^d))$ [9].[1]

---

[*] The second author is supported by the DFG research project GELO. We both thank M. Bojańczyk, Ch. Broadbent, and M. Lohrey for helpful discussions and comments.

[1] We define $\exp_0(n) := n$ and $\exp_{k+1}(n) := \exp(\exp_k(n))$ for any natural number k.

K. Chatterjee and J. Sgall (Eds.): MFCS 2013, LNCS 8087, pp. 528–539, 2013.
© Springer-Verlag Berlin Heidelberg 2013

In the setting of classical pushdown automata it is well known that restricting the stack alphabet to one single symbol, i.e., reducing the pushdown storage to a counter, often makes solving algorithmic problems easier. For instance, control state reachability for pushdown automata is **P**-complete whereas it is **NSPACE**($\log(n)$)-complete for counter automata. Then again, results from counter automata raise new insights to the pushdown case by providing algorithmic lower bounds and important subclasses of accepted languages separating different classes of complexity. In this paper we lift this idea to the higher-order setting by investigating reachability problems for higher-order counter automata (HOCA), i.e., HOPA over a one-element stack alphabet. Analogously to counter automata, we introduce level k HOCA in two variants: with or without 0-tests. Throughout this paper, we write k-HOCA$^-$ for the variant without 0-tests and k-HOCA$^+$ for the variant with 0-tests. Transferring our results' constructions back to HOPA will then allow to answer a recent open question [17,14].

To our knowledge, the only existing publication on HOCA is by Slaats [18]. She proved that $(k + 1)$-HOCA$^+$ can simulate level k pushdown automata (abbreviated k-HOPA). In fact, even $(k + 1)$-HOCA$^-$ simulate k-HOPA. Slaats conjectured that $L(\text{k-HOCA}^+) \subsetneq L(\text{k-HOPA})$ where $L(X)$ denotes the languages accepted by automata of type X. We can confirm this conjecture by combining the proof ideas of Engelfriet [7] with our main result on control-state reachability for HOCA in Theorems 13 and 14: control state reachability on k-HOCA$^+$ is complete for **DSPACE**($\bigcup_{d \in \mathbb{N}} \exp_{k-2}(n^d)$) and control state reachability on k-HOCA$^-$ is complete for **DTIME**($\bigcup_{d \in \mathbb{N}} \exp_{k-2}(n^d)$). These results are obtained by adapting a proof strategy relying on reductions to bounded space storage automata originally stated for HOPA by Engelfriet [7]. His main tool are auxiliary **SPACE**($b(n)$) P^k automata where P^k denotes the storage type of a k-fold nested pushdown (see Section 2 for a precise definition). Such a (two-way) automaton has an additional storage of type P^k, and a Turing machine worktape with space $b(n)$. His main technical result shows a trade off between the space bound b and the number of iterated pushdowns k. Roughly speaking, exponentially more space allows to reduce the number of nestings of pushdowns by one. Similarly, at the cost of another level of pushdown, one can trade alternation against nondeterminism. Here, we also restate reachability on k-HOCA$^+$ as a membership problem on alternating auxiliary SPACE($\exp_{k-3}(n)$) $\mathcal{Z}+$ automata (where $\mathcal{Z}+$ is the new storage type of a counter with 0-test). For our **DSPACE**($\bigcup_{d \in \mathbb{N}} \exp_{k-2}(n^d)$)-hardness proof we provide a reduction of **DSPACE**($\bigcup_{d \in \mathbb{N}} \exp(\exp_{k-3}(n^d))$) to alternating auxiliary **SPACE**($\exp_{k-3}(n)$) $\mathcal{Z}+$ automata that is inspired by Jancar and Sawa's **PSPACE**-completeness proof for the non-emptiness of alternating automata [11]. For containment we adapt the proof of Engelfriet [7] and show that membership for alternating auxiliary **SPACE**($\exp_{k-3}(n)$) $\mathcal{Z}+$ automata can be reduced to alternating reachability on counter automata of size $\exp_{k-2}(n)$, where n is the size of the original input, which is known to be in **DSPACE**($\bigcup_{d \in \mathbb{N}} \exp_{k-2}(n^d)$) (cf. [8]).

For the case of k-HOCA$^-$ the hardness follows directly from the hardness of reachability for level $(k-1)$ pushdown automata and the fact that the latter can be simulated by k-HOCA$^-$. For containment in **DTIME**$(\bigcup_{d \in \mathbb{N}} \exp_{k-2}(n^d))$ the mentioned machinery of Engelfriet reduces the problem to the case $k = 2$.

The proof that control-state reachability on 2-HOCA$^-$ is in **P** is implied by Theorem 5 which proves a stronger result: both the global regular forward and backward reachability problems for 2-HOCA$^-$ are **P**-complete. The backward reachability problem asks, given a regular set C of configurations, for a (regular) description of all configurations that allow to reach one in C. This set is typically denoted as pre$^*(C)$. Note that there is no canonical way of defining a regular set of configurations of 2-HOCA$^-$. We are aware of at least three possible notions: regularity via 2-store automata [2], via sequences of pushdown-operations [4], and via encoding in regular sets of trees. We stick to the latter, and use the encoding of configurations as binary trees introduced in [12]: We call a set C of configurations regular if the set of encodings of configurations $\{E(c) \mid c \in C\}$ is a regular set of trees (where E denotes the encoding function from [12]). Note that the other two notions of regularity are both strictly weaker (with respect to expressive power) than the notion of regularity we use here. Nevertheless, our result does not carry over to these other notions of regularity as they admit more succinct representations of certain sets of configurations. See [10] for details.

Besides computing pre$^*(C)$ in polynomial time our algorithm also allows to compute the reachable configurations post$^*(C)$ in polynomial time. Thus, 2-HOCA$^-$ subsumes the well-known class of pushdown systems [1] while still possessing the same good complexity with respect to reachability problems.

Due to the lack of space, detailed formal proofs are deferred to a long version of this article [10].

2   Formal Model of Higher-Order Counters

2.1   Storage Types and Automata

An elegant way for defining HOCA and HOPA is the use of storage types and operators on these (following [7]). For simplicity, we restrict ourselves to what Engelfriet calls *finitely encoded* storage types.

Definition 1. *For X some set, we call a function $t : X \to \{true, false\}$ an X-test and a partial function $f : X \to X$ an X-operation.*
A storage type is a tuple $\mathcal{S} = (X, T, F, x_0)$ where X is the set of $\mathcal{S}$-configurations, $x_0 \in X$ the initial $\mathcal{S}$-configuration, T a finite set of X-tests and F a finite set of X-operations containing the identity on X, i.e., $\mathrm{id}_X \in F$.

Let us fix some finite alphabet Σ with a distinguished symbol $\perp \in \Sigma$. Let $\mathcal{P}_\Sigma = (X, T, F, x_0)$ be the *pushdown storage type* where $X = \Sigma^+$, $x_0 = \perp$, $T = \{\mathrm{top}_\sigma \mid \sigma \in \Sigma\}$ with $\mathrm{top}_\sigma(w) = true$ if $w \in \Sigma^* \sigma$, and $F = \{\mathrm{push}_\sigma \mid \sigma \in \Sigma\} \cup \{\mathrm{pop}, \mathrm{id}\}$ with $\mathrm{id} = \mathrm{id}_X$, $\mathrm{push}_\sigma(w) = w\sigma$ for all $w \in X$, and $\mathrm{pop}(w\sigma) = w$ for all $w \in \Sigma^+$ and $\sigma \in \Sigma$ and $\mathrm{pop}(\sigma)$ undefined for all $\sigma \in \Sigma$. Hence, $\mathcal{P}_\Sigma$ represents a classical pushdown stack over the alphabet Σ. We write $\mathcal{P}$ for $\mathcal{P}_{\{\perp,0,1\}}$.

We define the storage type *counter without 0-test* $\mathcal{Z} = \mathcal{P}_{\{\bot\}}$, which is the pushdown storage over a unary pushdown alphabet. We define the storage type *counter with 0-test* $\mathcal{Z}+$ exactly like $\mathcal{Z}$ but we add the test *empty?* to the set of tests where $empty?(x) = true$ if $x = \bot$ (the plus in $\mathcal{Z}+$ stands for "with 0-test"). In other words, *empty?* returns false iff the operation pop is applicable.

Definition 2. *For a storage type* $\mathcal{S} = (X, T, F, x_0)$ *we define an* $\mathcal{S}$ *automaton as a tuple* $\mathcal{A} = (Q, q_0, q_f, \Delta)$ *where as usual* Q *is a finite set of states with initial state* q_0 *and final state* q_f *and* Δ *is the transition relation. The difference to a usual automaton is the definition of* Δ *by* $\Delta = Q \times \{true, false\}^T \times Q \times F$.

For $q \in Q$ and $x \in X$, a transition $\delta = (q, R, p, f)$ is applicable to the *configuration* (q, x) if $f(x)$ is defined and if for each test $t \in T$ we have $R(t) = t(x)$, i.e., the result of the storage-tests on the storage configuration x agree with the test results required by the transition δ. If δ is applicable, application of δ leads to the configuration $(p, f(x))$. The notions of a run, the accepted language, etc. are now all defined as expected.

The Pushdown Operator We also consider $\mathcal{P}_\Sigma$ as an operator on other storage types as follows. Given a storage type $\mathcal{S} = (X, T, F, x_0)$ let the storage type *pushdown of* $\mathcal{S}$ be $\mathcal{P}_\Sigma(\mathcal{S}) = (X', T', F', x_0')$ where $X' = (\Sigma \times X)^+$, $x_0' = (\bot, x_0)$, $T' = \{top_\sigma \mid \sigma \in \Sigma\} \cup \{test(t) \mid t \in T\}$, $F' = \{push_{\gamma, f} \mid \gamma \in \Sigma, f \in F\} \cup \{stay_f \mid f \in F\} \cup \{pop\}$, and where for all $x' = \beta(\sigma, x)$, $\beta \in (\Sigma \times X)^*$, $\sigma \in \Sigma$, $x \in X$ it holds that

- $top_\tau(x') = (\tau = \sigma)$,
- $test(t)(x') = t(x)$,
- $push_{\tau, f}(x') = \beta(\sigma, x)(\tau, f(x))$
  if f is defined on x (and undefined otherwise),
- $stay_f(x') = \beta(\sigma, f(x))$
  if f is defined on x (and undefined otherwise), and
- $pop(x') = \beta$ if β is nonempty (and undefined otherwise).

Note that $stay_{id_X} = id_{X'}$ whence F' contains the identity. As for storages, we define the operator $\mathcal{P}$ to be the operator $\mathcal{P}_{\{\bot,0,1\}}$.

2.2   HOPA, HOCA, and Their Reachability Problems

We can define the iterative application of the operator $\mathcal{P}$ on some storage $\mathcal{S}$ as follows: let $\mathcal{P}^0(\mathcal{S}) = \mathcal{S}$ and $\mathcal{P}^{k+1}(\mathcal{S}) = \mathcal{P}(\mathcal{P}^k(\mathcal{S}))$. A *level k higher-order pushdown automaton* is a $\mathcal{P}^{k-1}(\mathcal{P})$ automaton. We abbreviate the class of all these automata with k-HOPA. A *level k higher-order counter automaton with zero-test* is a $\mathcal{P}^{k-1}(\mathcal{Z}+)$ automaton and k-HOCA$^+$ denotes the corresponding class.[2]. Similarly, k-HOCA$^-$ denotes the class of *level k higher-order counter automata without zero-test* which is the class of $\mathcal{P}^{k-1}(\mathcal{Z})$ automata. Obviously,

---

[2] A priori our definition of k-HOCA$^+$ results in a stronger automaton model than that used by Slaats. In fact, both models are equivalent (cf. [10]).

for any level k it holds that $L(\text{k-HOCA}^-) \subseteq L(\text{k-HOCA}^+) \subseteq L(\text{k-HOPA})$ where $L(X)$ denotes the languages accepted by automata of type X.

We next define the reachability problems which we study in this paper.

Definition 3. *Given an S automaton and one of its control states $q \in Q$, then the control state reachability problem asks whether there is a configuration (q, x) that is reachable from (q_0, x_0) where $x \in X$ is an arbitrary S-configuration.*

Assuming a notion of regularity for sets of S configurations (and hence for sets of configurations of S automata), we can also define a global variant of the control state reachability problem.

Definition 4. *Given an S automaton $\mathcal{A}$ and a regular set of configurations C, the regular backwards reachability problem demands a description of the set of configurations from which there is a path to some configuration $c \in C$.*

Analogously, the regular forward reachability problem asks for a description of the set of configurations reachable from a given regular set C. In the following section, we consider the regular backwards (and forwards) reachability problem for the class of 2-HOCA$^-$ only.

3   Regular Reachability for 2-HOCA$^-$

The goal of this section is to prove the following theorem extending a known result on regular reachability on pushdown systems to 2-HOCA$^-$:

Theorem 5. *Reg. backwards/forwards reachability on 2-HOCA$^-$ is **P**-complete.*

3.1   Returns, Loops, and Control State Reachability

Proving Theorem 5 is based on the "returns-&-loops" construction for 2-HOPA of [12]. As a first step, we consider the simpler case of control-state reachability:

Proposition 6. *Control state reachability for 2-HOCA$^-$ is **P**-complete.*

In [12] it has been shown that certain runs, so-called *loops* and *returns*, are the building blocks of any run of a 2-HOPA in the sense that solving a reachability problem amounts to deciding whether certain loops and returns exist. Here, we analyse these notions more precisely in the context of 2-HOCA$^-$ in order to derive a polynomial control state reachability algorithm. Using this algorithm we can then also solve the regular backwards reachability problem efficiently.

For this section, we fix a $\mathcal{P}(\mathcal{Z})$-automaton $\mathcal{A} = (Q, q_0, F, \Delta)$. Recall that the $\mathcal{P}(\mathcal{Z})$-configurations of $\mathcal{A}$ are elements of $(\Sigma \times \{\bot\}^+)^+$. We identify $\bot^{m+1}$ with the natural number m and the set of storage configurations with $(\Sigma \times \mathbb{N})^+$.

Definition 7. *Let $s \in (\Sigma \times \mathbb{N})^+$, $t \in \Sigma \times \mathbb{N}$ and $q, q' \in Q$ be states of $\mathcal{A}$. A return of $\mathcal{A}$ from (q, st) to (q', s) is a run r from (q, st) to (q', s) such that except for the final configuration no configuration of r is in $Q \times \{s\}$.*

Let $s \in (\Sigma \times \mathbb{N})^*$, $t \in \Sigma \times \mathbb{N}$. A loop of $\mathcal{A}$ from (q, st) to (q', st) is a run r from (q, st) to (q', st) such that no configuration of r is in $Q \times \{s\}$.

One of the underlying reasons why control state reachability for pushdown systems can be efficiently solved is the fact that it is always possible to reach a certain state without increasing the pushdown by more than polynomially many elements. In the following, we prove an analogue of this fact for $\mathcal{P}(\mathcal{Z})$. For a given configuration, if there is a return or loop starting in this configuration, then this return or loop can be realised without increasing the (level 2) pushdown more than polynomially. This is due to the monotonic behaviour of $\mathcal{Z}$: given a $\mathcal{Z}$ configuration x, if we can apply a sequence φ of transitions to x then we can apply φ to all bigger configurations, i.e., to any configuration of the form $\mathsf{push}_1^n(x)$. Note that this depends on the fact that $\mathcal{Z}$ contains only trivial tests (the test $\mathsf{top}_\perp$ always returns true). In contrast, for $\mathcal{Z}+$, if φ applies a couple of pop operations and then tests for zero and performs a transition, then this is not applicable to a bigger counter because the 0-test would now fail.

For a $\mathcal{P}(\mathcal{Z})$ configuration $x = (\sigma_1, n_1)(\sigma_2, n_2)\ldots(\sigma_m, n_m)$, let $|x| = m$ be its *height*. Let r be some run starting in (q, x) for some $q \in Q$. The run r *increases the height by at most k* if $|x'| \leq |x| + k$ for all configurations (q', x') of r.

Definition 8. *Let $s \in (\{\perp\} \times \mathbb{N})^+$. We write $\mathsf{ret}_k(s)$ and $\mathsf{lp}_k(s)$, resp., for the set of pairs of initial and final control states of returns or loops starting in s and increasing the height by at most k. We write $\mathsf{ret}_\infty(s)$ and $\mathsf{lp}_\infty(s)$, resp., for the union of all $\mathsf{ret}_k(sw)$ or $\mathsf{lp}_k(s)$.*

The existence of a return (or loop) starting in sw (or $s'w$) (with $s \in (\{\perp\} \times \mathbb{N})^+, s' \in (\{\perp\} \times \mathbb{N})^*$ and $w \in \{\perp\} \times \mathbb{N}$) does not depend on the concrete choice of s or s'. Thus, we also write $\mathsf{ret}_k(w)$ for $\mathsf{ret}_k(sw)$ and $\mathsf{lp}_k(w)$ for $\mathsf{lp}_k(s'w)$.

By induction on the length of a run, we first prove that $\mathcal{P}(\mathcal{Z})$ is *monotone* in the following sense: let $s \in (\Sigma \times \mathbb{N})^*, t = (\sigma, n) \in \Sigma \times \mathbb{N}$, $q, q' \in Q$ and r a run starting in (q, st) and ending in state q'. If the topmost counter of each configuration of r is at least m, then for each $n' \geq n - m$ there is a run r' starting in $(q, s(\sigma, n'))$ and performing exactly the same transitions as r. In particular, for all $k \in \mathbb{N} \cup \{\infty\}$, $\sigma \in \Sigma$ and $m_1 \leq m_2 \in \mathbb{N}$, $\mathsf{ret}_k((\sigma, m_1)) \subseteq \mathsf{ret}_k((\sigma, m_2))$ and $\mathsf{lp}_k((\sigma, m_1)) \subseteq \mathsf{lp}_k((\sigma, m_2))$.

We next show that the sequence $(\mathsf{ret}_k((\sigma, m)))_{m \in \mathbb{N}}$ stabilises at $m = |\Sigma||Q|^2$. From this we conclude that $\mathsf{ret}_\infty = \mathsf{ret}_{|\Sigma|^2|Q|^4}$, i.e., in order to realise a return with arbitrary fixed initial and final configuration, we do not have to increase the height by more than $|\Sigma|^2|Q|^4$ (if there is such a return at all).

Lemma 9. *For $k \in \mathbb{N} \cup \{\infty\}$, $\sigma \in \Sigma$, $m \geq |\Sigma||Q|^2$, and $m' \geq 2 \cdot |\Sigma||Q|^2$, we have $\mathsf{ret}_k((\sigma, m)) = \mathsf{ret}_k((\sigma, |\Sigma||Q|^2))$ and $\mathsf{lp}_\infty((\sigma, m')) = \mathsf{lp}_\infty((\sigma, 2 \cdot |\Sigma||Q|^2))$.*

The proof uses the fact that we can find an $m' \leq |\Sigma||Q|^2$ with $\mathsf{ret}_k((\sigma, m')) = \mathsf{ret}_k(\sigma, m'+1)$ for all σ by the pigeonhole-principle. Using monotonicity of $\mathcal{P}(\mathcal{Z})$ we conclude that $\mathsf{ret}_k(\sigma, m') = \mathsf{ret}_k(\sigma, m)$ for all $m \geq m'$. A similar application of the pigeonhole-principle shows that there is a $k \leq |\Sigma|^2 \cdot |Q|^4$ such that $\mathsf{ret}_k((\sigma, i)) = \mathsf{ret}_{k+1}((\sigma, i))$ for all σ and all $i \leq |\Sigma||Q|^2$ (or equivalently for all $i \in \mathbb{N}$). By induction on $k' \geq k$ we show that $\mathsf{ret}_{k'} = \mathsf{ret}_k$ because any subreturn that increases the height by $k + 1$ can be replaced by a subreturn that only increases the height by k. Thus, we obtain the following lemma.

```
GeneratePDA(A, A):
Input: 2-HOCA⁻ A = (Q, q₀, Δ) over Σ, matrix A = (a_{σ,p,q})_{(σ,p,q)∈Σ×Q²} over ℕ ∪ {∞}
Output: 1-HOPA A' simulating A

1  k₀ := |Σ|² · |Q|⁴;  h₀ := |Σ| · |Q|²;  Δ' := ∅
2  foreach δ ∈ Δ:
3      if δ == (q, (σ, ⊥), stay_pop, p):
4          foreach i in {0, ..., h₀}:  Δ' := Δ' ∪ {((q, σ), ⊥ᵢ, pop, (p, σ)), ((q, σ), , ⊥_∞, pop, (p, σ))}
5      elseif δ == (q, (σ, ⊥), stay_{push_⊥}, p)
6          Δ' := Δ' ∪ {((q, σ), ⊥_∞, push_{⊥_∞}, (p, σ))} ∪ {((q, σ), ⊥_{h₀}, push_{⊥_∞}, (p, σ))}
7          foreach i in {0, ..., h₀ − 1}}:  Δ' := Δ' ∪ {((q, σ), ⊥ᵢ, push_{⊥_{i+1}}, (p, σ))}
8      elseif δ == (q, (σ, ⊥), push_{τ,id}, p)
9          foreach r in Q such that a_{τ,p,r} ≠ ∞:
10             foreach i in {a_{τ,p,r}, a_{τ,p,r} + 1, ..., h₀} ∪ {∞}:  Δ' := Δ' ∪ {((q, σ), ⊥ᵢ, id, (r, σ))}
11 A' := (Q × Σ, (q₀, ⊥), Δ')
12 return A'
```

Fig. 1. 2-HOCA⁻ to 1-HOPA Reduction Algorithm

Lemma 10. *For all $i \in \mathbb{N}$ and $\sigma \in \Sigma$, we have $\mathsf{ret}_\infty((\sigma, i)) = \mathsf{ret}_{|\Sigma|^2 \cdot |Q|^4}((\sigma, i))$ and $\mathsf{lp}_\infty = \mathsf{lp}_{|\Sigma|^2 |Q|^4 + 1}$.*

We now can prove that control-state reachability on 2-HOCA⁻ is **P**-complete.

Proof (of Proposition 6). Since 2-HOCA⁻ can trivially simulate pushdown automata, hardness follows from the analogous hardness result for pushdown automata. Containment in **P** uses the following ideas:

1. We assume that the input $(\mathcal{A}, q)$ satisfies that q is reachable in $\mathcal{A}$ iff $(q, (\bot, 0))$ is reachable and that $\mathcal{A}$ only uses instructions of the forms pop, $\mathsf{push}_{\sigma,\mathsf{id}}$, and stay_f. Given any 2-HOCA⁻ $\mathcal{A}'$ and a state q, it is straightforward to construct (in polynomial time) a 2-HOCA⁻ $\mathcal{A}$ that satisfies this condition such that q is reachable in $\mathcal{A}'$ iff it is reachable in $\mathcal{A}$.
2. Recall that $\mathsf{ret}_\infty(w) = \mathsf{ret}_{k_0}(w)$ for $k_0 = |\Sigma|^2 \cdot |Q|^4$ and for all $w \in \Sigma \times \mathbb{N}$. Set $h_0 = |\Sigma| \cdot |Q|^2$. We want to compute a table $(a_{\sigma,p,q})_{\sigma,p,q \in \Sigma \times Q^2}$ with values in $\{\infty, 0, 1, 2, \ldots, h_0\}$ such that $a_{\sigma,p,q} = \min\{i \mid (p, q) \in \mathsf{ret}_{k_0}((\sigma, i))\}$ (where we set $\min\{\emptyset\} = \infty$). Due to Lemmas 9 and 10 such a table represents ret_∞ in the sense that $(p, q) \in \mathsf{ret}_\infty((\sigma, i))$ iff $i \geq a_{\sigma,p,q}$.
3. With the help of the table $(a_{\sigma,p,q})_{(\sigma,p,q) \in \Sigma \times Q^2}$ we compute in polynomial time a $\mathcal{P}$ automaton $\mathcal{A}_\infty$ which executes the same level 1 transitions as $\mathcal{A}$ and simulates loops of $\mathcal{A}$ in the following sense: if there is a loop of $\mathcal{A}$ starting in $(q, (\sigma, i))$ performing first a $\mathsf{push}_{\tau,\mathsf{id}}$ operation and then performing a return with final state p, we allow $\mathcal{A}'$ to perform an id-transition from $(q, (\sigma, i))$ to $(p, (\sigma, i))$. This new system basically keeps track of the height of the pushdown up to h_0 by using a pushdown alphabet $\{\bot_0, \ldots, \bot_{h_0}, \bot_\infty\}$ where the topmost symbol of the pushdown is $\bot_i$ iff the height of the pushdown is i (where ∞ stands for values above h_0). After this change of pushdown alphabet, the additional id-transitions are easily computable from the table $(a_{\sigma,p,q})_{(\sigma,p,q) \in \Sigma \times Q^2}$. The resulting system has size $O(h_0^2 \cdot (|\mathcal{S}| + 1))$, i.e., is polynomial in the original system $\mathcal{A}$.
4. Using [1], check for reachability of q in the pushdown automaton $\mathcal{A}_\infty$.

```
ReachHOCA-(𝒜,q_f):
Input: 2-HOCA⁻ 𝒜 = (Q,q₀,Δ) over Σ, q_f ∈ Q
Output: whether q_f is reachable in 𝒜

 1  k₀ := |Σ|² · |Q|⁴; h₀ :=|Σ| · |Q²|;
 2  foreach (σ,p,q) in Σ × Q²: a_{σ,p,q} := ∞
 3  for k = 1,2,...,k₀:
 4      𝒜_k := GeneratePDA(𝒜,(a_{σ,p,q})_{(σ,p,q)∈Σ×Q²})
 5      foreach (r,(τ,⊥),pop,q) in Δ and (σ,p) in Σ × Q:
 6          for i = h₀,h₀ - 1,...,1,0:
 7              if ReachPDA(𝒜_k,((p,σ),i),(r,τ)): a'_{σ,p,q} := i
 8      foreach (σ,p,q) in Σ × Q²: a_{σ,p,q} :=a'_{σ,p,q}
 9  𝒜_∞ :=GeneratePDA(𝒜,(a_{σ,p,q})_{(σ,p,q)∈Σ×Q²})
10  if Reach(𝒜_∞,((q₀,⊥),0),(q_f,⊥)): return true else return false
```

Fig. 2. Reachability on 2-HOCA⁻ Algorithm 2

In fact, for step 2 we already use a variant of steps 3 and 4: we compute $\mathsf{ret}_\infty = \mathsf{ret}_{|\Sigma|^2|Q|^4}$ by induction starting with ret_0. If we remove all level 2 operations from $\mathcal{A}$ and store the topmost level 2 stack-symbol in the control state we obtain a pushdown automaton $\mathcal{B}$ such that $(q,q') \in \mathsf{ret}_0(\sigma,k)$ (w.r.t. $\mathcal{A}$) iff there is a transition $(p,(\sigma,\bot),\mathsf{pop},q')$ of $\mathcal{A}$ and the control state (p,σ) is reachable from $((p,\sigma),k)$ in $\mathcal{B}$. Thus, the results of polynomially many reachability queries for $\mathcal{B}$ determine the table for ret_0. Similarly, we can use the table of ret_i to compute the table of ret_{i+1} as follows. A return extending the height of the pushdown by $i+1$ decomposes into parts that do not increase the height at all and parts that perform a $\mathsf{push}_{\tau,\mathsf{id}}$ followed by a return increasing the height by at most i. Using the table for ret_i we can easily enrich $\mathcal{B}$ by id-transitions that simulate such push operations followed by returns increasing the height by at most i. Again, determining whether $(q,q') \in \mathsf{ret}_{i+1}(\sigma,k)$ reduces to one reachability query on this enriched $\mathcal{B}$ for each pop-transition of $\mathcal{A}$.

With these ideas in mind, it is straightforward to check that algorithm ReachHOCA- in Figure 2 (using algorithm GeneratePDA of Figure 1 as subroutine for step 3) solves the reachability problem for 2-HOCA⁻ (of the form described in step 1) in polynomial time. In this algorithm, ReachPDA $(\mathcal{A}',c,q)$ refers to the classical polynomial time algorithm that determines whether in the (level 1) pushdown automaton $\mathcal{A}'$ state q is reachable when starting in configuration c; a transition $(q,(\sigma,\tau),f,p)$ refers to a transition from state q to state p applying operation f that is executable if the (level 2) test top_σ and the (level 1) test $\mathit{test}(\mathsf{top}_\tau)$ both succeed. □

3.2 Regular Reachability

In order to define regular sets of configurations, we recall the encoding of 2-HOPA configurations as trees from [12]. Let $p = (\sigma_1,v_1)(\sigma_2,v_2)\ldots(\sigma_m,v_m) \in \mathcal{P}(\mathcal{Z})$. If $v_1 = 0$, we set $p_l = \emptyset$ and $p_r = (\sigma_2,v_2)\ldots(\sigma_m,v_m)$. Otherwise, there is a maximal $1 \leq j \leq m$ such that $v_1, \ldots, v_j \geq 1$ and we set $p_l = (\sigma_1, v_1 - 1) \ldots (\sigma_j, v_j - 1)$ and $p_r = (\sigma_{j+1}, v_{j+1}) \ldots (\sigma_m, v_m)$ if $j < m$ and $p_r = \emptyset$ if $j = m$. The *tree-encoding* E of p is given as follows:

$$E(p) = \begin{cases} \emptyset & \text{if } p = \emptyset \\ \bot(\sigma_1, E(p_r)) & \text{if } p = (\sigma_1, 0)p_r \\ \bot(E(p_l), E(p_r)) & \text{otherwise,} \end{cases}$$

where $\bot(t_1, t_2)$ is the tree with root $\bot$ whose left subtree is t_1 and whose right subtree is t_2. For a configuration $c = (q, p)$ we define $E(c)$ to be the tree $q(E(p), \emptyset)$. The picture beside the definition of E shows the encoding of the configuration $(q, (a, 2)(a, 2)(a, 0)(b, 1))$. Note that for each element (σ, i) of p, there is a path to a leaf l which is labelled by σ such that the path to l contains $i + 2$ left successors. Moreover, the inorder traversal of the tree induces an order of the leaves which corresponds to the left-to-right order of the elements of p. We call a set C of configurations *regular* if the set $\{E(c) \mid c \in C\}$ is a regular set of trees.

E turns the reachability predicate on 2-HOCA$^-$ into a tree-automatic relation [12], i.e., for a given 2-HOCA$^-$ $\mathcal{A}$, there is a tree-automaton $\mathcal{T}_{\mathcal{A}}$ accepting the convolution of $E(c_1)$ and $E(c_2)$ for 2-HOCA$^-$ configurations c_1 and c_2 iff there is a run of $\mathcal{A}$ from c_1 to c_2. This allows to solve the regular backwards reachability problem as follows. On input a 2-HOCA$^-$ and a tree automaton $\mathcal{T}$ recognising a regular set C of configurations, we first compute the tree-automaton $\mathcal{T}_{\mathcal{A}}$. Then using a simple product construction of $\mathcal{T}_{\mathcal{A}}$ and $\mathcal{T}$ and projection, we obtain an automaton $\mathcal{T}_{\text{pre}}$ which accepts pre$^*(C) = \{E(c) \mid \exists c' \in C$ and a run from c to $c'\}$. The key issue for the complexity of this construction is the computation of $\mathcal{T}_{\mathcal{A}}$ from $\mathcal{A}$. The explicit construction of $\mathcal{T}_{\mathcal{A}}$ in [12] involves an exponential blow-up. In this construction the blow-up is only caused by a part of $\mathcal{T}_{\mathcal{A}}$ that computes $\text{ret}_\infty(\sigma, m)$ for each $\sigma \in \Sigma$ on input a path whose labels form the word $\bot^m$. Thus, we can exhibit the following consequence.

Corollary 11 ([12]). *Given a 2-HOCA$^-$ $\mathcal{A}$ with state set Q, we can compute the tree automaton $\mathcal{T}_{\mathcal{A}}$ in $\mathbf{P}$, if we can compute from $\mathcal{A}$ in $\mathbf{P}$ a deterministic word automaton $\mathcal{T}'$ with state set $Q' \subseteq \prod_{\sigma \in \Sigma} (2^{Q \times Q})^2$ such that for all $m \in \mathbb{N}$ the state of $\mathcal{T}'$ on input $\bot^m$ is $(\text{ret}_\infty(\sigma, m), \text{lp}_\infty(\sigma, m))_{\sigma \in \Sigma}$.*

Thus, the following lemma completes the proof of Theorem 5.

Lemma 12. *Let $\mathcal{A}$ be a 2-HOCA$^-$ with state set Q. We can compute in polynomial time a deterministic finite word automaton $\mathcal{A}'$ with state set Q' of size at most $2 \cdot (|\Sigma| \cdot |Q|^2 + 1)$ such that $\mathcal{A}'$ is in state $(\text{ret}_\infty((\sigma, n)), \text{lp}_\infty((\sigma, n)))_{\sigma \in \Sigma}$ after reading $\bot^n$ for every $n \in \mathbb{N}$.*

Proof. Let $n_0 = 2 \cdot |\Sigma| \cdot |Q|^2$. Recall algorithm ReachHOCA- of Figure 2. In this polynomial time algorithm we computed a matrix $A = (a_{\sigma, p, q})_{(\sigma, p, q) \in \Sigma \times Q^2}$ representing ret_∞ and a pushdown automaton $\mathcal{A}_\infty$ (of level 1) simulating $\mathcal{A}$ in the sense that $\mathcal{A}_\infty$ reaches a configuration $((q, \sigma)p)$ for a pushdown p of height n if and only if $\mathcal{A}$ reaches $(q, (\sigma, n))$. It is sufficient to describe a polynomial time algorithm that computes $M_i = (\text{ret}_\infty((\sigma, n)), \text{lp}_\infty((\sigma, n)))_{\sigma \in \Sigma}$ for all $n \leq n_0$. $\mathcal{A}'$

is then the automaton with state set $\{M_i \mid i \leq n_0\}$, transitions from M_i to M_{i+1} for each $i < n_0$ and a transition from M_{n_0} to M_{n_0}. The correctness of this construction follows from Lemma 9.

Let us now describe how to compute M_i in polynomial time. Since $\mathcal{A}_\infty$ simulates $\mathcal{A}$ correctly, there is a loop from $(q, (\sigma, i))$ to $(q', (\sigma, i))$ of $\mathcal{A}$ if and only if there is a run of $\mathcal{A}_\infty$ from $((q, \sigma), p_i)$ to $((q', \sigma), p_i)$ for $p_i = \bot_0 \bot_1 \ldots \bot_i$ (where we identify $\bot_j$ with $\bot_\infty$ for all $j > h_0$). Thus, we can compute the loop part of M_i by n_0 many calls to an algorithm for reachability on pushdown systems. Note that $(p, q) \in \mathsf{ret}_\infty((\sigma, i))$ with respect to $\mathcal{A}$ if there is a state r and some $\tau \in \Sigma$ such that $(r, \tau, \mathsf{pop}, q)$ is a transition of $\mathcal{A}$ and (r, τ) is reachable in $\mathcal{A}_\infty$ from $((q, \sigma), i)$. Thus, with a loop over all transitions of $\mathcal{A}$ we reduce the computation of the returns component of M_i to polynomially many control state reachability problems on a pushdown system. $\qquad\square$

4 Reachability for k-HOCA⁻ and k-HOCA⁺

Using slight adaptations of Engelfriet's seminal paper [7], we can lift the result on reachability for 2-HOCA⁻ to reachability for k-HOCA⁻ (cf. [10]).

Theorem 13. *For $k \geq 2$, the control state reachability problem for k-HOCA⁻ is complete for $\mathbf{DTIME}(\bigcup_{d \in \mathbb{N}} \exp_{k-2}(n^d))$. For $k \geq 1$, the alternating control state reachability problem for k-HOCA⁻ is complete for $\mathbf{DTIME}(\bigcup_{d \in \mathbb{N}} \exp_{k-1}(n^d))$.*

Hardness follows from the hardness of control state reachability for $(k-1)$-HOPA [7] and the trivial fact that the storage type $\mathcal{P}^{k-1}$ of $(k-1)$-HOPA can be trivially simulated by the storage type $\mathcal{P}^{k-1}(\mathcal{Z})$ of k-HOCA⁻. Containment for the first claim is proved by induction on k (the base case $k = 2$ has been proved in the previous section). For $k \geq 3$, we use Lemma 7.11, Theorems 2.2 and 2.4 from [7] and reduce reachability of k-HOCA⁻ to reachability on (exponentially bigger) $(k-1)$-HOCA⁻. For the second claim, we adapt Engelfriet's Lemma 7.11 to a version for the setting of alternating automata (instead of nondeterministic automata) and use his Theorems 2.2. and 2.4 in order to show equivalence (up to logspace reductions) of alternating reachability for $(k-1)$-HOCA⁻ and reachability for k-HOCA⁻.

We can also reduce reachability for k-HOCA⁺ to reachability for $(k-1)$-fold exponentially bigger 1-HOCA⁺. Completeness for $\mathbf{NSPACE}(\log(n))$ of reachability for 1-HOCA⁺ (cf. [8]) yields the upper bounds for reachability for k-HOCA⁺. The corresponding lower bounds follow by applications of Engelfriet's theorems and an adaptation of the $\mathbf{PSPACE}$-hardness proof for emptiness of alternating finite automata by Jancar and Sawa [11].

Theorem 14. *For $k \geq 2$, (alternating) control state reachability for k-HOCA⁺ is complete for $(\mathbf{DSPACE}(\bigcup_{d \in \mathbb{N}} \exp_{k-1}(n^d)))$ $\mathbf{DSPACE}(\bigcup_{d \in \mathbb{N}} \exp_{k-2}(n^d))$.*

5 Back to HOPS: Applications to Languages

Engelfriet [7] also discovered a close connection between the complexity of the control state reachability problem for a class of automata and the class of languages recognised by this class. We restate a slight extension (cf. [10]) of these results and use them to confirm Slaat's conjecture from [18].

Proposition 15. *Let S_1 and S_2 be storage types and C_1, C_2 complexity classes such that $C_1 \subsetneq C_2$. If control state reachability for nondeterministic S_i automata is complete for C_i, then there is a deterministic S_2 automaton accepting some language L such that no nondeterministic S_1-automaton accepts L.*

In fact, Engelfriet's proof can be used to derive a separating language. For a storage type $S = (X, T, F, x_0)$, we define the *language of valid storage sequences* $\mathsf{VAL}(S)$ as follows. For each test $t \in T$ and $r \in \{true, false\}$ we set $t_r := \mathsf{id} \!\restriction_{\{x \in X \mid t(x) = r\}}$ and set $\Sigma = F \cup \{t_r \mid t \in T, r \in \{true, false\}\}$. For $s \in \Sigma^*$ such that $s = a_1 \ldots a_n$, and $x \in X$ we write $s(x)$ for $a_n(a_{n-1}(\ldots a_1(x) \ldots))$. We define $\mathsf{VAL}(S) = \{s \in \Sigma^* \mid s(x_0) \text{ is defined}\}$.

If the previous proposition separates the languages of S_2 automata from those of S_1 automata, then it follows from the proof that $\mathsf{VAL}(S_2)$ is not accepted by any S_1 automato (cf. [10]).

Corollary 16. *If $\mathbf{DTIME}(\bigcup_{d \in \mathbb{N}} \exp_k(n^d)) \subsetneq \mathbf{DSPACE}(\bigcup_{d \in \mathbb{N}} \exp_k(n^d)) \subsetneq \mathbf{DTIME}(\bigcup_{d \in \mathbb{N}} \exp_{k+1}(n^d))$, then $L((k-1)\text{-HOPA}) \subsetneq L(k\text{-HOCA}^-) \subsetneq L(k\text{-HOCA}^+) \subsetneq L(k\text{-HOPA})$.*

The crucial underlying construction detail of the proof of Proposition 15 is quite hidden within the details of Engelfriet's technical and long paper. Its usefulness in other contexts — e.g., for higher-order pushdowns or counters — has been overseen so far. Here we give another application to collapsible pushdown automata: reachability for collapsible pushdown automata of level k is $\mathbf{DSPACE}(\exp_{k-1}(n))$-complete (cf. [3]). Thus, Proposition 15 trivially shows that the language of valid level $(k+1)$ collapsible pushdown storage sequences separates the collapsible pushdown languages of level $k+1$ from those of level k. This answers a question asked by several experts in this field (cf. [17,14]). In fact, [17] uses a long and technical construction to prove the weaker result that there are more level $2k$ collapsible pushdown languages than level k collapsible pushdown languages. From Proposition 15 one also easily derives the level-by-level strictness of the collapsible pushdown tree hierarchy and the collapsible pushdown graph hierarchy (cf. [13,14]).

6 Future Work

Our result on regular reachability gives hope that also complexity results on model checking for logics like the μ-calculus extend from pushdown automata to 2-HOCA. 2-HOCA$^-$ probably is a generalisation of pushdown automata that retains the good complexity results for basic algorithmic questions. It is also interesting whether the result on regular reachability extends to the different notions of regularity for k-HOCA mentioned in the introduction. HOCA also can be seen as a new formalism in the context of register machines as currently used in the verification of concurrent systems. HOCA allow to store pushdown-like structures of register values and positive results on model checking HOCA could be transferred to verification questions in this concurrent setting.

References

1. Bouajjani, A., Esparza, J., Maler, O.: Reachability analysis of pushdown automata: Application to model-checking. In: Mazurkiewicz, A., Winkowski, J. (eds.) CONCUR 1997. LNCS, vol. 1243, pp. 135–150. Springer, Heidelberg (1997)
2. Bouajjani, A., Meyer, A.: Symbolic reachability analysis of higher-order context-free processes. In: Lodaya, K., Mahajan, M. (eds.) FSTTCS 2004. LNCS, vol. 3328, pp. 135–147. Springer, Heidelberg (2004)
3. Broadbent, C., Carayol, A., Hague, M., Serre, O.: A saturation method for collapsible pushdown systems. In: Czumaj, A., Mehlhorn, K., Pitts, A., Wattenhofer, R. (eds.) ICALP 2012, Part II. LNCS, vol. 7392, pp. 165–176. Springer, Heidelberg (2012)
4. Carayol, A.: Regular sets of higher-order pushdown stacks. In: Jedrzejowicz, J., Szepietowski, A. (eds.) MFCS 2005. LNCS, vol. 3618, pp. 168–179. Springer, Heidelberg (2005)
5. Carayol, A., Wöhrle, S.: The caucal hierarchy of infinite graphs in terms of logic and higher-order pushdown automata. In: Pandya, P.K., Radhakrishnan, J. (eds.) FSTTCS 2003. LNCS, vol. 2914, pp. 112–123. Springer, Heidelberg (2003)
6. Caucal, D.: On infinite terms having a decidable monadic theory. In: Diks, K., Rytter, W. (eds.) MFCS 2002. LNCS, vol. 2420, pp. 165–176. Springer, Heidelberg (2002)
7. Engelfriet, J.: Iterated stack automata and complexity classes. Inf. Comput. 95(1), 21–75 (1991)
8. Göller, S.: Reachability on prefix-recognizable graphs. Inf. Process. Lett. 108(2), 71–74 (2008)
9. Hague, M., Ong, C.-H.L.: Symbolic backwards-reachability analysis for higher-order pushdown systems. LMCS 4(4) (2008)
10. Heußner, A., Kartzow, A.: Reachability in higher-order-counters. CoRR, arxiv:1306.1069 (2013), http://arxiv.org/abs/1306.1069
11. Jancar, P., Sawa, Z.: A note on emptiness for alternating finite automata with a one-letter alphabet. Inf. Process. Lett. 104(5), 164–167 (2007)
12. Kartzow, A.: Collapsible pushdown graphs of level 2 are tree-automatic. Logical Methods in Computer Science 9(1) (2013)
13. Kartzow, A., Parys, P.: Strictness of the collapsible pushdown hierarchy. In: Rovan, B., Sassone, V., Widmayer, P. (eds.) MFCS 2012. LNCS, vol. 7464, pp. 566–577. Springer, Heidelberg (2012)
14. Kobayashi, N.: Pumping by typing. To appear in Proc. LICS (2013)
15. Maslov, A.N.: The hierarchy of indexed languages of an arbitrary level. Sov. Math., Dokl. 15, 1170–1174 (1974)
16. Maslov, A.N.: Multilevel stack automata. Problems of Information Transmission 12, 38–43 (1976)
17. Parys, P.: Variants of collapsible pushdown systems. In: Proc. of CSL 2012. LIPIcs, vol. 16, pp. 500–515 (2012)
18. Slaats, M.: Infinite regular games in the higher-order pushdown and the parametrized setting. PhD thesis, RWTH Aachen (2012)

Length-Increasing Reductions
for PSPACE-Completeness

John M. Hitchcock[1,*] and A. Pavan[2,**]

[1] Department of Computer Science, University of Wyoming
jhitchco@cs.uwyo.edu
[2] Department of Computer Science, Iowa State University
pavan@cs.iastate.edu

Abstract. Polynomial-time many-one reductions provide the standard notion of completeness for complexity classes. However, as first explicated by Berman and Hartmanis in their work on the isomorphism conjecture, all natural complete problems are actually complete under reductions with stronger properties. We study the length-increasing property and show under various computational hardness assumptions that all PSPACE-complete problems are complete via length-increasing reductions that are computable with a small amount of nonuniform advice.

If there is a problem in PSPACE that requires exponential time, then polynomial size advice suffices to give li-reductions to all PSPACE-complete sets. Under the stronger assumption that linear space requires exponential-size NP-oracle circuits, we reduce the advice to logarithmic size. Our proofs make use of pseudorandom generators, hardness versus randomness tradeoffs, and worst-case to average-case hardness reductions.

Keywords: computational complexity, completeness, length-increasing reductions, PSPACE.

1 Introduction

Completeness is arguably the single most important notion in computational complexity theory. Many natural problems that arise in practice turn out be complete for appropriate complexity classes. Informally, a set A is complete for a class $\mathcal{C}$ if A belongs to $\mathcal{C}$ and every set from $\mathcal{C}$ "polynomial-time reduces" to A. In his seminal paper, Cook [Coo71] used *Turing reductions* to define completeness. However, Karp [Kar72] used a much more restrictive notion, *many-one reductions*, to define completeness. Since then polynomial-time many-one reductions have been considered as the most natural reductions to define completeness.

It has been observed that most problems remain complete under more stringent notions of reducibility. Perhaps the most restrictive notion of a polynomial-time reduction is that of *isomorphic reduction*. Two sets A and B are p-*isomorphic*

---

* This research was supported in part by NSF grants 0652601 and 0917417.
** This research was supported in part by NSF grant 0916797.

K. Chatterjee and J. Sgall (Eds.): MFCS 2013, LNCS 8087, pp. 540–550, 2013.

if there exists a polynomial-time computable, one-one, onto, and polynomial-time invertible reduction from A to B. Berman and Hartmanis [BH77] observed that all known natural NP-complete sets are indeed p-isomorphic and this led to their famous "isomorphism conjecture"—all NP-complete sets are p-isomorphic to SAT. Berman and Hartmanis characterized isomorphism in terms of one-one, length-increasing reductions. They showed that two sets A and B are p-isomorphic if they are reducible to each other via one-one, polynomial-time invertible *length-increasing* reductions. We write "li-reduction" as an abbreviation for "length-increasing reduction." Thus the isomorphism conjecture is equivalent to the following statement: All NP-complete sets are complete via one-one, polynomial-time invertible li-reductions.

Though the original isomorphism conjecture concerns the class NP, a similar conjecture can be formulated for classes such as E, NE, and PSPACE. In spite of many years of research we do not have concrete evidence for or against the isomorphism conjecture for any complexity class. This has led researchers to ask weaker questions such as: Do complete sets for a class remain complete under one-one reductions? Do complete sets for a class remain complete under li-reductions? Even these weaker questions are not completely resolved and we only know of some partial answers.

Berman [Ber77] showed that all complete sets for E are complete under one-one, li-reductions. Ganesan and Homer [GH92] showed that all NE-complete sets are complete under one-one reductions. For quite sometime, there had been no progress on NP and the first major result for NP is due to Agrawal [Agr02]. He showed that if one-way permutations exist, then all NP-complete sets are complete via one-one, P/poly-computable li-reductions. Since then there have been several results of this nature. Hitchcock and Pavan [HP07] showed that if NP does not have p-measure zero, then all NP-complete sets are complete via P/poly li-reductions. Buhrman, Hescott, Homer, and Torenvliet [BHHT10] improved this result to show that under the same hypothesis, NP-complete sets are complete via li-reductions that use a logarithmic amount of advice. Next, Agrawal and Watanabe [AW09] showed that if regular one-way functions exist, then NP-complete sets are complete via one-one, P/poly li-reductions. Most recently, Gu, Hitchcock, and Pavan [GHP12] showed that if NP contains a language that requires time $2^{n^{\Omega(1)}}$ at almost all lengths, then NP-complete sets are complete via P/poly li-reductions. All of the known results till date concern the complexity classes NP, E, and NE.

In this paper, we consider the question of whether PSPACE-complete sets are complete via li-reductions. It should be noted that the proofs of many of the aforementioned results go through if one replaces NP with PSPACE. For example, Agrawal's proof shows that if one-way permutations exist, then all complete sets for PSPACE are complete via, one-one, P/poly li-reductions. Similarly, Hitchcock and Pavan's proof shows that if PSPACE does not have p-measure zero, then PSPACE-complete sets are complete via P/poly li-reductions. However, Gu, Hitchcock, and Pavan's proof does not go through if one replaces NP with PSPACE.

In this paper we establish new results regarding PSPACE-complete sets. Using ideas from [GHP12], we first give evidence that PSPACE-complete sets are complete via non uniform, li-reductions. Our first main result is the following.

Theorem I. If PSPACE contains a language that requires 2^{n^ϵ} time at almost all lengths, then PSPACE-complete sets are complete via li-reductions that use a polynomial amount of advice.

We note that the hypothesis used in this result is a *worst-case hardness* hypothesis (as opposed to average-case or almost-everywhere hardness hypotheses used in the works of [Agr02, HP07, BHHT10]). Next we address the question of whether we can eliminate or reduce the amount of nonuniformity used. We establish two sets of results. Our first result on this shows that nonuniformity in the reductions can be traded for nondetermisnism. We show that if NP contains a language that requires $2^{n^{\Omega(1)}}$ time at almost all lengths, then PSPACE-complete sets are complete via strong nondeterministic (SNP) li-reductions.

Next we show that using stronger hypotheses the amount of nonuniformity can be reduced. Our second main contribution is the following.

Theorem II. If there is a language in linear space that requires exponential size NP-oracle circuits, then PSPACE-complete sets are complete via li-reductions that use a logarithmic amount of advice.

The proof of this theorem is nonstandard. All known proofs that establish length-increasing completeness are of the following form: Say A is a complete language and we wish to prove that it is complete via li-reductions. All known proofs first define an intermediate language S and show that a standard complete language (such as SAT or K) length-increasing reduces to S, and there is a length-increasing reduction from S to A. We note that this approach may not work for our case (see the discussion after the statement of Theorem 3.4). Our proof proceeds by constructing two intermediate languages S_1 and S_2. We show both S_1 and S_2 length-increasing reduce to A. Our final length-increasing reduction from K to A goes via S_1 on some strings, and via S_2 on other strings. We use tools from pseudorandomness and hardness amplification to establish this result.

The following table compares some of the main results of this paper.

class	hardness assumption	li-reduction type
PSPACE	$2^{n^{\Omega(1)}}$ time	P/poly
NP	$2^{n^{\Omega(1)}}$ time	SNP
LINSPACE	$2^{\Omega(n)}$ NP-oracle circuits	P/log
PSPACE E	$2^{n^{\Omega(1)}}$ circuits $2^{\Omega(n)}$ NP-oracle circuits	P/log

The interpretation of a line in the table is that if the class (or pair of classes for the last line) satisfies the hardness assumption, then PSPACE-complete sets are complete under li-reductions of the stated type.

2 Preliminaries

Let $\mathcal{H}$ be a class of length bound functions mapping $\mathbb{N} \to \mathbb{N}$. A function $f : \Sigma^* \to \Sigma^*$ is P/$\mathcal{H}$-computable if there exist a polynomial-time computable $g : \Sigma^* \times \Sigma^* \to \Sigma^*$ and an $l(n) \in \mathcal{H}$ so that for every n, there is an advice string $a_n \in \Sigma^{\leq l(n)}$ such that for all $x \in \Sigma^n$, $f(x) = g(x, a_n)$. We will use the length bound classes poly = $\{l : l(n) = n^{O(1)}\}$ and log = $\{l : l(n) = O(\log n)\}$.

Given a language L, $L^{=n}$ denotes the set of strings of length n that belong to L. For a language L, we denote the characteristic function of L with L itself. That is, $L(x)$ is 1 if $x \in L$, otherwise $L(x)$ equals 0. Given two languages A and B, we say that A and B are *infinitely often equivalent*, $A =_{io} B$, if for infinitely many n, $A^{=n} = B^{=n}$. Given a complexity class $\mathcal{C}$, we define $_{io}\mathcal{C}$ as

$$_{io}\mathcal{C} = \{A \mid \exists B \in \mathcal{C}, A =_{io} B\}.$$

In this paper we will use strong nondeterministic reductions [AM77]. A language A is SNP-*reducible* to a language B if there is a polynomial-time bounded nondeterministic machine M such that for every x the following holds:

- Every path of $M(x)$ outputs a string y or outputs a special symbol $\perp$. Different paths of $M(x)$ may output different strings.
- If a path outputs a string y, then $x \in A \Leftrightarrow y \in B$.

We say that an SNP reduction is length-increasing if the length of every output (excluding $\perp$) is greater than the length of the input.

For a Boolean function $f : \Sigma^n \to \Sigma$, $CC(f)$ is the smallest number s such that there is circuit of size s that computes f, and $CC^{NP}(f)$ is the smallest number s such that there is a size s, NP-oracle circuit that computes f. The Boolean function f is (s, ϵ)-hard if for every circuit of size at most s, $\Pr_{x \in \Sigma^n}[C(x) \neq f(x)] \geq \epsilon$.

Definition 2.1. *A pseudorandom generator (PRG) family is a collection of functions* $G = \{G_n : \Sigma^{m(n)} \to \Sigma^n\}$ *such that* G_n *is uniformly computable in time* $2^{O(m(n))}$ *and for every circuit of* C *of size* $O(n)$,

$$\left| \Pr_{x \in \Sigma^n}[C(x) = 1] - \Pr_{y \in \{0,1\}^{m(n)}}[C(G_n(y)) = 1] \right| \leq \frac{1}{n}.$$

A pseudorandom generator is secure against NP-oracle circuits *if the above statement holds when the circuits have access to an* NP-oracle.

There are many results that show that the existence of hard functions in exponential time implies PRGs exist. We will use the following.

Theorem 2.2 ([KvM02]). *If there is a language A in* E *and an $\epsilon > 0$ such that $CC^{NP}(A_n) \geq 2^{\epsilon n}$ for all sufficiently large n, then there is a constant k and a PRG family $G = \{G_n : \Sigma^{k \log n} \to \Sigma^n\}$ that is secure against NP-oracle circuits.*

3   PSPACE-Complete Sets

We will first prove that if PSPACE contains a worst-case hard language, then PSPACE-complete sets are complete via P/poly li-reductions. We will use ideas from [GHP12]. As noted before, the proof of the analogous result in [GHP12] does not go through if we replace NP with PSPACE. This is because that proof uses the fact that NP contains complete languages that are disjunctively self-reducible. Since every disjunctively self-reducible language is in NP, we cannot hope that PSPACE has a disjunctively self-reducible complete set unless NP equals PSPACE. To get around this problem, we will use the fact that PSPACE is closed under complementation.

Theorem 3.1. *If there is language L in PSPACE that is not in $_\text{io}$DTIME(2^{n^ϵ}) for some $\epsilon > 0$, then all PSPACE-complete sets are complete via P/poly li-reductions.*

Proof. Let A be a PSPACE-complete set that can be decided in time 2^{n^k}, and let K be the standard PSPACE-complete set that can be decided in time 2^{cn}, for some constants k and c. We define the following intermediate language S, where $\delta = \frac{\epsilon}{ck}$.

$$S = \left\{ \langle x, y \rangle \mid |y| = |x|^\delta, L(x) \oplus K(y) = 1 \right\}$$

Since S is in PSPACE, there is a many-one reduction f from S to A.
    Let

$$T_n = \left\{ x \in \Sigma^n \mid x \in L, \forall y \in \Sigma^{n^\delta}, |f(\langle x, y \rangle)| > n^\delta \right\}.$$

We will first show that T_n is not an empty set.

Lemma 3.2. *For all but finitely many n, $T_n \neq \varnothing$.*

Proof. Suppose there exist infinitely many n for which $T_n = \emptyset$. We will exhibit an algorithm that decides L correctly in time 2^{n^ϵ} for every n at which $T_n = \emptyset$. Consider the following algorithm for L.

1. Input x, $|x| = n$.
2. Cycle through all y of length n^δ and find a y such that $|f(\langle x, y \rangle)| \leq n^\delta$. If no such y is found reject x.
3. Suppose such a y is found. Compute $K(y)$ and $A(f(\langle x, y \rangle))$.
4. Accept x if and only if $A(f(\langle x, y \rangle)) \oplus K(y) = 1$.

Consider a length n at which $T_n = \emptyset$. Let x be an input of length n. We first consider the case when the above algorithm finds a y in Step 2. Since f is many-one reduction from S to A, we have $L(x) \oplus K(y) = A(f(\langle x, y \rangle))$. Thus $L(x) = K(y) \oplus A(f(\langle x, y \rangle))$. Therefore the algorithm is correct in this case. Now consider the case when the algorithm does not find a y. Since $T_n = \emptyset$, for every x (of length n) in L, it must be the case that the length of $f(\langle x, y \rangle)$ is at most n^δ for some y of length n^δ. Thus if the algorithm does not find a y in Step 2, then it must be the case that $x \notin L$. In this case, the algorithm correctly rejects x. Therefore the algorithm is correct on all strings of length n.

The time taken find a y in Step 2 is bounded by 2^{2n^δ}. The time taken to compute $K(y)$ is 2^{cn^δ}. The time taken to compute $A(f(\langle x, y \rangle))$ is at most 2^{m^k}, where $m = |f(\langle x, y \rangle)|$. Since the length of $f(\langle x, y \rangle)$ is at most n^δ and $\delta = \epsilon/ck$, the total time taken by the above algorithm is bounded by 2^{n^ϵ}.

Thus if $T_n = \emptyset$ for infinitely many n, then L is in $_{\text{io}}\text{DTIME}(2^{n^\epsilon})$ and this contradicts the hardness of L. $\square$

We will now describe the P/poly many-one reduction from K to A. Let z_n be the lexicographically smallest string from T_n. It exists because of the previous lemma. On input y of length n^δ, the reduction outputs $f(\langle z_n, y \rangle)$. Since $z_n \in L$, $y \in K$ if and only if $\langle z_n, y \rangle$ is in S. Since f is a many-one reduction from S to A, this is a many-one reduction from K to A. By the definition of z_n, we have that the length of $f(\langle z_n, y \rangle)$ is bigger than n^δ. Since y is of length n^δ, this is a li-reduction from K to A. The advice for the reduction is z_n. Thus, this is a P/poly reduction. $\square$

Theorem 3.3. *If there is language L in NP that is not in $_{\text{io}}\text{DTIME}(2^{n^\epsilon})$ for some $\epsilon > 0$, then all PSPACE-complete sets are complete via SNP li-reductions.*

The proof of Theorem 3.3 uses the same setup as Theorem 3.1. Consider S and T_n as before. We have that T_n is not empty for all but finitely many lengths. The reduction will use nondeterminism to find string in T_n. Let y be an input of length n^δ. Nondeterministically guess a string z of length n and verify that such that z is in L and $|f(\langle z, y \rangle)| > n^\delta$. If the verification is successful, then output $f(\langle z, y \rangle)$. Otherwise, output $\perp$. Since T_n is not empty, there exist at least one path that guesses a z from T_n and on this path the reduction is correct. Note that any path that fails to guess a $z \in T_n$ will output $\perp$. Thus there is no path on which the reduction outputs a wrong answer. Thus S is SNP-complete via li-reductions.

We now show how to reduce the number of advice bits from polynomial to logarithmic with a stronger hypothesis. We will show that if the worst-case NP-oracle circuit complexity of LINSPACE is $2^{\Omega(n)}$, then PSPACE-complete sets are complete via li-reductions that are P/log-computable. We will first assume that PSPACE has a language that is hard on average, and then use a known worst-case to average-case connection for PSPACE.

Theorem 3.4. *Suppose there is a language L in LINSPACE such that for every n, L is $(2^{\epsilon n}, 3/8)$-hard for NP-oracle circuits, then PSPACE-complete sets are complete via P/log li-reductions.*

Before we present the proof, we will mention the idea behind the proof. Our goal is to proceed as in the proof of Theorem 3.1. Consider T_n—in the proof of Theorem 3.1, we have shown that T_n is not empty. Suppose, we could show a stronger claim and establish that T_n contains many (say $> 3/4$ fraction) strings. Then a randomly chosen string will be a good advice with high probability. If LINSPACE is hard on average, then E is also hard on average and pseudorandom generators exist. Thus we can derandomize the process of "randomly picking a string from T_n" and instead generate a small list of strings such that at least one string from this list belongs to T_n. Now, a small advices suffices to identify the good string from T_n.

However, this idea does not work for two reasons. Note that every string in T_n must be in L. Since L is in PSPACE, this places T_n in PSPACE. We would like to derandomize the process of "randomly picking a string from a language in PSPACE." For this to work, we need a pseudorandom generator that is secure against PSPACE-oracle circuits. For this, we need a language that is hard for PSPACE-oracle circuits. However, the hard language L that is guaranteed by our hypothesis is in linear space, and thus cannot be hard for PSPACE. The second reason is that it is not clear that T_n will contain a $3/4$ fraction of strings from Σ^n. Since L is hard on average for circuits, L contains roughly $1/2$-fraction of strings from Σ^n (at every length n). Since T_n is a subset of L, we cannot hope that the size of T_n will be bigger than $\frac{3}{4}2^n$.

We overcome these difficulties by considering two intermediate sets S_1 and S_2 instead of one intermediate set S. Say f and g are many-one reductions from S_1 and S_2 to the complete set A. Then we define T_n as the set of all strings x from Σ^n such that for every y, the length of $f(\langle x, y \rangle)$ and $g(\langle x, y \rangle)$ are both large enough. This will place T_n in coNP and we can pseudorandomly pick strings from T_n provided we have a pseudorandom generator that is secure against NP-oracle circuits. Depending on the string that we picked, we will either use reduction f or reduction g. Now, we present the details.

Proof. Let A be a PSPACE-complete language that can be decided in time 2^{n^k}, and let K be the standard complete language for PSPACE. Observe that K can be decided in time 2^{cn} for some constant $c > 0$. Let $\delta = \epsilon/k$. Consider the following two languages

$$S_1 = \left\{ \langle x, y \rangle \;\middle|\; K(x) \oplus L(y) = 1, |x| = \frac{\epsilon}{2c}|y| \right\},$$

$$S_2 = \left\{ \langle x, y \rangle \;\middle|\; K(x) \oplus L(y) = 0, |x| = \frac{\epsilon}{2c}|y| \right\}.$$

Both languages are in PSPACE, thus there is a many-one reduction f from S_1 to A and many-one reduction g from S_2 to A. We first show that these reductions must be honest for most strings.

Claim 3.5. *Under our hardness assumption of L, there is a polynomial time algorithm $\mathcal{A}$ such that for all but finitely many m, $\mathcal{A}(1^m)$ outputs polynomially many strings $y_1, y_2, \cdots y_t$ of length $n = \frac{2cn}{\epsilon}$, such that for some y_i $1 \leq i \leq t$, and for every $x \in \{0,1\}^m$, the lengths of both $f(\langle x, y_i \rangle)$ and $g(\langle x, y_i \rangle)$ are at least n^δ.*

Assuming that Claim 3.5 holds, we complete the proof of the theorem. We will describe a honest reduction h from K to A. Given a length m, let $y_1, \cdots, y_t$ be the strings output by the algorithm $\mathcal{A}$. By the lemma, there is y_i such that for every x the lengths of both $f(\langle x, y_i \rangle)$ and $g(\langle x, y_i \rangle)$ are at least n^δ. The reduction gets i and $L(y_i)$ as advice. Note that the length is the advice is $O(\log m)$.

Let x be a string of length m. The reduction first computes the list $y_1, \cdots, y_t$. If the $L(y_i) = 0$, then $h(x) = f(\langle x, y_i \rangle)$, else $h(x) = g(\langle x, y_i \rangle)$.

Thus h is $P/O(\log n)$ computable. If $y_i \notin L$, then $x \in K$ if and only if $\langle x, y_i \rangle \in S_1$. Similarly if $y_i \in L$, then $x \in K$ if and only if $\langle x, y_i \rangle \in S_2$. Since f is a reduction from S_1 to A and g is a reduction S_2 to A, h is a valid reduction from K to A. Since the lengths of both $f(\langle x, y_i \rangle)$ and $g(\langle x, y_i \rangle)$ are at least n^δ, h is an honest reduction from K to A. Since K is paddable, there is a length-increasing $P/O(\log n)$-reduction from K to A. This, together with the forthcoming proofs of Claims 3.6 and 3.5, complete the proof of Theorem 3.4. □

To prove Claim 3.5, we need the following result.

Claim 3.6. *Let*

$$T_n = \left\{ y \in \{0,1\}^n \mid \forall x \in \{0,1\}^{\frac{\epsilon n}{2c}}, |f(\langle x, y \rangle)| \geq n^\delta, \text{ and } |g(\langle x, y \rangle)| \geq n^\delta \right\}.$$

For all but finitely many n, $|T_n| \geq \frac{3}{4} 2^n$.

Proof. Suppose not. There exist infinitely many n for which T_n has at most $\frac{3}{4} 2^n$ strings. We will show that this implies L must be not be average-case hard at infinitely many lengths. Consider the following algorithm for L.

1. Input $y, |y| = n$.
2. Cycle through all strings of length $\frac{\epsilon n}{2c}$ to find a string x such that at least one of $f(\langle x, y \rangle)$ or $g(\langle x, y \rangle)$ has length less than n^δ.
3. If no such string is found output $\perp$.
4. If $|f(\langle x, y \rangle)| \leq n^\delta$, then output $A(f(\langle x, y \rangle)) \oplus K(x)$.
5. If $|g(\langle x, y \rangle)| \leq n^\delta$, then output $A(g(\langle x, y \rangle)) \oplus K(x)$.

Consider a length n for which the cardinality of T_n is less than $\frac{3}{4} 2^n$. We will first show that the above algorithm correctly solves L on $1/4$ fraction of strings from $\{0,1\}^n$.

A string y does not belong to T_n, if there is a string x of length $\frac{\epsilon n}{2c}$ such that at least one of the strings $f(\langle x, y \rangle)$ or $g(\langle x, y \rangle)$ has length less than n^δ. For all such string y the above algorithm halts in either Step 4 or in Step 5. It is clear that the decision made by the algorithm in these steps is correct. Thus if $y \notin T_n$, then the above algorithm correctly decide the membership of y in L. Since the size of T_n is at most $\frac{3}{4} 2^n$ for many strings, the above algorithm correctly decides L on at least $1/4$ fraction of strings at length n.

The running time of the above algorithm can be bounded as follows. It takes $2^{\frac{n\epsilon}{2c}}$ time to search for x. Computing $K(x)$ takes at most $2^{\frac{n\epsilon}{2}}$ time. If $|f(\langle x, y \rangle)| < n^\delta$, computing $A(f(\langle x, y \rangle))$ takes at most 2^{n^ϵ} time. Thus Step 4 take at most $O(2^{\frac{\epsilon n}{2}})$ time. Similarly Step 5 also takes at most $O(2^{\frac{\epsilon n}{2}})$ time. Thus the running time of the above algorithm is bounded by $O(2^{\frac{\epsilon n}{2}})$.

Observe that the above algorithm never errs. On any string y it either outputs $\perp$ or correctly decides L, and for at least $1/4$ fraction of the strings the algorithm does not output $\perp$. By providing one bit of advice, we can make the algorithm to correctly decide L on at least $5/8$ fraction of inputs from Σ^n.

We can convert this modified algorithm into a family of circuits of size at most $2^{\epsilon n}$. If the size of T_n is less than $\frac{3}{4}2^n$ for infinitely many n, then this circuit family correctly computes L on at least 5/8 fraction of strings from $\{0,1\}^n$ for infinitely many n. This contradicts the hardness of L. □

Now we return to the proof of Claim 3.5.

Proof of Claim 3.5. In the following we fix m and so n. Recall that $n = \frac{2cm}{\epsilon}$. There is a polynomial p such that the computation of both f and g on strings of form $\langle x,y \rangle$, $|y| = n$, $|x| = m$ is bounded by $p(n)$. Let $r = p^2(n)$. Since LINSPACE is hard on average for $2^{\epsilon n}$-size NP-oracle circuits, and LINSPACE $\subseteq$ E, by Theorem 2.2 there is a PRG family G_r that maps $d \log r$ bits to r bits. The algorithm $\mathcal{A}$ on input 1^m behaves as follows: For each $d \log r$ bit string u compute $G_r(u)$ and output its n-bit prefix. This generates at most r^d strings. Since r is a polynomial in m, the number of strings output are polynomial in m.

We have to show that there exists a string y_i from the output of $\mathcal{A}(1^m)$ such that for every $x \in \{0,1\}^m$, the lengths of both $f(\langle x,y_i \rangle)$ and $g(\langle x,y_i \rangle)$ are bigger than n^δ. Suppose not. Consider the following algorithm $\mathcal{B}$ that has SAT as oracle.

Given a string of length r as input, let y be its n-bit prefix. By making queries to the SAT find if there is a string x of length m such that one of $f(\langle x,y \rangle)$ or $g(\langle x,y \rangle)$ have length less than n^δ. If no such x is found accept, else reject.

This algorithm runs in time $p(n)$, and so can be converted into a circuit C of size at most r. By our assumption,

$$\Pr_{z \in \{0,1\}^{d \log r}} [C(G_r(z)) = 1] = 0$$

However, by Claim 3.6

$$\Pr_{z \in \{0,1\}^r} [C(z) = 1] \geq 3/4$$

This contradicts the fact that G is a pseudorandom generator against NP-oracle circuits.   □ ***Claim 3.5***

We can weaken the average-case hardness assumption in the above theorem to a worst-case hardness assumption. It is known that if LINSPACE requires $2^{\epsilon n}$-size NP-circuits at every length, there is a language in LINSPACE that is $(2^{\epsilon' n}, 3/8)$ hard for NP-oracle circuits at all lengths of the form t^2 [IW97, KvM02]. The proof of Theorem 3.4 requires average-case hardness of a language L at all lengths. However, the proof can be easily modified to work even when the language is average-case hard only at lengths of the form t^2. Thus we have the following theorem.

Theorem 3.7. *Suppose there is a language L in* LINSPACE *such that for every n, the worst-case NP-oracle circuit complexity of L is $2^{\epsilon n}$. Then all* PSPACE-*complete sets are complete via* P/ log *li-reductions.*

We conclude with an improvement of Theorem 3.7. Consider the proof of Theorem 3.4. The hardness assumption "LINSPACE is $(2^{\epsilon n}, 3/8)$ hard for NP-oracle circuits" is used at two places. First in the proof of Claim 3.6. Note that the proof of this lemma only needs a weaker assumption, namely "PSPACE is $(2^{\epsilon n}, 3/8)$ hard for circuits." By a very slight modification of the proof, we can further weaken the hypothesis needed to establish Claim 3.6. Consider the following definitions of S_1 and S_2.

$$S_1 = \left\{ \langle x, y \rangle \;\middle|\; K(x) \oplus L(y) = 1, |x| = \frac{|y|^\epsilon}{2c} \right\},$$

$$S_2 = \left\{ \langle x, y \rangle \;\middle|\; K(x) \oplus L(y) = 0, |x| = \frac{|y|^\epsilon}{2c} \right\}.$$

We define T_n as

$$T_n = \left\{ y \in \{0,1\}^n \;\middle|\; \forall x \in \{0,1\}^{\frac{n^\epsilon}{2c}}, |f(\langle x, y \rangle)| \geq n^\delta, \text{ and } |g(\langle x, y \rangle)| \geq n^\delta \right\}.$$

With this definition of S_1, S_2, and T_n, the proof proceeds exactly as before, except that to establish Claim 3.6, we only need that "PSPACE is $(2^{n^\epsilon}, 3/8)$ hard for circuits". Again, it is known that if PSPACE is does not have 2^{n^ϵ}-size circuits, then PSPACE is $(2^{n^\epsilon}, 3/8)$-hard for circuits.

The second place where the hardness of LINSPACE is used is in the proof of Claim 3.5. Note that the proof of this claim goes through if we merely have the assumption "E has a language with $2^{\epsilon n}$-size NP-oracle circuit complexity". These observations yield the following improvement.

Theorem 3.8. *Suppose there is a language L in PSPACE such that for every n, the worst-case circuit complexity of L is 2^{n^ϵ} for some $\epsilon > 0$. Further assume that E has a language whose worst-case NP-oracle circuit complexity is $2^{\delta n}$ for some $\delta > 0$. All PSPACE-complete sets are complete via P/ log li-reductions.*

References

[Agr02]   Agrawal, M.: Pseudo-random generators and structure of complete degrees. In: Proceedings of the Seventeenth Annual IEEE Conference on Computational Complexity, pp. 139–147. IEEE Computer Society (2002)

[AM77]   Adleman, L., Manders, K.: Reducibility, randomness, and intractability. In: Proceedings of the 9th ACM Symposium on Theory of Computing, pp. 151–163 (1977)

[AW09]   Agrawal, M., Watanabe, O.: One-way functions and the isomorphism conjecture. Technical Report TR09-019, Electronic Colloquium on Computational Complexity (2009)

[Ber77]   Berman, L.: Polynomial Reducibilities and Complete Sets. PhD thesis, Cornell University (1977)

[BH77]   Berman, L., Hartmanis, J.: On isomorphism and density of NP and other complete sets. SIAM Journal on Computing 6(2), 305–322 (1977)

[BHHT10]   Buhrman, H., Hescott, B., Homer, S., Torenvliet, L.: Non-uniform reductions. Theory of Computing Systems 47(2), 317–341 (2010)

[Coo71]   Cook, S.A.: The complexity of theorem proving procedures. In: Proceedings of the Third ACM Symposium on the Theory of Computing, pp. 151–158 (1971)

[GH92]   Ganesan, K., Homer, S.: Complete problems and strong polynomial re-
           ducibilities. SIAM Journal on Computing 21(4), 733–742 (1992)
[GHP12]   Gu, X., Hitchcock, J.M., Pavan, A.: Collapsing and separating complete
           notions under worst-case and average-case hypotheses. Theory of Comput-
           ing Systems 51(2), 248–265 (2012)
[HP07]   Hitchcock, J.M., Pavan, A.: Comparing reductions to NP-complete sets.
           Information and Computation 205(5), 694–706 (2007)
[IW97]   Impagliazzo, R., Wigderson, A.: P = BPP if E requires exponential cir-
           cuits: Derandomizing the XOR lemma. In: Proceedings of the 29th Sym-
           posium on Theory of Computing, pp. 220–229 (1997)
[Kar72]   Karp, R.M.: Reducibility among combinatorial problems. In: Miller, R.E.,
           Thatcher, J.W. (eds.) Complexity of Computer Computations, pp. 85–104.
           Plenum Press (1972)
[KvM02]   Klivans, A., van Melkebeek, D.: Graph nonisomorphism has subexponen-
           tial size proofs unless the polynomial-time hierarchy collapses. SIAM Jour-
           nal on Computing 31(5), 1501–1526 (2002)

Improved Complexity Results on k-Coloring P_t-Free Graphs

Shenwei Huang

School of Computing Science
Simon Fraser University, Burnaby B.C., V5A 1S6, Canada
shenweih@sfu.ca

Abstract. A graph is H-free if it does not contain an induced subgraph isomorphic to H. We denote by P_k the path on k vertices. In this paper, we prove that 4-COLORING is NP-complete for P_7-free graphs, and that 5-COLORING is NP-complete for P_6-free graphs. The second result is the first NP-completeness shown for any k-COLORING of P_6-free graphs. These two results improve two previously best results and almost complete the classification of complexity of k-COLORING P_t-free graphs for $k \geq 4$ and $t \geq 1$, leaving as the only missing case 4-COLORING P_6-free graphs. Our NP-completeness results use a general framework, which we show is not sufficient to prove the NP-completeness of 4-COLORING P_6-free graphs. We expect that 4-COLORING is polynomial solvable for P_6-free graphs.

1 Introduction

We consider computational complexity issues related to vertex coloring problems restricted to P_k-free graphs. It is well known that the usual k-COLORING problem is NP-complete for any fixed $k \geq 3$. Therefore, there has been considerable interest in studying its complexity when restricted to certain graph classes. One of the most remarkable results in this respect is that k-COLORING is polynomially solvable for perfect graphs. More information on this classical result and related work on coloring problems restricted to graph classes can be found in several surveys, e.g, [24, 25].

We continue the study of k-COLORING problem for P_t-free graphs. This problem has been given wide attention in recent years and much progress has been made through substantial efforts by different groups of researchers [4–6, 9, 13, 16, 19–21, 23, 26]. We summarize these results and explain our new results below.

We refer to [3] for standard graph theory terminology and [11] for terminology on computational complexity. Let $G = (V, E)$ be a graph and $\mathcal{H}$ be a set of graphs. We say that G is $\mathcal{H}$-free if G does not contain any graph $H \in \mathcal{H}$ as an induced subgraph. In particular, if $\mathcal{H} = \{H\}$ or $\mathcal{H} = \{H_1, H_2\}$, we simply say that G is H-free or (H_1, H_2)-free. Given any positive integer t, let P_t and C_t be the path and cycle on t vertices, respectively. A *linear forest* is a disjoint union of paths. We denote by $G + H$ the disjoint union of two graphs G and H.

K. Chatterjee and J. Sgall (Eds.): MFCS 2013, LNCS 8087, pp. 551–558, 2013.

We denote the complement of G by $\bar{G}$. The neighborhood of a vertex x in G is denoted by $N_G(x)$, or simply $N(x)$ if the context is clear. The *girth* of a graph G is the length of the shortest cycle.

A *k-coloring* of a graph $G = (V, E)$ is a mapping $\phi : V \rightarrow \{1, 2, \ldots, k\}$ such that $\phi(u) \neq \phi(v)$ whenever $uv \in E$. The value $\phi(u)$ is usually referred to as the *color* of u under ϕ. We say G is *k-colorable* if G has a k-coloring. The problem k-COLORING asks if an input graph admits an k-coloring. The k-LIST COLORING problem asks if an input graph G with lists $L(v) \subseteq \{1, 2, \ldots, k\}$, $v \in V(G)$, has a coloring ϕ that *respects* the lists, i.e., $\phi(v) \in L(v)$ for each $v \in V(G)$.

In the *pre-coloring extension of k-coloring* we assume that (a possible empty) subset $W \subseteq V$ of G is pre-colored with $\phi_W : W \rightarrow \{1, 2, \ldots, k\}$ and the question is whether we can extend ϕ_W to a k-coloring of G. We denote the problem of pre-coloring extension of k-coloring by k-PrExt . Note that k-COLORING is a special case of k-PrExt, which in turn is a special case of k-LIST COLORING.

Kamiński and Lozin [19] showed that, for any fixed $k \geq 3$, the k-COLORING problem is NP-complete for the class of graphs of girth at least g for any fixed $g \geq 3$. Their result has the following immediate consequence.

Theorem 1 ([19]). *For any $k \geq 3$, the k-COLORING problem is NP-complete for the class of H-free graphs whenever H contains a cycle.*

Holyer [17] showed that 3-COLORING is NP-complete for line graphs. Later, Leven and Galil [22] extended this result by showing that k-COLORING is also NP-complete for line graphs for $k \geq 4$. Because line graphs are claw-free, these two results together have the following consequence.

Theorem 2 ([17, 22]). *For any $k \geq 3$, the k-COLORING problem is NP-complete for the class of H-free graphs whenever H is a forest with a vertex of degree at least 3.*

Due to Theorems 1 and 2, only the case in which H is a linear forest remains. In this paper we focus on the case where H is a path. The k-COLORING problem is trivial for P_t-free graphs when $t \leq 3$. The first non-trivial case is P_4-free graphs. It is well known that P_4-free graphs (also called *cographs*) are perfect and therefore can be colored optimally in polynomial time by Grötschel et al. [15]. Alternatively, one can color cographs using the *cotree representation* of a cograph, see, e.g., [24]. Hoàng et al. [16] developed an elegant recursive algorithm showing that the k-COLORING problem can be solved in polynomial time for P_5-free graphs for any fixed k.

Woeginger and Sgall [26] proved that 5-COLORING is NP-complete for P_8-free graphs and 4-COLORING is NP-complete for P_{12}-free graphs. Later, Le et al. [21] proved that 4-COLORING is NP-complete for P_9-free graphs. The sharpest results so far are due to Broersma et al. [4, 6].

Theorem 3 ([6]). *4-COLORING is NP-complete for P_8-free graphs and 4-PrExt is NP-complete for P_7-free graphs.*

Theorem 4 ([4]). *6-COLORING is NP-complete for P_7-free graphs and 5-PrExt is NP-complete for P_6-free graphs.*

In this paper we strengthen these NP-completeness results. We prove that 5-COLORING is NP-complete for P_6-free graphs and that 4-COLORING is NP-complete for P_7-free graphs. We shall develop a novel general framework of reduction and prove both results simultaneously in Section 2. This leaves the k-COLORING problem for P_t-free graphs unsolved only for $k = 4$ and $t = 6$, except for 3-COLORING. (The complexity status of 3-COLORING P_t-free graphs for $t \geq 7$ is open. It is even unknown whether there exists a fixed integer $t \geq 7$ such that 3-COLORING P_t-free graphs is NP-complete.) We shall explain that our use of the reduction framework is tight in the sense that the framework is not sufficient to prove the NP-completeness of 4-COLORING for P_6-free graphs. Finally, we give some related remarks in Section 3.

2   The NP-Completeness Results

We begin this section by pointing out an error in the proof of NP-completeness of 6-COLORING P_7-free graphs [4]. In this paragraph we follow the notation of Broersma et al. [4]. They used a reduction from 3-SAT to the problem of 6-COLORING for P_7-free graphs. In [4], the authors constructed a graph G_I for an arbitrary instance I of 3-SAT in such a way that I is satisfiable if and only if G_I is 6-colorable. Furthermore, they claimed that G_I is P_7-free. Unfortunately, the last claim is not true in general. Here is one counterexample. Suppose I is an instance of 3-SAT which contains only one clause $C_1 = x_1 \vee \bar{x}_2 \vee x_3$. Then $\bar{x}_1 y_1 b_{11} d_1 b_{13} y_3 \bar{x}_3$ is an induced P_7 in the graph G_I from [4].

Next we shall prove our main results.

Theorem 5. *5-COLORING is NP-complete for P_6-free graphs.*

Theorem 6. *4-COLORING is NP-complete for P_7-free graphs.*

Instead of giving two independent proofs for Theorems 5 and 6, we provide a unified framework. The *chromatic number* of a graph G, denoted by $\chi(G)$, is the minimum positive integer k such that G is k-colorable. The *clique number* of a graph G, denoted by $\omega(G)$, is the maximum size of a clique in G. A graph G is called k-*critical* if $\chi(G) = k$ and $\chi(G - v) < k$ for any vertex v in G. We call a k-critical graph G *nice* if G contains three independent vertices $\{c_1, c_2, c_3\}$ such that $\omega(G - \{c_1, c_2, c_3\}) = \omega(G) = k - 1$. We point out that nice critical graphs do exist. For instance, any odd cycle of length at least 7 with any its three independent vertices is a nice 3-critical graph.

Let I be any 3-SAT instance with variables $X = \{x_1, x_2, \ldots, x_n\}$ and clauses $\mathcal{C} = \{C_1, C_2, \ldots, C_m\}$, and let H be a nice k-critical graph with three specified independent vertices $\{c_1, c_2, c_3\}$. We construct the graph G_I as follows.

- Introduce for each variable x_i a *variable component* T_i which is isomorphic to K_2, labeled by $x_i \bar{x}_i$. Call these vertices X-*type*.

- Introduce for each variable x_i a vertex d_i. Call these vertices D-*type*.
- Introduce for each clause $C_j = y_{i_1} \vee y_{i_2} \vee y_{i_3}$ a *clause component* H_j which is isomorphic to H, where y_{i_t} is either x_{i_t} or $\bar{x}_{i_t}$. Denoted three specified independent vertices in H_j by $c_{i_t j}$ for $t = 1, 2, 3$. Call $c_{i_t j}$ C-*type* and all remaining vertices U-*type*.

For any C-type vertex c_{ij} we call x_i or $\bar{x}_i$ its *corresponding literal vertex*, depending on whether $x_i \in C_j$ or $\bar{x}_i \in C_j$.

- Connect each U-type vertex to each D-type and each X-type vertices.
- Connect each C-type vertex c_{ij} to d_i and its corresponding literal vertex.

Lemma 1. *Let H be a nice k-critical graph. Suppose G_I is the graph constructed from H and a 3-SAT instance I. Then I is satisfiable if and only if G_I is $(k+1)$-colorable.*

Proof. We first assume that I is satisfiable and let σ be a truth assignment satisfying each clause C_j. Then we define a mapping $\phi : V(G) \to \{1, 2, \ldots, k+1\}$ as follows.

- Let $\phi(d_i) := k + 1$ for each i.
- If $\sigma(x_i)$ is TRUE, then $\phi(x_i) := k + 1$ and $\phi(\bar{x}_i) := k$. Otherwise, let $\phi(x_i) =: k$ and $\phi(\bar{x}_i) =: k + 1$.
- Let $C_j = y_{i_1} \vee y_{i_2} \vee y_{i_3}$ be any clause in I. Since σ satisfies C_j, at least one literal in C_j, say y_{i_t} ($t \in \{1, 2, 3\}$), is TRUE. Then the corresponding literal vertex of $c_{i_t j}$ receives the same color as d_{i_t}. Therefore, we are allowed to color $c_{i_t j}$ with color k. In other words, we let $\phi(c_{i_t j}) := k$.
- Since $H_j = H$ is k-critical, $H_j - c_{i_t j}$ has a $(k-1)$-coloring $\phi_j : V(H_j - c_{i_t j}) \to \{1, 2, \ldots, k - 1\}$. Let $\phi =: \phi_j$ on $H_j - c_{i_t j}$.

It is easy to check that ϕ is indeed a $(k + 1)$-coloring of G_I.

Conversely, suppose ϕ is a $(k + 1)$-coloring of G_I. Since $H_1 = H$ is a nice k-critical graph, the largest clique of U-type vertices in H_1 has size $k - 1$. Let R_1 be such a clique. Note that $\omega(G_I) = k+1$ and $R = R_1 \cup T_1$ is a clique of size $k+1$. Therefore, any two vertices in R receive different colors in any $(k + 1)$-coloring of G_I. Without loss of generality, we may assume $\{\phi(x_1), \phi(\bar{x}_1)\} = \{k, k + 1\}$. Because every U-type vertex is adjacent to every X-type and D-type vertex, we have the following three properties of ϕ.

(P1) $\{\phi(x_i), \phi(\bar{x}_i)\} = \{k, k + 1\}$ for each i.
(P2) $\phi(d_i) \in \{k, k + 1\}$ for each i.
(P3) $\phi(u) \in \{1, 2, \ldots, k - 1\}$ for each U-type vertex.

Next we construct a truth assignment σ as follows.

- Set $\sigma(x_i)$ to be TRUE if $\phi(x_i) = \phi(d_i)$ and FALSE otherwise.

It follows from (P1) and (P2) that σ is a truth assignment. Suppose σ does not satisfy $C_j = y_{i_1} \vee y_{i_2} \vee y_{i_3}$. Equivalently, $\sigma(y_{i_t})$ is FALSE for each $t = 1, 2, 3$. It follows from our definition of σ that the corresponding literal vertex of $c_{i_t j}$ receives a different color from the color of d_{i_t} under ϕ. Hence, $\phi(c_{i_t j}) \notin \{k, k+1\}$ for $t = 1, 2, 3$ and this implies that ϕ is a $(k - 1)$-coloring of $H_j = H$ by (P3). This contradicts the fact that $\chi(H) = k$. $\square$

Lemma 2. *Let H be a nice k-critical graph. Suppose G_I is the graph constructed from H and a 3-SAT instance I. If H is P_t-free where $t \geq 6$, then G_I is P_t-free as well.*

Proof. Suppose $P = P_t$ is an induced path with $t \geq 6$ in G_I. We first prove the following claim.

Claim A. *P contains no U-type vertex.*

Proof of Claim A. Suppose that u is a U-type vertex on P that lies in some clause component H_j. For any vertex x on P we denote by x^- and x^+ the left and right neighbor of x on P, respectively. Let us first consider the case when u is the left endvertex of P. If u^+ belongs to H_j, then $P \subseteq H_j$, since u is adjacent to all X-type and D-type vertices and P is induced. This contradicts the fact that H is P_t-free. Hence, u^+ is either X-type or D-type. Note that u^{++} must be C-type or U-type. In the former case we conclude that u^{+++} is U-type since C-type vertices are independent. Hence, $|P| \leq 3$ and this is a contradiction. In the latter case we have $|P| \leq 4$ for the same reason. Note that $|P| = 4$ only if P follows the pattern $U(X \cup D)UC$, namely the first vertex of P is U-type, the second vertex of P is X-type or D-type, and so on. Next we consider the case that u has two neighbors on P.

Case 1. Both u^- and u^+ belong to H_j. In this case $P \subseteq H_j$ and this contradicts the fact that $H = H_j$ is P_t-free.

Case 2. $u^- \in H_j$ but $u^+ \notin H_j$. Then u^+ is either X-type or D-type. Since each U-type vertex is adjacent to each X-type and D-type vertex, u^- is a C-type vertex and hence it is an endvertex of P. Now $|P| \leq 2 + 4 - 1 = 5$.

Case 3. Neither u^- nor u^+ belongs to H_j. Now both u^- and u^+ are X-type or D-type. Since each U-type vertex is adjacent to each X-type and D-type vertex, $P \cap U = \{u\}$ and $P \cap (X \cup D) = \{u^-, u^+\}$. Hence, $|P| \leq 5$. ($|P| = 5$ only if P follows the pattern $C(X \cup D)U(X \cup D)C$). □

Let C_i (resp. $\bar{C}_i$) be the set of C-type vertices that connect to x_i (resp. $\bar{x}_i$). Let $G_i = G[\{T_i \cup \{d_i\} \cup C_i \cup \bar{C}_i\}]$. Note that $G - U$ is disjoint union of G_i, $i = 1, 2, \ldots, n$. By Claim A, $P \subseteq G_i$ for some i. Let P' be a sub-path of P of order 6 . Since $C_i \cup \bar{C}_i$ is independent, $|P' \cap (C_i \cup \bar{C}_i)| \leq 3$. Hence, $|P' \cap (C_i \cup \bar{C}_i)| = 3$ and thus $\{d_i, x_i, \bar{x}_i\} \subseteq P'$. This contradicts the fact that P' is induced since d_i has three C-type neighbors on P'. □

Due to Lemmas 1 and 2, the following theorem follows.

Theorem 7. *Let $t \geq 6$ be an fixed integer. Then k-COLORING is NP-complete for P_t-free graphs whenever there exists a P_t-free nice $(k-1)$-critical graph.* □

Proof of Theorems 5 and 6. Let H_1 be the graph shown in Fig. 1 and let $H_2 = C_7$ be the 7-cycle. It is easy to check that H_1 is a P_6-free nice 4-critical graph and that H_2 is a P_7-free nice 3-critical graph. Applying Theorem 7 with $H = H_i$ $(i = 1, 2)$ will complete our proof. □

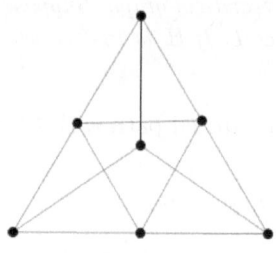

Fig. 1. H_1

We remark that Theorem 7 is not sufficient to prove the NP-completeness of 4-COLORING P_6-free graphs. In fact, there is no P_6-free nice 3-critical graph. Suppose H is P_6-free nice 3-critical graphs with $\{c_1, c_2, c_3\}$ being independent. Since the only 3-critical graphs are odd cycles and H is P_6-free, H must be C_5. But this contradicts the fact that C_5 contains at most two independent vertices.

3   Concluding Remarks

We have proved that 4-COLORING is NP-complete for P_7-free graphs, and that 5-COLORING is NP-complete for P_6-free graphs. These two results improve Theorems 3 and 4 obtained by Broersma et al. [4, 6]. We have used a reduction from 3-SAT and establish a general framework. The construction and the proof are simpler than those in previous papers. As pointed out above, however, they do not apply to 4-COLORING P_6-free graphs.

For graph H with at most five vertices, Golovach et al. [12] completed the dichotomy classification for 4-COLORING H-free graphs. The classification states that 4-COLORING is polynomially solvable for H-free graphs when H is a linear forest and is NP-complete otherwise. Note that linear forests on at most five vertices are all induced subgraph of P_6. Thus, all the polynomial cases from [12] are for subclasses of P_6-free graphs.

These results suggest the following conjecture.

Conjecture 1. 4-COLORING can be solved in polynomial time for P_6-free graphs.

Golovach et al. [14] showed that 4-LIST COLORING is NP-complete for P_6-free graphs. So far, there is no jump in complexity known for any value of t for the problems k-COLORING and its variants k-PrExt and k-LIST COLORING restricted to P_t-free graphs. If Conjecture 1 were true, then this would give a different picture.

In [18] we also prove that 4-COLORING (P_6, C_4)-free graphs is solvable in polynomial time. This also follows from a more general result of Golovach, Pauslusma and Song who proved that for all fixed integers k, t, r, s, the k-COLORING problem can be solved in polynomial time on $(K_{r,s}, P_t)$-free graphs. This result will appear in the journal version of the conference paper Golovach et al. [13].

Its proof follows from a recent result of Atminas et al. [1] who proved that for any given integer r and t there exists an integer $q(r,t)$ such that if a graph G has treewidth at least $q(r,t)$, then it either contains an induced P_t or a $K_{r,r}$ as a (not necessarily induced) subgraph. By an algorithm of Bodlaender [2] one can test in linear time if the treewidth of a graph G is at most $q(r,t)$. If so, we can solve k-COLORING by using a theorem of Courcelle et al. [8]. Otherwise, assuming that G is $(K_{r,s}, P_t)$-free, G must have a large balanced complete bipartite graph B as a subgraph. Applying Ramsey theorem on both partition classes of B then yields that G is not k-colorable. This produces a linear time algorithm for k-COLORING $(K_{r,s}, P_t)$-free graphs; however the constants due to tree decompositions and Ramsey's theorem are huge, and our algorithm from [18] which runs in $O(n^5)$ time may be more practical for up to a fairly large input size n.

This also suggests a new research direction, namely classifying the complexity of k-COLORING (P_t, C_l)-free graphs for every integer combination of k, l and t. Since k-COLORING is NP-complete for P_t-free graphs for even small k and t, say Theorems 5 and 6, it would be nice to know whether or not forbidding short induced cycles makes problem easier. As pointed out above, forbidding C_4 does make problem easier. In contrast, one recent result of Golovach et al. [13] showed that 4-COLORING is NP-complete for (P_{164}, C_3)-free graphs.

Acknowledgement. The author is grateful to his supervisor Pavol Hell for many useful conversations related to these results and helpful suggestions for improving the presentation of the paper. The author is also grateful to Daniel Paulusma for communicating the result from the journal version of Golovach et al. [13].

References

1. Atminas, A., Lozin, V.V., Razgon, I.: Linear time algorithm for computing a small biclique in graphs without long induced paths. In: Fomin, F.V., Kaski, P. (eds.) SWAT 2012. LNCS, vol. 7357, pp. 142–152. Springer, Heidelberg (2012)
2. Bodlaender, H.L.: On linear time minor tests with depth-first search. J. Algorithms 14(1), 1–23 (1993)
3. Bondy, J.A., Murty, U.S.R.: Graph Theory. Springer Graduate Texts in Mathematics, vol. 244 (2008)
4. Broersma, H.J., Fomin, F.V., Golovach, P.A., Paulusma, D.: Three complexity results on coloring P_k-free graphs. European Journal of Combinatorics (2012) (in press)
5. Broersma, H.J., Golovach, P.A., Paulusma, D., Song, J.: Determining the chromatic number of triangle-free $2P_3$-free graphs in polynomial time. Theoret. Comput. Sci. 423, 1–10 (2012)
6. Broersma, H.J., Golovach, P.A., Paulusma, D., Song, J.: Updating the complexity status of coloring graphs without a fixed induced learn forest. Theoret. Comput. Sci. 414, 9–19 (2012)
7. Chudnovsky, M., Cornuéjols, G., Liu, X., Seymour, P., Vušković, K.: Recognizing Berge graphs. Combinatorica 25, 143–187 (2005)

8. Courcelle, B., Makowsky, J.A., Rotics, U.: Linear Time Solvable Optimization Problems on Graphs of Bounded Clique Width. In: Hromkovič, J., Sýkora, O. (eds.) WG 1998. LNCS, vol. 1517, pp. 1–16. Springer, Heidelberg (1998)

9. Dabrowski, K., Lozin, V.V., Raman, R., Ries, B.: Colouring vertices of triangle-free graphs without forests. Discrete Math. 312, 1372–1385 (2012)

10. Edwards, K.: The complexity of coloring problems on dense graphs. Theoret. Comput. Sci. 43, 337–343 (1986)

11. Garey, M.R., Johnson, D.S.: Computers and Intractability: A Guide to the Theory of NP-Completeness. Freeman San Faranciso (1979)

12. Golovach, P.A., Paulusma, D., Song, J.: 4-coloring H-free graphs when H is small. Discrete Applied Mathematics (2012) (in press)

13. Golovach, P.A., Paulusma, D., Song, J.: Coloring graphs without short cycles and long induced paths. In: Owe, O., Steffen, M., Telle, J.A. (eds.) FCT 2011. LNCS, vol. 6914, pp. 193–204. Springer, Heidelberg (2011)

14. Golovach, P.A., Paulusma, D., Song, J.: Closing complexity gaps for coloring problems on H-free graphs. In: Chao, K.-M., Hsu, T.-s., Lee, D.-T. (eds.) ISAAC 2012. LNCS, vol. 7676, pp. 14–23. Springer, Heidelberg (2012)

15. Grötschel, M., Lovász, L., Schrijver, A.: Polynomial algorithms for perfect graphs. Ann. Discrete Math. 21, 325–356 (1984), Topics on Perfect Graphs

16. Hoàng, C.T., Kamiński, M., Lozin, V.V., Sawada, J., Shu, X.: Deciding k-colorability of P_5-free graphs in polynomial time. Algorithmica 57, 74–81 (2010)

17. Holyer, I.: The NP-completeness of edge coloring. SIAM J. Comput. 10, 718–720 (1981)

18. Huang, S.H.: Improved complexity results on k-coloring P_t-free graphs. arXiv: 0700726 [cs.CC] (2013)

19. Kamiński, M., Lozin, V.V.: Coloring edges and vertices of graphs without short or long cycles. Contrib. Discrete. Mah. 2, 61–66 (2007)

20. Král, D., Kratochvíl, J., Tuza, Z., Woeginger, G.J.: Complexity of coloring graphs without forbidden induced subgraphs. In: Brandstädt, A., Le, V.B. (eds.) WG 2001. LNCS, vol. 2204, pp. 254–262. Springer, Heidelberg (2001)

21. Le, V.B., Randerath, B., Schiermeyer, I.: On the complexity of 4-coloring graphs without long induced paths. Theoret. Comput. Sci. 389, 330–335 (2007)

22. Leven, D., Galil, Z.: NP-completeness of finding the chromatic index of regular graphs. J. Algorithm 4, 35–44 (1983)

23. Randerath, B., Schiermeyer, I.: 3-Colorability $\in$ P for P_6-free graphs. Discrete Appl. Math. 136, 299–313 (2004)

24. Randerath, B., Schiermeyer, I.: Vertex colouring and fibidden subgraphs-a survey. Graphs Combin. 20, 1–40 (2004)

25. Tuza, Z.: Graph colorings with local restrictions-a survey. Discuss. Math. Graph Theory 17, 161–228 (1997)

26. Woeginger, G.J., Sgall, J.: The complexity of coloring graphs without long induced paths. Acta Cybernet. 15, 107–117 (2001)

A Polychromatic Ramsey Theory for Ordinals

Martin Huschenbett[1] and Jiamou Liu[2]

[1] Institut für Theoretische Informatik
Technische Universität Ilmenau, Germany
martin.huschenbett@tu-ilmenau.de
[2] School of Computing and Mathematical Sciences
Auckland University of Technology, New Zealand
jiamou.liu@aut.ac.nz

Abstract The Ramsey degree of an ordinal α is the least number n such that any colouring of the edges of the complete graph on α using finitely many colours contains an n-chromatic clique of order type α. The Ramsey degree exists for any ordinal $\alpha < \omega^\omega$. We provide an explicit expression for computing the Ramsey degree given α. We further establish a version of this result for automatic structures. In this version the ordinal and the colouring are presentable by finite automata and the clique is additionally required to be regular. The corresponding automatic Ramsey degree turns out to be greater than the set theoretic Ramsey degree. Finally, we demonstrate that a version for computable structures fails.

1 Introduction

The (countably) infinite Ramsey's theorem states that any edge colouring of a countably infinite complete graph admits a complete monochromatic infinite clique. If we arrange the nodes in this graph into a well-ordering of order type ω, Ramsey's theorem guarantees the existence of a subordering of order type ω such that all pairs of its elements have the same colour. More specifically, by a *standard partition* of a set A we mean a partition of all 2-element subsets (or edges) of A into a finite number k of classes, where $k \geq 1$. A *homogeneous set* with respect to a standard partition of A is a subset $B \subseteq A$ such that all edges of B belong to one class of the partition. If α and β are ordinals, one writes $\alpha \to (\beta)$ for the fact that whenever $(A; \leq)$ has order type α, any standard partition of A admits a homogeneous subset B such that the suborder $(B; \leq \restriction B)$ has order type β. Ramsey's theorem can thus be stated as $\omega \to (\omega)$.

A question then arises as to whether one can extend the above statement to larger ordinals. Erdős and Rado gave a negative answer to this question for countable ordinals: For any countable well-ordering L there is a partition of edges of L such that any infinite homogeneous subset of L has order type ω [6]. Hence for any countable ordinal α, $\alpha \nrightarrow (\omega + 1)$. This result is the start of a vast amount of works on the partition relations of ordinals, which has become a central notion of combinatorial set theory [5].

K. Chatterjee and J. Sgall (Eds.): MFCS 2013, LNCS 8087, pp. 559–570, 2013.
© Springer-Verlag Berlin Heidelberg 2013

Since the 1970s, there has been another well-established extension to the Ramsey's theorem. The goal is to investigate the effective content of the homogeneous sets in computable standard partitions. Recall that a structure is computable if its domain as well as its atomic functions and predicates are decidable by Turing machines. Specker showed that the original statement in Ramsey's theorem cannot be made effective: there exists a computable standard partition of a computable copy of ω such that no infinite homogeneous set is computable [17]. Jockusch then showed that the infinite homogeneous sets of standard partitions of ω do not even necessarily belong to Σ_2. On the other hand, infinite homogeneous sets are guaranteed to exist in Π_2 [8].

More recently, attention has been given to automatic structures. These are structures that are defined in a similar way as computable structures except the "Turing machines" in the definition is replaced by "finite automata". Hence automatic structures form a subclass of computable structures. A main line of research in the study of automatic structures is to understand automaticity in classical theorems. Here, as opposed to computable structures for which numerous classical results (such as König's lemma and Ramsey's theorem) fail in the computable case, the automatic counterparts of these theorems hold. For example, Rubin proved that in any automatic standard partition of an infinite regular language there exists necessarily homogeneous sets that are recognisable by finite automata [15]. This result suggests that it makes sense to build a Ramsey theory on automatic ordinals.

We mention here that a standard partition can be viewed as a *colouring* function that maps the set of edges to a finite number of colours. The homogeneous sets mentioned above are thus "monochromatic". In this paper we consider colourings that consist of more than two colours and "ℓ-chromatic" subsets for some bounded number ℓ, that is, subsets whose edges are coloured by no more than ℓ colours. It is natural to ask the following: Let α be an ordinal.

1. Is there a number $\ell \in \mathbb{N}$ such that any edge colouring of α contains an ℓ-chromatic subset of order type α?
2. If such a number ℓ in the above question exists, how large must it be?

We call the least number ℓ that satisfies the first question the *Ramsey degree* of the ordinal α. Williams in [19, Theorem 7.2.7] showed that the Ramsey degree exists for any ordinal ω^n where $n \in \mathbb{N}$. Here we further provide a formula for computing the Ramsey degree of an arbitrary $\alpha < \omega^\omega$; see Theorem 3.3.

We then explore the same questions as above restricting to copies of ordinals and colourings that are finite-automata presentable, and regular ℓ-chromatic sets. For any ordinal $\alpha < \omega^\omega$, we show that the corresponding automatic Ramsey degree exists for α and give an explicit expression for computing it. The automatic Ramsey degree of α turns out to be strictly greater than its Ramsey degree if $\omega^2 \leq \alpha < \omega^\omega$. A by-product of our investigation is a similar result on automatic complete bipartite graphs, where each bipartition is an ordinal. Finally we briefly present a negative answer to the computable version of the above questions: For any $k \geq 1$, there is a computable edge colouring of the natural numbers using $k + 1$ colours that does not admit any infinite k-chromatic computably enumerable subsets.

Related works. The notion of Ramsey degrees used here has appeared in different forms in the literature. The paper [18] contains several results discussing similar notions. The result that motivated our study is F. Galvin's unpublished theorem on rationals: For any edge colouring of η, the order type of rationals, there must be a 2-chromatic sub-copy of η [7]. Pouzet and Sauer obtained a very similar result on the random graphs [12]. See [11] for an introduction on the automatic version of Ramsey's theorem.

Paper organisation. Section 2 introduces necessary background in Ramsey theory and automatic structure. Section 3 and Section 4 discusses Ramsey degrees in the general case and the automatic case respectively. Section 5 presents the computable case. Finally Section 6 discusses open problems.

2 Preliminaries

Throughout the whole paper, $\mathbb{N}$ denotes the natural numbers $1, 2, 3, \ldots$ and $\mathbb{N}_0$ denotes $\mathbb{N} \cup \{0\}$. We use the interval notation $[i, j]$ for the set $\{i, i+1, \ldots, j\}$.

Well-orderings and ordinals. A well-ordering is a linear ordering $(V; \leq_V)$ with no infinite descending chains. For details on basic notions and results on well-orderings and ordinals the reader is referred to [14]. We view sets also as well-ordered sets, i.e., a set V also denotes a well-ordering $(V, \leq_V)$. By "V has order type α" we mean "$(V; \leq_V)$ has order type α". By $U + V$ we mean the sum of the well-orderings $(U; \leq_U) + (V; \leq_V)$. If $U \subseteq V$ then we assume the ordering on U is the same as the ordering on V restricted to U.

Let $n \geq 0$. We view $\mathbb{N}^n$ as a well-ordered set using the order defined by $(x_0, \ldots, x_{n-1}) <_{\mathbb{N}^n} (y_0, \ldots, y_{n-1})$ if there is an $i \in [0, n-1]$ with $x_i \neq y_i$ and the least such i satisfies $x_i < y_i$. Then $\mathbb{N}^n$ has order type ω^n and is regarded as the *canonical representation* of ω^n. As we consider no other orders on $\mathbb{N}^n$ besides $\leq_{\mathbb{N}^n}$, we usually omit the subscript $\mathbb{N}^n$ from $\leq_{\mathbb{N}^n}$.

It is well-known that any ordinal $\alpha < \omega^\omega$ can be uniquely written in its *Cantor normal form* $\alpha = \omega^{n_1} + \omega^{n_2} + \cdots + \omega^{n_r}$ with $r \geq 0$ and $n_1 \geq n_2 \geq \cdots \geq n_r \geq 0$.

Finite automata and semigroups. We assume some familiarity with the basic concepts of (algebraic) automata theory (cf. [4]). Let Σ be an alphabet. (Nondeterministic) finite automata (over Σ) and their languages are defined as usual.

A *semigroup* is a set S equipped with an associative binary multiplication. Examples include the set Σ^* with concatenation and the direct product of finitely many semigroups. A *(semigroup) morphism* is a map between two semigroups which preserves multiplication. The Myhill-Nerode theorem states that a language $L \subseteq \Sigma^*$ is regular if, and only if, there is a morphism $h \colon \Sigma^* \to S$ into a finite semigroup S which *recognises* L, i.e., $L = h^{-1}(T)$ for some $T \subseteq S$. This theorem is effective in both directions, i.e., one can compute a morphism recognising L from a finite automaton recognising L and vice versa. For any finite number $L_1, \ldots, L_n \subseteq \Sigma^*$ of regular languages there exists a morphism into a finite semigroup which simultaneously recognises all the L_i.

An element $s \in S$ of a semigroup S is *idempotent* if $s^2 = s$. An *idempotency exponent* of S is a number $K \geq 1$ such that s^K is idempotent for all $s \in S$. Whenever S is finite, any multiple of $|S|!$ is an idempotency exponent of S.

Automatic structures. To recognise n-ary relations on Σ^*, we use finite automata which synchronously process n input tapes in parallel. Formally, let $\diamond \notin \Sigma$ be an additional padding symbol and $\Sigma_\diamond = \Sigma \cup \{\diamond\}$. The *convolution* of a tuple $\bar{u} = (u_0, \ldots, u_{n-1}) \in (\Sigma^*)^n$ is the word $\otimes \bar{u} \in (\Sigma_\diamond^n)^*$ of length $\max\{|u_0|, \ldots, |u_{n-1}|\}$ whose k^{th} symbol is $(\sigma_0, \ldots, \sigma_{n-1})$, where σ_i is the kth symbol of u_i if $k \leq |u_i|$, and $\diamond$ otherwise. An n-ary relation $R \subseteq (\Sigma^*)^n$ is *automatic* if its *convolution* $\otimes R = \{ \otimes \bar{u} \mid \bar{u} \in R \}$ is a regular subset of $(\Sigma_\diamond^n)^*$.

A *relational structure* $\mathcal{A} = (A; R_1, \ldots, R_k)$ consists of a set A, its *domain*, and relations $R_1, \ldots, R_k$ on A. A structure $\mathcal{A}$ is *automatic* if A is a regular language (over some alphabet Σ) and the relations R_i are automatic. In this situation, an *automatic presentation* of $\mathcal{A}$ is a tuple of finite automata recognising A and the $\otimes R_i$, respectively. We denote by AUT the class of all automatic structures, which actually includes all regular languages. The main motivation for investigating automatic structures is the decidability of their first-order theories (cf. [9,1]).

Theorem 2.1 (Khoussainov, Nerode [9]). *Every first-order definable relation R on an automatic structure $\mathcal{A}$ is automatic and one can compute a finite automaton recognising R from an automatic presentation of $\mathcal{A}$ and a first-order formula defining R. In particular, the first-order theory of $\mathcal{A}$ is decidable.*

A well-ordering A is automatic in the sense above if A is a regular language and $\leq_A$ an automatic relation. Automatic well-orderings are a well studied subject.

Theorem 2.2 (Delhommé [3]). *There is an automatic well-ordering of type α if, and only if, $\alpha < \omega^\omega$.*

Theorem 2.3 (Khoussainov et. al [10]). *Given an automatic presentation of a well-ordering, one can compute the Cantor normal form of its order type.*

3   Ramsey Relations and Ramsey Degrees

3.1   Ordinal Ramsey Relation

We use $[V]^2$ to denote the set of all 2-element subsets of a set V. For convenience we view $[V]^2$ as the irreflexive and symmetrical relation $\{ (x, y) \in V^2 \mid x \neq y \}$. It is customary to view standard partitions as colourings, which is the notion we adopt in this paper. Let α be an ordinal. An α-*colouring* is a function $C \colon [V]^2 \to Q$ where V is a set of order type α and Q is a finite set of *colours*. When α is clear from context we simply call C a *colouring (on V)*. Let $X \subseteq V$. We use $C(X)$ to denote the set $\{\, C(A) \mid A \in [X]^2 \,\}$. Let $D \subseteq Q$. The set X is D-*chromatic* w.r.t. C if $C(X) \subseteq D$. In this case, we also say that X is $|D|$-*chromatic*.

Let α, β be two ordinals, $k \in \mathbb{N}$, and $\ell \in \mathbb{N}_0$. The *ordinal Ramsey relation* is written as $\alpha \to (\beta)_{k,\ell}$ and denotes the fact that any α-colouring $C \colon [V]^2 \to [1, k]$ admits an ℓ-chromatic subset $X \subseteq V$ of order type β. We are interested in the Ramsey degrees of ordinals, which is defined below.

Definition 3.1. *Let $\alpha < \omega^\omega$ be an ordinal. The least $\ell \in \mathbb{N}_0$ such that $\alpha \to (\alpha)_{k,\ell}$ for all $k \in \mathbb{N}$ is called the* Ramsey degree *of α and denoted by $d_R(\alpha)$.*

The countably infinite case of Ramsey's theorem states that $d_R(\omega) = 1$ [13]. Williams in his book [19, Theorem 7.2.7] proved the following result, which extends Ramsey's theorem to ordinals ω^n where $n \in \mathbb{N}$.

Theorem 3.2 ([19, Theorem 7.2.7]). *For any $n \in \mathbb{N}$ there is an $\ell \in \mathbb{N}_0$ such that*
$$\forall k \in \mathbb{N} : \omega^n \to (\omega^n)_{k,\ell} .$$

The proof of Theorem 3.2 from [19] does not provide us the value of $d_R(\omega^n)$, which is presented in the following theorem.

Theorem 3.3. *For all ordinals $\alpha < \omega^\omega$, we have*
$$d_R(\alpha) = \sum_{1 \le i \le r} \sum_{1 \le j \le n_i} \binom{2j-1}{j} + \sum_{1 \le i < j \le r} \binom{n_i + n_j}{n_i} ,$$
where $\alpha = \omega^{n_1} + \cdots + \omega^{n_r}$ with $r \ge 0$ and $n_1 \ge \cdots \ge n_r \ge 0$ is the Cantor normal form of α.

Let $\mathcal{C}$ be a class of colourings and $\mathcal{D}$ a class of sets. We write
$$(\alpha : \mathcal{C}) \to (\beta : \mathcal{D})_{k,\ell}$$

if any α-colouring $C \colon [V]^2 \to [1, k]$ in $\mathcal{C}$ admits an ℓ-chromatic set $X \subseteq V$ of order type β such that $X \in \mathcal{D}$. We say that a colouring $C \colon [A]^2 \to Q$ is *automatic* if the well-ordering A is automatic and the relation $C^{-1}(q)$ is automatic for each $q \in Q$. The following is an automatic version of Ramsey's theorem.

Theorem 3.4 (Rubin [15]). *Let $k \in \mathbb{N}$ be a number. We have*
$$(\omega : \mathsf{AUT}) \to (\omega : \mathsf{AUT})_{k,1} .$$

This theorem is effective in the following sense: Given an automatic presentation of an ω-colouring on A, one can compute a finite automaton recognising a monochromatic, regular set $X \subseteq A$ of order type ω. A main goal of the paper is to extend Theorem 3.4 by presenting an automatic version of Theorem 3.3.

3.2   Bipartite Ordinal Ramsey Relation

As part of our investigation we also introduce a bipartite analogue of the Ramsey relation on ordinals. Let α and β be ordinals. A *bipartite (α, β)-colouring* is a function $C \colon U \times V \to Q$ where U and V have respectively order types α and β and Q is a finite set of colours. When α and β are clear we simply call C a *bipartite colouring* (on (U, V)).

Let $C \colon U \times V \to Q$ be a bipartite colouring. We write $(X, Y) \subseteq (U, V)$ to denote the fact that $X \subseteq U$ and $Y \subseteq V$. Let $\ell \in \mathbb{N}_0$. A pair $(X, Y) \subseteq (U, V)$ is *ℓ-chromatic* w.r.t. C if $|C(X \times Y)| \leq \ell$. We say that the pair (X, Y) has *order type* (γ, δ) if X and Y have order type γ and δ, respectively.

Let $\alpha, \beta, \gamma, \delta$ be ordinals, $k \in \mathbb{N}$, and $\ell \in \mathbb{N}_0$. The *bipartite ordinal Ramsey relation* is written as $(\alpha, \beta) \to (\gamma, \delta)_{k,\ell}$ and denotes the fact that any (α, β)-colouring $C \colon U \times V \to [1, k]$ admits an ℓ-chromatic pair $(X, Y) \subseteq (U, V)$ of order type (γ, δ). The finite version of Ramsey theory on complete bipartite graphs has been well studied; see [2] for example. Here we study the bipartite ordinal Ramsey relation when the ordinals involved are ω^n where $n \in \mathbb{N}$. We define bipartite Ramsey degrees as follows.

Definition 3.5. *Let $m, n \geq 0$. The least $\ell \in \mathbb{N}$ such that $(\omega^m, \omega^n) \to (\omega^m, \omega^n)_{k,\ell}$ for all $k \in \mathbb{N}$ is called the Ramsey degree of (ω^m, ω^n) and denoted by $d_R(\omega^m, \omega^n)$.*

The next theorem presents the value of the Ramsey degree in the bipartite case.

Theorem 3.6. *For all $m, n \geq 0$, we have*

$$d_R(\omega^m, \omega^n) = \binom{m+n}{m}.$$

In the following we generalise the above notion to specific classes of bipartite colourings. Let $\mathcal{C}$ be a class of colourings and $\mathcal{D}$ a class of sets. We write

$$(\alpha, \beta : \mathcal{C}) \to (\gamma, \delta : \mathcal{D})_{k,\ell}$$

for the fact that any (α, β)-colouring $C \colon U \times V \to [1, k]$ in $\mathcal{C}$ admits an ℓ-chromatic pair of sets $(X, Y) \subseteq (U, V)$ of order type (γ, δ) and $X, Y \in \mathcal{D}$.

4   The Automatic Case

In this section, we investigate an automatic analogue of the Ramsey degree from the previous section. The highlight is Theorem 4.7 which states that this degree exists for each ordinal $\alpha < \omega^\omega$ and provides a formula to compute its value.

Definition 4.1. *Let $\alpha < \omega^\omega$ be an ordinal. If there exists an $\ell \in \mathbb{N}_0$ such that $(\alpha : \mathsf{AUT}) \to (\alpha : \mathsf{AUT})_{k,\ell}$ for all $k \in \mathbb{N}$, the least such ℓ is called the automatic Ramsey degree of α and denoted by $d_{R,\mathsf{AUT}}(\alpha)$.*

Similarly, we define the *automatic (bipartite) Ramsey degree* $d_{R,\mathsf{AUT}}(\alpha, \beta)$ for ordinals $\alpha, \beta < \omega^\omega$.

4.1   Automatic Well-Orderings of Type ω^n

Our main tool in the investigation of the automatic Ramsey degree is Theorem 4.2 below which roughly states that every automatic well-ordering of type ω^n contains a simple automatic subordering of the same order type.

We call a map $f\colon \mathbb{N}^n \to \Sigma^*$ *presentable* if there are $u_0, \dots, u_{n-1}, u_n \in \Sigma^*$ and $v_0, \dots, v_{n-1} \in \Sigma^+$ such that

$$f(x_0, \dots, x_{n-1}) = u_0 v_0^{x_0 - 1} u_1 v_1^{x_1 - 1} \cdots u_{n-1} v_{n-1}^{x_{n-1}-1} u_n$$

for all $\bar{x} \in \mathbb{N}^n$. The tuple $(u_0, v_0, \dots, u_{n-1}, v_{n-1}, u_n)$ is called *presentation* of f. If there exists a $K \geq 1$ such that $|u_i| = |v_i| = K$ for $0 \leq i < n$ and $|u_n| \leq K$, we say that f is *(K-)uniformly presentable* and speak of a *(K-)uniform presentation*.

Theorem 4.2. *Let $n \geq 0$. For every automatic well-ordering A of type ω^n there exists a uniformly presentable embedding $f\colon \mathbb{N}^n \hookrightarrow A$.*

Proof. Let $A \subseteq \Sigma^*$ be an automatic well-ordering of type ω^n. We first show the existence of a (possibly non-uniformly) presentable embedding $f'\colon \mathbb{N}^n \hookrightarrow A$ by induction on n. The claim is trivial for $n = 0$. Therefore, we assume $n \geq 1$.

We define an equivalence relation $\sim$ on A by $u \sim v$ if there is *no* $(n-1)$-limit point $w \in A$ with $u < w \leq v$. For any $u \in A$ the $\sim$-class $[u]$ of u is an interval of A with order type ω^{n-1}. The set $P = \{\min[u] \mid u \in A\}$ is a system of representatives w.r.t. $\sim$ which has order type ω. Thus, $A = \sum_{u \in P}[u]$ is the unique representation of A as an ω-sum of copies of ω^{n-1}. For each $m \in \mathbb{N}_0$ the set of m-limit points of A is first-order definable in A. Thus, the relation $\sim$ and the set P are also first-order definable in A and hence automatic by Theorem 2.1.

We further define a binary relation R on Σ^* by

$$R = \left\{ (u, v) \in P \times \Sigma^* \mid |u| = |v| \text{ and } [u] \cap v\Sigma^* \text{ has order type } \omega^{n-1} \right\}.$$

Since any finite partition of a well-ordering of type ω^{n-1} contains a part of order type ω^{n-1}, for every $u \in P$ there exists a $v \in \Sigma^*$ such that $(u, v) \in R$. In addition, R is automatic as it is first-order definable in the automatic structure $(\Sigma^*; A, \leq_A, \sim, P, \equiv, \preceq)$, where $\equiv$ and $\preceq$ are the same-length and prefix relations, respectively. Similarly, $[u] \cap v\Sigma^*$ is regular for all $(u, v) \in R$.

Since $\otimes R$ is an infinite regular set and due to a pumping argument, there are words $p, q, r, \tilde{p}, \tilde{q}, \tilde{r} \in \Sigma^*$ with $|p| = |\tilde{p}|$ and $|q| = |\tilde{q}| > |r| = |\tilde{r}|$ such that $(pq^x r, \tilde{p}\tilde{q}^x \tilde{r}) \in R$ for each $x \geq 0$. Let η be a morphism into a finite semigroup S which simultaneously recognises $\leq_A$ and $\sim$. Pick an idempotency exponent $M \geq 1$ of S. We define presentable maps $g\colon \mathbb{N} \to P$ and $\tilde{g}\colon \mathbb{N} \to \Sigma^*$ by

$$g(x) = pq^{M \cdot (2x-1)} r \qquad \text{and} \qquad \tilde{g}(x) = \tilde{p}\tilde{q}^{M \cdot (2x-1)} \tilde{r}.$$

Using the idempotency property of M, we obtain $\eta(g(x) \otimes g(y)) = \eta(g(1) \otimes g(2))$ for all $x, y \in \mathbb{N}$ with $x < y$. This implies that g is an embedding $g\colon \mathbb{N} \hookrightarrow P$.

For every $x \in \mathbb{N}$ the regular set $B_x = [g(x)] \cap \tilde{g}(x)\Sigma^* \subseteq A$ has order type ω^{n-1}. We turn the regular set $Z \subseteq \Sigma^*$ with $B_1 = \tilde{g}(1)Z$ into an automatic well-ordering of type ω^{n-1} by defining $u \leq_Z v$ if $\tilde{g}(1)u \leq_A \tilde{g}(1)v$. Using the idempotency property of M once more yields that for each $x \in \mathbb{N}$ the map $i_x \colon Z \to B_x$ with $i_x(u) = \tilde{g}(x)u$ is an isomorphism between well-orderings.

By the induction hypothesis, there is a presentable embedding $h \colon \mathbb{N}^{n-1} \hookrightarrow Z$. The map $f' \colon \mathbb{N}^n \to \Sigma^*$ defined by

$$f'(x_0, \ldots, x_{n-1}) = \tilde{g}(x_0)h(x_1, \ldots, x_{n-1})$$

is a presentable embedding $f' \colon \mathbb{N}^n \hookrightarrow A$. This completes the induction.

Finally, let $(u_0, v_0, \ldots, u_{n-1}, v_{n-1}, u_n)$ be a presentation of f'. Pick a $K \geq 1$ which is divisible by each $|v_i|$, say $K = K_i \cdot |v_i|$, and satisfies $K \geq |u_0| + \cdots + |u_n|$. Then the map $f \colon \mathbb{N}^n \to A$ defined by

$$f(x_0, \ldots, x_{n-1}) = f'(K_0 x_0 + 1, \ldots, K_{n-1} x_{n-1} + 1)$$

can be shown to be a K-uniformly presentable embedding $f \colon \mathbb{N}^n \hookrightarrow A$. $\square$

4.2   The Automatic Ramsey Degree of ω^n

In this section, we apply Theorem 4.2 to determine the exact value of $d_{R,\mathsf{AUT}}(\omega^n)$ for each $n \geq 0$. In order to expresses these values, we need the following variation of binomial coefficients.

Definition 4.3. *For all $n, k \in \mathbb{N}_0$ with $k \leq n$ we define $\left\langle {n \atop k} \right\rangle \in \mathbb{N}$ as follows:*

(1) $\left\langle {n \atop k} \right\rangle = 1$ *if $k = 0$ or $k = n$,*
(2) $\left\langle {n \atop k} \right\rangle = \left\langle {n-1 \atop k-1} \right\rangle + \left\langle {n-1 \atop k} \right\rangle$ *if $0 < k < n$ and $2k \neq n$, and*
(3) $\left\langle {n \atop k} \right\rangle = \left\langle {n-1 \atop k-1} \right\rangle + \left\langle {n-1 \atop k} \right\rangle + \left\langle {n-2 \atop k-1} \right\rangle$ *if $0 < k < n$ and $2k = n$.*

Notice that $\binom{n}{k} \leq \left\langle {n \atop k} \right\rangle$ for all $k \leq n$. This inequality is strict whenever $0 < k < n$.

For the rest of this section, we fix some $n \geq 0$ and consider the alphabet $[n] = \{0, 1, \ldots, n-1\}$. The *lexicographic order* on $[n]^*$ w.r.t. the *reverse order* on $[n]$ is denoted by $\leq_{\mathrm{lex}}$. Whenever we use the alphabet $[n]_\diamond$, we identify the $\diamond$-symbol with n. For $\bar{x} \in \mathbb{N}^n$ we define

$$\langle \bar{x} \rangle = 0^{x_0} 1^{x_1} \cdots (n-1)^{x_{n-1}} \in [n]^* .$$

The set

$$\langle \mathbb{N}^n \rangle = \{ \langle \bar{x} \rangle \mid \bar{x} \in \mathbb{N}^n \} = 0^+ 1^+ \cdots (n-1)^+ \subseteq [n]^*$$

ordered by (the restriction of) $\leq_{\mathrm{lex}}$ is an automatic well-ordering of type ω^n. The map $\langle \cdot \rangle$ is the unique isomorphism (of well-orderings) between $\mathbb{N}^n$ and $\langle \mathbb{N}^n \rangle$.

For all $\bar{x}, \bar{y} \in \mathbb{N}^n$ the convolution $\langle \bar{x} \rangle \otimes \langle \bar{y} \rangle$ can be uniquely factorised as $\sigma_1^{e_1} \cdots \sigma_k^{e_k}$ with $k \geq 0$, $\sigma_1, \ldots, \sigma_k \in [n]_\diamond^2$ pairwise distinct, and $e_1, \ldots, e_k \geq 1$. In this situation, the sequence $p(\bar{x}, \bar{y}) = \sigma_1 \ldots \sigma_k (n, n)$ is a path through the 2-dimensional grid from $(0,0)$ to (n,n) using only steps $(0,1)$, $(1,0)$, and $(1,1)$. We call such sequences n-*paths*. We have $\bar{x} < \bar{y}$ if, and only if, $p(\bar{x}, \bar{y})$ contains a step different from $(1,1)$ and the first such is a $(1,0)$-step. We call n-paths with this latter property *lower* n-*paths*. An n-path is *restricted* if the $(1,1)$-step is used only on the main diagonal of the grid. There are precisely $\sum_{i=1}^{n} \left\langle {2i-1 \atop i} \right\rangle$ restricted lower n-paths.

Theorem 4.4. *Let $n \geq 0$. The automatic Ramsey degree $d_{R,\mathsf{AUT}}(\omega^n)$ exists and is given by*

$$d_{R,\mathsf{AUT}}(\omega^n) = \sum_{i=1}^{n} \left\langle \begin{matrix} 2i-1 \\ i \end{matrix} \right\rangle.$$

Proof. We prove existence and upper bound separately from the lower bound.

Existence and upper bound. Let $C \colon [A]^2 \to Q$ be an automatic ω^n-colouring. By Theorem 4.2, there exists a uniformly presentable embedding $f \colon \mathbb{N}^n \hookrightarrow A$. Consider the ω^n-colouring $D \colon [\langle \mathbb{N}^n \rangle]^2 \to Q$ with $D(\langle \bar{x} \rangle, \langle \bar{y} \rangle) = C(f(\bar{x}), f(\bar{y}))$. Due to the *uniform* presentability of f, D is automatic as well. Let η be a morphism into a finite semigroup S which simultaneously recognises all the $D^{-1}(q)$ and pick an idempotency exponent $M \geq 1$ of S. The set

$$X_M = \{ \bar{x} \in \mathbb{N}^n \mid \forall i \in [0, n-1] \colon x_i \equiv M \pmod{nM} \}$$

has order type ω^n and two useful properties for all $\bar{x}, \bar{y} \in X_M$ with $\bar{x} < \bar{y}$:

(1) The n-path $p(\bar{x}, \bar{y}) = \sigma_1 \ldots \sigma_k \, (n, n)$ is a restricted lower n-path.
(2) In the definition of $p(\bar{x}, \bar{y})$ above, each e_i is divisible by M. Thus, the idempotency properties of M imply $\eta(\langle \bar{x} \rangle \otimes \langle \bar{y} \rangle) = \eta(\sigma_1^M \cdots \sigma_k^M)$, i.e., $p(\bar{x}, \bar{y})$ determines $\eta(\langle \bar{x} \rangle \otimes \langle \bar{y} \rangle)$ and in turn also $C(f(\bar{x}), f(\bar{y})) = D(\langle \bar{x} \rangle, \langle \bar{y} \rangle)$.

Consequently, the regular set $f(X_M) \subseteq A$ is $\sum_{i=1}^{n} \left\langle \begin{smallmatrix} 2i-1 \\ i \end{smallmatrix} \right\rangle$-chromatic in C.

Lower bound. Let Q be the set of restricted lower n-paths. The sets

$$X_1 = \{ \bar{x} \in \mathbb{N}^n \mid \forall i \in [0, n-1] \colon x_i \equiv 1 \pmod{n} \}$$

and $\langle X_1 \rangle \subseteq \langle \mathbb{N}^n \rangle$ have order type ω^n. Like in property (1) of X_M above, we have $p(\bar{x}, \bar{y}) \in Q$ for all $\bar{x}, \bar{y} \in X_1$ with $\bar{x} < \bar{y}$. The ω^n-colouring $C \colon [\langle X_1 \rangle]^2 \to Q$ with $C(\langle \bar{x} \rangle, \langle \bar{y} \rangle) = p(\bar{x}, \bar{y})$ for $\bar{x} < \bar{y}$ is automatic. Since $|Q| = \sum_{i=1}^{n} \left\langle \begin{smallmatrix} 2i-1 \\ i \end{smallmatrix} \right\rangle$, it remains to show that for any regular subset $B \subseteq \langle X_1 \rangle$ of order type ω^n and all $\pi \in Q$ there are $u, v \in B$ with $u < v$ and $C(u, v) = \pi$.

Therefore, consider such X and π. By Theorem 4.2, there exists a uniformly presentable embedding $f \colon \mathbb{N}^n \hookrightarrow B$. Moreover, there are $\bar{x}, \bar{y} \in \mathbb{N}^n$ with $\bar{x} < \bar{y}$ and $\pi = (\langle \bar{x} \rangle \otimes \langle \bar{y} \rangle)(n, n)$. Finally, one can show that with $\bar{1} = (1, \ldots, 1) \in \mathbb{N}^n$ we have $C(f(n \cdot \bar{x} + \bar{1}), f(n \cdot \bar{y} + \bar{1})) = \pi$. $\square$

Remark 4.5. The remark after Definition 4.3 implies that $d_R(\omega^n) < d_{R,\mathsf{AUT}}(\omega^n)$ for $n \geq 2$. This is caused by the following reasons. In the proof of the lower bound above, you can find a non-regular subset $B \subseteq \langle X_1 \rangle$ such that $C(B)$ is the set of restricted lower n-paths in which the $(1,1)$-steps form an initial segment. There are precisely $d_R(\omega^n)$ such n-paths. However, for regular sets B you cannot avoid using $(1,1)$-steps after other steps as they provide more structure.

Using the same techniques, one can show a bipartite analogue of Theorem 4.4.

Theorem 4.6. *Let $m, n \geq 0$. The automatic Ramsey degree $d_{R,\mathsf{AUT}}(\omega^m, \omega^n)$ exists and is given by*

$$d_{R,\mathsf{AUT}}(\omega^m, \omega^n) = \binom{m+n}{m}.$$

4.3   The Automatic Ramsey Degree of Arbitrary Ordinals $\alpha < \omega^\omega$

Theorem 4.7. *Let $\alpha < \omega^\omega$ be an ordinal and $\alpha = \omega^{n_1} + \cdots + \omega^{n_r}$ with $r \geq 0$ and $n_1 \geq \cdots \geq n_r \geq 0$ its Cantor normal form. The automatic Ramsey degree $d_{R,\text{AUT}}(\alpha)$ exists and is given by*

$$d_{R,\text{AUT}}(\alpha) = \sum_{i=1}^{r} d_{R,\text{AUT}}(\omega^{n_i}) + \sum_{i=1}^{r} \sum_{j=i+1}^{r} d_{R,\text{AUT}}(\omega^{n_i}, \omega^{n_j}).$$

Proof. Let $\mu(\alpha)$ denote the sum on RHS above. Again, we prove existence/upper bound and lower bound separately.

Existence and upper bound. Let C be an automatic α-colouring on A. There is a unique decomposition $A = A_1 + \ldots + A_r$ such that each A_i has order type ω^{n_i}. All the A_i are regular. We construct regular subsets $B_{i,r} \subseteq \cdots \subseteq B_{i,1} \subseteq A_i$ of order type ω^{n_i} for $i = 1, \ldots, r$ in several stages. For $i = 1, \ldots, r$ choose $B_{i,1} \subseteq A_i$ by Theorem 4.4 such that $|C(B_{i,1})| \leq d_{R,\text{AUT}}(\omega^{n_i})$. For $i = 1, \ldots, r$ and $j = i + 1, \ldots, r$ choose $B_{i,j} \subseteq B_{i,j-1}$ and $B_{j,i+1} \subseteq B_{j,i}$ by Theorem 4.6 such that $|C(B_{i,j} \times B_{j,i+1})| \leq d_{R,\text{AUT}}(\omega^{n_i}, \omega^{n_j})$. Finally, the set $B = B_{1,r} + \cdots + B_{r,r} \subseteq A$ is regular, has order type α, and satisfies

$$|C(B)| \leq \sum_{i=1}^{r} |C(B_{i,r})| + \sum_{i=1}^{r} \sum_{j=i+1}^{r} |C(B_{i,r} \times B_{j,r})| \leq \mu(\alpha).$$

Lower bound. For $1 \leq i \leq r$ let $C_i \colon [A_i]^2 \to Q_i$ be (a slight modification of) the automatic ω^{n_i}-colouring proving the lower bound on $d_{R,\text{AUT}}(\omega^{n_i})$. Due to (the actual proof of) Theorem 4.4, for $1 \leq i < j \leq r$ there exists an automatic $(\omega^{n_i}, \omega^{n_j})$-colouring $C_{i,j} \colon A_i \times A_j \to Q_{i,j}$ which shows the lower bound on $d_{R,\text{AUT}}(\omega^{n_i}, \omega^{n_j})$. W.l.o.g. all the sets Q_i and $Q_{i,j}$ are mutually disjoint. Thus, their union Q has size $\mu(\alpha)$. The well-ordering $A = A_1 + \ldots + A_r$ is automatic and has type α. We define an automatic α-colouring $C \colon [A]^2 \to Q$ by $C(u, v) = C_i(u, v)$ if there is an i such that $u, v \in A_i$ and $C(u, v) = C_{i,j}(u, v)$ if there are $i < j$ such that $u \in A_i$ and $v \in A_j$. For every regular subset $B \subseteq A$ with order type α all the sets $A_i \cap B$ are regular and have order type ω^{n_i}. Thus, $C(B) = Q$ and hence $|C(B)| = \mu(\alpha)$. $\square$

Since all constructions employed throughout this section are effective, we obtain the following result which states that Theorem 4.7 is effective.

Theorem 4.8. *Given an automatic presentation of a colouring $C \colon [A]^2 \to Q$, one can compute the following:*

(1) $d_{R,\text{AUT}}(\alpha)$, where α is the order type of A,
(2) a subset $D \subseteq Q$ of size at most $d_{R,\text{AUT}}(\alpha)$, and
(3) a finite automaton recognising a D-chromatic subset $B \subseteq A$ of order type α.

5   The Computable Case

The reader can find the needed notions of computability theory in [16]. We call a colouring $C\colon [V]^2 \to F$ computable if $(V; \leq_V)$ is a computable ordinal and for each $i \in F$, the preimage $C^{-1}(i)$ is a computable set. Let COMP be the class of computable colourings.

Theorem 5.1 (Specker [17]). *For any $k \in \mathbb{N}$, $(\omega : \mathsf{COMP}) \nrightarrow (\omega : \mathsf{COMP})_{k,1}$.*

It is therefore a natural question whether the polychromatic version of Ramsey's theorem holds in the computable case. We remark in this section that polychromatic Ramsey's theorem also fails for computable colourings. Recall that Σ_1 denotes the class of computably enumerable sets.

Theorem 5.2. *For any $k \in \mathbb{N}$, we have $(\omega : \mathsf{COMP}) \nrightarrow (\omega : \Sigma_1)_{k+1,k}$.*

The proof is conceptually similar to Jockusch's proof of Theorem 5.1 in [8]. For this, one needs the following notion.

Definition 5.3. *A set $A \subseteq \mathbb{N}$ is* bi-immune *if it is infinite and neither A nor $\mathbb{N} \setminus A$ contains an infinite Σ_1 subset. A k-immune set partition is a partition $\mathbb{N} = A_1 \cup \cdots \cup A_k$ of the natural numbers such that each A_i is bi-immune.*

The proof of Proposition 5.4 uses a standard priority argument with finite injury.

Proposition 5.4. *For every $k \in \mathbb{N}$, there exists a k-immune set partition $A_1 \cup \cdots \cup A_k$ where each A_i is a Δ_2 set.*

Proof (of Theorem 5.2). Take a k-immune set partition $A_1 \cup \cdots \cup A_k$ where each A_i is Δ_2 as stipulated by Proposition 5.4. By the limit lemma each set A_i is limit computable. In other words there is a computable set $X_i \subseteq \mathbb{N}^2$ such that $\exists t \forall s \geq t : X_i(x,s) = A_i(x)$ for all $x \in \mathbb{N}$. We define a colouring $C\colon [\mathbb{N}]^2 \to [1,k]$ such that

$$C(x,s) = \begin{cases} \min\{i \mid X_i(x,s)\} & \text{if } \exists i\colon X_i(x,s); \\ 1 & \text{otherwise.} \end{cases}$$

Take any $(k-1)$-chromatic infinite set $H \subseteq \mathbb{N}$. Assume there is $x \in H \cap A_i$ for $i \in [1,k]$. There is some $t > x$ such that $\forall s \geq t : X_i(x,s)$. In particular there is some $y \in H$ such that $y > t$. This means that $C(x,y) = i \in C(H)$. We conclude that for some $i \in [1,k]$, $H \cap A_i = \emptyset$. However this means that $H \subseteq \mathbb{N} \setminus A_i$ and cannot be a Σ_1 set.   □

6   Final Remarks

This paper presents an explicit expression for computing the Ramsey degree of ordinals $\alpha < \omega^\omega$ and establishes the automatic version of this result. Below, we present some questions that came up but remained unanswered.

(1) Provided that an automatic α-colouring admits a regular, D-chromatic set of order type α, can one compute a finite automaton recognising such a set?

(2) Extend Jockusch's theorems [8, Corollary 3.2 and Theorem 4.2] to ordinals $\alpha < \omega^\omega$. In other words, does there exist an n such that Π_n contains a $d_R(\alpha)$-chromatic set of order type α in any computable α-colouring?

(3) Does Galvin's result on the rationals mentioned in Section 1 hold in the automatic case?

References

1. Blumensath, A., Grädel, E.: Automatic structures. In: LICS 2000, pp. 51–62. IEEE Computer Society (2000)

2. Conlon, D.: A new upper bound for the bipartite Ramsey problem. Journal of Graph Theory 58(4), 351–356 (2008)

3. Delhommé, C.: Automaticité des ordinaux et des graphes homogènes. Comptes Rendus Mathematique 339(1), 5–10 (2004)

4. Eilenberg, S.: Automata, Languages, and Machines. Pure and Applied Mathematics, vol. 58. Academic Press (1974)

5. Erdős, P., Hajnal, A., Máté, A., Rado, R.: Combinatorial set theory: Partition relations for cardinals. Studies in Logic, vol. 106. North-Holland (1984)

6. Erdős, P., Rado, R.: A partition calculus in set theory. Bull. Amer. Math. Soc. 62(5), 427–489 (1956)

7. Hajnal, A., Komjáth, P.: A strongly non-Ramsey order type. Combinatorica 17(3), 363–367 (1997)

8. Jockusch, C.G.: Ramsey's theorem and recursion theory. Journal of Symbolic Logic 37(2), 268–280 (1972)

9. Khoussainov, B., Nerode, A.: Automatic presentations of structures. In: Leivant, D. (ed.) LCC 1994. LNCS, vol. 960, pp. 367–392. Springer, Heidelberg (1995)

10. Khoussainov, B., Rubin, S., Stephan, F.: Automatic linear orders and trees. ACM Trans. Comp. Logic 6(4), 675–700 (2005)

11. Kuske, D.: (Un)countable and (non)effective versions of Ramsey's theorem. Contemporary Mathematics, vol. 558, pp. 467–487. Am. Math. Soc. (2011)

12. Pouzet, M., Sauer, N.: Edge partitions of the Rado graph. Combinatorica 16(4), 505–520 (1996)

13. Ramsey, F.P.: On a problem of formal logic. Proc. London Math. Soc. 30, 264–286 (1930)

14. Rosenstein, J.G.: Linear Orderings. Academic Press (1982)

15. Rubin, S.: Automata presenting structures: A survey of the finite string case. Bulletin of Symbolic Logic 14(2), 169–209 (2008)

16. Soare, R.: Recursively Enumerable Sets and Degrees: A study of computable functions and computably generated sets. Perspectives in Mathematical Logic. Springer (1987)

17. Specker, E.: Ramsey's theorem does not hold in recursive set theory. In: Logic Colloqium 1969. Studies in Logic and the Foundations of Mathematics, vol. 61, pp. 439–442. North-Holland (1971)

18. Todorcevic, S.: Some partitions of three-dimensional combinatorial cubes. Journal of Combinatorial Theory, Ser. A 68(2), 410–437 (1994)

19. Williams, N.: Combinatorial Set Theory. Elsevier (1977)

Detecting Regularities on Grammar-Compressed Strings

Tomohiro I[1,2], Wataru Matsubara[3], Kouji Shimohira[1],
Shunsuke Inenaga[1], Hideo Bannai[1], Masayuki Takeda[1],
Kazuyuki Narisawa[3], and Ayumi Shinohara[3]

[1] Department of Informatics, Kyushu University, Japan
{tomohiro.i,inenaga,bannai,takeda}@inf.kyushu-u.ac.jp
[2] Japan Society for the Promotion of Science (JSPS)
[3] Graduate School of Information Sciences, Tohoku University, Japan
{narisawa,ayumi}@ecei.tohoku.ac.jp

Abstract. We solve the problems of detecting and counting various forms of regularities in a string represented as a Straight Line Program (SLP). Given an SLP of size n that represents a string s of length N, our algorithm computes all runs and squares in s in $O(n^3 h)$ time and $O(n^2)$ space, where h is the height of the derivation tree of the SLP. We also show an algorithm to compute all gapped-palindromes in $O(n^3 h + gnh \log N)$ time and $O(n^2)$ space, where g is the length of the gap. The key technique of the above solution also allows us to compute the periods and covers of the string in $O(n^2 h)$ time and $O(nh(n + \log^2 N))$ time, respectively.

1 Introduction

Finding regularities such as squares, runs, and palindromes in strings, is a fundamental and important problem in stringology with various applications, and many efficient algorithms have been proposed (e.g., [13,6,1,7,14,2,10,9]). See also [5] for a survey.

In this paper, we consider the problem of detecting regularities in a string s of length N that is given in a compressed form, namely, as a straight line program (SLP), which is essentially a context free grammar in the Chomsky normal form that derives only s. Our model of computation is the word RAM: We shall assume that the computer word size is at least $\lceil \log_2 N \rceil$, and hence, standard operations on values representing lengths and positions of string s can be manipulated in constant time. Space complexities will be determined by the number of computer words (not bits).

Given an SLP whose size is n and the height of its derivation tree is h, Bannai et al. [3] showed how to test whether the string s is square-free or not, in $O(n^3 h \log N)$ time and $O(n^2)$ space. Independently, Khvorost [8] presented an algorithm for computing a compact representation of all squares in s in $O(n^3 h \log^2 N)$ time and $O(n^2)$ space. Matsubara et al. [15] showed that a compact representation of all maximal palindromes occurring in the string s can

K. Chatterjee and J. Sgall (Eds.): MFCS 2013, LNCS 8087, pp. 571–582, 2013.

be computed in $O(n^3 h)$ time and $O(n^2)$ space. Note that the length N of the decompressed string s can be as large as $O(2^n)$ in the worst case. Therefore, in such cases these algorithms are more efficient than *any* algorithm that work on uncompressed strings.

In this paper we present the following extension and improvements to the above work, namely,

1. an $O(n^3 h)$-time $O(n^2)$-space algorithm for computing a compact representation of squares and runs;
2. an $O(n^3 h + gnh \log N)$-time $O(n^2)$-space algorithm for computing a compact representation of palindromes with a gap (spacer) of length g.

We remark that our algorithms can easily be extended to count the number of squares, runs, and gapped palindromes in the same time and space complexities.

Note that Result 1 improves on the work by Khvorost [8] which requires $O(n^3 h \log^2 N)$ time and $O(n^2)$ space. The key to the improvement is our new technique of Section 3.3 called *approximate doubling*, which we believe is of independent interest. In fact, using the approximate doubling technique, one can improve the time complexity of the algorithms of Lifshits [11] to compute the periods and covers of a string given as an SLP, in $O(n^2 h)$ time and $O(nh(n + \log^2 N))$ time, respectively.

If we allow no gaps in palindromes (i.e., if we set $g = 0$), then Result 2 implies that we can compute a compact representation of all maximal palindromes in $O(n^3 h)$ time and $O(n^2)$ space. Hence, Result 2 can be seen as a generalization of the work by Matsubara et al. [15] with the same efficiency.

2   Preliminaries

2.1   Strings

Let Σ be the alphabet, so an element of Σ^* is called a string. For string $s = xyz$, x is called a prefix, y is called a substring, and z is called a suffix of s, respectively. The length of string s is denoted by $|s|$. The empty string ε is a string of length 0, that is, $|\varepsilon| = 0$. For $1 \leq i \leq |s|$, $s[i]$ denotes the i-th character of s. For $1 \leq i \leq j \leq |s|$, $s[i..j]$ denotes the substring of s that begins at position i and ends at position j. For any string s, let s^R denote the reversed string of s, that is, $s^R = s[|s|] \cdots s[2]s[1]$. For any strings s and u, let $lcp(s, u)$ (resp. $lcs(s, u)$) denote the length of the longest common prefix (resp. suffix) of s and u.

We say that string s has a *period* c ($0 < c \leq |s|$) if $s[i] = s[i + c]$ for any $1 \leq i \leq |s| - c$. For a period c of s, we denote $s = u^q$, where u is the prefix of s of length c and $q = \frac{|s|}{c}$. For convenience, let $u^0 = \varepsilon$. If $q \geq 2$, $s = u^q$ is called a *repetition* with root u and period $|u|$. Also, we say that s is *primitive* if there is no string u and integer $k > 1$ such that $s = u^k$. If s is primitive, then s^2 is called a *square*.

We denote a repetition in a string s by a triple $\langle b, e, c \rangle$ such that $s[b..e]$ is a repetition with period c. A repetition $\langle b, e, c \rangle$ in s is called a *run* (or *maximal periodicity* in [12]) if c is the smallest period of $s[b..e]$ and the substring cannot be extended to the left nor to the right with the same period, namely neither $s[b-1..e]$ nor $s[b..e+1]$ has period c. Note that for any run $\langle b, e, c\rangle$ in s, every substring of length $2c$ in $s[b..e]$ is a square. Let $Run(s)$ denote the set of all runs in s.

A string s is said to be a *palindrome* if $s = s^R$. A string s is said to be a *gapped* palindrome if $s = xux^R$ for some string $u \in \Sigma^*$. Note that u may or may not be a palindrome. The prefix x (resp. suffix x^R) of xux^R is called the *left arm* (resp. *right arm*) of gapped palindrome xuu^R. If $|u| = g$, then xux^R is said to be a g-gapped palindrome. We denote a *maximal g-gapped palindrome* in a string s by a pair $\langle b, e\rangle_g$ such that $s[b..e]$ is a g-gapped palindrome and $s[b-1..e+1]$ is not. Let $gPals(s)$ denote the set of all maximal g-gapped palindromes in s.

Given a text string $s \in \Sigma^+$ and a pattern string $p \in \Sigma^+$, we say that p occurs at position i ($1 \le i \le |s| - |p| + 1$) iff $s[i..i + |p| - 1] = p$. Let $Occ(s, p)$ denote the set of positions where p occurs in s. For a pair of integers $1 \le b \le e$, $[b, e] = \{b, b+1, \ldots, e\}$ is called an *interval*.

Lemma 1 ([16]). *For any strings $s, p \in \Sigma^+$ and any interval $[b, e]$ with $1 \le b \le e \le b + |p|$, $Occ(s, p) \cap [b, e]$ forms a single arithmetic progression if $Occ(s, p) \cap [b, e] \ne \emptyset$.*

2.2   Straight-Line Programs

A *straight-line program* (*SLP*) $\mathcal{S}$ of size n is a set of productions $\mathcal{S} = \{X_i \to expr_i\}_{i=1}^n$, where each X_i is a distinct variable and each $expr_i$ is either $expr_i = X_\ell X_r$ ($1 \le \ell, r < i$), or $expr_i = a$ for some $a \in \Sigma$. Note that X_n derives only a single string and, therefore, we view the SLP as a compressed representation of the string s that is derived from the variable X_n. Recall that the length N of the string s can be as large as $O(2^n)$. However, it is always the case that $n \ge \log N$. For any variable X_i, let $val(X_i)$ denote the string that is derived from variable X_i. Therefore, $val(X_n) = s$. When it is not confusing, we identify X_i with the string represented by X_i.

Let T_i denote the derivation tree of a variable X_i of an SLP $\mathcal{S}$. The derivation tree of $\mathcal{S}$ is T_n. Let $height(X_i)$ denote the height of the derivation tree T_i of X_i and $height(\mathcal{S}) = height(X_n)$. We associate each leaf of T_i with the corresponding position of the string $val(X_i)$. For any node z of the derivation tree T_i, let ℓ_z be the number of leaves to the left of the subtree rooted at z in T_i. The position of the node z in T_i is $\ell_z + 1$.

Let $[u, v]$ be any integer interval with $1 \le u \le v \le |val(X_i)|$. We say that the interval $[u, v]$ *crosses the boundary* of node z in T_i, if the lowest common ancestor of the leaves u and v in T_i is z. We also say that the interval $[u, v]$ *touches the boundary* of node z in T_i, if either $[u - 1, v]$ or $[u, v + 1]$ crosses the boundary of z in T_i. Assume $p = w[u..u + |p| - 1]$ and interval $[u, u + |p| - 1]$ crosses or touches the boundary of node z in T_i. When z is labeled by X_j, then we also say that the occurrence of p starting at position u in $val(X_i)$ crosses or touches the boundary of X_j.

Lemma 2 ([4]). *Given an SLP S of size n describing string w of length N, we can pre-process S in $O(n)$ time and space to answer the following queries in $O(\log N)$ time:*

- *Given a position u with $1 \leq u \leq N$, answer the character $w[u]$.*
- *Given an interval $[u,v]$ with $1 \leq u \leq v \leq N$, answer the node z the interval $[u,v]$ crosses, the label X_i of z, and the position of z in $T_S = T_n$.*

For any production $X_i \to X_\ell X_r$ and a string p, let $Occ^\xi(X_i, p)$ be the set of occurrences of p which begin in X_ℓ and end in X_r. Let S and T be SLPs of sizes n and m, respectively. Let the AP-table for S and T be an $n \times m$ table such that for any pair of variables $X \in S$ and $Y \in T$ the table stores $Occ^\xi(X,Y) = Occ^\xi(X, val(Y))$. It follows from Lemma 1 that $Occ^\xi(X,Y)$ forms a single arithmetic progression which requires $O(1)$ space, and hence the AP-table can be represented in $O(nm)$ space.

Lemma 3 ([11]). *Given two SLPs S and T of sizes n and m, respectively, the AP-table for S and T can be computed in $O(nmh)$ time and $O(nm)$ space, where $h = height(S)$.*

Lemma 4 ([11], local search (LS)). *Let S and T be SLPs that describe strings s and p, respectively. Using AP-table for S and T, we can compute, given any position b and constant $\alpha > 0$, $Occ(s,p) \cap [b, b+\alpha|p|]$ as a form of at most $\lceil \alpha \rceil$ arithmetic progressions in $O(h)$ time, where $h = height(S)$.*

Note that, given any $1 \leq i \leq j \leq |s|$, we are able to build an SLP of size $O(n)$ that generates substring $s[i..j]$ in $O(n)$ time. Hence, by computing the AP-table for S and the new SLP, we can conduct the local search LS operation on substring $s[i..j]$ in $O(n^2 h)$ time.

For any variable X_i of S and positions $1 \leq k_1, k_2 \leq |X_i|$, we define the "right-right" longest common extension query by

$$\mathsf{LCE}(X_i, k_1, k_2) = lcp(X_i[k_1..|X_i|], X_i[k_2..|X_i|]).$$

Using a technique of [16] in conjunction with Lemma 3, it is possible to answer the query in $O(n^2 h)$ time for each pair of positions, with no pre-processing. We will later show our new algorithm which, after $O(n^2 h)$-time pre-processing, answers to the LCE query for any pair of positions in $O(h \log N)$ time.

3   Finding Runs

In this section we propose an $O(n^3 h)$-time and $O(n^2)$-space algorithm to compute $O(n \log N)$-size representation of all runs in a text s of length N represented by SLP $S = \{X_i \to expr_i\}_{i=1}^n$ of height h.

For each production $X_i \to X_{\ell(i)} X_{r(i)}$ with $i \leq n$, we consider the set $Runs^\xi(X_i)$ of runs which touch or cross the boundary of X_i and are completed in X_i, i.e., those that are not prefixes nor suffixes of X_i. Formally,

$$Runs^\xi(X_i) = \{\langle b, e, c \rangle \in Run(X_i) \mid 1 \leq b - 1 \leq |X_{\ell(i)}| < e + 1 \leq |X_i|\}.$$

It is known that for any interval $[b, e]$ with $1 \leq b \leq e \leq |s|$, there exists a unique occurrence of a variable X_i in the derivation tree of SLP, such that the interval $[b, e]$ crosses the boundary of X_i. Also, wherever X_i appears in the derivation tree, the runs in $Runs^\xi(X_i)$ occur in s with some appropriate offset, and these occurrences of the runs are never contained in $Runs^\xi(X_j)$ with any other variable X_j with $j \neq i$. Hence, by computing $Runs^\xi(X_i)$ for all variables X_i with $i \leq n$, we can essentially compute all runs of s that are not prefixes nor suffixes of s. In order to detect prefix/suffix runs of s, it is sufficient to consider two auxiliary variables $X_{n+1} \to X_\$ X_n$ and $X_{n+2} \to X_{n+1} X_{\$'}$, where $X_\$$ and $X_{\$'}$ respectively derive special characters \$ and \$' that are not in s and $\$ \neq \$'$. Hence, the problem of computing the runs from an SLP $\mathcal{S}$ reduces to computing $Runs^\xi(X_i)$ for all variables X_i with $i \leq n + 2$.

Our algorithm is based on the divide-and-conquer method used in [3] and also [8], which detect squares crossing the boundary of each variable X_i. Roughly speaking, in order to detect such squares we take some substrings of $val(X_i)$ as *seeds* each of which is in charge of distinct squares, and for each seed we detect squares by using LS and LCE constant times. There is a difference between [3] and [8] in how the seeds are taken, and ours is rather based on that in [3]. In the next subsection, we briefly describe our basic algorithm which runs in $O(n^3 h \log N)$ time.

3.1 Basic Algorithm

Consider runs in $Runs^\xi(X_i)$ with $X_i \to X_\ell X_r$. Since a run in $Runs^\xi(X_i)$ contains a square which touches or crosses the boundary of X_i, our algorithm finds a run by first finding such a square, and then computing the maximal extension of its period to the left and right of its occurrence.

Each square ww that we want to find in X_i can be divided by its length and how it relates to the boundary of X_i. When $|w| > 1$, there exists $1 \leq t < \log |val(X_i)|$ such that $2^t \leq |w| < 2^{t+1}$ and there are four cases (see also Fig. 1); (1) $|w_\ell| \geq \frac{3}{2}|w|$, (2) $\frac{3}{2}|w| > |w_\ell| \geq |w|$, (3) $|w| > |w_\ell| \geq \frac{1}{2}|w|$, (4) $\frac{1}{2}|w| > |w_\ell|$, where w_ℓ is a prefix of ww which is also a suffix of $val(X_\ell)$.

The point is that in any case we can take a substring p of length 2^{t-1} of s which touches the boundary of X_i, and is completely contained in w. By using p as a seed we can detect runs by the following steps:

Step 1: Conduct local search of p in an "appropriate range" of X_i, and find a copy p' $(= p)$ of p.

Step 2: Compute the length *plen* of the longest common prefix to the right of p and p', and the length *slen* of the longest common suffix to the left of p and p', then check that $plen + slen \geq d - |p|$, where d is the distance between the beginning positions of p and p'.

Notice that Step 2 actually computes maximal extension of the repetition.

Since $d = |w|$, it is sufficient to conduct local search in the range satisfying $2^t \leq d < 2^{t+1}$, namely, the width of the interval for local search is smaller than

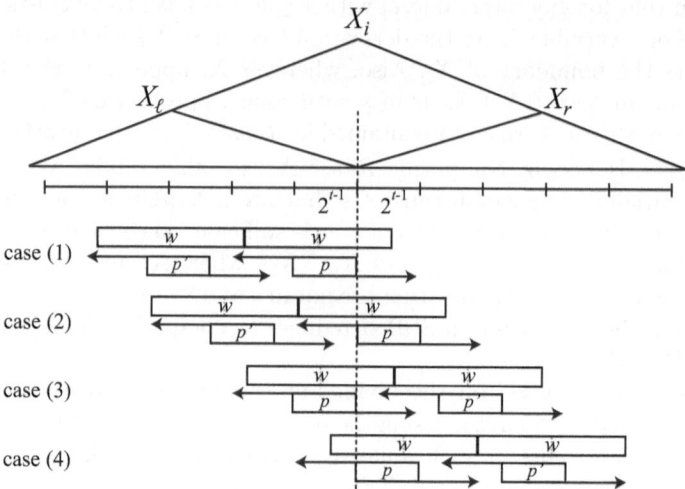

Fig. 1. The left arrows represent the longest common suffix between the left substrings immediately to the left of p and p'. The right arrows represent the longest common prefix between the substrings immediately to the right of p and p'.

$2|p|$, and all occurrences of p' are represented by at most two arithmetic progressions. Although exponentially many runs can be represented by an arithmetic progression, its periodicity enables us to efficiently detect all of them, by using LCE only constant times, and they are encoded in $O(1)$ space. We omit the details due to the lack of space but the employed techniques are essentially the same as in [8].

By varying t from 1 to $\log N$, we can obtain an $O(\log N)$-size compact representation of $Runs^\xi(X_i)$ in $O(n^2 h \log N)$ time. More precisely, we get a list of $O(\log N)$ quintuplets $\langle \delta_1, \delta_2, \delta_3, c, k \rangle$ such that the union of sets $\bigcup_{j=0}^{k-1} \langle \delta_1 - cj, \delta_2 + cj, \delta_3 + cj \rangle$ for all elements of the list equals to $Runs^\xi(X_i)$ without duplicates. By applying the above procedure to all the n variables, we can obtain an $O(n \log N)$-size compact representation of all runs in s in $O(n^3 h \log N)$ time. The total space requirement is $O(n^2)$, since we need $O(n^2)$ space at each step of the algorithm.

In order to improve the running time of the algorithm to $O(n^3 h)$, we will use new techniques of the two following subsections.

3.2 Longest Common Extension

In this subsection we propose a more efficient algorithm for LCE queries.

Lemma 5. *We can pre-process an SLP $\mathcal{S}$ of size n and height h in $O(n^2 h)$ time and $O(n^2)$ space, so that given any variable X_i and positions $1 \le k_1, k_2 \le |X_i|$, $\mathsf{LCE}(X_i, k_1, k_2)$ is answered in $O(h \log N)$ time.*

To compute $\mathsf{LCE}(X_i, k_1, k_2)$ we will use the following function: For an SLP $\mathcal{S} = \{X_i \rightarrow expr_i\}_{i=1}^n$, let Match be a function such that

$$\mathsf{Match}(X_i, X_j, k) = \begin{cases} \text{true} & \text{if } k \in Occ(X_i, X_j), \\ \text{false} & \text{if } k \notin Occ(X_i, X_j). \end{cases}$$

Lemma 6. *We can pre-process a given SLP $\mathcal{S}$ of size n and height h in $O(n^2 h)$ time and $O(n^2)$ space so that the query $\mathsf{Match}(X_i, X_j, k)$ is answered in $O(\log N)$ time.*

Proof. We apply Lemma 2 to *every* variable X_i of $\mathcal{S}$, so that the queries of Lemma 2 are answered in $O(\log N)$ time on the derivation tree T_i of each variable X_i of $\mathcal{S}$. Since there are n variables in $\mathcal{S}$, this takes a total of $O(n^2)$ time and space. We also apply Lemma 3 to $\mathcal{S}$, which takes $O(n^2 h)$ time and $O(n^2)$ space. Hence the pre-processing takes a total of $O(n^2 h)$ time and $O(n^2)$ space.

To answer the query $\mathsf{Match}(X_i, X_j, k)$, we first find the node of T_i the interval $[k, k+|X_j|-1]$ crosses, its label X_q, and its position r in T_i. This takes $O(\log N)$ time using Lemma 2. Then we check in $O(1)$ time if $(k - r) \in Occ^\xi(X_q, X_j)$ or not, using the arithmetic progression stored in the AP-table. Thus the query is answered in $O(\log N)$ time. □

The following function will also be used in our algorithm: Let FirstMismatch be a function such that

$$\mathsf{FirstMismatch}(X_i, X_j, k) = \begin{cases} |lcp(X_i[k..|X_i|], X_j)| & \text{if } |X_i| - k + 1 \leq |X_j|, \\ \text{undefined} & \text{otherwise.} \end{cases}$$

Using Lemma 6 we can establish the following lemma.

Lemma 7. *We can pre-process a given SLP $\mathcal{S}$ of size n and height h in $O(n^2 h)$ time and $O(n^2)$ space so that the query $\mathsf{FirstMismatch}(X_i, X_j, k)$ is answered in $O(h \log N)$ time.*

We are ready to prove Lemma 5:

Proof. Consider to compute $\mathsf{LCE}(X_i, k_1, k_2)$. Without loss of generality, assume $k_1 \geq k_2$. Let z be the lca of the k_2-th and $(k_2 - k_1 + |X_i|)$-th leaves of the derivation tree T_i. Let P_ℓ be the path from z to the k_2-th leaf of the derivation tree T_i, and let L be the list of the right child of the nodes in P_ℓ sorted in increasing order of their position in T_i. The number of nodes in L is at most $height(X_i) \leq h$, and L can be computed in $O(height(X_i)) = O(h)$ time. Let P_r be the path from z to the $(k_2 - k_1 + |X_i|)$-th leaf of the derivation tree T_i, and let R be the list of the left child of the nodes in P_r sorted in increasing order of their position in T_i. R can be computed in $O(h)$ time as well. Let $U = L \cup R = \{X_{u(1)}, X_{u(2)}, \ldots, X_{u(m)}\}$ be the list obtained by concatenating L and R. For each $X_{u(p)}$ in increasing order of $p = 1, 2, \ldots, m$, we perform query $\mathsf{Match}(X_i, X_{u(p)}, k_1 + \sum_{q=1}^{p-1} |X_{u(q)}|)$ until either finding the first variable

$X_{u(p')}$ for which the query returns false, or all the queries for $p = 1, \ldots, m$ have returned true. In the latter case, clearly $\mathsf{LCE}(X_i, k_1, k_2) = |X_i| - k_1 + 1$. In the former case, the first mismatch occurs between X_i and $X_{u(p')}$, and hence $\mathsf{LCE}(X_i, k_1, k_2) = \sum_{q'=1}^{p'-1} |X_{u(q')}| + \mathsf{FirstMismatch}(X_i, X_{u(p')}, k_1 + \sum_{q'=1}^{p'-1} |X_{u(q')}|)$.

Since U contains at most $2 \cdot height(X_i)$ variables, we perform $O(h)$ Match queries. We perform at most one FirstMismatch query. Thus, using Lemmas 6 and 7, we can compute $\mathsf{LCE}(X_i, k_1, k_2)$ in $O(h \log N)$ time after $O(n^2 h)$-time $O(n^2)$-space pre-processing. $\square$

We can use Lemma 5 to also compute "left-left", "left-right", and "right-left" longest common extensions on the uncompressed string $s = val(\mathcal{S})$: We can compute in $O(n)$ time an SLP $\mathcal{S}^R$ of size n which represents the reversed string s^R [15]. We then construct a new SLP $\mathcal{S}'$ of size $2n$ and height $h + 1$ by concatenating the last variables of $\mathcal{S}$ and $\mathcal{S}^R$, and apply Lemma 5 to $\mathcal{S}'$.

3.3   Approximate Doubling

Here we show how to reduce the number of AP-table computation required in Step 1 of the basic algorithm, from $O(\log N)$ to $O(1)$ times per variable.

Consider any production $X_i \to X_\ell X_r$. If we build a new SLP which contains variables that derive the prefixes of length 2^t of X_r for each $0 \le t < \log |X_r|$, we can obtain the AP-tables for X_i and all prefix seeds of X_r by computing the AP-table for X_i and the new SLP. However, it is uncertain if we can build such a new SLP of size $O(n)$ in a reasonable time or not. Here we notice that the lengths of the seeds do not have to be exactly doublings, i.e., the basic algorithm of Section 3.1 works fine as long as the following properties are fulfilled: (a) the ratio of the lengths for each pair of consecutive seeds is constant; (b) the whole string is covered by the $O(\log N)$ seeds [1]. We show in the next lemma that we can build an approximate doubling SLP of size $O(n)$ in $O(n)$ time.

Lemma 8. *Let $\mathcal{S} = \{X_i \to expr_i\}_{i=1}^n$ be an SLP that derives a string s. We can build in $O(n)$ time a new SLP $\mathcal{S}' = \{Y_i \to expr'_i\}_{i=1}^{n'}$ with $n' = O(n)$ and $height(\mathcal{S}') = O(height(\mathcal{S}))$, which derives s and contains $O(\log N)$ variables $Y_{a_1}, Y_{a_2}, \ldots, Y_{a_k}$ satisfying the following conditions:*

- *For any $1 \le j \le k$, Y_{a_j} derives a prefix of s, $|Y_{a_1}| = 1$ and $|Y_{a_k}| = |s|$.*
- *For any $1 \le j < k$, $|Y_{a_j}| < |Y_{a_{j+1}}| \le 2|Y_{a_j}|$.*

Proof. First, we copy the productions of $\mathcal{S}$ into $\mathcal{S}'$. Next we add productions needed for creating prefix variables $Y_{a_1}, Y_{a_2}, \ldots, Y_{a_k}$ in increasing order. When creating $Y_{a_{j+1}}$, we consider creating a variable Y_{b_j} that derives $s[|Y_{a_j}|+1..|Y_{a_{j+1}}|]$, i.e., $Y_{a_{j+1}} \to Y_{a_j} Y_{b_j}$, by traversing T_n and finding the sequence of nodes that represents the substring. Note that we have some degrees of freedom for creating Y_{b_j} since $|Y_{a_{j+1}}|$ is not fixed, and this helps us to keep the size of $\mathcal{S}'$ in $O(n)$. Let v_j and v_{j+1} be the nodes where the traversal for $Y_{b_{j+1}}$ begin and end, respectively.

---

[1]   A minor modification is that we conduct local search for a seed p at Step 1 with the range satisfying $2|p| \le d < 2|q|$, where q is the next longer seed of p.

We start from v_1 which is the leftmost node that derives $s[1]$. Suppose we have built prefix variables up to Y_{a_j} and now creating $Y_{a_{j+1}}$. At this moment we are at v_j. We move up to the node u_j such that u_j is the deepest node on the path from the root to v_j which contains position $2|Y_{a_j}|$, and move down from u_j towards position $2|Y_{a_j}|$. The traversal ends when we meet a node v_{j+1} which satisfies one of the following conditions; (1) the rightmost position of v_{j+1} is $2|Y_{a_j}|$, (2) v_{j+1} is labeled with X_i, and we have traversed another node labeled with X_i before.

- If Condition (1) holds, we let $|Y_{a_{j+1}}|$ be the rightmost position of v_{j+1}. It is clear that $|Y_{a_{j+1}}| = 2|Y_{a_j}|$.
- If Condition (1) does not hold but Condition (2) holds, we let $|Y_{a_{j+1}}|$ be the leftmost position of v_{j+1} minus 1. Since v_{j+1} contains position $2|Y_{a_j}|$, $|Y_{a_{j+1}}| < 2|Y_{a_j}|$. We remark that since X_i appears in $Y_{a_{j+1}}$, then $|Y_{a_{j+1}}| + |X_i| \le 2|Y_{a_{j+1}}|$, and therefore, we never move down v_{j+1} for creating prefix variables to follow.

We iterate the above procedures until we obtain a prefix variable $Y_{a_{k-1}}$ that satisfies $|s| \le 2|Y_{a_{k-1}}|$. We let u_k be the deepest node on the path from the root to v_{k-1} which contains position $|s|$, and let v_k be the right child of u_k. Since $|Y_{a_j}| < 2|Y_{a_{j+2}}|$ for any $1 \le j < k$, $k = O(\log N)$ holds.

We note that $val(Y_{b_j})$ can be represented by the concatenation of "inner" nodes attached to the path from v_j to v_{j+1}, and hence, the number of new variables needed for creating Y_{b_j} is bounded by the number of such nodes. Consider all the edges we have traversed in the derivation tree T_n of X_n. Each edge contributes to at most one new variable for Y_{b_j} for some $1 \le j < k$. Since each variable X_i in $\mathcal{S}$ is used constant times for moving down due to Condition (2), the number of the traversed edges as well as n' is $O(n)$. Also, it is easy to make the height of Y_{b_j} be $O(height(\mathcal{S}))$ for any $1 \le j < k$. Thus $O(height(\mathcal{S}')) = O(\log N + height(\mathcal{S})) = O(height(\mathcal{S}))$.   □

3.4   Improved Algorithm

Using Lemmas 5 and 8, we get the following theorem.

Theorem 1. *Given an SLP $\mathcal{S}$ of size n and height h that describes string s of length N, an $O(n \log N)$-size compact representation of all runs in s can be computed in $O(n^3 h)$ time and $O(n^2)$ working space.*

Proof. Using Lemma 5, we first pre-process $\mathcal{S}$ in $O(n^2 h)$ time so that any "right-right" or "left-left" LCE query can be answered in $O(h \log N)$ time. For each variable $X_i \rightarrow X_\ell X_r$, using Lemma 8, we build two temporary SLPs which have respectively approximately doubling suffix variables of X_ℓ and prefix variables of X_r, and compute two AP-tables for $\mathcal{S}$ and each of them in $O(n^2 h)$ time. For each of the $O(\log N)$ prefix/suffix variables, we use it as a seed and find all corresponding runs by using LS and LCE queries constant times. Hence the time complexity is $O(n^2 h + n(n^2 h + (h + h \log N) \log N)) = O(n^3 h)$. The space requirement is $O(n^2)$, the same as the basic algorithm.   □

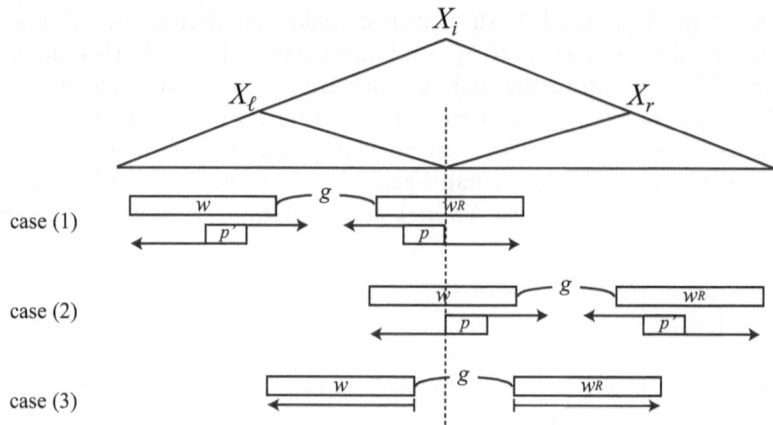

Fig. 2. Three groups of g-gapped palindromes to be found in X_i

4   Finding g-Gapped Palindromes

A similar strategy to finding runs on SLPs can be used for computing a compact representation of the set $gPals(s)$ of g-gapped palindromes from an SLP $\mathcal{S}$ that describes string s. As in the case of runs, we add two auxiliary variables $X_{n+1} \to X_\$ X_n$ and $X_{n+2} \to X_{n+1} X_{\$'}$. For each production $X_i \to X_\ell X_r$ with $i \le n + 2$, we consider the set $gPals^\xi(X_i)$ of g-gapped palindromes which touch or cross the boundary of X_i and are completed in X_i, i.e., those that are not prefixes nor suffixes of X_i. Formally,

$$gPals^\xi(X_i) = \{\langle b, e \rangle_g \in gPals(X_i) \mid 1 \le b - 1 \le |X_\ell| < e + 1 \le |X_i|\}.$$

Each g-gapped palindrome in X_i can be divided into three groups (see also Fig. 2); (1) its right arm crosses or touches with its right end the boundary of X_i, (2) its left arm crosses or touches with its left end the boundary of X_i, (3) the others.

For Case (3), for every $|X_\ell| - g + 1 \le j < |X_\ell|$ we check if $lcp(X_i[1..j]^R, X_i[j + g + 1..|X_i|]) > 0$ or not. From Lemma 5, it can be done in $O(gh \log N)$ time for any variable by using "left-right" LCE (excluding pre-processing time for LCE). Hence we can compute all such g-gapped palindromes for all productions in $O(n^2 h + gnh \log N)$ time, and clearly they can be stored in $O(ng)$ space.

For Case (1), let w_ℓ be the prefix of the right arm which is also a suffix of $val(X_\ell)$. We take approximately doubling suffixes of X_ℓ as seeds. Let p be the longest seed that is contained in w_ℓ. We can find g-gapped palindromes by the following steps:

Step 1: Conduct local search of $p' = p^R$ in an "appropriate range" of X_i and find it in the left arm of palindrome.

Step 2: Compute "right-left" LCE of p' and p, then check that the gap can be g. The outward maximal extension can be obtained by computing "left-right" LCE queries on the occurrences of p' and p.

As in the case of runs, for each seed, the length of the range where the local search is performed in Step 1 is only $O(|p|)$. Hence, the occurrences of p' can be represented by a constant number of arithmetic progressions. Also, we can obtain $O(1)$-space representation of g-gapped palindromes for each arithmetic progression representing overlapping occurrences of p', by using a constant number of LCE queries. Therefore, by processing $O(\log N)$ seeds for every variable X_i, we can compute in $O(n^2 h + n(n^2 h + (h + h \log N) \log N)) = O(n^3 h)$ time an $O(n \log N)$-size representation of all g-gapped palindromes for Case (1) in s.

In a symmetric way of Case (1), we can find all g-gapped palindromes for Case (2). Putting all together, we get the following theorem.

Theorem 2. *Given an SLP of size n and height h that describes string s of length N, and non-negative integer g, an $O(n \log N + ng)$-size compact representation of all g-gapped palindromes in s can be computed in $O(n^3 h + gnh \log N)$ time and $O(n^2)$ working space.*

5   Discussions

Let $\mathbb{R}$ and $\mathbb{G}$ denote the output compact representations of the runs and g-gapped palindromes of a given SLP $\mathcal{S}$, respectively, and let $|\mathbb{R}|$ and $|\mathbb{G}|$ denote their size. Here we show an application of $\mathbb{R}$ and $\mathbb{G}$; given any interval $[b, e]$ in s, we can count the number of runs and gapped palindromes in $s[b..e]$ in $O(n + |\mathbb{R}|)$ and $O(n + |\mathbb{G}|)$ time, respectively. We will describe only the case of runs, but a similar technique can be applied to gapped palindromes. As is described in Section 3.2, $s[b..e]$ can be represented by a sequence $U = (X_{u(1)}, X_{u(2)}, \ldots, X_{u(m)})$ of $O(h)$ variables of $\mathcal{S}$. Let $\mathcal{T}$ be the SLP obtained by concatenating the variables of U. There are three different types of runs in $\mathbb{R}$: (1) runs that are completely within the subtree rooted at one of the nodes of U; (2) runs that begin and end inside $[b, e]$ and cross or touch any border between consecutive nodes of U; (3) runs that begin and/or end outside $[b, e]$. Observe that the runs of types (2) and (3) cross or touch the boundary of one of the nodes in the path from the root to the b-th leaf of the derivation tree $T_{\mathcal{S}}$, or in the path from the root to the e-th leaf of $T_{\mathcal{S}}$. A run that begins outside $[b, e]$ is counted only if the suffix of the run that intersects $[b, e]$ has an exponent of at least 2. The symmetric variant applies to a run that ends outside $[b, e]$. Thus, the number of runs of types (2) and (3) can be counted in $O(n + 2|\mathbb{R}|)$ time. Since we can compute in a total of $O(n)$ time the number of nodes in the derivation tree of $\mathcal{T}$ that are labeled by X_i for all variables X_i, the number of runs of type (1) for all variables $X_{u(j)}$ can be counted in $O(n + |\mathbb{R}|)$ time. Noticing that runs are compact representation of squares, we can also count the number of occurrences of all squares in $s[b..e]$ in $O(n + |\mathbb{R}|)$ time by simple arithmetic operations.

The approximate doubling and LCE algorithms of Section 3 can be used as basis of other efficient algorithms on SLPs. For example, using approximate doubling, we can reduce the number of pairs of variables for which the AP-table has to be computed in the algorithms of Lifshits [11], which compute compact representations of all periods and covers of a string given as an SLP. As a result, we improve the time complexities from $O(n^2 h \log N)$ to $O(n^2 h)$ for periods, and from $O(n^2 h \log^2 N)$ to $O(nh(n + \log^2 N))$ for covers.

References

1. Apostolico, A., Breslauer, D.: An optimal $\mathcal{O}(\log \log N)$-time parallel algorithm for detecting all squares in a string. SIAM Journal on Computing 25(6), 1318–1331 (1996)
2. Apostolico, A., Breslauer, D., Galil, Z.: Parallel detection of all palindromes in a string. Theor. Comput. Sci. 141(1&2), 163–173 (1995)
3. Bannai, H., Gagie, T., I, T., Inenaga, S., Landau, G.M., Lewenstein, M.: An efficient algorithm to test square-freeness of strings compressed by straight-line programs. Inf. Process. Lett. 112(19), 711–714 (2012)
4. Bille, P., Landau, G.M., Raman, R., Sadakane, K., Satti, S.R., Weimann, O.: Random access to grammar-compressed strings. In: Proc. SODA 2011, pp. 373–389 (2011)
5. Crochemore, M., Ilie, L., Rytter, W.: Repetitions in strings: Algorithms and combinatorics. Theor. Comput. Sci. 410(50), 5227–5235 (2009)
6. Crochemore, M., Rytter, W.: Efficient parallel algorithms to test square-freeness and factorize strings. Information Processing Letters 38(2), 57–60 (1991)
7. Jansson, J., Peng, Z.: Online and dynamic recognition of squarefree strings. International Journal of Foundations of Computer Science 18(2), 401–414 (2007)
8. Khvorost, L.: Computing all squares in compressed texts. In: Proceedings of the 2nd Russian Finnish Symposium on Discrete Mathemtics, vol. 17, pp. 116–122 (2012)
9. Kolpakov, R., Kucherov, G.: Searching for gapped palindromes. Theor. Comput. Sci. 410(51), 5365–5373 (2009)
10. Kolpakov, R.M., Kucherov, G.: Finding maximal repetitions in a word in linear time. In: FOCS, pp. 596–604 (1999)
11. Lifshits, Y.: Processing compressed texts: A tractability border. In: Ma, B., Zhang, K. (eds.) CPM 2007. LNCS, vol. 4580, pp. 228–240. Springer, Heidelberg (2007)
12. Main, M.G.: Detecting leftmost maximal periodicities. Discrete Applied Mathematics 25(1-2), 145–153 (1989)
13. Main, M.G., Lorentz, R.J.: An $\mathcal{O}(n \log n)$ algorithm for finding all repetitions in a string. Journal of Algorithms 5(3), 422–432 (1984)
14. Manacher, G.K.: A new linear-time "on-line" algorithm for finding the smallest initial palindrome of a string. J. ACM 22(3), 346–351 (1975)
15. Matsubara, W., Inenaga, S., Ishino, A., Shinohara, A., Nakamura, T., Hashimoto, K.: Efficient algorithms to compute compressed longest common substrings and compressed palindromes. Theoretical Computer Science 410(8-10), 900–913 (2009)
16. Miyazaki, M., Shinohara, A., Takeda, M.: An improved pattern matching algorithm for strings in terms of straight-line programs. In: Hein, J., Apostolico, A. (eds.) CPM 1997. LNCS, vol. 1264, pp. 1–11. Springer, Heidelberg (1997)

Small Depth Proof Systems

Andreas Krebs[1], Nutan Limaye[2],
Meena Mahajan[3], and Karteek Sreenivasaiah[3]

[1] University of Tübingen, Germany
[2] Indian Institute of Technology, Bombay, India
[3] The Institute of Mathematical Sciences, Chennai, India

Abstract. A proof system for a language L is a function f such that
Range(f) is exactly L. In this paper, we look at proof systems from
a circuit complexity point of view and study proof systems that are
computationally very restricted. The restriction we study is: they can
be computed by bounded fanin circuits of constant depth (NC^0), or of
$O(\log \log n)$ depth but with $O(1)$ alternations (poly log AC^0). Each output bit depends on very few input bits; thus such proof systems correspond to a kind of local error-correction on a theorem-proof pair.

We identify exactly how much power we need for proof systems to
capture all regular languages. We show that all regular language have
poly log AC^0 proof systems, and from a previous result (Beyersdorff et
al, MFCS 2011, where NC^0 proof systems were first introduced), this is
tight. Our technique also shows that MAJ has poly log AC^0 proof system.

We explore the question of whether TAUT has NC^0 proof systems. Addressing this question about 2TAUT, and since 2TAUT is closely related
to reachability in graphs, we ask the same question about Reachability.
We show that both Undirected Reachability and Directed UnReachability have NC^0 proof systems, but Directed Reachability is still open.

In the context of how much power is needed for proof systems for
languages in NP, we observe that proof systems for a good fraction of
languages in NP do not need the full power of AC^0; they have SAC^0 or
$coSAC^0$ proof systems.

1 Introduction

Let f be any computable function mapping strings to strings. Then f can be
thought of as a proof system for the language $L = \text{range}(f)$ in the following sense:
to prove that a word x belongs to L, provide a word y that f maps to x. That is,
view y as a proof of the statement "$x \in L$", and computing $f(y)$ is then tanta-
mount to verifying the proof. From the perspective of computational complexity,
interesting proof systems are those functions that are efficiently computable and
have succinct proofs for all words in their range. If we use polynomial-time
computable as the notion of efficiency, and polynomial-size as the notion of suc-
cinctness, then NP is exactly the class of languages that have efficient proof
systems with succinct proofs. For instance, the coNP-complete language TAUT
has such proof systems if and only if NP equals coNP [1].

K. Chatterjee and J. Sgall (Eds.): MFCS 2013, LNCS 8087, pp. 583–594, 2013.
© Springer-Verlag Berlin Heidelberg 2013

Since we do not yet know whether or not NP equals co-NP, a reasonable question to ask is how much more computational power and/or non-succinctness is needed before we can show that TAUT has a proof system. For instance, allowing the verifier the power of randomized polynomial-time computation on polynomial-sized proofs characterizes the class MA; allowing quantum power characterizes the class QCMA; one could also allow the verifier access to some advice, yielding non-uniform classes; see for instance [2–5].

An even more interesting, and equally reasonable, approach is to ask: how much do we need to reduce the computational power of the verifier before we can formally establish that TAUT does not have a proof system within those bounds? This approach has seen a rich body of results, starting from the pathbreaking work of Cook and Reckhow [6]. The common theme in limiting the verifier's power is to limit the nature of proof verification, equivalently, the syntax of the proof; for example, proof systems based on resolution, Frege systems, and so on. See [7, 8] for excellent surveys on the topic.

Instead of restricting the proof syntax, if we only restrict the computational power of the verifier, it is not immediately obvious that we get anywhere. This is because it is already known that NP is characterised by succinct proof systems with extremely weak verifiers, namely AC^0 verifiers. Recall that in AC^0 we cannot even check if a binary string has an odd number of 1s [9, 10]. But an AC^0 computation can verify that a given assignment satisfies a Boolean formula. Nonetheless, one can look for verifiers even weaker than AC^0; this kind of study was initiated in [11] where NC^0 proof systems were investigated. In an NC^0 proof system, each output bit depends on just $O(1)$ bits of the input, so to enumerate L as the range of an NC^0 function f, f must be able to do highly local corrections to the alleged proof while maintaining the global property that the output word belongs to L. Unlike with locally-decodable error-correcting codes, the correction here must be deterministic and always correct. This becomes so restrictive that even some very simple languages, that are regular and in AC^0, do not have such proof systems, even allowing non-uniformity. And yet there is an NP-complete language that has a uniform NC^0 proof system (See [12]). (This should not really be that surprising, because it is known that in NC^0 we can compute various cryptographic primitives.) So the class of languages with NC^0 proof systems slices vertically across complexity classes. It is still not known whether TAUT has a (possibly non-uniform) NC^0 proof system. Figure 1 shows the relationships between classes of languages with proof systems of the specified kind. (Solid arrows denote proper inclusion, dotted lines denotes incomparability.)

The work in [11] shows that languages of varying complexity (complete for NC^1, P, NP) have uniform NC^0 proof systems, while the languages EXACT-OR, MAJ amongst others do not have even non-uniform NC^0 proof systems. It then focuses on regular languages, and shows that a large subclass of regular languages has uniform NC^0 proof systems. This work takes off from that point.

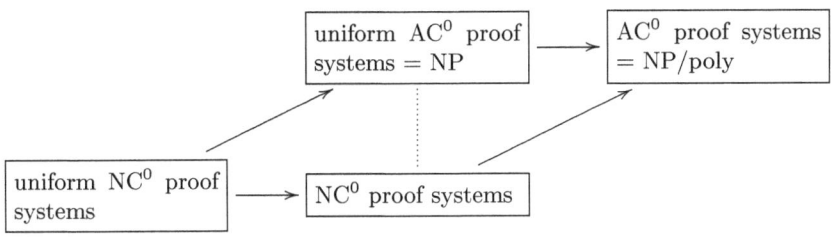

Fig. 1. Some constant-depth proof systems

Our Results

We address the question of exactly how much computational power is required to capture all regular languages via proof systems, and answer this question exactly. One of our main results (Theorem 3) is that every regular language has a proof system computable by a circuit with bounded fanin gates, depth $O(\log \log n)$, and $O(1)$ alternations. Equivalently, the proof system is computable by an AC^0 circuit where each gate has fanin $(\log n)^{O(1)}$; we refer to the class of such circuits as poly log AC^0 circuits. By the result of [11], EXACT-OR requires depth $\Omega(\log \log n)$, so (upto constant multiplicative factors) this is tight. Our proof technique also generalises to show that MAJ has poly log AC^0 proof systems (Theorem 4).

The most intriguing question here, posed in [11], is to characterize the regular languages that have NC^0 proof systems. We state a conjecture for this characterization; the conjecture throws up more questions regarding decidability of some properties of regular languages.

We believe that TAUT does not have AC^0 proof systems because otherwise NP = coNP (See [1]). As a weaker step, can we at least prove that it does not have NC^0 proof systems? Although it seems that this should be possible, we have not yet succeeded. So we ask the same question about 2TAUT, which is in NL, and hence may well have an NC^0 proof system. The standard NL algorithm for 2TAUT is via a reduction to REACH. So it is interesting to ask – does REACH have an NC^0 proof system? We do not know yet. However in our other main result, we show that undirected REACH, a language complete for L, has an NC^0 proof system (Theorem 5). Our construction relies on a careful decomposition of even-degrees-only graphs (established in the proof of Theorem 6) that may be of independent interest. We also show that directed unreachability has an NC^0 proof system (Theorem 7).

Finally, we observe that Graph Isomorphism does not have NC^0 proof systems. We also note that for every language L in NP, the language $(\{1\} \cdot L \cdot \{0\}) \cup 0^* \cup 1^*$ has both SAC^0 and $coSAC^0$ proof systems (Theorem 8).

2   Preliminaries

Unless otherwise stated, we consider only bounded fanin circuits over $\vee, \wedge, \neg$.

Definition 1 ([11]). *A circuit family* $\{C_n\}_{n>0}$ *is a proof system for a language* L *if there is a function* $m : \mathbb{N} \longrightarrow \mathbb{N}$ *such that for each* n *where* $L^{=n} \neq \emptyset$,

1. C_n *has* $m(n)$ *inputs and* n *outputs,*
2. *for each* $y \in L^{=n}$, *there is an* $x \in \{0,1\}^{m(n)}$ *such that* $C_n(x) = y$ *(complete-ness),*
3. *for each* $x \in \{0,1\}^{m(n)}$, $C_n(x) \in L^{=n}$ *(soundness).*

Note that the parameter n for C_n is the number of output bits, not input bits. NC^0 proof systems are proof systems as above where the circuit has $O(1)$ depth. The definition implies that the circuits are of linear size. AC^0 proof systems are proof systems as above where the circuit C_n has $O(\log n)$ depth but $O(1)$ alternations between gate types. Equivalently, they are proof systems as above of $n^{O(1)}$ size with unbounded fanin gates and depth $O(1)$.

Proposition 1 ([11]). *A regular language* L *satisfying any of the following has an* NC^0 *proof system:*

1. L *has a strict star-free expression (built from* ϵ, a, *and* Σ^*, *using concate-nation and union).*
2. L *is accepted by an automaton with a universally reachable absorbing final state.*
3. L *is accepted by a strongly connected automaton.*

Proposition 2 ([11])

1. *Proof systems for* MAJ *need* $\omega(1)$ *depth.*
2. *Proof systems for* EXACT-COUNT$_k^n$ *and* $\neg$TH$_{k+1}^n$ *need* $\Omega(\log(\log n - \log k))$ *depth. In particular, proof systems for* EXACT-OR *and for* EXACT-OR $\cup\, 0^*$ *need* $\Omega(\log \log n)$ *depth.*

3   Proof Systems for Regular Languages

We first explore the extent to which the structure of regular languages can be used to construct NC^0 proof systems. At the base level, we know that all finite languages have NC^0 proof systems. Building regular expressions involves unions, concatenation, and Kleene closure. And the resulting class of regular languages is also closed under many more operations. A natural idea is to somehow use the structure of the syntactic monoid (equivalently, the unique minimal deter-ministic automaton) to decide whether or not a regular language has an NC^0 proof system, and if so, to build one. Unfortunately, this idea collapses at once: the languages EXACT-OR and TH$_2$ have the same syntactic monoid; by Propo-sition 2, EXACT-OR has no NC^0 proof system; and by Proposition 1 TH$_2$ has such a proof system.

The next idea is to use the structure of a well-chosen (nondeterministic) au-tomaton for the language to build a proof system; Proposition 1 does exactly this. It describes two possible structures that can be used. However, one is sub-sumed in the other; see Observation 1 below.

Observation 1. *Let L be accepted by an automaton with a universally reachable absorbing final state. Then L is accepted by a strongly connected automaton.*

The above observation can be seen by adding ϵ moves from a universally reachable absorbing final state to every other state in the automaton accepting L to get a new automaton that accepts the same language and is strongly connected.

A small generalisation beyond strongly connected automata is automata with exactly two strongly connected components. However, the automaton for EXACT-OR is like this, so even with this small extension, we can no longer construct NC^0 proof systems. (In fact, we need as much as $\Omega(\log \log n)$ depth.)

Finite languages do not have strongly connected automata. But they are strict star-free and hence have NC^0 proof systems. Strict star-free expressions lack non-trivial Kleene closure. What can we say about their Kleene closure? It turns out that for any regular language, not just a strict-star-free one, the Kleene closure has an NC^0 proof system.

Theorem 2. *If L is regular, then L^* has an NC^0 proof system.*

Proof. Let M be an automaton accepting L, with no useless states. Adding ϵ moves from every final state to the start state q_0, and adding q_0 to the set of final states, gives an automaton M' for L^*. Now M' is strongly connected, so Proposition 1 gives the NC^0 proof system. □

Based on the above discussion and known (counter-) examples, we conjecture the following characterization. The structure implies the proof system, but the converse seems hard to prove.

Conjecture 1. Let L be a regular language. The following are equivalent:

1. L has an NC^0 proof system.
2. For some finite k, $L = \bigcup_{i=1}^{k} u_i \cdot L_i \cdot v_i$, where each u_i, v_i is a finite word, and each L_i is a regular language accepted by some strongly connected automaton.

An interesting question arising from this is whether the following languages are decidable:

$$\text{REG-SCC} = \left\{ M \mid \begin{array}{l} M \text{ is a finite-state automaton; } L(M) \text{ is accepted} \\ \text{by some strongly connected finite automaton} \end{array} \right\}$$

$$\text{REG-NC}^0\text{-PS} = \left\{ M \mid \begin{array}{l} M \text{ is a finite-state automaton; } L(M) \text{ has an } NC^0 \\ \text{proof system} \end{array} \right\}$$

(Instead of a finite-state automaton, the input language could be described in any form that guarantees that it is a regular language.)

We now establish one of our main results. NC^0 is the restriction of AC^0 where the fanin of each gate is bounded by a constant. By putting a fanin bound that is $\omega(1)$ but $o(n^c)$ for every constant c ("sub-polynomial"), we obtain intermediate classes. In particular, restricting the fanin of each gate to be at most poly $\log n$ gives the class that we call poly $\log AC^0$ lying between NC^0 and AC^0. We show that it is large enough to have proof systems for all regular languages. As mentioned earlier, Proposition 2 implies that this upper bound is tight.

Theorem 3. *Every regular language has a* poly log AC^0 *proof system.*

Proof (Sketch). We give a high-level description of the proof idea. A complete formal proof can be found in the full version of this article.

Consider a regular language L, and fix any automaton A for it. Assume for simplicity that A has a single accepting state. For a word $a = a_1 \ldots a_n$, a proof of $a \in L$ is a sequence of states $q_0, q_1, \ldots, q_n$ that is allegedly encountered on some accepting run ρ of A on x. Suppose we are given a and such a sequence. We can check local consistency ($\delta(q_{i-1}, a_i) = q_i$) and output a if all checks succeed. But this requires checking all n positions, implying $\Omega(\log n)$ depth. To circumvent this, we provide additional information in the proof. Represent the interval $(0, n]$ as a binary tree T where

1. the root corresponds to the interval $(0, n] = \{1, 2, \ldots, n\}$,
2. a node corresponding to interval $(i, j]$ has children corresponding to intervals $(i, \lceil \frac{i+j}{2} \rceil]$ (left child) and $(\lceil \frac{i+j}{2} \rceil, j]$ (right child), and
3. a node corresponding to interval $(k - 1, k]$ for $k \in [n]$ is a leaf.

We call this the interval tree. For each interval $(i, j]$ in T, we provide a pair of states $\langle u, v \rangle$; these are intended to be the states q_i and q_j in the alleged accepting run ρ. (Note that the state sequence on ρ itself is now supposed to be specified at the leaves of T.)

We use this additional information to self-correct the proof. Each leaf $(k-1, k]$ has $\theta(\log n)$ ancestors in T. To decide the bit at the kth position of the output, the proof system will look at all ancestors of $(k - 1, k]$ in the interval tree. It will find the lowest ancestor – some $(i, j]$ with $i < k \le j$ – such that $(i, j]$ and all its ancestors are locally consistent (in a sense that we define precisely). It then uses information at this ancestor (which could be the node $(k - 1, k]$ itself) to decide whether to output a_k or some other bit at position k. A complete formal proof can be found in the full version of this article. □

The above proof also works for branching programs (BPs) which have width $O(\log n^{O(1)})$ and are "structured" in the following sense:

Definition 2. *A BP for length-n inputs is* **structured** *if it satisfies the following:*

1. *It is layered: vertices are partitioned into $n+1$ layers $V_0, \ldots, V_n$ and all edges are between adjacent layers $E \subseteq \cup_i(V_{i-1} \times V_i)$.*
2. *Each layer has the same size $w = |V_i|$, the width of the BP. (This is not critical; we can let $w = \max |V_i|$.)*
3. *There is a permutation $\sigma \in S_n$ such that for $i \in [n]$, all edges in $V_{i-1} \times V_i$ read $x_{\sigma(i)}$ or $\overline{x_{\sigma(i)}}$.*

Potentially, this is much bigger than the class of languages accepted by non-uniform finite-state automata. The proof of Theorem 3, as given in the full version of this article, is stated in a way that clearly extends to polylog width structured BPs.

The language MAJ has constant-width branching programs, but these are not structured in the sense above. It can be shown that a structured BP for MAJ must have width $\Omega(n)$ (a family of growing automata M_n for MAJ, where M_n is guaranteed to be correct only on $\{0, 1\}^n$, must have $1 + n/2$ states in M_n). Nonetheless, the idea from the proof of Theorem 3 can be used to give a poly log AC^0 proof system for MAJ and for all threshold languages TH_k^n of strings with at least k 1s.

Theorem 4. *The threshold languages* TH_k^n *have* poly log AC^0 *proof systems.*

4   2TAUT, Reachability and NC^0 Proof Systems

In this section, we first look at the language Undirected Reachability, which is known to be in (and complete for) L ([13]). Intuitively, the property of connectivity is a global one. However, viewing it from a different angle gives us a way to construct an NC^0 proof system for it under the standard adjacency matrix encoding (i.e., our proof system will output adjacency matrices of all graphs that have a path between s and t, and of no other graphs). In the process, we give an NC^0 proof system for the set of all undirected graphs that are a union of edge-disjoint cycles.

Define the following languages:

$$uSTConn = \left\{ A \in \{0,1\}^{n \times n} \middle| \begin{array}{l} A \text{ is the adjacency matrix of an undirected graph } G \\ \text{where vertices } s = 1, \ t = n \text{ are in the same con-} \\ \text{nected component.} \end{array} \right\}$$

$$Cycles = \left\{ A \in \{0,1\}^{n \times n} \middle| \begin{array}{l} A \text{ is the adjacency matrix of an undirected graph} \\ G = (V, E) \text{ where } E \text{ is the union of edge-disjoint} \\ \text{simple cycles.} \end{array} \right\}$$

(For simplicity, we will say $G \in uSTConn$ or $G \in Cycles$ instead of referring to the adjacency matrices.)

Theorem 5. *The language* uSTConn *has an* NC^0 *proof system.*

Proof. We will need an addition operation on graphs: $G_1 \oplus G_2$ denotes the graph obtained by adding the corresponding adjacency matrices modulo 2. We also need a notion of upward closure: For any language A, UpClose(A) is the language $B = \{y : \exists x \in A, |x| = |y|, \forall i, x_i = 1 \implies y_i = 1\}$. In particular, if A is a collection of graphs, then B is the collection of super-graphs obtained by adding edges. Note that (undirected) reachability is monotone and hence UpClose(uSTConn) = uSTConn.

Let $L_1 = \{G = G_1 \oplus (s, t) | G_1 \in Cycles\}$ and $L_2 = $ UpClose(L_1). We show:

1. $L_2 = $ uSTConn.
2. If L_1 has an NC^0 proof system, then L_2 has an NC^0 proof system.

3. If CYCLES has an NC^0 proof system, then L_1 has an NC^0 proof system.
4. CYCLES has an NC^0 proof system.

Proof of 1: We show that $L_1 \subseteq$ USTCONN $\subseteq L_2$. Then applying upward closure, $L_2 = \text{UpClose}(L_1) \subseteq \text{UpClose}(\text{USTCONN}) = \text{USTCONN} \subseteq \text{UpClose}(L_2) = L_2$.

$L_1 \subseteq$ USTCONN: Any graph $G \in L_1$ looks like $G = H \oplus (s,t)$, where $H \in$ CYCLES. If $(s,t) \notin H$, then $(s,t) \in G$ and we are done. If $(s,t) \in H$, then s and t lie on a cycle C and hence removing the (s,t) edge will still leave s and t connected by a path $C \setminus \{(s,t)\}$.

USTCONN $\subseteq L_2$: Let $G \in$ USTCONN. Let ρ be an s-t path in G. Let $H = (V,E)$ be a graph such that $E = $ edges in ρ. Then, $G \in \text{UpClose}(\{H\})$. We can write H as $H' \oplus (s,t)$ where $H' = H \oplus (s,t) = \rho \cup (s,t)$; hence $H' \in$ CYCLES. Hence $H \in L_1$, and so $G \in L_2$.

Proof of 2: We show a more general construction for monotone properties, and then use it for USTCONN. The following lemma states that for any monotone function f, constructing a proof system for a language that sits in between Minterms(f) and $f^{-1}(1)$ suffices to get a proof system for $f^{-1}(1)$.

Lemma 1. *Let $f : \{0,1\}^* \longrightarrow \{0,1\}$ be a monotone boolean function and let $L = f^{-1}(1)$. Let L' be a subset of L that contains all the minterms of f. If L' has a proof system of depth d, size s and a alternations, then L has a proof system of depth $d+1$, size $s+n$ and at most $a+1$ alternations.*

Proof. Let C be a proof circuit for L' that takes input string x. We construct a proof system for L using C and asking another input string $y \in \{0,1\}^n$. The i'th output bit of our proof system is $C(x)_i \vee y_i$. $\square$

Now note that Minterms(USTCONN) is exactly the set of graphs where the edge set is a simple s-t path. We have seen that $L_1 \subseteq$ USTCONN. As above, we can see that $H \in$ Minterms(USTCONN) $\implies H \oplus (s,t) \in$ CYCLES $\implies H \in L_1$. Statement 2 now follows from Lemma 1.

Proof of 3: Let A be the adjacency matrix output by the the NC^0 proof system for CYCLES. The proof system for L_1 outputs A' such that $A'[s,t] = \overline{A[s,t]}$ and rest of A' is same as A.

Proof of 4: This is of independent interest, and is proved in theorem 6 below. This completes the proof of theorem 5. $\square$

We now construct NC^0 proof systems for the language CYCLES.

Theorem 6. *The language CYCLES has an NC^0 proof system.*

Proof. The idea is to find a set of triangles $T \subseteq$ CYCLES such that:

1. Every graph in CYCLES can be generated using triangles from T. i.e.,

$$\text{CYCLES} \subseteq \text{Span}(T) \triangleq \left\{ \sum_{i=1}^{|T|} a_i t_i \mid \forall i, a_i \in \{0,1\}, t_i \in T \right\}$$

2. Every graph generated from triangles in T is in CYCLES; $\text{Span}(T) \subseteq$ CYCLES.
3. $\forall u, v \in [n]$, the edge (u, v) is contained in at most 6 triangles in T.

Once we find such a set T, then our proof system asks as input the coefficients a_i which indicate the linear combination needed to generate a graph in CYCLES. An edge e is present in the output if, among the triangles that contain e, an odd number of them have coefficient set to 1 in the input. By property 3, each output edge needs to see only constant many input bits and hence the circuit we build is NC^0. We will now find and describe T in detail.

Let the vertices of the graph be numbered from 1 to n. Define the length of an edge (i, j) as $|i - j|$. A triple $\langle i, j, k \rangle$ denotes the set of triangles on vertices (u, v, w) where $|u - v| = i$, $|v - w| = j$, and $|u - w| = k$. We now define the set

$$T = \bigcup_{i=1}^{n/2} \langle i, i, 2i \rangle \cup \langle i, i+1, 2i+1 \rangle$$

Observation It can be seen that $|T| \leq \frac{3}{2}n^2$. This is linear in the length of the output, which has $\binom{n}{2}$ independent bits.

We now show that T satisfies all properties listed earlier.

T **satisfies property 3:** Take any edge $e = (u, v)$. Let its length be $l = |u - v|$. e can either be the longest edge in a triangle or one of the two shorter ones. If l is even, then e can be the longest edge for only 1 triangle in T and can be a shorter edge in at most 4 triangles in T. If l is odd, then e can be the longest edge for at most 2 triangles in T and can be a shorter edge in at most 4 triangles. Hence, any edge is contained in at most 6 triangles. T **satisfies property 2:** To see this, note first that $T \subseteq$ CYCLES. Next, observe the following closure property of cycles:

Lemma 2. *For any $G_1, G_2 \in$ CYCLES, the graph $G_1 \oplus G_2 \in$ CYCLES.*

Proof. A well-known fact about connected graphs is that they are Eulerian if and only if every vertex has even degree. The analogue for general (not necessarily connected) graphs is Veblen's theorem [14], which states that $G \in$ CYCLES if and only if every vertex in G has even degree.

Using this, we see that if for $i \in [2]$, $G_i \in$ CYCLES and if we add the adjacency matrices modulo 2, then degrees of vertices remain even and so the resulting graph is also in CYCLES. $\square$

It follows that $\text{Span}(T) \subseteq$ CYCLES.

T **satisfies property 1:** We will show that any graph $G \in$ CYCLES can be written as a linear combination of triangles in T. Define, for a graph G, the parameter $d(G) = (l, m)$ where l is the length of the longest edge in G and m is the number of edges in G that have length l. For graphs $G_1, G_2 \in$ CYCLES, with $d(G_1) = (l_1, m_1)$ and $d(G_2) = (l_2, m_2)$, we say $d(G_1) < d(G_2)$ if and only if either $l_1 < l_2$ holds or $l_1 = l_2$ and $m_1 < m_2$. Note that for any graph $G \in$ CYCLES with $d(G) = (l, m)$, $l \geq 2$.

Claim. Let $G \in$ CYCLES. If $d(G) = (2, 1)$, then $G \in T$.

Proof. It is easy to see that G has to be a triangle with edge lengths $1, 1$ and 2. All such triangles are contained in T by definition. $\square$

Lemma 3. *For every $G \in$ CYCLES with $d(G) = (l, m)$, either $G \in T$ or there is a $t \in T$, and $H \in$ CYCLES such that $G = H \oplus t$ and $d(H) < d(G)$.*

Proof. If $G \in T$, then we are done. So now consider the case when $G \notin T$:

Let e be a longest edge in G. Let C be the cycle which contains e. Pick $t \in T$ such that e is the longest edge in t. G can be written as $H \oplus t$ where $H = G \oplus t$. From Lemma 2 and since $T \subseteq$ CYCLES, we know that $H \in$ CYCLES. Let t have the edges e, e_1, e_2. Any edge present in both G and t will not be present in H. Since $e \in G \cap t$, $e \notin H$. Length of e_1 and e_2 are both less than l since e was the longest edge in t. Hence the number of times an edge of length l appears in H is reduced by 1 and the new edges added(if any) to H (namely e_1 and e_2) have length less then l. Hence if $m > 1$, then $d(H) = (l, m - 1) < d(G)$. If $m = 1$, then $d(H) = (l', m')$ for some m' and $l' < l$, and hence $d(H) < d(G)$. $\square$

By repeatedly applying Lemma 3, we can obtain the exact combination of triangles from T that can be used to give any $G \in$ CYCLES. A more formal proof will proceed by induction on the parameter $d(G)$ and each application of Lemma 3 gives a graph H with a $d(H) < d(G)$ and hence allows for the induction hypothesis to be applied. The base case of the induction is given by Lemma 4. Hence T satisifes property 1.

Since T satisfies all three properties, we obtain an NC^0 proof system for CYCLES, proving the theorem. $\square$

The above proof does not work for directed REACH. However, we can show that directed un-reachability can be captured by NC^0 proof systems.

Theorem 7. *The language* UNREACH *defined below has an NC^0 proof system under the standard adjacency matrix encoding.*

$$\text{UNREACH} = \left\{ A \in \{0, 1\}^{n \times n} \middle| \begin{array}{l} A \text{ is the adjacency matrix of a directed graph } G \\ \text{with no path from } s = 1 \text{ to } t = n. \end{array} \right\}$$

Proof. As proof, we take as input an adjacency matrix A and an n-bit vector X with $X(s) = 1$ and $X(t) = 0$ hardwired. Intuitively, X is like a characteristic vector that represents all vertices that can be reached by s.

The adjacency matrix B output by our proof system is:

$$B[i, j] = \begin{cases} 1 & \text{if } A[i, j] = 1 \text{ and it is not the case that } X(i) = 1 \text{ and } X(j) = 0, \\ 0 & \text{otherwise} \end{cases}$$

Soundness: No matter what A is, X describes an s, t cut since $X(s) = 1$ and $X(t) = 0$. So any gaph output by the proof system will not have a path from s to t.

Completeness: For any $G \in$ UNREACH, use the adjacency matrix of G as A and give input X such that $X(v) = 1$ for a vertex v if and only if v is reachable from s. $\square$

5   Discussion

In this section, we discuss certain observations that we have made and remark on related problems.

We know that any language in NP has AC^0 proof systems. Srikanth Srinivasan recently showed why proof system for characteristic functions of constant rate, linear distance binary error correcting codes require $\Omega(\log n)$ depth. Codes that can be computed efficiently and achieve these parameters are known (See for eg Justesen Code [15]). Thus AC^0 seems necessary. However, we note that proof systems for a big fragment of NP do not require the full power of AC^0:

Theorem 8. *Let L be any language in* NP.

1. *If L contains* 0^*, *then L has a proof system where negations appear only at leaf level,* $\wedge$ *gates have unbounded fanin,* $\vee$ *gates have* $O(1)$ *fanin, and the depth is* $O(1)$. *That is, L has a* coSAC^0 *proof system.*
2. *If L contains* 1^*, *then L has a proof system where negations appear only at leaf level,* $\vee$ *gates have unbounded fanin,* $\wedge$ *gates have* $O(1)$ *fanin, and the depth is* $O(1)$. *That is, L has an* SAC^0 *proof system.*
3. *The language* $(\{1\} \cdot L \cdot \{0\}) \cup 0^* \cup 1^*$ *has both* SAC^0 *and* coSAC^0 *proof systems.*

Using Lemma 1 and the known lower bound for MAJ from [11], we can show that the following languages have no NC^0 proof systems:

Lemma 4. *The following languages do not have* NC^0 *proof systems.*

1. EXMAJ, *consisting of strings x with exactly* $\lceil |x|/2 \rceil$ *1s.*
2. $L = \{xy \mid x, y \in \{0,1\}^*, |x| = |y|, |x|_1 = |y|_1\}$.
3. GI $= \{G_1, G_2 \mid \text{Graph } G_1 \text{ is isomorphic to graph } G_2\}$.
   Here we assume that G_1 *and* G_2 *are specified via their 0-1 adjacency matrices, and that 1s on the diagonal are allowed (the graphs may have self-loops).*

For MAJ, we have given a proof system with $O(\log \log n)$ depth (and $O(1)$ alternations), and it is known from [11] that $\omega(1)$ depth is needed. Can this gap between the upper and lower bounds be closed?

Can we generalize the idea we use in Theorem 5 and apply it to other languages? In particular, can we obtain good upper bounds using this technique for the language of *s-t* connected directed graphs? From the results of [11] and this paper, we know languages complete for NC^1, L, P and NP with NC^0 proof systems. A proof system for REACH would bring NL into this list.

Our construction from Theorem 3 can be generalized to work for languages accepted by growing-monoids or growing-non-uniform-automata with poly-log growth rate (see eg [16]). Can we obtain good upper bounds for linearly growing automata?

In [17], proof systems computable in DLOGTIME are investigated. The techniques used there seem quite different from those that work for small-depth circuits, especially poly log AC^0. Though in both cases each output bit can depend on at most poly $\log n$ input bits, the circuit can pick an arbitrary set of poly $\log n$ bits whereas a DLOGTIME proof system needs to write the index of each bit on the index tape using up $\log n$ time.

References

1. Cook, S.A.: The complexity of theorem-proving procedures. In: Proceedings of the Annual ACM Symposium on Theory of Computing, pp. 151–158 (1971)
2. Hirsch, E.A.: Optimal acceptors and optimal proof systems. In: Kratochvíl, J., Li, A., Fiala, J., Kolman, P. (eds.) TAMC 2010. LNCS, vol. 6108, pp. 28–39. Springer, Heidelberg (2010)
3. Hirsch, E.A., Itsykson, D.: On optimal heuristic randomized semidecision procedures, with application to proof complexity. In: Proceedings of 27th International Symposium on Theoretical Aspects of Computer Science, STACS, pp. 453–464 (2010)
4. Cook, S.A., Krajíček, J.: Consequences of the provability of NP ⊆ P/poly. Journal of Symbolic Logic 72(4), 1353–1371 (2007)
5. Pudlák, P.: Quantum deduction rules. Ann. Pure Appl. Logic 157(1), 16–29 (2009), See also ECCC TR07-032
6. Cook, S.A., Reckhow, R.A.: The relative efficiency of propositional proof systems. Journal of Symbolic Logic 44(1), 36–50 (1979)
7. Beame, P., Pitassi, T.: Propositional proof complexity: Past, present, and future. In: Current Trends in Theoretical Computer Science, pp. 42–70. World Scientific (2001)
8. Segerlind, N.: The complexity of propositional proofs. Bulletin of Symbolic Logic 13(4), 417–481 (2007)
9. Furst, M.L., Saxe, J.B., Sipser, M.: Parity, circuits, and the polynomial-time hierarchy. Mathematical Systems Theory 17(1), 13–27 (1984)
10. Håstad, J.: Almost optimal lower bounds for small depth circuits. In: Proceedings of the 18th Annual ACM Symposium on Theory of Computing STOC, pp. 6–20 (1986)
11. Beyersdorff, O., Datta, S., Krebs, A., Mahajan, M., Scharfenberger-Fabian, G., Sreenivasaiah, K., Thomas, M., Vollmer, H.: Verifying proofs in constant depth. ACM Transactions on Computation Theory (to appear, 2013) See also ECCC TR012-79. A preliminary version appeared in [18]
12. Cryan, M., Miltersen, P.B.: On pseudorandom generators in NC^0. In: Sgall, J., Pultr, A., Kolman, P. (eds.) MFCS 2001. LNCS, vol. 2136, pp. 272–284. Springer, Heidelberg (2001)
13. Reingold, O.: Undirected st-connectivity in log-space. In: Proceedings of the Thirty-Seventh Annual ACM Symposium on Theory of Computing STOC, pp. 376–385. ACM, New York (2005)
14. Veblen, O.: An application of modular equations in analysis situs. Annals of Mathematics 14(1/4), 86–94 (1912)
15. Justesen, J.: Class of constructive asymptotically good algebraic codes. IEEE Transactions on Information Theory 18(5), 652–656 (1972)
16. Bedard, F., Lemieux, F., McKenzie, P.: Extensions to Barrington's M-program model. Theoretical Computer Science 107, 31–61 (1993)
17. Krebs, A., Limaye, N.: Dlogtime-proof systems. Electronic Colloquium on Computational Complexity (ECCC) 19, 186 (2012)
18. Beyersdorff, O., Datta, S., Mahajan, M., Scharfenberger-Fabian, G., Sreenivasaiah, K., Thomas, M., Vollmer, H.: Verifying proofs in constant depth. In: Murlak, F., Sankowski, P. (eds.) MFCS 2011. LNCS, vol. 6907, pp. 84–95. Springer, Heidelberg (2011)

Reversibility of Computations in Graph-Walking Automata*

Michal Kunc[1] and Alexander Okhotin[2]

[1] Department of Mathematics and Statistics,
Masaryk University, Brno, Czech Republic
kunc@math.muni.cz
[2] Department of Mathematics and Statistics, University of Turku, Finland
alexander.okhotin@utu.fi

Abstract. The paper proposes a general notation for deterministic automata traversing finite undirected structures: the graph-walking automata. This abstract notion covers such models as two-way finite automata, including their multi-tape and multi-head variants, tree-walking automata and their extension with pebbles, picture-walking automata, space-bounded Turing machines, etc. It is then demonstrated that every graph-walking automaton can be transformed to an equivalent reversible graph-walking automaton, so that every step of its computation is logically reversible. This is done with a linear blow-up in the number of states, where the linear factor depends on the degree of graphs being traversed. The construction directly applies to all basic models covered by this abstract notion.

1 Introduction

Logical reversibility of computations is an important property of computational devices in general, which can be regarded as a stronger form of determinism. Informally, a machine is reversible, if, given its configuration, one can always uniquely determine its configuration at the previous step. This property is particularly relevant to the physics of computation, as irreversible computations incur energy dissipation [18]. It is known from Lecerf [20] and Bennett [3] that every Turing machine can be simulated by a reversible Turing machine. Later, the time and space cost of reversibility was analyzed in the works of Bennett [4], Crescenzi and Papadimitriou [10], Lange et al. [19] and Buhrman et al. [8]. A line of research on reversibility in high-level programming languages was initiated by Abramsky [1]. Reversibility in cellular automata also has a long history of research, presented in surveys by Toffoli and Margolus [26] and by Kari [15]. In the domain of finite automata, the reversible subclass of one-way deterministic finite automata (1DFAs) defines a proper subfamily of regular languages [23]. On the other hand, every regular language is accepted by a reversible two-way finite automaton (2DFA): as shown by Kondacs and Watrous [16], every n-state 1DFA can be simulated by a $2n$-state reversible 2DFA.

---

* Supported by ESF project CZ.1.07/2.3.00/20.0051 "Algebraic Methods in Quantum Logic", and by Academy of Finland project 257857.

K. Chatterjee and J. Sgall (Eds.): MFCS 2013, LNCS 8087, pp. 595–606, 2013.
© Springer-Verlag Berlin Heidelberg 2013

One of the most evident consequences of reversibility is that a reversible automaton halts on every input (provided that the state-space is bounded). The property of halting on all inputs has received attention on its own. For time-bounded and space-bounded Turing machines, halting can be ensured by explicitly counting the number of steps, as done by Hopcroft and Ullman [14]. A different method for transforming a space-bounded Turing machine to an equivalent halting machine operating within the same space bounds was proposed by Sipser [24], and his approach essentially means constructing a reversible machine, though reversibility was not considered as such. In particular, Sipser [24] sketched a transformation of an n-state 2DFA to an $O(n^2)$-state halting 2DFA (which is actually reversible), and also mentioned the possibility of an improved transformation that yields $O(n)$ states, where the multiplicative factor depends upon the size of the alphabet. The fact that Sipser's idea produces reversible automata was noticed and used by Lange et al. [19] to establish the equivalence of deterministic space $s(n)$ to reversible space $s(n)$. Next, Kondacs and Watrous [16] distilled the construction of Lange et al. [19] into the mathematical essence of constructing reversible 2DFAs. A similar construction for making a 2DFA halt on any input was later devised by Geffert et al. [13], who have amalgamated an independently discovered method of Kondacs and Watrous [16] with a pre-processing step. For tree-walking automata (TWA), a variant of Sipser's [24] construction was used by Muscholl et al. [22] to transform an n-state automaton to an $O(n^2)$-state halting automaton.

The above results apply to various models that recognize input structures by traversing them: such are the 2DFAs that walk over input strings, and the TWAs walking over input trees. More generally, these results apply to such models as deterministic space-bounded Turing machines, which have extra memory at their disposal, but the amount of memory is bounded by a function of the size of the input. What do these models have in common? They are equipped with a fixed finite-state control, as well as with a finite space of memory configurations determined by the input data, and with a fixed finite set of operations on this memory. A machine of such a type is defined by a transition table, which instructs it to apply a memory operation and to change its internal state, depending on the current state and the currently observed data stored in the memory.

This paper proposes a general notation for such computational models: the *graph-walking automata* (GWA). In this setting, the space of memory configurations is regarded as an input graph, where each node is a memory configuration, labelled by the data observed by the machine in this position, and the operations on the memory become labels of the edges. Then a graph-walking automaton traverses an input graph using a finite-state control and a transition function with a finite domain, that is, at any moment the automaton observes one of finitely many possibilities. The definitions assume the following conditions on the original models, which accordingly translate to graph-walking automata; these assumptions are necessary to transform deterministic machines to reversible ones:

1. Every elementary operation on the memory has an opposite elementary operation that undoes its effect. For instance, in a 2DFA, the operation of moving

the head to the left can be undone by moving the head to the right. In terms of graphs, this means that *input graphs are undirected*, and each edge has its end-points labelled by two opposite direction symbols, representing traversal of this edge in both directions.

2. The space of memory configurations on each given input object is finite, and it functionally depends on the input data. For graph-walking automata, this means that *input graphs are finite*. Though, in general, reversible computation is possible in devices with unbounded memory, the methods investigated in this paper depend upon this restriction.

3. The automaton can test whether the current memory configuration is the initial configuration. In a graph-walking automaton, this means that the initial node, where the computation begins, has a distinguished label.

Besides the aforementioned 2DFAs, TWAs and space-bounded Turing machines, graph-walking automata cover such generalizations as multi-head automata, automata with pebbles, etc.

The goal of this paper is to deal with the reversibility of computations on the general level, as represented by the model of graph-walking automata. The main results of this paper are transformations from automata of the general form to the *returning automata*, which may accept only in the initial node, and from returning to *reversible automata*. Both transformations rely on the same effective construction, which generalizes the method of Kondacs and Watrous [16], while the origins of the latter can be traced to the general idea due to Sipser [24]. The constructions involve only a linear blow-up in the number of states. Both results apply to every concrete model of computation representable as GWAs.

Investigating further properties of graph-walking automata is proposed as a worthy subject for future research. Models of this kind date back to *automata in labyrinths*, introduced by Shannon and later studied by numerous authors as a model of graph exploration by an agent following the edges of an undirected graph. This line of research has evolved into a thriving field of algorithms for searching and automatic mapping of graphs, which is surveyed in the recent paper by Fraigniaud et al. [12]. Other important models defining families of graphs are graph-rewriting systems and monadic second-order logic on graphs researched by Courcelle [9], and graph tilings studied by Thomas [25].

2 Graph-Walking Automata

The automata studied in this paper walk over finite undirected graphs, in which every edge can be traversed in both directions. The directions are identified by labels attached to both ends of an edge. These labels belong to a finite set of *directions* D, with a bijective operation $-: D \to D$ representing *opposite directions*. If a graph models the memory, the directions represent elementary operations on this memory, and the existence of opposite directions means that every elementary operation on the memory can be reversed by applying its opposite.

Definition 1. *A signature S consists of*

- *a finite set of directions D;*
- *a bijective operation $-: D \to D$, satisfying $-(-d) = d$ for all $d \in D$;*
- *a finite set Σ of possible labels of nodes of the graph;*
- *a non-empty subset $\Sigma_0 \subseteq \Sigma$ of labels allowed in the initial node;*
- *a set $D_a \subseteq D$ of directions for every $a \in \Sigma$.*

In each graph over $\mathcal{S}$, every node labelled with $a \in \Sigma$ must be of degree $|D_a|$, with the incident edges corresponding to the elements of D_a.

Definition 2. *A graph over the signature $\mathcal{S}$ is a quadruple $(V, v_0, +, \lambda)$, where*

- *V is a finite set of nodes;*
- *$v_0 \in V$ is the initial node;*
- *$+: V \times D \to V$ is a partial mapping, satisfying the following condition of invertibility by opposite directions: for every $v \in V$ and $d \in D$, if $v + d$ is defined, then $(v + d) + (-d)$ is defined too and $(v + d) + (-d) = v$. In the following, $v - d$ denotes $v + (-d)$;*
- *the total mapping $\lambda: V \to \Sigma$ is a labelling of nodes, such that for all $v \in V$,*
  (i) $d \in D_{\lambda(v)}$ if and only if $v + d$ is defined,
  (ii) $\lambda(v) \in \Sigma_0$ if and only if $v = v_0$.

Definition 3. *A deterministic graph-walking automaton (GWA) over a signature $\mathcal{S} = (D, -, \Sigma, \Sigma_0, (D_a)_{a \in \Sigma})$ is a quadruple $\mathcal{A} = (Q, q_0, \delta, F)$, where*

- *Q is a finite set of internal states,*
- *$q_0 \in Q$ is the initial state,*
- *$F \subseteq Q \times \Sigma$ is a set of acceptance conditions, and*
- *$\delta: (Q \times \Sigma) \setminus F \to Q \times D$ is a partial transition function, with $\delta(q, a) \in Q \times D_a$ for all a and q where it is defined.*

Given a graph $(V, v_0, +, \lambda)$, the automaton begins its computation in the state q_0, observing the node v_0. At each step of the computation, with the automaton in a state $q \in Q$ observing a node v, the automaton looks up the transition table δ for q and the label of v. If $\delta(q, \lambda(v))$ is defined as (q', d), the automaton enters the state q' and moves to the node $v + d$. If $\delta(q, \lambda(v))$ is undefined, then the automaton accepts the graph if $(q, \lambda(v)) \in F$ and rejects otherwise.

The two most well-known special cases of GWAs are the 2DFAs, which walk over path graphs, and TWAs operating on trees.

Example 1. A two-way deterministic finite automaton (2DFA) operating on a tape delimited by a left-end marker $\vdash$ and a right-end marker $\dashv$, with the tape alphabet Γ, is a graph-walking automaton operating on graphs over the signature $\mathcal{S}$ with $D = \{+1, -1\}$, $\Sigma = \Gamma \cup \{\vdash, \dashv\}$, $\Sigma_0 = \{\vdash\}$, $D_\vdash = \{+1\}$, $D_\dashv = \{-1\}$ and $D_a = \{+1, -1\}$ for all $a \in \Gamma$.

All connected graphs over this signature are path graphs, containing one instance of each end-marker and an arbitrary number of symbols from Γ in between.

For an input string $w = a_1 \ldots a_n$, with $n \geqslant 0$, the corresponding graph has the set of nodes $V = \{0, 1, \ldots, n, n+1\}$ representing positions on the tape, with $v_0 = 0$ and with $v + d$ defined as the sum of integers. These nodes are labelled as follows: $\lambda(0) = \vdash$, $\lambda(n+1) = \dashv$ and $\lambda(i) = a_i$ for all $i \in \{1, \ldots, n\}$.

Consider *tree-walking automata*, defined by Aho and Ullman [2, Sect. VI] and later studied by Bojańczyk and Colcombet [6,7]. Given an input binary tree, a tree-walking automaton moves over it, scanning one node at a time. At each step of its computation, it may either go down to any of the sons of the current node or up to its father. Furthermore, in any node except the root, the automaton is invested with the knowledge of whether this node is the first son or the second son [6]. Traversal of trees by these automata can be described using directions of the form "go down to the i-th son" and the opposite "go up from the i-th son to its father".

In the notation of graph-walking automata, the knowledge of the number of the current node among its siblings is given in its label: for each label a, the set of valid directions D_a contains exactly one upward direction and all downward directions. Furthermore, by analogy with 2DFAs, the input trees of tree-walking automata shall have end-markers attached to the root and to all leaves; in both cases, these markers allow a better readable definition.

Example 2. A tree-walking automaton on k-ary trees uses the set of directions $D = \{+1, +2, \ldots, +k, -1, -2, \ldots, -k\}$, with $-(+i) = -i$, where positive directions point to children and negative ones to fathers. Trees are graphs labelled with symbols in $\Sigma = \{\top, \bot_1, \ldots, \bot_k\} \cup \Gamma$, where the top marker $\top$ with $D_\top = \{+1\}$ is the label of the root v_0 (and accordingly, $\Sigma_0 = \{\top\}$), while each i-th bottom marker $\bot_i$ with $D_{\bot_i} = \{-i\}$ is a label for leaves. Elements of the set Γ are used to label internal nodes of the tree, so that for each $a \in \Gamma$ there exists $i \in \{1, \ldots, k\}$ with $D_a = \{-i, +1, \ldots, +k\}$, which means that every node labelled by a is the i-th child of its father.

In general, consider any computational device recognizing input objects of any kind, which has a fixed number of internal states and employs auxiliary memory holding such data as the positions of reading heads and the contents of any additional data structures. Assume that for each fixed input, the total space of possible memory configurations of the device and the structure of admissible transitions between these configurations are known in advance. The set of memory configurations with the structure of transitions forms a *graph of memory configurations*, which can be presented in the notation assumed in this paper by taking *elementary operations on the memory* as directions. The label attached to the currently observed node represents the information on the memory configuration available to the original device, such as the contents of cells observed by heads; along with its internal state, this is all the data it can use to determine its next move. Thus the device is represented as a graph-walking automaton.

As an example of such a representation, consider 2DFAs equipped with multiple reading heads, which can independently move over the same input tape: the *multi-head automata*.

Example 3. A k-head 2DFA with a tape alphabet Γ is described by a graph-walking automaton as follows. Its memory configuration contains the positions of all k heads on the tape. The set of directions is $D = \{-1, 0, +1\}^k \setminus \{0\}^k$, where a direction $(s_1, \ldots, s_k)$ with $s_i \in \{-1, 0, +1\}$ indicates that each i-th head is to be moved in the direction s_i. Each label in $\Sigma = \left(\Gamma \cup \{\vdash, \dashv\}\right)^k$ contains all the data observed by the automaton in a given memory configuration: this is a k-tuple of symbols scanned by all heads. There is a unique initial label corresponding to all heads parked at the left-end marker, that is, $\Sigma_0 = \{(\vdash, \ldots, \vdash)\}$. For each node label $(s_1, \ldots, s_k) \in \Sigma$, the set of directions $D_{(s_1, \ldots, s_k)}$ contains all k-tuples $(d_1, \ldots, d_k) \in \{-1, 0, +1\}^k$, where $d_i \neq -1$ if $s_i = \vdash$ and $d_i \neq +1$ if $s_i = \dashv$; the latter conditions disallow moving any heads beyond either end-marker.

The automaton operates on graphs of the following form. For each input string $a_1 \ldots a_n \in \Gamma^*$, let $a_0 = \vdash$ and $a_{n+1} = \dashv$ for uniformity. Then the set of nodes of the graph is a discrete k-dimensional cube $V = \{0, 1, \ldots, n, n+1\}^k$, with each node $(i_1, \ldots, i_k) \in V$ labelled with $(a_{i_1}, \ldots, a_{i_k}) \in \Sigma$. The initial node is $v_0 = (0, \ldots, 0)$, labelled with $(\vdash, \ldots, \vdash)$.

The graphs representing memory configurations of k-head 2DFAs, as described in Example 3, are not all connected graphs over the given signature. If edges are connected differently than in a grid of the form given above, the resulting graph no longer corresponds to the space of configurations of a k-head 2DFA on any input. However, on the subset of graphs of the intended form, a GWA defined in Example 3 correctly represents the behaviour of a k-head 2DFA.

Several other models of computation can be described by GWAs in a similar way. Consider *two-way finite automata with pebbles*, introduced by Blum and Hewitt [5]: these are 2DFAs equipped with a fixed number of pebbles, which may be dispensed at or collected from the currently visited cell. When such automata are represented as GWAs, the currently visited node of a graph represents the positions of the head and pebbles, while the label encodes the symbol observed by the head, together with the information on which pebbles are currently placed, and which of them are placed at the observed cell. This model can be extended to *tree-walking automata with pebbles*, first considered by Engelfriet and Hoogeboom [11] and subsequently studied by Muscholl et al. [22]. All these models can be further extended to have multiple reading heads, to work over multidimensional arrays (such as the 4DFAs of Blum and Hewitt [5]), etc., and each case can be described by an appropriate kind of GWAs operating over graphs that encode the space of memory configurations of the desired automata.

Typical models that cannot be described as automata walking on undirected graphs are those, which cannot immediately return to the previous configuration after any operation. Such are the 1DFAs [23] or pushdown automata [17].

3   Reversibility and Related Notions

The definition of logical reversibility for graph-walking automata is comprised of several conditions, and the first condition is that each state is accessed from a unique direction.

Definition 4. *A graph-walking automaton is called* direction-determinate, *if every state is reachable from a unique direction, that is, there exists a partial function* $d: Q \to D$, *such that* $\delta(q, a) = (q', d')$ *implies* $d' = d(q')$. *As the direction is always known, the notation for the transition function can be simplified as follows: for each* $a \in \Sigma$, *let* $\delta_a: Q \to Q$ *be a partial function defined by* $\delta_a(p) = q$ *if* $\delta(p, a) = (q, d(q))$.

A GWA can be made direction-determinate by storing the last used direction in its state.

Lemma 1. *For every graph-walking automaton with a set of states* Q *and a set of directions* D, *there exists a direction-determinate automaton with the set of states* $Q \times D$, *which recognizes the same set of graphs.*

Another subclass of automata requires returning to the initial node after acceptance.

Definition 5. *A graph-walking automaton is called* returning, *if it has* $F \subseteq Q \times \Sigma_0$, *that is, if it accepts only at the initial node.*

For each computational model mentioned in Section 2, returning after acceptance is straightforward: a 2DFA moves its head to the left, a 2DFA with pebbles picks up all its pebbles, a space-bounded Turing machine erases its work tape, etc. However, for graphs of the general form, finding a way back to the initial node from the place where the acceptance decision was reached is not a trivial task. This paper defines a transformation to a returning automaton, which finds the initial node by backtracking the accepting computation.

Theorem 1. *For every direction-determinate graph-walking automaton with* n *states, there exists a direction-determinate returning graph-walking automaton with* $3n$ *states recognizing the same set of graphs.*

For every direction-determinate graph-walking automaton, consider the inverses of transition functions by all labels, $\delta_a^{-1}: Q \to 2^Q$ for $a \in \Sigma$, defined by $\delta_a^{-1}(q) = \{ p \mid \delta_a(p) = q \}$. Given a configuration of a direction-determinate automaton, one can always determine the direction d, from which the automaton came to the current node v at the previous step; and if the function $\delta_{\lambda(v-d)}$ is furthermore injective, then the state at the previous step is also known, and hence the configuration at the previous step is uniquely determined. This leads to the following definition of automata, whose computations can be uniquely reconstructed from their final configurations:

Definition 6. *A direction-determinate graph-walking automaton is* reversible, *if*

   i. *every partial function* δ_a *is injective, that is,* $|\delta_a^{-1}(q)| \leqslant 1$ *for all* $a \in \Sigma$ *and* $q \in Q$, *and*

   ii. *the automaton is returning, and for each* $a_0 \in \Sigma_0$, *there exists at most one state* q, *such that* $(q, a_0) \in F$ *(this state is denoted by* $q_{\mathrm{acc}}^{a_0}$*).*

The second condition ensures that if an input graph $(V, v_0, +, \lambda)$ is accepted, then it is accepted in the configuration $(q_{\text{acc}}^{\lambda(v_0)}, v_0)$. Therefore, this assumed accepting computation can be traced back, beginning from its final configuration, until either the initial configuration (q_0, v_0) is reached (which means that the automaton accepts), or a configuration without predecessors is encountered (then the automaton does not accept this graph). This reverse computation can be carried out by another reversible GWA.

Lemma 2. *On each finite input graph* $(V, v_0, +, \lambda)$, *a reversible graph-walking automaton beginning in an arbitrary configuration* $(\widehat{q}, \widehat{v})$ *either halts after finitely many steps, or returns to the configuration* $(\widehat{q}, \widehat{v})$ *and loops indefinitely.*

The second case in Lemma 2 allows a reversible automaton to be non-halting, if its initial configuration can be re-entered. This possibility may be ruled out by disallowing any transitions leading to the initial state. Another imperfection of reversible automata is that while they may accept only in a single designated configuration, there are no limitations on where they may reject. Thus, backtracking a rejecting computation is not possible, because it is not known where it ends. The below strengthened definition additionally requires rejection to take place in a unique configuration, analogous to the accepting configuration.

Definition 7. *A strongly reversible automaton is a reversible automaton* $\mathcal{A} = (Q, q_0, \delta, F)$ *with non-reenterable initial state, which additionally satisfies the following conditions:*

iii. *for every non-initial label* $a \in \Sigma \setminus \Sigma_0$, *the partial function* δ_a *is a bijection from* $\{ p \in Q \mid -d(p) \in D_a \}$ *to* $\{ q \in Q \mid d(q) \in D_a \}$,

iv. *for each initial label* $a_0 \in \Sigma_0$, *there is at most one designated rejecting state* $q_{\text{rej}}^{a_0} \in Q$, *for which neither* $\delta_{a_0}(q_{\text{rej}}^{a_0})$ *is defined, nor* $(q_{\text{rej}}^{a_0}, a_0)$ *is in* F,

v. *for all* $a_0 \in \Sigma_0$ *and for all states* $q \in Q \setminus \{q_{\text{acc}}^{a_0}, q_{\text{rej}}^{a_0}\}$, $\delta_{a_0}(q)$ *is defined if and only if* $-d(q) \in D_{a_0}$ *or* $q = q_0$.

The requirement on the range of δ_a, with $a \notin \Sigma_0$, in condition (iii) means that if a-labelled nodes have a direction $d \in D_a$ for reaching a state $q \in Q$, then there is a state $p \in Q$, in which this direction d can be used to get to q. The requirement on the domain of δ_a means that this function is defined precisely for those states p, which can be possibly entered in a-labelled nodes, that is, for such states p, that the direction for entering p leads to these nodes. This in particular implies that whenever a computation of a strongly reversible automaton enters a configuration (p, v) with $\lambda(v) \notin \Sigma_0$ (that is, $v \neq v_0$), the next step of the computation is defined and the automaton cannot halt in this configuration. Similarly, condition (v) ensures that the computation cannot halt in the initial node, unless it reaches either the corresponding accepting state $q_{\text{acc}}^{a_0}$ or the corresponding rejecting state $q_{\text{rej}}^{a_0}$. Because the initial state of a strongly reversible automaton is not re-enterable, Lemma 2 guarantees that its computation beginning in the initial configuration always halts, with its head scanning the initial node, and either in the accepting state or in the rejecting state.

Lemma 3. *For every finite input graph $(V, v_0, +, \lambda)$, a strongly reversible graph-walking automaton, starting in the initial configuration, either accepts in the configuration $(q_{\text{acc}}^{\lambda(v_0)}, v_0)$ or rejects in the configuration $(q_{\text{rej}}^{\lambda(v_0)}, v_0)$.*

The transformation of a deterministic automaton to a reversible one developed in this paper ensures this strongest form of reversibility.

Theorem 2. *For every direction-determinate returning graph-walking automaton with n states, there exists a strongly reversible graph-walking automaton with $2n + 1$ states recognizing the same set of graphs.*

Theorems 1–2 and Lemma 1 together imply the following transformation:

Corollary 1. *For every graph-walking automaton with n states and d directions, there exists a strongly reversible automaton with $6dn + 1$ states recognizing the same set of graphs.*

4   Reversible Simulation of Irreversible Automata

The fundamental construction behind all results of this paper is the following reversible simulation of an arbitrary deterministic graph-walking automaton.

Lemma 4. *For every direction-determinate automaton $\mathcal{A} = (Q, q_0, \delta, F)$ there exists a reversible automaton $\mathcal{B} = (\overrightarrow{Q} \cup [Q], \delta', F')$ without an initial state, where $\overrightarrow{Q} = \{ \overrightarrow{q} \mid q \in Q \}$ and $[Q] = \{ [q] \mid q \in Q \}$ are disjoint copies of Q, with the corresponding directions $d'(\overrightarrow{q}) = d(q)$ and $d'([q]) = -d(q)$, and with acceptance conditions $F' = \{ ([\delta_{a_0}(q_0)], a_0) \mid a_0 \in \Sigma_0, \, \delta_{a_0}(q_0) \text{ is defined} \}$, which has the following property: For every graph $(V, v_0, +, \lambda)$, its node $\widehat{v} \in V$ and a state $\widehat{q} \in Q$ of the original automaton, for which $(\widehat{q}, \lambda(\widehat{v})) \in F$ and $-d(\widehat{q}) \in D_{\lambda(\widehat{v})}$, the computation of $\mathcal{B}$ beginning in the configuration $([\widehat{q}], \widehat{v} - d(\widehat{q}))$,*

- *accepts in the configuration $([\delta_{\lambda(v_0)}(q_0)], v_0)$, if $(\widehat{q}, \widehat{v}) \neq (q_0, v_0)$ and $\mathcal{A}$ accepts this graph in the configuration $(\widehat{q}, \widehat{v})$, as shown in Figure 1 (case 1).*
- *rejects in $(\overrightarrow{\widehat{q}}, \widehat{v})$, otherwise (see Figure 1, case 2).*

Proof (the overall idea). As per Sipser's [24] general approach, the automaton $\mathcal{B}$ searches through the tree of the computations of $\mathcal{A}$ leading to the configuration $(\widehat{q}, \widehat{v})$, until it finds the initial configuration of $\mathcal{A}$ or until it verifies that the initial configuration is not in the tree. While searching, it remembers *a single state* of $\mathcal{A}$, as well as *one bit* of information indicating the current direction of search: a state $[q] \in [Q]$ means tracing the computation in reverse, while in a state $\overrightarrow{q} \in \overrightarrow{Q}$ the computation of $\mathcal{A}$ is simulated forward[1].

---

[1] To compare, Sipser [24], followed by Muscholl et al. [22], has the simulating automaton remember *two states* of the original automaton, leading to a quadratic size blowup, while Morita's [21] simulation remembers a state and a symbol.

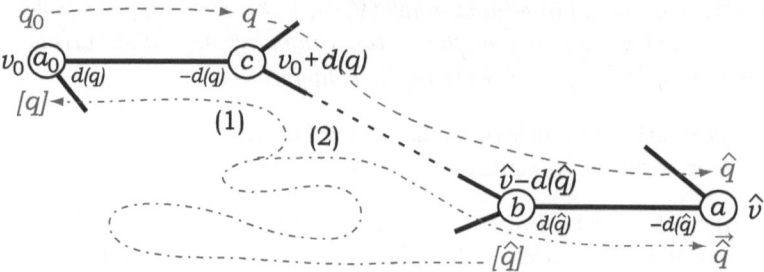

Fig. 1. Reversible GWA $\mathcal{B}$ in Lemma 4 checking whether $\mathcal{A}$ accepts in $(\widehat{q}, \widehat{v})$:
(1) if so, accept in $([q], v_0)$, where $q = \delta_{a_0}(q_0)$; (2) otherwise, reject in $(\overrightarrow{q}, \widehat{v})$

Whenever $\mathcal{B}$ reaches a state $[q]$ in a node v, this means that the computation of $\mathcal{A}$, beginning in the state q with the head in the neighbouring node $v + d(q)$, eventually leads to the configuration $(\widehat{q}, \widehat{v})$. In this way, the backward computation traces the state and the position of the head in a forward computation, but the state and the position are always out of synchronization by one step. When the automaton switches to forward simulation, and reaches a state $\overrightarrow{q}$, its head position is synchronized with the state, and this represents the original automaton's being in the state q, observing the same node.

The proofs of both theorems follow from this lemma. In the proof of Theorem 1, an arbitrary direction-determinate GWA $\mathcal{A}$ is transformed to a returning direction-determinate GWA, which operates as follows: first it simulates $\mathcal{A}$ until it accepts, and then backtracks the accepting computation of $\mathcal{A}$ to its initial configuration, using the reversible automaton constructed from $\mathcal{A}$ according to Lemma 4. If $\mathcal{A}$ rejects or loops, the constructed automaton will reject or loop in the same way, as it will never reach the backtracking stage.

In the proof of Theorem 2, a given returning direction-determinate automaton $\mathcal{A}$ is simulated by a reversible automaton $\mathcal{B}$ of Lemma 4.

5   Application to Various Types of Automata

The aim of this section is to revisit several models of computation represented as GWAs in Section 2, and apply the results of this paper to each of them.

Proposition 1. *Each n-state 2DFA has an equivalent $(4n + 3)$-state strongly reversible 2DFA.*

Indeed, for 2DFAs, the set of directions $D = \{-1, +1\}$ is a two-element set, and hence the transformation to direction-determinate duplicates the number of states. In order to make a direction-determinate 2DFA returning, it is sufficient to add one extra state, in which the automaton will move the head to the left-end marker after it decides to accept. Applying Theorem 2 to the resulting automaton gives a strongly reversible 2DFA with $4n + 3$ states.

In the case of 2DFAs, Theorem 2 is essentially a generalization of the construction by Kondacs and Watrous [16] from 1DFAs to direction-determinate 2DFAs. The transformation of an n-state 2DFA to a 2DFA with $4n + \text{const}$ states that halts on every input, presented by Geffert et al. [13], most likely results in the same reversible automaton as constructed in Proposition 1, but both main steps of the construction are amalgamated into one. Thus, the two-step transformation proving Proposition 1 explains the construction given by Geffert et al. [13].

Turning to tree-walking automata, Muscholl et al. [22] proved that an n-state TWA can be transformed to a halting TWA with $O(n^2)$ states, using another implementation of Sipser's method [24]. This can be now improved as follows.

Proposition 2. *Any n-state TWA over k-ary trees can be transformed to a $(4kn + 2k + 1)$-state strongly reversible TWA.*

Here the transformation to direction-determinate multiplies the number of states by $|D| = 2k$. Parking the head after acceptance generally requires only one extra state, in which the automaton will go up to the root. However, in order to keep the resulting automaton direction-determinate, one has to use k extra states $q_{\text{return}}^1, \ldots, q_{\text{return}}^k$ with $d(q_{\text{return}}^i) = -i$. Reversibility is ensured by Theorem 2, which produces $2(2kn + k) + 1$ states, as stated.

The next model are the multi-head automata, for which Morita [21] proved that an n-state k-head 2DFA can be transformed to a reversible k-head 2DFA with $O(n)$ states, where the constant factor depends both on k and on the alphabet. The general results of this paper imply a transformation with the constant factor independent of the alphabet.

Proposition 3. *Any n-state k-head 2DFA can be transformed to a $(2(3^k-1)n + 2k + 1)$-state strongly reversible k-head 2DFA.*

Since there are $3^k - 1$ directions, the transformation to direction-determinate automaton incurs a $(3^k - 1)$-times blowup. Adding k extra states to park all k heads after acceptance produces an automaton with $(3^k - 1)n + k$ states, to which Theorem 2 is applied.

In the full paper, it is similarly shown how to transform an n-state Turing machine operating in marked space $s(\ell)$, with an m-symbol work alphabet, to a $(6(m^2 - m + 4)n + 6m + 16)$-state reversible Turing machine of the same kind. There are also transformations of an n-state 4DFA to a $(8n + 9)$-state strongly reversible 4DFA, and of an n-state k-pebble 2DFA to a $((4k+4)n + 2k + 5)$-state strongly reversible k-pebble 2DFA. The list of such results can be continued further, by representing various models of computation with a bounded graph of memory configurations as graph-walking automata, and then applying the general theorems of this paper.

References

1. Abramsky, S.: A structural approach to reversible computation. Theoretical Computer Science 347(3), 441–464 (2005)
2. Aho, A.V., Ullman, J.D.: Translations on a context free grammar. Information and Control 19(5), 439–475 (1971)

3. Bennett, C.H.: Logical reversibility of computation. IBM Journal of Research and Development 17(6), 525–532 (1973)
4. Bennett, C.H.: Time/space trade-offs for reversible computation. SIAM Journal on Computing 81, 766–776 (1989)
5. Blum, M., Hewitt, C.: Automata on a 2-dimensional tape. In: SWAT 1967, pp. 155–160 (1967)
6. Bojańczyk, M., Colcombet, T.: Tree-walking automata cannot be determinized. Theoretical Computer Science 350(2-3), 164–173 (2006)
7. Bojańczyk, M., Colcombet, T.: Tree-walking automata do not recognize all regular languages. SIAM Journal on Computing 38(2), 658–701 (2008)
8. Buhrman, H., Tromp, J., Vitányi, P.: Time and space bounds for reversible simulation. Journal of Physics A: Mathematical and General 34(35), 6821–6830 (2001)
9. Courcelle, B.: Graph rewriting: An algebraic and logic approach. In: Handbook of Theoretical Computer Science, vol. B, pp. 193–242 (1990)
10. Crescenzi, P., Papadimitriou, C.H.: Reversible simulation of space-bounded computations. Theoretical Computer Science 143(1), 159–165 (1995)
11. Engelfriet, J., Hoogeboom, H.J.: Tree-walking pebble automata. Jewels are Forever, Contributions on Theoretical Computer Science in Honor of Arto Salomaa, 72–83 (1999)
12. Fraigniaud, P., Ilcinkas, D., Peer, G., Pelc, A., Peleg, D.: Graph exploration by a finite automaton. Theoretical Computer Science 345(2-3), 331–344 (2005)
13. Geffert, V., Mereghetti, C., Pighizzini, G.: Complementing two-way finite automata. Information and Computation 205(8), 1173–1187 (2007)
14. Hopcroft, J.E., Ullman, J.D.: Some results on tape bounded Turing machines. Journal of the ACM 16, 168–177 (1967)
15. Kari, J.: Reversible cellular automata. In: De Felice, C., Restivo, A. (eds.) DLT 2005. LNCS, vol. 3572, pp. 57–68. Springer, Heidelberg (2005)
16. Kondacs, A., Watrous, J.: On the power of quantum finite state automata. In: FOCS 1997, pp. 66–75 (1997)
17. Kutrib, M., Malcher, A.: Reversible pushdown automata. Journal of Computer and System Sciences 78(6), 1814–1827 (2012)
18. Landauer, R.: Irreversibility and heat generation in the computing process. IBM Journal of Research and Development 5(3), 183–191 (1961)
19. Lange, K.-J., McKenzie, P., Tapp, A.: Reversible space equals deterministic space. Journal of Computer and System Sciences 60(2), 354–367 (2000)
20. Lecerf, Y.: Machines de Turing réversibles. Comptes Rendus de l'Académie des Sciences 257, 2597–2600 (1963)
21. Morita, K.: A deterministic two-way multi-head finite automaton can be converted into a reversible one with the same number of heads. In: Glück, R., Yokoyama, T. (eds.) RC 2012. LNCS, vol. 7581, pp. 29–43. Springer, Heidelberg (2013)
22. Muscholl, A., Samuelides, M., Segoufin, L.: Complementing deterministic tree-walking automata. Information Processing Letters 99(1), 33–39 (2006)
23. Pin, J.-É.: On the languages accepted by finite reversible automata. In: Ottmann, T. (ed.) ICALP 1987. LNCS, vol. 267, pp. 237–249. Springer, Heidelberg (1987)
24. Sipser, M.: Halting space-bounded computations. Theoretical Computer Science 10(3), 335–338 (1980)
25. Thomas, W.: On logics, tilings, and automata. In: Leach Albert, J., Monien, B., Rodríguez-Artalejo, M. (eds.) ICALP 1991. LNCS, vol. 510, pp. 441–454. Springer, Heidelberg (1991)
26. Toffoli, T., Margolus, N.H.: Invertible cellular automata: A review. Physica D: Nonlinear Phenomena 45(1-3), 229–253 (1990)

Prime Languages

Orna Kupferman and Jonathan Mosheiff

School of Engineering and Computer Science, The Hebrew University, Jerusalem, Israel

Abstract. We say that a deterministic finite automaton (DFA) $\mathcal{A}$ is *composite* if there are DFAs $\mathcal{A}_1, \ldots, \mathcal{A}_t$ such that $L(\mathcal{A}) = \bigcap_{i=1}^{t} L(\mathcal{A}_i)$ and the index of every $\mathcal{A}_i$ is strictly smaller than the index of $\mathcal{A}$. Otherwise, $\mathcal{A}$ is *prime*. We study the problem of deciding whether a given DFA is composite, the number of DFAs required in a decomposition, methods to prove primality, and structural properties of DFAs that make the problem simpler or are retained in a decomposition.

1 Introduction

Compositionality is a well motivated and studied notion in computer science [2]. By decomposing a problem into several smaller problems, it is possible not only to increase parallelism, but also to sometimes handle inputs that are otherwise intractable. A major challenge is to identify problems and instances that can be decomposed.

Consider for example the LTL model-checking problem [9]. Given a system $\mathcal{S}$ and a specification ψ, checking whether all the computations of $\mathcal{S}$ satisfy ψ can be done in time linear in $\mathcal{S}$ and exponential in ψ. If ψ is a conjunction of smaller specifications, say $\psi = \varphi_1 \wedge \cdots \wedge \varphi_t$, then it is possible to check instead whether $\mathcal{S}$ satisfies each of the φ_i's.[1] Not all problems allow for easy decomposition. For example, if we wish to synthesize a transducer that realizes the specification ψ above, it is not clear how to use the decomposition of ψ into its conjuncts. In particular, it is not clear how to compose a transducer that realizes ψ from t transducers that realize $\varphi_1, \ldots, \varphi_t$ [6]. In the automata-theoretic approach to formal verification, we use automata in order to model systems and their specifications. A natural question then is whether we can decompose a given automaton $\mathcal{A}$ into smaller automata $\mathcal{A}_1, \ldots, \mathcal{A}_t$ such that $L(\mathcal{A}) = \bigcap_{i=1}^{t} L(\mathcal{A}_i)$. Then, for example, we can reduce checking $L(\mathcal{S}) \subseteq L(\mathcal{A})$ to checking whether $L(\mathcal{S}) \subseteq L(\mathcal{A}_i)$ for all $1 \leq i \leq t$.

The automata used for reasoning about systems and their specifications are typically nondeterministic automata on infinite words [11]. As it turns out, however, the question of automata decomposition is open already for the basic model of deterministic automata on finite words (DFAs). Studying DFAs also suggests a very clean mathematical approach, as each regular language has a canonical minimal DFA recognizing it. Researchers have developed a helpful algebraic approach for DFAs that offers some very interesting results on DFAs and their decomposition. To the best of our knowledge, however, the basic question of decomposing a DFA into smaller DFAs is still open.

---

[1] The ability to decompose the specification causes the systems, which are much bigger than the sub-specifications, to be the computational bottleneck in the model-checking problem. Thus, a different big challenge is to decompose the system [10].

K. Chatterjee and J. Sgall (Eds.): MFCS 2013, LNCS 8087, pp. 607–618, 2013.
© Springer-Verlag Berlin Heidelberg 2013

In the algebraic approach, a DFA $\mathcal{A}$ is matched with a *monoid* $M(\mathcal{A})$. The members of $M(\mathcal{A})$ are the actions of words in Σ^* on the states of $\mathcal{A}$. That is, each member is associated with a word and corresponds to the states-to-states transition function induced by the word. In particular, ϵ corresponds to the identity element. A DFA is called a *permutation DFA* if its monoid consists of permutations. A DFA is called a *reset DFA* if its monoid consists of constant functions and the identity function. The algebraic approach is used in [5] in order to show that every DFA $\mathcal{A}$ can be presented as a *wreath product* of reset DFAs and permutation DFAs, whose algebraic structure is simpler than that of $\mathcal{A}$. A wreath product of a sequence $\mathcal{A}_1, \mathcal{A}_2 \dots, \mathcal{A}_t$ of DFAs is a cascade in which the transition function of each DFA $\mathcal{A}_i$ may depend on the state of $\mathcal{A}_i$ as well as the states of the DFAs preceding $\mathcal{A}_i$ in the sequence.

The algebraic approach is based on a syntactic congruence between words in Σ^*: given a regular language $L \subseteq \Sigma^*$, we have that $x \sim_L y$, for $x, y \in \Sigma^*$, if for every $w, z \in \Sigma^*$, it holds that $w \cdot x \cdot z \in L$ iff $w \cdot y \cdot z \in L$. Thus, the congruence refers to extensions of words from both right and left. In the context of minimization, which motivates the practical study of decomposition, one is interested in *right congruence*. There, $x \sim_L y$ iff for all words $z \in \Sigma^*$, we have that $x \cdot z \in L$ iff $y \cdot z \in L$. By the Myhill-Nerode theorem [7,8], the equivalence classes of $\sim_L$ constitute the state space of a minimal canonical DFA for L. The number of equivalence classes is referred to as the *index* of L. We say that a language $L \subseteq \Sigma^*$ is *composite* if there are languages $L_1, \dots, L_t$ such that $L = \bigcap_{i=1}^t L_t$ and the index of L_i, for all $1 \leq i \leq t$, is strictly smaller than the index of L. Otherwise, we say that L is *prime*. The definitions applies also to DFAs, referring to the languages they recognize.

For example, for Σ with $|\Sigma| > 1$ and $w \in \Sigma^*$, let $L_w = \{w\}$. Clearly, the index of L_w is $|w| + 2$. We claim that if w contains at least two different letters, then L_w is composite. To see this, we show we can express L_w as the intersection of two DFAs of index at most $|w| + 1$. Let σ be some letter in w, and let m be its number of occurrences in w. By the condition on w, we have that $1 \leq m < |w|$. It is easy to see that L_w is the intersection of the language w^*, whose index is $|w| + 1$, and the language of all words in which σ appears exactly m times, whose index is $m + 2 \leq |w| + 1$. On the other hand, if w consists of a single letter, then L_w is prime. One of our goals in this work is to develop techniques for proving primality.

The decomposition of L_w described above is of *width* 2; that is, it has two factors. The case of decompositions of width 2 was studied in [3], where the question of whether one may need wider decompositions was left open. We answer the question positively; that is, we present a language that does not have a decomposition of width 2 but has one of width 3. For compositions of width 2, the question of deciding whether a given DFA is composite is clearly in NP, as one can guess the two factors. In the general case, the only bound we have on the width is exponential, which follows from the bound on the size of the underlying DFAs. This bound suggests an EXPSPACE algorithm for deciding whether a given DFA is composite.

Consider a DFA $\mathcal{A}$. We define the *roof* of $\mathcal{A}$ as the intersection of all languages $L(\mathcal{B})$, where $\mathcal{B}$ is a DFA such that $L(\mathcal{A}) \subseteq L(\mathcal{B})$ and the index of $\mathcal{B}$ is smaller than that of $\mathcal{A}$. Thus, the roof of $\mathcal{A}$ is the minimal (with respect to containment) language that can be defined as an intersection of DFAs whose language contain the language of $\mathcal{A}$ and whose index is smaller than the index of $\mathcal{A}$. Accordingly, $\mathcal{A}$ is composite iff $L(\mathcal{A}) = roof(\mathcal{A})$. We use roofs in order to study primality further. In particular, if $\mathcal{A}$ is prime then there exists a word $w \in roof(\mathcal{A}) \setminus L(\mathcal{A})$. The word w is called a *primality witness* for $\mathcal{A}$. Indeed, $\mathcal{A}$ is prime iff it has a primality witness.

Let us go back to the language L_w from the example above. We wish to prove that when $w = \sigma^n$ for some letter σ, then L_w is prime. Let $l > n$ be a natural number such that for every $p \leq n + 1$ it holds that $n \equiv l \mod p$. The existence of l is guaranteed by the Chinese remainder theorem. In the paper, we prove that σ^l is a primality witness for L_w and conclude that L_w is prime. We use the notion of a primality witness to prove the primality of additional, more involved, families of languages.

We then turn to study structural properties of composite and prime DFAs. Each DFA $\mathcal{A}$ induces a directed graph $G_\mathcal{A}$. We study the relation between the structure of $G_\mathcal{A}$ and the primality of $\mathcal{A}$. We identify cases, for example co-safety languages whose DFA contains one rejecting strongly connected component from which an accepting sink is reachable, where primality (with a short primality witness) is guaranteed. We also study structural properties that can be retained in a decomposition and prove, for example, that a composite strongly connected DFA can be decomposed into strongly connected DFAs.

Recall that a decomposition of a DFA $\mathcal{A}$ consists of DFAs that contain the language of $\mathcal{A}$ and are still of a smaller index. A simple way to get a DFA with the above properties is by merging states of $\mathcal{A}$. A *simple decomposition* of $\mathcal{A}$ is a decomposition in which each of the underlying DFAs is a result of merging states of $\mathcal{A}$. Simple decompositions have also been studied in [4] in the context of sequential machines. It follows from [3] that some DFAs have a decomposition of width 2 yet do not have a simple decomposition. We characterize simple decompositions and show that the problem of deciding whether a given DFA has a simple decomposition is in PTIME.

Finally, we develop an algebraic view of DFA primality. As [5], our approach is based on the transition monoid of $\mathcal{A}$. First, we show that once we fix the set of accepting states, the question of primality of a DFA $\mathcal{A}$ depends only on $\mathcal{A}$'s transition monoid, rather than its transition function or alphabet. We then focus on permutation DFAs. Given a permutation DFA $\mathcal{A}$ we construct a new DFA, termed the *monoid DFA*, such that compositionally of $\mathcal{A}$ can be reduced to simple-compositionality of its monoid DFA. Driven by observations about monoid DFAs, we show a PSPACE algorithm for deciding the primality of $\mathcal{A}$. We also show that composite permutation DFAs can be decomposed into permutation DFAs.

Due to lack of space, many examples and proofs are omitted. They can be found in the full version, in the authors' URLs.

2   Preliminaries

A *deterministic finite automaton* (DFA) is a 5-tuple $\mathcal{A} = \langle Q, \Sigma, q_0, \delta, F \rangle$, where Q is a finite set of states, Σ is a finite non-empty alphabet, $\delta : Q \times \Sigma \to Q$ is a transition function, $q_0 \in Q$ is an initial state, and $F \subseteq Q$ is a set of accepting states. For $q \in Q$, we use $\mathcal{A}^q$ to denote the DFA $\mathcal{A}$ with q as the initial state. That is, $\mathcal{A}^q = \langle Q, \Sigma, q, \delta, F \rangle$. We extend δ to words in the expected way, thus $\delta : Q \times \Sigma^* \to Q$ is defined recursively by $\delta(q, \epsilon) = q$ and $\delta(q, w_1 w_2 \cdots w_n) = \delta(\delta(q, w_1 w_2 \cdots w_{n-1}), w_n)$.

The *run* of $\mathcal{A}$ on a word $w = w_1 \ldots w_n$ is the sequence of states $s_0, s_1 \ldots s_n$ such that $s_0 = q_0$ and for each $1 \leq i \leq n$ it holds that $\delta(s_{i-1}, w_i) = s_i$. Note that $s_n = \delta(q_0, w)$. The DFA $\mathcal{A}$ *accepts* w iff $\delta(q_0, w) \in F$. Otherwise, $\mathcal{A}$ *rejects* w. The set of words accepted by $\mathcal{A}$ is denoted $L(\mathcal{A})$ and is called the *language of* $\mathcal{A}$. We say that $\mathcal{A}$ *recognizes* $L(\mathcal{A})$. A language recognized by some DFA is called a *regular language*.

A DFA $\mathcal{A}$ is *minimal* if every DFA $\mathcal{B}$ that has less states than $\mathcal{A}$ satisfies $L(\mathcal{B}) \neq L(\mathcal{A})$. Every regular language L has a single (up to DFA isomorphism) minimal DFA $\mathcal{A}$ such that $L(\mathcal{A}) = L$. The index of L, denoted $ind(L)$, is the size of the minimal DFA recognizing L.

Consider a language $L \subseteq \Sigma^*$. The *Myhill-Nerode relation* relative to L, denoted $\sim_L$, is a binary relation on Σ^* defined as follows: For $x, y \in \Sigma^*$, we say that $x \sim_L y$ if for every $z \in \Sigma^*$ it holds that $x \cdot z \in \Sigma^*$ iff $y \cdot z \in \Sigma^*$. Note that $\sim_L$ is an equivalence relation. It is known that L is regular iff $\sim_L$ has a finite number of equivalence classes. The number of these equivalence classes is equal to $ind(L)$.

Definition 1. [DFA decomposition] *Consider a DFA $\mathcal{A}$. For $k \in \mathbb{N}$, we say that $\mathcal{A}$ is k-decomposable if there exist DFAs $\mathcal{A}_1, \ldots, \mathcal{A}_t$ such that for all $1 \leq i \leq t$ it holds that $ind(\mathcal{A}_i) \leq k$ and $\bigcap_{i=1}^{t} L(\mathcal{A}_i) = L(\mathcal{A})$. The DFAs are then a k-decomposition of $\mathcal{A}$. The* depth *of $\mathcal{A}$, denoted $depth(\mathcal{A})$, is the minimal k such that $\mathcal{A}$ is k-decomposable.*

Obviously, every DFA $\mathcal{A}$ is $ind(\mathcal{A})$-decomposable. The question is whether a decomposition of $\mathcal{A}$ can involve DFAs of a strictly smaller index. Formally, we have the following.

Definition 2. [Composite and Prime DFAs] *A DFA $\mathcal{A}$ is* composite *if $depth(\mathcal{A}) < ind(\mathcal{A})$. Otherwise, $\mathcal{A}$ is* prime.

We identify a regular language with its minimal DFA. Thus, we talk also about a regular language being k-decomposable or composite, referring to its minimal DFA. Similarly, for a DFA $\mathcal{A}$, we refer to $ind(L(\mathcal{A}))$ as $ind(\mathcal{A})$.

Example 1. Let $\Sigma = \{a\}$ and $L_k = (a^k)^*$. We show that if k is not a prime power, then L_k is composite. Clearly, $ind(L_k) = k$. If k is not a prime power, there exist $2 \leq p, q < k$ such that p and q are coprime and $p \cdot q = k$. It then holds that $L_k = L_p \cap L_q$. Since $ind(L_p) < k$ and $ind(L_q) < k$, it follows that L_k is composite. □

Let $\mathcal{A}$ be a DFA. We define $\alpha(\mathcal{A}) = \{\mathcal{B} : \mathcal{B}$ is a minimal DFA such that $L(\mathcal{A}) \subseteq L(\mathcal{B})$ and $ind(\mathcal{B}) < ind(\mathcal{A})\}$. That is, $\alpha(\mathcal{A})$ is the set of DFAs that contain $\mathcal{A}$ and have an index smaller than the index of $\mathcal{A}$. The *roof* of $\mathcal{A}$ is the intersection of the languages of all DFAs in $\alpha(\mathcal{A})$. Thus, $roof(\mathcal{A}) = \bigcap_{\mathcal{B} \in \alpha(\mathcal{A})} L(\mathcal{B})$. Clearly, $L(\mathcal{A}) \subseteq roof(\mathcal{A})$. Also, if $\mathcal{A}$ is composite, then $\alpha(L)$ is an $(ind(\mathcal{A}) - 1)$-decomposition of $\mathcal{A}$. We thus have the following.

Theorem 1. *A DFA $\mathcal{A}$ is prime iff $L(\mathcal{A}) \neq roof(\mathcal{A})$, unless $L(\mathcal{A}) = \Sigma^*$.*

The PRIME-DFA problem is to decide, given a DFA $\mathcal{A}$, whether $\mathcal{A}$ is prime. A more general problem is, given a DFA $\mathcal{A}$, to compute $depth(\mathcal{A})$. We now prove an upper bound on the complexity of PRIME-DFA. We first need the following lemma.

Lemma 1. *Let $\mathcal{A}$ be a DFA and let $n = ind(\mathcal{A})$. Then, $|\alpha(\mathcal{A})| = 2^{O(n \cdot |\Sigma| \cdot \log n)}$ and $ind(roof(\mathcal{A})) \leq 2^{2^{O(n \cdot |\Sigma| \cdot \log n)}}$.*

Combining Theorem 1 and Lemma 1, we have the following.

Theorem 2. *The* PRIME-DFA *problem is in* EXPSPACE.

We note that the only lower bound for the problem is NLOGSPACE, by a reduction from reachability.

3   Primality Witnesses

Recall that a minimal DFA $\mathcal{A}$ is prime iff $roof(\mathcal{A}) \not\subseteq L(\mathcal{A})$. We define a *primality witness* for $\mathcal{A}$ as a word in $roof(\mathcal{A}) \setminus L(\mathcal{A})$. Clearly, a DFA $\mathcal{A}$ is prime iff $\mathcal{A}$ has a primality witness.

Let $\mathcal{A}$ be a DFA. By the above, we can prove that $\mathcal{A}$ is prime by pointing to a primality witness for L. Recall the language $L_k = (a^k)^*$ from Example 1. We show that the condition given there is necessary, thus if k is a prime power, then L_k is prime. Let $p, r \in \mathbb{N} \cup \{0\}$ be such that p is a prime and $k = p^r$. Since $w_k = a^{(p+1)p^{r-1}}$ is a primality witness for L_k, we can conclude that L_k is prime.

The bound on the size of $roof(\mathcal{A})$ from Lemma 1 implies the following.

Proposition 1. *A prime DFA has a primality witness of length doubly exponential.*

Proposition 1 implies a naive algorithm for PRIME-DFA: Given an input DFA $\mathcal{A}$, the algorithm proceeds by going over all words $w \in \Sigma^*$ of length at most $2^{2^{O(n \cdot |\Sigma| \cdot \log n)}}$, and checking, for each $\mathcal{B} \in \alpha(\mathcal{A})$, whether $w \in L(\mathcal{B})$. While the algorithm is naive, it suggests that if we strengthen Proposition 1 to give a polynomial bound on the length of minimal primality witnesses, we would have a PSPACE algorithm for PRIME-DFA. The question of whether such a polynomial bound exists is currently open.

The following examples introduce more involved families of prime languages.

Example 2. For $n \in \mathbb{N}$, let $K_n = \{ww : w \in \Sigma^n\}$ and $L_n = comp(K_n^*)$; that is, $L_n = \Sigma^* \setminus K_n^*$. Let w_n be a concatenation of all words of the form ss for $s \in \Sigma^n$ in some arbitrary order. Note that $w_n \notin L_n$. It can be shown that w_n is a primality witness for L_n. Hence, L_n is prime and a witness of length polynomial in $ind(L_n)$ exists.   □

Example 3. Consider words $s = s_1 \cdots s_m$ and $w = w_1 \cdots w_t$, both over Σ. If there exists an increasing sequence of indices $1 \leq i_1 < i_2 < \cdot < i_t \leq m$ such that for each $1 \leq j \leq t$ it holds that $w_j = s_{i_j}$, we say that w is a *subsequence* of s. If w is a subsequence of s and $w \neq s$, then we say that w is a *proper subsequence* of s.

For, $w \in \Sigma^*$, let $L_w = \{s \in \Sigma^* : w$ is a subsequence of $s\}$. In the full version, we show that L_w is prime via a primality witness of length $2 \cdot ind(L_w)$.   □

4   The Width of a Decomposition

Languages that can be decomposed into two factors have been studied in [3], where the question of whether one may need more than two factors was left open. In this section we answer the question positively. Formally, we have the following. Let $\mathcal{A}$ be a DFA. If there exist DFAs $\mathcal{A}_1, \mathcal{A}_2 \ldots \mathcal{A}_m \in \alpha(\mathcal{A})$ such that $\mathcal{A} = \bigcap_{i=1}^{m} L(\mathcal{A}_i)$, we say that $\mathcal{A}$ is m-factors composite. Assume that $\mathcal{A}$ is composite. Then, $width(L)$ is defined as the minimal m such that $\mathcal{A}$ is m-factors composite. Clearly, for every composite $\mathcal{A}$, it holds that $width(\mathcal{A}) \geq 2$. The question left open in [3] is whether there exists a composite $\mathcal{A}$ such that $width(L) > 2$. Such a language is presented in the following example.

Example 4. Let $\Sigma = \{a, b, c\}$ and let L be the language of prefixes of words in $c^*.(a^+.b^+.c^+)^*$. In the full version we show that L is composite with $width(L) = 3$. Also, it can be verified by a case-by-case analysis that L is not 2-factor composite. $\square$

Example 4 motivates us to conjecture that the width of composite languages is unbounded. That is, that width induces a strong hierarchy on the set of composite languages.

Given a composite $L \subseteq \Sigma^*$, we wish to provide an upper bound on $width(L)$. We conjecture that there exists a polynomial f such that $width(L) \leq f(ind(L))$ for some polynomial f. If this is true, the algorithm given in the proof of Theorem 2 can be improved to a PSPACE algorithm by going over all subsets of $D \subseteq \alpha(\mathcal{A})$ such that $|D| \leq f(ind(\mathcal{A}))$, and checking for each such D whether $\bigcap_{\mathcal{B} \in D} L(\mathcal{B}) = L(\mathcal{A})$.

5   Structural Properties

Consider a minimal DFA $\mathcal{A}$ and a DFA $\mathcal{B} \in \alpha(\mathcal{A})$. Recall that $L(\mathcal{A}) \subseteq L(\mathcal{B})$ and $ind(\mathcal{B}) < ind(\mathcal{A})$. Thus, intuitively, in $\mathcal{B}$, fewer states have to accept more words. In this section we examine whether this requirement on $\mathcal{B}$ can be of help in reasoning about possible decompositions.

The DFA $\mathcal{A} = \langle Q, \Sigma, q_0, \delta, F \rangle$ induces a directed graph $G_\mathcal{A} = \langle Q, E \rangle$, where $E = \{(q, q') : \exists \sigma \in \Sigma \text{ such that } \delta(q, \sigma) = q'\}$. The strongly connected components (SCCs) of $G_\mathcal{A}$ are called the SCCs of $\mathcal{A}$. We refer to the directed acyclic graph (DAG) induced by the SCCs of $G_\mathcal{A}$ as the SCC DAG of $\mathcal{A}$. A $leaf$ in $G_\mathcal{A}$ is a SCC that is a sink in this DAG. A DFA $\mathcal{A}$ is said to be $strongly\ connected$ if it consists of a single SCC.

Let $\mathcal{A} = \langle Q, \Sigma, q_0, \delta, F \rangle$ and $\mathcal{B} = \langle S, \Sigma, s_0, \eta, G \rangle$ be DFAs. Let $q \in Q$ and $s \in S$. If there exists a word $w \in \Sigma^*$ such that $\delta(q_0, w) = q$ and $\eta(s_0, w) = s$, then we say that q $touches$ s, denoted $q \sim s$. Obviously, this is a symmetric relation.

Lemma 2. *Let $\mathcal{A}$ and $\mathcal{B}$ be DFAs such that $L(\mathcal{A}) \subseteq L(\mathcal{B})$ and let q and s be states of $\mathcal{A}$ and $\mathcal{B}$, respectively, such that $q \sim s$. Then, $L(\mathcal{A}^q) \subseteq L(\mathcal{B}^s)$.*

For each $s \in S$, consider the subset of Q consisting of the states that touch s. Recall that $|S| < |Q|$. Intuitively, if one attempts to design $\mathcal{B}$ so that $L(\mathcal{B})$ over-approximates $L(\mathcal{A})$ as tightly as possible, one would try to avoid, as much as possible, having states in S that touch more than one state in Q. However, by the pigeonhole principle, there must be a state $s \in S$ that touches more than one state in Q. The following lemma provides a stronger statement: There must exist a non-empty set $Q' \subseteq Q$ relative to which the DFA $\mathcal{B}$ is "confused" when attempting to imitate $\mathcal{A}$.

Lemma 3. *Let $\mathcal{A} = \langle Q, \Sigma, q_0, \delta, F \rangle$ and $\mathcal{B} = \langle S, \Sigma, s_0, \eta, G \rangle$ be minimal DFAs such that $\mathcal{B} \in \alpha(\mathcal{A})$. Then, there exists a non empty set $Q' \subseteq Q$ such that for every $q_1 \in Q'$ and $s \in S$ with $q_1 \sim s$, there exists $q_2 \in Q'$ such that $q_1 \neq q_2$ and $q_2 \sim s$.*

We are going to use Lemma 3 in our study of primality of classes of DFAs. We start with safe and co-safe DFAs.

Let $\mathcal{A} = \langle Q, \Sigma, q_0, \delta, F \rangle$ be a minimal DFA such that $L(\mathcal{A}) \neq \Sigma^*$. It is easy to see that $L(\mathcal{A})$ is *co-safety* [1] iff $\mathcal{A}$ has a single accepting state s, which is an accepting sink. Obviously, the singleton $\{s\}$ is a SCC of $\mathcal{A}$. If $Q \setminus \{s\}$ is a SCC, we say that $\mathcal{A}$ is a *simple co-safety* DFA. For example, it is not hard to see that for all $w \in \Sigma^*$, if $|\Sigma| > 2$, then the language L_w of all words that have w as a subword is such that the minimal DFA for L_w is simple co-safe.

Theorem 3. *Every simple co-safe DFA is prime with a primality witness of polynomial length.*

Our main result about the structural properties of composite DFAs shows that strong connectivity can be carried over to the DFAs in the decomposition. Intuitively, let $\mathcal{A}_i$ be a member of a decomposition of $\mathcal{A}$, and let $\mathcal{B}_i$ be a DFA induced by a leaf of the SCC graph of $\mathcal{A}_i$. We can replace $\mathcal{A}_i$ by $\mathcal{B}_i$ and still get a valid decomposition of $\mathcal{A}$. Formally, we have the following.

Theorem 4. *Let $\mathcal{A}$ be a strongly connected composite DFA. Then, $\mathcal{A}$ can be decomposed using only strongly connected DFAs as factors.*

We find the result surprising, as strong connectivity significantly restricts the over-approximating DFAs in the decomposition.

6 Simple Decompositions

Consider the task of decomposing a DFA $\mathcal{A}$. A natural approach is to build the factors of $\mathcal{A}$ by merging states of $\mathcal{A}$ into equivalence classes in such a manner that the transition function of $\mathcal{A}$ respects the partition of its states into equivalence classes. The result of such a construction is a DFA $\mathcal{B}$ that contains $\mathcal{A}$ and still has fewer states. If $\mathcal{A}$ has a decomposition into factors constructed by this approach, we say that $\mathcal{A}$ is *simply-composite*. In this section, we formally define the concept of a simple decomposition and investigate its properties. In particular, we show that it is computationally easy to check whether a given DFA is simply-composite.

Consider DFAs $\mathcal{A} = \langle Q, \Sigma, q_0, \delta, F \rangle$ and $\mathcal{B} = \langle S, \Sigma, s_0, \eta, G \rangle$. We say that $\mathcal{B}$ is an *abstraction* of $\mathcal{A}$ if for every $q \in Q$ there exists a single $s \in S$ such that $q \sim s$. An abstraction $\mathcal{B}$ of $\mathcal{A}$ is called a *miser abstraction of* $\mathcal{A}$ if $L(\mathcal{A}) \subseteq L(\mathcal{B})$, and the set G of $\mathcal{B}$'s accepting states cannot be reduced retaining the containment. That is, for all $s \in G$, the DFA $\mathcal{B}_s = \langle S, \Sigma, s_0, \eta, G \setminus \{s\} \rangle$ is such that $L(\mathcal{A}) \not\subseteq L(\mathcal{B}_s)$. It is not hard to see that $L(\mathcal{A}) \subseteq L(\mathcal{B})$ iff for every $q \in F$ and $s \in S$ such that $q \sim s$, it holds that $s \in G$. The above suggests a simple criterion for fixing the set of accepting states required for an abstraction to be miser.

Simple decompositions of a DFA consists of miser abstractions. Let $L \subseteq \Sigma^*$. Clearly, if L is simply-composite, then it is composite. The opposite is not necessarily true. For example, while the singleton language $\{ab\}$ is composite, one can go over all the abstractions of its 4-state DFA and verify that $\mathcal{A}$ is not simply-composite. Consider a DFA

$\mathcal{A} = \langle Q, \Sigma, q_0, \delta, F \rangle$ and $t \in \mathbb{N}$. We use $\gamma_t(\mathcal{A})$ to denote the set of all miser abstractions of $\mathcal{A}$ with index at most t. Let $ceiling_t(\mathcal{A}) = \bigcap_{\mathcal{B} \in \gamma_t(\mathcal{A})} L(\mathcal{B})$. Note that $\mathcal{A}$ is simply-decomposable iff $L(\mathcal{A}) = ceiling_t(\mathcal{A})$ for $t = ind(\mathcal{A}) - 1$.

Our next goal is an algorithm that decides whether a given DFA is simply-composite. For $t \in \mathbb{N}$, a t-*partition* of Q is a set $P = \{Q_1, \ldots, Q_{t'}\}$ of $t' \le t$ nonempty and pairwise disjoint subsets of Q whose union is Q. For $q \in Q$, we use $[q]_P$ to refer to the set Q_i such that $q \in Q_i$.

Definition 3. *Let $\mathcal{A} = \langle Q, \Sigma, q_0, \delta, F \rangle$ be a DFA. A t-partition P of Q is a good t-partition of $\mathcal{A}$ if δ respects P. That is, for every $q \in Q$, $\sigma \in \Sigma$, and $q' \in [q]_P$, we have that $\delta(q', \sigma) \in [\delta(q, \sigma)]_P$.*

A good t-partition of $\mathcal{A}$ induces a miser abstraction of it. In the other direction, each abstraction of $\mathcal{A}$ with index at most t induces a t-partition of $\mathcal{A}$. Formally, we have the following.

Lemma 4. *There is a one-to-one correspondence between the DFAs in $\gamma_t(\mathcal{A})$ and the good t-partitions of $\mathcal{A}$.*

Let $\mathcal{A} = \langle Q, \Sigma, q_0, \delta, F \rangle$ be a DFA and let $q \in Q \setminus F$. Let P be a good partition of $\mathcal{A}$. If $[q]_P \cap F = \emptyset$, we say that P is a *q-excluding partition.*

Lemma 5. *Let $\mathcal{A} = \langle Q, \Sigma, q_0, \delta, F \rangle$ be a DFA such that every state of $\mathcal{A}$ is reachable and $F \ne Q$. Then, the DFA $\mathcal{A}$ is t-simply-decomposable iff for every state $q \in Q \setminus F$ there exists a q-excluding t-partition.*

For two partitions P and P' of Q, we say that P' is a *refinement* of P if for every $R \in P$ there exist sets $R'_1 \ldots R'_s \in P'$ such that $R = \bigcup_{i=1}^{s} R'_i$. Let $q, q' \in Q$ be such that $q \ne q'$. Let P be a good partition of $\mathcal{A}$. If $[q]_P = [q']_P$, we say that P *joins q and q'.*

It is not hard to see that good partitions are closed under intersection. That is, if P and P' are good partitions, so is the partition that contains the intersections of their members. Thus, there exists a unique good partition of $\mathcal{A}$, denoted $P_{q,q'}$ such that if P is a good partition of $\mathcal{A}$ that joins q and q', then $P_{q,q'}$ is a refinement of P. The partition $P_{q,q'}$ can be found by merging the two states q and q' and then merging only the pairs of states that must be merged in order for the generated partition to be good. Hence, the following holds.

Lemma 6. *Given a DFA $\mathcal{A} = \langle Q, \Sigma, q_0, \delta, F \rangle$ and $q, q' \in Q$ such that $q \ne q'$, the partition $P_{q,q'}$ can be computed in polynomial time.*

By Lemma 5, we can decide whether $\mathcal{A}$ is simply-composite, by deciding, for every rejecting state q of $\mathcal{A}$, whether there exists a q-excluding $(ind(\mathcal{A}) - 1)$-partition. This can be checked by going over all partitions of the form $P_{s,s'}$ for some states s and s'. By Lemma 6, the latter can be done in polynomial time. We can thus conclude with the following.

Theorem 5. *Given a DFA $\mathcal{A}$, it is possible to decide in polynomial time whether $\mathcal{A}$ is simply-composite.*

7   Algebraic Approach

In this section we use and develop concepts from the algebraic approach to automata in order to study the DFA primality problem.

Definitions and Notations.   A *semigroup* is a set S together with an associative binary operation $\cdot : S \times S \to S$. A *monoid* is a semigroup S with an *identity element* $e \in S$ such that for every $x \in S$ it holds that $e \cdot x = x \cdot e = x$. Let S be a monoid with an identity element $e \in S$ and let $A \subseteq S$. Let A^* be the smallest monoid such that $A \subseteq A^*$ and $e \in A^*$. If $A^* = S$ we say that A *generates* S.

Consider a DFA $\mathcal{A} = \langle Q, \Sigma, q_0, \delta, F \rangle$. For $w \in \Sigma^*$, let $\delta_w : Q \to Q$ be such that for every $q \in Q$, we have $\delta_w(q) = \delta(q, w)$. For $w_1, w_2 \in \Sigma^*$, the composition of δ_{w_1} and δ_{w_2} is, as expected, the operation $\delta_{w_2 \cdot w_1} : Q \to Q$ with $\delta_{w_2 \cdot w_1}(q) = \delta(q, w_2 \cdot w_1) = \delta_{w_1}(\delta_{w_2}(q))$. The set $\{\delta_w : w \in \Sigma^*\}$, equipped with the composition binary operation, is a monoid called the *transition monoid of* $\mathcal{A}$, denoted $M(\mathcal{A})$. Its identity element is δ_ϵ, denoted id. Note that $M(\mathcal{A})$ is generated by $\{\delta_\sigma : \sigma \in \Sigma\}$.

A Monoid-Driven Characterization of Primality.   The following theorem and its corollary show that in order to decide whether a DFA is composite, we only need to know its state set, set of accepting states, and transition monoid. Thus, interestingly, changing the transition function or even the alphabet does not affect the composability of a DFA as long as the transition monoid remains the same.

Theorem 6.   *Let $\mathcal{A}$ and $\mathcal{A}'$ be two DFAs with the same set of states, initial state, and set of accepting states. If $M(\mathcal{A}) = M(\mathcal{A}')$, then $depth(\mathcal{A}) = depth(\mathcal{A}')$.*

Let Σ and Σ' be the alphabets of $\mathcal{A}$ and $\mathcal{A}'$, respectively. The fact $M(\mathcal{A}) \subseteq M(\mathcal{A}')$ enables us to "encode" every letter in Σ by a word in Σ' that acts the same way on the set of states. By expanding this encoding, we can encode every word over Σ by a word over Σ'. In particular, a primality witness for $\mathcal{A}$ is encoded into a primality witness for $\mathcal{A}'$.

Theorem 6 suggests that we can relate the properties of a DFAs transition monoid to the question of its primality. In the next section we do so for the family of *permutation DFAs*.

Permutation DFAs.   Let $\mathcal{A} = \langle Q, \Sigma, q_0, \delta, F \rangle$ be a DFA. If for every $\sigma \in \Sigma$, it holds that δ_σ is a permutation, then $\mathcal{A}$ is called a *permutation DFA*. Equivalently, $\mathcal{A}$ is a permutation DFA if the monoid $M(\mathcal{A})$ is a group. It is easy to verify that the two definitions are indeed equivalent.

Note that a permutation DFA is strongly connected, unless it has unreachable states. From here on, we assume that all permutation DFAs we refer to are strongly connected.

Example 5. [The discrete cube DFA]: Let $n \in \mathbb{N}$. Recall that $\mathbb{Z}_2^n = (0, 1)^n$. Consider the DFA $\mathcal{A}_n = \langle \mathbb{Z}_2^n, \mathbb{Z}_2^n, 0, \delta, \mathbb{Z}_2^n \setminus \{0\} \rangle$, where $\delta(x, y) = x + y$. The language of $\mathcal{A}_n$ is the set of all words $w_1 \ldots w_m$ with $w_i \in \mathbb{Z}_2^n$ and $\sum_{i=1}^m w_i \neq 0$. It is easy to see that $\mathcal{A}_n$ is a permutation DFA of index 2^n and that it is minimal. In the full version we show that $\mathcal{A}_n$ is prime. $\square$

We start working towards an analogue of Theorem 4 for permutation DFAs. Thus, our goal is to show that a composite permutation DFA can be decomposed using only permutation DFAs as factors. We first need some notations.

Let $\mathcal{A} = \langle Q, \Sigma, q_0, \delta, F \rangle$ be a DFA and let $f \in M(\mathcal{A})$. The *degree* of f, denoted $\deg(f)$, is $|f(Q)|$. The *degree* of $\mathcal{A}$ is $\deg(\mathcal{A}) = \min\{\deg(f) : f \in M(\mathcal{A})\}$. Let $\mathcal{A}_i$ be a factor in a decomposition of $\mathcal{A}$ and let $w \in \Sigma^*$ be a word such that $\deg(\delta_w) = \deg(\mathcal{A}_i)$. Using simple observations about degrees, we can define a DFA $\mathcal{B}_i$ that is similar to $\mathcal{A}_i$ except that each letter σ acts in $\mathcal{B}_i$ as the word $\sigma \cdot w$ acts in $\mathcal{A}_i$. It can then be shown that $\mathcal{B}_i$ is a permutation DFA, and that it can replace $\mathcal{A}_i$ in the decomposition of $\mathcal{A}$. Hence the following theorem.

Theorem 7. *Let $\mathcal{A}$ be a permutation minimal DFA and let $t \in \mathbb{N}$. If $\mathcal{A}$ is t-decomposable then it can be t-decomposed using only permutation DFAs as factors.*

Beyond its theoretical interest, Theorem 7 implies that when checking the primality of a permutation DFA, we may consider only permutation DFAs as candidate factors. We now use this result in order to develop a more efficient algorithm for deciding the primality of permutation DFAs. We first need some observations on permutation DFAs.

Let $\mathcal{A} = \langle Q, \Sigma, q_0, \delta, F \rangle$ be a permutation DFA. We say that $\mathcal{A}$ is *inverse-closed* if for every $\sigma \in \Sigma$ there exists a letter, denoted $\sigma^{-1} \in \Sigma$, such that $\delta_{\sigma^{-1}} = (\delta_\sigma)^{-1}$. When $\mathcal{A}$ is not inverse-closed, we can consider the *inverse-closure* of $\mathcal{A}$, which is the DFA $\mathcal{A}' = \langle Q, \Sigma', q_0, \delta', F \rangle$, where $\Sigma' = \Sigma \cup \{\sigma^{-1} : \sigma \in \Sigma\}$ and $\delta'(q, \tau)$ is $\delta(q, \tau)$ if $\tau \in \Sigma$, and is $(\delta_\sigma)^{-1}(q)$ if $\tau = \sigma^{-1}$ for some $\sigma \in \Sigma$.

Recall that $\mathcal{A}$ is a permutation DFA, and thus $(\delta_\sigma)^{-1}$ is well-defined.

Lemma 7. *Let $\mathcal{A}$ be a permutation DFA and let $\mathcal{A}'$ be the inverse-closure of $\mathcal{A}$. Then, $depth(\mathcal{A}) = depth(\mathcal{A}')$.*

From now on we assume that the permutation DFAs that we need to decompose are inverse-closed. By Lemma 7 we do not lose generality doing so.

Given an alphabet Σ, we use F_Σ to denote the group generated by $\Sigma \cup \{\sigma^{-1} : \sigma \in \Sigma\}$ with the only relations being $\sigma \cdot \sigma^{-1} = \sigma^{-1} \cdot \sigma = \epsilon$ for every $\sigma \in \Sigma$. We note that the group F_Σ is also known as the *free group* over Σ.

Given a permutation DFA $\mathcal{A} = \langle Q, \Sigma, q_0, \delta, F \rangle$, we can think of F_Σ as a group acting on Q. Let $\tau \in \Sigma \cup \{\sigma^{-1} : \sigma \in \Sigma\}$. We describe the action of τ on Q by means of a function $\delta_\tau : Q \to Q$ defined as follows. If $\tau = \sigma \in \Sigma$, then $\delta_\tau(q) = \delta_\sigma(q)$. If $\tau = \sigma^{-1}$ with $\sigma \in \Sigma$, then $\delta_\tau(q) = \delta_\sigma^{-1}(q)$. Let $w \in F_\Sigma$ be such that $w = \tau_1 \cdots \tau_m$. The action of w on Q is then $\delta_w = \delta_{\tau_m} \cdots \delta_{\tau_1}$.

The *stabilizer subgroup* of q is the group $G(q) = \{w \in F_\Sigma : \delta_w(q_0) = q_0\}$. Let $w \in F_\Sigma$. The set $G(q) \cdot w$ is called a *right coset* of $G(q)$. The *index of $G(q)$ in F_Σ*, denoted $[F_\Sigma : G(q)]$, is defined as the number of right cosets of $G(q)$.

Fix Σ and let G be a subgroup of F_Σ such that $[F_\Sigma : G] = n$. Let C be the set of right cosets of G and let $D \subseteq C$. The pair $\langle G, D \rangle$ induces the DFA $\mathcal{A} = \langle C, \Sigma, G, \delta, D \rangle$, where $\delta(\sigma, G \cdot w) = G \cdot w \cdot \sigma$. Note that $\mathcal{A}$ is well defined, that it is a permutation DFA, and that $ind(\mathcal{A}) = n$. Consider a word $w \in \Sigma^*$. Note that $w \in L(\mathcal{A})$ iff the coset $G \cdot w$ is a member of D. Finally, note that the stabilizer set of the initial state of $\mathcal{A}$ is equal to G. Accordingly, we have the following.

Lemma 8. *There is a one to one correspondence between permutation DFAs and pairs* $\langle G, D \rangle$*, where* G *is a subgroup of* F_Σ *of the DFA's index and* D *is a set of right cosets of* G *in* F_Σ.

Let G be a group and let H be a subgroup of G. The subgroup H is said to be a *normal subgroup* of G if for every $g \in G$ it holds that $g \cdot H \cdot g^{-1} = H$.

Let $\mathcal{A} = \langle Q, \Sigma, q_0, \delta, F \rangle$ be a permutation DFA. Consider the subgroup $G(q_0)$. If $G(q_0)$ is a normal subgroup of F_Σ, we say that $\mathcal{A}$ is a *normal DFA*. It can be shown that $\mathcal{A}$ is normal iff for every $q, q' \in Q$, it holds that $G(q) = G(q')$. Equivalently, $\mathcal{A}$ is normal iff for every $w \in F_\Sigma$ such that δ_w has a fixed point, it holds that δ_w is the identity function on Q.

Lemma 9. *Let* $\mathcal{A} = \langle Q, \Sigma, q_0, \delta, F \rangle$ *be a normal permutation DFA and let* $\mathcal{B} = \langle S, \Sigma, s_0, \eta, G \rangle$ *be a permutation DFA. Let* $q_1, q_2 \in Q$ *and* $s_1, s_2 \in S$ *be such that* $q_1 \sim s_1$*,* $q_1 \sim s_2$*, and* $q_2 \sim s_1$*. Then,* $q_2 \sim s_2$.

Lemma 10. *Let* $\mathcal{A}$ *be a normal permutation DFA and* $\mathcal{B}$ *be a permutation DFA such that* $L(\mathcal{A}) \subseteq L(\mathcal{B})$*. Then, there exists a permutation DFA* $\mathcal{C}$ *such that* $L(\mathcal{A}) \subseteq L(\mathcal{C}) \subseteq L(\mathcal{B})$*,* $\mathcal{C}$ *is an abstraction of* $\mathcal{A}$*, and* $ind(\mathcal{C}) \le ind(\mathcal{B})$.

Theorem 8. *Let* $\mathcal{A}$ *be a normal t-decomposable permutation DFA. Then* $\mathcal{A}$ *is t-simply-decomposable.*

Let $\mathcal{A} = \langle Q, \Sigma, q_0, \delta, F \rangle$ be a DFA. We denote the *monoid DFA* of $\mathcal{A}$ by $\mathcal{A}_M = \langle M(\mathcal{A}), \Sigma, id, \delta_M, F_M \rangle$, with $\delta_M(f, w) = \delta_w \cdot f$ and $F_M = \{f \in M(\mathcal{A}) : f(q_0) \in F\}$. Simple observations about the monoid DFA show that $\mathcal{A}_M$ is normal and has the same depth as $\mathcal{A}$, meeting our goal

Let $\mathcal{A} = \langle Q, \Sigma, q_0, \delta, F \rangle$ be a permutation DFA. For each $q \in Q$, let $G(q)$ be the stabilizer subgroup of q in $\mathcal{A}$. Let $\pi \in M(\mathcal{A})$ and let $G_M(\pi)$ be the stabilizer subgroup of π in $\mathcal{A}_M$. It holds that $G_M(\pi) = \bigcap_{q \in Q} G(q)$.

The following theorem shows an immediate use of the transition monoid to prove the primality of a family of languages. Note that Example 5 is a special case of this more general theorem.

Theorem 9. *Let* $\mathcal{A} = \langle Q, \Sigma, q_0, \delta, F \rangle$ *be a permutation DFA. If* $|F| = ind(\mathcal{A}) - 1$*, then* $\mathcal{A}$ *is prime.*

Consider a permutation DFA $\mathcal{A}$ of index n. Let $\mathcal{A}_M$ be its monoid DFA. By the above, instead of checking whether $\mathcal{A}$ is prime, we can check whether $\mathcal{A}_M$ is $(n-1)$-decomposable. Since $\mathcal{A}_M$ is normal, this is equivalent to asking whether $\mathcal{A}_M$ is $(n-1)$-simply-decomposable. As $\mathcal{A}_M$ is of size at most exponential in n, this check can be done in PSPACE. We thus have the following.

Theorem 10. *Deciding the primality of permutation DFA can be done in* PSPACE.

8   Discussion

The motivation for this work has been compositional methods for LTL model checking and synthesis. Much to our surprise, we have realized that even the basic problem of

DFA decomposition was still open. Not less surprising has been the big gap between the EXPSPACE and NLOGSPACE upper and lower bounds for the primality problem. This work described our struggle and partial success with the problem and this gap. While the general case is still open, we managed to develop some intuitions and tools that we believe to be interesting and useful, to develop helpful primality-related theory, to identify easy cases, and to develop an algebraic approach to the problem.

Our future work involves both further investigations of the theory and tools developed here and a study of richer settings of decomposition. In the first front, we seek results that bound (from both below and above) the length of a primality witness or bound the width of decompositions. In the second front, we study richer types of automata, mainly automata on infinite words (as with nondeterministic automata, an additional challenge in this setting is the lack of a canonical minimal automaton), as well as richer definitions of decomposition. In particular, we are interested in union-intersection decompositions, where one may apply not only intersection but also take the union of the underlying automata. While it is easy to dualize our results for the case of union-only decomposition, mixing union and intersection results is a strictly stronger notion. Some of our results, for example the decomposition of permutation DFAs, apply also in this stronger notion.

References

1. Alpern, B., Schneider, F.B.: Recognizing safety and liveness. Distributed Computing 2, 117–126 (1987)
2. de Roever, W.-P., Langmaack, H., Pnueli, A. (eds.): COMPOS 1997. LNCS, vol. 1536. Springer, Heidelberg (1998)
3. Gazi, P.: Parallel decompositions of finite automata. Master's thesis, Comenius University, Bratislava, Slovakia (2006)
4. Hartmanis, J., Stearns, R.E.: Algebraic structure theory of sequential machines. Prentice-Hall international series in applied mathematics. Prentice-Hall (1966)
5. Krohn, K., Rhodes, J.: Algebraic theory of machines. i. prime decomposition theorem for finite semigroups and machines. Transactions of the American Mathematical Society 116, 450–464 (1965)
6. Kupferman, O., Piterman, N., Vardi, M.Y.: Safraless compositional synthesis. In: Ball, T., Jones, R.B. (eds.) CAV 2006. LNCS, vol. 4144, pp. 31–44. Springer, Heidelberg (2006)
7. Myhill, J.: Finite automata and the representation of events. Technical Report WADD TR-57-624, pp. 112–137, Wright Patterson AFB, Ohio (1957)
8. Nerode, A.: Linear automaton transformations. Proceedings of the American Mathematical Society 9(4), 541–544 (1958)
9. Pnueli, A.: The temporal semantics of concurrent programs. TCS 13, 45–60 (1981)
10. Pnueli, A.: Applications of temporal logic to the specification and verification of reactive systems: A survey of current trends. In: Rozenberg, G., de Bakker, J.W., de Roever, W.-P. (eds.) Current Trends in Concurrency. LNCS, vol. 224, pp. 510–584. Springer, Heidelberg (1986)
11. Vardi, M.Y., Wolper, P.: Reasoning about infinite computations. I&C 115(1), 1–37 (1994)

Logical Aspects of the Lexicographic Order on 1-Counter Languages

Dietrich Kuske

TU Ilmenau, Germany

Abstract. We prove two results to the effect that 1-counter languages give rise to the full complexity of context-free and even recursively enumerable languages: (1) There are pairs of disjoint deterministic one-counter languages whose union, ordered lexicographically, has an undecidable Σ_3-theory and, alternatively, true arithmetic can be reduced to its first-order theory. (2) It is undecidable whether the union of two disjoint deterministic 1-counter languages, ordered lexicographically, is dense.

In several aspects, these results cannot be sharpened any further: (a) they do not hold for single deterministic 1-counter languages [Cau02, Cau03], (b) they do not hold for the Σ_2-theory (Corollary 1.2), and (c) the first-order theory can always be reduced to true arithmetic (since these linear orders are computable structures).

1   Introduction

A natural structure on a language is provided by its lexicographic order $\leqslant_{\text{lex}}$ (provided the alphabet is linearly ordered). Up to isomorphism, every countable linear order arises this way, although the language has to be arbitrarily complicated. There has been quite some work on the structure of these linear orders depending on the complexity of the language (measured, e.g., in the Chomsky hierarchy). We mention a few of these results:

For every *regular* language L, the first-order theory of $(L, \leqslant_{\text{lex}})$ is decidable - this can be derived from Rabin's theorem [Rab69], from Büchi's theorem [Büc62, Theorem 4], or from the fact that $(L, \leqslant_{\text{lex}})$ is an automatic structure [KN95]. Since the decision procedure is even uniform in the regular language, the set of regular languages L with $(L, \leqslant_{\text{lex}}) \cong (\mathbb{Q}, \leqslant)$ is decidable (alternatively, this follows from the fact that the isomorphism problem for regular languages is decidable [Tho86, LM13]). For every ordinal $\alpha < \omega^\omega$, there exists a regular language L with $(L, \leqslant_{\text{lex}}) \cong \alpha$ and no larger ordinals can be presented this way [Del04].

For *context-free* languages, the following has been shown: There is a context-free language L such that the first-order theory of $(L, \leqslant_{\text{lex}})$ is undecidable (this language is even the disjoint union of three deterministic context-free languages) [CÉ12a]. For a deterministic context-free language K, the monadic second order theory of $(K, \leqslant_{\text{lex}})$ is decidable [Cau02, Cau03]; hence there is no

K. Chatterjee and J. Sgall (Eds.): MFCS 2013, LNCS 8087, pp. 619–630, 2013.

such language K with $(L, \leqslant_{\mathrm{lex}}) \cong (K, \leqslant_{\mathrm{lex}})$. Furthermore, the set of context-free languages L with $(L, \leqslant_{\mathrm{lex}}) \cong (\mathbb{Q}, \leqslant)$ is undecidable [Ési11]. Consequently, the isomorphism problem is undecidable and, as shown in [KLL11], even Σ_1^1-complete for deterministic context-free languages. For every ordinal $\alpha < \omega^{\omega^\omega}$, there exists a deterministic context-free language L with $(L, \leqslant_{\mathrm{lex}}) \cong \alpha$ and no larger ordinals can be presented this way [BÉ10].

It follows that the undecidability results also hold for recursively enumerable languages, the representable ordinals are those properly below ω_1^{CK} (by the definition of ω_1^{CK}).

In this paper, we study similar problems for 1-counter languages, i.e., a class of languages in between regular and context-free languages. Interest in these languages revived recently in particular in the verification community (see, e.g., [GHW12, BGc13]).

First, we present some ordinals by 1-counter languages:

Theorem 1.1. *For every ordinal $\alpha < \omega^{\omega^2}$, there is a deterministic real-time 1-counter language L such that $(L, \leqslant_{\mathrm{lex}}) \cong \alpha$.*

Already from this result it follows that 1-counter languages generate more linear orders than regular languages. I conjecture that no ordinal $\geqslant \omega^{\omega^2}$ can be represented by 1-counter languages. Recall that the bound for context-free languages is precisely ω^{ω^ω}. Hence, if the conjecture is true, the class of linear orders of 1-counter languages is properly in between those of regular and of context-free languages.

Then we turn to the main concern of this paper: the first-order theory of these linear orders. To set the scene, we show that, for every linear order $\mathcal{L}$, there exists a regular language L such that $\mathcal{L}$ and $(L, \leqslant_{\mathrm{lex}})$ cannot be distinguished by $\exists^* \forall^*$-sentences (Theorem 4.3). By the decidability for regular languages mentioned above, we obtain the following:

Corollary 1.2. *Let $\mathcal{L}$ be a linear order. Then the $\exists^* \forall^*$-theory of $\mathcal{L}$ is decidable.*

It is not difficult to construct a computable linear order whose $\exists^* \forall \exists$-theory is undecidable. Somewhat surprisingly, there is even a 1-counter language with (almost) this property:

Theorem 1.3. *There are two disjoint deterministic real-time 1-counter languages L_1 and L_2 such that the $\exists^* \forall^3 \exists$-theory of $(L_1 \cup L_2, \leqslant_{\mathrm{lex}})$ is undecidable.*

As a byproduct of the proof of this theorem, we obtain that the isomorphism problem is undecidable. More precisely, we show the following.

Theorem 1.4. *There is no algorithm that, given two disjoint deterministic real-time 1-counter languages L_1 and L_2, decides whether $(L_1 \cup L_2, \leqslant_{\mathrm{lex}}) \cong (\mathbb{Q}, \leqslant)$.*

Note that isomorphism to $(\mathbb{Q}, \leqslant)$ is expressible by a $\forall^* \exists$-formula. Hence, the decidability from Cor. 1.2 is not uniform for 1-counter languages.

The very basic idea of the proof of this and all the other undecidability results regarding 1-counter languages is the well known fact that the set of halting computations of a two-counter machine M is the *intersection* of two deterministic 1-counter languages. The language $L_1 \cup L_2$ from the theorem then is the *union* of these two 1-counter languages and a regular language. The regular language is, basically, chosen in such a way that the computations of the two-counter machine correspond to the intervals of length 2 in the linear order $\mathcal{L} = (L_1 \cup L_2, \leqslant_{\text{lex}})$ – hence, $\mathcal{L} \cong (\mathbb{Q}, \leqslant)$ if and only if the two-counter machine M has no halting computation.

For a 1-counter language L, the linear order $\mathcal{L} = (L, \leqslant_{\text{lex}})$ is a computable structure. Therefore, the first-order theory of $\mathcal{L}$ can be reduced to that of $(\mathbb{N}, +, \cdot)$, i.e., to true arithmetic. We construct a 1-counter language L such that also true arithmetic can be reduced to the first-order theory of $(L, \leqslant_{\text{lex}})$:

Theorem 1.5. *There are two disjoint deterministic real-time 1-counter languages L_1 and L_2 such that true arithmetic can be reduced to the first-order theory of $(L_1 \cup L_2, \leqslant_{\text{lex}})$.*

The proof of this theorem starts with the construction of a decidable language L such that true arithmetic can be reduced to the first-order theory of $(L, \leqslant_{\text{lex}})$ (this linear order is even scattered, cf. Theorem 7.1). In a second step, we build L_1 and L_2 as in the theorem and interpret $(L, \leqslant_{\text{lex}})$ in the linear order $(L_1 \cup L_2, \leqslant_{\text{lex}})$.

Summary. This study demonstrates that 1-counter languages realize the full complexity of computable languages when ordered lexicographically. This sharpens recent results by Ésik [Ési11] and Carayol and Ésik [CÉ12a] considerably. These undecidability results are pushed to the limit since they neither hold for deterministic 1-counter languages [Cau02, Cau03] nor for the Σ_2-theory (Cor. 1.2) nor can their first-order theory be harder than true arithmetic.

2   Preliminaries

2.1   Notions for the Main Results

Languages and Automata. A *deterministic 1-counter automaton* is a deterministic pushdown automaton with a single stack symbol (apart from the bottom-of-stack symbol) whose acceptance condition is given by a set of accepting states. It is *real-time* if it does not have any ε-transitions, i.e., consumes a letter of the input with every transition. A language L is a *deterministic real-time 1-counter language* if it can be accepted by some deterministic real-time 1-counter automaton.

Let Σ be some set and $\leqslant$ a linear order on Σ. Then $\leqslant_{\text{lex}}$ denotes the *lexicographic order* on Σ^*: $u \leqslant_{\text{lex}} v$ if u is a prefix of v or $u = xay$ and $v = xbz$ with $x, z, y \in \Sigma^*$ and $a, b \in \Sigma$ such that $a < b$.

Logic. Fix the set $\{v_i \mid i \geqslant 0\}$ of elementary variables. If x and y are elementary variables, then $x \leqslant y$, $x < y$, and $x = y$ are atomic formulas. Complex formulas can be built as usual using Boolean connectives and the quantifiers $\exists$ and $\forall$. A formula without free variables is a *sentence*. Let $\mathcal{L}$ be some linear order and φ a sentence. Then $\mathcal{L} \models \varphi$ denotes that $\mathcal{L}$ satisfies φ.

Example 2.1. The formula $x \lessdot y$ abbreviates $x < y \wedge \neg \exists z (x < z \wedge z < y)$. It expresses that y is the direct successor of x.

A formula φ is in *prenex normal form* if it is of the form

$$Q_1 x_1 \, Q_2 x_2 \, \ldots \, Q_m x_m \, \alpha$$

where α is a Boolean combination of atomic formulas and $Q_1, \ldots, Q_m \in \{\exists, \forall\}$. If $H \subseteq \{\exists, \forall\}^*$, then we say "$\varphi$ belongs to the class H" if $Q_1 Q_2 \ldots Q_n \in H$. In particular, Σ_2 arises from the language $H = \exists^* \forall^*$ and Σ_3 from $\exists^* \forall^* \exists^*$.

For a linear order $\mathcal{L}$, the *theory of* $\mathcal{L}$ is the set of sentences that are satisfied in $\mathcal{L}$. The *H-theory* is the set of sentences from the class H that hold true in $\mathcal{L}$.

2.2   Notions for the Proofs

2-Counter Machines. Let Δ be some alphabet. A *2-counter machine* is a tuple $M = (I_1, I_2, \ldots, I_m)$ where every I_j is of one of the following forms:

(1) halt
(2) $x_c := x_c + 1$; goto ℓ
(3) if $x_c = 0$ then goto ℓ_1 else $x_c := x_c - 1$; goto ℓ_2 endif
(4) read$((\ell_\sigma)_{\sigma \in \Delta})$

where $c \in \{1, 2\}$ and $1 \leqslant \ell, \ell_1, \ell_2, \ell_\sigma \leqslant m$ for $\sigma \in \Delta$. A 2-counter machine is *inputless* if all its instructions are of the form (1), (2), or (3).

These machines accept a word if the machine reaches the instruction halt after reading the whole input from left to right. The meaning of the instructions (1), (2) and (3) should be clear, in particular, they do not consume any letter of the input. Differently, when executing the instruction from (4), the machine reads the next input symbol σ and the computation continues with instruction I_{ℓ_σ}.

Usually, the semantics of a 2-counter machine is defined in terms of sequences of configurations. In the context of this paper, it is more convenient to define it in terms of sequences of atomic actions. These atomic actions are: incrementation of counter c denoted $+_c$, decrementation of counter c denoted $-_c$, successful test for emptiness of counter c denoted $\mathbf{0}_c$, and reading of a letter $a \in \Delta$ denoted a. Therefore, computations of a 2-counter machine will be words over the alphabet $\Delta_2 = \Delta \cup \{+_1, +_2, -_1, -_2, \mathbf{0}_1, \mathbf{0}_2\}$.

To qualify as computation, a word over Δ_2 has to satisfy three conditions that we define next.

A word $w = a_0 a_1 \ldots a_{n-1}$ with $a_0, a_1, \ldots, a_{n-1} \in \Delta_2$ *conforms to the control flow of* M if there are $\ell_0, \ell_1, \ldots, \ell_n \in \{1, 2, \ldots, m\}$ such that, for all $1 \leqslant i < n$ and $c \in \{1, 2\}$, the following hold

- $\ell_0 = 1$
- If $a_i = +_c$, then $I_{\ell_i} = (x_c := x_c + 1;$ goto $\ell_{i+1})$.
- If $a_i = -_c$, then $I_{\ell_i} = ($if $x_c = 0$ then goto ℓ else $x_c := x_c - 1;$ goto ℓ_{i+1} endif$)$
  for some $1 \leqslant \ell \leqslant m$.

- If $a_i = \mathbf{0}_c$, then $I_{\ell_i} = ($if $x_c = 0$ then goto ℓ_{i+1} else $x_c := x_c - 1;$ goto ℓ endif$)$ for some $1 \leqslant \ell \leqslant m$.
- If $a_i \in \Delta$, then $I_{\ell_i} = \mathrm{read}((\ell'_\sigma)_{\sigma \in \Delta})$ with $\ell'_{a_i} = \ell_{i+1}$.
- $I_{\ell_n} = \mathrm{halt}$.

Note that the set of words that conform to the control flow of M form a regular language and that a deterministic finite automaton accepting this language can be computed from M.

A word $w \in \Delta_2^*$ *conforms to the counter conditions of counter c from* $(n_1, n_2) \in \mathbb{N}^2$ if

- $n_c + |u|_{+_c} \leqslant |u|_{-_c}$ for all prefixes u of w and
- $n_c + |u|_{+_c} = |u|_{-_c}$ for all prefixes $u\mathbf{0}_c$ of w

where $|u|_a$ denotes the number of occurrences of the letter a in the word u. The idea is that, when started with value n_c in counter c, the value of the counter after executing the sequence of atomic actions u equals $n_c + |u|_{+_c} - |u|_{-_c}$. Hence the first condition expresses that the counters will always hold non-negative integers. The second condition expresses that the emptiness test of counter c is successful whenever it is claimed to be successful.

The set of words that conform to the counter conditions of counter c from $(n_1, n_2) \in \mathbb{N}^2$ form a deterministic real-time 1-counter language (that only depends on Δ, but not on M).

A word $w \in \Delta_2^*$ is a *halting* or *accepting computation of M from* $(n_1, n_2) \in \mathbb{N}^2$ if it conforms to the control flow and to both counter conditions from (n_1, n_2).

The following easy observation is central to the proofs of this paper:

> The set of accepting computations of a 2-counter machine M is the intersection of two deterministic real-time 1-counter languages that can be computed from M.

(But our results are concerned with disjoint unions as opposed to intersections.)

A word $w \in \Delta^*$ is accepted by M if it is the projection to Δ of an accepting computation of M from $(0,0)$. We denote the language of M, i.e., the set of all words accepted by M, by $L(M)$. For an inputless 2-counter machine, we also define the *halting set*: $H(M)$ is the set of pairs $(n_1, n_2) \in \mathbb{N}^2$ such that there is an accepting computation of M from (n_1, n_2).

Theorem 2.2 (Minsky [Min61])

(1) From a Turing machine accepting the language $L \subseteq \Sigma^*$ and a letter $\$ \notin \Sigma$, one can compute a 2-counter machine M with $L(M) = L\$$.

(2) From a Turing machine accepting a set $A \subseteq \mathbb{N}$, one can compute an inputless 2-counter machine M with $H(M) = \{(2^n \cdot m, 0) \mid n \in A, m \in \mathbb{N}$ odd$\}$.

Recall that the function $n \mapsto 2^n$ cannot be computed by any inputless 2-counter machine [Sch72].

Linear Orders. We write ω for $(\mathbb{N}, \leqslant)$, ζ for $(\mathbb{Z}, \leqslant)$, η for $(\mathbb{Q}, \leqslant)$, and (for $n \in \mathbb{N}$), $\mathbf{n}$ stands for $(\{1, 2, \ldots, n\}, \leqslant)$.

Let $(I, \leqslant_I)$ be a linear order and, for $i \in I$, let $(L_i, \leqslant_i)$ be linear orders. Set $L = \bigcup_{i \in I} L_i \times \{i\}$ and, for $(x, i), (y, j) \in L$, set $(x, i) \leqslant (y, j)$ if $i <_I j$ or $i = j$ and $x \leqslant_i y$. Then $(L, \leqslant)$ is a linear order denoted $\sum_{i \in (I, \leqslant_I)} (L_i, \leqslant_i) = (L, \leqslant)$. Intuitively, one obtains $\sum_{i \in (I, \leqslant_I)} (L_i, \leqslant_i)$ by replacing every element i of $(I, \leqslant_I)$ by a copy of the linear order $(L_i, \leqslant_i)$.

If $(I, \leqslant_I) = \mathbf{2}$, we write $(L_1, \leqslant_1) + (L_2, \leqslant_2)$ for $\sum_{i \in \mathbf{2}} (L_i, \leqslant_i)$, i.e., for the concatenation of the two linear orders. Furthermore, if $(K, \leqslant_K) = (L_i, \leqslant_i)$ for all $i, j \in I$, then we write

$$(K, \leqslant_K) \cdot (I, \leqslant_I) = \sum_{i \in (I, \leqslant_I)} (L_i, \leqslant_i).$$

Since every countable dense linear order without endpoints is isomorphic to η [Can97] (see also [Ros82]), one gets the following:

(A) $\eta + \eta \cong \eta \cong \eta + \mathbf{1} + \eta$, but $\eta + \mathbf{2} + \eta \not\cong \eta$.
(B) $\eta \cdot \mathcal{L} \cong \eta$ for all non-empty countable linear orders $\mathcal{L}$.

Below we will refer to these statements as (A) and (B), resp.

3   Ordinals (Proof of Theorem 1.1)

First consider the ordinal ω^ω which is the set of all tuples of natural numbers with the length-lexicographic order $\leqslant_{\text{llex}}$. We use the deterministic real-time 1-counter language $L_1 = \bigcup_{n \geqslant 1} c^n (b^* a)^n$ (note that the number of occurrences of c is arbitrary and equals that of a). With $a < b < c$, the mapping

$$(n_1, n_2, \ldots, n_m) \mapsto c^m b^{n_1} a b^{n_2} \ldots b^{n_m} a$$

proves $(L_1, \leqslant_{\text{lex}}) \cong (\mathbb{N}^+, \leqslant_{\text{llex}}) \cong \omega^\omega$.

Next, $(\omega^\omega)^k$ is the set of k-tuples of elements of ω^ω ordered lexicographically. Hence $(L_k, \leqslant_{\text{lex}}) \cong (\omega^\omega)^k$ for the deterministic real-time 1-counter language

$$L_k = L_1^k = \left(\bigcup_{n \geqslant 1} c^n (b^* a)^n \right)^k.$$

Finally, let $\alpha < \omega^{\omega^2}$. Then there exists $k \geqslant 1$ with $\alpha < (\omega^\omega)^k$. Therefore, we find $u \in L_k$ with $\alpha \cong (\{v \in L_k \mid v <_{\text{lex}} u\}, \leqslant_{\text{lex}})$. Since the language $\{v \in \{a, b, c\}^* \mid v <_{\text{lex}} u\}$ is regular, the intersection L of this language with L_k is a deterministic real-time 1-counter language with $\alpha \cong (L, \leqslant_{\text{lex}})$. This finishes the proof of Theorem 1.1.

Comments and Open Questions

Let L be a context-free language. Then the Hausdorff rank of $(L, \leqslant_{\text{lex}})$ (see [Ros82] for the definition) is properly below ω^ω [CÉ12b]. Hence, context-free languages can only represent ordinals properly below ω^{ω^ω} [BÉ10]. Furthermore, any such ordinal can be represented by a deterministic context-free language [BÉ10].

In view of these results, I pose the following two questions:

1. Is there a 1-counter language L such that $(L, \leqslant_{\text{lex}}) \cong \omega^{\omega^2}$?
2. Is there a 1-counter language L such that the Hausdorff rank of $(L, \leqslant_{\text{lex}})$ is at least ω^2?

4 Σ_2-Theories of Arbitrary Linear Orders (Proof of Cor. 1.2)

The central concept here is that of the *FC-class*[1] of an element x of a linear order $\mathcal{L}$: this is the set of elements y of $\mathcal{L}$ such that the interval with endpoints x and y is finite. Typical properties expressible in Σ_2 concern the relative order of FC-classes of size at least k. One such example is "there is an FC-class of size $\geqslant 3$ followed by some FC-class of size $\geqslant 2$":

$$\exists x_1, x_2, x_3, y_1, y_2 \, \forall z \left(\begin{array}{l} x_1 < x_2 < x_3 < y_1 < y_2 \\ \wedge \bigwedge_{1 \leqslant i \leqslant 2} \neg(x_i < z < x_{i+1}) \\ \wedge \neg(y_1 < z < y_2) \end{array} \right)$$

Having this in mind, the following two lemmas are natural. ($\mathcal{L}_1 \equiv_{\Sigma_2} \mathcal{L}_2$ expresses that $\mathcal{L}_1$ and $\mathcal{L}_2$ cannot be distinguished by Σ_2-formulas.)

Lemma 4.1. *Let $\mathcal{L}_1$ and $\mathcal{L}_2$ be infinite linear orders. Then $\mathcal{L}_1 \equiv_{\Sigma_2} \mathcal{L}_2$ if one of the statements (1), (2), or (3) holds:*

(1) $\mathcal{L}_1$ *and* $\mathcal{L}_2$ *have no endpoints and the size of FC-classes in* $\mathcal{L}_1$ *and in* $\mathcal{L}_2$ *is not finitely bounded.*

(2) $\mathcal{L}_1$ *and* $\mathcal{L}_2$ *have minimal elements, the FC-classes of these minimal elements are infinite, and* $\mathcal{L}_1$ *and* $\mathcal{L}_2$ *do not have maximal elements.*

(3) $\mathcal{L}_1$ *and* $\mathcal{L}_2$ *have infinitely many FC-classes of size N and no FC-classes of size $> N$. Furthermore,* $\mathcal{L}_1$ *and* $\mathcal{L}_2$ *do not have endpoints.*

Proof. We only consider the case (1). Note that every Σ_2-sentence is a disjunction of sentences of the form

$$\Phi = \exists x_1, \ldots, x_k \, (x_1 < x_2 < \cdots < x_k \wedge \forall x_{k+1}, \ldots, x_{k+\ell} \, \Psi) \tag{1}$$

where Ψ is quantifier-free. Hence it suffices to show that $\mathcal{L}_1$ and $\mathcal{L}_2$ cannot be distinguished by sentences of this form. So assume $\mathcal{L}_1 \models \Phi$, i.e., there are $a_1 < a_2 < \cdots < a_k$ in $\mathcal{L}_1$ such that

$$(\mathcal{L}_1, a_1, \ldots, a_k) \models \forall x_{k+1}, \ldots, x_{k+\ell} \, \Psi . \tag{2}$$

---

[1] "FC" stands for "finite condensation", see [Ros82].

Since the size of FC-classes in $\mathcal{L}_2$ is not bounded, there are elements $b_1 \lessdot b_2 \lessdot \cdots \lessdot b_k$ in $\mathcal{L}_2$.

Now let $b_{k+1}, \ldots, b_{k+\ell} \in \mathcal{L}_2$ be arbitrary. Then, for all $1 \leqslant i \leqslant \ell$, we have $b_{k+i} \in \{b_1, \ldots, b_k\}$, $b_{k+i} < b_1$, or $b_{k+i} > b_k$. Since $\mathcal{L}_1$ has no endpoints, there are elements $a_{k+1}, \ldots, a_{k+\ell} \in \mathcal{L}$ such that

$$a_i < a_j \iff b_i < b_j \text{ for all } i, j \in \{1, 2, \ldots, k+\ell\}. \tag{3}$$

From (2), we infer

$$(\mathcal{L}_1, a_1, \ldots, a_k, a_{k+1}, \ldots, a_{k+\ell}) \models \Psi. \tag{4}$$

Since Ψ is quantifier-free, (3) and (4) imply

$$(\mathcal{L}_2, b_1, \ldots, b_k, b_{k+1}, \ldots, b_{k+\ell}) \models \Psi.$$

Since $b_{k+1}, \ldots, b_{k+\ell} \in \mathcal{L}_2$ are arbitrary and since $b_1 < b_2 < \cdots < b_k$, we have $\mathcal{L}_2 \models \Phi$.   □

Applying a similar proof (or analysing the proof of the Theorem by Feferman-Vaught [Hod93, Thm. 9.6.2]), one also obtains

Lemma 4.2. *Let $\mathcal{L}'_i$ and $\mathcal{L}''_i$ be linear orders with $\mathcal{L}'_1 \equiv_{\Sigma_2} \mathcal{L}'_2$ and $\mathcal{L}''_1 \equiv_{\Sigma_2} \mathcal{L}''_2$. Then $\mathcal{L}'_1 + \mathcal{L}'_2 \equiv_{\Sigma_2} \mathcal{L}''_1 + \mathcal{L}''_2$.*

Using these two lemmas, we next prove that every linear order is Σ_2-equivalent to a regular language ordered lexicographically.

Theorem 4.3. *Let $\mathcal{L}$ be an infinite linear order. There exists a regular language L such that $\mathcal{L} \equiv_{\Sigma_2} (L, \leqslant_{\mathrm{lex}})$.*

Proof. In this sketch of proof, we only consider the case of linear orders without endpoints.

If there is no finite upper bound for the size of FC-classes in $\mathcal{L}$, Lemma 4.1(1) ensures $\mathcal{L} \equiv_{\Sigma_2} \zeta \cong (0^+ 1 \cup 1^+, \leqslant_{\mathrm{lex}})$.

Now let $N \in \mathbb{N}$ be the maximal size of an FC-class in $\mathcal{L}$. We prove the claim by induction on N. If $N = 1$, then $\mathcal{L} \cong \eta \cong (\{0, 1\}^* 1, \leqslant_{\mathrm{lex}})$.

In case $N > 1$, we distinguish two cases. If there are infinitely many FC-classes of size N, then Lemma 4.1(3) implies

$$\mathcal{L} \equiv_{\Sigma_2} \mathbf{N} \cdot \eta \cong (\{N, N+1\}^* (N+1) \{0, \ldots, N-1\}, \leqslant_{\mathrm{lex}}).$$

Now suppose there are $n \in \mathbb{N}$ FC-classes of size N. Then there are infinite linear orders $\mathcal{L}_i$ for $i \in \{0, 1, \ldots, n\}$ such that $\mathcal{L} = \mathcal{L}_0 + \sum_{1 \leqslant i \leqslant n} (\mathbf{N} + \mathcal{L}_i)$ and the induction hypothesis is applicable to the linear orders $\mathcal{L}_i$.   □

Now Cor. 1.2 follows immediately since the first-order theory of $(L, \leqslant_{\mathrm{lex}})$ is decidable for every regular language L [Rab69].

5 An Undecidable Theory (Proof of Theorem 1.3)

The plan is as follows: By Theorem 2.2(2), there is an inputless 2-counter machine U with an undecidable halting set $H(U) \subseteq \mathbb{N} \times \{0\}$. Let $A = \{n \in \mathbb{N} \mid (n,0) \in H(U)\}$. We will construct, from U, two disjoint deterministic real-time 1-counter languages L_1 and L_2 and reduce A to the $\exists^* \forall^3 \exists$-theory of $(L_1 \cup L_2, \leqslant_{\mathrm{lex}})$.

5.1 Encoding of A in a Linear Order $\mathcal{L}$

For $n \in A$ let $\lambda_n = \eta + 2 + \eta$ and, for $a \in \mathbb{N} \setminus A$, set $\lambda_n = \eta$. Note that $\lambda_n \models \exists x, y \colon x \lessdot y$ if and only if $n \in A$. Furthermore, set

$$\mathcal{L} = \sum_{n \in \omega} (3 + \lambda_n).$$

Note that $n \in A$ if and only if, in $\mathcal{L}$, there is an interval of length 2 between the intervals of length 3 number $n + 1$ and $n + 2$. In other words, $n \in A$ if and only if the linear order $\mathcal{L}$ satisfies the following formula:

$$\exists_{0 \leqslant i \leqslant n+1} x_i, y_i, z_i \left(\begin{array}{l} \bigwedge_{0 \leqslant i \leqslant n+1} x_i \lessdot y_i \lessdot z_i \\ \wedge \bigwedge_{0 \leqslant i < n+1} z_i \lessdot x_{i+1} \\ \wedge \forall x, y, z \, (x \lessdot y \lessdot z \to \bigvee_{0 \leqslant i \leqslant n+1} y = y_i \vee y_{n+1} \lessdot y) \\ \wedge \exists x, y \, (z_n \lessdot x \lessdot y \lessdot x_{n+1}) \end{array} \right)$$

The subformula $x \lessdot y \lessdot z$ in the premise of the third line hides an existential quantifier; whence the whole formula is logically equivalent to a $\exists^* \forall^3 \exists$-formula φ_n. Since φ_n can be computed from n, we reduced A to the $\exists^* \forall^3 \exists$-theory of the linear order $\mathcal{L}$. In particular, the $\exists^* \forall^3 \exists$-theory of $\mathcal{L}$ is undecidable.

5.2 Representation of $\mathcal{L}$ by 1-Counter Languages

We consider the base language $\mathrm{BL} = +_1^* \# \Delta_2^+ \#$. Let $+_1^n \# v \# \in \mathrm{BL}$. The idea is that $+_1^n$ "preloads" the first counter to n and that then v is a computation of the 2-counter machine U that starts in $(n, 0)$.

On the language BL, we define unary relations R_1 and R_2. Let $c \in \{1, 2\}$ and $u = +_1^n \# v \# \in \mathrm{BL}$. Then $u \in R_c$ if and only if

- v conforms to the control flow of the 2-counter machine U and
- u conforms to the counter conditions of counter c.

Later, we will use that R_1 and R_2 both are deterministic real-time 1-counter languages. We have the following:

1. Suppose $u = +_1^n \# v \# \in \mathrm{BL}$ belongs to $R_1 \cap R_2$. Then v conforms to the control flow of U and u satisfies all counter conditions. Hence v is a halting computation from $(n, 0)$ implying $n \in A$.
2. Conversely, let $n \in A$. Then there exists a (uniquely determined) halting computation v from $(n, 0)$. Hence $+_1^n \# v \# \in R_1 \cap R_2$.

We now define a new language

$$K = +_1^* \#\{s_1, s_2, s_3\} \cup BL(\{0,1\}^*1 \cup \{2,3\}^*3) \cup R_1\, m_1 \cup R_2\, m_2 \,.$$

We order the alphabet Γ of this language in such a way that

$$\# < s_1 < s_2 < s_3 < 0 < 1 < m_1 < m_2 < 2 < 3 < \Delta_2 \,.$$

For a word $x \in \Gamma^*$, let $K(x) = K \cap x\Gamma^*$ denote the set of extensions of x that belong to the language K. Then $(K(x), \leqslant_{\text{lex}})$ is an interval in $(K, \leqslant_{\text{lex}})$ (or empty). Since $\# <_{\text{lex}} +_1\# <_{\text{lex}} +_1^2\# <_{\text{lex}} \dots$ and $K \subseteq +_1^*\#\Gamma^*$, we get

$$(K, \leqslant_{\text{lex}}) \cong \sum_{n \in \omega} (K(+_1^n\#), \leqslant_{\text{lex}}) \,.$$

Let $u = +_1^n\#$. Then $us_1 <_{\text{lex}} us_2 <_{\text{lex}} us_3$ are the three minimal elements of $(K(u), \leqslant_{\text{lex}})$, i.e., we have $(K(u), \leqslant_{\text{lex}}) \cong \mathbf{3} + (K'(u), \leqslant_{\text{lex}})$ where $K'(u) = K(u) \setminus \{us_1, us_2, us_3\}$. Next let $v \in \Delta_2^+\#$. Then

$$K(uv) = uv\{0,1\}^*1 \cup \{uvm_i \mid uv \in R_i, i \in \{1,2\}\} \cup uv\{2,3\}^*3 \,.$$

With $h \in \{0,1,2\}$ the number of sets R_1 and R_2 that uv belongs to, we get

$$(K(uv), \leqslant_{\text{lex}}) \;\cong\; (uv\{0,1\}^*1, \leqslant_{\text{lex}}) + h + (uv\{2,3\}^*3, \leqslant_{\text{lex}}) \cong \eta + h + \eta$$

$$\overset{\text{by (A)}}{\cong} \begin{cases} \eta + \mathbf{2} + \eta & \text{if } h = 2, \text{ i.e., } uv \in R_1 \cap R_2 \text{ and} \\ \eta & \text{otherwise.} \end{cases}$$

Since there is at most one word $v \in \Delta_2^+\#$ such that $uv \in R_1 \cap R_2$, we get

$$(K'(u), \leqslant_{\text{lex}}) \cong \sum_{v \in (\Delta_2^+\#, \leqslant_{\text{lex}})} (K(uv), \leqslant_{\text{lex}}) \overset{\text{by (B)}}{\cong} \begin{cases} \eta + \mathbf{2} + \eta & \text{if } n \in A \text{ and} \\ \eta & \text{otherwise.} \end{cases}$$

In other words, $(K(u), \leqslant_{\text{lex}}) \cong \mathbf{3} + (K'(u), \leqslant_{\text{lex}}) \cong \mathbf{3} + \lambda_n$ for all $n \in \mathbb{N}$. Thus,

$$\mathcal{L} = \sum_{n \in \omega} (\mathbf{3} + \lambda_n) \cong \sum_{n \in \omega} (K(+_1^n\#), \leqslant_{\text{lex}}) \cong (K, \leqslant_{\text{lex}}) \,.$$

Finally note that $L_1 = R_1 m_1$ and

$$L_2 = +_1^*\#\{s_1, s_2, s_3\} \cup BL(\{0,1\}^*1 \cup \{2,3\}^*3) \cup R_2 m_2$$

are deterministic real-time 1-counter languages. Since $K = L_1 \cup L_2$, this finishes the proof of Theorem 1.3 from the introduction.

6   The Density Problem (Proof of Theorem 1.4)

Theorem 1.4 is a consequence of the following result that we demonstrate by an adaptation of the above proof.

Theorem 6.1. *There are two disjoint deterministic real-time 1-counter languages L_1 and L_2 such that the set of regular languages L with $((L_1 \cap L) \cup (L_2 \cap L), \leqslant_{\text{lex}}) \cong \eta$ is undecidable.*

Proof. Let L_1 and L_2 be the languages from the previous section. For $n \in \mathbb{N}$, the language $L(n) = \{x \in \Gamma^* \mid +_1^n \# s_3 <_{\text{lex}} x <_{\text{lex}} +_1^{n+1} \#\}$ is regular. Since $(L_1 \cup L_2) \cap L(n) = K'(+_1^n \#)$, we have $((L_1 \cup L_2) \cap L(n), \leqslant_{\text{lex}}) \cong \eta$ if and only if $n \notin A$. Since A is undecidable, this proves Theorem 6.1. $\qquad\square$

Now Theorem 1.4 follows from Theorem 6.1 since the set of deterministic real-time 1-counter languages over Γ is effectively closed under the intersection with regular languages.

7   A Non-arithmetical Theory (Proof of Theorem 1.5)

While the official definition of arithmetical properties is a bit technical, here we can use a result by Tarski [Tar36] saying that a set is non-arithmetical provided true arithmetic can be reduced to it.

The starting point of this proof is the following result:

Theorem 7.1. *There exists a decidable language $L \subseteq \{01, 11\}00$ such that the first-order theory of $(L, \leqslant_{\text{lex}})$ is not arithmetical. Even more, η does not embed into $(L, \leqslant_{\text{lex}})$ (i.e., $(L, \leqslant_{\text{lex}})$ is scattered).*

It remains to encode the computable linear order $(L, \leqslant_{\text{lex}})$ into a 1-counter language. By Theorem 2.2(1), there exists a 2-counter machine M with $L(M) = L\$$. For $c \in \{1, 2\}$, let R_c denote the set of words over

$$\Delta_2 = \{0, 1, \$, +_1, +_2, -_1, -_2, \mathbf{0}_1, \mathbf{0}_2\}$$

that conform to the control flow of M and to the counter condition of counter c. Then $R_1 \cap R_2$ is the set of accepting computations of the 2-counter machine M.

Now observe the following:

1. If $w \in R_1 \cap R_2$, then the projection to $\{0, 1, \$\}$ belongs to $L\$$.
2. If $u \in L$, then there exists a uniquely determined word in $R_1 \cap R_2$ whose projection to $\{0, 1, \$\}$ equals $u\$$.
3. Due to the determinism of the 2-counter machine M, the above correspondence of words from $L\$$ and $R_1 \cap R_2$ is order-preserving, i.e., $(L, \leqslant_{\text{lex}}) \cong (L\$, \leqslant_{\text{lex}}) \cong (R_1 \cap R_2, \leqslant_{\text{lex}})$ (the first isomorphism holds since L is prefix-free).

Now let $K = \Delta_2^* \$ \cup R_1 \$ \mathbf{m}_1 \cup R_2 \$ \mathbf{m}_2$ and order the letters of the alphabet of K by $\mathbf{m}_1 < \mathbf{m}_2 < 0 < 1 < +_1 < +_2 < -_1 < -_2 < \mathbf{0}_1 < \mathbf{0}_2 < \$$. Then the elements of $R_1 \cap R_2$ correspond to the intervals of length 3 in $(K, \leqslant_{\text{lex}})$. Therefore, $(L, \leqslant_{\text{lex}}) \cong (R_1 \cap R_2, \leqslant_{\text{lex}})$ can be interpreted in $(K, \leqslant_{\text{lex}})$. Hence the first-order theory of $(L, \leqslant_{\text{lex}})$ can be reduced to that of $(K, \leqslant_{\text{lex}})$ which is therefore non-arithmetical. Since K is the disjoint union of the two deterministic real-time 1-counter languages $R_1 \$ \mathbf{m}_1$ and $\Delta_2^* \$ \cup R_2 \$ \mathbf{m}_2$, this finishes the proof of Theorem 1.5.

References

[BÉ10]   Bloom, S.L., Ésik, Z.: Algebraic ordinals. Fundamenta Informaticae 99(4), 384–407 (2010)

[BGc13]   Böhm, S., Göller, S., Jančar, P.: Equivalence of deterministic one-counter automata is NL-complete. In: STOC 2013 (to appear, 2013)

[Büc62]   Büchi, J.R.: On a decision method in restricted second order arithmetics. In: Nagel, E., et al. (eds.) Proc. Intern. Congress on Logic, Methodology and Philosophy of Science, pp. 1–11. Stanford University Press, Stanford (1962)

[Can97]   Cantor, G.: Beiträge zur Begründung der transfiniten Mengenlehre, II. Math. Annalen 49, 207–246 (1897)

[Cau02]   Caucal, D.: On infinite terms having a decidable monadic theory. In: Diks, K., Rytter, W. (eds.) MFCS 2002. LNCS, vol. 2420, pp. 165–176. Springer, Heidelberg (2002)

[Cau03]   Caucal, D.: On infinite transition graphs having a decidable monadic theory. Theoretical Computer Science 290(1), 79–115 (2003)

[CÉ12a]   Carayol, A., Ésik, Z.: A context-free linear ordering with an undecidable first-order theory. In: Baeten, J.C.M., Ball, T., de Boer, F.S. (eds.) TCS 2012. LNCS, vol. 7604, pp. 104–118. Springer, Heidelberg (2012)

[CÉ12b]   Carayol, A., Ésik, Z.: The FC-rank of a context-free language. arXiv:1202.6275 (2012)

[Del04]   Delhommé, C.: Automaticité des ordinaux et des graphes homogènes. C. R. Acad. Sci. Paris, Ser. I 339, 5–10 (2004)

[Ési11]   Ésik, Z.: An undecidable property of context-free linear orders. Inform. Processing Letters 111(3), 107–109 (2011)

[GHW12] Göller, S., Haase, C., Ouaknine, J., Worrell, J.: Branching-time model checking of parametric one-counter automata. In: Birkedal, L. (ed.) FOSSACS 2012. LNCS, vol. 7213, pp. 406–420. Springer, Heidelberg (2012)

[Hod93]   Hodges, W.: Model Theory. Cambridge University Press (1993)

[KLL11]   Kuske, D., Liu, J., Lohrey, M.: The isomorphism problem on classes of automatic structures with transitive relations. Transactions of the AMS (2011) (accepted)

[KN95]   Khoussainov, B., Nerode, A.: Automatic presentations of structures. In: Leivant, D. (ed.) LCC 1994. LNCS, vol. 960, pp. 367–392. Springer, Heidelberg (1995)

[LM13]   Lohrey, M., Mathissen, C.: Isomorphism of regular trees and words. Information and Computation 224, 71–105 (2013)

[Min61]   Minsky, M.: Recursive unsolvability of Post's problem of 'tag' and other topics in theory of Turing machines. Annals of Mathematics 74(3), 437–455 (1961)

[Rab69]   Rabin, M.O.: Decidability of second-order theories and automata on infinite trees. Trans. Amer. Math. Soc. 141, 1–35 (1969)

[Ros82]   Rosenstein, J.G.: Linear Orderings. Academic Press (1982)

[Sch72]   Schroeppel, R.: A two counter machine cannot calculate 2^N. Artificial Intelligence Memo 257, Massachusetts Institute of Technology, A.I. Laboratory (1972)

[Tar36]   Tarski, A.: Der Wahrheitsbegriff in den formalisierten Sprachen. Studia Philosophica I, 261–405 (1935/1936)

[Tho86]   Thomas, W.: On frontiers of regular trees. RAIRO – Theoretical Informatics 20(4), 371–381 (1986)

Helly Circular-Arc Graph Isomorphism
Is in Logspace

Johannes Köbler, Sebastian Kuhnert*, and Oleg Verbitsky**

Humboldt-Universität zu Berlin, Institut für Informatik

Abstract. We present logspace algorithms for the canonical labeling problem and the representation problem of Helly circular-arc (HCA) graphs. The first step is a reduction to canonical labeling and representation of interval intersection matrices. In a second step, the Δ trees employed in McConnell's linear time representation algorithm for interval matrices are adapted to the logspace setting and endowed with additional information to allow canonization. As a consequence, the isomorphism and recognition problems for HCA graphs turn out to be logspace complete.

1 Introduction

A graph G is *circular-arc* if each vertex $v \in V(G)$ can be assigned an arc $\rho(v)$ on a circle such that two vertices are adjacent if and only if their arcs intersect. We call any such assignment ρ a *circular-arc representation* of G and the arc system $\rho(G) = \{\rho(v) \mid v \in V(G)\}$ a *circular-arc model* of G. G is *Helly circular-arc* (HCA) if G has a representation ρ such that the arcs of the vertices in every clique C of G have non-empty intersection. We call such a ρ an *HCA representation* and $\rho(G)$ an *HCA model* of G. In this article, we solve the *canonical representation problem* for HCA graphs in logspace. That is, we give a logspace algorithm that computes for any given HCA graph G an HCA representation ρ_G such that isomorphic HCA graphs G and H receive identical HCA models $\rho_G(G) = \rho_H(H)$. If the input graph G is not HCA, the algorithm will detect this.

Previous results. HCA graphs were introduced by Gavril under the name of Θ circular-arc graphs [Gav74]. Gavril gave an $O(n^3)$ time representation algorithm for HCA graphs. Hsu improved this to $O(nm)$ [Hsu95]. Recently, Joeris et al. gave a linear time algorithm [JLM+11]. Chen gave a parallel AC^2 algorithm [Che96]. The fastest known isomorphism algorithm for HCA graphs is due to Curtis et al. and works in linear time [CLM+13]. Note that, though a logspace algorithm can take time bounded by a polynomial of high degree, the logspace solvability implies that the problem can be solved even in logarithmic time by a CRCW PRAM with polynomially many processors.

For the special case of interval graphs (which are easily seen to be HCA), the linear time algorithms by Booth and Lueker for recognition [BL76] and

---

* Supported by DFG grant KO 1053/7–1.
** Supported by DFG grant VE 652/1–1. On leave from the Institute for Applied Problems of Mechanics and Mathematics, Lviv, Ukraine.

K. Chatterjee and J. Sgall (Eds.): MFCS 2013, LNCS 8087, pp. 631–642, 2013.

isomorphism [LB79] have been known for many decades. Recently, these have been supplemented with a logspace algorithm for canonical representation of interval graphs [KKL+11].

Generalizing these results to the class of all circular-arc graphs remains a challenging problem. While the representation problem for this class is solved in linear time by McConnell [McC03], no polynomial-time isomorphism test for circular-arc graphs is currently known (see the discussion in [CLM+13], where a counterexample to the correctness of Hsu's $O(nm)$ time isomorphism algorithm [Hsu95] is given). The history of the isomorphism problem for circular-arc graphs is surveyed in more detail by Uehara [Ueh13].

This motivates the persistent interest in isomorphism algorithms for subclasses of circular-arc graphs. Besides HCA graphs, mainly proper circular-arc graphs and concave-round graphs have been studied. The isomorphism problem for these two classes can be solved in linear time [LSS08, CLM+13] and in logspace [KKV12].

Overview of our results. Our logspace algorithm for canonical representation of HCA graphs proceeds in several steps; see Fig. 1.

Hsu observed that the structure of certain circular-arc graphs G allows to prescribe the intersection structure of each pair of arcs in a circular-arc representation of G as di (disjoint), cd (contained), cs (contains), cc (circle cover), and ov (overlap) [Hsu95]. We store this information in the *neighborhood matrix* λ_G of G (for more details see Section 2).

The motivation for switching to the matrix λ_G is that flipping the arc of a vertex (i.e., exchanging its two start and end points) can be mimicked in λ_G by substituting some of its entries (details are given in Section 3). We show how to identify a subset $X \subseteq V(G)$ such that flipping the arcs of all vertices in X results in a matrix $\lambda_G^{(X)}$ that can be realized by an interval system. We choose X as an inclusion-maximal clique of G that is the common neighborhood of two vertices, and prove that at least one such clique can be found in logspace.

This gives a Turing reduction of the (canonical) representation problem of Helly circular-arc graphs to that of interval matrices: Flipped vertices are marked with a color, and in the representation returned by the oracle their intervals are flipped back to give a Helly circular-arc representation of both λ_G and G.

Our logspace algorithm for computing a representation of a given interval matrix is described in Section 4. McConnell gave a linear time algorithm for this problem as part of his representation algorithm for circular-arc graphs [McC03].

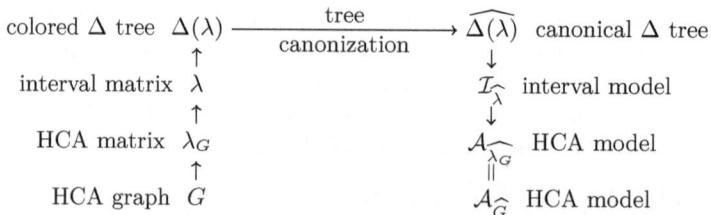

Fig. 1. Overview of the canonical representation algorithm for HCA graphs

He introduced the Δ *tree* of an interval matrix to capture all possible interval representations. Our key contribution here is to compute the Δ tree in logspace.

In Section 5, we show how to compute *canonical* representations of interval matrices. This is a significant extension of McConnell's algorithm, which only deals with representation. We implement this step as a reduction to colored tree canonization, which can be solved in logspace using Lindell's algorithm [Lin92].

To save space, some proofs have been omitted; they can be found in [KKV13].

2   Preliminaries

A *circular-arc system* $\mathcal{A}$ is a set of non-empty arcs on a circle. An *interval system* $\mathcal{I}$ is a set of non-empty intervals on a line. Equivalently, we can define an interval system as a circular-arc system $\mathcal{I}$ having the special property that there is at least one point on the circle that is not covered by any arc of $\mathcal{I}$. A set system $\mathcal{S}$ has the *Helly property* if every subsystem $\mathcal{S}' \subseteq \mathcal{S}$ with non-empty pairwise intersections has a non-empty overall intersection, i.e., $(\forall A, B \in \mathcal{S}' : A \cap B \neq \varnothing) \Rightarrow \bigcap_{A \in \mathcal{S}'} A \neq \varnothing$. It is easy to see that every interval system has the Helly property, but that there are non-Helly circular-arc systems; see Figure 2 (a) for an example. To keep notation concise, we use *CA* as a shorthand for circular-arc and *HCA* as an abbreviation of Helly circular-arc.

Two sets A and B *intersect* if $A \cap B \neq \varnothing$. They *overlap* (written $A \between B$) if additionally $A \setminus B \neq \varnothing$ and $B \setminus A \neq \varnothing$.

Given a set system $\mathcal{S}$, its *intersection graph* $\mathbb{I}(\mathcal{S})$ has one vertex for each set $A \in \mathcal{S}$, and two nodes $A, B \in \mathcal{S}$ are adjacent if and only if $A \cap B \neq \varnothing$. A graph G is a *CA graph* if there is a CA system $\mathcal{A}$ such that $G \cong \mathbb{I}(\mathcal{A})$. In this case, $\mathcal{A}$ is called a *CA model* of G, and an isomorphism $\rho \colon V(G) \to \mathcal{A}$ from G to $\mathbb{I}(\mathcal{A})$ is called a *CA representation* of G. HCA graphs and interval graphs are defined analogously, and so are their respective models and representations.

Given a graph G and $v \in V(G)$, let $N_G[v]$ denote the *closed neighborhood* of v, i.e., the set of vertices with distance at most 1 from v. The *common neighborhood* of two vertices $u, v \in V(G)$ is $N_G[u, v] = N_G[u] \cap N_G[v]$. If G is understood from the context, the index will be omitted. A vertex $v \in V(G)$ is *universal* if $N[v] = V(G)$. Two vertices $u, v \in V(G)$ are *twins* if $N[u] = N[v]$. A *twin class* is an inclusion-maximal set $U \subseteq V(G)$ such that all pairs of vertices in U are twins.

Let $\mu = (\mu_{i,j})_{i \neq j \in V}$ be a quadratic matrix. We call the elements of V the *vertices* of μ and we assume that V is linearly ordered. Another quadratic matrix $\lambda = (\lambda_{i,j})_{i \neq j \in U}$ is *isomorphic* to μ (written $\lambda \cong \mu$) if there is a bijection $\sigma \colon U \to V$ such that $\lambda_{i,j} = \mu_{\sigma(i), \sigma(j)}$ for all $i \neq j \in U$. Note that two graphs are isomorphic if and only if their adjacency matrices are isomorphic.

An *intersection matrix* is a matrix $\mu = (\mu_{u,v})_{u \neq v \in V}$ with entries $\mu_{u,v} \in \{\mathtt{di}, \mathtt{cs}, \mathtt{cd}, \mathtt{cc}, \mathtt{ov}\}$ that satisfies (a) $\mu_{u,v} = \mathtt{cd} \Leftrightarrow \mu_{v,u} = \mathtt{cs}$ and (b) $\mu_{u,v} = \mu_{v,u}$ in all other cases. Our interest is in intersection matrices that describe the intersection types between the arcs of a CA system. The following notation was introduced in [LS09].

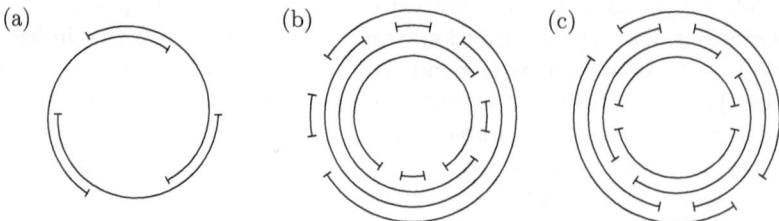

Fig. 2. (a) A non-HCA model of the HCA graph K_3. (b) Let G_n denote the split graph on $n + n$ vertices consisting of an n-clique C and a set S of n independent vertices, which are connected by the bipartite complement of a perfect matching between C and S. Every G_n is HCA; the figure shows an HCA model of G_4. Note that G_n has exactly $n + 1$ maxcliques, each of size n, and the maxclique C cannot be described as intersection or difference of less than n neighborhoods. (c) The complement graph H_n of n independent edges is CA. It has 2^n maxcliques C_i, each containing exactly one endpoint of each edge in $\overline{H_n}$. Since the common neighborhood of fewer than n vertices of H_n contains both endpoints of at least one edge in $\overline{H_n}$, no maxclique C_i can be described in this way. The figure shows a CA model of H_4.

Definition 2.1. *Let $\mathcal{A}$ be a CA system such that no single arc $C \in \mathcal{A}$ covers the whole circle and the endpoints of all arcs $C \in \mathcal{A}$ are pairwise distinct. The intersection matrix $\mu_{\mathcal{A}} = (\mu_{A,B})_{A \neq B \in \mathcal{A}}$ of $\mathcal{A}$ is defined by the entries*

$$\mu_{A,B} := \begin{cases} \text{di} & \text{if } A \cap B = \varnothing; \\ \text{cd} & \text{if } A \subsetneq B; \\ \text{cs} & \text{if } A \supsetneq B; \\ \text{cc} & \text{if } A \between B \text{ and } A \text{ and } B \text{ jointly cover the circle}; \\ \text{ov} & \text{if } A \between B \text{ but } A \text{ and } B \text{ do not jointly cover the circle.} \end{cases}$$

The intersection matrix $\mu_{\mathcal{I}}$ of an interval system $\mathcal{I}$ with pairwise distinct endpoints is defined similarly, using only the entries di, cd, cs *and* ov *(for $A \between B$). A matrix μ is a CA matrix if there is a CA system $\mathcal{A}$ such that $\mu \cong \mu_{\mathcal{A}}$. HCA matrices and interval matrices are defined analogously.*

Definition 2.2. *Given a graph G, its* neighborhood matrix *$\lambda_G = (\lambda_{u,v})_{u \neq v \in V(G)}$ is defined by the entries*

$$\lambda_{u,v} := \begin{cases} \text{di} & \text{if } \{u, v\} \notin E(G); \\ \text{cd} & \text{if } N[u] \subsetneq N[v]; \\ \text{cs} & \text{if } N[u] \supsetneq N[v]; \\ \text{cc} & \text{if } N[u] \between N[v], N[u] \cup N[v] = V, \\ & \text{and } \forall w \in N[u] \setminus N[v] : N[w] \subset N[u], \\ & \text{and } \forall w \in N[v] \setminus N[u] : N[w] \subset N[v]; \\ \text{ov} & \text{otherwise.} \end{cases}$$

Note that λ_G can be viewed as an augmented adjacency matrix, as 0 entries correspond to di and 1 entries are subdivided into four different categories.

The *underlying graph* of an intersection matrix $\mu = (\mu_{u,v})_{u \neq v \in V}$ is denoted by G_μ and consists of the vertices V and the edges $\{\{u, v\} \mid \mu_{u,v} \neq \texttt{di}\}$.

Following [Hsu95], we call a CA representation $\rho \colon V(G) \to \mathcal{A}$ of G *normalized* if ρ is an isomorphism between the neighborhood matrix λ_G and the CA matrix $\mu_\mathcal{A}$. Hsu provides an algorithm that transforms any CA representation of a CA graph with certain properties into a normalized representation, obtaining the following.

Lemma 2.3 ([Hsu95]). *Any CA graph G without twins and universal vertices has a normalized CA representation.*

All normalized CA representations have a property that is called *stable* by Joeris et al. who prove that every stable CA representation of an HCA graph G yields an HCA model [JLM+11, Theorem 4.1]. This implies the following.

Lemma 2.4. *Any normalized CA representation of an HCA graph G without twins and universal vertices provides an HCA model for G.*

Lemma 2.5. *There is a logspace reduction from the (canonical) HCA representation problem for HCA graphs G to the (canonical) CA representation problem of vertex-colored HCA matrices.*

Proof sketch. First consider the case that G is twin-free and has no universal vertex. Compute the neighborhood matrix λ_G. By Lemma 2.3, λ_G admits a normalized CA representation ρ. Any such ρ is Helly by Lemma 2.4, and easily seen to be also a HCA representation of G. If ρ is canonical for (colored) matrices, it is also canonical for (colored) graphs, as $G \cong H$ is equivalent to $\lambda_G \cong \lambda_H$.

It remains to observe that each twin class can be represented by a single vertex that is colored with the size of the twin class, and that universal vertices can be ignored if arcs covering the circle are added for them in the end.   □

3   Transforming HCA Matrices into Interval Matrices

In this section we describe a logspace reduction of the (canonical) CA representation problem for HCA matrices to the (canonical) representation problem of interval matrices. Note that it suffices for our purposes to obtain *any* (not necessarily Helly) representation: Lemma 2.4 implies that the representation is Helly if the HCA matrix is the neighborhood matrix of some HCA graph G.

Following McConnell, we can transform a CA system $\mathcal{A}$ into an interval system $\mathcal{A}^{(X)} = \{C \in \mathcal{A} \mid x \notin C\} \cup \{\tilde{C} \mid C \in \mathcal{A}, x \in C\}$ by choosing any point x on the circle that is different from all endpoints of $\mathcal{A}$ and flipping all the arcs in the set $X = \{C \in \mathcal{A} \mid x \in C\}$. *Flipping* an arc C just means that we replace it with the arc $\tilde{C}$ having the same endpoints as C but covering the opposite part of the circle. McConnell observed that flipping arcs of $\mathcal{A}$ corresponds to the replacements in the CA matrix $\mu_\mathcal{A}$ (cf. Definition 2.1) that are given in Table 1. Denote the result of flipping a subset X of the vertices of a CA matrix λ as $\lambda^{(X)}$. Note that $\lambda^{(X)}$ will become an interval matrix if exactly the arcs that contain a point x are flipped.

Table 1. The effect of flipping arcs of a CA system $\mathcal{A}$ on the entries of its CA matrix $\mu_\mathcal{A} = (\mu_{A,B})_{A \neq B \in \mathcal{A}}$

$\mu_{A,B}$	di cd cs cc ov
$\mu_{A,\bar{B}}$	cs cc di cd ov
$\mu_{\bar{A},B}$	cd di cc cs ov
$\mu_{\bar{A},\bar{B}}$	cc cs cd di ov

The rules described in Table 1 can be applied to any CA matrix $\mu = (\mu_{i,j})_{i \neq j \in V}$. To ensure that the resulting matrix $\mu^{(X)}$ is interval, a suitable vertex set $X \subseteq V$ has to be used. If we don't have a CA representation of μ, the set X has to be identified only from the structure of μ. If μ is HCA, the underlying graph G_μ is also HCA. The following fact implies that any inclusion-maximal clique (*max-clique* for short) C of G_μ can be used as X in this case.

Fact 3.1 *Let $\rho\colon V(G) \to \mathcal{A}$ be any HCA representation of a graph G and let C be any maxclique of G. Then there is a point x in the HCA model $\mathcal{A}$ such that no arc has x as its endpoint and $\{\rho(v) \mid v \in C\} = \{A \in \mathcal{A} \mid x \in A\}$.*

Proof. As C is a clique, the arcs in $\rho(C) = \{\rho(v) \mid v \in C\}$ intersect pairwise. As $\mathcal{A}$ is Helly, $\bigcap_{v \in C} \rho(v)$ is non-empty. By maximality of C, no further arc can contain any point x in this intersection. $\qquad\square$

In an interval graph, all maxcliques can be characterized as the common neighborhood of two vertices. This property was used in [KKL$^+$11] to reduce the canonical representation problem of interval graphs to that of interval hypergraphs. The same approach is not possible for HCA graphs, as they may contain maxcliques that cannot be characterized as the intersection or difference of constantly many neighborhoods; see Fig. 2 (b) for an example. However, at least one maxclique can be found in this way.

Theorem 3.2. *Let G be an HCA graph. Then there are $u, v \in V(G)$ (possibly $u = v$) such that $N[u, v]$ is a maxclique.*

We remark that general CA graphs do not necessarily have such a maxclique, see Fig. 2 (c) for an example.

Proof. Let $\lambda_G = (\lambda_{u,v})_{u \neq v \in V(G)}$ be the neighborhood matrix and $\rho\colon V(G) \to \mathcal{A}$ a normalized HCA representation of G. In order to find two vertices $u, v \in V(G)$ such that $N[u, v]$ is a maxclique, we start with an arbitrary vertex v such that there is no vertex w with $\lambda_{v,w} = \mathsf{cs}$ (i.e., $\not\exists w : N[w] \subsetneq N[v]$). Note that there cannot be a vertex w' with $\lambda_{v,w'} = \mathsf{cc}$, since this would imply that there is a vertex $w \in N[v] \setminus N[w']$ (because $N[w'] \not\subseteq N[v]$) with $N[w] \subsetneq N[v]$ (we can rule out equality because $w' \in N[v] \setminus N[u']$).

In case there is no vertex w with $\lambda_{v,w} = \mathsf{ov}$, $N[v]$ is a maxclique. This follows since $\lambda_{v,w} = \mathsf{cd}$ for all $w \in N(v)$ and hence, for all $w, w' \in N[v]$ it holds that $w \in N[v] \subseteq N[w']$.

Otherwise, we choose a vertex $u \in N[v]$ with $\lambda_{v,u} = \mathsf{ov}$, such that $N[u, v]$ is minimal w.r.t. inclusion and claim that $N[u, v]$ is a maxclique. In order to derive a contradiction assume that there exist $w, w' \in N[u,v]$ such that $w \notin N[w']$. If $\lambda_{v,w} = \mathsf{cd}$ (or $\lambda_{v,w'} = \mathsf{cd}$) then it follows that $w' \in N[v] \subseteq N[w]$ (or $w \in N[v] \subseteq N[w']$), a contradiction.

If $\lambda_{v,w} = \lambda_{v,w'} = \mathsf{ov}$ then $\rho(w) \cap \rho(w') = \varnothing$ and $\rho(w) \between \rho(v) \between \rho(w')$. Since $\rho(u) \between \rho(v)$ it follows that $\rho(u)$ overlaps $\rho(v)$ from the same side as one of $\rho(w)$ and $\rho(w')$, say $\rho(w)$. Because of $w' \in N[u,v] \setminus N[w]$ and the Helly property, it follows that $\rho(u) \cap \rho(v) \cap \rho(w') \neq \varnothing$ but $\rho(w) \cap \rho(v) \cap \rho(w') = \varnothing$, implying that $\rho(v) \cap \rho(w) \subseteq \rho(v) \cap \rho(u)$. Using again the Helly property, it now follows for any $x \in N[w,v]$ that $\rho(v) \cap \rho(w) \cap \rho(x) \neq \varnothing$ which in turn implies that $\rho(v) \cap \rho(u) \cap \rho(x) \neq \varnothing$. Hence, we get the inclusion $N[w,v] \subseteq N[u,v]$, contradicting the choice of u, since $w' \in N[u,v] \setminus N[w,v]$. □

Theorem 3.3. *The (canonical) CA representation problem for vertex-colored HCA matrices can be reduced in logspace to the (canonical) representation problem for vertex-colored interval matrices.*

Proof sketch. Given an HCA matrix $\mu = (\mu_{u,v})_{u,v \in V}$, the algorithm works as follows.

1. Find all pairs $u, v \in V$ such that $N[u,v]$ is a maxclique in G_μ (allowing $u = v$). By Theorem 3.2 at least one such pair exists. Denote the set of all maxcliques that are found in this way by $\mathcal{M}$.
2. For each $M \in \mathcal{M}$: Compute the interval matrix $\mu^{(M)}$ and mark the flipped vertices with a new color. Compute a (canonical) interval representation of $\mu^{(M)}$ and flip back all colored arcs, obtaining a CA representation $\rho_{\mu,M}$ of μ.
3. Among the $\rho_{\mu,M}$ computed in the previous step, choose ρ_μ as one that results in a lexicographically least CA model $\rho_{\mu,M}(\mu)$. Output ρ_μ.

It is not hard to see that ρ_μ is canonical if the interval representation of $\mu^{(M)}$ is canonical. □

4   Finding Representations of Interval Matrices in Logspace

McConnell [McC03] showed how to find interval representations of interval matrices in linear time. In this section, we apply some of his techniques to solve this task in logspace.

Given an intersection matrix $\lambda = (\lambda_{u,v})_{u \neq v \in V}$, define $G_{\mathsf{ov,di}}$ as the undirected graph on the vertex set V with edges $\{u,v\}$ for each pair with $\lambda_{u,v} \in \{\mathsf{ov}, \mathsf{di}\}$. Similarly, define D_{cd} (resp. D_{cs}) as the directed graph on V with arrows (u,v) for each pair with $\lambda_{u,v} = \mathsf{cd}$ (resp. $\lambda_{u,v} = \mathsf{cs}$).

A *transitive orientation* of an undirected graph is an assignment of directions to all edges such that the resulting set of arrows is transitive. An *interval orientation* of an intersection matrix λ is a transitive orientation $D_{\mathsf{ov,di}}$ of $G_{\mathsf{ov,di}}$ that remains transitive when restricted to G_{di} and that satisfies

$$\lambda_{u,v} = \mathsf{di} \wedge \lambda_{u,w} = \lambda_{v,w} = \mathsf{ov} \Rightarrow \text{either } (u,w) \in D_{\mathsf{ov,di}} \text{ or } (v,w) \in D_{\mathsf{ov,di}} \quad (1)$$

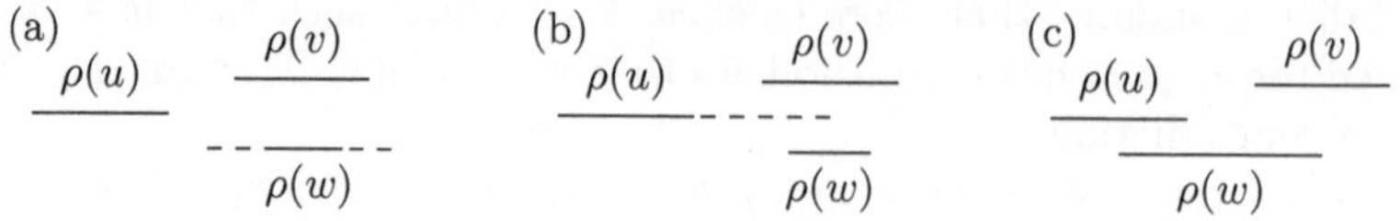

(a) $\rho(u)$ $\rho(v)$ (b) $\rho(u)$ $\rho(v)$ (c) $\rho(u)$ $\rho(v)$

$\rho(w)$ $\rho(w)$ $\rho(w)$

Fig. 3. In all three cases of Definition 4.2 there is no way to place $\rho(u)$ between $\rho(v)$ and $\rho(w)$

The last condition requires that if w stays in overlap relation with two disjoint vertices u, v, then w has to be arranged in between u and v. Any interval representation ρ of λ induces an interval orientation of λ: An edge $\{u, v\}$ of $G_{\mathrm{ov,di}}$ is oriented as (u, v) if and only if $\rho(u) < \rho(v)$, i.e., if the interval $\rho(u)$ starts left of the interval $\rho(v)$. The following lemma shows the converse, implying that interval orientations are in 1-1 correspondence with interval representations (provided that we fix the set of endpoints as $\{0, \ldots, 2n - 1\}$).

Lemma 4.1. *Let λ be an interval matrix, and let $D_{\mathrm{ov,di}}$ be an interval orientation of λ. Then there exists an interval representation ρ of λ that induces the interval orientation $D_{\mathrm{ov,di}}$. Moreover, ρ is computable in logspace on input λ and $D_{\mathrm{ov,di}}$.*

Proof sketch. Let $\lambda = (\lambda_{u,v})_{u \neq v \in V}$. To obtain ρ, order the left endpoints according to $D_{\mathrm{ov,di}} \cup D_{\mathrm{cs}}$ and the right endpoints according to $D_{\mathrm{ov,di}} \cup D_{\mathrm{cd}}$. Interleave these two linear orders such that the relationships in λ are obeyed. □

By Lemma 4.1 it suffices to compute an interval orientation $D_{\mathrm{ov,di}}$ of a given interval matrix λ to get an interval representation of λ.

Definition 4.2 (cf. [McC03, Definition 6.3]). *Let λ be an intersection matrix, and let $\{u, v\}$ and $\{u, w\}$ be edges in $G_{\mathrm{ov,di}}$. The binary relation Δ contains the entries $(u, v)\Delta(u, w)$ and $(v, u)\Delta(w, u)$ if one of the following holds:*
(a) $\lambda_{u,v} = \lambda_{u,w} = \mathrm{di}$ and $\lambda_{v,w} \neq \mathrm{di}$
(b) $\lambda_{u,v}, \lambda_{u,w} \in \{\mathrm{ov}, \mathrm{di}\}$ and $\lambda_{v,w} \in \{\mathrm{cd}, \mathrm{cs}\}$
(c) $\lambda_{u,v} = \mathrm{di}$ and $\lambda_{u,w} = \lambda_{v,w} = \mathrm{ov}$

If any of these three condition holds true, then in any interval representation ρ of λ, the intervals $\rho(v)$ and $\rho(w)$ must be on the same side of $\rho(u)$; see Fig. 3. In other words, any interval orientation $D_{\mathrm{ov,di}}$ of λ must contain (u, v) if and only if it contains (u, w). This is the rationale for the following definition: Δ *implication classes* are the equivalence classes of the symmetric transitive closure of Δ. The union of a Δ implication class and its transpose is called Δ *color class* and can be viewed as a set of (undirected) edges in $G_{\mathrm{ov,di}}$.

Lemma 4.3 ([McC03, Theorem 6.4]). *Each interval orientation of λ contains exactly one Δ implication class from each Δ color class.*

This implies that (u, v) and (v, u) cannot be in the same Δ implication class. However, not any selection of one Δ implication class from each Δ color class yields an interval orientation of λ. To find a valid selection, we need to consider the Δ *tree* of λ.

A *module* of a matrix $\lambda = (\lambda_{u,v})_{u \neq v \in V}$ is a subset $U \subseteq V$ that is not distinguished by any vertex outside U, i.e., for any $u \neq v \in U$ and $w \in V \setminus U$ it holds $\lambda_{u,w} = \lambda_{v,w}$ and $\lambda_{w,u} = \lambda_{w,v}$. McConnell [McC03] calls a module U of an intersection matrix λ a Δ *module*, if it is a clique in the corresponding intersection graph (i.e., $\lambda_{u,v} \neq$ di for all $u \neq v \in U$) or if there is no $v \in V \setminus U$ such that $\lambda_{v,u} =$ ov for all $u \in U$. The Δ modules of an intersection matrix form a tree decomposable family [McC03, Definition 6.7 and Theorem 6.9]. The resulting decomposition tree, i.e., the transitive reduction of the containment relation among strong Δ modules U (i.e., U does not overlap any other Δ module), is called Δ *tree* of λ. The leaves of the Δ tree are trivial modules consisting of single vertices. An inner node in the Δ tree is called *degenerate* if taking the union of any of its children gives a Δ module, and *prime* otherwise. If U is an inner node in the Δ tree and $W_1, \ldots, W_k$ are its children, the *quotient of λ at U* is the submatrix $\lambda[U]$ of λ on the vertices $W = \{w_1, \ldots, w_k\}$ with $w_i \in W_i$. As the W_i are disjoint modules, $\lambda[U]$ does not depend on the actual choice of the w_i. In the quotient matrix of a degenerate node, its children are either in pairwise ov, in pairwise di, or in pairwise cd/cs relation [McC03, 110]. Hence, the inner nodes of the Δ tree can be classified as prime, disjoint, overlap or containment nodes.

The following results from [McC03] show that the Δ tree provides a compact representation of all possible interval orientations of λ.

Lemma 4.4 ([McC03, Lemma 6.14]). *The set of vertices spanned by a Δ color class in an interval intersection matrix λ is a Δ module of λ.*

Lemma 4.5 ([McC03, Theorem 6.15]). *A set of edges of $G_{\text{ov,di}}$ is a Δ color class if and only if it is the set of edges of $G_{\text{ov,di}}$ connecting all children of a prime node or a pair of children of a degenerate node in the Δ tree.*

Lemma 4.6 ([McC03, Theorem 6.19]). *Any acyclic union of Δ implication classes gives an interval orientation of λ.*

The next lemma reduces the problem of computing an interval orientation of λ to the problem of computing interval orientations of the quotient matrices of the inner nodes of the Δ tree.

Lemma 4.7. *Let λ be an intersection matrix, and let $U_1, \ldots, U_c$ be the inner nodes of its Δ tree. Any sequence of interval orientations $D_1, \ldots, D_c$ for the quotient matrices $\lambda[U_1], \ldots, \lambda[U_c]$ induces an interval orientation D of λ, which can be computed in logspace.*

As soon as we have the Δ tree, it's very easy to compute interval orientations for the quotient matrices corresponding to its inner nodes U. If U is prime, we can take any of the two implication classes of the color class connecting all its children. If U is degenerate of type overlap or disjoint, any linear ordering of its children provides an interval orientation for its quotient matrix. Finally, if U is of type containment, no edges have to be oriented.

Theorem 4.8. *The Δ implication classes, the Δ color classes, and the Δ tree of a given intersection matrix λ can be computed in logspace.*

By combining Theorem 4.8 with Lemma 4.1 we obtain the following result.

Corollary 4.9. *Given an intersection matrix λ, an interval representation for it can be computed in logspace.*

5   Finding Canonical Representations for Interval Matrices

In this section, we describe a logspace algorithm for computing a canonical representation of a given interval matrix λ. The main task is to choose between the different possible interval orientations of the quotient matrices corresponding to the inner nodes of the Δ tree. By providing the Δ tree with additional information we can reduce this task to (colored) tree canonization.

Lemma 5.1. *Given an interval matrix λ and its Δ tree T', for each inner node U of T' the following can be computed in logspace:*
- *The quotient $\lambda[U]$ of λ at U.*
- *All possible interval models of $\lambda[U]$ (either only one, or two that are the reverse of each other).*
- *For each interval model M_U of $\lambda[U]$, the possible correspondences of the children of U to the intervals in M_U. This can either be arbitrary, a fixed mapping or one of two fixed mappings.*

Definition 5.2. *Given an intersection matrix λ, the colored Δ tree $\mathbb{T}(\lambda)$ has the same nodes as the Δ tree (i.e., the strong Δ modules of λ that are not overlapped by another Δ module), plus three additional nodes $\mathrm{lo}_U, \mathrm{mi}_U, \mathrm{hi}_U$ for each inner node U that admits exactly two assignments of its children to the interval model of its quotient matrix (cf. Lemma 5.1); these nodes are inserted between U and its children. Each Δ tree node U receives a tuple (p_U, M_U) as color, where M_U is the interval model of the quotient $\lambda[U]$ given by Lemma 5.1 (if there are two different models, take the smaller one), and p_U is the position of U among the children of its parent: If U is the root or if the parent of U admits an arbitrary mapping of its children to its quotient intervals, let $p_U = 0$. If the parent of U has a fixed assignment of its children to its intervals, let p_U be the position of the interval corresponding to U among the other intervals. If the parent of U allows two assignments of its children, let $p_{U,1}$ and $p_{U,2}$ be the positions of U under the two assignments, respectively. If $p_{U,1} < p_{U,2}$, make U a child of lo_U and define $p_U = (p_{U,1}, p_{U,2})$; if $p_{U,1} = p_{U,2}$, make U a child of mi_U and define $p_U = (p_{U,1}, p_{U,2})$; if $p_{U,1} > p_{U,2}$, make U a child of hi_U and define $p_U = (p_{U,2}, p_{U,1})$. Finally, color all lo_U and hi_U nodes with 0 and all mi_U nodes with 1.*

By Theorem 4.8 and Lemma 5.1, $\mathbb{T}(\lambda)$ can be computed in logspace.

Lemma 5.3. *If λ and λ' are isomorphic interval matrices, then $\mathbb{T}(\lambda) \cong \mathbb{T}(\lambda')$.*

Lemma 5.4. *Let λ be an interval matrix. Given an isomorphic copy T' of $\mathbb{T}(\lambda)$, an isomorphic copy λ' of λ (that depends only on T') can be computed in logspace. When also given an isomorphism $\ell \colon \mathbb{T}(\lambda) \to T'$, an isomorphism $\varphi \colon \lambda \to \lambda'$ can be computed within the same space bound.*

Theorem 5.5. *The canonical representation problem for interval matrices can be solved in logspace.*

Proof. The algorithm works as follows:
1. Compute the Δ tree of λ (see Theorem 4.8).
2. Compute interval models of the quotient matrices at the nodes of the Δ tree to obtain the colored Δ tree $\mathbb{T}(\lambda)$ (see Lemma 5.1 and Definition 5.2).
3. Compute a canonical labeling of $\mathbb{T}(\lambda)$ and use the algorithm of Lemma 5.4 to compute a canonical copy λ' of λ and a canonical labeling φ of λ.
4. Compute the Δ tree of λ' and interval orderings for the quotient matrices at its inner nodes (in fact, the information from $\mathbb{T}(\lambda)$ can be reused; only the assignment of children needs to be revisited). Combine these orientations into one for the whole matrix (see Lemma 4.7) and convert it into an interval representation ρ' of λ' (see Lemma 4.1). Combined with the canonical labeling φ of λ, this results in an interval representation $\rho = \rho' \circ \varphi$ of λ.

Note that λ' depends only on the canon of $\mathbb{T}(\lambda)$, so $\lambda_1 \cong \lambda_2$ implies $\lambda_1' = \lambda_2'$. As ρ' depends only on λ', the resulting interval model $\rho(\lambda) = \rho'(\lambda')$ is canonical. $\square$

6   Conclusion

Our algorithms also allow recognition of HCA graphs: If the input graph does not belong to this class, either one of the steps will fail (e.g. finding a suitable maxclique M), or the resulting arcs will not be a representation of G (which can easily be checked), or the resulting arcs are not Helly. The latter can be checked in logspace using [JLM+11, Theorem 3.1].

We remark that by combining Theorem 3.3 and Corollary 4.9 we already get a logspace algorithm that computes for any given HCA graph G an HCA representation of G. Since any HCA representation of G allows to compute *all* maxcliques in logspace, we can reduce the canonical representation problem of HCA graphs to that of CA hypergraphs $\mathcal{H}_G$: the vertex set of $\mathcal{H}_G$ consists of all maxcliques of G and for each vertex $v \in V(G)$, $\mathcal{H}_G$ contains a hyperedge consisting of all maxcliques that contain v. It is known that a graph G is HCA if and only if $\mathcal{H}_G$ is a CA hypergraph [Gav74]. Moreover, the hypergraph $\mathcal{H}_G$ provides a canonical HCA model for G, if we order its maxcliques by a canonical circular ordering. Hence an alternative canonical representation algorithm for HCA graphs can be obtained by using the algorithm for computing a canonical CA model of $\mathcal{H}_G$ given in [KKV12]. However, we believe that finding canonical representations of interval matrices is of independent interest, as these allow additional constraints on the structure of the intervals compared to interval graphs. For a different kind of constraint, namely prescribing the lengths of pairwise intersections (and optionally interval lengths), both logspace and $O(nm)$ time (resp. linear time) algorithms are known [KKW12].

References

[BL76]   Booth, K.S., Lueker, G.S.: Testing for the consecutive ones property, in-
           terval graphs, and graph planarity using PQ-tree algorithms. J. Comput.
           Syst. Sci. 13(3), 335–379 (1976)
[Che96]   Chen, L.: Graph isomorphism and identification matrices: Parallel algo-
           rithms. Trans. Paral. Distrib. Syst. 7(3), 308–319 (1996)
[CLM+13]   Curtis, A.R., Lin, M.C., McConnell, R.M., Nussbaum, Y., Soulignac, F.J.,
           Spinrad, J.P., Szwarcfiter, J.L.: Isomorphism of graph classes related to the
           circular-ones property. Discrete Math. Theor. Comp. Sci. 15(1), 157–182
           (2013)
[Gav74]   Gavril, F.: Algorithms on circular-arc graphs. Networks 4, 357–369 (1974)
[Hsu95]   Hsu, W.L.: $O(m \cdot n)$ algorithms for the recognition and isomorphism prob-
           lems on circular-arc graphs. SIAM J. Comput. 24(3), 411–439 (1995)
[JLM+11]   Joeris, B.L., Lin, M.C., McConnell, R.M., Spinrad, J.P., Szwarcfiter, J.L.:
           Linear time recognition of helly circular-arc models and graphs. Algorith-
           mica 59(2), 215–239 (2011)
[KKL+11]   Köbler, J., Kuhnert, S., Laubner, B., Verbitsky, O.: Interval graphs: Canon-
           ical representations in logspace. SIAM J. Comput. 40(5), 1292–1315 (2011)
[KKV12]   Köbler, J., Kuhnert, S., Verbitsky, O.: Solving the canonical representation
           and star system problems for proper circular-arc graphs in logspace. In:
           D'Souza, D., Kavitha, T., Radhakrishnan, J. (eds.) Proc. 32nd FSTTCS,
           Dagstuhl, Leibniz-Zentrum für Informatik. LIPIcs, vol. 18, pp. 387–399
           (2012)
[KKV13]   Köbler, J., Kuhnert, S., Verbitsky, O.: Helly circular-arc graph isomor-
           phism is in logspace. Electr. Colloq. Comput. Complexity (2013), TR13-
           074 http://eccc.hpi-web.de/report/2013/074/
[KKW12]   Köbler, J., Kuhnert, S., Watanabe, O.: Interval graph representation with
           given interval and intersection lengths. In: Chao, K.-M., Hsu, T.-S., Lee, D.-
           T. (eds.) ISAAC 2012. LNCS, vol. 7676, pp. 517–526. Springer, Heidelberg
           (2012)
[LB79]   Lueker, G.S., Booth, K.S.: A linear time algorithm for deciding interval
           graph isomorphism. J. ACM 26(2), 183–195 (1979)
[Lin92]   Lindell, S.: A logspace algorithm for tree canonization. In: Proc. 24th
           STOC, pp. 400–404 (1992)
[LSS08]   Lin, M.C., Soulignac, F.J., Szwarcfiter, J.L.: A simple linear time algo-
           rithm for the isomorphism problem on proper circular-arc graphs. In: Gud-
           mundsson, J. (ed.) SWAT 2008. LNCS, vol. 5124, pp. 355–366. Springer,
           Heidelberg (2008)
[LS09]   Lin, M.C., Szwarcfiter, J.L.: Characterizations and recognition of circular-
           arc graphs and subclasses: A survey. Discrete Math. 309(18), 5618–5635
           (2009)
[McC03]   McConnell, R.M.: Linear-time recognition of circular-arc graphs. Algorith-
           mica 37(2), 93–147 (2003)
[Ueh13]   Uehara, R.: Tractabilities and intractabilities on geometric intersection
           graphs. Algorithms 6(1), 60–83 (2013)

Zeno, Hercules and the Hydra: Downward Rational Termination Is Ackermannian

Ranko Lazić[1], Joël Ouaknine[2], and James Worrell[2]

[1] Department of Computer Science, University of Warwick, UK
[2] Department of Computer Science, University of Oxford, UK

Abstract. Metric temporal logic (MTL) is one of the most prominent specification formalisms for real-time systems. Over infinite timed words, full MTL is undecidable, but satisfiability for its safety fragment was proved decidable several years ago [18]. The problem is also known to be equivalent to a fair termination problem for a class of channel machines with insertion errors. However, the complexity has remained elusive, except for a non-elementary lower bound. Via another equivalent problem, namely termination for a class of rational relations, we show that satisfiability for safety MTL is not primitive recursive, yet is Ackermannian, i.e., among the simplest non-primitive recursive problems. This is surprising since decidability was originally established using Higman's Lemma, suggesting a much higher non-multiply recursive complexity.

1 Introduction

Metric temporal logic (MTL) is one of the most popular approaches for extending temporal logic to the real-time setting. MTL extends linear temporal logic by constraining the temporal operators with intervals of real numbers. For example, the formula $\Diamond_{[3,4]}\varphi$ means that φ will hold within 3 to 4 time units in the future. There are two main semantic paradigms for MTL: continuous (state-based) and pointwise (event-based)—cf. [3,12]. In the former, an execution of a system is modelled by a flow which maps each point in time to the state propositions that are true at that moment. In the latter, one records only a countable sequence of events, corresponding to instantaneous changes in the state of the system. In this paper we interpret MTL over the pointwise semantics[1] and assume that time is *dense* (arbitrarily many events can happen in a single time unit) but *non-Zeno* (only finitely many events can occur in a single time unit).

Over the past few years, the theory of *well-structured transition systems* has been used to obtain decidability results for MTL. Well-structured transition systems are a general class of infinite-state systems for which certain verification problems, such as reachability and termination, are decidable; see [9] for a comprehensive survey. In [19] satisfiability and model checking for MTL were shown to be decidable by reduction to the reachability problem for a class of well-structured transition systems. Likewise, for a syntactically defined fragment of

---

[1] Note that it follows from the thesis work of Henzinger [11] that safety MTL satisfiability is undecidable over the continuous semantics.

K. Chatterjee and J. Sgall (Eds.): MFCS 2013, LNCS 8087, pp. 643–654, 2013.

MTL that expresses safety properties, called *safety MTL*, model checking and satisfiability were shown decidable over infinite timed words by reduction to the termination problem on well-structured transition systems [18].

Extracting well-structured systems from MTL formulas relies on Higman's Lemma, which states that over a finite alphabet the subword order is a well-quasi order. Analysis of termination arguments that use Higman's Lemma has been applied to bound the complexity of reachability in lossy channel systems and insertion (or gainy) channel systems: two classes of well-structured systems that arise naturally in the modelling of communication over faulty media. For the reachability and termination problem in lossy channel systems, an upper bound in level $\mathfrak{F}_{\omega^\omega}$ of the fast-growing hierarchy was obtained in [7]. (Recall that $\mathfrak{F}_{<\omega}$ comprises the primitive recursive functions, Ackermann's function lies in $\mathfrak{F}_\omega$, while $\mathfrak{F}_{\omega^\omega}$ contains the first non-multiply recursive function.) The same paper also shows that neither problem lies in a lower level of the hierarchy and observes that both lower and upper bounds carry over to MTL satisfiability over *finite* words and to reachability in insertion channel systems, among many other problems.[2] An upper bound in $\mathfrak{F}_{\omega^\omega}$ for safety MTL satisfiability has also been sketched in [21] using related techniques.

Meanwhile, complexity lower bounds for safety MTL have been obtained utilising a correspondence with the termination problem for insertion channel systems. In [4] it is shown that termination for insertion channel machines with emptiness tests is primitive recursive, though non-elementary.[3] This result is used to give a non-elementary lower bound in $\mathfrak{F}_3$ for the satisfiability problem for safety MTL. An improved lower bound in $\mathfrak{F}_4$ is given in [13], again via insertion channel machines, but still leaving a considerable gap with the above-mentioned $\mathfrak{F}_{\omega^\omega}$ upper bound. This gap was highlighted recently in [14].

The key to determining the precise complexity of satisfiability for safety MTL is to study a refined version of the termination problem for channel machines—namely the *fair termination problem*. Roughly speaking, an infinite computation of an insertion channel machine is *fair* if every message that is written to the channel is eventually consumed—and not continuously preempted by insertion errors. (In the translation between channel machines and MTL, fairness corresponds in a precise sense to the *non-Zenoness* assumption.) We obtain lower and upper complexity bounds for this problem that are *Ackermannian*, i.e., that lie in level $\mathfrak{F}_\omega$ of the fast-growing hierarchy. These bounds also apply to safety MTL satisfiability, finally closing the above-mentioned complexity gap.

Unlike [4], we consider channel machines with a single channel. In [4], without the hypothesis of fairness, the termination problem was shown to be non-elementary in the number of channels. On the other hand, fair termination is already undecidable if there are two channels. But with a single channel fair

---

[2] Incidentally, the *model-checking* problem over infinite timed words for safety MTL against timed automata can also be shown to have complexity precisely in $\mathfrak{F}_{\omega^\omega}$, following arguments presented in [19] together with the results of [7].

[3] In the presence of insertion errors, read-transitions can always be taken, so the channel is redundant unless there is an extra hypothesis, such as emptiness tests.

termination is non-primitive recursive in the size of the channel alphabet. In common with [4] we find that termination for insertion channels has a lower complexity than termination for lossy channel systems or reachability for either type of system, neither of which is multiply recursive.

Our technical development is carried out in a slightly more abstract framework than insertion channel systems. We study the termination problem for well-structured transition systems whose states are words over a given alphabet, and whose transition relation is a rational relation that is (downwards) compatible with the subword order. (This is similar to the basic framework of regular model checking [2], but with the additonal hypothesis of monotonicity.)

To obtain an Ackermannian upper bound, we associate a *Hydra battle* with each finite computation of such a system. For our purposes, a Hydra battle is a sequence of 'flat' regular expressions that express assertions about states in the computation. Each regular expression can be seen as arising from its precedessor by a process of truncation (by the sword of Hercules) and regeneration. Our Hydra correspond to the classical tree Hydra of Kirby and Paris [15] via a natural correspondence between flat regular expressions and trees of height 2.

The basic pattern for proving our lower bound result is a standard one, namely to reduce from the halting problem for Ackermannianly bounded Turing machines by simulating their computations. However, in contrast to the common approach in the literature, in which a large function and its inverse are computed weakly before and after the simulation respectively (cf. e.g. [7,22,14]), we bootstrap a counter that can count accurately to an Ackermannian bound even in the presence of insertion errors. The bootstrapping involves extending Stockmeyer's yardstick construction, which reaches beyond the elementary functions, to surpass all primitive recursive ones.

2   Preliminaries

2.1   Fast Growing Hierarchy

We define an initial segment of the fast growing hierarchy [16] of computable functions by following the presentation of Figueira et al. [8].

For each $k \in \mathbb{N}$, class $\mathfrak{F}_k$ is the closure under substitution and limited recursion of constant, sum and projection functions, and F_n functions for $n \leq k$. The latter are defined so that F_0 is the successor function, and each F_{n+1} is computed by iterating F_n:

$$F_0(x) = x + 1 \qquad\qquad F_{n+1}(x) = F_n^{x+1}(x)$$

The following are a few simple observations:

- $\mathfrak{F}_0 = \mathfrak{F}_1$ contains all linear functions, like $\lambda x.x + 3$ or $\lambda x.2x$;
- $\mathfrak{F}_2$ contains all elementary functions, like $\lambda x.2^{2^x}$;
- $\mathfrak{F}_3$ contains all tetration functions, like $\lambda x.\underbrace{2^{2^{\cdot^{\cdot^{2}}}}}_{x}$.

The hierarchy is strict for $k \geq 1$, i.e., $\mathfrak{F}_k \subsetneq \mathfrak{F}_{k+1}$, because $F_{k+1} \notin \mathfrak{F}_k$. Also, for each $k \geq 1$ and $f \in \mathfrak{F}_k$, there exists $p \geq 1$ such that F_k^p majorises f, i.e., $f(x_1, \ldots, x_n) < F_k^p(\max(x_1, \ldots, x_n))$ for all $x_1, \ldots, x_n$ [16, Theorem 2.10].

The union $\bigcup_k \mathfrak{F}_k$ is the class of all primitive recursive functions, while F_ω defined by $F_\omega(x) = F_x(x)$ is an Ackermann-like non-primitive recursive function; we call Ackermannian such functions that lie in $\mathfrak{F}_\omega \setminus \bigcup_k \mathfrak{F}_k$.

We remark that, following this pattern for successor and limit ordinals, the hierarchy can be continued up to level ω^ω. The union $\bigcup_{\alpha < \omega^\omega} \mathfrak{F}_\alpha$ is the class of all multiply recursive functions, and the non-multiply recursive functions in $\mathfrak{F}_{\omega^\omega}$ have been called 'hyper-Ackermannian'.

2.2   Finite Transducers

We work with normalised transducers with ϵ-transitions, whose input and output alphabets are the same. They are tuples of the form $\langle Q, \Sigma, \delta, I, F \rangle$, where Q is a finite set of states, Σ is a finite alphabet, $\delta \subseteq Q \times (\Sigma \cup \{\epsilon\}) \times (\Sigma \cup \{\epsilon\}) \times Q$ is a transition relation, and $I, F \subseteq Q$ are sets of initial and final states respectively. We write transitions as $q \xrightarrow{a|a'} q'$, which can be thought of as reading a from the input word (if $a \in \Sigma$) and writing a' to the output word (if $a' \in \Sigma$).

For a transducer $\mathcal{T}$ as above, we say that τ is a *transduction* iff it is a path $q_0 \xrightarrow{a_1|a_1'} q_1 \cdots \xrightarrow{a_n|a_n'} q_n$ where q_0 is initial and q_n is final, and we write $\mathrm{In}(\tau)$ and $\mathrm{Out}(\tau)$ for the words $a_1 \ldots a_n$ and $a_1' \ldots a_n'$ respectively. The relation of $\mathcal{T}$ is then $\mathrm{R}(\mathcal{T}) = \{\langle \mathrm{In}(\tau), \mathrm{Out}(\tau) \rangle : \tau \text{ is a transduction of } \mathcal{T}\}$. The transducers recognise exactly rational relations between Σ^* and Σ^* (cf. e.g. [20, Chapter IV]).

A *computation* of a transducer $\mathcal{T}$ from a word w_1 is a finite or infinite sequence of words $w_1, w_2, \ldots$ such that $w_1 \mathrm{R}(\mathcal{T}) w_2 \mathrm{R}(\mathcal{T}) \cdots$.

If q and q' are states of a transducer $\mathcal{T}$, we write $\mathcal{T}(q, q')$ for the transducer obtained from $\mathcal{T}$ by making q the only initial state and q' the only final state.

2.3   Composing Transducers

We write $\fatsemi$ for relational composition, as well as for its counterpart in terms of transducers. Recalling a standard definition of the latter operation, given two transducers $\mathcal{T}_1 = \langle Q_1, \Sigma, \delta_1, I_1, F_1 \rangle$ and $\mathcal{T}_2 = \langle Q_2, \Sigma, \delta_2, I_2, F_2 \rangle$, the transition relation of their composition $\mathcal{T}_1 \fatsemi \mathcal{T}_2 = \langle Q_1 \times Q_2, \Sigma, \delta, I_1 \times I_2, F_1 \times F_2 \rangle$ is defined so that every output of $\mathcal{T}_1$ must be consumed by an input of $\mathcal{T}_2$:

$$\langle q_1, q_2 \rangle \xrightarrow{a|a'} \langle q_1', q_2' \rangle \text{ iff } \begin{cases} q_1 \xrightarrow{a|\epsilon} q_1' \text{ and } a' = \epsilon \text{ and } q_2 = q_2', \text{ or} \\ q_1 \xrightarrow{a|a''} q_1' \text{ and } q_2 \xrightarrow{a''|a'} q_2' \text{ for some } a'' \in \Sigma, \text{ or} \\ q_1 = q_1' \text{ and } a = \epsilon \text{ and } q_2 \xrightarrow{\epsilon|a'} q_2'. \end{cases}$$

We then have $\mathrm{R}(\mathcal{T}_1 \fatsemi \mathcal{T}_2) = \mathrm{R}(\mathcal{T}_1) \fatsemi \mathrm{R}(\mathcal{T}_2)$.

2.4   Downwards Monotone Transducers

Given an alphabet Σ, we write $\sqsubseteq$ for the subword ordering on Σ^*, i.e., $w \sqsubseteq w'$ iff w' can be obtained from w by a number of insertions of letters. The *downward closure* of a subset L of Σ^*, i.e., $\{w \mid \exists w'. w \sqsubseteq w' \wedge w' \in L\}$, is denoted by $\downarrow L$.

We say that a relation R on Σ^* is *downwards monotone* iff, whenever $w_1 \, R \, w_2$, every replacement of w_1 by a subword w_1' can be matched on the right-hand side of R, i.e., $\forall w_1, w_2, w_1'. \, w_1 \, R \, w_2 \wedge w_1' \sqsubseteq w_1 \Rightarrow \exists w_2'. \, w_1' \, R \, w_2' \wedge w_2' \sqsubseteq w_2$. Note that this is the same notion as downward compatibility of R with respect to $\sqsubseteq$ in the theory of well-structured transition systems [9].

A transducer $\mathcal{T}$ is *downwards monotone* iff its relation $\mathrm{R}(\mathcal{T})$ has the property.

Proposition 1. *Composing transducers preserves downward monotonicity.*

We leave open the decidability of whether a given transducer is downward monotone. We note however that this problem is at least as hard as the regular Post embedding problem (PEP^{reg}) [6], and therefore not multiply recursive [7].

2.5   Downward Rational Termination

The principal problem we study is whether a given downwards monotone transducer terminates from a given word:

> Given a downwards monotone transducer $\mathcal{T}$ and a word w_1 over its alphabet, is every computation of $\mathcal{T}$ from w_1 finite?

We remark that the standard rational termination problem, i.e., without the assumption of downward monotonicity, is undecidable. Indeed, it is straightforward to compute a transducer that recognises the one-step relation between configurations of a given Turing machine.

Another closely related problem is gainy rational termination (also called increasing rational termination [14]):

> Given a transducer $\mathcal{T}$ and a word w_1 over its alphabet, is every computation of $\mathcal{T}_{\sqsubseteq}$ from w_1 finite?

Here $\mathcal{T}_{\sqsubseteq} = \sqsubseteq \, \mathbin{\fatsemi} \, \mathcal{T} \, \mathbin{\fatsemi} \, \sqsubseteq$, where $\sqsubseteq$ on the right-hand side denotes a transducer whose relation is the subword ordering over the alphabet Σ of $\mathcal{T}$:

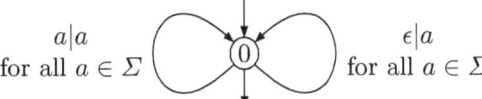

$$
\begin{matrix}
a|a & & \epsilon|a \\
\text{for all } a \in \Sigma & & \text{for all } a \in \Sigma
\end{matrix}
$$

Thus, $\mathcal{T}_{\sqsubseteq}$ can be thought of as a 'faulty' version of $\mathcal{T}$ that may gain arbitrary letters in both input and output words, i.e., suffers from 'insertion errors'.

By observing that $\mathcal{T}_{\sqsubseteq}$ is downwards monotone for every transducer $\mathcal{T}$, gainy rational termination reduces to downward rational termination. Conversely, for a downwards monotone transducer $\mathcal{T}$, it is easy to see that $\mathcal{T}$ has an infinite computation from w_1 iff the same is true of $\mathcal{T}_{\sqsubseteq}$.

3   Upper Bound

We obtain an Ackermannian upper bound for downward rational termination by proving that, given an instance $\mathcal{T}, w_1$ of the problem, there is an Ackermannianly large positive integer $N(\mathcal{T}, w_1)$ such that if $\mathcal{T}$ terminates from w_1 then all its computations from w_1 have lengths bounded by $N(\mathcal{T}, w_1)$.

At the heart of the proof, there is an analysis of computations of $\mathcal{T}$ from w_1 in terms of how frequently they contain words that belong to certain regular languages. A trivial case is when the regular language consists of all words over the alphabet of $\mathcal{T}$, for which the frequency is 1. More interestingly, our central lemma (Lemma 6) shows that, assuming that the frequency of the language of a regular expression E in a computation of length N is u^{-1} and that N is sufficiently large in terms of u, either some segment of the computation can be pumped to produce an infinite computation, or E can be refined to some E' whose frequency is some smaller u'^{-1}. The notion of refinement of the regular expressions is such that only finitely many successive refinements are ever possible, and so if $\mathcal{T}$ terminates from w_1 then repeated applications of the lemma must stop because N is not sufficiently large. Moreover, the refinements of the regular expressions and the decreases in their frequences observe certain bounds (that depend on $\mathcal{T}$, but not on w_1 or N), which together with the preceding reasoning enables us to obtain a global bound on the lengths of all the computations (provided that $\mathcal{T}$ terminates from w_1).

Before the central lemma, we have two lemmas that are about pumpability of computation segments, and its connection with the regular expressions and their refinements. Leading to the main result, we have another two lemmas, which are concerned with bounding the sequences of regular expressions and frequences that can arise from repeated applications of the central lemma, and consequences of those bounds for the lengths of computations. However, we first introduce the class of regular expressions used and the notion of refinement, as well as a useful class of auxiliary transducers.

3.1   Flat Regular Expressions (FRE)

A prominent role in the sequel is played by the following subclass of the simple regular expressions of Abdulla et al. [1]: we say that a regular expression over an alphabet Σ is *flat* iff it is of the form $\Delta_1^* d_1 \Delta_2^* d_2 \cdots \Delta_K^*$ with $K \geq 1$, $\Delta_1, \ldots, \Delta_K \subseteq \Sigma$ and $d_1, \ldots, d_{K-1} \in \Sigma \cup \{\epsilon\}$.

For such a regular expression E, let: the *length* of E be K; the *height* of E be $\max_{i=1}^{K} |\Delta_i|$. If $l \in \mathbb{N}^+$, let us say that E is *l-refined* by E' iff E' can be obtained from E by replacing some Δ_i^* with an FRE $E^\dagger$ over Δ_i such that:

- the length of $E^\dagger$ is at most l, and
- the height of $E^\dagger$ is strictly less than $|\Delta_i|$, i.e., each set in $E^\dagger$ is strictly in Δ_i.

In that case, E' is also an FRE over Σ, of length at most $K + l - 1$. When each set in $E^\dagger$ has size $|\Delta_i| - 1$, we call the refinement *maximal*.

For E still as above, let $\mathcal{I}_E$ denote an identity transducer on the downward closure of the language of E as follows:

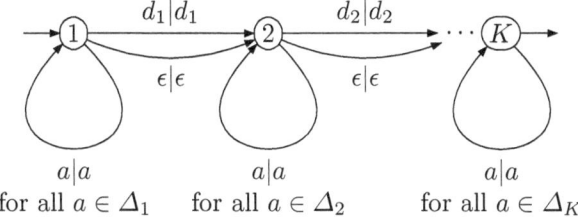

Indeed, $R(\mathcal{I}_E) = \{\langle w, w \rangle \; : \; w \in {\downarrow}L(E)\}$, so $\mathcal{I}_E$ is downwards monotone.

3.2 Pumpable Transductions

Since finite sequences of consecutive transductions can be seen as single transductions of composite transducers, it suffices to consider pumpability of transductions instead of considering it for computation segments. The notion we define applies to transductions between words in the language of an FRE $E = \Delta_1^* d_1 \cdots \Delta_K^*$, and essentially requires that, for all i, while reading the portion of the input word in Δ_i^*, the transduction visits a part of the transducer that is able to consume any word in Δ_i^*. The composition with the identity transducer is a technical tool to ensure that traversing different paths in the state-transition graph still produces words that conform to E.

Definition 2. *If $\mathcal{T}$ is a downwards monotone transducer and $E = \Delta_1^* d_1 \cdots \Delta_K^*$ is a FRE over an alphabet Σ, and τ is a transduction of composite transducer $\mathcal{T}\,{}_{\S}\,\mathcal{I}_E$ such that $\mathrm{In}(\tau) \in L(E)$, let us say that τ is pumpable iff it can be factored as $s_1 \xrightarrow{\tau_1} s_1' \xrightarrow{d_1|e_1} s_2 \xrightarrow{\tau_2} s_2' \xrightarrow{d_2|e_2} \cdots s_K \xrightarrow{\tau_K} s_K'$, where, for each $i \in \{1, \ldots, K\}$, $\Delta_i^* \subseteq {\downarrow}\mathrm{dom}(R((\mathcal{T}\,{}_{\S}\,\mathcal{I}_E)(s_i, s_i')))$.*

Lemma 3. *If $\mathcal{T}$ is downwards monotone, and $\mathcal{T}\,{}_{\S}\,\mathcal{I}_E$ has a transduction τ such that $\mathrm{In}(\tau) \in L(E)$ and which is pumpable, then $\mathcal{T}$ has an infinite computation from any word in ${\downarrow}L(E)$.*

The following is a 'pumping lemma': roughly, if a transduction from E to E is such that its input word is not in the language of any 'short' refinement of E, then it is pumpable. Here 'short' amounts to a bound which is the product of the length of E and the size of the transducer's state space.

Lemma 4. *Suppose that: $\mathcal{T}$ is a downwards monotone transducer with set of states Q and alphabet Σ; E is a FRE over Σ, of length K; τ is a transduction of $\mathcal{T}\,{}_{\S}\,\mathcal{I}_E$ such that $\mathrm{In}(\tau) \in L(E)$. Then either τ is pumpable, or $\mathrm{In}(\tau) \in L(E')$ for some $K|Q|$-refinement E' of E.*

3.3 Sword of Hercules

Our central lemma, assuming that γ is a computation of length N from w_1 of a downwards monotone $\mathcal{T}$ which terminates from w_1, can be applied repeatedly to γ to yield some sequence $\langle E_0, u_0 \rangle, \langle E_1, u_1 \rangle, \ldots$ of pairs of FREs and positive integers, as long as N is sufficiently large. For each h, there are at least $\lfloor N/u_h \rfloor$ occurrences in γ of words from the language of E_h. Moreover, each E_{h+1} refines E_h, and the length of E_{h+1} as well as u_{h+1} are bounded by elementary functions of: the number of states of $\mathcal{T}$, the length and height of E_h, and u_h. Recalling the notion of refinement, each application of the lemma can be thought of as a strike of Hercules on the FRE E_h, after which the latter has a Hydra-like response: although some component of the form Δ_h^* is removed from E_h, it is replaced in E_{h+1} by some FRE $E_h^\dagger$. The height of $E_h^\dagger$, however, must be strictly smaller than the size of Δ_h, but the bound on its length grows with every strike.

Definition 5. *For $\alpha \in (0,1]$, let us say that a regular expression E is α-frequent in a sequence of words $w_1, \ldots, w_N$ iff there exists $J \in \{1, \ldots, N\}$ of size $\lfloor N\alpha \rfloor$ such that $w_j \in \mathrm{L}(E)$ for all $j \in J$.*

Lemma 6. *Suppose that $\gamma = w_1, \ldots, w_N$ is a computation of a downwards monotone transducer $\mathcal{T}$ with set of states Q and alphabet Σ, and that $\mathcal{T}$ terminates from w_1. If an FRE E over Σ and $u \in \mathbb{N}^+$ are such that $N \geq 16u^2$ and E is u^{-1}-frequent in γ, then there exists a $K|Q|^{4u}$-refinement E' of E which is u'^{-1}-frequent in γ, where $K = \mathrm{len}(E)$, $H = \mathrm{hgt}(E)$, and $u' = 16u^2 K(H+1)^{2K|Q|^{4u}}$.*

3.4 Slaying the Hydra

The next two lemmas show that every sequence of pairs of FREs and positive integers that can arise from repeated applications of Lemma 6 is finite, i.e., Hercules always defeats the Hydra eventually, and that if $N \geq 16u^2$ for every u in such a sequence and $\mathcal{T}$ terminates from w_1, then $\mathcal{T}$ cannot have a computation from w_1 of length N. Moreover, from the single-step bounds in Lemma 6, we establish a bound for each pair in terms of $|Q|$, $|\Sigma|$ and the distance from the initial pair $\langle \Sigma^*, 1 \rangle$, where Q and Σ are the state space and the alphabet of $\mathcal{T}$.

We first define a directed graph which contains every sequence that Lemma 6 can yield. To show that every path that starts from $\langle \Sigma^*, 1 \rangle$ is finite, we also introduce a measure on FREs E over Σ in terms of $|\Sigma|$-tuples of natural numbers. The latter records, for each $s \in \{1, \ldots, |\Sigma|\}$, how many sets of size s occur in E.

We say that a sequence $y_0, y_1, \ldots$ of tuples in some $\mathbb{N}^k$ is *bad* iff there do not exist $i < j$ such that $y_i \leq y_j$, where $\leq$ is the pointwise ordering. We recall that, by Dickson's Lemma, $\leq$ is a well-quasi ordering on $\mathbb{N}^k$, i.e., there is no infinite bad sequence. Hence, the finiteness of every path from $\langle \Sigma^*, 1 \rangle$ follows once we show that every corresponding sequence of measures in $\mathbb{N}^{|\Sigma|}$ is bad.

Definition 7. *Given a set of states Q and an alphabet Σ, let $\Upsilon_{Q,\Sigma}$ be the graph:*

- *the vertices are pairs $\langle E, u \rangle$ where E is an FRE over Σ and $u \in \mathbb{N}^+$;*
- *there is an edge from $\langle E, u \rangle$ to $\langle E', u' \rangle$ iff E' is a $K|Q|^{4u}$-refinement of E and $u' = 16u^2 K(H+1)^{2K|Q|^{4u}}$, where $K = \mathrm{len}(E)$ and $H = \mathrm{hgt}(E)$.*

Definition 8. *For $E = \Delta_1^* d_1 \cdots \Delta_K^*$ an FRE over Σ and $s \in \{0, \ldots, |\Sigma|\}$, let $Y_s(E) = |\{i : i \in \{1, \ldots, K\}$ and $|\Delta_i| = s\}|$.*

Lemma 9. *Suppose that Q is a set of states, Σ is an alphabet, and $\langle E_0, u_0 \rangle \to \langle E_1, u_1 \rangle \to \ldots$ is a path from $\langle \Sigma^*, 1 \rangle$ in $\Upsilon_{Q, \Sigma}$. Then $\langle Y_1(E_0), \ldots, Y_{|\Sigma|}(E_0) \rangle$, $\langle Y_1(E_1), \ldots, Y_{|\Sigma|}(E_1) \rangle$, $\ldots$ is a bad sequence, and letting $f(u) = 16u^{3+2u^{1+4u}}$, we have $\sum_{s=0}^{|\Sigma|} Y_s(E_h), u_h < f^{h + \max(|Q|, |\Sigma|)}(2)$ for all h.*

Lemma 10. *Suppose that: $\mathcal{T}$ is a downwards monotone transducer with set of states Q and alphabet Σ; $\mathcal{T}$ terminates from w_1; $N \geq 16u^2$ for all vertices $\langle E, u \rangle$ that are reachable from $\langle \Sigma^*, 1 \rangle$ in $\Upsilon_{Q, \Sigma}$. Then $\mathcal{T}$ does not have a computation from w_1 of length N.*

3.5   Main Result

Given the preceding lemmas, it remains to do two things. The first is to show that, in every graph $\Upsilon_{Q, \Sigma}$, the positive integers in all vertices that are reachable from $\langle \Sigma^*, 1 \rangle$ are bounded by an Ackermannian function of $|Q|$ and $|\Sigma|$. Although the vertices and edges of $\Upsilon_{Q, \Sigma}$ can be encoded using the classical Hydra trees of Kirby and Paris [15], we do not require the full generality of the latter, but are able to obtain an Ackermannian bound using Lemma 9 and recent results of Figueira et al. [8] on lengths of bad sequences of tuples of natural numbers.

Writing $N(|Q|, |\Sigma|)$ for the obtained bound, it then remains to argue that a computation of $\mathcal{T}$ from w_1 can be non-deterministically guessed and checked in Ackermannian time or space, but that can be done by a straightforward non-deterministic algorithm that explores the state-transition graph of the iterated transducer $\mathcal{T}^{N(|Q|, |\Sigma|)-1}$ on the fly.

Theorem 11. *Termination for a downwards monotone transducer $\mathcal{T}$ with set of states Q and alphabet Σ, from a word w_1 over Σ, is decidable by an algorithm whose complexity is bounded by an Ackermannian function. For fixed $|\Sigma|$, the bound is in $\mathfrak{F}_{|\Sigma|+2}$.*

Proof. From Lemma 9, for every path $\langle E_0, u_0 \rangle \to \langle E_1, u_1 \rangle \to \ldots$ from vertex $\langle \Sigma^*, 1 \rangle$ in $\Upsilon_{Q, \Sigma}$, the sequence $\langle Y_1(E_0), \ldots, Y_{|\Sigma|}(E_0) \rangle$, $\langle Y_1(E_1), \ldots, Y_{|\Sigma|}(E_1) \rangle$, $\ldots$ in $\mathbb{N}^{|\Sigma|}$ is bad, and for all h, $\max(Y_1(E_h), \ldots, Y_{|\Sigma|}(E_h)) < f^{h + \max(|Q|, |\Sigma|)}(2)$, i.e., in the terminology of Figueira et al. [8], the sequence is $\max(|Q|, |\Sigma|)$-controlled by the function $g(h) = f^h(2)$. Since f is in class $\mathfrak{F}_2$ of the fast growing hierarchy, we have that g belongs to $\mathfrak{F}_3$. Also, g is monotone and satisfies $g(h) \geq \max(1, h)$ for all h, and we can assume that $|\Sigma| \geq 1$. Hence, [8, Proposition 5.2] applies and gives us a function $M_s(t)$ such that M_s is in $\mathfrak{F}_{s+2}$ for each $s \geq 1$, and the length of $\langle Y_1(E_0), \ldots, Y_{|\Sigma|}(E_0) \rangle$, $\langle Y_1(E_1), \ldots, Y_{|\Sigma|}(E_1) \rangle$, $\ldots$ is at most $M_{|\Sigma|}(\max(|Q|, |\Sigma|))$.

Since the distance of each $\langle E, u \rangle$ reachable from $\langle \Sigma^*, 1 \rangle$ in $\Upsilon_{Q, \Sigma}$ is at most $M_{|\Sigma|}(\max(|Q|, |\Sigma|)) - 1$, we have by Lemma 9 that $N(|Q|, |\Sigma|) \geq 16u^2$, where

$$N(k, s) = 16(g(M_s(\max(k, s)) - 1 + \max(k, s)))^2.$$

Therefore, by Lemma 10, $\mathcal{T}$ terminates from w_1 iff it does not have a computation from w_1 of length $N(|Q|, |\Sigma|)$.

We conclude that termination of $\mathcal{T}$ from w_1 is decidable by guessing and checking an $N(|Q|, |\Sigma|)$-long computation of $\mathcal{T}$ from w_1, which is equivalent to guessing and checking a transduction of the iterated transducer $\mathcal{T}^{N(|Q|,|\Sigma|)-1}$ from w_1. It follows that space $O(N(|Q|, |\Sigma|) \times (\log |Q| + \log |\Sigma|) + \log |w_1|)$ is sufficient for a non-deterministic algorithm.

Recalling that $M_{|\Sigma|}$ is in $\mathfrak{F}_{|\Sigma|+2}$ and that g is in $\mathfrak{F}_3 \subseteq \mathfrak{F}_{|\Sigma|+2}$, we have that $N(|Q|, |\Sigma|)$ as a function of $|Q|$ is also in $\mathfrak{F}_{|\Sigma|+2}$. Therefore, as a function of the combined size of $\mathcal{T}$ and w_1, the non-deterministic space bound is in $\mathfrak{F}_{|\Sigma|+2}$ when $|\Sigma|$ is fixed, and in $\mathfrak{F}_\omega$ in general. Since the classes involved of the fast growing hierarchy are closed under squaring and exponentiation, the same coarse classifications apply to consequent deterministic space and time bounds.   □

4   Lower Bound

We use the following variant of the fast growing functions F_k, which give rise to the *Ackermann hierarchy* (cf. e.g. [10]):

$$A_1(x) = 2x \qquad\qquad A_{k+1}(x) = A_k^x(1), \text{ for } k \geq 1.$$

For example, A_2 is exactly exponentiation of 2, and A_3 is exactly tetration of 2. One can check that, for all $k, p \geq 1$, there exists $x_{k,p} \geq 0$ such that, for all $x \geq x_{k,p}$, we have $A_k(x) > F_{k-1}^p(x)$; hence $A_k \notin \mathfrak{F}_{k-1}$ if $k \geq 2$ by [16, Theorem 2.10]. Conversely, $A_k(x) \leq F_k(x)$ for all $k \geq 1$ and $x \geq 0$, so $A_k \in \mathfrak{F}_k$.

To obtain our lower bound result, we provide a construction of 'dependent counter programs' $D_1, D_2, \ldots$ such that each D_{k+1} is computable from D_k in logarithmic space. For every k, D_k consists of routines for basic counter operations (initialisation, increment, decrement, zero testing, maximum testing), and is dependent in the sense that it may operate on an as yet unspecified counter by calling the latter's operations as subroutines. Moreover, D_k is closely related to the A_k function above: provided C is a counter program that reliably implements a counter bounded by N (in the sense that transducers that correspond to its routines compute correctly, even if insertion errors are possible), then $D_k[C]$ reliably implements a counter bounded by $A_k(N)$. Given a Turing machine of size K, we then use $D_K[C]$ with C reliable up to K to build a transducer that reliably simulates $A_K(K)$ steps of the machine (in the presence of insertion errors), and diverges iff the machine halts.

Theorem 12. *Given a deterministic Turing machine $\mathcal{M}$ of size K, we have that a transducer $\mathcal{T}(\mathcal{M})$ and a word w_1, over an alphabet of linear size, are computable in elementary time, such that $\mathcal{M}$ halts within time $A_K(K)$ iff $\mathcal{T}(\mathcal{M})_\sqsubseteq$ does not terminate from w_1.*

5   Safety MTL Satisfiability

We now show that the satisfiability problem for the safety fragment of MTL is inter-reducible with the termination problem for gainy transducers (equivalently, for downwards monotone transducers, cf. Sect. 2.5), thus improving the best known upper and lower bounds for the former. This reduction relies on results in the literature concerning *insertion channel machines (ICMs)*—a model that is very closely related to gainy transducers.

The formulas of MTL are built over a set of atomic events Σ using monotone Boolean connectives and time-constrained versions of the *next* operator $\bigcirc$, *until* operator $\mathcal{U}$, and the *dual until* operator $\tilde{\mathcal{U}}$:

$$\varphi ::= \top \mid \bot \mid \varphi_1 \wedge \varphi_2 \mid \varphi_1 \vee \varphi_2 \mid a \mid \bigcirc_I \varphi \mid \varphi_1 \mathcal{U}_I \varphi_2 \mid \varphi_1 \tilde{\mathcal{U}}_I \varphi_2,$$

where $a \in \Sigma$ and $I \subseteq \mathbb{R}_{\geq 0}$ is an interval with endpoints in $\mathbb{N} \cup \{\infty\}$.

A *timed word* over alphabet Σ is a pair $\rho = \langle \sigma, \tau \rangle$, where σ is an infinite word over Σ and τ is an infinite sequence of non-negative reals that is strictly increasing and unbounded (i.e., *non-Zeno*). The *satisfiability problem* for MTL asks whether a given formula is satisfied by some timed word. This problem was shown undecidable in [17], motivating the introduction of the sub-logic safety MTL in [18]. Safety MTL is the fragment of MTL obtained by requiring that the interval I in each until operator $\mathcal{U}_I$ have finite length. Thus safety MTL allows bounded eventualities, such as $\Diamond_{(0,1)}\varphi$, but not unbounded eventualites, such as $\Diamond_{(0,\infty)}\varphi$. The satisfiability problem for safety MTL was shown to be decidable in [18] by an argument involving Higman's Lemma. It was later observed that this argument yields an upper bound in level $\mathfrak{F}_{\omega^\omega}$ of the fast-growing hierarchy [21]. A non-elementary lower bound (in $\mathfrak{F}_3$) is given in [4] and an improved lower bound in $\mathfrak{F}_4$ is given in [13].

Theorems 11 and 12 yield upper and lower bounds for safety MTL satisfiability that are both in $\mathfrak{F}_\omega$ through four reductions:

where an ICM is a finite-state automaton acting on an unbounded channel that is subject to insertion errors, and their *fair termination problem* ask whether there is no infinite computation which is *fair*, i.e., in which every message written to the channel is eventually read.

The reductions (i) and (ii) are almost immediate and require only logarithmic space. Details for the reduction (iii), which can also be done in logarithmic space, can be found in [4]. The most complex reduction is (iv): it is doubly exponential, and its details are available in [13, Proposition 5.27], which builds on a translation from MTL to channel machines in [5].

References

1. Abdulla, P.A., Collomb-Annichini, A., Bouajjani, A., Jonsson, B.: Using forward reachability analysis for verification of lossy channel systems. Formal Meth. Sys. Des. 25(1), 39–65 (2004)

2. Abdulla, P.A., Jonsson, B., Nilsson, M., Saksena, M.: A survey of regular model checking. In: Gardner, P., Yoshida, N. (eds.) CONCUR 2004. LNCS, vol. 3170, pp. 35–48. Springer, Heidelberg (2004)

3. Alur, R., Henzinger, T.A.: Back to the future: Towards a theory of timed regular languages. In: FOCS, pp. 177–186. IEEE Comput. Soc. (1992)

4. Bouyer, P., Markey, N., Ouaknine, J., Schnoebelen, P., Worrell, J.: On termination and invariance for faulty channel machines. Formal Asp. Comput. 24(4-6), 595–607 (2012)

5. Bouyer, P., Markey, N., Ouaknine, J., Worrell, J.: The cost of punctuality. In: LICS, pp. 109–120. IEEE Comput. Soc. (2007)

6. Chambart, P., Schnoebelen, P.: Post embedding problem is not primitive recursive, with applications to channel systems. In: Arvind, V., Prasad, S. (eds.) FSTTCS 2007. LNCS, vol. 4855, pp. 265–276. Springer, Heidelberg (2007)

7. Chambart, P., Schnoebelen, P.: The ordinal recursive complexity of lossy channel systems. In: LICS, pp. 205–216. IEEE Comput. Soc. (2008)

8. Figueira, D., Figueira, S., Schmitz, S., Schnoebelen, P.: Ackermannian and primitive-recursive bounds with Dickson's Lemma. In: LICS, pp. 269–278. IEEE Comput. Soc. (2011), ext. ver. in CoRR, vol. abs/1007.2989

9. Finkel, A., Schnoebelen, P.: Well-structured transition systems everywhere! Theor. Comput. Sci. 256(1-2), 63–92 (2001)

10. Friedman, H.: Long finite sequences. J. Combin. Theor., Ser. A 95(1), 102–144 (2001)

11. Henzinger, T.A.: The Temporal Specification and Verification of Real-Time Systems. Ph.D. thesis, Stanford University (1991), tech. rep. STAN-CS-91-1380

12. Henzinger, T.A.: It's about time: Real-time logics reviewed. In: Sangiorgi, D., de Simone, R. (eds.) CONCUR 1998. LNCS, vol. 1466, pp. 439–454. Springer, Heidelberg (1998)

13. Jenkins, M.: Synthesis and Alternating Automata over Real Time. D.Phil. thesis, Oxford University (2012)

14. Karandikar, P., Schmitz, S.: The parametric ordinal-recursive complexity of Post embedding problems. In: Pfenning, F. (ed.) FOSSACS 2013. LNCS, vol. 7794, pp. 273–288. Springer, Heidelberg (2013)

15. Kirby, L., Paris, J.: Accessible independence results for Peano arithmetic. Bull. London Math. Soc. 14(4), 285–293 (1982)

16. Löb, M., Wainer, S.: Hierarchies of number-theoretic functions. I. Arch. Math. Log. 13(1), 39–51 (1970)

17. Ouaknine, J., Worrell, J.: On metric temporal logic and faulty Turing machines. In: Aceto, L., Ingólfsdóttir, A. (eds.) FOSSACS 2006. LNCS, vol. 3921, pp. 217–230. Springer, Heidelberg (2006)

18. Ouaknine, J., Worrell, J.: Safety metric temporal logic is fully decidable. In: Hermanns, H., Palsberg, J. (eds.) TACAS 2006. LNCS, vol. 3920, pp. 411–425. Springer, Heidelberg (2006)

19. Ouaknine, J., Worrell, J.: On the decidability of metric temporal logic over finite words. Log. Meth. Comput. Sci. 3(1) (2007)

20. Sakarovitch, J.: Elements of Automata Theory. Camb. Univ. Press (2009)

21. Schmitz, S.: Scientific report - ESF Short Visit to Oxford University (2012)

22. Schnoebelen, P.: Revisiting Ackermann-hardness for lossy counter machines and reset Petri nets. In: Hliněný, P., Kučera, A. (eds.) MFCS 2010. LNCS, vol. 6281, pp. 616–628. Springer, Heidelberg (2010)

Strong Completeness for Markovian Logics

Dexter Kozen[1], Radu Mardare[2], and Prakash Panangaden[3]

[1] Computer Science Department, Cornell University, Ithaca, New York, USA
[2] Department of Computer Science, Aalborg University, Denmark
[3] School of Computer Science, McGill University, Montreal, Quebec, Canada

Abstract. In this paper we present Hilbert-style axiomatizations for three logics for reasoning about continuous-space Markov processes (MPs): (i) a logic for MPs defined for probability distributions on measurable state spaces, (ii) a logic for MPs defined for sub-probability distributions and (iii) a logic defined for arbitrary distributions. These logics are not compact so one needs infinitary rules in order to obtain strong completeness results.

We propose a new infinitary rule that replaces the so-called Countable Additivity Rule (CAR) currently used in the literature to address the problem of proving strong completeness for these and similar logics. Unlike the CAR, our rule has a countable set of instances; consequently it allows us to apply the Rasiowa-Sikorski lemma for establishing strong completeness. Our proof method is novel and it can be used for other logics as well.

1 Introduction

Markov processes (MPs) are standard models used for abstracting and reasoning about complex natural and man-made systems in order to handle either a lack of knowledge or inherent randomness. There are various levels of abstraction that one can consider in the definition of Markov processes: (i) the state space can be modeled by using particular types of structures that can vary from discrete finite spaces to topological or measurable spaces; (ii) the indeterminacy can be modeled by using probability or sub-probability distributions over the state space to describe the probability of transitions or by assuming exponentially distributed random variables to characterize the time durations between transitions.

To specify properties of Markov processes, the natural logic is a simple modal logic in which bounds on probabilities enter into the modalities. This logic can be stripped down to a very spartan core —just the modalities and finite conjunction— and still characterize bisimulation for labeled Markov processes [5,6]. It is therefore tempting to understand this logic from a proof theoretic perspective. Recent papers [4,11,19] have established complete proof systems and prove finite model properties for similar logics. Goldblatt in [11] presents a proof-theoretic analysis of the logic of T-coalgebras, where T is any polynomial functor constructed from a standard monad on the category of measurable spaces. He proves that the semantic consequence relation over T-coalgebras is equal to the least deducibility relation that satisfies Lindenbaum's lemma, which states that any consistent set of formulas can be extended to a maximally consistent set. In other words, this consequence relation is equal to the least of all deducibility relations

K. Chatterjee and J. Sgall (Eds.): MFCS 2013, LNCS 8087, pp. 655–666, 2013.
© Springer-Verlag Berlin Heidelberg 2013

if and only if that least deducibility relation satisfies Lindenbaum's lemma. These logics are not compact and for proving the aforementioned results in [11] it is used a powerful infinitary axiom scheme named the Countable Additivity Rule (CAR). In [21] Zhou and Ying prove that such a logic is not strongly-complete in the absence of CAR.

A feature of CAR is that it has an uncountable set of instances. This fact makes it difficult to prove that maximally consistent sets exist for such logics and consequently, in the papers concerned with the strong completeness of the modal logics for Harsanyi type spaces [19,21] or for Markov processes [4,15] it had to be assumed that consistent sets can be extended to maximally consistent sets. The completeness theorems cited are contingent on this assumption.

In this paper we reconsider the axiomatizations of the deducibility relations for three modal logics for Markov processes. The first one refers to what we call probabilistic Markov processes (PMPs), which are Markov processes defined by probability distribution over the state space. The second one is a modal logic for subprobabilistic Markov processes (SMPs), which are Markov processes defined for sub-probability distributions, and the third one is defined for what we call general Markov processes (GMPs), which are Markov processes defined for arbitrary distributions, usually these are interpreted as rates.

We propose a new infinitary axiom schema to replace CAR. Unlike CAR, our axiom has a countable set of instances. This fact allows us to invoke the Rasiowa-Sikorski Lemma and prove the strong completeness theorem via a canonical models construction *without needing to assume that consistent sets can be enlarged to maximal consistent sets* (Lindenbaum's lemma). In fact Lindenbaum's lemma can be directly proven from the Rasiowa-Sikorski lemma.

The Rasiowa-Sikorski lemma is a model-theoretic result that exploits a topological result known as Baire category theorem and the Stone duality for boolean algebras with operators. Applied to logics, the Rasiowa-Sikorski lemma states that given a multimodal logic (possibly involving an infinite set of modalities) for which the provability relation admits an axiomatization such that the set of instances of the infinitary proof rules (if any) is countable, then for any consistent formula φ, there exists a maximally-consistent set of formulas containing φ. Since we manage to replace CAR with an infinitary rule having a countable set of instances, we can apply this result and prove strong completeness for each of the three logics.

The contribution of this paper consists in the novelty of the proof method for strong completeness. We have used already these types of techniques in [13], where we proved a Stone duality for PMPs. That result implies strong completeness for the logic for PMPs but the logical aspects were not spelled out in that paper; there we concentrated on the algebraic versions of the logic and proved a duality theorem. In this paper we spell out the completeness theorem explicitly and, in addition, demonstrate that, in fact, the proof method can be used for the other two logics as well. Though these logics are superficially very similar, the axiomatizations are different and none of the completeness theorems follow directly from the others.

We also show that the infinitary axiom needed to replace CAR can be obtained by a lifting of the so-called archimedean axioms already present in the old versions of these axiomatizations. We are confident that similar results can be obtained for the general case of the measurable polynomial functors on the category of measurable spaces considered in [11], but we do not have such a result yet.

2 Background

Let $\mathbb{Q}_0 = \mathbb{Q} \cap [0, 1]$, $\mathbb{Q}^+ = \mathbb{Q} \cap [0, \infty)$, $\mathbb{R}_0 = \mathbb{R} \cap [0, 1]$, and $\mathbb{R}^+ = \mathbb{R} \cap [0, \infty)$.

2.1 Measurable Spaces and Measures

In this section we introduce a few concepts and results from measure theory that we will find useful. For more details, we refer the reader to [3,8].

Let M be an arbitrary nonempty set.

A *field* (of sets) over M is a boolean algebra of subsets of M under the usual set-theoretic boolean operations. A *σ-algebra* (also called a *σ-field*) over M is a field of sets over M closed under countable union. The tuple (M, Σ) where Σ is a σ-algebra over M, is called a *measurable space* and the elements of Σ *measurable sets*.

If $\Omega \subseteq 2^M$, the *σ-algebra generated by Ω*, denoted $\sigma(\Omega)$, is the smallest σ-algebra containing Ω. Every topological space has a natural σ-algebra associated with it, namely the one generated by the open sets. This is called the *Borel algebra* of the space, and the measurable sets are called *Borel sets*.

Given two measurable spaces (M, Σ) and (N, Ω), a function $f : M \to N$ is *measurable* if $f^{-1}(T) \in \Sigma$ for all $T \in \Omega$. We use $[\![M \to N]\!]$ to denote the family of measurable functions from (M, Σ) to (N, Ω).

A nonnegative real-valued set function μ is *finitely additive* if $\mu(A \cup B) = \mu(A) + \mu(B)$ whenever $A \cap B = \emptyset$. We say that μ is *countably subadditive* if $\mu(\bigcup_i A_i) \leq \sum_i \mu(A_i)$ for a countable family of measurable sets, and we say that μ is *countably additive* if $\mu(\cup_i A_i) = \sum_i \mu(A_i)$ for a countable *pairwise-disjoint* family of measurable sets. Finite additivity implies monotonicity and countable additivity implies certain continuity properties; see the references for precise statements.

Given a measurable space (M, Σ), a countably additive set function $\mu : \Sigma \to \mathbb{R}^+$ is a *measure* on (M, Σ). A measure $\mu : \Sigma \to \mathbb{R}_0$ is a *subprobability measure*. Thus, for a subprobability measure $\mu(M) \leq 1$; if in addition $\mu(M) = 1$, μ is a *probability measure*. We use $\Delta(M, \Sigma)$, $\Pi(M, \Sigma)$ and $\Pi^*(M, \Sigma)$ to denote the set of measures, probability measures and subprobability measures on (M, Σ) respectively.

We view $\Delta(M, \Sigma)$ as a measurable space by defining the σ-algebra generated by the sets $\{\mu \in \Delta(M, \Sigma) \mid \mu(S) \geq r\}$ for $S \in \Sigma$ and $r \in \mathbb{R}^+$. This is the least σ-algebra on $\Delta(M, \Sigma)$ such that all maps $\mu \mapsto \mu(S) : \Delta(M, \Sigma) \to \mathbb{R}^+$ for $S \in \Sigma$ are measurable, where the set of positive reals is endowed with the σ-algebra generated by all rational intervals, i.e. the Borel σ-algebra. Similarly, $\Pi(M, \Sigma)$ and $\Pi^*(M, \Sigma)$ can be viewed as measurable spaces by defining the σ-algebras generated by the sets $\{\mu \in \Pi(M, \Sigma) \mid \mu(S) \geq r\}$ and $\{\mu \in \Pi^*(M, \Sigma) \mid \mu(S) \geq r\}$ respectively, defined for $S \in \Sigma$ and $r \in [0, 1]$.

The next theorem is a key tool in our constructions.

Theorem 1. *[Theorem 11.3 of [3]] Let $\mathcal{F} \subseteq 2^M$ be a field of sets. Let $\mu : \mathcal{F} \to \mathbb{R}^+$ be finitely additive and countably subadditive. Then μ extends uniquely to a measure on $\sigma(\mathcal{F})$.*

2.2 Analytic Spaces

Recall that a topological space is said to be *separable* if it contains a countable dense subset and *second countable* if its topology has a countable base. Second countability implies separability, but not vice versa; however, the two concepts coincide for metric spaces. A *Polish space* is the topological space underlying a complete separable metric space. An *analytic space* is a continuous image of a Polish space in a Polish space.

Analytic spaces enjoy remarkable properties that were crucial in proving the logical characterization of bisimulation [6,16]. We note that the completeness theorems proved in [4,15,21] were established for Markov processes defined on analytic spaces.

2.3 The Baire Category Theorem

The Baire category theorem is a topological result with important applications in logic. It is used to prove the Rasiowa-Sikorski lemma [17,10] which is crucial for this paper.

A subset D of a topological space X is *dense* if its closure $\overline{D}$ is all of X. Equivalently, a dense set is one intersecting every nonempty open set. A set $N \subseteq X$ is *nowhere dense* if every nonempty open set contains a nonempty open subset disjoint from N. A set is *of the first category* or *meager* if it is a countable union of nowhere dense sets.

A *Baire space* is one in which the intersection of countably many dense open sets is dense. It follows from these definitions that the complement of a first category set is dense in any Baire space. Baire originally proved that the real line is a Baire space. More generally, every Polish space is Baire and every locally compact Hausdorff space is Baire. For us, the relevant version is the following special case: every compact Hausdorff space is Baire.

Definition 2. *Let $\mathcal{B}$ be a boolean algebra and let $T \subseteq \mathcal{B}$ be such that T has a greatest lower bound $\bigwedge T$ in $\mathcal{B}$. An ultrafilter (maximal filter) U is said to respect T if $T \subseteq U$ implies that $\bigwedge T \in U$. If $\mathcal{T}$ is a family of subsets of $\mathcal{B}$, we say that an ultrafilter U respects $\mathcal{T}$ if it respects every member of $\mathcal{T}$.*

Theorem 3 (Rasiowa–Sikorski lemma [17]). *For any boolean algebra $\mathcal{B}$ and any countable family $\mathcal{T}$ of subsets of $\mathcal{B}$, each member of which has a meet in $\mathcal{B}$, and for any nonzero $x \in \mathcal{B}$, there exists an ultrafilter in $\mathcal{B}$ that contains x and respects $\mathcal{T}$.*

3   Markov Processes

In this section we introduce three classes of models of probabilistic systems with a continuous state space: (i) *probabilistic Markov processes* (PMPs), (ii) *subprobabilistic Markov processes* (SMPs) and (iii) *general Markov processes* (GMPs). The first two classes contain the systems for which the transition from a state to a measurable set of states is characterized by its probability. The third class represents the systems with continuous-time transitions, i.e., the probability of a transition from a state to a measurable set of states depends on time. In earlier papers, they were called *labeled* Markov processes to emphasize the fact that there were multiple possible actions, but here we will suppress the labels, as they do not contribute any relevant structure for our results.

Definition 4 (Markov process). *Given an analytic space* (M, Σ),

- *a* probabilistic Markov process *is a measurable mapping* $\theta \in \llbracket M \to \Pi(M, \Sigma) \rrbracket$;
- *a* subprobabilistic Markov process *is a measurable mapping* $\theta \in \llbracket M \to \Pi^*(M, \Sigma) \rrbracket$;
- *a* general Markov process *is a measurable mapping* $\theta \in \llbracket M \to \Delta(M, \Sigma) \rrbracket$.

In what follows we identify a Markov process with the tuple $\mathcal{M} = (M, \Sigma, \theta)$; M is called the *support set*, denoted by $\mathrm{supp}(\mathcal{M})$, and θ is called the *transition function*.

If $\mathcal{M} = (M, \Sigma, \theta)$ is a (probabilistic/subprobabilistic/general) Markov process, then for $m \in M$, $\theta(m)$ is a (probabilistic/subprobabilistic/general) measure on the state space (M, Σ). If $\mathcal{M}$ is a PMP or a SMP, the value $\theta(m)(N)$ for $N \in \Sigma$ represents the probability of a transition from m to a state in N; otherwise, if $\mathcal{M}$ is a GMP, then $\theta(m)$ is a measure on the state space and the value $\theta(m)(N) \in \mathbb{R}^+$ represents the rate of an exponentially distributed random variable that characterizes the transition from m to a state in N.

The condition that θ is measurable is equivalent to the condition that for fixed $N \in \Sigma$, the function $m \mapsto \theta(m)(N)$ is measurable (see e.g. Proposition 2.9 of [7]).

4 Markovian Logics

Markovian logics are multi-modal logics for semantics based on the three classes of Markov processes introduced in the previous section. They have been introduced and studied in various contexts [1,2,14,12,19,9,4,15]. In addition to the boolean operators, these logics are equipped with modal operators of type L_r for rational numbers r that are used to approximate the numerical labels of the transitions. Intuitively, the formula $L_r \varphi$ is satisfied by $m \in \mathcal{M}$ whenever the probability/rate of a transition from m to a state satisfying the logical property φ is at least r.

In this paper we study three Markovian logics: the probabilistic Markovian logic (PML), the subprobabilistic Markovian logic (SML) and the general Markovian logic (GML); they are interpreted on PMPs, SMPs and GMPs respectively. Despite their apparent similarities, we have found it necessary to treat these logics separately because of subtle technical differences that make a uniform treatment difficult.

4.1 Syntax and Semantics

Definition 5. *Given a countable set $\mathcal{P}$ of atomic propositions, the grammars below define the sets of formulas $\mathcal{L}(\Pi)$ of probabilistic and subprobabilistic Markovian logic and $\mathcal{L}(\Delta)$ of general Markovian logic*

$$\mathcal{L}(\Pi): \quad \varphi ::= p \mid \neg\varphi \mid \varphi \wedge \varphi \mid L_r\varphi, \quad \text{for arbitrary } p \in \mathcal{P} \text{ and } r \in \mathbb{Q}_0$$

$$\mathcal{L}(\Delta): \quad \varphi ::= p \mid \neg\varphi \mid \varphi \wedge \varphi \mid L_r\varphi, \quad \text{for arbitrary } p \in \mathcal{P} \text{ and } r \in \mathbb{Q}^+$$

For each of these logics we assume that the usual boolean operators $\top, \bot, \vee, \to$ are available as derived constructs as well as the additional derived operator

$$L_{r_1 \cdots r_n}\varphi = L_{r_1} \cdots L_{r_n}\varphi$$

defined for $r_1, \ldots, r_n \in \mathbb{Q}_0$ for $\mathcal{L}(\Pi)$ and for $r_1, \ldots, r_n \in \mathbb{R}^+$ for $\mathcal{L}(\Delta)$.

To differentiate the probabilistic and the subprobabilistic logics, which have the same syntax but different semantics, we denote in what follows by $\mathcal{L}(\Pi^*)$ the logic interpreted on subprobabilistic distributions and we use $\mathcal{L}(\Pi)$ to refer to the logic interpreted on probabilistic distributions.

In what follows we define *en masse* the semantics for three logics using a generic $\mathcal{L}$ that ranges over the set $\{\mathcal{L}(\Pi), \mathcal{L}(\Pi^*), \mathcal{L}(\Delta)\}$. However, each of the following concepts has to be properly interpreted in each case. Let $\mathcal{M} = (M, \Sigma, \theta)$ be a PMP when we consider $\mathcal{L} = \mathcal{L}(\Pi)$, an SMP when we consider $\mathcal{L} = \mathcal{L}(\Pi^*)$ and an GMP when we consider $\mathcal{L} = \mathcal{L}(\Delta)$. Let $m \in M$ be an arbitrary state and $i : M \to 2^P$ an arbitrary interpretation function for the atomic propositions. The semantics of the three logics is defined as follows.

- $\mathcal{M}, m, i \vDash p$ iff $p \in i(m)$,
- $\mathcal{M}, m, i \vDash \varphi \wedge \psi$ iff $\mathcal{M}, m, i \vDash \varphi$ and $\mathcal{M}, m, i \vDash \psi$,
- $\mathcal{M}, m, i \vDash \neg\varphi$ iff not $\mathcal{M}, m, i \vDash \varphi$.
- $\mathcal{M}, m, i \vDash L_r\varphi$ iff $\theta(m)(\llbracket\varphi\rrbracket_{\mathcal{M}}^i) \geq r$,
  where $\llbracket\varphi\rrbracket_{\mathcal{M}}^i = \{m \in M \mid \mathcal{M}, m, i \vDash \varphi\}$.

For the last clause to make sense, $\llbracket\varphi\rrbracket_{\mathcal{M}}^i$ must be measurable. This is guaranteed, for each of the three types of Markov process, by the fact that θ is a measurable mapping between the measurable space of states and the measurable space of probabilistic/subprobabilistic/general distributions (see e.g. [4] for a complete proof).

Given $\mathcal{M} = (M, \Sigma, \theta)$ and i, we say that $m \in M$ *satisfies* φ if $\mathcal{M}, m, i \vDash \varphi$. We write $\mathcal{M}, m, i \nvDash \varphi$ if it is not the case that $\mathcal{M}, m, i \vDash \varphi$; and we write $\mathcal{M}, m, i \vDash \Phi$ if $\mathcal{M}, m, i \vDash \varphi$ for all $\varphi \in \Phi$. We write $\Phi \vDash \varphi$ if $\mathcal{M}, m, i \vDash \varphi$ whenever $\mathcal{M}, m, i \vDash \Phi$. A formula or set of formulas is *satisfiable* if there exist an MP $\mathcal{M}$, an interpretation function i for M and $m \in \mathrm{supp}(\mathcal{M})$ that satisfies it. We say that φ is *valid* and write $\vDash \varphi$, if $\neg\varphi$ is not satisfiable.

In what follows, when we have to differentiate between the three semantics, we will use indexes: $\vDash_{\Pi}$ will be used for PML, $\vDash_{\Pi^*}$ for SML and $\vDash_{\Delta}$ for GML.

4.2   Hilbert-Style Axiomatizations

We now present Hilbert-style axiomatic systems for the three logics. These axiomatic systems are meant to include the axioms of propositional logic; we do not write propositional axioms explicitly. In the next section we prove that these system are *strongly complete* for their semantics, meaning that an arbitrary formula φ can be proven from an arbitrary set Φ of formulae if and only if the models of Φ are also models of φ.

As we did for the semantics, we introduce the concepts related to the provability *en masse*. However, they have a specific meaning for each logic and depend directly of each particular provability relation.

As usual, for an arbitrary formula φ, $\vdash \varphi$ denotes the fact that φ is an axiom or a theorem in the system. If Φ is a set of formulas, we write $\Phi \vdash \varphi$ and say that Φ *derives* φ if φ is provable from the axioms and the extra assumptions Φ; we implicitly assume that the provability relation is adapted for the infinitary proofs allowed by the axiomatic systems. A formula or set of formulas is *consistent* if it cannot derive $\bot$. We say that Φ is *maximally consistent* if it is consistent and it has no proper consistent extensions.

When we have to differentiate between the three provability relations, we will use indexes: $\vdash_\Pi$ will be used for PML, $\vdash_{\Pi^*}$ for SML and $\vdash_\Lambda$ for GML.

The axiomatic system of PML is listed in Table 1. The axioms and the rules are stated for arbitrary $\varphi, \psi \in \mathcal{L}$ and arbitrary $r, s, r_1, .., r_k \in \mathbb{Q}_0$ for $k \geq 0$.

Table 1. The axioms of $\mathcal{L}(\Pi)$

(A1): $\vdash_\Pi L_0\varphi$

(A2): $\vdash_\Pi L_r\top$

(A3): $\vdash_\Pi L_r\varphi \rightarrow \neg L_s\neg\varphi, \quad r + s > 1$

(A4): $\vdash_\Pi L_r(\varphi \wedge \psi) \wedge L_s(\varphi \wedge \neg\psi) \rightarrow L_{r+s}\varphi, \quad r + s \leq 1$

(A5): $\vdash_\Pi \neg L_r(\varphi \wedge \psi) \wedge \neg L_s(\varphi \wedge \neg\psi) \rightarrow \neg L_{r+s}\varphi, \quad r + s \leq 1$

(R1): $\dfrac{\vdash_\Pi \varphi \rightarrow \psi}{\vdash_\Pi L_r\varphi \rightarrow L_r\psi}$

(R2): $\{L_{r_1\cdots r_k r}\psi \mid r < s\} \vdash_\Pi L_{r_1\cdots r_k s}\psi$

A similar axiomatic system was studied in [19,20]. The novelty of our axiomatization is the rule (R2). In [19], for proving the strongly completeness of the axiomatic system, Lindenbaum's lemma is assumed as a meta-axiom and instead of (R2), the rules in Table 2 are used, stated for arbitrary $\varphi \in \mathcal{L}(\Pi)$ and arbitrary set $\Phi \subseteq \mathcal{L}(\Pi)$ closed under conjunction, where $L_r\Phi = \{L_r\psi \mid \psi \in \Phi\}$.

Table 2. Zhou's rules of $\mathcal{L}(\Pi)$

(R2'): $\{L_r\psi \mid r < s\} \vdash_\Pi L_s\psi$

(R2"): $\dfrac{\Phi \vdash_\Pi \varphi}{L_r\Phi \vdash_\Pi L_r\varphi}$

While (R2') is an instance of (R2), (R2") is much stronger and it has an uncountable set of instances. This makes the proof of the existence of the maximally consistent sets difficult. In our case that fact that (R2) has countably many instances means that the existence of maximally consistent sets is guaranteed by the Rasiowa-Sikorski lemma.

Before introducing the axiomatization of SML, notice that (A3) guarantees that the semantics must use distributions bounded by 1 while (A2) guarantees that these are probability distributions.

The axiomatic system of SML is listed in Table 3. The axioms and the rules are stated for arbitrary $\varphi, \psi \in \mathcal{L}$ and arbitrary $r, s, r_1, .., r_k \in \mathbb{Q}_0$ for $k \geq 0$.

Notice the difference between this system and the previous one. For SML the axiom (A2) is not sound anymore, since for a subprobability distribution the measure of the entire space can be smaller than 1. However, (A2') which replaces (A2) is also sound for PML.

The axiomatic system of GML is listed in Table 4. The axioms and the rules are stated for arbitrary $\varphi, \psi \in \mathcal{L}$ and arbitrary $r, s, r_1, .., r_k \in \mathbb{Q}^+$ for $k \geq 0$.

The difference with respect to the axiomatic system of SML is that the axiom (A3) is not sound anymore. Moreover, the indexes of the modal operator can be any positive rational, meaning that these axioms have more instances than the corresponding ones in the other two systems. Since the semantics does not allow infinite measures, rule (R3)

Table 3. The axioms of $\mathcal{L}(\Pi^*)$

(A1): $\vdash_{\Pi^*} L_0\varphi$
(A2'): $\vdash_{\Pi^*} L_r\bot \to \bot$
(A3): $\vdash_{\Pi^*} L_r\varphi \to \neg L_s\neg\varphi,\ \ r+s>1$
(A4): $\vdash_{\Pi^*} L_r(\varphi \wedge \psi) \wedge L_s(\varphi \wedge \neg\psi) \to L_{r+s}\varphi,\ \ r+s\le 1$
(A5): $\vdash_{\Pi^*} \neg L_r(\varphi \wedge \psi) \wedge \neg L_s(\varphi \wedge \neg\psi) \to \neg L_{r+s}\varphi,\ \ r+s\le 1$
(R1): $\dfrac{\vdash_{\Pi^*} \varphi \to \psi}{\vdash_{\Pi^*} L_r\varphi \to L_r\psi}$
(R2): $\{L_{r_1\cdots r_k r}\psi \mid r < s\} \vdash_{\Pi^*} L_{r_1\cdots r_k s}\psi$

Table 4. The axioms of $\mathcal{L}(\Delta)$

(A1): $\vdash_\Delta L_0\varphi$
(A2'): $\vdash_\Delta L_r\bot \to \bot$
(A4): $\vdash_\Delta L_r(\varphi \wedge \psi) \wedge L_s(\varphi \wedge \neg\psi) \to L_{r+s}\varphi,\ \ r+s\le 1$
(A5): $\vdash_\Delta \neg L_r(\varphi \wedge \psi) \wedge \neg L_s(\varphi \wedge \neg\psi) \to \neg L_{r+s}\varphi,\ \ r+s\le 1$
(R1): $\dfrac{\vdash_\Delta \varphi \to \psi}{\vdash_\Delta L_r\varphi \to L_r\psi}$
(R2): $\{L_{r_1\cdots r_k r}\psi \mid r < s\} \vdash_\Delta L_{r_1\cdots r_k s}\psi$
(R3): $\{L_{r_1\cdots r_k r}\psi \mid r \in \mathbb{Q}^+\} \vdash_\Delta L_{r_1\cdots r_k}\bot$

guarantees that divergent sequences of modalities prefixing some formula generates an inconsistent set of formulas.

Strong completeness for this logic was proven in [15] where, as in the case of Zhou's completeness for PML, Lindenbaum's lemma is postulated and the rules (R2') and (R2'') are involved.

The next theorem states the soundness of the axioms of the three logics for their corresponding semantics

Theorem 6. *[Soundness]*

1. *The axiomatization of PML is sound for the PMPs semantics, i.e.,*

   for any $\varphi \in \mathcal{L}(\Pi), \vdash_\Pi \varphi$ implies $\vDash_\Pi \varphi$.

2. *The axiomatization of SML is sound for the SMPs semantics, i.e.,*

   for any $\varphi \in \mathcal{L}(\Pi^*), \vdash_{\Pi^*} \varphi$ implies $\vDash_{\Pi^*} \varphi$.

3. *The axiomatization of GML is sound for the GMPs semantics, i.e.,*

   for any $\varphi \in \mathcal{L}(\Delta), \vdash_\Delta \varphi$ implies $\vDash_\Delta \varphi$.

4.3 Canonical Models

In this section we construct canonical models for the three logics. The canonical model for a logic $\mathcal{L} \in \{\mathcal{L}(\Pi), \mathcal{L}(\Pi^*), \mathcal{L}(\Delta)\}$ is a Markov processes $\mathcal{M}_\mathcal{L} = (\mathcal{U}_\mathcal{L}, \Sigma_\mathcal{L}, \theta_\mathcal{L})$ having the set $\mathcal{U}_\mathcal{L}$ of $\mathcal{L}$-maximally consistent sets of formulas as the state space and satisfying the property that for any $\varphi \in \mathcal{L}$ and $u \in \mathcal{U}_{\mathcal{L}}$, $\mathcal{M}_{\mathcal{L}}, u, i_{\mathcal{L}} \vDash \varphi$ iff $\varphi \in u$, where $i_{\mathcal{L}}$ is an appropriate interpretation function.

In order to complete such a construction, we have to:

- prove that $\mathcal{U}_{\mathcal{L}} \neq \varnothing$ for each $\mathcal{L} \in \{\mathcal{L}(\Pi), \mathcal{L}(\Pi^*), \mathcal{L}(\Delta)\}$;
- define $\Sigma_{\mathcal{L}}$ such that $(\mathcal{U}_{\mathcal{L}}, \Sigma_{\mathcal{L}})$ is an analytic space;
- define a measure $\theta_{\mathcal{L}}$ on $(\mathcal{U}_{\mathcal{L}}, \Sigma_{\mathcal{L}})$;
- define an interpretation function $i_{\mathcal{L}}$ such that for any $\varphi \in \mathcal{L}$, $[\![\varphi]\!]_{\mathcal{M}_{\mathcal{L}}}^{i_{\mathcal{L}}} \in \Sigma_{\mathcal{L}}$;
- and prove the Truth Lemma stating that $\mathcal{M}_{\mathcal{L}}, u, i_{\mathcal{L}} \vDash \varphi$ iff $\varphi \in u$.

Lemma 7. *For $\mathcal{L} \in \{\mathcal{L}(\Pi), \mathcal{L}(\Pi^*), \mathcal{L}(\Delta)\}$ with the proof systems previously defined, the set $\mathcal{U}_{\mathcal{L}}$ of $\mathcal{L}$-maximally consistent sets is nonempty.*

Proof. Note that in each case $\mathcal{L}$ forms a boolean algebra and the instances of all the axioms and rules define a countable family of subsets of $\mathcal{L}$, each member of which has the meet in $\mathcal{L}$; in particular the instances of (R2) define the subsets $\{L_{r_1 \cdots r_k r} \psi \mid r < s\}$ of $\mathcal{L}$ each having the meet $L_{r_1 \cdots r_k s} \psi \in \mathcal{L}$. Observe also that a set $u \subseteq \mathcal{L}$ is a $\mathcal{L}$-maximally consistent set iff it is a boolean ultrafilter that respects all the instances of the axioms of $\mathcal{L}$. Consequently, the Rasiowa-Sikorski Lemma guarantees that $\mathcal{U}_{\mathcal{L}} \neq \varnothing$.

Let $(\![\mathcal{L}]\!) = \{(\![\varphi]\!) \mid \varphi \in \mathcal{L}\}$, where $(\![\varphi]\!) = \{u \in \mathcal{U}_{\mathcal{L}} \mid \varphi \in u\}$. Using this, we define $\Sigma_{\mathcal{L}} = \sigma((\![\mathcal{L}]\!))$. The space $(\mathcal{U}_{\mathcal{L}}, \Sigma_{\mathcal{L}})$ is an analytic space for each $\mathcal{L} \in \{\mathcal{L}(\Pi), \mathcal{L}(\Pi^*), \mathcal{L}(\Delta)\}$. The proof for the probabilistic case —which is decidedly non-trivial— can be found in [13] and works similarly for the other two cases.

The next step in our construction is to define an appropriate measure $\theta_{\mathcal{L}}$ on $(\mathcal{U}_{\mathcal{L}}, \Sigma_{\mathcal{L}})$. To do this we prove the following lemma.

Lemma 8. *1. For arbitrary $u \in \mathcal{U}_{\mathcal{L}(\Pi)}$ and $\varphi \in \mathcal{L}(\Pi)$, or $u \in \mathcal{U}_{\mathcal{L}(\Pi^*)}$ and $\varphi \in \mathcal{L}(\Pi^*)$,*

$$x_u^\varphi = \sup\{r \in \mathbb{Q}_0 \mid L_r \varphi \in u\} = \inf\{r \in \mathbb{Q}_0 \mid \neg L_r \varphi \in u\}.$$

Moreover, if $x_u^\varphi \in \mathbb{Q}$, then $L_{x_u^\varphi} \varphi \in u$.

2. For arbitrary $u \in \mathcal{U}_{\mathcal{L}(\Delta)}$ and $\varphi \in \mathcal{L}(\Delta)$,

$$x_u^\varphi = \sup\{r \in \mathbb{Q}^+ \mid L_r \varphi \in u\} = \inf\{r \in \mathbb{Q}^+ \mid \neg L_r \varphi \in u\} \in \mathbb{R}^+.$$

Moreover, if $x_u^\varphi \in \mathbb{Q}$, then $L_{x_u^\varphi} \varphi \in u$.

The previous lemma allows us to define, for each $\mathcal{L} \in \{\mathcal{L}(\Pi), \mathcal{L}(\Pi^*), \mathcal{L}(\Delta)\}$ and arbitrary $u \in \mathcal{U}_{\mathcal{L}}, \varphi \in \mathcal{L}$,

$$\theta_{\mathcal{L}}(u)((\![\varphi]\!)) = \sup\{r \in \mathbb{Q}^+ \mid L_r \varphi \in u\}.$$

Obviously, $\theta_{\mathcal{L}}(u)$ is a set function defined on the field $(\![\mathcal{L}]\!)$ and Theorem 1 ensures us that it can be uniquely extended to a measure on $\Sigma_{\mathcal{L}}$ if it is finitely additive and countable subadditive on $(\![\mathcal{L}]\!)$. This is what we prove next.

Lemma 9. *For all $u \in \mathcal{U}_{\mathcal{L}}$, the function $\theta_{\mathcal{L}}(u)$ is finitely additive.*

Now we prove that the function is also countable subadditive and this is a central result of the paper where we make use of (R2). In related papers, to prove a similar result a so-called countable additivity axiom was used. This is an infinitary axiom with uncountable instances [11,21,15].

The main technical lemma that lies at the heart of the construction is proved in our previous paper on Stone duality [13] for the probabilistic case. The proof can be similarly done for the other two cases.

Lemma 10. *For $u \in \mathcal{U}_{\mathcal{L}}$, the function $\theta_{\mathcal{L}}(u)$ is countably subadditive.*

The previous lemmas guarantees that $\theta_{\mathcal{L}}$ can be extended to a measure on $(\mathcal{U}_{\mathcal{L}}, \Sigma_{\mathcal{L}})$. From the construction we also obtain that $\theta_{\mathcal{L}(\Pi)}$ is a probabilistic measure and $\theta_{\mathcal{L}(\Pi^*)}$ is a subprobabilistic measure.

Theorem 11 (Canonical models)

1. $\mathcal{M}_{\mathcal{L}(\Pi)} = (\mathcal{U}_{\mathcal{L}(\Pi)}, \Sigma_{\mathcal{L}(\Pi)})$ *is a probabilistic Markov process;*
2. $\mathcal{M}_{\mathcal{L}(\Pi^*)} = (\mathcal{U}_{\mathcal{L}(\Pi^*)}, \Sigma_{\mathcal{L}(\Pi^*)})$ *is a subprobabilistic Markov process;*
3. $\mathcal{M}_{\mathcal{L}(\Delta)} = (\mathcal{U}_{\mathcal{L}(\Delta)}, \Sigma_{\mathcal{L}(\Delta)})$ *is a general Markov process.*

Proof. In the generic case we only need to verify that $\theta_{\mathcal{L}}$ is a measurable function. Let $\varphi \in \mathcal{L}$, and $r \in [0,1]$ for $\mathcal{L}(\Pi)$ and $\mathcal{L}(\Pi^*)$ and $r \in \mathbb{R}^+$ for $\mathcal{L}(\Delta)$. Consider $(r_i)_i \subseteq \mathbb{Q}$ an increasing sequence with supremum r. Let $X = \{\mu \in \Pi(\mathcal{U}_{\mathcal{L}(\Pi)}, \Sigma_{\mathcal{L}(\Pi)}) \mid \mu(\langle\!\langle\varphi\rangle\!\rangle) \geq r\}$ for $\mathcal{L}(\Pi)$, $X = \{\mu \in \Pi^*(\mathcal{U}_{\mathcal{L}(\Pi^*)}, \Sigma_{\mathcal{L}(\Pi^*)}) \mid \mu(\langle\!\langle\varphi\rangle\!\rangle) \geq r\}$ for $\mathcal{L}(\Pi^*)$ and $X = \{\mu \in \Delta(\mathcal{U}_{\mathcal{L}(\Delta)}, \Sigma_{\mathcal{L}(\Delta)}) \mid \mu(\langle\!\langle\varphi\rangle\!\rangle) \geq r\}$ for $\mathcal{L}(\Delta)$. It suffices to prove, in each case, that $\theta_{\mathcal{L}}^{-1}(X) \in \Sigma_{\mathcal{L}}$. But

$$\theta_{\mathcal{L}}^{-1}(X) = \{u \in \mathcal{U}_{\mathcal{L}} \mid \theta(u)(\langle\!\langle\varphi\rangle\!\rangle) \geq r\} = \bigcap_i \{u \in \mathcal{U}_{\mathcal{L}} \mid \theta_{\mathcal{L}}(u)(\langle\!\langle\varphi\rangle\!\rangle) \geq r_i\} = \bigcap_i \langle\!\langle L_{r_i}\varphi\rangle\!\rangle \in \Sigma_{\mathcal{L}}.$$

We define an interpretation function $i_{\mathcal{L}}$ for arbitrary $u \in \mathcal{U}_{\mathcal{L}}$ by $i_{\mathcal{L}}(u) = u \cap \mathcal{P}$. Now we are ready to prove the Truth Lemma.

Lemma 12 (Truth Lemma). *For $\mathcal{L} \in \{\mathcal{L}(\Pi), \mathcal{L}(\Pi^*), \mathcal{L}(\Delta)\}$, $\Phi \subseteq \mathcal{L}$ and $u \in \mathcal{U}_{\mathcal{L}}$,*

$$\mathcal{M}_{\mathcal{L}}, u, i_{\mathcal{L}} \vDash \Phi \text{ iff } \Phi \subseteq u.$$

Proof. It is sufficient to prove inductively that for any $\varphi \in \mathcal{L}$, $\mathcal{M}_{\mathcal{L}}, u, i_{\mathcal{L}} \vDash \varphi$ iff $\varphi \in u$. The case $\varphi \in \mathcal{P}$ and the boolean cases are trivial.

The case $\varphi = L_r\psi$: ($\Longrightarrow$) Suppose that $\mathcal{M}_{\mathcal{L}}, u, i_{\mathcal{L}} \vDash \varphi$ and $\varphi \notin u$. Hence $\neg\varphi \in u$. Let $x_u^\varphi = \inf\{r \in \mathbb{Q} \mid \neg L_r\psi \in u\}$. Then, from $\neg L_r\psi \in u$, we obtain $r \geq x_u^\varphi$. But $\mathcal{M}_{\mathcal{L}}, u, i_{\mathcal{L}} \vDash L_r\psi$ is equivalent with $\theta_{\mathcal{L}}(u)(\langle\!\langle\psi\rangle\!\rangle) \geq r$, i.e. $x_u^\varphi \geq r$. Hence, $x_u^\varphi = r \in \mathbb{Q}$ and Lemma 8 implies $L_{x_u^\varphi}\varphi \in u$, i.e., $\varphi \in u$ - contradiction.

($\Longleftarrow$) If $L_r\psi \in u$, then $r \leq x_u^\varphi$, i.e., $r \leq \theta_{\mathcal{L}}(u)(\langle\!\langle\psi\rangle\!\rangle)$. Hence, $\mathcal{M}_{\mathcal{L}}, u, i_{\mathcal{L}} \vDash L_r\psi$.

The Truth Lemma allows us to prove strong completeness for all three logics.

Theorem 13 (Completeness). *For $\mathcal{L} \in \{\mathcal{L}(\Pi), \mathcal{L}(\Pi^*), \mathcal{L}(\Delta)\}$, $\Phi \subseteq \mathcal{L}$ and $\varphi \in \mathcal{L}$*

$$\Phi \vDash \varphi \;\; \text{iff} \;\; \Phi \vdash \varphi.$$

Proof. ($\Longleftarrow$) This is a consequence of soundness, Theorem 6.
($\Longrightarrow$) If Φ is inconsistent, the statement is trivially true. Suppose that Φ is consistent, and let $u \in \mathcal{U}_{\mathcal{L}}$ be an arbitrary maximally consistent set. We have that $\Phi \subseteq u$ iff $\mathcal{U}_{\mathcal{L}}, u, i_{\mathcal{L}} \vDash \Phi$ (from Truth Lemma). But if $\mathcal{U}_{\mathcal{L}}, u, i_{\mathcal{L}} \vDash \Phi$, since $\Phi \vDash \varphi$, we obtain that $\mathcal{U}_{\mathcal{L}}, u, i_{\mathcal{L}} \vDash \varphi$. Applying again the truth lemma we get $\varphi \in u$. Consequently, for an arbitrary maximally-consistent set $u \in \mathcal{U}_{\mathcal{L}}$, $\Phi \subseteq u$ implies $\varphi \in u$. Hence, $\Phi \vdash \varphi$.

5   Conclusions and Related Work

The most closely related work to ours is the work of Goldblatt [10] on the role of the Baire category theorem in completeness proofs, and his work on deduction systems for coalgebras [11]. The main difference between his work and ours is that we have replaced the Countable additivity Rule (CAR) that he uses, with a different infinitary axiom that has only countably many instances. Goldblatt uses CAR in order to show countable additivity of the measures that he defines; this is where we have been able to use of the Rasiowa–Sikorski lemma. As far as we know this is a new idea. Furthermore, Goldblatt's results are contingent on the assumption that consistent sets can be expanded to maximally consistent sets; we have essentially proved this fact for our logics.

Regarding the completeness proofs for Markovian logics, the results for probabilistic case were proved by Zhou in [19] and for the general case by Mardare-Cardelli-Larsen in [15]. In these papers the strong completeness is solved using CAR.

In this paper, we have used some of the results of our earlier paper [13] to show that we can obtain strong completeness theorems for three types of Markov processes and their related logics. The main technical lemmas about measure theory are in [13] but the canonical model constructions which use those facts are in the present paper. That paper focussed on algebra and duality whereas the present paper is primarily about logic and can be read independently.

A very tempting future research project is to extend these completeness theorems to the entire class of systems described as coalgebras of polynomial functors described by Goldblatt [11]. It is possible that the results of Pattinson and Schröder [18] will be useful for this.

Though the focus of the present paper has been on probabilistic systems and Markovian logics, the techniques may well apply to any non-compact modal logic. We are investigating whether there is a general way of introducing an infinitary axiom that will allow us to mimic the techniques of the present paper.

Acknowledgments. We would like to thank Ernst-Erich Doberkat, Rob Goldblatt, Jean Goubault-Larrecq, Larry Moss and Chunlai Zhou for helpful discussions.

Panangaden's research was supported by an NSERC grant. Mardare's research was supported by VKR Center of Excellence MT-LAB and by the Sino-Danish Basic Research Center IDEA4CPS.

References

1. Aumann, R.: Interactive epistemology I: knowledge. International Journal of Game Theory 28, 263–300 (1999)
2. Aumann, R.: Interactive epistemology II: probability. International Journal of Game Theory 28, 301–314 (1999)
3. Billingsley, P.: Probability and Measure. Wiley-Interscience (1995)
4. Cardelli, L., Larsen, K.G., Mardare, R.: Continuous markovian logic - from complete axiomatization to the metric space of formulas. In: CSL, pp. 144–158 (2011)
5. Desharnais, J., Edalat, A., Panangaden, P.: A logical characterization of bisimulation for labelled Markov processes. In: Proceedings of the 13th IEEE Symposium On Logic In Computer Science, Indianapolis, pp. 478–489. IEEE Press (June 1998)
6. Desharnais, J., Edalat, A., Panangaden, P.: Bisimulation for labeled Markov processes. Information and Computation 179(2), 163–193 (2002)
7. Doberkat, E.-E.: Stochastic Relations. Foundations for Markov Transition Systems. Chapman and Hall, New York (2007)
8. Dudley, R.M.: Real Analysis and Probability. Wadsworth and Brookes/Cole (1989)
9. Fagin, R., Halpern, J.Y.: Reasoning about knowledge and probability. Journal of the ACM 41(2), 340–367 (1994)
10. Goldblatt, R.: On the role of the Baire category theorem in the foundations of logic. Journal of Symbolic Logic, 412–422 (1985)
11. Goldblatt, R.: Deduction systems for coalgebras over measurable spaces. Journal of Logic and Computation 20(5), 1069–1100 (2010)
12. Heifetz, A., Mongin, P.: Probability logic for type spaces. Games and Economic Behavior 35(1-2), 31–53 (2001)
13. Kozen, D., Larsen, K.G., Mardare, R., Panangaden, P.: Stone duality for markov processes. In: Proceedings of the 28th Annual IEEE Symposium on Logic in Computer Science: LICS 2013. IEEE Computer Society (2013)
14. Larsen, K.G., Skou, A.: Bisimulation through probablistic testing. Information and Computation 94, 1–28 (1991)
15. Mardare, R., Cardelli, L., Larsen, K.G.: Continuous markovian logics - axiomatization and quantified metatheory. Logical Methods in Computer Science 8(4) (2012)
16. Panangaden, P.: Labelled Markov Processes. Imperial College Press (2009)
17. Rasiowa, H., Sikorski, R.: A proof of the completeness theorem of gödel. Fund. Math. 37, 193–200 (1950)
18. Schröder, L., Pattinson, D.: Modular algorithms for heterogeneous modal logics. In: Arge, L., Cachin, C., Jurdziński, T., Tarlecki, A. (eds.) ICALP 2007. LNCS, vol. 4596, pp. 459–471. Springer, Heidelberg (2007)
19. Zhou, C.: A complete deductive system for probability logic with application to Harsanyi type spaces. PhD thesis, Indiana University (2007)
20. Zhou, C.: Probability logic of finitely additive beliefs. J. Logic, Language and Information 19(3), 247–282 (2010)
21. Zhou, C., Ying, M.: Approximating Markov processes through filtration. Theoretical Computer Science 446, 75–97 (2012)

Arithmetic Branching Programs with Memory

Stefan Mengel*

Institute of Mathematics
University of Paderborn
D-33098 Paderborn, Germany

Abstract. We extend the well known characterization of the arithmetic circuit class VP_{ws} as the class of polynomials computed by polynomial size arithmetic branching programs to other complexity classes. In order to do so we add additional memory to the computation of branching programs to make them more expressive. We show that allowing different types of memory in branching programs increases the computational power even for constant width programs. In particular, this leads to very natural and robust characterizations of VP and VNP by branching programs with memory.

1 Introduction

Arithmetic Branching Programs (ABPs) are a well studied model of computation in algebraic complexity: They were already used in the VNP-completeness proof of the permanent by Valiant [14] and have since then contributed to the understanding of arithmetic circuit complexity (see e.g. [10,7]). The computational power of ABPs is well understood: They are equivalent to both skew and weakly skew arithmetic circuits and thus capture the determinant, matrix power and other natural problems from linear algebra [12,8]. The complexity of bounded width ABPs is also well understood: In analogy to Barrington's Theorem [1], Ben-Or and Cleve [2] proved that polynomial size ABPs of bounded width are equivalent to arithmetic formulas.

We modify ABPs by giving them memory during their computations and ask how this changes their computational power. There are several different motivations for doing this: We define branching programs with stacks, that are an adaption of the nondeterministic auxiliary pushdown automaton (NAuxPDA) model to the arithmetic circuit model. The NAuxPDA-characterization of LOGCFL has been very successful in the study of this class and has contributed greatly to its understanding. We give a characterization of VP—a class that is well known for its apparent lack of natural non-circuit characterizations. Our characterization also has some similarity to results in the Boolean setting in which graph connectivity problems on edge-labeled graphs that are similar to our ABPs with stacks were shown to be complete for LOGCFL [11,15]. One motivation for adapting these results to the arithmetic circuit setting is the hope that one can apply techniques from the NAuxPDA setting to arithmetic circuits.

---

* Partially supported by DFG grants BU 1371/2-2 and BU 1371/3-1.

K. Chatterjee and J. Sgall (Eds.): MFCS 2013, LNCS 8087, pp. 667–678, 2013.

Another motivation is that our modified branching programs in different settings give various very similar characterizations of different arithmetic circuit classes. This allows us to give a new perspective on problems like VP vs. VP_{ws}, VP vs. VNP that are classical question from arithmetic circuit complexity. This is similar to the motivation that Kintali [6] had for studying similar graph connectivity problems for the Boolean setting.

Finally, all modifications we make to ABPs are straightforward and natural. The basic question is the following: ABPs are in a certain sense a memoryless model of computation. At each point of time during the computation we do not have any information about the history of the computation sofar apart from the state we are in. So what happens if we allow memory during the computation? Intuitively, the computational power should increase, and we will see that it indeed does (under standard complexity assumptions of course). How do different types of memory compare? What is the role of the width of the branching programs if we allow memory? In this paper we will answer several of these questions.

The structure of the paper is a follows: After some preliminaries we start off with ABPs that may use a stack during their computation. We show that they characterize VP even when restricted to bounded width. Next we consider ABPs with random access memory and show that they characterize VNP. Due to lack of space, several proofs are omitted and can be found in the full version of this paper [9].

2   Preliminaries

2.1   Arithmetic Circuits

We briefly recall the relevant definitions from arithmetic circuit complexity. A more thorough introduction into arithmetic circuit classes can be found in the book by Bürgisser [5]. Newer insights into the nature of VP and especially of VP_{ws} are presented in the excellent paper of Malod and Portier [8].

An *arithmetic circuit* over a field $\mathbb{F}$ is a labeled directed acyclic graph (DAG) consisting of vertices or *gates* with indegree or *fanin* 0 or 2. The gates with fanin 0 are called input gates and are labeled with constants from $\mathbb{F}$ or variables $X_1, X_2, \ldots$. The gates with fanin 2 are called computation gates and are labeled with $\times$ or $+$.

The polynomial computed by an arithmetic circuit is defined in the obvious way: An input gates computes the value of its label, a computation gate computes the product or the sum of its childrens' values, respectively. We assume that a circuit has only one sink which we call the output gate. We say that the polynomial computed by the circuit is the polynomial computed by the output gate. The *size* of an arithmetic circuit is the number of gates. The *depth* of a circuit is the length of the longest path from an input gate to the output gate in the circuit.

We also consider circuits in which the $+$-gates may have unbounded fanin. We call these circuits *semi-unbounded circuits*. Observe that in semi-unbounded circuits ×-gates still have fanin 2. A circuit is called *multiplicatively disjoint* if for each ×-gate v the subcircuits that have the children of v as output-gates are disjoint. A circuit is called *skew*, if for all of its ×-gates one of the children is an input gate.

We call a sequence (f_n) of multivariate polynomials a family of polynomials or *polynomial family*. We say that a polynomial family is of polynomial degree, if there is a univariate polynomial p such that $\deg(f_n) \le p(n)$ for each n. VP is defined as the class of polynomial families of polynomial degree computed by families of polynomial size arithmetic circuits. We will use the following well known characterizations of VP.

Theorem 1. *([13,8]) Let (f_n) be a family of polynomials. The following statements are equivalent:*

1. $(f_n) \in$ VP
2. (f_n) *is computed by a family of multiplicatively disjoint polynomial size circuits.*
3. (f_n) *is computed by a family of semi-unbounded circuits of logarithmic depth and polynomial size.*

VP$_e$ is defined analogously to VP with the circuits restricted to trees. By a classical result of Brent [3], VP$_e$ can equivalently be defined as the class of polynomial families computed by arithmetic circuits of depth $O(\log(n))$. VP$_{ws}$ is defined as the class of families of polynomials computed by families of skew circuits of polynomial size[1]. Finally, a family (f_n) of polynomials is defined to be in VNP, if there is a family $(g_n) \in$ VP and a polynomial p such that $f_n(X) = \sum_{e \in \{0,1\}^{p(n)}} g_n(e, X)$ for all n where X denotes the vector $(X_1, \dots, X_{q(n)})$ for some polynomial q.

A polynomial f is called a *projection* of g (symbol: $f \le g$), if there are values $a_i \in \mathbb{F} \cup \{X_1, X_2, \dots\}$ such that $f(X) = g(a_1, \dots, a_q)$. A family (f_n) of polynomials is called a *p-projection* of (g_n) (symbol: $(f_n) \le_p (g_n)$), if there is a polynomial r such that $f_n \le g_{r(n)}$ for all n. As usual we say that (g_n) is hard for an arithmetic circuit class $\mathcal{C}$ if for every $(f_n) \in \mathcal{C}$ we have $(f_n) \le_p (g_n)$. If further $(g_n) \in \mathcal{C}$ we say that (g_n) is $\mathcal{C}$-complete.

The following criterion by Valiant [14] (see also [5, Prop. 2.20]) for containment in VNP is often helpful:

Lemma 1 (Valiant's criterion). *Let $\phi : \{0,1\}^* \to \mathbb{N}$ be a function in #P/poly. Then the family (f_n) of polynomials defined by $f_n = \sum_{e \in \{0,1\}^n} \phi(e) \prod_{i=1}^n X_i^{e_i}$ is in VNP.*

---

[1] The "ws" in VP$_{ws}$ stands for "weakly skew". The reason for this notation is that VP$_{ws}$ can also be defined by weakly skew circuits which, in contrast to what their name suggests, are equivalent to skew circuits [12,8]. Since we will not consider weakly skew circuits in this paper, we will not introduce them but nevertheless stick to the usual notation VP$_{ws}$ for the complexity class.

2.2   Arithmetic Branching Programs

The second common model of computation in arithmetic circuit complexity are arithmetic branching programs.

Definition 1. *An arithmetic branching program (ABP) G is a DAG with two vertices s and t and an edge labeling $w : E \to \mathbb{F} \cup \{X_1, X_2, \dots\}$. A path $P = v_1 v_2 \dots v_r$ in G has the weight $w(P) := \prod_{i=1}^{r-1} w(v_i v_{i+1})$. Let v and u be two vertices in G, then we define $f_{v,u} = \sum_P w(P)$, where the sum is over all v-u-paths P. The ABP G computes the polynomial $f_G = f_{s,t}$. The size of G is the number of vertices of G.*

Toda and Malod and Portier proved the following theorem:

Theorem 2. *([12,8]) We have $(f_n) \in \mathsf{VP}_{ws}$, iff (f_n) is computed by a family of polynomial size ABPs.*

Definition 2. *An ABP of width k is an ABP in which all vertices are organized into layers $L_i, i \in \mathbb{N}$, there are only edges from layer L_i to L_{i+1} and the number of vertices in each layer L_i is at most k.*

The computational power of ABPs of constant width was settled by Ben-Or and Cleve:

Theorem 3. *([2]) $(f_n) \in \mathsf{VP}_e$, iff (f_n) is computed by a family of polynomial size ABPs of constant width.*

3   Stack Branching Programs

3.1   Definition

Let S be a set called *symbol set*. For a symbol $s \in S$ we define two *stack operations*: $push(s)$ and $pop(s)$. Additionally we define the stack operation *nop* without any arguments. A *sequence of stack operations* on S is a sequence $op_1 op_2 \dots op_r$, where either $op_i = \bar{op}_i(s_i)$ for $\bar{op}_i \in \{push, pop\}$ and $s_i \in S$ or $op_i = nop$. *Realizable sequences* of stack operations are defined inductively:

- The empty sequence is realizable.
- If P is a realizable sequence of stack operations, then the sequence $push(s)$ $P pop(s)$ is realizable for all $s \in S$. Furthermore, $nop\, P$ and $P\, nop$ are realizable sequences.
- If P and Q are realizable sequences of stack operations, then PQ is a realizable sequence.

Definition 3. *A stack branching program (SBP) G is an ABP with an additional edge labeling $\sigma : E \to \{op(s) \mid op \in \{push, pop\}, s \in S\} \cup \{nop\}$. A path $P = v_1 v_2 \dots v_r$ in G has the sequence of stack operations $\sigma(P) := \sigma(v_1 v_2) \sigma(v_2 v_3) \dots \sigma(v_{r-1} v_r)$. If $\sigma(P)$ is realizable we call P a stack-realizable path. The SBP G computes the polynomial $f_G = \sum_P w(P)$, where the sum is over all stack-realizable s-t-paths P.*

It is helpful to interpret the stack operations as operations on a real stack that happen along a path through G. On an edge uv with the stack operation $\sigma(uv) = push(s)$ we simply push s onto the stack. If uv has the stack operation $\sigma(uv) = pop(s)$ we pop the top symbol of the stack. If it is s we continue the path, but if it is different from s the path is not stack realizable and we abort it. nop stands for "no operation" and thus as this name suggests the stack is not changed on edges labelled with nop. Realizable paths are exactly the paths on which we can go from s to t in this way without aborting while starting and ending with an empty stack.

To ease notation we sometimes call edges e with $\sigma(e) = push(s)$ for an $s \in S$ simply $push$-edges. pop-edges and nop-edges are defined in the obvious analogous way.

It will sometimes be convenient to consider only SBPs that have no nop-edges. The following easy proposition shows that this is not a restriction.

Proposition 1. *Let G be an SBP of size s. There is an SBP G' of size $O(s^2)$ such that $f_G = f_{G'}$ and G' does not contain any nop-edges. If G is layered with width k, then G' is layered, too, and has width at most k^2.*

3.2   Characterizing VP

In this section we show that stack branching programs of polynomial size characterize VP.

Theorem 4. *$(f_n) \in$ VP, iff (f_n) is computed by a family of polynomial size SBPs.*

We prove the two directions of Theorem 4 in two steps.

Lemma 2. *If (f_n) is computed by a family of polynomial size SBPs, then $(f_n) \in$ VP.*

Proof. Let (G_n) be a family of SBPs computing (f_n), of size at most $p(n)$ for a polynomial p. Observe that $\deg(f_n) \le p(n)$, so we only have to show that we can compute the f_n by polynomial size circuits C_n.

Let $G = G_n$ be an SBP with m vertices, source s and sink t. The construction of $C = C_n$ uses the following basic observation: Every stack-realizable path P of length i between two vertices v and u can be uniquely decomposed in the following way. There are vertices $a, b, c \in V(G)$ and a symbol $s \in S$ such that there are edges va and bc with $\sigma(va) = push(s)$ and $\sigma(bc) = pop(s)$. Furthermore there are stack-realizable paths P_{ab} from a to b and P_{cu} from c to u such that $length(P_{ab}) + length(P_{cu}) = i - 2$ and $P = vaP_{ab}bcP_{cu}$. The paths P_{ab} and P_{cu} may be empty. We define $w(u, v, i) := \sum_P w(P)$ where the sum is over all stack-realizable s-t-paths of length i.

We now show that the values $w(v, u, i)$ can be computed efficiently with a straightforward dynamic programming approach. First observe that $w(v, u, i) = 0$ for odd i. For $i = 0$ we set $w(v, u, 0) = 0$ for $v \neq u$ and $w(v, v, 0) = 1$. For even $i > 0$ we get

$$w(v, u, i) = \sum_{a,b,c,j,s} w(v, a)w(a, b, j)w(b, c)w(c, u, i - j - 2),$$

where the sum is over all $s \in S$, all $j \leq i - 2$ and all a, b, c such that $\sigma(va) = push(s)$ and $\sigma(bc) = pop(s)$. With this recursion formula we can compute all $w(v, u, i)$ with a polynomial number of arithmetic operations. Having computed all $w(v, u, i)$ we get $f_G = \sum_{i \in [m]} w(s, t, i)$. □

The more involved direction of the proof of Theorem 4 will be the reverse direction. To prove it it will be convenient to slightly relax our model of computation. A *relaxed SBP* G is an SBP where the underlying directed graph is not necessarily acyclic. To make use of cyclicity, in a relaxed SBP G, we do not consider paths but *walks*, i.e., vertices and edges of G may be visited several times. *Realizable walks* are defined completely analogously to realizable paths. Also the weight $w(P)$ of a walk is defined in the obvious way. Clearly, we cannot define the polynomial computed by a relaxed ABP by summing over the weight of all realizable walks, because there may be infinitely many of them since they may be arbitrarily long. Hence, we define for each pair u, w of vertices and for each integer m the polynomial $f_{u,v,m} := \sum_P w(P)$, where the sum is over all stack-realizable u-v-walks P in G that have length m. Furthermore, we say that for each m the relaxed SBP G computes the polynomial $f_{G,m} := f_{s,t,m}$.

The connection to SBPs is given by the following straight-forward lemma.

Lemma 3. *Let G be a relaxed SBP and $m \in \mathbb{N}$. Then for each m there is an SBP G'_m of size $m|G|$ that computes $f_{G,m}$.*

Proof. The idea is to unwind the computation of the relaxed SBP into m layers. Let $G = (V, E, w, \sigma)$, then for each $v \in V$ the SBP G' has m copies $\{v_1, \ldots, v_m\}$. For each $uv \in E$ the SBP G' had the edges $u_i v_{i+1}$ for $i \in [m-1]$ with weight $w(u_i v_{i+1}) := w(uv)$ and stack operation $\sigma(u_i v_{i+1}) := \sigma(uv)$. This completes the construction of G'.

Clearly, G' indeed computes $f_{G,m}$ and has size $m|G|$. □

Proposition 2. *Let C be a multiplicatively disjoint arithmetic circuit. For each $v \in V$ we denote by C_v the subcircuit of C with output v and we denote by f_v the polynomial computed by C_v. Then there is a relaxed SBP $G = (V, E, w, \sigma)$ of size at most $2|C|(|C| + 1) + 3(|C|)$ such that for each $v \in V$ there is a pair $v_-, v_+ \in V$ and an integer $m_v \leq 4|C_v|$ with*

- $f_v = f_{v_-, v_+, m_v}$, *and*
- *there is no stack-realizable walk from v_- to v_+ in G that is shorter than m_v.*

Proof. We construct G iteratively along a topological order of C by adding new vertices and edges, starting from the relaxed SBP with empty vertex set. We distinguish three cases:

Case 1: Let first v be an input of C with label X. We add two new vertices v_-, v_+ to G and the edge $v_- v_+$ with weigth $w(v_- v_+) := X$ and stack-operation $\sigma(v_- v_+) := nop$. Furthermore, $m_v := 1$. Clearly, none of the polynomials computed before change and the size of the relaxed SBP grows only by 2. Thus all statements of the proposition are fulfilled.

Case 2: Let now v be an addition gate with children u, w. By induction G contains vertices u_-, u_+, w_-, w_+ and there are m_u, m_v such that $f_{u_-, u_+, m_u} = f_u$

and $f_{w_-,w_+,m_w} = f_w$. Assume w.l.o.g. $m_u \geq m_w$. We add the new vertices v_-, v_+, v_s, v_t to G. We further add the edges v_-u_-, v_-v_s, v_tw_-, u_+v_+ and w_+v_+. Moreover, we connect v_s an v_t by a path of length $m_u - m_w$ whose inner vertices are also new. All edges we add get weight 1. Furthermore, we set the stack symbol operations $\sigma(v_-u_-) := push(vu)$, $\sigma(u_+v_+) := pop(vu)$, $\sigma(v_-v_s) := push(vw)$ and $\sigma(w_+v_+) := pop(vw)$ for new stack symbols vu and vw. All other edges we added are nop-edges. Finally, set $m_v := m_u + 2$.

Let us check that G computes the correct polynomials. First observe that the edges we added do not allow any new walks between old vertices, so we still compute all old polynomials by induction. Thus we only have to consider the realizable v_--v_+-walks of length m_v. Each of these either starts with the edge v_-u_- or the edge v_-v_s. In the first case, because of the stack symbols the walk must end with the edge u_+v_+. Thus the realizable v_-v_+-walks of length m_v that start with v_-u_- contribute exactly the same weight as the realizable u_--u_+-walks of length m_u. Hence, these weights add up to f_u by induction. Every v_-v_+-walks of length m_v that start with v_-v_s first makes $m_u - m_w$ unweighted steps to w_- and ends with the edge w_+v_+. Thus, these walks contribute exactly the same as the stackrealizable w_--w_+ walks of length $m_v - 2 - (m_u - m_w) = m_w$, so they contribute f_w. Combining all walks we get $f_{v_-,v_+,m_v} = f_u + f_w = f_v$ as desired.

We have that every realizable walk from u_+ to u_- has length at least m_u, and thus there is no realizable v_--v_+-walk starting with v_-u_- that is shorter than $m_u + 2 = m_v$. Moreover, since the realizable w_--w_+-walks have length at least m_w, the realizable paths starting with v_-w_- have length at least $m_w + (m_u - m_w) + 2 = m_u + 2 = m_w$. Thus there is no realizable v_--v_+-walk of length less that m_v.

We have $m_v = m_u + 2 \leq 4|C_u| + 2 \leq 4|C_v|$ where the first inequality is by induction and the second inequality follows from the fact that v is not contained in C_u and thus $|C_v| > |C_u|$. To see the bound on $|G|$ let s be the size of G before adding the new edges and vertices. By induction $s \leq 2(|C_v| - 1)(|C_v| - 1 + 1) + 3(|C_v| - 1)$. We have added $2 + m_u - m_v + 1$ vertices and thus G has now size $s + 3 + m_u - m_v \leq s + 3 + m_u$. But we have $m_u \leq 4|C_u| \leq 4|C_v|$ and thus the number of vertices in G is at most $2(|C_v| - 1)|C_v| + 3(|C_v| - 1) + 3 + 4|C_v| \leq 2|C_v|(|C_v| + 1) + 3|C_v|$. This completes the case that v is an addition gate.

Case 3: Let now v be a multiplication gate with children u, w. As before, G already contains u_-, u_+, w_-, w_+ and there are m_u, m_v with the desired properties. We add three vertices v_-, v_+ and v_* and the edges v_-u_-, u_+v_*, v_*w_- and w_+v_+ all with weight 1. The new edges have the stack symbols $\sigma(v_-u_-) := push(vu)$, $\sigma(u_+v_*) := pop(vu)$, $\sigma(v_*w_-) := push(vw)$ and $\sigma(w_+v_+) := pop(vw)$ for new stack symbols vu and vw. Finally, set $m_v := m_u + m_w + 4$.

Clearly, no stack-realizable walk between any pair of old vertices can traverse v_-, v_+ or v_* and thus these walks still compute the same polynomials as before. Thus we only have to analyse the v_--v_+-walks of length m_v in G. Let P be such a walk. Because of the stack symbols vu and vw the walk P must have the structure $P = v_-u_-P_1u_+v_*w_-P_2w_+v_+$ where P_1 and P_2 are a stack-realizable u_--u_+-walk and a stack-realizable w_--w_+-walk, respectively. The walk P is of length m_v and thus P_1 and P_2 must have the combined length $m_u + m_w$. But by induction P_1 must at least have length m_u and P_2 must have at least length m_w, so it follows that P_1 has length exactly m_u and P_2 has length exactly m_w. The walks P_1 and P_2 are independent and thus we have $f_{v_-,v_+,m_v} = f_{u_-,u_+,m_u} f_{w_-,w_+,m_w} = f_u f_w$ as desired.

The circuit C is multiplicatively disjoint and thus we have $|C_v| = |C_u| + |C_w| + 1$. It follows that $m_v = m_u + m_w + 4 \leq 4|C_u| + 4|C_w| + 4 = 4|C_v|$ where we get the inequality by induction. The relaxed SBP has grown only by 3 vertices which gives the bound on the size of G. This completes the proof for the case that v is an addition gate and hence the proof of the lemma.   □

Now the second direction of Theorem 4 is a straightforward combination of Lemma 3 and Proposition 2.

Lemma 4. *Every family* $(f_n) \in$ VP *can be computed by a family of SBPs of polynomial size.*

3.3   Width Reduction

In this section we show that, unlike for ordinary ABPs, bounding the width of SBPs does not decrease the computational power.

Lemma 5. *Every family* $(f_n) \in$ VP *can be computed by a SBP of width* 2 *with the stack symbol set* $\{0, 1\}$.

Proof. The idea of the proof is to start from the characterization of VP by SBPs from Theorem 4. We use the stack to remember which edge will be used next on a realizable path through the branching program. We will show how this can be done with width 2 SBPs with a bigger stack symbol size. In a second step we will seee how to reduce the stack symbol set to $\{0, 1\}$.

So let (G_n) be a family of SBPs. Fix n and let $G := G_n$ with vertex set V and edge set E. Furthermore, let w be the weight function, σ the stack operation labeling and S the stack symbol of G. Let s and t be the source and the sink of the SBP G. We assume without loss of generality that s has one single outgoing edge e_s. Furthermore t is only entered by one *nop*-edge e_t with weight 1. We will construct a new SBP G' with weight function w' and stack operation labeling σ'. G' will have stack symbol set $S \cup E$. For each edge e with a successor edge e' the SBP G' contains a gadget $G_{e,e'}$. The vertex set of $G_{e,e'}$ is $\{v_{e,e'}^1, v_{e,e'}^2, v_{e,e'}^3, v_{e,e'}^4, v_{e,e'}^5, v_{e,e'}^6\}$. These vertices are connected to a DAG by the edges $\{v_{e,e'}^1 v_{e,e'}^2, v_{e,e'}^1 v_{e,e'}^3, v_{e,e'}^2 v_{e,e'}^4, v_{e,e'}^3 v_{e,e'}^5, v_{e,e'}^4 v_{e,e'}^6, v_{e,e'}^5 v_{e,e'}^6\}$. All these edges have weight 1 except for $v_{e,e'}^2 v_{e,e'}^4$ which we give the weight $w'(v_{e,e'}^2 v_{e,e'}^4) := w(e)$. We call $v_{e,e'}^2 v_{e,e'}^4$ the *weighted edge* of $G_{e,e'}$. Furthermore we set $\sigma(v_{e,e'}^1 v_{e,e'}^2) := pop(e)$, $\sigma(v_{e,e'}^2 v_{e,e'}^4) := \sigma(e)$, $\sigma(v_{e,e'}^4 v_{e,e'}^6) := push(e')$. All other edges are *nop*-edges.

Now choose an order $\leq_E$ of E such that for each pair $uv, vw \in E$, the edge uv comes before vw. This order can be iteratively constructed from a topological order $\leq_V$ of V: For each vertex v along $\leq_V$ iteratively add the edges entering v to $\leq_E$ as the new maximum. From $\leq_E$ we construct an order $\leq_G$ of the gadgets $G_{e,e'}$ by defining

$$G_{e_1,e_2} \leq_G G_{e_3,e_4} \leftrightarrow e_1 < e_3 \vee (e_1 = e_3 \wedge e_2 < e_4).$$

We now connect the gadgets along the order $\leq_G$ in the following way: Let G_{e_1,e_2} and G_{e_3,e_4} be two successors in $\leq_G$. We connect v_{e_1,e_2}^6 to v_{e_3,e_4}^1 by a nop-edge of weight 1. Let $G_{e,e'}$ be the minimum of $\leq_G$. We add a new vertex s and the edge $sv_{e,e'}^1$ with weigth 1 and stack operation $\sigma(sv_{e,e'}^1) := push(e_s)$ where e_s is the single outgoing edge of s in G. Let now $G_{e,e'}$ be the maximum gadget in $\leq_G$. We add a new vertex t and the edge $v_{e,e'}t$ with weight 1 and stack operation $pop(e_t)$. This concludes the construction of G'.

It is easy to see that G' has indeed width 2. Thus we only need to show that G and G' compute the same polynomial. This will follow directly from the following claim:

Claim 1. *There is a bijection π between the stack-realizable paths in G and G'. Furthermore $w(P) := w'(\pi(P))$ for each stack-realizable path in G.*

In a final step we now reduce the stack symbol size to $\{0,1\}$ in a straightforward way. Let $\ell := \lceil \log(|S \cup E|) \rceil$, then each stack symbol s can be encoded into a $\{0,1\}$-string $\mu(s)$ of length ℓ. Now we substitute each edge e of G' by a path P_e of length ℓ. If $\sigma'(e) = push(s)$ we the edges along P_e are $push$-edges, too, that push $\mu(e)$ onto the stack. If $\sigma'(e) = pop(s)$ we pop $\mu(s)$ in reverse order along P_e. If e is a nop-edge, all edges of P_e are nop-edges, too. Finally, we give one of the edges in P_e the weight $w'(e)$, while all other edges get weight 1. Doing this for all edges, it is easy to see that the resulting SBP computes the same polynomial as G'. Furthermore, its width is 2. $\qquad\square$

4   Random Access Memory

4.1   Definition

We change the model of computation by allowing random access memory instead of a stack. We still work over a symbol set S like for SBPs but we introduce three *random access memory operations*: The operation *write* and *delete* take an argument $s \in S$ while the operation *nop* again takes no argument. Let $op(s)$ be a random access memory operation with $op \in \{write, delete\}$ and $P = op_1 op_2 \ldots op_r$ a sequence of memory operations. By $occ(P, op(s))$ we denote the number of occurences of $op(s)$ in P. We call a sequence P realizable if for all symbols $s \in S$ we have that $occ(P, write(s)) = occ(P, delete(s))$ and for all prefixes P' of P we have $occ(P', write(s)) \geq occ(P', delete(s))$ for all $s \in S$.

Intuitively, the random access memory operations do the following: $write(s)$ writes the symbol s into the random access memory. If s is already there it adds it another time. The operation $delete(s)$ deletes one occurrence of the symbol s from the memory if there is one. Otherwise an error occurs. nop is the "no operation" operation again like for SBPs. A sequence of operations is realizable if no error occurs during the deletions, and moreover starting from empty memory the memory is empty again after the sequence of operations.

Definition 4. *A* random access branching program *(RABP) G is an ABP with an additional edge labeling $\sigma : E \rightarrow \{op(s) \mid op \in \{write, delete\}, s \in S\} \cup \{nop\}$. A path $P = v_1v_2 \ldots v_r$ in G has the sequence of random access memory operations $\sigma(P) := \sigma(v_1v_2)\sigma(v_2v_3) \ldots \sigma(v_{r-1}v_r)$. If $\sigma(P)$ is realizable we call P a random-access-realizable path. The RABP G computes the polynomial $f_G = \sum_P w(P)$, where the sum is over all random-access-realizable s-t-paths P.*

Proposition 3. *Let G be an RABP of size s. There is an SBP G' of size $O(s^2)$ such that $f_G = f_{G'}$ and G' does not contain any nop-edges. If G is layered with width k, then G' is layered, too, and has width at most k^2.*

4.2   Characterizing VNP

Intuitively random access on the memory allows us more fine-grained control over the paths in the branching program that contribute to the computation. While in SBPs nearly all of the memory content is hidden, in RABPs we have access to the complete memory at all times. This makes RABPs more expressive than SBPs which is formalized in the following theorem.

Theorem 5. *If (f_n) is computed by a family of polynomial size RABPs, then $(f_n) \in$ VNP. Moreover, for every family $(f_n) \in$ VNP there is a family of width 2 RABPs of polynomial size computing (f_n).*

Proof. The upper bound is easy to see with Valiant's criterion (Lemma 1) and the fact that checking if a path through a RABP is realizable is certainly in P.

To show the lower bound we consider the dominating-set polynomial for a graph $G = (V, E)$ defined as $DSP_G(X_1, \ldots, X_n) := \sum_D \prod_{v \in D} X_v$, where the sum is over all dominating sets D in G.

Proposition 4. *There is a family (G_n) of graphs such that the resulting family (DSP_{G_n}) of polynomials is VNP-complete.*

We will show that for a graph $G = (V, E)$ with n vertices there is a RABP of size $n^{O(1)}$ and width 2 that computes $DSP_G(X_1, \ldots, X_n)$. The RABP works in two stages. The symbol set of the RABP will be V. In a first stage it iteratively selects vertices v and writes v and all of its neightbors into the memory. In a second stage it checks that each vertex v was written at least once into the memory, i.e., either v or one of its neighbors was chosen in the first phase. Thus the set of chosen vertices must have been a dominating set.

So fix a graph G and set $w(v) = X_v$ for each $v \in V$. For each vertex v with neighbors $v_1, \ldots, v_d$ we construct a gadget G_v as shown in Figure 1. We call the path through G_v with the edges that have memory operations the *choosing path*. Now for each vertex v we construct a second gadget G'_v that is shown in

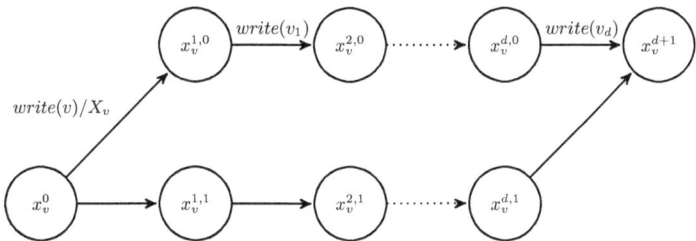

Fig. 1. The gadget G_v. Let v be a vertex with neighbors $v_1, \ldots, v_d$. The weight of $x_v^0 x_v^{1,0}$ is X_v while all other edges have weight 1. G_v has two paths. Every realizable path that traverses G_v on the upper path writes v and all of its neighbors into the memory. This path has weight X_v. Realizable paths through the upper path do not change the memory in G_v and have a weight weight contribution of 1 in G_v.

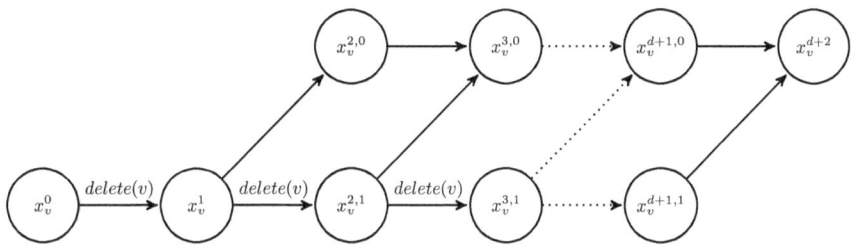

Fig. 2. The gadget G_v'. Let d be the degree of v, then G_v' has $d + 3$ layers. All edges have weight 1. The edges connecting vertices in the lower level have operation $delete(v)$ while all other edges have no memory operation. Every realizable path through G_v' has weight 1 and deletes between 1 and $d + 1$ occurences of the symbol v from memory.

Figure 2. Choose an order on the vertices. For each non-maximal vertex v in the order with successor u, we connect the sink of G_v to the source of G_u and the sink of G_v' to the source of G_u' with a *nop*-edge of weight 1. Finally, let x be the maximal vertex in the order and y the minimal vertex. Connect the sink of G_x to the source of G_y' again by a *nop*-edge of weight 1.

We claim that G' computes DSP_G. To see this, define the weight of a vertex set D in G to be $w(S) := \prod_{v \in S} X_v$. The following claim and the observation that G' has width 2, complete the proof.

Claim 2. *There is a bijection π between dominating sets in G and random-access-realizable paths in G' such that for each dominating set D in G we have $w(D) := w(\pi(D))$.*   □

Acknowledgements. The author would like to thank Sébastien Tavenas for pointing out an error in an earlier proof of Lemma 4. The corrected proof presented in this paper is the result of discussions with him and Pascal Koiran. The author is very thankful for this contribution. Furthermore, the author is grateful to Guillaume Malod who gave very helpful feedback on a draft of this paper.

Finally, the author would like to thank Peter Bürgisser and Meena Mahajan for encouraging him to write up these results as a paper.

References

1. Barrington, D.A.: Bounded-width polynomial-size branching programs recognize exactly those languages in NC1. Journal of Computer and System Sciences 38(1), 150–164 (1989)
2. Ben-Or, M., Cleve, R.: Computing algebraic formulas using a constant number of registers. SIAM J. Comput. 21(1), 54–58 (1992)
3. Brent, R.P.: The complexity of multiple-precision arithmetic. In: Brent, R.P., Andersson, R.S. (eds.) The Complexity of Computational Problem Solving, pp. 126–165. Univ. of Queensland Press (1976)
4. Briquel, I., Koiran, P.: A dichotomy theorem for polynomial evaluation. In: Královič, R., Niwiński, D. (eds.) MFCS 2009. LNCS, vol. 5734, pp. 187–198. Springer, Heidelberg (2009)
5. Bürgisser, P.: Completeness and reduction in algebraic complexity theory. Springer (2000)
6. Kintali, S.: Realizable paths and the NL vs L problem. Electronic Colloquium on Computational Complexity (ECCC) 17, 158 (2010)
7. Koiran, P.: Arithmetic circuits: The chasm at depth four gets wider. Theor. Comput. Sci. 448, 56–65 (2012)
8. Malod, G., Portier, N.: Characterizing Valiant's algebraic complexity classes. J. Complexity 24(1), 16–38 (2008)
9. Mengel, S.: Arithmetic Branching Programs with Memory, arXiv:1303.1969 (2013)
10. Nisan, N.: Lower bounds for non-commutative computation. In: Proceedings of the Twenty-Third Annual ACM Symposium on Theory of Computing, p. 418. ACM (1991)
11. Skyum, S., Valiant, L.G.: A complexity theory based on boolean algebra. J. ACM 32(2), 484–502 (1985)
12. Toda, S.: Classes of arithmetic circuits capturing the complexity of computing the determinant. IEICE Transactions on Information and Systems 75(1), 116–124 (1992)
13. Valiant, L.G., Skyum, S., Berkowitz, S., Rackoff, C.: Fast parallel computation of polynomials using few processors. SIAM J. Comput. 12(4), 641–644 (1983)
14. Valiant, L.G.: Completeness classes in algebra. In: Proceedings of the Eleventh Annual ACM Symposium on Theory of Computing, pp. 249–261. ACM (1979)
15. Weber, V., Schwentick, T.: Dynamic complexity theory revisited. Theory Comput. Syst. 40(4), 355–377 (2007)

Subexponential Algorithm for d-Cluster Edge Deletion: Exception or Rule?

Neeldhara Misra[1], Fahad Panolan[2], and Saket Saurabh[2]

[1] Indian Institute of Science, Bangalore, India
neeldhara@csa.iisc.ernet.in
[2] The Institute of Mathematical Sciences, Chennai, India
{fahad,saket}@imsc.res.in

Abstract. The correlation clustering problem is a fundamental problem in both theory and practice, and it involves identifying clusters of objects in a data set based on their similarity. A traditional modeling of this question as a graph theoretic problem involves associating vertices with data points and indicating similarity by adjacency. Clusters then correspond to cliques in the graph. The resulting optimization problem, CLUSTER EDITING (and several variants) are very well-studied algorithmically. In many situations, however, translating clusters to cliques can be somewhat restrictive. A more flexible notion would be that of a structure where the vertices are mutually "not too far apart", without necessarily being adjacent.

One such generalization is realized by structures called s-clubs, which are graphs of diameter at most s. In this work, we study the question of finding a set of at most k edges whose removal leaves us with a graph whose components are s-clubs. Recently, it has been shown that unless Exponential Time Hypothesis fail (ETH) fails CLUSTER EDITING (whose components are 1-clubs) does not admit sub-exponential time algorithm [*STACS, 2013*]. That is, there is no algorithm solving the problem in time $2^{o(k)}n^{O(1)}$. However, surprisingly they show that when the number of cliques in the output graph is restricted to d, then the problem can be solved in time $O(2^{O(\sqrt{dk})} + m + n)$. We show that this sub-exponential time algorithm for the fixed number of cliques is rather an exception than a rule. Our first result shows that assuming the ETH, there is no algorithm solving the s-CLUB CLUSTER EDGE DELETION problem in time $2^{o(k)}n^{O(1)}$. We show, further, that even the problem of deleting edges to obtain a graph with d s-clubs cannot be solved in time $2^{o(k)}n^{O(1)}$ for any fixed s, $d \geq 2$. This is a radical contrast from the situation established for cliques, where sub-exponential algorithms are known.

Keywords: subexponential algorithms, s-clubs, cluster edge deletion, ETH-hardness.

1 Introduction

The correlation clustering problem involves identifying clusters of objects in a data set based on their similarity. A traditional way of posing this as a graph

K. Chatterjee and J. Sgall (Eds.): MFCS 2013, LNCS 8087, pp. 679–690, 2013.
© Springer-Verlag Berlin Heidelberg 2013

theoretic question involves associating vertices with data points and indicating similarity by adjacency. In this setting, the natural notion of a cluster would correspond to a *clique*, a set of mutually adjacent vertices. Thus, we call a graph G a *cluster graph* if every connected component of G is a complete graph. The task of identifying clusters can now be viewed as an optimization problem. In particular, a subset $F \subseteq E$ is called a cluster edge deletion set if $G \backslash F = (V, E \backslash F)$ is a cluster. On the other hand, if for some $F \subset V \times V$, $G \Delta F = (V, E \Delta F)$ is a cluster, then F is called cluster editing set. (Here $E \Delta F$ is the symmetric difference between E and F.) In the CLUSTER EDGE DELETION (CLUSTER EDITING) problem, we are given a graph G and an integer k, and we want to check whether there exists a cluster edge deletion set (cluster editing set), F of size at most k.

The complexity of CLUSTER EDGE DELETION and CLUSTER EDITING is well-understood. The problems are NP-complete and admit constant-factor approximation algorithms. On the other hand, they are also known to be APX-hard. Further, it has been recently shown that Cluster Editing cannot be solved in time $2^{o(k)} n^{O(1)}$ unless the Exponential Time Hypothesis (ETH) fails [7,3]. This led the authors of [3] to consider the question of editing at most k edges to obtain a graph with at most d clusters. This variant continues to be well motivated in several practical settings, where the number of clusters corresponds to an external constraint. With the restriction on the number of clusters in place, there is good news, as [3] describes an algorithm that solves the problem in time $O(2^{O(\sqrt{dk})} + m + n)$.

So far, we have considered the clustering problem in the graph theoretic context using cliques as a natural means for modeling the notion of a cluster. This effectively restricts us to a binary notion of similarity, in that a pair of data points are either similar or not, and we would like to maximize similarities within a cluster and minimize non-similarities across clusters. In many situations, however, this translation can be somewhat severe. A more flexible notion would be that of a structure where the vertices are mutually "not too far apart", without necessarily being adjacent. Additionally, note that cliques are also a popular choice for modeling highly correlated or connected substructures in applications. Given that cliques impose a very strict connectivity requirement, this modeling suffers from being overly restrictive.

A natural generalization of the notion of cliques would be along the lines of small-diameter graphs. These structures are called *clubs* and have been proposed as a more reasonable measure of connectivity and correlation. Formally, note that the complete graphs can be thought of as graphs of diameter one. A *s*-club is a graph of diameter at most s, and note that cliques are exactly 1-clubs. The notion of *s*-clubs was introduced in [1]. The *s*-club concept was defined in the context of social sciences [1], and it has recently been used in the analysis of social [11] and biological networks. In [5,6,12] parameterized studies of finding *s*-clubs were undertaken. It is worth to mention that several other generalizations of cliques such as *s*-cliques and s-plexes [4] and the related notion of clustering into these graphs have been studied in literature before.

The immediate question that arises in the context of clustering is the s-CLUB CLUSTER EDGE DELETION problem: is there a set of at most k edges whose removal leaves us with a graph whose components are s-clubs? It is known that the problem is NP-complete for $s = 2$, and there is an algorithm that solves the problem in $O(2.74^k n^{O(1)})$ [8]. It is natural to ask if the problem admits a sub-exponential algorithm. Our first result shows that assuming the ETH, the answer is in the negative:

Theorem 1. 2-CLUB CLUSTER EDGE DELETION *cannot be solved in time* $2^{o(k)}n^{O(1)}$, *unless ETH fails.*

In the setting of cliques, it was useful to consider the question with the additional dimension of the number of clusters: if we demanded deletion into at most d clusters, then the problem turned out to admit a sub-exponential algorithm. It is therefore natural to consider the corresponding question in the s-club setting: can we identify at most k edges whose removal leaves us with at most d s-clubs? It turns out that the slightest generalization of the cluster editing problem makes the problem significantly harder in the context of sub-exponential algorithms. In particular, we show:

Theorem 2. s-CLUB d-CLUSTER EDGE DELETION *for $s \geq 2$ and $d \geq 2$ cannot be solved in time* $2^{o(k)}n^{O(1)}$, *unless ETH fails.*

Our Theorem 2 shows that the sub-exponential algorithm in the case of 1-CLUB d-CLUSTER EDGE DELETION is rather an exception. All our results are obtained by reductions from 3-CNFSAT. The Exponential Time Hypothesis states that there is no algorithm that solves 3-CNFSAT in time $2^{o(m+n)}$ time (via sparsification). Our reductions produce instances where the size of the solution depends linearly on $(m + n)$. We refer to recent survey of Lokshtanov et al. [9] for a detailed discussions on ETH and to the books [2,10] for an introduction to the area of parameterized complexity.

Organization of the paper. In Section 2 we establish the notation and state the problems formally. In Sections 3 and 4, we prove Theorems 1 and 2, respectively. The proof of Theorem 2 is split into three cases, namely $s = 2$, $s = 3$, and $s \geq 4$.

2   Preliminaries

Graphs. For a finite set V, a pair $G = (V, E)$ such that $E \subseteq V^2$ is a graph on V. The elements of V are called *vertices*, while pairs of vertices (u, v) such that $(u, v) \in E$ are called *edges*. In the following, let $G = (V, E)$ and $G' = (V', E')$ be graphs, and $U \subseteq V$ some subset of vertices of G. Let G' be a subgraph of G. If E' contains all the edges $\{u, v\} \in E$ with $u, v \in V'$, then G' is an *induced subgraph* of G, *induced by* V'. For any set of vertices $U \subseteq V$, $G[U]$ denotes the subgraph of G induced by U. For $v \in V$, $N(v) = \{u \mid (u, v) \in E\}$ and $N[v] = N(v) \cup \{v\}$. For $U \subseteq V$, $N(U) = \left(\bigcup_{u \in U} N(u)\right) \setminus U$.

The *distance* between vertices u, v of G is the length of a shortest path from u to v in G; if no such path exists, the distance is defined to be ∞. The *diameter* of G is the greatest distance between any two vertices in G. A graph G is said to be *connected* if there is a path in G from every vertex of G to every other vertex of G. If $U \subseteq V$ and $G[U]$ is connected, then U itself is said to be connected in G. A subset of vertices U is said to induce a s-club if $G[U]$ has diameter at most s, or in other words, the distance between every pair of vertices in U is at most s in $G[U]$. A graph is said to be a s-club cluster if every connected component of the graph induces a s-club.

Satisfiability. Let P be an arbitrary set, whose elements we shall refer to as *variables*. It will be convenient to assume that P is a countably infinite set. The set of formulas over P is inductively defined to be the smallest set of expressions such that: (a) Each variable in the set P is a formula, (b) $(\neg\alpha)$ is a formula whenever α is, and (c) $(\alpha \,\square\, \beta)$ is a formula whenever α and β are formulas and $\square$ is one of the binary connectives $\wedge$, $\vee$.

We denote by $\mathcal{F}(P)$ the set of all formulas over P. An valuation or an assignment of P is a function $v : P \rightarrow \{0, 1\}$, which may be extended to a function $\bar{v} : \mathcal{F}(P) \rightarrow \{0, 1\}$, as follows. For each variable x in the set P, $\bar{v}(x) = v(x)$. Further, $\bar{v}(\neg\alpha) = 1 - \bar{v}(\alpha)$, $\bar{v}(\alpha \wedge \beta) = min\{\bar{v}(\alpha), \bar{v}(\beta)\}$, and $\bar{v}(\alpha \vee \beta) = max\{\bar{v}(\alpha), \bar{v}(\beta)\}$.

A formula is in *conjunctive normal form* (CNF) if it is a conjunction of clauses, where a clause is a disjunction of literals. Every propositional formula can be converted into an equivalent formula that is in CNF. The question of *satisfiability* is whether, given a formula α, there exists a valuation v such that $v(\alpha) = 1$. This is one of the most well-studied NP-complete problems. The problem continues to be NP-complete if the formula is offered in CNF even when every clause has no more than three variables.

Notice that given a 3-SAT formula, we may preprocess it effectively to ensure that each variable appears at least twice: at least once in a positive literal and at least once in a negative one. This is because any variable that appears only positively (respectively, negatively) can be assigned 1 (respectively, 0) by a satisfying assignment without loss of generality. Similarly, we assume that any variable will not appear both positively and negatively in a clause, because such a clause can be removed from the formula without affecting the satisfiability of the formula. Finally, we may assume that each clause of ϕ consists of exactly three literals by a standard padding argument using dummy variables. We say that a 3-CNF formula is standardized if it satisfies all three properties above. In our discussions, we work with standardized formulas.

The problems we study in this work are the following:

---

s-CLUB CLUSTER EDGE DELETION

Instance: An undirected graph $G = (V, E)$ and a positive integer k.
Problem: Does there exist $E' \subseteq E$ with $|E'| \leq k$ such that $G \setminus E'$ is an s-club cluster?

---

s-CLUB d-CLUSTER EDGE DELETION

Instance: An undirected graph $G = (V, E)$ and a positive integer k.
Problem: Does there exist $E' \subseteq E$ with $|E'| \leq k$ such that $G \setminus E'$ is an
s-club cluster containing d components?

---

3   2-Club Cluster Edge Deletion

In this section we will show that 2-CLUB CLUSTER EDGE DELETION cannot
be solved in $2^{o(k)} n^{O(1)}$ unless ETH fails. To show this result we will give a
reduction from 3-SAT to 2-CLUB CLUSTER EDGE DELETION. More precisely,
from an instance ϕ with m clauses and n variables, of 3-SAT, we will construct
an instance (G, k) of 2-CLUB CLUSTER EDGE DELETION with the property that
ϕ is satisfiable iff (G, k) is an YES instance, where $k = O(m + n)$.

Lemma 1. $(\star)^1$ *Let $G = (V, E)$ be an undirected graph. Let $X \subseteq V$ such that
$G[X]$ is a clique, $\forall x, y \in X, N[x] = N[y]$ and $G[N(X)]$ is a clique. Then there
exist an optimum solution F to 2-CLUB CLUSTER EDGE DELETION such that
X is contained in a single component in $G \setminus F$.*

Lemma 2. $(\star)$ *There exists a polynomial-time algorithm that, given a 3-CNF
formula ϕ with n variables and m clauses, constructs a 2-CLUB CLUSTER EDGE
DELETION instance (G, k) such that (i) ϕ is satisfiable if and only if (G, k) is a
YES-INSTANCE, AND (ii) $k = O(n + m)$.*

Theorem 3. *2-CLUB CLUSTER EDGE DELETION cannot be solved in time
$2^{o(k)} n^{O(1)}$, unless ETH fails.*

4 s-Club d-Cluster Edge Deletion

In this section, we show the hardness of s-CLUB d-CLUSTER EDGE DELETION
for all $s \geq 2$. The results are divided into three parts. First, we demonstrate
a reduction from 3-SAT to 2-CLUB 2-CLUSTER EDGE DELETION. With minor
modifications, we show that this reduction works for the problem of edge deletion
into two 3-clubs. For $s \geq 4$, we show a general reduction from 3-SAT to s-CLUB
2-CLUSTER EDGE DELETION. The construction in the first reduction serves as a
basis for the general reduction, but we note that the finer details involve several
nuances. We also note that the problem of deleting into two s-clubs easily reduces
to the problem of deleting into d s-clubs.

4.1   2-Club 2-Cluster Edge Deletion

In this section we will show that 2-CLUB 2-CLUSTER EDGE DELETION cannot
be solved in $2^{o(k)} n^{O(1)}$ unless ETH fails. To this end, we will give a reduction

---

[1] The proofs of Lemmas marked with a $\star$ will appear in the full version of the paper.

from 3-SAT to 2-CLUB 2-CLUSTER EDGE DELETION. More precisely, based on an instance ϕ of 3-SAT with m clauses and n variables, we will construct an instance (G, k) of 2-CLUB 2-CLUSTER EDGE DELETION with the property that ϕ is satisfiable if and only if (G, k) is an YES instance, where $k = O(m + n)$.

Lemma 3. *There exists a polynomial-time algorithm that, given a 3-CNF formula ϕ with n variables and m clauses, constructs a 2-CLUB 2-CLUSTER EDGE DELETION instance (G, k) such that (i) ϕ is satisfiable iff (G, k) is a YES instance, and (ii) $k = O(m + n)$.*

Proof. Let ϕ be a standardized 3-CNF formula with m clauses and n variables. Let $C_1, C_2, \ldots, C_m$ be the clauses and $x_1, x_2, \ldots, x_n$ be the variables.

Construction. We construct a graph $G = (V, E)$ based on ϕ as follows. The graph G contains two clause gadgets $\mathcal{C}_1$ and $\mathcal{C}_2$, two connection gadgets K_1 and K_2, two selection gadgets S_1 and S_2, one variable gadget $\mathcal{V}$ and four global vertices $\{p, p', g_1, g_2\}$. The clause gadget $\mathcal{C}_1$ contains m vertices $c_1, c_2, \ldots, c_m$ and there are no edges within $\mathcal{C}_1$. Similarly, $\mathcal{C}_2$ contains m vertices $c'_1, c'_2, \ldots, c'_m$ and there are no edges within $\mathcal{C}_2$. The variable gadget $\mathcal{V}$ contains $2n$ vertices, one for each literal. Let these vertices be named $x_1, x_2, \ldots, x_n$ and $\overline{x_1}, \overline{x_2}, \ldots, \overline{x_n}$. The connection gadgets K_1 and K_2 are cliques of size $k + 2$. The selection gadget S_1 contains n vertices $a_1, a_2, \ldots, a_n$ and no edges within S_1. Similarly, S_2 contains n vertices $b_1, b_2, \ldots, b_n$ and no edges within S_2.

For each $1 \leq i, j \leq m$ we add an edge (c_i, c'_j) if $i \neq j$. For any literal $l = x_i$ or $l = \overline{x_i}$, and for every clause C_i that contains l, we add the edges (c_i, l) and (c'_i, l). For each $1 \leq i, j \leq n$ we add an edge (a_i, b_j) if $i \neq j$. For every pair of literals x_i and $\overline{x_i}$, add the edges (x_i, a_i), (x_i, b_i), $(\overline{x_i}, a_i)$ and $(\overline{x_i}, b_i)$. Also, add all possible edges between: K_1 and g_2; K_2 and g_2; K_1 and S_1; K_2 and S_2; g_1 and $\mathcal{V}$; g_2 and $\mathcal{V}$; p and $\mathcal{C}_1$; p' and $\mathcal{C}_2$. Finally, add the edges (g_1, p) and (g_1, p'). (See Fig. 1.) We set $k = 4(m + n)$.

Completeness. Let ϕ be satisfiable, and $f : \{x_1, \ldots, x_n\} \to \{0, 1\}$ be a satisfying assignment. Now we construct the edge deletion set $F \subseteq E(G)$ as follows. For each $1 \leq i \leq n$, if $f(x_i) = 1$, then include in F the edges between x_i and S_b ($b = 1, 2$), the edges between $\overline{x_i}$ and $\mathcal{C}_b$ (for $b = 1, 2$), and the edges $(\overline{x_i}, g_1), (x_i, g_2)$. On the other hand, if $f(x_i) = 0$, the we include in F the edges between $\overline{x_i}$ and S_b ($b = 1, 2$), edges between x_i and $\mathcal{C}_b$ (for $b = 1, 2$), and the edges $(\overline{x_i}, g_2), (x_i, g_1)$.

Note that the number of edges in F which are between $\mathcal{C}_b$ (for $b = 1, 2$) and $\mathcal{V}$ is at most $4m$. This is because every vertex $c \in \mathcal{C}_1 \cup \mathcal{C}_2$ has three neighbors in $\mathcal{V}$, of which F picks at most two (since the choice of F is based on a satisfying assignment f). The number of edges in F which are between S_i (for $i = 1, 2$) and $\mathcal{V}$, is clearly $2n$ and the number of edges in F which are between $\mathcal{V}$ and $\{g_1, g_2\}$, is also $2n$. So we have that $|F| \leq 4(m + n)$, as desired.

Now, we need to show that $G[E \setminus F]$ consists of two components which are 2-clubs. For $b = 0, 1$, let V_b denote the set of vertices corresponding to literals that evaluate to b under the assignment f. It is easy to see that K_1, K_2, S_1, S_2, g_2

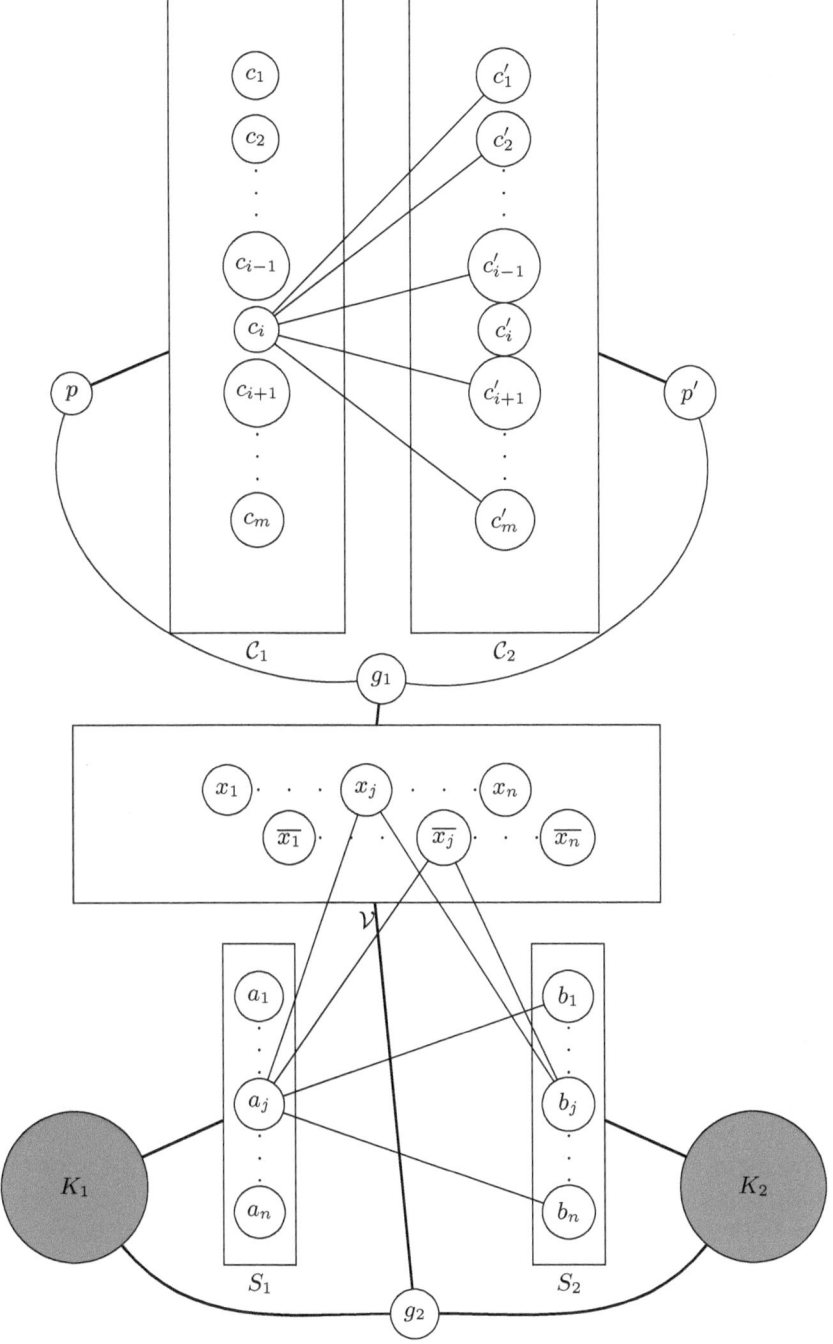

Fig. 1. Graph G constructed from ϕ. Vertices in a gadget which colored gray are completely connected. Edges between the clause gadgets and the variable gadget are not drawn in the figure. Thick lines are used to represent all possible edges between two sets of vertices.

and V_0 form a connected component (call it G_0). Also $\mathcal{C}_1, \mathcal{C}_2, p, p', g_1$ and V_1 form a connected component (call it G_1). We now argue that G_0 and G_1 are 2-clubs.

Any pair of vertices except (a_i, b_i) for all $1 \le i \le n$, in the graph induced on S_1, S_2, K_1, K_2, g_2 are at a distance at most two. Since either x_i or $\overline{x_i}$ is in G_0, the distance between a_i and b_i is 2. Since each vertex $y \in G_0$ that corresponds to a literal is adjacent to g_2, the distance between y and vertices in K_b (for $b = 1, 2$) is two in G_0. Finally, since y is adjacent to one vertex from S_1 and one vertex from S_2 (say a_i, b_i), y is at a distance at most two from any vertex in S_1, S_2 (recall that a_i is adjacent to b_j for all $i \ne j$). Since all the vertices in $\mathcal{V} \cap G_0$ has a common neighbor g_2 in G_0, these vertices are also at a distance two from each other in G_0. Hence G_0 is a 2-club (see Table 1).

Table 1. G_0 is a 2-club

	S_1	S_2	K_1	K_2	V_0	g_2
S_1	2 (S_2)	2 (V_0)	1	2 (S_2)	2 (S_2)	2 (K_1)
S_2		2 (S_1)	2 (S_1)	1	2 (S_1)	2 (K_2)
K_1			1	2 (g_2)	2 (S_1)	1
K_2				1	2 (S_2)	1
V_0					2 (g_2)	1
g_2						0

Table 2. G_1 is a 2-club

	$\mathcal{C}_1$	$\mathcal{C}_2$	p	p'	V_1	g_1
$\mathcal{C}_1$	2 ($\mathcal{C}_2$)	2 (V_1)	1	2 ($\mathcal{C}_2$)	2 ($\mathcal{C}_2$)	2 (p)
$\mathcal{C}_2$		2 ($\mathcal{C}_1$)	2 ($\mathcal{C}_1$)	1	2 ($\mathcal{C}_1$)	2 (p')
p			0	2 (g_1)	2 ($\mathcal{C}_1$)	1
p'				0	2 ($\mathcal{C}_2$)	1
V_1					2 (g_1)	1
g_1						0

Now consider G_1. Again any pair of vertices except (c_i, c_i') for all $1 \le i \le m$, in the graph induced on $\mathcal{C}_1, \mathcal{C}_2, p, p', g_1$ are at a distance at most two. Since f is a satisfying assignment, for all $1 \le i \le m$ there exists a literal from the clause C_i that is set to 1. Therefore, for each $1 \le i \le m$, vertices c_i, c_i' has a common neighbor in G_1. Using arguments similar to the case of G_0, we can show that all vertices $\mathcal{V} \cap G_1$ are at a distance of at most two from all other vertices in G_1. Hence G_1 is a 2-club (see Table 2 for details).

Soundness. Suppose (G, k) is an Yes instance of 2-Club 2-Cluster Edge Deletion. Let $F \subseteq E(G)$ is the edge deletion set. Let G_a, G_b be the two connected components in $G \setminus F$. We first claim that, without loss of generality, $(K_1 \cup K_2 \cup S_1 \cup S_2 \cup g_2) \subset G_a$. Since K_1 induces a clique of size $k + 2$, no set of at most k edges will disconnect K_1. Thus, the vertices of K_1 will belong to one of the two connected components in $G \setminus F$. Without loss of generality, let $K_1 \subseteq G_a$. Since the number of edges between g_2 and K_1, between any vertex in S_1 and K_1 is $k + 2$, $\{g_2\} \cup S_1 \subset G_a$. By similar arguments $\{K_2 \cup S_2 \cup g_2\}$ will belong to the same component. Hence $\{K_1 \cup S_1 \cup g_2 \cup K_2 \cup S_2\} \subset G_a$.

Notice that $N(K_1) = \{a_1, \ldots, a_n, g_2\}$. Consider any $v \in \mathcal{C}_1 \cup \mathcal{C}_2 \cup \{p, p', g_1\}$. It is easily checked that for all $1 \le i \le n$, $a_i \notin N(v)$, and therefore, $N(v) \cap N(K_1) = \emptyset$. This implies that in G, the vertices of $\mathcal{C}_1 \cup \mathcal{C}_2 \cup \{p, p', g_1\}$ are at a distance more than two from K_1, and therefore, $\mathcal{C}_1 \cup \mathcal{C}_2 \cup \{p, p', g_1\} \subset G_b$.

Observe that for each $1 \leq i \leq m$ c_i and c_i' are at a distance more than two in the graph induced on $\mathcal{C}_1 \cup \mathcal{C}_2 \cup \{p, p', g_1\}$. Also, for each $1 \leq i \leq n$, a_i, b_i are at a distance more than 2 in the graph induced on K_1, K_2, S_1, S_2, g_2. Therefore, these vertices can be made closer only via vertices in $\mathcal{V}$. In particular, for each $1 \leq i \leq n$, a_i, b_i are at a distance of at most two in G_a, at least one of x_i or $\overline{x_i}$ belongs to G_a, or equivalently, at most one of x_i and $\overline{x_i}$ belongs to G_b. Whenever either literal associated with x_i belongs to G_b, we define $f(x_i)$ as follows: $f(x_i) = 1$ if $x_i \in G_b$ and $f(x_i) = 0$ if $\overline{x_i} \in G_b$. If $x_i, \overline{x_i} \in G_a$, then let $f(x_i) = 1$ (the setting is arbitrary). Now we show that the f thus defined is a satisfying assignment. Consider any clause C_j. Since G_b is a 2-club, there exists a vertex y from $\mathcal{V}$ which is a common neighbor of c_j and c_j'. By the definition of f, we have that $f(l_y) = 1$, where l_y is the literal corresponding to the vertex y. So f is a satisfying assignment for ϕ.   □

It is now easy to see that 2-CLUB d-CLUSTER EDGE DELETION cannot be solved in time $2^{o(k)}n^{O(1)}$ unless ETH fails, for any $d \geq 2$. We would reduce from 3-CNF SAT as described in the proof of Lemma 3, and add $d - 2$ disjoint cliques of size $k + 2$ each to the reduced graph. With this, we have shown the following theorem.

Theorem 4. 2-CLUB d-CLUSTER EDGE DELETION *for $d \geq 2$, cannot be solved in time $2^{o(k)}n^{O(1)}$, unless ETH fails.*

4.2   3-Club 2-Cluster Edge Deletion

In this section we will show that 3-CLUB 2-CLUSTER EDGE DELETION cannot be solved in $2^{o(k)}n^{O(1)}$ unless ETH fails. The proof is a slight modification of the construction described in the proof of Lemma 3.

Lemma 4. ($\star$) *There exists a polynomial-time algorithm that, given a 3-CNF formula ϕ with n variables and m clauses, constructs a 3-CLUB 2-CLUSTER EDGE DELETION instance (G, k) such that (i) ϕ is satisfiable iff (G, k) is a YES instance, and (ii) $k = O(m + n)$.*

Theorem 5. 3-CLUB d-CLUSTER EDGE DELETION *for $d \geq 2$, cannot be solved in time $2^{o(k)}n^{O(1)}$, unless ETH fails.*

4.3 s-Club d-Cluster Edge Deletion

We now present a general reduction: for all $s \geq 4$, we show that s-CLUB d-CLUSTER EDGE DELETION cannot be solved in time $2^{o(k)}n^{O(1)}$ unless the ETH fails.

Lemma 5. *There exists a polynomial-time algorithm that, given a 3-CNF formula ϕ with n variables and m clauses, constructs a s-CLUB 2-CLUSTER EDGE DELETION instance (G, k) for $s \geq 4$ such that (i) ϕ is satisfiable iff (G, k) is a YES instance, and (ii) $k = O(m + n)$.*

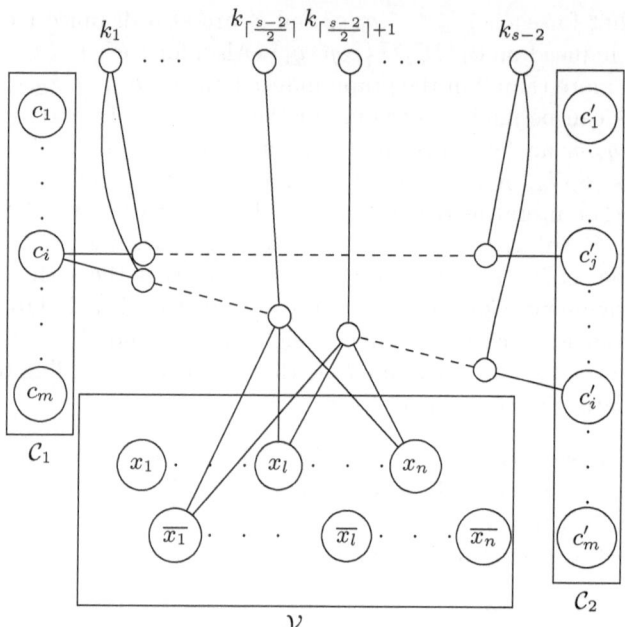

Fig. 2. Example of connection within clause gadget and between clause gadget and variable gadget with $C_i = \overline{x_1} \vee x_l \vee x_n$

Proof. Let ϕ be a standardized 3-CNF formula with m clauses and n variables. Let $C_1, C_2, \ldots, C_m$ be the clauses and $x_1, x_2, \ldots, x_n$ be the variables.

Construction. We construct a graph $G = (V, E)$ based on ϕ as follows. The graph G contains two clause gadgets $\mathcal{C}_1$ and $\mathcal{C}_2$, s connection gadgets $K_1, \ldots, K_s$, two selection gadgets S_1 and S_2, one variable gadget $\mathcal{V}$ and vertices $\{k_1, \ldots, k_{s-2}\}$. The clause gadget $\mathcal{C}_1$ contains m vertices $c_1, c_2, \ldots, c_m$ and there are no edges within $\mathcal{C}_1$. Similarly, $\mathcal{C}_2$ contains m vertices $c'_1, c'_2, \ldots, c'_m$ and there are no edges within $\mathcal{C}_2$. The variable gadget $\mathcal{V}$ contains $2n$ vertices, one for each literal. Let these vertices be named $x_1, x_2, \ldots, x_n$ and $\overline{x_1}, \overline{x_2}, \ldots, \overline{x_n}$. The connection gadgets $\{K_i\}_{i=0}^{s}$ are cliques of size $k + 2$. The selection gadget S_1 contains n vertices $a_1, a_2, \ldots, a_n$ and no edges within S_1. Similarly, S_2 contains n vertices $b_1, b_2, \ldots, b_n$ and no edges within S_2.

For each $1 \le i, j \le m$ we add an edge (c_i, c'_j). For each $1 \le i, j \le m$ subdivide the edge (c_i, c'_j) $s-2$ times and let the new vertices be named $t_{ij}(1), \ldots, t_{ij}(s-2)$. Let T denote the set of these newly introduced subdivision vertices. For each $1 \le i \le m$ delete the edge $(t_{ii}(\lceil \frac{s-2}{2} \rceil), t_{ii}(\lceil \frac{s-2}{2} \rceil + 1))$. Further, add the edges $(k_l, t_{ij}(l))$ for all $1 \le i, j \le m$, $1 \le l \le s - 2$.

If a clause C_i contains a literal x_j then we add two edges $(t_{ii}(\lceil \frac{s-2}{2} \rceil), x_j)$ and $(t_{ii}(\lceil \frac{s-2}{2} \rceil + 1), x_j)$. See Fig 2 for a sketch of the clause gadgets and its connection with the variable gadget as described so far.

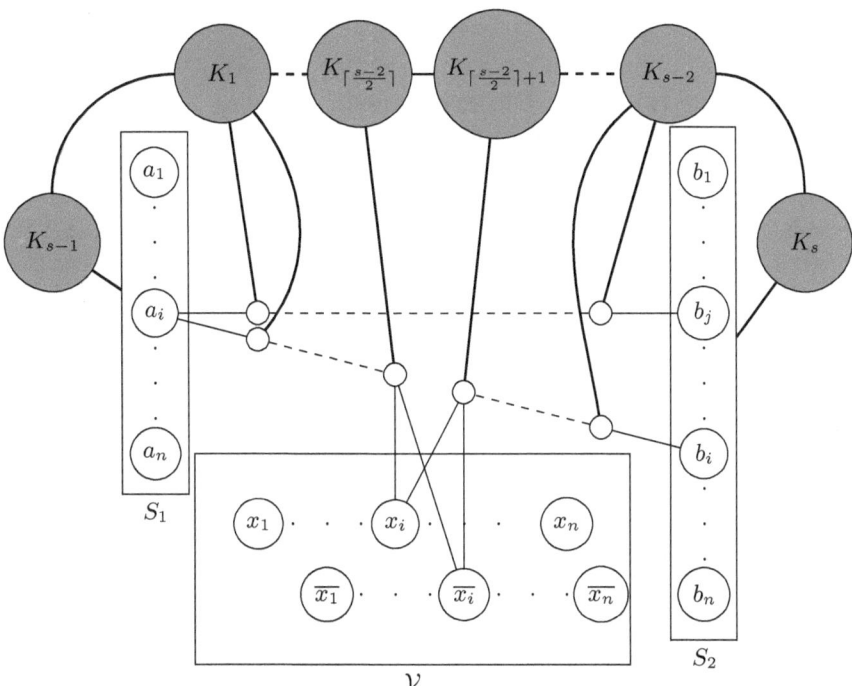

Fig. 3. Example of connection within selection gadget and between selection gadget and variable gadget. Thick lines are used to represent all possible edges between two sets of vertices.

We now perform an analogous construction between the selection gadgets. For each $1 \leq i, j \leq n$ we add an edge (a_i, b_j). For each $1 \leq i, j \leq n$ subdivide the edge (a_i, b_j) $s - 2$ times and let the new vertices be named $u_{ij}(1), \ldots, u_{ij}(s - 2)$. In this case, let U denote the set of these newly introduced subdivision vertices. For each $1 \leq i \leq n$ delete the edge $(u_{ii}(\lceil \frac{s-2}{2} \rceil), u_{ii}(\lceil \frac{s-2}{2} \rceil + 1))$. For each $1 \leq j \leq n$ add edges $(u_{jj}(\lceil \frac{s-2}{2} \rceil), x_j), (u_{jj}(\lceil \frac{s-2}{2} \rceil + 1), x_j), (u_{jj}(\lceil \frac{s-2}{2} \rceil), \overline{x_j})$ and $(u_{jj}(\lceil \frac{s-2}{2} \rceil), \overline{x_j})$. We add all possible edges between K_l and $t_{ij}(l)$ for all $1 \leq i, j \leq n$, $1 \leq l \leq s - 2$. Finally, we add all possible edges between K_i and K_{i+1} for all $1 \leq i \leq s - 3$. We add all possible edges between K_{s-2} and K_s, between K_{s-1} and K_1, between K_{s-1} and S_1, between K_s and S_2. Fig. 3 shows the selection gadget and its connection with variable gadget. We set $k = 4m+2n$. This concludes the description of the construction. Due to space constraints, we defer the proof of correctness to the full version of the paper. □

Theorem 6. s-Club d-Cluster Edge Deletion *for $s \geq 4$ and $d \geq 2$ cannot be solved in time $2^{o(k)}n^{O(1)}$, unless ETH fails.*

5   Conclusions

In this work, we established that assuming the ETH, there is no algorithm solving the s-CLUB CLUSTER EDGE DELETION question in time $2^{o(k)}n^{O(1)}$. We also showed that even the problem of deleting edges to obtain a graph with d s-clusters cannot be solved in time $2^{o(k)}n^{O(1)}$ for any $s \geq 2$.

In the context of cluster editing, the exact and approximation results are consistent, in that the general Cluster Editing problem is APX-hard, and does not admit a sub-exponential algorithm unless the ETH fails. On the other hand, the problem of deleting into a sub-linear number of cliques allows for both a sub-exponential algorithm and a PTAS. A natural direction would be to pursue the approximation of these problems so as to either establish or disprove a similar connection in the context of s-clubs.

References

1. Alba, R.D.: A graph-theoretic definition of a sociometric clique. Journal of Mathematical Sociology 3, 113–126 (1973)
2. Flum, J., Grohe, M.: Parameterized Complexity Theory. Texts in Theoretical Computer Science. An EATCS Series. Springer-Verlag New York, Inc., Secaucus (2006)
3. Fomin, F.V., Kratsch, S., Pilipczuk, M., Pilipczuk, M., Villanger, Y.: Subexponential fixed-parameter tractability of cluster editing. To appear in STACS 2013, abs/1112.4419 (2013)
4. Guo, J., Komusiewicz, C., Niedermeier, R., Uhlmann, J.: A more relaxed model for graph-based data clustering: s-plex cluster editing. SIAM J. Discrete Math. 24(4), 1662–1683 (2010)
5. Hartung, S., Komusiewicz, C., Nichterlein, A.: Parameterized algorithmics and computational experiments for finding 2-clubs. In: Thilikos, D.M., Woeginger, G.J. (eds.) IPEC 2012. LNCS, vol. 7535, pp. 231–241. Springer, Heidelberg (2012)
6. Hartung, S., Komusiewicz, C., Nichterlein, A.: On structural parameterizations for the 2-club problem. In: van Emde Boas, P., Groen, F.C.A., Italiano, G.F., Nawrocki, J., Sack, H. (eds.) SOFSEM 2013. LNCS, vol. 7741, pp. 233–243. Springer, Heidelberg (2013)
7. Komusiewicz, C.: Parameterized Algorithmics for Network Analysis: Clustering & Querying. PhD thesis, Technische Universität Berlin, Berlin, Germany (2011)
8. Liu, H., Zhang, P., Zhu, D.: On editing graphs into 2-club clusters. In: Snoeyink, J., Lu, P., Su, K., Wang, L. (eds.) FAW-AAIM 2012. LNCS, vol. 7285, pp. 235–246. Springer, Heidelberg (2012)
9. Lokshtanov, D., Marx, D., Saurabh, S.: Lower bounds based on the exponential time hypothesis. Bulletin of the EATCS 105, 41–72 (2011)
10. Niedermeier, R.: Invitation to Fixed Parameter Algorithms. Oxford Lecture Series in Mathematics and Its Applications. Oxford University Press, USA (2006)
11. Pasupuleti, S.: Detection of protein complexes in protein interaction networks using n-clubs. In: Marchiori, E., Moore, J.H. (eds.) EvoBIO 2008. LNCS, vol. 4973, pp. 153–164. Springer, Heidelberg (2008)
12. Schäfer, A., Komusiewicz, C., Moser, H., Niedermeier, R.: Parameterized computational complexity of finding small-diameter subgraphs. Optimization Letters 6(5), 883–891 (2012)

Unlimited Decidability of Distributed Synthesis with Limited Missing Knowledge*

Anca Muscholl[1] and Sven Schewe[2]

[1] LaBRI, CNRS, Universiy of Bordeaux, France
[2] University of Liverpool, UK

Abstract. We study the problem of controller synthesis for distributed systems modelled by Zielonka automata. While the decidability of this problem in full generality remains open and challenging, we suggest here to seek controllers from a parametrised family: we are interested in controllers that ensure frequent communication between the processes, where frequency is determined by the parameter. We show that this restricted controller synthesis version is affordable for synthesis standards: fixing the parameter, the problem is EXPTIME-complete.

1 Introduction

The synthesis problem has a long tradition that goes back to Church's solvability problem [4], which asks for devices that generate output streams from input streams, such that a given specification is met. Synthesis of sequential systems has been thoroughly studied and driven various results for infinite 2-player games (see [15] for a survey).

Synthesis of *distributed systems* has a bad reputation. Many possible variants of distributed synthesis could be considered. But in the best known and most studied one, initiated by Pnueli and Rosner [26], the problem is undecidable in general – cf. also variations [18] and generalisations [22,9] thereof. These models extend Church's formulation to a fixed architecture of *synchronously* communicating processes that exchange messages through one-slot communication channels. Undecidability in this setting comes mainly from *partial information*: architectures (the communication topology) restrict the flow of information about the global system state. Synthesis in a given architecture is decidable, iff this partial knowledge defines a preorder on the processes [9]. The complexity of the decision problem is non-elementary in the number of quotients. When extended to asynchronous communication with one-slot channels, only systems where a single process needs to be synthesised remain decidable [30].

We use here a different synchronisation model, based on shared variables and known as Zielonka automata [31]. In this model, processes that execute shared actions get full information about the states of the processes with whom they synchronise. Therefore, partial information is reduced to concurrency: the only missing knowledge that a process might have concerns those events that happen concurrently. Partial information in this model is therefore minimalistic, in the sense that it is not driven by the specification or the architecture. As a consequence, establishing the (un)decidability of distributed synthesis in this setting has proven to be challenging and remains open. We know, however, of some non-trivial cases where the problem is decidable. The first one

---

* The research was supported through a visiting professorship by the University of Bordeaux and by the Engineering and Physical Sciences Research Council grant EP/H046623/1.

K. Chatterjee and J. Sgall (Eds.): MFCS 2013, LNCS 8087, pp. 691–703, 2013.
© Springer-Verlag Berlin Heidelberg 2013

[23] imposes a bound on the missing knowledge of a process concerning the evolution of other processes. This restriction mainly says that every event in the system may have only a bounded number of concurrent events. In this setting, the distributed game can be reduced to a 2-player game with complete information. The proof of [23] actually uses Rabin's theorem about decidability of monadic second-order logic over infinite trees. The second decidability result is based on a restriction on the distributed alphabet of actions [11], which needs to be a co-graph, and it applies to global reachability conditions. More recently, it has been shown that distributed synthesis with local reachability conditions is decidable under the assumption that the synchronisation graph is acyclic [13]. The exact complexity is non-elementary in the depth of the synchronisation tree. For instance, it is EXPTIME for trees of depth 1, as for architectures involving one server and several clients. The decidability proof also involves a reduction to a 2-player game.

The complexity of distributed synthesis with shared variables is therefore forbiddingly high, unless the class of strategies under consideration is restricted. The reason for this high complexity is, once again, the partial knowledge a process has about other processes. In the acyclic case studied in [13], partial knowledge is hierarchical. This resembles the situation from Pnueli and Rosner's setting [26,18,22,9], and similarly increases the complexity by one exponent for each additional level of the hierarchy.

With this observation in mind, we reconsider the result of [23] and restrict the class of strategies in such a way, that missing knowledge is uniformly limited. The restriction on strategies is very similar to the notion of N-communicating plants used in [23] to show decidability of monadic second-order logic over the event structure associated with the plant. The main differences are that (1) we do not require that the N-communicating restriction is made explicit in the plant, but more liberally look for strategies that impose N-communication on the controlled system, (2) the bound N applies only to synchronisation events: there is no limitation of local actions, and (3) the winning condition is local on each process. The first condition above is reminiscent of the bounded-context restriction used in model-checking [27], where local computations are unrestricted and only context-switches are limited. To keep the presentation simple, we do not consider divergent infinite plays, where two disjoint groups of processes can synchronise infinitely often in parallel (our result can be adapted to include this case).

Our main result is that the existence of distributed strategies for a system described by a Zielonka automaton $\mathcal{A}$ and a fixed bound N is exponential in the size of $\mathcal{A}$ and doubly exponential in N. If N is fixed, then the problem is EXPTIME-complete.

Related Work. The restriction to solutions that obey various bounds in synthesis [10,17,8,6,2] has been inspired by similar restrictions in model-checking, e.g., in bounded model-checking [3] and model-checking with bounded context switches [27,1].

The first two bounds used in synthesis were bounds on the size of the model [10,17] and bounds on the number of rejecting states [19,10] in emptiness equivalent determinisation procedures from universal Co-Büchi automata to deterministic Büchi [19] and safety [10] automata. The latter approach has been implemented by different groups [8,6], while the first has been extended to quantitative specification languages [2], as well as to restrictions on the size of symbolic representations of implementations [7,20]. The implementations of genetic synthesis algorithms in [16] is of the same kind, as the fitness functions used effectively restrict the size of the synthesised programs.

2 Zielonka Automata

Informally, Zielonka automata are parallel compositions of finite-state processes that synchronise on shared actions. There is no global clock, so between two synchronisations, two processes can perform a different number of actions. Because of this, Zielonka automata are also called asynchronous automata.

A Zielonka automaton has a (fixed) assignment of actions to sets of processes. A *distributed action alphabet* on a finite set $\mathbb{P}$ of processes is a pair (Σ, dom), where Σ is the finite set of *actions* and $dom : \Sigma \to (2^{\mathbb{P}} \setminus \emptyset)$ is the *location function*. The location $dom(a)$ of an action $a \in \Sigma$ comprises all processes that synchronise in order to perform a. Similar to other classical synchronisation mechanisms, e.g., CCS-like rendez-vous or Petri net transitions, executing a shared action is only possible if the states of all processes in $dom(a)$ allow to execute a. In addition, the execution of a shared action allows to "broadcast" some information between its processes: for instance, an action shared between processes p and q may produce a swap between the states of p and q. Related concepts are used in multithreaded programming, where atomic instructions like compare-and-swap (CAS) allow to exchange values between two processes.

A (deterministic) *Zielonka automaton* $\mathcal{A} = \langle \{S_p\}_{p \in \mathbb{P}}, s_{in}, \{\delta_a\}_{a \in \Sigma}, F \rangle$ is given by

- a finite set S_p of (local) states for every process p,
- the initial state $s_{in} \in \prod_{p \in \mathbb{P}} S_p$, a set $F \subseteq \prod_{p \in \mathbb{P}} S_p$ of accepting states, and
- a partial transition function $\delta_a : \prod_{p \in dom(a)} S_p \dashrightarrow \prod_{p \in dom(a)} S_p$ for every action $a \in \Sigma$, acting on tuples of states of processes in $dom(a)$.

For convenience, we abbreviate a tuple $(s_p)_{p \in P}$ of local states by s_P, where $P \subseteq \mathbb{P}$. We also refer to S_p as the set of *p-states* and of $\prod_{p \in \mathbb{P}} S_p$ as *global states*.

A Zielonka automaton can be seen as a sequential automaton with the state set $S = \prod_{p \in \mathbb{P}} S_p$ and transitions $s \xrightarrow{a} s'$ if $(s_{dom(a)}, s'_{dom(a)}) \in \delta_a$, and $s_{\mathbb{P} \setminus dom(a)} = s'_{\mathbb{P} \setminus dom(a)}$. By $L(\mathcal{A})$ we denote the language of this sequential automaton.

This definition has an important consequence. The location mapping dom defines in a natural way an independence relation $I \subseteq \Sigma \times \Sigma$: two actions $a, b \in \Sigma$ are independent (written as $(a, b) \in I$) if the processes they involve are disjoint, that is, if $dom(a) \cap dom(b) = \emptyset$. Note that the order of execution of two independent actions $(a, b) \in I$ in a Zielonka automaton is irrelevant, they can be executed as a, b, or b, a – or even concurrently. More generally, we can consider the congruence $\sim_I$ on Σ^* generated by I, and observe that, whenever $u \sim_I v$, the state reached from the initial state on u and v, respectively, is the same. Hence, $u \in L(\mathcal{A})$ if, and only if, $v \in L(\mathcal{A})$. We denote u, v as *trace-equivalent* whenever $u \sim_I v$ (and write $u \sim v$ for simplicity).

The idea of describing concurrency by an independence relation on actions was introduced in the late seventies by Mazurkiewicz [24] (see also [5]). An equivalence class $[w]$ of $\sim$ is called a Mazurkiewicz *trace*, it can be viewed as a labelled pomset. We will often refer to a trace using just a word w instead of writing $[w]$. As we have observed $L(\mathcal{A})$ is a sum of such equivalence classes. In other words, the language of a Zielonka automaton is *trace-closed*.

Actions a with $|dom(a)| = 1$ are called *local*, and Σ^{loc} is the set of local actions. If $|dom(a)| > 1$ then a is called *synchronisation action*, and Σ^{sync} is the set of such actions. Actions from $\Sigma_p = \{a \in \Sigma \mid p \in dom(a)\}$ are denoted as *p-actions*. We write $\Sigma_p^{sync} = \Sigma^{sync} \cap \Sigma_p$ and $\Sigma_p^{loc} = \Sigma^{loc} \cap \Sigma_p$. For $u \in \Sigma^*$ we write $state_p(u)$ for the *p*-state reached by $\mathcal{A}$ on u.

Example 1. Consider the example automaton with processes $P_1, \ldots, P_n, S_1, \ldots, S_m$ as shown in Figure 1. Here, processes $S_1, \ldots, S_m$ are backup servers and each of the processes P_i loops on a sequence of internal actions, followed by backup actions on some server. We abstract this by the following distributed alphabet: ℓ_i, b_i are local actions of P_i (i.e., $dom(\ell_i) = dom(b_i) = \{P_i\}$), where b_i denotes a backup request on P_i, and $s_{i,k}$ is a shared (backup) action with $dom(s_{i,k}) = \{P_i, S_k\}$. Action $s_{i,k}$ is enabled if P_i is in state 1_i. Actions $\ell_i, b_i, s_{i,k}$ ($k = 1, \ldots, m$) are P_i-actions, and $\Sigma^{sync} = \{s_{i,k} \mid i, k\}$. Note also that s_{i_1,k_1}, s_{i_2,k_2} are independent iff $i_1 \neq i_2$ and $k_1 \neq k_2$.

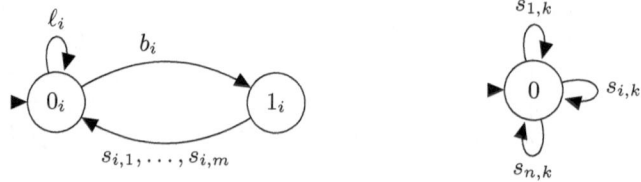

Fig. 1. Example Zielonka automaton; Process P_i on the left and server S_k on the right

A major result about Zielonka automata is stated in the theorem below. Note that it is one of the few examples of synthesis of (*closed*) distributed systems.

Theorem 1. *[31] Let dom : $\Sigma \to (2^{\mathbb{P}} \setminus \{\emptyset\})$ be a distribution of letters. If a language $L \subseteq \Sigma^*$ is regular and trace-closed, then there is a deterministic Zielonka automaton recognising L. Its size is exponential in the number of processes and polynomial in the size of the minimal automaton for L [12].*

3   Distributed Control and Games

The synthesis problem considered here was proposed in [23]. It can be viewed as a distributed instantiation of supervisory control, as considered in the framework of Ramadge and Wonham [28]. In supervisory control, one is given a plant $\mathcal{A}$, together with a partition $\Sigma = \Sigma^{sys} \,\dot\cup\, \Sigma^{env}$ of Σ into controllable actions Σ^{sys} and uncontrollable actions Σ^{env}. As in [23,13] we assume that all uncontrollable actions are local, $\Sigma^{env} \subseteq \Sigma^{loc}$. The goal is to synthesise a controller $\mathcal{C}$, which is a device that never blocks uncontrollable actions. The controlled plant is then the product of $\mathcal{A}$ and $\mathcal{C}$, and it needs to satisfy additional conditions like safety, reachability, or parity conditions.

We will work with the game description of the controller problem, and start by illustrating it on an example.

Example 2. Reconsider the automaton from Figure 1 and assume that local actions are uncontrollable, whereas synchronisation actions are controllable. In this model, the environment decides whether a process P_i continues to use local transitions or needs backup, while the system is in charge of deciding each time, on which server(s) the backup is made. One possible objective of the control strategy could be to achieve a balanced use of the servers, which avoids using certain servers more often than others. A round-robin strategy on each process P_i, e.g., that asks each time for backup on the next server, guarantees that a server S_k can be "behind" the other servers by at most $O(nm)$ backup actions.

The game formulation refers to a game between the distributed *system* and *local environments*, one for each process. A Zielonka automaton $\mathcal{A}$ defines a game arena, with plays corresponding to initial runs. Since $\mathcal{A}$ is deterministic, we can view a play x as a word from $L(\mathcal{A})$ – or a trace, since $L(\mathcal{A})$ is trace-closed. Let $Plays(\mathcal{A})$ denote the set of traces associated with words from $L(\mathcal{A})$.

A strategy for the system will be a collection of individual strategies for each process. The important notion here is the view each process has of the global state of the system. Intuitively, this is the part of the current play that the process could see or learn about from other processes by synchronising with them. Formally, the p-view of a play x, denoted $view_p(u)$, is the smallest trace $[v]$ such that $u \sim vy$ and y contains no action from Σ_p. We write $Plays_p(\mathcal{A})$ for the set of plays that are p-views: $Plays_p(\mathcal{A}) = \{view_p(u) \mid u \in Plays(\mathcal{A})\}$.

A *strategy for a process* p is a function $\sigma_p : Plays_p(\mathcal{A}) \to 2^{\Sigma_p^{sys}}$, where $\Sigma_p^{sys} = \{a \in \Sigma^{sys} \mid p \in dom(a)\}$. We require in addition, for every $u \in Plays_p(\mathcal{A})$, that $\sigma_p(u)$ is a subset of the set enabled(s_p) of actions that are enabled in $s_p = state_p(u)$. A *strategy* is a family of strategies $\{\sigma_p\}_{p \in \mathbb{P}}$, one for each process.

The set of plays respecting a strategy $\sigma = \{\sigma_p\}_{p \in \mathbb{P}}$, denoted $Plays(\mathcal{A}, \sigma)$, is the smallest set that contains the empty play ε and that satisfies, for every $u \in Plays(\mathcal{A}, \sigma)$, the following two conditions: (1) if $a \in \Sigma^{env}$ and $ua \in Plays(\mathcal{A})$, then ua is in $Plays(\mathcal{A}, \sigma)$, and (2) if $a \in \Sigma^{sys}$ and $ua \in Plays(\mathcal{A})$, then $ua \in Plays(\mathcal{A}, \sigma)$ provided that $a \in \sigma_p(view_p(u))$ *for all* $p \in dom(a)$. So this definition says that actions of the environment are always possible / enabled, whereas actions of the system are possible only if they are allowed by the strategies of all involved processes.

Before defining winning (control) strategies, we need to introduce infinite plays that are consistent with a given strategy σ. Such plays can be viewed as (infinite) traces associated with infinite initial runs of $\mathcal{A}$ that satisfy both conditions of the definition of $Plays(\mathcal{A}, \sigma)$. The precise definition is very intuitive when using pomsets, here we just give an example: the infinite play $a^\omega b^\omega$ is the set of all ω-words with infinitely many as and infinitely many bs. We write $Plays^\infty(\mathcal{A}, \sigma)$ for the set of such *finite or infinite* plays. A play from $Plays^\infty(\mathcal{A}, \sigma)$ is also denoted as σ-*play*. A play $u \in Plays^\infty(\mathcal{A}, \sigma)$ is called *maximal*, if there is no action c such that $uc \in Plays^\infty(\mathcal{A}, \sigma)$.

Winning conditions. In analogy to regular 2-player games, winning conditions in these games can be provided by regular, trace-closed languages [23]. In this paper, we consider simpler conditions, namely *local parity* conditions, because we are interested in the game complexity and do not want to add the specification as an extra parameter.

Our system $\mathcal{A}$ is thus a deterministic Zielonka automaton with local states, coloured by integers from $[k] = \{0, \ldots, k-1\}$: let $\mathcal{A} = \langle (S_p)_{p \in \mathbb{P}}, (\delta_a)_{a \in \Sigma}, s^0, \chi \rangle$, $\chi : \bigcup_{p \in \mathbb{P}} S_p \to [k]$. A maximal play $u \in Plays^\infty(\mathcal{A}, \sigma)$ is winning, if the following holds for every process p. Write $view_p(u)$ as $u_0 u_1 \ldots$, for $u_0, u_1, \ldots$ such that, for every n, we have that $view_p(u_0 \cdots u_n) = u_0 \cdots u_n$ and either u_n is empty or it has only one p-action (which is the last one). Then we require that $\liminf_{n \to \infty} \chi(state_p(u_0 \cdots u_n))$ is even. Equivalently, if $view_p(u)$ is infinite, then the local parity condition should hold, and if $view_p(u)$ is finite, then the colour of the last state reached by p needs to be even. A strategy σ is winning, if every maximal play in $Plays^\infty(\mathcal{A}, \sigma)$ is winning. Maximality is a sort of fairness condition for such automata. Requiring infinite plays as in [23] is also possible, but it does not guarantee fairness for each process. A more refined notion of fairness can be found in [14].

Remark 1. The decidability of the existence of a winning distributed control strategy for systems modelled by Zielonka automata is an open problem. It is worth noting that slight modifications of the problem statement lead to undecidability. First, if one uses regular, but *not trace-closed* specifications, then the problem is known to be undecidable (see e.g. [21]). Second, if the individual strategy σ_p only depends only on the local history of process p (i.e., $\sigma_p : (\Sigma_p)^* \to 2^{\Sigma_p^{sys}}$), then the problem is again undecidable [21]. In both cases, undecidability stems from the *restricted* partial knowledge of the processes.

4 Resuming Local Behaviour

Recall that $\Sigma^{env} \subseteq \Sigma^{loc}$, i.e., environment actions are local. As shown in this section, this allows to summarise local behaviour, such that one can reason about distributed strategies only w.r.t. synchronisation actions.

Lemma 1. *Let $\mathcal{A}$ be a Zielonka automaton. If there is a winning control strategy $\sigma = (\sigma_p)_{p \in \mathbb{P}}$ for $\mathcal{A}$, then there also exists a winning one that satisfies, for every process $p \in \mathbb{P}$ and every play $t \in Plays_p(\mathcal{A}, \sigma)$, either $\sigma_p(t) = \{a\}$ for some $a \in \Sigma_p^{loc} \cap \Sigma^{sys}$, or $\sigma_p(t) \subseteq \Sigma_p^{sync}$. In addition, $\sigma_p(t) = \emptyset$ if $enabled(state_p(t)) \cap \Sigma_p^{env} \neq \emptyset$.*

The proof exploits that, when both local and synchronisation actions are enabled, disabling the synchronisation actions reduces the set of plays, but they are still winning.

Definition 1. *Fix some process p. A local p-play is a word from $(\Sigma_p^{loc})^*$. A p-context is a play from $Plays_p(\mathcal{A})$ that ends with an action from Σ_p^{sync} (unless it is empty).*
Given a distributed strategy $(\sigma_p)_{p \in \mathbb{P}}$, we associate with a p-context u a local strategy from u: this is the mapping $\sigma_p[u] : (\Sigma_p^{loc})^* \to 2^{\Sigma_p}$ defined as

$$\sigma_p[u](x) := \sigma_p(ux) \qquad \text{for all } x \in (\Sigma_p^{loc})^*.$$

We assume in the following that $\sigma = (\sigma_p)_{p \in \mathbb{P}}$ satisfies Lemma 1, thus $\sigma_p[u] : (\Sigma_p^{loc})^* \to (\Sigma_p^{loc} \cap \Sigma^{sys}) \cup 2^{\Sigma_p^{sync}}$. We are interested in the configurations that result after a *maximal* local run of process p from a given p-context u with $s_p = state_p(u)$. We define:

$$Sync^\sigma(p, u) = \{(s'_p, A, c) \mid \exists x \in (\Sigma_p^{loc})^* : s'_p = state_p(ux), A = \sigma_p(ux) \subseteq \Sigma_p^{sync},$$

$$enabled(s'_p) \cap \Sigma_p^{env} = \emptyset, \text{ and the minimal colour seen on } s_p \xrightarrow{x} s'_p \text{ is } c\}.$$

A local strategy $\sigma_p[u]$ is called *simple* if, for every $(s'_p, A, c), (s''_p, A', c') \in Sync^\sigma(p, u)$, we have that $s'_p = s''_p$ implies $A = A'$. In this case $Sync^\sigma(p, u)$ is a partial mapping $Sync^\sigma(p, u) : S_p \dashrightarrow 2^{\Sigma_p^{sync}} \times 2^{[k]}$.

A local strategy $\sigma_p[u]$ from context u is computable with memory M if $\sigma_p[u](x)$ can be computed from $state_p(ux)$ using an additional *finite* memory M. In this case, $\sigma_p[u]$ is a mapping from $S_p \times M$ to $(\Sigma_p^{loc} \cap \Sigma^{sys}) \cup 2^{\Sigma_p^{sys}}$.

Lemma 2. *If there is a winning control strategy $\sigma = (\sigma_p)_{p \in \mathbb{P}}$ for $\mathcal{A}$ with a local parity condition with k colours, then there is also a winning one, say $\tau = (\tau_p)_{p \in \mathbb{P}}$, where, for each process p and every p-context u, the local strategy $\tau_p[u]$ is simple, computable with memory of size k, and such that every infinite (and thus local) $\tau_p[u]$-play satisfies the parity condition for process p.*

The proof exploits that, for every s'_p that can occur at the end of a run, one can select a triple (s'_p, A, c) with worst color among the elements of $Sync^\sigma(p, u)$ and then change the decision for each such end-point s'_p to A. It is then easy to turn the resulting simple local strategy into one, where the decision is only based on the state and the minimal colour that occurred so far.

We denote local strategies $\tau_p[u]$ as in Lemma 2, as *good* strategies. In Section 6 we will compose good strategies, and we therefore define their *outcomes*.

Definition 2. *Let u be a p-context. The outcome of a simple strategy $\tau_p[u]$ is a partial mapping $f : S_p \dashrightarrow 2^{\Sigma_p^{sync}} \times [k]$ that satisfies the following side constraints:*

1. *f and $Sync^\tau(p, u)$ have the same domain, and*
2. *for each state s_p in the domain of f: if $Sync^\tau(p, u)(s_p) = (A, C)$ for some $A \subseteq \Sigma_p^{sync}$ and $C \subseteq [k]$ then $f(s_p) = (A, c)$ where:*
   - *either $C \nsubseteq 2\mathbb{N}$, c is odd and $c \leq d$ for every odd colour $d \in C$, or*
   - *$C \subseteq 2\mathbb{N}$, c is even, and $c \geq \max(C)$.*

Remark 2. Note that we can test, for given $s_p \in S_p$ and partial mapping $f : S_p \dashrightarrow 2^{\Sigma_p^{sync}} \times [k]$, whether f is the outcome of a good local strategy from state s_p. The test amounts to solving a 2-player game with parity condition on infinite plays. Finite plays are won if the last state, say t_p, is in the domain of f. In addition, if $f(t_p) = (A, c)$ for some A, then t_p can be reached only with even minimal colours $d \leq c$ if c is even. If c is odd, and t_p is reached with odd minimal colour d, then $d \geq c$. The condition on colours can be checked using additional memory k.

5 Well-Informed Strategies

We start by defining the distributed strategies we are interested in. They are very similar to the notion of N-communicating plants used in [23], with two exceptions. First, our bound N applies only to synchronisation actions. That is, there is no limitation on local actions. Second, our definition implies that infinite plays are non-divergent, which is a restriction that we impose only for simplifying the presentation (see also Remark 4).

Definition 3. *Let $N > 0$ be an integer. A strategy $\sigma = (\sigma_p)_{p\in\mathbb{P}}$ is called N-informed if, for every play $u \in Plays^\infty(\mathcal{A}, \sigma)$ such that $u = u'av$ with $a \in \Sigma^{sync}$, $v \in \Sigma^\infty$ and $dom(a) \cap dom(v) = \emptyset$, it holds that v has at most N actions from Σ^{sync}.*

The round-robin strategy mentioned in Example 2 is N-informed with $N \in O(nm)$. Note that Lemma 2 preserves N-informedness, since only local strategies are modified and Def. 3 refers only to synchronisation actions. By abuse of notations, we will call a sequence from $(\Sigma^{sync})^\infty$ N-*informed*, if it satisfies the above definition.

Let us fix some total order $<$ on Σ^{sync}. A sequence $u \in (\Sigma^{sync})^*$ is said to be in lexicographic normal form (w.r.t. $<$) if there is no trace-equivalent sequence $u' \sim u$ such that $u = vbw$, $u' = vaw'$ with $a < b$. We denote by $\mathsf{lnf}(u)$ the trace-equivalent sequence $v \sim u$ that is in lexicographic normal form. Sequences in lexicographic normal form build a regular set: a sequence $u \in (\Sigma^{sync})^*$ is *not* in lexicographic normal form iff, for some $x, y, z \in (\Sigma^{sync})^*$:

$$u = xbyaz \quad \text{with } dom(a) \cap dom(by) = \emptyset \text{ and } a < b. \tag{1}$$

Let $\mathfrak{s} = \max_{p\in\mathbb{P}} |S_p|$, $\mathfrak{c} = \max_{p\in\mathbb{P}} |\Sigma_p^{sync}|$, $\mathfrak{p} = |\mathbb{P}|$, and recall that k is the number of colours. A deterministic safety automaton of size $O(\mathfrak{c} \cdot 2^{\mathfrak{p}})$ exists that accepts the set of sequences in lexicographic normal form. This automaton records, for every $a \in \Sigma^{sync}$, the set of processes in whose view the last a occurs.

The next lemma considers how the lexicographic normal form changes when extending a sequence from $(\Sigma^{sync})^*$, showing that only a bounded suffix is modified. The lemma is essentially the same as Lemma 3 in [23]:

Lemma 3. *[23] Let $u \in (\Sigma^{sync})^*$ be an N-informed sequence and $a \in \Sigma^{sync}$. Then $\mathsf{lnf}(u) = zx$ and $\mathsf{lnf}(u \cdot a) = zy$ for some x, y, z with $|x| \leq N$.*

The next lemma makes the statement of Lemma 3 more precise:

Lemma 4. *Let $u, v \in (\Sigma^{sync})^*$ be N-informed and $p \in \mathbb{P}$ such that $u = view_p(v)$ and both u, v are in lexicographic normal form. Then we can write $u = zx$ and $v = zy$, with $y = y_0 x_1 y_1 x_2 \cdots y_{m-1} x_m y_m$ for some $m \leq \mathfrak{c} \cdot \mathfrak{p}$, such that (1) $|x| \leq N$, (2) $x = x_1 \cdots x_m$, and (3) $dom(y_i) \cap dom(x_{i+1} \cdots x_m) = \emptyset$ for every $i < m$ hold.*

6 Strategy Trees

Let $\Omega = \prod_{p\in\mathbb{P}} \Omega_p$, where Ω_p is a set of tuples (s_p, f, b), where f is outcome of some good local p-strategy τ from state s_p and $b \in \{0, 1\}$ says whether τ allows infinite (local) plays from s_p. Let $\Delta = \bigcup_{a\in\Sigma^{sync}} \delta_a$. A *strategy tree* is an infinite tree with directions $\Gamma = \Sigma^{sync} \times \Delta$ and nodes labelled by elements of $\Omega \cup \{\bot\}$. Note that $|\Gamma| \leq \mathfrak{c} \cdot |\mathcal{A}|$. A node in the tree is identified with the sequence from Γ^* labelling the path from the root to that node. We require for every pair of nodes $u, u \cdot \langle a, d \rangle \in \Gamma^*$:

1. $d \in \delta_a$, and
2. if the labels of u and $u \cdot \langle a, d \rangle$ are ω' and ω, resp., then $\omega'_q = \omega_q$ for all $q \notin dom(a)$.

For $u \in \Gamma^*$, we denote by $state_p(u)$ the state of process p after u. This is namely the state that occurs in the transition d of the last pair $\langle a, d \rangle$ with $p \in dom(a)$.

Nodes in a strategy tree that do not correspond to *realisable* summarised plays, are called *sink* nodes and are labelled by $\bot$. A node $u \cdot \langle a, d \rangle$ with label ω' is a sink, but u is not, if, and only if, (1) either the action a is not allowed by $\omega_{dom(a)}$, where ω is the label of u, (2) or the a-transition d does not result from $\omega_{dom(a)}$, (3) or $\omega'_p \neq (t_p, *, *)$ for some $p \in dom(a)$ with $t_p = state_p(u \cdot \langle a, d \rangle)$.

Remark 3. If we fix local good strategies for every pair of local states and outcomes from $S_p \times \Omega_p$, then every initial path $\omega_0, \langle a_0, d_0 \rangle, \omega_1, \langle a_1, d_1 \rangle \ldots$ in a strategy tree such that all $\omega_i \neq \bot$ can be "expanded" in a natural way to a set of plays from $Plays(\mathcal{A})$.

Lemma 5. *A deterministic safety automaton with $O(\mathfrak{s}^{\mathfrak{p}})$ states can check that the label $\bot$ correctly identifies sink nodes.*

A strategy tree is *winning* if all plays that are expansions of maximal initial paths of the tree, are winning. Checking consistency of the labels ω requires to include in the label of each node $u \in \Gamma^*$ a bit that reflects whether or not the projection of u on Σ^{sync} is in lexicographic normal form. As mentioned in Section 5, a deterministic safety automaton with $O(\mathfrak{c} \cdot 2^{\mathfrak{p}})$ states can check that this labelling is correct. In the following we will focus on non-sink nodes in lexicographic normal form. We will refer to them as *normalised* nodes.

We will need to ensure that a strategy tree has a consistent node labelling. The next definition tells when two nodes $u, u' \in \Gamma^*$ correspond to the same summarised play.

Definition 4. *Two nodes $u, u' \in \Gamma^*$ are called* play-equivalent *if the following hold:*

1. *The projections of u, u' onto Σ^{sync} are trace-equivalent.*
2. *For all $a \in \Sigma^{sync}$ and $k \geq 0$: suppose that $u = u_1 \langle a, d \rangle u_2$ and $u' = u'_1 \langle a, d' \rangle u'_2$, with $\langle a, d \rangle, \langle a, d' \rangle$ being the k-th occurrence of a in u and u', respectively. Assume also that u_1 and u'_1 is labelled by ω and ω', respectively. Then we require that*

$$d = d' \qquad and \qquad \omega_{dom(a)} = \omega'_{dom(a)} .$$

We ensure that the strategy tree is labelled consistently by local strategies by comparing normalised nodes that are play-equivalent. By abuse of notation, we write $view_p(u)$ for the p-view of $u \in \Gamma^*$.

Definition 5. *A strategy tree is* labelled consistently *if, for all normalised non-sink nodes $u, v \in \Gamma^*$ and every process p such that u and $view_p(v)$ are play-equivalent, it holds that $\omega_p = \omega'_p$, where ω and ω' are the labels of u and v, respectively.*

Informally, the strategy tree is labelled consistently if the choice of the next outcomes $\omega_{dom(a)}$ after some synchronisation $a \in \Sigma^{sync}$ depends only on the history associated with the views of processes $p \in dom(a)$ after a.

Lemma 6. *Every good control strategy (cf. Lemma 2) maps to a consistently labelled strategy tree. Conversely, from every consistently labelled strategy tree we can construct a good control strategy.*

Lemma 7. *An alternating safety automaton with $O(|\Gamma|^N \cdot N \cdot \mathfrak{s} \cdot \mathfrak{c} \cdot k)$ states can check that a strategy tree associated with an N-informed strategy is labelled consistently.*

Proof. Let $u, v \in \Gamma^*$ be as in Definition 5, so in particular normalised, non-sink, and such that u and $view_p(v)$ are play-equivalent. We can apply Lemma 4 to (the projections on Σ^{sync} of) u, v. Thus, we can write $u = zx$, $v = zy$ with $y = y_0 x_1 y_1 x_2 \cdots y_{m-1} x_m y_m$ for some $m \le \mathfrak{c} \cdot \mathfrak{p}$, such that (1) $|x| \le N$, (2) $x = x_1 \cdots x_m$, and (3) $dom(y_i) \cap dom(x_{i+1} \cdots x_m) = \emptyset$ for every $i < m$ hold.

An alternating (reachability) tree automaton can check that some $u, v \in \Gamma^*$ as above *do not* satisfy $\omega_p = \omega'_p$, with ω, ω' the labels of u, v. The automaton first guesses (and moves to) node z. It then guesses $x \in \Gamma^{\le N}$ and two directions where to proceed; it also guesses the difference between the labels of zx and zy, e.g., a state from S_p, an action from Σ_p^{sync}, and some colour. In the first direction it checks that a path labelled by x ends with a p-label consistent with the guessed difference. In the second direction, it checks that the path is of the form $y_0 x_1 y_1 x_2 \cdots y_{m-1} x_m$, with $x = x_1 \cdots x_m$ and $dom(y_i) \cap dom(x_{i+1} \cdots x_m) = \emptyset$ for every $i < m$, and that it ends with a p-label consistent with the guessed difference.

Note that we do not need to remember the intermediate labels ω on the path x, because we can look for a shortest u that witnesses the inconsistency. Then we can assume that u and $view_p(v)$ are play-equivalent (and not only trace-equivalent). The alternating automaton has $O(|\Gamma|^N \cdot N \cdot \mathfrak{s} \cdot \mathfrak{c} \cdot k)$ states, $|\Gamma|^N \cdot N$ for matching x inside $y_0 x_1 y_1 x_2 \cdots y_{m-1} x_m$ and $\mathfrak{s} \cdot \mathfrak{c} \cdot k$ for the guessed difference between node labels.

Lemma 8. *A deterministic safety automaton with $O(\mathfrak{c} \cdot \mathfrak{p}! \cdot 2^{\mathfrak{p}} \cdot (N+1)^{\mathfrak{p}})$ states can check that a strategy tree corresponds to an N-informed control strategy.*

Proof. The state records, at each node u and for each $u = u_1 \langle a, d \rangle u_2$, how many synchronisation actions in u are concurrent to this a (up to N), and the set of processes in the causal future of a in u. Note that, if $u = u_1 \langle a, d \rangle u_2 = u'_1 \langle a, d' \rangle u'_2$ with $|u_1| < |u'_1|$, then the set of processes in the causal future of $\langle a, d \rangle$ is a superset of the set of processes in the future of $\langle a, d' \rangle$. In addition, we need to count, for each $a \in \Sigma^{sync}$, the length of $u \setminus view_{dom(a)}(u)$ (up to $N+1$). Thus, $O(\mathfrak{c} \cdot \mathfrak{p}! \cdot 2^{\mathfrak{p}} \cdot (N+1)^{\mathfrak{p}})$ states suffice.

Lemma 9. *For a given consistently labelled strategy tree for an N-informed strategy, a universal Co-Büchi automaton with $1 + \mathfrak{p}(k+2)$ states can check that each run satisfies the parity condition.*

The proof idea is to construct an automaton that rejects if it can guess, for some process p, a path where the minimal colour occurring infinitely often is an odd colour o. It can guess a point where no lower colour than o occurs and verify (1) this (safety) and (2) that o occurs infinitely often (Co-Büchi). To test the corner case of a process p being scheduled finitely often, the automaton can guess a point where p is not scheduled again and verify (1) this and (2) that it might end in a state with odd colour (both safety).

Remark 4. As mentioned, we consider strategies that produce only non-divergent plays. A divergent play is a play where, whenever a synchronisation event a has more than N synchronisation events $b_1, \ldots, b_M$ in parallel, then the processes of a and those of the

b_i are henceforth separated. It is easy to extend Lemma 8 to the divergent case, we merely need to check N-informedness on non-divergent plays. Extending the winning condition (Lemma 9) requires more care, since we need to consider processes that are scheduled finitely often and to show that a play is maximal for them.

Summing up, we showed how to check the following properties of strategy trees:

1. Sink nodes are identified correctly: deterministic safety automaton $\mathcal{B}_1$ with $O(|S|) = O(\mathfrak{s}^\mathfrak{p})$ states.
2. Normalised nodes are identified correctly: deterministic safety automaton $\mathcal{B}_2$ with $O(\mathfrak{c} \cdot 2^\mathfrak{p})$ states.
3. Strategy tree is labelled consistently: alternating safety automaton $\mathcal{B}_3$ with $O((\mathfrak{c} \cdot |\mathcal{A}|)^N \cdot N \cdot \mathfrak{s} \cdot \mathfrak{c} \cdot k)$ states (Lemma 7).
4. Control strategy is N-informed: deterministic safety automaton $\mathcal{B}_4$ with $O(\mathfrak{c} \cdot \mathfrak{p}! \cdot 2^\mathfrak{p} \cdot (N+1)^\mathfrak{p})$ states (Lemma 8).
5. The parity condition is satisfied: universal Co-Büchi automaton $\mathcal{B}_5$ with $O(\mathfrak{p} \cdot k)$ states (Lemma 9).

Theorem 2. *Given a Zielonka automaton $\mathcal{A}$ with local parity condition and an integer N, the existence of a winning N-informed control strategy can be decided in time doubly exponential in N and exponential in $\mathcal{A}$. For fixed N, the problem is* EXPTIME-*complete. The same bounds apply to the construction of a winning strategy (if it exists).*

The proof exploits the correspondence between N-informed control strategies and trees accepted by the intersection of $\mathcal{B}_1$ through $\mathcal{B}_5$. We intersect them in two steps. Invoking the simulation theorem [25], we first construct a nondeterministic parity automaton $\mathcal{B}_5'$, which is language equivalent to $\mathcal{B}_5$, with polynomially many colours and exponentially many states in the states of $\mathcal{B}_5$. We likewise construct a nondeterministic safety automaton $\mathcal{B}_3'$, which is language equivalent to and exponential in $\mathcal{B}_3$. We can then intersect $\mathcal{B}_1, \mathcal{B}_2, \mathcal{B}_3', \mathcal{B}_4$, and $\mathcal{B}_5'$ to a nondeterministic parity automaton $\mathcal{B}$ with the same colours as $\mathcal{B}_5'$, whose states are the product states of these five automata.

The emptiness of $\mathcal{B}$ can be checked (and a control strategy constructed) by solving the resulting emptiness parity game, which is polynomial in the number of states, and exponential only in the number of colours [29]. This provides the claimed complexity.

References

1. Atig, M.F., Bouajjani, A., Qadeer, S.: Context-bounded analysis for concurrent programs with dynamic creation of threads. Logical Methods in Computer Science 7(4) (2011)
2. Bertrand, N., Fearnley, J., Schewe, S.: Bounded satisfiability for PCTL. In: Proc. of CSL 2012, pp. 92–106 (2012)
3. Biere, A., Cimatti, A., Clarke, E.M., Strichman, O., Zhu, Y.: Bounded model checking. Advances in Computers 58, 118–149 (2003)
4. Church, A.: Logic, arithmetics, and automata. In: Proceedings of the International Congress of Mathematicians, pp. 23–35 (1962)
5. Diekert, V., Rozenberg, G. (eds.): The Book of Traces. World Scientific (1995)
6. Ehlers, R.: Symbolic bounded synthesis. In: Touili, T., Cook, B., Jackson, P. (eds.) CAV 2010. LNCS, vol. 6174, pp. 365–379. Springer, Heidelberg (2010)

7. Fearnley, J., Peled, D., Schewe, S.: Synthesis of succinct systems. In: Chakraborty, S., Mukund, M. (eds.) ATVA 2012. LNCS, vol. 7561, pp. 208–222. Springer, Heidelberg (2012)
8. Filiot, E., Jin, N., Raskin, J.-F.: An antichain algorithm for LTL realizability. In: Bouajjani, A., Maler, O. (eds.) CAV 2009. LNCS, vol. 5643, pp. 263–277. Springer, Heidelberg (2009)
9. Finkbeiner, B., Schewe, S.: Uniform distributed synthesis. In: Proc. of LICS 2005, pp. 321–330 (2005)
10. Finkbeiner, B., Schewe, S.: Bounded synthesis. International Journal on Software Tools for Technology Transfer, online-first:1–12 (2012)
11. Gastin, P., Lerman, B., Zeitoun, M.: Distributed games with causal memory are decidable for series-parallel systems. In: Lodaya, K., Mahajan, M. (eds.) FSTTCS 2004. LNCS, vol. 3328, pp. 275–286. Springer, Heidelberg (2004)
12. Genest, B., Gimbert, H., Muscholl, A., Walukiewicz, I.: Optimal Zielonka-type construction of deterministic asynchronous automata. In: Abramsky, S., Gavoille, C., Kirchner, C., Meyer auf der Heide, F., Spirakis, P.G. (eds.) ICALP 2010. LNCS, vol. 6199, pp. 52–63. Springer, Heidelberg (2010)
13. Genest, B., Gimbert, H., Muscholl, A., Walukiewicz, I.: Asynchronous games over tree architectures. In: Fomin, F.V., Freivalds, R., Kwiatkowska, M., Peleg, D. (eds.) ICALP 2013, Part II. LNCS, vol. 7966, pp. 275–286. Springer, Heidelberg (2013)
14. Gastin, P., Sznajder, N.: Fair Synthesis for Asynchronous Distributed Systems. ACM Transactions on Computational Logic (2013)
15. Grädel, E., Thomas, W., Wilke, T. (eds.): Automata, Logics, and Infinite Games. LNCS, vol. 2500. Springer, Heidelberg (2002)
16. Katz, G., Peled, D.: Model checking-based genetic programming with an application to mutual exclusion. In: Ramakrishnan, C.R., Rehof, J. (eds.) TACAS 2008. LNCS, vol. 4963, pp. 141–156. Springer, Heidelberg (2008)
17. Kupferman, O., Lustig, Y., Vardi, M.Y., Yannakakis, M.: Temporal synthesis for bounded systems and environments. In: Proc. of STACS 2011, pp. 615–626 (2011)
18. Kupferman, O., Vardi, M.Y.: Synthesizing distributed systems. In: Proc. of LICS 2001, pp. 389–398 (2001)
19. Kupferman, O., Vardi, M.Y.: Safraless decision procedures. In: Proc. of FOCS 2005, pp. 531–540 (2005)
20. Madhusudan, P.: Synthesizing Reactive Programs. In: Proc. of CSL, pp. 428–442 (2011)
21. Madhusudan, P., Thiagarajan, P.S.: A decidable class of asynchronous distributed controllers. In: Brim, L., Jančar, P., Křetínský, M., Kučera, A. (eds.) CONCUR 2002. LNCS, vol. 2421, pp. 145–160. Springer, Heidelberg (2002)
22. Madhusudan, P., Thiagarajan, P.S.: Distributed controller synthesis for local specifications. In: Orejas, F., Spirakis, P.G., van Leeuwen, J. (eds.) ICALP 2001. LNCS, vol. 2076, pp. 396–407. Springer, Heidelberg (2001)
23. Madhusudan, P., Thiagarajan, P.S., Yang, S.: The MSO theory of connectedly communicating processes. In: Sarukkai, S., Sen, S. (eds.) FSTTCS 2005. LNCS, vol. 3821, pp. 201–212. Springer, Heidelberg (2005)
24. Mazurkiewicz, A.: Concurrent program schemes and their interpretations. DAIMI Rep. PB 78, Aarhus University, Aarhus (1977)
25. Muller, D.E., Schupp, P.E.: Simulating alternating tree automata by nondeterministic automata: New results and new proofs of the theorems of Rabin, McNaughton and Safra. Theoretical Computer Science 141, 69–107 (1995)
26. Pnueli, A., Rosner, R.: Distributed reactive systems are hard to synthesize. In: Proc. of FOCS 1990, pp. 746–757 (1990)

27. Qadeer, S., Rehof, J.: Context-Bounded Model Checking of Concurrent Software. In: Halbwachs, N., Zuck, L.D. (eds.) TACAS 2005. LNCS, vol. 3440, pp. 93–107. Springer, Heidelberg (2005)

28. Ramadge, P.J.G., Wonham, W.M.: The control of discrete event systems. Proceedings of the IEEE 77(2), 81–98 (1989)

29. Schewe, S.: Solving parity games in big steps. In: Arvind, V., Prasad, S. (eds.) FSTTCS 2007. LNCS, vol. 4855, pp. 449–460. Springer, Heidelberg (2007)

30. Schewe, S., Finkbeiner, B.: Synthesis of asynchronous systems. In: Puebla, G. (ed.) LOPSTR 2006. LNCS, vol. 4407, pp. 127–142. Springer, Heidelberg (2007)

31. Zielonka, W.: Notes on finite asynchronous automata. RAIRO–Theoretical Informatics and Applications 21, 99–135 (1987)

Revisiting Space in Proof Complexity: Treewidth and Pathwidth

Moritz Müller[1] and Stefan Szeider[2]

[1] Kurt Gödel Research Center, University of Vienna, Vienna, Austria
[2] Institute of Information Systems, Vienna University of Technology, Vienna, Austria

Abstract. So-called *ordered* variants of the classical notions of pathwidth and treewidth are introduced and proposed as proof theoretically meaningful complexity measures for the directed acyclic graphs underlying proofs. The ordered pathwidth of a proof is shown to be roughly the same as its formula space. Length-space lower bounds for $R(k)$-refutations are generalized to arbitrary infinity axioms and strengthened in that the space measure is relaxed to ordered treewidth.

1 Introduction

Proof complexity seeks to show that certain propositional contradictions do not admit short refutations in certain propositional refutation systems; here, short means polynomial in the size of the contradiction refuted. It is well-known and easy to see [7] that NP $\neq$ coNP if and only no propositional refutation system (in a sufficiently abstract sense) has short refutations of all contradictions. The so-called program of Cook-Reckhow asks to establish superpolynomial lower bounds for natural refutation systems. This can be interpreted as a bottom-up approach to the hypothesis NP $\neq$ coNP.

Of special interest are Resolution-based refutation systems and meaningful contradictions expressing combinatorial principles in some natural way. Common instances of the latter are given by propositional translations of first-order formulas, and in particular of *infinity axioms* (cf. Section 7).

By a Resolution-based refutation system we mean Krajíček's systems $R(k)$ [15] for k a constant or log, and their treelike versions $R^*(k)$. $R(1)$ is the same as Resolution and $R(k)$ is a straightforward generalization of Resolution operating with k-DNFs instead of clauses, i.e., cutting on conjunctions of k literals instead of single literals. From a practical perspective this special interest derives from the fact that SAT solvers are based on such systems. From a more theoretical perspective the special interest derives from the fact that lower bounds for these systems are prerequisite for understanding independence from bounded arithmetic [14]. The systems line up in a hierarchy with respect to p-simulation [7], denoted $\leq_p$:

$$R^*(1) \leq_p R^*(2) \leq_p \cdots R^*(\log) \leq_p R(1) \leq_p R(2) \leq_p \cdots R(\log). \quad (R)$$

Besides *proof length* the most popular complexity measure of proofs is *proof space* (formula-space or clause-space) as introduced by Esteban and Toran [10]. Intuitively, a space 100 refutation of a set Γ of clauses, say in Resolution, is one that can be presented as follows.

K. Chatterjee and J. Sgall (Eds.): MFCS 2013, LNCS 8087, pp. 704–716, 2013.

A teacher is in class equipped with a blackboard containing up to 100 formulas. The teacher starts from the empty blackboard and finally arrives at one containing the empty clause. The blackboard can be altered by either writing down an axiom from Γ, or by wiping out some formula, or by deriving a new clause from clauses currently written on the blackboard by means of the Resolution rule.

A motivation for studying the space of refutations is to understand memory requirements for SAT solvers [5]. The hierarchy (R) of Resolution-based proof systems is not only strict with respect to length (indeed, no $\leq_p$ can be reversed; see [23] for a survey) but it is also strict with respect to space [9,5]. A sequence of work established lower bounds on space for $R(k)$ refutations, and especially for translations of infinity axioms [10,1,9]. Resolution lower bounds on space follow from lower bounds on width [10,2] but not the other way around [19].

Ben-Sasson showed that size and space cannot in general be simultaneously optimized [4], laying the ground for various so-called *length-space trade-off* results. An exponential length-space trade-off states that there exists a sequence of contradictions that have short Resolution refutations in small space while refutations in somewhat smaller space require exponential length (length-space lower bound). Ben-Sasson and Nordström found such sequences for various settings for the qualifications "small" and "somewhat smaller", e.g., for $O(n)$ versus $o(n/\log n)$. Moreover, they managed to extend the length-space lower bound to $R(k)$ for constant k when taking the $(k+1)$th root of the qualification "somewhat smaller." The contradictions are substitution instances of pebbling contradictions. What Nordström and Ben-Sasson showed is how to transfer trade-off results for pebbling games to Resolution proofs. We refer to the survey [5] for more information. The wording trade-off has to be taken with some care in that the upper bounds are claimed only for the very special contradictions constructed. In this paper we shall focus on the lower bound part of trade-offs.

This paper. We revisit refutation space by means of natural invariants of the refutation DAG, using variants of the notions of *pathwidth* and *treewidth* which play an important role in Robertson and Seymour's graph minors project and have evolved as very successful and ubiquitously used complexity measures (see, for instance, Bodlaender's survey [6]). We introduce *ordered* variants of these graph width measures that, in contrast to earlier adaptions of the width notions to digraphs [3], allow us to distinguish between DAGs. Our notions are well-motivated from a graph theoretic point of view; for example on DAGs, ordered pathwidth coincides with a straightforward variant of the vertex separation number [6] adapted to DAGs (Proposition 1). We show that the notions have proof theoretic sense: Resolution refutations of minimal ordered pathwidth are just input Resolution refutations (Theorem 1), and those of minimal ordered treewidth are just the treelike ones. More importantly, we show that ordered pathwidth is roughly the same as refutation space (Theorem 2). Conceptually, these results allow to rethink space as a measure of how far a Resolution proof is from being an input Resolution refutation.

This gives interest to *ordered treewidth*, a notion that relaxes ordered pathwidth in much the same way as treewidth relaxes pathwidth. Ordered treewidth of a refutation can be interpreted as measuring how far a refutation is from being treelike . We also propose an interpretation of ordered treewidth in terms of space, using the following two player game, that continues the metaphor above.

> Imagine a student visits the teacher in her office asking her to explain the proof. The teacher has a blackboard potentially containing up to 100 formulas and writes the empty clause on it. The student asks how to prove it. The teacher produces a length ≤ 100 proof from Γ plus some additional axioms. The student chooses one of these additional axioms and asks how to prove it. And so on. The game ends when the teacher comes up with a proof using no additional axioms.

The new graph invariants also provide the means for making progress with respect to the already mentioned *length-space lower bounds* from Ben-Sasson and Nordström [5]. Our main technical result (Theorem 4) is a lower bound on length and ordered treewidth for $R(k)$-refutations of infinity axioms in general. This makes progress with respect to the known length-space lower bounds in that it applies to infinity axioms in general, and thereby to a large class of formulas having a natural meaning. It relaxes the refutation space measure (i.e., ordered pathwidth) to ordered treewidth, and it gives nontrivial lower bounds for all $R(k)$ simultaneously, and for $R(\log)$. The latter feature overcomes a bottleneck in constructions from [5] which give good lower bounds for $R(k)$ with constant k but become trivial for $R(\log)$.

Proof idea. The proof for our general lower bound follows the adversary type argument of [16] against treelike $R(\log)$ refutations of translations of infinity axioms. One uses restrictions that describe finite parts of some infinite model of the infinity axiom. Starting with the empty restriction, first choose a node as in Spira's theorem, namely one that splits the refutation tree into two subtrees of size at most 2/3 of total. In case no extension of the current restriction satisfies the formula at the chosen node, recurse to the subtree rooted at this node and stick with the current restriction. Otherwise, delete this subtree and recurse with a "small" restriction satisfying the formula at the chosen node. The invariant maintained is a proof of a formula "forced false" from axioms plus some formulas "forced true." If the proof has length S, this process reaches a constant size proof after $O(\log S)$ steps. If S is not too large, it is argued that the final restriction can be further extended to force all remaining axioms true and a contradiction is reached. The proof of our lower bound proceeds similarly but by recursion on a tree decomposition of the refutation. To make sense of this idea we show that we can always find a tree decomposition whose underlying tree is binary (to find a Spira type split node) and whose size is linear in the size of the refutation (Lemma 3). Further care is needed to ensure that the partial tree decompositions during the recursion are decompositions of refutations with similar properties as the invariant described above (Lemma 4).

Carrying this out requires some work. This extended abstract sketches the proof and states the main lemmas needed.

2 Preliminaries

Digraphs. We consider *directed graphs (digraphs*, for short) without self-loops and denote the set of vertices and the set of directed edges of a digraph D by $V(D)$ and $E(D)$, respectively. If $(u, v) \in E(D)$, then u is a *predecessor* of v and v a *successor* of u. An *ancestor* of $v \in V(D)$ is a vertex w such that there is a directed path from w to v in D; we understand that there is a directed path of length 0 from any vertex to itself. The *in-degree (out-degree) of* v is the number of its predecessors (successors). The in-degree (out-degree) of D is the maximal in-degree (out-degree) over all vertices. Vertices of in-degree 0 are *sources*, vertices of out-degree 0 are *sinks*. An *(induced) subdigraph* of D is a digraph $D[X]$ induced on a nonempty $X \subseteq V(D)$; if $V(D) \setminus X$ is nonempty, we write $D - X$ for $D[V(D) \setminus X]$. The graph $\underline{D}$ *underlying* a digraph D has the sane vertices as D and as edges the symmetric closure of $E(D)$. In general, a *graph* is a digraph D with symmetric $E(D)$. A *DAG* is a directed acyclic graph (i.e., a digraph without directed cycles), and a *tree* is a DAG T with a unique sink r_T called *root* such that for every $v \in V(T)$ there is exactly one directed path from v to r_T. We shall refer to vertices in a tree as *nodes*. The *subtree* T_t *rooted at* $t \in V(T)$ is the subtree of T induced on the set of ancestors of t in T; it has root $r_{T_t} = t$.

Propositional Logic. A *literal* is a propositional variable X or its negation $\neg X$; for a literal ℓ we let $\neg \ell$ denote $\neg X$, if $\ell = X$, and X, if $\ell = \neg X$. A *(k-)term* is a set of (at most k) literals. A *(k-)DNF* is a set of *(k-)*terms. The empty DNF is denoted by 0 and the empty term by 1. A *clause* is a 1-DNF. An *assignment* is a function from the propositional variables into $\{0, 1\}$. A *restriction* ρ is a finite partial assignment. For a restriction or assignment ρ and a term t we let $t \upharpoonright \rho$ be 0 if t contains a literal falsified by ρ (in the usual sense) and otherwise the subterm obtained by deleting all literals satisfied by ρ. For a DNF D we let $D \upharpoonright \rho := \bigcup_{t \in D} \{t \upharpoonright \rho\}$ if this DNF does not contain 1, and otherwise $D \upharpoonright \rho := 1$. Note, if ρ is defined on all variables appearing in D then $D \upharpoonright \rho$ equals the truth value of D under ρ.

Definition 1. A *(k-)DNF proof* is a pair $(D, (F_v)_{v \in V(D)})$ where D is a DAG with a unique sink and in which every vertex has at most two predecessors, and F_v is a *(k-)*DNF for every $v \in V(D)$. The proof is said to be *of* F if $F = F_v$ for v the sink of D, and *from* Γ if $F_v \in \Gamma$ for all sources v of D. It is said to be *treelike* if D is a tree. Proofs of 0 are *refutations*. The *length* of the proof is $|V(D)|$. A *refutation system* is a set of refutations.

Usually one requires refutation systems to satisfy certain further properties like soundness or completeness or being polynomial time decidable (cf. [7]).

Definition 2. A proof $(D, (F_v)_{v \in V(D)})$ is *sound (strongly sound)* if for every inner vertex $v \in V(D)$ and every assignment (respectively, every restriction) ρ we have $F_v \upharpoonright \rho = 1$ whenever $F_u \upharpoonright \rho = 1$ for all predecessors u of v in D.

One easily checks that this definitions of strong soundness generalizes the one given in [24] for DNF proofs. We consider the following rules of inference, namely *weakening*, *introduction of conjunction* and *cut*:

$$\frac{D}{D \cup \{t\}} \qquad \frac{D \cup \{t\} \quad D' \cup \{t'\}}{D \cup D' \cup \{t \cup t'\}} \qquad \frac{D \cup \{t\} \quad D' \cup D''}{D \cup D'},$$

where D, D', D'' are DNFs, t, t' are terms and in the cut rule we assume $\emptyset \neq D'' \subseteq \{\{\neg \ell\} \mid \ell \in t\}$. A k-DNF proof $(D, (F_v)_{v \in V(D)})$ is an $R(k)$-proof if for every inner vertex v with predecessors u, w the formula F_v is obtained from F_u and F_w by one of the three rules above. An $R(k)$-proof is an $R(\log)$-*proof* if its length is at least 2^k. An $R(1)$-proof is a *Resolution* proof. The refutation system consisting of all $R(k)$-refutations ($R(\log)$-refutations) is denoted $R(k)$ ($R(\log)$).

Remark 1. $R(k)$ is strongly sound. We have completeness in the sense that for every k-DNF F implied by some set Γ of k-DNFs, there is an $R(k)$-proof of F from Γ plus some additional 'axioms' of the form $(X \lor \neg X)$, i.e., $\{\{X\}, \{\neg X\}\}$. $R(k)$ is refutation-complete in the sense that no such axioms are needed in case $F = 0$. If one adds a new rule allowing to infer such an axiom from any formula, then the system ceases to be strongly sound.

First-Order Logic and Propositional Translation. A *vocabulary* is a finite set τ of relation and function symbols, each with an associated *arity*; function symbols of arity 0 are *constants*. The *arity* of τ is the maximum arity of one of its symbols. τ-*terms* are first-order variables $x, y, z \ldots$ or of the form $f t_1 \cdots t_r$ where $t_1, \ldots, t_r$ are again τ-terms and $f \in \tau$ is a function symbol of arity r. τ-*atoms* are of the form $t_1 = t_2$ or $R t_1 \cdots t_r$ where $t_1, t_2, \ldots, t_r$ are τ-terms and $R \in \tau$ is a relation symbol of arity r. τ-*formulas* are built from τ-atoms using $\land, \lor, \neg$ and existential and universal quantifiers $\exists x, \forall x$. For a tuple of first-order variables $\bar{x}$ we write $\varphi(\bar{x})$ for a τ-formula φ to indicate that the free variables of φ are among the components of $\bar{x}$. A τ-*sentence* is a τ-formula without free variables. A τ-*structure* M consists of a nonempty set, its *universe*, that we also denote by M and for every, say, r-ary relation symbol $R \in \tau$ (function symbol $f \in \tau$) an *interpretation* $R^M \subseteq M^r$ ($f^M : M^r \to M$); we identify the interpretation of a constant with its unique value.

Recall that the *spectrum* of a first-order sentence φ is the set of those naturals $n \geq 1$ such that φ has a model of cardinality n. Skolemization and elementary formula manipulation allows to compute from every first-order sentence ψ a sentence φ with the same spectrum of the form

$$\forall \bar{x} \bigwedge_{i \in I} C_i(\bar{x}), \tag{1}$$

where I is a nonempty finite set, the C_is are first-order clauses (disjunctions of atoms and negated atoms) whose atoms have the form $R\bar{y}$ or $f\bar{y} = z$.

Following Paris and Wilkie [20] we define for every natural $n \geq 1$ a set $\langle \varphi \rangle_n$ of clauses that is satisfied exactly by those assignments that describe a model of φ with universe $[n] := \{0, 1 \ldots, n-1\}$.

Let τ denote the vocabulary of φ. We use as propositional variables $R\bar{a}, f\bar{a} = b$ where $r \in \mathbb{N}, \bar{a} \in [n]^r, b \in [n]$, R is an r-ary relation symbol in τ and f is an r-ary function symbol in τ. For $i \in I$ and $\bar{a} \in [n]^{|\bar{x}|}$ substitute $\bar{a}$ for $\bar{x}$ in $C_i(\bar{x})$; this transforms every literal into a propositional literal or into an expression of the form $a = a'$ or $\neg a = a'$ where a, a' are components of $\bar{a}$; the propositional clause $\langle C_i(\bar{a}) \rangle$ is $\{1\}$ if one of these expressions is "true" in the obvious sense; otherwise $\langle C_i(\bar{a}) \rangle$ is the set of propositional literals (of the form $R\bar{a}, f\bar{a} = b$) obtained by the substitution. Then $\langle \varphi \rangle_n$ is the set of the clauses $\langle C_i(\bar{a}) \rangle$ obtained this way plus the *functionality clauses* $\{\{f\bar{a} = b\} \mid b \in [n]\}, \{\{\neg f\bar{a} = b\}, \{\neg f\bar{a} = b'\}\}$ for $f \in \tau$ an r-ary function symbol, $\bar{a} \in [n]^r$ and distinct $b, b' \in [n]$.

It should be clear that the assignments that satisfy the functional clauses bijectively correspond to τ-structures on $[n]$; moreover, such an assignment satisfies $\langle \varphi \rangle_n$ if and only if the corresponding τ-structure models φ. In particular, $\langle \varphi \rangle_n$ is unsatisfiable if and only if n is not in the spectrum of φ.

3   Width Notions for DAGs

3.1   Treewidth and Pathwidth

Let G be graph. A *tree decomposition* of G is a pair (T, χ) where T is a tree and $\chi : V(T) \to 2^{V(G)}$ is a mapping such that the following three conditions hold:

(a) $V(G) \subseteq \bigcup_{t \in V(T)} \chi(t)$;
(b) $E(G) \subseteq \bigcup_{t \in V(T)} (\chi(t) \times \chi(t))$;
(c) for every $v \in V(G)$ the set $\{t \in V(T) \mid v \in \chi(t)\}$ is connected in $\underline{T}$.

Recall, $\underline{T}$ is the graph underlying T. The *width* of a tree decomposition (T, χ) is the maximum $|\chi(t)| - 1$ over all $t \in V(T)$. The *treewidth* $tw(G)$ of G is the minimum width over all its tree decompositions. A *path decomposition* is a tree decomposition (T, χ) where T is a (directed) path. The *pathwidth* $pw(G)$ of a graph G is the minimum width over all its path decompositions.

Let (T, χ) be a tree decomposition of a graph G. We say that a vertex $v \in V(G)$ is *introduced* at $t \in V(T)$ if $v \in \chi(t)$ but $v \notin \chi(t')$ for any predecessor t' of t. Similarly, we say that v is *forgotten* at $t \in V(T)$ if $v \in \chi(t)$ and either $t = r_T$ or $v \notin \chi(t')$ for the successor t' of t. Note that every vertex $v \in V(G)$ is introduced at at least one tree node (by condition (a)) and forgotten at exactly one tree node (by condition (c)). In a path decomposition every vertex is introduced at exactly one tree node.

The same definitions apply literally to digraphs, so we can also speak of tree and path decompositions of digraphs. Consequently, treewidth and pathwidth of a digraph equal the treewidth and pathwidth of the digraph's underlying graph, respectively. Thus the direction of edges is completely irrelevant for the treewidth or pathwidth of a digraph. For some considerations, however, one needs the direction of edges to be reflected in the decomposition and the associated width measure. For example [11] introduces the notion of *directed treewidth*, and it is known that every DAG has directed treewidth 1. We introduce new width measures that can distinguish between DAGs.

3.2   Ordered Treewidth and Ordered Pathwidth

Although we shall be mainly interested in DAGs, we give the definitions and some first observations generally for digraphs.

Definition 3. A tree decomposition (T, χ) of a digraph D is *ordered* if the following condition holds:

(d) for every directed edge $(u, v) \in E(D)$ and every $t \in V(t)$ where v is introduced, $u \in \chi(t)$.

As above, we define the *ordered treewidth* $otw(D)$ of D as the minimum width over all ordered tree decompositions of D, and the *ordered pathwidth* $opw(D)$ of D as the minimum width over all ordered path decompositions of D.

The ordered width measures are different from their classical counterparts:

Remark 2. For every digraph D, $otw(D)$ is at least the in-degree of D.

We say that a class $\mathcal{C}$ of digraphs has *bounded ordered pathwidth* if there is a constant $w \in \mathbb{N}$ such that every digraph in $\mathcal{C}$ has ordered pathwidth at most w; we say $\mathcal{C}$ has *unbounded ordered pathwidth* if it does not have bounded ordered pathwidth. We use a similar mode of speech for the other width notions.

Example 1.   1. The ordered treewidth of a tree is its in-degree.
2. A directed path with at least one edge has ordered pathwidth 1.
3. The class of full binary trees (with edges directed towards the root) has unbounded ordered pathwidth and bounded ordered treewidth.
4. The class of full binary trees with all edges reversed (edges directed away from the root) has unbounded ordered treewidth and bounded treewidth.

We have the following analogues of well-known results.

Definition 4. (T, χ) is *succinct* if every node forgets some vertex.

Lemma 1. *Every digraph D has a succinct ordered tree decomposition of width $otw(D)$, and a succinct ordered path decomposition of width $opw(D)$.*

Lemma 2. *A succinct ordered tree decomposition of a digraph D has at most $|V(D)|$ many nodes.*

Based on these lemmas we get the first of two lemmas, key to carry out the proof sketch from the Introduction.

Lemma 3. *For every digraph D there exists an ordered tree decomposition (T, χ) of width $otw(D)$ where T has in-degree at most 2 and $|V(T)| < 2|V(D)|$.*

The second key lemma is the following.

Definition 5. A subtree T' of a tree T is *complete in T* if for every node of T' either all or none of its predecessors in T are in $V(T')$.

Lemma 4. *Let (T, χ) be an ordered tree decomposition of a digraph D, let T' be a subtree of T and set $\chi' := \chi \upharpoonright V(T')$. Assume $\bigcup_{t' \in V(T')} \chi(t') \neq \emptyset$ and set $D' := D[\bigcup_{t' \in V(T')} \chi(t')]$. Then*

1. *(T', χ') is an ordered tree decomposition of D';*
2. *if T' is complete in T, then there exists for every edge $(u, v) \in E(D)$ with $u \notin V(D')$ and $v \in V(D')$ a leaf t of T' which is not a leaf of T such that $v \in \chi(t)$.*

3.3 Vertex Separation Numbers

Recall, the vertex separation number vsn of a graph G is defined as the minimum $s \in \mathbb{N}$ taken over all linear orders $\leq$ of $V(G)$ such that for all $v \in V(G)$ there are at most s many vertices $\leq v$ with an edge to some vertex $> v$. It is known [12] that $vsn(G) = pw(G)$ for all graphs G.

For a DAG D it seems natural to take the minimum not over all linear orders on $V(D)$ but only over those embedding D. We call the resulting number $ovsn(D)$ the *ordered vertex separation number* of D and show the following result.

Proposition 1. $opw(D) = ovsn(D)$ *for every DAG D.*

4 Resolution Proofs of Minimal Width

Recall, the ordered treewidth of a proof containing an application of the cut rule is at least 2 (Remark 2). Clearly, when talking about the ordered pathwidth or ordered treewidth of a proof we mean the ordered pathwidth or ordered treewidth of its underlying DAG. Recall that a Resolution refutation of Γ is called *input* if it contains only applications of the cut rule and each such application has at least one premiss in Γ.

Theorem 1. *Let ℓ be a natural and Γ a set of clauses.*

1. *There is a Resolution refutation of Γ of ordered pathwidth at most 2 and length at most ℓ if and only if there is an input Resolution refutation of Γ of length at most ℓ.*
2. *If there is a Resolution refutation of Γ of ordered treewidth at most 2 and length at most ℓ, then there is a treelike Resolution refutation of Γ of length at most 3ℓ.*

This result allows us to think of ordered pathwidth (ordered treewidth) as a measure of how far a refutation is from being input (treelike). Concerning a converse of (2), recall that treelike refutations have ordered treewidth 2 (cf. Example 1 (1)).

5 Proof Space

Let $k, w, \ell > 0$ be naturals, F a k-DNF and Γ a set of k-DNFs.

5.1 Ordered Pathwidth Is Proof Space

In the Introduction we informally explained a bounded space proof by a sequence of blackboards. Formally, we follow [10] and define a *space w $R(k)$-proof of F from Γ* to be a finite sequence $(\mathbb{B}_0, \dots, \mathbb{B}_{\ell-1})$ of sets $\mathbb{B}_i$ of k-DNFs each of cardinality at most w such that $\mathbb{B}_0 = \emptyset$ and $F \in \mathbb{B}_{\ell-1}$ and for all $0 < i < \ell$ there is a formula G such that

(B1) $\mathbb{B}_i = \mathbb{B}_{i-1} \cup \{G\}$ and $G \in \Gamma$, or
(B2) $\mathbb{B}_i = \mathbb{B}_{i-1} \cup \{G\}$ and G is derived from at most two formulas in $\mathbb{B}_{i-1}$ by one application of some inference rule of $R(k)$, or
(B3) $\mathbb{B}_i = \mathbb{B}_{i-1} \setminus \{G\}$.

The space measure above is known as "formula space" or, in case $k = 1$, as "clause space." This is roughly the same as ordered pathwidth:

Theorem 2.   *1. If there is a space w $R(k)$-proof of F from Γ of length ℓ, then there is an $R(k)$-proof of F from Γ of length $< \ell$ and ordered pathwidth $< w$.*
  2. If there is a $R(k)$-proof of F from Γ of length $< \ell$ and ordered pathwidth $< w$, then there is a space w $R(k)$-proof of F from Γ of length at most $2w\ell$.

Combining with Theorem 1 (1) this result allows to think of the space of a Resolution refutation as a measure of how far it is from being input.

5.2   Ordered Treewidth as Interactive Proof Space

The conversation of a teacher with her student described informally in the Introduction is described more formally by a game $\Pi_w^k(\Gamma, F)$ between two players called *Student* and *Teacher* on the following game graph.

Its vertices are partitioned into *Student positions* and *Teacher positions*, the former are $R(k)$-proofs of length at most w and the latter are k-DNFs. Its directed edges are from a k-DNF to a length $\leq w$ proof of it, and from a proof to a label of one of its sources which is outside Γ. In particular, precisely the proofs from Γ are sinks. The *initial* position is the Teacher position F. Paths starting at the initial position are *plays*.

A *strategy for Teacher* (in $\Pi_w^k(\Gamma, F)$) is a function that maps plays ending in a Teacher position to a successor of this position; it is *positional* in case this value depends only on the Teacher position reached by the play. A play is *conform* to the strategy if every Student position in it is the value of the strategy on the initial segment of the play up to it. The strategy is *(ℓ-)winning* if all plays conform to it are finite (of length at most $2\ell - 1$, i.e., Teacher wins making $\leq \ell$ moves).

Remark 3.   The game $\Pi_w^k(\Gamma, F)$ can be seen as a parity game, so it is memory-less determined; in particular, if a winning strategy for Teacher exists, then so does a positional one [18].

Proposition 2.   *If there is an ℓ-winning strategy for Teacher in $\Pi_w^k(\Gamma, F)$, then there is also a positional one.*

This is verified by standard arguments (cf. [18]) and eases the proof of the next theorem.

Theorem 3.   *There is an ℓ-winning strategy for Teacher in $\Pi_w^k(\Gamma, F)$ if and only if there is an $R(k)$-proof of F from Γ with an ordered tree-decomposition of width at $< w$ and height $< \ell$.*

Remark 4.   Assume Γ is a set of clauses. By Theorem 3 Teacher wins $\Pi_w^k(\Gamma, 0)$ if and only if there is an $R(k)$-refutation of Γ. This is equivalent to there being a treelike Resolution refutation, so equivalent to Teacher winning $\Pi_3^1(\Gamma, 0)$ (Theorem 1) . Thus, the parameters k and w only matter when taking into account how fast the Teacher can win, that is, when considering ℓ-winning strategies.

6 Lower Bounds

Theorem 4. *Let φ be a first-order τ-sentence of the form (1) that has an infinite model. Let r be the maximal arity of some function symbol in τ and assume $r \geq 1$. Then there exists a real $c_\varphi > 0$ such that for every natural $n \geq 1$ and every natural $k \geq 1$, every strongly sound k-DNF refutation $(D, (F_v)_{v \in V(D)})$ of $\langle \varphi \rangle_n$ satisfies*

$$k \cdot otw(D) \cdot \log |V(D)| > c_\varphi \cdot n^{1/r}.$$

Remark 5. The assumption that $r \geq 1$ does not exclude interesting cases. If $r = 0$, all function symbols of τ are constants. In an infinite model of φ every nonempty set containing the interpretations of these constants carries a submodel which too models φ (being universal). Hence, the spectrum of φ is co-finite, so all but finitely many translations $\langle \varphi \rangle_n$ are satisfiable and have no sound refutations at all.

Proof (sketch). Let M be an infinite model of φ. Define a *condition* to be a pair (κ, λ) of partial injections $\kappa \subseteq \lambda$ from $[n]$ into M such that the image of λ contains all values of functions $f^M, f \in \tau$, taken on the image of κ. A restriction $\rho = \rho(\kappa, \lambda)$ associated with such a condition evaluates a propositional variable $R\bar{a}$ or $f\bar{a} = b$ according to the truth value of $\kappa(\bar{a}) \in R^M$ or $\lambda^{-1}(f^M(\kappa(\bar{a}))) = b$. An *extension* of such a restriction is one associated with a condition (κ', λ') such that $\kappa \subseteq \kappa', \lambda \subseteq \lambda'$. Given a refutation $(D, (F_v)_{v \in V(D)})$ of $\langle \varphi \rangle_n$, choose a tree decomposition (T_0, χ) of width $w := otw(D)$ according Lemma 3. Iteratively move to subtrees T complete in T_0 and restrictions ρ as sketched in the introduction: choose a split node t and distinguish cases as to whether ρ can be extended to some ρ' so that $F_v \restriction \rho' = 1$ for all $v \in \chi(t)$ or not. The invariant maintained is that for every extension ρ'' of ρ there exists $v \in \chi(r_T)$, the current root bag, such that $F_v \restriction \rho'' \neq 1$; further, Lemma 4 ensures that $(T, \chi \restriction V(T))$ decomposes a proof from axioms $\langle \varphi \rangle_n$ plus some formulas restricting to 1 under ρ.

In the first case the extension ρ' of the current $\rho = \rho(\kappa, \lambda)$ needs to force true at most one k-term per k-DNF $F_v, v \in \chi(t)$. To find such ρ' one needs to extend κ to at most $O(k \cdot w)$ new elements from $[n]$. Such a "small" extension can be found provided sufficiently many elements from $[n]$ are left, i.e., still outside the domain of λ. After $O(\log |V(T)|) \leq O(\log |V(D)|)$ iterations a constant size subtree is found and one needs sufficiently many elements still left to extend the current ρ once more to some ρ'' evaluating all $O(w)$ many still appearing clauses from $\langle \varphi \rangle_n$. These then restrict to 1 because our restrictions cannot falsify them. By the invariant above one reaches a contradiction to strong soundness. $\square$

This proof has the following corollary. It generalizes lower bounds on space (recall Theorem 2) known for particular infinity axioms (cf. Introduction).

Corollary 1. *Let φ be a first-order τ-sentence of the form (1) that has an infinite model. Let r be the maximal arity of some function symbol in τ and assume $r \geq 1$. Then there exists a real $c_\varphi > 0$ such that for every natural $n \geq 1$ and every natural $k \geq 1$, every strongly sound k-DNF refutation $(D, (F_v)_{v \in V(D)})$ of $\langle \varphi \rangle_n$ satisfies*

$$k \cdot opw(D) > c_\varphi \cdot n^{1/r}.$$

7 Infinity Axioms

An *infinity axiom* is a first-order sentence φ of the form (1) that does not have finite models but does have an infinite model. Note that in this case all propositional translations $\langle\varphi\rangle_n, n \geq 1$, are contradictory. Strong lower bounds on the length of refutations of these principles are known for the treelike systems [16,22,8]. One also knows, however, some few short DAG-like refutations:

Example 2. The *least number principle* is formulated using a unary function symbol f and a binary relation symbol $<$:

$$lnp := \forall xyz(\neg x < x \wedge (\neg x < y \vee \neg y < z \vee x < z) \wedge f(x) < x).$$

Stålmarck [25] gave polynomial length Resolution refutations of $\langle lnp \rangle_n$.

Example 3. The *very weak pigeonhole principle* states that n^2 pigeons cannot fly injectively into n holes. This principle can be formulated as a first-order infinity axiom *wphp* using a binary function symbol f:

$$\forall xx'yy'z((\neg fxx' = z \vee \neg fyy' = z \vee x = y) \wedge (\neg fxx' = z \vee \neg fyy' = z \vee x' = y')).$$

For $\langle wphp \rangle_n$ one knows a $2^{\Omega(n/(\log n)^2)}$ lower bound in Resolution [21] and a quasipolynomial upper bound in $R(\log)$ [17].

We note that short DAG-like refutations of translations of infinity axioms need to be far from being treelike in that they require unbounded ordered treewidth.

Corollary 2. *Let φ be as in Theorem 4.*

1. *Length $\leq 2^{n^{o(1)}}$ $R(\log)$-refutations of $\langle\varphi\rangle_n$ have ordered treewidth $\geq n^{\Omega(1)}$.*
2. *Ordered treewidth $\leq n^{o(1)}$ $R(\log)$-refutations of $\langle\varphi\rangle_n$ have length $\geq 2^{n^{\Omega(1)}}$.*

Specifically for the above two examples we can say the following.

Corollary 3. *1. Polynomial length $R(100)$-refutations of $\langle lnp \rangle_n$ have ordered treewidth at least $\Omega(n/\log n)$.*
2. *Quasipolynomial length $R(\log)$-refutations of $\langle wphp \rangle_n$ have ordered treewidth at least $\Omega(n^{0.4})$.*

8 Conclusion

In this paper we have revisited proof complexity using the graph invariants ordered treewidth and ordered pathwidth. Whereas the first corresponds to ordinary proof space, the latter gives rise to a notion of interactive proof space, which can be described in terms of a student-teacher game. These graph invariants provide the means for length-space lower bounds for $R(k)$-refutations that apply to a large class of formulas having a natural meaning (infinity axioms). It relaxes the refutation space measure (i.e., ordered pathwidth) to ordered treewidth and applies to $R(\log)$.

Acknowledgements. The restrictions used in the proof of Theorem 4 come from unpublished work of Albert Atserias, Sergi Oliva and the first author. We thank Albert Atserias and Sergi Oliva for their kind allowance to use them here. The first author thanks the FWF (Austrian Science Fund) for its support through Project P 24654 N25. The second author thanks the ERC (European Research Council) for its support through Project COMPLEX REASON 239962.

References

1. Alekhnovich, M., Ben-Sasson, E., Razborov, A.A., Wigderson, A.: Space complexity in propositional calculus. SIAM J. Comput. 31(4), 1184–1211 (2002)
2. Atserias, A., Dalmau, V.: A combinatorial characterization of resolution width. J. of Computer and System Sciences 74(3), 323–334 (2008)
3. Bang-Jensen, J., Gutin, G.: Digraphs, 2nd edn. Springer Monographs in Mathematics. Springer-Verlag London Ltd., London (2009)
4. Ben-Sasson, E.: Size-space tradeoffs for resolution. SIAM J. Comput. 38(6), 2511–2525 (2009)
5. Ben-Sasson, E., Nordström, J.: Understanding space in proof complexity: Separations and trade-offs via substitutions. ECCC 17, 125 (2010)
6. Bodlaender, H.L.: A partial k-arboretum of graphs with bounded treewidth. Theoretical Computer Science 209(1-2), 1–45 (1998)
7. Cook, S., Reckhow, R.: The relative efficiency of propositional proof systems. The Journal of Symbolic Logic 44, 36–50 (1979)
8. Dantchev, S., Riis, S.: On relativisation and complexity gap for resolution-based proof systems. In: Baaz, M., Makowsky, J.A. (eds.) CSL 2003. LNCS, vol. 2803, pp. 142–154. Springer, Heidelberg (2003)
9. Esteban, J.L., Galesi, N., Messner, J.: On the complexity of resolution with bounded conjunctions. Theoretical Computer Science 321(2-3), 347–370 (2004)
10. Esteban, J.L., Torán, J.: Space bounds for resolution. Information and Computation 171(1), 84–97 (2001)
11. Johnson, T., Robertson, N., Seymour, P.D., Thomas, R.: Directed tree-width. Journal of Combinatorial Theory, Series B 82(1), 138–154 (2001)
12. Kinnersley, N.G.: The vertex separation number of a graph equals its path-width. Information Processing Letters 42(6), 345–350 (1992)
13. Kleine Büning, H., Lettman, T.: Propositional logic: deduction and algorithms. Cambridge University Press, Cambridge (1999)
14. Krajíček, J.: Bounded arithmetic, propositional logic, and complexity theory. Encyclopedia of Mathematics and its Applications, vol. 60. Cambridge Univ. Press (1995)
15. Krajíček, J.: On the weak pigeonhole principle. Fund. Math. 170(1-2), 123–140 (2001), Dedicated to the memory of Jerzy Łoś
16. Krajíček, J.: Combinatorics of first order structures and propositional proof systems. Archive for Mathematical Logic 43(4), 427–441 (2004)
17. Maciel, A., Pitassi, T., Woods, A.R.: A new proof of the weak pigeonhole principle. J. of Computer and System Sciences 64(4), 843–872 (2002)
18. Mazala, R.: 2 Infinite games. In: Grädel, E., Thomas, W., Wilke, T. (eds.) Automata, Logics, and Infinite Games. LNCS, vol. 2500, pp. 23–38. Springer, Heidelberg (2002)
19. Nordström, J.: Narrow proofs be spacious: separating space and width in resolution. SIAM J. Comput. 39(1), 59–121 (2009)

20. Paris, J., Wilkie, A.: Counting problems in bounded arithmetic. In: Methods in Mathematical Logic. LNM, vol. 1130, pp. 317–340 (1985)
21. Razborov, A.: Resolution lower bounds for the weak functional pigeonhole principle. Theoretical Computer Science 303(1), 233–243 (2003)
22. Riis, S.: A complexity gap for tree resolution. Comput. Compl. 10(3), 179–209 (2001)
23. Segerlind, N.: The complexity of propositional proofs. Bull. of Symbolic Logic 13(4), 417–481 (2007)
24. Segerlind, N., Buss, S., Impagliazzo, R.: A switching lemma for small restrictions and lower bounds for k-DNF resolution. SIAM J. Comput. 33(5), 1171–1200 (2004)
25. Stålmarck, G.: Short resolution proofs for a sequence of tricky formulas. Acta Informatica 33(3), 277–280 (1996)

Space-Efficient Parallel Algorithms
for Combinatorial Search Problems*

Andrea Pietracaprina[1], Geppino Pucci[1],
Francesco Silvestri[1], and Fabio Vandin[2]

[1] University of Padova, Dip. Ingegneria dell'Informazione, Padova, Italy
{capri,geppo,silvest1}@dei.unipd.it
[2] Brown University, Computer Science Dept., Providence, RI, USA
vandinfa@cs.brown.edu

Abstract. We present space-efficient parallel strategies for two fundamental combinatorial search problems, namely, *backtrack search* and *branch-and-bound*, both involving the visit of an n-node tree of height h under the assumption that a node can be accessed only through its father or its children. For both problems we propose efficient algorithms that run on a distributed-memory machine with p processors. For backtrack search, we give a deterministic algorithm running in $O\left(n/p + h\log p\right)$ time, and a Las Vegas algorithm requiring optimal $O\left(n/p + h\right)$ time, with high probability. Building on the backtrack search algorithm, we also derive a Las Vegas algorithm for branch-and-bound which runs in $O\left((n/p + h\log p\log n)h\log n\right)$ time, with high probability. A remarkable feature of our algorithms is the use of only constant space per processor, which constitutes a significant improvement upon previously known algorithms whose space requirements per processor depend on the (possibly huge) tree to be explored ($\Omega\left(h\right)$ for backtrack search and $\Omega\left(n/p\right)$ for branch-and-bound).

1   Introduction

The exact solution of a combinatorial (optimization) problem is often computed through the systematic exploration of a tree-structured solution space, where internal nodes correspond to partial solutions (growing progressively more refined as the depth increases) and leaves correspond to feasible solutions. A suitable algorithmic template used to study this type of problems (originally proposed in [1]) is the exploration of a tree $\mathcal{T}$ under the constraints that: (i) only the tree root is initially known; (ii) the structure, size and height of the tree are unknown; and (iii) a tree node can be accessed if it is the root of the tree or if either its father or one of its children is available.

In the paper, we focus on two important instantiations of the above template. The *backtrack search problem* [2] requires to explore the entire tree $\mathcal{T}$ starting from its root r, so to enumerate all solutions corresponding to the leaves.

---

* This work was supported, in part, by the University of Padova Projects *STPD08JA32* and *CPDA121378*. F. Vandin was also supported by NSF grant IIS-1247581.

K. Chatterjee and J. Sgall (Eds.): MFCS 2013, LNCS 8087, pp. 717–728, 2013.

In the *branch-and-bound problem*, each tree node is associated to a cost, and costs satisfy the min-heap order property, so that the cost of an internal node is a lower bound to the cost of the solutions corresponding to the leaves of its subtree. The objective here is to determine the leaf associated with the solution of minimum cost. We define n and h to be, respectively, the number of nodes and the height of the tree to be explored. It is important to remark that in the branch-and-bound problem, the nodes that must necessarily be explored are only those whose cost is less than or equal to the cost of the solution to be determined. These nodes form a subtree $\mathcal{T}^*$ of $\mathcal{T}$ and in this case n and h refer to $\mathcal{T}^*$. Assuming that a node is explored in constant time, it is easy to see that the solution to the above problems requires $\Omega(n)$ time, on a sequential machine, and $\Omega(n/p + h)$ time on a p-processor parallel machine.

Due to the elevated computational requirements of search problems, many parallel algorithms have been proposed in literature that speed-up the execution by evenly distributing the computation among the available processing units. All these studies have focused mainly on reducing the running time while the resulting memory requirements (expressed as a function of the number of nodes to be stored locally at each processor) may depend on the tree parameters. However, the search space of combinatorial problems can be huge, hence it is fundamental to design algorithms which exploit parallelism to speed up execution and yet need a small amount of memory per processor, possibly independent of the tree parameters. Reducing space requirements allows for a better exploitation of the memory hierarchy and enables the use of cheap distributed-memory parallel platforms where each processing units is endowed with limited memory.

Previous Work. Parallel algorithms for backtrack search have been studied in a number of different parallel models. Randomized algorithms have been developed for the complete network [2,3] and the butterfly network [4], which require optimal $\Theta(n/p + h)$ node explorations (ignoring the overhead due to manipulations of local data structures). The work of Herley et al. [5] gives a deterministic algorithm running in $O\left((n/p + h)(\log\log\log p)^2\right)$ time on a p-processor COMMON CRCW PRAM. While the algorithm in [2] performs depth-first explorations of subtrees locally at each processor requiring $\Omega(h)$ space per processor, the other algorithms mostly concentrate on balancing the load of node explorations among the available processors but may require $\Omega(n/p)$ space per processor.

In [2] an $\Theta(n/p + h)$-time randomized algorithm for branch-and-bound is also provided for the complete network. In [6,7] Herley et al. show that a parallelization of the heap-selection algorithm of [8] gives, respectively, a deterministic algorithm running in time $O\left(n/p + h\log^2(np)\right)$ on an EREW-PRAM, and one running in time $O\left((n/p + h\log^4 p)\log\log p\right)$ on the Optically Connected Parallel Computer (OCPC), a weak variant of the complete network [9]. All of these works adopt a best-first like strategy, hence they may need $\Omega(n/p)$ space per processor. In [10] deterministic algorithms for both backtrack search and branch-and-bound are given which run in $O\left(\sqrt{nh}\log n\right)$ time on an n-node mesh with constant space per processor. However, any straightforward implementation of these algorithms on a p-processor machine, with $p < n$, would still require $\Omega\left(n/p\right)$ space per processor. Karp et al. [1] describe sequential algorithms for the branch-and-bound problem featuring a range of space-time tradeoffs. The minimum space they attain is $O\left(\sqrt{\log n}\right)$ in time $O\left(n2^{O(\sqrt{\log n})}\right)$[1]. Some papers (see [11] and references therein) describe sequential and parallel algorithms for branch-and-bound with limited space, which interleave depth-first and breadth-first strategies, but provide no analytical guarantee on the running time.

Our Contribution. In this paper, we present space-efficient parallel algorithms for the backtrack search and branch-and-bound problems. The algorithms are designed for a p-processor distributed-memory message-passing system similar to the one employed in [2], where in one time step each processor can perform $O\left(1\right)$ local operations and send/receive a message of $O\left(1\right)$ words to/from another arbitrary processor. In case $x > 1$ messages are sent to the same processor in one step, we make the restrictive assumption that none of these messages is delivered (as in the OCPC model [9,12]). Consistently with most previous works, we assume that a memory word is sufficient to store a tree node, and, as in [1], we also assume that, given a tree node, a processor can generate any one of its children or its father in $O\left(1\right)$ steps and $O\left(1\right)$ space. We let $P_0, P_1, \ldots, P_{p-1}$ denote the processors in our system.

For the backtrack search problem we develop a deterministic algorithm which runs in $O\left(n/p + h\log p\right)$ time, and a Las Vegas randomized algorithm which runs in optimal $\Theta\left(n/p + h\right)$ time with high probability, if $p = O\left(n/\log n\right)$. Both algorithms require only constant space per processor and are based on a nontrivial lazy implementation of the work-distribution strategy featured in the backtrack search algorithm by [2], whose exact implementation requires $\Omega\left(h\right)$ space per processor. By using the deterministic backtrack search algorithm as a subroutine, we develop a Las Vegas randomized algorithm for the branch-and-bound problem which runs in $O\left((n/p + h\log p\log n)h\log n\right)$ time with high probability, using again constant space per processor.

To the best of our knowledge, our backtrack search algorithms are the first to achieve (quasi) optimal time using constant space per processor, which constitutes a significant improvement upon the aforementioned previous works. As for the branch-and-bound algorithm, while its running time may deviate substantially from the trivial lower bound, for search spaces not too deep and sufficiently high parallelism, it achieves sublinear time using constant space per processor. For instance, if $h = O\left(n^\epsilon\right)$ and $p = \Theta\left(n^{1-\epsilon}\right)$, with $0 < \epsilon < 1/2$, the algorithm runs in $O\left(n^{2\epsilon}\mathrm{polylog}(n)\right)$ time, with high probability, using $\Theta\left(n^{1-\epsilon}\right)$ aggregate space. Again, to the best of our knowledge, ours is the first algorithm achieving sublinear running time using sublinear (aggregate) space, thus providing evidence that branch-and-bound can be parallelized in a space-efficient way.

For simplicity, our results are presented assuming that the tree $\mathcal{T}$ to be explored is binary and that each internal node has both left and right children. The same results extend to the case of d-ary trees, with $d = \Theta\left(1\right)$, and to trees

---

[1] The authors claim a constant-space randomized algorithm running in $O\left(n^{1+\epsilon}\right)$ time which, however, disregards the nonconstant space required by the recursion stack.

that allow an internal node to have only one child. (The details will be provided in the full version of this extended abstract.)

The rest of the paper is organized as follows. In Sec. 2 we first present a generic strategy for parallel backtrack search and then instantiate this strategy to derive our deterministic and randomized algorithms. In Sec. 3 we describe the randomized parallel algorithm for branch-and-bound. Sec. 4 concludes with some final remarks and open problems. Due to space constraints, we refer to [13] for omitted proofs.

2   Space-Efficient Backtrack Search

In this section we describe two parallel algorithms, a deterministic algorithm and a Las Vegas randomized algorithm, for the backtrack search problem. Both algorithms implement the same generic strategy described in Sec. 2.1 and require constant space per processor. The deterministic implementation of the generic strategy (Sec. 2.2) requires global synchronization, while the randomized one (Sec. 2.3) avoids explicit global synchronization.

2.1   Generic Strategy

The main idea behind our generic strategy moves along the same lines as the backtrack search algorithm of [2], where at each time a processor is either *idle* or *busy* exploring a certain subtree of $\mathcal{T}$ in a depth-first fashion. The computation evolves as a sequence of *epochs*, where each epoch consists of three consecutive phases of fixed durations: (1) a *traversal phase*, where each busy processor continues the depth-first exploration of its assigned subtree; (2) a *pairing phase*, where busy processors are matched with distinct idle processors; and (3) a *donation phase*, where each busy processor P_i that was paired with an idle processor P_j in the preceding phase, attempts to entrust a portion of its assigned subtree to P_j which becomes in charge of the exploration of this portion.

In [2] it is shown that the best progress towards completion is achieved by letting a busy processor donate the *topmost* unexplored right subtree of the subtree it is currently exploring. A straightforward implementation of this donation rule requires that a busy processor either stores a list of up to $\Theta(h)$ nodes, or, at each donation, traverses up to $\Theta(h)$ nodes in order to retrieve the subtree to be donated, thus incurring a large overhead. As anticipated in the introduction, our algorithm features a lazy implementation of this strategy which uses constant space per processor and incurs only a small time overhead.

We now describe in more detail how the three phases of an epoch are performed. At any time, a busy processor P_i maintains the following information, which can be stored in constant space:

- r_i: the root of its assigned subtree;
- v_i: the last node reached by the processor in the depth-first exploration of its assigned subtree;
- $d_i \in \{\texttt{left}, \texttt{right}, \texttt{parent}\}$: a *direction flag* identifying the next node (left child, right child, or parent node, respectively) to be touched after v_i;

- (t_i, q_i): a pair of nodes that are used to identify a portion of the subtree to donate to an idle processor; in particular, t_i is a node on the path from r_i to v_i (r_i and v_i included), while q_i is either the right child of r_i or is undefined. We refer to the path from t_i up to r_i as the *tail* associated with processor P_i, and define the tail's length as the number of edges it comprises.

At the beginning of the first epoch, only processor P_0 is busy and its variables are initialized as follows: r_0 is set to the root of the tree $\mathcal{T}$ to be explored; $v_0 = t_0 = r_0$; q_0 is set to the right child of r_0; and $d_0 = \texttt{left}$. Consider now an arbitrary epoch. (In the description below, Δ_t, Δ_p, and Δ_d denote suitable values which will be fixed by the analysis.)

Traversal Phase. Each busy processor P_i advances of at most Δ_t steps in the depth-first exploration of the subtree rooted at r_i, starting from v_i and proceeding in the direction indicated by d_i. Variables v_i and d_i are suitably updated at each step. During the exploration, if the processor touches r_i and moves to its right child w, it sets r_i to w and q_i to w's right child. Also, t_i is updated when either the processor touches t_i and moves to its father u, or when $t_i = r_i$ and r_i is updated. Variable t_i is set to u in the former case, and to the new value of r_i in the latter. P_i finishes the exploration of its assigned subtree and becomes idle when $v_i = r_i$ and $d_i = \texttt{parent}$.

Pairing Phase. Busy and idle processors are paired in preparation of the subsequent donation phase. The phase runs for Δ_p steps. Different pairing mechanisms are employed by the deterministic and the randomized algorithm, as described in detail in the respective sections.

Donation Phase. Consider a busy processor P_i that has been paired to an idle processor P_j. Two types of donations from P_i to P_j are possible, namely a *quick donation* or a *slow donation*, depending on the status of q_i. As we will see, a quick donation always starts and terminates within the same epoch, assigning a subtree to explore to P_j, while a slow donation may span several epochs and may even fail to assign a subtree to P_j.

If q_i is defined (i.e., it is the right child of r_i) a quick donation occurs. In this case, P_i donates to P_j the subtree rooted in q_i and P_i keeps the subtree rooted at the left child of r_i for exploration. Then, P_j sets r_j, v_j and t_j all equal to q_i, d_j to $\texttt{left}$, and q_j to the right child of q_i. Instead, P_i sets r_i and t_i to the left child of r_i, q_i to undefined, and keeps v_i, d_i to their current values. Note that quick donation coincides with the donation strategy in [2].

If q_i is undefined, a slow donation is performed where the tail associated with P_i is climbed upwards to identify an unexplored subtree of the one rooted at r_i, which is then donated to P_j. To amortize the cost of tail climbing, P_i attempts to donate a subtree rooted at the right child of a node located in the middle of the tail, so to halve the length of the residual tail that P_i has to climb in future slow donations. This halving is crucial for reducing the running time.

Let us see in more detail how a slow donation is accomplished. Initially P_i verifies if a new tail must be created. This happens if $t_i = r_i$ and $v_i \neq r_i$. In this case, a *tail creation* is performed by setting $t_i = v_i$. Then, two cases are possible depending on the tail length.

Case 1: tail length ≤ 1. If t_i is the left child of r_i, then P_i donates to P_j the subtree rooted at the right child of r_i, performing the same steps of a quick donation. Otherwise, if $t_i = r_i$ (i.e., the tail length is 0) or t_i is the right child of r_i, then P_i must have already explored the left subtree of r_i. In this case, no donation is performed and, since the current root is no longer needed, P_i sets r_i, v_i and t_i to the right child of r_i, and d_i to `left`. Note that in all cases, the level of the root of the subtree assigned to P_i increases by 1.

Case 2: tail length > 1. First, processor P_i identifies the middle node m_i of the tail by backtracking twice from t_i to r_i. Let ℓ_i be the parent of m_i: P_i seeks the node u_i along the path from ℓ_i to r_i which is closest to r_i and is the left child of its parent z_i. If u_i is found, all nodes in the path from z_i (excluded) to r_i (included) are unnecessary to complete the exploration of the subtree rooted at r_i, since they and their left subtrees have already been explored. Therefore r_i is set to z_i and q_i to the right child of z_i. If instead, no such u_i is found, then all nodes in the path from ℓ_i (excluded) to r_i (included) are discarded, r_i is set to ℓ_i and q_i is left undefined. Finally, P_i donates to P_j the (partially explored) subtree rooted at m_i. Namely, P_j sets $r_j = m_i$, $v_j = v_i$, $d_j = d_i$, $t_j = t_i$, and sets q_j as undefined. Instead, P_i continues exploring the tree rooted at r_i setting both v_i and t_i to ℓ_i and d_i to `right`, if m_i is the left child of ℓ_i, or to `parent` if m_i is the right child of ℓ_i. Note that the level of the root of the subtree assigned to P_j is always greater than the level of the root of the subtree assigned to P_i. Moreover, the level of the root of the subtree assigned to P_i either increases or remains unchanged. In this latter case, however, q_i can be set during the tail traversal so that the next donation of P_i will be a quick donation.

The donation phase runs for Δ_d steps, where we assume Δ_d to be greater than or equal to the maximum between the time for a quick donation and the time for Case 1 of a slow donation. However, for efficiency reasons, Δ_d cannot be chosen large enough to perform entirely Case 2 of a slow donation, since its duration is proportional to the tail length, which may be rather large. In this case, if P_i does not conclude the donation in Δ_d steps, it saves its state (requiring constant space) at the end of the donation phase and resumes the computation in the donation phase of the subsequent epoch, in which it maintains the pairing with P_j and refuses any novel pairing. If t_i changes in the subsequent traversal phase, the state is updated accordingly: namely, if t_i is set to its father, the tail length is updated and, if needed, m_i is moved to its father. Also, if the tail length becomes at most one, the slow donation switches from Case 2 to Case 1.

It is easy to check that the above algorithms touches all the nodes in the tree $\mathcal{T}$, therefore solving the backtrack search problem.

2.2 Deterministic Algorithm

In the deterministic algorithm each pairing phase is performed through a prefix-like computation that finds a maximal matching between idle processors and busy processors; such computation requires $\Theta(\log p)$ parallel time. For this algorithm we set $\Delta_p, \Delta_d = \Theta\left(\log p\right)$, and $\Delta_t = \Delta_d/\kappa$, for a suitable constant κ defined in the proof. We call an epoch *full* if at the last step of its traversal phase at least $p/2$ processors are busy, and we call it *non-full* otherwise.

Lemma 1. *The total number of parallel steps in full epochs is $O\left(n/p\right)$.*

Proof. Since each node is touched at most 3 times in a traversal phase (after descending from the parent, after exploring the left subtree, and after exploring the right subtree), the total number of times nodes are touched is $O\left(n\right)$. The lemma follows by observing that in a full epoch $\Theta\left(p\right)$ processors touch $\Theta\left(\log p\right)$ nodes each, and that the epoch runs in $O\left(\log p\right)$ parallel steps.   □

Consider an arbitrary leaf q of $\mathcal{T}$. Now, we bound the number of parallel steps before q is touched. Observe that after all leaves have been touched, the algorithm terminates in $O\left(h + \log p\right)$ additional parallel steps, when all busy processors have gone back to the roots of their assigned subtrees. In each epoch, we define the *special processor* of q as the processor exploring the subtree containing q with the deepest root; note that there is a unique special processor in any epoch. When the special processor S performs a donation to a processor P_j, then for the subsequent epoch either S remains the special processor or P_j becomes the special processor.

We refer to non-full epochs as *donating* or *preparing* depending on the status of the special processor of q. Namely, a non-full epoch is donating if the special processor S completes a donation in the epoch, while it is preparing if S is involved in Case 2 of a slow donation and, at the end of the epoch, it has not finished to execute all operations prescribed by this type of donation. Note that, before q is touched, any non-full epoch is always either donating or preparing.

Lemma 2. *The total number of parallel steps in donating epochs before leaf q is touched is $O\left(h \log p\right)$.*

Proof. We claim that the level of the root of the subtree explored by special processor S increases by at least one after at most two donating epochs. If a quick donation, or Case 1 of slow donation is performed by S, then the claim is verified. Suppose S is involved in Case 2 of a slow donation. Let $S = P_i$ and let P_j be the processor paired to P_i. If after the donation P_i remains the special processor and the root r_i of its subtree is unchanged, then during the slow donation q_i has been set and hence the next donation of the special processor is a quick donation. In all other cases, the level root of the special processor is increased after the donation. Thus, the claim is proved. Since the height of the tree to be explored is h, there are $O\left(h\right)$ donating epochs and the total number of parallel steps in donating epochs is $O\left(h \log p\right)$.   □

We now bound the total number of parallel steps in preparing epochs.

Lemma 3. *The total number of parallel steps in preparing epochs before leaf q is touched is $O\left(n/p + h \log p\right)$.*

Proof. Consider the time interval from the beginning of the algorithm until leaf q is explored. Clearly, at any time within this interval a special processor is defined. We partition this interval into *eras* delimited by subsequent donation phases in which tail creations are performed by the special processor. (Recall that a processor P_i creates a tail in the donation phase of an epoch whenever q_i is undefined, $t_i = r_i$ and $v_i \neq r_i$: then the tail is created by setting $t_i = v_i$.) More precisely, for $i \geq 1$, the i-th era begins at the donation phase of the i-th tail creation, and ends right before the donation phase of the $(i+1)$-st tail creation (or the end of the interval if there is no such tail creation). Observe that the beginning of the interval does not coincide with the beginning of the first era, however no preparing epochs occur before the first tail creation. Note that an era may involve more than one donation from the special processor, and that all preparing epochs in the same era work on segments of the tail whose creation defines the beginning of the era. We denote with Φ the number of eras and with $\phi_i \geq 1$ the number of slow donations in the i-th era, for each $1 \leq i \leq \Phi$.

Let T_i^j be the number of distinct nodes that the special processor touches by walking up a subtree to prepare the j-th slow donation of the i-th era, with $1 \leq i \leq \Phi$ and $1 \leq j \leq \phi_i$ (nodes can be touched in both donating and preparing epochs). Since a slow donation splits the tail in half, we have that $T_i^{j+1} \leq T_i^j/2$ for all $1 \leq j \leq \phi_i$. Since the number of steps in preparing epochs for one slow donation is at most a constant factor the tail length, the total time spent in preparing epochs is $\sum_{i=1}^{\Phi} \sum_{j=1}^{\phi_i} cT_i^j \leq 2c \sum_{i=1}^{\Phi} T_i^1$, where $c \geq 1$ is a suitable constant. Clearly, we have $T_1^1 \leq h$.

Consider an arbitrary era $i \geq 2$. A node u in the tail of the era has been touched for the first time in a traversal phase of an era $\ell < i$. Note that $\ell = i - 1$ since if it was $\ell < i - 1$, u would have been part of a tail created in an era before the i-th one and it is easy to verify that tails of different eras are disjoint. Therefore the number of nodes touched (walking upward in the tree) in the preparing epochs for the first donation of era i is bounded by the node touched in the traversal phases of era $i - 1$, which can be partitioned in three (disjoint) sets:

- the nodes touched for the first time in traversal phases of full epochs in era $i - 1$; we denote the number of such nodes as E_i;
- the nodes touched for the first time in the traversal phases of donating epochs in era $i - 1$; we denote the number of such nodes as D_i;
- the nodes touched for the first time in the traversal phases of preparing epochs in era $i - 1$; we denote the number of such nodes as C_i.

Thus we have $\sum_{i=2}^{\Phi} T_i^1 \leq \sum_{i=2}^{\Phi} E_i + \sum_{i=2}^{\Phi} D_i + \sum_{i=2}^{\Phi} C_i$. By Lemma 1 we have $\sum_{i=2}^{\Phi} E_i = O(n/p)$, while by Lemma 2 it follows that $\sum_{i=2}^{\Phi} D_i = O(h \log p)$. We now only need to bound $\sum_{i=2}^{\Phi} C_i$. Remember that C_i is the number of nodes touched in the preparing epochs of the i-th era that have been touched for the first time in the traversal phases of preparing epochs of the $(i-1)$-st era. Consider the second era: in order to bound C_2, we need to bound the number of nodes that have been touched in the traversal phases of epochs in the first era. Since cT_1^j is an upper bound to the time required for preparing the j-th donation in the first era, and since the number of nodes visited in the traversal phase of a preparing epoch is at most a factor $1/\kappa$ the time of the respective donation phase, for a suitable constant κ (i.e., $\Delta_t = \Delta_d/\kappa$), we have

$C_2 \leq \sum_{j=1}^{\phi_i} cT_1^j/\kappa \leq T_1^1/2 \leq h/2$ by setting $\kappa = 2c$. In general, for era $i > 2$ we have: $C_i \leq \sum_{j=1}^{\phi_i} cT_{i-1}^j/\kappa \leq T_{i-1}^1/2 \leq (E_{i-1} + D_{i-1} + C_{i-1})/2$. Then, by unrolling the above inequality, we get

$$C_i \leq \frac{1}{2}E_{i-1} + \frac{1}{4}E_{i-2} + \ldots + \frac{1}{2^{i-2}}E_2 + \frac{1}{2}D_{i-1} + \frac{1}{4}D_{i-2} + \ldots + \frac{1}{2^{i-2}}D_2 + \frac{1}{2^{i-1}}h.$$

Therefore, by summing up among all eras, we have

$$\sum_{i=1}^{\Phi} C_i \leq \sum_{i=1}^{\Phi} \left[\frac{h}{2^i} + \sum_{j=1}^{\infty} \frac{E_i}{2^j} + \sum_{j=1}^{\infty} \frac{D_i}{2^j} \right] \leq h + \sum_{i=1}^{\Phi} (E_i + D_i) = O\left(\frac{n}{p} + h \log p \right).$$

As already noticed, the number of steps in preparing epochs is proportional to the number of nodes touched in such epochs, and this establishes the result. □

By combining the above three lemmas, we obtain the following theorem.

Theorem 1. *The deterministic algorithm for backtrack search completes in $O(n/p + h \log p)$ parallel steps and constant space per processor.*

2.3 Randomized Algorithm

In the randomized algorithm, the durations of the traversal and of the pairing phase are set to a constant (i.e., $\Delta_d, \Delta_p = O(1)$), and the duration of a donation phase is set to $\Delta_t = \Delta_d/\kappa$, for a suitable constant κ. While the traversal phase and the donation phase are as described in Sec. 2.1 and are the same as in the deterministic algorithm, the pairing phase is implemented differently as follows. In a first step, each idle processor sends a pairing request to a random processor; in a second step, a busy processor P_i that has received a pairing request from (idle) processor P_j, sends a message to P_j to establish the pairing. Note that the communication model described in Sec. 1 guarantees that each busy processor receives at most one pairing request in the first step. The analysis of the randomized algorithm combines elements of the analysis of the above deterministic algorithm and the one for the randomized backtrack search algorithm in [2].

Theorem 2. *The randomized algorithm completes in $O(n/p + h)$ parallel steps with probability at least $1 - ne^{-n/(4p)}$, and requires constant space per processor. In particular, if $n/\ln n \geq 4(1 + c)p$ for a constant $c > 0$, then the probability is at least $1 - n^{-c}$.*

3 Space-Efficient Branch-and-Bound

In this section we present a Las Vegas algorithm for the branch-and-bound problem, which requires to explore a heap-ordered (binary) tree $\mathcal{T}$ starting from the root to find the minimum-cost leaf. For simplicity, we assume that all node costs are distinct. The algorithm implements, in a parallel setting, a simplified version of the sequential space-efficient strategy proposed in [1] which is based on the solution of a suitable selection problem. Specifically, the strategy reduces the branch-and-bound problem to the problem of finding the node with the n-th smallest cost, for exponentially increasing values of n. We first give the algorithm for this latter selection problem, using the deterministic backtrack algorithm of the previous section as a subroutine; then we describe the Las-Vegas branch-and-bound algorithm.

Selection. Let $\mathcal{T}$ be an infinite binary tree whose nodes are associated with distinct costs satisfying the min-heap order property, and let $c(u)$ denote the cost associated with a node u. Consider the following *selection problem*: given an integer n and the root r of $\mathcal{T}$, find the node u_n with the n-th smallest cost $c(u_n)$. We denote by $\mathcal{T}_c$ the subtree of $\mathcal{T}$ containing all nodes of cost less than or equal to a value c. Clearly, $\mathcal{T}_{c(u_n)}$ contains exactly n nodes, and we let h denote its height. We say that a node is *good* if its cost is not larger than $c(u_n)$.

Suppose we want to determine whether a node u is good. We explore $\mathcal{T}_{c(u)}$ using the deterministic backtrack algorithm and counting, at the end of each epoch, how many nodes have been touched for the first time. The visit finishes as soon as the subtree $\mathcal{T}_{c(u)}$ is completely visited or the count becomes larger than n. Node u is flagged good only in the former case. We have:

Lemma 4. *Determining whether a node u is good can be accomplished in time $O\left(n/p + h\log p\right)$ using constant space per processor.*

Consider a subtree $\mathcal{T}'$ of $\mathcal{T}$ with n nodes and height h, and suppose that some nodes of $\mathcal{T}'$ are marked as *distinguished*. Our selection algorithm makes use of a subroutine to efficiently pick a node uniformly at random among the *distinguished* ones of $\mathcal{T}'$. To this purpose, we use reservoir sampling [14], which allows to sample an element uniformly at random from a data stream of unknown size in constant space. Specifically, $\mathcal{T}'$ is explored using backtrack search. During the exploration, each processor counts the number of distinguished nodes it touches for the first time, and picks one of them uniformly at random through reservoir sampling. The final random node is obtained from the p selected ones in $\log p$ rounds, by discarding half of the nodes at each round, as follows. For $0 \leq k < p$, let q_k^0 be the number of nodes counted by processor P_k in the backtrack search. In the i-th round, processor $P_{2^i j}$, with $0 \leq i < \log p$ and $0 \leq j < p/2^i$, replaces its selected node with the node selected by $P_{2^i(j+1)-1}$ with probability $q_{2^i(j+1)}^i/(q_{2^i j}^i + q_{2^i(j+1)}^i)$, and sets $q_{2^i j}^{i+1}$ to $q_{2^i j}^i + q_{2^i(j+1)}^i$. After the last round, the distinguished node held by P_0 is returned. We have:

Lemma 5. *Selecting a node uniformly at random from a set of distinguished nodes in a subtree $\mathcal{T}'$ of $\mathcal{T}$ with n nodes and height h can be accomplished in time $O\left(n/p + h\log p\right)$, with high probability, using constant space per processor.*

We are now ready to describe the parallel selection algorithm, which works in epochs. In the i-th epoch, the algorithm starts with a lower bound L_i (initially $L_1 = -\infty$) on the n-th smallest cost $c(u_n)$ and ends with a new lower bound $L_{i+1} > L_i$ computed by exploring the set F_i consisting of the children in $\mathcal{T}$ of the leaves in $\mathcal{T}_{L_i}$. More in details, L_{i+1} is set to the largest cost of a good node in F_i (note that there exists at least one good node otherwise $\mathcal{T}_{L_i}$ would contain n nodes). The algorithm ends when there are exactly n nodes in $\mathcal{T}_{L_i}$, and the one we seek, namely u_n, is the node with the largest cost among these. The largest good node in F_i is computed by a binary search using random splitters as suggested in [1]. The algorithm iteratively updates two values X_L^i and X_U^i, which represent lower and upper bounds on the largest cost of a good node in F_i, until $X_L^i = X_U^i$. Initially, we set $X_L^i = L_i$ and $X_U^i = +\infty$. The two values are updated as follows: by using the strategy analyzed in Lemma 5, the algorithm selects a node u, called *random splitter*, uniformly at random among those in F_i with cost in the range $[X_L^i, X_U^i]$ (which are the distinguished nodes). Then, by using the strategy analyzed in Lemma 4, the algorithm verifies if u is good: if this is the case, then X_L^i is set to $c(u)$, otherwise X_U^i is set to $c(u)$.

Theorem 3. *The n-th smallest element in a heap-ordered binary tree $\mathcal{T}$ can be selected in time $O((n/p + h \log p)h \log n)$, with high probability, and constant space per processor.*

We note that by using the randomized backtrack search algorithm, the complexity of the selection algorithm can be slightly improved. However, this complicates the analysis and we postpone such an improvement to the full version.

Branch-and-Bound. The algorithm we propose for this problem consists of a number of iterations. In the i-th iteration, with $i \geq 1$, the above selection algorithm is employed to determine the node with the 2^i-th smallest cost in the tree. Let c_i be the cost of such a node. Then, backtrack search is performed on $\mathcal{T}_{c_i}$ to assess whether its nodes include a leaf of the original tree and, if so, return the one with minimum cost. If no leaf of the original tree belongs to $\mathcal{T}_{c_i}$ then the algorithm proceeds to the next iteration. Clearly, the algorithm terminates after $O(\log n)$ iterations. The following corollary is easily established.

Corollary 1. *The branch-and-bound algorithm requires $O((n/p + h \log p \log n) h \log n)$ parallel steps, with high probability, and constant space per processor.*

4 Conclusions

We presented the first time-efficient combinatorial parallel search strategies which work in constant space per processor. For backtrack search, the time of our deterministic algorithm comes within a factor $O(\log p)$ from optimal, while our randomized algorithm is time-optimal. Building on backtrack search, we provided a randomized algorithm for the more difficult branch-and-bound problem, which requires constant space per processor and whose time is an $O(h \, \text{polylog}(n))$ factor away from optimal.

While our results for backtrack search show that the nonconstant space per processor required by previous algorithms is not necessary to achieve optimal running time, our result for branch-and-bound still leaves a gap open, and more work is needed to ascertain whether better space-time tradeoffs can be established. However, the reduction in space obtained by our branch-and-bound strategy could be crucial for enabling the solution of large instances, where n is huge but $\Omega\left(n/p\right)$ space per processor cannot be tolerated. The study of space-time tradeoffs is crucial for novel computational models such as MapReduce, suitable for cluster and cloud computing [15]. However, algorithms for combinatorial search strategies on such new models deserve further investigations.

As in [1], our algorithms assume that the father of a tree node can be accessed in constant time, but this feature may be hard to implement in certain application contexts, especially for branch-and-bound. However, our algorithm can be adapted so to avoid the use of this feature by increasing the space requirements of each processor to $\Theta\left(h\right)$. We remark that even with this additional overhead, the space required by our branch-and-bound algorithm is still considerably smaller, for most parameter values, than that of the state-of-the-art algorithm of [2], where $\Theta\left(n/p\right)$ space per processor may be needed.

References

1. Karp, R.M., Saks, M., Wigderson, A.: On a search problem related to branch-and-bound procedures. In: Proc. of 27th FOCS, pp. 19–28. IEEE (1986)
2. Karp, R.M., Zhang, Y.: Randomized parallel algorithms for backtrack search and branch-and-bound computation. J. ACM 40, 765–789 (1993)
3. Liu, P., Aiello, W., Bhatt, S.: An atomic model for message-passing. In: Proc. of 5th ACM SPAA, pp. 154–163. ACM (1993)
4. Ranade, A.: Optimal speedup for backtrack search on a butterfly network. In: Proc. of 3rd ACM SPAA, pp. 40–48. ACM (1991)
5. Herley, K.T., Pietracaprina, A., Pucci, G.: Deterministic parallel backtrack search. Theor. Comput. Sci. 270, 309–324 (2002)
6. Herley, K.T., Pietracaprina, A., Pucci, G.: Fast deterministic parallel branch-and-bound. Parallel Processing Letters 9, 325–333 (1999)
7. Herley, K.T., Pietracaprina, A., Pucci, G.: Deterministic branch-and-bound on distributed memory machines. Int. J. Found. Comput. Sci. 10, 391–404 (1999)
8. Frederickson, G.N.: The information theory bound is tight for selection in a heap. In: Proc. of 22nd ACM STOC, pp. 26–33. ACM (1990)
9. Anderson, R., Miller, G.: Optical communication for pointer based algorithms. Technical Report CRI-88-14, CS Department, Univ. South. California (1988)
10. Kaklamanis, C., Persiano, G.: Branch-and-bound and backtrack search on mesh-connected arrays of processors. Theory of Comp. Syst. 27, 471–489 (1994)
11. Mahapatra, N.R., Dutt, S.: Sequential and parallel branch-and-bound search under limited-memory constraints. In: Par. Proc. Disc. Prob., vol. 106, pp. 139–158 (1999)
12. Goldberg, L., Jerrum, M., Leighton, F., Rao, S.: Doubly logarithmic communication algorithms for optical-communication parallel computers. SIAM J. Comput. 26, 1100–1119 (1997)
13. Pietracaprina, A., Pucci, G., Silvestri, F., Vandin, F.: Space-efficient parallel algorithms for combinatorial search problems (2013), arXiv:abs/1305.3828
14. Vitter, J.S.: Random sampling with a reservoir. ACM Trans. Math. Softw. 11, 37–57 (1985)
15. Pietracaprina, A., Pucci, G., Riondato, M., Silvestri, F., Upfal, E.: Space-round tradeoffs for MapReduce computations. In: Proc. 26th ACM ICS, pp. 235–244 (2012)

Separating Regular Languages by Piecewise Testable and Unambiguous Languages*

Thomas Place, Lorijn van Rooijen, and Marc Zeitoun

LaBRI, Univ. Bordeaux & CNRS UMR 5800
351 cours de la Libération, 33405 Talence Cedex, France
{tplace,lvanrooi,mz}@labri.fr

Abstract. Separation is a classical problem asking whether, given two sets belonging to some class, it is possible to separate them by a set from another class. We discuss the separation problem for regular languages. We give a PTIME algorithm to check whether two given regular languages are separable by a piecewise testable language, that is, whether a $\mathcal{B}\Sigma_1(<)$ sentence can witness that the languages are disjoint. The proof refines an algebraic argument from Almeida and the third author. When separation is possible, we also express a separator by saturating one of the original languages by a suitable congruence. Following the same line, we show that one can as well decide whether two regular languages can be separated by an unambiguous language, albeit with a higher complexity.

1 Introduction

Separation is a classical notion in mathematics and computer science. In general, one says that two structures L_1, L_2 from a class $\mathcal{C}$ are *separable* by a structure L if $L_1 \subseteq L$ and $L_2 \cap L = \varnothing$. In this case, L is called a *separator*. In separation problems, the separator L is required to belong to a given class Sep. The problem asks whether two disjoint elements L_1, L_2 of $\mathcal{C}$ can always be separated by an element of the class Sep. In the case that disjoint elements of $\mathcal{C}$ cannot always be separated by an element of Sep, several natural questions arise:

(1) given elements L_1, L_2 in $\mathcal{C}$, can we decide whether a separator exists in Sep?
(2) if so, what is the complexity of this decision problem?
(3) can we, in addition, compute a separator, and what is the complexity?

In this context, it is known for example that separation of two context-free languages by a regular one is undecidable [9].

Separating Regular Languages. This paper looks at separation problems for the class $\mathcal{C}$ of regular languages, and for classes Sep closed under complement. Under this last condition, a separation algorithm for Sep entails an algorithm for deciding membership in Sep, *i.e.*, membership reduces to separability. Indeed, membership in Sep can be checked by testing whether the input language is Sep-separable from its complement.

---

* Supported by the Agence Nationale de la Recherche ANR 2010 BLAN 0202 01 FREC.

K. Chatterjee and J. Sgall (Eds.): MFCS 2013, LNCS 8087, pp. 729–740, 2013.

Conversely, while finding a decidable characterization for Sep already requires a deep understanding of the subclass, the search for separation algorithms is intrinsically more difficult. Indeed, powerful tools are available to decide membership in Sep: one normally makes use of a recognizing device of the input language, *viz.* its syntactic monoid. A famous result along these lines is Schützenberger's Theorem [14], which states that a language is definable in first-order logic if and only if its syntactic monoid is aperiodic, a property one can easily decide.

Now for a separation algorithm, the question is whether the input languages are *sufficiently different*, from the point of view of the subclass Sep, to allow this to be witnessed by an element of Sep. Note that we cannot use standard methods on the recognizing devices, as was the case for the membership problem. We now have to decide whether there *exists* a recognition device of the given type that separates the input: we do not have it in hand, nor its syntactic monoid. An even harder question then is to actually construct the so-called separator in Sep.

Contributions. In this paper, we study this problem for two subclasses of the regular languages: piecewise testable languages and unambiguous languages.

Piecewise testable languages are languages that can be described by the presence or absence of scattered subwords up to a certain size within the words. Equivalently, these are the languages definable using $\mathcal{B}\Sigma_1(<)$ formulas, *i.e.*, first-order logic formulas that are boolean combinations of $\Sigma_1(<)$ formulas. A $\Sigma_1(<)$ formula is a first-order formula with a quantifier prefix $\exists^*$, followed by a quantifier-free formula. A well-known result about piecewise testable languages is Simon's Theorem [16], which states that a regular language is piecewise testable if and only if its syntactic monoid is $\mathcal{J}$-trivial. This property yields a decision procedure to check whether a language is piecewise testable, refined by Stern into a PTIME algorithm [18], of which the complexity has been improved by Trahtman [21].

The second class that we consider is the class of unambiguous languages, *i.e.*, languages defined by unambiguous products. This class has been given many equivalent characterizations [19]. For example, these are the $\mathrm{FO}^2(<)$-definable languages, that is, languages that can be defined in first-order logic using only two variables. Equivalently, this is the class $\Delta_2(<)$ of languages that are definable by a first-order formula with a quantifier prefix $\exists^*\forall^*$ and simultaneously by a first-order formula with a quantifier prefix $\forall^*\exists^*$. Note that consequently, all piecewise testable languages are $\mathrm{FO}^2(<)$-definable. It has been shown in [10] for $\Delta_2(<)$, and in [20] for $\mathrm{FO}^2(<)$ that these are exactly the languages whose syntactic monoid belongs to the decidable class DA.

There is a common difficulty in the separation problems for these two classes. *A priori*, it is not known up to which level one should proceed in refining the candidate separators to be able to answer the question of separability. For piecewise testable languages, this refinement basically means increasing the size of the considered subwords. For unambiguous languages, it means increasing the size of the unambiguous products. For both of these classes, we are able to compute, from the two input languages, a number that suffices for this purpose. This entails decidability of the separability problem for both classes.

In both cases, we obtain a better complexity bound to answer the decision problem starting from NFAs: we show that two languages are separable if and only if the corresponding automata contain certain forbidden patterns of the same type. We prove that for piecewise testable languages this property can be decided in polynomial time wrt. the size of the automata and of the alphabet. For unambiguous languages this can be done in exponential space.

Related Work. The classes of piecewise testable and unambiguous languages are varieties of regular languages. For such varieties, there is a generic connection found by Almeida [1] between profinite semigroup theory and the separation problem: Almeida has shown that two regular languages over A are separable by a language of a variety $A^*\mathcal{V}$ if and only if the topological closures of these two languages inside a profinite semigroup, depending only on $A^*\mathcal{V}$, intersect. Note that this theory does not give any information about how to actually *construct* the separator, in case two languages are separable. To turn Almeida's result into an algorithm deciding separability, we should compute representations of these topological closures, and test for emptiness of intersections of such closures.

So far, these problems have no generic answer and have been studied in an algebraic context for a small number of specific varieties. Deciding whether the closures of two regular languages intersect is equivalent to computing the so-called 2-pointlike sets of a finite semigroup wrt. the considered variety, see [1]. This question has been answered positively for the varieties of finite group languages [4,12], piecewise testable languages [3,2], star-free languages [8,7], and a few other varieties, but it was left open for unambiguous languages.

A general issue is that the topological closures may not be describable by a finite device. However, for piecewise testable languages, the approach of [3] builds an automaton over an extended alphabet, of exponential size wrt. the original alphabet, recognizing the closure of a regular language. The algorithm is polynomial wrt. the size of the original automaton (the construction was presented for deterministic automata but also works for nondeterministic ones). These automata admit the usual construction for intersection and can be checked for emptiness in NLOGSPACE. This yields an algorithm which, from two NFAs, decides separability by a piecewise testable language in time polynomial in the number of states of the NFAs, and exponential in the size of the original alphabet.

Our proof for separability by piecewise testable languages follows the same pattern as the method described above. A significant improvement is that we show that non-separability is witnessed by paths of the same shape in both automata, which yields an algorithm providing better complexity: it runs in polynomial time in both the size of the automata *and* in the size of the alphabet. Also, we do not make use of the theory of profinite semigroups: we work only with elementary concepts. We have described this algorithm in [13]. Furthermore, we show how to compute from the input languages an index that suffices to separate them. We use the same technique for unambiguous languages. Recently, Czerwinski et. al. [6] also provided a PTIME algorithm for deciding separability by piecewise testable languages, but do not provide the computation of such an index.

Due to space constraints, some proofs only appear in the full version of this paper.

2 Preliminaries

We fix a finite alphabet $A = \{a_1, \ldots, a_m\}$. We denote by A^* the free monoid over A. The empty word is denoted by ε. For a word $u \in A^*$, the smallest $B \subseteq A$ such that $u \in B^*$ is called the *alphabet* of u and is denoted by $\mathsf{alph}(u)$.

Separability. Given languages L, L_1, L_2, we say that L *separates* L_1 from L_2 if

$$L_1 \subseteq L \text{ and } L_2 \cap L = \varnothing.$$

Given a class Sep of languages, we say that the pair (L_1, L_2) is Sep-*separable* if some language $L \in \mathsf{Sep}$ separates L_1 from L_2. Since all classes we consider are closed under complement, (L_1, L_2) is Sep-separable if and only if (L_2, L_1) is, in which case we simply say that L_1 and L_2 are Sep-separable.

We are interested in two classes Sep of separators: the class of piecewise testable languages, and the class of unambiguous languages.

Piecewise Testable Languages. We say that a word u is a *piece* of v, if

$$u = b_1 \cdots b_k, \text{ where } b_1, \ldots, b_k \in A, \quad \text{and } v \in A^* b_1 A^* \cdots A^* b_k A^*.$$

For instance, ab is a piece of $bb\underline{a}c\underline{c}\underline{b}a$. The *size* of a piece is its number of letters. A language $L \subseteq A^*$ is *piecewise testable* if there exists $\kappa \in \mathbb{N}$ such that membership of w in L only depends on the pieces of size up to κ occurring in w. We write $w \sim_\kappa w'$ when w and w' have the same pieces of size up to κ. Clearly, $\sim_\kappa$ is a congruence of finite index. Therefore, a language is piecewise testable if and only if it is a union of $\sim_\kappa$-classes for some $\kappa \in \mathbb{N}$. In this case, the language is said to be of *index* κ. It is easy to see that a language is piecewise testable if and only if it is a finite boolean combination of languages of the form $A^* b_1 A^* \cdots A^* b_k A^*$.

Piecewise testable languages are languages definable by $\mathcal{B}\Sigma_1(<)$ formulas, that is, boolean combinations of first-order formulas of the form:

$$\exists x_1 \ldots \exists x_n \; \varphi(x_1, \ldots, x_n),$$

where φ is quantifier-free. For instance, $A^* b_1 A^* \cdots A^* b_k A^*$ is defined by the formula $\exists x_1 \ldots \exists x_k \left[\bigwedge_{i<k}(x_i < x_{i+1}) \wedge \bigwedge_{i \leqslant k} b_i(x_i) \right]$, where the first-order variables $x_1, \ldots, x_k$ are interpreted as positions, and where $b(x)$ is the predicate testing that position x carries letter b.

We denote by $PT[\kappa]$ the class of all piecewise testable languages of index κ or less, and by $PT = \bigcup_\kappa PT[\kappa]$ the class of all piecewise testable languages. Given $L \subseteq A^*$ and $\kappa \in \mathbb{N}$, the smallest $PT[\kappa]$-language containing L is

$$[L]_{\sim_\kappa} = \{ w \in A^* \mid \exists u \in L \text{ and } u \sim_\kappa w \}.$$

In general however, there is no smallest PT-language containing a given language.

Unambiguous Languages. A product $L = B_0^* a_1 B_1^* \cdots B_{k-1}^* a_k B_k^*$ is called *unambiguous* if every word of L admits exactly one factorization witnessing its membership in L. The number k is called the *size* of the product. An *unambiguous language* is a finite disjoint union of unambiguous products. Observe that unambiguous languages are connected to piecewise testable languages. Indeed, it was proved in [15] that the class of unambiguous languages is closed under boolean operations. Moreover, languages of the form $A^* b_1 A^* \cdots A^* b_k A^*$ are unambiguous, witnessed by the product $(A \setminus \{b_1\})^* b_1 (A \setminus \{b_2\})^* \cdots (A \setminus \{b_k\})^* b_k A^*$. Therefore, piecewise testable languages form a subclass of the unambiguous languages.

Many equivalent characterizations for unambiguous languages have been found [19]. From a logical point of view, unambiguous languages are exactly the languages definable by an $FO^2(<)$ formula [20]. Here, $FO^2(<)$ denotes the two-variable restriction of first-order logic. Another logical characterization which further illustrates the link with piecewise testable languages (*i.e.*, $\mathcal{B}\Sigma_1(<)$-definable languages) is $\Delta_2(<)$. A $\Sigma_2(<)$ formula is a first-order formula of the form:

$$\exists x_1 \ldots \exists x_n \ \forall y_1 \ldots \forall y_m \ \varphi(x_1, \ldots, x_n, y_1, \ldots, y_m),$$

where φ is quantifier-free. A language is $\Delta_2(<)$-*definable* if it can be defined both by a $\Sigma_2(<)$ formula and the negation of a $\Sigma_2(<)$ formula. It has been proven in [10] that a language is unambiguous if and only if it is $\Delta_2(<)$-definable.

For two words w, w', we write, $w \cong_\kappa w'$ if w, w' belong to the same unambiguous products of size κ or less. We denote by $UL[\kappa]$ the class of all languages that are unions of $\cong_\kappa$-classes, and we let $UL = \bigcup_\kappa UL[\kappa]$. Since unambiguous languages are closed under boolean operations, UL is the class of all unambiguous languages. Given $L \subseteq A^*$ and $\kappa \in \mathbb{N}$, the smallest $UL[\kappa]$-language containing L is

$$[L]_{\cong_\kappa} = \{w \in A^* \mid \exists u \in L \text{ and } u \cong_\kappa w\}.$$

Again, in general there is no smallest UL-language containing a given language.

Automata. A *nondeterministic finite automaton* (NFA) over A is denoted by a tuple $\mathcal{A} = (Q, A, I, F, \delta)$, where Q is the set of states, $I \subseteq Q$ the set of initial states, $F \subseteq Q$ the set of final states and $\delta \subseteq Q \times A \times Q$ the transition relation. The *size* of an automaton is its number of states plus its number of transitions. We denote by $L(\mathcal{A})$ the language of words accepted by $\mathcal{A}$. Given a word $u \in A^*$, a subset B of A and two states p, q of $\mathcal{A}$, we denote

- by $p \xrightarrow{u} q$ a path from state p to state q labeled u,
- by $p \xrightarrow{\subseteq B} q$ a path from p to q of which all transitions are labeled over B,
- by $p \xrightarrow{=B} q$ a path from p to q of which all transitions are labeled over B, with the additional demand that every letter of B occurs at least once along it.

Given a state p, we denote by $\mathsf{scc}(p, \mathcal{A})$ the strongly connected component of p in $\mathcal{A}$ (that is, the set of states that are reachable from p and from which p can be reached), and by $\mathsf{alph\_scc}(p, \mathcal{A})$ the set of labels of all transitions occurring in this strongly connected component. Finally, we define the restriction of $\mathcal{A}$ to a subalphabet $B \subseteq A$ by $\mathcal{A} \restriction_B \overset{\text{def}}{=} (Q, B, I, F, \delta \cap (Q \times B \times Q))$.

3 Separation by Piecewise Testable Languages

Since $PT[\kappa] \subset PT$, $PT[\kappa]$-separability implies PT-separability. Furthermore, for a fixed κ, it is obviously decidable whether two languages L_1 and L_2 are $PT[\kappa]$-

separable: there is a finite number of $PT[\kappa]$ languages over A, and for each of them, one can test whether it separates L_1 and L_2. The difficulty for deciding whether L_1 and L_2 are PT-separable is to effectively compute a witness $\kappa = \kappa(L_1, L_2)$, i.e., such that L_1 and L_2 are PT-separable if and only if they are $PT[\kappa]$-separable. Actually, we show that PT-separability is decidable, by different arguments:

(1.a) We give a necessary and sufficient condition on NFAs recognizing L_1 and L_2, in terms of forbidden patterns, to test whether L_1 and L_2 are PT-separable.

(1.b) We give a polynomial time algorithm to check this condition.

(2) We compute $\kappa \in \mathbb{N}$ from L_1, L_2, such that PT-separability and $PT[\kappa]$-separability are equivalent for L_1 and L_2. Hence, if the PTIME algorithm answers that L_1 and L_2 are PT-separable, then $[L_1]_{\sim\kappa}$ is a valid PT-separator.

Let us first introduce some terminology to explain the necessary and sufficient condition on NFAs. Let $\mathcal{A}$ be an NFA over A. For $u_0, \ldots, u_p \in A^*$ and nonempty subalphabets $B_1, \ldots, B_p \subseteq A$, let $\boldsymbol{u} = (u_0, \ldots, u_p)$ and $\boldsymbol{B} = (B_1, \ldots, B_p)$. A $(\boldsymbol{u}, \boldsymbol{B})$-path in $\mathcal{A}$ is a successful path (leading from the initial state to a final state of $\mathcal{A}$), of the form shown in Fig. 1.

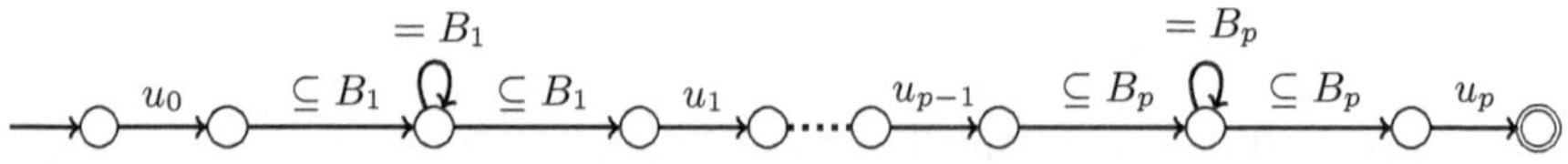

Fig. 1. A $(\boldsymbol{u}, \boldsymbol{B})$-path

Recall that edges denote sequences of transitions (see section *Automata*, p. 733). Therefore, if $\mathcal{A}$ has a $(\boldsymbol{u}, \boldsymbol{B})$-path, then $L(\mathcal{A})$ contains a language of the form $u_0(x_1 y_1^* z_1) u_1 \cdots u_{p-1}(x_p y_p^* z_p) u_p$, where $\mathsf{alph}(x_i) \cup \mathsf{alph}(z_i) \subseteq \mathsf{alph}(y_i) = B_i$.

Given NFAs $\mathcal{A}_1$ and $\mathcal{A}_2$, a pair $(\boldsymbol{u}, \boldsymbol{B})$ is a *witness of non PT-separability* for $(\mathcal{A}_1, \mathcal{A}_2)$ if there is a $(\boldsymbol{u}, \boldsymbol{B})$-path in both $\mathcal{A}_1$ and $\mathcal{A}_2$. For instance in Fig. 2, $\boldsymbol{u} = (\varepsilon, c, \varepsilon)$ and $\boldsymbol{B} = (\{a, b\}, \{a\})$ define such a witness of non PT-separability.

Fig. 2. A witness of non PT-separability for $(\mathcal{A}_1, \mathcal{A}_2)$: $\boldsymbol{u} = (\varepsilon, c, \varepsilon)$, $\boldsymbol{B} = (\{a, b\}, \{a\})$

We are now ready to state our main result regarding PT-separability.

Theorem 1. *Let $\mathcal{A}_1$ and $\mathcal{A}_2$ be two NFAs over A. Let $L_1 = L(\mathcal{A}_1)$ and $L_2 = L(\mathcal{A}_2)$. Let k_1, k_2 be the number of states of $\mathcal{A}_1$ resp. $\mathcal{A}_2$. Define $p = \max(k_1, k_2) + 1$ and $\kappa = p|A|2^{2^{|A|}|A|(p|A|+1)}$. Then the following conditions are equivalent:*

(1) L_1 and L_2 are PT-separable.
(2) L_1 and L_2 are PT[κ]-separable.
(3) The language $[L_1]_{\sim_\kappa}$ separates L_1 from L_2.
(4) There is no witness of non PT-separability in $(\mathcal{A}_1, \mathcal{A}_2)$.

Condition (2) yields an algorithm to test PT-separability of regular languages. Indeed, one can effectively compute all piecewise testable languages of index κ (of which there are finitely many), and for each of them, one can test whether it separates L_1 and L_2. Before proving Theorem 1, we show that Condition (4) can be tested in polynomial time (and hence, PT-separability is PTIME decidable).

Proposition 2. *Given two NFAs $\mathcal{A}_1$ and $\mathcal{A}_2$, one can determine whether there exists a witness of non PT-separability in $(\mathcal{A}_1, \mathcal{A}_2)$ in polynomial time wrt. the sizes of $\mathcal{A}_1$ and $\mathcal{A}_2$, and the size of the alphabet.*

Proof. Let us first show that the following problem is in PTIME: given states p_1, q_1, r_1 of $\mathcal{A}_1$ and p_2, q_2, r_2 of $\mathcal{A}_2$, determine whether there exist a nonempty $B \subseteq A$ and paths $p_i \xrightarrow{\subseteq B} q_i \xrightarrow{(=B)} q_i \xrightarrow{\subseteq B} r_i$ in $\mathcal{A}_i$ for both $i = 1, 2$.

To do so, we compute a decreasing sequence $(C_i)_i$ of alphabets overapproximating the greatest alphabet B that can be chosen for labeling the loops around q_1 and q_2. Note that if there exists such an alphabet B, it should be contained in

$$C_1 \stackrel{\text{def}}{=} \mathsf{alph\_scc}(q_1, \mathcal{A}_1) \cap \mathsf{alph\_scc}(q_2, \mathcal{A}_2).$$

Using Tarjan's algorithm to compute strongly connected components in linear time, one can compute C_1 in linear time as well. Then, we restrict the automata to alphabet C_1, and we repeat the process to obtain the sequence $(C_i)_i$:

$$C_{i+1} \stackrel{\text{def}}{=} \mathsf{alph\_scc}(q_1, \mathcal{A}_1 \upharpoonright_{C_i}) \cap \mathsf{alph\_scc}(q_2, \mathcal{A}_2 \upharpoonright_{C_i}).$$

After a finite number n of iterations, we obtain $C_n = C_{n+1}$. Note that $n \leqslant |\mathsf{alph}(\mathcal{A}_1) \cap \mathsf{alph}(\mathcal{A}_2)| \leqslant |A|$. If $C_n = \varnothing$, then there exists no nonempty B for which there is an $(= B)$-loop around both q_1 and q_2. If $C_n \neq \varnothing$, then it is the maximal nonempty alphabet B such that there are $(= B)$-loops around q_1 in $\mathcal{A}_1$ and q_2 in $\mathcal{A}_2$. It then remains to determine whether there exist paths $p_1 \xrightarrow{\subseteq B} q_1 \xrightarrow{\subseteq B} r_1$ and $p_2 \xrightarrow{\subseteq B} q_2 \xrightarrow{\subseteq B} r_2$, which can be performed in linear time.

To sum up, since the number n of iterations such that $C_n = C_{n+1}$ is bounded by $|A|$, and since each computation is linear wrt. the size of $\mathcal{A}_1$ and $\mathcal{A}_2$, one can decide in PTIME wrt. to both $|A|$ and these sizes whether such a pair of paths occurs.

Now we build from $\mathcal{A}_1$ and $\mathcal{A}_2$ two new automata $\tilde{\mathcal{A}}_1$ and $\tilde{\mathcal{A}}_2$ as follows. The procedure first initializes $\tilde{\mathcal{A}}_i$ as a copy of $\mathcal{A}_i$. Denote by Q_i the state set of $\mathcal{A}_i$. For each 4-uple $\tau = (p_1, r_1, p_2, r_2) \in Q_1^2 \times Q_2^2$ such that there exist $B \neq \varnothing$, two states $q_1 \in Q_1, q_2 \in Q_2$ and paths $p_i \xrightarrow{\subseteq B} q_i \xrightarrow{=B} q_i \xrightarrow{\subseteq B} r_i$ in $\mathcal{A}_i$ both for $i = 1$ and $i = 2$, we add in both $\tilde{\mathcal{A}}_1$ and $\tilde{\mathcal{A}}_2$ a new letter a_τ to the alphabet, and "summary" transitions $p_1 \xrightarrow{a_\tau} r_1$ and $p_2 \xrightarrow{a_\tau} r_2$. Since there is a polynomial number of tuples

$(p_1, q_1, r_1, p_2, q_2, r_2)$, the above shows that computing these new transitions can be performed in PTIME. So, computing $\tilde{\mathcal{A}}_1$ and $\tilde{\mathcal{A}}_2$ can be done in PTIME.

By construction, there exists some pair $(\boldsymbol{u}, \boldsymbol{B})$ such that $\mathcal{A}_1$ and $\mathcal{A}_2$ both have a $(\boldsymbol{u}, \boldsymbol{B})$-path if and only if $L(\tilde{\mathcal{A}}_1) \cap L(\tilde{\mathcal{A}}_2) \neq \varnothing$. Since both $\tilde{\mathcal{A}}_1$ and $\tilde{\mathcal{A}}_2$ can be built in PTIME, this can be decided in polynomial time as well. □

The following is an immediate consequence of Theorem 1 and Proposition 2.

Corollary 3. *Given two NFAs, one can determine in polynomial time, with respect to the number of states and the size of the alphabet, whether the languages recognized by these NFAs are PT-separable.* □

In the rest of the section, we sketch the proof of Theorem 1. The implications $(3) \Longleftrightarrow (2) \Longrightarrow (1)$ are obvious. To show $(1) \Longrightarrow (2)$, we introduce some terminology. Let us fix an arbitrary order $a_1 < \cdots < a_m$ on A.

(p, B)**-patterns.** Let $B = \{b_1, \ldots, b_r\} \subseteq A$ with $b_1 < \cdots < b_r$, and let $p \in \mathbb{N}$. We say that a word $w \in A^*$ is a (p, B)-*pattern* if $w \in (B^* b_1 B^* \cdots B^* b_r B^*)^p$. The number p is called the *power* of w. For example, set $B = \{a, b, c\}$ with $a < b < c$. The word $bb\underline{aaba}b\underline{cca}c\underline{ba}ba\underline{ca}$ is a $(2, B)$-pattern but not a $(3, B)$-pattern.

ℓ**-templates.** An ℓ-*template* is a sequence $T = t_1, \ldots, t_\ell$ of length ℓ, such that every t_i is either a letter or a nonempty subset of the alphabet A. The main idea behind ℓ-templates is that they yield decompositions of words that can be detected using pieces and provide a suitable decomposition for pumping. Unfortunately, not all ℓ-templates are actually detectable. Because of this we restrict ourselves to a special case of ℓ-templates. An ℓ-template is said to be *unambiguous* if all pairs t_i, t_{i+1} are either two letters, two incomparable sets or a set and a letter that is not included in the set. For example, $T = a, \{b, c\}, d, \{a\}$ is unambiguous, while $T' = \underline{b}, \{\underline{b}, c\}, d, \{a\}$ and $T'' = a, \{b, \underline{c}\}, \{\underline{c}\}, \{a\}$ are not.

p**-implementations.** A word $w \in A^*$ is a p-*implementation* of an ℓ-template $T = t_1, \ldots, t_\ell$ if $w = w_1 \cdots w_\ell$ and for all i either $t_i = w_i \in A$ or $t_i = B \subseteq A$, $w_i \in B^*$ and w_i is a (p, B)-pattern. For example, $abccbbcbdaaaa = a.(\underline{bccbbc}b).d.(\underline{aa}aa)$ is a 2-implementation of the 4-template $T = a, \{b, c\}, d, \{a\}$, since $\underline{bccbbc}b$ is a $(2, \{b, c\})$-pattern and $\underline{aa}aa$ is a $(2, \{a\})$-pattern.

We now prove $(1) \Longrightarrow (2)$ by contraposition: we show that if $w_1 \in L_1, w_2 \in L_2$ are such that $w_1 \sim_\kappa w_2$, then for any h, one can build $v_1 \in L_1$ and $v_2 \in L_2$ such that $v_1 \sim_h v_2$. Therefore, non-$PT[\kappa]$-separability entails non-PT-separability.

Lemma 4. *From regular languages L_1, L_2, we can compute $p \in \mathbb{N}$ such that whenever L_1 and L_2 both contain p-implementations of the same ℓ-template T, then L_1 and L_2 are not PT-separable.*

Proof. Let p be greater than the number of states of NFAs recognizing L_1, L_2. Let w_1, w_2 be p-implementations of an ℓ-template $T = t_1, \ldots, t_\ell$. Fix $h \in \mathbb{N}$. Whenever t_i is a set B, the corresponding factors in w_1, w_2 are (p, B)-patterns. By choice of p, these factors can be pumped into (h, B)-patterns in $v_1 \in L_1$ and $v_2 \in L_2$, respectively. It is then easy to check that $v_1 \sim_h v_2$. Hence, L_1, L_2 are not $PT[h]$-separable. Since h is arbitrary, L_1, L_2 are not PT-separable. □

It remains to prove that if two words contain the same pieces of a large enough size κ, they are both p-implementations of a common unambiguous ℓ-template, where p is the number introduced in Lemma 4. We split the proof in two parts. We begin by proving that it is enough to look for ℓ-templates for a bounded ℓ.

Lemma 5. *Let $p \in \mathbb{N}$. Every word is the p-implementation of some unambiguous N_A-template, for $N_A = 2^{2^{|A|}|A|(p|A|+1)}$.*

Proof. We first get rid of the unambiguity condition. Any ambiguous ℓ-template T can be reduced to an unambiguous ℓ'-template T' with $\ell' < \ell$ by merging the ambiguities. It is then straightforward to reduce any p-implementation of T into a p-implementation of T'. Therefore, it suffices to prove that every word is the p-implementation of some (possibly ambiguous) N_A-template.

The choice of N_A comes from Erdös-Szekeres' upper bound of Ramsey numbers. Indeed, a complete graph with edges labeled over $c = 2^{|A|}$ colors, there exists a complete monochromatic subgraph of size $m = p|A| + 1$ provided the graph has at least 2^{mc} vertices (see [5] for a short proof that this bound suffices).

Observe that a word is always the p-implementation of the ℓ-template which is just the sequence of its letters. Therefore, in order to complete our proof, it suffices to prove that if a word is the p-implementation of some ℓ-template T with $\ell > N_A$, then it is also the p-implementation of an ℓ'-template with $\ell' < \ell$.

Fix a word w, and assume that w is the p-implementation of some ℓ-template $T = t_1, \ldots, t_\ell$ with $\ell > N_A$. By definition, we get a decomposition $w = w_1 \cdots w_\ell$. We construct a complete graph Γ with vertices $\{0, \ldots, \ell\}$ and edges labeled by subsets of A. For all $i < j$, we set $\mathsf{alph}(w_{i+1} \cdots w_j)$ as the label of the edge (i, j). Since Γ has more than $\ell > N_A$ vertices, by definition of N_A there exists a complete monochromatic subgraph with $p|A| + 1$ vertices $\{i_1, \ldots, i_{p|A|+1}\}$. Let B be the color of the edges of this monochromatic subgraph. Let $w' = w_{i_1+1} \cdots w_{i_{p|A|+1}}$. By construction, w' is the concatenation of $p|A| \geqslant p$ words with alphabet exactly B. Hence w' is a (p, B)-pattern. It follows that w is a p-implementation of the ℓ'-template $t_1, \ldots, t_{i_1}, B, t_{i_{p|A|+2}}, \ldots, t_\ell$ with $\ell' = \ell - p|A| + 1$. Hence $\ell' < \ell$ (except for the trivial case $p = |A| = 1$).   □

The next lemma proves that once ℓ and p are fixed, given w it is possible to describe by pieces ℓ-templates that w p-implements, as long as they are unambiguous.

Lemma 6. *Let $\ell, p \in \mathbb{N}$. From p and ℓ, we can compute κ such that for every pair of words $w \sim_\kappa w'$ and every unambiguous ℓ-template T, w' is a p-implementation of T whenever w is a $(p + 1)$-implementation of T.*   □

We finish the proof of the implication (1) $\Longrightarrow$ (2) by assembling the results. Let p be greater than the number of states of NFAs recognizing L_1 and L_2, as introduced in the proof of Lemma 4. Let N_A be as introduced in Lemma 5 for $p + 1$, and let $\kappa = |A|(p + 1)N_A$ be as introduced in Lemma 6. Fix $h > \kappa$ and assume that we have $w_1 \in L_1$ and $w_2 \in L_2$ such that $w_1 \sim_\kappa w_2$. By Lemma 5, w_1 is the $(p + 1)$-implementation of some unambiguous N_A-template T. Moreover, it follows from Lemma 6 that w_2 is a p-implementation of T. By Lemma 4, we finally obtain that L_1 and L_2 are not PT-separable.

The implication (1) $\Longrightarrow$ (4) of Theorem 1 is easy to show by contraposition, see [13, Lemma 2]. The remaining implication (4) $\Longrightarrow$ (1) can be shown using Lemma 6. For a direct proof, see [13, Lemma 3], where the key for getting a forbidden pattern out of two non-separable languages is to extract a suitable p-implementation using Simon's Factorization Forest Theorem [17].

4   Separation by Unambiguous Languages

This section is devoted to proving that UL-separability is a decidable property. Again, the result is twofold. Using an argument that is analogous to property (2) of Theorem 1 in Section 3, we prove that given L_1, L_2, it is possible to compute a number κ such that L_1, L_2 are UL-separable if and only if they are $UL[\kappa]$-separable. It is then possible to test separability by using a brute-force approach that tests all languages in $UL[\kappa]$.

The second part of our theorem is an algorithm providing only a 'yes/no' answer, but running in exponential space. This algorithm is more complicated than the one of Section 3. In this case, we cannot search for a witness of non-separability directly on the NFAs of the languages. A precomputation is needed. We present the algorithm before stating our main theorem.

UL-intersection. Let $\mathcal{A}_1 = (Q_1, A, I_1, F_1, \delta_1)$, $\mathcal{A}_2 = (Q_2, A, I_2, F_2, \delta_2)$ be NFAs. The purpose of our precomputation is to associate to all 4-uples $(q_1, r_1, q_2, r_2) \in Q_1^2 \times Q_2^2$ a set $\alpha(q_1, r_1, q_2, r_2)$ of subalphabets. Intuitively, $B \in \alpha(q_1, r_1, q_2, r_2)$ if, for all $\kappa \in \mathbb{N}$, there are two words w_1, w_2 such that

(1) $B = \mathsf{alph}(w_1) = \mathsf{alph}(w_2)$,
(2) $q_1 \xrightarrow{w_1} r_1$, and $q_2 \xrightarrow{w_2} r_2$,
(3) $w_1 \cong_\kappa w_2$.

The precomputation of $\alpha : Q_1^2 \times Q_2^2 \to 2^{2^A}$ is performed via a fixpoint algorithm. For all $(q_1, r_1, q_2, r_2) \in Q_1^2 \times Q_2^2$, we initially set $\alpha(q_1, r_1, q_2, r_2) = \{\{a\} \mid q_1 \xrightarrow{a} r_1 \text{ and } q_2 \xrightarrow{a} r_2\}$. The sets are then saturated with the following two operations:

(1) When $\alpha(p_1, q_1, p_2, q_2) = B$ and $\alpha(q_1, r_1, q_2, r_2) = C$, then add $B \cup C$ to $\alpha(p_1, r_1, p_2, r_2)$.
(2) When $B \in \alpha(q_1, q_1, q_2, q_2) \cap \alpha(r_1, r_1, r_2, r_2)$ and there exist words $w_1, w_2 \in B^*$ such that $q_1 \xrightarrow{w_1} r_1$ and $q_2 \xrightarrow{w_2} r_2$ then add B to $\alpha(q_1, r_1, q_2, r_2)$.

Since every set $\alpha(q_1, r_1, q_2, r_2)$ only grows with respect to inclusion, and is bounded from above by 2^A, the computation terminates. It is straightforward to see that α can be computed in EXPSPACE using a fixpoint algorithm. Finally, we say that L_1, L_2 have *empty UL-intersection* if $\alpha(q_1, r_1, q_2, r_2) = \varnothing$ for all $q_1, q_2 \in I_1, I_2$ and $r_1, r_2 \in F_1, F_2$. We now state the main theorem of this section.

Theorem 7. *Let $\mathcal{A}_1$ and $\mathcal{A}_2$ be two NFAs over alphabet A. Let $L_1 = L(\mathcal{A}_1)$ and $L_2 = L(\mathcal{A}_2)$. Let k_1, k_2 be the number of states of $\mathcal{A}_1$, resp. $\mathcal{A}_2$. Define $\kappa = (2k_1 k_2 + 1)(|A| + 1)^2$. Then the following conditions are equivalent:*

(1) L_1 and L_2 are UL-separable.
(2) L_1 and L_2 are $UL[\kappa]$-separable.
(3) The language $[L_1]_{\cong_\kappa}$ separates L_1 from L_2.
(4) L_1, L_2 have empty UL-intersection.

As in the previous section, Conditions (2) and (4) yield algorithms for testing whether two languages are separable. Moreover, it can be shown that empty UL-intersection can be tested in PSPACE from α. Therefore, we get the following corollary.

Corollary 8. *It is decidable whether two regular languages can be separated by an unambiguous language. Moreover, this can be done in* EXPSPACE *in the size of the NFAs recognizing the languages.*

Observe that by definition of $UL[\kappa]$, the bound κ is defined in terms of unambiguous products. A rephrasing of the theorem would be: there exists a separator iff there exists one defined by a boolean combination of unambiguous products of size κ. It turns out that the same κ also works for $\mathrm{FO}^2(<)$, *i.e.*, there exists a separator iff there exists one defined by an $\mathrm{FO}^2(<)$-formula of quantifier rank κ. This can be proved by minor adjustements to the proof of Theorem 7.

The proof of Theorem 7 is inspired from techniques used in [11] and relies heavily on the notion of (p, B)-patterns. It works by induction on the size of the alphabet. There are two non-trivial implications: (1) $\Longrightarrow$ (4) and (4) $\Longrightarrow$ (3). We now provide an insight into the most difficult one, *i.e.*, (4) $\Longrightarrow$ (3). The following proposition is used to prove this.

Proposition 9. *Let* $B \subseteq A$ *and* $\kappa = (2k_1 k_2 + 1)(|B| + 1)^2$. *For all pairs of words* $w_1 \cong_\kappa w_2$ *such that* $B = \mathsf{alph}(w_1) = \mathsf{alph}(w_2)$ *and all pairs of states* $(q_1, r_1) \in Q_1^2$ *and* $(q_2, r_2) \in Q_2^2$ *such that* $q_1 \xrightarrow{w_1} r_1$ *and* $q_2 \xrightarrow{w_2} r_2$, *we have* $B \in \alpha(q_1, r_1, q_2, r_2)$.

Observe that a consequence of Proposition 9 is that as soon as there exists $w_1 \in L_1, w_2 \in L_2$ such that $w_1 \cong_\kappa w_2$ (*i.e.*, $[L_1]_{\cong_\kappa}$ is not a separator), there exists a witness of nonempty UL-intersection. This is the contrapositive of (4) $\Longrightarrow$ (3).

5   Conclusion

We proved separation results for both piecewise testable and unambiguous languages. Both results provide a means to decide separability. In the PT case, we even prove that this can be done in PTIME. Moreover, in both cases we give an insight on the actual separator by providing a bound on its size, should it exist.

There remain several interesting questions in this field. First, one could consider other subclasses of regular languages, the most interesting one being full first-order logic. Separability by first-order logic has already been proven to be decidable using semigroup theory [7]. However, this approach is difficult to understand, and it yields a costly algorithm that only provides a yes/no answer, without insight about a possible separator. Another question is to get tight complexity bounds. For unambiguous languages for instance, it is likely that our

EXPSPACE upper bound can be improved, and even for piecewise testable languages, we do not know any tight bounds.

A final observation is that right now, we have no general approach and are bound to use *ad-hoc* techniques for each subclass. An interesting direction would be to invent a general framework that is suitable for this problem in the same way that monoids are a suitable framework for decidable characterizations.

References

1. Almeida, J.: Some algorithmic problems for pseudovarieties. Publ. Math. Debrecen 54(suppl.), 531–552 (1999); Automata and formal languages, VIII (Salgótarján, 1996)
2. Almeida, J., Costa, J.C., Zeitoun, M.: Pointlike sets with respect to R and J. J. Pure Appl. Algebra 212(3), 486–499 (2008)
3. Almeida, J., Zeitoun, M.: The pseudovariety **J** is hyperdecidable. RAIRO Inform. Théor. Appl. 31(5), 457–482 (1997)
4. Ash, C.J.: Inevitable graphs: a proof of the type II conjecture and some related decision procedures. Internat. J. Algebra Comput. 1, 127–146 (1991)
5. Bacher, R.: An easy upper bound for Ramsey numbers. HAL, 00763927
6. Czerwiński, W., Martens, W., Masopust, T.: Efficient separability of regular languages by subsequences and suffixes. In: Fomin, F.V., Freivalds, R., Kwiatkowska, M., Peleg, D. (eds.) ICALP 2013, Part II. LNCS, vol. 7966, pp. 150–161. Springer, Heidelberg (2013)
7. Henckell, K.: Pointlike sets: the finest aperiodic cover of a finite semigroup. J. Pure Appl. Algebra 55(1-2), 85–126 (1988)
8. Henckell, K., Rhodes, J., Steinberg, B.: Aperiodic pointlikes and beyond. IJAC 20(2), 287–305 (2010)
9. Hunt III, H.B.: Decidability of grammar problems. J. ACM 29(2), 429–447 (1982)
10. Pin, J.-E., Weil, P.: Polynomial closure and unambiguous product. Theory of Computing Systems 30(4), 383–422 (1997)
11. Place, T., Segoufin, L.: Deciding definability in $FO_2(<_h, <_v)$ on trees. Journal Version (to appear, 2013)
12. Ribes, L., Zalesskiĭ, P.A.: On the profinite topology on a free group. Bull. London Math. Soc. 25, 37–43 (1993)
13. van Rooijen, L., Zeitoun, M.: The separation problem for regular languages by piecewise testable languages (2013), http://arxiv.org/abs/1303.2143
14. Schützenberger, M.: On finite monoids having only trivial subgroups. Information and Control 8(2), 190–194 (1965)
15. Schützenberger, M.: Sur le produit de concaténation non ambigu. Semigroup Forum 13, 47–75 (1976)
16. Simon, I.: Piecewise testable events. In: Brakhage, H. (ed.) GI-Fachtagung 1975. LNCS, vol. 33, pp. 214–222. Springer, Heidelberg (1975)
17. Simon, I.: Factorization forests of finite height. Th. Comp. Sci. 72(1), 65–94 (1990)
18. Stern, J.: Complexity of some problems from the theory of automata. Information and Control 66(3), 163–176 (1985)
19. Tesson, P., Therien, D.: Diamonds are forever: The variety DA. In: Semigroups, Algorithms, Automata and Languages, pp. 475–500. World Scientific (2002)
20. Thérien, D., Wilke, T.: Over words, two variables are as powerful as one quantifier alternation. In: Proc. of STOC 1998, pp. 234–240. ACM (1998)
21. Trahtman, A.N.: Piecewise and local threshold testability of DFA. In: Freivalds, R. (ed.) FCT 2001. LNCS, vol. 2138, pp. 347–358. Springer, Heidelberg (2001)

An Unusual Temporal Logic

Alexander Rabinovich

The Blavatnik School of Computer Science, Tel Aviv University

Abstract. Kamp's theorem states that the temporal logic with modalities Until and Since has the same expressive power as the First-Order Monadic Logic of Order (FOMLO) over Real and Natural time flows. Kamp notes that there are expressions which deserve to be regarded as tense operators but are not representable within FOMLO. The words 'mostly' and 'usually' are examples of such expressions. We propose a formalization of 'usually' as a generalized Mostowski quantifier and prove an analog of Kamp's theorem.

1 Introduction

Temporal Logic (*TL*), introduced to Computer Science by Pnueli in [5], is a convenient framework for reasoning about "reactive" systems. This has made temporal logics a popular subject in the Computer Science community, enjoying extensive research in the past 40 years. In *TL* we describe basic system properties by *atomic propositions* that hold at some points in time, but not at others. More complex properties are expressed by formulas built from the atoms using Boolean connectives and *Modalities* (temporal connectives): A k-place modality M transforms statements $\varphi_1, \ldots, \varphi_k$ possibly on 'past' or 'future' points to a statement $M(\varphi_1, \ldots, \varphi_k)$ on the 'present' point t_0. The rule to determine the truth of a statement $M(\varphi_1, \ldots, \varphi_k)$ at t_0 is called a *truth table* of M. The choice of particular modalities with their truth tables yields different temporal logics. A temporal logic with modalities $M_1, \ldots, M_k$ is denoted by $TL(M_1, \ldots, M_k)$.

The simplest example is the one place modality $\Diamond P$ saying: "P holds some time in the future." Its truth table is formalized by $\varphi_\Diamond(x_0, X) := \exists x(x > x_0 \wedge P(x))$. This is a formula of the First-Order Monadic Logic of Order (*FOMLO*) - a fundamental formalism in Mathematical Logic where formulas are built using atomic propositions $P(x)$, atomic relations between elements $x_1 = x_2$, $x_1 < x_2$, Boolean connectives and first-order quantifiers $\exists x$ and $\forall x$. Two more natural modalities are the modalities Until (*"Until"*) and Since (*"Since"*). XUntilY means that X will hold from now until a time in the future when Y will hold. XSinceY means that Y was true at some point of time in the past and since that point X was true until (not necessarily including) now.

The main *canonical*, linear time intended models are the non-negative integers $\omega := \langle \mathbb{N}, < \rangle$ for discrete time and the reals $\langle \mathbb{R}, < \rangle$ for continuous time.

K. Chatterjee and J. Sgall (Eds.): MFCS 2013, LNCS 8087, pp. 741–752, 2013.
© Springer-Verlag Berlin Heidelberg 2013

Kamp's theorem [3] states that the temporal logic with modalities Until and Since and *FOMLO* have the same expressive power over the above two linear time canonical[1] models. After explaining his main theorem, Kamp writes:

> This still leaves open the question whether *all* English tense operators are representable in a language like *TL*. .... One easily verifies that indeed a very large number of expressions which are naturally classified as tense operators because of their function have first order definable tenses as their meanings. Yet there are expressions which deserve to be regarded as tense operators but which are nonetheless not representable within *TL*. The words 'mostly' and 'usually' are examples of such expressions. The impossibility of representing these particular expressions stems from the fact that their meanings involve a measure on time in an essential manner.

In this paper we suggest a formalization of "usually" over the standard discrete time $\omega := (\mathbb{N}, <)$ and prove a generalization of Kamp's theorem.

Here are three natural possibilities to formalize "P is unusual."

1. If P is finite.
2. If $\limsup_{n \to \infty} \dfrac{\text{the cardinality of } P \cap [0,n]}{n} = 0$.
3. If $\sum \frac{1}{p_i + 1}$ finite, where $p_0 < p_1 < \cdots < p_i < \cdots$ is the enumeration of the elements of P.

$P \subseteq \mathbb{N}$ is usual if its complement is unusual. Note that "P is finite" is definable in *FOMLO* (over ω); however, formalizations (2)-(3) of "P is unusual" are not first-order definable.

A. Mostowski [4] initiated a study of so-called generalized quantifiers. Generalized quantifiers are now standard equipment in the toolboxes of both logicians and linguists.

The first-order logic with a (unary) generalized quantifier Q is obtained by extending the syntax of first-order logic by the rule if φ is a formula then $(Qx)\varphi$ is a formula. A (unary) generalized quantifier Q in a structure $\mathcal{M}$ is defined as a set Q of subsets of the domain of $\mathcal{M}$. The corresponding semantical clause for $(Qx)\varphi$ is $\mathcal{M} \models (Qx)\varphi(x, \overline{b})$ if $\{a \mid \mathcal{M} \models \varphi(a, \overline{b})\}$ is in Q.

For a family of subsets Q of $\mathbb{N}$, we define a temporal modality $\langle Q \rangle$ as follows: $\langle Q \rangle \varphi$ holds iff the set of points where φ holds is in Q.

Each of the above formalizations of unusual has the following properties:

1. If $P_1 \in Q$ and $P_2 \subseteq P_1$ then $P_2 \in Q$, i.e., if P_1 is unusual and $P_2 \subseteq P_1$, then so is P_2.
2. If $P_1, P_2 \in Q$ then $P_1 \cup P_2 \in Q$, i.e., if both P_i are unusual then their union is also unusual.
3. If $P_1 \in Q$ and P_2 is finite then $P_1 \cup P_2 \in Q$, i.e., if a finite set is added to an unusual event then the new set is still unusual.

---

[1] The technical notion which unifies $\langle \mathbb{N}, \ < \rangle$ and $\langle \mathbb{R}, < \rangle$ is Dedekind completeness.

Our main theorem states that for every family $\mathcal{Q}$ of subsets of $\mathbb{N}$ with properties (1)-(3) the temporal logic with modalities Until and Since and $\langle Q \rangle$ is expressively equivalent over $\omega := (\mathbb{N}, <)$ to the extension of *FOMLO* by the generalized quantifier Q. Moreover, our meaning preserving translations between these logics are computable and independent of $\mathcal{Q}$.

The rest of the paper is organized as follows. In Section 2 we recall the definitions of the monadic logic, the temporal logics and state Kamp's theorem. In Section 3 we provide a formalization of unusual as a Mostowski generalized quantifier and state our main result. In Section 4 we prove the main theorem. This proof is based on our simple proof of Kamp's theorem [6]. The proof of one proposition is postponed to Section 5. Section 6 states further results and open questions.

2 Kamp's Theorem

In this section we recall the definitions of the first-order monadic logic of order, the temporal logics and state Kamp's theorem.

Fix a set Σ of *atoms*. We use $P, R, S \ldots$ to denote members of Σ. The syntax and semantics of both logics are defined below with respect to such Σ.

First-Order Monadic Logic of Order. In the context of *FOMLO*, the atoms of Σ are referred to (and used) as *unary predicate symbols*. Formulas are built using these symbols, plus two binary relation symbols: $<$ and $=$, and a set of first-order variables (denoted: $x, y, z, \ldots$). Formulas are defined by the grammar:

$$\varphi ::= \quad x < y \mid x = y \mid P(x) \mid \neg\varphi_1 \mid \varphi_1 \vee \varphi_2 \mid \varphi_1 \wedge \varphi_2 \mid \exists x \varphi_1 \mid \forall x \varphi_1$$

where $P \in \Sigma$. We will also use the standard abbreviated notation for **bounded quantifiers**, e.g., $(\exists x)_{>z}(\ldots)$ denotes $\exists x((x > z) \wedge (\ldots))$, and $(\forall x)^{<z}(\ldots)$ denotes $\forall x((x < z) \rightarrow (\ldots))$, and $((\forall x)^{<z_2}_{>z_1}(\ldots)$ denotes $\forall x((z_1 < x < z_2) \rightarrow (\ldots))$, etc.

Semantics. Formulas are interpreted over *labeled linear orders* which are called *chains*. A Σ-**chain** is a triplet $\mathcal{M} = (\mathcal{T}, <, \mathcal{I})$ where $\mathcal{T}$ is a set - the *domain* of the chain, $<$ is a linear order relation on $\mathcal{T}$, and $\mathcal{I} : \Sigma \rightarrow \mathcal{P}(\mathcal{T})$ is the *interpretation* of Σ (where $\mathcal{P}$ is the powerset notation). We use the standard notation $\mathcal{M}, t_1, t_2, \ldots, t_n \models \varphi(x_1, x_2, \ldots, x_n)$ to indicate that the formula φ with free variables among $x_1, \ldots, x_n$ is satisfiable in $\mathcal{M}$ when x_i are interpreted as elements t_i of $\mathcal{M}$. For atomic $P(x)$ this is defined by: $\mathcal{M}, t \models P(x)$ iff $t \in \mathcal{I}(P)$; the semantics of $<, =, \neg, \wedge, \vee, \exists$ and $\forall$ is defined in a standard way.

Temporal Logics. In the context of temporal logics the atoms of Σ are used as *atomic propositions* (also called *propositional atoms*). Formulas are built using these atoms, and a set (finite or infinite) B of **modality names**, where a non-negative integer **arity** is associated with each $\mathsf{M} \in B$. The syntax of TL with the **basis** B over the signature Σ, denoted by **TL(B)**, is defined by the grammar:

$$F ::= \quad P \mid \neg F_1 \mid F_1 \vee F_2 \mid F_1 \wedge F_2 \mid \mathsf{M}(F_1, F_2, \ldots, F_n)$$

where $P \in \Sigma$ and $\mathsf{M} \in B$ is an n-place modality. As usual, **True** denotes $P \vee \neg P$ and **False** denotes $P \wedge \neg P$.

Semantics. The semantics defines when a temporal formula holds at a *time-point* (or *moment* or element of the domain) in a chain.

The semantics of each n-place modality $\mathsf{M} \in B$ is defined by a 'rule' specifying how the set of moments where $\mathsf{M}(F_1, \ldots, F_n)$ holds (in a given structure) is determined by the n sets of moments where each of the formulas F_i holds. Such a 'rule' for M is formally specified (over time flow $(\mathcal{T}, <)$), by an operator $\mathcal{O}_\mathsf{M}$: $(\mathcal{P}(\mathcal{T}))^n \longrightarrow \mathcal{P}(\mathcal{T})$, which assigns to each n tuples of subsets of $\mathcal{T}$ a subset of $\mathcal{T}$.

The semantics of $TL(B)$ formulas is then defined inductively: Given a structure $\mathcal{M} = (\mathcal{T}, <, \mathcal{I})$ and a moment $t \in \mathcal{M}$ (read $t \in \mathcal{M}$ as $t \in \mathcal{T}$), define when a formula F **holds** in $\mathcal{M}$ at t - notation: $\mathcal{M}, t \models F$ - as follows:
- $\mathcal{M}, t \models P$ iff $t \in \mathcal{I}(P)$ for any propositional atom P.
- $\mathcal{M}, t \models F \vee G$ iff $\mathcal{M}, t \models F$ or $\mathcal{M}, t \models G$; similarly ("pointwise") for $\wedge$, $\neg$.
- $\mathcal{M}, t \models \mathsf{M}(F_1, \ldots, F_n)$ iff $t \in \mathcal{O}_\mathsf{M}(T_1, \ldots, T_n)$ where $\mathsf{M} \in B$ is an n-place modality, $F_1, \ldots, F_n$ are formulas and $T_i := \{s \in \mathcal{T} : \mathcal{M}, s \models F_i\}$.

Truth tables. Practically, most standard modalities studied in the literature can be specified in *FOMLO*: A *FOMLO* formula $\varphi(x, P_1, \ldots, P_n)$ (with a single free first-order variable x and with n predicate symbols P_i) is called an n-**place first-order truth table**. Such a truth table φ **defines** an n-ary modality M whose semantics is given by an operator $\mathcal{O}_\mathsf{M}$ such that for any time flow $(\mathcal{T}, <)$, for any $T_1, \ldots, T_n \subseteq \mathcal{T}$ and for any structure $\mathcal{M} = (\mathcal{T}, <, \mathcal{I})$ where $\mathcal{I}(P_i) = T_i$:

$$\mathcal{O}_M(T_1, \ldots, T_n) = \{t \in \mathcal{T} : \mathcal{M}, t \models \varphi(x, P_1, \ldots, P_n)\}$$

Example 2.1. *Below are truth-table definitions for the well known "**Eventually**", the (binary) **strict**-Until and **strict**-Since of [3,1].*

- $\Diamond$ (*"**Eventually**"*) is defined by: $\varphi_\Diamond(x, P) := (\exists x')_{>x} P(x')$
- Until *is defined by* : $\varphi_{\mathrm{Until}}(x, P_1, P_2) := (\exists x')_{>x}(P_2(x') \wedge (\forall y)^{\leq x'}_{\leq x} P_1(y))$
- Since *is defined by*: $\varphi_{\mathrm{Since}}(x, P_1, P_2) := (\exists x')^{<x}(P_2(x') \wedge (\forall y)^{<x}_{>x'} P_1(y))$

Example 2.2 (Modality $\langle \mathcal{Q} \rangle$). *Let $\mathcal{Q}$ be a family of subsets of the domain $\mathcal{T}$ of a structure $\mathcal{M}$. We can define a unary modality $\langle \mathcal{Q} \rangle$ by the operator*

- $\mathcal{O}(T_1) := \begin{cases} \mathcal{T} & \text{if } T_1 \in \mathcal{Q} \\ \emptyset & \text{otherwise.} \end{cases}$

In the next section we will formalize "usually" by special families of subsets of $\mathbb{N}$. It is clear that there are $\mathcal{Q}$ such that the corresponding modality $\langle \mathcal{Q} \rangle$ has no first-order truth table.

Kamp's Theorem. Equivalence between temporal and monadic formulas is naturally defined: F is equivalent to $\varphi(x)$ over a class $\mathcal{C}$ of structures iff for any $\mathcal{M} \in \mathcal{C}$ and $t \in \mathcal{M}$: $\mathcal{M}, t \models F \Leftrightarrow \mathcal{M}, t \models \varphi(x)$. If $\mathcal{C}$ is the class of all chains, we will say that F is equivalent to φ.

A linear order $(T, <)$ is *Dedekind complete* if every non-empty subset (of the domain) which has an upper bound has a least upper bound. The canonical linear time models $\omega := (\mathbb{N}, <)$ and $(\mathbb{R}, <)$ are Dedekind complete, while the order of the rationals is not Dedekind complete. A chain is Dedekind complete if its underlying linear order is Dedekind complete.

The fundamental theorem of Kamp's states that $TL(\mathsf{Until}, \mathsf{Since})$ is expressively equivalent to $FOMLO$ over Dedekind complete chains.

Theorem 2.3 (Kamp [3]).   *1. Given any $TL(\mathsf{Until}, \mathsf{Since})$ formula A there is a FOMLO formula $\varphi_A(x)$ which is equivalent to A over all chains.*
2. Given any FOMLO formula $\varphi(x)$ with one free variable, there is a $TL(\mathsf{Until}, \mathsf{Since})$ formula which is equivalent to φ over Dedekind complete chains.

3   An Unusual Quantifier and Modality

3.1   Generalized Quantifier

The syntax of the first-order logic with a unary generalized quantifier Q (notation $FO[Q]$) is obtained by extending the usual first-order syntax by the new quantifier.

The formulas of $FO[Q]$ are built by the usual formation rules and the following new (variable-binding) formation rule:
- if x is a variable and φ is a formula of $FO[Q]$, then so is $(Qx)\varphi$, and Qx binds all free occurrences of x in φ.

The semantics of $FO[Q]$ is provided by enriching the domain of first-order structures with a set $\mathcal{Q}$ of subsets of its domain and extending the usual definition of satisfaction by a clause for $(Qx)\varphi$:

$$\mathcal{M}, b_1, \ldots, b_n \models (Qx)\varphi(x, y_1, \ldots, y_n) \text{ if } \{a \mid \mathcal{M}, a, b_1, \ldots, b_n \models \varphi(x, y_1, \ldots, y_n)\} \text{ is in } \mathcal{Q}.$$

$FOMLO[Q]$ denotes the extension of $FOMLO$ by a generalized quantifier Q.

For a generalized quantifier Q we also introduce **modality** $\langle Q \rangle$, defined by $\mathcal{M}, t \models \langle Q \rangle \varphi$ iff $\{a \mid \mathcal{M}, a \models \varphi\} \in \mathcal{Q}$. Note that if $\mathcal{M}, t \models \langle Q \rangle \varphi$, then $\mathcal{M}, t' \models \langle Q \rangle \varphi$ for every t'.

3.2   Usual and Unusual over $\mathbb{N}$

Let us start with some intuitive requirements on unusual sets. If P never happens (respectively, always holds) then P is unusual (respectively, is not unusual). If P_1 is unusual and $P_2 \subseteq P_1$, then P_2 is unusual. If both P_1 and P_2 are unusual, then their union is also unusual. It is also natural to require that a finite subset of an infinite set is unusual. These lead to the following definition.

Given a set X, an **unusual** family on X is a set $\mathcal{Q}$ consisting of subsets of X such that

1. $\emptyset \in \mathcal{Q}$ and $X \notin \mathcal{Q}$.
2. If A and B are subsets of X, A is a subset of B, and B is an element of $\mathcal{Q}$, then A is also an element of $\mathcal{Q}$.

3. If A and B are elements of $\mathcal{Q}$, then so is the union of A and B.
4. If A is finite then A is in $\mathcal{Q}$.

In model theory, an ideal $\mathcal{Q}$ on a set X is a family of subsets of X which satisfies (1)-(3). A filter is a dual notion to an ideal. Hence, a family $\mathcal{Q}$ of subsets of X is a filter, if it is a non-empty proper subset of $\mathcal{P}(X)$ and it is closed under superset and finite intersection. In model theory ideals (respectively, filters) are considered as families of small (respectively, big) subsets of X.

Several collections of "small" subsets of $\mathbb{N}$ are presented below:
1. $\mathcal{Q}_1 := \{P \mid P \text{ is finite}\}$.
2. $\mathcal{Q}_2 := \{P \mid \limsup_{n \to \infty} \dfrac{\text{the cardinality of } P \cap [0,n]}{n} = 0\}$.
3. Van der Waerden ideal is the family $\{P \mid P \text{ does not contain an arithmetic progression of arbitrary length}\}$.

Let $p_0 < p_1 < \cdots < p_i < \cdots$ be the enumeration of the elements of P.

4. $\mathcal{Q}_4 := \{P \mid \sum \frac{1}{p_i+1} \text{ is finite }\}$.
5. P is 1-sparse if for every n there is N such that $[m, m+n]$ contains at most one element from P for every $m > N$.
6. P is 1-thin if $\lim_{n \to \infty} \frac{p_n}{p_{n+1}} = 0$.
7. P is almost 1-thin if $\limsup_{n \to \infty} \frac{p_n}{p_{n+1}} < 1$.

Note that 1-sparse (respectively, 1-thin, or almost 1-thin) sets are not closed under union. Hence, these families of sets are not ideals.

A set is sparse (respectively, thin or almost thin) if it is finite or a finite union of 1-sparse (respectively, 1-thin, or almost 1-thin) sets. The family of sparse (respectively, thin or almost thin) is an ideal.

The families defined in examples (1)-(4), as well as the families of sparse, thin and almost thin sets are unusual. The family $\{P \mid \text{the set of even elements of } P \text{ is finite}\}$ is also unusual.

A generalized **quantifier** Q is **unusual** if its corresponding family of subsets of $\mathbb{N}$ is unusual. Dually, we say that a family $\mathcal{Q}$ of sets is **usual** if $\{\mathbb{N} \setminus P \mid P \in \mathcal{Q}\}$ is unusual. The corresponding quantifier and modality are *usual*. The next Lemma states some immediate equivalences:

Lemma 3.1. *If Q is an unusual quantifier. Then:*

1. $(Qx)(\varphi_1 \vee \varphi_2)$ *is equivalent to* $((Qx)\varphi_1) \wedge (Qx)\varphi_2$.
2. $(Qx)(\varphi \wedge x < z)$ *is equivalent to True.*
3. *If x does not occur free in φ, then $(Qx)(\varphi \wedge \psi)$ is equivalent to $\neg\varphi \vee (Qx)\psi$,*
4. *Assume that x does not occur free in φ. Then $(Qx)\varphi$ is equivalent to $\neg\varphi$.*

3.3   Expressive Equivalence

Theorem 3.2 (Main). *Let $\mathcal{Q}$ be an unusual family of subsets of $\mathbb{N}$. Let Q and $\langle Q \rangle$ be the corresponding generalized quantifier and modality. Then*

1. *Given any $TL(\text{Until}, \text{Since}, \langle Q \rangle)$ formula A there is a $FOMLO[Q]$ formula $\varphi_A(x)$ which is equivalent (over ω-chains) to A.*

2. *Given any FOMLO[Q] formula $\varphi(x)$ with one free variable, there is a TL(Until, Since, $\langle Q \rangle$) formula A_φ which is equivalent (over ω-chains) to φ.*

Moreover, φ_A and A_φ are computable from φ and A and independent of $\mathcal{Q}$.

The meaning preserving translation from $TL(\text{Until}, \text{Since}, \langle Q \rangle)$ to $FOMLO[Q]$ is easily obtained by structural induction. The main technical contribution of our paper is a proof of Theorem 3.2 (2). The proof is constructive. An algorithm which for every $FOMLO[Q]$ formula $\varphi(x)$ constructs a $TL(\text{Until}, \text{Since}, \langle Q \rangle)$ formula which is equivalent to φ is easily extracted from our proof.

4   Proof of the Main Theorem

First, we introduce $\overrightarrow{\exists} \forall$ formulas which are instances of the Decomposition formulas of [2,6].

Definition 4.1 ($\overrightarrow{\exists} \forall$-formulas). *Let Σ be a set of monadic predicate names. An $\overrightarrow{\exists} \forall$-formula over Σ is a formula of the form:*

$$\psi(z_0, \ldots, z_m) := \exists x_n \ldots \exists x_1 \exists x_0$$

$$\left(\bigwedge_{k=0}^{m} z_k = x_{i_k} \right) \wedge (x_n > x_{n-1} > \cdots > x_1 > x_0) \quad \text{``ordering of x_i and z_j''}$$

$$\wedge \bigwedge_{j=0}^{n} \alpha_j(x_j) \quad \text{``Each α_j holds at x_j''}$$

$$\wedge \bigwedge_{j=1}^{n} [(\forall y)_{>x_{j-1}}^{<x_j} \beta_j(y)] \quad \text{``Each β_j holds along (x_{j-1}, x_j)''}$$

$$\wedge (\forall y)_{>x_n} \beta_{n+1}(y) \quad \text{``β_{n+1} holds everywhere after x_n''}$$

$$\wedge (\forall y)^{<x_0} \beta_0(y) \quad \text{``β_0 holds everywhere before x_0''}$$

with a prefix of $n+1$ existential quantifiers and with all α_j, β_j quantifier free formulas with one variable over Σ. (ψ has $m+1$ free variables $z_0, \ldots, z_m$ and $m+1 \le n+1$ existential quantifiers are dummy and are introduced just in order to simplify notations.)

Definition 4.2 ($\vee \overrightarrow{\exists} \forall$-formulas). *A formula is a $\vee \overrightarrow{\exists} \forall$ formula if it is equivalent to a disjunction of $\overrightarrow{\exists} \forall$-formulas.*

The set of $\vee \overrightarrow{\exists} \forall$ formulas is closed under disjunction, conjunction and existential quantification. The set of $\vee \overrightarrow{\exists} \forall$ formulas is not closed under negation. However, the negation of a $\vee \overrightarrow{\exists} \forall$ formula is equivalent to a $\vee \overrightarrow{\exists} \forall$ formula in the expansion of chains by all $TL(\text{Until}, \text{Since})$ definable predicates (see Proposition 4.7).

The next definition plays a major role in the proof of Kamp's theorem [2,6].

Definition 4.3. *Let $\mathcal{M}$ be a Σ chain and $\mathcal{L}$ be a temporal logic. We denote by $\mathcal{L}[\Sigma]$ the set of unary predicate names $\Sigma \cup \{A \mid A$ is an $\mathcal{L}$-formula over $\Sigma \}$. The canonical $\mathcal{L}$-expansion of $\mathcal{M}$ is an expansion of $\mathcal{M}$ to an $\mathcal{L}[\Sigma]$-chain, where each predicate name $A \in \mathcal{L}[\Sigma]$ is interpreted as $\{a \in \mathcal{M} \mid \mathcal{M}, a \models A\}^2$. We say that first-order formulas in the signature $\mathcal{L}[\Sigma] \cup \{<\}$ are equivalent over $\mathcal{M}$ (respectively, over a class of Σ-chains $\mathcal{C}$) if they are equivalent in the canonical expansion of $\mathcal{M}$ (in the canonical expansion of every $\mathcal{M} \in \mathcal{C}$).*

Note that if A is a $\mathcal{L}$ formula over $\mathcal{L}[\Sigma]$ predicates, then it is equivalent to a $\mathcal{L}$ formula over Σ, and hence to an atomic formula in the canonical $\mathcal{L}$-expansions.

The $\overrightarrow{\exists}\forall$ and $\vee\overrightarrow{\exists}\forall$ formulas are defined as previously, but now they can use as atoms $\mathcal{L}$ definable predicates.

The next Proposition was proved in [6].

Proposition 4.4. *Let $\mathcal{L}$ be a temporal logic which contains modalities* Until *and* Since*. Every FOMLO formula is equivalent (over the canonical $\mathcal{L}$ expansions of ω-chains) to a disjunction of $\overrightarrow{\exists}\forall$-formulas.*

The $\vee\overrightarrow{\exists}\forall$ formulas with one free variable can be easily translated to temporal formulas.

Proposition 4.5 (From $\vee\overrightarrow{\exists}\forall$-formulas to $\mathcal{L}$ formulas). *If $\mathcal{L}$ contains modalities* Until *and* Since*, then every $\vee\overrightarrow{\exists}\forall$ formula with one free variable is equivalent (over the canonical $\mathcal{L}$-expansions) to an $\mathcal{L}$ formula.*

The proof of the next proposition is postponed to Sect. 5.

Proposition 4.6. *(Closure under unusual quantifier) Let Q be an unusual quantifier on $\mathbb{N}$ and $\mathcal{L}$ be a temporal logic which contains modalities* Until*,* Since *and $\langle Q \rangle$. If ψ is an $\overrightarrow{\exists}\forall$-formula, then $(Qx)\psi$ is equivalent (over the canonical $\mathcal{L}$ expansions of ω-chains) to a disjunction of $\overrightarrow{\exists}\forall$-formulas.*

As a consequence we obtain:

Proposition 4.7. *Let Q be an unusual quantifier on $\mathbb{N}$ and $\mathcal{L}$ be a temporal logic which contains modalities* Until*,* Since *and $\langle Q \rangle$. Every FOMLO[Q] is equivalent (over the canonical $\mathcal{L}$ expansions of ω-chains) to a disjunction of $\overrightarrow{\exists}\forall$-formulas.*

Now, we are ready to prove the unusual version of Kamp's Theorem:

Theorem 4.8. *Let Q be an unusual quantifier on $\mathbb{N}$. For every FOMLO[Q] formula $\varphi(x)$ with a single free variable, there is a $TL(\mathsf{Until}, \mathsf{Since}, \langle Q \rangle)$ formula which is equivalent (on ω-chains) to φ.*

Proof. By Proposition 4.7, $\varphi(x)$ is equivalent to a disjunction of $\overrightarrow{\exists}\forall$ formulas. By Proposition 4.5, an $\overrightarrow{\exists}\forall$ formula is equivalent to a $TL(\mathsf{Until}, \mathsf{Since}, \langle Q \rangle)$ formula. Hence, $\varphi(x)$ is equivalent to a $TL(\mathsf{Until}, \mathsf{Since}, \langle Q \rangle)$ formula. $\square$

This completes the proof of our main theorem, except for Proposition 4.6 which is proved in the next section.

---

^{2} We often use "$a \in \mathcal{M}$" instead of "a is an element of the domain of $\mathcal{M}$"

5   Proof of Proposition 4.6

In this section we say that "formulas are equivalent in a chain $\mathcal{M}$" instead of "formulas are equivalent in the canonical $\mathcal{L}$-expansion of $\mathcal{M}$." We also say that "formulas are equivalent" instead of "formulas are equivalent in the canonical $\mathcal{L}$-expansions of chains over ω."

If ψ has at most one free variable then, by Proposition 4.5, ψ is equivalent to a $TL(\mathsf{Until}, \mathsf{Since}, \langle Q \rangle)$ formula A. Hence, $(Qx)\psi$ is equivalent to a temporal logic formula $\langle Q \rangle A$.

Let $\psi(z_0, \ldots, z_m)$ be an $\overset{\rightarrow}{\exists}\forall$-formula as in Definition 4.1 with $m \geq 1$. W.l.o.g. assume that $\psi \to \wedge_{i=0}^{m-1} z_i < z_{i+1}$.

If x is not free in ψ then, by Lemma 3.1, $(Qx)\psi$ is equivalent to a $\neg\psi$ and hence to a $\vee\overset{\rightarrow}{\exists}\forall$ formula by Proposition 4.4.

If $x \in \{z_0, \ldots, z_{m-1}\}$, then there are at most finitely many x which satisfy ψ, therefore $(Qx)\psi$ is equivalent to True.

If x is z_m then ψ is equivalent to the conjunction of an $\overset{\rightarrow}{\exists}\forall$-formula $\psi_1(z_0, \ldots, z_{m-1})$ and an $\overset{\rightarrow}{\exists}\forall$-formula $\psi_2(z_{m-1}, z_m)$ with two free variables z_{m-1} and z_m. By Lemma 3.1, $(Qz_m)\psi$ is equivalent to $\neg\psi_1 \vee (Qz_m)\psi_2(z_{m-1}, z_m)$. By Proposition 4.4, it is sufficient to show that $(Qz_m)\psi_2(z_{m-1}, z_m)$ is equivalent to a $\vee\overset{\rightarrow}{\exists}\forall$ formula.

It is easy to show that any $\overset{\rightarrow}{\exists}\forall$ formula with the free variables z_0, z_1 is equivalent to a formula of the following form:

$$\exists x_0 \ldots \exists x_n [(z_0 = x_0 < \cdots < x_n = z_1) \wedge \bigwedge_{j=0}^{n} \alpha_j(x_j) \wedge \bigwedge_{j=1}^{n} (\forall y)_{>x_{j-1}}^{<x_j} \beta_j(y)] \quad (1)$$

where α_i, β_i are quantifier free.

Therefore, to complete our proof it is sufficient to prove the following lemma:

Lemma 5.1. *Let $\psi(z_0, z_1)$ be a formula as in (1). Then $(Qz_1)\psi$ is equivalent to a $\vee\overset{\rightarrow}{\exists}\forall$ formula.*

In the rest of this section we prove Lemma 5.1. Our proof is organized as follows. In Lemma 5.3 we prove an instance of Lemma 5.1 where all β_i are equivalent to True. Then we derive a more general instance (Corollary 5.5) where $\beta_1(x)$ holds for all $x > z_0$. Finally, in Lemma 5.6(2) we prove the full version of Lemma 5.1. First, we introduce some helpful notations.

Notation 5.2. *We use the abbreviated notation $[\alpha_0, \beta_1 \ldots, \alpha_{n-1}, \beta_n \alpha_n](z_0, z_1)$ for the $\overset{\rightarrow}{\exists}\forall$-formula as in (1).*

In this notation Lemma 5.1 can be rephrased as $(Qz_1)[\alpha_0, \beta_1 \ldots, \alpha_{n-1}, \beta_n \alpha_n]$ (z_0, z_1) is equivalent to a $\vee\overset{\rightarrow}{\exists}\forall$ formula.

We start with the instance of Lemma 5.1 where all β_i are True.

Lemma 5.3. $(Qz_1)\exists x_0 \exists x_1 \ldots \exists x_n (z_0 = x_0 < x_1 < \cdots < x_n = z_1) \wedge \bigwedge_{i=0}^n P_i(x_i)$ *is equivalent to a* $\vee \overset{\rightarrow}{\exists} \forall$ *formula.*

Proof. This formula is equivalent to the disjunction of $(Qz_1)P_n(z_1)$ and $\neg \exists x_0 \exists x_1 \ldots \exists x_{n-1} (z_0 = x_0 < x_1 < \cdots < x_{n-1}) \wedge \bigwedge_{i=0}^{n-1} P_i(x_i)$. The first disjunct is equivalent to $\langle Q \rangle P_n$. The second disjunct is equivalent to a $\vee \overset{\rightarrow}{\exists} \forall$ formula by Proposition 4.4. Hence, this formula is equivalent to a $\vee \overset{\rightarrow}{\exists} \forall$ formula. $\square$

The next Lemma does not deal with generalized quantifiers.

Lemma 5.4. $((\forall y)_{>z_0} \beta_1) \wedge [\alpha_0, \beta_1, \alpha_1, \beta_2, \ldots, \alpha_{n-1}, \beta_n, \alpha_n](z_0, z_1)$ *is equivalent to* $((\forall y)_{>z_0} \beta_1) \wedge \exists x_0 \exists x_1 \ldots \exists x_n (z_0 = x_0 < x_1 < \cdots < x_n = z_1) \wedge \bigwedge_{i=0}^n \alpha_i'(x_i)$, *where* α_i' *are atoms.*

As a consequence we obtain:

Corollary 5.5.

Let $\psi(z_0, z_1)$ *be* $((\forall y)_{>z_0} \beta_1(y)) \wedge [\alpha_0, \beta_1, \alpha_1, \beta_2, \ldots, \alpha_{n-1}, \beta_n, \alpha_n](z_0, z_1)$. *Then* $(Qz_1)\psi$ *is equivalent to a* $\vee \overset{\rightarrow}{\exists} \forall$ *formula.*

Proof. Immediately by Lemmas 3.1(3), 5.3, and 5.4. $\square$

Now we are ready to prove Lemma 5.1, i.e., $(Qz_1)[\alpha_0, \beta_1 \ldots, \beta_{n-1}, \alpha_{n-1}, \beta_n, \alpha_n]$ (z_0, z_1) is equivalent to a $\vee \overset{\rightarrow}{\exists} \forall$ formula.

Lemma 5.6.

1. Let $\psi(z_0, z_1)$ be $((\exists y)_{>z_0} \neg \beta_1(y)) \wedge [\alpha_0, \beta_1, \alpha_1, \beta_2, \ldots, \alpha_{n-1}, \beta_n, \alpha_n](z_0, z_1)$. Then $(Qz_1)\psi$ is equivalent to a $\vee \overset{\rightarrow}{\exists} \forall$ formula.
2. $(Qz_1)[\alpha_0, \beta_1, \alpha_1, \beta_2, \ldots, \alpha_{n-1}, \beta_n, \alpha_n](z_0, z_1)$ is equivalent to a $\vee \overset{\rightarrow}{\exists} \forall$ formula..

Proof. We prove (1) and (2) simultaneously by induction on n. Observe that A is equivalent to $(((\exists y)_{>z_0} \neg \beta_1(y)) \wedge A) \vee (((\forall y)_{>z_0} \beta_1(y)) \wedge A)$. Hence, if (1) holds for n, then by Corollary 5.5, Lemma 3.1(1) and the closure of $\vee \overset{\rightarrow}{\exists} \forall$ formulas under conjunction we obtain that (2) holds for n. Therefore, for the inductive step it is sufficient to prove that if (1) and (2) hold for n then (2) holds for $n+1$.

Note that $(\exists y)_{>z_0} \neg \beta_1(y)$ implies that there is at most one z such that $[\alpha_0, \beta_1, \alpha_1](z_0, z)$ and $\neg(\exists y)_{>z}[\alpha_0, \beta_1, \alpha_1](z_0, y)$.

If there is no such z, then $(Qz_1)\psi$ is equivalent to True.

So, we assume that there is a unique such z. It is definable by the formula

$$def(z_0, z) := [\alpha_0, \beta_1, \alpha_1](z_0, z) \wedge \neg(\exists y)_{>z}[\alpha_0, \beta_1, \alpha_1](z_0, y). \tag{2}$$

It is sufficient to show that $(\exists z)_{>z_0} def(z) \wedge (Qz_1)[\alpha_0, \beta_1, \alpha_1, \ldots, \beta_{n+1}, \alpha_{n+1}](z_0, z_1)$ is equivalent to a $\vee \overset{\rightarrow}{\exists} \forall$ formula ψ'. Then $(Qz_1)\psi$ is equivalent to $(\forall y)_{>z_0} \beta_1(y) \vee (\neg \exists z\, def) \vee (\exists z\, def \wedge \psi')$, and by Proposition 4.4, to a $\vee \overset{\rightarrow}{\exists} \forall$ formula.

We prove this by induction on n. The *basis* is trivial.

Inductive step $n \mapsto n+1$. Define:

$$
\begin{array}{ll}
A_i^-(z_0, z) := [\alpha_0, \beta_1, \ldots, \beta_i, \alpha_i](z_0, z) & i = 1, \ldots, n \\
A_i^+(z, z_1) := [\alpha_i, \beta_{i+1}, \ldots \beta_{n+1}\alpha_{n+1}](z, z_1) & i = 1, \ldots, n \\
A_i(z_0, z, z_1) := A_i^-(z_0, z) \wedge A_i^+(z, z_1) & i = 1, \ldots, n \\
B_i^-(z_0, z) := [\alpha_0\beta_1, \ldots, \beta_{i-1}, \alpha_{i-1}, \beta_i, \beta_i](z_0, z) & i = 1, \ldots, n+1 \\
B_i^+(z, z_1) := [\beta_i, \beta_i, \alpha_i\beta_{i+1}\alpha_{i+1}, \ldots, \beta_{n+1}, \alpha_{n+1}](z, z_1) & i = 1, \ldots, n+1 \\
B_i(z_0, z, z_1) := B_i^-(z_0, z) \wedge B_i^+(z, z_1) & i = 1, \ldots, n+1
\end{array}
$$

If the interval (z_0, z_1) is non-empty, these definitions imply

$$
[\alpha_0, \beta_1, \alpha_1, \ldots, \beta_{n+1}, \alpha_{n+1}](z_0, z_1) \Leftrightarrow (\forall z)_{>z_0}^{<z_1} \big(\bigvee_{i=1}^n A_i \vee \bigvee_{i=1}^{n+1} B_i \big)
$$

$$
[\alpha_0, \beta_1, \alpha_1, \ldots, \beta_{n+1}, \alpha_{n+1}](z_0, z_1) \Leftrightarrow (\exists z)_{>z_0}^{<z_1} \big(\bigvee_{i=1}^n A_i \vee \bigvee_{i=1}^{n+1} B_i \big)
$$

Hence, for every $\varphi(z_0, z)$:

$$
((\exists z)_{>z_0}^{<z_1} \varphi(z_0, z)) \wedge [\alpha_0, \beta_1, \alpha_1, \ldots, \beta_{n+1}, \alpha_{n+1}](z_0, z_1)
$$

is equivalent to $(\exists z)_{>z_0}^{<z_1} \big(\varphi(z_0, z) \wedge (\bigvee_{i=1}^n A_i \vee \bigvee_{i=1}^{n+1} B_i) \big)$. In particular,

$$
\begin{array}{c}
(\exists z)_{>z_0}^{<z_1} def(z_0, z) \wedge [\alpha_0, \beta_1, \alpha_1, \ldots, \beta_{n+1}, \alpha_{n+1}](z_0, z_1) \\
\text{is equivalent to} \\
(\exists z)_{>z_0}^{<z_1} \big(def(z_0, z) \wedge (\bigvee_{i=1}^n A_i \vee \bigvee_{i=1}^{n+1} B_i) \big),
\end{array} \tag{3}
$$

where *def* was defined in equation (2). To proceed we use the following simple properties of the unusual quantifier:

Lemma 5.7. *Assume that z_1 does not occur free in φ, and $\exists! z \varphi$. Then*

1. $(Qz_1)(\exists z)^{<z_1}(\varphi \wedge C)$ *is equivalent to* $(\exists z)(\varphi \wedge (Qz_1)C)$
2. $(Qz_1)\exists z(\varphi \wedge \bigvee C_i)$ *is equivalent to* $\bigwedge \exists z(\varphi \wedge (Qz_1)C_i)$

Now $(\exists z)_{>z_0} def(z_0, z) \wedge (Qz_1)[\alpha_0, \beta_1, \alpha_1, \ldots, \beta_{n+1}, \alpha_{n+1}](z_0, z_1)$ is equivalent (by Lemma 5.7(1)) to $(Qz_1)(\exists z)_{>z_0}^{<z_1} def(z_0, z) \wedge [\alpha_0, \beta_1, \alpha_1, \ldots, \beta_{n+1}, \alpha_{n+1}](z_0, z_1)$ is equivalent, by (3), to $(Qz_1)\big((\exists z)_{>z_0}^{<z_1} def(z_0, z) \wedge (\bigvee_{i=1}^n A_i \vee \bigvee_{i=1}^{n+1} B_i)\big)$ is equivalent (by Lemma 5.7(2)) to

$$
\big(\bigwedge_{i=1}^n (\exists z)_{>z_0}^{<z_1} def(z_0, z) \wedge (Qz_1)A_i \big) \wedge \big(\bigwedge_{i=1}^{n+1} (\exists z)_{>z_0}^{<z_1} def(z_0, z) \wedge (Qz_1)B_i \big) \tag{4}
$$

We are going to show that $(Qz_1)A_i$ $(i = 1, \ldots, n)$ and $(Qz_1)B_i$ $(i = 2, \ldots, n+1)$, and $(\exists z)_{>z_0}^{<z_1} def(z_0, z) \wedge (Qz_1)B_1$ are equivalent to $\vee \overrightarrow{\exists} \forall$ formulas and therefore, by Proposition 4.4, we obtain that (4) is equivalent to a $\vee \overrightarrow{\exists} \forall$ formula.

Recall that $A_i := A_i^-(z_0, z) \wedge A_i^+(z, z_1)$ and $B_i := B_i^-(z_0, z) \wedge B_i^+(z, z_1)$. By Lemma 3.1(3), we obtain that $(Qz_1)A_i$ is equivalent to $\neg A_i^- \vee (Qz_1)A_i^+$. By the inductive assumption $(Qz_1)A_i^+$ is equivalent to a $\vee \overrightarrow{\exists} \forall$ formula for $i = 1, \ldots, n$. Hence, by Proposition 4.4, $(Qz_1)A_i$ is equivalent to a $\vee \overrightarrow{\exists} \forall$ formula. Similar arguments show that $(Qz_1)B_i$ is equivalent to a $\vee \overrightarrow{\exists} \forall$ formula for $i = 2, \ldots, n+1$.

Finally, $def(z_0, z)$ implies that there is no $x > z$ such that $\alpha_1(x)$ and β_i holds on $[z, x)$. Therefore, B_1^+ is equivalent to False and $(Qz_1)B_1^+$ is equivalent to True. Hence, $(\exists z)_{\geq z_0}^{< z_1} def(z_0, z) \wedge (Qz_1)B_1$ is equivalent to a $\vee \overrightarrow{\exists} \forall$ formula $(\exists z)_{\geq z_0}^{< z_1} def(z_0, z)$.

This completes our proof of Lemma 5.1 and of Proposition 4.6.   □

6   Further Results and Open Questions

We provided a natural interpretation of usual/unusual over $\mathbb{N}$ and proved an analog of Kamp's theorem. We can consider several unusual quantifiers $\mathcal{Q}_1, \ldots \mathcal{Q}_k$ and prove that $FOMLO[Q_1, \ldots, Q_k]$ and $TL(\mathsf{Until}, \mathsf{Since}, \langle Q_1 \rangle, \ldots, \langle Q_k \rangle)$ have the same expressive power over ω. Our result can be easily extended to the time domain of integers; however, in this case we have to require that if $\mathcal{Q}$ is a family of unusual sets over integers and $P \in \mathcal{Q}$, then neither $(-\infty, k]$ nor $[k, \infty)$ is a subset of P. It is open how to formalize "usually/unusually" over the reals.

Standard notions of "fairness" are based on the ideal of finite sets. For example, strong fairness is formalized as: if P_1 occurs infinitely often, then P_2 occurs infinitely often. It is natural to base fairness on an unusual modality $\langle Q \rangle$, and define Q-fairness as $Fair_Q(P_1, P_2) := \langle Q \rangle P_2 \to \langle Q \rangle P_1$. More general notions of "fairness" can be introduced by using several unusual quantifiers; e.g., $Fair_{Q_1, Q_2}(P_1, P_2) := \langle Q_2 \rangle P_2 \to \langle Q_1 \rangle P_1$.

Unfortunately, in our extension a phrase like "It is unusual that the weather is sunny when it rains" is not expressible, and further extensions are needed to express such a binary unusual modality.

We can show that under each of the seven interpretations of unusual described in Section 3.2, the problem whether a $TL(\mathsf{Until}, \mathsf{Since}, \langle Q \rangle)$ formula is satisfiable is PSPACE-complete. Moreover, the interpretations (2)-(7) of unusual give the same set of satisfiable $TL(\mathsf{Until}, \mathsf{Since}, \langle Q \rangle)$ formulas.

References

1. Gabbay, D., Hodkinson, I., Reynolds, M.: Temporal logic: Mathematical Foundations and Computational Aspects. Oxford University Press (1994)
2. Gabbay, D., Pnueli, A., Shelah, S., Stavi, J.: On the Temporal Analysis of Fairness. In: POPL 1980, pp. 163–173 (1980)
3. Kamp, H.W.: Tense logic and the theory of linear order. Phd thesis, University of California, Los Angeles (1968)
4. Mostowski, A.: On a Generalization of Quantifiers. Fund. Math. 44, 12–36 (1957)
5. Pnueli, A.: The temporal logic of programs. In: Proc. IEEE 18th Annu. Symp. on Found. Comput. Sci., New York, pp. 46–57 (1977)
6. Rabinovich, A.: A proof of Kamp's theorem. In: CSL 2012, pp. 516–527 (2012)

A More Efficient Simulation Algorithm on Kripke Structures

Francesco Ranzato

Dipartimento di Matematica, University of Padova, Italy

Abstract. A number of algorithms for computing the simulation preorder (and equivalence) on Kripke structures are available. Let Σ denote the state space, $\rightarrow$ the transition relation and P_{sim} the partition of Σ induced by simulation equivalence. While some algorithms are designed to reach the best space bounds, whose dominating additive term is $|P_{\mathrm{sim}}|^2$, other algorithms are devised to attain the best time complexity $O(|P_{\mathrm{sim}}||\rightarrow|)$. We present a novel simulation algorithm which is both space and time efficient: it runs in $O(|P_{\mathrm{sim}}|^2 \log |P_{\mathrm{sim}}| + |\Sigma| \log |\Sigma|)$ space and $O(|P_{\mathrm{sim}}||\rightarrow| \log |\Sigma|)$ time. Our simulation algorithm thus reaches the best space bounds while closely approaching the best time complexity.

1 Introduction

The simulation preorder is a fundamental behavioral relation widely used in process algebra for establishing system correctness and in model checking as a suitable abstraction for reducing the size of state spaces. The problem of efficiently computing the simulation preorder (and consequently simulation equivalence) on finite Kripke structures has been thoroughly investigated and generated a number of simulation algorithms. Both time and space complexities play an important role in simulation algorithms, since in several applications, especially in model checking, memory requirements may become a serious bottleneck as the input transition system grows.

Consider a finite Kripke structure where Σ denotes the state space, $\rightarrow$ the transition relation and P_{sim} the partition of Σ induced by simulation equivalence. The best simulation algorithms are those by, in chronological order, Gentilini, Piazza and Policriti (GPP) [3] (subsequently corrected in [4]), Ranzato and Tapparo (RT) [10,12], Markovski (Mar) [8], Cécé (Space-Céc and Time-Céc) [2]. The simulation algorithms GPP and RT are designed for Kripke structures, while Space-Céc, Time-Céc and Mar are for more general labeled transition systems. Their space and time complexities are summarized in the following table.

Algorithm	Space complexity	Time complexity
Space-Céc [2]	$O(|P_{\mathrm{sim}}|^2 + |\rightarrow| \log |\rightarrow|)$	$O(|P_{\mathrm{sim}}|^2|\rightarrow|)$
Time-Céc [2]	$O(|P_{\mathrm{sim}}||\Sigma| \log |\Sigma| + |\rightarrow| \log |\rightarrow|)$	$O(|P_{\mathrm{sim}}||\rightarrow|)$
GPP [3]	$O(|P_{\mathrm{sim}}|^2 \log |P_{\mathrm{sim}}| + |\Sigma| \log |\Sigma|)$	$O(|P_{\mathrm{sim}}|^2|\rightarrow|)$
Mar [8]	$O((|\Sigma| + |P_{\mathrm{sim}}|^2) \log |P_{\mathrm{sim}}|)$	$O(|\rightarrow| + |P_{\mathrm{sim}}||\Sigma| + |P_{\mathrm{sim}}|^3)$
RT [12]	$O(|P_{\mathrm{sim}}||\Sigma| \log |\Sigma|)$	$O(|P_{\mathrm{sim}}||\rightarrow|)$
ESim (this paper)	$O(|P_{\mathrm{sim}}|^2 \log |P_{\mathrm{sim}}| + |\Sigma| \log |\Sigma|)$	$O(|P_{\mathrm{sim}}||\rightarrow| \log |\Sigma|)$

K. Chatterjee and J. Sgall (Eds.): MFCS 2013, LNCS 8087, pp. 753–764, 2013.

We remark that all the above space bounds are bit space complexities, i.e., the word size is a single bit. Let us also remark that both articles [3,4] state that the bit space complexity of GPP is in $O(|P_{\text{sim}}|^2 + |\Sigma| \log |P_{\text{sim}}|)$. However, as observed also in [2], this is not precise. In fact, the algorithm GPP [3, Section 4, p. 98] assumes that the states belonging to some block are stored as a doubly linked list, and this entails a bit space complexity in $O(|\Sigma| \log |\Sigma|)$. Furthermore, GPP uses Henzinger, Henzinger and Kopke [5] simulation algorithm (HKK) as a subroutine, whose bit space complexity is in $O(|\Sigma|^2 \log |\Sigma|)$, which is called on a Kripke structure where states are blocks of the current partition. The bit space complexity of GPP must therefore include an additive term $|P_{\text{sim}}|^2 \log |P_{\text{sim}}|$ and therefore results to be $O(|P_{\text{sim}}|^2 \log |P_{\text{sim}}| + |\Sigma| \log |\Sigma|)$. It is worth observing that a space complexity in $O(|P_{\text{sim}}|^2 + |\Sigma| \log |P_{\text{sim}}|)$ can be considered optimal for a simulation algorithm, since this is of the same order as the size of the output, which needs $|P_{\text{sim}}|^2$ space for storing the simulation preorder as a partial order on simulation equivalence classes and $|\Sigma| \log |P_{\text{sim}}|$ space for storing the simulation equivalence class for any state. Hence, the bit space complexities of GPP and Space-Céc can be considered quasi-optimal. As far as time complexity is concerned, the algorithms RT and Time-Céc both feature the best time bound $O(|P_{\text{sim}}||{\rightarrow}|)$.

We present here a novel space and time Efficient Simulation algorithm, called ESim, which features a time complexity in $O(|P_{\text{sim}}||{\rightarrow}| \log |\Sigma|)$ and a bit space complexity in $O(|P_{\text{sim}}|^2 \log |P_{\text{sim}}| + |\Sigma| \log |\Sigma|)$. Thus, ESim reaches the best space bound of GPP and significantly improves the GPP time bound $O(|P_{\text{sim}}|^2|{\rightarrow}|)$ by replacing a multiplicative factor $|P_{\text{sim}}|$ with $\log |\Sigma|$. Furthermore, ESim significantly improves the RT space bound $O(|P_{\text{sim}}||\Sigma| \log |\Sigma|)$ and closely approaches the best time bound $O(|P_{\text{sim}}||{\rightarrow}|)$ of RT and Time-Céc.

ESim is a partition refinement algorithm, meaning that it maintains and iteratively refines a so-called partition-relation pair $\langle P, \trianglelefteq \rangle$, where P is a partition of Σ that over-approximates the final simulation partition P_{sim}, while $\trianglelefteq$ is a binary relation over P which overapproximates the final simulation preorder. ESim relies on the following three main points, which in particular allow to attain the above complexity bounds.

(1) Two distinct notions of partition and relation stability for a partition-relation pair are introduced. Accordingly, at a logical level, ESim is designed as a partition refinement algorithm which iteratively performs two clearly distinct refinement steps: the refinement of the current partition P which splits some blocks of P and the refinement of the relation $\trianglelefteq$ which removes some pairs of blocks from $\trianglelefteq$.

(2) ESim exploits a logical characterization of partition refiners, i.e. blocks of P that allow to split the current partition P, which admits an efficient implementation.

(3) ESim only relies on data structures, like lists and matrices, that are indexed on and contain blocks of the current partition P. The hard task here is to devise efficient ways to keep updated these partition-based data structures along the iterations of ESim. We show that this can be done efficiently, in particular by resorting to Hopcroft's "process the smaller half" principle [7] when updating a crucial data structure after a partition split.

Due to lack of space, some auxiliary algorithms and the proofs of all the results are omitted.

2   Background

Notation. If $R \subseteq \Sigma \times \Sigma$ is any relation and $X \subseteq \Sigma$ then $R(X) \triangleq \{x' \in \Sigma \mid \exists x \in X. (x, x') \in R\}$. Recall that R is a preorder relation when it is reflexive and transitive. If f is a function defined on $\wp(\Sigma)$ and $x \in \Sigma$ then we often write $f(x)$ to mean $f(\{x\})$. Part(Σ) denotes the set of partitions of Σ. If $P \in$ Part(Σ), $s \in \Sigma$ and $S \subseteq \Sigma$ then $P(s)$ denotes the block of P that contains s while $P(S) = \cup_{s \in S} P(s)$. Part$(\Sigma)$ is endowed with the standard partial order $\preceq$: $P_1 \preceq P_2$, i.e. P_2 is coarser than P_1, iff for any $s \in \Sigma$, $P_1(s) \subseteq P_2(s)$. If $P_1 \preceq P_2$ and $B \in P_1$ then $P_2(B)$ is a block of P_2 which is also denoted by parent$_{P_2}(B)$. For a given nonempty subset $S \subseteq \Sigma$ called splitter, we denote by $Split(P, S)$ the partition obtained from P by replacing each block $B \in P$ with $B \cap S$ and $B \smallsetminus S$, where we also allow no splitting, namely $Split(P, S) = P$ (this happens exactly when $P(S) = S$).

Simulation Preorder and Equivalence. A transition system $(\Sigma, \rightarrow)$ consists of a set Σ of states and of a transition relation $\rightarrow \subseteq \Sigma \times \Sigma$. Given a set AP of atoms (of some specification language), a Kripke structure (KS) $\mathcal{K} = (\Sigma, \rightarrow, \ell)$ over AP consists of a transition system $(\Sigma, \rightarrow)$ together with a state labeling function $\ell : \Sigma \rightarrow \wp(AP)$. The state partition induced by ℓ is denoted by $P_\ell \triangleq \{\{s' \in \Sigma \mid \ell(s) = \ell(s')\} \mid s \in \Sigma\}$. The predecessor/successor transformers pre, post $: \wp(\Sigma) \rightarrow \wp(\Sigma)$ are defined as usual: $\text{pre}(T) \triangleq \{s \in \Sigma \mid \exists t \in T. s \rightarrow t\}$ and $\text{post}(S) \triangleq \{t \in \Sigma \mid \exists s \in S. s \rightarrow t\}$. If $S_1, S_2 \subseteq \Sigma$ then $S_1 \rightarrow^{\exists} S_2$ iff there exist $s_1 \in S_1$ and $s_2 \in S_2$ such that $s_1 \rightarrow s_2$.

A relation $R \subseteq \Sigma \times \Sigma$ is a simulation on a Kripke structure $(\Sigma, \rightarrow, \ell)$ if for any $s, s' \in \Sigma$, if $s' \in R(s)$ then:

(A) $\ell(s) = \ell(s')$;
(B) for any $t \in \Sigma$ such that $s \rightarrow t$, there exists $t' \in \Sigma$ such that $s' \rightarrow t'$ and $t' \in R(t)$.

Given $s, t \in \Sigma$, t simulates s, denoted by $s \leq t$, if there exists a simulation relation R such that $t \in R(s)$. It turns out that the largest simulation on a given KS exists, is a preorder relation called simulation preorder and is denoted by R_{sim}. Thus, for any $s, t \in \Sigma$, $s \leq t$ iff $(s, t) \in R_{\text{sim}}$. Simulation equivalence R_{simeq} is the symmetric reduction of R_{sim}, namely $R_{\text{simeq}} \triangleq R_{\text{sim}} \cap R_{\text{sim}}^{-1}$, so that $(s, t) \in R_{\text{simeq}}$ iff $s \leq t$ and $t \leq s$. $P_{\text{sim}} \in$ Part(Σ) denotes the partition corresponding to the equivalence R_{simeq} and is called the simulation partition.

3   Logical Simulation Algorithm

A partition-relation pair $\mathcal{P} = \langle P, \trianglelefteq \rangle$, PR for short, is a state partition $P \in$ Part(Σ) together with a binary relation $\trianglelefteq \subseteq P \times P$ between blocks of P. We write $B \lhd C$ when $B \trianglelefteq C$ and $B \neq C$ and $(B', C') \trianglelefteq (B, C)$ when $B' \trianglelefteq B$ and $C' \trianglelefteq C$. When $\trianglelefteq$ is a preorder/partial order then $\mathcal{P}$ is called, respectively, a preorder/partial order PR.

PRs allow to represent symbolically, i.e. through state partitions, a relation between states. A relation $R \subseteq \Sigma \times \Sigma$ induces a PR PR$(R) = \langle P, \trianglelefteq \rangle$ defined as follows:

$$\forall s \in \Sigma. P(s) \triangleq \{t \in \Sigma \mid R(s) = R(t)\}; \quad \forall s, t \in \Sigma. P(s) \trianglelefteq P(t) \text{ iff } t \in R(s).$$

It is easy to note that if R is a preorder then $\mathrm{PR}(R)$ is a partial order PR. On the other hand, a PR $\mathcal{P} = \langle P, \trianglelefteq \rangle$ induces the following relation $\mathrm{Rel}(\mathcal{P}) \subseteq \Sigma \times \Sigma$:

$$(s, t) \in \mathrm{Rel}(\mathcal{P}) \; \Leftrightarrow \; P(s) \trianglelefteq P(t).$$

Here, if $\mathcal{P}$ is a preorder PR then $\mathrm{Rel}(\mathcal{P})$ is clearly a preorder.

A PR $\mathcal{P} = \langle P, \trianglelefteq \rangle$ is defined to be a simulation PR on a KS $\mathcal{K}$ when $\mathrm{Rel}(\mathcal{P})$ is a simulation on $\mathcal{K}$, namely when $\mathcal{P}$ represents a simulation relation between states. Hence, if $\mathcal{P}$ is a simulation PR and $P(s) = P(t)$ then s and t are simulation equivalent, while if $P(s) \trianglelefteq P(t)$ then t simulates s.

Given a PR $\mathcal{P} = \langle P, \trianglelefteq \rangle$, the map $\mu_{\mathcal{P}} : \wp(\Sigma) \to \wp(\Sigma)$ is defined as follows:

$$\text{for any } X \in \wp(\Sigma), \; \mu_{\mathcal{P}}(X) \triangleq \mathrm{Rel}(\mathcal{P})(X) = \cup\{C \in P \mid \exists s \in X. \, P(s) \trianglelefteq C\}.$$

Note that, for any $s \in \Sigma$, $\mu_{\mathcal{P}}(s) = \mu_{\mathcal{P}}(P(s)) = \cup\{C \in P \mid P(s) \trianglelefteq C\}$. For preorder PRs, this map allows us to characterize the property of being a simulation PR as follows.

Theorem 3.1. *Let $\mathcal{P} = \langle P, \trianglelefteq \rangle$ be a preorder PR. Then, $\mathcal{P}$ is a simulation iff*

(i) *if $B \trianglelefteq C$, $b \in B$ and $c \in C$ then $\ell(b) = \ell(c)$;*
(ii) *if $B{\to}^{\exists}C$ and $B \trianglelefteq D$ then $D{\to}^{\exists}\mu_{\mathcal{P}}(C)$;*
(iii) *for any $C \in P$, $P = Split(P, \mathrm{pre}(\mu_{\mathcal{P}}(C)))$.*

By Theorem 3.1, assuming that condition (i) holds, there are two possible reasons for a PR $\mathcal{P} = \langle P, \trianglelefteq \rangle$ for not being a simulation:

(1) There exist $B, C, D \in P$ such that $B{\to}^{\exists}C$, $B \trianglelefteq D$, but $D \not\to^{\exists}\mu_{\mathcal{P}}(C)$; in this case we say that the block C is a *relation refiner* for $\mathcal{P}$.
(2) There exist $B, C \in P$ such that $B \cap \mathrm{pre}(\mu_{\mathcal{P}}(C)) \neq \varnothing$ and $B \smallsetminus \mathrm{pre}(\mu_{\mathcal{P}}(C)) \neq \varnothing$; in this case we say that the block C is a *partition refiner* for $\mathcal{P}$.

We therefore define $\mathrm{RRefiner}(\mathcal{P})$ and $\mathrm{PRefiner}(\mathcal{P})$ as the sets of blocks of P that are, respectively, relation and partition refiners for $\mathcal{P}$. Accordingly, $\mathcal{P}$ is defined to be relation or partition *stable* when, respectively, $\mathrm{RRefiner}(\mathcal{P}) = \varnothing$ or $\mathrm{PRefiner}(\mathcal{P}) = \varnothing$. Then, Theorem 3.1 can be read as follows: $\mathcal{P}$ is a simulation iff $\mathcal{P}$ satisfies condition (i) and is both relation and partition stable.

If $C \in \mathrm{PRefiner}(\mathcal{P})$ then P is first refined to $P' \triangleq Split(P, \mathrm{pre}(\mu_{\mathcal{P}}(C)))$, i.e. P is split w.r.t. the splitter $S = \mathrm{pre}(\mu_{\mathcal{P}}(C))$. Accordingly, the relation $\trianglelefteq$ on P is transformed into the following relation $\trianglelefteq'$ defined on P':

$$\trianglelefteq' \; \triangleq \; \{(D, E) \in P' \times P' \mid \mathrm{parent}_P(D) \trianglelefteq \mathrm{parent}_P(E)\} \tag{$\dagger$}$$

Hence, two blocks D and E of the refined partition P' are related by $\trianglelefteq'$ if their parent blocks $\mathrm{parent}_P(D)$ and $\mathrm{parent}_P(E)$ in P were related by $\trianglelefteq$. Hence, if $\mathcal{P}' = \langle P', \trianglelefteq' \rangle$ then for all $D \in P'$, we have that $\mu_{\mathcal{P}'}(D) = \mu_{\mathcal{P}}(\mathrm{parent}_P(D))$. We will show that this refinement of $\langle P, \trianglelefteq \rangle$ is correct because if $B \in P$ is split into $B \smallsetminus S$ and $B \cap S$ then all the states in $B \smallsetminus S$ are not simulation equivalent to all the states in $B \cap S$. Note that if $B \in P$ has been split into $B \cap S$ and $B \smallsetminus S$ then both $B \cap S \trianglelefteq' B \smallsetminus S$ and $B \smallsetminus S \trianglelefteq' B \cap S$ hold, and consequently $\mathcal{P}'$ becomes relation unstable.

---

1 ESim(**PR** $\langle P, \trianglelefteq \rangle$)
2   *Initialize*(); *PStabilize*(); **bool** PStable := *RStabilize*(); **bool** RStable := **tt**;
3   **while** ¬(PStable & RStable) **do**
4   **if** ¬ PStable **then** {RStable := *PStabilize*(); PStable := **tt**;}
5   **if** ¬ RStable **then** {PStable := *RStabilize*(); RStable := **tt**;}

6 **bool** *PStabilize*()
7   P_{old} := P;
8   **while** $\exists C \in \text{PRefiner}(\mathcal{P})$ **do**
9      $S := \text{pre}(\mu_{\mathcal{P}}(C))$; $P := Split(S)$;
10   **forall** $(D, E) \in P \times P$ **do** $D \trianglelefteq E := \text{parent}_P(D) \trianglelefteq \text{parent}_P(E)$;
11   **return** $(P = P_{\text{old}})$;

12 **bool** *RStabilize*()
13   // Precondition: PStable = **tt**
14   $\trianglelefteq_{\text{old}} := \trianglelefteq$; Delete := $\varnothing$;
15   **while** $\exists C \in \text{RRefiner}(\mathcal{P})$ **do**
16   Delete := Delete $\cup \{(B, D) \in P \times P \mid B \trianglelefteq D, B{\to}^{\exists}C, D \not\to^{\exists}\mu_{\mathcal{P}}(C)\}$;
17   $\trianglelefteq := \trianglelefteq \setminus$ Delete;
18   **return** $(\trianglelefteq = \trianglelefteq_{\text{old}})$;

Fig. 1. Logical Simulation Algorithm

On the other hand, if $\mathcal{P}$ is partition stable and $C \in \text{RRefiner}(\mathcal{P})$ then we will show that $\trianglelefteq$ can be safely refined to the following relation $\trianglelefteq'$:

$$\trianglelefteq' \triangleq \trianglelefteq \setminus \{(B, D) \in P \times P \mid B{\to}^{\exists}C, B \trianglelefteq D, D \not\to^{\exists}\mu_{\mathcal{P}}(C)\}$$
$$= \{(B, D) \in P \times P \mid B \trianglelefteq D, \big(B{\to}^{\exists}C \Rightarrow D{\to}^{\exists}\mu_{\mathcal{P}}(C)\big)\} \quad (\ddagger)$$

because if $(B, D) \in \trianglelefteq \setminus \trianglelefteq'$ then all the states in D cannot simulate all the states in B.

The above facts lead us to design a basic simulation algorithm ESim described in Figure 1. ESim maintains a PR $\mathcal{P} = \langle P, \trianglelefteq \rangle$, which initially is $\langle P_\ell, \text{id} \rangle$ and is iteratively refined as follows:

PStabilize(): If $\langle P, \trianglelefteq \rangle$ is not partition stable then the partition P is split for $\text{pre}(\mu_{\mathcal{P}}(C))$ as long as a partition refiner C for $\mathcal{P}$ exists, and when this happens the relation $\trianglelefteq$ is transformed to $\trianglelefteq'$ as defined by (†); at the end of this process, we obtain a PR $\mathcal{P}' = \langle P', \trianglelefteq' \rangle$ which is partition stable and if P has been actually refined, i.e. $P' \prec P$ then the current PR $\mathcal{P}'$ becomes relation unstable.

RStabilize(): If $\langle P, \trianglelefteq \rangle$ is not relation stable then the relation $\trianglelefteq$ is refined to $\trianglelefteq'$ as described by (‡) as long as a relation refiner for $\mathcal{P}$ exists; hence, at the end of this refinement process $\langle P, \trianglelefteq' \rangle$ becomes relation stable but possibly partition unstable.

Moreover, the following properties of the current PR of ESim hold.

Lemma 3.2. *In any run of* ESim, *the following two conditions hold:*

(i) *If PStabilize() is called on a partial order PR $\langle P, \unlhd \rangle$ then at the exit we obtain a PR $\langle P', \unlhd' \rangle$ which is a preorder.*

(ii) *If RStabilize() is called on a preorder PR $\langle P, \unlhd \rangle$ then at the exit we obtain a PR $\langle P, \unlhd' \rangle$ which is a partial order.*

The main loop of ESim terminates when the current PR $\langle P, \unlhd \rangle$ becomes both partition and relation stable. By the above Lemma 3.2, the output PR $\mathcal{P}$ of ESim is a partial order, and hence a preorder, so that Theorem 3.1 can be applied to $\mathcal{P}$ which then results to be a simulation PR. It turns out that this algorithm is correct, meaning that the output PR $\mathcal{P}$ actually represents the simulation preorder.

Theorem 3.3 (Correctness). *Let Σ be finite. ESim is correct, i.e., ESim terminates on any input and if $\langle P, \unlhd \rangle$ is the output PR of ESim on input $\langle P_\ell, \mathrm{id} \rangle$ then for any $s, t \in \Sigma$, $s \leq t \Leftrightarrow P(s) \unlhd P(t)$.*

4 Efficient Implementation

4.1 Data Structures

ESim is implemented by relying on the following data structures.

States: A state s is represented by a record that contains the list $\mathrm{post}(s)$ of its successors, a pointer s.block to the block $P(s)$ that contains s and a boolean flag used for marking purposes. The whole state space Σ is represented as a doubly linked list of states. $\{\mathrm{post}(s)\}_{s \in \Sigma}$ therefore represents the input transition system.

Partition: The states of any block B of the current partition P are consecutive in the list Σ, so that B is represented by two pointers begin and end: B.begin is the first state of B in Σ and B.end is the successor of the last state of B in Σ, i.e., $B = [B.\text{begin}, B.\text{end}[$. Moreover, B stores a boolean flag B.intersection and a block pointer B.brother whose meanings are as follows: after a call to $Split(P, S)$ for splitting P w.r.t. a set of states S, if $B_1 = B \cap S$ and $B_2 = B \smallsetminus S$, for some $B \in P$ that has been split by S then B_1.intersection = **tt** and B_2.intersection = **ff**, while B_1.brother points to B_2 and B_2.brother points to B_1. If instead B has not been split by S then B.intersection = **null** and B.brother = **null**. Also, any block B stores in $Rem(B)$ a list of blocks of P, which is used by $RStabilize()$, and in B.preE the list of blocks $C \in P$ such that $C {\rightarrow}^{\exists} B$. Finally, any block B stores in B.size the size of B, in B.count an integer counter bounded by $|P|$ which is used by $PStabilize()$ and a pair of boolean flags used for marking purposes. The current partition P is stored as a doubly linked list of blocks.

Relation: The current relation $\unlhd$ on P is stored as a resizable $|P| \times |P|$ boolean matrix. Recall that insert operations in a resizable array (whose capacity is doubled as needed) take amortized constant time and that a resizable matrix (or table) can be implemented as a resizable array of resizable arrays. The boolean matrix $\unlhd$ is resized by adding a new entry to $\unlhd$, namely a new row and a new column, for any block B that is split into two new blocks $B \smallsetminus S$ and $B \cap S$. The old entry B becomes the entry for the new block $B \smallsetminus S$ while the new entry is used for the new block $B \cap S$.

```
1  bool PStabilize()
2      list⟨Block⟩ split := ∅;
3      while (C := FindPRefiner()) ≠ null) do
4      │   list⟨State⟩ S := preμ(C); split := Split(S); updateRel(split);
5      └   updateBCount(split); updatePreE(); updateCount(split); updateRem(split);
6      return (split = ∅);

7  Block FindPRefiner()
8      forall B ∈ P do
9      │   list⟨Block⟩ p := Post(B);
10     └   forall C ∈ p do  if (Count(B, C) = 1) then return C;
11     return null;

12 list⟨Block⟩ Post(Block B)
13     list⟨Block⟩ p := ∅;
14     forall b ∈ B, do
15     │   forall c ∈ post(b) do
16     │   │   Block  C := c.block;
17     │   │   if unmarked1(C) then  {mark1(C); C.count = 0; p.append(C);}
18     │   └   if unmarked2(C) then  {mark2(C); C.count++;}
19     └   forall C ∈ p do  unmark2(C);
20     forall C ∈ p do  {unmark1(C); if (C.count = B.size) then p.remove(C);}
21     return p;
```

Fig. 2. *PStabilize()* Algorithm

Auxiliary Data Structures: We store and maintain a resizable boolean matrix BCount and a resizable integer matrix Count, both indexed over P, whose meanings are as follows:

$$\text{BCount}(B, C) \triangleq \begin{cases} 1 \text{ if } B\rightarrow^\exists C \\ 0 \text{ if } B \nrightarrow^\exists C \end{cases} \quad \text{Count}(B, C) \triangleq \sum_{E \trianglerighteq C}\text{BCount}(B, E).$$

Hence, $\text{Count}(B, C)$ stores the number of blocks E such that $C \trianglelefteq E$ and $B\rightarrow^\exists E$. The table Count allows to implement the test $B \nrightarrow^\exists \text{pre}(\mu_{\mathcal{P}}(C))$ in constant time as $\text{Count}(B, C) = 0$.

The data structures BCount, preE, Count and *Rem* are initialized by a function *Initialize()* at line 2 of ESim, which is here omitted.

4.2 Partition Stability

Our implementation of ESim will exploit the following logical characterization of partition refiners.

Theorem 4.1. *Let $\langle P, \trianglelefteq \rangle$ be a partial order PR. Then, $\text{PRefiner}(\langle P, \trianglelefteq \rangle) \neq \emptyset$ iff there exist $B, C \in P$ such that the following three conditions hold:*

(i) $B \rightarrow^{\exists} C$;
(ii) *for any $C' \in P$, if $C \lhd C'$ then $B \not\rightarrow^{\exists} C'$;*
(iii) $B \nsubseteq \mathrm{pre}(C)$.

Notice that this characterization of partition refiners requires that the current PR is a partial order relation and, by Lemma 3.2, for any call to *PStabilize*(), this is actually guaranteed by the ESim algorithm.

The algorithm in Figure 2 is an implementation of the *PStabilize*() function that relies on Theorem 4.1 and on the above data structures. The function *FindPRefiner*() implements the conditions of Theorem 4.1: it returns a partition refiner for the current PR $\mathcal{P} = \langle P, \unlhd \rangle$ when this exists, otherwise it returns a null pointer. Given a block $B \in P$, the function *Post*(B) returns a list of blocks $C \in P$ that satisfy conditions (i) and (iii) of Theorem 4.1, i.e., those blocks C such that $B \rightarrow^{\exists} C$ and $B \nsubseteq \mathrm{pre}(C)$. This is accomplished through the counter C.count that at the exit of the for-loop at lines 14-19 in Figure 2 stores the number of states in B having (at least) an outgoing transition to C, i.e., C.count $= |B \cap \mathrm{pre}(C)|$. Hence, we have that:

$$B \rightarrow^{\exists} C \text{ and } B \nsubseteq \mathrm{pre}(C) \iff 1 \leq C.\text{count} < B.\text{size}.$$

Then, for any candidate partition refiner $C \in Post(B)$, it remains to check condition (ii) of Theorem 4.1. This condition is checked in *FindPRefiner*() by testing whether $\mathrm{Count}(B, C) = 1$: this is correct because $\mathrm{Count}(B, C) \geq 1$ holds since $C \in Post(B)$ and therefore $B \rightarrow^{\exists} C$, so that

$$\mathrm{Count}(B, C) = 1 \text{ iff } \forall C' \in P.C \lhd C' \Rightarrow B \not\rightarrow^{\exists} C'.$$

Hence, if $\mathrm{Count}(B, C) = 1$ holds at line 10 of *FindPRefiner*(), by Theorem 4.1, C is a partition refiner. Once a partition refiner C has been returned by *Post*(B), *PStabilize*() splits the current partition P w.r.t. the splitter $S = \mathrm{pre}(\mu_{\mathcal{P}}(C))$ by calling the function *Split*(S), updates the relation $\unlhd$ as defined by equation (†) in Section 3 by calling *updateRel*(), updates the data structures BCount, preE, Count and *Rem*, and then check again whether a partition refiner exists. At the exit of the main while-loop of *PStabilize*(), the current PR $\langle P, \unlhd \rangle$ is partition stable.

PStabilize() calls the functions preμ() and *Split*() that are here omitted. Recall that the states of a block B of P are consecutive in the list of states Σ, so that B is represented as $B = [B.\text{begin}, B.\text{end}[$. The implementation of *Split*(S) is quite standard (see e.g. [12]): this is based on a linear scan of the states in S and for each state in S performs some constant time operations. Hence, *Split*(S) takes $O(|S|)$ time. Also, *Split*(S) returns the list *split* of blocks $B \smallsetminus S$ such that $\varnothing \subsetneq B \smallsetminus S \subsetneq B$ (i.e., B.intersection $=$ **ff**). Let us remark that a call *Split*(S) may affect the ordering of the states in the list Σ because states are moved from old blocks to newly generated blocks.

We will show that the overall time complexity of *PStabilize*() along a whole run of ESim is in $O(|P_{\mathrm{sim}}||\!\rightarrow\!|)$.

4.3   Updating Data Structures

In the function *PStabilize*(), after calling *Split*(S), firstly we need to update the boolean matrix that stores the relation $\unlhd$ in accordance with definition (†) in Section 3. After that, since both P and $\trianglelefteq$ are changed we need to update the data structures BCount, preE, Count and *Rem*. We omit the implementations of the functions *updateRel*(), *updateBCount*(), *updatePreE*() and *updateRem*(), which are quite straightforward.

The function *updateCount*() is in Figure 3 and deserves special care in order to design a time efficient implementation. The core of the *updateCount*() algorithm follows Hopcroft's "process the smaller half" principle [7] for updating the integer matrix Count. Let P' be the partition which is obtained by splitting the partition P w.r.t. the splitter S. Let B be a block of P that has been split into $B \cap S$ and $B \setminus S$. Thus, we need to update Count$(B \cap S, C)$ and Count$(B \setminus S, C)$ for any $C \in P'$ by knowing Count$(B, \text{parent}_P(C))$. Let us first observe that after lines 4-6 of *updateCount*(), we have that for any $B, C \in P'$, Count$(B, C) = $ Count$(\text{parent}_P(B), \text{parent}_P(C))$. Let X be the block in $\{B \cap S, B \setminus S\}$ with the smaller size, and let Z be the other block, so that $|X| \leq |B|/2$ and $|X| + |Z| = |B|$. Let C be any block in P'. We set Count(X, C) to 0, while Count(Z, C) is left unchanged, namely Count$(Z, C) = $ Count(B, C). We can correctly update both Count(Z, C) and Count(X, C) by just scanning all the outgoing transitions from X. In fact, if $x \in X$, $x{\to}y$ and the block $P(y)$ is scanned for the first time then for all $C \trianglelefteq P(y)$, Count(X, C) is incremented by 1, while if $Z \not\to^{\exists} P(y)$, i.e. BCount$(Z, P(y)) = 0$, then Count(Z, C) is decremented by 1. The correctness of this procedure goes as follows:

(1) At the end, Count(X, C) is clearly correct because its value has been re-computed from scratch.

(2) At the end, Count(Z, C) is correct because Count(Z, C) initially stores the value Count(B, C), and if there exists some block D such that $C \trianglelefteq D$, $B{\to}^{\exists}D$ whereas $Z \not\to^{\exists} D$ — this is correctly implemented at line 19 as BCount$(Z, D) = 0$, since the date structure BCount is up to date — then necessarily $X{\to}^{\exists}D$, because B has been split into X and Z, so that $D = P(y)$ for some $y \in \text{post}(X)$, namely D has been taken into account by some increment Count(X, C)++ and consequently Count(Z, C) is decremented by 1 at line 19.

Moreover, if some block $D \in P' \setminus \{B \cap S, B \setminus S\}$ is such that both $D{\to}^{\exists}X$ and $D{\to}^{\exists}Z$ hold then for all the blocks $C \in P$ such that $C \trianglelefteq X$ (or, equivalently, $C \trianglelefteq Z$), we need to increment Count(D, C) by 1. This is done at lines 21-22 by relying on the updated date structures preE and BCount.

Let us observe that the time complexity of a single call of *updateCount*(*split*) is

$$|P|\Big(|split| + \sum_{X \in split}\big(|\{(x,y) \mid x \in X, y \in \Sigma, x{\to}y\}| + |\{(X, D) \mid D \in P, X{\to}^{\exists}D\}|\big)\Big).$$

Hence, let us calculate the overall time complexity of *updateCount*(). If X and X' are two blocks that are scanned in two different calls of *updateCount* and $X' \subseteq X$ then $|X'| \leq |X|/2$. Consequently, any transition $x{\to}y$ at line 16 and $D{\to}^{\exists}X$ at line 21 can be scanned in some call of *updateCount*() at most $\log_2 |\Sigma|$ times. Thus, the overall time complexity of *updateCount*() is in $O(|P_{\text{sim}}||{\to}| \log |\Sigma|)$.

4.4   Relation Stability

The basic procedure *RStabilize*() in Figure 1 is implemented by the algorithm in Figure 4. Let $\mathcal{P}^{\text{in}} = \langle P, \trianglelefteq^{\text{in}} \rangle$ be the current PR when calling *RStabilize*(). For each

```
 1  //  Precondition: BCount and preE are updated with the current PR
 2  updateCount(list⟨Block⟩ split)
 3      forall B ∈ split do addNewEntry(B) in matrix Count;
 4      forall B ∈ P, C ∈ split do
 5      │   if (B.intersection = tt) then  Count(B, C) := Count(B.brother, C.brother);
 6      └   else Count(B, C) := Count(B, C.brother);

 7      forall C ∈ P, B ∈ split do
 8      └   if (C.intersection = ff) then  Count(B, C) := Count(B.brother, C);

 9      forall C ∈ P do unmark(C);
10      forall B ∈ split do
11      │   //  Update Count(B, ·) and Count(B.brother, ·)
12      │   Block X, Z;
13      │   if (B.size ≤ B.brother.size) then {X := B; Z := B.brother;}
14      │   else {X := B.brother;  Z := B;}
15      │   forall C ∈ P do {Count(X, C) := 0; /* Count(Z, C) := Count(B, C); */}
16      │   forall x ∈ X, y ∈ post(x) such that unmarked(y.block) do
17      │   │   mark(y.block);
18      │   │   forall C ∈ P such that C ⊴ y.block do
19      │   └   └   Count(X,C)++;  if (BCount(Z, y.block) = 0) then Count(Z,C)--;

20      │   //  For all D ∉ {B, B.brother}, updateCount(D, ·)
21      │   forall D ∈ X.preE such that (D ≠ X & D ≠ Z & BCount(D, Z) = 1) do
22      └   └   forall C ∈ P such that C ⊴ X do Count(D, C)++;
```

Fig. 3. *updateCount()* function

relation refiner $C \in P$, *RStabilize()* must iteratively refine the initial relation $\trianglelefteq^{\text{in}}$ in accordance with equation (‡) in Section 3. Hence, if $B \rightarrow^\exists C$, $B \trianglelefteq D$ and $D \not\rightarrow^\exists \mu_{\mathcal{P}^{\text{in}}}(C)$, the entry $B \trianglelefteq D$ of the boolean matrix that represents the relation $\trianglelefteq$ must be set to **ff**. Thus, the idea is to store and incrementally maintain for each block $C \in P$ a list $Rem(C)$ of blocks $D \in P$ such that: (A) If C is a relation refiner for $\mathcal{P}^{\text{in}}$ then $Rem(C) \neq \varnothing$; (B) If $D \in Rem(C)$ then necessarily $D \not\rightarrow^\exists \mu_{\mathcal{P}}^{\text{in}}(C)$. It turns out that C is a relation refiner for $\mathcal{P}^{\text{in}}$ iff there exist blocks B and D such that $B \rightarrow^\exists C$, $D \in Rem(C)$ and $B \trianglelefteq D$. Hence, the set of blocks $Rem(C)$ is reminiscent of the set of states *remove(s)* used in Henzinger et al.'s [5] simulation algorithm, since each pair (B, D) which must be removed from the relation $\trianglelefteq$ is such that $D \in Rem(C)$, for some block C.

Initially, namely at the first call of *RStabilize()* by ESim, $Rem(C)$ is set by the function *Initialize()* to $\{D \in P \mid D \rightarrow^\exists \Sigma, D \not\rightarrow^\exists \mu_{\mathcal{P}}(C)\}$. Hence, *RStabilize()* scans all the blocks in the current partition P and selects those blocks C such that $Rem(C) \neq \varnothing$, which are therefore candidate to be relation refiners. Then, by scanning all the blocks $B \in C.\text{preE}$ and $D \in Rem(C)$, if $B \trianglelefteq D$ holds then the entry $B \trianglelefteq D$ must be set to **ff**. However, the removal of the pair (B, D) from the current relation $\trianglelefteq$ may affect the function $\mu_{\mathcal{P}}$. This is avoided by making a copy *oldRem(C)* of all the $Rem(C)$'s

```
1  bool RStabilize()
2    // μ_P^in := μ_P;
3    forall C ∈ P do {oldRem(C) := Rem(C); Rem(C) = ∅; }
4    bool Removed := ff;
5    forall C ∈ P such that oldRem(C) ≠ ∅ do
6    | // Invariant (Inv): ∀C ∈ P. Rem(C) = {D ∈ P | D →^∃ μ_P^in(C), D ≁^∃ μ_P(C)}
7    | forall B ∈ C.preE, D ∈ oldRem(C) such that (B ⊴ D) do
8    |  | B ⊴ D := ff; Removed := tt;
9    |  | // update Count and Rem
10   |  | forall F ∈ D.preE do
11   |  |  | Count(F, B) := Count(F, B) − 1;
12   |  |  | if (Count(F, B) = 0 & Rem(B) = ∅) then Rem(B).append(F);
13   return ¬Removed;
```

Fig. 4. $RStabilize()$ Algorithm

at the beginning of $RStabilize()$ and then using this copy. During the main for-loop of $RStabilize()$, $Rem(C)$ must satisfy the following invariant property:

$$\text{(Inv): } \forall C \in P. \, Rem(C) = \{D \in P \mid D \to^\exists \mu_P^{in}(C), \, D \not\sim^\exists \mu_P(C)\}.$$

This means that at the beginning of $RStabilize()$, any $Rem(C)$ is set to empty, and after the removal of a pair (B, D) from $\trianglelefteq$, since $\mu_P(B)$ has changed, we need: (i) to update the matrix Count, for all the entries (F, B) where $F \to^\exists D$, and (ii) to check if there is some block F such that $F \not\sim^\exists \mu_P(B)$, because any such F must be added to $Rem(B)$ in order to maintain the invariant property (Inv).

4.5 Complexity

The time complexity of the algorithm ESim relies on the following key properties:

(1) The overall number of partition refiners found by ESim is in $O(|P_{sim}|)$. Moreover, the overall number of newly generated blocks by the splitting operations performed by calling $Split(S)$ at line 4 of $PStabilize()$ is in $O(|P_{sim}|)$. In fact, let $\{P_i\}_{i \in [0,n]}$ be the sequence of different partitions computed by ESim where P_0 is the initial partition P_ℓ, P_n is the final partition P_{sim} and for all $i \in [1, n]$, P_i is the partition after the i-th call to $Split(S)$, so that $P_i \prec P_{i-1}$. The number of new blocks which are produced by a call $Split(S)$ that refines P_i to P_{i+1} is $2(|P_{i+1}| - |P_i|)$. Thus, the overall number of newly generated blocks is $\sum_{i=1}^n 2(|P_i| - |P_{i-1}|) = 2(|P_{sim}| - |P_\ell|) \in O(|P_{sim}|)$.

(2) The invariant (Inv) of the sets $Rem(C)$ guarantees the following property: if C_1 and C_2 are two blocks that are selected by the for-loop at line 5 of $RStabilize()$ in two different calls of $RStabilize()$, and $C_2 \subseteq C_1$ (possibly $C_1 = C_2$) then $(\cup Rem(C_1)) \cap (\cup Rem(C_2)) = \emptyset$.

Theorem 4.2 (Complexity). ESim *runs in* $O(|P_{sim}|^2 \log |P_{sim}| + |\Sigma| \log |\Sigma|)$-*space and* $O(|P_{sim}||\to| \log |\Sigma|)$-*time.*

5   Further Work

We see a couple of interesting avenues for further work. A first natural question arises: can the time complexity of ESim be further improved and reaches the time complexity of RT? This would require to eliminate the multiplicative factor $\log |\Sigma|$ from the time complexity of ESim and, presently, this seems to us quite hard to achieve. More in general, it would be interesting to investigate whether some lower space and time bounds can be stated for the simulation preorder problem. Secondly, ESim is designed for Kripke structures. While an adaptation of a simulation algorithm from Kripke structures to labeled transition systems (LTSs) can be conceptually simple, unfortunately such a shift may lead to some loss in both space and time complexities, as argued in [2]. We mention the works [1,6] and [8] that provide simulation algorithms for LTSs by adapting, respectively, RT and GPP. It is thus worth investigating whether and how ESim can be efficiently adapted to work with LTSs.

Acknowledgements. We acknowledge the contribution of Francesco Tapparo to a preliminary stage of this research which was informally presented in [11]. This work was partially supported by Microsoft Research SEIF 2013 Award and by the University of Padova under the project BECOM.

References

1. Abdulla, P.A., Bouajjani, A., Holík, L., Kaati, L., Vojnar, T.: Computing simulations over tree automata. In: Ramakrishnan, C.R., Rehof, J. (eds.) TACAS 2008. LNCS, vol. 4963, pp. 93–108. Springer, Heidelberg (2008)
2. Cécé, G.: Three simulation algorithms for labelled transition systems. Preprint cs.arXiv:1301.1638, arXiv.org (2013)
3. Gentilini, R., Piazza, C., Policriti, A.: From bisimulation to simulation: coarsest partition problems. J. Automated Reasoning 31(1), 73–103 (2003)
4. van Glabbeek, R., Ploeger, B.: Correcting a space-efficient simulation algorithm. In: Gupta, A., Malik, S. (eds.) CAV 2008. LNCS, vol. 5123, pp. 517–529. Springer, Heidelberg (2008)
5. Henzinger, M.R., Henzinger, T.A., Kopke, P.W.: Computing simulations on finite and infinite graphs. In: Proc. 36th IEEE FOCS, pp. 453–462 (1995)
6. Holík, L., Šimáček, J.: Optimizing an LTS-simulation algorithm. Computing and Informatics 29(6+), 1337–1348 (2010)
7. Hopcroft, J.E.: A $n \log n$ algorithm for minimizing states in a finite automaton. In: Kohavi, Z., Paz, A. (eds.) Theory of Machines and Computations, pp. 189–196. Academic Press (1971)
8. Markovski, J.: Saving time in a space-efficient simulation algorithm. In: Proc. 11th Int. Conf. on Quality Software, pp. 244–251. IEEE (2011)
9. Paige, R., Tarjan, R.E.: Three partition refinement algorithms. SIAM J. Comput. 16(6), 973–989 (1987)
10. Ranzato, F., Tapparo, F.: A new efficient simulation equivalence algorithm. In: Proc. 22nd Annual IEEE Symp. on Logic in Computer Science, LICS 2007, pp. 171–180 (2007)
11. Ranzato, F., Tapparo, F.: A time and space efficient simulation algorithm. Short talk at 24th Annual IEEE Symposium on Logic in Computer Science, LICS 2009 (2009)
12. Ranzato, F., Tapparo, F.: An efficient simulation algorithm based on abstract interpretation. Inf. Comput. 208(1), 1–22 (2010)

A Planarity Test via Construction Sequences

Jens M. Schmidt

Max Planck Institute for Informatics, Germany

Abstract. Linear-time algorithms for testing the planarity of a graph are well known for over 35 years. However, these algorithms are quite involved and recent publications still try to give simpler linear-time tests. We give a conceptually simple reduction from planarity testing to the problem of computing a certain construction of a 3-connected graph. This implies a linear-time planarity test. Our approach is radically different from all previous linear-time planarity tests; as key concept, we maintain a planar embedding that is 3-connected at each point in time. The algorithm computes a planar embedding if the input graph is planar and a Kuratowski-subdivision otherwise.

1 Introduction

Testing the planarity of a graph is a fundamental algorithmic problem that has initiated significant contributions to data structures and the design of algorithms in the past. Although optimal linear-time algorithms for this problem are known for over 35 years [13,3], they are involved and recent publications still try to give simpler linear-time algorithms [4,6,9,11,25].

We give a linear-time planarity test that is based on a conceptually very simple reduction to the problem of computing a certain construction C of a 3-connected graph G (we will give a precise definition of C in Section 3). The existence of a similar construction has also been used by Kelmans [14] and Thomassen [27] to give a short proof of Kuratowski's Theorem. Although their proof itself is constructive (in the sense that it gives a polynomial-time planarity test) and received much attention in graph theory due to its simplicity, it has not been utilized algorithmically. We give the first linear-time planarity test that captures this proof scheme. Our hope is that this new approach will lead to simple planarity tests, just as the same concept led to simple proofs of Kuratowski's Theorem.

Currently, the fastest algorithm known for computing C achieves a linear running time [24], but is quite involved. For that reason, our reduction does not qualify to be regarded as a simple planarity test yet. However, every simplification made for computing C will immediately result in a simpler linear-time planarity test. In fact, much less is needed, as our reduction relies only on the part of the construction C until a first non-planar graph occurs; thus, one may assume planarity for computing the necessary part of C. In the case that a

K. Chatterjee and J. Sgall (Eds.): MFCS 2013, LNCS 8087, pp. 765–776, 2013.
© Springer-Verlag Berlin Heidelberg 2013

quadratic running time is allowed, a very simple algorithm that computes C (and, thus, planarity) is known [23].

Recent planarity tests like [4,6,9,11,25] maintain a planar embedding at each step, where all steps either add paths/edges (*path addition method*) or vertices (*vertex addition method*) to the embedding; for a thorough survey on planarity tests, we refer to Patrignani [22]. In our algorithm, each step will essentially add an edge, possibly after subdividing one or two edges in advance. Unlike all previous linear-time planarity tests, we maintain a planar embedding that is always 3-connected. This is a key concept for the following reason. The 3-connectivity constraint fixes the planar embedding (up to flipping), which will allow to test efficiently whether the addition of a next edge e preserves planarity.

A well-known connection between 3-connectivity and planarity is that both can be characterized by conditions on *segments* (or *components*) of cycles [30]. In fact, the first linear-time tests on planarity and 3-connectivity [12,13] due to Hopcroft and Tarjan use such conditions. A detailed exposition of the connection was given later by Vo and Williamson [29,30,33], respectively, with an emphasis on explaining the algorithms in [12,13]. Nevertheless, the precise interplay between 3-connectivity and planarity and its algorithmic consequences are still far from understood; e. g., the known connection suggests to ask whether there is a general approach that combines linear-time 3-connectivity and planarity tests. Our reduction makes a first step towards such a general approach. The combining element is C; on the one hand, C proves a graph to be 3-connected, on the other hand, C provides a unique embedding as long as the constructed graphs are planar, which allows to check planarity efficiently.

A planarity test is *certifying* in the sense of [16] if its *yes/no*-output is augmented with a planar embedding if the input graph is planar and a Kuratowski-subdivision otherwise. The first two linear-time planarity tests of Hopcroft and Tarjan [13] and Booth and Lueker [3] did not give a planar embedding for planar input graphs. Mehlhorn and Mutzel [18] and Chiba, Nishizeki, Abe and Ozawa [7] extended these tests to compute a planar embedding in the same asymptotic running time. The algorithm presented here is certifying.

2   Preliminaries

We use standard graph-theoretic terminology from [2]. Let $G = (V, E)$ be a simple finite graph with $n := |V|$ and $m := |E|$. Multiedges do not matter for planarity and can be removed in advance by performing two bucket sorts on the endpoints of edges in E.

A vertex whose deletion increases the number of connected components is called a *cut vertex*. A graph G is *biconnected* if it is connected and contains no cut vertex. A *biconnected component* of a graph G is a maximal biconnected subgraph of G. A pair of vertices whose deletion disconnects a biconnected graph is called a *separation pair*. A biconnected graph is *triconnected* if it contains no separation pair.

A *(straight-line) planar embedding* of a graph $G = (V, E)$ is an injective function $\pi \colon V \to \mathbb{R}^2$ such that for any two distinct edges ab and cd the straight line segments $\overline{\pi(a)\pi(b)}$ and $\overline{\pi(c)\pi(d)}$ are internally disjoint (i.e., they may only intersect at their endpoints). Two embeddings Emb_1 and Emb_2 of the same planar graph are *(combinatorially) different* if there is a vertex v such that the cyclic order of edges around v in Emb_1 and Emb_2 is different.

A *subdivision* of a graph G (a *G-subdivision*) is a graph obtained by replacing the edges of G with internally disjoint paths of length at least one. Triconnected graphs and their subdivisions have the following property, which we will use throughout this paper.

Lemma 1 (Whitney [32], Thm. 1.1 in [20]). *Every subdivision of a triconnected graph has a unique planar embedding (up to flipping).*

The *triconnected components* of a graph G are obtained by the following recursive process on every biconnected component H of G: As long as there is a separation pair $\{x, y\}$ in H, we split H into two subgraphs H_1 and H_2 that partition $E(H)$ and have only x and y in common, followed by adding the edge $e = xy$ to both H_1 and H_2. We refer to [10] for a precise definition of this process. The edge e that was added to H_1 (respectively, H_2) is called the *virtual edge* of H_1 (H_2) and can be seen as a replacement of the graph H_2 (H_1) in this decomposition.

The graphs resulting from this process are either sets of three parallel edges (*triple-bonds*), triangles or simple triconnected graphs. To obtain the triconnected components of G, triple-bonds containing a common virtual edge are successively merged to maximal sets of parallel edges (*bonds*); similarly, triangles containing a common virtual edge are successively merged to maximal cycles (*polygons*). Thus, a triconnected component of G is either a bond, a polygon, or a simple triconnected graph. The triconnected components form a tree, which is called *SPQR-tree* of G [10].

It is well-known that a graph G is planar if and only if all its biconnected components are planar [13]. A similar result holds for the triconnected components of G: If G is planar, all triconnected components of G are planar, as every triconnected component is a minor of G. Conversely, if all triconnected components of a graph G are planar, we can successively merge the planar embeddings of two triconnected components containing the same virtual edge to a bigger planar embedding [31, Lemma 6.2.6], and obtain a planar embedding for G in linear time. This gives the following result.

Lemma 2 ([17]). *A graph is planar if and only if all its triconnected components are planar.*

As bonds and cycles are planar, planarity has only to be checked for simple triconnected graphs. The triconnected components can be computed in linear time [10,12]. Although this is not a trivial task, reliable implementations are publicly available [21] and future steps will benefit greatly from this.

3   Constructions of Triconnected Graphs

With the above arguments we can assume that the input graph G is simple and triconnected. We will make use of a special construction of triconnected graphs due to Barnette and Grünbaum [1].

Definition 1. *Let G be a simple triconnected graph with $n \geq 4$. We define the following operations on G (all vertices and edges are assumed to be distinct; see Figure 1).*

(a) Add an edge xy between two non-adjacent vertices x and y.
(b) Subdivide an edge ab by a vertex x and add the edge xy for a vertex $y \notin \{a, b\}$.
(c) Subdivide two non-parallel edges e and f by vertices x and y, respectively, and add the edge xy (note that e and f may intersect in one vertex).
(d) Add a new vertex x and join it to exactly three old vertices a, b and c.

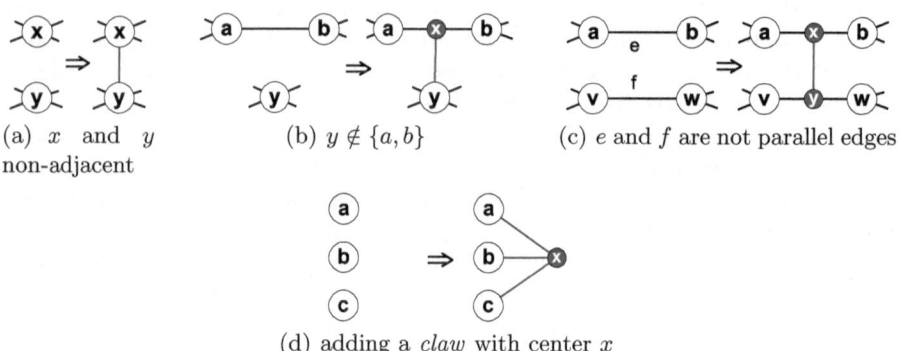

(a) x and y (b) $y \notin \{a, b\}$ (c) e and f are not parallel edges
non-adjacent

(d) adding a *claw* with center x

Fig. 1. Operations on triconnected graphs

Operations 1a-c correspond to *adding edges* (the *added* edge is xy) while Operation 1d corresponds to *adding a claw* (i. e., $K_{1,3}$) with a designated *center vertex* x. The *attachments* of an operation O on G are the vertices and edges in G involved in the operation, i. e., the attachments of Operations 1a–d are $\{x, y\}$, $\{ab, y\}$, $\{ab, vw\}$ and $\{a, b, c\}$, respectively. Let *suppressing* a vertex x with exactly two non-adjacent neighbors y and z be the operation of deleting x and adding the edge yz.

Applying any of the Operations 1a–d to G generates a graph that is simple and triconnected again. A classical result of Barnette and Grünbaum [1] and Tutte [28] characterizes the triconnected graphs in terms of the first three operations.

Theorem 1 ([1,28]). *A simple graph G with $n \geq 4$ is triconnected if and only if G can be constructed from K_4 using only operations of Types 1a–c.*

For testing planarity, we will use the following slightly modified construction. It restricts operations of Type 1a to be at the end of the construction; in order to achieve this, we allow to use additional operations of Type 1d. Having all Type 1a operations at the end of the construction will allow for an easier efficient data structure in the planarity test.

Theorem 2 ([23]). *A simple graph G with $n \geq 4$ is triconnected if and only if G can be constructed from K_4 using only operations of Types 1a–d such that all operations of Type 1a are applied last.*

A *construction sequence* C of G is a sequence of operations that constructs G from K_4 precisely as stated in Theorem 2. Note that any edge that was added by an Operation 1a in C will not be subdivided by later operations.

A construction sequence has a space complexity that is linear in the size of G by using a labeling scheme on vertices and edges that essentially assigns a new label to one half of an edge e after e was subdivided by an operation [23]. The labeling scheme allows additionally for a constant-time access to the edges and vertices that are involved in an operation O, i. e., to the edge e that is added by O and to the vertices and edges on which the endpoints of e lie.

Recently, it was shown that a construction C' of G as stated in Theorem 1 can be computed in linear time [24]. The algorithm is certifying and hence a reliable implementation has already been made publicly available [19]. A construction sequence C can be obtained from C' by a simple linear-time transformation as pointed out in [23].

4   The Planarity Test

We can assume that the input graph G is simple and triconnected. Observe that if $n \leq 3$, G is planar. Assume $n \geq 4$. Let C be a construction sequence of G.

The planarity test starts with the (unique) planar embedding of K_4 and computes iteratively a planar embedding for the graph that is obtained from the next operation O in C if possible. The following lemma characterizes under which conditions an operation O in C preserves planarity.

Lemma 3. *Let H be a planar embedding of a simple triconnected graph on at least 4 vertices and let H' be the graph that is obtained from H by applying an operation O of Type 1a–d. Then H' is planar if and only if the attachments of O are part of one face f of H.*

Proof. $\Leftarrow$: Clearly, subdividing edges in the facial cycle of f preserves planarity and so does the addition of an edge or a claw inside f.

$\Rightarrow$: Assume to the contrary that not all attachments of O are contained in one face of H. Note that when O is of Type 1d it is possible that every two of the three attachments are contained in a face of H, respectively. We will show a contradiction to the unique embedding of H. Let Emb be the planar embedding that is obtained from the planar embedding of H' by reversing Operation O,

i. e., by deleting the added edge in H' and suppressing all vertices of degree two if O is of Type 1a–c, and by deleting the center vertex of the added claw in H' if O is of Type 1d. Then Emb and H embed the same simple triconnected graph, but are combinatorially different, as Emb has a face containing all attachments of O, while H has no such face by assumption. This contradicts Lemma 1.   □

If all attachments of O are in one face f, applying O gives a planar embedding. If all operations in C satisfy this condition, we obtain a planar embedding of G. Otherwise, let H be the graph obtained from the first operation in C that does not satisfy the condition of Lemma 3. Then H is non-planar and G must be non-planar, as G contains a subdivision of H as a subgraph. We will show how to extract this subdivision in linear time in the next section. Lemma 3 suggests the following Algorithm 1.

Algorithm 1. PlanarityTest(G)   ▷ G simple and triconnected with $n \geq 4$

1: compute a construction sequence $C = O_1, \ldots, O_k$ of G
2: initialize the (unique) planar embedding H of K_4
3: **for** $i = 1$ **to** k **do**
4:     **if** all attachments of O_i are in one face f of H **then**   ▷ planar
5:         apply O_i to H by adding the edge or claw inside f
6:     **else**   ▷ non-planar
7:         compute a Kuratowski-subdivision

It remains to discuss how the condition in Lemma 3 can be checked efficiently for every operation in C.

A *plane st-graph* is an embedding of a planar directed acyclic graph with exactly one source s and exactly one sink t such that s and t are contained in the external face of the embedding. It is well-known that every biconnected planar graph can be oriented and drawn as plane st-graph (see, e. g., [5, Lemmas 1+2]). In every step of Algorithm 1, the planar embedding H is triconnected and thus biconnected. To check the condition in Lemma 3 efficiently, we will maintain H as plane st-graph and use a data structure that is able to answer queries whether edges and vertices are contained in the same face of H in amortized constant time.

We modify a data structure due to Djidjev [8, Lemma 3.1], which runs on a standard word-RAM. The original data structure maintains a plane st-graph H in which the incoming and the outgoing edges for any vertex x appear consecutively around x; hence, the boundary of each face f in H consists of two oriented paths from a common start vertex (the *source* of f) to a common end vertex (the *sink* of f); see [26]. Note that every vertex is source or sink of at least one face, as H has minimum degree 3. Additionally, every vertex $x \notin \{s, t\}$ is contained in exactly two faces for which x is neither source nor sink; we call these faces the *left* and the *right* face of x, respectively (see Figure 2). The data structure maintains pointers to the source and sink for each face in H and a pointer from

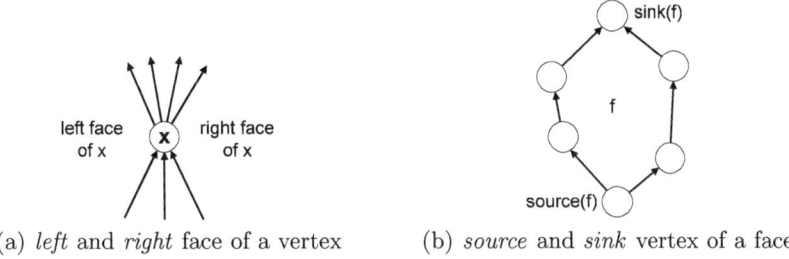

(a) *left* and *right* face of a vertex   (b) *source* and *sink* vertex of a face

Fig. 2. Plane *st*-graphs

each vertex $x \notin \{s, t\}$ to its left and right face. The following two queries for triconnected graphs H are supported by performing simple tests along the above pointer structure.

(1) Given a vertex a and an edge b of H, output a face of H that contains a and b or report that there is none.
(2) Given two vertices a and b of H such that a is a source or sink of at most 11 faces, output a face of H that contains a and b or report that there is none.

Each of these queries takes worst-case time $O(1)$. Note that the only query for which we cannot expect a constant running time using the above pointer structure would be a query where a and b are source or sink vertices of an unbounded number of faces. That is why query (2) assumes only a constant number of such faces. We augment the data structure by the following query type and show that each such query can be computed in worst-case time $O(1)$.

(3) Given three vertices a, b and c of H, output a face of H that contains a, b and c or report that there is none.

We can compute the set F of all left and right faces of the vertices a, b and c in constant time; note that F contains at most 6 faces. If there is a face f in H containing a, b and c, at least one vertex in $\{a, b, c\}$ is neither source nor sink of f, which implies that f must be in F. For a query (3), it therefore suffices to test each face $f \in F$ for containing a, b and c, respectively. A vertex v is contained in f if and only if v is either source or sink of f, which can be checked in time $O(1)$, or one of the remaining vertices in f, which can be checked in time $O(1)$ by testing whether f is the left or right face of v.

The data structure additionally supports each of the following modifications to H in amortized time $O(1)$ and maintains a plane *st*-graph after every modification.

(4) Subdivide an edge.
(5) Given two non-adjacent vertices a and b and a face f of H that contains a and b, add the edge ab inside f.

Clearly, K_4 can be embedded as a plane st-graph and we initialize H with this embedding. Every operation O of Type 1a–d can be converted into at most three of the modifications (4) and (5). E. g., we can add a claw having its attachments $\{a, b, c\}$ in a common face by consecutively inserting the edge ab, subdividing ab with a new vertex x and adding the edge xc. For operations O of Type 1b–d, the condition in Lemma 3 can be checked in constant time by one query (1) or one query (3).

It only remains to show how we can check the condition in Lemma 3 if O is of Type 1a. According to Theorem 2, all operations in C that follow O will be of Type 1a, which implies that H is a spanning subgraph of G. In other words, each of the remaining operations in C adds only an edge that will not be subdivided afterwards. Hence, the order in which these remaining edges E' are added does not matter.

We use a trick similar as in [8, Lemma 3.2] and add the remaining edges E' in an order such that each added edge has an endpoint that is the sink or source of at most 11 faces. If we know this order, we can use query (2) to ensure that every step can be computed in constant time.

In order to compute this order, we maintain an auxiliary graph H^A whose vertex set consists of all vertices in $V(H)$ that are incident to an edge in E'. There is an edge between two vertices a and b in H^A if a and b are source and sink vertices of the same face. Note that H^A may have parallel edges. We construct H^A in linear time when the first operation of Type 1a in C is encountered; after every modification (5), H^A can be updated in time $O(1)$, as each face f stores a pointer to its source and sink.

As H^A is planar and has at most two parallel edges between every two vertices (as H is simple and triconnected), it contains at most $6|V(H^A)| - 12$ edges. Hence, there is at least one vertex of degree at most 11 in H^A. We note that the degree bound 6 proposed in [8, Lemma 3.2] should also be 11, as the auxiliary graph used there is not necessarily simple.

Before the first operation of Type 1a in C is applied to H, we construct a list $Small$ of all vertices in H^A having degree at most 11 in linear time; again, this list is easy to maintain under modifications (5) in time $O(1)$. Now we just choose successively a vertex $v \in Small$ and an edge $e = vw$ in E' and perform modification (5) with v and w if v and w have been reported to be in the same face. This allows to check the condition in Lemma 3 for each of the remaining edges in E' in constant time using query (2). We conclude the following theorem.

Theorem 3. *The planarity test Algorithm 1 can be implemented in linear time.*

5   Extensions

A *Kuratowski-subdivision* is a subdivision of either a $K_{3,3}$ or of a K_5 and proves every graph that contains it to be non-planar. We show how a Kuratowski-subdivision can be computed if an operation O is encountered that has not all attachment vertices on one face in H. The computation follows in parts the arguments given in the short proof of Kuratowski's Theorem in [27]. The fact that the Kuratowski-subdivision is computed in a triconnected component does not matter; it is straight-forward to get a corresponding Kuratowski-subdivision in the input graph by reversing the splits that were done to obtain the triconnected components.

We first recall planarity-related terminology.

Definition 2. *For a cycle C in a graph G, let a C-component be either an edge $e \notin C$ with both endpoints in C or a connected component of $G \setminus V(C)$ together with all edges that join the component to C and all endpoints of these edges. The vertices of attachment of a C-component H are the vertices in $H \cap C$.*

Two C-components H_1 and H_2 *avoid each other* if C contains two vertices u and v such that H_1 has all vertices of attachment on one path in C from u to v and H_2 has all vertices of attachment on the other path in C from u to v.

Two C-components *overlap* if they do not avoid each other. Let two C-components H_1 and H_2 be C-*equivalent* if $H_1 \cap C = H_2 \cap C$ and this set contains exactly three vertices. Let H_1 and H_2 be *skew* if C contains four distinct vertices x_1, x_2, x_3 and x_4 in cyclic order such that x_1 and x_3 are in H_1 and x_2 and x_4 are in H_2. We will need the following basic fact about C-components.

Lemma 4 ([27]). *Two C-components overlap if and only if they are either skew or C-equivalent.*

Now we are prepared to compute a Kuratowski-subdivision.

Lemma 5. *Let H be a planar embedding of a simple triconnected graph on at least 4 vertices and let O be an operation of Type 1a–d on H whose attachments are not all contained in one face of H. Then the graph H' that is obtained from H by applying O contains a subdivision of K_5 or $K_{3,3}$, and this subdivision can be computed in linear time.*

Proof. First assume that O adds a claw and every two of the three attachments $\{a, b, c\}$ of O are contained in a face of H, respectively; we call these three faces f_1, f_2 and f_3. Let J be a closed Jordan curve in $f_1 \cup f_2 \cup f_3$ that intersects H exactly at a, b and c. Since a, b and c are not all contained in a face of H, there are vertices v_{in} and v_{out} strictly inside and strictly outside J, respectively. Since H is 3-connected, we can compute from v_{in} and v_{out} three internally vertex-disjoint paths to $\{a, b, c\}$, respectively, by performing one depth first search. Adding the claw of O to these paths gives a $K_{3,3}$-subdivision in H'.

The only remaining case is that O has at least two attachments a and b that are not contained in one face of H; note that a and b may be edges. As $H \setminus a$ is 2-connected, it contains a cycle C that is the boundary of the face which contains a in its interior. By assumption, $b \notin C$. Let H_a and H_b be the C-components of H containing a and b, respectively. By definition of C, H_a is the only C-component in the interior of C.

We show that H_a and H_b overlap. Assume the contrary. Then H_b has two vertices of attachment u and v such that H_a has all vertices of attachment on one path $P_a \subset C$ from u to v and H_b has all vertices of attachment on the other path $P_b \subset C$ from u to v. If a is a vertex, a and b are in different components of $H \setminus \{u, v\}$, since H is a planar embedding. This contradicts H to be triconnected. Otherwise, a is an edge (which will be subdivided by O) and $H_a = a$. Then, as H is simple, P_a has length at least two, which implies that an inner vertex in P_a is in a different component of $H \setminus \{u, v\}$ than b. This contradicts H to be triconnected. Thus, H_a and H_b overlap.

According to Lemma 4, H_a and H_b are either skew or C-equivalent. The cycle C, H_a and H_b can be easily computed in linear time. Deciding whether H_a and H_b are skew and computing the vertices x_1, x_2, x_3 and x_4 on C, whose existence defines this property amounts to one traversal along C. If a is an edge, subdivide a and let a' be the new vertex of degree two; otherwise let $a' = a$. Define b' accordingly. Due to Menger's Theorem, there are either two or three internally disjoint paths from a' to C in H_a (and from b' to C in H_b), depending on whether H_a (H_b) is an edge. These paths can be computed by a depth-first search that starts with the desired vertex.

If H_a and H_b are skew, we compute two of these paths in H_a that end at x_1 and x_3, respectively, and two in H_b that end at x_2 and x_4, respectively. Taking the union of these four paths, C and T forms a $K_{3,3}$-subdivision, where T is either the added edge of O or the path of length two from a to b if O adds a claw. If H_a and H_b are C-equivalent, the union of the three paths in H_a and H_b, respectively, C and T gives a K_5-subdivision.   □

We remark that our algorithm can be easily extended to output always a $K_{3,3}$-subdivision in linear time when the input graph G is 3-connected, non-planar and $G \neq K_5$. This is based on the following variant of Kuratowski's Theorem for triconnected graphs.

Lemma 6 ([15]). *A simple triconnected graph $G \neq K_5$ is planar if and only if G does not contain a $K_{3,3}$-subdivision.*

Note that we get a K_5-subdivision K only in the case that O adds a claw. The desired $K_{3,3}$-subdivision can then be obtained from K by rerouting one of the paths of K that ends at a to the center vertex of the claw.

Open Questions. The most immediate question is whether there is a simple linear-time algorithm that computes the construction sequence C of a triconnected graph. This would immediately imply a simple linear-time planarity test. As argued before, one may even assume planarity to find such a sequence. Further, it seems possible that such an algorithm, or the existing one in [24], can be extended to compute the triconnected components of the input graph, similarly as in the triconnectivity test of Hopcroft and Tarjan [12]. This would subsume the computation of C and the preprocessing of the graph into triconnected components. The proposed new algorithmic approach to planarity testing might also allow to recognize other subclasses of planar graphs efficiently (e.g., planar graphs that contain no subdivision of $K_5 - e$ or of W_4).

References

1. Barnette, D.W., Grünbaum, B.: On Steinitz's theorem concerning convex 3-polytopes and on some properties of 3-connected graphs. In: Many Facets of Graph Theory, pp. 27–40 (1969)
2. Bondy, J.A., Murty, U.S.R.: Graph Theory. Springer (2008)
3. Booth, K.S., Lueker, G.S.: Testing for the consecutive ones property, interval graphs, and graph planarity using PQ-Tree algorithms. J. Comput. Syst. Sci. 13, 335–379 (1976)
4. Boyer, J.M., Myrvold, W.J.: On the cutting edge: Simplified $O(n)$ planarity by edge addition. Journal of Graph Algorithms an Applications 8(3), 241–273 (2004)
5. Brandes, U.: Eager st-Ordering. In: Möhring, R.H., Raman, R. (eds.) ESA 2002. LNCS, vol. 2461, pp. 247–256. Springer, Heidelberg (2002)
6. Brandes, U.: The left-right planarity test (2009) (manuscript submitted for publication)
7. Chiba, N., Nishizeki, T., Abe, S., Ozawa, T.: A linear algorithm for embedding planar graphs using PQ-Trees. J. Comput. Syst. Sci. 30(1), 54–76 (1985)
8. Djidjev, H.N.: A linear-time algorithm for finding a maximal planar subgraph. SIAM J. Discrete Math. 20(2), 444–462 (2006)
9. de Fraysseix, H., de Mendez, P.O., Rosenstiehl, P.: Trémaux Trees and planarity. Int. J. Found. Comput. Sci. 17(5), 1017–1030 (2006)
10. Gutwenger, C., Mutzel, P.: A linear time implementation of SPQR-trees. In: Marks, J. (ed.) GD 2000. LNCS, vol. 1984, pp. 77–90. Springer, Heidelberg (2001)
11. Haeupler, B., Tarjan, R.E.: Planarity algorithms via PQ-Trees (extended abstract). Electronic Notes in Discrete Mathematics 31, 143–149 (2008)
12. Hopcroft, J.E., Tarjan, R.E.: Dividing a graph into triconnected components. SIAM J. Comput. 2(3), 135–158 (1973)
13. Hopcroft, J.E., Tarjan, R.E.: Efficient planarity testing. J. ACM 21(4), 549–568 (1974)
14. Kelmans, A.K.: A new planarity criterion for 3-connected graphs. Journal of Graph Theory 5, 259–267 (1981)
15. Liebers, A.: Planarizing graphs — A survey and annotated bibliography. J. Graph Algorithms Appl. 5(1) (2001)
16. McConnell, R.M., Mehlhorn, K., Näher, S., Schweitzer, P.: Certifying algorithms. Computer Science Review 5(2), 119–161 (2011)
17. McLane, S.: A structural characterization of planar combinatorial graphs. Duke Mathematical Journal 3, 466–472 (1937)
18. Mehlhorn, K., Mutzel, P.: On the embedding phase of the Hopcroft and Tarjan planarity testing algorithm. Algorithmica 16(2), 233–242 (1996)
19. Neumann, A.: Implementation of Schmidt's algorithm for certifying triconnectivity testing. Master's thesis, Universität des Saarlandes and Graduate School of CS, Germany (2011)
20. Nishizeki, T., Chiba, N.: Planar graphs: Theory and algorithms. North-Holland (1988)
21. OGDF - The Open Graph Drawing Framework (June 2013), http://www.ogdf.net
22. Patrignani, M.: Planarity testing and embedding. In: Tamassia, R. (ed.) Handbook of Graph Drawing and Visualization. CRC Press (to appear)
23. Schmidt, J.M.: Construction sequences and certifying 3-connectedness. In: Proceedings of the 27th Symposium on Theoretical Aspects of Computer Science, STACS 2010, pp. 633–644 (2010)

24. Schmidt, J.M.: Certifying 3-connectivity in linear time. In: Czumaj, A., Mehlhorn, K., Pitts, A., Wattenhofer, R. (eds.) ICALP 2012, Part I. LNCS, vol. 7391, pp. 786–797. Springer, Heidelberg (2012)
25. Shih, W.K., Hsu, W.L.: A new planarity test. Theor. Comput. Sci. 223, 179–191 (1999)
26. Tamassia, R., Tollis, I.G.: A unified approach to visibility representation of planar graphs. Discrete & Computational Geometry 1, 321–341 (1986)
27. Thomassen, C.: Kuratowski's theorem. Journal of Graph Theory 5(3), 225–241 (1981)
28. Tutte, W.T.: Connectivity in graphs. In: Mathematical Expositions, vol. 15. University of Toronto Press (1966)
29. Vo, K.-P.: Finding triconnected components of graphs. Linear and Multilinear Algebra 13, 143–165 (1983)
30. Vo, K.-P.: Segment graphs, depth-first cycle bases, 3-connectivity, and planarity of graphs. Linear and Multilinear Algebra 13, 119–141 (1983)
31. West, D.B.: Introduction to Graph Theory. Prentice Hall (2001)
32. Whitney, H.: Congruent graphs and the connectivity of graphs. American Journal of Mathematics 54(1), 150–168 (1932)
33. Williamson, S.G.: Embedding graphs in the plane — algorithmic aspects. In: Combinatorial Mathematics, Optimal Designs and Their Applications. Annals of Discrete Mathematics, vol. 6, pp. 349–384. Elsevier (1980)

Feasible Combinatorial Matrix Theory

Ariel Germán Fernández and Michael Soltys

McMaster University
Hamilton, Canada
{fernanag,soltys}@mcmaster.ca

Abstract. We give the first, as far as we know, feasible proof of König's Min-Max Theorem (KMM), a fundamental result in combinatorial matrix theory, and we show the equivalence of KMM to various Min-Max principles, with proofs of low complexity.

Keywords: Proof Complexity, Min-Max principle, **LA**, **VTC**^{0}.

1 Introduction

König's Mini-Max Theorem (KMM) is a cornerstone result in Combinatorial Matrix Theory. We give the first, as far as we know, feasible proof of KMM, and we show that it is equivalent to a host of other theorems: Menger's, Hall's, and Dilworth's, with the equivalence provable in low complexity.

The standard textbook proof of KMM given in [BR91], can be formalized with Π_2^B reasoning. On the other hand, our approach yields a Σ_1^B proof. We use the theory of Bounded Arithmetic **LA**, introduced by [SC04].

Let A be an $n \times m$ 0-1 matrix, i.e., a matrix with entries in $\{0, 1\}$. A *line* is a row or column of A; given an entry A_{ij} of A, we say that a line *covers* that entry if this line is either row i or column j. KMM states that the minimum number of lines that cover all of the 1s in A is equal to the maximum number of 1s in A with no two of the 1s on the same line.

LA is a first-order theory, of three sorts: indices, ring elements, and matrices. It formalizes basic index manipulations, as well as ring properties, and has a matrix constructor. The details can be found in [SC04]. While **LA** allows for bounded index quantification and arbitrary matrix quantification, its induction is restricted to be over formulas without matrix quantifiers, i.e., over $\Sigma_0^B = \Pi_0^B$ formulas. On the other hand, $\exists$**LA** allows Σ_1^B induction. When the underlying ring is $\mathbb{Z}$, the theorems of **LA** translate into TC^0-Frege while the theorems of $\exists$**LA** translate into extended Frege, [SC04, §6.5].

It follows more or less directly that our **LA** results can also be formalized in the theory **VTC**^{0} (and vice versa), defined in [CN10, pg. 283]. The reason is that the function ΣA is exactly Buss' function Numones(A) ([Bus86] and [Bus90, pg. 6]), i.e., the function that counts the number of 1s in A, and **TC**^{0} is the **AC**^{0} closure of Numones, [CN10, Proposition IX.3.1]. On the other hand, our $\exists$**LA** results can also be formalized in **V**^{1}, defined in [CN10, pg. 133].

K. Chatterjee and J. Sgall (Eds.): MFCS 2013, LNCS 8087, pp. 777–788, 2013.

Recently [LC12] formalized the proof of correctness of the Hungarian algorithm, which is an algorithm based on KMM.

The language of **LA** is well suited to express concepts in combinatorial matrix theory. For example, we say that the matrix α is a *cover* of a matrix A with the predicate:

$$\text{Cover}(A, \alpha) := \forall i, j \le r(A)(A(i, j) = 1 \to \alpha(1, i) = 1 \lor \alpha(2, j) = 1) \quad (1)$$

We use $r(A)$ and $c(A)$ to denote the rows and columns of a matrix A. We abbreviate $r(A) \le n \land c(A) \le n$ with $|A| \le n$. The matrix α keeps track of the lines that cover A; it does so with two rows: the top row keeps track of the horizontal lines, and the bottom row keeps track of the vertical line. The condition ensures that any 1 in A is covered by some line stipulated in α.

We say that β is a *selection* of A with the predicate $\text{Select}(A, \beta)$ defined as the conjunction of

$$\forall i, j \le r(A)(\beta(i, j) = 1 \to A(i, j) = 1),$$

which asserts that β is a selection of 1s from A, and

$$\forall k \le r(A)(\beta(i, j) = 1 \to \beta(i, k) = 0 \land \beta(k, j) = 0)),$$

which asserts that no two of those 1s are in the same row or column.

We are interested in a minimum cover (as few 1s in α as possible) and a maximum selection (as many 1s in β as possible). The following two predicates express that α is a minimum cover and β a maximum selection.

$$\text{MinCover}(A, \alpha) := \text{Cover}(A, \alpha) \land \forall \alpha' \le c(\alpha)(\text{Cover}(A, \alpha') \to \Sigma\alpha' \ge \Sigma\alpha)$$

$$\text{MaxSelect}(A, \beta) := \text{Select}(A, \beta) \land \forall \beta' \le r(\beta)(\text{Select}(A, \beta') \to \Sigma\beta' \le \Sigma\beta)$$

Clearly MinCover and MaxSelect are Π_1^B formulas. We can now state KMM:

$$\text{MinCover}(A, \alpha) \land \text{MaxSelect}(A, \beta) \to \Sigma\alpha = \Sigma\beta \quad (2)$$

Note that (2) is a Σ_1^B formula. The reason is that in prenex form, the universal matrix quantifiers in MinCover and MaxSelect become existential as we pull them out of the implication; they are also bounded.

Given a matrix A, its *n-th principal minor* consists of A with the first $r(A) - n$ rows deleted, and the first $c(A) - n$ columns deleted. For instance, for a square matrix A, when $n = |A|$, the n-th submatrix is just A, and when $n = 1$, then n-th submatrix is just $[A_{|A|,|A|}]$, i.e., the matrix consisting of just the lower-right entry. Let $A[n]$ denote the n-th principal minor, and note that $A[n]$ can be expressed as follows in the language of **LA**: $\lambda ij\langle n, n, e(A, r(A) - n + i, c(A) - n + j)\rangle$.

Let $\text{KMM}(A, n)$ assert that formula (2) holds for the n-th submatrix of A. More precisely, $\text{KMM}(A, n)$ is the prenex form of (2) with A replaced by $A[n]$. Thus, $\text{KMM}(A, n)$ is a Σ_1^B formula. Let $l_A = \Sigma\alpha$ where $\text{MinCover}(A, \alpha)$, and $o_A = \Sigma\beta$ where $\text{MaxSelect}(A, \beta)$. It can be stated with a Σ_0^B predicate that a matrix P is a permutation matrix. That is,

$$\text{Perm}(P) := (\forall i \le |P|\exists j \le |P|P_{ij} = 1) \land (\forall i, j \ne k \le |P|(P_{ij} = 0 \lor P_{ik} = 0)).$$

2 Feasible Proof of KMM

We prove the main theorem with a sequence of Lemmas.

Theorem 1. $\exists$**LA** *proves König's Min-Max (KMM) Theorem.*

We prove KMM for any matrix A by induction on the principal minors of A.

Lemma 2. $\exists$**LA** $\vdash \forall n \mathrm{KMM}(A, n)$.

Recall that the predicate $\mathrm{KMM}(A, n)$ has been defined in the last paragraph of the previous section. Showing $\forall n \mathrm{KMM}(A, n)$ is enough to prove KMM for A since letting $n = |A|$ we obtain $A[n] = A$.

We start by showing the following technical Lemma which states that l_A and o_A are invariant under permutations of rows and columns.

Lemma 3. *Given a matrix A, and given any permutation matrix P, we have that* **LA** $\vdash l_{PA} = l_{AP} = l_A$ *and* **LA** $\vdash o_{PA} = o_{AP} = o_A$.

Proof. **LA** shows that if we reorder the rows or columns (or both) of a given matrix A, then the new matrix, call it A', where $A' = PA$ or $A' = AP$, has the same size minimum cover and the same size maximum selection. Of course, we can reorder both rows and columns by applying the statement twice: $A' = PA$ and $A'' = A'Q = PAQ$.

LA proves $\mathrm{Cover}(A, \alpha) \to \mathrm{Cover}(A', \alpha')$ and $\mathrm{Select}(A, \beta) \to \mathrm{Select}(A', \beta')$, where A' is defined as in the above paragraph, and α' is the same as α, except the first row of α is now reordered by the same permutation P that multiplied A on the left (and the second row of α is reordered if P multiplied A on the right). The matrix β is even easier to compute, as $\beta' = P\beta$ if $A' = PA$, and $\beta' = \beta P$ if $A' = AP$. It follows from P being a permutation matrix that $\Sigma\alpha = \Sigma\alpha'$ and $\Sigma\beta = \Sigma\beta'$: we can show by **LA** induction on the size of matrices that if X' is the result of rearranging X (i.e., $X' = PXQ$, where P, Q are permutation matrices), then $\Sigma X = \Sigma X'$. We do so first on X consisting of a single row, by induction on the length of the row. Then we take the single row as the basis case for induction over the number of rows of a general X.

It is clear that given A', the cover α' has been adjusted appropriately; same for the selection β'. We can prove it formally in **LA** by contradiction: suppose some 1 in A' is not covered in α'; then the same 1 in A would not be covered by α. For the selections, note that reordering rows and/or columns we maintain the property of being a selection: we can again prove this formally in **LA** by contradiction: if β' has two 1s on the same line, then so would β.

The last thing to show is that **LA** proves $\mathrm{MinCover}(A, \alpha) \to \mathrm{MinCover}(A', \alpha')$ $\mathrm{MaxSelect}(A, \beta) \to \mathrm{MaxSelect}(A', \beta')$. If the right-hand side does not hold, we would get that the left-hand side does not hold by applying the inverse of the permutation matrix. □

We are going to prove Lemma 2 by induction on n, breaking it down into Claims 4 and 7.

Claim 4. LA ⊢ $o_A \leq l_A$.

Proof. Given a covering of A consisting of l_A lines, we know that every 1 we pick for a maximal selection of 1s has to be on one of the lines of the covering. We also know that we cannot pick more than one 1 from each line. Thus, the number of lines in the covering provide an upper bound on the size of such selection, giving us $o_A \leq l_A$.

We can formalize this argument in **LA** as follows: let $\mathcal{A}$ be an $l_A \times o_A$ matrix whose rows represent the lines of the covering, and whose columns represent the 1s no two on the same line. Let $\mathcal{A}(i,j) = 1 \iff$ the line labeled with i covers the 1 labeled with j. Then,

$$o_A = c(\mathcal{A}) \leq \Sigma\mathcal{A} \tag{a}$$
$$= \Sigma_i(\Sigma\lambda pq\langle 1, c(\mathcal{A}), \mathcal{A}(i,q)\rangle) \tag{b}$$
$$\leq \Sigma_i 1 = r(\mathcal{A}) = l_A, \tag{c}$$

where (a) can be shown by induction on the number of columns of $\mathcal{A}$ which has the condition that each column contains at least one 1 (i.e., each 1 from the selection must be covered by some line); (b) follows from the fact that we can add all the entries in a matrix by rows; and (c) can be shown by induction on the number of rows of $\mathcal{A}$ which has the condition that each row contains at most one 1 (i.e., no two 1s from the selection can be on the same line). □

Note that in the proof of Claim 4 we implicitly show the Pigeonhole Principle (PHP). We showed that if we have a set of n items $\{i_1, i_2, \ldots, i_n\}$ and a second set of m items $\{j_1, j_2, \ldots, j_m\}$, and we represent the matching by A as follows: $A(p,q) = 1 \iff i_p \mapsto j_q$, then injectivity means that each column of A has at most one 1. Thus:

$$n \leq \Sigma A = \Sigma_i(\text{col } i \text{ of } A) \leq \Sigma_i 1 \leq m.$$

This is to be expected as we already mentioned that **LA** over $\mathbb{Z}$ corresponds to $\mathbf{VTC}^0$, which proves PHP.

Bondy's Theorem states that for any $n \times n$ 0-1 matrix whose rows are distinct, we can always delete a column so that the remaining $n \times (n-1)$ matrix still has n distinct rows. [CN10, §IX.3.8] investigate the connection between Bondy's Theorem (BONDY) and PHP, and they show that $\mathbf{V}^0 \vdash$ BONDY $\leftrightarrow$ PHP. It would be interesting to know if $\mathbf{V}^0 \vdash$ KMM $\leftrightarrow$ PHP.

As Claim 4 shows, **LA** is sufficient to prove $o_A \leq l_A$; on the other hand, we seem to require the stronger theory ∃**LA** (which is **LA** with induction over Σ_1^B formulas) in order to prove the other direction of the inequality. We start with the following definition.

Definition 5. *We say that an $n \times n$ 0-1 matrix has the* diagonal property *if for each diagonal entry (i,i) of A, either $A_{ii} = 1$, or $\forall j \geq i[A_{ij} = 0 \wedge A_{ji} = 0]$.*

Claim 6. *Given any matrix A, **LA** proves that there exist permutation matrices P, Q such that PAQ has the diagonal property.*

Proof. We construct P, Q inductively on $n = |A|$. Let the *i-th layer of A* consist of the following entries of A: A_{ij}, for $j = i, \ldots, n$ and A_{ji} for $j = i + 1, \ldots, n$. Thus, the first layer consists of the first row and column of A, and the n-th layer (also the last layer), is just A_{nn}. We transform A by layers, $i = 1, 2, 3, \ldots$. At step i, let A' be the result of having dealt already with the first $i - 1$ layers. If $A'_{ii} = 1$ move to the next layer, $i + 1$. Otherwise, find a 1 in layer i of A'. If there is no 1, also move on to the next layer, $i + 1$. If there is a 1, permute it from position $A_{ij'}$, $j' \in \{i, \ldots, n\}$ to A'_{ii}, or from position $A_{j'i}$, $j' \in \{i + 1, \ldots, n\}$. Note that such a permutation does not disturb the work done in the previous layers; that is, if A'_{kk}, $k < i$, was a 1, it continues being a 1, and if it was not a 1, then there are no 1s in layer k of A'. Note that each layer can be computed independently of the others.   □

Claim 7. $\exists \mathbf{LA} \vdash o_A \geq l_A$.

Proof. Let

$$A = \left[\begin{array}{c|c} a & R \\ \hline S & M \end{array} \right], \tag{3}$$

where a is the top-left entry, and M the principal sub-matrix of A, and R (resp. S) is $1 \times (n - 1)$ (resp. $(n - 1) \times 1$).

By Claim 6 we can ensure that A has the diagonal property, which simplifies the analysis of the cases. Indeed, from the diagonal property we know that one of the following two cases is true:

Case 1. $a = 1$

Case 2. a, R, S consist entirely of zeros

In the second case, $o_A \geq l_A$ follows directly from the induction hypothesis, $o_M \geq l_M$, as $o_A = o_M \geq l_M = l_A$. Thus, it is the first case, $a = 1$, that is interesting. The first case, in turn, can be broken up into two subcases: $l_M = n-1$ and $l_M < n - 1$.

Subcase (1-a) $l_M = n - 1$

By induction hypothesis, $o_M \geq l_M = n - 1$. We also have that $a = 1$, and a is in position $(1, 1)$, and hence no matter what subset of 1s is selected from M, none of them lie on the same line as a. Therefore, $o_A \geq o_M + 1$. Since $o_M \geq n-1$, $o_A \geq n$, and since we can *always* cover A with n lines, we have that $n \geq l_A$, and so $o_A \geq l_A$.

Subcase (1-b) $l_M < n - 1$

Let A and M be as in (3), and let α_M be a set of lines of M, i.e., α_M consists of rows $i_1, i_2, \ldots, i_k$, and columns $j_1, j_2, \ldots, j_\ell$. The *extension* of α_M to A, denoted $\hat{\alpha}_M$, is simply the set of rows $i_1 + 1, i_2 + 1, \ldots, i_k + 1$, and the set of columns $j_1 + 1, j_2 + 1, \ldots, j_\ell + 1$.

We say that a minimal cover α_A is proper if it does not consists entirely of all the rows or of all the columns of A; that is, α_A is proper if it is minimal, i.e., $|\alpha_A| = l_A$, and each row of α_A has at least one zero. If $l_M < n - 1$, then we know that α_A has a proper cover, as we can always cover A with $\hat{\alpha}_M$ plus the first row and column of A.

Let α_A be a proper minimal cover of A, and let P, Q be two permutations that place all the rows of the cover in the initial position, and place all the columns of the cover in the initial position—Figure 1 illustrates this.

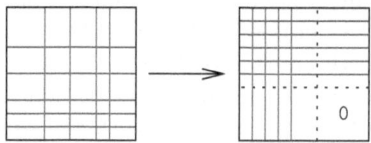

Fig. 1. Permuting the rows and columns of the cover to be in initial positions

Now suppose that α_A consists of e rows and f columns (in the diagram, e horizontal lines and f vertical lines). Clearly $l_A = e + f$. The rearranging of A produces four quadrants; the lower-right quadrant, of size $(|A| - f) \times (|A| - e)$, consists entirely of zeros (since no lines cross it), and since α_A is proper, we know that it is not empty. The upper-right quadrant is of size $e \times (|A| - f)$, and it cannot be covered by fewer than e lines. The lower-left quadrant is of size $(|A| - e) \times f$ and cannot be covered by fewer than f lines.

Claim 8. $\exists$**LA** *shows that if X is an $e \times h$ matrix, and $l_X = e$, then $o_X \geq e$.*

Proof. We state the claim formally as follows:

$$[\forall \alpha \leq r(A)\mathrm{Cover}(A, \alpha) \to \Sigma\alpha \geq r(A)] \to [\exists\beta \leq r(A)\mathrm{Select}(A, \beta) \wedge \Sigma\beta \geq r(A)]$$

and we prove it by induction on the number of rows of A. To this end, let A_n denote the first n rows of A, so that $A_{r(A)} = A$. We now prove the Σ_1^B formula:

$$\exists\alpha, \beta \leq n \left[(\mathrm{Cover}(A_n, \alpha) \wedge \Sigma\alpha < n) \vee (\mathrm{Select}(A_n, \beta) \wedge \Sigma\beta \geq n)\right],$$

which is equivalent to the formula above it for $n = r(A)$. The claim holds for $n = 1$, as in that case we have a single row, which is either zero and hence has a cover of size 0, or the row has a 1, in which case we can select it. For the induction step, suppose the claim holds for $n = k$. Suppose that any cover for A_{k+1} requires $k + 1$ rows. Then, A_k requires k rows (for otherwise, a cover of A_k of size $< k$ plus row $k + 1$ would give a cover of size $\leq k$ of A_{k+1}, which is a contradiction). By IH, A_k has a selection of size at least k.

Let $\mathcal{S} = \{(1, \ell_1), (2, \ell_2), \ldots, (k, \ell_k)\}$ be a selection from A_k. Let $C_\mathcal{S}$ be the set of k vertical lines going through $\mathcal{S}$. Consider row $k + 1$; we know that this row cannot be empty. If there is a 1 in row $k+1$ not covered by $C_\mathcal{S}$, then select that 1. Otherwise, suppose that there are $p > 0$ 1s in row $k + 1$; label their columns as $c_1, c_2, \ldots, c_p$. Let r_i be the row with the unique 1 in $\mathcal{S}$ such that $\ell_{r_i} = c_i$.

Let $\rho_i = \{(k + 1, c_i), (r_i, c_i), (r_i, x_1), (y_1, x_1), (y_1, x_2), \ldots, (a, b)\}$, so that each position has a 1 in A_{k+1}, and in particular (a, b) corresponds to a 1 not covered by $C_\mathcal{S}$. Then, ρ_i describes a re-arrangement of the selection. A ρ_i with (a, b) not covered by $C_\mathcal{S}$ must exist. See Figure 2 for an illustration.   □

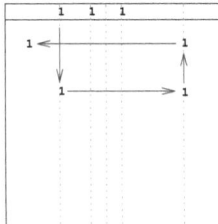

Fig. 2. ρ_1 consisting of five positions

Since the size of selections is invariant under permutations, it follows that $o_A \geq e + f = l_A$. □

As an aside, we present a recursive algorithm for computing minimal covers. It would be interesting to know if it has a polytime proof of correctness. First convert A into diagonal form.

Case 1. If $a = 0$ (so $R = S = 0$, by the diagonal form of A), then $l_A = l_M$, and proceed to compute α_M; output $\hat{\alpha}_A$.

Case 2. If $a \neq 0$, we first examine R to see if the matrix M', consisting of the columns of M minus those columns of M which correspond to 1s in R, has a cover of size $l_M - \Sigma R$ (of course, if $l_M < \Sigma R$, then the answer is "no").

If the answer is "yes", compute the minimal cover of M', $\alpha_{M'}$. Then let α_M be the cover of M consisting of the lines in $\alpha_{M'}$ properly renamed to account for the deletion of columns that transformed M into M', plus the columns of M corresponding the 1s in R. Let $\alpha_A = \hat{\alpha}_A \cup \{$1st column of $A\}$.

If the answer is "no", repeat the same with S: check whether M' has a cover of size $l_M - \Sigma S$. If the answer is "yes" then $\alpha_A = \hat{\alpha}_A \cup \{$1st row of $A\}$.

If the answer is "no", then compute any minimal cover for M, extend it to A, and add the first row and column of A; this results in a cover for A.

3   Equivalence of Various Min-Max Principles

Theorem 9. *The theory* **LA** *proves the equivalence of KMM, Menger's, Hall's and Dilworth's Theorems.*

3.1   Menger's Theorem

Given a graph $G = (V, E)$, an x, y-*path* in G is a sequence of distinct vertices $v_1, v_2, \ldots, v_n$ such that $x = v_1$ and $y = v_n$ and for all $1 \leq i < n, (v_i, v_{i+1}) \in E$. The vertices $\{v_2, \ldots, v_{n-1}\}$ are called *internal vertices*; we say that two x, y-paths are *internally disjoint* if they do not have internal vertices in common. We also say that $S \subseteq V$ is an x, y-*cut* if there is no path from x to y in the graph $G' = (V - S, E')$, where E' is the subset of those edges in E which have no end-point in S.

Let $\kappa(x, y)$ represent the size of the smallest x, y-cut, and let $\lambda(x, y)$ represent the size of the largest set of pairwise internally disjoint x, y-paths. Menger's theorem states that for any graph $G = (V, E)$, if $x, y \in V$ and $(x, y) \notin E$, then the minimum size of an x, y-cut equals the maximum number of pairwise internally disjoint x, y-paths. That is, $\kappa(x, y) = \lambda(x, y)$. Menger's Theorem is of course the familiar Min-Cut Max-Flow Theorem where all edges have capacity 1. For more details on Menger's Theorem turn to [Men27, Gör00, Pym69].

Let β be a matrix that encodes disjoint paths; the rows of β correspond to the paths, and the columns to the vertices of G, where $\beta(i, j) = 1$ if path i contains vertex j. The disjointness can be stated by insisting that each column has at most one 1. Let γ be a $1 \times |V|$ matrix that encodes a cut in the natural way. Maximality and minimality can be expressed as in the KMM Theorem. We leave the details to the reader:

$$\text{Menger}(A) := \text{MaxDisj}(A, x, y, \beta) \wedge \text{MinCut}(A, x, y, \gamma) \to \Sigma\beta = \Sigma\gamma \quad (4)$$

Lemma 10. LA $\cup$ Menger $\vdash$ KMM.

Proof. Consider a bipartite graph $G = (V_0 \cup V_1, E)$, where $E \subseteq V_0 \times V_1$. Let A be the adjacency matrix for G where $A(i, j) = 1$ iff $i \in V_0$ and $j \in V_1$ and $(i, j) \in E$. We now extend G to $G_{x,y}$ by adding two new vertices, x and y, and edges $\{(x, v) : v \in V_1\}$, denoted "red edges", and edges $\{(y, v) : y \in V_0\}$, denoted "green edges."

The adjacency matrix $A_{x,y}$ of $G_{x,y}$ is of size $(|A| + 1) \times (|A| + 1)$ and:

$$A_{x,y}(i, j) = \begin{cases} A(i, j) & \text{for } 1 \le i, j \le |A| \\ 1 & \text{one } \{i, j\} \text{ equals } |A| + 1 \\ 0 & \text{both } \{i, j\} \text{ equal } |A| + 1 \end{cases}$$

i.e., $\lambda ij \langle r(A) + 1, c(A) + 1, \text{cond}(1 \le i, j \le |A|, A(i, j), \text{cond}(i = j = |A| + 1, 0, 1)) \rangle$.

As the graphs related to Menger's Theorem are not bipartite, we convert $A_{x,y}$ to a non-bipartite graph A' as follows:

$$A' = \begin{bmatrix} 0 & A_{x,y} \\ A_{x,y}^T & 0 \end{bmatrix},$$

Let G' be the non-bipartite graph represented by A'. We now finish the proof of the Lemma with a sequence of claims.

Claim 11. LA *proves that if there is a cut in G' of size k, then there is a cut in G' if size k that only cuts the red/green edges, i.e., only those edges that are adjacent to either x or y.*

Proof. Suppose that a black edge is part of a cut. Every x, y-path crosses from V_0 to V_1, and taking off one black edge can only block one x, y-path; the same path is blocked by taking off the corresponding red or green edge. $\square$

Claim 12. LA *proves the following two:*

1. *G has a matching of size k* $\iff$ *G' has k disjoint x, y-paths.*
2. *G has a vertex cover of size k* $\iff$ *G' has an x, y-cut of size k.*

Claim 12 follows directly from Claim 11. On the other hand, the direct consequence of Claim 12 is that the size of a maximum matching in G equals the size of a maximum set of disjoint x, y-paths in G'; and the size of the minimum vertex cover in G equals the size of the minimum x, y-cut in G'. All this is provable in **LA**. This ends the proof of Lemma 10 because by Menger's Theorem, the size of the maximum set of disjoint x, y-paths in G' equals the size of the minimum x, y-cut in G'. Therefore, the size of the maximum matching in G equals the size of the minimum vertex cover in G. $\qquad\square$

Lemma 13. LA $\cup$ KMM $\vdash$ Menger.

Proof. Each path in β must have at least one vertex in the cut γ and no vertex of γ can be in more than one path in β, hence $\lambda \leq \kappa$. The proof of this is identical to the proof of Claim 4.

Thus, it remains to show, using KMM, that $\lambda \geq \kappa$. The proof of this is inspired by [Aha83]; we assume that G is directed, but a simple construction gives us the undirected case as well. Let $A = \{u \in V : (x, u) \in E\}$ and let $B = \{v \in V : (v, y) \in E\}$. Let $X = V - (A \cup B)$, and also split every vertex $v \in V$ into two vertices v', v''. We now construct a new bipartite graph Γ where the two sides are given by $A' \cup X'$ and $B'' \cup X''$, and where the edges are given by $\{(u', v'') : (u, v) \in E\} \cup \{(x', x'') : x \in X\}$. By KMM there is a matching M and a cover C in Γ of the same size. We let $\mathcal{P}$ be the set of paths $\{x_1, x_2, \ldots, x_k\}$ such that $(x_i', x_{i+1}'') \in M$, and we let $\mathcal{S}$ be a cut consisting of $\{v \in V : v', v'' \in C \text{ or } v' \in A' \cap C \text{ or } v'' \in B'' \cap C\}$. **LA** can prove that $\mathcal{P}$ is a set of disjoint paths, and $\mathcal{S}$ is a cut, and $|\mathcal{P}| \geq |\mathcal{S}|$. This is enough to prove the lemma as: $\lambda \geq |\mathcal{P}| \geq |\mathcal{S}| \geq \kappa$. $\qquad\square$

3.2 Hall's Theorem

Let $S_1, S_2, \ldots, S_n$ be n subsets of a given set M. Let D be a set of n elements of M, $D = \{a_1, a_2, \ldots, a_n\}$, such that $a_i \in S_i$ for each $i = 1, 2, \ldots, n$. Then D is said to be a *system of distinct representative* (SDR) for the subsets $S_1, S_2, \ldots, S_n$.

If the subsets $S_1, S_2, \ldots, S_n$ have an SDR, then any k of the sets must contain between them at least k elements. The converse proposition is the combinatorial theorem of P. Hall: suppose that for any $k = 1, 2, \ldots, n$, any $S_{i_1} \cup S_{i_2} \cup \cdots \cup S_{i_k}$ contains at least k elements of M; we call this the *union property*. Then there exists an SDR for these subsets. See [Hal87, EW49, HV50] for more on Hall's theorem.

We formalize Hall's theorem in **LA** with an adjacency matrix A such that the rows of A represent the sets S_i, and the columns of A represent the indices of the elements in M, i.e., the columns are labeled with $[n] = \{1, 2, \ldots, n\}$, and

$A(i, j) = 1 \iff j \in S_i$. Let SDR($A$) be the following Σ_1^B formula which states that A has a system of distinct representatives:

$$\text{SDR}(A) := (\exists P \leq n)(\forall i \leq n)(AP)_{ii} = 1 \tag{5}$$

The next predicate is a Π_2^B formula stating the union property:

$$\text{UnionProp}(A) := (\forall P \leq n \forall k \leq n \exists Q \leq n)[\Sigma \lambda pq \langle 1, k, (PAQ)_{pp} \rangle = k] \tag{6}$$

Therefore, we can state Hall's theorem as a Σ_2^B formula:

$$\text{Hall}(A) := \text{UnionProp}(A) \rightarrow \text{SDR}(A) \tag{7}$$

Lemma 14. LA $\cup$ KMM $\vdash$ Hall.

Proof. Let A be a 0-1 sets/elements incidence matrix of size $n \times n$. Assume that we have UnionProp(A); our goal is to show in **LA**, using KMM, that SDR(A) holds.

Since by Claim 6, every matrix can be put in a diagonal form, using the fact that we have UnionProp(A), it follows that we can find $P, Q \leq n$ such that $\forall k \leq n (PAQ)_{kk} = 1$. Thus we need n lines to cover all the 1s, but by KMM there exists a selection of n 1s no two on the same line, hence $o_A = n$.

But this means that the maximal selection of 1s, no two on the same line, constitutes a permutation matrix P (since A is $n \times n$, and we have n 1s, no two on the same line). Note that AP^T has all ones on the diagonal, and this in turn implies SDR(A). □

Lemma 15. LA $\cup$ Hall $\vdash$ KMM.

Proof. Suppose that we have MinCover(A, α) and MaxSelect(A, β); we want to conclude that $\Sigma \alpha = \Sigma \beta$ using Hall's Theorem.

As usual, let $l_A = \Sigma \alpha$ and $o_A = \Sigma \beta$, and by Claim 4 we already have that **LA** $\vdash o_A \leq l_A$. We now show in **LA** that $o_A \geq l_A$ using Hall's Theorem.

Suppose that the minimum number of lines that cover all the 1s of A consists of e rows and f columns, so that $l_A = e + f$. Both l_A and o_A are invariant under permutations of the rows and the columns of A (Lemma 3), and so we reorder the rows and columns of A so that these e rows and f columns are the initial rows and columns of A',

$$A' = \begin{bmatrix} A_1 & A_2 \\ A_3 & A_4 \end{bmatrix},$$

where A_1 is of size $e \times f$. Now, we shall work with the term rank of A_2 and A_3 in order to show that $o_A \geq l_A$. More precisely, we will show that the maximum number of 1s, no two on the same line, in A_2 is e, while in A_3 it is f.

Let us consider A_2 as an incidence matrix for subsets $S_1, S_2, \ldots, S_e$ of a universe of size $|A| - f$, and A_3^t (which is the transpose of A_3) as an incidence matrix for subsets $S_1', S_2', \ldots, S_f'$ of a universe of size $|A| - e$. It is not difficult to prove that UnionProp(A_2) and UnionProp(A_3^t) holds (and can be proven in **LA**; this is left to the reader), which in turn implies SDR(A_2) and SDR(A_3^t), resp., by Hall's Theorem. But the system of distinct representative of A_2 (resp. A_3^t) implies that $o_{A_2} \geq e$ (resp. $o_{A_3^t} = o_{A_3} \geq f$), and since $o_A \geq o_{A_2} + o_{A_3}$, this yields that $o_A \geq e + f = l_A$. □

3.3 Dilworth's Theorem

Let $\mathcal{P}$ be a *finite partially ordered set* or *poset* (we use a "script $\mathcal{P}$" in order to distinguish it from permutation matrices, denoted with P). We say that $a, b \in \mathcal{P}$ are *comparable elements* if either $a < b$ or $b < a$. A subset C of $\mathcal{P}$ is a *chain* if any two distinct elements of C are comparable. A subset S of $\mathcal{P}$ is an *anti-chain* (also called an *independent set*) if no two elements of S are comparable.

We want to partition a poset into chains; a poset with an anti-chain of size k cannot be partitioned into fewer than k chains, because any two elements of the anti-chain must be in a different partition. Dilworth's Theorem states that the maximum size of an anti-chain equals the minimum number of chains needed to partition $\mathcal{P}$. For more on Dilworth's Theorem see [Dil50, Per63].

In order to formalize Dilworth's theorem in **LA**, we represent finite posets $\mathcal{P} = (X = \{x_1, x_2, \ldots, x_n\}, <)$ with an incidence matrix $A = A_{\mathcal{P}}$ of size $|X| \times |X|$, which expresses the relation $<$ as follows: $A(i, j) = 1 \iff x_i < x_j$. For more material regarding formalizing posets see [Sol11]. Let $1 \times n$ α encode a chain:

$$\text{Chain}(A, \alpha) := (\forall i \neq j \leq n)[\alpha(i) = \alpha(j) = 1 \to A(i, j) = 1 \vee A(j, i) = 1]. \quad (8)$$

In a similar fashion we define an anti-chain β; the only difference is that the succedent of the implication expresses the opposite: $A(i, j) = 0 \wedge A(j, i) = 0$.

Dilworth(A) can be stated as:

$$\text{MinChain}(A, \alpha) \wedge \text{MaxAntiChain}(A, \beta) \to \Sigma\alpha = \Sigma\beta, \quad (9)$$

where MinChain and MaxAntiChain are defined in the same style as the predicates expressing the other theorems. Note that (9) also requires a statement that A encodes a poset, that is, $A(i, i) = 1$, $A(i, j) = 1 \to A(j, i) = 1$, and $A(i, j) \wedge A(j, k) \to A(i, k)$.

Lemma 16. LA $\cup$ KMM $\vdash$ Dilworth

Proof. Suppose that MinChain(A, α) and MaxAntiChain(A, β); we want to use **LA** reasoning and KMM in order to show that $\Sigma\alpha = \Sigma\beta$.

As usual we define a matrix A' whose rows are labeled by the chains in β, and whose columns are labeled by the elements of the poset. As there cannot be more chains than elements in the poset, it follows that the number of rows of A' is bounded by $|A|$ (while the number of columns is exactly $|A|$). The proof of this is similar to the proof of Claim 4.

We have that $A'(i, j) = 1 \iff$ chain i contains element j. Clearly each column contains at least one 1, as β is a partition of the poset. On the other hand, rows may contain more than one 1, as in general chains may have more than one element.

Note that a maximal selection of 1s, no two on the same line, corresponds naturally to a maximal anti-chain; such a selection picks one 1 from each line, and so its size is the number of rows of A'. By KMM, $\Sigma\alpha = o_{A'} = l_{A'} = r(A') = \Sigma\beta$, where $r(A')$ is the number of rows of A'. $\square$

Lemma 17. LA ∪ Dilworth ⊢ KMM

Proof. It is in fact easier to show that that **LA** ∪ Dilworth ⊢ Hall, and since by Lemma 15 we have that **LA** ∪ Hall ⊢ KMM, we will be done.

Assume that we have 0-1 sets/elements $n \times n$ matrix A, and that we have UnionProp(A); our goal is to show in **LA**, using Dilworth, that SDR(A) holds.

Let $S_1, S_2, \ldots, S_n$ be subsets of $[n]$ where $n = |A|$. We define a partial order $\mathcal{P}$ based on A; the universe of $\mathcal{P}$ is $X = \{S_1, S_2, \ldots, S_n\} \cup [n]$. The relation $<_{\mathcal{P}}$ is defined as follows: $i <_{\mathcal{P}} S_j \iff A(i,j) = 1$. Note that the the maximum size of an anti-chain in $\mathcal{P}$ is n. The $[n]$ form an anti-chain of length n, and we cannot add any of the S_j, as some $i \in S_j$, and hence $i <_{\mathcal{P}} S_j$.

By Dilworth we can partition $\mathcal{P}$ into n chains, where each of the chains has two elements $\{i, S_j\}$, giving the set of distinct representatives, and hence SDR(A). □

References

[Aha83] Aharoni, R.: Menger's theorem for graphs containing no infinite paths. European Journal of Combinatorics 4, 201–204 (1983)

[BR91] Brualdi, R.A., Ryser, H.J.: Combinatorial Matrix Theory. Cambridge University Press (1991)

[Bus86] Buss, S.R.: Bounded Arithmetic. Bibliopolis, Naples (1986)

[Bus90] Buss, S.R.: Axiomatizations and conservations results for fragments of Bounded Arithmetic. AMS Contemporary Mathematics 106, 57–84 (1990)

[CN10] Cook, S.A., Nguyen, P.: Logical Foundations of Proof Complexity. Cambridge Univeristy Press (2010)

[Dil50] Dilworth, R.P.: A decomposition theorem for partially ordered sets. Annals of Mathematics 51(1), 161–166 (1950)

[EW49] Everett, C.J., Whaples, G.: Representations of sequences of sets. American Journal of Mathematics 71(2), 287–293 (1949)

[Gör00] Göring, F.: Short proof of Menger's theorem. Discrete Mathematics 219, 295–296 (2000)

[Hal87] Hall, P.: On representatives of subsets. In: Gessel, I., Rota, G.-C. (eds.) Classic Papers in Combinatorics. Modern Birkhäuser Classics, pp. 58–62. Birkhäuser, Boston (1987)

[HV50] Halmos, P.R., Vaughan, H.E.: The marriage problem. American Journal of Mathematics 72(1), 214–215 (1950)

[LC12] Lê, D.T.M., Cook, S.A.: Formalizing randomized matching algorithms. Logical Methods in Computer Science 8, 1–25 (2012)

[Men27] Menger, K.: Zur allgemeinen kurventheorie. Fund. Math. 10, 95–115 (1927)

[Per63] Perles, M.A.: A proof of dilworth's decomposition theorem for partially ordered sets. Israel Journal of Mathematics 1, 105–107 (1963)

[Pym69] Pym, J.S.: A proof of Menger's theorem. Monatshefte für Mathematik 73(1), 81–83 (1969)

[SC04] Soltys, M., Cook, S.: The proof complexity of linear algebra. Annals of Pure and Applied Logic 130(1-3), 207–275 (2004)

[Sol11] Soltys, M.: Feasible proofs of Szpilrajn's theorem: a proof-complexity framework for concurrent automata. Journal of Automata, Languages and Combinatorics (JALC) 16(1), 27–38 (2011)

Approximation Algorithms for Generalized Plant Location

Alexander Souza

apixxo AG
alex.souza@apixxo.com

Abstract. We consider the following GENERALIZED PLANT LOCATION problem: There are m possible plant locations and n customers. Each customer j has a demand d_j of some utility. A plant i can be constructed at cost c_i. Serving a customer j with plant i incurs cost $s_{i,j}$ and yields that the plant produces $u_{i,j}$ units of the demanded utility. The goal is to serve all demands at minimal total cost. In BUDGETED PLANT LOCATION each customer has a budget of b_j and serving this customer with plant i charges the budget an amount $a_{i,j}$.

We give a unified randomized algorithm which is $(4+\varepsilon)\cdot\ln n$-approximate for both versions of the problem for any $\varepsilon > 0$. This result is best possible up to a constant factor. In the budgeted version, we will violate the budgets by factors of at most $2 \cdot (4+\varepsilon) \cdot \ln n$. Our approach is based on LP-relaxations of the problems strengthened by additional KNAPSACK COVER inequalities. This allows us to round the LP by rounding some "large" variables deterministically and the other "small" ones randomly.

1 Introduction

In this paper, we consider generalizations of the classical the PLANT LOCATION problem and give a unified approach for these. GENERALIZED PLANT LOCATION means the following problem: There are m possible plants and n customers. Customer j, say, has a *demand* of d_j units of some utility. Each plant i can be *constructed* at cost c_i. If a constructed plant i *serves* a customer j, it produces $u_{i,j}$ units of the *utility* at *service cost* $s_{i,j}$. The goal is to serve the demand of each customer at minimal total cost. In BUDGETED PLANT LOCATION, each customer j additionally has a *budget* b_j. If a constructed plant i serves a customer j, the budget is charged an amount of $a_{i,j}$ and we require that the total charge can not be more than b_j.

In the classical PLANT LOCATION problem we have $d_j = 1$ and $u_{i,j} = 1$ for all i and j. That is, in this version it is only required that each customer is served by at least one facility. In our version, we have demands d_j and customer-dependent utility production $u_{i,j}$.

PLANT LOCATION is a prominent problem in operations research and has numerous applications since many economic decisions involve selecting and placing facilities to serve certain demands. Examples include manufacturing plants, storage facilities, depots, warehouses, libraries, fire stations and so on. The budgeted version reflects the willingness of customers to pay for some utility.

K. Chatterjee and J. Sgall (Eds.): MFCS 2013, LNCS 8087, pp. 789–800, 2013.

The PLANT LOCATION problem is NP-hard and can thus not be solved optimally in polynomial time, unless P = NP. Instead, we are interested in approximation algorithms. Recall that an algorithm is called ρ-*approximation* if the objective value of its returned solution is always within ρ times the optimal value. A straightforward reduction from SET COVER shows that PLANT LOCATION is inapproximable within $(1 - o(1)) \cdot \ln n$, under mild complexity theoretic assumptions, see Feige [9].

As explained below, we give a unified randomized algorithm which is $O(\log n)$-approximate for both versions of the problem. However, in the budgeted version, we will violate the budget by a factor of at most $O(\log n)$. Our approach is based on LP-relaxations of the problems strengthened by additional valid inequalities, so-called KNAPSACK COVER inequalities introduced by Carr et al. [7].

Related Work. There is a large body of literature on the PLANT LOCATION problem for practical models, exact methods, and approximation algorithms for special cases, especially SET COVER and FACILITY LOCATION. To the best of our knowledge, there are no approximation algorithms known for GENERALIZED PLANT LOCATION and BUDGETED PLANT LOCATION. For a survey of approaches for "simple" versions of the PLANT LOCATION problem, spanning from heuristics to exact methods, we refer to Krarup and Pruzan [13]. For an extensive treatment we also refer to Mirchandani and Francis [15].

Observe that we obtain the SET COVER problem for $d_j = 1$, $s_{i,j} = 0$ and $u_{i,j} \in \{0, 1\}$. Chvatal [8] proved that there is an H_n-approximate GREEDY algorithm for this problem. Here $H_n = \sum_{k=1}^{n} 1/k$ denotes the n-th *Harmonic number* and it is well known that $H_n = \ln n + O(1)$. As stated already, since SET COVER can not be approximated better than $(1 - o(1)) \cdot \ln n$ [9], this is essentially best possible, unless P = NP. The PLANT LOCATION problem, i.e., with $d_j = 1$ and $u_{i,j} = 1$, is also called the DISCRETE MEDIAN problem. Hochbaum [11] gave a (best-possible) H_n-approximation algorithm with running time $O(n^2 m)$ for this and related problems. A version of PLANT LOCATION with demands d_j in which the constraints $\sum_{i=1}^{m} y_{i,j} \geq d_j$ hold for *integer* $y_{i,j}$ was treated by Bar-Ilan et al. [4]. They gave an $O(\log n + \log w_{\max})$-approximation algorithm (even for a budgeted version), where $w_{\max}$ is the largest input parameter. The PLANT LOCATION problem in which the $s_{i,j}$ satisfy the triangle inequality $s_{i,j} \leq s_{i,j'} + s_{i',j'} + s_{i',j}$ for all i, i' and j, j' is called the FACILITY LOCATION problem. This problem can be approximated within small constant factors. The currently best known bound is 1.52 due to Mahdian et al. [14].

As described shortly, the KNAPSACK COVER inequalities [7] allow us to derive a randomized algorithm for GENERALIZED PLANT LOCATION. These valid inequalities have earlier proved useful for strengthening LP-relaxations of covering problems. In specific, in the GENERAL COVERING problem one is asked to minimize $c^{\top} x$ subject to $Ax \geq b$ and $x \leq d$, where $x \in \mathbb{N}^m$ and the entries of $A \in \mathbb{R}^{n \times m}$, $b \in \mathbb{R}^n$, $c \in \mathbb{R}^m$, and $d \in \mathbb{R}^m$ are non-negative. In a seminal paper Carr et al. [7] introduced the KNAPSACK COVER valid inequalities, which enabled them to prove strong bounds on the integrality gaps of this and related problems. Kolliopoulos and Young [12] gave a $O(\log n)$-approximation algorithm, by giving an LP-formulation with additional KNAPSACK COVER inequalities and a randomized algorithm, which is then derandomized with the method of conditional probabilities.

Moreover, the inequalities were used for obtaining an $O(1)$-approximation algorithm for the GENERALIZED MIN-SUM SET COVER problem by Bansal et al. [2]. In this problem, there is a universe of elements and a collection of sets given, where each set S has an associated covering requirement of $k(S)$. The goal is to output an ordering of the elements such that the total cover time of all the sets is minimized, where the cover time of a set S is the first time when $k(S)$ elements from S have been output.

In the GENERAL SCHEDULING problem there is one machine and n jobs given, where job j has a size p_j and a time-dependent weight function $w_j(t)$. The objective is to schedule the jobs as to minimize the total weight. Special cases include TOTAL WEIGHTED TARDINESS SCHEDULING and TOTAL WEIGHTED FLOW TIME SCHEDULING. Bansal and Pruhs [3] used KNAPSACK COVER inequalities and a sophisticated rounding scheme to derive an approximation algorithm with ratio $O(\log \log n \max_j p_j)$.

In the GENERALIZED CACHING problem, there is a cache with size k and pages with arbitrary sizes and fetching cost given. The goal is to serve a sequence of requests to pages that must be made available in cache at minimal total fetching cost. Bansal et al. [1] used the inequalities [7] to give an $O(\log^2 k)$-approximate algorithm.

Our Contribution. As explained above, to the best of our knowledge, there is no approximation algorithm known for the GENERALIZED PLANT LOCATION problem. However, there are approximations for special cases and relaxations, e.g., [8,11,14,4]. We close this gap by giving a randomized expected $(4+\varepsilon) \cdot \ln n$-approximation algorithm for any $\varepsilon > 0$ for the general case. By inapproximability of SET COVER [9] this is best possible up to a constant factor, unless $\mathsf{P} = \mathsf{NP}$. The two main technical ingredients of our algorithm are:

(I) We give an LP-relaxation of the GENERALIZED PLANT LOCATION problem strengthened by KNAPSACK COVER valid inequalities [7]. As stated above, these valid inequalities have proved useful for several other (covering) problems, e.g., [12,2,3,1,6]. This strengthened formulation allows us to round a fractional solution randomly thus yielding logarithmic integrality gap.

(II) The second tool we use (for the analysis of the rounding) is Bernstein's inequality [5], which is a Chernoff-type bound on the concentration of measure of sums of independent random variables. A property of this bound is that it depends on the variance of the random sum under investigation and that the absolute values of the summands can be arbitrary constants. The dependence on the variance is useful: We use the KNAPSACK COVER inequalities to derive a rounding scheme in which some "large" variables are rounded deterministically and the other "small" ones are rounded randomly. For the latter we can prove that the variance of the associated sums is not "too large" and we derive a sufficiently strong bound on the concentration of measure.

We obtain the algorithm as follows: In Section 2 we consider the case $n = 1$, in which the GENERALIZED PLANT LOCATION problem becomes the KNAPSACK COVER problem. As a first step, we give a randomized rounding algorithm for this special case. In Section 3, we observe that the GENERALIZED PLANT LOCATION problem can be seen as n simultaneous KNAPSACK COVER problems with the additional requirement of constructing plants. By solving all of these problems feasibly with high probability we obtain which plant shall serve which customer. Having these decisions made, we construct a plant if it serves some customer. With the help of the ingredients (I) and (II) we are able to prove that the algorithm is expected $(4 + \varepsilon) \cdot \ln n$-approximate. In Section 4, we turn our attention to BUDGETED PLANT LOCATION. We find that the same rounding algorithm, which simply uses an adjusted LP-relaxation, also produces an expected $(4 + \varepsilon) \cdot \ln n$-approximate solution, which violates the budgets by factors of at most $2 \cdot (4 + \varepsilon) \cdot \ln n$ with high probability.

Preliminaries. The possible plants are indexed by the set $P = \{1, \ldots, m\}$ and the customers are indexed by $C = \{1, \ldots, n\}$. We are given a *demand vector* $d = (d_1, \ldots, d_j, \ldots, d_n)$ with $d_j \geq 0$ for all $j \in C$. Furthermore, we are given a *cost vector* $c = (c_1, \ldots, c_i, \ldots, c_m)$ and *constructing* a plant $i \in P$ incurs cost $c_i \geq 0$. For each plant $i \in P$ there is a *service cost vector* $s_i = (s_{i,1}, \ldots, s_{i,j}, \ldots, s_{i,n})$. A constructed plant i can *serve* a customer j at cost $s_{i,j}$. Each plant $i \in P$ is associated a *utility vector* $u_i = (u_{i,1}, \ldots, u_{i,j}, \ldots, u_{i,n})$ with $u_{i,j} \geq 0$ for all $j \in C$. We may assume that $\sum_{i \in P} u_{i,j} \geq d_j$ for all $j \in C$, since the problem is otherwise clearly infeasible. If plant i serves customer j, $u_{i,j}$ units of the demanded utility are produced.

Our goal is to decide which plants to construct and which plants shall serve which customer in order to cover the demand at minimal total cost. More precisely, introducing the variables $x_i \in \{0, 1\}$ of a vector $x = (x_1, \ldots, x_i, \ldots, x_m)$ that indicate if plant i is constructed and the variables $y_{i,j} \in \{0, 1\}$ of a vector $y = (y_{1,1}, \ldots, y_{i,j}, \ldots, y_{m,n})$ that indicate if plant i serves customer j, we define GENERALIZED PLANT LOCATION:

$$\text{minimize} \quad \sum_{i=1}^{m} c_i x_i + \sum_{i=1}^{m} \sum_{j=1}^{n} s_{i,j} y_{i,j} \tag{1}$$

$$\text{subject to} \quad \sum_{i=1}^{m} u_{i,j} y_{i,j} \geq d_j \qquad j = 1, \ldots, n, \tag{2}$$

$$x_i - y_{i,j} \geq 0 \qquad i = 1, \ldots, m, j = 1, \ldots, n, \tag{3}$$

$$x_i \in \{0, 1\} \qquad i = 1, \ldots, m, \tag{4}$$

$$y_{i,j} \in \{0, 1\} \qquad i = 1, \ldots, m, j = 1, \ldots, n. \tag{5}$$

Furthermore, in the case of BUDGETED PLANT LOCATION, each customer $j \in C$ has a *budget* $b_j \geq 0$ gathered in a vector $b = (b_1, \ldots, b_j, \ldots, b_n)$. For each plant $i \in P$, there is a vector $a_i = (a_{i,1}, \ldots, a_{i,j}, \ldots, a_{i,n})$ given, where $a_{i,j} \geq 0$ for all $i \in P$ and $j \in C$. A plant i which serves customer j charges $a_{i,j}$ on the budget b_j. Now we have the additional constraints that

$$\sum_{i=1}^{m} a_{i,j} y_{i,j} \leq b_j \qquad j = 1, \ldots, n.$$

As a shorthand we write $\mathrm{cost}(x, y) = \sum_{i=1}^{m} c_i x_i + \sum_{i=1}^{m} \sum_{j=1}^{n} s_{i,j} y_{i,j}$.

2 Knapsack Cover

Using similar notation, in the KNAPSACK COVER problem we are given a knapsack with *demand* $d \geq 0$ and items $P = \{1, \ldots, m\}$. Item $i \in P$ has *utility* $u_i \geq 0$ and *cost* $c_i \geq 0$. Find a subset of items with total utility covering the demand at minimal total cost. That is, KNAPSACK COVER is the problem:

$$\text{minimize} \qquad \sum_{i=1}^{m} c_i x_i \tag{6}$$

$$\text{subject to} \qquad \sum_{i=1}^{m} u_i x_i \geq d, \tag{7}$$

$$x_i \in \{0, 1\} \qquad i = 1, \ldots, m. \tag{8}$$

We write $\mathrm{cost}(x) = \sum_{i=1}^{m} c_i x_i$. Using relaxed variables $x_i \in [0, 1]$ instead of the $x_i \in \{0, 1\}$ yields an arbitrarily large integrality gap, see [7]. For any set $Q \subseteq P$ define

$$d(Q) = d - \sum_{i \in Q} u_i \quad \text{and} \quad u_i(Q) = \min\{u_i, d(Q)\}$$

as the *residual demand* $d(Q)$ and *residual utility* $u_i(Q)$, respectively. An alternative formulation with bounded integrality gap due to [7] is:

$$\text{minimize} \qquad \sum_{i=1}^{m} c_i x_i \tag{9}$$

$$\text{subject to} \qquad \sum_{i \in P-Q} u_i(Q) x_i \geq d(Q) \qquad \text{for all } Q \subseteq P, \tag{10}$$

$$x_i \in [0, 1] \qquad i = 1, \ldots, m. \tag{11}$$

The constraint $\sum_{i \in P-Q} u_i(Q) x_i \geq d(Q)$ states that, even if we choose the items in Q, the remaining items $P - Q$ must still cover the residual demand $d(Q)$.

Carr et al. [7] proved that the integrality gap of this relaxation is 2. We are not aware of an algorithm that solves this relaxation exactly. However, Carr et al. [7] define the following type of solution, which is sufficient for our purpose and which can be found in polynomial time with the ELLIPSOID algorithm. For $c > 0$, call a vector $\bar{x}$ a c-relaxed solution for (9) if $\text{cost}(\bar{x}) \leq \text{cost}(x)$, where x is an optimal (fractional) solution for (9), and $\bar{x}$ satisfies (11) and the KNAPSACK COVER inequalities (10) for the set $Q = \{i \in P : \bar{x}_i \geq c\}$. A c-relaxed solution can be found in polynomial time, because the separation problem of the ELLIPSOID algorithm can be solved in polynomial time: Given a solution candidate $\bar{x}$ already satisfying (11), we can compute the set Q in linear time and check if (10) is also satisfied for this particular set. If so, we can stop, if not, we have found a violated constraint, as needed by the ELLIPSOID algorithm.

Algorithm 2.1. ROUNDKC

Input. $c \in (0, 1]$.
Output. $X = (X_1, \dots, X_m) \in \{0, 1\}^m$.

Step 1. Find a c-relaxed solution for (9) $\bar{x} = (\bar{x}_1, \dots, \bar{x}_m) \in [0, 1]^m$.
Step 2. For $i = 1, \dots, m$ let $C_i \sim \text{Uni}(0, c)$ and let

$$
X_i = \begin{cases} 1 & \text{if } \bar{x}_i \geq C_i, \\ 0 & \text{otherwise.} \end{cases}
$$

Step 3. Return $X = (X_1, \dots, X_m)$.

Theorem 1. *Let* $c \in (0, 1]$. *The algorithm* ROUNDKC(c) *runs in polynomial time and returns an expected* $1/c$-*approximate solution, which is feasible for* KNAPSACK COVER *with probability at least* $1 - 2\exp(-(1 - c)^2/4c)$.

The proof of Theorem 1 can be found below and uses the following intermediate result.

Lemma 1. *Let* $c \in (0, 1]$. *With probability at least* $1 - 2\exp(-(1 - c)^2/4c)$, ROUNDKC(c) *returns a feasible solution for* KNAPSACK COVER.

Proof. The proof uses Bernstein's inequality [5], which is stated in a version taken from [10] below.

Theorem 2 (Bernstein [5]). *Let* $Z_1, \dots, Z_m$ *be independent random variables with* $\mathbb{E}[Z_i] = 0$ *and* $|Z_i| \leq \delta$. *For* $Z = \sum_{i=1}^m Z_i$ *and* $\lambda > 0$ *it holds that*

$$
\Pr[|Z| > \lambda] \leq 2\exp\left(-\frac{1}{2}\frac{\lambda^2}{\text{Var}[Z] + \lambda\delta}\right).
$$

We have to show that $U = \sum_{i=1}^{m} u_i X_i \geq d$ with probability at least $1 - 2\exp(-(1-1/c)^2/4c)$, or equivalently $\Pr[U < d] \leq 2\exp(-(1-1/c)^2/4c)$. First observe that the items $i \in Q \subseteq P$ with $\bar{x}_i \geq c$ are rounded to one deterministically, i.e., $X_i = 1$. Now, if $\sum_{i \in Q} u_i X_i \geq d$, there is nothing to show as the solution is already feasible. Thus we assume $\sum_{i \in Q} u_i X_i < d$. For the set Q, since $\bar{x}$ is c-relaxed, we have the constraint

$$\sum_{i \in P-Q} u_i(Q)\bar{x}_i \geq d(Q).$$

For the items $i \in P - Q$ with $\bar{x}_i \leq c$ we have

$$\mathbb{E}[X_i] = \Pr[X_i = 1] = \Pr[\bar{x}_i \geq C_i] = \bar{x}_i/c.$$

Therefore, by defining $U(Q) = \sum_{i \in P-Q} u_i(Q)X_i$ we have

$$\mathbb{E}[U(Q)] = \sum_{i \in P-Q} u_i(Q)\mathbb{E}[X_i] = \frac{1}{c} \cdot \sum_{i \in P-Q} u_i(Q)\bar{x}_i \geq \frac{1}{c} \cdot d(Q).$$

Define $Z_i = u_i(Q)(X_i - \mathbb{E}[X_i])$ for all $i \in P - Q$ and $Z(Q) = \sum_{i \in P-Q} Z_i = U(Q) - \mathbb{E}[U(Q)]$. Observe that the Z_i are independent, $\mathbb{E}[Z_i] = 0$, and $|Z_i| \leq \max_{i \in P-Q} u_i(Q) =: \delta$. Furthermore, by independence, we have $\mathrm{Var}[Z(Q)] = \mathrm{Var}[U(Q)] = \sum_{i \in P-Q} u_i^2(Q)\mathrm{Var}[X_i] \leq \delta \cdot \mathbb{E}[U(Q)]$, since $\mathrm{Var}[X_i] = 1/c \cdot \bar{x}_i(1 - 1/c \cdot \bar{x}_i) \leq 1/c \cdot \bar{x}_i = \mathbb{E}[X_i]$. Using $\mathbb{E}[U(Q)] \geq 1/c \cdot d(Q) \geq 1/c \cdot \delta$ and Bernstein's inequality yields

$$\begin{aligned}
\Pr[U(Q) < d(Q)] &\leq \Pr[U(Q) < c \cdot \mathbb{E}[U(Q)]] \\
&= \Pr[\mathbb{E}[U(Q)] - U(Q) > \mathbb{E}[U(Q)](1-c)] \\
&\leq \Pr[|Z(Q)| > \mathbb{E}[U(Q)](1-c)] \\
&\leq 2\exp\left(-\frac{1}{2}\frac{\mathbb{E}[U(Q)]^2(1-c)^2}{\mathrm{Var}[Z(Q)] + \mathbb{E}[U(Q)](1-c)\delta}\right) \\
&\leq 2\exp\left(-\frac{(1-c)^2}{4c}\right).
\end{aligned}$$

With $X_i = 1$ for $i \in Q$ we find

$$\begin{aligned}
\Pr[U < d] &= \Pr\left[\sum_{i \in P-Q} u_i X_i < d - \sum_{i \in Q} u_i\right] \\
&\leq \Pr\left[\sum_{i \in P-Q} u_i(Q)X_i < d - \sum_{i \in Q} u_i\right] \\
&= \Pr[U(Q) < d(Q)] \leq 2\exp\left(-\frac{(1-c)^2}{4c}\right)
\end{aligned}$$

completing the proof.

Proof (Proof of Theorem 1). Using the ELLIPSOID algorithm for solving the fractional relaxation of KNAPSACK COVER, ROUNDKC(c) runs in polynomial time. The probability that X is feasible is stated in Lemma 1. Let $\bar{x}$ be the c-relaxed solution found by the algorithm and let x^* be an optimal solution for KNAPSACK COVER. We clearly have $\text{cost}(\bar{x}) \leq \text{cost}(x^*)$.

Let $Q = \{i \in P : \bar{x}_i \geq c\}$. Recall that variables X_i with $i \in Q$ are rounded to one deterministically. Hence $X_i = 1 = c/c \leq \bar{x}_i/c$. For the variables X_i with $i \in P - Q$ we have $X_i = 1$ with probability equal to $\bar{x}_i/c$. Therefore, for any X found by the algorithm we have

$$\mathbb{E}\left[\text{cost}(X)\right] = \mathbb{E}\left[\sum_{i=1}^{m} c_i X_i\right] = \sum_{i \in Q} c_i \mathbb{E}\left[X_i\right] + \sum_{i \in P-Q} c_i \mathbb{E}\left[X_i\right]$$

$$\leq \sum_{i \in Q} \frac{1}{c} \cdot c_i \bar{x}_i + \sum_{i \in P-Q} \frac{1}{c} \cdot c_i \bar{x}_i$$

$$= \frac{1}{c} \cdot \text{cost}(\bar{x}) \leq \frac{1}{c} \cdot \text{cost}(x^*)$$

as claimed.

3   Generalized Plant Location

Here we give an LP relaxation of GENERALIZED PLANT LOCATION, which also uses KNAPSACK COVER inequalities. For any $Q \subseteq P$ and customer j define

$$d_j(Q) = d_j - \sum_{i \in Q} u_{i,j} \quad \text{and} \quad u_{i,j}(Q) = \min\{u_{i,j}, d_j(Q)\}$$

as the *residual demand* $d_j(Q)$ and *residual utility* $u_{i,j}(Q)$ for customer j, respectively. This yields the following relaxation for GENERALIZED PLANT LOCATION:

$$\text{minimize} \quad \sum_{i=1}^{m} c_i x_i + \sum_{i=1}^{m} \sum_{j=1}^{n} s_{i,j} y_{i,j} \tag{12}$$

$$\text{subject to} \quad \sum_{i=1}^{m} u_{i,j}(Q) y_{i,j} \geq d_j(Q) \qquad \text{for all } Q \subseteq P, j = 1, \ldots, n, \tag{13}$$

$$x_i - y_{i,j} \geq 0 \qquad i = 1, \ldots, m, j = 1, \ldots, n, \tag{14}$$

$$x_i \in [0,1] \qquad i = 1, \ldots, m, \tag{15}$$

$$y_{i,j} \in [0,1] \qquad i = 1, \ldots, m, j = 1, \ldots, n. \tag{16}$$

Consider the algorithm ROUNDPL. It is important to note that it uses the random variables $C_1, \ldots, C_m$, i.e., *one* variable per potential plant for the *whole* set of customers. For $c > 0$, call a vector $(\bar{x}, \bar{y})$ a *c-relaxed solution* for (12) if $\text{cost}(\bar{x}, \bar{y}) \leq \text{cost}(x, y)$, where (x, y) is a (fractional) optimum solution for (12),

---

Algorithm 3.1. ROUNDPL

Input. $c \in (0, 1]$.

Output. $X = (X_1, \ldots, X_m) \in \{0, 1\}^m$, $Y = (Y_{1,1}, \ldots, Y_{m,n}) \in \{0, 1\}^{m \times n}$.

Step 1. Find a c-relaxed solution for (12)

$$\bar{x} = (\bar{x}_1, \ldots, \bar{x}_m) \in [0, 1]^m, \bar{y} = (\bar{y}_{1,1}, \ldots, \bar{y}_{m,n}) \in [0, 1]^{m \times n}.$$

Step 2. For $i = 1, \ldots, m$ let $C_i \sim \mathrm{Uni}\,(0, c)$.

Step 3. For $i = 1, \ldots, m$ and $j = 1, \ldots, n$ let

$$Y_{i,j} = \begin{cases} 1 & \text{if } \bar{y}_{i,j} \geq C_i, \\ 0 & \text{otherwise.} \end{cases}$$

Step 4. For $i = 1, \ldots, m$ let $X_i = \max_{j=1,\ldots,n} Y_{i,j}$.

Step 5. Return $X = (X_1, \ldots, X_m)$, $Y = (Y_{1,1}, \ldots, Y_{m,n})$.

---

and $(\bar{x}, \bar{y})$ satisfies (14), (15), (16), and for each j the KNAPSACK COVER inequalities (13) for the set $Q = \{i \in P : \bar{y}_{i,j} \geq c\}$. A c-relaxed solution can be found in polynomial time with the ELLIPSOID algorithm, see Section 2.

Theorem 3. *Let $c \in (0, 1]$. The algorithm* ROUNDPL(c) *runs in polynomial time and returns an expected $1/c$-approximate solution, which is feasible for* GENERALIZED PLANT LOCATION *with probability at least $1 - 2n \exp(-(1 - c)^2/4c)$.*

Proof. By ignoring the x_i for the moment consider the following problem:

$$\text{minimize} \quad \sum_{i=1}^{m} \sum_{j=1}^{n} s_{i,j} y_{i,j} \tag{17}$$

$$\text{subject to} \quad \sum_{i=1}^{m} u_{i,j}(Q) y_{i,j} \geq d_j(Q) \qquad \text{for all } Q \subseteq P, j = 1, \ldots, n, \tag{18}$$

$$y_{i,j} \in \{0, 1\} \qquad i = 1, \ldots, m, j = 1, \ldots, n. \tag{19}$$

Observe that this problem consists of n many KNAPSACK COVER problems – one for each customer. Let $\bar{y}$ be a c-relaxed solution over the variables $\bar{y}_{i,j} \in [0, 1]$. Let $C_i \sim \mathrm{Uni}\,(0, c)$ and round the $Y_{i,j}$ as given in the algorithm.

Let the variable $I_j \in \{0, 1\}$ indicate if the variables $Y_{i,j}$ with $i \in P$ are *infeasible* for the KNAPSACK COVER problem of customer j. Thus $I = \sum_{j=1}^{n} I_j$ counts the number of infeasibly solved problems. We have $\Pr[I > 0] \leq \mathbb{E}[I] = \sum_{j=1}^{n} \mathbb{E}[I_j] = \sum_{j=1}^{n} \Pr[I_j = 1] \leq 2n \exp(-(1 - c)^2/4c)$.

For each *fixed* customer j, the $Y_{i,j}$ are independent by construction. Thus Lemma 1 also applies to each individual KNAPSACK COVER problem and we have the bound $\Pr[I_j = 1] \leq 2 \exp(-(1 - c)^2/4c)$. As there are n such problems, the union bound gives that the vector Y is feasible for the whole problem with probability at least $1 - 2n \exp(-(1 - c)^2/4c)$.

For any fixed j let $Q = \{i \in P : \bar{y}_{i,j} \geq c\}$. Thus the $Y_{i,j} = 1 = c/c \leq \bar{y}_{i,j}/c$ deterministically for $i \in Q$ and $Y_{i,j} = 1$ with probability $\bar{y}_{i,j}/c$ for $i \in P - Q$. Hence $\mathbb{E}\left[\sum_{i=1}^{m} s_{i,j}Y_{i,j}\right] = \sum_{i=1}^{m} s_{i,j}\mathbb{E}\left[Y_{i,j}\right] \leq \frac{1}{c} \cdot \sum_{i=1}^{m} s_{i,j}\bar{y}_{i,j}$.

Now return to the formulation of the problem including the x_i. Let $(\bar{x}, \bar{y})$ be the c-relaxed solution found by ROUNDPL(c). Let $\bar{y}_{i,\max} = \max_{j=1,\dots,n} \bar{y}_{i,j}$. Since there is *one* random variable C_i per plant i, we have

$$\mathbb{E}\left[X_i\right] = \Pr\left[X_i = 1\right] = \Pr\left[\bar{y}_{i,\max} \geq C_i\right] = \begin{cases} 1 \leq \bar{y}_{i,\max}/c & \text{if } \bar{y}_{i,\max} > c, \\ \bar{y}_{i,\max}/c & \text{if } \bar{y}_{i,\max} \leq c. \end{cases}$$

By the constraints $x_i - y_{i,j} \geq 0$ we have $\bar{y}_{i,\max} \leq \bar{x}_i$.

Let (x^*, y^*) be an optimum solution for the GENERALIZED PLANT LOCATION problem. We have that $\text{cost}(\bar{x}, \bar{y}) \leq \text{cost}(x^*, y^*)$. Now we calculate and obtain

$$\mathbb{E}\left[\text{cost}(X, Y)\right] = \mathbb{E}\left[\sum_{i=1}^{m} c_i X_i + \sum_{i=1}^{m}\sum_{j=1}^{n} s_{i,j}Y_{i,j}\right]$$

$$= \sum_{i=1}^{m} c_i \mathbb{E}\left[X_i\right] + \sum_{j=1}^{n} \mathbb{E}\left[\sum_{i=1}^{m} s_{i,j}Y_{i,j}\right]$$

$$\leq \frac{1}{c} \cdot \sum_{i=1}^{m} c_i \bar{x}_i + \frac{1}{c} \cdot \sum_{j=1}^{n}\sum_{i=1}^{m} s_{i,j}\bar{y}_{i,j}$$

$$= \frac{1}{c} \cdot \text{cost}(\bar{x}, \bar{y}) \leq \frac{1}{c} \cdot \text{cost}(x^*, y^*)$$

as claimed.

The notion *high probability* refers to probability converging to one as n tends to infinity.

Corollary 1. *Choose* $c = 1/((4 + \varepsilon) \cdot \ln n)$ *for any* $\varepsilon > 0$ *and* ROUNDPL(c) *is expected* $(4 + \varepsilon) \cdot \ln n$-*approximate and feasible with high probability.*

4   Budgeted Plant Location

Here we add to the GENERALIZED PLANT LOCATION problem the constraints that

$$\sum_{i=1}^{m} a_{i,j} y_{i,j} \leq b_j \qquad j = 1, \dots, n, \tag{20}$$

where the $a_{i,j} \geq 0$ are arbitrary coefficients and b_j is the *budget* of customer j. The resulting problem is called BUDGETED PLANT LOCATION and it is not hard to see that it is already NP-complete to even decide if there exists a *feasible* solution. To overcome this difficulty we will consider a relaxation and allow that these constraints be violated somewhat. More precisely, a vector (x, y) is

λ-*feasible* if $\sum_{i=1}^{m} a_{i,j} y_{i,j} \leq \lambda b_j$ for all $j \in C$ and the remaining constraints are satisfied. Define a *c-relaxed solution* analogously as above.

We define the algorithm ROUNDBPL by adjusting ROUNDPL in the following manner: Firstly, we add the constraints (20) to the linear programming relaxation of the problem without budgets. Secondly, we introduce the constraint $y_{i,j} = 0$ for all $i \in P$ and $j \in C$ for which $a_{i,j} > b_j$. This requirement is valid since we can clearly not have any such $y_{i,j} = 1$ in any feasible integral solution. In the sequel we will assume that this relaxation admits a feasible solution, since there is clearly no integral feasible solution otherwise. The fractional solution is then rounded as in the version without budgets by using the parameter c and the random variables $C_i \sim \mathrm{Uni}\,(0, c)$ for $i = 1, \ldots, m$.

Theorem 4. *Let $c \in (0, 1]$. The algorithm* ROUNDBPL(c) *runs in polynomial time and returns an expected $1/c$-approximate solution, which is $2/c$-feasible for* BUDGETED PLANT LOCATION *with probability at least $1 - 4n \exp(-(1-c)^2/4c)$.*

Proof. Let $j \in C$ and define $A_j = \sum_{i=1}^{m} a_{i,j} Y_{i,j}$, where the $Y_{i,j}$ are the respective random variables in the algorithm. We show that $\Pr\,[A_j > 2b_j/c] \leq 2\exp\,(-1/4c)$.

Then, repeating the proof of Theorem 3 yields that the probability that the returned solution is $1/c$-approximate and $2/c$-feasible is at least $1 - 4n \exp(-(1-c)^2/4c)$ as claimed.

Let $(\bar{x}, \bar{y})$ be the c-relaxed solution determined by ROUNDBPL(c). For any $i \in P$ and $j \in C$ we have that $\bar{y}_{i,j} > 0$ implies $a_{i,j} \leq b_j$ by construction. This property helps bounding the variance of A_j. Notice that for any fixed $j \in C$ the $Y_{i,j}$ are independent. Furthermore notice that $\mathbb{E}\,[Y_{i,j}] = \Pr\,[Y_{i,j} = 1] \leq \bar{y}_{i,j}/c$ by construction of the algorithm. Then we find

$$\mathrm{Var}\,[A_j] = \mathrm{Var}\left[\sum_{i=1}^{m} a_{i,j} Y_{i,j}\right] = \sum_{i=1}^{m} a_{i,j}^2 \mathrm{Var}\,[Y_{i,j}]$$

$$\leq b_j \cdot \sum_{i=1}^{m} a_{i,j} \mathbb{E}\,[Y_{i,j}] \leq b_j \cdot \sum_{i=1}^{m} a_{i,j} \frac{\bar{y}_{i,j}}{c} \leq \frac{b_j^2}{c},$$

where we have used $\mathbb{E}\,[A_j] \leq \sum_{i=1}^{m} a_{i,j} \bar{y}_{i,j}/c \leq b_j/c$ by feasibility of $(\bar{x}, \bar{y})$.

Now, using this property again and Bernstein's inequality we have

$$\Pr\,[A_j > 2b_j/c] = \Pr\,[A_j - \mathbb{E}\,[A_j] > 2b_j/c - \mathbb{E}\,[A_j]] \leq \Pr\,[A_j - \mathbb{E}\,[A_j] > b_j/c]$$

$$\leq 2\exp\left(-\frac{1}{2}\frac{b_j^2}{c^2(\mathrm{Var}\,[A_j] + b_j^2/c)}\right) \leq 2\exp\left(-\frac{1}{4c}\right)$$

and the result follows.

Corollary 2. *Choose $c = 1/((4+\varepsilon) \cdot \ln n)$ for any $\varepsilon > 0$ and* ROUNDBPL(c) *is expected $(4+\varepsilon) \cdot \ln n$-approximate and $2 \cdot (4+\varepsilon) \cdot \ln n$-feasible with high probability.*

Acknowledgement. We are grateful to Nikhil Bansal, firstly for bringing the KNAPSACK COVER inequalities to our attention and secondly for sharing an early version of [3].

References

1. Bansal, N., Buchbinder, N., Naor, J.: Randomized competitive algorithms for generalized caching. In: Proceedings of the 40th Annual ACM Symposium on Theory of Computing, STOC 2008, pp. 235–244 (2008)
2. Bansal, N., Gupta, A., Krishnaswamy, R.: A constant factor approximation algorithm for generalized min-sum set cover. In: Proceedings of the 21st ACM-SIAM Symposium on Discrete Algorithms, SODA 2010, pp. 1539–1545 (2010)
3. Bansal, N., Pruhs, K.: The geometry of scheduling. In: Proceedings of the 51st Annual Symposium on Foundations of Computer Science, FOCS 2010, pp. 407–414 (2010)
4. Bar-Ilan, J., Kortsarz, G., Peleg, D.: Generalized submodular cover problems and applications. Theoretical Computer Science 250, 179–200 (2001)
5. Bernstein, S.N.: On a modification of chebyshev's inequality and of the error formula of laplace. Ann. Sci. Inst. Sav. Ukraine, Sect. Math. 1 4(5) (1924)
6. Carnes, T., Shmoys, D.: Primal-dual schema for capacitated covering problems. In: 13th MPS Conference on Integer Programming and Combinatorial Optimization, IPCO 2008, pp. 288–302 (2008)
7. Carr, R.D., Fleischer, L.K., Leung, V.J., Phillips, C.A.: Strengthening integrality gaps for capacitated network design and covering problems. In: Proceedings of the 6th ACM-SIAM Symposium on Discrete Algorithms, SODA 2000, pp. 106–115 (2000)
8. Chvatal, V.: A greedy heuristic for the set-covering problem. Mathematics of Operations Research 4(3), 233–235 (1979)
9. Feige, U.: A threshold of $\ln n$ for approximating set cover. Journal of the ACM 45, 634–652 (1998)
10. Hazewinkel, M.: Encyclopaedia of Mathematics (2002), http://eom.springer.de
11. Hochbaum, D.S.: Heuristics for the fixed cost median problem. Mathematical Programming 22, 148–162 (1982)
12. Kolliopoulos, S.G., Young, N.E.: Tight approximation results for general covering integer programs. In: Proceedings of the 42nd Annual Symposium on Foundations of Computer Science, FOCS 2001, pp. 522–528 (2001)
13. Krarup, J., Pruzan, P.M.: The simple plant location problem: Survey and synthesis. European Journal of Operational Research 12, 36–81 (1983)
14. Mahdian, M., Ye, Y., Zhang, J.: Approximation algorithms for metric facility location problems. SIAM Journal on Computing 36(2), 411–432 (2006)
15. Mirchandani, P.B., Francis, R.L.: Discrete Location Theory. John Wiley and Sons (1990)

Hardness of Classically Simulating Quantum Circuits with Unbounded Toffoli and Fan-Out Gates

Yasuhiro Takahashi[1], Takeshi Yamazaki[2], and Kazuyuki Tanaka[2]

[1] NTT Communication Science Laboratories, NTT Corporation,
Atsugi 243-0198, Japan
`takahashi.yasuhiro@lab.ntt.co.jp`
[2] Mathematical Institute, Tohoku University,
Sendai 980-8578, Japan
`{yamazaki,tanaka}@math.tohoku.ac.jp`

Abstract. We study the classical simulatability of constant-depth quantum circuits followed by only one single-qubit measurement, where they consist of universal gates on at most two qubits and additional gates on an unbounded number of qubits. First, we consider unbounded Toffoli gates as additional gates and deal with the weak simulation, i.e., sampling the output probability distribution. We show that there exists a constant-depth quantum circuit with only one unbounded Toffoli gate that is not weakly simulatable, unless $\mathsf{BQP} \subseteq \mathsf{PostBPP} \cap \mathsf{AM}$. Then, we consider unbounded fan-out gates as additional gates and deal with the strong simulation, i.e., computing the output probability. We show that there exists a constant-depth quantum circuit with only two unbounded fan-out gates that is not strongly simulatable, unless $\mathsf{P} = \mathsf{PP}$. These results are in contrast to the fact that any constant-depth quantum circuit without additional gates is strongly and weakly simulatable.

1   Introduction and Summary of Results

An important problem in quantum information processing is to understand the difference between the computational power of a quantum computer and that of a classical computer. For considering this problem, it is known to be useful to study the classical simulatability of quantum computation processes. In this context, the above difference can be found even in rather simple computation processes [14,7,2,5,11], such as constant-depth polynomial-size quantum circuits. There is great interest in studying the classical simulatability of such simple computation processes since this is particularly useful for identifying the source of the computational power of a quantum computer.

We study the classical simulatability of constant-depth polynomial-size quantum circuits. In 2004, Terhal et al. provided evidence for the hardness of classically simulating such circuits followed by polynomially many single-qubit measurements, where they consist of universal gates on at most two qubits [14]. Subsequently, some authors provided (or mentioned) another evidence [7,2,5].

K. Chatterjee and J. Sgall (Eds.): MFCS 2013, LNCS 8087, pp. 801–812, 2013.
© Springer-Verlag Berlin Heidelberg 2013

An important assumption in these arguments is that the number of measurements is polynomial in the length of the input [14,7,5]. In fact, for example, any constant-depth quantum circuit followed by only one single-qubit measurement is efficiently simulatable classically [14]. Even in this simplest output setting, however, it has not been known how the use of gates on an unbounded number of qubits affects the classical simulatability of constant-depth quantum circuits.

We focus on quantum circuits in the simplest output setting, where they consist of universal gates on at most two qubits (Hadamard, $\pi/8$, and CNOT gates [12]) and additional gates on an unbounded number of qubits, more concretely, unbounded Toffoli and fan-out gates. An unbounded Toffoli gate on $n+1$ qubits computes the AND of n inputs. An unbounded fan-out gate on $n+1$ qubits makes n copies of a classical source bit. When $n = 1$, the gate is a CNOT gate. The main reason for adopting these gates as elementary gates is that they are a generalization of the classical ones assumed to be elementary gates for studying the computational power of small-depth classical circuits. Moreover, the study of an unbounded fan-out gate in the context of the classical simulatability complements previous studies of the gate showing that it is very powerful [13].

We deal with the strong and weak simulations [14,9,5,10,11]. The strong simulation of a quantum circuit means that, when an input to the circuit and its output are specified, the probability of obtaining the output can be efficiently computed classically. The weak simulation means that the output probability distribution of the circuit can be efficiently sampled classically. The strong simulation implies the weak simulation [14,5,10]. The error setting in the weak simulation is different from Terhal et al.'s efficient simulation [14] in that the error in the weak simulation is not a multiple of the output probability. Our setting seems more natural than the previous multiplicative one.

First, we consider constant-depth quantum circuits with unbounded Toffoli gates and their weak simulatability. We provide evidence for the hardness of weakly (and thus strongly) simulating a $\mathsf{QNC}^0_{t,1}$ circuit, which is a constant-depth quantum circuit with only one unbounded Toffoli gate:

Theorem 1. *There exists a $\mathsf{QNC}^0_{t,1}$ circuit that is not weakly simulatable, unless* $\mathsf{BQP} \subseteq \mathsf{PostBPP} \cap \mathsf{AM}$.

It is considered unlikely that $\mathsf{BQP} \subseteq \mathsf{PostBPP} \cap \mathsf{AM}$ since this (or even a weaker containment, such as $\mathsf{BQP} \subseteq \mathsf{PostBPP}$) would imply that BQP is contained in the polynomial hierarchy, which is considered unlikely [1]. Theorem 1 shows a boundary between classical and quantum computation: any constant-depth quantum circuit without additional gates on an unbounded number of qubits is strongly and weakly simulatable, but such a circuit with only one unbounded Toffoli gate is not strongly or weakly simulatable (under a plausible assumption).

To prove Theorem 1, we first show that, if any $\mathsf{QNC}^0_{t,1}$ circuit is weakly simulatable, then $\mathsf{BQP} \subseteq \mathsf{PostBPP}$. To do this, we parallelize a quantum circuit for $L \in \mathsf{BQP}$ by Fenner et al.'s method [7] and obtain a QNC^0 circuit, which is a constant-depth quantum circuit without gates on an unbounded number of qubits. The QNC^0 circuit has polynomially many postselection qubits that have to be measured to relate a measurement on the output qubit of the circuit to L. Using an unbounded Toffoli gate, we regard the postselection qubits and the output qubit as "one" new output qubit in two ways and construct two $\mathsf{QNC}^0_{t,1}$ circuits. Their weak simulations yield a $\mathsf{PostBPP}$ algorithm for L. We then deal with the containment $\mathsf{BQP} \subseteq \mathsf{AM}$ based on Terhal et al.'s argument in terms of the efficient simulation [14]. Since the number of measurements in their argument is polynomial and the efficient simulation is different from the weak simulation as described above, the argument does not work directly. We modify the argument and make an error analysis using one of the two $\mathsf{QNC}^0_{t,1}$ circuits.

As shown in [9], there exists a weakly simulatable polynomial-size quantum circuit that is not strongly simulatable (under a plausible assumption). Based on the idea of the proof of Theorem 1, we show the difference between the strong and weak simulatability in a simpler setting: there exists a weakly simulatable $\mathsf{QNC}^0_{t,1}$ circuit that is not strongly simulatable, unless $\mathsf{P} = \mathsf{PP}$. This contributes to our understanding not only of the classical simulatability of a $\mathsf{QNC}^0_{t,1}$ circuit but also of the notions of the strong and weak simulatability.

Then, we consider constant-depth quantum circuits with unbounded fan-out gates and their strong simulatability. The OR circuit in [13] allows us to replace an unbounded Toffoli gate in Theorem 1 with polynomially many unbounded fan-out gates. Thus, there exists a constant-depth quantum circuit with poly-nomially many unbounded fan-out gates that is not strongly (or weakly) simu-latable (under a plausible assumption). We provide evidence for the hardness of strongly simulating a simpler circuit, more concretely, a $\mathsf{QNC}^0_{f,2}$ circuit, which is a constant-depth quantum circuit with only two unbounded fan-out gates:

Theorem 2. *There exists a* $\mathsf{QNC}^0_{f,2}$ *circuit that is not strongly simulatable, un-less* $\mathsf{P} = \mathsf{PP}$.

It is considered unlikely that $\mathsf{P} = \mathsf{PP}$, which would imply the collapse of the polynomial hierarchy. As in the case of Theorem 1, Theorem 2 shows a bound-ary: any constant-depth quantum circuit without additional gates is strongly simulatable, but such a circuit with only two unbounded fan-out gates is not strongly simulatable (under a plausible assumption).

Our idea for showing Theorem 2 is to use the Hadamard test [11], more precisely, to parallelize it by two unbounded fan-out gates. For a QNC^0 circuit, the parallelized Hadamard test is a $\mathsf{QNC}^0_{f,2}$ circuit and allows us to show that, if any $\mathsf{QNC}^0_{f,2}$ circuit is strongly simulatable, there exists a polynomial-time deterministic classical algorithm for computing a matrix element of a QNC^0 circuit with exponential precision. This algorithm can be transformed into the one for computing a matrix element of a polynomial-size quantum circuit with exponential precision by Fenner et al.'s method [7] for parallelizing quantum circuits. This implies that $\mathsf{P} = \mathsf{PP}$ [11] and thus Theorem 2.

More generally, based on the idea of the proof, we characterize the relationship $\mathsf{P} = \mathsf{PP}$ using the strong simulatability of the parallelized Hadamard test for a QNC^0 circuit, which is a $\mathsf{QNC}^0_{f,2}$ circuit. This contributes to our understanding of the strong simulatability of such a circuit in the sense that the hardness of its strong simulation is exactly evaluated. Moreover, this is interesting in that the simple quantum computation process characterizes the classical relationship.

2 Preliminaries

We use the standard notation for quantum states and the standard diagrams for quantum circuits [12]. A quantum circuit consists of elementary gates. Our elementary gates are Hadamard gates H, $\pi/8$ gates $T = Z(\pi/4)$, CNOT gates, unbounded Toffoli gates, and unbounded fan-out gates, where

$$H = \frac{1}{\sqrt{2}}\begin{pmatrix} 1 & 1 \\ 1 & -1 \end{pmatrix}, \; Z(\theta) = \begin{pmatrix} 1 & 0 \\ 0 & e^{i\theta} \end{pmatrix}$$

for any $\theta \in \mathbb{R}$. We denote T^4 and HT^4H as Z and X, respectively. A Toffoli gate on $k+1$ qubits implements the quantum operation defined as

$$\left(\bigotimes_{j=0}^{k-1}|x_j\rangle\right)|y\rangle \mapsto \left(\bigotimes_{j=0}^{k-1}|x_j\rangle\right)|y \oplus \bigwedge_{j=0}^{k-1} x_j\rangle,$$

where $x_j, y \in \{0,1\}$, $k \geq 2$, and $\oplus$ denotes addition modulo 2. The first k input qubits, i.e., the qubits in state $\bigotimes_{j=0}^{k-1}|x_j\rangle$, are called the control qubits. A Toffoli gate on three qubits is simply called a Toffoli gate. A fan-out gate on $k+1$ qubits implements the quantum operation defined as

$$|y\rangle\bigotimes_{j=0}^{k-1}|x_j\rangle \mapsto |y\rangle\bigotimes_{j=0}^{k-1}|x_j \oplus y\rangle,$$

where $y, x_j \in \{0,1\}$ and $k \geq 1$. The first input qubit, i.e., the qubit in state $|y\rangle$, is called the control qubit. When $k = 1$, a fan-out gate is a CNOT gate. When a Toffoli gate or a fan-out gate is applied on an unbounded number of qubits, it is called an unbounded Toffoli gate or an unbounded fan-out gate, respectively.

The complexity measures of a quantum circuit are its size and depth. The size of a quantum circuit is defined as the total size of all elementary gates in it, where the size of an elementary gate is defined as the number of qubits affected by the gate. The depth of a quantum circuit is defined as follows. Input qubits are considered to have depth 0. For each gate G, the depth of G is equal to 1 plus the maximum depth of a gate on which G depends. The depth of a quantum circuit is defined as the maximum depth of a gate in it. Intuitively, the depth is the number of layers in the circuit, where a layer consists of gates that can be applied in parallel. A quantum circuit can use ancillary qubits initialized to $|0\rangle$. It is not required to reset the states to $|0\rangle$ at the end of the computation.

We deal with a uniform family of polynomial-size quantum circuits $\{C_n\}_{n\geq 1}$. Each C_n is a quantum circuit with n input qubits and poly(n) ancillary qubits. A symbol denoting a quantum circuit, such as C_n, also denotes its matrix representation. When a classical output is obtained from C_n, the circuit is followed by only one measurement in the computational basis, i.e., a Z-measurement, on a specified qubit called the output qubit. The output qubit in this paper is one of the ancillary qubits. The uniformity means that there exists a polynomial-time deterministic classical algorithm for computing the function $1^n \mapsto \overline{C_n}$, where $\overline{C_n}$ is the encoding of the description of C_n. For simplicity, we denote $\{C_n\}_{n \geq 1}$ as C_n (and thus call $\{C_n\}_{n \geq 1}$ a circuit). Let C_n be a constant-depth polynomial-size quantum circuit. In general, the number of gates on an unbounded number of qubits in C_n is poly(n). In particular, when the number of such gates in C_n is one and the gate is an unbounded Toffoli gate, we call C_n a $\mathsf{QNC}^0_{t,1}$ circuit. When the number is two and the gates are unbounded fan-out gates, we call C_n a $\mathsf{QNC}^0_{f,2}$ circuit. When C_n has no such gates, we call it a QNC^0 circuit.

The strong and weak simulatability is defined as follows [9,5,10,11]:

Definition 1. *Let C_n be a polynomial-size quantum circuit with n input qubits and* poly(n) *ancillary qubits including the output qubit. For any $x \in \{0,1\}^*$ of length n and $y \in \{0,1\}$, let* $\Pr[C_n(x) = y]$ *be the probability of obtaining y by a Z-measurement on the output qubit of C_n with the input state $|x\rangle$.*

- *C_n is strongly simulatable if, for any polynomial p, there exists a polynomial-time deterministic classical algorithm A such that, for any $x \in \{0,1\}^*$ of length n and $y \in \{0,1\}$, $|A(x,y) - \Pr[C_n(x) = y]| \leq 1/2^{p(n)}$.*
- *C_n is weakly simulatable if, for any polynomial p, there exists a polynomial-time probabilistic classical algorithm A such that, for any $x \in \{0,1\}^*$ of length n and $y \in \{0,1\}$, $|\Pr[A(x) = y] - \Pr[C_n(x) = y]| \leq 1/2^{p(n)}$.*

A strongly simulatable quantum circuit is weakly simulatable [14,5,10]. The weak simulation is different from Terhal et al.'s efficient simulation [14] in that the error in the weak simulation, which is the right hand of the above inequality, is not a multiple of $\Pr[C_n(x) = y]$. In other words, the error in the weak simulation is an absolute one, but that in the efficient simulation is a one relative to $\Pr[C_n(x) = y]$. In this sense, the error setting in the weak simulation seems more natural. We note that any QNC^0 circuit followed by only one single-qubit measurement is strongly (and thus weakly) simulatable [14].

The complexity classes we deal with are defined as follows [12,1,5]:

Definition 2. *Let $L \subseteq \{0,1\}^*$.*

- *$L \in \mathsf{BQP}$ if there exists a polynomial-size quantum circuit C_n with n input qubits and* poly(n) *ancillary qubits including the output qubit such that, for any $x \in \{0,1\}^*$ of length n,*
  - *if $x \in L$, $\Pr[C_n(x) = 1] \geq 2/3$,*
  - *if $x \notin L$, $\Pr[C_n(x) = 1] \leq 1/3$.*
- *$L \in \mathsf{PostBPP}$ if there exists a polynomial-time probabilistic classical algorithm A that, for any $x \in \{0,1\}^*$, outputs $A(x)$, post$(x) \in \{0,1\}$ such that*
  - *$\Pr[\text{post}(x) = 0] > 0$,*
  - *if $x \in L$, $\Pr[A(x) = 1|\text{post}(x) = 0] \geq 2/3$,*
  - *if $x \notin L$, $\Pr[A(x) = 1|\text{post}(x) = 0] \leq 1/3$.*

We note that $\mathsf{PostBPP}$ is equal to $\mathsf{BPP}_{\mathsf{path}}$ defined in [8]. The constants $2/3$ and $1/3$ in the definitions can be replaced with $1/2 + \varepsilon$ and $1/2 - \varepsilon$, respectively, for any constant $0 < \varepsilon < 1/2$ [12,5]. We also deal with the well-known complexity classes P, AM, and PP [3]. Moreover, we deal with the function classes FP and #P [3]: FP is the class of functions for which there exists a polynomial-time deterministic classical algorithm and #P is the class of functions counting the number of solutions to polynomial-time decidable relations.

We frequently use a quantum circuit obtained by Fenner et al.'s method [7]. The existence of the circuit (combined with a constant-depth polynomial-size quantum circuit for permuting qubits) can be described as follows:

Lemma 1. *For any polynomial-size quantum circuit C_n with n input qubits and a ancillary qubits, there exists a QNC^0 circuit D_n with n input qubits and $a+b$ ancillary qubits such that b is even, $b = O(\mathrm{size}(C_n))$, and, for any $x \in \{0,1\}^*$ of length n,*

$$D_n|x\rangle|0\rangle^{\otimes(a+b)} = \frac{1}{\sqrt{2^b}}(C_n|x\rangle|0\rangle^{\otimes a})|0\rangle^{\otimes b} + \sum_{y\in\{0,1\}^b\backslash\{0^b\}} \alpha_y|\psi_y\rangle|y\rangle,$$

where $\mathrm{size}(C_n)$ is the polynomial representing the size of C_n, $\alpha_y \in \mathbb{C}$, and $|\psi_y\rangle$ is an $(n+a)$-qubit state.

The b new ancillary qubits are called the postselection qubits.

3   Circuit with One Unbounded Toffoli Gate

3.1   Proof of Theorem 1

A Toffoli gate outputs 1 if and only if the states of the control qubits are $|11\rangle$. Combining the gate with an X gate, we can obtain a circuit that outputs 1 if and only if the states of the control qubits are $|10\rangle$. We call it a (1,0)-Toffoli gate. We define a (0,0)-Toffoli gate similarly. Moreover, combining an unbounded Toffoli gate with X gates, we can obtain (a reversible version of) an OR gate with unbounded fan-in. Using these gates, we first show the following lemma:

Lemma 2. *If any $\mathsf{QNC}^0_{t,1}$ circuit is weakly simulatable, then $\mathsf{BQP} \subseteq \mathsf{PostBPP}$.*

Proof. We assume that any $\mathsf{QNC}^0_{t,1}$ circuit is weakly simulatable. Let $L \in \mathsf{BQP}$. There exists a polynomial-size quantum circuit C_n with n input qubits and a ancillary qubits including the output qubit such that, for any $x \in \{0,1\}^*$ of length n, $\Pr[C_n(x) = 1] \geq 2/3$ if $x \in L$ and $\Pr[C_n(x) = 1] \leq 1/3$ if $x \notin L$. By Lemma 1, there exists a QNC^0 circuit D_n with n input qubits and $a+b$ ancillary qubits including the output qubit such that, for any $x \in \{0,1\}^*$ of length n, $\Pr[D_n(x) = 1|\mathrm{post}_n(x) = 0^b] \geq 2/3$ if $x \in L$ and $\Pr[D_n(x) = 1|\mathrm{post}_n(x) = 0^b] \leq 1/3$ if $x \notin L$, where $b = O(\mathrm{size}(C_n))$, "$\mathrm{post}_n(x) = 0^b$" means that all results of Z-measurements on the postselection qubits are 0, and $\Pr[\mathrm{post}_n(x) = 0^b] = 1/2^b$. This implies that, for any $x \in \{0,1\}^*$ of length n,

- if $x \in L$, $\Pr[D_n(x) = 1 \& \mathrm{post}_n(x) = 0^b] \geq \frac{2}{3} \cdot \frac{1}{2^b}$,
- if $x \notin L$, $\Pr[D_n(x) = 1 \& \mathrm{post}_n(x) = 0^b] \leq \frac{1}{3} \cdot \frac{1}{2^b}$.

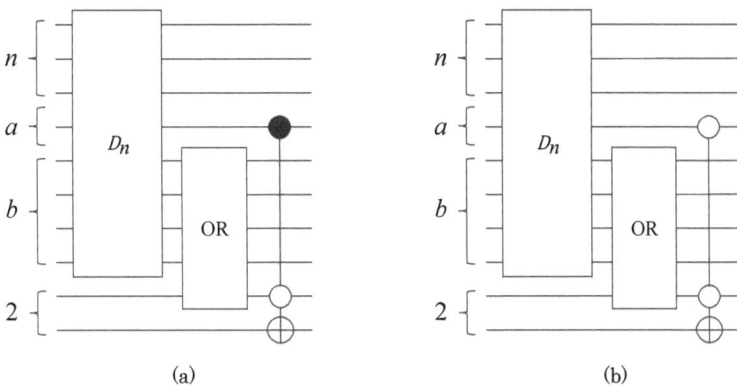

Fig. 1. (a) The circuit E_n. (b) The circuit F_n.

We define a quantum circuit E_n as follows, where it has n input qubits, $a + b$ ancillary qubits for D_n, and two new ancillary qubits including the output qubit for E_n:

1. Apply D_n on the n input qubits and $a + b$ ancillary qubits.
2. Apply the OR gate (with unbounded fan-in) on the b postselection qubits and one of the two new ancillary qubits that is not the output qubit for E_n. The output of the gate is written into the new ancillary qubit.
3. Apply the $(1,0)$-Toffoli gate on the output qubit for D_n and the two new ancillary qubits, where the output qubit for D_n is used as the control qubit that is in state $|1\rangle$ when the gate outputs 1. The output of the gate is written into the output qubit for E_n.

We also define a quantum circuit F_n similarly to E_n except that the $(1,0)$-Toffoli gate in E_n is replaced with the $(0,0)$-Toffoli gate. The circuits E_n and F_n are depicted in Fig. 1 (a) and (b), respectively, where the bottom qubits are the output qubits. A Toffoli gate can be decomposed exactly into a constant-depth constant-size quantum circuit consisting of H, T, and CNOT gates with no ancillary qubits [12]. Since D_n is a QNC^0 circuit, E_n and F_n are $\mathsf{QNC}^0_{t,1}$ circuits.

The OR gate in Step 2 reduces the b postselection qubits to one new postselection qubit, which is the new ancillary qubit that is not the output qubit for E_n. Moreover, the state of the output qubit for E_n is $|1\rangle$ if and only if the state of the output qubit for D_n and that of the new postselection qubit are $|1\rangle$ and $|0\rangle$, respectively. Thus, for any $x \in \{0,1\}^*$ of length n, $\Pr[E_n(x) = 1] = \Pr[D_n(x) = 1 \& \mathrm{post}_n(x) = 0^b]$. Similarly, $\Pr[F_n(x) = 1] = \Pr[D_n(x) = 0 \& \mathrm{post}_n(x) = 0^b]$. Since $\Pr[\mathrm{post}_n(x) = 0^b] = 1/2^b$, $\Pr[E_n(x) = 1] + \Pr[F_n(x) = 1] = 1/2^b$.

As described above, E_n and F_n are $\mathsf{QNC}^0_{t,1}$ circuits. Thus, by the assumption, there exist polynomial-time probabilistic classical algorithms A and B such that, for any $x \in \{0,1\}^*$ of length n, $|\Pr[A(x) = 1] - \Pr[E_n(x) = 1]| \leq 1/2^{b+6}$, $|\Pr[B(x) = 1] - \Pr[F_n(x) = 1]| \leq 1/2^{b+6}$. We define a polynomial-time probabilistic classical algorithm G as follows, where the input is $x \in \{0,1\}^*$:

1. Choose $r \in \{0, 1\}$ uniformly at random.
2. (a) If $r = 1$, compute $A(x)$.
     i. If $A(x) = 1$, set $post(x) = 0$ and $G(x) = 1$.
     ii. If $A(x) = 0$, set $post(x) = 1$ and $G(x) = 1$.
   (b) If $r = 0$, compute $B(x)$.
     i. If $B(x) = 1$, set $post(x) = 0$ and $G(x) = 0$.
     ii. If $B(x) = 0$, set $post(x) = 1$ and $G(x) = 0$.

Using the algorithm G, we can show that $L \in$ PostBPP. □

We then deal with the containment $\mathsf{BQP} \subseteq \mathsf{AM}$ based on Terhal et al.'s argument [14]. The argument uses a set of results of internal coin tosses in the classical simulation of (a parallelized version of) a quantum circuit for $L \in \mathsf{BQP}$, where the number of elements of the set on input x is large when $x \in L$ and the number is small when $x \notin L$. Even if the difference of the numbers is somewhat small, the Goldwasser-Sipser set lower bound protocol can decide whether the number is large or small, which implies that $L \in \mathsf{AM}$. More precisely, the key part of the argument is to show that, under an assumption about the classical simulatability of a QNC^0 circuit, any $L \in \mathsf{BQP}$ has the following property, which we call the property $\mathcal{P}$: there exist a constant $0 < \varepsilon < 1/3$, polynomials m, q ($m \geq q$), and a family of sets $\{S_x\}_{x \in \{0,1\}^*}$ such that, for any $x \in \{0, 1\}^*$ of length n, $S_x \subseteq \{0, 1\}^{m(n)}$, $|S_x| \geq (1 - \varepsilon) \cdot \frac{2}{3} \cdot 2^{q(n)}$ if $x \in L$ and $|S_x| \leq (1 + \varepsilon) \cdot \frac{1}{3} \cdot 2^{q(n)}$ if $x \notin L$, where the problem of deciding whether a bit string of length $m(n)$ is in S_x has a polynomial-time deterministic classical algorithm. If L has the property $\mathcal{P}$, the Goldwasser-Sipser protocol implies that $L \in \mathsf{AM}$ [14,3] as described above. Thus, to show Theorem 1, it suffices to show the following lemma:

Lemma 3. *If any $\mathsf{QNC}^0_{t,1}$ circuit is weakly simulatable, then any $L \in \mathsf{BQP}$ has the property $\mathcal{P}$.*

Proof. We assume that any $\mathsf{QNC}^0_{t,1}$ circuit is weakly simulatable. Let $L \in \mathsf{BQP}$. As in the proof of Lemma 2, there exists a QNC^0 circuit D_n such that, for any $x \in \{0, 1\}^*$ of length n,

- if $x \in L$, $\Pr[D_n(x) = 1 \& post_n(x) = 0^b] \geq \frac{2}{3} \cdot \frac{1}{2^b}$,
- if $x \notin L$, $\Pr[D_n(x) = 1 \& post_n(x) = 0^b] \leq \frac{1}{3} \cdot \frac{1}{2^b}$.

We define a quantum circuit E_n as in the proof of Lemma 2. It holds that, for any $x \in \{0, 1\}^*$ of length n, $\Pr[E_n(x) = 1] = \Pr[D_n(x) = 1 \& post_n(x) = 0^b]$.

We fix a polynomial p satisfying $(3 \cdot 2^b)/2^p < 1/10$. Since E_n is a $\mathsf{QNC}^0_{t,1}$ circuit, by the assumption, there exists a polynomial-time probabilistic classical algorithm A such that, for any $x \in \{0, 1\}^*$ of length n, $|\Pr[A(x) = 1] - \Pr[E_n(x) = 1]| \leq 1/2^{p(n)}$. More concretely, there exist such an algorithm A and a polynomial m such that, for any $x \in \{0, 1\}^*$ of length n, the above inequality holds, where $\Pr[A(x) = 1] = |\{r \in \{0, 1\}^{m(n)} | A_r(x) = 1\}|/2^{m(n)}$ and A_r is A with the result of its internal coin tosses r. We note that A with a fixed r can be regarded as a polynomial-time deterministic classical algorithm. We can choose m satisfying $m \geq b$. Let $\varepsilon = 1/4$, $q = m - b$, and $S_x = \{r \in \{0, 1\}^{m(n)} | A_r(x) = 1\}$ for any $x \in \{0, 1\}^*$ of length n. This implies that L has the property $\mathcal{P}$. □

3.2 Difference between the Strong and Weak Simulatability

We consider a polynomial-time computable function $f = \{f_n\}_{n \geq 1}$, where $f_n :$ $\{0,1\}^n \to \{0,1\}$. That is, there exists a uniform family of polynomial-size classical circuits such that it computes f. For simplicity, we denote $\{f_n\}_{n \geq 1}$ as f_n. For any f_n, there exists a polynomial-size quantum (in fact, classical reversible) circuit C_n^f with n input qubits and $a \geq 1$ ancillary qubits including the output qubit such that C_n^f consists only of Toffoli and X gates and implements the quantum operation $|y\rangle|0\rangle^{\otimes a} \mapsto |y\rangle|f_n(y)\rangle|0\rangle^{\otimes(a-1)}$, where $y \in \{0,1\}^n$. The Hadamard-Toffoli circuit for f_n, which we call HT_n^f, is defined as follows [9], where it has n input qubits and a ancillary qubits including the output qubit: apply H gates on the input qubits, then apply C_n^f on the $n + a$ qubits by using the n input qubits as the input qubits for C_n^f. It can be shown that, for any f_n, HT_n^f is weakly simulatable and that, unless $\mathsf{FP} = \#\mathsf{P}$, that is, unless $\mathsf{P} = \mathsf{PP}$ [3], there exists an f_n such that HT_n^f is not strongly simulatable [9].

By Lemma 1, for any HT_n^f with n input qubits and a ancillary qubits, there exists a QNC^0 circuit D_n^f with n input qubits and $a + b$ ancillary qubits. As in the proof of Lemma 2, we consider a $\mathsf{QNC}_{t,1}^0$ circuit E_n^f, which is defined similarly to E_n except that D_n in E_n is replaced with D_n^f. By the construction of E_n^f and Lemma 1, $\Pr[E_n^f(x) = 1] = \#_n^f(1)/2^{n+b}$ and $\Pr[E_n^f(x) = 0] = (1 - 1/2^b) + \#_n^f(0)/2^{n+b}$, where $\#_n^f(c) = |\{y \in \{0,1\}^n | f_n(y) = c\}|$ for any $c \in \{0,1\}$. Based on this analysis, we can define a polynomial-time probabilistic classical algorithm for weakly simulating E_n^f. On the other hand, we can show that a polynomial-time deterministic classical algorithm for strongly simulating E_n^f yields such an algorithm for computing $\#_n^f(1)$ (and $\#_n^f(0)$). This implies the following lemma:

Lemma 4. *The following statements hold:*

(1). For any polynomial-time computable function f_n, E_n^f is weakly simulatable.
(2). If, for any polynomial-time computable function f_n, E_n^f is strongly simulatable, then $\mathsf{P} = \mathsf{PP}$.

Lemma 4 immediately implies that there exists a weakly simulatable $\mathsf{QNC}_{t,1}^0$ circuit that is not strongly simulatable, unless $\mathsf{P} = \mathsf{PP}$.

4 Circuit with Two Unbounded Fan-Out Gates

Let C_n be a polynomial-size quantum circuit with n input qubits and a ancillary qubits. The Hadamard test for C_n is the well-known quantum circuit that relates its output probability to the real or imaginary part of the matrix element $\langle 0|^{\otimes(n+a)} C_n |0\rangle^{\otimes(n+a)}$ [11]. It has n input qubits and $a + 1$ ancillary qubits including the output qubit. The circuit is defined as follows: apply an H gate on the output qubit, then apply the controlled version of C_n on the $n + a + 1$ qubits by using the output qubit as the control qubit, and then apply an H gate on the output qubit. For example, let C_3 be the quantum circuit depicted in Fig. 2 (a). In this case, $a = 0$ and the Hadamard test for the circuit is depicted in Fig. 2 (b), where the top qubit is the output qubit.

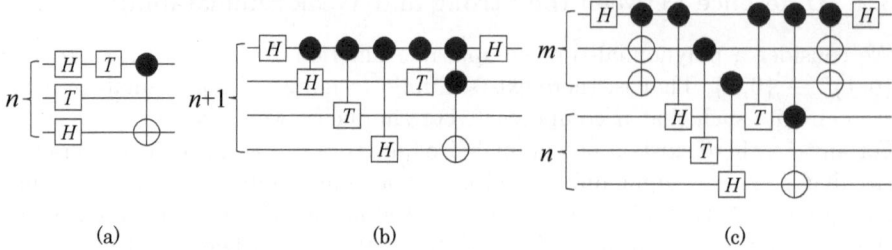

Fig. 2. (a) An example of a quantum circuit C_3 with $a = 0$. (b) The Hadamard test for C_3 in (a). (c) The parallelized Hadamard test for C_3 in (a) and $U_1 = I$. In this case, $m = 3$. The gate next to H is an unbounded fan-out gate, where the top qubit is the control qubit.

A key circuit for showing Theorem 2 is a parallelized version of the Hadamard test. It relates its output probability to the matrix element as the standard Hadamard test. Moreover, the depth of the parallelized version of the Hadamard test for a QNC^0 circuit is constant, in contrast to the fact that, in general, the depth of the standard Hadamard test for a QNC^0 circuit is polynomial in the length of the input. To describe the parallelized version, let C_n be a polynomial-size quantum circuit with n input qubits and a ancillary qubits, U_1 be a single-qubit unitary gate, and m be the maximum number of gates included in a layer in C_n. We define a quantum circuit, which we call the parallelized Hadamard test for C_n and U_1, as follows, where it has n input qubits, a ancillary qubits for C_n, and m new ancillary qubits including the output qubit:

1. Apply an H gate on the output qubit.
2. Apply an unbounded fan-out gate on the m new ancillary qubits, where the output qubit is used as the control qubit.
3. Apply the controlled version of C_n, where the m new ancillary qubits are used as the control qubits. The gates of the controlled version of C_n are arranged so that, if the original gates in C_n are in a layer, their controlled versions are also in a layer in this new circuit.
4. This step is the same as Step 2.
5. This step is the same as Step 1.
6. Apply the U_1 gate on the output qubit.

The parallelized Hadamard test for the circuit in Fig. 2 (a) and $U_1 = I$ is depicted in Fig. 2 (c), where the top qubit is the output qubit and $m = 3$.

The parallelized Hadamard test is described by the gates that are not our elementary gates. Fortunately, we can decompose such gates exactly into constant-depth constant-size quantum circuits using our elementary gates as shown in [4]. Moreover, a direct calculation shows that the output probabilities of the parallelized Hadamard test for C_n are related to the real and imaginary parts of $\langle x|\langle 0|^{\otimes a} C_n|x\rangle|0\rangle^{\otimes a}$ as follows, where $\mathrm{Re}(z)$ and $\mathrm{Im}(z)$ are the real and imaginary parts of z, respectively, for any $z \in \mathbb{C}$:

Lemma 5. *For any* QNC^0 *circuit* C_n *(with n input qubits and a ancillary qubits) and single-qubit unitary gate U_1 generated by a constant number of H and T, there exists a $\mathsf{QNC}^0_{f,2}$ circuit D_n that implements the same operation exactly as the parallelized Hadamard test for C_n and U_1. Moreover, for any $x \in \{0,1\}^*$ of length n, when the input state is $|x\rangle$ and $U_1 = I$, it holds that $\Pr[D_n(x) = 0] = (1 + \mathrm{Re}(\langle x|\langle 0|^{\otimes a}C_n|x\rangle|0\rangle^{\otimes a}))/2$. Similarly, when $U_1 = HT^2$, it holds that $\Pr[D_n(x) = 0] = (1 + \mathrm{Im}(\langle x|\langle 0|^{\otimes a}C_n|x\rangle|0\rangle^{\otimes a}))/2$.*

In the following, the parallelized Hadamard test represents the $\mathsf{QNC}^0_{f,2}$ circuit consisting only of our elementary gates. Roughly speaking, Lemma 5 means that the existence of a polynomial-time deterministic classical algorithm for computing the output probability of the parallelized Hadamard test for C_n with exponential precision is equivalent to that of a polynomial-time deterministic classical algorithm for computing $\langle x|\langle 0|^{\otimes a}C_n|x\rangle|0\rangle^{\otimes a}$ with exponential precision. This implies the following lemma:

Lemma 6. *The following statements are equivalent:*

(1). *For any QNC^0 circuit C_n and single-qubit unitary gate $U_1 \in \{I, HT^2\}$, the parallelized Hadamard test for C_n and U_1 is strongly simulatable.*

(2). *For any QNC^0 circuit C_n, there exists a polynomial-time deterministic classical algorithm for computing $\langle x|\langle 0|^{\otimes a}C_n|x\rangle|0\rangle^{\otimes a}$ with exponential precision. More precisely, for any polynomial p, there exists a polynomial-time deterministic classical algorithm A such that, for any $x \in \{0,1\}^*$ of length n, $|A(x) - \mathrm{Re}(\langle x|\langle 0|^{\otimes a}C_n|x\rangle|0\rangle^{\otimes a})| \leq 1/2^{p(n)}$. Moreover, such an algorithm exists also for computing the imaginary part.*

To show Theorem 2, we need the following lemma, which is a simple consequence of Lemma 1 and the results in [6,11]:

Lemma 7. *The following statements are equivalent:*

(1). *For any QNC^0 circuit C_n, there exists a polynomial-time deterministic classical algorithm for computing $\langle x|\langle 0|^{\otimes a}C_n|x\rangle|0\rangle^{\otimes a}$ with exponential precision.*

(2). *For any polynomial-size quantum circuit C_n, there exists a polynomial-time deterministic classical algorithm for computing $\langle x|\langle 0|^{\otimes a}C_n|x\rangle|0\rangle^{\otimes a}$ with exponential precision.*

(3). $\mathsf{P} = \mathsf{PP}$.

Lemmas 6 and 7 immediately imply the characterization of the relationship $\mathsf{P} = \mathsf{PP}$. That is, the following statements are equivalent:

(1). For any QNC^0 circuit C_n and single-qubit unitary gate $U_1 \in \{I, HT^2\}$, the parallelized Hadamard test for C_n and U_1 is strongly simulatable.

(2). $\mathsf{P} = \mathsf{PP}$.

As shown in Lemma 5, the parallelized Hadamard test for any QNC^0 circuit C_n and $U_1 \in \{I, HT^2\}$ is a $\mathsf{QNC}^0_{f,2}$ circuit. Thus, this characterization implies Theorem 2.

5   Open Problems

Interesting challenges would be to further investigate the relationships between the classical simulatability of constant-depth quantum circuits and complexity classes. We give two examples of such problems:

- Can we provide stronger evidence? For example, can we show that, if any $\mathsf{QNC}^0_{t,1}$ circuit (followed by only one single-qubit measurement) is weakly simulatable, then $\mathsf{P} = \mathsf{PP}$?
- Can we provide evidence for the hardness as in Theorems 1 and 2 when we consider the error $1/\mathrm{poly}(n)$ in place of $1/2^{p(n)}$ in the classical simulations?

References

1. Aaronson, S.: BQP and the polynomial hierarchy. In: Proceedings of the 42nd ACM Symposium on Theory of Computing, STOC, pp. 141–150 (2010)
2. Aaronson, S., Arkhipov, A.: The computational complexity of linear optics. In: Proceedings of the 43rd ACM Symposium on Theory of Computing, STOC, pp. 333–342 (2011)
3. Arora, S., Barak, M.: Computational Complexity: A Modern Approach. Cambridge University Press (2009)
4. Bera, D., Fenner, S., Green, F., Homer, S.: Efficient universal quantum circuits. Quantum Information and Computation 10(1&2), 16–27 (2010)
5. Bremner, M.J., Jozsa, R., Shepherd, D.J.: Classical simulation of commuting quantum computations implies collapse of the polynomial hierarchy. Proceedings of the Royal Society A 467, 459–472 (2011)
6. Dawson, C.M., Hines, A.P., Mortimer, D., Haselgrove, H.L., Nielsen, M.A., Osborne, T.J.: Quantum computing and polynomial equations over the finite field $\mathbb{Z}_2$. Quantum Information and Computation 5(2), 102–112 (2005)
7. Fenner, S., Green, F., Homer, S., Zhang, Y.: Bounds on the power of constant-depth quantum circuits. In: Liśkiewicz, M., Reischuk, R. (eds.) FCT 2005. LNCS, vol. 3623, pp. 44–55. Springer, Heidelberg (2005)
8. Han, Y., Hemaspaandra, L.A., Thierauf, T.: Threshold computation and cryptographic security. SIAM Journal on Computing 26(1), 59–78 (1997)
9. van den Nest, M.: Classical simulation of quantum computation, the Gottesman-Knill theorem, and slightly beyond. Quantum Information and Computation 10(3&4), 258–271 (2010)
10. van den Nest, M.: Simulating quantum computers with probabilistic methods. Quantum Information and Computation 11(9&10), 784–812 (2011)
11. Ni, X., van den Nest, M.: Commuting quantum circuits: efficient classical simulations versus hardness results. Quantum Information and Computation 13(1&2), 54–72 (2013)
12. Nielsen, M.A., Chuang, I.L.: Quantum Computation and Quantum Information. Cambridge University Press (2000)
13. Takahashi, Y., Tani, S.: Collapse of the hierarchy of constant-depth exact quantum circuits. In: Proceedings of the 28th IEEE Conference on Computational Complexity, CCC, pp. 168–178 (2013)
14. Terhal, B.M., DiVincenzo, D.P.: Adaptive quantum computation, constant-depth quantum circuits and Arthur-Merlin games. Quantum Information and Computation 4(2), 134–145 (2004)

Improved Bounds for Reduction
to Depth 4 and Depth 3

Sébastien Tavenas

LIP⋆, École Normale Supérieure de Lyon
sebastien.tavenas@ens-lyon.fr

Abstract. Koiran [7] showed that if an n-variate polynomial of degree d (with $d = n^{O(1)}$) is computed by a circuit of size s, then it is also computed by a homogeneous circuit of depth four and of size $2^{O(\sqrt{d}\log(d)\log(s))}$. Using this result, Gupta, Kamath, Kayal and Saptharishi [6] gave an $\exp\left(O\left(\sqrt{d\log(d)\log(n)\log(s)}\right)\right)$ upper bound for the size of the smallest depth three circuit computing an n-variate polynomial of degree $d = n^{O(1)}$ given by a circuit of size s.

We improve here Koiran's bound. Indeed, we show that if we reduce an arithmetic circuit to depth four, then the size becomes $\exp\left(O\left(\sqrt{d\log(ds)\log(n)}\right)\right)$. Mimicking the proof in [6], it also implies the same upper bound for depth three circuits.

This new bound is not far from optimal in the sense that Gupta, Kamath, Kayal and Saptharishi [5] also showed a $2^{\Omega(\sqrt{d})}$ lower bound for the size of homogeneous depth four circuits such that gates at the bottom have fan-in at most $\sqrt{d}$. Finally, we show that this last lower bound also holds if the fan-in is at least $\sqrt{d}$.

1 Introduction

Agrawal and Vinay proved [1] that if an n-variate polynomial f of degree $d = O(n)$ has a circuit of size $2^{o(d+d\log(\frac{n}{d}))}$, then f can also be computed by a depth-four circuit ($\sum\prod\sum\prod$) of size $2^{o(d+d\log(\frac{n}{d}))}$. This result shows that for proving arithmetic circuit lower bounds or black-box derandomization of identity testing, the case of depth four arithmetic circuit is the general case in a certain sense. This result arose after other ones on parallelization. Valiant, Skyum, Berkowitz and Rackoff [9] proved that if a size-s depth-d circuit computes a polynomial of degree d, then this polynomial can also be computed by a circuit of depth $O(\log(d)\log(s))$ and of size bounded by a polynomial in s. Some years later, Allender, Jiao, Mahajan and Vinay [2] showed that this parallelization could be done uniformly. Their method for parallelization is reused in [1] and will be the basis for the parallelization in this paper.

Agrawal and Vinay's result only deals with polynomials of sub-exponential complexity. But if the hypothesis is strengthened, it is possible to get a stronger

---

⋆ UMR 5668 ENS Lyon - CNRS - UCBL - INRIA, Université de Lyon.

K. Chatterjee and J. Sgall (Eds.): MFCS 2013, LNCS 8087, pp. 813–824, 2013.

conclusion. Indeed, Koiran [7] showed that if the circuit at the beginning is of size s, then it can be computed by a homogeneous depth-four circuit of size $2^{O(\sqrt{d}\log(d)\log(s))}$. For example, if the permanent family is computed by a polynomial size circuit (i.e., of size n^c), then it is computed by a depth-four circuit of size $2^{O(\sqrt{n}\log^2(n))}$. These results appear as an interesting approach to lower bounds: if one finds a $2^{\omega(\sqrt{n}\log^2(n))}$ lower bound on the size of depth-4 circuits computing the permanent, then it will imply that there are no polynomial size circuits for the permanent. The interest of this approach is confirmed by Gupta, Kamath, Kayal and Saptharishi's recent result [5]. They showed that if a homogeneous $\sum\prod\sum\prod$ circuit where the bottom fan-in is bounded by t computes the permanent of a matrix of size $n \times n$, then its size is $2^{\Omega(\frac{n}{t})}$. In a recent paper [6], the same authors improve the upper bound by transforming n-variate circuits of size s and depth d (with $d = n^{O(1)}$) into depth-3 circuits of size $\exp\left(O(\sqrt{d\log s\log n\log d})\right)$, moreover if the input is a branching program (and not a circuit), the upper bound becomes $\exp\left(O(\sqrt{d\log s\log n})\right)$. In particular, this result gives a depth-3 circuit of size $2^{O(\sqrt{n}\log n)}$ computing the determinant of a matrix $n \times n$. Nevertheless, the depth-3 circuit they get is not homogeneous, and uses intermediate gates which compute polynomials of very high degree.

In this paper we improve Koiran's bound. We show that a circuit of size s can be parallelized homogeneously in depth 4 and in size $\exp\left(O\left(\sqrt{d\log(ds)\log(n)}\right)\right)$ such that the fan-in of each multiplication gate is bounded by $O\left(\sqrt{d\frac{\log ds}{\log n}}\right)$. We can notice that as $n \leq s$, the result implies Koiran's bound and is generally better (in the case where $d, s = n^{\Theta(1)}$, Koiran's bound is $2^{O(\sqrt{n}\log^2 n)}$ while the new bound is $2^{O(\sqrt{n}\log n)}$). It implies that a $2^{\omega(\sqrt{n}\log(n))}$ lower bound for depth-4 circuits computing the permanent gives a super-polynomial lower bound for general circuits computing the permanent. Moreover, using this result in Gupta, Kamath, Kayal and Saptharishi's proof instead of Koiran's result slightly improves the depth-3 upper bound. An n-variate circuit of size s and depth d is computed by a depth-3 circuit of size $\exp\left(O(\sqrt{d\log(ds)\log n})\right)$. So, we get the same bound for the reduction at depth 3 starting from an arithmetic circuit as from an arithmetic branching program. Finally in Section 6, we show, by a counting argument, that if a homogeneous $\sum\prod\sum\prod$ circuit where the bottom fan-in is lower-bounded by t computes the permanent (or the determinant) of a matrix of size $n \times n$, then its size is $2^{\Omega(t\log n)}$.

2   Arithmetic Circuits

We give here a brief introduction to arithmetic circuits theory. The reader can find more detailed information in [10,3,8,4]. In this theory, we measure the complexity of polynomial functions using arithmetic circuits.

Definition 1. *An arithmetic circuit is a finite acyclic directed graph with vertices of in-degree 0 or more and exactly one vertex of out-degree 0. Vertices of* *in-degree 0 are called inputs and labeled by a constant or a variable. The other vertices are labeled by × or + (or sometimes by ⊙ in this paper) and called computation gates (the in-degree of these gates will be also called the fan-in). The vertex of out-degree 0 is called the output. The vertices of a circuit are commonly called gates and its edges arrows. Finally, we call a formula, an arithmetic circuit such that the underlying graph is a tree.*

Each gate of a circuit computes a polynomial (defined by induction). The polynomial computed by a circuit corresponds to the polynomial computed by the output of this circuit. For a gate α, we denote $[\alpha]$ the polynomial computed by this gate. A circuit is called *homogeneous* is all its gates compute homogeneous polynomials. In fact, for some proofs, we will use circuits with several outputs (each one corresponds to an out-degree 0 gate). A ⊙-gate corresponds to a multiplication-by-a-scalar gate. The fan-in of such a gate will be always 2 and at least one of its inputs corresponds to a constant (We will give a syntactic restriction just after the next definition).

Definition 2. *The size of a circuit is its number of gates. The depth is the maximal length of a directed path from an input to an output. The degree of a gate is defined recursively: any variable input is of degree 1, constant inputs are of degree 0, the degree of a + or ⊙-gate is the maximum of the incoming degrees and the degree of a × -gate is the sum of the incoming degrees.*

We can now put a restriction for the ⊙-gates. For each one of these gates, one of its child has to be of degree 0.

For a given circuit we will consider graphs called parse trees. A parse tree corresponds, in the spirit, to the computation of one particular monomial.

Definition 3. *The set of parse trees of a circuit C is defined by induction on its size:*

- *If C is of size 1 it has only one parse tree, itself.*
- *If the output gate of C is a +-gate whose arguments are the gates $\alpha_1, \ldots, \alpha_k$, then the parse trees of C are obtained by taking, for an arbitrary $i \leq k$, a parse tree of the sub-circuit rooted in α_i and the arrow from α_i to the output.*
- *If the output gate of C is a × -gate or an ⊙-gate whose arguments are the gates $\alpha_1, \ldots, \alpha_k$, the parse trees of C are obtained by taking disjoint copies of parse tree of the sub-circuits rooted in α_i for all $i \leq k$ and the arrows from all α_i to the output.*

The polynomial computed by a circuit C becomes the sum of the monomials computed by the parse trees of C.

We will use some convenient notations which are defined in [6]. A depth-4 circuit such that gates are multiplication gates at level one and three and addition gates at levels two and four are denoted $\sum \prod \sum \prod$ circuits. Furthermore, a $\sum \prod^{[\alpha]} \sum \prod^{[\beta]}$ circuit is a $\sum \prod \sum \prod$ circuit such that the fan-in of the multiplication gates at level 3 is bounded by α, and the fan-in of the multiplication gates at level 1 is bounded by β. For example, a $\sum \prod^{[\alpha]} \sum \prod^{[\beta]}$ circuit computes a polynomial of the form:

$$\sum_{i=1}^{t} \prod_{j=1}^{a_i} \sum_{k=1}^{u_{i,j}} \prod_{l=1}^{b_{i,j,k}} x_{i,j,k,l}$$

where $a_i \le \alpha$, $b_{i,j,k} \le \beta$.

3   Upper Bounds

Here, we state the main theorem in this paper.

Theorem 1. *Let f be an n-variate polynomial computed by a circuit of size s and of degree d. Then f is computed by a $\sum \prod \sum \prod$ circuit C of size $2^{O\left(\sqrt{d \log(ds) \log n}\right)}$. Furthermore, if f is homogeneous, it will be also the case for C.*

The previous theorem can be directly applied for the permanent.

Theorem 2. *If the $n \times n$ permanent is computed by a circuit of size polynomial in n, then it is also computed by a $\sum \prod \sum \prod$ circuit of size $2^{O(\sqrt{n} \log(n))}$.*

In their paper [6], Gupta, Kamath, Kayal and Saptharishi used the previous $2^{\sqrt{d} \log^2(s)}$ bound [7] for parallelizing at depth 3. They showed that:

Proposition 1 (Theorem 1.1 in [6]). *Let $f(x) \in \mathbb{Q}[x_1, \ldots, x_n]$ be an n-variate polynomial of degree $d = n^{O(1)}$ computed by an arithmetic circuit of size s. Then it can also be computed by a $\sum \prod \sum$ circuit of size $2^{O(\sqrt{d \log n \log s \log d})}$.*

In fact, their proof is divided into three parts. First they transform circuits into depth-4 circuits, then they transform depth-4 circuits into depth-5 circuits using only sum and exponentiation gates. And finally they transform these last circuits into depth-3 circuits. Using Theorem 1 instead of Theorem 4.1 in their paper improves the first part of their proof. That implies a small improvement of Theorem 1.1 in [6]:

Corollary 1. *Let $f(x) \in \mathbb{Q}[x_1, \ldots, x_n]$ be an n-variate polynomial of degree $d = n^{O(1)}$ computed by an arithmetic circuit of size s. Then it can also be computed by a $\sum \prod \sum$ circuit of size $2^{O(\sqrt{d \log n \log s})}$.*

4   Useful Propositions

For proving Theorem 1, we will need the following propositions.
    The next result is folklore. A proof can be found in [2].

Proposition 2. *If f is a degree-d polynomial computed by a $\{+, \times\}$-circuit C of size s such that the fan-in of each $+$-gate is unbounded and the fan-in of each $\times$-gate is bounded by 2, then there exists a circuit $\tilde{C}$ of size $s(d+1)^2$ with $d+1$ outputs $O_0, O_1, \ldots, O_d$ such that:*

- *the fan-in of each $+$-gate is unbounded,*
- *the fan-in of each $\times$-gate is bounded by 2,*
- *for each i, the gate O_i computes the homogeneous part of f of degree i,*
- *$\tilde{C}$ is homogeneous,*
- *the degree of each gate of $\tilde{C}$ equals the degree of the polynomial computed by this gate.*

We define $\times$-*balanced* $\{\times, +, \odot\}$-circuits.

Definition 4. *A $\{\times, +, \odot\}$-circuit C is called $\times$-balanced if and only if all the following properties are verified:*

- *the fan-in of each $\times$-gate is at most 5,*
- *the fan-in of each $+$-gate is unbounded,*
- *the fan-in of each $\odot$-gate is at most 2,*
- *for each $\times$-gate α, each one of its arguments is of degree at most half of the degree of α.*

The last condition can not be true for the multiplication by a scalar. It is the reason, we introduced the operator $\odot$.

The next proposition which is implicitly a first result of parallelization is almost the same result that we can find in Section 2 in [1] or in Theorem 2.7 in [8]. We give a proof in appendix.

Proposition 3. *Let f be a homogeneous degree-d polynomial computed by a size-s circuit $\tilde{C}$ defined as in the conclusion of Proposition 2. Then f is computed by a homogeneous $\times$-balanced $\{\times, +, \odot\}$-circuit of size $s^6 + s^4 + 1$ and of degree d.*

Agrawal and Vinay already noticed that Valiant, Skyum, Berkowitz and Rackoff's famous result [9] is a direct corollary of this proposition.

Corollary 2. *Let f be a polynomial of degree d computed by a circuit of size s. Then f is computed by a $\{+, \times\}$-circuit of size $(sd)^{O(1)}$ and of depth $O(\log(s)\log(d))$ where each $+$ and $\times$-gate is of fan-in 2.*

5   Proof of Theorem 1

For realizing the reduction to depth four, Koiran begins by transforming the circuit into an equivalent arithmetic branching program. Then, he parallelizes the branching program, and finally comes back to the circuits. The problem with this strategy is that the transformation from circuits to branching programs requires an increase in the size of our object. If the circuit is of size s, our new branching program is of size $s^{\log(d)}$. Here, the approach is to directly parallelize the circuit without using arithmetic branching programs in intermediate steps.

The idea is to split the circuit into two parts: gates of degree lower than $\sqrt{d}$ and gates of larger degree. Furthermore, a circuit such that the degree of each gate is bounded by $\sqrt{d}$ computes a degree-$\sqrt{d}$ polynomial and so can be written as a sum of at most $s^{O(\sqrt{d})}$ monomials. Then, if each part of our circuit computes polynomials of degrees bounded by $\sqrt{d}$, we just have to get the two depth-2 circuits and connect them together. The main difficulty comes from the fact it is not always true that the sub-circuit obtained by the gates of degree larger than $\sqrt{d}$ is of degree smaller than $\sqrt{d}$. For example, for the comb graph with $n-1$ ×-gates and n variable inputs:

$$x_1 \cdot (x_2 \cdot (x_3 \cdot (\ldots)))$$

the degree of the first part is $\sqrt{n}$, but the degree of the second one is $n - \sqrt{n}$.

In fact, following ideas from [6], we are going to cut not exactly at level $\sqrt{d}$. It will give a sharper result.

Lemma 1. *Let f be a homogeneous n-variate polynomial of degree d computed by a homogeneous ×-balanced $\{\times, +, \odot\}$-circuit C of size σ. Then f is computed by a homogeneous $\sum \prod^{[15a]} \sum \prod^{[\frac{d}{a}]}$ circuit of size $1 + \binom{\sigma+15a}{15a} + \sigma + \sigma\binom{n+\frac{d}{a}}{\frac{d}{a}} + n$ for any positive constant a smaller than d.*

To get nicer expressions, we will use the following consequence of Stirling's formula: (A proof appears in [1])

Lemma 2.

$$\binom{k+l}{l} = 2^{O\left(l + l \log \frac{k}{l}\right)}$$

First, let us see how Lemma 1 implies Theorem 1.

Proof (Proof of Theorem 1). Let f be an n-variate polynomial computing by a circuit of size s and degree d. Let $\tilde{C}$ be the homogeneous circuit for the polynomial that we get by Proposition 2. The circuit $\tilde{C}$ is of size $t = s(d+1)^2$ and computes all polynomials $f_0, \ldots, f_d$ where f_i is the homogeneous part of f of degree i. Then for each $i \le d$, there exists a homogeneous ×-balanced circuit C of size $\sigma = t^6 + t^4 + 1$ computing f_i. We apply Lemma 1 for the circuit C with $a = \sqrt{d \frac{\log n}{\log \sigma}}$. Using Lemma 2 we get a homogeneous $\sum \prod \sum \prod$ circuit of size $1 + \binom{\sigma+15a}{15a} + \sigma + \sigma\binom{n+\frac{d}{a}}{\frac{d}{a}} + n = 2^{O\left(\sqrt{d \log \sigma \log n}\right)}$. At the end, we just have to add together homogeneous parts f_i. As $\sigma = O(s^6 d^{12})$, it gives a $2^{O\left(\sqrt{d \log(ds) \log n}\right)}$ upper bound for the size.

Remark 1. Choosing the easier assignment $a = \sqrt{d}$ gives a $2^{O\left(\sqrt{d} \log(ds)\right)}$ upper bound.

Proving Lemma 1 will complete the proof.

Proof (Proof of Lemma 1). We define circuits C_1 and C_2 as follows. C_1 is the circuit we get by keeping only gates of C of degree $< \frac{d}{a}$. Circuit C_2 is made up of the remaining gates (i.e., those of degree $\geq \frac{d}{a}$) and of the inputs of these gates. These inputs are the only gates which belong both in C_1 and in C_2.

Each gate α of C_1 has degree at most $\frac{d}{a}$, so computes a polynomial of degree at most $\frac{d}{a}$. By homogeneity of C, the polynomial computed in α is homogeneous. Consequently, α is a homogeneous sum of at most $\binom{n+\frac{d}{a}}{\frac{d}{a}}$ monomials, and so, can be computed by a homogeneous depth-2 circuit of size $1 + \binom{n+\frac{d}{a}}{\frac{d}{a}} + n$ (The "1" encodes the +-gate, the "n" encodes the input gates, and the remainder encodes the $\times$-gates).

We are going to show now that the degree of C_2 is bounded by $15a$.

Let δ be the degree of C_2. There exists a degree-δ monomial m in C_2. Let T be a parse tree computing m.

We partition the set of $\times$-gates of T into 3 sets:

- $\mathcal{G}_0 = \{\alpha \in T | \alpha$ is a $\times$-gate and all children of α are leaves of $T\}$
- $\mathcal{G}_1 = \{\alpha \in T | \alpha$ is a $\times$-gate and exactly one child of α is not a leaf$\}$
- $\mathcal{G}_2 = \{\alpha \in T | \alpha$ is a $\times$-gate and at least two children of α are not leaves$\}$.

Then, if we consider the sub-tree S of T with only gates of C_2, then $\mathcal{G}_0$ are leaves of S, $\mathcal{G}_1$ are internal vertices of fan-in 1 and $\mathcal{G}_2$ are internal vertices of fan-in at least 2.

The proof is in two parts. First we upperbound the size of the sets $\mathcal{G}_0$, $\mathcal{G}_1$ and $\mathcal{G}_2$. Then, we upperbound the degree of m.

In C, the degree of "m" is at least the sum of the degrees of the gates of $\mathcal{G}_0$ (since two of these gates can not appear on the same path). Each one of these gates is in C_2, so is of degree at least $\frac{d}{a}$ in C. As m is of degree at most d in C, it means that the number of gates in $\mathcal{G}_0$ is at most a.

In C, the degree of "m" is at least the sum of the degrees of the leaves directly connected to a gate of $\mathcal{G}_1$. For each gate α of $\mathcal{G}_1$, exactly one of its inputs β is in C_2, hence of degree at least $\frac{d}{a}$ in C. By Proposition 3, the degree of α is at least two times the degree of β, it yields that the sum of degrees of inputs of α which are in C_1 is also at least $\frac{d}{a}$. Then, the number of vertices in $\mathcal{G}_1$ is at most a.

Finally, in a tree, the number of leaves is larger than the number of vertices of fan-in at least 2. Then in S, we get that:

$$|\mathcal{G}_2| \leq |\mathcal{G}_0| \leq a.$$

In C_2, the degree of the monomial m is the number of leaves labelled by a non-constant leaf in T. We match each leaf with the first $\times$-gate which is connected to it. As in T, the fan-in of the $\times$-gates is bounded by 5, the fan-in of the +-gates is bounded by 1 and each $\odot$-gates add only one constant input, then the number of variable leaves connected to a particular $\times$-gate is at most 5. So the number of leaves in T is at most:

$$5 \times (|\mathcal{G}_0| + |\mathcal{G}_1| + |\mathcal{G}_2|) \leq 15a.$$

This proves that the degree of C_2 is at most $15a$. Then, the number of inputs of C_2 is bounded by the number of gates in C_1 and so in C (which is σ). So, there exists a depth-2 circuit which compute C_2, of size $1 + \binom{\sigma + 15a}{15a} + \sigma$ with as inputs the gates of C_1.

Consequently, each polynomial f_i can be computed by a homogeneous $\sum \prod^{[a]} \sum \prod^{[\frac{d}{a}]}$ circuit of size at most $1 + \binom{\sigma + 15a}{15a} + \sigma + \sigma\binom{n + \frac{d}{a}}{\frac{d}{a}} + n$.

6   A Lower Bound

In [5], it was proved that if a homogeneous depth-four circuit computing PERM_n has its bottom fan-in bounded by t, then the size of the circuit is at least $2^{\Omega(\frac{n}{t})}$. But what happens if bottom multiplication gates all have a large fan-in? We show that this implies a similar lower bound for the size of the circuit:

Theorem 3. *If C is a homogeneous $\sum \prod \sum \prod$ circuit which computes PERM_n (or DET_n) such that the fan-in of each bottom multiplication gate is at least t, then the size of C is at least $2^{\Omega(t \log(n))}$.*

Our approach is only based on counting the number of monomials. We begin by some definitions.

Definition 5. *For a multivariate polynomial $f(\mathbf{x}) = \sum_{i=1}^{m_f} a_i \mathbf{x_i}$, we will denote $\mathcal{M}_f$ the set $\{ \mathbf{x_i} \mid \mathbf{x_i} \text{ is a monomial of } f \}$. If E is a set of polynomials, we also define $\mathcal{M}_E = \bigcup_{f \in E} \mathcal{M}_f$.*

We can notice $\mathcal{M}_{\text{PERM}_n} = \{ x_{1,\sigma(1)} \cdots x_{n,\sigma(n)} \mid \sigma \in \mathfrak{S}_n \}$. So, $|\mathcal{M}_{\text{PERM}_n}| = n!$.

Definition 6. *Let E be a set of polynomials. Let us denote*

$$E^+ = \{ f_1 + \ldots + f_m \mid m \in \mathbb{N} \text{ and } \forall i \le m, f_i \in E \}$$
$$\text{and } E^{\times k} = \{ f_1 \times \ldots \times f_m \mid m \le k \text{ and } \forall i \le m, f_i \in E \}$$

Lemma 3. *Let E be a set of polynomials. Then,*

$$\mathcal{M}_{E^+} = \mathcal{M}_E \text{ and } |\mathcal{M}_{E^{\times s}}| \le (|\mathcal{M}_E| + 1)^s.$$

Proof. If $\mathbf{x}$ is a monomial in $\mathcal{M}_{E^+}$, it means there exist polynomials $f_1, \ldots, f_m$ in E such that $\mathbf{x}$ is a monomial of $f_1 + \ldots + f_m$. Then there exists $i \le m$ such that $\mathbf{x}$ is a monomial of f_i and so $\mathbf{x}$ is an element of $\mathcal{M}_E$. Hence $\mathcal{M}_{E^+} \subseteq \mathcal{M}_E$. Moreover, as $E \subseteq E^+$, we get $\mathcal{M}_E \subseteq \mathcal{M}_{E^+}$.

Moreover, if $\mathbf{x}$ is a monomial in $\mathcal{M}_{E^{\times s}}$, it means there exist polynomials $f_1, \ldots, f_m$ in E such that $\mathbf{x}$ is a monomial of $f_1 \times \ldots \times f_m$ with $m \le s$. It implies that $\mathbf{x} \in \{ \mathbf{x_1} \times \ldots \times \mathbf{x_m} \mid m \le s \text{ and } \mathbf{x_i} \in \mathcal{M}_E \}$. That is to say, $\mathbf{x} \in \{ \mathbf{x_1} \times \ldots \times \mathbf{x_s} \mid \text{ and } \mathbf{x_i} \in (\mathcal{M}_E \cup \{1\}) \}$. It proves the lemma.

Let C be a $\sum \prod \sum \prod$ circuit. The gates of the circuit are layered into five levels. Inputs are at level 0, multiplication gates at levels 1 and 3 and addition gates at levels 2 and 4. For each level i, let us denote s_i the number of gates at this level, t_i an upper bound on the fan-in of these gates and E_i the set of polynomials computed at this level.

Lemma 4. *Any* $\sum \prod \sum \prod$ *circuit that computes* PERM_n *(or* DET_n*) such that the fan-in of the multiplication gates at level 3 is bounded by* v *must have size* $\exp\left[\Omega\left(\frac{n}{v}\log(n)\right)\right]$.

Proof. We notice that the hypothesis in the lemma about the bound of the fan-in just states that $t_3 \leq v$.

The polynomials in E_1 are just monomials. So, $|\mathcal{M}_{E_1}| \leq s_1$. We have:

$$E_4 \subseteq E_3^+, \quad E_3 \subseteq E_2^{\times t_3} \text{ and } E_2 \subseteq E_1^+.$$

Then by Lemma 3,

$$|\mathcal{M}_{E_4}| \leq (s_1 + 1)^{t_3} \leq (s_1 + 1)^v.$$

However, as PERM_n is an element of E_4, we also have:

$$|\mathcal{M}_{E_4}| \geq |\mathcal{M}_{\text{PERM}_n}| = n!.$$

So, $s_1 \geq (n!)^{\frac{1}{v}} - 1 = 2^{\Omega\left(\frac{n}{v}\log(n)\right)}$

The result of this lemma directly implies Theorem 3.

Proof (Proof of Theorem 3). Let C be a homogeneous $\sum \prod \sum \prod$ circuit which computes PERM_n (or DET_n) such that the fan-in of each bottom gate is at least t. It implies that the degree of each gate at level 1 and 2 is at least t. As the circuit is homogeneous, the degree of a gate at level 3 is upperbounded by n and lowerbounded by t times the number of inputs of this gate. Consequently, in C, the fan-in of the multiplication gates at level 3 is bounded by $\frac{n}{t}$. Then Lemma 4 implies the theorem.

In fact, for computing the determinant, we can also notice that the fan-in of multiplication gates in the depth-four circuits that we get either in [7] or here in Section 5, is linear in $\sqrt{n}$. It implies that in this case, the bounds are tight.

Corollary 3. *If C is a $\sum \prod \sum \prod$ circuit which computes DET_n such that the fan-in of each bottom multiplication gate is $\Omega(\sqrt{n})$ or such that the fan-in of each multiplication gate of level 3 is $O(\sqrt{n})$, then the minimal size of C is $2^{\Theta(\sqrt{n}\log(n))}$.*

Proof. Koiran's result [7] implies that there exist depth-four circuits for DET_n of size $2^{O(\sqrt{n}\log n)}$ such that all multiplication gates have fan-in bounded by $O(\sqrt{n})$. For the lowerbound, the case where the bottom fan-in is lowerbounded by $\Omega(\sqrt{n})$ is given by Theorem 3. The case where the fan-in of gates of level 3 is bounded by $O(\sqrt{n})$ is given by Lemma 4.

Consequently, it would be an interesting question to know the lower bound on the size of an homogeneous circuit computing DET_n. In [5] the authors show that if the circuit is such that the fan-in of bottom gates is bounded by $O(\sqrt{n})$, then the size is $2^{\Omega(\sqrt{n})}$. Here, we show that if all bottom fan-in are lowerbounded by $\Omega(\sqrt{n})$, then the size is $2^{\Omega(\sqrt{n}\log n)}$. What happens if in the circuit, there are some bottom gates with a large fan-in and some bottom gates with a small fan-in?

Question 1. Is it true that if $\mathcal{C}$ is a homogeneous depth-four circuit which computes DET_n then the size of $\mathcal{C}$ is at least $2^{\Omega(\sqrt{n})}$?

Acknowledgments. The author thanks Pascal Koiran for helpful discussions and comments on this work.

References

1. Agrawal, M., Vinay, V.: Arithmetic circuits: A chasm at depth four. In: Proceedings-Annual Symposium on Foundations of Computer Science, pp. 67–75 (2008)
2. Allender, E., Jiao, J., Mahajan, M., Vinay, V.: Non-commutative arithmetic circuits: depth reduction and size lower bounds. Theoretical Computer Science 209(1-2), 47–86 (1998)
3. Bürgisser, P.: Completeness and Reduction in Algebraic Complexity Theory. Algorithms and Computation in Mathematics, vol. 7. Springer (2000)
4. Chen, X., Kayal, N., Wigderson, A.: Partial Derivatives in Arithmetic Complexity and Beyond. Foundations and Trends in Theoretical Computer Science (2011)
5. Gupta, A., Kamath, P., Kayal, N., Saptharishi, R.: Approaching the chasm at depth four. In: Proceedings of the Conference on Computational Complexity, CCC (2013)
6. Gupta, A., Kamath, P., Kayal, N., Saptharishi, R.: Arithmetic circuits: A chasm at depth three. Electronic Colloquium on Computational Complexity (2013)
7. Koiran, P.: Arithmetic circuits: The chasm at depth four gets wider. Theoretical Computer Science 448, 56–65 (2012)
8. Shpilka, A., Yehudayoff, A.: Arithmetic circuits: A survey of recent results and open questions. Foundations and Trends in Theoretical Computer Science, vol. 5 (2010)
9. Valiant, L., Skyum, S., Berkowitz, S., Rackoff, C.: Fast parallel computation of polynomials using few processors. SIAM Journal on Computing 12(4), 641–644 (1983)
10. von zur Gathen, J.: Feasible arithmetic computations: Valiant's hypothesis. Journal of Symbolic Computation 4(2), 137–172 (1987)

Appendix: Proof of Proposition 3

Let f be a homogeneous polynomial computed by a circuit $\tilde{C}$ of size s like in the proposition. First, we can delete the "calculus with constants". To do that, we just have to replace recursively each gate such that all entries are constants by the constant value of this gate. Then, by homogeneity, constants can not be entries of a +-gate. Then, for each ×-gate such that one entry is a constant, we replace the ×-gate by a scalar ⊙-gate. We can notice that this transformation does not increase the size of the circuit. Second, we can reorder the children of the ×-gates and of the ⊙-gates so as to for each one of these gates, the degree of the rightmost child is larger or equals the degree of the other child. We get a circuit C_1 of size s.

We define now a new circuit C_2 which satisfies the criteria of the proposition. For each pair of gates α and β in C_1, we define the gate $(\alpha; \beta)$ in C_2 as follows:

- If β is a leaf, then $[(\alpha; \beta)]$ equals the sum of the parse trees rooted in α such that β appears in the rightmost path (ie, β is the leaf of the rightmost path).
- If β is not a leaf, then $[(\alpha; \beta)]$ equals the sum of the parse trees rooted in α such that β appears in the rightmost path and where the subcircuit rooted in β is deleted. That is as if we replace the gate β by the input 1 in the rightmost path and we compute $[(\alpha; \beta)]$ with $\beta = 1$ a leaf.

We notice here that it is easy to get the polynomial computed by the gate α: $[\alpha] = \sum_{l \text{ leaf}} [(\alpha; l)]$.

Now, we show how one can compute the value of the gates $(\alpha; \beta)$.

- If β does not appear on the rightmost path of a parse tree rooted in α, then $(\alpha; \beta) = 0$.
- If α is a leaf, then $(\alpha; \alpha) = \alpha$ and else $(\alpha; \alpha) = 1$.
- Otherwise α and β are two different gates and α is not a leaf. If α is a +-gate, then $[(\alpha; \beta)]$ is simply the sum of all $[(\alpha', \beta)]$, where α' is a child of α.
- If α is a $\odot$-gate, then one child is a constant c and the other child is a gate α'. Then $(\alpha; \beta)$ is simply the scalar operation $[(\alpha; \beta)] = [(c; c)] \odot [(\alpha'; \beta)]$.
- If α is a $\times$-gate. There are two cases.
  - First case: β is a leaf. Then $\deg(\alpha) > \deg(\beta) = 1$. On each rightmost path ending on β of a parse tree rooted in α, there exists exactly one $\times$-gate γ and its right child on this path γ_r such that:

$$\deg(\gamma) > \deg(\alpha)/2 \geq \deg(\gamma_r). \tag{1}$$

Conversely, we notice that for each gate γ satisfying (1), if $[(\alpha; \gamma)]$ and $[(\gamma_r; \beta)]$ are not zero, then γ is on a rightmost path from α to β. Then,

$$[(\alpha; \beta)] = \sum_{l \text{ leaf, } \gamma \text{ $\times$-gate verifying (1)}} [(\alpha; \gamma)][(\gamma_l; l)][(\gamma_r; \beta)].$$

One can notice that $\deg(\alpha; \beta) = \deg(\alpha)$. Using (1):

$$\deg(\alpha; \gamma) = \deg(\alpha) - \deg(\gamma) < \deg(\alpha)/2$$
$$\deg(\gamma_r; \beta) = \deg(\gamma_r) \leq \deg(\alpha)/2$$
$$\deg(\gamma_l; l) = \deg(\gamma_l) \leq \deg(\gamma_r) \leq \deg(\alpha)/2.$$

Consequently, $[(\alpha; \beta)]$ is computed by a depth-2 circuit of size at most $s^2 + 1$: a +-gate where each child is a $\times$-gate of fan-in 3. Each child of these $\times$-gates is of degree at most the half of the degree of the $\times$-gate.
  - Second case: β is not a leaf. Then there exists on every rightmost paths rooted in α a $\times$-gate γ and its child on this path γ_r such that:

$$\deg(\gamma) \geq (\deg(\alpha) + \deg(\beta))/2 > \deg(\gamma_r). \tag{2}$$

Then by the same argument,

$$[(\alpha; \beta)] = \sum_{l \text{ leaf, } \gamma \text{ $\times$-gate verifying (2)}} [(\alpha; \gamma)][(\gamma_l; l)][(\gamma_r; \beta)]. \tag{3}$$

We have this time with (2):

$$\deg(\alpha;\beta) = \deg(\alpha) - \deg(\beta)$$
$$\deg(\alpha;\gamma) = \deg(\alpha) - \deg(\gamma) \leq (\deg(\alpha) - \deg(\beta))/2$$
$$\deg(\gamma_r;\beta) = \deg(\gamma_r) < (\deg(\alpha) - \deg(\beta))/2.$$

The problem here is that the degree of $(\gamma_l; l)$ could be larger than the average of the degrees of α and β. If γ_l is of degree at most 1 (and so exactly 1) and if the degree of $(\alpha;\beta)$ is also 1, then $\gamma = \alpha$ (they are the same gate) and $(\gamma_r;\beta)$ is of degree 0 and computes a constant c_γ. Hence,

$$[(\alpha;\beta)] = \sum_{l \text{ leaf, } \gamma \text{ }\times\text{-gate verifying (2)}} [c_\gamma] \odot [(\gamma_l; l)].$$

Now, if the degree of γ_l is again 1 but if $(\alpha;\beta)$ is of degree at least 2, then the computation of the gate $(\alpha;\beta)$ by the formula (3) works (ie., the degree of $(\gamma_l; l)$ is smaller than half of the degree of $(\alpha;\beta)$). Otherwise, the degree of γ_l is at least 2 and at most $\deg(\alpha;\beta)$. As l is a leaf, we can apply the first case (even if γ_l is not a $\times$-gate). There exists also on every rightmost paths rooted in γ_l a $\times$-gate μ and its child on this path μ_r such that:

$$\deg(\mu) > \deg(\gamma_l)/2 \geq \deg(\mu_r). \tag{4}$$

Then,

$$[(\alpha;\beta)] = \sum_{l_1,l_2,\gamma,\mu} [(\alpha;\gamma)][(\gamma_r;\beta)][(\gamma_l;\mu)][(\mu_l;l_2)][(\mu_r;l_1)] \tag{5}$$

where the sum is taken over all l_1, l_2 leaves, γ $\times$-gate verifying (2) and μ $\times$-gate verifying (4).

The degrees of the gates $(\gamma_l;\mu)$, $(\mu_l;l_2)$ and $(\mu_r;l_1)$ are bounded by half of the degree of γ_l. Hence, $[(\alpha;\beta)]$ is computed by a depth-2 size-$s^4 + 1$ circuit. The $\times$-gates are of fan-in bounded by 5 and the degree of their children is bounded by half their degree.

Consequently, for each gates α and β in C_1, the gate $(\alpha;\beta)$ is computed in C_2 by a sub-circuit of size at most $s^4 + 1$. At the end we get a circuit of size at most $s^6 + s^2$ which computes all gates $(\alpha;\beta)$. Finally, f is computed by a circuit of size bounded by $s^6 + s^2 + 1$.

That proves the proposition.

Parameterized Algorithms for Module Motif

Meirav Zehavi

Department of Computer Science, Technion - Israel Institute of Technology,
Haifa 32000, Israel
meizeh@cs.technion.ac.il

Abstract. Module Motif is a pattern matching problem that was introduced in the context of biological networks. Informally, given a multiset
of colors P and a graph H whose nodes have sets of colors, it asks if P
occurs in a module of H (i.e. in a set of nodes that have the same neighborhood outside the set). We present three parameterized algorithms for
this problem that measure similarity between matched colors and handle
deletions and insertions of colors to P. We observe that the running time
of two of them might be essentially tight and prove that the problem is
unlikely to admit a polynomial kernel.

Keywords: parameterized algorithm, module motif, pattern matching.

1 Introduction

Graph Motif is an important problem in the analysis of biological networks that
has received considerable attention since its introduction by Lacroix et al. [16]
(see [1, 2, 4, 5, 7, 10–13, 15, 19–21]). Informally, given a multiset of colors P and
a graph H whose nodes have sets of colors, it asks if P occurs in a subtree of H.

A module M of a graph $H = (V, E)$ is a subset of V s.t. $\forall u, v \in M, \forall x \notin
M : \{v, x\} \in E$ iff $\{u, x\} \in E$ [8] (i.e. it is a set of nodes that have the same
neighborhood outside the set). Rizzi et al. [21] replace the connectivity constraint
of Graph Motif with modularity and thus define Module Motif. They present
biological justifications for considering this replacement.

Module Motif

- Input: A set of colors C, a multiset P of colors from C, a graph $H = (V, E)$
  and $Col : V \to 2^C$.
- Decide if there is a module M of H and $m : M \to C$ s.t.
  1. $\forall v \in M : m(v) \in Col(v)$.
  2. $\forall c \in C : c$ occurs in P exactly $|\{v \in M : m(v) = c\}|$ times.

In the limited case of the problem we have that $|Col(v)| = 1$ for all $v \in V$. We
denote $MaxCol = \max_{v \in V}\{|Col(v)|\}$.

We use the common O^* and $\widetilde{O}$ notation to hide factors polynomial and poly-
logarithmic in the input size respectively. As in [21], we denote $k = |P|$.

K. Chatterjee and J. Sgall (Eds.): MFCS 2013, LNCS 8087, pp. 825–836, 2013.
© Springer-Verlag Berlin Heidelberg 2013

Rizzi et al. [21] prove that the limited case of the problem is NP-complete even if P is a set and H is a collection of paths of size 3. They denote by c the number of different colors in P and give an $O^*((k(2^c)^k)^{k+1}c^k)$ time algorithm for the problem, which is not satisfying for practical issues. They also give an $O^*(2^k)$ time and space algorithm for the limited case. They use a modular tree decomposition [24] and dynamic programming.

Rizzi et al. [21] leave the handling of deletions and insertions of colors to P as an open problem. Several Graph Motif algorithms handle deletions and insertions (see e.g. [19]), and we shall handle them in a similar manner. In biological networks such as protein-protein interaction networks and metabolic networks, we can measure the similarity between the elements that the colors represent (see e.g. [18, 23]) and thus assess the relevance of a solution to Module Motif. This approach is used in the alignment query problem [19], which is another pattern matching problem that was introduced in the context of biological networks. Thus we define the following generalization of Module Motif:

General Module Motif

- Input: A set of colors C, a multiset P of colors from C, a graph $H = (V, E)$, $col : V \rightarrow C$, $Col : V \rightarrow 2^C$, $\Delta : C \times C \rightarrow \mathbb{R}$, $D \in \mathbb{N}^0$, $I \in \mathbb{N}^0$ and $S \in \mathbb{R}$.
- Decide if there is a module M of H, $U \subseteq M$ and $m : U \rightarrow C$ s.t.
  1. $\forall v \in U : m(v) \in Col(v)$.
  2. $\forall c \in C : c$ occurs in P at least $|\{v \in U : m(v) = c\}|$ times.
  3. $|U| = k - D = |M| - I$.
  4. $\sum_{v \in U} \Delta(col(v), m(v)) \geq S$.

The function col assigns to each node a color that represents the element that the node represents, and Col assigns to each node a set of colors that represent elements that are similar to the element that the node represents. For example, in protein-protein interaction networks $col(v)$ is the color that represents the protein that v represents, and $Col(v)$ is the set of colors that represent proteins that are homologous to the protein that v represents. Δ is symmetric, and D, I and S stand for *Deletions*, *Insertions* and *Score*, respectively.

Conditions 1 and 2 are similar to those of Module Motif. Condition 3 states that we delete D occurrences from P and add I occurrences to P. Condition 4 states that the score of the solution is at least S. We denote the occurrences of colors in P by $p_1, p_2, \ldots, p_k$, the color of an occurrence p_i by $col(p_i)$ and $S^+ = \max\{S, 1\}$. We assume WLOG that $D \leq k$, $I \leq |V|$ and $MaxCol \leq \min\{|C|, k\}$.

Module Motif is the special case of this problem in which $(\forall c, c' \in C : \Delta(c, c') = 0)$ and $I = D = S = 0$.

Fixed-parameter algorithms [17] are an approach to solve NP-hard problems by confining the combinatorial explosion to a parameter t. More precisely, a problem is fixed-parameter tractable with respect to a parameter t if an instance of size n can be solved in $O^*(f(t))$ time for some function f. We shall consider the parameters k and $k - D$.

Our algorithms use modular tree decompositions (see Section 2). Section 3 presents an $O^*(2^k)$ time General Module Motif algorithm that uses dynamic programming. In Section 4 we use it and improved color coding [14] to design an $O^*(4.314^{k-D})$ time General Module Motif algorithm. Section 5 presents an $O^*(2^{k-D}S^+)$ time algorithm for General Module Motif where S and $\Delta(c, c')$ are nonnegative integers $\forall c, c' \in C$. It uses the algebraic framework of Björklund et al. [3]. We get an $O^*(2^k)$ time and $O^*(1)$ space Module Motif algorithm, which improves the $O^*((k(2^c)^k)^{k+1}c^k)$ time algorithm of Rizzi et al. [21]. In Section 6 we observe that some of our results might be essentially tight. Finally, in Section 7, we prove that Module Motif is unlikely to admit a polynomial kernel.

Due to space constraints, some of the proofs are omitted.

2   Sets of Disjoint Sets Instead of Modules

In this section we show that we can focus our attention on finding a certain subset of a set of disjoint sets instead of a certain module of a graph.

A modular tree decomposition of a graph $H = (V, E)$ is a linear-sized representation of all its modules. It includes a rooted tree $T = (V_T, E_T)$, a function $f : V_T \rightarrow 2^V$ and a function $g : V_T \rightarrow \{0, 1\}$. A formal definition appears in [24]. In this paper we are only interested in its following properties (which are also used in [21]):

1. M is a module of H iff there is a node $v \in V_T$ s.t. $M = f(v)$ or $(g(v) = 1$ and there is a subset U of the set of children of v s.t. $M = \bigcup_{u \in U} f(u))$.
2. Every $v, u \in V_T$ that have the same father satisfy $f(v) \cap f(u) = \emptyset$.
3. $|V_T| \leq 2|V| - 1$.

Now we define a problem whose algorithms (which we design in Sections 3 and 5) and modular tree decompositions will help us solve General Module Motif.

Set Motif

- Input: A set of colors C, a multiset P of colors from C, a set $\mathcal{A}$ of disjoint sets, $col : \bigcup \mathcal{A} \rightarrow C$, $Col : \bigcup \mathcal{A} \rightarrow 2^C$, $\Delta : C \times C \rightarrow \mathbb{R}$, $D \in \mathbb{N}^0$, $I \in \mathbb{N}^0$ and $S \in \mathbb{R}$.
- Decide if there is $\widetilde{\mathcal{A}} \subseteq \mathcal{A}$, $U \subseteq \bigcup \widetilde{\mathcal{A}}$ and $m : U \rightarrow C$ s.t.
    1. $\forall a \in U : m(a) \in Col(a)$.
    2. $\forall c \in C : c$ occurs in P at least $|\{a \in U : m(a) = c\}|$ times.
    3. $|U| = k - D = |\bigcup \widetilde{\mathcal{A}}| - I$.
    4. $\sum_{a \in U} \Delta(col(a), m(a)) \geq S$.

We denote the sets of $\mathcal{A}$ by $A_1, A_2, \ldots, A_{|\mathcal{A}|}$. For each $A_i \in \mathcal{A}$, we denote its elements by $a_1^i, a_2^i, \ldots, a_{|A_i|}^i$.

Let SetALG$(C, P, \mathcal{A}, col, Col, \Delta, D, I, S)$ be a Set Motif algorithm that uses $t(MaxCol, k, |\bigcup \mathcal{A}|, D, I, S)$ time and $s(MaxCol, k, |\bigcup \mathcal{A}|, D, I, S)$ space. Next we present a procedure that solves General Module Motif by using a modular tree decomposition and SetALG.

ModuleALG$(C, P, H = (V, E), col, Col, \Delta, D, I, S)$

1. Compute a modular tree decomposition $(T = (V_T, E_T), f, g)$ of H in $O(|V|^2)$ time and $O(|V|)$ space (e.g. use the algorithm of Tedder et al. [24]).
2. $\forall v \in V_T$:
   If SetALG$(C, P, \{f(v)\}, col, Col, \Delta, D, I, S)$ accepts or $(g(v) = 1$ and SetALG$(C, P, \{f(u) : u$ is a child of $v\}, col, Col, \Delta, D, I, S)$ accepts): Accept.
3. Reject.

Property 2 of a modular tree decomposition implies that $\forall v \in V_T$, $\{f(u) : u$ is a child of $v\}$ is a set of disjoint subsets of V. Thus each call of SetALG is legal and uses $O(t(MaxCol, k, |V|, D, I, S))$ time and $O(s(MaxCol, k, |V|, D, I, S))$ space. By Property 3 of a modular tree decomposition, we get that Step 2 uses $O(|V| t(MaxCol, k, |V|, D, I, S))$ time and $O(s(MaxCol, k, |V|, D, I, S))$ space.

Property 1 of a modular tree decomposition implies that if (M, U, m) is a solution, then there is $v \in V_T$ s.t. $(\{f(v)\}, U, m)$ is a solution to Set Motif whose input is $(C, P, \{f(v)\}, col, Col, \Delta, D, I, S)$ or $(g(v) = 1$ and $(\{f(u) : u \in M\}, U, m)$ is a solution to Set Motif whose input is $(C, P, \{f(u) : u$ is a child of $v\}, col, Col, \Delta, D, I, S))$, and if there is $v \in V_T$ s.t. $(\widetilde{\mathcal{A}}, U, m)$ is a solution to Set Motif whose input is $(C, P, \{f(v)\}, col, Col, \Delta, D, I, S)$ or $(g(v) = 1$ and it is a solution to Set Motif whose input is $(C, P, \{f(u) : u$ is a child of $v\}, col, Col, \Delta, D, I, S))$, then $(\bigcup \widetilde{\mathcal{A}}, U, m)$ is a solution.

We get the following lemma:

Lemma 1. *ModuleALG solves General Module Motif in* $O(|V| t(MaxCol, k, |V|, D, I, S) + |V|^2)$ *time and* $O(s(MaxCol, k, |V|, D, I, S) + |V|)$ *space.*

3   An $O^*(2^k)$-Time Algorithm

We present an $O^*(2^k)$ time and space Set Motif algorithm that uses dynamic programming. By Lemma 1, we thus get an $O^*(2^k)$ time and space General Module Motif algorithm, which we denote by ALG3.

First we define the partial solutions that we consider in our computation.

Definition 1. *Given a multiset* $\widetilde{P} \subseteq P$ *s.t.* $|\widetilde{P}| \leq k - D$, $1 \leq i \leq |\mathcal{A}|$, $1 \leq j \leq |A_i|$ *and* $0 \leq ins \leq I$, $Sol(\widetilde{P}, i, j, ins)$ *denotes the set of tuples* $(\widetilde{\mathcal{A}}, U, m)$ *s.t.* $\widetilde{\mathcal{A}} \subseteq \{A_1, A_2, \dots, A_{i-1}, \{a_1^i, a_2^i, \dots, a_j^i\}\}$, $U \subseteq \bigcup \widetilde{\mathcal{A}}$, $m : U \to C$ *and*

1. $\forall a \in U : m(a) \in Col(a)$.
2. $\forall c \in C : c$ *occurs in* $\widetilde{P}$ *exactly* $|\{a \in U : m(a) = c\}|$ *times.*
3. $|U| = |\widetilde{P}| = |\bigcup \widetilde{\mathcal{A}}| - ins$.

We use two matrices:

1. M has a cell $[\widetilde{P}, i, j, ins]$ for every multiset $\widetilde{P} \subseteq P$ s.t. $|\widetilde{P}| \leq k - D$, $1 \leq i \leq |\mathcal{A}|$, $1 \leq j \leq |A_i|$ and $0 \leq ins \leq I$.
2. N has a cell $[\widetilde{P}, i, ins]$ for every multiset $\widetilde{P} \subseteq P$ s.t. $|\widetilde{P}| \leq k - D$, $1 \leq i \leq |\mathcal{A}|$ and $0 \leq ins \leq I$.

The cells of the matrices hold the following scores:

1. $M[\widetilde{P}, i, j, ins] =$
   $\max_{(\widetilde{A}, U, m) \in Sol(\widetilde{P}, i, j, ins) \text{ s.t. } \{a_1^i, a_2^i, \ldots, a_j^i\} \in \widetilde{A}} \{\sum_{a \in U} \Delta(col(a), m(a))\}.$
2. $N[\widetilde{P}, i, ins] = \max_{(\widetilde{A}, U, m) \in Sol(\widetilde{P}, i, |A_i|, ins)} \{\sum_{a \in U} \Delta(col(a), m(a))\}.$

Lemma 2. *The cells of M and N can be computed in $O(2^k MaxCol|\bigcup A|I)$ time and $O(2^k I)$ space by using dynamic programming.*

After we compute the cells, we accept iff $\max_{\widetilde{P} \subseteq P \text{ s.t. } |\widetilde{P}| = k - D} \{N[\widetilde{P}, |A|, I]\}$ $\geq S$, and the correctness immediately follows from the definition of Set Motif. We get the following theorem:

Theorem 1. *ALG3 solves General Module Motif in $O(2^k MaxCol|V|^2 I)$ time and $O(2^k I + |V|)$ space.*

4   An $O^*(4.314^{k-D})$-Time Algorithm

We use improved color coding [14] and ALG3 (see Section 3) to design an $O^*(4.314^{k-D})$ time and $O^*(2.463^{k-D})$ space General Module Motif algorithm.

The idea of the algorithm is to introduce a new multiset of colors that represents P and whose size is a function of $k - D$. We reduce the size from k to a function of $k - D$ by allowing each occurrence in the new multiset to represent several occurrences in P. Then we call ALG3 with the new multiset and get a parameterized algorithm whose parameter is $k - D$.

Let $P^* = \{p_1^*, p_2^*, \ldots, p_{1.3(k-D)}^*\}$ be a set of $1.3(k - D)$ new colors. For each $v \in V$, define $col^*(v) = c_v^*$, where c_v^* is a new color. Define $C^* = P^* \cup \{c_v^* : v \in V\}$.

Given $f : P \to P^*$, we define $Col_f : V \to 2^{C^*}$ as follows $\forall v \in V$:

- $Col_f(v) = \{f(p_i) : p_i \in P, col(p_i) \in Col(v)\}.$

We also define a symmetric $\Delta_f : C^* \times C^* \to \mathbb{R}$ as follows $\forall c, d \in C^*$:

1. If $c = c_v^*$ and $d \in Col_f(v)$ for a node $v \in V$:
   $\Delta_f(c, d) = \Delta_f(d, c) = \max_{p_i \in P \text{ s.t. } f(p_i) = d \wedge col(p_i) \in Col(v)} \Delta(col(v), col(p_i)).$
2. Else: $\Delta_f(c, d) = \Delta_f(d, c) = 0.$

Now we present the algorithm:

ALG4$(C, P, H = (V, E), col, Col, \Delta, D, I, S)$:

1. Repeat 1.752^{k-D} times:
   (a) $\forall p_i \in P$, independently and uniformly at random assign a color from P^*. Denote the resulting function by $f : P \to P^*$.
   (b) If ALG3$(C^*, P^*, H, col^*, Col_f, \Delta_f, 0.3(k - D), I, S)$ accepts: Accept.
2. Reject.

Next we prove the correctness of ALG4.

Observation 1. *If there is $f : P \to P^*$ s.t. (M, U, m^*) is a solution to $(C^*, P^*, H, col^*, Col_f, \Delta_f, 0.3(k - D), I, S)$, then there is a solution to the input instance.*

Proof. $\forall v \in U$, we define $m(v) = col(p_i)$ for some $p_i \in P$ s.t. $f(p_i) = m^*(v)$, $col(p_i) \in Col(v)$ and $\Delta_f(col^*(v), f(p_i)) = \Delta(col(v), col(p_i))$ (our definitions of Col_f and Δ_f imply that such a p_i exists). Clearly M is a module of H, $U \subseteq M$ and $m : U \to C$. Moreover, they fulfill the following conditions:

1. Let $v \in U$. We have that $m(v) \in Col(v)$.
2. Let $c \in C$. If c does not occur in P, then $|\{v \in U : m(v) = c\}| = 0$. If there are two different $v, u \in U$ s.t. we defined both $m(v)$ and $m(u)$ as c because of the same occurrence p_i of c in P, then $|\{v \in U : m^*(v) = f(p_i)\}| \geq 2$, which is a contradiction (since $f(p_i)$ occurs once in P^*). Thus the number of occurrences of c in P is at least $|\{v \in U : m(v) = c\}|$.
3. $|U| = |P^*| - 0.3(k - D) = k - D$.
4. $|U| = |M| - I$.
5. $\sum_{v \in U} \Delta(col(v), m(v)) = \sum_{v \in U} \Delta_f(col^*(v), m^*(v)) \geq S$.

Thus (M, U, m) is a solution to the input instance.   □

Observation 2. *If there is a solution (M, U, m) to the input instance, then with probability at least $1 - 1/e$, there is an iteration where we choose $f : P \to P^*$ s.t. there is a solution to $(C^*, P^*, H, col^*, Col_f, \Delta_f, 0.3(k - D), I, S)$.*

Proof. Each $c \in C$ occurs in P at least $|\{v \in U : m(v) = c\}|$ times. Thus $\forall c \in C$ we can denote some set of $|\{v \in U : m(v) = c\}|$ occurrences of c in P by $Occ(c)$.

We denote by F the set of functions $f : P \to P^*$ that satisfy $\forall p_i, p_j \in \bigcup_{c \in C} Occ(c)$ s.t. $i \neq j$, $f(p_i) \neq f(p_j)$. Note that $|\bigcup_{c \in C} Occ(c)| = k - D$.

Given sets A, $B \subseteq A$ and C s.t. $1.3|B| = |C|$, Hüffner et al. [14] prove that if we repeat $1.752^{|B|}$ times the step

– $\forall a \in A$, independently and uniformly at random assign an element from C.

then with probability at least $1 - 1/e$, there is a step where we assign to each element in B a different element from C.

We get that with probability at least $1 - 1/e$, there is an iteration where we choose $f \in F$. Next consider an iteration that corresponds to such a f.

For each $v \in U$, choose a different occurrence p_i of $m(v)$ in $Occ(m(v))$ and denote $m^*(v) = f(p_i)$. Clearly M is a module of H, $U \subseteq M$ and $m^* : U \to C^*$. Moreover, they fulfill the following conditions:

1. Let $v \in U$. $m(v) \in Col(v)$, and thus $m^*(v) \in \{f(p_i) : p_i \in P, col(p_i) \in Col(v)\}$. Therefore $m^*(v) \in Col_f(v)$.
2. Let $c \in C^*$. If $c \notin P^*$, then $|\{v \in U : m^*(v) = c\}| = 0$, and if $c \in P^*$, then by our choice of f, $|\{v \in U : m^*(v) = c\}| \leq 1$.
3. $|U| = k - D = |P^*| - 0.3(k - D)$.
4. $|U| = |M| - I$.
5. $\sum_{v \in U} \Delta_f(col^*(v), m^*(v)) \geq \sum_{v \in U} \Delta(col(v), m(v)) \geq S$.

Thus (M, U, m^*) is a solution to $(C^*, P^*, H, col^*, Col_f, \Delta_f, 0.3(k-D), I, S)$.   □

The time and space complexities of Step 1b are $O(2^{1.3(k-D)}(k-D)|V|^2 I)$ and $O(2^{1.3(k-D)} I + |V|)$ respectively. Since we repeat it 1.752^{k-D} times, we get that ALG4 uses $O(4.314^{k-D} k |V|^2 I)$ time and $O(2.463^{k-D} I + |V|)$ space.

We get the following theorem:

Theorem 2. *ALG4 solves General Module Motif. It has one-sided error and uses $O(4.314^{k-D} k |V|^2 I)$ time and $O(2.463^{k-D} I + |V|)$ space.*

5   An $O^*(2^{k-D} S^+)$-Time Algorithm

We present an $O^*(2^{k-D} S^+)$ time and $O^*(S^+)$ space algorithm for Set Motif where S and $\Delta(c, c')$ are nonnegative integers $\forall c, c' \in C$. By Lemma 1, we thus get an $O^*(2^{k-D} S^+)$ time and $O^*(S^+)$ space algorithm for General Module Motif where S and $\Delta(c, c')$ are nonnegative integers $\forall c, c' \in C$, which we denote by ALG5. Since in Module Motif, $(\forall c, c' \in C : \Delta(c, c') = 0)$ and $D = I = S = 0$, we get an $O^*(2^k)$ time and $O^*(1)$ space Module Motif algorithm.

We use the algebraic framework of Björklund et al. [3]. We express our parameterized problem by associating monomials with potential solutions. A correct solution is associated with a unique monomial, and a monomial which is not associated with a correct solution is associated with an even number of potential solutions. Having a polynomial which is the sum of such monomials, we need to determine whether it has a monomial whose coefficient is odd.

Given $i \in \mathbb{N}$, we denote $[i] = \{1, 2, \ldots, i\}$.

5.1   Potential Solutions

First we define the potential solutions (PS stands for Potential Solutions).

Definition 2. *Given $0 \le size \le k - D$, $1 \le i \le |\mathcal{A}|$, $1 \le j \le |A_i|$, $0 \le ins \le I$ and $sco \in \mathbb{N}^0$, $PS(size, i, j, ins, sco)$ is the set of tuples $(\widetilde{\mathcal{A}}, U, m, l)$ s.t. $\widetilde{\mathcal{A}} \subseteq \{A_1, A_2, \ldots, A_{i-1}, \{a_1^i, a_2^i, \ldots, a_j^i\}\}$, $U \subseteq \bigcup \widetilde{\mathcal{A}}$, $m : U \to P$, $l : U \to [k - D]$ and*

1. $\forall a \in U : col(m(a)) \in Col(a)$.
2. $|U| = size = |\bigcup \widetilde{\mathcal{A}}| - ins$.
3. $\sum_{a \in U} \Delta(col(a), col(m(a))) = sco$.

We denote:

1. $Bij = \bigcup_{s \in \mathbb{N}^0 \text{ s.t. } s \ge S} \{(\widetilde{\mathcal{A}}, U, m, l) \in PS(k-D, |\mathcal{A}|, |A_{|\mathcal{A}|}|, I, s) : l \text{ is bijective}\}$.
2. $BijM = \{(\widetilde{\mathcal{A}}, U, m, l) \in Bij : m \text{ is injective}\}$.

Observation 3. *The input instance has a solution iff $BijM \ne \emptyset$.*

5.2   Associating Monomials with Potential Solutions

We introduce an indeterminate $x(A_i)$ for all $A_i \in \mathcal{A}$, an indeterminate $y(a,p)$ for all $a \in \bigcup \mathcal{A}$ and $p \in P$, and an indeterminate $z(p,i)$ for all $p \in P$ and $i \in [k-D]$. We order them arbitrarily as $q_1, q_2, \ldots, q_r$ where $r = |\mathcal{A}| + k(|\bigcup \mathcal{A}| + k - D)$.

Definition 3. *Given* $(\widetilde{\mathcal{A}}, U, m, l) \in PS(size, i, j, ins, sco)$, *its monomial is*

$$mon(\widetilde{\mathcal{A}}, U, m, l) = \prod_{A_i \in \widetilde{\mathcal{A}}} x(A_i) \prod_{a \in U} y(a, m(a)) z(m(a), l(a)).$$

Observation 4. *If* $(\widetilde{\mathcal{A}}, U, m, l) \in BijM$ *and* $(\widetilde{\mathcal{A}}', U', m', l') \in Bij \setminus \{(\widetilde{\mathcal{A}}, U, m, l)\}$, *then* $mon(\widetilde{\mathcal{A}}, U, m, l) \neq mon(\widetilde{\mathcal{A}}', U', m', l')$.

Observation 5. *There is a fixed-point-free involution (i.e. a permutation that is its own inverse)* $f : Bij \setminus BijM \to Bij \setminus BijM$ *s.t.* $mon(\widetilde{\mathcal{A}}, U, m, l) = mon(f(\widetilde{\mathcal{A}}, U, m, l))$ *for all* $(\widetilde{\mathcal{A}}, U, m, l) \in Bij \setminus BijM$.

5.3   Evaluating the Sum of the Monomials

Denote $POL(q_1, q_2, \ldots, q_r) = \sum_{(\widetilde{\mathcal{A}}, U, m, l) \in Bij} mon(\widetilde{\mathcal{A}}, U, m, l)$. Observations 3, 4 and 5 imply that the input instance has a solution iff POL has a monomial with an odd coefficient. We evaluate the polynomial over the finite field $\mathbb{F}_q$ (i.e. the finite field of order q), where $q = 2^{\lceil \log_2(e(3k+I)) \rceil}$. Since this field has characteristic 2, we get the following observation:

Observation 6. *The input instance has a solution iff* $POL \not\equiv 0$.

Denote the image of a function $l : U \to [k-D]$ by $l(U)$.
   Given $L \subseteq [k-D]$, denote:

1. $PS_L = \{(\widetilde{\mathcal{A}}, U, m, l) \in \bigcup_{s \in \mathbb{N}^0 \text{ s.t. } s \geq S} PS(k-D, |\mathcal{A}|, |\mathcal{A}_{|\mathcal{A}|}|, I, s) : l(U) \subseteq L\}$.
2. $POL_L(q_1, q_2, \ldots, q_r) = \sum_{(\widetilde{\mathcal{A}}, U, m, l) \in PS_L} mon(\widetilde{\mathcal{A}}, U, m, l)$.

By the inclusion-exclusion principle and since $\mathbb{F}_q$ has characteristic 2, we get the following observation:

Observation 7. $POL(q_1, q_2, \ldots, q_r) = \sum_{L \subseteq [k-D]} POL_L(q_1, q_2, \ldots, q_r)$.

Lemma 3. *Given* $L \subseteq [k-D]$ *and* $b_1, b_2, \ldots, b_r \in \mathbb{F}_q$, $POL_L(b_1, b_2, \ldots, b_r)$ *can be evaluated in* $\widetilde{O}(S^+ \log S^+ k | \bigcup \mathcal{A} | I)$ *time and* $\widetilde{O}(S^+ k I)$ *space by using dynamic programming.*

5.4   The Algorithm

SetALG5$(C, P, \mathcal{A}, col, Col, \Delta, D, I, S)$

1. Select $b_1, b_2, \ldots, b_r \in \mathbb{F}_q$ independently and uniformly at random.
2. $SUM \Leftarrow 0$.

3. For all $L \subseteq [k - D]$: $SUM \Leftarrow SUM + POL_L(b_1, b_2, \ldots, b_r)$.
4. Accept iff $SUM \neq 0$.

The proof of the following lemma appears in [22]:

Lemma 4. *Let $p(x_1, x_2, \ldots, x_n)$ be a nonzero polynomial of total degree at most d over the finite field F. Then, for $b_1, b_2, \ldots, b_n \in F$ selected independently and uniformly at random: $Pr(p(b_1, b_2, \ldots, b_n) \neq 0) \geq 1 - d/|F|$.*

By Observations 6 and 7 and Lemmas 3 and 4 (note that the degree of POL is at most $3k + I$), we have that:

Lemma 5. *SetALG5 solves Set Motif where $S \in \mathbb{N}^0$ and $\Delta(c, c') \in \mathbb{N}^0$ for all $c, c' \in C$. It has one-sided error and uses $\widetilde{O}(2^{k-D}S^+ \log S^+ k |\bigcup \mathcal{A}| I)$ time and $\widetilde{O}(k(S^+ I + |\bigcup \mathcal{A}|))$ space.*

We get the following theorem:

Theorem 3. *ALG5 solves General Module Motif where $S \in \mathbb{N}^0$ and $\Delta(c, c') \in \mathbb{N}^0$ for all $c, c' \in C$. It has one-sided error and uses $\widetilde{O}(2^{k-D}S^+ \log S^+ k |V|^2 I)$ time and $\widetilde{O}(k(S^+ I + |V|))$ space.*

6 The Tightness of the Results

In this section we observe that further improvement on the running time of the algorithms we have presented in Sections 3 and 5 is substantially harder.

Set Cover

– Input: $t \in \mathbb{N}^0$, a set of sets $\mathcal{S} = \{S_1, S_2, \ldots, S_m\}$ and $A = \bigcup \mathcal{S}$ where $|A| = n$.
– Decide if there is $\widetilde{\mathcal{S}} \subseteq \mathcal{S}$ s.t. $|\widetilde{\mathcal{S}}| = t$ and $A = \bigcup \widetilde{\mathcal{S}}$.

We assume WLOG that $2 \leq t \leq m$.

We prove that for any $\epsilon > 0$, there is $\epsilon' > 0$ s.t. the existence of an $O^*((2-\epsilon)^k)$ time algorithm for Module Motif even if P is a set and H is a collection of paths implies an $O^*((2 - \epsilon')^n)$ time Set Cover algorithm. Thus the existence of an $O^*((2-\epsilon)^k)$ General Module Motif algorithm or an $O^*((2-\epsilon)^{k-D}S^+)$ algorithm for General Module Motif where $S \in \mathbb{N}^0$ and $\Delta(c, c') \in \mathbb{N}^0$ for all $c, c' \in C$ implies an $O^*((2 - \epsilon')^n)$ time Set Cover algorithm.

Björklund et al. [4] use this approach to claim that further improvement on the running time of their Graph Motif algorithm is substantially harder. Set Cover is a well-known problem researched for decades, what suggests that an $O^*((2 - \epsilon)^n)$ time algorithm for it, if possible at all, would be a major breakthrough in the field. The nonexistence of such an algorithm has already been used as an assumption for proving hardness results [9].

Theorem 4. *Let ALG6 be an $O^*((2 - \epsilon)^k)$ time algorithm for Module Motif where P is a set and H is a collection of paths. Then there is $\epsilon' > 0$ s.t. there is an $O^*((2 - \epsilon')^n)$ time Set Cover algorithm.*

Proof. Consider the following algorithm:

SetCoverALG$(t, \mathcal{S} = \{S_1, S_2, \ldots, S_m\}, A)$

1. Construct an instance of Module Motif where P is a set and H is a collection of paths as follows:
   (a) $C = P = A \cup \{c_1, c_2, \ldots, c_t\}$. Note that $k = n + t$.
   (b) $V = \{v_{i,r,j} : 1 \leq i \leq m, 1 \leq r \leq |S_i|, 1 \leq j \leq r\} \cup$
       $\{u_{i,r} : 1 \leq i \leq m, 0 \leq r \leq |S_i|\}$.
   (c) $E = \{\{v_{i,r,j}, v_{i,r,j+1}\} : 1 \leq i \leq m, 1 \leq r \leq |S_i|, 1 \leq j \leq r - 1\} \cup$
       $\{\{v_{i,r,r}, u_{i,r}\} : 1 \leq i \leq m, 1 \leq r \leq |S_i|\}$.
   (d) $\forall v_{i,r,j} \in V : Col(v_{i,r,j}) = S_i$.
   (e) $\forall u_{i,r} \in V : Col(u_{i,r}) = \{c_1, c_2, \ldots, c_t\}$.
2. Accept iff ALG6$(C, P, H = (V, E), Col)$ accepts.

Lemma 6. *SetCoverALG solves Set Cover.*

Since ALG6 runs in $O^*((2-\epsilon)^k)$ time, we get that SetCoverALG runs in $O^*((2-\epsilon)^{n+t})$ time. Cygan et al. [9] prove that the existence of an $O^*((2 - \epsilon)^{n+t})$ time Set Cover algorithm implies that there is $\epsilon' > 0$ s.t. there is an $O^*((2 - \epsilon')^n)$ time Set Cover algorithm, and thus we get the theorem.   □

7   No Polynomial Kernel

We denote by LMMS the **L**imited case of **M**odule **M**otif where P is a **S**et and the parameter is k. We prove that even LMMS is unlikely to admit a polynomial kernel (i.e. there is no $O^*(1)$ time algorithm that for every input of this problem, its output is an input of this problem that has a solution iff the original input has a solution and whose size is polynomial in k).

Given an input IN of a parameterized problem, we denote by $p(IN)$ the parameter of IN (e.g. the parameter of the input $(C, P, H, col, Col, D, I, S)$ of General Module Motif that we have considered in Section 3 is $k = |P|$). The unparameterized version of a parameterized problem L is $\widetilde{L} = \{x\#^{p(x)} : x$ is an input of L that has a solution$\}$ where $\#$ is a symbol that does not appear in L [6]. We also need the following definition of Bodlaender et al. [6]:

Definition 4. *A parameterized problem L is compositional if it has a compositional algorithm, which is an algorithm whose input is a tuple $(x_1, x_2, \ldots, x_t)$ of inputs of L s.t. $p(x_1) = p(x_2) = \ldots = p(x_t)$, runs in time polynomial in $\sum_{1 \leq i \leq t} |x_i| + p(x_1)$, and outputs an input y of L s.t. (y has a solution iff $\exists 1 \leq i \leq t$ s.t. x_i has a solution) and $p(y)$ is polynomial in $p(x_1)$.*

The proof of the following theorem appears in [6]:

Theorem 5. *If a compositional parameterized problem whose unparameterized version is NP-complete has a polynomial kernel, then $NP \subseteq coNP/Poly$.*

Rizzi et al. [21] prove that LMMS is NP-complete. Assume that there is an $O^*(1)$ time algorithm ALG7 for the unparameterized version of LMMS. Given an input x of LMMS, we can call ALG7 on $x\#^{p(x)}$ and answer the same. Since $|x\#^{p(x)}| = O(|x|)$, we thus get an $O^*(1)$ time LMMS algorithm and have a contradiction. We get the following observation:

Observation 8. *The unparameterized version of LMMS is NP-complete.*

Consider the following algorithm:

CompALG$((C_1, P_1, H_1 = (V_1, E_1), Col_1), \ldots, (C_t, P_t, H_t = (V_t, E_t), Col_t))$:

1. $C \Leftarrow P_1 \cup \{c^*\}$ where c^* is a new color.
2. $V \Leftarrow V_1 \cup V_2 \cup \ldots \cup V_t \cup \{v_1^*, v_2^*, \ldots, v_t^*\}$ where each v_i^* is a new node.
3. $E \Leftarrow E_1 \cup E_2 \cup \ldots \cup E_t \cup \{\{v_i^*, v\} : 1 \leq i \leq t, v \in V_i\}$.
4. For $i = 1, 2, \ldots, t$, let f_i be some bijective function $f_i : P_i \to P_1$.
5. Define $Col : V \to C$ as follows for $i = 1, 2, \ldots, t$:
   (a) $\forall v \in V_i$ s.t. $Col_i(v) = \{c\}$ for $c \in P_i$: $Col(v) = \{f_i(c)\}$.
   (b) $\forall v \in V_i$ s.t. $Col_i(v) = \{c\}$ for $c \notin P_i$: $Col(v) = \{c^*\}$.
   (c) $Col(v_i^*) = \{c^*\}$.
6. Return $(C, P_1, H = (V, E), Col)$.

Lemma 7. *CompALG is a compositional algorithm for LMMS.*

Theorem 5, Observation 8 and Lemma 7 imply the following theorem, which states that LMMs is unlikely to admit a polynomial kernel:

Theorem 6. *If LMMS admits a polynomial kernel, then $NP \subseteq coNP/Poly$.*

References

1. Ambalath, A.M., Balasundaram, R., Rao H., C., Koppula, V., Misra, N., Philip, G., Ramanujan, M.S.: On the kernelization complexity of colorful motifs. In: Raman, V., Saurabh, S. (eds.) IPEC 2010. LNCS, vol. 6478, pp. 14–25. Springer, Heidelberg (2010)
2. Betzler, N., Bevern, R., Fellows, M.R., Komusiewicz, C., Niedermeier, R.: Parameterized algorithmics for finding connected motifs in biological networks. IEEE/ACM Trans. Comput. Biol. Bioinform. 8(5), 1296–1308 (2011)
3. Björklund, A., Husfeldt, T., Kaski, P., Koivisto, M.: Narrow sieves for parameterized paths and packings. CoRR abs/1007.1161 (2010)
4. Björklund, A., Kaski, P., Kowalik, L.: Probably optimal graph motifs. In: Proc. STACS, pp. 20–31 (2013)
5. Blin, G., Sikora, F., Vialette, S.: Gramofone: a cytoscape plugin for querying motifs without topology in protein-protein interactions networks. In: Proc. BICoB, pp. 38–43 (2010)
6. Bodlaender, H.L., Downey, R.G., Fellows, M.R., Hermelin, D.: On problems without polynomial kernels. J. Comput. Syst. Sci. 75(8), 423–434 (2009)
7. Bruckner, S., Hüffner, F., Karp, R.M., Shamir, R., Sharan, R.: Topology-free querying of protein interaction networks. In: Batzoglou, S. (ed.) RECOMB 2009. LNCS, vol. 5541, pp. 74–89. Springer, Heidelberg (2009)

8. Chein, M., Habib, M., Maurer, M.C.: Partitive hypergraphs. Discrete Mathematics 37(1), 35–50 (1981)
9. Cygan, M., Dell, H., Lokshtanov, D., Marx, D., Nederlof, J., Okamoto, Y., Paturi, R., Saurabh, S., Wahlstrom, M.: On problems as hard as cnf-sat. In: Proc. CCC, pp. 74–84 (2012)
10. Dondi, R., Fertin, G., Vialette, S.: Weak pattern matching in colored graphs: minimizing the number of connected components. In: Proc. ICTCS, pp. 27–38 (2007)
11. Dondi, R., Fertin, G., Vialette, S.: Finding approximate and constrained motifs in graphs. In: Giancarlo, R., Manzini, G. (eds.) CPM 2011. LNCS, vol. 6661, pp. 388–401. Springer, Heidelberg (2011)
12. Fellows, M.R., Fertin, G., Hermelin, D., Vialette, S.: Upper and lower bounds for finding connected motifs in vertex-colored graphs. J. Comput. Syst. Sci. 77(4), 799–811 (2011)
13. Guillemot, S., Sikora, F.: Finding and counting vertex-colored subtrees. In: Hliněný, P., Kučera, A. (eds.) MFCS 2010. LNCS, vol. 6281, pp. 405–416. Springer, Heidelberg (2010)
14. Hüffner, F., Wernicke, S., Zichner, T.: Algorithm engineering for color-coding with applications to signaling pathway detection. Algorithmica 52(2), 114–132 (2008)
15. Koutis, I.: Constrained multilinear detection for faster functional motif discovery. Inf. Process. Lett. 112(22), 889–892 (2012)
16. Lacroix, V., Fernandes, C.G., Sagot, M.F.: Motif search in graphs: Application to metabolic networks. IEEE/ACM Trans. Comput. Biol. Bioinform. 3(4), 360–368 (2006)
17. Niedermeier, R.: Invitation to fixed-parameter algorithms. Oxford University Press (2006)
18. Pinter, R.Y., Rokhlenko, O., Yeger-Lotem, E., Ziv-Ukelson, M.: Alignment of metabolic pathways. Bioinformatics 21(16), 3401–3408 (2005)
19. Pinter, R.Y., Zehavi, M.: Algorithms for topology-free and alignment queries. Technion Technical Reports CS-2012-12 (2012)
20. Pinter, R.Y., Zehavi, M.: Partial information network queries. In: Proc. IWOCA (to appear, 2013)
21. Rizzi, R., Sikora, F.: Some results on more flexible versions of graph motif. In: Hirsch, E.A., Karhumäki, J., Lepistö, A., Prilutskii, M. (eds.) CSR 2012. LNCS, vol. 7353, pp. 278–289. Springer, Heidelberg (2012)
22. Schwartz, J.T.: Fast probabilistic algorithms for verification of polynomial identities. J. Assoc. Comput. Mach. 27(4), 701–717 (1980)
23. Shlomi, T., Segal, D., Ruppin, E., Sharan, R.: Qpath: a method for querying pathways in a protein-protein interaction networks. BMC Bioinform. 7, 199 (2006)
24. Tedder, M., Corneil, D., Habib, M., Paul, C.: Simpler linear-time modular decomposition via recursive factorizing permutations. In: Aceto, L., Damgård, I., Goldberg, L.A., Halldórsson, M.M., Ingólfsdóttir, A., Walukiewicz, I. (eds.) ICALP 2008, Part I. LNCS, vol. 5125, pp. 634–645. Springer, Heidelberg (2008)

On the Quantifier-Free Dynamic Complexity of Reachability

Thomas Zeume and Thomas Schwentick

TU Dortmund University
Germany

Abstract. The dynamic complexity of the reachability query is studied in the dynamic complexity framework of Patnaik and Immerman, restricted to quantifier-free update formulas.

It is shown that, with this restriction, the reachability query cannot be dynamically maintained, neither with binary auxiliary relations nor with unary auxiliary functions, and that ternary auxiliary relations are more powerful with respect to graph queries than binary auxiliary relations. Further results are obtained including more inexpressibility results for reachability in a different setting, inexpressibility results for some other queries and normal forms for quantifier-free update programs.

1 Introduction

In modern data management scenarios, data is subject to frequent changes. In order to avoid costly re-computations from scratch after each small update, one can try to (re-)use auxiliary data structures that has been already computed before to keep the information about the data up-to-date. However, the auxiliary data structures need to be updated dynamically whenever the data changes.

The descriptive dynamic complexity framework (short: dynamic complexity) introduced by Patnaik and Immerman [1] models this setting. It was mainly inspired by relational databases. For a relational database subject to change, auxiliary relations are maintained with the intention to help answering a query Q. When an update to the database, an insertion or deletion of a tuple, occurs, every auxiliary relation is updated through a first-order query that can refer to the database as well as to the auxiliary relations. A particular auxiliary relation shall always represent the answer to Q. The class of all queries maintainable in this way, and thus also in the core of SQL, is called DynFO.

Beyond query or view maintenance in databases we consider it an important goal to understand the dynamic complexity of fundamental algorithmic problems. Reachability in directed graphs is the most intensely investigated problem in dynamic complexity (and also much studied in dynamic algorithms and other dynamic contexts) and the main query studied in this paper. It is one of the simplest inherently recursive queries and thus serves as a kind of drosophila in the study of the dynamic maintainability of recursive queries by non-recursive means. It can be maintained with first-order update formulas supplemented by

K. Chatterjee and J. Sgall (Eds.): MFCS 2013, LNCS 8087, pp. 837–848, 2013.

counting quantifiers on general graphs [2] and with plain first-order update formulas on both acyclic graphs and undirected graphs [1]. However, it is not known whether Reachability on general graphs is maintainable with first-order updates. This is one of the major open questions in dynamic complexity.

All attempts to show that Reachability *cannot* be maintained in DYNFO have failed. In fact, there are no general inexpressibility results for DYNFO at all.[1] This seems to be due to a lack of understanding of the underlying mechanisms of DYNFO. To improve the understanding of dynamic complexity, mainly two kinds of restrictions of DYNFO have been studied: (1) limiting the information content of the auxiliary data by restricting the arity of auxiliary relations and functions and (2) reducing the amount of quantification in update formulas.

A study of bounded arity auxiliary relations was started in [3] and it was shown that unary auxiliary relations are not sufficient to maintain the reachability query with first-order updates. Further inexpressibility results for unary auxiliary relations were shown and an arity hierarchy for auxiliary relations was established. However, to separate level k from higher levels, database relations of arity larger than k were used. Thus the hierarchy has not yet been established for queries on graphs. In [4] it was shown that unary auxiliary relations are not sufficient to maintain Reachability for update formulas of any local logic. The proofs strongly use the "static" weakness of local logics and do not fully exploit the dynamic setting, as they only require update sequences of constant length.

The second line of research was initiated by Hesse [5]. He invented and studied the class DYNPROP of queries maintainable with quantifier-free update formulas. He proved that Reachability on deterministic graphs (i.e. graphs of unary functions) can be maintained with quantifier-free first-order update formulas.

There is still no proof that Reachability on general graphs cannot be maintained in DYNPROP. However, *some* inexpressibility results for DYNPROP have been shown in [6]: the alternating reachability query (on graphs with ∧- and ∨-nodes) is not maintainable in DYNPROP. Furthermore, on strings, DYNPROP exactly contains the regular languages (as Boolean queries on strings).

Contributions. The high-level goal of this paper is to achieve a better understanding of the dynamic maintainability of Reachability and dynamic complexity in general. Our main result is that the reachability query cannot be dynamically maintained by quantifier-free updates with binary auxiliary relations. This result is weaker than that of [3] in terms of the logic (quantifier-free vs. general first-order) but it is stronger with respect to the information content of the auxiliary data (binary vs. unary). We establish a strict hierarchy between DYNPROP for unary, binary and ternary auxiliary relations (this is still open for DYNFO).

We further show that Reachability is not maintainable with unary auxiliary *functions* (plus unary auxiliary relations). Although unary functions provide less information content than binary relations, they offer a very weak form of

---

[1] Of course, a query maintainable in DYNFO can be evaluated in polynomial time and thus queries that cannot be evaluated in polynomial time cannot be maintained in DYNFO either.

quantification in the sense that more elements of the domain can be taken into account by update formulas.

All these results hold in the setting of Patnaik and Immerman where update sequences start from an empty database as well as in the setting that starts from an arbitrary database, where the auxiliary data is initialized by an arbitrary function. We show that if, in the latter setting, the initialization mapping is permutation-invariant, quantifier-free updates cannot maintain Reachability even with auxiliary functions and relations of arbitrary arity. Intuitively a permutation-invariant initialization mapping maps isomorphic databases to isomorphic auxiliary data. A particular case of permutation-invariant initialization mappings, studied in [7], is when the initialization is specified by logical formulas. In this case, lower bounds for first-order update formulas have been obtained for several problems [7].

We transfer many of our inexpressibility results to the k-CLIQUE query for fixed $k \geq 3$ and the k-COL query for fixed $k \geq 2$.

Finally, we show two normal form results: every query in DYNPROP is already maintainable with negation-free quantifier-free formulas only as well as with conjunctive quantifier-free formulas only. Thus, one approach to inexpressibility proofs could be to use these syntactically restricted update formulas.

Related Work. We already mentioned the related work that is most closely related to our results. As said before, the reachability query has been studied in various dynamic frameworks, one of which is the Cell Probe model. In the Cell Probe model, one aims for lower bounds for the number of memory accesses of a RAM machine for static and dynamic problems. For dynamic reachability, lower bounds of order $\log n$ have been proved [8].

Outline. In Section 2 we fix our notation and in Section 3 we define our dynamic setting more precisely. The lower bound results for Reachability are presented in Section 4 (for auxiliary relations) and in Section 5 (for auxiliary functions). In Section 6 we transfer the lower bounds to other queries. Finally, we establish the two normal forms for DYNPROP in Section 7. Due to the space limit, most proofs are only available in the full version of the paper [9].

2   Preliminaries

A *domain* is a finite set. A *schema (or signature)* τ consists of a set τ_{rel} of relation symbols and a set τ_{const} of constant symbols together with an arity function $\text{Ar} : \tau_{\text{rel}} \mapsto \mathbb{N}$. A *database* $\mathcal{D}$ of schema τ with domain D is a mapping that assigns to every relation symbol $R \in \tau_{\text{rel}}$ a relation of arity $\text{Ar}(R)$ over D and to every constant symbol $c \in \tau_{\text{const}}$ an element (called *constant*) from D.

A τ-*structure* $\mathcal{S}$ is a pair $(D, \mathcal{D})$ where $\mathcal{D}$ is a database with schema τ and domain D. Sometimes we omit the schema when it is clear from the context. If $\mathcal{S}$ is a structure over domain D and D' is a subset of D that contains all constants of $\mathcal{S}$, then the substructure of $\mathcal{S}$ induced by D' is denoted by $\mathcal{S} \upharpoonright D'$. For two structures $\mathcal{S}$ and $\mathcal{T}$ we write $\mathcal{S} \simeq_\pi \mathcal{T}$ if $\mathcal{S}$ and $\mathcal{T}$ are isomorphic via π.

The k-ary atomic type[2] $\langle S, \vec{a} \rangle$ of a tuple $\vec{a} = (a_1, \ldots, a_k)$ over D with respect to a τ-structure S is the set of all atomic formulas $\varphi(\vec{x})$ with $\vec{x} = (x_1, \ldots, x_k)$ for which $\varphi(\vec{a})$ holds in S, where $\varphi(\vec{a})$ is short for the substitution of $\vec{x}$ by $\vec{a}$ in φ. We note that the atomic formulas can use constant symbols.

An *s-t-graph* is a graph $G = (V, E)$ with two distinguished nodes s and t. A *k-layered s-t-graph* G is a directed graph (V, E) in which $V - \{s, t\}$ is partitioned into k layers $A_1, \ldots, A_k$ such that every edge is from s to A_1, from A_k to t or from A_i to A_{i+1}, for some $i \in \{1, \ldots, k-1\}$. The *s-t-reachability query* s-t-REACH is a Boolean query that is true for an *s-t-graph* G, if and only if t can be reached from s in G.

3   Dynamic Queries and Programs

The following presentation follows [10] and [11]. For a more formal introduction, see [9].

A *dynamic instance* of the *s-t*-reachability query is a pair (G, α), where G is an *s-t*-graph and α is a sequence of changes to G, i.e. a sequence of edge insertions and deletions. The dynamic *s-t*-reachability query DYN(s-t-REACH) yields as result the relation that is obtained by first applying the updates from α to G and then evaluating the *s-t*-reachability query on the resulting graph. This setting extends to general databases and other queries in a straightforward way.

The database resulting from applying an update δ to a database $\mathcal{D}$ is denoted by $\delta(\mathcal{D})$. The result $\alpha(\mathcal{D})$ of applying a sequence of updates $\alpha = \delta_1 \ldots \delta_m$ to a database $\mathcal{D}$ is defined by $\alpha(\mathcal{D}) \overset{\text{def}}{=} \delta_m(\ldots (\delta_1(\mathcal{D})) \ldots)$.

Dynamic programs, to be defined next, consist of an initialization mechanism and an update program. The former yields, for every database $\mathcal{D}$ an initial state of P with initial auxiliary data (and possibly with further built-in data). The latter defines the new state, for each update in α. Built-in data never changes. In general, built-in data can be "simulated" by auxiliary data yet this does not (seem to) hold for all of the restricted kinds of auxiliary data studied in this paper.

A *dynamic schema* is a triple $(\tau_{\text{in}}, \tau_{\text{aux}}, \tau_{\text{bi}})$ of schemas of the input database, the built-in database and the auxiliary database, respectively. We always let $\tau \overset{\text{def}}{=} \tau_{\text{in}} \cup \tau_{\text{aux}} \cup \tau_{\text{bi}}$.

Definition 1. (Update program) An *update program* P over dynamic schema $(\tau_{\text{in}}, \tau_{\text{aux}}, \tau_{\text{bi}})$ is a set of first-order formulas (called *update formulas* in the following) that contains, for every $R \in \tau_{\text{aux}}$ and every $\delta \in \{\text{INS}_S, \text{DEL}_S\}$ with $S \in \tau_{\text{in}}$, an update formula $\phi_\delta^R(x_1, \ldots, x_l; y_1, \ldots, y_m)$ over the schema τ where l is the arity of S and m is the arity of R.

A *program state* S over dynamic schema $(\tau_{\text{in}}, \tau_{\text{aux}}, \tau_{\text{bi}})$ is a structure $(D, \mathcal{I}, \mathcal{A}, \mathcal{B})$ where D is the domain, $\mathcal{I}$ is a database over the input schema (the *current*

---

[2] As we only consider atomic types in this paper, we will often simply say type instead of atomic type.

database), $\mathcal{A}$ is a database over the auxiliary schema (the *auxiliary database*) and $\mathcal{B}$ is a database over the built-in schema (the *built-in database*).

The semantics of update programs is as follows. For an update $\delta(\vec{a})$ and program state $\mathcal{S} = (D, \mathcal{I}, \mathcal{A}, \mathcal{B})$ we denote by $P_\delta(\mathcal{S})$ the state $(D, \delta(\mathcal{I}), \mathcal{A}', \mathcal{B})$, where $\mathcal{A}'$ consists of relations $R' \stackrel{\text{def}}{=} \{\vec{b} \mid \mathcal{S} \models \phi_\delta^R(\vec{a}; \vec{b})\}$. The effect $P_\alpha(\mathcal{S})$ of an update sequence $\alpha = \delta_1 \ldots \delta_m$ to a state $\mathcal{S}$ is the state $P_{\delta_m}(\ldots(P_{\delta_1}(\mathcal{S}))\ldots)$.

Definition 2. (Dynamic program) A *dynamic program* is a triple (P, INIT, Q), where P is an update program over some dynamic schema $(\tau_{\text{in}}, \tau_{\text{aux}}, \tau_{\text{bi}})$, the tuple $\text{INIT} = (\text{INIT}_{\text{aux}}, \text{INIT}_{\text{bi}})$ consists of a function INIT_{aux} that maps τ_{in}-databases to τ_{aux}-databases and a function INIT_{bi} that maps domains to τ_{bi}-databases, and $Q \in \tau_{\text{aux}}$ is a designated *query symbol*.

A dynamic program $\mathcal{P} = (P, \text{INIT}, Q)$ *maintains* a dynamic query $\text{DYN}(\mathcal{Q})$ if, for every dynamic instance $(\mathcal{D}, \alpha)$, the relation $\mathcal{Q}(\alpha(\mathcal{D}))$ coincides with the query relation $Q^{\mathcal{S}}$ in the state $\mathcal{S} = P_\alpha(\mathcal{S}_{\text{INIT}}(\mathcal{D}))$, where $\mathcal{S}_{\text{INIT}}(\mathcal{D})$ is the initial state, i.e. $\mathcal{S}_{\text{INIT}}(\mathcal{D}) \stackrel{\text{def}}{=} (D, \mathcal{D}, \text{INIT}_{\text{aux}}(\mathcal{D}), \text{INIT}_{\text{bi}}(\mathcal{D}))$.

Several dynamic settings and restrictions of dynamic programs have been studied in the literature [1, 12, 7, 11]. Possible parameters are, for instance:

- the logic in which update formulas are expressed;
- whether in dynamic instances $(\mathcal{D}, \alpha)$, the initial database $\mathcal{D}$ is always empty;
- whether the initialization mapping INIT_{aux} is *permutation-invariant* (short: *invariant*), that is, whether $\pi(\text{INIT}_{\text{aux}}(\mathcal{D})) = \text{INIT}_{\text{aux}}(\pi(\mathcal{D}))$ holds, for every database $\mathcal{D}$ and permutation π of the domain; and
- whether there are any built-in relations at all.

Definition 3. (DYNFO, DYNPROP) DYNFO is the class of all dynamic queries maintainable by dynamic programs with first-order update formulas. DYNPROP is the subclass of DYNFO, where update formulas do not use quantifiers. A dynamic program is k-*ary* if the arity of its auxiliary relation symbols[3] is at most k. By k-ary DYNPROP (resp. DYNFO) we refer to dynamic queries that can be maintained with k-ary dynamic programs.

Thus in our basic setting the initialization mappings can be arbitrary. We will explicitly state when we relax this most general setting. Figure 1 illustrates the relationships between the various settings for the initialization. From now on we restrict our attention to quantifier-free update programs. Next, we give a non-trivial example for such a program.

Example 1. We provide a DYNPROP-program P for the dynamic variant of the Boolean query NONEMPTYSET, where, for a unary relation U subject to insertions and deletions of elements, one asks whether U is empty. It illustrates a technique to maintain lists with quantifier-free dynamic programs, introduced in [11, Proposition 4.5], which is used in some of our upper bounds.

---

[3] We note that this restriction does not apply to the built-in relations.

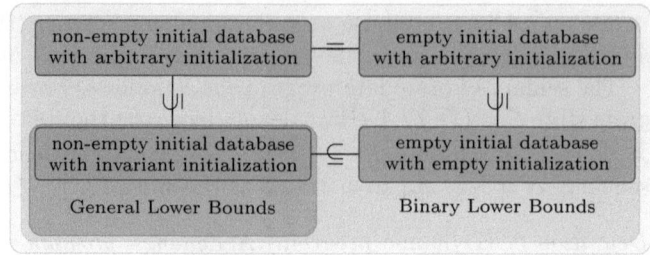

Fig. 1. Initializations considered in literature and lower bounds obtained in this paper for quantifier-free updates.

The program P is over auxiliary schema $\tau_{\text{aux}} = \{Q, \text{FIRST}, \text{LAST}, \text{LIST}\}$, where Q is the query bit (i.e. a 0-ary relation symbol), FIRST and LAST are unary relation symbols, and LIST is a binary relation symbol. The idea is to store in a program state $\mathcal{S}$ a list of all elements currently in U. The list structure is stored in the binary relation $\text{LIST}^{\mathcal{S}}$ such that $\text{LIST}^{\mathcal{S}}(a, b)$ holds for all elements a and b that are adjacent in the list. The first and last element of the list are stored in $\text{FIRST}^{\mathcal{S}}$ and $\text{LAST}^{\mathcal{S}}$, respectively. We note that the order in which the elements of U are stored in the list depends on the order in which they are inserted into the set.

For a given instance of NONEMPTYSET the initialization mapping initializes the auxiliary relations accordingly.

The update formulas for insertions are as follows:

$$\phi_{\text{INS}}^{\text{FIRST}}(a; x) \stackrel{\text{def}}{=} (\neg Q \wedge a = x) \vee (Q \wedge \text{FIRST}(x)) \qquad \phi_{\text{INS}}^{\text{LAST}}(a; x) \stackrel{\text{def}}{=} a = x$$

$$\phi_{\text{INS}}^{\text{LIST}}(a; x, y) \stackrel{\text{def}}{=} \text{LIST}(x, y) \vee (\text{LAST}(x) \wedge a = y) \qquad \phi_{\text{INS}}^{Q}(a) \stackrel{\text{def}}{=} \top$$

For deletions we only exhibit the update formula for LIST, the others are similar.

$$\phi_{\text{DEL}}^{\text{LIST}}(a; x, y) \stackrel{\text{def}}{=} x \neq a \wedge y \neq a \wedge \big(\text{LIST}(x, y) \vee (\text{LIST}(x, a) \wedge \text{LIST}(a, y))\big)$$

4   Lower Bounds for Dynamic Reachability

In this section we prove lower bounds for the maintainability of the dynamic s-t-reachability query $\text{DYN}(s\text{-}t\text{-REACH})$.

The proofs use the following tool which is a slight variation of Lemma 1 from [11]. The intuition is as follows. When updating an auxiliary tuple $\vec{c}$ after an insertion or deletion of a tuple $\vec{d}$, a quantifier-free update formula has access to $\vec{c}$, $\vec{d}$, and the constants only. Thus, if a sequence of updates changes only tuples from a substructure $\mathcal{S}'$ of $\mathcal{S}$, the auxiliary data of $\mathcal{S}'$ is not affected by information outside $\mathcal{S}'$. In particular, two isomorphic substructures $\mathcal{S}'$ and $\mathcal{T}'$ should remain isomorphic, when corresponding updates are applied to them.

We formalize the notion of corresponding updates as follows. Let π be an isomorphism from a structure $\mathcal{S}$ to a structure $\mathcal{T}$. Two updates $\delta(\vec{a})$ on $\mathcal{S}$ and $\delta(\vec{b})$ on $\mathcal{T}$ are said to be π-*respecting* if $\vec{b} = \pi(\vec{a})$. Two sequences $\alpha = \delta_1 \cdots \delta_m$ and $\beta = \delta_1' \cdots \delta_m'$ of updates respect π if, for every $i \leq m$, δ_i and δ_i' are π-respecting.

Lemma 2 (Substructure Lemma). *Let $\mathcal{P}$ be a DynProp program and $\mathcal{S}$ and $\mathcal{T}$ states of $\mathcal{P}$ with domains S and T, respectively. Further, let $S' \subseteq S$ and $T' \subseteq T$ such that $\mathcal{S} \upharpoonright S'$ and $\mathcal{T} \upharpoonright T'$ are isomorphic via π. Then $P_\alpha(\mathcal{S}) \upharpoonright S'$ and $P_\beta(\mathcal{T}) \upharpoonright T'$ are isomorphic via π for all π-respecting update sequences α, β on S' and T'.*

The Substructure Lemma can be applied along the following lines to prove that Dyn(s-t-Reach) cannot be maintained in some settings with quantifier-free updates. Towards a contradiction, assume that there is a quantifier-free program $\mathcal{P} = (P, \text{Init}, Q)$ that maintains Dyn(s-t-Reach). Then, find
- two states $\mathcal{S}$ and $\mathcal{T}$ occurring as states[4] of $\mathcal{P}$ with current graphs $G_\mathcal{S}$ and $G_\mathcal{T}$;
- substructures $\mathcal{S}'$ and $\mathcal{T}'$ of $\mathcal{S}$ and $\mathcal{T}$ isomorphic via π; and
- two π-respecting update sequences α and β such that $\alpha(G_\mathcal{S})$ is in s-t-Reach and $\beta(G_\mathcal{T})$ is not in s-t-Reach.

This yields the desired contradiction, since Q has the same value in $P_\alpha(\mathcal{S})$ and $P_\beta(\mathcal{T})$ by the Substructure Lemma.

How such states $\mathcal{S}$ and $\mathcal{T}$ can be obtained depends on the particular setting. Yet, Ramsey's Theorem and Higman's Lemma often prove to be useful for this task. Next, we present the variants of these theorems used in our proofs. We refer to [9] and [13, Proposition 2.5, page 3] for proofs.

Theorem 3 (Ramsey's Theorem for Structures). *For every schema τ and all natural numbers k and n there exists a number $R_{\tau,k}(n)$ such that, for every τ-structure $\mathcal{S}$ with domain A of size $R_{\tau,k}(n)$, every $\vec{d} \in A^k$ and every order $\prec$ on A, there is a subset B of A of size n with $B \cap \vec{d} = \emptyset$, such that, for every l, the type of $(\vec{a}, \vec{d})$ in $\mathcal{S}$ is the same, for all $\prec$-ordered l-tuples $\vec{a}$ over B.*

A word u is a *subsequence* of a word v, in symbols $u \sqsubseteq v$, if $u = u_1 \ldots u_k$ and $v = v_0 u_1 v_1 \ldots v_{k-1} u_k v_k$ for some words $u_1, \ldots, u_k$ and $v_0, \ldots, v_k$.

Theorem 4 (Higman's Lemma). *For every alphabet of size c and function $g : \mathbb{N} \to \mathbb{N}$ there is a natural number $H(c)$ such that in every sequence $(w_i)_{1 \leq i \leq H(c)}$ of $H(c)$ many words with $|w_i| \leq g(i)$ there are l and k with $l < k$ and $w_l \sqsubseteq w_k$.*

Before turning towards lower bounds for arbitrary initialization, we state a lower bound for the restricted setting of invariant initialization. Intuitively lower bounds in this setting can be obtained easier because invariant initialization cannot generate complex initial auxiliary structures such as lists from simple-structured input databases.

Theorem 5. *Dyn(s-t-Reach) cannot be maintained in DynProp with invariant initialization mapping and empty built-in schema. This holds even for 1-layered s-t-graphs.*

---

[4] I.e. $\mathcal{S} = \mathcal{P}_\delta(\mathcal{S}_{\text{Init}}(G))$ for some s-t-graph G, and likewise for $\mathcal{T}$.

4.1   A Binary Lower Bound

As already mentioned in the introduction, the proof that $\textsc{Dyn}(s\text{-}t\text{-}\textsc{Reach})$ is not in unary $\textsc{DynFO}$ in [3] uses constant-length update sequences, and is mainly an application of a locality-based static lower bound for monadic second order logic. This technique does not seem to generalize to binary $\textsc{DynFO}$. We prove the first unmaintainability result for $\textsc{Dyn}(s\text{-}t\text{-}\textsc{Reach})$ with respect to binary auxiliary relations.

Theorem 6. $\textsc{Dyn}(s\text{-}t\text{-}\textsc{Reach})$ *is not in binary* $\textsc{DynProp}$.

The proof of Theorem 6 will actually show that binary $\textsc{DynProp}$ cannot even maintain $\textsc{Dyn}(s\text{-}t\text{-}\textsc{Reach})$ on 2-layered $s\text{-}t$-graphs. These restricted graphs will then help us to separate binary $\textsc{DynProp}$ from ternary $\textsc{DynProp}$. This separation shows that the lower bound technique for binary $\textsc{DynProp}$ does not immediately transfer to ternary $\textsc{DynProp}$. At the moment we do not know whether it is possible to adapt the technique to full $\textsc{DynProp}$.

The following notion of homogeneous sets is used in the proof of Theorem 6. Let $\mathcal{S}$ be a structure of some schema τ and A, B disjoint subsets of the domain of $\mathcal{S}$. We say that B is $A\text{-}\prec$-*homogeneous up to arity* m, if for every $l \leq m$, all tuples $(a, \vec{b})$, where $a \in A$ and $\vec{b}$ is an $\prec$-ordered l-tuple over B, have the same type. We may drop the order $\prec$ from the notation if it is clear from the context, and we may drop A if $A = \emptyset$. We observe that if the maximal arity of τ is m and B is A-homogeneous up to arity m, then B is A-homogeneous up to arity m' for every m'. In this case we simply say B is A-*homogeneous*.

Lemma 7. *For every schema* τ *and natural number* n, *there is a natural number* $R_\tau^{hom}(n)$ *such that for any two disjoint subsets* A, B *of the domain of a* τ-*structure* $\mathcal{S}$ *with* $|A|, |B| \geq R_\tau^{hom}(n)$, *there are subsets* $A' \subseteq A$ *and* $B' \subseteq B$ *such that* $|A'|, |B'| = n$ *and* B' *is* A'-*homogeneous in* $\mathcal{S}$.

PROOF (OF THEOREM 6).   Let us assume, towards a contradiction, that the dynamic program $(P, \textsc{Init}, Q)$ over schema $\tau = (\tau_{\text{in}}, \tau_{\text{aux}}, \tau_{\text{bi}})$ with binary τ_{aux} maintains the dynamic $s\text{-}t$-reachability query for 2-layered $s\text{-}t$-graphs. We choose numbers n, n_1, n_2 and n_3 such that n_3 is sufficiently large with respect to τ, n_2 is sufficiently large with respect to n_3, n_2 is sufficiently large with respect to n_1 and n is sufficiently large with respect to n_1.

Let $G = (V, E)$ be a 2-layered $s\text{-}t$-graph with layers A, B, where A and B are both of size n and $E = \{(b, t) \mid b \in B\}$. Further, let $\mathcal{S} = (V, E, \mathcal{A}, \mathcal{B})$ be the state obtained by applying $\textsc{Init}$ to G.

We will first choose homogeneous subsets. By Lemma 7 and because n is sufficiently large, there are subsets A_1 and B_1 such that $|A_1| = |B_1| = n_1$ and B_1 is $A_1\text{-}\prec$-homogeneous in $\mathcal{S}$, for some order $\prec$. Next, let A_2 and B_2 be arbitrarily chosen subsets of A_1 and B_1, respectively, of size $|B_2| = n_2$ and $|A_2| = 2^{|B_2|}$, respectively. We note that B_2 is still A_2-homogeneous. In particular, B_2 is still A_2-homogeneous with respect to schema τ_{bi}. We associate with every subset $X \subseteq B_2$ a unique vertex a_X from A_2 in an arbitrary fashion.

Now, we define the update sequence α as follows.

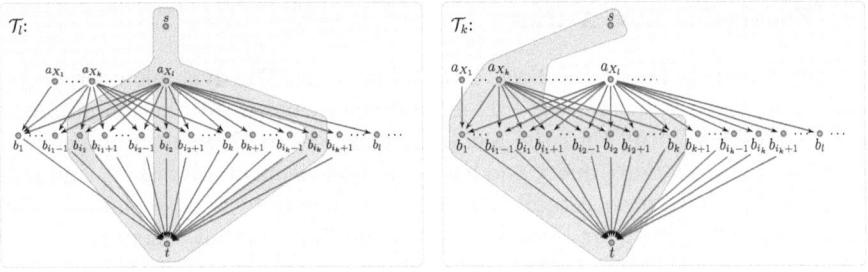

Fig. 2. The structure $\mathcal{S}'$ from the proof of Theorem 6. The isomorphic substructures $\mathcal{T}_k$ and $\mathcal{T}_l$ are highlighted in blue.

(α) For every subset X of B_2 and every $b \in X$ insert an edge (a_X, b), in some arbitrarily chosen order.

Let $\mathcal{S}' \stackrel{\text{def}}{=} (V, E', \mathcal{A}', \mathcal{B})$ be the state of $\mathcal{P}$ after applying α to $\mathcal{S}$, i.e. $\mathcal{S}' = P_\alpha(\mathcal{S})$. We observe that the built-in data has not changed, but the auxiliary data might have changed. In particular, B_2 is not necessarily A_2-homogeneous with respect to schema τ_{aux} in state $\mathcal{S}'$.

Our plan is to exhibit two sets X, X' such that $X \subsetneq X' \subseteq B_2$ such that the restriction of $\mathcal{S}'$ to $\{s, t, a_{X'}\} \cup X'$ contains an isomorphic copy of $\mathcal{S}'$ restricted to $\{s, t, a_X\} \cup X$. Then the Substructure Lemma will easily give us a contradiction.

By Ramsey's Theorem and because $|B_2|$ is sufficiently large with respect to n_2, there is a subset $B_3 \subseteq B_2$ of size n_3 such that B_3 is $\prec$-homogeneous in $\mathcal{S}'$. Let $b_1 \prec \ldots \prec b_{n_3}$ be an enumeration of the elements of B_3 and let $X_i \stackrel{\text{def}}{=} \{b_1, \ldots, b_i\}$, for every $i \in \{1, \ldots, n_3\}$.

Let $\mathcal{S}'_i$ denote the restriction of $\mathcal{S}'$ to $X_i \cup \{s, t, a_{X_i}\}$. For every i, we construct a word w_i of length i, that has a letter for every node in X_i and captures all relevant information about those nodes in $\mathcal{S}'_i$. More precisely, $w_i \stackrel{\text{def}}{=} \sigma_i^1 \cdots \sigma_i^i$, where for every i and j, σ_i^j is the binary type of (a_{X_i}, b_j).

Since B_3 is sufficiently large with respect to τ_{aux}, it follows, by Higman's Lemma, that there are k and l such that $k < l$ and $w_k \sqsubseteq w_l$, that is $w_k = \sigma_k^1 \sigma_k^2 \ldots \sigma_k^k = \sigma_l^{i_1} \sigma_l^{i_2} \ldots \sigma_l^{i_k}$ for suitable numbers $i_1 < \ldots < i_k$. Let $\vec{b} \stackrel{\text{def}}{=} (b_1, \ldots, b_k)$ and $\vec{b}' \stackrel{\text{def}}{=} (b_{i_1}, \ldots, b_{i_k})$. Further, let $\mathcal{T}_k \stackrel{\text{def}}{=} \mathcal{S}'_k \upharpoonright T_k$ where $T_k = \{s, t, a_{X_k}\} \cup \vec{b}$, and $\mathcal{T}_l \stackrel{\text{def}}{=} \mathcal{S}'_l \upharpoonright T_l$ where $T_l \stackrel{\text{def}}{=} \{s, t, a_{X_l}\} \cup \vec{b}'$. We refer to Figure 2 for an illustration of the substructures $\mathcal{T}_k$ and $\mathcal{T}_l$ of $\mathcal{S}'$.

It can be shown that $\mathcal{T}_k \simeq_\pi \mathcal{T}_l$, where π is the isomorphism that maps s and t to themselves, a_{X_k} to a_{X_l} and b_j to b_{i_j} for every $j \in \{1, \ldots, k\}$.

Thus, by the Substructure Lemma, application of the following two update sequences to $\mathcal{S}'$ results in the same query result:

(β_1) Deleting edges $(a_{X_k}, b_1), \ldots, (a_{X_k}, b_k)$ and adding an edge (s, a_{X_k}).
(β_2) Deleting edges $(a_{X_l}, b_{i_1}), \ldots, (a_{X_l}, b_{i_k})$ and adding an edge (s, a_{X_l}).

However, applying β_1 yields a graph in which t is not reachable from s, whereas by applying β_2 a graph is obtained in which t is reachable from s. This is the desired contradiction. $\qquad\square$

4.2 Separating Low Arities

An arity hierarchy for DYNFO was established in [3]. The dynamic queries $\mathcal{Q}_{k+1}$ used to separate k-ary and $(k + 1)$-ary DYNFO can already be maintained in $(k + 1)$-ary DYNPROP, thus the hierarchy transfers to DYNPROP immediately. However, $\mathcal{Q}_{k+1}$ is a k-ary query and has an input schema of arity $6k+1$ (improved to $3k+1$ in [14]). Here we establish a strict arity hierarchy between unary, binary and ternary DYNPROP for Boolean queries and binary input schemas.

We use the problems s-t-TWOPATH, where one asks whether there is a path of length two from s to t in a given s-t-graph G, and the problem s-TWOPATH where one asks whether there is *any* path of length 2 starting in s.

Proposition 8. *The dynamic query* DYN(s-t-TWOPATH) *is in binary* DYNPROP, *but not in unary* DYNPROP.

Proposition 9. *The dynamic query* DYN(s-TWOPATH) *is in ternary* DYNPROP, *but not in binary* DYNPROP.

5 Lower Bounds with Auxiliary Functions

In this section we consider the extension of the quantifier-free update formalism by auxiliary functions. Recall that DYNPROP-update formulas have access only to the inserted or deleted tuple $\vec{a}$ and the currently updated tuple of an auxiliary relation $\vec{b}$. When auxiliary functions are allowed in update formulas, further elements of the structure can be accessed by function application. This can be seen as adding weak quantification to quantifier-free formulas. The class of dynamic queries that can be maintained with quantifier-free update formulas and auxiliary functions is denoted DYNQF.

DYNQF is strictly more expressive than DYNPROP. E.g., it contains all Dyck languages, among other non-regular languages [6]. Further, undirected reachability can be maintained in DYNQF with built-in relations [5].

Lists can be represented by unary functions in a straightforward way. Therefore, it is not surprising that the upper bound of Proposition 8 already holds for unary DYNPROP with unary built-in functions.

Proposition 10. DYN(s-t-REACH) *on 1-layered s-t-graphs can be maintained in unary* DYNPROP *with unary built-in functions.*

Yet, unary DYNQF cannot maintain the reachability query. Also Theorem 5 can be extended to quantifier-free programs with auxiliary functions.

Theorem 11. DYN(s-t-REACH) *is not in unary* DYNQF.

Theorem 12. DYN(s-t-REACH) *cannot be maintained in* DYNQF *with invariant initialization mapping and empty built-in schema. This holds even for 1-layered s-t-graphs.*

6   Lower Bounds for Other Dynamic Queries

Lower bounds for the dynamic variants of the k-CLIQUE and k-COL problems (where k is fixed) can be established via reductions to the dynamic s-t-reachability query for shallow graphs.

Proposition 13. *The dynamic query* DYN(k-CLIQUE), *for $k \geq 3$, and the dynamic query* DYN(k-COL), *for $k \geq 2$, are not in binary* DYNPROP.

Proposition 14. *The dynamic query* DYN(k-CLIQUE), *for $k \geq 3$, and the dynamic query* DYN(k-COL), *for $k \geq 2$, cannot be maintained in* DYNQF *with invariant initialization mapping.*

7   Normal Forms for Dynamic Programs

In this section, we give normal forms for dynamic programs. The study of normal forms has a long tradition in logics. Normal forms are often helpful in proofs based on the structure of formulas and yield insights for the construction of algorithms.

A formula is *negation-free* if it does not use negation at all. A formula is *conjunctive* if it is a conjunction of (positive or negated) literals. A dynamic program is negation-free (conjunctive, respectively) if all its update formulas are negation-free (conjunctive, respectively). Two dynamic programs $\mathcal{P}$ and $\mathcal{P}'$ are equivalent, if they maintain the same query. The results in this section allow arbitrary initialization but no auxiliary functions. The first theorem is a straightforward generalization of Theorem 6.6 from [5] which states this observation for a subclass of DYNPROP.

Theorem 15. *(a) Every* DYNFO-*program   has   an   equivalent   negation-free* DYNFO-*program.*
(b) Every DYNPROP-*program has an equivalent negation-free* DYNPROP-*program.*

Theorem 16. *Every* DYNPROP-*program   has   an   equivalent   conjunctive* DYNPROP-*program.*

8   Future Work

The question whether Reachability is maintainable with first-order updates remains one of the major open questions in dynamic complexity. Proving that Reachability cannot be maintained with quantifier-free updates with arbitrary auxiliary data seems to be a worthwhile intermediate goal, but it appears nontrivial as well.

We contributed to the intermediate goal by giving a first lower bound for binary auxiliary relations. Whether the strictness of the arity hierarchy for DYNPROP extends beyond arity three is another open question.

For (full) first-order updates a major challenge is the development of lower bound tools. Current techniques are in some sense not fully dynamic: either results from static descriptive complexity are applied to constant-length update sequences; or non-constant but very regular update sequences are used. In the latter case, the updates do not depend on previous changes to the auxiliary data (as, e.g., in [7] and in this paper). Finding techniques that adapt to changes could be a good starting point.

The normal forms obtained for DYNPROP give hope that some fragments of DYNFO collapse. Therefore, we plan to study normal forms for DYNFO extensively. One interesting question being which fragments of DYNFO can be captured by a conjunctive query normal form.

Acknowledgement. We thank Ahmet Kara and Martin Schuster for careful proofreading. We acknowledge the financial support by the German DFG under grant SCHW 678/6-1.

References

[1] Patnaik, S., Immerman, N.: Dyn-FO: A parallel, dynamic complexity class. In: PODS, pp. 210–221. ACM Press (1994)

[2] Hesse, W.: The dynamic complexity of transitive closure is in DynTC^0. In: Van den Bussche, J., Vianu, V. (eds.) ICDT 2001. LNCS, vol. 1973, pp. 234–247. Springer, Heidelberg (2000)

[3] Dong, G., Su, J.: Arity bounds in first-order incremental evaluation and definition of polynomial time database queries. J. Comput. Syst. Sci. 57(3), 289–308 (1998)

[4] Dong, G., Libkin, L., Wong, L.: Incremental recomputation in local languages. Inf. Comput. 181(2), 88–98 (2003)

[5] Hesse, W.: Dynamic Computational Complexity. PhD thesis, University of Massachusetts Amherst (2003)

[6] Gelade, W., Marquardt, M., Schwentick, T.: The dynamic complexity of formal languages. In: STACS, pp. 481–492 (2009)

[7] Grädel, E., Siebertz, S.: Dynamic definability. In: ICDT, 236–248 (2012)

[8] Patrascu, M., Demaine, E.D.: Lower bounds for dynamic connectivity. In: Babai, L. (ed.) STOC, pp. 546–553. ACM (2004)

[9] Zeume, T., Schwentick, T.: On the quantifier-free dynamic complexity of reachability. CoRR abs/1306.3056 (2013), http://arxiv.org/abs/1306.3056

[10] Weber, V., Schwentick, T.: Dynamic complexity theory revisited. Theory Comput. Syst. 40(4), 355–377 (2007)

[11] Gelade, W., Marquardt, M., Schwentick, T.: The dynamic complexity of formal languages. ACM Trans. Comput. Log. 13(3), 19 (2012)

[12] Etessami, K.: Dynamic tree isomorphism via first-order updates. In: PODS, pp. 235–243. ACM Press (1998)

[13] Schmitz, S., Schnoebelen, P.: Multiply-recursive upper bounds with Higman's lemma. In: Aceto, L., Henzinger, M., Sgall, J. (eds.) ICALP 2011, Part II. LNCS, vol. 6756, pp. 441–452. Springer, Heidelberg (2011)

[14] Dong, G., Zhang, L.: Separating auxiliary arity hierarchy of first-order incremental evaluation systems using (3k+1)-ary input relations. Int. J. Found. Comput. Sci. 11(4), 573–578 (2000)

Erratum:
A Constant Factor Approximation for the Generalized Assignment Problem with Minimum Quantities and Unit Size Items

Marco Bender[1], Clemens Thielen[2], and Stephan Westphal[1]

[1] Institute for Numerical and Applied Mathematics, University of Goettingen,
Lotzestr. 16-18, D-37083 Goettingen, Germany
{m.bender,s.westphal}@math.uni-goettingen.de
[2] Department of Mathematics, University of Kaiserslautern,
Paul-Ehrlich-Str. 14, D-67663 Kaiserslautern, Germany
thielen@mathematik.uni-kl.de

K. Chatterjee and J. Sgall (Eds.): MFCS 2013, LNCS 8087, pp. 135–145, 2013.
© Springer-Verlag Berlin Heidelberg 2013

DOI 10.1007/978-3-642-40313-2_74

Abstract. We identify an error in the approximation algorithm for the generalized assignment problem with minimum quantities and unit size items in [1]. We show that the previously presented randomized algorithm can be slightly modified in order to obtain, for every $c \geq 2$, a randomized approximation algorithm with approximation ratio $((2c - 1) \cdot \frac{e}{e-1})$ that outputs a feasible solution with probability at least $1 - \min \left\{ \frac{1}{c}, \frac{e^{c-1}}{c^c} \right\}$.

In the analysis of the randomized approximation algorithm in [1], we implicitly assumed that the expected profit after a clean-up step can be calculated as the product of the expected profit before this step and the expected loss factor in this step, which is not valid since the random variables are correlated. In this note, we show how we can modify the algorithm slightly in order to obtain a correct approximation result. We refer to the original paper for details about the problem and the terminology.

Following the notation in [1], we denote by x^{IP} the binary vector obtained from the randomized rounding process. If we choose the scaling factor $\alpha = 1$, we have $\mathbb{E}(\mathrm{PROFIT}(x^{\mathrm{IP}})) = \mathrm{OPT_{LP}}$. In x^{IP}, at most one feasible packing is chosen for each bin, but it is possible that some items are assigned to multiple bins. Therefore, we perform two clean-up steps on x^{IP} as before. In the following, $c \geq 2$ is an arbitrary but fixed integer.

In the first step, we discard a subset of the bins opened in x^{IP} in order to ensure that the total number of places used in the bins is at most n. In the second step, we then replace copies of multiply assigned items by unassigned items in order to ensure that each item is packed into at most one bin.

When discarding a subset of the bins opened in x^{IP} during the first clean-up step, we distinguish two cases:

1. If $\#(\text{places used in } x^{\text{IP}}) < cn$, we can obtain a subset of the packings such that the total number of places used is at most n and the remaining profit is at least $\frac{1}{2c-1} \cdot \text{PROFIT}(x^{\text{IP}})$. This can be achieved by the technique used in the first clean-up step in [1] (choosing $k := c - 1$).
2. If $\#(\text{places used in } x^{\text{IP}}) \geq cn$, we discard no packings and leave the solution unchanged.

In both cases, the first clean-up step yields a solution with expected profit at least $\frac{1}{2c-1} \cdot \text{OPT}_{\text{LP}}$.

After the first clean-up step, some items might still be assigned to multiple bins. Therefore, in the second clean-up step, we remove each multiply assigned item from all bins but the one where it yields the highest profit in the solution obtained from the first step. The following well-known analysis shows that we lose at most a factor of $(1 - \frac{1}{e})$ in the total profit by this removal process:

We fix an item i and denote by y_{ij} the probability that item i is assigned to bin j after the first clean-up step. We assume without loss of generality that the bins are sorted by nonincreasing profit of item i, i.e., $p_{i1} \geq p_{i2} \geq \ldots \geq p_{im}$. The expected profit obtained from item i in the solution obtained after the second clean-up step is then given as

$$y_{i1}p_{i1} + (1 - y_{i1})y_{i2}p_{i2} + \ldots + \prod_{j=1}^{m-1}(1 - y_{ij})y_{im}p_{im}$$

$$\geq \left(1 - \left(1 - \frac{1}{m}\right)^m\right) \sum_{j=1}^{m} p_{ij}y_{ij}$$

$$\geq \left(1 - \frac{1}{e}\right) \sum_{j=1}^{m} p_{ij}y_{ij}.$$

Here, we used the arithmetic-geometric mean inequality and the fact that $(1 - 1/m)^m \leq e^{-1}$ for all $m \geq 1$. Since the total expected profit is the sum of the expected profits obtained from each item, we can perform this procedure separately for every item and, altogether, we obtain a solution with expected profit at least

$$\frac{1}{2c-1} \cdot \left(1 - \frac{1}{e}\right) \cdot \text{OPT}_{\text{LP}}.$$

After the removal of multiply assigned items, some of the bins that are opened may not be filled to their minimum quantities anymore.

In case 1, the total number of places used in the bins after the first clean-up step was no more than the total number n of items available. Hence, for each item i that was assigned to $l \geq 2$ bins, there must be $l - 1$ items that were not assigned to any bin after the first step. Thus, we can refill the $l - 1$ places vacated by deleting item i from all but one bin with items that were previously unassigned, and doing so for all multiply assigned items yields a feasible integral solution that respects the minimum quantities of the bins.

Although this is, in general, not possible in case 2 and our algorithm will end up with an infeasible solution in this case, we can bound the probability of this bad event. By definition of x^{IP}, we know that the expected number of places used in x^{IP} is at most n. Hence, Markov's inequality yields that

$$\Pr\left(\#(\text{places used in } x^{\mathrm{IP}}) \geq cn\right) \leq \frac{n}{cn} = \frac{1}{c}.$$

Alternatively, we can make use of the following Chernoff bound:

$$\Pr\left(\#(\text{places used in } x^{\mathrm{IP}}) \geq cn\right) < \frac{e^{c-1}}{c^c}$$

This yields sharper tail bounds for $c \geq 3$.

In total, for every integer $c \geq 2$, we obtain a randomized approximation algorithm that achieves an approximation ratio of

$$(2c - 1) \cdot \frac{e}{e - 1}$$

and outputs a feasible solution with probability at least

$$1 - \min\left\{\frac{1}{c}, \frac{e^{c-1}}{c^c}\right\}.$$

Acknowledgments

We would like to thank Alexander Souza for many helpful discussions and for pointing out the error in the original paper.

References

1. Bender, M., Thielen, C., Westphal, S.: A constant factor approximation for the generalized assignment problem with minimum quantities and unit size items. In: Proceedings of the 38th International Symposium on Mathematical Foundations of Computer Science (MFCS). (2013) 135–145

The original online version for this chapter can be found at
http://dx.doi.org/10.1007/978-3-642-40313-2_14

Author Index